VIEWING LIFE MATHEMATICALLY

A Pathway to Quantitative Literacy

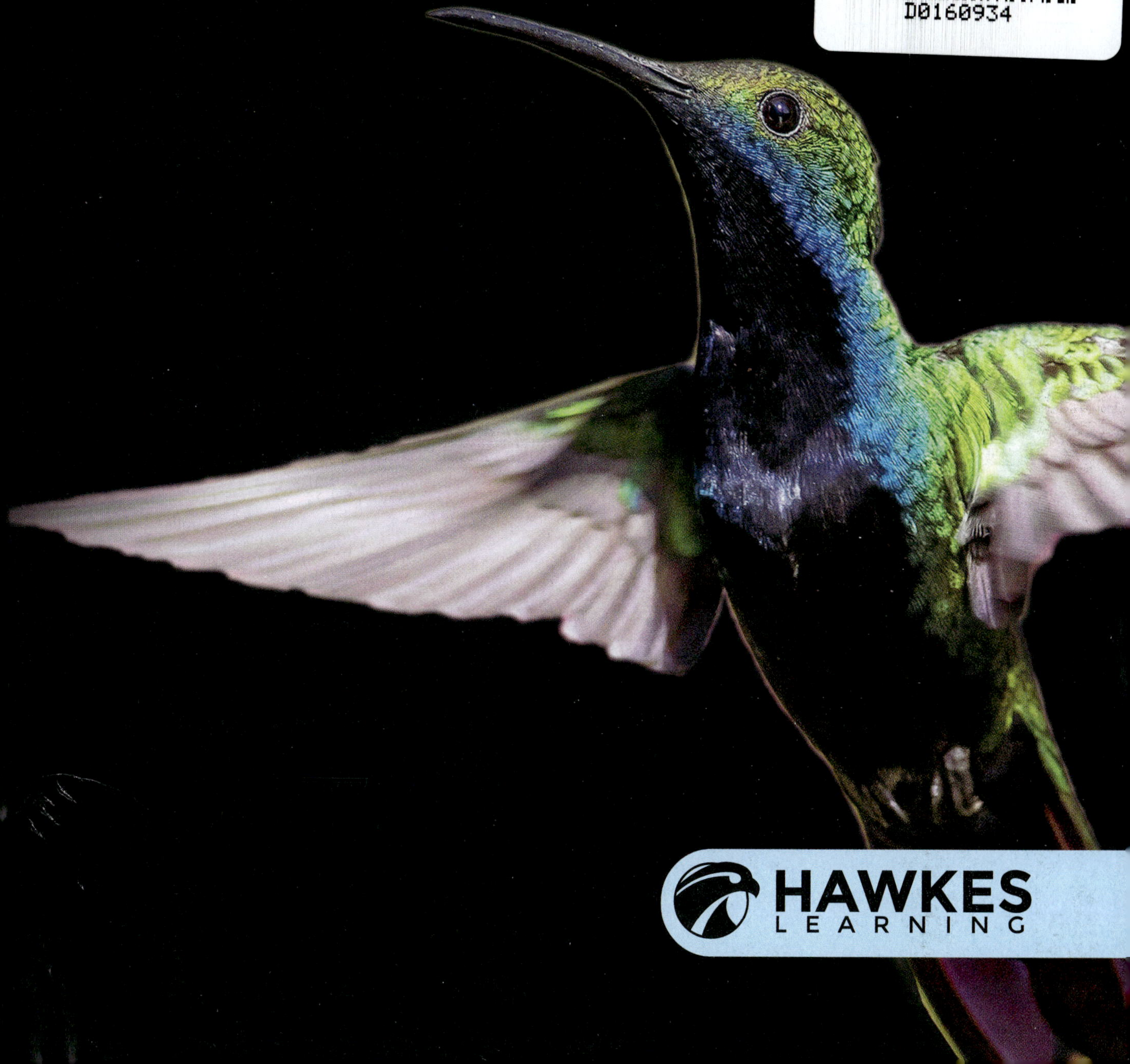

HAWKES LEARNING

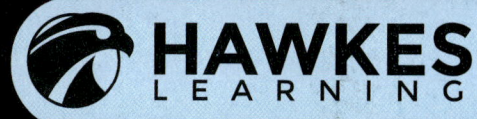

Barbara Miller

Trudy Tronco

Answer Key Assistance:
Quant Systems India Pvt. Ltd.

Editors:
Danielle C. Bess,
Allison Conger,
Jolie Even,
Robin Hendrix,
S. Rebecca Johnson,
Claudia Vance

Designers:
Lizbeth Mendoza,
Patrick Thompson,
Joel Travis

Cover Design:
Lizbeth Mendoza

Indexer:
Barbara Miller

Courseware Developers:
Allison Conger,
Jolie Even,
Adam Flaherty,
Kyle Gilstrap

Content Contributors:
Karen Crossin,
Muff Glover,
Lisa Hinton,
Silke Hunker,
Leonardo Pinheiro,
Kara Roché,
Joe A. Stickles, Jr.

Manager of Math Content Development:
Blair Dunivan

A division of Quant Systems, Inc.

546 Long Point Road
Mount Pleasant, SC 29464

Library of Congress Control Number   2022932250

Printed in the United States of America 🇺🇸

10 9 8 7 6 5 4 3 2 1

ISBN: 978-1-64277-491-7

Table of Contents

Chapter 1

Critical Thinking and Problem Solving

Chapter 2

Set Theory

Chapter 3

Logic

Chapter 4

Ratios, Percentages, Rates, and Proportionality

Chapter 5

Algebra: Equations, Inequalities, and Functions

Chapter 6

Finance

Chapter 7

Numeration and Measurement Systems

Chapter 8

Number Theory

Chapter 9

Geometry

Chapter 10

Probability

Chapter 11

Statistics

Chapter 12

Data Science

Chapter 13

Voting and Apportionment

Chapter 14

Graph Theory

Preface

Letter from the Author

It is my pleasure to introduce to you the second edition of *Viewing Life Mathematically*. Building from the first edition, I have endeavored to curate fourteen chapters that not only teach concepts but propel students to be curious about the world around us and how things relate to one another. When I first started writing mathematics textbooks, it was out of a desire to create a text that was both approachable and readable for all students. This mission to create a robust text that is presented with a conversational approach continues to be central to my writing philosophy.

While this may be the only college-level mathematics course some students take, it is my hope that this course will not be the end of their mathematical curiosity. Instead, I intend this course to serve as a springboard for students to see the ways in which mathematical thought pertains to their chosen discipline, no matter the field. To this end, I have consciously tried to concentrate the material around pivotal concepts that open doors to future mathematics.

Any modern quantitative reasoning text contains far more chapters than could reasonably be taught in any single semester. Consequently, rather than devoting entire chapters to topics such as sports or art, as in the first edition, I have interwoven those topics among the other chapters. In addition, I have expanded topics such as functions and mathematical growth, which are now contained in a new chapter on algebraic methods.

Completely new to this second edition is a chapter on data science. The study of data is no longer limited to statistical professionals; it penetrates every profession. As such, I am thrilled to introduce students to the exciting world of a data scientist—wrangling, analyzing, and storytelling.

In addition to these changes, this edition contains improved and expanded exercise sets with additional real-world data, new graphics, and updated examples that are of interest to the reader. When appropriate, I've also included the use of Excel as a calculation tool in examples. As one of the most sought-after skills for employers, I endeavored to equip students with as much proficiency in Excel as possible.

I hope you enjoy this new edition.

Best Regards,

Kim Denley

Special Features

Chapter and Section Openers

Each chapter begins with a list of sections to prepare students for the topics that will be covered. Objectives are listed at the beginning of each section, helping students to identify the most important concepts in the section and enabling students to focus their time and effort appropriately.

Margin Boxes

💬 Think Back

Think Backs are just-in-time information located in the margins of the text to remind students of important concepts that they may have forgotten from previous math courses or previous chapters in this course.

💬 Fun Fact

Fun Facts appear in the margin with additional information to pique student interest and make mathematics more relatable.

💬 Math Milestone

Math Milestones appear in the margins to give more information about mathematicians and the history of mathematics to help students connect further with the material.

💬 Helpful Hint

Helpful Hints appear in the margins of the text and provide students with extra insightful pieces of information to help them understand concepts more deeply and avoid common mistakes.

Chapter Introductions

An application of the chapter content is introduced on the first page of each chapter to inspire student interest and provide an understanding of how the topics to be studied are useful.

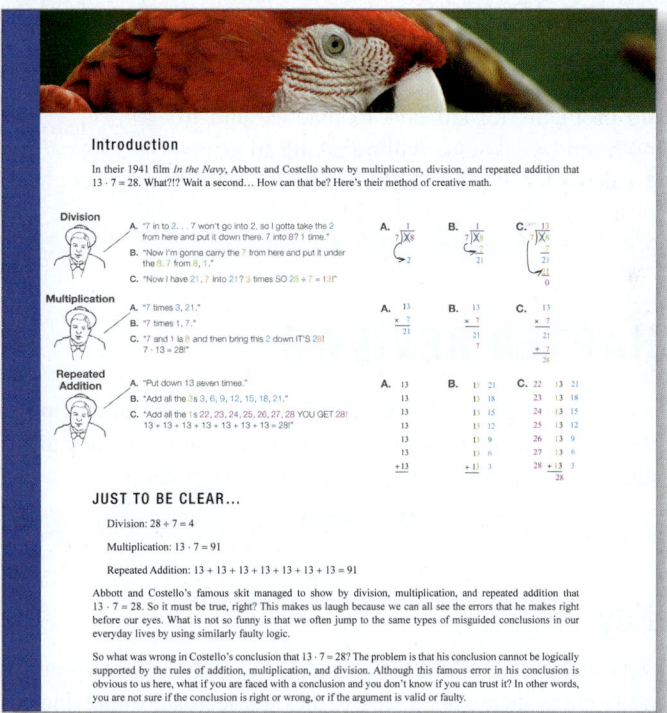

Examples

Examples are presented in a step-by-step manner that is easy for students to follow. Titles are given to alert students to the concept they will be learning throughout the example. Examples make use of tables, diagrams and graphs, and technology where applicable, giving clarification to the mathematical skill being presented.

✔ Skill Check

Skill Check questions allow students to test their knowledge as they progress through a chapter to ensure that they understand the material. Solutions to these problems are located at the end of the section in which they appear before the section exercises.

📈 TECH TIP

Tech Tips provide additional insights to help the student follow along with the TI-83/84 Plus graphing calculator and Microsoft Excel solution methods in the examples.

📖 DEFINITIONS AND FORMULAS

Definitions and formulas are presented in highly visible boxes for easy reference.

Section Exercises

Each section includes a variety of exercises to give students practice applying and reinforcing the skills they learned in the section. The exercises exhibit a wide range of difficulty levels and applications, often using real-world data.

SECTION PROJECTS

At the end of most sections, a Section Project provides students with the opportunity to put the concepts and skills learned in the section into practice.

Chapter Exercises

At the end of each chapter, a set of Chapter Exercises provides extra problems for students to practice and to identify strengths and weaknesses before taking an exam.

CHAPTER PROJECTS

At the end of each chapter, a Chapter Project provides a thoughtful activity to apply the concepts and skills learned in the chapter to real-world situations.

📋 CHAPTER REVIEWS

Each chapter features a Chapter Review that highlights important definitions, properties, processes, and formulas covered in the chapter. Material is listed by section in the order that it appears, allowing students to quickly find the information they need when studying for a test or doing homework assignments.

Answer Key

The Answer Key in the back of the book contains the answers for odd-numbered Section Exercises and Chapter Exercises. This allows students to check their work to ensure they are accurately applying the methods and skills that they have learned.

Content

Chapter 1: Critical Thinking and Problem Solving

For most students in college, thinking mathematically means thinking algebraically. This chapter aims to introduce thinking mathematically as creative problem solving through reasoning, processes, and techniques.

Chapter 2: Set Theory

From social media followers to pizza toppings; from understanding surveys to organizing a closet; and from factors to prime numbers—looking at the way items are grouped provides a solid mathematical foundation. This chapter uses a systematic approach to organizing and analyzing information and groups of items in order to solve problems involving sets. Formal set theory is introduced, including its definitions and notation, as a means to express, analyze, and manipulate categories and groupings.

Chapter 3: Logic

Logic is the basic structure of thinking mathematically. This chapter introduces formal mathematical logic, along with its definitions and notation. The ultimate goal of the chapter is for students to create valid logical arguments and critique invalid logical arguments.

Chapter 4: Ratios, Percentages, Rates, and Proportionality

Ratios, percentages, rates, and proportions are universal, from scale models and mixtures to retail sales and tipping. This chapter develops the skills for understanding and calculating these important values.

Chapter 5: Algebra: Equations, Inequalities, and Functions

Mathematical models use information about the past to create images of the future. In this chapter, methodology is developed to create, manipulate, and analyze models for everyday situations. The focus on real-world applications involves linear, quadratic, exponential, and logarithmic models along with an introduction to linear programming.

Chapter 6: Finance

Wise financial planning both for the short term and the long term is an important life skill. This chapter introduces the basic skills of mathematical finance for effective budgeting and long-term decision making. Topics include saving, investing, and borrowing money along with filing taxes and creating a budget.

Chapter 7: Numeration and Measurement Systems

Understanding the history of numeration systems, such as the Babylonian and Greek numeration systems, can help build an understanding and appreciation of the Hindu-Arabic system we use today. The exploration of ancient numeration systems leads into an investigation of different base number systems, such as binary and hexadecimal. The chapter wraps up with a review of the metric system and converting to and from the US customary system.

Chapter 8: Number Theory

From ISBNs to internet security, the properties of numbers play an often unnoticed role throughout modern life. This chapter provides a gentle introduction into the power of number theory, covering topics from prime and composite numbers to modular arithmetic, as well as applications of public-key encryption.

Chapter 9: Geometry

From planning a garden to determining the volume of a swimming pool, geometry is closely involved with everyday decision making. This chapter reviews the concepts of perimeter, area, volume, and surface area in the context of real-life situations. The chapter concludes with right triangles and trigonometry, introducing students to concepts such as angle of elevation or depression and the laws of sines and cosines.

Chapter 10: Probability

Little in life is certain. Behavioral economists suggest that many mistakes in everyday life stem from massively overestimating or underestimating chance. Being able to express and manipulate the uncertainties of random events is a crucial skill in life. This chapter introduces formal probability, including its definitions and notation, as a means to express, analyze, and calculate chance.

Chapter 11: Statistics

Being able to understand statistical concepts is an important part of mathematical literacy in every field of study at the college level. This chapter introduces the fundamental concepts behind collecting and analyzing statistical data.

Chapter 12: Data Science

A step beyond statistics, the field of data science involves cleaning up and analyzing large data sets, and then presenting findings. In this chapter, students will learn how to clean up large data sets using Excel and how to present their findings with a variety of visuals.

Chapter 13: Voting and Apportionment

Making the right choice can be challenging. This chapter introduces the mathematics behind the way individual preferences affect the decisions groups make and how to ensure fair outcomes. Topics include apportionment, voting methods, and voting paradoxes.

Chapter 14: Graph Theory

This chapter includes a thorough coverage of introductory topics to graph theory and how they relate to real-world applications. The structure of networks, including Facebook and the interstate highway system, provides powerful mediums to connect people. This chapter looks at the mathematics behind these networks.

Acknowledgements

Recognizing that truly good teaching and writing does not happen in a vacuum, but rather takes a team of talented individuals with different points of view to critique and hone the thoughts that appear on the page, I am grateful for those who partnered with me in this endeavor. To my extraordinary editing team—Barbara Miller, Rebecca Johnson, Allison Conger, Jolie Even, and Danielle Bess—you are a joy to work with and have made this edition exceptional! This work would only be a shadow of the current text without your input and patience. To Jim Hawkes, Marcel Prevuznak, Abby O'Leary, and Katherine Binder, thank you for your continued support. Thanks also to the numerous individuals at Hawkes who have put together a beautiful product both in print and digitally. To the reviewers, who helped not only shape the content, but also fine-tune the language, I am deeply appreciative.

Iana Anguelova, College of Charleston

Dr. Edna Bazik, National Louis University–Chicago

Dr. Shari Beck, Navarro College–Corsicana Campus

Jennifer Briney, MacMurray College

Joan Brown, Eastern New Mexico University

David Busch, Iowa Central Community College

John Callaghan, Gateway Community College

Dr. Judith Covington, Louisiana State University–Shreveport

Karen Crossin, George Mason University

Dr. John Dawson, Kirkwood Community College–Iowa City

Erik Degenhardt, UW Platteville Baraboo Sauk County

Dr. Sarah Duffin, Southern Utah University

Dr. Johnny Duke, Georgia Highlands College

Kevin Dyke, Georgia Highlands College

Professor Gilbert Eyabi, Anderson University

Ryan D. Fox, Belmont University

Kay Geving, Belmont University

Dr. Kim Harris, University of North Carolina–Charlotte

Carla Hill, Marist College

Dr. Peggy Hohensee, Fletcher Technical Community College/Kaplan University

Silke Hunker, Northeastern Junior College

Dr. Liz Jurisich, College of Charleston

Dr. Dorothy Kerzel, Mississippi University for Women

Denise A. Krueger, Mount St. Joseph University

Michael W. Lanstrum, Cuyahoga Community College

Dr. Chris Mattingly, Martin Methodist College

Holly McDow, Sandhills Community College

Tonya Meisner, University of Wisconsin–Marinette

Usha Midkiff, The University of Alabama

Andrew J. Miller, Belmont University

Janette Miller, University of Wisconsin–Sheboygan

Dr. Emily J. Olson, Millikin University

Kristi Peters, Enterprise State Community College

Dr. Traci M. Reed, St. Johns River State College

Robin Rufatto, Ball State University

Dr. Tsvetanka Sendova, Bennett College for Women

Jim Sheff, Spoon River College

Melody Shipley, North Central Missouri College

Lymeda Singleton, Texas A&M University–Commerce

Eugene Slason, Regis College

Joe A. Stickles, Jr., Millikin University

Paula R. Stickles, Millikin University

Joseph Szurek, University of Pittsburgh–Greensburg

John Thoo, Yuba College

Dr. Alfredo Vaquiax-Alvarado, Our Lady of the Lake University

John W. Weber, Iowa Central Community College

Dr. Catherine Whatley, York Technical College

Christy Wilson, South Arkansas Community College

Dr. Nancy Wyshinski, Trinity College

Amy Young, Navarro College

My sincere thanks go to Mike Hall, my co-author for the first edition. Without your partnership along the creation journey, this project would never have seen the light of day. Thank you also to all of you who have used that first edition and offered feedback. Your comments and suggestions have greatly shaped and improved this version and are welcome both now and in the future.

Lastly, to my husband, Tristan. Everything we do is a partnership, and this book is no different. You have a heart for education that is immense and encourages me; your vision, inspiration, and support truly propelled the book to be what it is. I love you.

Hawkes Learning: A Clear Path to Mastery

Hawkes' software employs an adaptive, competency-based approach to knowledge mastery supported by a user-friendly interface. The student-centric platform promotes positive active learning by adapting to each student's needs through algorithmically generated questions based on an individual learner's pace, skill, and knowledge level. The real-time adaptive feedback addresses errors immediately, so that students learn from their mistakes when they make them. For each topic, the Hawkes Learning path to content mastery engages students through three simple modes: Learn, Practice, and Certify.

Competency-based learning made simple in three steps:

 **LEARN** offers a multimedia-rich presentation of the lesson content. It includes instructional videos, interactive examples, and more.

 **PRACTICE** engages students with algorithmically generated questions and intelligent tutoring in an ungraded, penalty-free environment.

 **CERTIFY** requires students to demonstrate mastery of the material at a defined proficiency level without access to tutoring aids.

Support

If you have questions or comments, we can be contacted as follows:

24/7 Chat: chat.hawkeslearning.com

Phone: 1-800-426-9538

Email: support@hawkeslearning.com

Web: support.hawkeslearning.com

VIEWING LIFE
MATHEMATICALLY
A Pathway to Quantitative Literacy

2ND EDITION

CHAPTER 1

Critical Thinking and Problem Solving

■ SECTIONS

Introduction

Land navigation is essential in military training. As part of their preparation, new Marine Corps recruits are introduced to land navigation skills in boot camp. They go on to hone these skills in the School of Infantry. Using critical thinking, recruits must find their way through unfamiliar terrain, read maps, use a compass, and shoot an azimuth. In land navigation, to shoot an azimuth means to climb to a height, sight a celestial object or feature on the horizon in the direction you are traveling, and then adjust your compass to make sure you are moving in the correct direction. Mastering these skills can potentially mean the difference between life and death during future expeditions.

18th-Century Azimuthal Compass

Though all critical thinking does not have life-and-death consequences, being able to make logical and informed decisions helps us navigate through our lives more easily. Whether we are making a decision in a logical argument, deciding which freeway to take on the way to work, or maximizing our work output, mathematics plays a key role in the choices we make and the problems we solve. This chapter, along with the rest of the book, will focus on the use of mathematical concepts that can be used to analyze the world around us and, in turn, help us make informed decisions.

OBJECTIVES

1. Identify inductive reasoning and deductive reasoning.

2. Use inductive reasoning to reach conclusions.

3. Find counterexamples.

4. Use deductive reasoning to reach conclusions.

The ability to use math as a tool can separate you from many others in your chosen field. Just as carpenters work at a level above laborers partly because of their skill with tools, you can work at a higher level with the ability to formulate and solve number problems.—Ned Pelger

OBJECTIVE 1 Inductive Reasoning versus Deductive Reasoning

As we begin our voyage into *Viewing Life Mathematically*, we hope to show you both the beauty and the every-day value of being numerate; that is, mathematically competent. Being *numerate* is having the ability to think about and communicate with numbers, just as being *literate* is having the ability to function in the world of words. The beauty is that numeracy and literacy overlap more often than we think. Consider questions like the following.

- Did you get charged the correct amount when you purchased your shirt at the mall?

- Does your final average in chemistry reflect the work you did throughout the semester?

- How much time will be required to finish a project?

- Is it reasonable to say that 95% of Americans live a middle-class lifestyle?

Reading these questions requires literacy while answering them requires numeracy.

The skills that numeracy provide help us excel at solving problems—all kinds of problems. These problems include performing calculations, making estimations, observing patterns, testing conjectures, making investment decisions, predicting outcomes, and deciphering codes. You can see from the quote at the beginning of the lesson that Ned Pelger asserts this view on problem solving for those looking to excel in the construction field. The authors of the popular book *Super Crunchers* have the same opinion when talking about why shoppers should be savvy thinkers. They claim that consumers can no longer be asleep at the wheel while spending money since businesses are figuring out ways to treat price-oblivious consumers differently from the price-conscience consumers, encouraging them to spend more money. Living in the 21st century means that everyone's lives are steadily becoming more quantitative, and problem-solving situations that unfold each day are steadily requiring a broader familiarity with mathematical tools.

> **Fun Fact**
>
> Ian Ayres, author of *Super Crunchers*, is an American lawyer and economist. He is a professor at the Yale Law School and at the Yale School of Management.

Let's launch our exploration into thinking mathematically by considering the ways that we reason; in other words, how we form inferences or conclusions. As thinkers, we often draw conclusions using either **inductive** or **deductive reasoning**. The difference between the two lies in what information we begin with and the pathway to a conclusion. Inductive reasoning takes specific examples and makes a general conclusion. An example of inductive reasoning would be noticing that every time you walk outside you sneeze, and then deciding to always carry a tissue with you any time you venture outside. Inductive reasoning can be considered a *generalization*.

On the other hand, deductive reasoning begins with general statements that are commonly accepted as facts and arrives at specific conclusions by using logic. For example, suppose you believe that the best way to sell perfume is for people to smell it on others. Then deductive reasoning would lead you to give out free samples to as many people as possible as a marketing tactic.

Deductive reasoning produces stronger arguments than inductive reasoning. Using logical deductions, if you start with true statements, you can provide absolute proof of your conclusions.

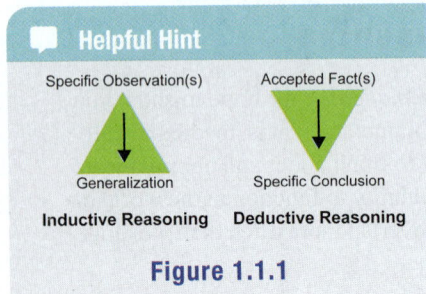

Helpful Hint

Figure 1.1.1

📖 INDUCTIVE AND DEDUCTIVE REASONING

Inductive Reasoning

Inductive reasoning is a line of argument that takes specific examples as its premise and then draws a general conclusion from these examples.

Deductive Reasoning

Deductive reasoning uses statements that are commonly accepted as facts to make a case for a specific conclusion.

Example 1.1.1

Identifying Inductive and Deductive Reasoning

Determine whether the following arguments use inductive or deductive reasoning.

a. In Boulder, Colorado, it snowed 16.2 inches during January 2019 and 16.3 inches during January 2020. Therefore, Boulder will receive at least 16 inches of snow every January.

b. All even numbers are divisible by two. The number 2,045,890 is even. Therefore, 2,045,890 is divisible by two.

c. To get a bachelor's degree at Georgia Southern University, a student must have 120 credit hours. James is about to graduate from Georgia Southern University with a bachelor's degree in engineering. Therefore, James has at least 120 credit hours on his transcript.

d. Colleagues in the office have determined that the optimal time to leave work is 4:42 p.m. based on traffic patterns they experienced over the past two weeks.

Solution

a. Notice that in this argument, the examples are specific to January 2019 and January 2020, and then a very general conclusion is made. The likelihood of it snowing about 16 inches each January is not very likely; nevertheless, the argument is inductive.

b. Here two mathematical facts are used: all even numbers are divisible by two and the number 2,045,890 is even. A specific conclusion is then drawn from these facts. Thus, this is an example of deductive reasoning.

c. The fact about the number of hours required to get a bachelor's degree is used to draw a specific conclusion. This is deductive reasoning.

d. Because the office workers are basing their conclusion on specific examples over the past two weeks, they are using inductive reasoning to draw a general conclusion.

OBJECTIVE 2 Inductive Reasoning

Realizing that patterns exist is the foundation of inductive reasoning. The process of exploring possible conclusions from these patterns often involves a bit of trial and error. This brings up an important point—mathematics is a process that is potentially full of missteps. More often than not, the first thing you try will be wrong; but there is no harm in making mistakes. Mistakes are often a first step on the road to discovering the right thing. For instance, when looking for patterns in the list of numbers 17, 12, 8, 5, …, you might notice that the numbers are getting smaller. Is that a pattern? Sure. But is it an interesting pattern, one that helps us determine the next number in the sequence? Probably not. This means we need to continue to look for more features of the pattern to create a more detailed conclusion.

One way to spot patterns in numbers is to look at the difference between two consecutive terms instead of just looking at the terms themselves. If a sequence is structured so that the difference between any two consecutive terms in a number sequence is the same (that is, consecutive terms have a common difference), the sequence is called an **arithmetic sequence**. If every pair of consecutive terms in a number sequence differs by a common multiple (that is, the quotient of consecutive terms is the same), the sequence is called a **geometric sequence**.

However, not all sequences are either arithmetic or geometric. Example 1.1.2 looks at techniques you can make use of when searching for the next number in a sequence. The more practice you have with these techniques, the easier it becomes to identify interesting and useful patterns.

Example 1.1.2

Using Inductive Reasoning

Use inductive reasoning to identify a pattern in each of the following sequences of numbers in order to establish the next term in the sequence.

a. 4, 9, 14, 19, ___

b. 2, 6, 18, 54, ___

c. 5, 6, 8, 11, ___

Solution

When identifying a pattern in a sequence of numbers, there is not a set method to follow. Instead, we use trial-and-error techniques such as looking for common differences, common multiples, or other computational operations.

a. In the sequence 4, 9, 14, 19, ___, let's begin by determining whether the difference between the consecutive terms is constant; that is, whether the difference is the same for each consecutive pairs of terms or if the difference between the consecutive pairs of terms varies.

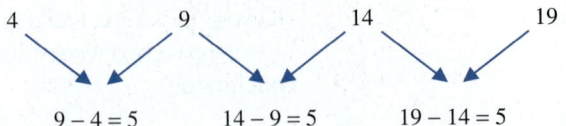

We can see here that there is a common difference between each number in the sequence, making it an arithmetic sequence. The common difference is 5 and the sequence is increasing. Thus, the next term in the sequence will be $19 + 5 = 24$.

b. The sequence 2, 6, 18, 54, __ does not have a common difference between terms like we had in part a. (Check this for yourself.) Since there is no common difference, we need another approach. Perhaps there is a common multiple between the consecutive pairs of terms. We can check this by finding the quotient of consecutive pairs of terms.

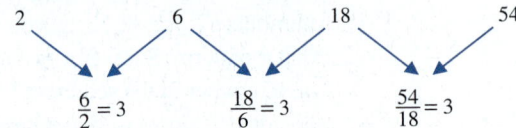

We can see that each successive number is the product of the previous number and 3. So $2 \cdot 3 = 6$, $6 \cdot 3 = 18$, $18 \cdot 3 = 54$, and so on. This means that the next term would be $54 \cdot 3 = 162$. Because the gap between each term is a constant multiple, this is a geometric sequence.

c. For the sequence, 5, 6, 8, 11, __ , there is no common difference or common ratio between the numbers. (Check this for yourself.) Thus, we need to ask ourselves, "What do we have to do to the first term in order to obtain the second?" In this case, we have to add 1 to 5 in order to get 6. For the next term, we see that the sum is $2 + 6 = 8$. We can now see a pattern developing that we need to investigate further. We added 1 to the first term and 2 to the second term. If we continue in this manner by adding 3 to the third term, 4 to the fourth term, and so on, does the pattern continue? For the next term, $8 + 3 = 11$, so indeed it does.

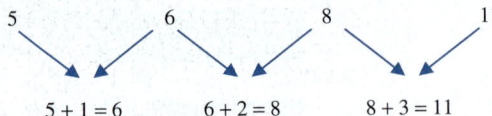

Thus, to find the next term in the sequence, we add 4 to 11 to get 15.

Example 1.1.3

Using Inductive Reasoning

A certain type of human cell reproduces in the following manner: 1 cell, 4 cells, 9 cells, 16 cells.

a. Determine the number of cells present on the next production of cells.

b. How many cells would be present on the nth production cycle?

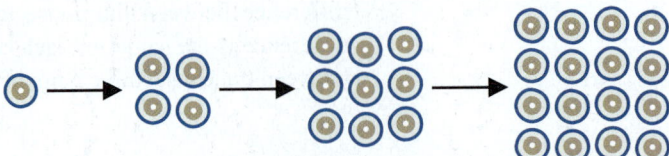

Figure 1.1.2

✔ Skill Check 1.1.1

Consider the patterns in Figure 1.1.3. The first requires 16 line segments to join all of the dots, the second requires 28 line segments, and the third requires 40 line segments. How many line segments would be needed for the 8th pattern? The nth pattern?

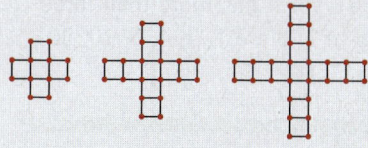

Figure 1.1.3

Solution

a. The first iteration contains 1 cell and the second iteration contains 4 cells, or 2^2. The third and fourth iterations contain $3^2 = 9$ cells and $4^2 = 16$ cells, respectively. These numbers should look familiar to you. They are the squares of the natural numbers. So, the next iteration in the sequence will contain $5^2 = 25$ cells. (Notice that $1^2 = 1$, so the pattern holds.)

b. We've already established that at each production (1st, 2nd, 3rd, etc.) we square the production number. So on the nth production cycle, we will square n; that is, there will be n^2 cells.

OBJECTIVE 3 Counterexamples

Although we use inductive reasoning often, there are a few drawbacks to this type of reasoning. First, since we are using a limited number of specific examples (or a limited amount of evidence) to draw generalized conclusions, we can only speculate that our conclusion is true. This is because the conclusion is based only on the information and examples we have. For instance, after dining out in your local area, you may conclude that Friday night is a good night to be a waiter because it seems that more people dine out on Fridays than any other night of the week. Unless you can consider every restaurant that exists, on every night of the week, you cannot truly be sure that this statement is true.

With inductive reasoning, there is always the possibility that new evidence will be discovered that will prove the conclusion to be incorrect. This leads to the second drawback. If we can find even one situation that does not satisfy our conclusion from inductive reasoning, then the argument is *invalid*. A contradictory example like this is called a **counterexample**. It is important to note that one counterexample is enough to prove that a line of reasoning is false, but one positive example is never enough to prove that a line of reasoning is true. We'll study more about valid and invalid arguments in a later chapter.

📖 COUNTEREXAMPLE

A **counterexample** is an example that shows that a generalization or rule is false by satisfying the conditions of the generalization or rule, but not satisfying the conclusion.

Example 1.1.4

Finding Counterexamples

Consider each conclusion that was drawn using inductive reasoning. Show each conclusion is false by finding a counterexample.

a. You must have a degree in computer science to become wealthy in the computer tech industry.

b. If a sport uses a ball, then the ball is spherical.

c. If a number is a perfect square, then it has exactly three factors.

💬 Fun Fact

Between 1894 and 1900, scientists discovered six new elements, all of which they believed had the property that they could not combine with any other element. These became known as "noble gasses." The conclusion that "noble gases cannot form compounds" became generally accepted by chemists as fact until 1962 when a compound was made for the first time from one of the gasses. The creation of this compound proved that the inductive reasoning behind noble gasses and their ability to form compounds was false.

Solution

✔ **Skill Check 1.1.2**

Which of the following is a counterexample to the following argument?

If the product of two numbers is positive, then their sum is positive.

a. 5 and 5

b. –2 and 5

c. –8 and –8

a. For this example, we need to look no further than two of the most famous people in the computer tech industry—Steve Jobs (cofounder of Apple Computers) and Bill Gates (cofounder of Microsoft). Neither Jobs nor Gates has a degree in computer science, yet both became quite wealthy in the computer tech industry. Hence, we have found a counterexample to our conclusion, and the argument is false.

b. Football is a sport that uses a ball, but the ball is not a sphere; it is oblong. Thus, the argument is false.

c. The number 100 is a perfect square ($10^2 = 100$), yet it has more than three factors, making the argument false. The factors of 100 are 1, 2, 4, 5, 10, 20, 25, 50, and 100.

OBJECTIVE 4 Deductive Reasoning

Inductive reasoning is of great significance in *suggesting* plausible conclusions. However, with deductive reasoning we can be certain that the conclusions drawn are reliable. In contrast to inductive reasoning, deductive reasoning takes statements that are commonly accepted as facts and uses logic to make a case for a certain conclusion. In this manner, the argument is said to be *valid* since the conclusion follows logically from the beginning statements. Deductive reasoning usually follows steps. First there is an assumption or two, which then leads to a series of logical inferences. One of the most famous examples of deductive reasoning is from the Greek philosopher Aristotle:

All men are mortal. Socrates is a man. Therefore, Socrates is mortal.

Helpful Hint

Mathematical statements proved by deductive reasoning are called **theorems**.

As long as the two assumptions are based on accurate information, the outcome of this type of conclusion is true. However, it's important to recognize that deductive reasoning may sometimes lead to a false conclusion if one or both of the beginning statements are false. That is why we have placed an emphasis on starting with statements or assumptions that are regarded as true. The truth of the facts guarantees the truth of the conclusion that is reached through deductive reasoning.

Example 1.1.5

Using Deductive Reasoning

Use deductive reasoning to choose the correct conclusion for each argument.

a. My state requires all lawyers pass the bar exam to practice legally. If I do not pass the bar, then I _____

   1. will not be able to represent someone legally in my state.

   2. will have to take the exam again.

b. The tutoring center at my college is offering free exam reviews to students. I am a student and I plan on reviewing for my sports management exam, so I

_____

   1. will pass my exam.

   2. will not have to pay anything for this service.

c. My boss said the person with the highest sales would get a promotion at the end of the year. I generated the highest sales, so I _____

   1. am anticipating my promotion.

   2. am the highest earner in my workplace.

Solution

a. Since the state requires all practicing lawyers to pass the bar, it is impossible to practice law legally without first passing the bar exam. Although one might consider taking the exam again, it is not a definite conclusion. Consequently, the correct answer is conclusion 1.

b. The tutoring center is offering free reviews to students. I satisfy all of the criteria to be eligible for that free tutoring. Neither of the premises ensure that I pass the exam. So the correct answer is conclusion 2.

c. Since my boss promised a promotion to the best salesman, my having the highest sales would lead me to conclusion 1, if my boss keeps their word. However, neither premise guarantees that I earn the most, ruling out conclusion 2. Thus, the correct answer is conclusion 1.

There are many popular games that use deductive reasoning. One of those is Sudoku, a Japanese puzzle game based on the logical placement of numbers. The goal of Sudoku is to fill a 9 x 9 grid with numbers so that each row, column, and 3 x 3 section contain all of the digits between 1 and 9, inclusive. The only catch is, no digit can be repeated in any row, column, or 3 x 3 section. In the Sudoku grid in Figure 1.1.4, consider which number must go in the highlighted square, where the first column and fourth row intersect.

6		7	1		5		9	
5	9	4		2		1	6	
8		3	9	4				
			6		4			
3	6					4		
2	4	8			7			
				6	9		3	
7			5		1	2		
9	8	2	4		3	6	5	

Figure 1.1.4: Sudoku Grid Partially Filled In

Notice that in the first column, only the digits 1 and 4 are missing. Using deductive reasoning, we know that one of those numbers (1 or 4) must then go in the highlighted square. If we look across the fourth row, we can see that the row already contains the digits 6 and 4. In order to not have duplicates, deductive reasoning tells us we must rule out the number 4. Thus, the only number that can go in the square is the number 1. Example 1.1.6 looks at solving a portion of a Sudoku puzzle using deductive reasoning.

Example 1.1.6

Using Deductive Reasoning

Use deductive reasoning to complete the first three rows of the Sudoku puzzle.

4	2	7	5		8	9	1	3
9	1	5	3	4			8	7
6	8		1	9	7	2		4
7	6			1	9		3	8
1		2	6	8	5	4	7	9
5	9	8	7		4	1	6	
3			8	5	1	7		
8	7	1	9		6	3	4	5
2	5	6	4	7	3		9	1

Figure 1.1.5

Solution

When completing a Sudoku puzzle, there are often many options of where to begin. We'll start in the top left-hand 3 x 3 section. Notice that in that section, there is only one number missing: the number 3. So this means that 3 must go in the intersection of the third row and third column, as shown in Figure 1.1.6.

4	2	7	5		8	9	1	3
9	1	5	3	4			8	7
6	8	**3**	1	9	7	2		4
7	6			1	9		3	8
1		2	6	8	5	4	7	9
5	9	8	7		4	1	6	
3			8	5	1	7		
8	7	1	9		6	3	4	5
2	5	6	4	7	3		9	1

Figure 1.1.6

Notice that this leaves only one number missing in the 3rd row: the number 5. So we can fill in that square with a 5, as shown in Figure 1.1.7.

4	2	7	5		8	9	1	3
9	1	5	3	4			8	7
6	8	3	1	9	7	2	5	4
7	6			1	9		3	8
1		2	6	8	5	4	7	9
5	9	8	7		4	1	6	
3			8	5	1	7		
8	7	1	9		6	3	4	5
2	5	6	4	7	3		9	1

Figure 1.1.7

Next, notice that the 6th column is only missing one number: the number 2.

4	2	7	5		8	9	1	3
9	1	5	3	4	2		8	7
6	8	3	1	9	7	2	5	4
7	6			1	9		3	8
1		2	6	8	5	4	7	9
5	9	8	7		4	1	6	
3			8	5	1	7		
8	7	1	9		6	3	4	5
2	5	6	4	7	3		9	1

Figure 1.1.8

When we place that number, as shown in Figure 1.1.8, we know the 1st and 2nd rows have every value except 6, thus the last two missing squares must both be 6s. The first three rows are now complete in Figure 1.1.9.

4	2	7	5	6	8	9	1	3
9	1	5	3	4	2	6	8	7
6	8	3	1	9	7	2	5	4
7	6			1	9		3	8
1		2	6	8	5	4	7	9
5	9	8	7		4	1	6	
3			8	5	1	7		
8	7	1	9		6	3	4	5
2	5	6	4	7	3		9	1

Figure 1.1.9

Verify for yourself that each of the top three rows and the top three 3 x 3 boxes have the values 1 through 9. We leave the remainder for you to complete now that you have the idea of how to solve a Sudoku puzzle.

To illustrate the difference between inductive and deductive reasoning, Example 1.1.7 considers a process and evaluates the conclusion from both reasoning perspectives.

Example 1.1.7

Using Inductive versus Deductive Reasoning

Consider the following process and its conclusion.

Choosing a positive integer, multiply it by 2, then add 1 to the product.

The result will be an odd number.

a. Evaluate this conclusion as an inductive argument.

b. Evaluate this conclusion as a deductive argument.

Solution

a. To evaluate the conclusion with an inductive perspective, we will begin by looking at some specific examples using the numbers 3, 6, and 8. Table 1.1.1 shows the results for each number.

Table 1.1.1

Positive Integer	Arithmetic	Result
3	$3 \cdot 2 + 1$	7
6	$6 \cdot 2 + 1$	13
8	$8 \cdot 2 + 1$	17

We can see that choosing a positive integer, multiplying it by 2, and adding 1 to the product does result in an odd number in each of these three examples, regardless of whether we choose an even or and odd positive integer. Since the conclusion (getting an odd number as the result) is a generalization based on three specific examples, we have used inductive reasoning to show that the conclusion is likely correct.

b. To evaluate the conclusion using deductive reasoning, we will introduce the variable x to represent any positive integer we might use for the process. The first step is to multiply the integer by 2, so we now have $2x$. Then we are to add 1, so our number can now be expressed as $2x + 1$.

Using deductive reasoning, we can apply two mathematical facts:

1. Any number multiplied by 2 is always an even number.

2. When you add 1 to an even number, you will always get an odd number.

Therefore, we can assert that given any positive integer x, when x is multiplied by 2 and then added to 1, the result will be an odd number. We can now use this general statement (which we have just proven to be true) and apply it to specific numbers.

Skill Check Answers
1. $100; 4 + 12n$ **2. c.** -8 and -8

1.1   Exercises

1. _____ reasoning starts with specific examples or observations and comes to a general conclusion.

2. _____ reasoning starts with accepted facts and comes to a specific conclusion.

3. True or False: If every pair in a sequence of terms differs by a multiple of two, then the sequence is an arithmetic sequence.

4. True or False: If a single counterexample to a line of reasoning is found, then the line of reasoning is false.

💡 PRACTICE

Identify whether each statement is an example of inductive or deductive reasoning.

5. The sides of all squares are proportional. The three quadrilaterals shown are all squares. Therefore all of their sides are proportional.

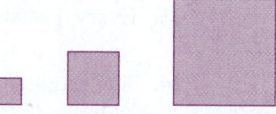

6. If Jessica plays basketball and makes 7 out of every 12 free throws, then she should make 35 out of every 60 free throws.

7. I have an 8:00 a.m. math class on Tuesdays and Thursdays. Each class day, I leave for class in my car at 7:30 a.m. Every day that the drive to campus takes 15 minutes, I arrive to class on time. Therefore, if I leave for class at 7:30 a.m. today and the drive to campus takes 15 minutes, I will be on time.

8. My grade on the first test in Quantitative Reasoning was 85, so I will make a B in the course.

9. If you live in New York, you are a resident of the United States.

10. All squares are rectangles, and all rectangles have four sides. Therefore all squares have four sides.

11. All known planets travel about the sun in elliptical orbits; therefore all planets travel about the sun in elliptical orbits.

12. I spent $76.54 on groceries last week and $77.23 on groceries this week, so I will spend less than $80 on groceries next week.

Find the missing terms of each sequence and determine if the sequence is arithmetic, geometric, or neither. If the sequence is arithmetic, state the common difference; if the sequence is geometric, state the common ratio.

13. 1, 3, 5, 7, ____ , ____ , ____

14. 1, 5, 9, 13, ____ , ____ , ____

15. 15, 10, 5, 0, ____ , ____ , ____

16. 5, 10, 20, 40, ____ , ____ , ___

17. $1, \dfrac{1}{2}, \dfrac{1}{4}, \dfrac{1}{8},$ ____ , ____ , ____

18. 3, 5, 8, 12, 17, ____ , ____ , __

19. 10, 100, 1000, 10,000, ____ , ____ , ____

20. 1, 3, 4, 7, 11, 18,

21.

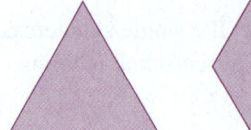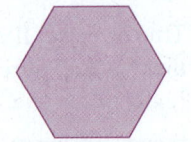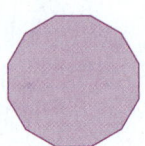

22.

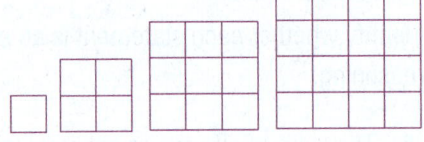

Find a counterexample to each statement.

23. Every Tuesday is an even day of the month.

24. The product of two numbers is always even.

25. If the difference between two numbers is even, then the numbers are both even.

26. If a geometric figure has 4 sides, then the figure is a square.

27. If $a > b$ and $a > c$, then $b > c$.

28. If the sum of two numbers is odd, then both numbers are odd.

Use deductive reasoning to choose the correct conclusion for each argument.

29. A grocery store requires a membership card to receive exclusive discounts on groceries. I do not have a store membership card, so I ____.

    a. cannot buy groceries from this store.

    b. cannot receive the exclusive discounts on groceries from this store.

30. A local animal shelter only allows residents of the county it is located in to adopt animals from the shelter. Julian is a resistant of the county the shelter is located in, so Julian ____.

    a. is allowed to adopt an animal from the shelter.

    b. will adopt a puppy from the shelter.

31. Chicken that is baked in an oven needs to reach an internal temperature of 165 °F to be considered fully cooked. A piece of chicken that was baked in the oven has an internal temperature of 150 °F, so the piece of chicken ___.

   a. needs to go back into the oven.

   b. is not fully cooked.

32. An album receives gold status after attaining the equivalent of 500,000 sales. A musician's album passed the equivalent of 500,000 sales last week, so the album ___.

   a. will receive gold status.

   b. was well received.

Figurate numbers are numbers that can be represented by any regular geometric figure. (A regular geometric figure is a figure such as a triangle, pentagon, or hexagon where all sides have the same length.)

33. Triangular Numbers: Triangular numbers are numbers that can be represented by an equilateral triangle. The first three triangular numbers, 1, 3, and 6, are shown in the figure. Find the next two triangular numbers and represent them with dots in the shape of a triangle.

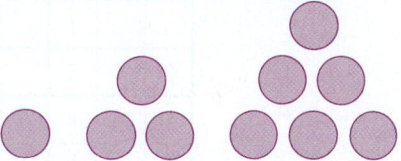

34. Square Numbers: Square numbers are numbers that can be represented by a square. We discussed the first five square numbers, 1, 4, 9, 16, and 25, in Examples 2 and 5. Determine the 6th and 7th square numbers and represent them using squares in the shape of a square.

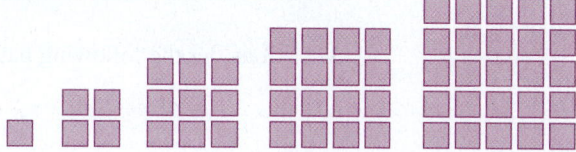

Complete the following Sudoku puzzles.

35.

9	3		5	6	8	1	2	
	2	6	7	1		5	9	3
	5	7	9		3	4	6	
	7	8	1	5	9	3	4	6
6	4	1	3	8		2	5	9
3	9		6	4		7	8	1
5	6	3	4		1	8	7	2
	8		2	3	5	6	1	
4	1	2					3	5

36.

	5				9	3		6
7		9		5	6	8		1
4	6	8	3	7	1			5
	8	7		2	4		5	9
		1		6	3	4	2	8
2	4	6		9		7		3
9	1	4	6	3	7	5	8	2
	2	5	9		8	1	3	7
			5	1	2	9		4

✏ WRITING & THINKING

37. Consider the following pattern.

$$1 \cdot 12 = 12$$
$$11 \cdot 12 = 132$$
$$111 \cdot 12 = 1332$$
$$1111 \cdot 12 = 13,332$$

a. Based on this pattern, predict the product $111{,}111 \cdot 12$.

b. Check your prediction by finding the product.

c. What type of reasoning did you employ?

38. Consider the following information.

Multiplication Patterns		
Multiplication	**Repeated Addition**	**Sum**
$4 \cdot -2$	$-2 + (-2) + (-2) + (-2)$	-8
$3 \cdot -7$	$-7 + (-7) + (-7)$	-21
$5 \cdot -6$	$-6 + (-6) + (-6) + (-6) + (-6)$	-30

What can you conclude about the sign of the final sum based on multiplication?

1.1 PROJECT

THE LOGIC OF NONOGRAMS

In Section 1.1, you learned about solving problems using logic and you explored a variety of situations and a logic puzzle. Another type of logic puzzle is the nonogram. Nonograms are picture logic puzzles and were created by Non Ishida in 1987. A beginner level 6 x 5 nonogram is shown here.

The numbers across the top and along the side indicate how many squares in the columns or rows should be shaded to solve the puzzle. For instance, the 5 in the first row indicates that 5 consecutive squares should be shaded in that row. A space between numbers indicates that there is a gap of *at least* one square between the consecutive shaded squares. For instance, 1 1 indicates that there are two single shaded squares that are separated by at least one empty square. When a nonogram is solved, the solution creates a picture.

There are several ways to approach solving these puzzles, but we'll only cover a few in this project. As you answer the questions, shade any squares you are confident about in the puzzle. If you know a cell will *not* be shaded, you may place an X in that square.

1. While we can start anywhere with solving this puzzle, we'll start with the largest numbers. The third column indicates that 5 consecutive squares should be shaded. Can we guarantee any of the squares in this column will be shaded? Explain your reasoning.

2. The top row indicates that 5 of the 6 squares in the row should be shaded. Since the 5 shaded squares are consecutive, which squares can we guarantee will be shaded? Explain your reasoning.

3. The second row indicates that there are three single squares shaded with at least one empty square between each. What is the minimum number of consecutive squares that are needed to complete the indicated pattern? With the information we have so far, which squares in this row can we guarantee will be shaded? Explain your reasoning.

4. Continue solving the puzzle, explaining your steps as you go. (**Hint:** Look at the column with 4 shaded squares next.)

5. Describe the shape created by the shaded solution.

6. Now that you know the basics of solving nonograms, solve one of the following slightly harder puzzles. As you work, describe the logic you used to determine which squares are shaded. Then, describe the shape created by the shaded solution.

Option 1

						1	1					
					1	3	1	1				
		1			5	1	1	1	3	1		
	4	4	8	2	1	1	1	1	1	1	1	7
1 1 5												
3 1 1												
1 1 1 2 1												
3 1 1 1 1												
2 1 1 1 1												
2 1 1 1												
2 3 1												
2 1												
7												

Option 2

	1		1		1				
	2	4	6	7	3	3	3	7	1
2 2									
3									
2 1 2									
4 1									
6									
6									
6									
1 1									
1 1									

1. Estimate values using given information.

Enrico Fermi was an Italian American Nobel winning physicist who created the world's first nuclear reactor. Born in Rome, he later moved to New York and then on to Chicago. Fermi was known as an inspiring teacher and his method of getting quick approximations became informally known as the *Fermi method*.

Figure 1.2.1

OBJECTIVE 1 Estimating

To estimate means to make an approximate calculation or to roughly judge the value of a quantity. An estimate is not just any "shot in the dark" guess; it is a guess based on the best information that you have, aimed at being as accurate as possible. Easy, right? Sometimes it is, like when estimating your grocery bill or how long it will take to complete a journey. But sometimes we face questions of accurately estimating when a volcano will erupt, if a war will break out, or how many people will order a particular dish at a restaurant on a given night. These types of estimations are not as straightforward.

One of the most famous estimators was the physicist Enrico Fermi. He is legendary for asking his students at the University of Chicago to estimate the number of piano tuners in Chicago using only what they already knew. Without cellphones, the internet, or even a physical phone book, he encouraged his students to see their capacity for finding an answer to the question with little or no facts. The solution goes something like this:

1. Estimate Chicago's population.

2. Estimate the number of households in Chicago.

3. Guess how many households have pianos.

4. Estimate how often pianos are tuned.

5. Estimate how long it takes to tune a piano.

6. Estimate how many pianos a tuner can tune in a week, and *voila*!

The key is to unravel the original question into more questions that explore the details of what would need to be true for the event to happen. If you want to know the number of piano tuners, then knowing the number of pianos would be helpful; and to know the number of pianos, knowing the size of the population is a good start. The reason for Fermi's question was not that he wanted to explore whether opening a piano tuning business would be lucrative in Chicago nor was it simply curiosity. Instead, he asked the question to help his students hone their estimating skills so they could employ them in much more serious scenarios later. Fermi was famous for this type of back-of-the-envelope calculation, which was more than just a guess. As he watched the first atom bomb explosion from 10 miles away, he threw a handful of torn paper bits into the air and watched how far the shock moved them. From their motion, he estimated the energy of the explosion to be over ten kilotons of TNT. Months later, the actual blast turned out to be 18.6 kilotons—Fermi had made a reasonably accurate estimation within moments of the explosion!

You might be asking yourself, since we have the internet at our fingertips, why bother estimating? The reason for estimating is that there are still things that we (or the internet) cannot know exactly but we need a rough estimate for. For instance, consider life insurance companies. They are in the business of estimating disability and death probabilities for individuals, which is something you can't find on the internet. This doesn't mean that they know precisely when a person will die, but based on a person's profile, they can estimate how long that person is likely to live. The insurance companies that are good at estimating these unknown values have been in business a long time and the companies that were not so good have gone bankrupt!

Fermi would also encourage us to continually ask ourselves if the answers we find are reasonable. Think about the moment when your waiter brings your bill for the evening's meal. How do you know the waiter correctly calculated the total amount due? Usually, a computer handles the actual addition, but a human is responsible for the input of the items into the computer. This is where errors tend to happen. Was everything you ordered on the ticket? Are there charges for items you did not order? If you have an estimated total in mind for your bill before the ticket is brought, you'll be ready to catch any potential errors.

Fortunately, the ability to make good estimates based on foresight is not a "you have it or you don't" talent. As Philip Tetlock and Dan Gardner explain in their book *Superforecasting: The Art and Science of Prediction*, estimating (or foresight) is a skill to be learned and one that sharpens with practice. "Foresight isn't a mysterious gift bestowed at birth. It is the product of particular ways of thinking, of gathering information, of updating beliefs. These habits of thought can be learned and cultivated by any intelligent, thoughtful, determined person."

As we begin the estimating process, one of the first things to decide is how precise we want the estimate to be. If we are estimating a food bill, rounding to the nearest dollar or ten dollars is probably sufficiently detailed. However, if we are estimating the cost of building a new home, rounding to the nearest ten thousand dollars might be adequate. In fact, contractors often suggest estimating with a 10% margin, to account for unexpected issues that may arise.

Estimating also involves rounding numbers in calculations. We can round numbers before any calculations are done or wait and round at the end of all the calculations. Keeping in mind the particular situation, it's best to choose the method where the numbers are easiest to work with. For instance, when making purchases, it might actually be better to round up at the beginning and overestimate the total price rather than underestimate it and be short on money. However, overestimating the amount of income you'll earn from working six hours of overtime next week could be disappointing on payday.

Let's pause and note that the accuracy of some estimations are more significant than the accuracy of other estimations. If you are ordering a catered meal for a wedding reception, getting an accurate estimate of attendance is important so that no one is sitting without a plate of food. However, incorrectly estimating how many people attend a pop-up free concert is likely to have relatively minor consequences.

Another tip for estimating is to think about the *order of magnitude* of any answer you expect to get. This is often referred to as a "ballpark answer." For instance, when estimating the amount of people attending a concert, to judge the order of magnitude is to consider whether the answer to "how many people will attend" is likely to be in the tens, hundreds, or the hundred-thousands. While this type of thinking may seem simple to some people, it is often surprising how unnatural it feels for many people. Narrowing down an answer to a ballpark figure can help put some perspective on the calculation—estimating now doesn't feel as overwhelming as choosing from any possible number.

Recently, a friend of mine observed exactly this unexpected phenomenon. My friend was considering accepting a job in a new city and wanted to compare the cost of living given their new proposed salary package. Since they were expecting their first child, determining the cost of having a baby at the new local hospital was of importance to them. Could they afford it? Had they saved enough? When they inquired at the hospital, they couldn't get anyone to commit to answering the question, "What does it cost to have a baby at your hospital?" Most of the hospital staff focused on quoting the precise cost, which of course varies widely. So, my friend was told, "It depends." One employee offered, "If there are complications the cost goes up." But up from what initial amount? Others said, "Your insurance will determine the cost." What my friend really wanted to know was the ballpark figure that they should expecting—was it $500, $5000, or $50,000? When they rephrased their request as wanting an order of magnitude, they immediately got an answer: "Oh, an average, births usually cost around $22,000 here." This reply helped my friend plan accordingly.

Determining an order of magnitude often depends a lot on a person's individual knowledge of the subject in question. If you are interested in space, estimating the distance to the moon might come easily for you. However, if you have no formerly learned context about a topic, then guessing any information can be quite a hurdle. The trick to determining order of magnitude is to try to compare what you don't know to something that you do know.

Pulling from past knowledge, thinking about order of magnitude, breaking questions into smaller ones, and rounding are all tools to sharpen our estimating skills. However, it's important to remember that when we estimate, there isn't one correct answer or method.

≔ TIPS FOR ESTIMATING

1. Start with something you know about the topic or the closest thing you know to the topic.

2. Have a rough order of magnitude. For example, will it take 5 days, 5 weeks, 5 months, or 5 years?

3. Use past experiences to find context.

4. Break the original question into simpler, more manageable questions.

5. Round numbers to make calculations easier.

6. Sample a small area and scale it up by multiplying.

Example 1.2.1

Estimating Order of Magnitude

Determine which piece of information would be most helpful for you to know in order to make each estimation. Then estimate an appropriate order of magnitude.

a. Estimate the depth of the deepest known point of the ocean.

1. The height of the tallest mountain

2. The radius of Earth

3. The distance a boat has to travel to cross each ocean

4. The distance of the longest river

b. Estimate the weight of an empty commercial passenger jet.

1. The length of a commercial passenger jet

2. The weight of an empty school bus

3. The density of steel

4. The weight of an empty passenger train car

Solution

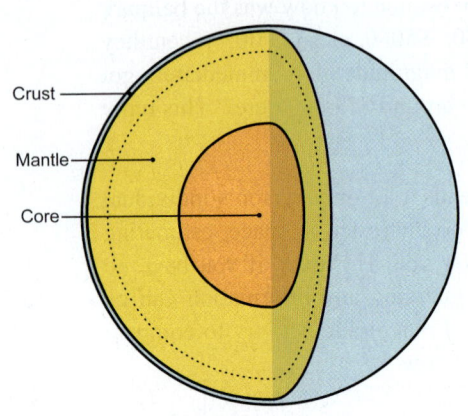

Figure 1.2.2

a. When we think about what information might be most useful, we can eliminate some of the answers right away. Knowing the distance of the longest river will not provide any context for the depth of the ocean. The same is true for the distance across each ocean—it could be that the ocean is very wide, but shallow. So knowing the values for options 3 or 4 would not be very useful. From science class, you might remember that the ocean, even though very deep, only accounts for a very small portion of Earth's overall composition. Knowing the radius of Earth might help you, a small amount, if you had some other facts like the depth of the other layers of Earth, as shown in Figure 1.2.2. However, suppose you knew the height of the tallest mountain on Earth, Mount Everest. We could reasonably guess that the ocean might be as deep as the highest mountain.

So knowing that Mt. Everest is about 29,000 feet high gives us a good magnitude of scale. We could guess that the deepest part of the ocean is somewhere in the tens of thousands of feet deep. The actual deepest part of the ocean is 36,161 feet. So we indeed have chosen the right magnitude of scale.

b. Again, we are looking for the most helpful piece of information for an estimate of scale. Even though we may not be familiar with all of the given choices, we might need to just make an educated guess. Looking through the list of facts, the least useful seems to be length of the commercial jet. Although that gives us a sense of scale, without other facts, it doesn't help us with the weight of the jet. Thinking about the construction of the jet compared to that of a train, we can determine that a passenger train car is the wrong type of comparison. (A train is built with much heavier materials than a jet.) That leaves us with the empty school bus and the density of steel. You might find the density of steel to be useful, but with that fact alone we have no sense of the dimensions of the aircraft. If we think about an empty school bus, one might guess that its construction makeup is the closest to the jet's.

We can guess that one passenger jet can hold approximately the same number of people as four school buses. If we knew the weight of the bus, we could multiply that by four to have the magnitude of scale. An empty school bus weighs about 25,000 pounds. So our guess would be that an empty passenger jet weighs in the hundred-thousand-pound range. In actuality, an empty passenger jet weighs 90,000 pounds on average. Again, we had a very good estimate!

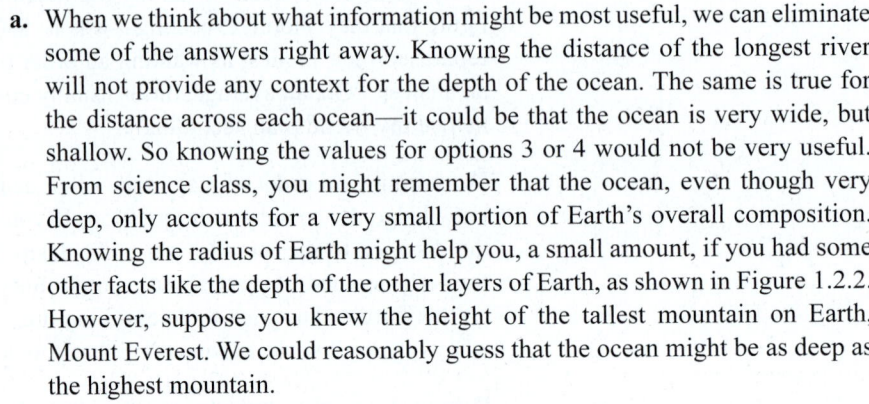

✔ **Skill Check 1.2.1**

Choose the best order of magnitude.

How many countries belong to OPEC (Organization of the Petroleum Exporting Countries)?

1. Single Digits

2. Tens

3. Hundreds

4. Thousands

Example 1.2.2

Using Past Experience in Estimation

A public university is expanding its campus by building a new Mathematics and Computer Science Center. Estimate the cost of the new center by using the cost specifications of the most recently constructed building on campus, the library. The library had 20,000 square feet of assigned space, with a construction cost of $6,000,000. The new center needs to contain 40,000 square feet of assigned space.

Solution

The problem asks for an estimate of the building cost for the new center based on previous figures. State-funded universities build all academic buildings at roughly the same level of quality. Since we are given the construction cost of the first building, using deductive reasoning we can reasonably conclude that the new center will cost approximately the same amount per square foot. Because we are doubling the square footage of assigned space in the new center, we will need to double the construction cost. Using the information we are given, a good estimate for the cost of the new Mathematics and Computer Science Center is $12,000,000.

Example 1.2.3

Breaking Down the Question

Suppose you are planning a wedding reception for 100 people. This is your first time planning such an event. Estimate the cost of the reception.

Solution

If we are estimating the cost of the reception, then we need to account for all of the possible expenses we might encounter and add the approximate expenses to find an estimate for the total. There are several factors that go into the cost of the reception. Breaking the overall cost into smaller segments by listing as many of these as possible is a great place to start.

How many cost factors can you name? If you're not in the reception business, you might seek the help of an online planner or other resource so that you don't leave anything out. Suppose an internet search for "planning a reception checklist" produced the following factors.

- Venue cost
- Food
- Décor
- Entertainment

First of all, you need a venue big enough to comfortably hold 100 people. Since this cost varies based on location, it is best to get a few rough estimates—again, these estimates can be obtained online. According to one online report, couples spend between $3000 and $11,000 on average for their wedding reception venue. Since all we need is an estimate, we can take the middle of these two numbers as our average: $\dfrac{\$3000 + \$11,000}{2} = \$7000$. However, keep in mind that some venues include the cost of food in the total cost of the venue, while other venues do not include the cost of food and may still cost tens of thousands of dollars.

Food for a reception is usually based on a *per person* amount. As a result, changing the number of guests could drastically change the total cost of the food. Wedding-cost estimator sites give an average cost for 100 people in the range of $45 per person, which is $4500 in total for 100 people. Again, keeping in mind that we're looking for an estimate and not an exact figure, this will help keep us moving forward on finding the estimate and keep us from being bogged down in finding the least expensive price per head.

Decorations can be anything from very simple to spectacular! However, the average reported cost is in the $600 to $800 range for reception decorations and flower arrangements. So we'll take the average again and use $700 for our estimate.

Finally, assuming entertainment is a desired feature of the reception, we'll need to make an initial decision based on some quick research. Having a DJ or single musician ranges from $300 to $700 for the reception, while having a live band ranges from $900 to $1600. It might seem like averaging the two of these would give a good general idea, but stop and think for a minute. An average of the lowest and highest amounts would give us an amount of $950. This amount is either far too much for the first category, or just barely enough for the higher category. So instead of being off in either direction, we will simply choose one of the categories and mention it in our estimation. For now, choosing the less expensive option of having a DJ means that the average entertainment cost is $500.

Now we're prepared to combine all of the individual estimates in order to arrive at a total estimate for a wedding reception.

Venue cost	$7000
Food	$4500
Décor	$700
DJ	$500
Total	$12,700

So a reasonable estimate for the cost of a wedding reception could be $12,700.

✔ **Skill Check 1.2.2**

Using the estimation prices given in Example 1.2.3, what is the estimated cost for the same reception if the number of guests is only 50? Or 200 (assuming that all venues in our previous calculations could accommodate 200 people)? Think carefully about how you might arrive at the estimates.

Example 1.2.4

Estimating by Rounding

The average adult has between 4500 mL to 5700 mL of blood in their body. When you donate blood, approximately 473 mL is taken. On average, the local blood bank center collects 43 blood donations per day.

a. Estimate the amount of donated blood the center collects in one week.

b. Estimate the amount of donated blood the center collects in a year.

Solution

First, notice that there is information in the question that we don't need to use. Although interesting, the amount of blood the average adult has in their body does not help us estimate the amount of blood donated at a blood center. What we are concerned with is the amount of blood that is taken (473 mL) and how many donors there are (43).

a. We'll round each of these numbers so that the multiplication is easier. Remember that when we round, we are making a trade-off for accuracy. The more we round, the less accurate our answer will be. We could round the blood amount to 480 mL for greater precision or round to 500 mL for an easier calculation. The same is true for the 43 donors. We can round down to 40 for a minimum estimate or up to 50, which is likely an overestimate. As a compromise to increase accuracy of the estimate, let's round the blood amount up to 500 mL and the donor count down to 40. We will also assume that the center collects blood all 7 days of the week. We estimate the amount the center will collect per week as follows.

$$\frac{500 \text{ mL}}{1 \text{ donation}} \cdot \frac{40 \text{ donation}}{1 \text{ day}} \cdot \frac{7 \text{ days}}{1 \text{ week}} = 140,000 \text{ mL per week}$$

b. We can scale up the estimate we found in part a. to guess the amount of blood the center collects in a year. Since there are 52 weeks in a year, we can multiply by 140,000 mL to estimate that the center collects.

$$\frac{140,000 \text{ mL}}{1 \text{ week}} \cdot \frac{52 \text{ weeks}}{1 \text{ year}} = 7,280,000 \text{ mL per year}$$

Notice that both estimates assume the blood center collects blood every day of the year—even on holidays—and has roughly the same intake of donors per day. It is unlikely for that to be the case, so we could say that both estimates are upper estimates for the amount of blood the center collects. However, both answers are certainly in the right magnitude of scale.

💬 **Think Back**

Only consider the digit directly to the right of the place you are rounding to. Round down for digits less than 5. Round up for all other digits.

✔ **Skill Check 1.2.3**

Estimate how many people ride New York City subway system every year if approximately 4.3 million people ride it every day.

Example 1.2.5

Estimating by Scaling

At most concerts, the floor area immediately in front of the arena stage is designated as standing room only. Using an aerial camera view, a local news crew counted 27 people in an area that was approximately $\frac{1}{11}$ of this portion of the floor. How could the news station estimate the number of people in the standing-room-only section?

Solution

A common technique to estimate crowd size is to divide the area of people into equal sections and then determine the number of people in one of those sections. Once this value is known, the amount of people in one section is multiplied by the total number of sections. In this case, the news crew counted 27 people in one section and estimated that there were 11 sections. We can multiply the two numbers together to get an estimate for the size of the crowd in front of the stage.

$$27 \cdot 11 = 297$$

Thus, the news could reasonably report there were approximately 300 people in the standing-room-only section.

One final thought on estimating. It turns out that humans are better at estimating as a group than we are as individuals. If a group of people each make an estimate that is not influenced by the others in the group, then the average of their estimates is more accurate than any single estimate—even an estimate by an expert in the field. This phenomenon is called the *wisdom of the crowd*. Francis Galton is credited with first noting this phenomenon in 1907 when he published an article entitled *Vox Populi* in the March 7th issue of *Nature*, a weekly journal of science. At the West of England Fat Stock and Poultry Exhibition, he realized that the average of all the entries in a "guess the weight of the ox" competition was amazingly accurate. It was better than most individual guesses and those of alleged cattle experts. Although there still is controversy over his calculations, the core of his conclusion was the beginning of the study of *collective intelligence*.

Scholars agree that the strength in collective intelligence lies in diversity. Having opinions from different angles pushes the answer so that it comes closer to the truth. For example, consider a group of students who were asked to estimate the number of stories in the tallest building in Dubai. The individual guesses ranged from 150 floors to 250 floors. The actual height is 209 floors, so the average of their guesses turned out to be pretty accurate. But where did their estimations come from? The group that guessed on the lower side seemed to use the reasoning that although it was likely taller than the Empire State Building, which they knew to be around 100 floors, it was a beach area and people liked unobstructed views of the water. However, the group that predicted more floors reasoned that Dubai was a very wealthy community and would have huge new skyscrapers to show off their wealth. The students made different guesses based on their vantage points, but when combined, their estimates were very close to the truth.

Fun Fact

The Burj Khalifa, in Dubai, stands as the tallest skyscraper in the world with a height of 2722 feet, just over half a mile! It took the crown in 2010, beating out Taipei 101, the previous title holder from 2004–2009, which stands 1667 feet tall.

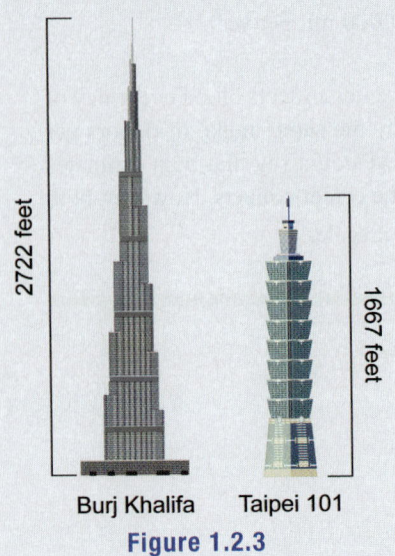

2722 feet

1667 feet

Burj Khalifa   Taipei 101

Figure 1.2.3

Example 1.2.6

Using the Wisdom of the Crowd

Thirty people were asked to estimate the attendance at an outdoor community concert to the nearest 10. Their estimates are shown in Table 1.2.1. Use the crowd-sourced data to make an estimate of the community concert attendance.

Table 1.2.1: Community Concert Attendance Estimates

370	210	310	270	300	260
270	220	270	320	220	330
240	280	230	300	310	200
210	260	240	250	260	300
200	260	310	340	350	290

Solution

One method of making an estimate could be to look for the number that was most commonly guessed. However, the wisdom of the crowds says that the average of many estimates is better than a single estimate. So instead of using a single number, we can add all of the estimates together and divide by the number of estimates (30).

The sum of all of the estimates is 8180. Dividing by 30, we have the following.

$$\frac{8180}{30} \approx 272.7$$

Thus, we should estimate that there were approximately 273 people at the community concert.

Skill Check Answers

1. Tens (In 2021, there were 13 member countries) **2.** Only the estimation for food will change; $10,450 for 50 guests and $17,200 for 200 guests **3.** Approximately 1.6 billion people

1.2   Exercises

✔ CONCEPT CHECK

1. The values 10 minutes, 10 hours, and 10 days are examples of
   _____.

2. Breaking the original question into _____ questions can make estimating easier.

3. True or False: Collective intelligence is used to make a more accurate estimate than a single person can make.

4. True or False: When making an estimate, you should never round the values in your calculations.

💡 PRACTICE

Determine the most appropriate order of magnitude for each situation.

5. Student enrollment at a university

6. Number of hamburgers sold worldwide every day

7. Number of flavors of ice cream sold at a local ice cream stand

8. Number of seats in a movie theater

For each estimation, determine which piece of information would be most helpful. Explain your reasoning.

9. Estimate the average gestation period of an Asian elephant.

   a. The average lifespan of an Asian elephant

   b. The average weight of newborn Asian elephant

   c. The average number of elephants born each year

   d. The average weight of a newborn human

10. Estimate the number of pounds of food served for lunch at a local elementary school each day.

   a. The average weight of a sandwich

   b. The average weight of the students

   c. The number of students who attend the elementary school

   d. The length of the lunch period

11. Estimate the total number of hours employees of a certain company spend in meetings per year.

   a. The number of employees who work in the office

   b. The number of meeting rooms in the office

   c. The average number of minutes employees spend in the breakroom each day

   d. The average number of vacation days employees get per year

12. Estimate the length, in minutes, of the audiobook recording of this year's best-selling fiction novel.

    a. The number of chapters in the novel

    b. The number of five star reviews the novel received

    c. The average amount of time it takes to read a single page out loud

    d. The average length of the audiobook recording of the top five best-selling fiction novels from the past five years

🚀 APPLICATIONS

Estimate the indicated cost.

13. A city is planning a new 3-acre park that will include a playground, a walking trail, and several picnic shelters. Previous parks the city constructed that were similar in amenities averaged 5-acres in size with cost an average of $900,000 to create. Estimate the cost to create the new park.

14. The Ziegler family plans to create a cement patio in their backyard to use for picnics and grilling. They'd like the patio to be a rectangle that is 18 feet by 20 feet. The previous year, they had a cement patio poured onto the side entrance of their house that was 7 feet by 4 feet, which cost $250. Estimate the cost of the new backyard patio. Round your answer to the nearest dollar.

15. A family of four plans to go on a seven-night vacation next summer. The previous summer, they went to a similar location for a four-night trip and spent $4500 total. Estimate the cost of a seven-night vacation.

16. The Wild Flower Foundation is planning their annual Bee the Difference fundraising dinner. The previous dinners had a maximum of 100 seats available and cost on average $8500 to host (including venue, food, and entertainment). This year, the foundation wants to increase the seating capacity to 175. Estimate the budget they should expect for this year's fundraising dinner.

Break each situation down into smaller parts.

17. You are hosting a birthday party for an 8-year-old child. A total of 20 children plus their parents will attend. You need to estimate the total cost of the party.

18. You are building a new house that needs three bedrooms, two bathrooms, and an attached garage. You need to estimate the total cost to build the house.

19. You and a friend are opening a restaurant. You need to estimate the upfront costs to run the restaurant for 6 months.

20. You are planning a destination wedding and will cover the travel expenses for members of the bridal party. You need to estimate the cost of the wedding.

Solve.

21. A local café sells beverages and pastries and is open for 8 hours each day. On average, customers spend $6.73 per order. The café averages 24 orders per hour.

    a. Estimate the amount of money made by the café in one day.

    b. Estimate the amount of money made by the café in one week.

22. A new video streaming service claims to have over 700,000 hours of video available for streaming. Each day, an average of 47,882 subscribers use the streaming service for approximately 126 minutes each.

    a. Estimate the number of hours of video that is streamed each day.

    b. Estimate the number of hours of video that is streamed each month.

23. A local food bank takes donations of nonperishable foods every day of the year. The average donation weighs 6.25 pounds, and the food bank receives approximately 32 donations per day.

    a. Estimate how many pounds of donated food the food bank receives per month.

    b. Estimate how many pounds of donated food the food bank receives per year.

24. A taco truck is open for business during a 4-hour window each day. They fulfill an average of 46 orders per hour and each order has an average of 12 tacos.

    a. Estimate the number of tacos that are served each day.

    b. Estimate the number of tacos that are served each month.

25. A farmer is estimating his expected crop yield for the season. He has a farm that is 250 acres, and he planted corn on 100 acres. He planted 31,000 seeds per acre and expects 95% of the seeds to germinate and grow into full stalks. Based on previous years, the farmer expects an average of one ear of corn per stalk. Estimate the number of ears of corn per acre and then estimate the expected corn crop yield for the season.

26. A local news network is estimating the number of people attending a parade based on aerial footage. A single block of the parade route is determined to hold 173 people. The parade route is 24 blocks long. Estimate the number of people who attended the parade.

27. A bakery held a contest for people to guess how many sprinkles are on the top of a cake. The winner would receive a $50 gift card. You determine that the cake is a half-sheet, meaning it measures 18-by-13 inches. You count that there are 31 sprinkles on a section of cake that is approximately one square inch in area. Based on this information, how many sprinkles would you guess are on the cake?

28. An editor is asked to give an estimate for the cost to edit a chapter of a book. To create the estimate, she needs to first determine the number of words in the chapter. The chapter is 56 pages long and the first two pages of the chapter each have an average of 472 words. How many words should she estimate are in the chapter?

29. Twenty people are asked to estimate the number of tacos sold by a local taco stand per day, rounded to the nearest ten. Their estimates are shown in the following table. Use the crowd-sourced data to make an estimate of the number of tacos sold per day by the local taco stand.

840	890	840	810	870
730	810	720	790	770
870	870	800	830	760
760	830	820	780	800

30. Twenty-five people are asked to estimate the number of people who attend a local corn festival per day, rounded to the nearest ten. Their estimates are shown in the following table. Use the crowd-sourced data to make an estimate of the number of corn-fest attendees per day.

810	960	870	1170	920
870	1030	860	1000	890
1160	920	1080	1060	1150
1070	1010	980	1120	820
1100	890	860	1100	1190

31. A contest is held to guess the number of jelly beans in a jar. The guesses of 20 people are provided in the following table. Use the crowd-sourced data to make an estimate of the total number of jelly beans in the jar.

282	312	328	331	311
373	341	328	284	308
277	353	277	389	275
385	280	286	374	277

32. Twenty-five college students are asked to estimate the number of minutes they think students spend on social media per day. Their estimates are shown in the following table. Use the crowd-sourced data to estimate the total number of minutes students spend on social media per day.

103	162	168	167	124
151	174	176	218	116
148	194	158	181	170
182	91	144	156	154
204	191	166	91	202

33. Suppose your project needs one project manager, two user-experience professionals, and three app developers. The rates for each of these team members are as follows.

    Project manager: $118.50 per hour

    User-experience professionals: $88 per hour

    App developers: $125 per hour

    The project requires the project manager for 100 hours, each user experience professional for 50 hours, and each app developer for 100 hours. Estimate the total cost of the project.

34. In his first week at college, Chase spent the following.

    Entertainment: $28

    Apparel: $24

    Travel (gas): $18

    If there are 12 weeks in the semester, estimate the amount of money Chase will spend on these things throughout the semester.

35. Trying to prepare a budget for his first few months in his new place, Darren has done research and found the following average costs per month: health insurance $183, phone/internet/cable $140, electric $179. If Darren will be earning $1700 per month, estimate the amount of money he will have after paying these bills each month.

36. Jane had a dream to complete an Ironman triathlon. The race consists of a 2.4-mile swim, a 112-mile bike ride, and a marathon run of 26.2 miles. In training, Jane can swim at an average pace of 1.9 mph. She rides her bike at an average of 20.8 mph, and runs at 4.1 mph. Estimate how long it will take her to complete the Ironman triathlon.

37. Because Elias is self-employed, he is advised to pay quarterly estimated federal tax payments since his income tax withholding will not fully cover next year's tax liability. Online advice instructs him to determine last year's tax, minus any withholding, and divide by four. Estimate the quarterly taxes Elias should pay if his total tax from last year was $23,741 and his total withholding was $7500.

38. Amelia is planning a road trip from Nashville, Tennessee, to Washington, D.C. The distance from Nashville to Washington is 651 miles and will take about 36 gallons of gas to complete.

    a. If gas costs an average of $3.745, estimate the fuel cost for the one-way trip.

    b. Use your answer for part **a.** to estimate the total cost of fuel for the round trip.

39. Debbie needs to purchase some printer paper and USB jump drives for the department where she works. Each ream of paper costs $4.55 and each pack of 4 jump drives costs $27.85. She needs nine reams of paper and six packages of jump drives. Approximate to the nearest dollar the amount of money Debbie will spend to make the purchase.

40. Marcel has decided to paint his new apartment. He estimates the cost of paint will be $385. Which estimate should Marcel use for the job so that his budget includes a contingency factor of 8% in addition to the cost of paint?

    a. $38

    b. $432

    c. $395

    d. $800

✏ WRITING & THINKING

41. Three students estimate the product $57 \cdot 159$. Explain which of the following estimations will be the most precise and why: $50 \cdot 100$, $60 \cdot 200$, $60 \cdot 160$.

1.2 PROJECT

ESTIMATING TUNERS AND TOILETS

The physicist Enrico Fermi once asked a class of students to estimate the number of piano tuners in Chicago. We'll walk through the same estimating process that he used to solve this problem. For each step, explain your thought process for the estimate or calculation. Do not look up any values during this project. Use your reasoning skills to estimate at each step along the way.

1. Suppose the city of Chicago has approximately 3 million people living in it. From this value, estimate how many households are in Chicago. (**Hint:** First estimate how many people are in a household.)

2. Estimate how many households in Chicago own a piano. Using this estimate, determine how many pianos are likely to exist in Chicago.

3. Next, estimate how often the average piano owner has their piano tuned per year.

4. Now, consider how many pianos a single piano tuner can tune per year. Assume that it takes an hour to an hour and a half to tune a piano. (**Hint:** Would the piano tuner work every day of the year? How much travel time might the tuner need between appointments?)

5. Use the estimates and educated guesses you've made so far to make one final estimate: how many piano tuners are in Chicago?

6. Now, use a similar thought process to estimate the number of restrooms the Pentagon has knowing the following information: the Pentagon has 7 floors and houses approximately 27,000 employees. Explain your entire thought process and how you arrived at the final number.

7. Compare your estimates to the estimates of other people in your class. Are the estimates similar? What assumptions might have resulted in answers that are not similar?

1.3 PROBLEM SOLVING: PROCESSES AND TECHNIQUES

OBJECTIVES

1. Use Pólya's problem-solving technique.

OBJECTIVE 1 Pólya's Problem-Solving Technique

A great discovery solves a great problem... Your problem may be modest; but if it challenges your curiosity and brings into play your inventive faculties, and if you solve it by your own means, you may experience the tension and enjoy the triumph of the discovery. —George Pólya, *How to Solve It*

George Pólya wrote numerous books on problem solving during his lifetime. His love for teaching and for nurturing the mind leaps off the pages of his work. Pólya believed that every one of his students could become a better problem solver and encouraged both teachers and students along this path, expecting that "having tasted the pleasure in mathematics, [one] will not forget it easily." One of Pólya's most famous and lasting contributions was a four-step process to problem solving. Although he focused his attention on the field of mathematics, the process can be employed in any situation or field. It consists of the following four steps.

Step 1: Understand the Problem

On its surface, this seem obvious, yet it is often the proverbial stumbling block for many people. Before you are able to solve a problem, you must first *understand* what is being asked. Pólya suggested asking questions to help identify the end goal.

- What is unknown?
- What are the data?
- Can I restate the problem?
- Do I understand all the words used in the problem?
- Can I draw a figure that helps?
- Is there enough information to find a solution?

To understand the problem means reading and rereading the problem to identifying pertinent information, discerning what is being asked, writing down important facts, and assigning names or variables to unknown quantities.

Step 2: Devise a Plan

Once you understand the problem, the next step is to find the connection between what is known and what is unknown, and then devise a method to determine the unknown. Pólya offers many strategies to devising a plan. You can look to problems that you have solved before (or related problems that could be useful) and use similar methods. You should look for any patterns in the data and consider different points of view. You should also ask questions to help propel you onward.

- Can you restate the problem?
- If you can't solve the entire problem, can you solve a part of it?
- Have you used all the data you were given?

Think Back

Recall that a variable is a representation of an unknown quantity or amount in a mathematical equation. We often solve equations for a given variable to determine its value.

The strategy or plan will vary since some problems may require multiple strategies and multiple attempts to find a solution. Here are some standard strategies to consider.

- Find a pattern.
- Adopt a new point of view.
- Solve a related problem.
- Work backward.
- Draw a picture.
- Estimate and test.
- Account for all possibilities.
- Use inductive reasoning.
- Use a variable.
- Make a list or table.
- Use trial and error.
- Solve a simpler problem.

Step 3: Carry Out the Plan

Carry out the plan set out in Step 2. Be persistent and carefully check each step. If the plan does not work, understanding why the plan fails may allow you to discover something new about the problem that you didn't understand before. This enables you to use that new information to devise a new plan. And take heart—this is the nature of mathematics. Failure is only a step toward success.

Step 4: Look Back

An essential part of problem solving is having the ability to be reflective in thought. Examine the solution you obtained. Ask yourself the following questions.

- Is my answer feasible and correct?
- What worked?
- What didn't work?
- Will the solution to this problem assist me the next time I see a similar problem?

Looking back with a critical eye on what you have done—what worked and what didn't work—helps you predict what strategy to use to solve future problems.

≔ PÓLYA'S PROBLEM-SOLVING TECHNIQUE

1. Understand the problem.
2. Devise a plan.
3. Carry out the plan.
4. Look back.

Remember that problem solving is a skill that improves with practice. The more you do it, the easier it becomes. Since there are no algorithms for problem solving, the process can seem challenging. Just dig in and know that we are all at different skill levels. A problem that might be difficult for one person may seem trivial for you. For example, if you were asked to find the product of 12 and 11, you would almost immediately arrive at a strategy that would allow you to find the solution. However, a student who is just learning the concept of multiplication might struggle quite a bit with a solution strategy. This is true at every level.

The following examples walk through a few different methods of problem solving.

Math Milestone

Hungarian mathematician George Pólya is best known for his book titled *How to Solve It*. Since its publication in 1945 it has been continuously in print and sold over one million copies.

Figure 1.3.1

Example 1.3.1

Working Backward

While three watchmen were guarding an orchard, a thief slipped in and stole some apples. On his way out, the thief met the three watchmen, one after another, and to each in turn he gave one-half of the apples he had, plus two more in addition to that. In this way, he managed to escape with one apple. How many apples had the thief stolen originally?

Solution

Step 1: Understand the Problem

After reading the problem, identify the pertinent information needed to solve the problem. We are told three important points: the thief had only one apple when he escaped the orchard, he met three watchmen, and he gave each of the three watchmen one-half of the apples in his possession, plus two more. We are asked to determine how many apples he stole originally.

Step 2: Devise a Plan

In this problem, let's try working backward from the ending value of one apple. If the thief gave away one-half of the apples plus two more per watchman, we need to add them back in for each watchman he met. Remember that when using a method of working backward, all of the arithmetic operations are done in reverse. We can use a table to help us keep track of the process.

Step 3: Carry Out the Plan

We have decided to work backward. Recall from the initial problem that the thief is dividing his apples in half and then subtracting two more from the number of apples in his possession each time he meets a watchman. To work backward, we must reverse the operations. So instead of subtracting 2 apples, we add 2. Instead of dividing the number of apples in half (that is, instead of dividing by 2), we multiply by 2. Knowing that the thief escaped with only 1 apple, we start with this value. The thief must have had $1 + 2 = 3$ apples, and then twice that amount, or 6 apples, when he met the final watchman. Using the same logic, we can continue to work backward to calculate the number of apples he had as he met each watchman in turn.

Table 1.3.1: Number of Apples

Phase of the Theft	Number of Apples
Escaped	1 apple
Met 3rd Watchman	(1 apple + 2 apples) · 2 = 6 apples
Met 2nd Watchman	(6 apples + 2 apples) · 2 = 16 apples
Met 1st Watchman	(16 apples + 2 apples) · 2 = 36 apples

Thus, the thief originally stole 36 apples.

Step 4: Look Back

The last step in the process requires that we consider the answer to determine if it is feasible. We can use our solution of 36 apples to see if the thief actually ends up with 1 apple on escape after meeting each of the three watchmen.

- He meets the 1st watchman and gives him one-half of his apples and 2 more.

$$36 \text{ apples} - \frac{1}{2}\left(36 \text{ apples}\right) - 2 \text{ apples} = 16 \text{ apples remaining}$$

- He meets the 2nd watchman and gives him one-half of his remaining apples and 2 more.

$$16 \text{ apples} - \frac{1}{2}\left(16 \text{ apples}\right) - 2 \text{ apples} = 6 \text{ apples remaining}$$

- He meets the 3rd watchman and gives him one-half of his remaining apples and 2 more.

$$6 \text{ apples} - \frac{1}{2}\left(6 \text{ apples}\right) - 2 \text{ apples} = 1 \text{ apple remaining}$$

When looking at our process, we should ask ourselves if what we have done makes sense and is feasible. Notice that we have done more than show our answer is feasible, we have shown that it works. So the answer to both of these is yes. We can have confidence in our answer that the thief originally stole 36 apples.

> **💬 Helpful Hint**
>
> Language we use casually every day may have precise meanings in mathematics—perhaps different from its everyday meanings. Here, for example, the following all have the same meaning in mathematics.
>
> Dividing in half
>
> Dividing in 2
>
> Dividing by 2
>
> Be careful: dividing by one-half is *not* the same thing.

Example 1.3.2

Using the Trial-and-Error Method

Arrange the numbers 1, 2, 3, 4, 5, and 6 in the circles of the given triangle so that the sum along each side is 12.

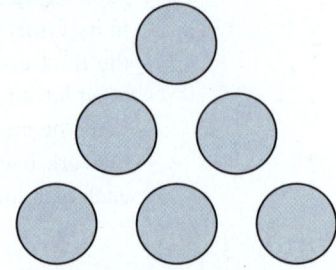

Figure 1.3.2

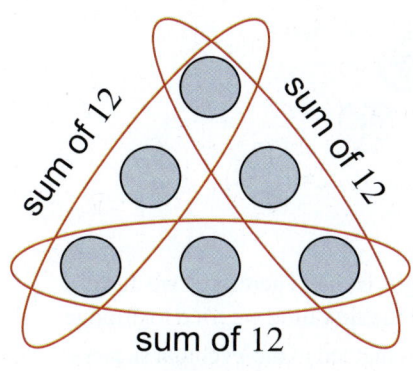

sum of 12

Figure 1.3.3

Solution

Step 1: Understand the Problem

Since we are given a figure to consider, understanding the problem must include understanding the figure. In this case, we are being asked to place each of the numbers 1 through 6 in the circles such that each side of the triangle has a sum of 12.

Step 2: Devise a Plan

Unless the answer is obvious to you, we most likely need to use the strategy of trial and error. Trial and error is an effective strategy for solving problems where you are given enough information to test your results immediately and the data is a manageable size. Trial and error is sometimes the best strategy for helping to determine where to begin in solving a problem, and it makes it possible to quickly eliminate erroneous solutions.

Step 3: Carry Out the Plan

Our first guess will be to place the numbers 1 through 6 in any order on the circles and add the numbers along each side to see what happens. We'll start by placing 1 at the top and then ordering the numbers clockwise around the triangle.

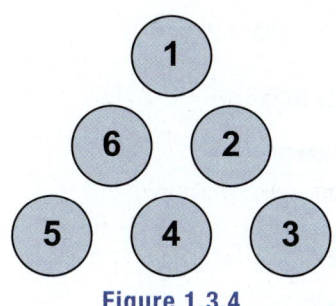

Figure 1.3.4

From our initial guess, we can see that the sum along two of the sides is 12, but not along all three sides.

$$1 + 2 + 3 = 6 \qquad 3 + 4 + 5 = 12 \qquad 5 + 6 + 1 = 12$$

We can use this information to formulate a better second guess. Consider the possibilities of acquiring a sum of 12 when using the numbers 1 through 6. The only possible sums of 12 using the numbers 1 through 6 are the following.

$$1 + 5 + 6 = 12$$
$$2 + 4 + 6 = 12$$
$$3 + 4 + 5 = 12$$

Looking at our three sums, we can see that the numbers 4, 5, and 6 are each used in two of the three sums, while the numbers 1, 2, and 3 are used only once. This indicates that the numbers 4, 5, and 6 should be placed in the corner circles of our triangle so that the numbers are used in two sums each.

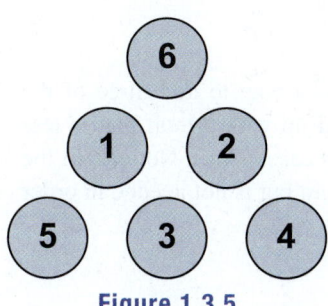

Figure 1.3.5

Step 4: Look Back

Now, we need to check to see that the sums of all of the sides are indeed 12. Now we can check our answer.

A magic square is a square arrangement of numbers in which the sums of all the rows, columns, and diagonals are equal. This magic square uses the numbers 2, 7, 12, 17, 22, 27, 32, 37, and 42 only once each. Determine the placement of the missing numbers in the given magic square.

17		7
12	22	
		27

Figure 1.3.7

$5 + 1 + 6 = 12$

$5 + 3 + 4 = 12$

$6 + 2 + 4 = 12$

Figure 1.3.6

So the sum of each side of the triangle is 12. Therefore, we have a solution. Notice that although we failed the first time we tried to arrange the numbers, we gained new information that led to our eventual success.

Example 1.3.3

Making a List

Stefan has four bills due this month. They are listed in Table 1.3.2.

Table 1.3.2: Stefan's Monthly Bills

Bill	Amount Owed
Water	$125
Internet	$80
Electricity	$170
Phone	$110
Total	**$485**

Stefan only has $350 available to pay his bills. Each company imposes a $25 late fee if the bill is not paid on time. Is it possible for Stefan to pay three of the four bills in order to incur only one late fee?

Solution

Step 1: Understand the Problem

The problem is asking us whether there is a way to add three of the values from the set of four values and end up with a total that is less than or equal to a particular number (in this case, $350). Notice that the amount of the late fee is given in the problem but is not needed in order to answer the question.

Step 2: Devise a Plan

To solve this problem, we can make a list of all the possible ways we can add up three of the four bills. We can then check the total to see if it is less than or equal to the amount Stefan has available. We will use a table here, but you can organize your list in any way you would like.

Magic squares are very ancient, dating back more than 400 years according to Chinese legend. They have been associated with mythical significance, have been the center of perfect harmony and balance, and have appeared in works of art.

Figure 1.3.8: *Melancholia I* by Albrecht Dürer

Step 3: Carry Out the Plan

There are four ways to group three bills at a time. Each of these combinations is listed in the leftmost column in Table 1.3.3.

After adding up the total of each of the possible groups of three, we see that the second row is the only one less than Stefan's available $350. This means that yes, it is possible for Stefan to cover three of his bills and only incur one late fee.

Table 1.3.3: Bill Combinations

Bills Covered	Sum	Total
Water, Internet, Electricity	$125 + $80 + $170	$375
Water, Internet, Phone	$125 + $80 + $110	$315
Water, Electricity, Phone	$125 + $170 + $110	$405
Internet, Electricity, Phone	$80 + $170 + $110	$360

Step 4: Look Back Looking back, we check for errors and see that we didn't make any. We also see that the largest bill, $170 for electricity, didn't make it into the final grouping. Perhaps we could have predicted that if any bill wouldn't be able to be covered, it would be the largest.

Example 1.3.4

Solving a Simpler Problem

Find the sum of the numbers 1 through 1000.

Solution

Step 1: Understand the Problem

In this problem, we are asked to add the numbers 1 through 1000 to find the total. In other words, we are asked to compute $1 + 2 + 3 + \ldots + 998 + 999 + 1000$. The sum will be quite large, and it could be tedious to solve this problem using only paper and pencil.

Step 2: Devise a Plan

One plan we could use is to get out some scratch paper and start calculating. Another strategy could be to use a calculator. We could also break 1000 into 10 groups of 100 numbers each and then find the total sum.

This problem is an example of a classic problem and solution developed by Carl Gauss over 200 years ago. Did he solve the problem using a calculator? Of course not! There were no calculators at the time.

While there are many ways to approach this problem, we are going to use the strategy of solving a simpler problem. Consider just the first few numbers, 1 through 10. If we find the sum of these, we get the following.

$$1 + 2 + 3 + 4 + 5 + 6 + 7 + 8 + 9 + 10 = 55 = 5 \cdot 11$$

How is this helpful? Well, if you notice from our simpler problem of 10 values, if we start with the outside numbers and work our way toward the middle, there are 5 "pairs" of numbers that add to 11.

1st	$1 + 10 = 11$
2nd	$2 + 9 = 11$
3rd	$3 + 8 = 11$
4th	$4 + 7 = 11$
5th	$5 + 6 = 11$

Using this knowledge, we should be able to carry out a similar plan and find a solution.

Step 3: Carry Out the PlanUsing the same type of strategy that we devised in Step 2, we can now carry out the plan to find the sum of the numbers 1 through 1000. Using the end numbers and working our way in, we can identify the "pairs" of numbers that add together to give 1001. Notice there will be $\dfrac{1000}{2} = 500$ pairs all together.

This means we will have 500 pairs of 1001. Using our calculators or pencil and paper, we can now obtain the final answer of $500 \cdot 1001 = 500,500$.

1st	$1 + 1000 = 1001$
2nd	$2 + 999 = 1001$
3rd	$3 + 998 = 1001$
⋮	⋮
500th	$500 + 501 = 1001$

Step 4: Look Back

We need to determine if the answer makes sense and is feasible. Have we committed any type of mathematical errors? Well, by solving the simpler problem, we have used deductive reasoning. We know the pattern we developed can always be followed as long as the numbers used in the sequence remain consistent. So we have indeed found the solution to be 500,500.

Fun Fact

Example 1.3.4 is based on a famous (yet unconfirmed) story about the mathematician Carl Friedrich Gauss. The story goes that as a child in primary school, after the young Gauss misbehaved, his teacher gave him a task to add a list of integers in arithmetic progression (1 to 100). To the teacher's astonishment, young Gauss produced the result in a matter of seconds.

Gauss found the solution by a method of realizing that pairwise addition of terms from opposite ends of the list yielded identical intermediate sums: $1 + 100 = 101$; $2 + 99 = 101$, $3 + 98 = 101$, and so on until $50 + 51 = 101$, for a total sum of $50 \cdot 101 = 5050$.

✔ Skill Check 1.3.2

Suppose you could spend $20 every minute of every day for a year. How much money would you spend in 365 days?

Example 1.3.5

Finding a Pattern

Rosa tracks her car's gas mileage with an app on her phone. Each time she fills up her tank, she records the number of miles driven since her last fill up and the number of gallons of gas replenished. The ten most recent entries from Rosa's log are shown in Table 1.3.4. Looking at the data, do you think Rosa's log is accurate or do you suspect that an entry may have a typo? If so, which one?

Table 1.3.4: Rosa's Mileage Log

Distance Traveled (Miles)	Gallons Used
320	13.2
278	11.7
334	13.4
313	13.3
180	7.6
297	12.2
311	7.4
294	12.6
323	13.1
308	12.9

Solution

Step 1: Understand the Problem

This question is asking us to see if any data values seem "out of bounds" compared to the others. We need to inspect the data we've been given, develop a theory about how accurate it is, and be able to justify our theory.

Step 2: Devise a Plan

We can inspect data in different ways. One option is to create a graph, which is particularly helpful when there's a lot of data. Another option is to look for patterns. Let's try that strategy here. Note that while Rosa's log tells us the miles driven and the gallons of gas used, it doesn't tell us the "miles per gallon" rate of each entry. As you likely know from personal experience, gas mileage varies based on factors such as weather and driver habits, but in general each car has an expected range. Let's determine the miles per gallon for each row and then inspect our updated table.

Step 3: Carry Out the Plan

Calculate the miles per gallon for each entry by dividing the miles driven by the number of gallons of gas purchased. Here's our updated table.

Table 1.3.5: Rosa's Mileage Log with Miles Per Gallon

Distance Traveled (Miles)	Gallons Used	MPG
320	13.2	24.24
278	11.7	23.76
334	13.4	24.93
313	13.3	23.53
180	7.6	23.68
297	12.2	24.34
311	7.4	42.03
294	12.6	23.33
323	13.1	24.66
308	12.9	23.88

Now we can look for patterns in the entries.

First, you might notice that 180 miles in the distance column is a smaller value than the other entries in that column. That stands out, but when we read the rest of the row, we see that fewer gallons of gas were purchased and the miles per gallon is similar to other entries.

Now you might notice that all of the entries in the miles per gallon column are similar (between 23 and 25 miles per gallon) except for one: the row with a distance of 311 miles, 7.4 gallons of gas, and 42.03 miles per gallon. That should grab our attention. Two rows above that row is an entry with 7.6 gallons of gas used for 180 miles. Compared to this entry with 7.4 gallons of gas used for 311 miles, that's a pretty big difference in miles traveled for a similar amount of gas.

While we can expect variation in gas mileage, it's suspicious that one entry reports gas mileage that is almost double the other entries. It is reasonable to suspect this entry has a typo.

Step 4: Look Back

This question asked us to develop a theory and not necessarily to conduct a formal calculation. By inspecting the data, we found that while most entries were similar to each other, one differed from the pattern. The higher gas mileage on the one entry is notable enough to raise suspicion, making our theory reasonable. (We will explore more formal ways of measuring "out of bounds" data in the Statistics chapter.)

Example 1.3.6

Drawing a Diagram

A baseball league is being created where each of the teams will play four games against every other team. There are five teams in the league: the Raiders, the Jackals, the Blazers, the Warriors, and the Eagles. Determine how many total games will be played.

Solution

Step 1: Understand the Problem

The problem asks us to consider five teams playing each other in a baseball league where each team will play every other team four times. This means that each team will play 16 games. By this logic, it would appear that the total number of games played will be 80, since each of the 5 teams will play 16 games. We need to actually solve the problem; however, before settling on an answer.

Step 2: Devise a Plan

As with other problems, there are many ways to approach a solution. For this example, we will use a diagram to solve the problem. Our plan is to let each team be represented by their team name and use arrows to connect each team to the other teams, where each connection represents the 4 games played between pairs of teams.

Step 3: Carry Out the Plan

Carrying out the plan here involves creating a visual representation of the problem. We begin with a figure that represents the games the Raiders will play against the Blazers (and consequently, the games the Blazers will play against the Raiders). Because each team plays each other four times, the line represents four games.

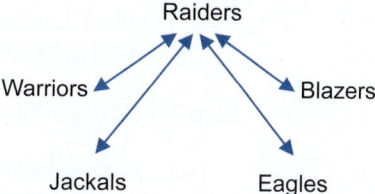

Next, we consider all games that will be played by the Raiders against all the other teams (and the games they all play against the Raiders).

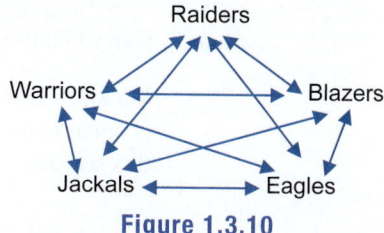

Figure 1.3.9: The 16 Games Played by the Raiders

Continuing in this manner, we arrive at all of the possible game matchups.

Figure 1.3.10

From the figure, we can see that there are 10 lines. Remembering that each line represents 4 games, we conclude that a total of 40 games will be played.

Step 4: Look Back

Recall from our understanding of the problem that we thought there might be 80 total games played. Why is there a difference in our actual solution? The answer lies in the fact that although the Raiders play the Blazers four times and the Blazers play the Raiders four times, there is a total of only four games between the two teams instead of the expected eight games. With our initial thinking, when each team plays another team, we counted that game as a game for both teams. So in our preliminary discussion of the problem, we made an error when we counted each game twice.

Example 1.3.7

Using a Variable

Devon has $100 to spend on a pair of shoes. The pair he wants costs $120. Ignoring sales tax, what is the minimum percentage off Devon would need so he can buy the shoes?

Solution

Step 1: Understand the Problem

We are being asked to find the percentage off the original price of the shoes that leads to a discounted price that is $100 or less. The original price is $120, so Devon needs a discount that is at least $20.

Step 2: Devise a Plan

A discount is typically a percentage of the sale price. We need a discount that is at least $20, so we need to find a percentage that when multiplied by the original price of $120 results in a value of at least $20. We can set up an inequality using a variable, say x, to represent the unknown percentage in decimal form.

Step 3: Carry Out the Plan

We know that we need to find the value of x where $120 multiplied by x is equal to a value that is greater than or equal to $20. We can represent this scenario with the inequality $120x \geq 20$.

$$120x \geq 20$$
$$x \geq \frac{20}{120}$$
$$x \geq \frac{1}{6} \approx 0.17$$

This tells us that $20 is approximately 17% of the original price. If the original price is reduced by at least 17%, Devon will be able to purchase the shoes.

Step 4: Look Back

We use the look back step to check the accuracy and feasibility of our solution. We can confirm the percent off we found falls into a reasonable range with some quick checks. We know that 10% of $120 would be savings of $12. Similarly, 20% of $120 would be $24. We needed to find the percentage off associated with savings of $20, and our answer of 17% falls between 10% and 20%, as expected.

✔ **Skill Check 1.3.3**

A number is multiplied by 8 and the product is added to 6. If the sum is 30, what is the number?

Skill Check Answers

1.

17	42	7
12	22	32
37	2	27

2. $10,512,000   **3.** 3

1.3 Exercises

✔ CONCEPT CHECK

1. True or False: When understanding a problem, it can be helpful to restate the problem using different phrasing.

2. True or False: When devising a plan to solve a problem, it can be helpful to look for patterns in the given data.

3. True or False: When carrying out the plan to solve a problem, if the plan you devised doesn't work, it's not worth creating a new plan and trying again.

4. True or False: It is useful to look back over your problem-solving process to help ensure that the conclusion you reached makes sense for the situation.

🚀 APPLICATIONS

5. A child snuck out of the house one evening and bought a bag of candy. His older sister caught him sneaking back into the house and said that if he handed over one-fourth of the candy, she wouldn't tell their parents. So he gave her the candy. His younger sister saw him sneaking back to his room and said if he gave her twenty percent of the candy, she wouldn't tell their parents, so he gave her some candy. His younger brother found him hiding the candy in the room they shared and said if he gave him half of the candy, he wouldn't tell their parents. So he gave him some candy. After the series of negotiations, the child had 6 pieces of candy to hide. How many pieces of candy were in the bag when he returned home?

6. An employee baked a fresh batch of cookies and brought them to work the next day. She told a coworker that she'd give him one-fourth of the cookies plus three if he went to a meeting in her place. He accepted. She then told another coworker that she'd give her one-third of the remaining cookies plus two if she created a PowerPoint presentation for a meeting later in the week. She accepted. The employee finally told another coworker that she'd give him three-fourths of the remaining cookies if he researched a new product line for her. He accepted. After these work negotiations, she had three cookies left. How many cookies did she take to work with her that day?

7. Complete the following magic square using the numbers 1 through 16, where the sum of every row, column, and diagonal is the same.

16	2		
5	11		8
	7	6	
4	14		1

8. Arrange the numbers 1, 2, 3, 4, 5, and 6 in the given circles so that the sum of the numbers of each side is equal to 9.

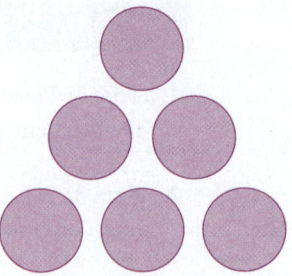

9. Arrange the numbers 1, 2, 3, 4, 5, and 6 in the given circles so that the sum of the numbers of each side is equal to 10.

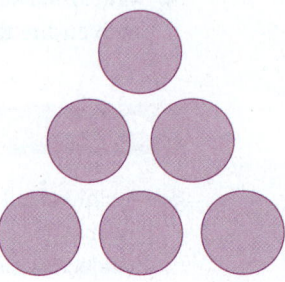

10. Place the digits 1, 2, 3, 4, and 5 in the circles so that the sums across (horizontally) and down (vertically) are equal.

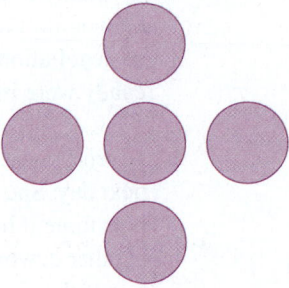

11. Arrange the numbers 1, 2, 3, 4, 5, 6, 7, 8, and 9 in the given circles so that the sum of the numbers of each side is equal to 19.

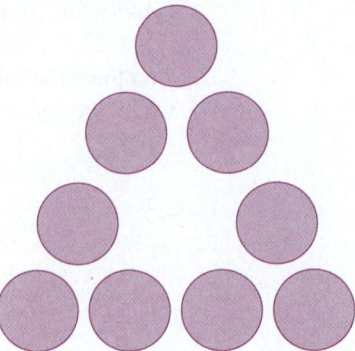

12. Arrange the numbers 1, 2, 3, 4, 5, 6, 7, 8, and 9 in the given circles so that the sum of the numbers of each side is equal to 23.

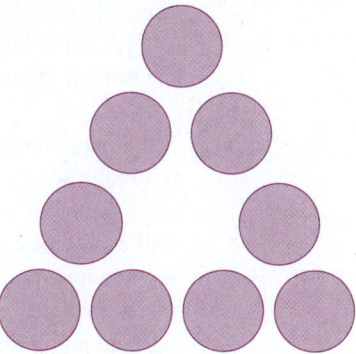

13. Place the numbers 1, 2, 3, 4, 5, 6, 7, 8, and 9 in the magic square so that the sum of the numbers in each row, column, and diagonal is equal to 15.

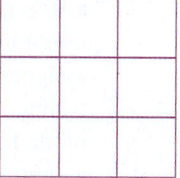

14. Use the digits 0, 1, 2, 3, 4, 5, 6, 7, and 8 to create another magic square so that the sum of the numbers in each row, column and diagonal is equal to 12.

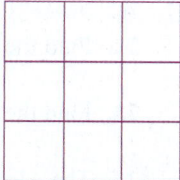

15. A rectangle has an area of 72 square inches. Its length and width are whole numbers.

    a. What are the possible dimensions of the rectangle?

    b. Which of those dimensions yield a rectangle with the smallest perimeter?

16. Carlos is placing an order online for curbside pickup and notices a promotion that if he spends $75 on his order (before tax), he will receive a $20 gift card for a future purchase. He needs to purchase a pack of LED lightbulbs for $32.99, a bottle of laundry detergent for $15.99, a throw blanket for $24.99, and a travel coffee mug for $17.99. Can he purchase three of the four items with this online order and use the gift card to purchase the remaining item? If so, which three of items should he purchase?

17. Marcus is considering which presents to buy for his son's birthday. He wants to limit the overall cost to $100, including tax, so he is aiming for the gift total to be no more than $90 before taxes. His son has the following toys on his wish list: fire station playset ($19.69), dinosaur expedition playset ($22.99), toy soldiers set ($35.99), construction truck set ($31.99), miniature racetrack with cars ($9.99), and a medieval castle ($85.89). If Marcus wants to purchase as many gifts as possible and stay within the set budget, which toys should he buy?

18. Emmet has five bills due this month: water & sewer ($95.50), internet ($75.45), electric ($115.79), gas ($63.98), and phone ($112.79). He currently only has $350 to put towards bills. Each company imposes at $20 late fee if the bill is not paid on time. It is possible for Emmet to pay four of the bills in order to incur only one late fee? If so, which bills should he pay?

19. Sofía is planning a weekend trip and wants to limit her spending to $500, not including travel expenses. She'll stay for two nights at a hotel that costs $115 per night (including taxes and fees), and she estimates her meals will cost a total of $120. The activities in the area that Sofía is considering include an aquarium ($58), an art museum ($25), a zipline tour ($75), and a wine tasting tour ($65). Can Sofía pick three of the activities and stay within her budget? If so, which three activities?

20. Azra has four assignments that are due by the end of the week and only has 20 hours to spend working on them. Any late assignments will incur a 10% grade deduction. She estimates that her math assignment will take 4 hours, her literature assignment will take 7.5 hours, her sociology assignment will take 8 hours, and her business ethics assignment will take 5 hours. All courses are equally important to her. Can Azra complete three of the four assignments and only incur one late penalty? If so, which three assignments?

21. Find the sum of the whole numbers from 1 to 900.

22. Find the sum of the even numbers from 1 to 500.

23. Find the sum of the odd numbers from 1 to 700.

24. The Alpha Zeta Math Club is having a pizza party. The president of the club decides to have fun cutting the pizza and challenges the members to cut a pizza into 11 pieces with only 4 straight cuts. Show a way that this could be done.

25. There are eight teams in the intramural basketball league. Each of the teams will play each other five times. What is the total number of games played in the league?

26. A local bowling league consists of seven teams and each team plays each other team three times. What is the total number of games played in the league?

27. At a birthday party, a child was permitted to cut his own cake to serve to the guests. The cake was round, and the child cut it into 16 pieces using only 5 cuts. (Each cut was vertical through the top of the cake.) Draw a diagram of how this was possible.

28. A farmer looks over a field and sees 37 heads and 98 feet. Some are goats, some are chickens. How many of each are there?

29. Jessica needs to mail a USB drive to her friend. She uses a combination of 41-cent stamps and 8-cent stamps to pay $1.71 in postage. How many of each stamp did Jessica use?

30. While inspecting tanks at an aquarium, a keeper counts 148 legs and 16 heads in a tank that holds octopuses and crabs, which have 8 and 10 legs, respectively. How many of each creature are there?

31. Ruth is sending a care package to her granddaughter, Sarah, who is in college. The care package contains two of Sarah's favorite cookies: fudge swirls, which come in packs of 12, and extra chocolate chunk, which come in packs of 5. The care package contains a total of 107 cookies. How many packs of each cookie type were included?

32. The product of two whole numbers is 196 and their sum is 35. What are the two numbers?

33. The product of two whole numbers is 902 and their sum is 63. What are the two numbers?

34. Twice the difference of a number and 1 is 4 more than that number. Find the number.

35. A number is multiplied by 8, and that product is added to 3. The sum is equal to the product of 5 and 7. Find the number.

✏ **WRITING & THINKING**

36. One of the most interesting and useful patterns in all of mathematics is called Pascal's triangle, as shown. It is named after the French mathematician Blaise Pascal (1623–1662) who showed that these numbers play an important role in the theory of probability.

$$
\begin{array}{c}
1 \\
1 \quad 1 \\
1 \quad 2 \quad 1 \\
1 \quad 3 \quad 3 \quad 1 \\
1 \quad 4 \quad 6 \quad 4 \quad 1 \\
1 \quad 5 \quad 10 \quad 10 \quad 5 \quad 1 \\
1 \quad 6 \quad 15 \quad 20 \quad 15 \quad 6 \quad 1
\end{array}
$$

a. Copy the array and fill in the next three rows of numbers using the pattern(s) that you notice.

b. Compute the sum of the elements in each row of Pascal's triangle for the first 6 rows.

c. Look for a pattern in the sums of the first 6 rows. See if you can predict the sum of the elements for row 7, then check by adding. Do the same for rows 8 and 9. State the general rule for the sum of the elements in words.

Chapter 1 Exercises

Identify whether each statement is an example of inductive or deductive reasoning.

1. An online store determines that sales are highest on Tuesday evenings based on sales data from the past two months.

2. In order to qualify for a $25 discount each month at Things and Stuff, you need to make three $75 purchases during the first three weeks of the month. Emilia made purchases of $85.53, $75.29, and $77.82 during the first three weeks of the month. So Emilia qualifies for the $25 discount.

Find a counterexample to each statement

3. The Super Bowl is played on the first Sunday in February.

4. The product of an even number and an odd number is odd.

5. If the sum of two numbers is odd, then both numbers are odd.

Find the missing terms of each sequence and determine if the sequence is arithmetic, geometric, or neither. If it is an arithmetic sequence, state the common difference; if it is a geometric sequence, state the common ratio.

6. 4, 7, 10, _____ , _____ , _____

7. 3, 6, 18, 36, _____ , _____ , _____

8. $\dfrac{1}{3}, \dfrac{1}{9}, \dfrac{1}{27}$, _____ , _____ , _____

Use inductive reasoning to predict the next line of each pattern. Complete the computations to verify.

9. Consider the following pattern.

$$1 \cdot 8 + 1 = 9$$
$$12 \cdot 8 + 2 = 98$$
$$123 \cdot 8 + 3 = 987$$
$$1234 \cdot 8 + 4 = 9876$$
$$12{,}345 \cdot 8 + 5 = 98{,}765$$
$$123{,}456 \cdot 8 + 6 = 987{,}654$$
$$1{,}234{,}567 \cdot 8 + 7 = 9{,}876{,}543$$

10. Consider the following pattern.

$$999{,}999 \cdot 1 = 999{,}999$$
$$999{,}999 \cdot 2 = 1{,}999{,}998$$
$$999{,}999 \cdot 3 = 2{,}999{,}997$$
$$999{,}999 \cdot 4 = 3{,}999{,}996$$

11. Determine the most appropriate order of magnitude for the daily temperature in degrees Fahrenheit

12. You are hosting a graduation party for your child who just graduated from high school. Your child wants to invite 25 friends, and you will also invite 50 family members. How could you break this situation down into smaller parts to estimate the total cost of the party?

13. A local newspaper is estimating the number of people attending a championship parade based on aerial footage. A single block of the parade route is determined to hold 142 people. The parade route is 32 blocks long. Estimate the number of people who attended the parade.

14. A pet store employee counts 23 heads and 68 feet on the animals in the store. The store currently only has kittens and parakeets left in the store. How many of each are there?

15. A box office sells tickets for a concert at two different prices: $35 and $55. The box office sold a total of 95 tickets at a value of $4325 today. How many of each of the different tickets did the box office sell?

16. The product of two whole numbers is 234 and their sum is 31. What are the two numbers?

17. Place the numbers 1 through 9 in the given circles so that the sum of the numbers on each side of the triangle is 20.

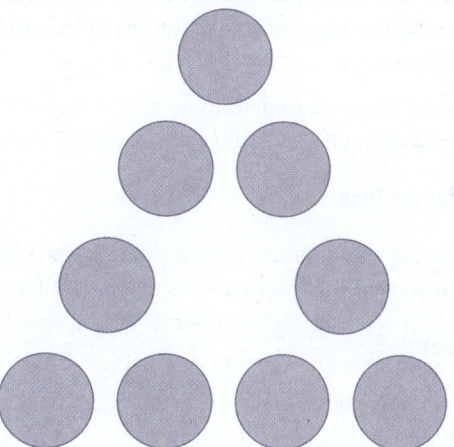

18. Find the sum of the whole numbers from 1 to 200.

19. How many line segments are there in each of these figures?

CHAPTER 1 PROJECT

Dream Plans

Owning your own home has always been part of the American Dream. But how feasible is building that home? In this project, you will use Pólya's problem-solving process to estimate the cost of building a house. Assume that you already own the land the house will be built on, and the neighborhood has the following building restrictions.

The house must meet the following parameters:

- Square footage: 1600 ft^2 to 2400 ft^2
- Bedrooms: 2 to 4
- Bathrooms: 1 to 3
- Other required rooms: kitchen, dining area, living room, laundry room

Step 1: Understand the Problem

1. The first step is to identify what is being asked of you for this project. Describe in your own words the goal of this project and the steps you will take to complete the project.

Step 2: Devise a Plan

2. List the different aspects of building a house from initial design to finalizations. Be sure to consider both physical labor required and materials to be purchased. (**Note:** You do not need to list specific costs, only the categories.)

3. Do some research to determine an estimated cost per square foot. Describe what considerations went into estimating the overall cost. Be sure to state where the house will be built.

Step 3: Carry Out the Plan

4. Determine the number of bedrooms, bathrooms, and total square footage of your home. Include the dimensions of the rooms and as much detail as possible.

Step 4: Look Back

5. Determine a rough estimate for the cost of the house designed in Step 3 and the cost per square foot from Step 2.

6. How does the estimated cost of your house compare with the real estate market in your city or town? What source(s) did you use to make the comparison?

7. Name at least two factors that might cause the cost of building your house to be more than you estimated. Explain how you can decrease the cost of building the house.

8. Name at least two things you learned in the process about building a house.

9. What would you change about your original house plan based on your findings?

CHAPTER 1 REVIEW

Section 1.1 Thinking Mathematically

Definitions

Inductive Reasoning: Inductive reasoning is a line of argument that takes specific examples as its premise and then draws a general conclusion from these examples.

Deductive Reasoning: Deductive reasoning uses statements that are commonly accepted as facts to make a case for a specific conclusion.

Counterexample: A **counterexample** is an example that shows that a generalization or rule is false by satisfying the conditions of the generalization or rule, but not satisfying the conclusion.

Section 1.2 Estimating and Evaluating

Procedure

Tips for Estimating

1. Start with something you know about the topic or the closest thing you know to the topic.
2. Have a rough order of magnitude. For example, will it take 5 days, 5 weeks, 5 months, or 5 years?
3. Use past experiences to find context.
4. Break the original question into simpler, more manageable questions.
5. Round numbers to make calculations easier.
6. Sample a small area and scale it up by multiplying.

Section 1.3 Problem Solving: Processes and Techniques

Procedure

Pólya's Problem-Solving Technique

1. Understand the problem.
2. Devise a plan.
3. Carry out the plan.
4. Look back.

CHAPTER 2

Set Theory

█ SECTIONS

Introduction

At a dinner party one night, one of the guests died from eating food that was poisoned. Oddly enough, no one else attending the dinner party got sick. The waitress who served the guests told the detective investigating the case that 14 dinner plates were served with shrimp, 14 dinner plates had lobster, and 14 dinner plates had steak. In addition, she told the detective that some plates had more than one meat. The detective decided to put the information in a diagram but forgot to indicate how many plates had a single type of meat. When the autopsy report came back, it stated that the victim only had one type of meat in his stomach. Does this diagram contain enough information to determine which type of meat the victim consumed?

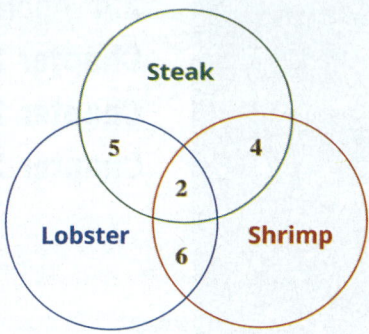

Diagrams like the one created by the detective can be used for a variety of reasons other than solving logic puzzles. Researchers analyzing survey results use these diagrams to breakdown known information, find overlaps, and determine exactly how many items fall into each potential category. Business managers use diagrams to compare and contrast jobs or projects. Therapists might use diagrams to show how aspects of psychology and well-being intersect. Even you and your friends could discover the perfect restaurant for dinner by knowing where group opinions and choices overlap so that everyone is happy.

In this chapter, we'll study relationships between sets and how to represent these with Venn diagrams. Then, we'll use Venn diagrams to perform survey analysis to help us make informed decisions in our day-to-day lives.

2.1 SET NOTATION

OBJECTIVES

1. Write sets in roster notation.

2. Determine whether sets are equal or equivalent.

3. Write sets in set-builder notation.

4. Identify empty sets.

5. Determine complements of sets.

OBJECTIVE 1 Roster Notation

A set is a Many that allows itself to be thought of as a One.
—Georg Cantor

Georg Cantor was a German mathematician whose work in the late 1800s is credited with starting the branch of mathematics called set theory. His work successfully established that sets had intrinsic value and needed to be studied.

Simply put, a **set** is a collection of objects. A flock of sheep, a litter of kittens, and an ensemble of musicians are examples of sets of things with living members. The alphabet is a set of letters, while a line is a set of points. We can also discuss sets of famous people, scary animals, or innovative architecture. However, notice that these last three set examples aren't as obviously definable. What classifies as a "scary animal" to one person, may not with another. Thus, in mathematics, we restrict our study of sets to sets that are clearly defined. In other words, if we are given the criteria for a set, we can impartially determine whether an object is a member of the set or not. For instance, the set of people who have been president of the United States is a well-defined set. However, the set of best movies in history is not a well-defined set because the term "best," just like the term "scary," is highly subjective from person to person.

> 📖 **SET**
>
> A **set** is a collection of objects made up of specified elements, or members.

Mathematical sets are commonly represented using capital letters, and the elements of the set are represented by lowercase letters. We can state that the member x is an element of a set G symbolically by writing $x \in G$ (read "x is an element of G"). If y is not an element of G, we write $y \notin G$ (read "y is not an element of G").

One of the most common ways to describe a set is to use **roster notation**, which is a list of all of the elements in the set. When a set is written in roster notation, the elements of the set are surrounded by braces and separated with commas. For example, sets A, B, and C are described using roster notation.

$$A = \{1, 2, x, y, z\}$$

$$B = \left\{\frac{1}{2}, \frac{2}{3}, \frac{3}{4}, \frac{4}{5}\right\}$$

$$C = \{\text{Gwen, King Charles spaniel, Zoë, taxi}\}$$

It's important to note that sets are collections where repetition and order are ignored. That is, every element of a set will be unique. For instance, the set M consisting of the letters of the word "Mississippi" will contain the four letters M, I, S, and P. Using roster notation, we write $M = \{M, I, S, P\}$. We do not list any element more than once when using roster notation.

Figure 2.1.1

📖 **ROSTER NOTATION**

Roster notation specifies the members of a set by listing all of the elements in the set, separated by commas and surrounded by curly braces.

As we've seen, elements of sets are not restricted to numbers. The elements in a set also do not need to have any obvious connection. In fact, the only feature that elements of a set might have in common is that they are all members of the same set, as in set C that was defined earlier: $C = \{$Gwen, King Charles spaniel, Zoë, taxi$\}$.

In set theory, sets themselves can be elements of other sets. Consider the set G, which is a set consisting of sets of paired numbers.

$$G = \{\{1, 2\}, \{3, 4\}, \{5, 6\}, \{7, 8\}, \{9, 10\}\}$$

Notice that the individual numbers 1 through 10 are not members of the set G, but the five sets of paired numbers $\{1, 2\}$, $\{3, 4\}$, $\{5, 6\}$, $\{7, 8\}$, and $\{9, 10\}$ are members of set G.

Example 2.1.1

Representing a Set with Roster Notation

Use roster notation to represent the following sets.

a. S is the set of states in the United States that begin with the letter M.

b. Given the sets $\{$blue, red, green$\}$, $\{$red, white, blue$\}$, $\{$blue, green, aqua$\}$, and $\{$red, white$\}$, let T be the set of sets that contain the element "red."

Solution

a. There are eight states in the United States that begin with the letter M.

$S = \{$Maine, Maryland, Massachusetts, Michigan, Minnesota, Mississippi, Missouri, Montana$\}$.

Note that although we listed the elements of S in alphabetical order for the ease of the reader, this is not necessary when using roster notation. The order of elements does not matter in a set.

b. Three of the sets listed contain the element "red."

$T = \{\{$blue, red, green$\}$, $\{$red, white, blue$\}$, $\{$red, white$\}\}$.

OBJECTIVE 2 Equal and Equivalent Sets

As we mentioned earlier, there is no concept of order in a set. Thus, set $X = \{$a, b, c, d$\}$ is the same as set $Y = \{$d, c, b, a$\}$. When two sets contain exactly the same elements, the sets are said to be **equal**, and we can write $X = Y$ to indicate set X is equal to set Y.

📖 EQUAL SETS

Two sets are said to be **equal** if they contain exactly the same elements. If sets A and B are equal, we write $A = B$.

Sets come in all sizes: some have infinitely many elements and some have a finite number of elements. A set having infinitely many elements means that a list of the elements would go on forever, whereas a finite set has a specific number of elements. Roster notation for an infinite set often includes a series of three dots, called an *ellipsis*, that indicates that the set continues on without ceasing. For instance, the set of positive integers can be written as $Z = \{1, 2, 3, 4,...\}$. An ellipsis can also be used to indicate when a certain pattern is continued within a list of set members. For instance, when describing the even numbers between 2 and 100, we could write $L = \{4, 6, 8,..., 98\}$.

In this text, we will focus most of our discussion on finite sets. Since finite sets have a specific number of elements, we can discuss their cardinal numbers. The **cardinality** of a finite set is the number of elements contained in the set; in other words, the cardinality indicates the size of the set. The cardinality of a set is usually denoted by two vertical lines around the name of the set. For example, if $A = \{a, b, c, d, e\}$, then $|A| = 5$, which is read "the cardinality of set A is 5."

> 💬 **Helpful Hint**
>
> The notation |A| can mean either the cardinality of set A or the absolute value of some number A. The meaning of the notation depends on context.

When two finite sets have the same cardinal number (that is, the same number of elements, no matter if the elements are of the same type or not), the sets are said to be **equivalent**. For instance, consider set $C = \{\alpha, \beta, \lambda\}$ and set $D = \{\neq, \geq, \leq\}$. Both sets contain 3 elements, so C is equivalent to D. Symbolically, we write $C \sim D$, which is read "C is equivalent to D."

📖 CARDINAL NUMBER AND EQUIVALENT SETS

Cardinal Number

The number of elements contained in a finite set is called the cardinal number of the set, or the **cardinality**. The cardinal number of set A is denoted by $|A|$.

> 💬 **Helpful Hint**
>
> The cardinal number may also be denoted by $n(A)$, $\overline{\overline{A}}$, card(A), or #A.

Equivalent Sets

Two sets are said to be **equivalent** if they have the same cardinal number. If set A is equivalent to set B, we write $A \sim B$.

Example 2.1.2

Determining Equal and Equivalent Sets

Determine if the given pairs of sets are equal, equivalent, or neither.

> 💬 **Helpful Hint**
>
> When describing set equality, $X = Y$ means the same as $Y = X$. The same holds true with equivalences.

a. $A = \{$Public Health, International Studies, Mechanical Engineering, Music Education, Political Science$\}$

   $B = \{$Tim, Gloria, Alan, Warren, Kalif$\}$

b. $X = \{2, 4, 6, 8, 10, 12, 14, 16, 18, 20\}$

   $Y = \{20, 18, 16, 14, 12, 10, 8, 6, 4, 2\}$

Solution

a. $|A| = 5$ and $|B| = 5$, therefore, the sets are equivalent to one another, and we can write $A \sim B$. However, because they do not contain the same elements, they are not equal to one another.

b. $|X| = 10$ and $|Y| = 10$. Even though the order of the elements is different, both contain the even numbers from 2 to 20. Therefore, the sets are equal, and we can write $X = Y$. When two sets are equal, they are also always equivalent.

OBJECTIVE 3 Set-Builder Notation

Describing a set using roster notation is very convenient for small sets, however it quickly loses its usefulness for big sets. For instance, consider the set of fractions between the numbers three and four. Because there are infinitely many fractions between any two whole numbers, it is impossible to list them out and it is not convenient to use an ellipsis to represent them. However, we can encompass the entire set by describing it. When we identify a set by stating the properties that each element must have, we refer to this as set-builder notation. Set-builder notation starts by stating a variable, then has a vertical line, followed by a rule that defines the variable. For instance, the set of fractions between the numbers three and four can be described by the following set-builder notation.

$$A = \{x \mid x \text{ is a fraction and } 3 < x < 4\}$$

This is read, "A is the set of all x such that x is a fraction and x is between three and four."

✔ Skill Check 2.1.1

Find the cardinality of W and Y. Determine if the sets are equal, equivalent, or neither.

$W = \{3, 6, 12, 24\}$ and
$Y = \{2, 4, 8, 16, 32\}$

💬 Helpful Hint

Not all numbers can be written as fractions. The irrational number π is a number between 3 and 4 that would not be a member of the set A.

💬 Think Back

$\mathbb{R}$ represents the set of real numbers, $\mathbb{N}$ represents the set of natural numbers, $\mathbb{Z}$ represents the set of integers, and $\mathbb{W}$ represents the set of whole numbers. All of these sets are infinite sets.

📖 SET-BUILDER NOTATION

Set-builder notation specifies the members of a set using a variable, a vertical separator, and a rule defining the elements, all surrounded by curly braces. It can be used when the members of the set all share certain properties. For instance, the set of all integers is represented by

$$\mathbb{Z} = \{n \mid n \text{ is an integer}\}$$

and is read, "$\mathbb{Z}$ is the set of all n such that n is an integer."

Example 2.1.3

Representing a Set with Set-Builder Notation

Use set-builder notation to represent Y, the set of all even integers.

Solution

💬 Helpful Hint

Set-builder notation may also use a colon instead of a vertical line as a separator. For example,

$Y = \{x : x \text{ is an even integer}\}$.

When we use set-builder notation, there is often more than one way to state the rule for the elements of the set. When you are choosing, it is best to be as precise and clear as you can be. In this case, the cleanest way to describe the even integers is to simply state that an element must be an even integer. Thus, we have the following.

$$Y = \{x \mid x \text{ is an even integer}\}$$

However, we could also describe the even integers as elements that are members of the set of integers *and* are even numbers.

$$Y = \{x \mid x \in \mathbb{Z} \text{ and } x \text{ is even}\}$$

Alternatively, we could describe the even integers as any integer multiplied by two.

$$Y = \{2x \mid x \text{ is an integer}\}$$

All of these notations are valid ways to describe the set of even integers.

✔ **Skill Check 2.1.2**

Represent the set J, which consists of all natural numbers less than 10, using both roster notation and set-builder notation.

OBJECTIVE 4 Empty Sets

There are a couple of special types of sets that we need to define. The first of these is the **empty set**. Think for a moment about the set A, which is the set of states that border Hawaii. Since Hawaii is a made up of islands that do not border any other state, A has no elements and thus is the empty set. The empty set is denoted symbolically by $\varnothing$, or sometimes $\{\ \}$. The cardinality of the empty set is 0.

> 📖 **EMPTY SET**
>
> The **empty set** is the set that contains no elements. If set A is empty, we write
>
> $$A = \varnothing \text{ or } A = \{\ \}.$$
>
> The cardinality of the empty set is 0.

The empty set should not be confused with the set $\{\varnothing\}$ or the set $\{0\}$, both of which contain a single element. The first is the set whose only element is the empty set and the second is the set containing the number 0, both of which have a cardinality of 1.

Example 2.1.4

Identifying the Empty Set

Determine if the following sets are the empty sets.

a. $A = \{x \mid x \text{ is a negative number less than } 100\}$

b. The set B of any state that contains the letter q in its name.

c. $C = \{\_\}$

Solution

✔ **Skill Check 2.1.3**

Determine if set Y is the empty set.

$Y = \{x \mid x \text{ is a US citizen who held the office of president before Barack Obama, and } x \text{ is a woman}\}$

a. Since it is the case that all negative numbers are less than 100, the set A is not empty. In fact, it is a set with infinitely many elements.

b. Since there are no states with the letter q in their name, this set is empty. That is, $B = \varnothing$.

c. The set C contains the underscore character, so it is not empty.

OBJECTIVE 5 Universal Sets and Complements

The second special set we will discuss is the **universal set**, U. The universal set is the set of all elements that are being considered in any particular situation. For instance, suppose you are buying a car. The universal set could be the set of all new cars available; it could be the set of all used cars available, or it might be both—the set of all new and used cars available. You can think of choosing a universal set as "setting the scene" for the elements that will be considered.

> 📖 **UNIVERSAL SET**
>
> The set of all elements being considered for any particular situation is called the **universal set** and is denoted by U.

The universal set is especially useful in that it allows us to talk about the elements that are being considered in a situation, but don't belong to a particular set. For instance, let's go back to our car example. Suppose the universal set is the set of all new and used cars that are available. Assume you are only interested in buying a car under \$25,000. Then the set of cars you can buy is $W = \{c \mid c \in U$ and c costs $< \$25,000\}$. Notice that not all cars in the universal set are included in W. For instance, a \$4,000,000 Lamborghini is not in set W, although it is in the universal set. The Lamborghini is in the **complement** of W, denoted W'. That is, the Lamborghini is an element of the universal set that is not in the set W. In our car example, $W' = \{c \mid c \in U$ and c costs $\geq \$25,000\}$.

> 💬 **Helpful Hint**
>
> The complement of A can be denoted by A', A^c, or $\overline{A}$.

> 📖 **COMPLEMENT**
>
> The **complement** of set A consists of all the elements in the given universal set that are not contained in A. The complement of A is denoted A'.

The set-builder notation for the complement of set A is $A' = \{x \mid x \in U, x \notin A\}$.

Consider for a moment the two special sets we've discussed: the universal set and the empty set. One contains all of the elements being considered and the other contains no elements. Hence, the universal set and the empty set will always be complements of one another.

$$U' = \varnothing \text{ and } \varnothing' = U$$

Example 2.1.5

Determining the Complement of a Set

According to a recent poll, approximately 30% of American adults have at least one tattoo.[1]

Let

$U = \{x \mid x$ is an American adult$\}$

$A = \{x \mid x$ is an American adult who has at least one tattoo$\}$

$B = \{x \mid x$ is an American adult who has exactly two tattoos$\}$

Write the complements of A and B using set-builder notation.

---

1   Chris Jackson, "More Americans Have Tattoos Today than Seven Years Ago," *Ipsos*, August 29, 2019, https://www.ipsos.com/en-us/news-polls/more-americans-have-tattoos-today.

Solution

Since A consists of American adults who have at least one tattoo, the complement of A consists of American adults who do *not* have at least one tattoo. We know that *at least* one means greater than or equal to one, so *does not have at least one* implies has less than one or simply no tattoos at all. We can write the complement of A as follows.

$A' = \{x \mid x \in U$ and x has no tattoos$\}$

This could also be written as $A' = \{x \mid x$ is in U and x has no tattoos$\}$. Note that based on the poll data, A' represents approximately 70% of American adults.

B is the set of American adults with exactly two tattoos. The complement of B includes all of the elements in our universal set that are not contained in B. This translates to American adults that do *not* have exactly two tattoos. The complement of B therefore contains adults with 0 tattoos, 1 tattoo, or 3 or more tattoos. We can write the complement of B in either of the following ways.

$B' = \{x \mid x$ is an American adult who does not have exactly two tattoos$\}$

Or

$B' = \{x \mid x \in U$ and x has either 0 tattoos, 1 tattoo, or 3 or more tattoos$\}$

> ✔ **Skill Check 2.1.4**
>
> Let
>
> $U = \{x \mid x$ is a book published in the US in 2021$\}$
>
> $A = \{x \mid x \in U$ and you read x as an e-book$\}$.
>
> 1. Find A'.
>
> 2. Is it possible for you to have read a book in A'?

Skill Check Answers

1. $|W| = 4$, $|Y| = 5$; W and Y are neither equal nor equivalent.
2. Roster notation: $J = \{1,2,3,4,5,6,7,8,9\}$; Set-builder notation: possible answers include $J = \{x \mid x$ is a natural number less than 10$\}$ and $J = \{x \mid x \in \mathbb{N}, x < 10\}$
3. Yes, $Y = \varnothing$.
4. $A' = \{x \mid x \in U$ and you did not read x as an e-book$\}$; Yes, but not as an e-book; perhaps as a printed or audio book.

2.1 Exercises

✔ CONCEPT CHECK

Determine whether each statement is true or false. If the statement is false, explain why.

1. It is always possible to list every element of a set using roster notation.

2. $\varnothing \sim \{0\}$

3. Let $U = \{$set of all students enrolled at Xavier University$\}$ and $A = \{x \mid x \in U$ and x is a student with less than 30 earned credit hours$\}$. Then $A' = \{x \mid x \in U$ and x is a student with at least 30 earned credit hours$\}$.

4. $0 \in \varnothing$

5. $x \in \{x, y, z\}$

6. $\{x\} \in \{x, y, z\}$

7. $|\varnothing| = 0$

8. Let $Y = \{$Tim, Emilia, Aleesa, Whit$\}$, then $|Y| = 4$.

9. Let $A = \{$Sounds, Angels, Ravens, Titans$\}$ and $B = \{201, 38, 46, 23\}$, then $A \sim B$.

🔆 PRACTICE

Write each set using roster notation.

10. A is the set of months of the year that have exactly 30 days.

11. B is the set of states whose names begin with the letter N.

12. C is the set of positive numbers smaller than 100 that have 2 digits that are the same.

13. D is the set of last names of people who teach this course at your university.

14. E is the set of planets in our solar system.

15. F is the set of weekdays.

16. $A = \{x \mid x$ is an odd number and $20 < x < 30\}$

17. $B = \{x \mid x \leq 15$ and x is a positive multiple of $3\}$

18. $C = \{x \mid 2x = 4\}$

19. $D = \{x \mid x$ is a state that shares a common border with Colorado$\}$

Determine the cardinal number of each set.

20. $W = \{3, 4, 5, 6, 7, 8, 9, 10, 0\}$

21. $X = \{x \mid x$ is a positive integer and $x \leq 10\}$

22. The empty set

23. $Y = \{x \mid x$ is a United States president, past or present$\}$

Determine if the given pairs of sets are equal, equivalent, or neither.

24. $A = \{1, 2, 3, 5, 7, 11, 13, 17, 19\}$

    $B = \{2, 4, 6, 10, 14, 22, 26, 34, 38\}$

25. $C = \{$banana, tomato, grape, eggplant, watermelon, pumpkin$\}$

    $D = \{$blackberry, raspberry, strawberry, sloeberry$\}$

26. $E = \{$c, o, n, s, t, r, u, e$\}$

    $F = \{$r, e, c, o, u, n, t, s$\}$

27. $G = \{$r, e, s, c, u, e, d$\}$

    $H = \{$s, e, c, u, r, e$\}$

Write each set using set-builder notation.

28. Let G be the set of whole numbers less than five.

29. Let H be the set of natural numbers less than or equal to 50.

30. Let U be the set of all the states in the United States of America and J be the set of all states that border an ocean.

31. Let U be the set of all students enrolled at a public school of higher education and K be the set of all collegiate athletes.

32. $M = \{2, 3, 4, 5, 6, 7\}$

33. $Q = \{3, 6, 9, 12, 15, 18, 21, \ldots\}$

34. $U = \{-10, -8, -6, -4, -2\}$

35. $P = \{1, 2, 4, 9, 16, 25, 36, \ldots\}$

Determine whether each set is the empty set.

36. $W = \{x \mid x$ is a positive integer less than 0$\}$

37. The set X is the set of months that have fewer than 30 days.

38. $Y = \{\varnothing\}$

39. $Z = \{x \mid x$ is divisible by 10$\}$

Use the set $A = \{$b, a, s, k, e, t$\}$ to solve each problem.

40. Find $|A|$.

41. If $U = \{$a, b, c, d, . . . , x, y, z$\}$, find A'.

42. If $U = \{$a, b, c, d, . . . , x, y, z$\}$, find $|A'|$.

43. If $U = \{$a, b, c, d, . . . , x, y, z, A, B, C, D, . . . , X, Y, Z$\}$, find A'.

44. If $U = \{a, b, c, d, \ldots, x, y, z, A, B, C, D, \ldots, X, Y, Z\}$, find $|A'|$.

Use the set $B = \{1, 2, 3, 4\}$ to solve each problem

45. Find $|B|$.

46. If $U = \{0, 1, 2, 3, 4, 5, 6, 7, 8, 9\}$, find B'.

47. If $U = \{0, 1, 2, 3, 4, 5, 6, 7, 8, 9\}$, find $|B'|$.

48. If $U = \{-9, -8, -7, -6, \ldots, 6, 7, 8, 9\}$, find B'.

49. If $U = \{-9, -8, -7, -6, \ldots, 6, 7, 8, 9\}$, find $|B'|$.

Use the given sets to answer each question.

　　$P = \{$lasagna, rotini, orzo, tortellini, penne$\}$

　　$Q = \{x \mid x$ is a pasta shape$\}$

　　$R = \{$penne, tortellini, orzo, rotini, lasagna$\}$

　　$S = \{$marinara, pesto, alfredo, Bolognese, carbonara$\}$

50. Is $P = Q$? Why or why not?

51. Is $P = R$? Why or why not?

52. Is $P = S$? Why or why not?

53. Are any of P, Q, R, and S equivalent? Explain.

Use the given sets to answer each question.

　　$A = \{$Business, Physics, Psychology, Kinesiology, Graphic Design, History$\}$

　　$B = \{$History, Graphic Design, Kinesiology, Physics, Business$\}$

　　$C = \{$Art History, Education, Nursing, Biology, Statistics$\}$

　　$D = \{x \mid x$ is a university major$\}$

54. Is $A = B$? Why or why not?

55. Is $B = C$? Why or why not?

56. Is $A = D$? Why or why not?

57. Are any of A, B, C, and D equivalent? Explain.

✎ **WRITING & THINKING**

58. Give an example of a set that cannot be represented using roster notation. Explain your reasoning.

2.1 PROJECT

THE DEDEKIND INN: THE STRANGE CASE OF SETS WITH LARGE CARDINALITY

In Section 2.1, you learned about cardinality and how it represents the number of elements in a set. In this activity, you will explore this idea a little further and investigate the cardinality of a very large set.

Let's start by refreshing our memory. Consider the set of colors of the rainbow, X.

$$X = \{\text{Red, Orange, Yellow, Green, Blue, Indigo, Violet}\}$$

Also consider the set of days of the week, Y.

$$Y = \{\text{Sunday, Monday, Tuesday, Wednesday, Thursday, Friday, Saturday}\}$$

1. What is the cardinality of set X?
2. What is the cardinality of set Y?

These two sets are said to be **equipotent** because they have the same cardinality; that is, we can match each element of set X with a unique element of set Y in what is called a *one-to-one correspondence*. Every element of X is matched to one unique element of Y and every element of Y has a unique match in X. No element in Y is matched to more than one element in X. Here is one possibility of the matching between the sets.

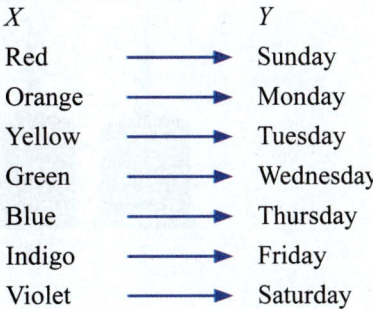

3. Find a different one-to-one correspondence between the two sets.

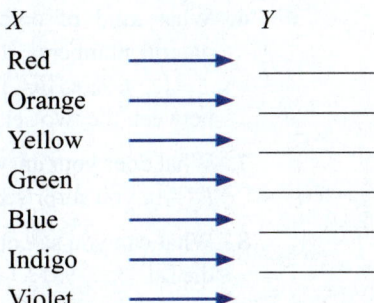

It may surprise you, but we use this idea all the time without even thinking about it! For instance, if a hotel has 100 rooms and they are all full, the hotel can't accept another guest. There is no one-to-one correspondence between the set of rooms (the cardinality of which is 100) and the set of potential guests (the cardinality of which is 101).

Now, imagine a hotel called The Dedekind Inn[2]. This is a strange place with a lot of rooms. In fact, there is one room for each natural number. There is a Room 1, a Room 2, a Room 3, and so on, never ending.

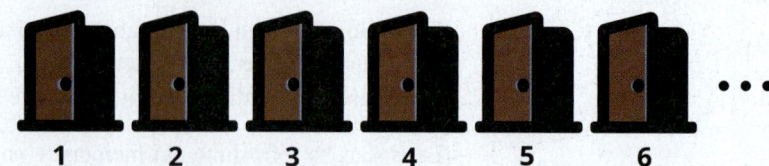

Suppose that the hotel is completely full, and a new guest arrives and requests a room.

4. Would the hotel be able to accommodate the new guest? Why or why not?

Imagine that now management asks every current guest to move to the next room over; that is, the guest in Room 1 moves to Room 2, the guest in Room 2 moves to Room 3, and so on.

5. After this shuffling of guests, which room is now open and ready to receive the new guest? Where did the extra room come from?

Now, suppose that management needs to empty half of the rooms for cleaning. They ask each guest to move to the room whose number is twice their current one.

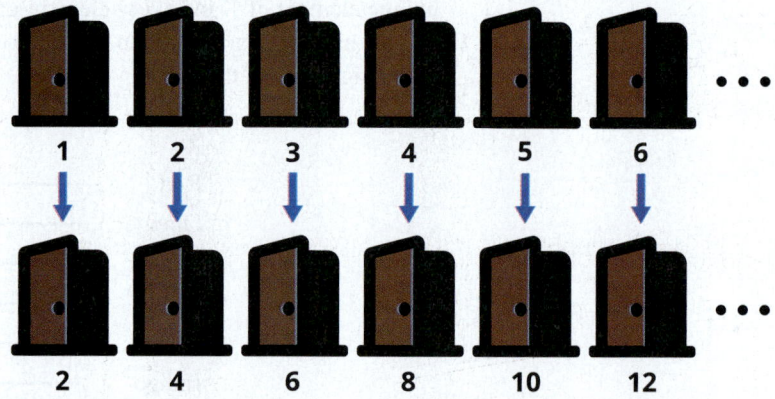

After this rearrangement, every odd room is empty and cleaning can be done in those rooms.

6. What kind of correspondence have you constructed between the set of natural numbers, $\mathbb{N} = \{1, 2, 3, 4, 5, ...\}$, and the set of positive even integers, $E = \{2, 4, 6, 8, 10, ...\}$? Create an expression to describe the correspondence between the two sets.

7. What does your answer in part 6 say about the cardinality of the two sets $\mathbb{N}$ and E? Are you surprised? Explain.

8. What can you say about the correspondence of the set of natural numbers and the set $\{3, 6, 9, 12, 15, ...\}$?

---

2   Julius Wilhelm Richard Dedekind (October 6, 1831 to February 12, 1916) was a German mathematician considered one of the fathers of modern set theory.

2.2 SUBSETS AND VENN DIAGRAMS

OBJECTIVE 1 Venn Diagrams

As we will see throughout this book, mathematics is about so much more than numbers; it is also about structures, mappings, and networks. We are familiar with ways to visualize numbers, whether as a collection of objects, a length, or a sum of money. Similarly, visualizing other mathematical properties such as sets, or data, or structures of ideas is not simply a way to be educational or informative, but also serves as a tool to help solve problems and answer questions.

Visualizing the relationships between sets can be done with the help of a **Venn diagram**. Now used in fields including set theory, logic, probability, statistics, and computer science, Venn diagrams were first introduced by British logician John Venn in the late 1800s. These diagrams have stood the test of time, partly because they are so intuitive and easy to understand. The Venn Diagram shown in Figure 2.2.1 doesn't take much explaining and can easily be used to compare and classify jobs as you begin your search for your dream job.

Figure 2.2.1: Dream Job Venn Diagram

A Venn diagram represents all possible relationships between a collection of sets. The sets are represented by circles (or ovals) contained within a rectangular region representing the universal set. Figure 2.2.2 shows a Venn diagram that represents the sets $A = \{1, 3, 4\}$ and $B = \{1, 2, 8, 10\}$ within the universal set $U = \{x \mid x$ is a whole number and $1 \leq x \leq 10\}$.

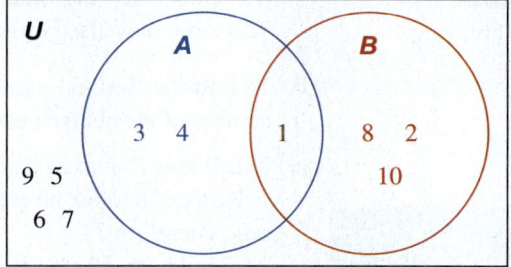

Figure 2.2.2: Venn Diagram

Notice how easy it is to quickly identify the numbers in each set and the numbers not contained in either set. The circles are each labeled with the names of the sets that they represent, and the elements of each set are listed within each circle.

Figure 2.2.3: Leonhard Euler

Figure 2.2.4: John Venn

Think Back

Sets *A* and *B* are equal if they contain exactly the same elements, denoted *A* = *B*. Sets *A* and *B* are equivalent if they have the same cardinal number, denoted *A* ~ *B*.

VENN DIAGRAM

A **Venn diagram** is a visualization of the relationships between a collection of sets. In a Venn diagram, the sets are represented by circles (or ovals) contained within a rectangular region that represents the universal set.

Example 2.2.1

Interpreting Venn Diagrams

The following Venn diagram represents the sets S, T, and V within the universal set $U = \{x \mid x \in \text{positive integers less than } 30\}$. The sets are defined as $S = \{x \mid x \text{ is divisible by } 11\}$, $T = \{x \mid x \text{ is a perfect square}\}$, and $V = \{x \mid x > 2 \text{ and } x \text{ is a perfect cube}\}$. Use the diagram to answer the following questions.

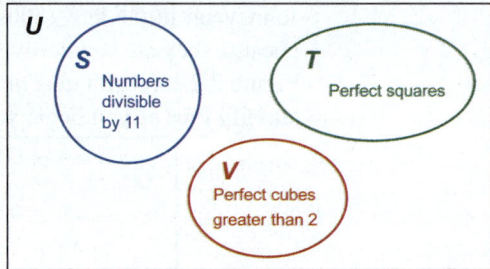

Figure 2.2.5

a. List the elements of the sets S, T, and V in roster notation.

b. Find $|S|$, $|T|$, and $|V|$.

c. Find T'.

d. Is $S = V$? Is $S \sim V$? Explain your answers.

Solution

a. The elements of each set in roster form are as follows.

$S = \{11, 22\}$

$T = \{1, 4, 9, 16, 25\}$

$V = \{8, 27\}$

Remember that the order in which the elements are listed is not important. Therefore, it is also correct to list the elements of each set in a different order.

b. To find the cardinal number of each set (S, T, and V), we need to count the number of elements in each set. Therefore, $|S| = 2$, $|T| = 5$, and $|V| = 2$.

c. Recall that T' contains all of the elements in the universal set that are not in T. Be careful to list all elements not in T and not just the elements in the other sets. Therefore $T' = \{2, 3, 5, 6, 7, 8, 10, 11, 12, 13, 14, 15, 17, 18, 19, 20, 21, 22, 23, 24, 26, 27, 28, 29\}$.

d. To have $S = V$, the two sets would need to have exactly the same elements. Since this is not the case, $S \neq V$. However, S is equivalent to V since both sets have the same cardinal number. So $S \sim V$.

Example 2.2.2

Interpreting Venn Diagrams

An increasing number of electric vehicles are on the roads in the United States. The number of charging locations available for these cars varies from state to state. The five states with the most public and private electric charging locations are California, New York, Florida, Texas, and Massachusetts.

Let

$U = \{x \mid x \in$ all public and private electric charging locations in the United States$\}$

$C = \{x \mid x \in$ all public and private electric charging locations in California$\}$

$N = \{x \mid x \in$ all public and private electric charging locations in New York$\}$

$F = \{x \mid x \in$ all public and private electric charging locations in Florida$\}$

$T = \{x \mid x \in$ all public and private electric charging locations in Texas$\}$

$M = \{x \mid x \in$ all public and private electric charging locations in Massachusetts$\}$

The following Venn diagram depicts the five states with the most electric charging locations. The cardinal number for each set is shown inside the appropriate circle.

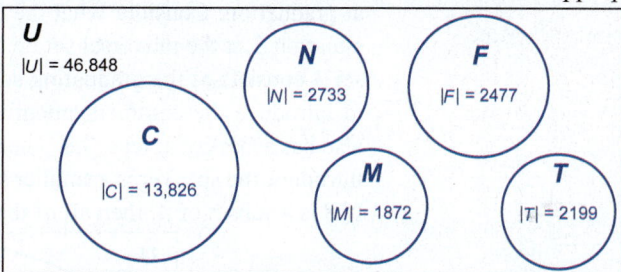

Figure 2.2.6

Use the Venn diagram to answer the following questions. Assume all questions refer to both public and private locations.

a. Which state has the most electric charging locations?

b. Which state has the second highest number of charging locations?

c. How many electric charging locations do these five states have all together?

d. How many charging locations are in the United States but are not in one of the five states in the diagram?

Solution

a. Using the Venn diagram, we can see that California has the most charging locations with 13,826 in the state.

b. Don't be fooled by the size of the circles in the diagram. Remember that the size of a circle is of no consequence in a Venn diagram. Looking at the size of the sets by their cardinality, we see that New York has the second highest number of charging locations with 2733 locations.

c. In order to determine how many combined charging locations there are in these five states, we need to add together the size of all five sets.

$$\text{Total Number of Charging Locations} = |C| + |F| + |N| + |T| + |M|$$
$$= 13{,}826 + 2477 + 2733 + 2199 + 1872 = 23{,}107$$

So the five states with the most electric charging locations have a combined total of 23,107.

d. To find the total number of electric charging locations in the United States that are not in one of these five states, we need to subtract the answer we found in part c. from the total number of charging locations in the universal set. We know the cardinal number of U is 46,848. Therefore, we have $46{,}848 - 23{,}107 = 23{,}741$ electric charging locations that are in the United States but are not in one of the top five states. This means that nearly half of all locations are in these five states!

✔ **Skill Check 2.2.1**

List the elements in A'.

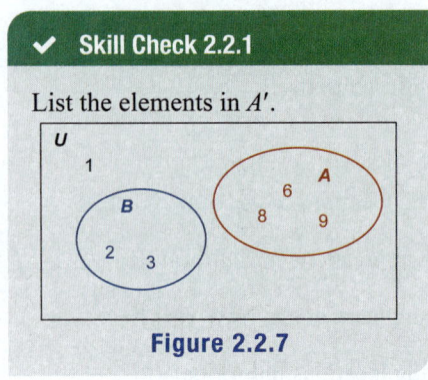

Figure 2.2.7

OBJECTIVE 2 Subsets

Let's consider a situation where the circles in a Venn diagram overlap in a particular way. Suppose that at State University, a graduating senior with a GPA of at least 3.8 is always chosen at random to introduce the commencement speaker at graduation. Consider what the Venn diagram would look like to illustrate this situation. Let the universal set be $U = \{$all students at State University$\}$. Then, let set A consists of the graduating seniors at State University. In order to be chosen to introduce the commencement speaker, you must be a graduating senior who also has a GPA of at least 3.8. The set of all possible students that can be chosen to introduce the speaker is a smaller set B contained within A, referred to as a **subset**. If B is a subset of A, then all of the elements of B are contained within A.

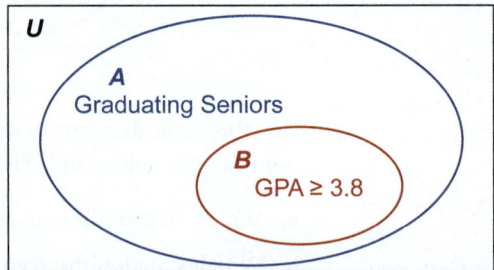

Figure 2.2.8: Venn Diagram of State University

📖 **SUBSET**

Let A and B be sets. The set B is a **subset** of A if every element of B is also an element of A. This is denoted by $B \subseteq A$.

Let's think about what constitutes a subset. Consider the set consisting of the five most downloaded apps in 2021, $X = \{$TikTok, YouTube, Facebook, HBO Max, Instagram$\}$. We could begin to list out all of the subsets, a few of which would be $\{$TikTok, YouTube$\}$, $\{$Instagram, HBO Max$\}$, and $\{$YouTube$\}$. But what about the set $\{$TikTok, YouTube, Facebook, HBO Max, Instagram$\}$, which is equal to X? Is this a subset of X? This set does meet the definition of a subset, since every element of this set is also a member of X. In fact, it will always be true that a set is a subset of itself. Now consider the empty set: is it a subset of X? Since the empty set has no elements, then it also meets the definition since every element in it is also in X; that is, you cannot find an element in the empty set which is not also in X.

If a set is not a subset of another set, the symbol $\not\subseteq$ is used to relate the two sets. For instance, the set $P = \{$Cash App, Zoom, Netflix$\}$ is not a subset of $X = \{$TikTok, YouTube, Facebook, HBO Max, Instagram$\}$, so we write $P \not\subseteq X$.

Example 2.2.3

Determining Subsets

The school orchestra is divided into four sections: strings, woodwinds, brass, and percussion. Let S represent the four instruments in the strings section, $S = \{$violin, viola, cello, double bass$\}$. List all of the subsets of S.

Solution

We know from our discussion that the empty set (which has 0 elements) and the entire set (which has 4 elements) will both be subsets of S. We'll need to find all possible subsets with 1 element, 2 elements, and 3 elements. We can use a table to help us list the subsets so that we don't miss any. Remember that the order of elements in a set is not important, so the set $\{$violin, cello$\}$ is the same as $\{$cello, violin$\}$. Be careful not to list any subsets more than once!

Table 2.2.1: Subsets of the String Section

Subsets with 0 Instruments	Subsets with 1 Instrument	Subsets with 2 Instruments	Subsets with 3 Instruments	Subsets with 4 Instruments
∅	{violin}	{violin, viola}	{violin, viola, cello}	{violin, viola, cello, double bass}
	{viola}	{violin, cello}	{violin, viola, double bass}	
	{cello}	{violin, double bass}	{violin, cello, double bass}	
	{double bass}	{viola, cello}	{viola, cello, double bass}	
		{viola, double bass}		
		{cello, double bass}		

In total there are 16 subsets of S.

Example 2.2.4

Drawing a Venn Diagram with Subsets

According to the Centers for Disease Control and Prevention (CDC), food allergies affect an estimated 8% of children in the United States[3]. Wheat, soy, and peanuts are some of the most common food allergens.

---

3   "Food Allergies," Centers for Disease Control, June 8, 2020, https://www.cdc.gov/healthyschools/foodallergies/index.htm.

Let

$U = \{x | x$ is a child in the United States$\}$

$W = \{x | x$ is a child in the United States who is allergic to wheat$\}$

$Y = \{x | x$ is a child in the United States who is allergic to wheat, soy, or peanuts$\}$

Draw a Venn diagram to represent the sets U, W, and Y.

Solution

Begin by drawing a rectangle representing the universal set of all children in the United States. This rectangle can be any size you like.

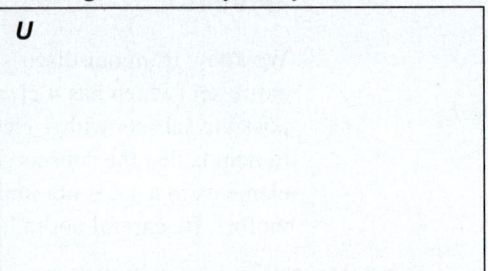

Figure 2.2.9: The Universal Set

Next, we need to decide how the sets W and Y are to be drawn. Notice the set Y contains all children in the United States with an allergy to one of three foods listed, while W is limited to children who are allergic to one of the three foods that define set Y. Therefore, $W \subseteq Y$ and the Venn diagram will look like the following.

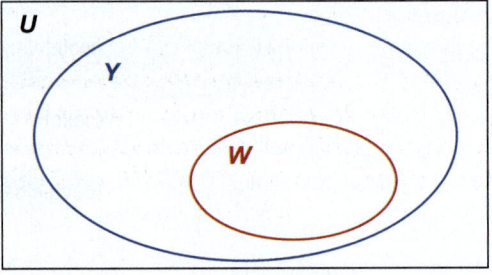

Figure 2.2.10

Although your diagram may look a little different than the one shown in Figure 2.2.10, it should resemble the structure of this one: the oval representing W should be completely contained within the oval representing Y.

> **✔ Skill Check 2.2.2**
>
> Draw a Venn diagram of the following.
>
> $U = \{x \mid x$ is a computer$\}$
>
> $A = \{x \mid x$ is an iPad$\}$
>
> $B = \{x \mid x$ is a tablet computer$\}$

OBJECTIVE 3 Proper Subsets

If set A is a subset of set B, but A is not equal to B, we say that A is a **proper subset** of B. In Example 2.2.4, notice that the set W is a proper subset of Y. This means that there is at least one element in Y (such as a child allergic to peanuts and wheat) who is not in W (children only allergic to wheat); that is, Y and W are not equal to each other. You can think of W as being "properly contained" within Y. When W is a proper subset of Y, we use the notation $W \subset Y$.

> **💬 Helpful Hint**
>
> $\not\subseteq$ = "not a subset of"
>
> $\not\subset$ = "not a proper subset of"

📖 **PROPER SUBSET**

When $B \subseteq A$, and A contains at least one element that is not contained in B, B is a **proper subset** of A and is denoted by $B \subset A$.

Example 2.2.5

Identifying Proper Subsets

Let $M = \{a, b, c, d, e, f\}$. Determine if the following sets are proper subsets of M.

$N = \{a, b, f\}$

$P = \{c\}$

$R = \{a, b, c, d, e, f\}$

$S = \{a, b, h\}$

Solution

Both sets N and P are proper subsets of M because each are subsets of M but are not equal to M. The set R is not a proper subset of M because it contains every element of M; that is, $R = M$. Set S is also not proper subset of M because S is not a subset at all. It contains the element h, which is not in M. That is, $S \nsubseteq M$.

Let's stop and consider the empty set as a proper subset. A proper subset is defined as any combination of elements within the set that does not include the entire set. Since the empty set is a combination of zero elements, it is a proper subset of all sets that include at least one element. However, the empty set cannot be a proper subset of itself by definition. In that same vein, every set is a subset of itself but not a proper subset of itself.

Example 2.2.6

Identifying Proper Subsets

Let $X = \{$Brazil, Vietnam, Columbia$\}$, which are the top three coffee-producing countries in the world (as measured in metric tons). List all the proper subsets of X.

Solution

All proper subsets of X must exclude at least one member of X. Since X has three members, each proper subset can have at most two elements. In fact, the proper subsets may contain two elements, one element, or no elements. A table listing out the possible proper subsets will help us keep track.

Table 2.2.2: Proper Subsets of *X*

Proper Subsets with 0 Elements	Proper Subsets with 1 Element	Proper Subsets with 2 Elements
∅	{Brazil}	{Brazil, Vietnam}
	{Vietnam}	{Brazil, Columbia}
	{Columbia}	{Vietnam, Columbia}

✔ **Skill Check 2.2.3**

List all of the proper subsets of the set {a, b, c, d}.

In total there are 7 proper subsets of *X*.

OBJECTIVE 4   Number of Subsets

In Example 2.2.6, we determined that the set *X*, which consists of 3 elements, has 7 proper subsets. However, it is important to note that the number of proper subsets of a set does not depend on what the individual elements in the set are. The elements could be 1, 2, and 3 instead of Brazil, Vietnam, and Columbia; the names of the elements are different, but the number of subsets will stay the same. In fact, any set with 3 elements will have 7 proper subsets. Although determining the number of proper subsets by listing them out seems easy enough when there are only 3 elements in the set, this method quickly becomes tedious with larger sets. Luckily, knowing the cardinal number of a set is enough to determine the number of subsets and proper subsets of a finite set.

> 📖 **NUMBER OF SUBSETS AND PROPER SUBSETS OF A SET**
>
> If the cardinal number of a set is n, then the set has 2^n subsets and $2^n - 1$ proper subsets.

Since the definition of a proper subset requires that we exclude at least one member of the set, the only subset that is not proper is the set itself. Therefore, the number of subsets and proper subsets differ by only one. We'll leave the verification of why there are 2^n subsets as an exercise.

Example 2.2.7

Determining the Number of Subsets

Dr. Williams is eating at a local buffet one afternoon and notices a sign that says

> "So many possibilities—you could spend a lifetime eating at our buffet and never have the same meal twice!"

He wonders how many different meals he could make from the 16 items on the buffet. He can have all, none, or some of the items. Determine the number of different meals he could make at the buffet.

Solution

When Dr. Williams makes a meal, there is no requirement on the number of food items he needs to have in a meal. For example, he could have a meal with just one item or a meal with all 16 items, or anything in between. Thus, each meal he chooses is a subset of the items from the buffet. The number of different meals that

Dr. Williams could put together is the number of subsets of the 16 items on the buffet. We can use the formula for the number of subsets to determine how many different meals he could make. Using the formula for the number of subsets, we have the following.

$$\text{Number of Plates} = 2^{16} = 65{,}536.$$

This means that if Dr. Williams ate at the buffet once a day every day, he could eat for almost 180 years without duplicating a meal. Certainly, it would seem the sign is not misleading.

Skill Check Answers

1. 1, 2, 3
2.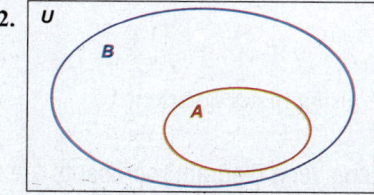
3. $\varnothing, \{a\}, \{b\}, \{c\}, \{d\}, \{a, b\}, \{a, c\}, \{a, d\}, \{b, c\}, \{b, d\}, \{c, d\}, \{a, b, c\},$
   $\{a, b, d\}, \{a, c, d\}, \{b, c, d\}$

2.2   Exercises

✔ CONCEPT CHECK

1. A _____ is a visualization of the relationships that exist between sets.

2. If set A is a subset of set B, that means every element of set ____ is also an element of set ____.

3. In a proper subset, there is at least one _____ of the set that is not in the subset.

4. True or False: The empty set is a subset of every set.

5. True or False: If set X is a subset of set Y and set X is also equivalent to set Y, then set X is not a proper subset of set Y.

Use the Venn diagram to solve each problem.

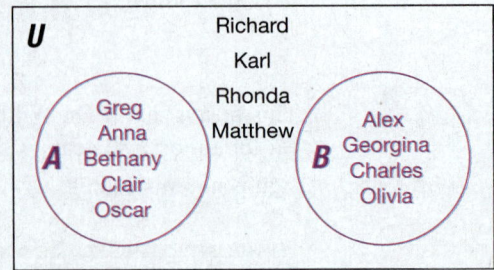

6. List *A* and *B* using roster notation.

7. Find *A'*.

8. List *U* using roster notation.

The following Venn diagram represents the sets *M*, *N*, and *P* within the universal set $U = \{x \mid x$ is a positive integer and $x \le 25\}$ The sets are defined as $M = \{x \mid x$ is a multiple of 5\}, $N = \{x \mid x$ is divisible by 12\}, and $P = \{x \mid x$ is a perfect squuare\}. Use the diagram to answer the following questions.

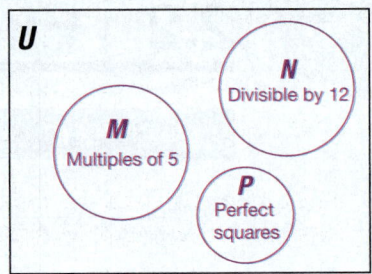

9. List the elements of sets *M*, *N*, and *P* in roster notation.

10. Find $|M|$, $|N|$, and $|P|$.

11. Find *M'*.

12. Is $M = P$? Is $M \sim P$? Explain your answers.

The following Venn diagram represents the sets Q, R, and S within the universal set $U = \{x \mid x$ is a positive integer and $x < 20\}$. The sets are defined as $Q = \{x \mid x$ is an even number$\}$, $R = \{x \mid x$ is divisible by 2$\}$, and $S = \{x \mid x$ is a multiple of 0$\}$. Use the diagram to answer the following questions.

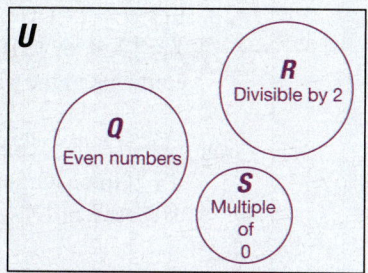

13. List the elements of sets Q, R, and S in roster notation.

14. Find $|Q|$, $|R|$, and $|S|$.

15. Find R'.

16. Is $Q = R$? Is $Q \sim R$? Explain your answers.

17. Let $A = \{0, 1, 2, 3, 4, 5\}$ and $B = \{5, 4, 3, 2, 1\}$. Is $B \subseteq A$?

18. Let $A = \{\{a, b\}, c, d, e\}$ and $B = \{a, b, c\}$. Is $B \subseteq A$?

19. The set B contains the names of three of the most expensive paintings ever sold. If $B = \{$*The Card Players* by Cézanne, *No. 5 1948* by Pollock, *Woman III* by de Kooning$\}$, list all the subsets of B.

20. Given $A = \{x \mid x \in$ positive whole numbers less than 100$\}$ and $B = \{x \mid x \in$ even non-negative integers less than 100$\}$, is either set a subset of the other? Explain.

List all subsets of the given set.

21. $L = \{$lemon, lime$\}$

22. $T = \{1, 2, 3\}$

23. $Q = \{$four, vier, cuatro, quatre$\}$

24. $W = \{$one$\}$

Draw each Venn diagram.

25. Let U consist of all artists. Draw a Venn diagram to represent the two sets violinist and musicians.

26. Let U consist of all four legged animals. Draw a Venn diagram to represent the two sets male dogs and female dogs.

27. Draw a separate Venn diagram to illustrate each of the following.

    a. $W \subseteq Y$

    b. $Y \subseteq W$

28. Let $U = \{x \mid x \in \mathbb{R}\}$, $W = \{x \mid x$ is a counting number less than $20\}$, and $Y = \{2, 4, 6, 8, 10\}$. Draw a Venn diagram to represent U, W, and Y with the elements in the proper regions.

29. Let U = {Red, Orange, Yellow, Green, Blue, Indigo, Violet}, A = {Green, Orange, Yellow}, and B = {Indigo, Violet}. Draw a Venn diagram to represent U, A, and B with the elements in the proper regions.

Solve.

30. Let P = {baklava, cannoli, eclair, polvorones, strudel}. Determine if the following sets are proper subsets of P.

    a. A = {baklava, cannoli, eclair}

    b. B = { }

    c. C = {baklava, strudel, cannoli, eclair, polvorones}

    d. D = {polvorones, pie, paxlava}

31. Let M = {c, a, l, i, p, e, r}. Determine if the following sets are proper subsets of M.

    a. A = {r, e, p, l, i, c, a}

    b. B = {r, e, c, a, p}

    c. C = {l, e, a, p}

    d. D = {r, a, i, s, e}

List all proper subsets of the given set.

32. The set C contains the names of the top three grossing films of all time as of Summer 2014, C = {*Avatar, Titanic, Marvel's The Avengers*}.

33. F = {2, 4, 6}

34. P = {pen, pencil}

35. W = {cloudy, rainy, sunny}

36. T = {10, 20, 30, 40}

37. L = {0}

Solve.

38. W contains The New York Times' top five fiction books for 2013. If W = {*The Goldfinch, Americanah, The Flamethrowers, Life After Life, Tenth of December*}, how many subsets does W contain? How many proper subsets?

39. Given that $B = \{ \text{♩}, \text{⚕}, \heartsuit, \text{☺}, \text{◎}, \text{☎}, \text{🚲}, \text{☺} \}$. How many subsets does B have? How many proper subsets does B have?

40. Determine the number of subsets contained in Y if $Y = \{ x \mid x \text{ is an odd positive integer and } x < 25 \}$.

41. A set has 32 subsets. How many elements are in the set?

42. A set has 127 proper subsets. How many elements are in the set?

43. The national fast food chain Wendy's advertises that there are 256 ways to personalize a Wendy's hamburger. How many condiments must Wendy's carry for their customer to have this many choices?

🚀 **APPLICATIONS**

Every year, thousands of deaths happen as the result of a drunk driver causing a car crash. The number of deaths varies state to state, with higher numbers occurring in states with higher populations. In 2019, the five states with the highest number of deaths due to drunk drivers were California, Florida, Georgia, North Carolina, and Texas. Let

$U = \{ x \mid x \in \text{death as the result of drunk driving in the United States} \}$

$C = \{ x \mid x \in \text{death as the result of drunk driving in California} \}$

$F = \{ x \mid x \in \text{death as the result of drunk driving in Florida} \}$

$G = \{ x \mid x \in \text{death as the result of drunk driving in Georgia} \}$

$N = \{ x \mid x \in \text{death as the result of drunk driving in North Carolina} \}$

$T = \{ x \mid x \in \text{death as the result of drunk driving in Texas} \}$

The following Venn diagram depicts the five states with the most deaths that occur as the result of drunk driving in 2019 within the universal set of all deaths that occurred as a result of drunk driving in the United States. The cardinal number for each set is shown inside the appropriate circle.

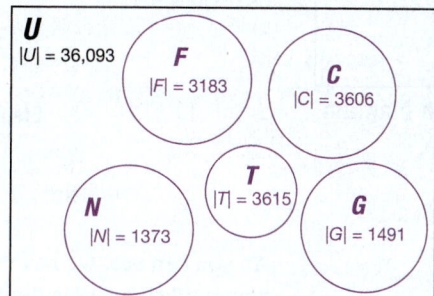

44. Which state had the most deaths caused by drunk driving in 2019?

45. Which state had the second highest number of deaths caused by drunk driving in 2019?

46. How many total deaths caused by drunk driving happened in these five states in 2019?

47. How many deaths caused by drunk driving happened in the United States in 2019 but did not happen in one of these five states?

✏ **WRITING & THINKING**

48. Explain why Red ⊄ {Red, Blue, Green}.

49. A new keyless car security system allows the owner to create a "handprint" to start the car by selecting up to five places on the six-space keypad on which to place their fingers. How many possible handprints can the user choose from if at least one space must be selected?

50. Pizza Jet promotes a special lifetime offer—once you order all possible pizza combinations from Pizza Jet, you can get free pizza for life! Pizza Jet offers a selection of 12 toppings (including cheese) for their pizzas. How many different pizzas must you order from Pizza Jet to qualify for their lifetime offer? Is this possible? Explain.

2.2 PROJECT

ONES, ZEROS, AND THE NUMBER OF REGIONS IN A VENN DIAGRAM

A computer chip is basically a collection of transistors, which are electronic components that work as switches. A transistor can be in one of two states: on or off. In computer engineering, the number 1 is used to represent the *on* state while 0 is used to represent the *off* state. Numbers composed only of zeros and ones are called binary numbers.

In this activity, you will investigate the relationship between binary numbers and the regions of a Venn diagram. Consider the one-set Venn diagram in Figure 1. We have labeled each region of the diagram by asking the question, "Does this region contain elements of set A?" If the answer is yes, we labeled the region with (1); if the answer is no, we labeled the region with (0). Table 1 shows a summary of the regions.

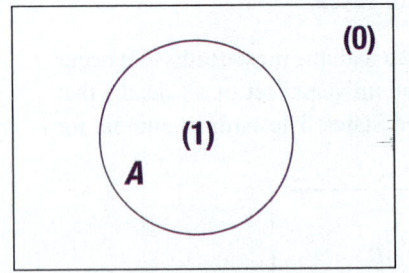

Figure 1: One-Set Venn Diagram

Table 1: Regions in Figure 1

Elements of Set A?	Region
Yes	(1)
No	(0)

As you can see, we have exactly 2 regions: (1) and (0). We are labeling the Venn diagram using single digit binary numbers. Let's use the same idea to label a two-set Venn diagram. (Note that some regions in the figures may end up containing no elements when actual sets are considered, but the diagrams take into account all possible regions when considering the relationship between a fixed number of sets.) The region in Figure 2 that contains no elements of set A but some of the elements of set B is labeled (10), and the region that contains no elements of set

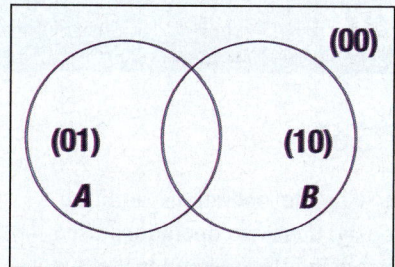

Figure 2: Two-Set Venn Diagram

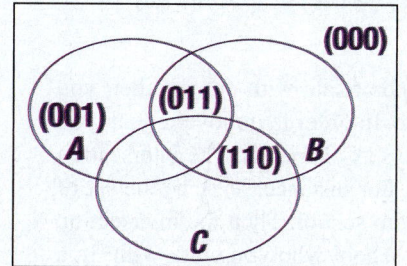

Figure 3: Three-Set Venn Diagram

B but some of the elements of set *A* is labeled (01). This time, we are labeling the Venn diagram using two-digit binary numbers.

Table 2: Regions in Figure 2

Elements of Set *B*?	Elements of Set *A*?	Region
No	No	(00)
No	Yes	(01)
Yes	No	(10)
Yes	Yes	

1. Write 2, which is the number of regions in Figure 1, as a power of 2. In other words, 2 raised to what power is equal to 2?

2. In Figure 2, what should be the label for the intersection of set *A* and set *B*, the region that contains elements that are both in set *A* and in set *B* at the same time?

3. How many regions are in Figure 2? Write this value as a power of 2.

4. Complete Table 3 and label the three-set Venn diagram in Figure 3 using the same process we used with the one-set Venn diagram and the two-set Venn diagram. For this Venn diagram, you will work with three-digit binary numbers.

Table 3: Regions in Figure 3

Elements of Set *C*?	Elements of Set *B*?	Elements of Set *A*?	Region
No	No	No	(000)
No	No	Yes	(001)
	Yes	No	(010)
	Yes	Yes	(011)
			(100)
Yes	No	Yes	
Yes	Yes	No	
Yes	Yes	Yes	

5. How many regions are there in Figure 3? Write the number as a power of 2.

6. How do you determine the number of regions in a Venn diagram for a fixed number of *n* sets?

7. Without drawing a diagram determine the number of regions in a four-set Venn diagram.

8. List all regions for a four-set Venn diagram using four-digit binary numbers. (**Hint:** Notice that in Tables 1, 2, and 3, if a region contains elements of set *A*, then the rightmost digit of the region is 1. Look for similar patterns.)

2.3 OPERATIONS WITH SETS

OBJECTIVES

1. Determine the intersection of sets.

2. Determine the union of sets.

3. Combine set operations.

4. Use De Morgan's Laws.

OBJECTIVE 1 Intersection of Sets

When we work with numbers, we are familiar with the operations addition, subtraction, multiplication, and division. Analogously, there are operations used to combine sets together to create other sets, but not using the operations that we use with numbers. When working with sets, the *set operations* are **intersection** and **union**. In this section, we explore the definitions of intersection and union, their properties, and their applications.

Suppose you are coordinating an event. You want to invite friends from work, friends from school, friends you know through working out, and friends of friends. Perhaps some of the people you wish to invite fall into more than one category. You could have a friend from work who also exercises with you. You could have friends of friends that also attend your school but aren't people you know. You could also have a friend who you want to invite that doesn't fit into any of the categories mentioned.

By looking at the way the different sets of friends relate with one another, you can determine how big your invitation list will be. In order to not over count, you can look for duplicates in the **intersection** of the sets of friends. The intersection of two sets is the elements that are in both sets. For instance, let X be the set of friends from work and Y be the set of friends from school. Then the intersection of X and Y are the friends who you know from school who you work with. In a Venn diagram, we represent an intersection with overlapping circles. Figure 2.3.1 shows the intersection of friends from work and school.

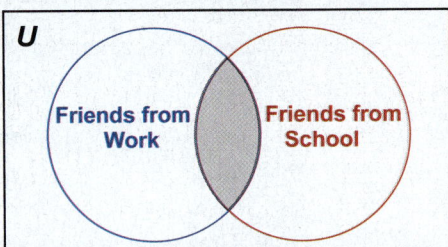

Figure 2.3.1: Intersection of Sets

📖 INTERSECTION

The **intersection** of two sets A and B is the set of all elements common to both A and B. We denote the intersection of A and B as $A \cap B = \{x \mid x \in A$ and $x \in B\}$.

Example 2.3.1

Determining the Intersection of Sets

a. Find the intersection of the sets $A = \{2, 3, 5, 7, 11, 13\}$ and $B = \{2, 4, 6, 8, 10, 12\}$.

b. Interpret the intersection of the sets in the Venn diagram in Figure 2.3.2.

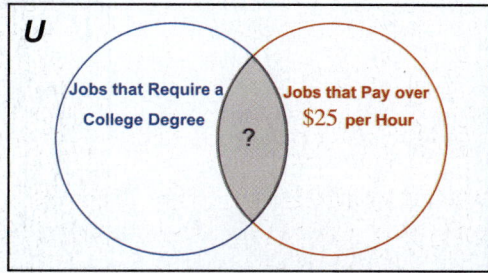

Figure 2.3.2

Solution

a. Since the intersection of two sets consists of all of the elements that appear in both sets at the same time, we can see that the intersection of A and B consists of only the number 2.

$$A \cap B = \{2, 3, 5, 7, 11, 13\} \cap \{2, 4, 6, 8, 10, 12\} = \{2\}$$

We can also use a Venn diagram to find the intersection.

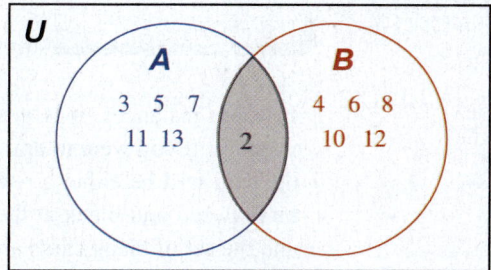

Figure 2.3.3

You may have noticed that A is a list of the first six prime numbers and B is the list of the first six positive even integers. As a reminder, 2 is the only prime number that is even!

b. The intersection of two sets is the set of elements that are common to both sets. In the given Venn diagram, we see a set representing jobs that require a college degree and a second set representing jobs that pay over $25 an hour. The intersection of these, denoted by the overlapping portion of the circles in the diagram, represents jobs that require a college degree *and* pay over $25 an hour.

> **💬 Helpful Hint**
>
> In a Venn diagram, any elements in the intersection of sets are only listed once. Likewise, we only list elements once when using the roster notation for an intersection.

As is the case in part b. of the previous example, it is sometimes impractical to list every element of a set in a Venn diagram. In such cases, we can display the number of elements in each region instead of listing the elements themselves. Example 2.3.2 illustrates this.

Example 2.3.2

Using a Venn Diagram to Find the Intersection of Sets

Harvard Law School has produced more Supreme Court Justices than any other law school. Use the Venn diagram below to interpret how many of the sitting Justices as of August 2021 graduated from Harvard Law School.

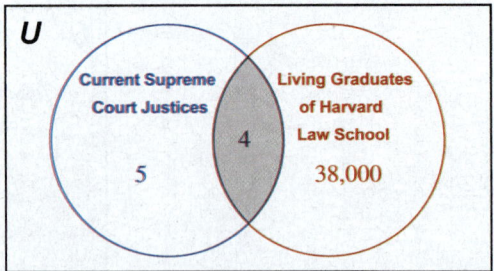

Figure 2.3.4

Solution

Using the Venn diagram, we see that there are more than 38,000 living alumni of Harvard Law School and that of the 9 Supreme Court Justices, 5 of them graduated from a law school other than Harvard. The overlapping portion in the middle of the diagram gives us our answer and tells us that 4 of the current Supreme Court Justices graduated from Harvard Law School. (If you're curious, the 4 justices are Breyer, Gorsuch, Kagan, and Roberts.)

> ✔ **Skill Check 2.3.1**
>
> $A = \{27, 111, 213\}$ and $B = \{213, 40, 61, 88, 210, 27\}$. Find $A \cap B$.

In some instances, it is possible that two sets have no common elements. For instance, if you were to draw a single card out of a standard deck of playing cards, the card will be either a red card or a black card. It is impossible for the card to be both red and black at the same time. In this case, we say the set of red cards and the set of black cards are **disjoint**; that is, they have no elements in common. When sets are disjoint, their intersection will always be the empty set.

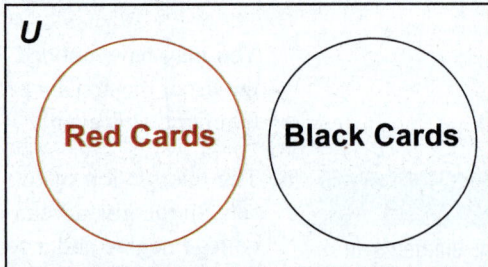

Figure 2.3.5: Disjoint Sets

> 💬 **Helpful Hint**
>
> Disjoint sets can also be represented by a Venn diagram where the overlap between the sets is empty, as shown below.
>
>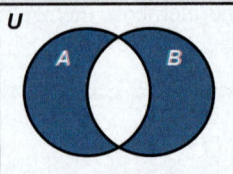
>
> **Figure 2.3.6**

📖 DISJOINT

Two sets A and B are **disjoint** if there are no elements in set A that are also contained in set B. Their intersection is the empty set, denoted by $A \cap B = \varnothing$.

Example 2.3.3

Identifying Disjoint Sets

Let $U = \{$all students$\}$, $A = \{$students with GPA $< 2.5\}$, and $B = \{$students with GPA $> 3.0\}$. Determine if sets A and B are disjoint and draw a Venn diagram to illustrate the relationship between sets A and B.

Solution

Since it would be impossible for a student to be in both set A (by having a GPA that is less than 2.5) and set B (by having a GPA greater than 3.0), sets A and B are disjoint.

In this Venn diagram, the universal set represents all students regardless of their grade point average. Because sets A and B are disjoint, there are no common GPAs between the two sets, which is illustrated by two circles not overlapping. Any student with a GPA between 2.5 and 3.0 is not represented in either set A or set B.

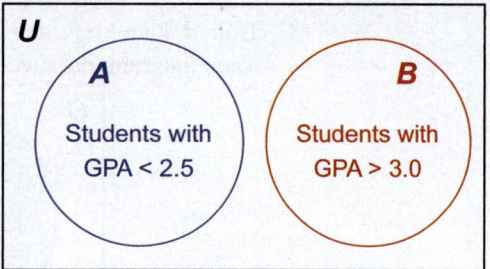

Figure 2.3.7

OBJECTIVE 2 Union of Sets

The second set operation is the **union** of sets. The union of two sets is a new set of elements that are in either of the original sets. For instance, if set A = {Paris, Amsterdam, New York} and set B = {Tokyo, Delhi, New York, London}, then the union of A and B is the set {Paris, Amsterdam, New York, Tokyo, Delhi, London}. Notice that the union includes every element in each set, but we only list a particular element once since repetition of elements is ignored.

📖 UNION

The **union** of two sets A and B is the set of all elements in A or in B. We denote the union of A and B as $A \cup B = \{x \mid x \in A \text{ or } x \in B\}$.

Example 2.3.4

Determining the Union of Sets

Sarah has lived in Ohio, Pennsylvania, and Michigan. Her friend Trevor has lived in Colorado, Ohio, and Georgia. Her friend Ron has lived in Washington and Idaho. Let the residential locations of each friend be represented by the following sets: S = {OH, PA, MI}, T = {CO, OH, GA}, and R = {WA, ID}. Find the following unions.

a. $S \cup T$

b. $S \cup R$

Solution

a. To find the union of sets S and T, we need to find the elements that appear in either set S or set T. The elements of set S are OH, PA, and MI, and the elements of set T are CO, OH, and GA. We simply combine the items together to create a new set to represent the union.

$$S \cup T = \{\text{OH,PA,MI}\} \cup \{\text{CO,OH,GA}\}$$
$$= \{\text{OH,PA,MI,CO,GA}\}$$

Note that any repeated elements only get listed once. Figure 2.3.8 shows the Venn diagram that gives us a visual illustration of the union.

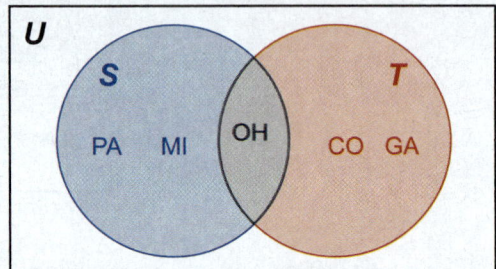

Figure 2.3.8

Note that $S \cap T = \{\text{OH}\}$. In other words, the state that Sarah and Trevor have both lived in is Ohio.

b. Finding the union of sets S and R is done in a similar manner.

$$S \cup R = \{\text{OH,PA,MI}\} \cup \{\text{WA,ID}\}$$
$$= \{\text{OH,PA,MI,WA,ID}\}$$

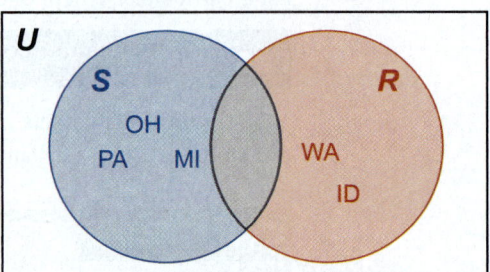

Figure 2.3.9

Using the Venn diagram, it is easy to see that the intersection of these two sets is empty because Sarah and Ron have not lived in any of the same states. Therefore, $S \cap R = \varnothing$ and sets S and R are disjoint.

> ✔ **Skill Check 2.3.2**
>
> Let U be students enrolled at MLK High School, $A = \{x \mid x \in$ student band$\}$ and $B = \{x \mid x \in$ student government$\}$. Find $A \cup B$.

Just as with other sets, we can calculate the cardinality of the union of two sets, denoted by $|A \cup B|$. Remember that when we find the union, we don't list an element more than once. Likewise, when we find the cardinality of the union, we need to be careful not to count elements more than once. That means we have to do a bit more than simply add the cardinal numbers of the sets together. We must also subtract off the number of elements that appear in the intersection of the sets. This principle is called the **inclusion-exclusion principle**.

📖 INCLUSION-EXCLUSION PRINCIPLE

The **inclusion-exclusion principle** states that the cardinality of the union of two sets A and B is calculated by adding the number of elements in set A to the number of elements in set B and subtracting the number of elements that appear in both sets.

$$|A \cup B| = |A| + |B| - |A \cap B|$$

Example 2.3.5

Applying the Inclusion-Exclusion Principle

Consider the scenario where a state senate is voting on a bill. The senate has 35 members, of which 15 are Democrats and 20 are Republicans. Five of the 35 members represent urban areas of the state, two of which are Democrats. The bill is expected to receive "Yes" votes from the Democrat members and from members who represent urban areas. Determine how many Yes votes the bill is expected to receive.

Solution

Let set A consist of Democrats and set B consist of members representing urban areas. The number of expected Yes votes is the number of elements in their union.

The Inclusion-Exclusion Principle tells us that $|A \cup B| = |A| + |B| - |A \cap B|$. In other words, the total number of Yes votes will be the total number of Democrat members plus the total number of members representing urban areas minus the number of members who appear in both of those groups.

We are told that two members are Democrats who represent urban areas; that is, $|A \cap B| = 2$. We also know $|A| = 15$ and $|B| = 5$. Using the formula, we have the following.

$$|A \cup B| = |A| + |B| - |A \cap B| = 15 + 5 - 2 = 18$$

Thus, the bill is expected to receive 18 Yes votes.

OBJECTIVE 3 Combining Operations

Helpful Hint

Recall that the complement of set A consists of all the elements in the given universal set that are not contained in A. The complement of A is denoted A'.

In addition to finding the intersection or the union of two sets, we can also combine operations involving intersections, unions, and complements. When working with numbers, when we want an operation to be performed first, we indicate this with parentheses. Likewise with sets, we always execute operations inside parentheses first. If there are nested parentheses, we work from the inside out. Then we determine complements. The operations of union, intersection, and complement are then done in order from left to right.

Example 2.3.6

Combining Intersection and Union

Let $U = \{a, b, c, d, e, \ldots, z\}$, $M = \{m, a, t, h\}$, $N = \{m, o, n, e, y\}$, and $K = \{i, n, v, e, s, t, o, r\}$. Find the following.

a. $M \cup (N \cap K)$

b. $M \cap (N \cup K)$

Solution

a. In order to find the solution when combining the operations of intersections and unions of sets, it might be best to describe the set first. The set $M \cup (N \cap K)$ is the set of all elements that are in M or are in the intersection of N and K.

$$M \cup (N \cap K) = \{m, a, t, h\} \cup (\{m, o, n, e, y\} \cap \{i, n, v, e, s, t, o, r\})$$

Just as in order of operations with numbers, we need to perform the operation in parentheses first. That is, find $N \cap K$ first.

$$N \cap K = \{m, o, n, e, y\} \cap \{i, n, v, e, s, t, o, r\} = \{o, n, e\}$$

Now, we can replace $N \cap K$ with the simplified set we just found and solve as follows.

$$M \cup (N \cap K) = \{m, a, t, h\} \cup \{o, n, e\}$$
$$= \{m, a, t, h, o, n, e\}$$

b. Similarly, we can find $M \cap (N \cup K)$ by performing the union inside the parentheses first and then the intersection with M.

$$M \cap (N \cup K) = \{m, a, t, h\} \cap (\{m, o, n, e, y\} \cup \{i, n, v, e, s, t, o, r\})$$
$$= \{m, a, t, h\} \cap \{m, o, n, e, y, i, v, s, t, r\}$$
$$= \{m, t\}$$

> ✔ **Skill Check 2.3.3**
>
> Let $M = \{m, a, t, h\}$, $N = \{m, o, n, e, y\}$, and $K = \{i, n, v, e, s, t, o, r\}$.
>
> Find $K \cap (M \cup N)$

Example 2.3.7

Combining Operations

Let $U = \{1, 2, 3, 4, 5, 6, 7, 8, 9, 10\}$, $A = \{1, 2, 3, 4, 5\}$, and $B = \{2, 4, 6, 8\}$. Find the following.

a. $(A \cup B)'$

b. $B' \cap A$

Solution

a. Notice that the complement sign is on the outside of the parentheses. That means that we need to perform the operation inside the parentheses first; that is, find the union of A and B.

$$A \cup B = \{1, 2, 3, 4, 5, 6, 8\}$$

Now, we take the complement of $A \cup B$.

$$(A \cup B)' = \{7, 9, 10\}$$

b. To find the intersection of these sets, we need to first find the complement of B.

$$B' = \{1, 3, 5, 7, 9, 10\}$$

Next, we take the intersection of B' and A.

$$B' \cap A = \{1, 3, 5\}$$

OBJECTIVE 4 De Morgan's Laws

One of the particulars of set operations is determining when to take the complement of a set, as we saw in the Example 2.3.7. We have to be careful not to assume that the expression $A' \cap B'$ results in the same set as the expression $(A \cap B)'$. Notice that the first expression says to find the complements of each set and then find their intersection. The second expression says to find the intersection of the sets and then take the complement. Let's look at a simplified example of the difference in the two.

Let $U = \{1, 2, 3\}$, $A = \{1, 2\}$, and $B = \{2, 3\}$. Then $A' = \{3\}$, $B' = \{1\}$, and $A \cap B = \{2\}$. Now we can find both $A' \cap B'$ and $(A \cap B)'$.

$$A' \cap B' = \{3\} \cap \{1\} = \varnothing$$

$$(A \cap B)' = \{1, 3\}$$

Thus, these operations do not produce the same set; that is, $A' \cap B' \neq (A \cap B)'$.

Take a closer look at what we got for $(A \cap B)'$. It is the set $\{1, 3\}$. Can you see another way to describe this set based on what we know? The set $\{1, 3\}$ is also equal to $A' \cup B'$. In fact, this will always be true, regardless of what sets A and B are. The complement of an intersection of two sets is equal to the union of the complements of the two sets.

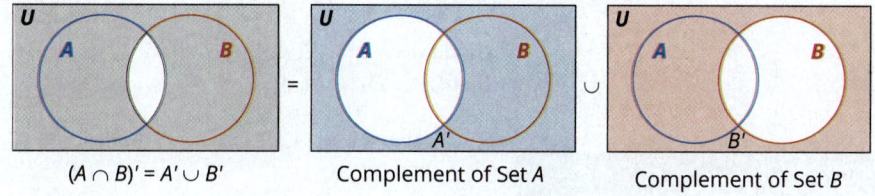

$(A \cap B)' = A' \cup B'$ Complement of Set A Complement of Set B

Figure 2.3.10

In the same manner, let's look at what happens with the complement of a union. Using the same sets, let's find $(A \cup B)'$ and $A' \cup B'$.

Let $U = \{1, 2, 3\}$, $A = \{1, 2\}$, and $B = \{2, 3\}$. Then $A' = \{3\}$, $B' = \{1\}$, and $A \cup B = \{1, 2, 3\}$. Thus, $A' \cup B' = \{3, 1\}$ and $(A \cup B)' = \varnothing$. Again, these operations do not produce the same results; that is, $(A \cup B)' \neq A' \cup B'$. However, $(A \cup B)'$ is equal to $A' \cap B'$, which in this case are both the empty set. (Check this for yourself.)

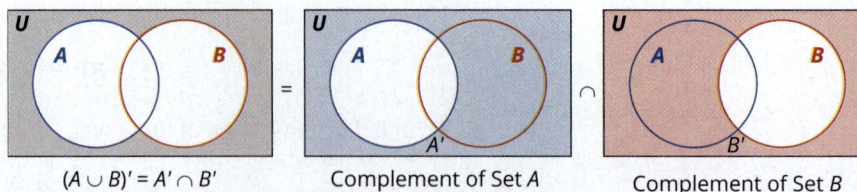

$(A \cup B)' = A' \cap B'$ Complement of Set A   Complement of Set B

Figure 2.3.11

These two rules are known as **De Morgan's Laws**, named after the British mathematician Augustus De Morgan.

> ### 📖 DE MORGAN'S LAWS
>
> Let A and B be sets. Then
>
> $$(A \cup B)' = A' \cap B'$$
>
> and
>
> $$(A \cap B)' = A' \cup B'.$$

Example 2.3.8

Using De Morgan's Laws

Let $U = \{x \mid x \in \text{positive integers less than or equal to } 10\}$. If set A represents even numbers that are less than or equal to 10 and set B represents prime numbers that are less than or equal to 10, then we can write the sets as follows.

$U = \{1, 2, 3, 4, 5, 6, 7, 8, 9, 10\}$

$A = \{2, 4, 6, 8, 10\}$

$B = \{2, 3, 5, 7\}$

Verify that $(A \cup B)' = A' \cap B'$.

Solution

Since $A \cup B = \{2, 4, 6, 8, 10\} \cup \{2, 3, 5, 7\} = \{2, 3, 4, 5, 6, 7, 8, 10\}$, we know that $(A \cup B)' = \{1, 9\}$.

Similarly, $A' = \{1, 3, 5, 7, 9\}$ and $B' = \{1, 4, 6, 8, 9, 10\}$. That means that $A' \cap B' = \{1, 9\}$.

Notice that $(A \cup B)' = A' \cap B'$. Hence, we have verified De Morgan's law for A and B.

✔ Skill Check 2.3.4

Given sets $U = \{a, b, c, d, e, f, g, h, i, j, k, l\}$, $A = \{l, i, k, e\}$, and $B = \{c, a, k, e\}$, verify $(A \cap B)' = A' \cup B'$.

Skill Check Answers

1. $A \cap B = \{27, 213\}$

2. $A \cup B = \{x \mid x \in \text{student band or } x \in \text{student government}\}$ 3. $\{e, n, o, t\}$

4. $A \cap B = \{k, e\}$, so $(A \cap B)' = \{a, b, c, d, f, g, h, i, j, l\}$.

$A' = \{a, b, c, d, f, g, h, j\}$ and $B' = \{b, d, f, g, h, i, j, l\}$, so $A' \cup B' = \{a, b, c, d, f, g, h, i, j, l\}$.

This gives that $(A \cap B)' = A' \cup B'$.

2.3 Exercises

✔ CONCEPT CHECK

1. The _____ of two sets is represented by the overlapping region in a Venn diagram.

2. The _____ of two sets contains all elements of the two sets.

3. De Morgan's Laws are used to find the _____ of the union or intersection of two sets.

4. True or False: If two sets are disjoint, then the cardinality of the union of the two sets is equal to the sum of the cardinality of each set.

💡 PRACTICE

Use set notation to represent each shaded region.

5.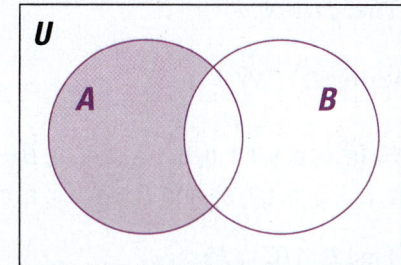

6.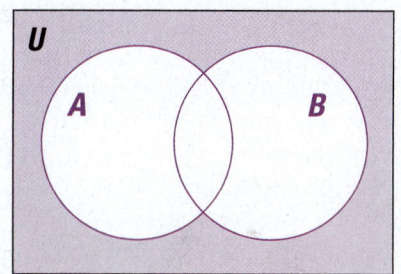

Show that each pair of sets is equal by drawing a Venn diagram of each set.

7. $A \cap B$ and $B \cap A$

8. $A \cup B$ and $B \cup A$

9. $(A \cap B) \cap C$ and $A \cap (B \cap C)$

10. $(A \cup B) \cup C$ and $A \cup (B \cup C)$

11. $A \cup \varnothing$ and A

Let $U = \{1, 2, 3, 4, 5, 6, 7, 8, 9, 10\}$, $A = \{2, 4, 6, 8, 10\}$, and $B = \{1, 4, 9\}$.

12. Find $A \cup B$.

13. Find $A \cap B$.

14. Find $|A \cup B|$.

15. Find $|A \cap B|$.

16. Verify $(A \cup B)' = A' \cap B'$.

Let $U = \{1, 2, 3, 5, 8, 13, 21, 34\}$, $X = \{1, 3, 5, 13, 21\}$, and $Y = \{3, 21\}$.

17. Find $X \cup Y$.

18. Find $X \cap Y$.

19. Find $|X \cup Y|$.

20. Find $|X \cap Y|$.

21. Verify $(X \cup Y)' = X' \cap Y'$.

Let $U = \{a, r, c, h, i, v, e, d\}$, $H = \{h, i, r, e, d\}$, and $R = \{r, i, c, h\}$.

22. Find $H \cup R$.

23. Find $H \cap R$.

24. Find $|H \cup R|$.

25. Find $|H \cap R|$.

26. Verify $(H \cap R)' = H' \cup R'$.

Let $U = \{e, d, u, c, a, t, i, o, n\}$, $D = \{d, a, n, c, e\}$, and $N = \{n, o, t, e\}$.

27. Find $D \cup N$.

28. Find $D \cap N$.

29. Find $|D \cup N|$.

30. Find $|D \cap N|$.

31. Verify $(D \cap N)' = D' \cup N'$.

Let $U = \{c, o, p, y, r, i, g, h, t, a, b, l, e\}$, $B = \{b, i, o, g, r, a, p, h, y\}$, $C = \{c, h, i, p, o, t, l, e\}$, and $P = \{p, i, r, a, c, y\}$.

32. Find $B \cap (C \cup P)$.

33. Find $B \cup (C \cap P)$.

34. Find $P \cup (B \cap C)$.

35. Verify $(C \cup P)' = C' \cap P'$.

Let $U = \{m, e, l, t, d, o, w, n, s\}$, $S = \{s, n, o, w, m, e, l, t\}$, $M = \{m, o, d, e, l\}$, and $T = \{t, o, w, e, d\}$.

36. Find $S \cap (M \cup T)$.

37. Find $S \cup (M \cap T)$.

38. Find $T \cup (S \cap M)$.

39. Verify $(M \cap T)' = M' \cup T'$.

🚀 **APPLICATIONS**

Use the Venn diagram to solve each problem.

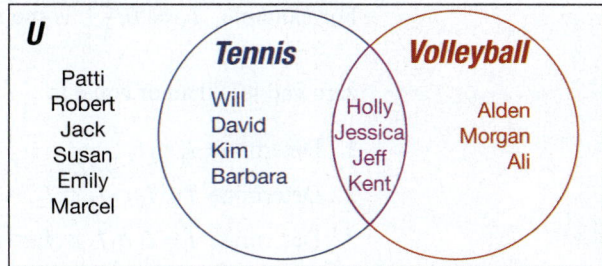

40. Which students played only tennis?

41. Determine which students played tennis or volleyball.

42. Determine which students played tennis and volleyball.

43. Find the number of students that play tennis or volleyball.

Solve.

44. A grocery store found that 275 of its customers use push carts to shop, 185 used a carry basket to shop, and that 145 used both a push cart and a carry basket. How many customers use only a push cart or a carry basket? Draw the Venn diagram.

45. Determine the number of playing cards in a standard deck that are red cards or face cards.

46. Determine the number of playing cards in a standard deck that are odd numbered cards or black cards.

2.3 PROJECT

EXPLORING INTERVALS: INTERSECTIONS AND UNIONS

In Section 2.3, you learned about the intersection and union of sets. In this activity, you will investigate intersections involving sets that cannot be described using roster notation.

Consider the set I_1 of all real numbers that are greater than or equal to 0 and less than or equal to 1. Using set-builder notation, we have $I_1 = \{x \mid 0 \leq x \leq 1\}$, which we can also represent using interval notation as $I_1 = [0,1]$ or graphically as shown in Figure 1.

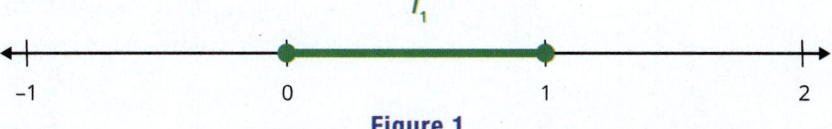

Figure 1

Now, for each natural number n, that is $n = 1, 2, 3, 4, \ldots$, we can think of the interval $I_n = \left[0, \dfrac{1}{n} \right]$.

For example, $I_3 = \left[0, \dfrac{1}{3} \right]$ is the set of all real numbers greater than or equal to zero and less than or equal to $\dfrac{1}{3}$.

1. Determine $I_1 \cap I_2$.

2. Determine $I = I_1 \cap I_2 \cap I_3$.

3. Determine $I = I_1 \cap I_2 \cap I_3 \cap I_4$.

This sequence of intervals is called *nested intervals*, where each set in the sequence is contained within the previous one. This means that the intersection of the nested intervals is equal to the smallest interval.

Let's see what happens if we keep going with taking intersection forever. Let's consider

$$I = I_1 \cap I_2 \cap I_3 \cap I_4 \cdots,$$

which is the intersection of all such intervals. If a positive number x is in I, then it has to be in every one of the intervals. Let's see if this is possible.

4. Find an interval of the form $\left[0, \dfrac{1}{n} \right]$ that does not contain the number 0.01.

5. Find an interval of the form $\left[0, \dfrac{1}{n} \right]$ that does not contain the number 0.0001.

No matter how small of a number you pick, there is always an interval in our list of nested intervals that does not contain the number you picked. We can conclude that no positive number can be in the intersection $I = I_1 \cap I_2 \cap I_3 \cap I_4 \cdots$.

6. In fact, there is exactly one number in the intersection $I = I_1 \cap I_2 \cap I_3 \cap I_4 \cdots$. What is the number in this intersection? Explain your reasoning.

7. What is the union $J = I_1 \cup I_2 \cup I_3 \cup I_4 \cdots$ equivalent to?

2.4 APPLICATIONS AND SURVEY ANALYSIS

OBJECTIVE 1 Two-Set Survey Analysis

Venn diagrams are often used when describing and analyzing information collected from surveys. At times, survey results can be somewhat confusing or even appear to be inconsistent at first glance. Take, for example, the following survey results from 64 adults about things they believed as a child.

- 45 believed that if they ate a watermelon seed, a watermelon would grow in their stomach.

- 49 believed that white cows made white milk and brown cows made chocolate milk.

- 42 believed both of these things.

Initially, the first two numbers might make it seem that there were more responses than there were people surveyed, since 45 + 49 is more than 64. However, when we consider the fact that there were 42 people who believed *both* things, we can resolve this apparent contradiction. By using a Venn diagram to help us categorize the responses in a meaningful way, we can make better sense of the results. Figure 2.4.1 shows how the results from the survey are less puzzling when displayed with a Venn diagram. The figure also highlights that 12 of the adults surveyed believed neither of the two things, even though we weren't specifically given that information.

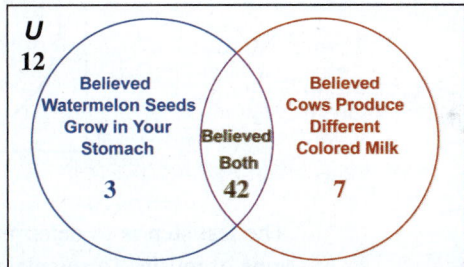

Figure 2.4.1: Venn Diagram of Survey Results

Example 2.4.1

Drawing a Venn Diagram for Survey Analysis

A survey that asked 500 donors of an artistic nonprofit organization about their musical choices showed that 350 of them listen to jazz, 300 listen to classical, and 200 listen to both. Draw a Venn diagram to illustrate this survey and determine how many donors surveyed don't listen to either jazz or classical music.

Solution

The Venn diagram will have two circles to represent each of the sets "Jazz" and "Classical." We know that the circles will overlap, since we also are told that some respondents listen to both. Figure 2.4.2 shows the skeleton Venn diagram without the number of responses filled in.

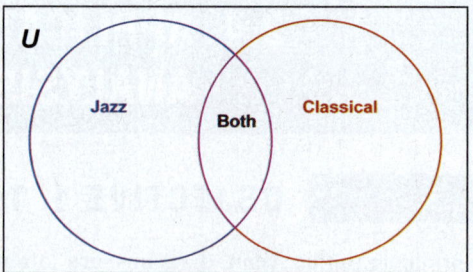

Figure 2.4.2

When determining the number of responses in each category, it is easiest to start with the intersection of the sets. In other words, determine how many people listened to both types of music. We are told that the number of donors that listen to jazz and classical music is 200. So this number will go in the intersection.

We must notice that each donor that likes both types of music are also counted in the categories for their respective types of music. For instance, although 350 donors listen to jazz, 200 of those listeners also listen to classical. Recall from Section 2.3 that the inclusion-exclusion principle, which states that we must subtract the donors that have been counted twice. Thus, 350 – 200 = 150 donors listen to jazz but not classical music. Similarly, 300 – 200 = 100 donors listen to classical music but not jazz. Putting these numbers in the diagram, we have the following.

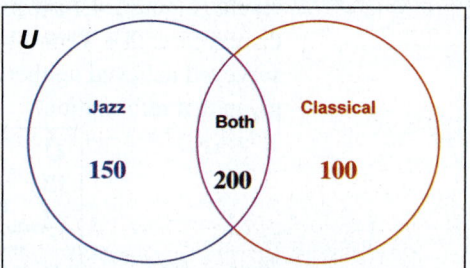

Figure 2.4.3

The last step is to determine how many donors surveyed do not listen to either type of music. To calculate this, we need to subtract the numbers in each region from the total number of people surveyed.

$$500 - 150 - 200 - 100 = 50$$

Thus, there are 50 donors who do not listen to either jazz or classical music. This number is placed on the outside of the circles. Figure 2.4.4 shows the completed Venn diagram.

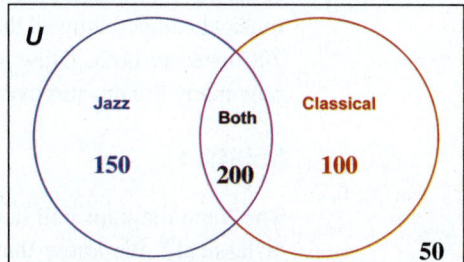

Figure 2.4.4: Completed Venn Diagram

✔ **Skill Check 2.4.1**

A survey of 400 customers at an ice cream shop showed that 225 customers like chocolate ice cream, 300 like vanilla, and 200 like both. Draw a Venn diagram to illustrate the survey results. How many customers responded that they didn't like either chocolate or vanilla ice cream?

OBJECTIVE 2 Three-Set Survey Analysis

When we use Venn diagrams, we are not limited in the number of sets that can be shown. However, as you might imagine, the more sets, the more complicated the diagram because of the possible intersections among the sets. For this reason, we will only consider Venn diagrams with up to three sets. For example, Figure 2.4.5 shows a Venn diagram of a group of tourists and the different languages that each speaks fluently.

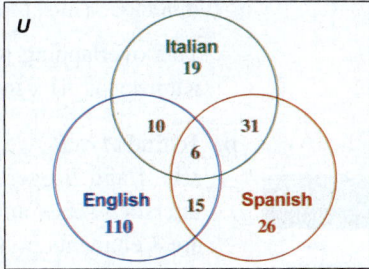

Figure 2.4.5

The Venn diagram shows that 110 tourists speak only English, 26 speak only Spanish, 19 speak only Italian, 10 speak English and Italian but not Spanish, 31 speak Italian and Spanish but not English, 15 speak English and Spanish but not Italian, and 6 speak all three languages. Whew, that's a lot of information in one diagram, but notice that it is easier to digest in the diagram than it is written out.

Example 2.4.2

Interpreting a Venn Diagram of Three Sets

The Heisman Trophy, created in 1935, is awarded every year to an outstanding college football player. The Venn diagram in Figure 2.4.7 tells us about some of the past winners and contains the number of elements that belong to the three sets A, B, and C. A is the set of winners whose position was that of running back, B is the set of winners from The Ohio State University (OSU), and C is the set of winners who were classified as juniors when they won.

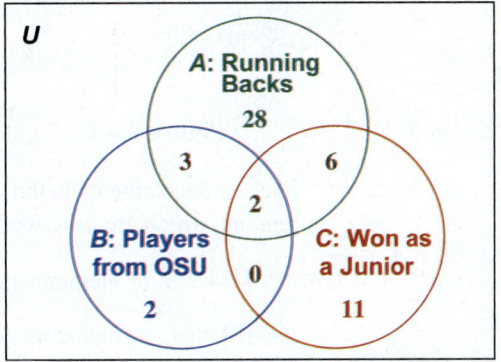

Figure 2.4.7

Use the information in the diagram to determine the following and describe what each represents.

a. $|A \cap B \cap C|$

b. $|A \cap B|$

Solution

a. To find $|A \cap B \cap C|$, we need to consider the intersection of all three sets A, B, and C. Working with three sets and their intersections, we need to be careful and make sure we are considering the proper intersection. This is represented by the triangular middle section of the Venn diagram. We are looking for the number of elements that are common to all three sets. We observe in the Venn diagram that there are two elements represented in the intersection of sets A, B, and C. Thus, $|A \cap B \cap C| = 2$.

This overlapping region represents the number of times the Heisman Trophy winner was in his junior year, a running back, and from The Ohio State University.

b. To find $|A \cap B|$, we need to determine the number of elements in the area where sets A and B overlap. Referring to the Venn diagram, we can see that set C intersects sets A and B as well. In the overlapping regions of sets A and B, there are 3 elements in $A \cap B$ but not in C and 2 elements in $A \cap B$ that are also in C. Therefore, $A \cap B$ contains $3 + 2 = 5$ elements, so $|A \cap B| = 5$.

This intersection of A and B represents occurrences where the Heisman Trophy winner was a running back and was from The Ohio State University.

> ✔ **Skill Check 2.4.2**
>
> Use the Venn diagram in Example 2.4.2 to find the following.
>
> **a.** $|A \cap C|$
>
> **b.** $|B \cap C|$

As illustrated in Example 2.4.2, a Venn diagram having three sets will always be divided into eight distinct regions: 7 regions representing the three sets and their possible intersections and 1 region outside of the sets representing all universal elements not in any of the sets. These regions are shown in Figure 2.4.8.

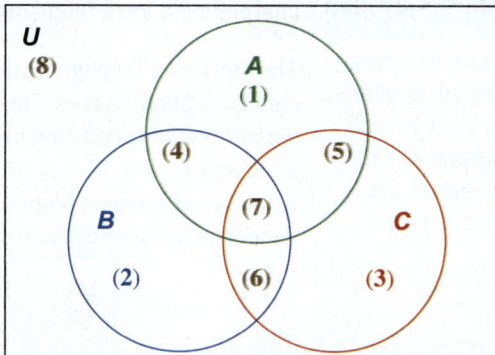

Figure 2.4.8: Eight Regions of a Venn Diagram with Three Sets

Each of the elements in the universal set may only be placed in one of these eight regions, which are described as follows.

(1) contains elements in set A only.

(2) contains elements in set B only.

(3) contains elements in set C only.

(4) contains elements in sets A and B but not C.

(5) contains elements in sets A and C but not B.

(6) contains elements in sets B and C but not A.

(7) contains elements in sets A and B and C.

(8) contains elements in the universal set but not sets A or B or C.

Much like other logic puzzles, there are many approaches to determining how many elements are in each region. One way is to begin with the innermost region—the intersection of all three sets, region (7)—and then work your way out to each of the other seven regions while making sure to not count any element more than once. Example 2.4.3 looks at this process of placing the correct number of elements in each region of the diagram.

Example 2.4.3

Constructing a Venn Diagram of Three Sets

Consider the universal set $U = \{a, b, c, ..., z\}$. Given subsets $A = \{a, e, i, o, u\}$, $B = \{a, b, c, d, e, f, g, h, i, j, k, l\}$, and $C = \{a, l, u, m, n, i\}$, draw a Venn diagram to represent the relationships between the sets.

Solution

Let's begin by noting what our sets are and how they relate to one another. Notice that the universal set consists of the letters in the English alphabet, set A consists of all of the vowels, set B contains the first 12 letters of the alphabet, and the elements of set C spell the word *alumni*. As we place the elements in the diagram, remember that each of the elements in the universal set may only be placed in one of the eight regions in the Venn Diagram, as shown in Figure 2.4.9.

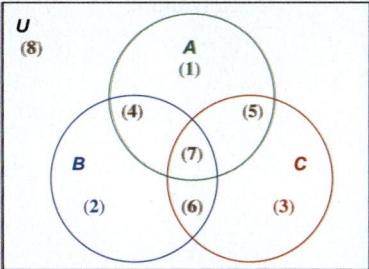

Figure 2.4.9: Regions of a Three-Set Venn Diagram

We'll begin with the intersection of all three sets, region (7). Comparing the elements of sets A, B, and C, the elements common to all three sets are the vowels a and i. In other words, $A \cap B \cap C = \{a, i\}$.

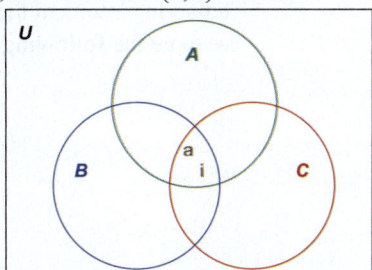

Figure 2.4.10: Region (7), $A \cap B \cap C$

Next, we'll determine where the three remaining elements of set A belong—the vowels e, o, and u.

When compared to sets B and C, the only vowel not in either B or C is o. So the only element in area (1) is o. That leaves us with placing e and u in their correct regions. Notice that $A \cap B = \{a, e, i\}$. We've already placed the elements a and i in region (7), which means e is the only element in region (4), the intersection of A and B but not C. Finally, $A \cap C = \{a, u, i\}$. Again, we've placed a and i in region (7), so u belongs in region (5), the intersection of A and C but not B.

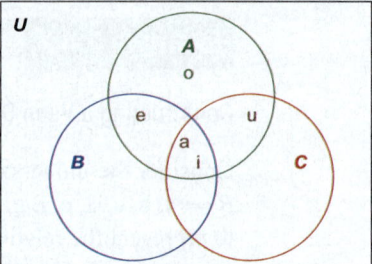

Figure 2.4.11: Placement of the Elements in Set A

Now that we've positioned all of the elements in set A, we turn our attention to set B, which consists of $\{a, b, c, d, e, f, g, h, i, j, k, l\}$. Let's look at region (6), which is the intersection of sets B and C only (not including set A). The intersection of sets B and C is $\{a, l, i\}$. However, since the vowels in set B have already been accounted for, we are only left with the element l for region (6). We then place the remaining elements of set B in region (2), which are the elements b, c, d, f, g, h, j, and k.

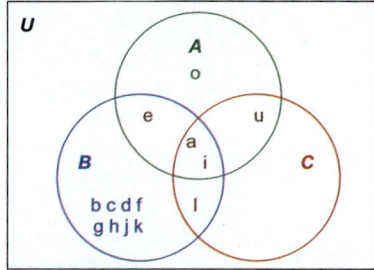

Figure 2.4.12: Placement of the Elements in Sets A and B

Now we turn our attention to set C. We can see that region (3) will contain the elements of C that we have not already placed—the elements m and n. Any remaining letters of the alphabet are placed in region (8) of the universal set. Thus, we have the following Venn diagram.

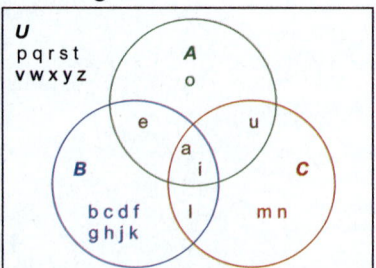

Figure 2.4.13: Venn Diagram Solution

Example 2.4.4

Constructing a Venn Diagram for Survey Analysis

A group of students majoring in international relations are polled on whether they have taken courses in any of three languages: French, German, and Russian. Every student took at least one of the languages. No student who took French also took Russian, but 39 who took French also took German. Eighty-four students who took German also took Russian. Altogether, 55 reported taking French, 141 reported taking German, and 92 reported taking Russian. Draw a Venn diagram to illustrate this survey and determine how many students were polled.

Solution

Since we are not given the individual elements, we will place the count of the number of elements in a region instead of listing the elements as we did before. Recall that when there are three sets of consideration, there are eight possible regions to place elements in. Each of the student responses may only be placed in one of the eight regions. The eight regions for our diagram represent students taking: (1) French only, (2) German only, (3) Russian only, (4) French and German but not Russian, (5) French and Russian but not German, (6) German and Russian but not French, (7) French and German and Russian, and (8) None. Figure 2.4.14 shows our skeleton diagram.

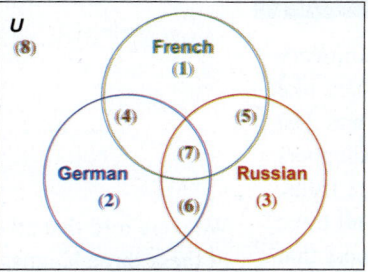

Figure 2.4.14

We will start with the intersection of all three sets. We are told that no student who took French also took Russian, which means that region (5) will have 0 elements. Notice that this also means that no student could have taken all three languages, so region (7) will also have 0 elements. This means that regions (7) and (5) will both have 0 in them.

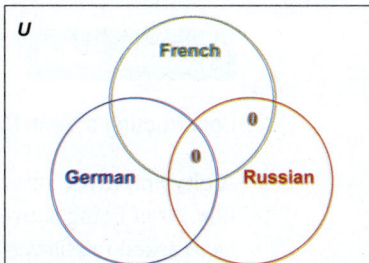

Figure 2.4.15: Regions (7) and (5) Filled In

That leaves two more intersection regions—regions (4) and (6). We were given the information about both of these intersections. We know that 39 students took French and German, so region (4) will have 39 elements. And 84 students took both German and Russian, so region (6) will have 84.

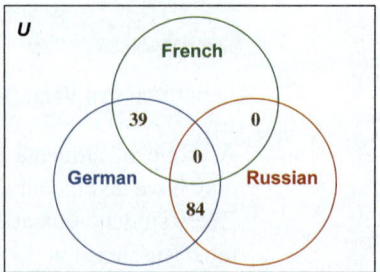

Figure 2.4.16: Regions (7), (6), (5), and (4) Filled In

Finally, we need to determine the number of students who took a single language in order to fill in regions (1), (2), and (3). Since we are given that 55 students took French, of which 39 also took German, we can remove the students that have been counted twice to determine that 55 − 39 = 16 students took only French. There were 141 students that took German, and of those students, 84 also took Russian and 39 also took French. Therefore, 141 − 84 − 39 = 18 students took only German. Lastly, 92 students reported taking Russian. Of those students, 84 also took German and 0 students also took French. Thus, 92 − 84 = 8 students took only Russian. Placing these in the diagram, we have the following.

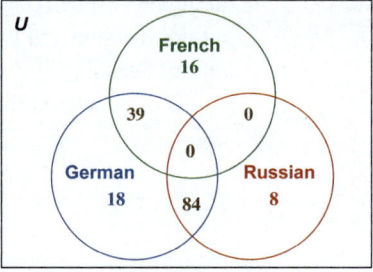

Figure 2.4.17

We were told that all students majoring in international relations took at least one of these three languages, which means there are no students in region (8). We can now add the regions together to determine how many students were polled.

$$16 + 39 + 18 + 84 + 8 = 165$$

Thus, there were 165 students majoring in international relations in the survey.

> ✔ **Skill Check 2.4.3**
>
> A survey of shoppers at a grocery store found that 225 shoppers like bananas, 198 like apples, and 180 like grapes. All of the shoppers like at least one of the three fruits. Twenty-five shoppers like all three fruits. There are 110 shoppers that like bananas and apples, 58 that like apples and grapes, and 55 that like bananas and grapes. Draw a Venn diagram to illustrate this poll and determine how many shoppers were in the survey.

Example 2.4.5

Constructing a Venn Diagram of Three Sets

A city employee surveyed 1000 residents to gauge interest in a possible new bus line. After being shown information about the proposed route, survey respondents were asked to answer Yes or No to the following questions.

1. Do you think the new bus line should be added?

2. Would you be willing to pay additional taxes to implement the new route?

3. Would you be likely to ride on the new route?

The survey responses are summarized in the table below. How many respondents responded No to all three questions?

Table 2.4.1: City Employee Survey Results

Number of Responses	Yes Votes
556	Question 1
262	Question 2
331	Question 3
257	Question 1 and 2
325	Question 1 and 3
152	Question 2 and 3
150	Question 1, 2, and 3

Solution

Once again, we need a Venn diagram with three sets. The circles will represent the Yes votes for each of the three questions. We start by looking at the intersection of all three sets—the respondents who voted Yes on all three questions. There are 150 respondents in this intersection.

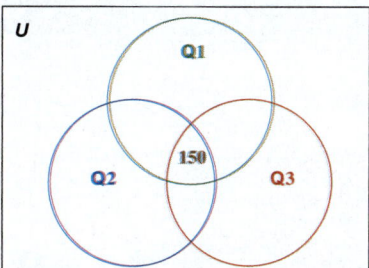

Figure 2.4.18: Answered Yes on All 3 Questions

Although tempting to simply look at the table and use the number of Yes answers given for each question, remember that we will overcount responses if we do that. Instead, we need to fill in the remaining intersections next. We know that 257 responded Yes to questions 1 and 2. Subtracting the 150 responses in the middle region, we have that $257 - 150 = 107$ responded Yes to questions 1 and 2, but not all three questions. Similarly, $325 - 150 = 175$ had positive responses for both questions 1 and 3 but not question 2. And finally, for questions 2 and 3, we have $152 - 150 = 2$ Yes answers for these two questions only. Figure 2.4.19 shows the diagram thus far.

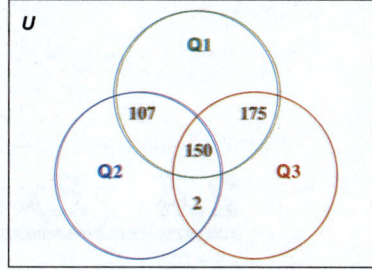

Figure 2.4.19

Next, we use subtraction for the remaining regions in the sets that are residents who answered Yes to only one question

Question 1: $556 - 107 - 150 - 175 = 124$

Question 2: $262 - 107 - 150 - 2 = 3$

Question 3: $331 - 175 - 150 - 2 = 4$

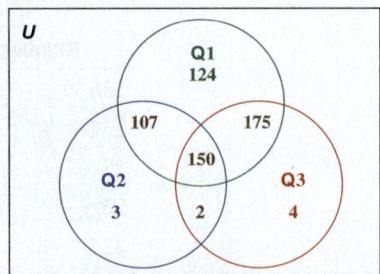

Figure 2.4.20

Finally, subtract all the numbers in the regions (1) through (7) from 1000 to determine how many residents answered No to all three questions.

$$1000 - 124 - 107 - 175 - 150 - 3 - 2 - 4 = 435$$

Figure 2.4.21 shows the completed Venn Diagram.

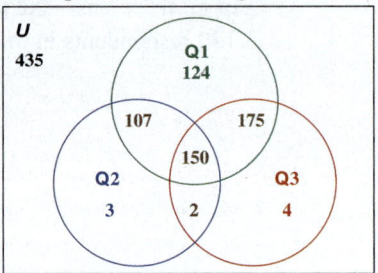

Figure 2.4.21

There were 435 residents who responded No to all three questions.

Skill Check Answers

1. 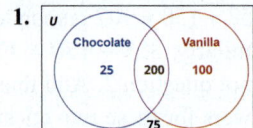   75 like neither ice cream.   **2. a.** $|A \cap C| = 8$ **b.** $|B \cap C| = 2$

3. 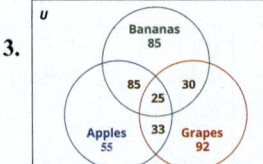   405 shoppers in the survey.

2.4   Exercises

✔ CONCEPT CHECK

1. A two-set Venn diagram has a total of _____ regions.

2. A three-set Venn diagram has a total of _____ regions.

3. True or False: Region (8) of a three-set Venn diagram represents the complement of the union of the three sets.

4. True or False: If three sets are disjoint, then regions (4), (5), and (6) will be empty and regions (1), (2), (3), and (7) will contain elements.

💡 PRACTICE

5. Construct a Venn diagram illustrating the following sets: A = {apple, orange, grape, peach}, B = {grape, banana, apple, kiwi}, and C = {kiwi, apple, peach, banana} if U = {apple, orange, peach, grape, banana, kiwi}.

6. Construct a Venn diagram illustrating the sets: $A = \{1, 2, 3, 4\}$, $B = \{2, 4, 6, 8, 10\}$, and $C = \{3, 4, 6\}$ if $U = \{1, 2, 3, 4, 5, 6, 7, 8, 9, 10\}$.

🚀 APPLICATIONS

7. A survey of 400 college freshmen showed that 200 drink soda, 300 drink juice, and 150 drink both. Draw a Venn diagram to determine how many students responded that they drink neither.

8. A survey of 350 students showed that 225 listen to rap and 200 listen to rock and 135 listen to both. Draw a Venn diagram to determine how many students responded that they listen to neither.

9. A researcher collecting data on 100 households finds that 47 have a DVD player; 52 have only streaming video, and 27 have both. Determine the answer to the following questions.

   a. How many do not have video streaming?

   b. How many have neither video streaming nor a DVD player?

   c. How many have a DVD player but not video streaming?

10. A survey of 600 workers yielded the following information: 417 belonged to the Auto Workers Union, 275 were Democrats, and 215 of the Auto Workers Union were Democrats.

   a. How many workers belonged to the Auto Workers Union or were Democrats?

   b. How many workers belonged to the Auto Workers Union but were not Democrats?

   c. How many workers were Democrats but did not belong to the Auto Workers Union?

   d. How many workers neither belonged to the Auto Workers Union nor were Democrats?

11. A three-course meal is served during the soft opening of a new restaurant. The three-course meal included an appetizer, the main course, and dessert. After the meal, diners were asked whether they enjoyed each course. The following Venn diagram summarizes the results.

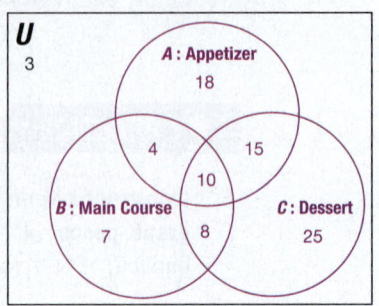

Use the information in the diagram to determine the following and describe what each represents.

a. $|(A \cup B \cup C)'|$

b. $|B \cup C|$

12. There are 43 students in the University Travel Club. They discovered that 17 members have visited Mexico, 10 have been to England, 28 have visited Canada, 8 have been to Mexico and Canada, 3 have only been to England, and 4 have only been to Mexico. No student has been to only England and Canada. Two students have been to all three countries. Some of the club members have not been to any of the three.

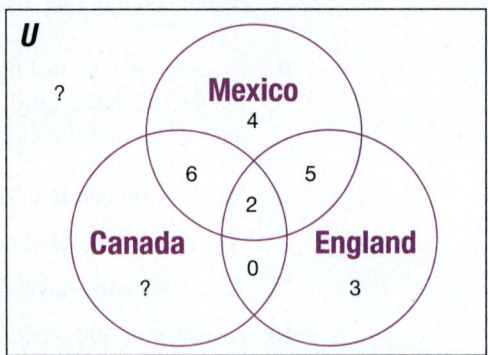

a. How many students have been to all three countries?

b. How many students have been only to Canada?

c. How many have been to Mexico or Canada but not England?

d. How many have been to none of the countries?

13. Students were polled on which of the three restaurants in the student center they have purchased food from in the past week. Out of those surveyed, 250 students made a purchase from Pizza Palace, 278 students made a purchase from Maing Wok, 226 students made a purchase from Salad Express, 87 students made a purchase from both Pizza Palace and Maing Wok, 57 students made a purchase from both Pizza Palace and Salad Expression, 95 students make a purchase from both Maing Wok and Salad Express, and 22 students made a purchase from all three restaurants. A total of 63 students replied that they had not made a purchase from any of the restaurants. Draw a Venn diagram to illustrate this survey and determine how many students were polled.

14. A variety show consisted of three acts: music, comedy, and magic. All audience members were asked which acts they enjoyed. Out of those surveyed, 105 enjoyed the music, 122 enjoyed the comedy, 95 enjoyed the magic, 50 enjoyed both the music and the comedy, 37 enjoyed both the music and the magic, 57 enjoyed both the comedy and the magic, and 15 enjoyed all three. Only 7 audience members did not enjoy any of the three acts. Assuming every audience member completed a survey, draw a Venn diagram to illustrate this survey and determine how many people were in the audience.

15. A survey of 125 freshman business students at a large university produced the following results:

> 35 read *Money*;
> 25 read *The Wall Street Journal*;
> 32 read *Fortune*;
> 21 read *Money* but not *The Wall Street Journal*;
> 11 read *The Wall Street Journal* and *Fortune*;
> 13 read *Money* and *Fortune*;
> 9 read all three.

Use this information to answer the following questions:
a. How many students read none of the publications?

b. How many read only *Fortune*?

c. How many students read *Money* and *The Wall Street Journal*, but not *Fortune*?

16. Taylor Swift, Drake, and BLACKPINK toured the United States. A large group of teenagers were surveyed about whether they went to any of the concerts and the following information was obtained: 825 saw Drake, 1033 saw Taylor Swift, 1247 saw BLACKPINK, 211 saw all three, 514 saw none, 240 saw only BLACKPINK, 677 saw BLACKPINK and Taylor Swift, and 201 saw Taylor Swift and Drake but not BLACKPINK.

a. What percent of the teenagers saw at least one concert?

b. What percent of the teenagers saw exactly one concert?

17. A book publishing company asked 400 random customers of a bookstore whether they purchased print books, e-books, or audiobooks in the past year. The results are summarized in the following table.

Bookstore Customer Book Format Preference	
Print Book	149
e-Book	281
Audiobook	215
Print Book and e-Book	98
Print Book and Audiobook	69
e-Book and Audiobook	153
Print Book, e-Book, and Audiobook	42

a. How many people purchased none of the three book formats in the past year?

b. How many people purchased e-books and audiobooks but not print books?

c. How many people purchased only print books?

18. An artist sent out a survey with her newsletter asking fans which type of merchandise they would be interested in buying. The options were art prints, T-shirts, and stickers. The results of the 326 fans who completed the survey are shown in the following table.

Merchandise Preference	
Art Prints	153
T-Shirts	121
Stickers	198
Art Prints and T-Shirts	37
Art Prints and Stickers	66
T-Shirts and Stickers	62
Art Prints, T-Shirts, and Stickers	12

a. How many fans indicated they would buy none of the options?

b. How many fans indicated they would buy art prints and stickers but not T-shirts?

c. How many fans indicated they would buy only T-shirts?

19. There are three types of blood antigens that determine blood type: A, B, and Rh+. An individual's blood type is determined by the specific combination of these antigens. In order to receive a blood transfusion, you can't receive blood from a donor who has an antigen that you don't have yourself. That means that people with AB+ blood can receive a transfusion from ANY donor, since they have all of the possible antigens. They can only donate to other people with AB+. People with O− blood (none of the antigens) can only receive type O− blood, since all other blood types have at least one of the antigens. However, they can donate their blood to anyone, since their blood does not have any of the antigens. A laboratory looked at blood samples for 200 patients and found the following information provided in the table. How many patients were classified as O−? Explain your reasoning.

Blood Antigen Survey Results	
Number of Samples	**Antigen in Blood**
80	A
36	B
82	Rh
10	A and B
62	A and Rh
22	B and Rh
4	A, B, and Rh

2.4 PROJECT

SET THEORY AND ALLOCATION OF RESOURCES

According to the job interviewing coaches Jeff & Mike *The Interview Guys*,

Analytical skill is the ability for an individual to solve complex issues by gathering and then analyzing the information that is available to them through a variety of other skills including critical thinking, research, and attention to detail.

In this activity, you will solve a seemingly complex allocation of resources problem by using a mathematical model involving Venn diagrams.

A medium-size tech company has to staff 3 distinct departments: Social Media Outreach (SM), Information Technology (IT), and Web Development (WD). The fast-paced and innovative environment at the company may require an employee to be part of more than one department.

The company has the following staffing requirements.

- The total number of employees in the three departments must be exactly 40.

- There must be exactly 16 employees in Information Technology and exactly 20 employees in Social Media Outreach.

- No employee can work in Information Technology and Social Media Outreach without also being a part of Web Development.

- There must be exactly 8 employees working in both Social Media Outreach and Web Development.

- Exactly 4 employees will be required to work in Web Development and Social Media Outreach but not in Information Technology.

- Exactly 2 employees work in Information Technology and Web Development but not in Social Media Outreach.

1. Draw a three-set Venn diagram representing the three departments and their overlap.

2. Determine the number of employees in each of the regions of the Venn diagram from part 1.

The company decides to add another department named Customer Experience (CE). This department can only share employees with Social Media Outreach.

3. Draw a four-set Venn diagram that models this situation.

4. After adding the Customer Experience department, the total number of employees increases from 40 to 50. Knowing that 6 people work in the CE department only, determine how many employees now work in CE and SM departments at the same time.

Chapter 2 Exercises

Determine whether each statement is true or false. If the statement is false, explain why.

1. $\{3\} \in \{1, 2, 3, 4, 5, 6\}$

2. $1 \in \{x \mid x \text{ is an integer}\}$

3. $\{3\} \subseteq \{1, 2, 3, 4, 5, 6\}$

4. $\{1\} \not\subset \{x \mid x \text{ is an integer}\}$

5. Let $A = \{\text{red, yellow, blue}\}$. Then $|A| = 3$.

6. Let $B = \{-2\}$. Then $|B| = 2$.

7. $|\varnothing| = 1$

8. $\left|\{\varnothing\}\right| = 1$

Write each set as indicated.

9. Let the set A consist of the even counting numbers less than 14. Write A using roster notation.

10. Use roster notation to write the set B that consists of the seasons of the year.

11. Use set-builder notation to write the set C that consists of the set of real numbers between 100 and 1000.

12. Use set-builder notation to write the set D that consists of the months of the year that have 30 days.

Use the given sets to solve each problem.

  $A = \{\text{Felix, Amber}\}$

  $B = \{\text{moral, social, civil}\}$

13. Find $|A|$ and $|B|$.

14. List all the subsets of A and subsets of B.

15. List all the proper subsets of A.

16. Is $A = B$? Why or why not?

17. Is $A \sim B$? Why or why not?

Use the given sets to solve each problem.

G = {I, II, III}

F = {love, joy, peace}

18. Find $|G|$ and $|F|$.

19. List all the subsets of G and subsets of F.

20. List all the proper subsets of F.   **21.** Is $G = F$? Why or why not?

22. Is $G \sim F$? Why or why not?

Determine the number of proper subsets of each set.

23. $\{\alpha, \beta, \chi, \delta, \varepsilon, \phi, \mu, \pi\}$ **24.** $\varnothing$

Draw a Venn diagram to illustrate each group of sets. A universal set is not given, so choose one that fits and define it.

25. Parents and their children   **26.** Sculptors and Artists

27. Cats and Dogs

28. $A = \{x \mid x \in \mathbb{R}\}$ and $B = \{x \mid x \text{ is an integer}\}$

Use the given sets to write each set in roster notation.

U = {1, 2, 3, 4, 5, 6, 7, 8, 9, 10}

A = {1, 2, 3, 4, 5}

B = {1, 3, 5, 7}

29. $A \cap B$ **30.** $A \cup B$

31. $A' \cap B$ **32.** $A' \cup B'$

33. $|A \cap B|$ **34.** $\left|(A \cup B)'\right|$

Use the given Venn diagram to write each set in roster notation.

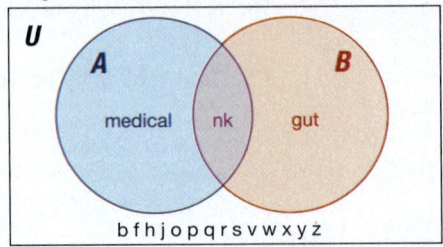

35. A **36.** B

37. $A \cap B$ **38.** $A \cup B$

39. $(A \cup B)'$ **40.** $(A \cap B)'$

41. U **42.** $|A \cup B|$

43. $|A \cap B|$ **44.** $|(A \cup B)'|$

Solve each problem.

45. A school gym teacher is trying to determine what sports students enjoy the most. She collected information on 250 students and found that 150 like volleyball, 110 like soccer, and 65 students like both.

    a. Draw a Venn diagram to represent the findings of the teacher.

    b. How many students like only volleyball?

    c. How many students like only soccer?

    d. How many students like neither soccer nor volleyball?

46. A camp counselor is planning activities for the summer and wants to know what campers would enjoy. He asks 650 campers and finds that 457 enjoy swimming, 250 enjoy tennis, 223 enjoy jogging. He finds that 176 enjoy swimming and tennis, 75 enjoy swimming and jogging, 105 enjoy tennis and jogging, and 45 enjoy all three.

    a. Draw a Venn diagram to represent the survey results.

    b. How many campers enjoy only swimming?

    c. How many campers enjoy only tennis?

    d. How many campers enjoy only jogging?

    e. How many campers enjoy only swimming and tennis?

    f. How many campers enjoy only tennis and jogging?

    g. How many campers enjoy only swimming and jogging?

    h. How many campers enjoy none of the three?

47. A study found that 25% of a certain population has blue eyes, 20% of the population has blonde hair, and 12% of the population has blonde hair and blue eyes. Estimate the percentage of the population that has blue eyes or blonde hair?

CHAPTER 2 PROJECT

Ready... SET... Go

Venn diagrams can be a useful way to display the results of a survey with multiple questions. At a glance of the diagram, you can get a general feel of the ways in which the experiences or knowledge of a group of people overlap. In this project, you will collect data in a survey and then use a Venn diagram to analyze the results.

Step 1: Design the Survey

Create a survey with 3 questions that are binary in their responses. In other words, each question on the survey has only 2 options to choose from. (This is easiest if you think of yes/no questions.) You may use the following questions or come up with some of your own. If you choose to make up questions on your own, remember that they must only have a choice of 2 responses.

- Were you born in this state? [**Note:** Specify which state.]

- Have you ever traveled to another country?

- Can you fluently speak another language besides English?

1. Create three questions if you choose not to use the ones above.

2. How many possible answer combinations are there? _____

Step 2: Collect and Organize the Data

3. Survey 50 people using your chosen questions and record their answers in a table similar to the one shown here. You will need to complete the table to include all of the possible combinations of responses.

Survey Results

Number of Responses	Born in This State?	Traveled to Another Country?	Fluently Speak Another Language?
	Yes	Yes	Yes
	Yes	Yes	No
	Yes	No	Yes
	⋮	⋮	⋮

Step 3: Display the Data

Now that you've collected and organized the data, it's time to display the results with a 3-circle Venn diagram like the one provided.

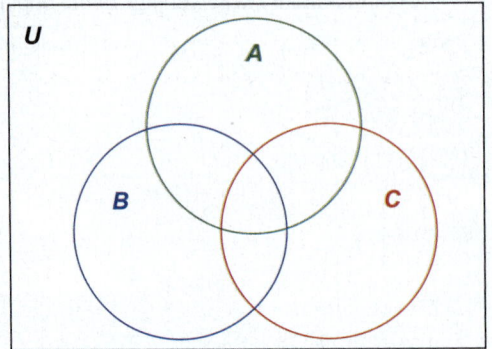

4. Begin by labeling the circles. Think about what type of labeling your Venn diagram will require. How will you label the circles?

5. As you begin to populate the circles with the response numbers, where should you start?

6. Finish populating your diagram with results.

Step 4: Analyze and Interpret the Results

Answer the following about your survey using complete sentences.

7. Using the Venn diagram you created, summarize your results.

8. What conclusions can you draw from the Venn diagram?

9. Why was it important that the questions only had a choice of 2 responses?

Step 5: Collect and Organize New Data

10. Now survey 50 different people using the same questions. Record your results in the same manner as above.

Step 6: Display

11. Using the same method as before, draw a second Venn diagram to display the results of the second survey.

Step 7: Analyze and Interpret

Answer the following about your survey.

12. Using the second Venn diagram, summarize the second set of responses that you gathered.

13. What conclusions can you draw from this Venn diagram?

Step 8: Compare

Answer the following questions using complete sentences in order to compare the diagrams.

14. Do the two Venn diagrams look the same? If the Venn diagrams are not the same, describe how they are different.

15. What do you think a Venn diagram would look like if it contained the results from all 100 people you surveyed combined?

16. Would you be able to draw any more conclusions with the data combined? Why or why not?

CHAPTER 2 REVIEW

Section 2.1 Set Notation

Definitions

Set: A **set** is a collection of objects made up of specified elements, or members.

Roster Method: Roster notation specifies the members of a set by listing all of the elements in the set, separated by commas and surrounded by curly braces.

Equal Sets: Two sets are said to be **equal** if they contain exactly the same elements. If sets A and B are equal, we write $A = B$.

Cardinal Number: The number of elements contained in a finite set is called the **cardinal number** of the set, or the **cardinality**. The cardinal number of set A is denoted by $|A|$.

Equivalent Sets: Two sets are said to be **equivalent** if they have the same cardinal number. If set A is equivalent to set B, we write $A \sim B$.

Set-Builder Notation: Set-builder notation specifies the members of a set using a variable, a vertical separator, and a rule defining the elements, all surrounded by curly braces. It can be used when the members of the set all share certain properties. For instance, the set of all integers is represented by

$$\mathbb{Z} = \{n \mid n \text{ is an integer}\}$$

and is read, "$\mathbb{Z}$ is the set of all n such that n is an integer."

Empty Set: The **empty set** is the set that contains no elements. If set A is empty, we write

$$A = \varnothing \text{ or } A = \{\ \}.$$

The cardinality of the empty set is 0.

Universal Set: The set of all elements being considered for any particular situation is called the **universal set** and is denoted by U.

Complement: The **complement** of set A consists of all the elements in the given universal set that are not contained in A. The complement of A is denoted A'.

Section 2.2 Subsets and Venn Diagrams

Definitions

Venn Diagram: A **Venn diagram** is a visualization of the relationships between a collection of sets. In a Venn diagram, the sets are represented by circles (or ovals) contained within a rectangular region that represents the universal set.

Subset: Let A and B be sets. The set B is a **subset** of A if every element of B is also an element of A. This is denoted by $B \subseteq A$.

Proper Subset: When $B \subseteq A$, and A contains at least one element that is not contained in B, B is a **proper subset** of A and is denoted by $B \subset A$.

Number of Subsets and Proper Subsets of a Set: If the cardinal number of a set is n, then the set has 2^n subsets and $2^n - 1$ proper subsets.

Section 2.3 Operations with Sets

Definitions

Intersection: The **intersection** of two sets A and B is the set of all elements common to both A and B. We denote the intersection of A and B as $A \cap B = \{x | x \in A \text{ and } x \in B\}$.

Disjoint: Two sets A and B are **disjoint** if there are no elements in set A that are also contained in set B. Their intersection is the empty set, denoted by $A \cap B = \varnothing$.

Union: The **union** of two sets A and B is the set of all elements in A or in B. We denote the union of A and B as $A \cup B = \{x | x \in A \text{ or } x \in B\}$.

Inclusion-Exclusion Principle: The **inclusion-exclusion principle** states that the cardinality of the union of two sets A and B is calculated by adding the number of elements in set A to the number of elements in set B and subtracting the number of elements that appear in both sets.

$$|A \cup B| = |A| + |B| - |A \cap B|$$

De Morgan's Laws: Let A and B be sets. Then

$$(A \cup B)' = A' \cap B'$$

and

$$(A \cap B)' = A' \cup B'.$$

Section 2.4 Applications and Survey Analysis

Definition

Eight Regions of a Venn Diagram with Three Sets: A Venn diagram having three sets will always be divided into eight distinct regions: 7 regions representing the three sets and their possible intersections and 1 region outside of the sets representing all universal elements not in any of the sets. These regions are shown in Figure 2.4.8.

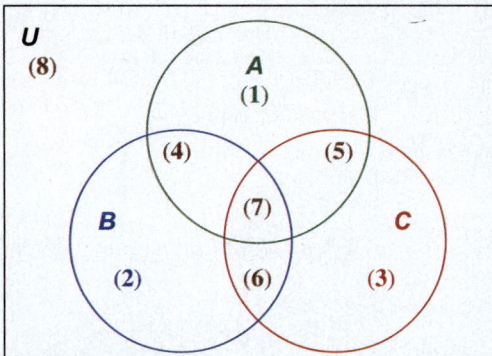

Figure 2.4.8: Eight Regions of a Venn Diagram with Three Sets

Each of the elements in the universal set may only be placed in one of these eight regions, which are described as follows.

(1) contains elements in set A only.

(2) contains elements in set B only.

(3) contains elements in set C only.

(4) contains elements in sets A and B but not C.

(5) contains elements in sets A and C but not B.

(6) contains elements in sets B and C but not A.

(7) contains elements in sets A and B and C.

(8) contains elements in the universal set but not sets A or B or C.

CHAPTER 3

Logic

🔖 SECTIONS

Introduction

In their 1941 film *In the Navy*, Abbott and Costello show by multiplication, division, and repeated addition that $13 \cdot 7 = 28$. What?!? Wait a second… How can that be? Here's their method of creative math.

Division

A. "7 in to 2. . . 7 won't go into 2, so I gotta take the 2 from here and put it down there. 7 into 8? 1 time."

B. "Now I'm gonna carry the 7 from here and put it under the 8. 7 from 8, 1."

C. "Now I have 21, 7 into 21? 3 times SO $28 \div 7 = 13$!"

A.
$$7\overline{)8} \quad \begin{matrix}1\\ \\ 2\end{matrix}$$

B.
$$7\overline{)8} \quad \begin{matrix}1\\ 7\\ 21\end{matrix}$$

C.
$$7\overline{)8} \quad \begin{matrix}13\\ 7\\ 21\\ 21\\ 0\end{matrix}$$

Multiplication

A. "7 times 3, 21."

B. "7 times 1, 7."

C. "7 and 1 is 8 and then bring this 2 down IT'S 28! $7 \cdot 13 = 28$!"

A.
$$\begin{matrix}13\\ \times\ 7\\ \hline 21\end{matrix}$$

B.
$$\begin{matrix}13\\ \times\ 7\\ \hline 21\\ 7\end{matrix}$$

C.
$$\begin{matrix}13\\ \times\ 7\\ \hline 21\\ +\ 7\\ \hline 28\end{matrix}$$

Repeated Addition

A. "Put down 13 seven times."

B. "Add all the 3s 3, 6, 9, 12, 15, 18, 21."

C. "Add all the 1s 22, 23, 24, 25, 26, 27, 28 YOU GET 28! $13 + 13 + 13 + 13 + 13 + 13 + 13 = 28$!"

A.
$$\begin{matrix}13\\13\\13\\13\\13\\13\\+13\end{matrix}$$

B.
13	21
13	18
13	15
13	12
13	9
13	6
+ 13	3

C.
22	13	21
23	13	18
24	13	15
25	13	12
26	13	9
27	13	6
28	+ 13	3
	28	

JUST TO BE CLEAR...

Division: $28 \div 7 = 4$

Multiplication: $13 \cdot 7 = 91$

Repeated Addition: $13 + 13 + 13 + 13 + 13 + 13 + 13 = 91$

Abbott and Costello's famous skit managed to show by division, multiplication, and repeated addition that $13 \cdot 7 = 28$. So it must be true, right? This makes us laugh because we can all see the errors that he makes right before our eyes. What is not so funny is that we often jump to the same types of misguided conclusions in our everyday lives by using similarly faulty logic.

So what was wrong in Costello's conclusion that $13 \cdot 7 = 28$? The problem is that his conclusion cannot be logically supported by the rules of addition, multiplication, and division. Although this famous error in his conclusion is obvious to us here, what if you are faced with a conclusion and you don't know if you can trust it? In other words, you are not sure if the conclusion is right or wrong, or if the argument is valid or faulty.

As we move through the chapter, not only will we show you ways to identify common errors in logic, like those in the Costello scene, we will also help you better understand how to make strong arguments for yourself.

3.1 LOGIC STATEMENTS AND THEIR NEGATIONS

OBJECTIVES

1. Identify mathematical statements.

2. Negate mathematical statements.

3. Translate compound statements into symbolic expressions.

OBJECTIVE 1 Mathematical Statements

Consider the following observation a student made in class one day.

Teacher: Before the end of the week, you will have a surprise quiz.

Student: That can never happen. Because, if you waited until Friday to give it, we would all realize it was coming and it would no longer be a surprise. So the quiz can't be given on Friday. Similarly, since it's not on Friday, it can't be on Thursday. If you waited until after Wednesday, we already know it can't be on Friday. So we would know it must be on Thursday and hence it would not be a surprise. We could make the same argument for Wednesday and Tuesday. So that means the quiz is today, Monday. We now know it is happening today, so it's not a surprise quiz and therefore you can't give it.

Although all his classmates were initially excited at the thought that they were not going to have a surprise quiz that week, they were surprised when the next morning they heard the words, "Clear your desks and take out a blank sheet of paper." Sadly, the student's logic did not hold true.

Lots of claims similar to those made by the hopeful student are made every day.

If you run for office, you will surely win.

9 out of 10 dentists choose this toothpaste.

It always rains the first day of school.

Not only does our ordinary English language encompass these types of statements, but it is also littered with opinions, sarcasm, riddles, commandments, and the list goes on. No wonder we often find it difficult to determine the validity of many of the things we hear day to day, just like the student's classmates. In these instances, we want to determine with certainty if statements are not only factually true, but also logically true. Mathematical logic provides us with a way to do just that. It provides a consistent framework in which to evaluate claims for logical truth.

All claims that we can logically evaluate through mathematics are made up of what are called statements. A mathematical **statement** is a complete sentence that asserts a claim that is either true or false, but not both at the same time. Any statement has exactly one of two possible **truth values** at any particular moment: true (T) or false (F). If a statement is true, then we say the truth value is true. If a statement is false, then its truth value is false. If it is not possible to assign a truth value to a sentence, then the sentence is not a mathematical statement.

> **💬 Helpful Hint**
>
> Mathematical logic statements are most commonly represented by lower case letters.

> **📖 STATEMENT**
>
> A mathematical **statement** is a complete sentence that asserts a claim which is either true or false, but not both at the same time.

The following sentences are all examples of statements. They are represented by italicized lowercase letters, as is the practice in mathematical logic.

a: The car is blue and the cat is black.

b: The number two is even.

c: Jeff Bezos is the wealthiest person in the world.

d: Charlie Brown was the first president of the United States.

These four sentences are indeed statements because they have a clear true or false value at any given time, even though we may not know which it is. For instance, sentence a might be true or false depending on which car and cat are being described. However, sentence *b* is always true and sentence *d* is always false. Regardless of their truth value, all four of the sentences are examples of mathematical statements. Decide for yourself if the truth value of statement *c* is either always true, always false, or dependent on something.

It has to be noted that not all complete sentences in the English language are statements. That is, some sentences don't assert a claim that is either true or false. Consider the following examples.

s: This is the best movie ever made.

t: Stop!

u: Are we going to the movies tonight?

v: Chair sees box cat telephone.

w: This sentence is false.

At first sight, some of these sentences might seem to vie for statement status. Note that just because they are labeled with a lowercase letter, we cannot assume that they are statements with truth values.

Let's consider the first sentence, *s*. Claiming a movie to be the best one ever made is a matter of opinion, and therefore the sentence cannot really be said to be true or false. Matters of opinion are not mathematically legitimate statements.

Neither the command "Stop!" nor the question "Are we going to the movies tonight?" make any claim to be tested and are therefore also not statements. As for sentence *v*, this is just a random and meaningless string of words and quite obviously not a statement.

That leaves us with example *w*, which is a little more perplexing. It looks like a statement, but is it true or false? From our definition, we know that to be a mathematical statement it must be one or the other, but not both at the same time. When reading the definition of a statement, you might have wondered how a sentence could be both true and false at the same time. To see how that might happen, let's suppose that sentence *w* is actually true. Then just as it asserts, the sentence must also be false. So now we have a sentence that is both true and false at the same time. On the other hand, what if we assume the sentence is false? Then it is false that "this sentence is false," so it is true. Somehow it is both true and false yet again! This type of sentence is an example of a **paradox**. Paradoxes are not statements because they have no truth value. At first, it might seem that this example is a little contrived, but once you start looking, you will see that paradoxes are more common than you think.

📖 **PARADOX**

A **paradox** is a sentence that contradicts itself and therefore has no single truth value. A paradox cannot be a mathematical statement.

Example 3.1.1

Identifying Mathematical Statements

Determine if the following sentences are statements.

    a: It is raining outside.

    b: Beaches are the most beautiful place to vacation.

    c: Today is Monday.

    d: Today is Monday and tomorrow is Friday.

    e: I lie all the time.

Solution

a is a statement because it can be assigned a truth value depending on the weather outside.

b is an opinion and therefore not a statement.

c is a statement since it can be either true or false, depending on the current day the statement is read.

d is a statement even though it is always false.

e is a paradox and not a statement since it contradicts itself and therefore has no truth value.

✔ **Skill Check 3.1.1**

Determine if the following sentences are mathematical statements.

m: Aretha Franklin was the first woman inducted into the Rock and Roll Hall of Fame.

n: A physician has the most stressful job in the world.

o: Any number multiplied by zero equals one.

OBJECTIVE 2 Negation

Sometimes, it is the case that we want the opposite truth value of a statement, or its **negation**. The negation of a statement is the logical opposite of that statement, or its denial. If a statement is true, its negation is always false. Likewise, if a statement is false, its negation is always true. Although there are a variety of different ways to express the idea of negation in the English language, mathematical negations of statements can be written with the symbol ~, read as "not." For instance, the negation of statement *a* is ~*a*.

📖 **NEGATION**

The **negation** of a statement is the logical opposite of that statement, or its denial. Negations always have the opposite truth value of the original statement. Negations are denoted by the symbol ~.

Consider the following statement.

e: 5 is a prime number.

Since 5 is actually a prime number, statement *e* is true. Therefore, $\sim e$ is false. The most common way to write a negation of a statement is to insert the word "not" into the statement. For instance, we could write $\sim e$ as "5 is not a prime number." However, there are many ways to negate statements with words. Here is another way to negate *e*.

$\sim e$: It is not true that 5 is a prime number.

Sometimes it takes a bit more thought to negate a statement in English. This is true when the statement contains words that are **quantifiers**, such as *all*, *some*, *none*, *no*, or *there is*. In the English language, we can negate the word *all* with *not all*. However, English does not allow us to put the word *not* in front of every quantifier. For example, it is incorrect to say "not some" or "not none." Instead, we use other ways to negate the words *some* and *none*. Table 3.1.1 gives us ways to negate these quantifiers.

Table 3.1.1: Negating Quantifiers

Quantifier	Negations
All are	Not all are; Some are not; At least one is not
Some are	None are
Some are not	All are
None are	There is at least one that is; Some are

Notice that the negation of *all* is not the word *none* (or *no*) since this does not give an opposite truth value to a statement. For instance, the statements "all leaves are green" and "no leaves are green" are both false and therefore cannot be negations of one another.

One last note about negations. In logic, when you negate the negation of a statement, you end up with the original statement.

$$\sim (\sim p) = p$$

The negative of the statement "this is blue" is "this is not blue." If we then negate the second statement, we have "this is not not blue." In other words, "this is blue," and we're back to where we started.

Example 3.1.2

Negating Mathematical Statements

Negate the following statements.

a: Melony is wearing a red raincoat.

b: The door is not closed.

c: None of the tourists brought raincoats.

d: I run less than Cara.

Solution

~**a**: Melony is not wearing a red raincoat.

~**b**: The door is not not closed.

> Instead of having two consecutive "nots" in a sentence, we more commonly say, "The door is closed." Notice that statement *b* is actually the negation of the statement "The door is closed." Thus, when we negate statement *b*, we are back to the original statement "The door is closed."

~**c**: Some of the tourists brought raincoats.

> We could also say, "At least one of the tourists brought a raincoat."

~**d**: I do not run less than Cara.

> Notice that we could also write "I run the same as or more than Cara." We need both parts since the opposite of "less than" is "more than or equal to."

> ✔ **Skill Check 3.1.2**
>
> Negate the following statement.
>
> Some of the students completed their assignments.

OBJECTIVE 3 Compound Statements

There is rarely a need to mathematically evaluate a single statement, such as the ones we have looked at so far, since they are inherently either true or false. More often, we need to evaluate the truth value of two or more statements combined together using connecting words such as *and*, *or*, or *implies*. We call these types of statements **compound statements**.

> 📖 **COMPOUND STATEMENTS**
>
> A **compound statement** is composed of two or more statements joined together by connective words such as *and*, *or*, or *implies*.

Conjunctions: *And* Compound Statements

Two or more statements combined together using the word *and* form a compound statement called a **conjunction**. To write this mathematically, we use the symbol ∧ between the lowercase letters representing each statement. The compound statement *c*, which is composed of the two statements *a* and *b*, can be written symbolically as $c = a \wedge b$.

Consider the following two statements *p* and *q*.

p: It is raining.

q: It is sunny outside.

We can combine these two simple statements using the connecting word *and* to form the following compound statement, *r*.

r: It is raining and it is sunny outside.

Symbolically, we write, $r = p \wedge q$.

CONJUNCTION

If *a* and *b* are statements, then "*a* and *b*" is a compound statement called a **conjunction**. The symbol ∧ is used to represent a conjunction.

Example 3.1.3

Writing Conjunctions Symbolically

Consider the following statements.

> *a*: Snow is falling.

> *b*: The sun is shining.

Write the following compound statements using logic symbols.

> *c*: Snow is falling and the sun is shining.

> *d*: The sun is shining and snow is not falling.

Solution

> *c* = Snow is falling and the sun is shining.

> = (Snow is falling) **and** (the sun is shining).

> = $a \wedge b$

> *d* = The sun is shining and snow is not falling.

> = (The sun is shining) **and** (snow is **not** falling).

> = $b \wedge \sim a$

Disjunctions: *Or* Compound Statements

A compound sentence that uses the connecting word *or* is called a **disjunction**. There are two ways to think about the word *or*. Consider the statement "Your painting is so inspirational that you are very talented or have had lots of training." You could interpret this statement to mean that you are either very talented or had lots of training, but not both. This is referred to as an **exclusive *or*** , meaning *one or the other, but not both.*

However, it could mean that both are true; in other words, you are talented and have had lots of training. This type of *or*, meaning either one or both of the options can be true, is known as the **inclusive *or*** and is what is used in mathematical logic. To represent the inclusive *or*, we use the symbol ∨.

DISJUNCTION

If *a* and *b* are statements, then "*a* or *b*" is a compound statement called a **disjunction**. The symbol ∨ is used to represent a disjunction.

Example 3.1.4

Writing Disjunctions Symbolically

Consider the following statements.

> p: He will go to the movies tonight.

> q: He will stay home to give the dog a bath tonight.

Write the following compound statements using logic symbols.

> r: He will go to the movies tonight or he will stay home to give the dog a bath tonight.

> s: He will not go to the movies tonight or he will not stay home to give the dog a bath tonight.

Solution

> r = He will go to the movies tonight or he will stay home to give the dog a bath tonight.

> = (He will go to the movies tonight) **or** (he will stay home to give the dog a bath tonight).

> = $p \vee q$

> s = He will not go to the movies tonight or he will not stay home to give the dog a bath tonight.

> = (He will **not** go to the movies tonight) **or** (he will **not** stay home to give the dog a bath tonight).

> = $\sim p \vee \sim q$

> **Helpful Hint**
>
> Mathematical logic always uses the inclusive *or*, meaning one or the other, or both. The inclusive *or* is represented by the symbol $\vee$.

Conditionals: *If-Then* Compound Statements

We can also combine simple statements into compound statements with implications. Two statements can be joined together using the sentence structure "if a, then b." We call this type of combination **conditional** (or *implication*). Both "if a, then b" and "a implies b" have identical meanings in the English language, but sometimes one sounds more natural than the other. Symbolically, we use $a \Rightarrow b$ to represent both of these statements. Often these are simply called "if-then" statements.

> **CONDITIONAL**
>
> If a and b are statements, then "if a, then b" is a compound statement called a **conditional**. The symbol $\Rightarrow$ is used to represent a conditional statement.

Helpful Hint

Here are a few of the many ways to convey $p \Rightarrow q$.

"p implies q"

"if p, then q"

"p is sufficient for q"

"q is necessary for p"

"p will lead to q"

"q if p"

"q whenever p"

"p only if q"

✔ **Skill Check 3.1.3**

Consider the following statements.

a: I am hungry.

b: I am tired.

c: I am in college.

Write the following three compound statements symbolically.

1. I am hungry and tired.

2. I am hungry or I am in college.

3. I am tired and not in college.

Example 3.1.5

Writing Conditional Statements Symbolically

Consider the following statements.

   s: The water temperature on Saturday is below 76.2 °F.

   t: You are allowed to wear a wetsuit in the triathlon.

Write the following compound statements using logic symbols.

   q: If the water temperature on Saturday is below 76.2 °F, then you are allowed to wear a wetsuit in the triathlon.

   r: You are not allowed to wear a wetsuit in the triathlon if the water temperature on Saturday is not below 76.2 °F.

Solution

   $q =$ If the water temperature on Saturday is below 76.2°F, then you are allowed to wear a wetsuit in the triathlon.

   $=$ **If** (the water temperature on Saturday is below 76.2°F), **then** (you are allowed to wear a wetsuit in the triathlon).

   $= s \Rightarrow t$

Be careful on the is last one. Notice which simple statement follows the "if" part of the condition. This is the part that needs to be written first symbolically.

   $r =$ You are not allowed to wear a wetsuit in the triathlon if the water temperature on Saturday is not below 76.2°F

   $=$ (You are **not** allowed to wear a wetsuit in the triathlon) **if** (the water temperature on Saturday is **not** below 76.2°F).

   $=$ **If** (the water temperature on Saturday is **not** below 76.2°F), **then** (you are **not** allowed to wear a wetsuit in the triathlon).

   $= {\sim}s \Rightarrow {\sim}t$

Biconditionals: *If-and-Only-If* Compound Statements

Finally, we can combine statements using a stronger if-then combination, called a **biconditional statement**. A biconditional statement is a compound statement formed using the words *if and only if* between the two simple statements. This type of statement means that *a* implies *b* and *b* implies *a*. We use $a \Leftrightarrow b$ to symbolically represent "*a* if and only if *b*."

 BICONDITIONAL

If *a* and *b* are statements, then "*a* if and only if *b*" is a compound statement called a **biconditional**. The symbol $\Leftrightarrow$ is used to represent a biconditional statement.

Example 3.1.6

Writing Biconditional Statements Symbolically

Consider the following statements.

    c: Octopuses have three hearts.

    d: The platypus does not have a stomach.

Write the following compound statements using logic symbols.

    e: The platypus does not have a stomach if and only if octopuses have three hearts.

    f: Octopuses do not have three hearts if and only if the platypus has a stomach.

Solution

    e = The platypus does not have a stomach if and only if octopuses have three hearts.

      = (The platypus does not have a stomach) **if and only if** (octopuses have three hearts).

      = $c \Leftrightarrow d$

    f = Octopuses do not have three hearts if and only if the platypus has a stomach.

      = (Octopuses do **not** have three hearts) **if and only if** (the platypus **has** a stomach).

      = $\sim c \Leftrightarrow \sim d$

Table 3.1.2 summarizes the logic symbols discussed in this section.

Table 3.1.2: Logic Symbols

	Symbol	Read
Negation	~	*Not*
Conjunction	∧	*And*
Disjunction	∨	*Or*
Conditional	⇒	*Implies*
Biconditional	⇔	*If And Only If*

Skill Check Answers

1. *m* and *o* are statements; *n* is not a statement. **2.** None of the students completed their assignments. **3.** $a \wedge b$, $a \vee c$, $b \wedge \sim c$

3.1 Exercises

1. A _____ is a statement that contradicts itself.

2. Negations of a statement has the _____ truth value as the original statement.

3. Conjunctions are compound statements that use the connective word _____, while disjunctions are compound statements that use the connective word _____.

4. True or False: Questions are considered mathematical statements.

5. True or False: The sentence "this statement is false" is an example of a paradox.

Decide if each sentence is a mathematical statement.

6. My computer is fast.

7. The car in front of me is turning around.

8. Who are you voting for in the election?

9. I always lie.

10. Running is fun.

11. Are you cold today?

12. I received 35 e-mails today, half of which ended up in my spam folder.

13. Get out of my room!

14. This computer's processor is at least 1.8 GHz.

15. Either we are going to the beach or the mountains for vacation this year.

16. No one goes to that mall anymore; it's too crowded.

17. I will go to the concert if and only if I can get front row seats.

Negate each statement.

18. Kelsey's website had more than 50,000 visits yesterday.

19. Three hundred twenty-nine people applied for the same job I did.

20. I did not get the job.

21. Austin slept until 8:00 a.m. this morning.

22. None of the Christmas tree lights are not working.

23. Every student is volunteering at the food pantry this year.

Using the letters given to represent simple statements and the proper logic connectives, express each compound statement in symbolic form.

24. It is not true that Miranda likes both art and history.

    m: Miranda likes art.

    n: Miranda likes history.

25. Miranda does not like art or Miranda does not like history.

    m: Miranda likes art.

    n: Miranda likes history.

26. If I don't eat meat, then I don't get sick.

    a: I don't eat meat.

    b: I get sick.

27. I don't get sick or I don't eat meat.

    a: I don't eat meat.

    b: I get sick.

28. I will bake cookies if and only if I can give them all away.

    a: I will bake cookies.

    b: I can give all the cookies away.

29. If Mrs. Walker is a teacher, then she is not a rocket scientist.

    t: Mrs. Walker is a teacher.

    r: Mrs. Walker is a rocket scientist.

30. If Mrs. Walker is not a rocket scientist, then she is a teacher.

    t: Mrs. Walker is a teacher.

    r: Mrs. Walker is a rocket scientist.

31. Right angles are formed by the lines if the lines are perpendicular.

    w: The lines are perpendicular.

    z: Right angles are formed by the lines.

32. The lines are not perpendicular or right angles are formed by the lines.

   w: The lines are perpendicular.

   z: Right angles are formed by the lines.

33. I will not adopt a puppy if and only if I have to move next month.

   a: I will adopt a puppy.

   b: I have to move next month.

Use the given simple statements to write each compound statement in words.

   a: Driving makes me smile.

   b: I grill more often than I bake.

   c: It is sunny.

34. $b \wedge a$	**35.** $a \vee c$	**36.** $c \wedge \sim b$
37. $c \Rightarrow b$	**38.** $\sim a \Leftrightarrow c$	**39.** $\sim c \Rightarrow \sim b$

Use the given simple statements to write each compound statement in words.

   p: My video reached 1000 views on YouTube.

   q: I have 1000 Facebook friends.

   r: The home page of my website has a bounce rate of less than 20%.

40. $p \wedge q$	**41.** $\sim q \Rightarrow \sim r$
42. $q \vee p$	**43.** $\sim r \Leftrightarrow \sim p$

Use the given statements to write each compound statement in words.

   p: The moon is full.

   q: I don't know if it's cloudy or bright outside.

   r: I've lost my glasses.

44. $p \wedge r$	**45.** $(\sim p) \vee q$	**46.** $\sim (q \wedge r)$
47. $p \Leftrightarrow \sim q$	**48.** $r \Rightarrow q$	**49.** $\sim r \Rightarrow \sim q$

3.1 PROJECT

SELF-REFERENCE, BARBERS, ALLIGATORS, AND PARADOXES

As you have learned in Section 3.1, not all sentences in English qualify as statements. There are several ways in which seemingly simple sentences can lead to head-scratching situations. In this project, we will consider three different sentences that do not qualify as statements.

First, consider a barber who operates under the following assumption, which is the barber paradox: The barber shaves all those, and those only, who do not shave themselves.

1. Suppose that Cristiano, who is not the barber, does not shave himself. Will the barber shave Cristiano?

2. Arjun, who is also not the barber, shaves himself every morning. Will the barber shave Arjun?

3. Does the barber shave himself? (**Hint**: Explore the consequences of answering "yes, the barber does shave himself" and "no, the barber does not shave himself.")

Now, imagine the admittedly bizarre situation where an alligator steals a child and promises the child's safe return if the child's mother can correctly guess what action the alligator will take with the child next: returning the child safely or not returning the child.

4. How would the alligator respond in the case the mother guesses that the child will not be returned?

Finally, recall the famous tale of Pinocchio, whose nose would grow every time he told a lie.

5. What happens if Pinocchio says, "My nose grows now"?

The issues you may have encountered while thinking about these situations have to do with the phenomenon of self-reference; these are *self-referential statements*.

6. Perform an internet search and find the definition of self-reference.

7. Explain how self-reference is working in each of the three examples we covered in this project to make them paradoxes.

3.2 TRUGH TABLES

OBJECTIVES

1. Construct truth tables for compound statements containing two simple statements.

2. Construct truth tables for compound statements containing more than two simple statements.

3. Determine whether compound statements are tautologies.

OBJECTIVE 1 Introduction to Truth Tables

As we stated in Section 3.1, mathematical logic deals in part with deciding the truth value of different types of statements. For instance, consider the statement a and its negation $\sim a$. If we let a be the statement "Today is Monday," then it is either true or false, depending on when you are reading this. If a is true, then its negation, $\sim a$ will be false. However, if a is false, then its negation will be true.

To help us keep track of the different possible truth values for a statement, we can construct a **truth table**. A truth table is a chart with rows and columns that systematically lists out each possible combination of truth values. Once fully complete, the truth table reveals the conditions under which a statement is true and when it is false. One of the simplest truth tables is that of the negation we just discussed. It is shown in Table 3.2.1.

Table 3.2.1: Negation Truth Table

a	$\sim a$
T	F
F	T

While the truth value of some statements is easy to determine, like that of a negation, others require more thought. If the statement we are considering is a compound statement, then its truth value will vary based on the truth values of the various component parts. For example, let d be the statement "It is sunny outside." The conjunction using a and d is given as follows.

$a \wedge d$: Today is Monday **and** it is sunny outside.

Is this statement true or false? To decide the truth of this compound statement for today, we need to consult the calendar to determine the day of the week while also checking the weather. Suppose the portion "Today is Monday" is true and the portion "It is sunny outside" is also true. Then the compound statement in which the two pieces are put together as one is obviously true as well. However, what truth value would the statement have if today is not Monday, but it is sunny outside? It can often get complicated to keep up with the truth values for compound statements, especially when they have more and more pieces to them. This is where truth tables come in handy.

Let's look at constructing the truth table for the conjunction we were just discussing—$a \wedge d$: Today is Monday **and** it is sunny outside. To construct a truth table, we need a column for each part of the compound statement. In this case, we will have three columns: two for the simple statements a and d, and one for the compound statement itself.

Table 3.2.2: Conjunction Truth Table Template

a	d	$a \wedge d$

Helpful Hint

As compound statements get more complicated, you will begin to see the natural progression of difficulty in the columns leading up to the last column, which always contains the entire compound statement.

We next list all the possible combinations that could occur for the truth values of the simple statements. For example, both a and d could be true, meaning that it actually could be Monday and sunny outside. Or a could be true while b is false. We fill in the table systematically to list the other possible combinations for a and d.

Table 3.2.3: Conjunction Truth Table

a	d	$a \wedge d$
T	T	
T	F	
F	T	
F	F	

In order to complete the last column of our truth table, we need to know when compound statements are true. Specifically in our example, we want to know when a conjunction is true and when it is false. In order for a conjunction to be true, each of the individual parts of the compound statement must be true; otherwise, it is false. So we can look across each row of the truth table and decide which truth value should be placed in the final column.

Table 3.2.4: Conjunction Truth Table

a	d	$a \wedge d$
T	T	T
T	F	F
F	T	F
F	F	F

Notice that only one row in the conjunction truth table is true—the first row where both a and d are both true. So now we are fully armed to answer the question "Is the statement $a \wedge d$, 'Today is Monday and it is sunny outside,' true or false?" The actual answer is "It depends." We can look at the table and see that there are four possible truth value combinations in our situation. A better question might be, "Under what circumstances is the compound statement $a \wedge d$, 'Today is Monday and it is sunny outside,' true?" We can see that it is true only when it is both Monday and it is sunny.

📖 **CONJUNCTION TRUTH VALUE**

If a and b are statements, then the **conjunction** "a and b" is true only when both a and b are true; otherwise, the conjunction is false.

Let's take a look at the truth values for the other three types of compound statements: disjunctions, conditionals, and biconditionals. If a compound statement is a disjunction (that is, it contains the word *or*), recall that we will assume that the disjunction is inclusive. That is to say that the statement is true if either of its components are true. So in contrast to a conjunction, where there is only one true value, a disjunction will always be true unless each of the simple statements are false. Table 3.2.5 shows the truth table for the disjunction $a \vee d$: Today is Monday or it is sunny outside.

Table 3.2.5: Disjunction Truth Table

a	d	$a \vee d$
T	T	T
T	F	T
F	T	T
F	F	F

Helpful Hint

The order of statements in a disjunction does not matter; that is, $p \vee q$ has the same truth value as $q \vee p$.

📖 **DISJUNCTION TRUTH VALUE**

If a and b are statements, then the **disjunction** "a or b" is always true unless a and b are both false.

Next is the conditional statement, where one statement implies another. A conditional is true in all cases except when p is true and q is false. The statement structure of "If a, then b" can be best described by thinking of a promise. If you do a, then I promise to do b. The only time the promise is false, or broken, is when you do a and I don't do b. Think of everything before the *implies* arrow as the *if* part of the statement, and everything after the arrow is the *then* statement. Table 3.2.6 shows the truth table for the conditional statement $a \Rightarrow d$: If today is Monday, then it is sunny outside.

Table 3.2.6: Conditional Truth Table

a	d	$a \Rightarrow d$
T	T	T
T	F	F
F	T	T
F	F	T

Helpful Hint

The order of statements in a conditional matters; that is, $p \Rightarrow q$ does not have the same truth value as $q \Rightarrow p$.

📖 **CONDITIONAL TRUTH VALUE**

If a and b are statements, then the **conditional** "if a, then b" is always true unless a is true and b is false.

Finally, a biconditional statement (if and only if) is true only when each component of the statement has the same truth value; that is, either both are true or both are false. Table 3.2.7 shows the truth table for the biconditional statement $a \Leftrightarrow b$: Today is Monday if and only if it is sunny outside.

Table 3.2.7: Biconditional Truth Table

a	d	$a \Leftrightarrow d$
T	T	T
T	F	F
F	T	F
F	F	T

Helpful Hint

The order of statements in a biconditional does not matter; that is, $p \Leftrightarrow q$ has the same truth value as $q \Leftrightarrow p$.

📖 BICONDITIONAL TRUTH VALUE

If a and b are statements, then the **biconditional** "a if and only if b" is true only if a and b have the same truth value; that is, either both are true or both are false.

Table 3.2.8: Summary of Logic Statements

	Notation	Read	Truth Value Rule
Negation	$\sim p$	not p	opposite of the truth value of p
Conjunction	$p \wedge q$	p and q	true only when both p and q are true
Disjunction	$p \vee q$	p or q	false only when both p and q are false
Conditional	$p \Rightarrow q$	if p, then q (or p implies q)	false only when p is true and q is false
Biconditional	$p \Leftrightarrow q$	p if and only if q	true when p and q have the same truth value

These are the basic building blocks to determining the truth values for more complex compound statements. We simply break down a statement and then apply these rules. Let's get our feet wet with the negation of a conjunction. Then we will move on to more involved scenarios.

Example 3.2.1

Constructing a Truth Table

Construct a truth table for the conjunction $\sim(a \wedge b)$ to determine when the statement is true and when it is false.

Solution

Notice that we have parentheses in this compound statement. Just like in algebra, parentheses are used to reduce ambiguity with logical symbols. If there are parentheses, we work from the inside out. So in our table we will first find $(a \wedge b)$, and then we will negate it. We need a column for a, a column for b, a column for the conjunction $a \wedge b$, and finally a column for the negation of the conjunction, $\sim(a \wedge b)$. So our truth table template looks like Table 3.2.9. Notice that the last column is the compound statement we were given.

Table 3.2.9: Truth Table for $\sim(a \wedge b)$ Template

a	b	$a \wedge b$	$\sim(a \wedge b)$

Begin by completing the first two columns of the table so that we have all four possible combinations of truth values for a and b.

Table 3.2.10: Truth Table for ~$(a \wedge b)$

a	b	$a \wedge b$	~$(a \wedge b)$
T	T		
T	F		
F	T		
F	F		

To fill in the third column, remember that the conjunction is true only when both parts are true. Referring to the first and second columns, we can fill in the third column.

Table 3.2.11: Truth Table for ~$(a \wedge b)$

a	b	$a \wedge b$	~$(a \wedge b)$
T	T	T	
T	F	F	
F	T	F	
F	F	F	

The negation has the opposite truth value of the original statement. So we can fill in the fourth column by changing the truth vales of the third column.

Table 3.2.12: Completed Truth Table for ~$(a \wedge b)$

a	b	$a \wedge b$	~$(a \wedge b)$
T	T	T	F
T	F	F	T
F	T	F	T
F	F	F	T

This truth table tells us that the negation of a conjunction statement is only false on one occasion: when both parts are true. Otherwise, the statement is true.

✔ **Skill Check 3.2.1**

Consider the following statements.

f: Arizona is the first state listed alphabetically.

g: Albany is the capital of New York.

Determine the truth value of each statement and then use Table 3.2.12 to determine if the statement ~$(f \wedge g)$ is true or false.

The usefulness of truth tables is that they are not dependent on the individual statements you are given. A truth table for the negation of a conjunction will always look like Table 3.2.12. If we are given two statements and asked to form this type of compound statement, we can know the truth value straightaway. For example, consider the following statements.

a: England is the largest country in the world by landmass.

b: England is the most populated country in the world.

The statement ~$(a \wedge b)$ translates to "It is not the case that England is the largest country in the world by landmass and England is the most populated country in the world." To determine the truth value of this negation, we do not need to construct another table. We simply need to know the truth values of each individual statement. In actuality, both of these statements are false. (Russia is the largest country by land mass and China is the most populated.) Table 3.2.12 tells us that when both individual statements are false—as in this case—the negation of their conjunction is true; that is, ~$(a \wedge b)$ is true.

OBJECTIVE 2 Truth Tables for More Complex Compound Statements

Let's try our hand at another example. This time we will consider a compound statement involving three simple statements. Before we begin, note that the tables we have made thus far don't actually have enough rows for all of the possible outcomes with three statements. For each individual statement within a compound statement, the number of rows required for the truth table increases by a factor of 2. For example, a compound statement with 4 simple statements needs $2 \cdot 2 \cdot 2 \cdot 2 = 2^4 = 16$ rows. In general, a truth table for a compound statement with n simple statements will have 2^n rows.

Example 3.2.2

Constructing a Truth Table Containing Three Simple Statements

Construct a truth table for the conjunction $(c \vee d) \Rightarrow {\sim} e$ to determine when the statement is true and when it is false.

Solution

To form the truth table for this compound statement, we need several columns. We need a column for each original statement c, d, and e, and we'll also need columns for the negation of e along with $c \vee d$ and the final statement $(c \vee d) \Rightarrow {\sim} e$. This means that the table will have six columns and $2^3 = 8$ rows. We can fill in the first three columns with all possible truth combinations for the simple statements as shown in Table 3.2.13.

Table 3.2.13: Truth Table for $(c \vee d) \Rightarrow {\sim} e$

c	d	e	${\sim} e$	$(c \vee d)$	$(c \vee d) \Rightarrow {\sim} e$
T	T	T			
T	T	F			
T	F	T			
T	F	F			
F	T	T			
F	T	F			
F	F	T			
F	F	F			

Helpful Hint

When constructing truth tables, the order of the columns is not crucial. Some tables show all of the simple statements of the compound statement first, while others follow a progression of the compound statement reading left to right.

Next, the fourth column contains the negation of e, which is simply the opposite truth value of the third column. We also know how to fill in the disjunction in the fifth column by referring to the columns for c and d. Recall that a disjunction is false only when both parts are false. Table 3.2.14 shows these columns filled in.

Table 3.2.14: Truth Table for $(c \lor d) \Rightarrow \sim e$

c	d	e	~e	(c ∨ d)	(c ∨ d) ⇒ ~e
T	T	T	F	T	
T	T	F	T	T	
T	F	T	F	T	
T	F	F	T	T	
F	T	T	F	T	
F	T	F	T	T	
F	F	T	F	F	
F	F	F	T	F	

Lastly, the final column contains the conditional statement "if $(c \lor d)$, then $\sim e$." As we fill in the final column, we will be focused on the columns that contain the "if a, then b" parts—the fourth and fifth columns. Remember that the only time a conditional statement is false is when the *if* part is true while the *then* part is false.

Table 3.2.15: Truth Table for $(c \lor d) \Rightarrow \sim e$

			"Then" ↓	"If" ↓	
c	d	e	~e	(c ∨ d)	(c ∨ d) ⇒ ~e
T	T	T	F	T	F
T	T	F	T	T	T
T	F	T	F	T	F
T	F	F	T	T	T
F	T	T	F	T	F
F	T	F	T	T	T
F	F	T	F	F	T
F	F	F	T	F	T

Thus, we know that the statement $(c \lor d) \Rightarrow \sim e$ will be false in three circumstances, while the remaining five circumstances have a truth value of true.

Example 3.2.3

Constructing a Truth Table from Words

Construct the truth table for the following compound statement and determine when the statement is true and when it is false.

If you're not making mistakes, then you're not doing anything.

—John Wooden

Solution

First, we need to identify and label the two simple statements of the compound sentence. The easiest way to do this is to write them down without using the negations. We'll add the negations later. So our two statements, *a* and *b*, are the following.

> *a*: You are making mistakes.

> *b*: You are doing something.

Notice that the word "anything" is used in the second part of the original statement. But in English, since we don't commonly say "you are doing anything," we will use the word "something" in our second simple statement.

Then, we can negate each of these two simple statements.

> ~*a*: You are not making mistakes.

> ~*b*: You are not doing anything.

Symbolically, our conditional statement is $\sim a \Rightarrow \sim b$. Notice that we have "not doing anything" as the negation of "doing something," which matches the original quote.

> If you are **not** making mistakes, **then** you are **not** doing something.

The truth table we need will contain columns for a, b, $\sim a$, $\sim b$, and $\sim a \Rightarrow \sim b$.

Table 3.2.16: Truth Table for $\sim a \Rightarrow \sim b$ Template

a	b	~a	~b	~a ⇒ ~b

Fill in the table from left to right as we have done so far. Remember that a conditional statement is false only when the *if* part is true and the *then* part is false. The completed truth table will look like Table 3.2.17.

Table 3.2.17: Truth Table for $\sim a \Rightarrow \sim b$

a	b	"If" ↓ ~a	"Then" ↓ ~b	~a ⇒ ~b
T	T	F	F	T
T	F	F	T	T
F	T	T	F	F
F	F	T	T	T

From the truth table, we can see that the compound statement is true in all but one of the cases: the statement is false when *you are not making mistakes* and *you are doing something.*

✓ Skill Check 3.2.2

Construct a truth table for the conditional statement in Example 3.2.3. This time let statements *a* and *b* be the following.

a: You are not making mistakes.

b: You are not doing anything.

Does your truth table have the same truth values as the table in Example 3.2.3? Should it?

OBJECTIVE 3 Tautologies

For some compound statements, the values in the last column of the truth table are all true. When this happens, the statement is called a **tautology**. A tautology will have a *true* truth value in all possible circumstances.

> 📖 **TAUTOLOGY**
>
> A **tautology** is a statement that is true in all possible circumstances.

The conditional statement from Example 3.2.3 does not represent a tautology because there is one instance where the statement is false. Let's take a look at a statement that *is* a tautology.

Example 3.2.4

Constructing a Truth Table for a Tautology

Construct the truth table for the following compound statement.

Next year, Imre can take physics or he cannot take physics.

Solution

Notice that this time, there is really only one simple statement. The second part is simply the negation of the first part. Thus, we have c and $\sim c$ as the following.

  c: Imre can take physics next year.

  $\sim c$: Imre cannot take physics next year.

Our entire compound statement can then be expressed symbolically by $c \vee \sim c$. The truth table will only need 3 columns: c, $\sim c$, and $c \vee \sim c$. Remember, a disjunction will only be false when both parts are false. Table 3.2.18 shows the completed truth table.

Table 3.2.18: Truth Table for $c \vee \sim c$

c	$\sim c$	$c \vee \sim c$
T	F	T
F	T	T

Hence, "Next year, Imre can take physics or he cannot take physics" is a tautology, since all truth values for the disjunction in the last column are true.

Example 3.2.5

Constructing a Truth Table for a Conditional Statement

Consider the following quote from the movie *The Hunger Games: Catching Fire*[1].

---

1   Simon Beaufoy and Michael Arndt. *The Hunger Games: Catching Fire*. Directed by Francis Lawrence. (Santa Monica, CA: Lionsgate, 2014), DVD.

Peeta Mellark to Katniss Everdeen:

"…if you can stop looking at me like I'm wounded, then I can quit acting like it. Then maybe we have a shot at being friends."

a. Construct a truth table for Peeta's conditional quote.

b. Determine if Peeta's statement is a tautology.

Solution

a. As we've stated before, English is a rich and complicated language. We need to be cautious about simply seeing the words *if…, then* and assuming it's a straightforward conditional statement without looking at the intent behind the statement. In the original quotation, the words *if…, then* do actually appear; however, the word *then* occurs again without a matching *if* to go with it. The *then* part of Peeta's conditional statement can be perceived as twofold. *If* Katniss changes her gaze, *then* two things will happen: Peeta's behavior will change and they have a shot at friendship.

"**…if** you can stop looking at me like I'm wounded, **then** I can quit acting like it, **and** (then) maybe we have a shot at being friends."

As a formal "if *a*, then *b*" compound statement, Peeta's quote involves 3 simple statements: *p*, *q*, and *r*.

p: You stop looking at me like I'm wounded.

q: I can quit acting like I'm wounded.

r: We have a shot at being friends.

The logical mathematical statement that follows from the quote is $p \Rightarrow (q \wedge r)$.

As we build the truth table for the logic statement, we need to include a column for each individual part that is contained in the final conditional statement. Also, remember to include enough rows for three simple statements; that is, $2 \cdot 2 \cdot 2 = 8$ rows.

The first part of the table should look like Table 3.2.19.

Table 3.2.19: Truth Table for $p \Rightarrow (q \wedge r)$

p	q	r	$q \wedge r$	$p \Rightarrow (q \wedge r)$
T	T	T		
T	T	F		
T	F	T		
T	F	F		
F	T	T		
F	T	F		
F	F	T		
F	F	F		

Next, complete the column for the conjunction (that is, the *and* statement). Remember that a conjunction is true only if both of the individual statements are true.

Table 3.2.20: Truth Table $p \Rightarrow (q \wedge r)$

p	q	r	q ∧ r	p ⇒ (q ∧ r)
T	T	T	T	
T	T	F	F	
T	F	T	F	
T	F	F	F	
F	T	T	T	
F	T	F	F	
F	F	T	F	
F	F	F	F	

Finally, fill in the conditional column using the columns that contain the *if* and *then* parts.

Table 3.2.21: Truth Table $p \Rightarrow (q \wedge r)$

If ↓				Then ↓
p	q	r	q ∧ r	p ⇒ (q ∧ r)
T	T	T	T	T
T	T	F	F	F
T	F	T	F	F
T	F	F	F	F
F	T	T	T	T
F	T	F	F	T
F	F	T	F	T
F	F	F	F	T

b. The truth table shows us that the statement would be true in all but 3 cases. Because the statement is not always true, it is not a tautology. The 3 cases in which Peeta's statement turns out to be false are as follows.

    a. TTF = Katniss <u>stops</u> looking at him like he's wounded; he <u>stops</u> acting like it, but they **don't** <u>have</u> a shot at being friends.

    b. TFT = Katniss <u>stops</u> looking at him like he's wounded; he <u>does</u> **not** stop acting like it, and they <u>do have</u> a shot at being friends.

    c. TFF = Katniss <u>stops</u> looking at him like he's wounded; but he <u>does</u> **not** stop acting like it, and they **don't** <u>have</u> a shot at being friends.

Skill Check Answers

1. f is false and g is true, so $\sim(f \wedge g)$ is true   **2.** No, the truth tables do not look exactly the same. The truth tables have the same meaning, but do not look identical since a and b are defined differently.

a	b	a ⇒ b
T	T	T
T	F	F
F	T	T
F	F	T

3.2 Exercises

✔ CONCEPT CHECK

1. Truth tables are used to keep track of different possible _____ for a mathematical statement.

2. A conjunction of two statements a and b is true when _____.

3. A disjunction of two statements a and b is true unless _____.

4. True or False: The statement "if today is Sunday, then I will eat a sundae" is false if today is Tuesday and I eat a Sunday.

5. True or False: The statement "the snow will melt if and only if the temperature is above freezing" is true if the snow doesn't melt and the temperature is below freezing.

💡 PRACTICE

Complete each truth table.

6. $a \wedge \sim b$

Truth Table

a	b	$\sim b$	$a \wedge \sim b$
T	T		
T	F		
F	T		
F	F		

7. $\sim w \vee \sim z$

Truth Table

w	z	$\sim w$	$\sim z$	$\sim w \vee \sim z$
T	T			
T	F			
F	T			
F	F			

8. $c \Rightarrow \sim d$

Truth Table

c	d	$\sim d$	$c \Rightarrow \sim d$
T	T		
T	F		
F	T		
F	F		

9. $m \Leftrightarrow \sim n$

Truth Table

m	n	$\sim n$	$m \Leftrightarrow \sim n$
T	T		
T	F		
F	T		
F	F		

10. $\sim(p \Rightarrow q)$

Truth Table

p	q	$p \Rightarrow q$	$\sim(p \Rightarrow q)$
T	T		
T	F		
F	T		
F	F		

11. $\sim(s \Leftrightarrow t)$

Truth Table

s	t	$s \Leftrightarrow t$	$\sim(s \Leftrightarrow t)$
T	T		
T	F		
F	T		
F	F		

12. $(a \vee b) \vee c$

Truth Table

a	b	c	$a \vee b$	$(a \vee b) \vee c$
T	T	T		
T	T	F		
T	F	T		
T	F	F		
F	T	T		
F	T	F		
F	F	T		
F	F	F		

13. $(p \vee r) \Rightarrow q$

Truth Table

p	r	q	$p \vee r$	$(p \vee r) \Rightarrow q$
T	T	T		
T	T	F		
T	F	T		
T	F	F		
F	T	T		
F	T	F		
F	F	T		
F	F	F		

14. $\left(\sim a \wedge \sim b\right) \wedge \sim c$

Truth Table

a	b	c	~a	~b	~c	~a∧~b	$\left(\sim a \wedge \sim b\right) \wedge \sim c$
T	T	T					
T	T	F					
T	F	T					
T	F	F					
F	T	T					
F	T	F					
F	F	T					
F	F	F					

15. $\left(m \wedge n\right) \Leftrightarrow r$

Truth Table

m	n	r	m∧n	$\left(m \wedge n\right) \Leftrightarrow r$
T	T	T		
T	T	F		
T	F	T		
T	F	F		
F	T	T		
F	T	F		
F	F	T		
F	F	F		

16. $\sim p \Rightarrow (q \vee r)$

Truth Table

p	q	r	~p	q∨r	$\sim p \Rightarrow (q \vee r)$
T	T	T			
T	T	F			
T	F	T			
T	F	F			
F	T	T			
F	T	F			
F	F	T			
F	F	F			

17. $(\sim b \Leftrightarrow c) \Rightarrow \sim a$

Truth Table

a	b	c	$\sim a$	$\sim b$	$\sim b \Leftrightarrow c$	$(\sim b \Leftrightarrow c) \Rightarrow \sim a$
T	T	T				
T	T	F				
T	F	T				
T	F	F				
F	T	T				
F	T	F				
F	F	T				
F	F	F				

Convert each compound statement into variables and then construct the truth table for the compound statement.

18. We will buy a new smartphone or we will buy a new computer.

19. I plan to go to the movies and go out to eat this weekend.

20. Ella is not on the dance team and Coleman is on the soccer team.

21. My calculator is not working correctly and I need to buy a new one if and only if it calculates that one plus one equals three.

22. If it rains and you don't put the convertible hood up, the inside of your convertible will get wet.

23. If Meg doesn't get gas, then her car will break down and she will miss her exam.

Use a truth table to determine if each statement is a tautology.

24. $(\sim p \vee \sim q) \vee (p \wedge q)$ **25.** $w \Rightarrow v$

26. $(m \wedge n) \vee (\sim n)$ **27.** $(a \wedge \sim b) \vee (a \Rightarrow b)$

Convert each compound statement into variables and create a truth table for the given statement. Then determine whether it is a tautology.

28. If I like the cake, then it is chocolate, or if the cake is chocolate, then I like the cake.

29. If the bird is blue or green, then the bird is not from here.

30. You wrote the love letter if and only if you liked her, or you did not write the love letter if and only if you did not like her.

31. The animal is a penguin implies it wears a bow tie if and only if the animal does not wear a bow tie implies it is not a penguin.

3.3 LOGICAL EQUIVALENCE AND DE MORGAN'S LAWS

OBJECTIVES

1. Determine the logical equivalence of statements.

2. Negate compound statements.

3. Write variations of conditional statements.

As we said at the beginning of the chapter, we encounter a barrage of assertions every day. The first two sections laid a foundation that we can use to evaluate these claims mathematically within the framework of logic. Of course, there are many ways to express ideas—and sometimes two things can be said very differently but actually express precisely the same thing. When evaluating statements, it is sometimes important to determine whether two statements are equal to one another or whether they are the complete opposite of one another. This section looks at ways to identify statements that fall into one of these categories. We'll begin with statements that are equal to one another and investigate what equal means when it comes to logical statements.

OBJECTIVE 1 Logical Equivalence

As we noted in Section 3.2, switching the order of the simple statements in a conjunction makes no difference to the truth value. That is to say, $p \wedge q$ has the same meaning as $q \wedge p$. Likewise, the disjunction $p \vee q$ has the same meaning as $q \vee p$. However, for statements, rather than being *equal*, we talk about *logical equivalence*. When statements have the same truth values in all corresponding circumstances, they are **logically equivalent**; that is, the statements are equal and mean the same thing. Thus, we say "$p \wedge q$ is logically equivalent to $q \wedge p$."

📖 LOGICALLY EQUIVALENT STATEMENTS

Logically equivalent statements are statements that have exactly the same truth values in all corresponding circumstances. Equivalence is denoted with the symbol $\equiv$.

For example, consider the two compound statements "If my cat is hungry, she will rub my leg" and "If my cat does not rub my leg, then she is not hungry." Are these statements logically equivalent? If they are, then they will have the same truth values when we look at their individual truth tables. Let's write each of the symbolically using the following simple statements.

a: My cat is hungry.

b: My cat will rub my leg.

The first statement, "If my cat is hungry, she will rub my leg," translates to $a \Rightarrow b$. The second statement, "If my cat does not rub my leg, then she is not hungry," translates to $\sim b \Rightarrow \sim a$. The following two tables show the truth tables for each of these conditional statements.

Table 3.3.1: Truth Table for $a \Rightarrow b$

a	b	$a \Rightarrow b$
T	T	T
T	F	F
F	T	T
F	F	T

Table 3.3.2: Truth Table for $\sim b \Rightarrow \sim a$

a	b	$\sim a$	$\sim b$	$\sim b \Rightarrow \sim a$
T	T	F	F	T
T	F	F	T	F
F	T	T	F	T
F	F	T	T	T

Notice that the truth values in the final column of each table are the same (T, F, T, T). This means that the two compound statements are logically equivalent; that is, they mean the same thing. We write this symbolically as $a \Rightarrow b \equiv \sim b \Rightarrow \sim a$. Notice that we have done more than show the logical equivalence of these specific statements about my cat. By showing that $a \Rightarrow b \equiv \sim b \Rightarrow \sim a$, we have shown that for any possible statements a and b, $a \Rightarrow b$ and $\sim b \Rightarrow \sim a$ express the same thing.

Consider the statement "If my cat rubs my leg, then she is hungry." Is this statement also logically equivalent to the original implication? Using the same simple statements, "If my cat rubs my leg, then she is hungry" translates to $b \Rightarrow a$. Again, we can use a truth table to help us.

Table 3.3.3: Truth Table for $a \Rightarrow b$ and $b \Rightarrow a$

a	b	$a \Rightarrow b$	$b \Rightarrow a$
T	T	T	T
T	F	F	T
F	T	T	F
F	F	T	T

You can see in Table 3.3.3 that the truth values on the second and third rows are not the same for $a \Rightarrow b$ and $b \Rightarrow a$. Thus, the statements are *not* logically equivalent to one another. We write this mathematically as $a \Rightarrow b \not\equiv b \Rightarrow a$. Thus, we know that unlike with conjunctions and disjunctions, switching the order of simple statements in a conditional statement does affect the truth value.

Example 3.3.1

Determining Logical Equivalence

Use a truth table to determine if p and $\sim(\sim p)$ are logically equivalent.

Solution

The truth table will contain three columns, one each for p, $\sim p$ and $\sim(\sim p)$. Remember that the negation of a truth value is its opposite. Table 3.3.4 contains the truth values of each.

Table 3.3.4: Truth Table for p, $\sim p$ and $\sim(\sim p)$

p	$\sim p$	$\sim(\sim p)$
T	F	T
F	T	F

We are interested in the first and last columns, p and $\sim(\sim p)$. Both contain true and then false. Thus, we know that p and its double negation are equivalent; in other words, $p \equiv \sim(\sim p)$. This ought to sound familiar to you. We occasionally use double negatives in our everyday speech. For instance, the statement "There is no way I'm not going to the party tonight" contains a double negative but is equivalent to the statement that doesn't contain a negative at all: "I am going to the party tonight."

Example 3.3.2

Determining Logical Equivalence

Use truth tables to show that $p \Rightarrow q \equiv \sim p \vee q$.

Solution

We are asked to show that the conditional statement $p \Rightarrow q$ is equivalent to the disjunction statement $\sim p \vee q$ using truth tables. If the final columns of each table contain the same truth vales, then the statements are logically equivalent. We'll construct the conditional truth table first. Remember that a conditional statement is false only when the *if* statement is true and the *then* statement is false. Thus, the truth table for the conditional looks like Table 3.3.5.

Table 3.3.5: Truth Table for $p \Rightarrow q$

p	q	$p \Rightarrow q$
T	T	T
T	F	F
F	T	T
F	F	T

The truth table for the second compound statement, $\sim p \vee q$, will have a column for $\sim p$ before the disjunction statement. Remember that a disjunction is only false when both statements are false. Table 3.3.6 shows the truth table for $\sim p \vee q$.

Helpful Hint

When comparing the final column of two truth tables to show equivalence, the order of the truth values for the simple statements must be the same in each table; that is, both tables must have the same first two columns.

Table 3.3.6: Truth Table for ~ $p \vee q$

p	q	~ p	~ $p \vee q$
T	T	F	T
T	F	F	F
F	T	T	T
F	F	T	T

In order for the two expressions to be equivalent, their truth values in the last column must be the same for all four combinations. We can see that both tables have T, F, T, T in the last column, proving that the statements are indeed equivalent.

Example 3.3.3

Determining Logical Equivalence

Determine if the following statements are logically equivalent to the given conditional statement.

If it rains on Saturday, we will not go to the ballgame.

1. It is raining on Saturday and we will not go to the ballgame.

2. It is not raining on Saturday or we will not go to the ballgame.

3. If we do go to the ballgame, then it is not raining on Saturday.

Solution

To determine if the statements are logically equivalent, we begin by writing the original statement in symbolic form. Let a be the statement "It rains on Saturday" and b be the statement "We will not go to the ballgame." Then the original statement can be written in the form $a \Rightarrow b$. Now we can write each of the choices in symbolic form and create a truth table to compare the truth values to the original statement.

1. "It is raining on Saturday and we will not go to the ballgame" translates to $a \wedge b$.

2. "It is not raining on Saturday or we will not go to the ballgame" translates to $\sim a \vee b$.

3. "If we do go to the ballgame, then it is not raining on Saturday" translates to $\sim b \Rightarrow \sim a$.

Next construct a truth table containing the original statement along with each of the alternative statements.

Table 3.3.7: Truth Table for $a \Rightarrow b$, $a \wedge b$, $\sim a \vee b$, and $\sim b \Rightarrow \sim a$

Original
↓

a	b	$\sim a$	$\sim b$	$a \Rightarrow b$	**1.** $a \wedge b$	**2.** $\sim a \vee b$	**3.** $\sim b \Rightarrow \sim a$
T	T	F	F	T	T	T	T
T	F	F	T	F	F	F	F
F	T	T	F	T	F	T	T
F	F	T	T	T	F	T	T

We can see from the truth table that statements 2 and 3 both have the same corresponding truth values as those for the original statement. Therefore, the original conditional, "If it rains on Saturday, we will not go to the ballgame," is logically equivalent to "It is not raining on Saturday or we will not go to the ballgame" and "If we do go to the ballgame, then it is not raining on Saturday."

OBJECTIVE 2 Negating Compound Statements

As these examples have demonstrated, it is important to know how to apply negations to complicated compound statements. Augustus De Morgan, an English mathematician and logician, formally defined two famous equivalent negations that show how to negate *and* statements and *or* statements. De Morgan's Laws show how a negation sign is "distributed" across compound statements.

> **📖 DE MORGAN'S LAWS**
>
> **1.** $\sim(p \wedge q) \equiv \sim p \vee \sim q$
>
> **2.** $\sim(p \vee q) \equiv \sim p \wedge \sim q$

> **💬 Helpful Hint**
>
> Notice that in De Morgan's Laws, one side of the equivalence has a negation sign on the outside of a set of the parentheses while the other side has no parentheses at all.

De Morgan observed that conjunctions become disjunctions and vice versa when they are negated. That is to say, the *and* ($\wedge$) becomes an *or* ($\vee$) and the *or* ($\vee$) becomes an *and* ($\wedge$) when they are negated. For example, the negation of the statement "I will mow the lawn *or* I will wash the car" is "I will *not* mow the lawn *and* I will *not* wash the car." We will leave the verification of De Morgan's laws by constructing their truth tables from scratch as an exercise at the end of this section.

Example 3.3.4

Applying De Morgan's Laws

Write the negations of the following statements using De Morgan's laws.

a. I will exercise or sleep in tomorrow.

b. Jack and Jill went up the hill.

Solution

a. To negate the disjunction statement, we negate both simple statements, then change the disjunction to a conjunction by using the word *and* instead of *or*.

I will not exercise and I will not sleep in tomorrow.

b. To negate the conjunction, we similarly negate each simple statement and use the word *or* instead of *and* to make the statement a disjunction.

Jack did not go up the hill or Jill did not go up the hill.

Because this original statement is such a familiar one, it might look like a simple statement instead of a compound statement and you might have found it hard to negate the statement correctly. Be careful not to say, "Neither Jack nor Jill went up the hill."

Example 3.3.5

Applying De Morgan's Laws

Write the negation of the following compound statement using De Morgan's laws.

The football team will win or they will be out of the tournament, and their ranking will improve.

Solution

Notice that this compound statement has an *or* statement as well as an *and* statement. We need to use De Morgan's laws twice to negate each part. To help us keep track of both, let's write the statement out symbolically.

Begin by labeling each part of the compound statement. Let

p = The football team will win.

q = They will be out of the tournament

r = Their ranking will improve.

Thus, our original statement is $(p \vee q) \wedge r$.

This is a conjunction where the first part is $(p \vee q)$ and the second part is r. De Morgan's laws say that its negation will be a disjunction made from the negations of these two parts. Thus, we have the following.

$$\sim((p \vee q) \wedge r) \equiv \sim(p \vee q) \vee \sim r$$

We can then use De Morgan's Laws again to negate $\sim(p \vee q)$.

$$\sim(p \vee q) \equiv \sim p \wedge \sim q$$

Putting these together, we have that the negation of the original statement is as follows.

$$\sim((p \vee q) \wedge r) \equiv (\sim p \wedge \sim q) \vee \sim r$$

Now, we can translate the symbols back into words, giving us the following.

The football team will not win and they will not be out of the tournament, or their ranking will not improve.

De Morgan showed that negating a conjunction forms a disjunction and vice versa. One might assume that negating a conditional statement is another conditional statement; however, the negation of a conditional statement is actually a compound statement. Think of the conditional statement as a promise and the negation as the broken promise. For example, consider the conditional statement "If you make an A in the class, then I'll give you a reward." In order for the opposite of this to happen, then the promise will need to be broken, so to speak. In other words, you made an A in the class, and I did not give you a reward.

📖 NEGATION OF CONDITIONAL STATEMENTS

$$\sim(p \Rightarrow q) \equiv p \wedge \sim q$$

Example 3.3.6

Negating a Conditional Statement

Write the negation of the following conditional statement.

If I go to Moss' Diner, then I get the triple stack pancakes.

Solution

The statement we are given is a conditional statement, $a \Rightarrow b$, where a is the statement "I go to Moss' Diner" and b is the statement "I get the triple stack pancakes." The negation of a conditional statement is a conjunction comprised of a along with $\sim b$. Thus, the negation of the original statement is as follows.

I go to Moss' Diner and I do not get the triple stack pancakes.

> ✔ **Skill Check 3.3.2**
>
> Negate the following statement.
>
> If the weather gets worse, then we will leave.

Example 3.3.7

Negating a Conditional Statement Symbolically

Negate the following conditional statement by using the rule of negation of conditional statements along with De Morgan's Laws. Remember that the solution will be a compound statement.

$$a \Rightarrow (c \wedge d)$$

Solution

The first step is to negate the conditional statement. We know that its negation will be a conjunction made from the first part (a) and the negation of the second part ($c \wedge d$). Thus, we have the following.

$$a \wedge \sim(c \wedge d)$$

We can then use De Morgan's Laws to negate the second half of the statement.

$$\sim(c \wedge d) \equiv \sim c \vee \sim d$$

Putting these together, we have that the negation of the original implication is $a \wedge (\sim c \vee \sim d)$.

$$\sim(a \Rightarrow (c \wedge d)) \equiv a \wedge (\sim c \vee \sim d)$$

You can see that using the negation rule along with De Morgan's Laws is far less time consuming than building truth tables for large conditional statements as we did earlier.

OBJECTIVE 3 Variations on Conditional Statements

Conditional statements play such an important role in logic that variations on a conditional statement and whether they are logically equivalent are important tools. Given a conditional statement, there are three related conditional statements that have special names: **converse**, **inverse**, and **contrapositive**.

Table 3.3.8: Variations on the Conditional Statement

Conditional Statement	Converse	Inverse	Contrapositive
If p, then q.	If q, then p.	If not p, then not q.	If not q, then not p.
$p \Rightarrow q$	$q \Rightarrow p$	$\sim p \Rightarrow \sim q$	$\sim q \Rightarrow \sim p$

Consider the statement "If you smile, the world smiles with you," which is a take on the song *The Whole World Smiles With You* by Louis Armstrong. If we let a be the statement "you smile" and b be the statement "the world smiles," the variations on this conditional statement are as follows.

Converse ($b \Rightarrow a$): If the world smiles, then you smile.

Inverse ($\sim a \Rightarrow \sim b$): If you do not smile, then the world does not smile.

Contrapositive ($\sim b \Rightarrow \sim a$): If the world does not smile, then you do not smile.

The relationship between a conditional statement and these variations can be seen by looking at the truth values shown in Table 3.3.9.

Table 3.3.9: Truth Table for a Conditional Statement and Its Variations

a	b	$\sim a$	$\sim b$	$a \Rightarrow b$	$b \Rightarrow a$	$\sim a \Rightarrow \sim b$	$\sim b \Rightarrow \sim a$
T	T	F	F	T	T	T	T
T	F	F	T	F	T	T	F
F	T	T	F	T	F	F	T
F	F	T	T	T	T	T	T

Notice that the conditional statement and its contrapositive have the same truth values, so they are logically equivalent; that is, $a \Rightarrow b \equiv \sim b \Rightarrow \sim a$. The same is true for the converse and inverse: $b \Rightarrow a \equiv \sim a \Rightarrow \sim b$.

Example 3.3.8

Writing Variations of a Conditional Statement

Write the converse, inverse, and contrapositive of the following statement.

If I save enough money, I will go on a vacation.

State which of these new conditional statements is equivalent to the original statement.

Solution

The statement we are given is a conditional statement, $a \Rightarrow b$, where a is the statement "I save enough money" and b is the statement "I will go on a vacation." Now we can write each of the variations.

Converse ($b \Rightarrow a$): If I go on vacation, then I saved enough money.

Inverse ($\sim a \Rightarrow \sim b$): If I do not save enough money, then I will not go on vacation.

Contrapositive ($\sim b \Rightarrow \sim a$): If I do not go on vacation, then I did not save enough money.

The contrapositive is logically equivalent to the original statement.

✔ Skill Check 3.3.3

Write the converse, inverse, and contrapositive of the following statement.

If I cannot find my phone, then it is in the car.

Skill Check Answers

1. Truth Table for a and $a \vee a$

a	a	a ∨ a
T	T	T
F	F	F

2. The weather gets worse and we will not leave.   **3.** Converse: If my phone is in the car, then I cannot find it. Inverse: If I can find my phone, then it is not in the car. Contrapositive: If my phone is not in the car, then I can find it.

3.3   Exercises

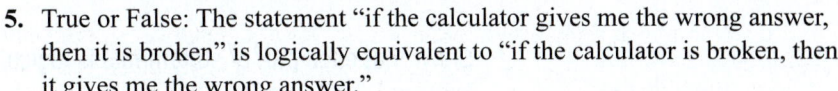

✔ CONCEPT CHECK

1. _____ statements have exactly the same truth values in all corresponding circumstances.

2. De Morgan's Laws define the negation of two _____.

3. The negation of a conditional statement is a _____.

4. True or False: The statement "if the dog is a good boy, then he gets a treat" is logically equivalent to "if the dog did not get a treat, then he is not a good boy."

5. True or False: The statement "if the calculator gives me the wrong answer, then it is broken" is logically equivalent to "if the calculator is broken, then it gives me the wrong answer."

🔵 PRACTICE

Complete each truth table for the given tautology and then decide which two statements within the compound statement are logically equivalent, if applicable.

6. $(p \Rightarrow \sim q) \Leftrightarrow \sim(p \wedge q)$

Truth Table

p	q	$\sim q$	$p \Rightarrow \sim q$	$p \wedge q$	$\sim(p \wedge q)$	$(p \Rightarrow \sim q) \Leftrightarrow \sim(p \wedge q)$
T	T					
T	F					
F	T					
F	F					

7. $\sim(p \Rightarrow q) \Rightarrow (p \vee q)$

Truth Table

p	q	$p \Rightarrow q$	$\sim(p \Rightarrow q)$	$p \vee q$	$\sim(p \Rightarrow q) \Rightarrow p \vee q$
T	T				
T	F				
F	T				
F	F				

8. $(\sim p \wedge \sim q) \Leftrightarrow \sim(p \vee q)$

Truth Table

p	q	$\sim p$	$\sim q$	$\sim p \wedge \sim q$	$p \vee q$	$\sim(p \vee q)$	$(\sim p \wedge \sim q) \Leftrightarrow \sim(p \vee q)$
T	T						
T	F						
F	T						
F	F						

9. $\left[p \wedge (\sim p \vee q) \right] \Leftrightarrow (p \wedge q)$

Truth Table

p	q	$\sim p$	$\sim p \vee q$	$p \wedge (\sim p \vee q)$	$p \wedge q$	$[p \wedge (\sim p \vee q)] \Leftrightarrow (p \wedge q)$
T	T					
T	F					
F	T					
F	F					

Show that each pair of statements is logically equivalent or explain why they are not.

10. **a.** $a \Rightarrow b$ **b.** $b \Rightarrow a$

11. **a.** $n \Rightarrow m$ **b.** $m \vee \sim n$

12. **a.** $p \wedge \sim q$ **b.** $\sim(\sim p \vee q)$

13. **a.** $r \wedge \sim s$ **b.** $\sim(r \Rightarrow s)$

14. **a.** $w \Rightarrow z$ **b.** $\sim w \Rightarrow \sim z$

15. **a.** $p \vee q$ **b.** $p \vee (q \wedge \sim p)$

Write the negation of the given compound statement using De Morgan's laws.

16. The band will go on tour, or the band will release a new album and film a new music video.

17. The dog will wear a bow tie and will not wear a hat, and the dog will have his picture taken.

De Morgan's Laws state that $\sim(p \wedge q) \equiv \sim p \vee \sim q$ and $\sim(p \vee q) \equiv \sim p \wedge \sim q$. Construct a truth table to show that each pair of compound statements is logically equivalent.

18. $\sim(p \wedge q)$ and $\sim p \vee \sim q$ 19. $\sim(p \vee q)$ and $\sim p \wedge \sim q$

Use De Morgan's Laws to write an equivalent statement without using parentheses.

20. $\sim(\sim p \wedge q)$ 21. $\sim(\sim p \vee \sim q)$

Use De Morgan's Laws to write an equivalent statement.

22. $\sim p \wedge q$ 23. $p \vee q$

Negate each conditional statement.

24. $\sim p \Rightarrow a$ 25. $a \Rightarrow (c \vee d)$

26. $(c \wedge d) \Rightarrow b$ 27. $(w \vee z) \Rightarrow (w \wedge z)$

28. If Brooke comes to my room before 2:00 p.m. then I do not finish my homework before 10:00 p.m.

29. My computer goes to sleep whenever I leave it alone for 15 minutes.

30. Having a ticket from the Sunday paper is sufficient to get a free ice cream at Dairy Dip.

31. Being in Charleston, South Carolina, implies that I am on Eastern Standard Time.

32. Consider the conditional statement "If the weather is cold enough, we will go ice skating." Determine if the following conditional statements are logically equivalent to the given conditional statement.

    a: If we go ice skating, then the weather is cold enough.

    b: If we do not go ice skating, then the weather is not cold enough.

    c: We will go ice skating if and only if the weather is cold enough.

33. Consider the conditional statement "If the puppy is not asleep, then the puppy is causing mischief." Determine if the following conditional statements are logically equivalent to the given conditional statement.

   a: If the puppy is not causing mischief, then the puppy is asleep

   b: If the puppy is asleep, the puppy is not causing mischief.

   c: The puppy is asleep or the puppy is causing mischief.

Use De Morgan's Laws to write an equivalent statement.

34. There is a space available in the 8:00 a.m. Biology class or I am not able to make the perfect schedule for next semester.

35. In Biology the nucleotide bases adenine and thymine pair together in DNA and the bases do not pair together in RNA.

Write the converse, inverse, contrapositive, and biconditional for each conditional statement.

36. If I go to the movies, then I will get popcorn.

37. Getting an 89 on the final is sufficient for me to get an A in biology.

38. It is dark whenever I get out of my last class.

39. Seeing a puppy will lead to me smiling.

Find each conditional variation.

40. Find the inverse of $p \Rightarrow q$.

41. Find the converse of $\sim q \Rightarrow p$.

42. Find the contrapositive of $\sim p \Rightarrow q$.

43. Find the inverse of $\sim q \Rightarrow \sim p$.

3.3 PROJECT

THE CASE OF DESCARTES

The French mathematician and philosopher René Descartes (1596–1650) famously wrote "cogito, ergo sum," which is Latin for the well-known philosophical statement, "I think, therefore I am." He wrote this sentence when studying a place of deep questioning in the human mind.

Descartes' work is of great importance in Western thinking, and it is widely studied in academia. To lighten their studies up a little, many mathematics and philosophy students are told the following joke.

> René Descartes is having dinner and has just finished his drink. The server asks him if he would like another. Descartes replies, "I do not think so," and disappears in a puff of smoke.

While a witty one, the joke's punch line may not be mathematically correct.

1. Rewrite Descartes' statement as a conditional (if-then) statement.

2. Let p be the statement "I think" and q be the statement "I am." Complete the truth table.

p	q	$\sim p$	$\sim q$	$p \Rightarrow q$	$\sim p \Rightarrow \sim q$
T	T				
T	F				
F	T				
F	F				

3. Write the conditional $\sim p \Rightarrow \sim q$ as a sentence in English.

4. Are $p \Rightarrow q$ and $\sim p \Rightarrow \sim q$ logically equivalent? Explain.

5. Use your previous answers to explain why the punch line for the joke does not really work.

OBJECTIVES

1. Identify the parts of an argument.

2. Determine whether an argument is valid.

3. Determine whether an argument is sound.

4. Identify fallacies.

OBJECTIVE 1 Arguments

Now we're ready to use the tools we've amassed so far in the chapter to analyze logical arguments. When we refer to an argument in logic, we are not talking about a quarrel or shouting match; instead, we are referring to the process of following a line of reasoning that is meant to support a particular belief or idea. In logic, the subject matter is irrelevant when determining if a conclusion is justified. So although it might be difficult, we have to take our feelings about the subject out of the picture when analyzing arguments.

Formally, a logical **argument** consists of a series of logical statements. Statements at the beginning of the argument are called **premises** (or *hypotheses* or *assumptions*) and are used to support a particular **conclusion**.

> ### 📖 LOGICAL ARGUMENTS
>
> **Argument**
>
> A logical **argument** consists of a series of logical statements.
>
> **Premise**
>
> A **premise** is a statement at the beginning of an argument used to support a particular conclusion.
>
> **Conclusion**
>
> A **conclusion** is the ending statement of an argument.

The terms *premise* and *conclusion* are relative terms. They suggest a specific role or function in an argument. Premises are offered up as the supporting evidence for the conclusion. No statement taken out of the context of the argument can be either a premise or a conclusion on its own; they work together. To illustrate this, consider one of the most famous logical arguments.

All men are mortal. Socrates is a man. Therefore, Socrates is mortal.

Premise 1 is "All men are mortal." Premise 2 is "Socrates is a man." These lead to the conclusion that "Socrates is mortal." Notice that if you considered each of these statements on its own, it could be either a premise or the conclusion for any argument, depending on the order of the statements. Having the statements together in context tells you which part of the argument they are.

In speech, we often leave out some of the implied words in an argument. When this happens, there are particular words that often indicate conclusions or premises are to follow. The following table gives an overview of some of the more common words or phrases you might come across in spoken arguments.

Table 3.4.1: Premise and Conclusion Indicators

Premise Indicators	Conclusion Indicators
as	accordingly
as indicated by	consequently
because	entails that
due to the fact that	hence
for	implies that
for the reason that	it must be that
given that	it follows that
if	so
in as much as	then
in that	therefore
may be concluded from	thus
may be inferred from	we may conclude that
seeing that	we may infer that
since	whence
the reason that	wherefore

Here's a rather lengthy, but fun, set of premises leading to a conclusion. Consider the following reply from Sherlock Holmes when he was questioned on how he knew at a glance that a man was a retired Marine sergeant in *A Study in Scarlet* by Sir Arthur Conan Doyle.[2]

> ...I could see a great blue anchor tattooed on the back of the fellow's hand. That smacked of the sea. He had a military carriage, however, and regulation side whiskers. There we have the marine. He was a man with some amount of self-importance and a certain air of command. You must have observed the way in which he held his head and swung his cane. A steady, respectable, middle-aged man, too, on the face of him—all facts which led me to believe that he had been a sergeant.

As you can see, Holmes gave a large set of premises to support his conclusion.

Example 3.4.1

Identifying Premises and Conclusions

Identify the premises and conclusion in each of the following arguments.

a. If we update the technology, then we will have a better product and fewer complaints. Our technology was updated last month, so we will receive fewer complaints this month.

b. The president of the United States must be younger than 40. Bruce Springsteen is president of the United States. Therefore, Bruce Springsteen must be younger than 40.

Solution

a. The premises in support of the conclusion come first in the argument. The conclusion is the last statement. Thus, we have the following.

Helpful Hint

If-then statements can be used in the premise of an argument. Be careful to correctly identify the conclusion in such cases.

---

2   Arthur Conan Doyle, *A Study in Scarlet*, London, Ward Lock & Co., 1888.

Premise 1: If we update the technology, then we will have a better product and fewer complaints.

Premise 2: Our technology was updated last month.

Conclusion: We will receive fewer complaints this month.

b. This argument has two premises and a conclusion.

Premise 1: The president of the United States must be younger than 40.

Premise 2: Bruce Springsteen is president of the United States.

Conclusion: Bruce Springsteen must be younger than 40.

✔ **Skill Check 3.4.1**

Identify the premises and conclusion in the following argument.

If the weather is warm, we will go the beach. We're at the beach, so it must be warm outside.

Let's stop and notice that the second argument in Example 3.4.1 doesn't sound right to us because none of the parts of the argument are true. The president of the United States does *not* have to be younger than 40; Bruce Springsteen is *not* the president of the United States; and Mr. Springsteen is *not* younger than 40. However, it is important here to separate the truth of the statements from the structure of the argument. As we move forward, we'll look at ways to determine if a conclusion can logically be drawn from the premises regardless of the truth of the statements.

OBJECTIVE 2 Valid Arguments

Even though people might not agree with an original premise (or the premise may indeed be false), an argument is said to be valid if the conclusion follows logically from the premise. That is, a **valid argument** is a deductive argument where the conclusion is always guaranteed from the premises. If the conclusion is not always guaranteed from the premises, then the argument is **invalid**. It is important to realize that having an argument that is valid implies no judgment on the truth of the premise. The validity of an argument is concerned only with the logical connection *between* the premise and the conclusion.

> 📖 **VALID AND INVALID ARGUMENTS**
>
> **Valid Argument**
>
> A **valid argument** is a deductive argument where the conclusion is always guaranteed from the premises.
>
> **Invalid Argument**
>
> An **invalid argument** is a deductive argument where the conclusion is *not* always guaranteed from the premises.

💬 **Think Back**

Deductive reasoning uses statements that are commonly accepted as facts to make a case for a certain conclusion.

For example, consider the following two arguments.

Valid Argument: Cutting out sweets from your diet makes you lose weight. He hasn't eaten sweets in months. So he must be losing weight.

Invalid Argument: Cutting out sweets from your diet makes you lose weight. He has been losing weight. So he must be cutting out sweets from his diet.

Let's look at why the second argument is invalid. The premises say that cutting out sweets makes you lose weight, and he lost weight. However, the conclusion that he must be cutting out sweets from his diet doesn't always follow since there are other reasons that he might have lost weight; for example, he started exercising or he's been ill. Thus, we cannot conclude that he has cut out sweets from his diet.

One method for determining the validity of an argument is to use truth tables. By writing the argument in symbolic form, we can determine if it's valid by looking at its truth table. The argument will have the form of an implication where the *if* portion is made up of the premises joined together with conjunctions and the *then* portion is the conclusion. If the implication is a tautology, then the argument is valid. If it is not, the argument is invalid.

☰ USING A TRUTH TABLE TO DETERMINE THE VALIDITY OF AN ARGUMENT

1. Express the argument in symbolic form as an *if-then* statement. The *if* portion will consist of the premises joined together with conjunctions. The *then* portion is the conclusion. For example, if there are n premises, the statement will be of the form

   $$[(\text{premise } 1) \wedge (\text{premise } 2) \wedge \ldots \wedge (\text{premise } n)] \Rightarrow \text{conclusion}.$$

2. Construct a truth table for the *if-then* statement.

3. If the implication is a tautology, then the argument is valid; otherwise, the argument is invalid.

Example 3.4.2

Determining the Validity of an Argument

Decide if the following argument is valid by using a truth table.

> If someone lives in the city of Phoenix, Arizona, then they live in the Mountain Standard Time Zone. Sebastian does not live in the Mountain Standard Time Zone. Therefore, Sebastian does not live in the city of Phoenix, Arizona.

Solution

We begin by identifying the premises and conclusion.

> Premise 1: If someone lives in the city of Phoenix, Arizona, then they live in the Mountain Standard Time Zone.

> Premise 2: Sebastian does not live in the Mountain Standard Time Zone.

> Conclusion: Sebastian does not live in the city of Phoenix, Arizona.

Next, we will represent the simple statements with letters so that we can write the argument symbolically.

> p: Someone lives in Phoenix, Arizona.

> q: They live in the Mountain Standard Time Zone.

💬 Think Back

A tautology is a statement that is true in all possible circumstances.

Therefore,

Premise 1 is $p \Rightarrow q$.

Premise 2 is $\sim q$.

Conclusion is $\sim p$.

To write the argument symbolically, we write a conditional statement where we join the premises together as a conjunction that implies the conclusion. Thus, the argument is written in the following way.

Conditional Argument

$$((p \Rightarrow q) \wedge \sim q) \Rightarrow \sim p$$

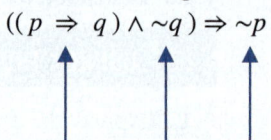

Premise 1   Premise 2   Conclusion

We can now construct the truth table for the argument.

Table 3.4.2: Truth Table for $((p \Rightarrow q) \wedge \sim q) \Rightarrow \sim p$

p	q	$p \Rightarrow q$	$\sim q$	$(p \Rightarrow q) \wedge \sim q$	$\sim p$	$((p \Rightarrow q) \wedge \sim q) \Rightarrow \sim p$
T	T	T	F	F	F	T
T	F	F	T	F	F	T
F	T	T	F	F	T	T
F	F	T	T	T	T	T

Notice that the final column is true in every case, meaning it is a tautology. Therefore, the argument is valid, which means that the conclusion follows from the premises.

Example 3.4.3

Determining the Validity of an Argument

Decide if the following argument is valid by using a truth table.

If you work hard, you will succeed. You succeeded, so you must have worked hard.

Solution

Begin by identifying the premises and conclusion.

Premise 1: If you work hard, you will succeed.

Premise 2: You succeeded.

Conclusion: You worked hard.

Next, we will represent the simple statements with letters so that we can write the argument symbolically.

a: You work hard.

b: You succeeded.

Therefore,

Premise 1 is $a \Rightarrow b$.

Premise 2 is b.

Conclusion is a.

Thus, the argument is written symbolically as follows.

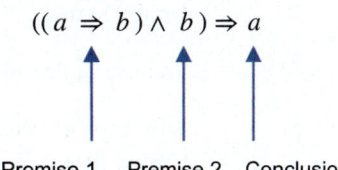

Next, construct the truth table for the argument.

Table 3.4.3: Truth Table for $((a \Rightarrow b) \wedge b) \Rightarrow a$

a	b	$a \Rightarrow b$	$(a \Rightarrow b) \wedge b$	$((a \Rightarrow b) \wedge b) \Rightarrow a$
T	T	T	T	T
T	F	F	F	T
F	T	T	T	F
F	F	T	F	T

Notice that the final column is not true in every case. Row three has a false truth value. Therefore, the argument is invalid, which means that the conclusion does not follow from the premises. In other words, your success was not necessarily because you worked hard.

OBJECTIVE 3 Sound Arguments

When faced with an argument that is logically valid but has a conclusion that is not true, the problem lies in the premise, not the argument. In these instances, there must be an error in the beginning statement; that is, the premises were faulty to begin with. The best arguments require a logically valid argument *and* true premises. That is, the arguments are **sound**. To have a sound argument is the most confident we can be.

📖 **SOUND ARGUMENT**

A **sound argument** is a valid argument that uses true premises.

Example 3.4.4

Determining If an Argument Is Sound

Use a truth table to determine if the following is a sound argument.

If the President of the United States is unable to carry out his or her constitutional role, then the Vice President becomes the President of the United States. In November 1963, President John F. Kennedy was fatally shot. Therefore, without an election, Vice President Lyndon Johnson became President of the United States.

Solution

Begin by identifying the premises and conclusion.

Premise 1: If the President of the United States is unable to carry out his or her constitutional role, then the Vice President becomes the President of the United States.

Premise 2: President John F. Kennedy was fatally shot.

Conclusion: Vice President Lyndon Johnson became President of the United States.

Next, represent the simple statements with letters.

p: The President of the United States is unable to carry out his or her constitutional role.

q: The Vice President becomes the President of the United States.

Therefore,

Premise 1 is $p \Rightarrow q$.

Premise 2 is p.

Conclusion is q.

Thus, the argument is represented by $((p \Rightarrow q) \land p) \Rightarrow q$ and its truth table is as follows.

Table 3.4.4: Truth Table for $((p \Rightarrow q) \land p) \Rightarrow q$

p	q	$p \Rightarrow q$	$(p \Rightarrow q) \land p$	$((p \Rightarrow q) \land p) \Rightarrow q$
T	T	T	T	T
T	F	F	F	T
F	T	T	F	T
F	F	T	F	T

You can see that the implication is a tautology, and thus valid.

For the argument to be sound, we must start with a true statement. Therefore, we'll only consider the first two rows of the table. We can confirm (via history books or the internet) that in 1963, President John F. Kennedy was fatally shot. In other words, statement p is true (the President was unable to carry out his constitutional role). In the same manner, we know that statement q is true in this argument because the Vice President at the time, Lyndon Johnson, became President of the United States. Since we are starting with true statements, and this is a valid argument, we know that this argument is also sound.

✔ **Skill Check 3.4.3**

Determine whether the following is a sound argument.

Marcy is very allergic to strawberries. Last week, she inadvertently drank some punch with strawberry puree in it. As a result, she had an allergic reaction and had to be taken to urgent care.

OBJECTIVE 4 Fallacies

Not all arguments make solid and compelling cases for their conclusions. Arguments often contain **fallacies**. Loosely speaking, a fallacy is a lapse in logic or an error in reasoning, often referred to as a faulty argument. It is this reasoning that leads to an invalid argument.

> 📖 **FALLACY**
>
> A **fallacy** is an error in reasoning that leads to an invalid argument.

If you search for "fallacy" on the internet, you will find dozens of categorizations of fallacies. Since our aim here is to have you think critically about logical arguments, and not memorize the names of every fallacy known, we will introduce a few of the most common fallacies you might run across. Evident by their Latin titles, people have been loose with their logic for many, many centuries.

As we consider each of the following eight fallacies, we want to point out that when we speak, arguments involve implications that are often not necessarily verbalized in the argument itself. Although, mathematically speaking, some of the scenarios we will look at are simply statements and not arguments at all because they omit the implied conclusion, we will consider such implications and note the unspoken conclusion (or premises) since this is often the case in the world in which we live.

1. *Post Hoc, Ergo Propter Hoc* (After This, Therefore Because of This)

   My wallet had money in it before you went to get gum from my purse.

   The brakes worked just fine before you drove the car.

   A black cat crossed my father's path. The next day, he had a heart attack.

This title comes from the Latin meaning "after this, therefore because (on account) of this." That is, since one thing happened first, it must have caused the other to happen, without any evidence at all. Formally, the argument is completed with a conclusion that it merely implies, as in these examples. The first statement implies that the person getting gum stole the money from the wallet in the purse. The second example implies that the driver damaged the car brakes. And the final example implies that the black cat caused the heart attack.

2. *Dicto Simpliciter* (Hasty Generalization)

   People from Chicago are just rude. We had terrible service at our restaurant and the hotel staff were so unhelpful.—A statement made on the first day of a vacation.

   I don't like coffee. It's too sweet.—A statement made after the first taste of a white chocolate mocha with an extra shot of syrup.

   I've decided to buy a lottery ticket every day this year. This is easy money.—A statement made after winning $50 on the first lottery ticket purchased.

We frequently make broad sweeping conclusions based on only a few occurrences. These hasty generalizations are often because the size of the sample is too small to make a reasoned conclusion. Recall that this type of reasoning is inductive and can be thrown off its course by one counterexample. If we generalize a rule based on very few occurrences, we are guilty of committing the fallacy of hasty generalization.

3. *Ad Hominem* (Personal Attack)

> *How can you expect anything good from such a bleeding-heart politician?*—The response from Candidate B after Candidate A outlined the five points of change he would make if elected.

> *Do we really want to believe what a wrestling fan has to say about our fine arts program?*—A school board member responding to a parent voicing concern over recent curriculum updates.

From the Latin meaning "to the man," the *ad hominem* fallacy refers to attacking someone's person, character, or motives, instead of addressing the premise of the argument. Notice that neither example addressed the issue, but instead attacked the person speaking. An attack on the person and not their argument simply changes the focus. Using the *ad hominem* attack does not prove or disprove the argument.

4. *Petitio Principii* (Circular Reasoning)

> *You should eat healthy meals. They're good for you.*

> *She's really happy right now because she's in a good mood.*

> *I think, therefore I am.*—René Descartes

Meaning to "beg the question," this form of fallacy is circular reasoning; that is, the premise and conclusion state the same thing. The speaker uses the conclusion that he is trying to prove as the evidence to prove it. The speaker just uses different wording to restate the same thing. As the audience, we often have a hard time dissecting this fallacy because there seems to be no obvious beginning or end to the argument.

Consider the argument "Tim Tebow is the most successful college quarterback this decade because he's the best we've ever seen." Do not be fooled by the insertion of the word "because" here. There is no supporting evidence following it. Both parts of the sentence are premises (or opinions, if you like) until they have some evidence to back them up. Of course, if you believe that *thinking* and *being* are not synonymous, then Descartes may not have been fallacious in his statement after all.

5. *Non Sequitur* (It Does Not Follow)

> *If you wear this outfit, you're sure to get a date.*

> *By giving to this overseas charity, you are neglecting the needs of your local community.*

> *She lives on Main Street. She must be very wealthy.*

Non sequitur comes from Latin, literally meaning "it does not follow." Arguments like this can be persuasive because they use conclusions that seem to be related to the premise, but actually aren't related if you think carefully about them. For instance, getting a date might have something to do with someone finding you attractive, which could potentially have something to do with how you are dressed. Giving to one charity could have something to do with giving to other charities as you have limited resources (but it has nothing to do with whether you neglect other charities). And it is not unreasonable to think that someone who lives in an area where property is expensive has the means to afford property there (but living somewhere doesn't mean you are paying to do so). In these types of arguments, we say a diversion or "red herring" was introduced into the argument. Political endorsements by famous people are often non sequiturs. In many cases, an endorsement from a celebrity is not enough evidence that the candidate will make a good leader.

Math Milestone

René Descartes was a famed philosopher who thought that science and mathematics could be used to explain everything in nature and wrote many works explaining his reasoning. At the age of 41, he published La Geométrie, which gave birth to Cartesian geometry that is widely studied today. Sadly, he died of pneumonia at the age of 54.

Figure 3.4.1

6. *Straw Man* (Exaggerated or Distorted View of the Opponent)

We can't have national health care. No one wants a government death panel deciding their loved one's fate.

I'm against extending tax cuts for the wealthy. Why should we be worried about Bill Gates losing a few pennies?

Like the straw man who is easily blown over, this fallacy involves the distortion of someone's ideas or beliefs so that they can easily be knocked down. The premise is mostly arguing on a poor representation of the truth. It's easy to make ourselves look good when we can effortlessly tear down someone else. This fallacy includes any attempt to "prove" an argument by overstating, exaggerating, or oversimplifying the arguments of the opposing side.

7. *False Dilemma* (Illusion of Limited Choice)

You're either part of the solution or part of the problem.—Eldridge Cleaver during his 1968 presidential campaign.

Either you can afford this new car or you can decide to walk around for the next year.

If you aren't for us, you're against us.

"Life isn't always black or white" would be a great thing to keep in mind when faced with this type of fallacy. As we well know, rarely are there ever simple solutions with only two choices. Yet the false dilemma fallacy relies on you thinking just that. It occurs when we build an argument around the assumption that only two possible choices are available. Users of this type of fallacy hope for you to ignore the idea that other options might exist in the dilemma.

8. *Argumentum ad Populum* (Appeal to the People)

9 out 10 doctors prefer ...

You should watch Reality Takes a Leap ... the number 1 show on television.

I have hundreds of letters supporting his good character. He couldn't have committed that crime. No one would believe it.

When we want to use the fact that the majority, or even just a large number, of people are doing something as the basis for supporting a conclusion, we are appealing to the "everybody does it" idea. The Latin meaning of *argumentum ad populum* is literally "argument to the people." This type of fallacy is used extensively in propaganda.

My young daughter once tried to convince me to get her a puppy by telling me, "Zoe has a new puppy, Mae Mae has a new puppy, and Isabella has a new puppy. Everyone is getting a new puppy." This is also a hasty generalization, or *dicto simpliciter*. Not everyone is getting a new puppy! My mother used to counter this line of argument with her own version of logic: "If everyone else jumped off a bridge, would you?"

Fun Fact

Eldridge Cleaver ran as a candidate in the 1968 presidential election under the Peace and Freedom Party. He received 36,571 votes, even though some states held that he did not meet the constitutional requirement for age and therefore could be excluded from the ballot. The Constitution requires that the US President be at least 35 but does not specify when he must have reached that age. Cleaver would not have been 35 until more than a year after the inauguration day in 1969.

Figure 3.4.2

Example 3.4.5

Identifying Fallacies

Identify the type of fallacy used in each of the following statements.

a. If public spending is not reduced, our economy will collapse.

b. Heather just had a new fuel pump put in her car. Forty-eight hours later, her car won't start. Heather's father thinks the mechanic must not have installed the fuel pump correctly.

Solution

a. This is a false dilemma. The argument makes it seem like there are only two options: either public spending decreases or the economy collapses. However, it is also possible that the economy can be strong while public spending remains the same.

b. This is *post hoc, ergo propter hoc.* There are many other things that could cause Heather's car to not start, including a lack of fuel! It is not unreasonable to consider that there might be a connection, but it is still a fallacy to assert that the one thing happening after the other is the *only* possible explanation.

This is in no way an exhaustive list of the fallacies that we encounter in our daily lives. However, it does give us a basis for building our critical-thinking skills when we are next confronted with a statement or argument that does not sit quite right with us. If nothing else, it has hopefully raised your awareness of the ways in which language and ideas can be manipulated.

Table 3.4.5: Summary of Fallacies

Type of Fallacy	Description
Post Hoc, Ergo Propter Hoc (After This, Therefore Because of This)	Because one thing happened first, it caused the other to happen.
Dicto Simpliciter (Hasty Generalization)	Making broad, sweeping generalizations based on a few specific occurrences.
Ad Hominem (Personal Attack)	Attacking someone's person, character, or motives instead of addressing the premise of the argument.
Petitio Principii (Circular Reasoning)	The premise and conclusion state the same thing.
Non Sequitur (It Does Not Follow)	The conclusion does not logically follow from the premise; it follows a diversion.
Straw Man (Exaggerated or Distorted View of the Opponent)	The distortion of someone's ideas or beliefs so that they can easily be knocked down.
False Dilemma (Illusion of Limited Choice)	Argument that rests on the assumption that there are only two choices as a solution.
Argumentum ad Populum (Appeal to the People)	Stating that the majority, or even just a large number, of people are doing something as the basis for supporting a conclusion.

Skill Check Answers
1. Premise 1: If the weather is warm, we will go the beach. Premise 2: We're at the beach. Conclusion: It is warm outside.   **2.** Invalid   **3.** The argument is sound.   **4.** *Non sequitur*

3.4   Exercises

✔ CONCEPT CHECK

1. In an argument, the _____ provides support for the argument and the _____ is the ending statement.

2. A _____ is a statement that is true in every single possible circumstance.

3. A _____ argument must have true premises.

4. True or False: An argument can be valid as long as there is a logical connection between the premise and conclusion, even if the conclusion seems like nonsense.

5. True or False: Fallacies are caused by a faulty line of reasoning to reach a conclusion.

💡 PRACTICE

Identify the premise and conclusion in each argument.

6. "If the owners truly cared about the game and the fans, they would lift the lockout and allow the season to begin on time while negotiations continue," NHLPA Executive Director Donald Fehr said.[3]

7. "Even if we stopped burning fossil fuels today, there is enough carbon dioxide in the atmosphere—and it is such a persistent, lasting gas—that temperatures will continue to rise for a few hundred years."[4]

8. The sea level is rising because melting glaciers and ice sheets are adding more water to the oceans.

9. Fast food obesity has strikingly increased in many countries because of the easy availability of fast food in the grocery shops, gas stations and dispensers everywhere. So it is difficult to escape from the lure of these delicious advertisements and showcases.[5]

10. The current US population is approximately 314 million, about half of which are males, so if 2 percent of the 157 million American men suffer from one of these severe disorders (schizophrenia, major depression, or psychopathy), this results in a figure of 3,140,000.[6]

---

3   BBC, http://www.bbc.co.uk
4   BBC, http://www.bbc.com
5   ygoy, http://www.ygoy.com
6   *The Chronicle of Higher Education*, http://www.chronicle.com

11. "If a man is struck down by a heart attack in the street, Americans will care for him whether or not he has insurance."[7]

12. An excerpt from the TV show *The West Wing*:[8]
Press secretary, C. J., says, ". . .*USA Today* asks you (the president) why you don't spend more time campaigning in Texas and you say because you don't look good in funny hats."

"It was *big hats*," Sam corrects her.

"The point is we got whomped in Texas," C. J. states.

"We got whomped in Texas twice," Josh adds.

"We got whomped in the primary and we got whomped in November," says C. J.

"I think I was there," replies the president.

"And it was avoidable. Sir."

"C. J. on your tombstone, it's going to read, 'Post hoc, ergo propter hoc.'"

13. A cartoon by Randy Glasbergen.[9]

14. "Dear Friend, a man who has studied law to its highest degree is a brilliant lawyer, for a brilliant lawyer has studied law to its highest degree." Oscar Wilde, *De Profundis*.

15. All potatoes have skin. I have skin. Therefore, I must be a potato.

16. During the fall of 2011, the movement "Occupy Wall Street" spread almost overnight into a nationwide/worldwide phenomenon. Many famous personalities voiced their opinions both for and against the movement. One film star claimed, " . . . they're a bunch of posers, propping up the same corporate giant they're trying to take down. I find it a little ironic that most of these kids not only own all the latest technology, but use it constantly while pretending to 'take a stand.' They're a major player in the consumerism that goes on in this country."

Determine whether each argument uses inductive or deductive reasoning, and whether the reasoning is valid.

17. The number pi (π) begins with 3.14159. Four is the only even digit when you write out the number π.

7   Fox News, http://www.foxnews.com
8   Sorkin, *The West Wing*
9   Courtesy of Randy Glasbergen

18. We have put up our Christmas tree the day after Thanksgiving every year for the last 5 years. A photograph from 2 years ago shows our Christmas tree in the background, so it couldn't have been taken before Thanksgiving.

19. We have put up our Christmas tree the day after Thanksgiving every year for the last 5 years. It's Black Friday today, so we must be putting up the Christmas tree.

20. All of my friends have smartphones. They must be the most popular type of cell phone.

21. In a survey of 200 adults, 5 out of 7 people said they carry smartphones as their cell phone. With 95% confidence, I can say that most people carry a smartphone as their cell phone.

Determine whether each argument is valid or invalid.

22. All pigs are pink. I am pink, therefore I must be a pig.

23. All professors at the university received a 3% raise this year. My dad is a psychology professor at the university, so he must have gotten a raise.

24. To renew a US passport for a child under 16, you need to demonstrate that he or she is a US citizen with either a US birth certificate or a valid US passport. I am renewing my daughter's passport, so I need her birth certificate.

25. Most people who live in California are Democrats. Stephanie lives in California, so she must be a Democrat.

26. All dogs bark. I have a dog. Therefore, my dog barks.

27. If it rains, I will need an umbrella. I do not need my umbrella so it must not be raining.

28. Most people who live in Kansas are Republicans. Jim lives in Kansas. Therefore, Jim is a Republican.

Determine the missing piece of information needed to make each argument valid.

29. I can only carry a bottle of liquid that is at most 3.4 fl oz in my carry-on luggage while flying. A larger bottle will be confiscated by airport security. My hand sanitizer got taken away when I went through airport security.

30. I was born on US soil. Therefore, I am a US citizen.

31. Whenever I play folk music, I tune my guitar to an open D tuning. This evening I am playing some folk music.

32. I am tired, so it must be late.

33. If you are not over 16, then you cannot apply for a driver's license in the state of Tennessee. Emma applied for a driver's license.

34. To get into the local state university, you must have a cumulative ACT score of at least 21. Ben's cumulative ACT score is less than 21.

35. If you buy a new car, then you cannot pay your student loan bills. You are able to pay your student loan bills.

Use a truth table to determine if the given argument is a sound argument.

36. If an album receives platinum certification, then at least 1,000,000 copies of the album were sold. Taylor Swift's album *Lover* received platinum status on September 27, 2019. So at least 1,000,000 copies of Taylor Swift's album *Lover* were sold.

    a. Determine the premises and conclusion of the argument.

    b. Write the argument using logic symbols.

    c. Use a truth table to determine whether the argument is a valid argument.

    d. Is the argument sound? Explain why or why not.

37. If a movie is extremely popular and financially successful, then it is considered a blockbuster. The 2009 movie *Avatar* was not a blockbuster. Therefore, *Avatar* was not popular nor financially successful.

    a. Determine the premises and conclusion of the argument.

    b. Write the argument using logic symbols.

    c. Use a truth table to determine whether the argument is a valid argument.

    d. Is the argument sound? Explain why or why not.

Identify the type of fallacy being used in each statement.

38. "Some people believe the answer to this problem is to wall off our economy from the world," he said this month in India, talking about migration of US jobs overseas. "I strongly disagree."

39. You must be an atheist. You never go to church on a Sunday morning.

40. On a discussion of the Treaty of Versailles in a history class, a student responds, "You said this happened five years before Hitler came to power. Why are you so fascinated with Hitler? Are you anti-Semetic?"

41. After hitting the side of the television, the picture on Emma's old television goes back into focus. Emma tells Karan that hitting the television fixed it.

42. Everyone drives over the speed limit on this road, so the city should raise the speed limit.

43. Can we really take his opinion on Star Wars seriously? He's a Star Trek fan!

44. It's healthy to take vitamins every day because they are good for you.

45. This cookie you made is burnt. You must be terrible at making cookies.

3.4 PROJECT

QUATERNIO TERMINORUM: ANOTHER INTERESTING TYPE OF FALLACY

In Section 3.4, you learned how to recognize sound deductive reasoning and explored a few different types of fallacies. In this activity, you will explore one more type of fallacy.

Consider this perfectly fine logical argument.

All mammals have bones.

All cats are mammals.

All cats have bones.

This type of argument is known as a syllogism, which means we have exactly three elements: two premises and one conclusion. This syllogism connects the three terms *mammals*, *bones*, and *cats* to reach a logical conclusion.

1. Identify the two premises and the conclusion in the mammal syllogism.

2. Construct your own syllogism using the three terms *eggs*, *birds*, and *parrots*.

If we use a fourth term in the conclusion that is not connected by the two premises, we get a fallacy. Consider the following argument.

All mammals have bones.

All cats are mammals.

All fish are mammals.

This is total nonsense, which stems from the fact that the term *fish* is introduced and is not related to either premise (that is, fish are not a type of mammal). The two premises are not enough to connect the four terms involved. This type of fallacy is known as a fallacy of four terms, or *quaternio terminorum*.

3. Create a fallacy of four terms using the premises you created in part 2.

Not all fallacies of this type introduce an obvious fourth term. Consider the feature of the English language that allows a single word to have different meanings depending on how it is used in a sentence. This can lead to a fallacy where multiple instances of the same word actually represent different terms. Here is an example.

Nobody is perfect.

I am a nobody.

I am perfect.

4. In this argument, a single word appears multiple times and seems to be a single term. Identify this term and explain how the term actually takes on two different meanings.

5. Write an argument that is a fallacy of four terms using three terms, where one of the terms has two different meanings. Explain what causes your argument fallacy. (**Hint**: If you are stuck, try to think of a common word that has more than one meaning, like *mouse*, *bat*, *light*, or *bright*.)

Chapter 3 Exercises

Negate each statement.

1. The puppy could not keep her eyes open after 10:00.

2. Jour means "the day" in French.

3. All houses have fireplaces.

4. Every senior that graduated had a job offer.

Construct the truth table for each compound statement.

5. $(a \vee {\sim} b) \Rightarrow c$

6. $w \wedge (x \wedge {\sim} y)$

7. If you and Kathy both go to the movies, then I will go as well.

8. I don't eat dessert and do not have a cup of coffee.

Write the converse, inverse, contrapositive, and biconditional for each conditional statement.

9. If my grade in this course is an A, then I can enroll in the next course.

10. If my car runs out of gas, then my car will not start.

Solve each problem.

11. Complete the truth table for $\left(p \wedge (p \Rightarrow q) \right) \Rightarrow q.$

Truth Table

p	q	$p \Rightarrow q$	$p \wedge (p \Rightarrow q)$	$\left(p \wedge (p \Rightarrow q) \right) \Rightarrow q$
T	T			
T	F			
F	T			
F	F			

12. If the bakery has fresh muffins, then I will buy some for my roommates. If I buy fresh muffins then my roommates will be happy. What can you deduce from these statements?

13. Which of the following is an example of inductive logic?

   a. If you liked all of Richie Rock's movies, then you will also like his latest movie.

   b. If there is an accident on the freeway, then we will be late for the play.

   c. If I get a raise next week, then I can buy a new car.

   d. I have to leave work early if the bank closes at 5:00 today.

Determine whether each arguments is valid or invalid.

14. Eagles have feathers.

Sparrows have feathers.

Ducks have feathers.

15. All birds have wings.

Ducks are birds.

Ducks have wings.

16. Some birds have feathers.

Sparrows are birds.

Sparrows have feathers.

17. All birds can fly.

Eagles are birds.

Eagles can fly.

Identify the premise and conclusion in each statement.

18. "If the Egyptian government had been doing what the IDF is doing during the Arab Spring, it would have been a very different picture," he said, referring to the activists who used social media to organize protests that toppled Egypt's Hosni Mubarak last year.[10]

19. "The most prominent advocates of global warming aren't scientists," said Heartland's president, Joseph Bast. "They are Charles Manson, a mass murderer; Fidel Castro, a tyrant; and Ted Kaczynski, the Unabomber. Global warming alarmists include Osama bin Laden and James J. Lee (who took hostages inside the headquarters of the Discovery Channel in 2010)."[11]

Identify each type of fallacy being used.

20. If you believe that a man who can cheat on his wife of 38 years will be honest with you on all levels, than you deserve what you get for voting for him.

21. "The only question is if Washington has the courage to seize this opportunity. If they do, this moment won't be remembered as a drawn-out partisan showdown over spending, but as the launching pad from which we began to sail into a healthier, more prosperous future."[12]

---

10   CNN, http://www.cnn.com
11   The Heartland Institute, http://www.heartland.org
12   CNN, http://www.cnn.com

CHAPTER 3 PROJECT

Keeping the Lights On

Electronic circuits are a major part of our everyday lives whether we are conscious of it or not. Alarm systems, computers, diagnostic medical equipment, Christmas lights, and even a clothes dryer all use logical circuit systems. The truth tables that we've been studying in this chapter actually have a direct use in designing these circuits. Let's take a look at how they correspond.

Recall that each variable in a truth table can either be *true* or *false* at any given time, but not both at the same time. The same is true of a switch. It can either be *on* or *off*, but not both, at any given time. In a digital circuit we associate *true* with *on* and *false* with *off*.

For instance, if we have a circuit variable that controls a light bulb, when the variable is true, the light is on, and vice versa, as shown in the following diagrams.

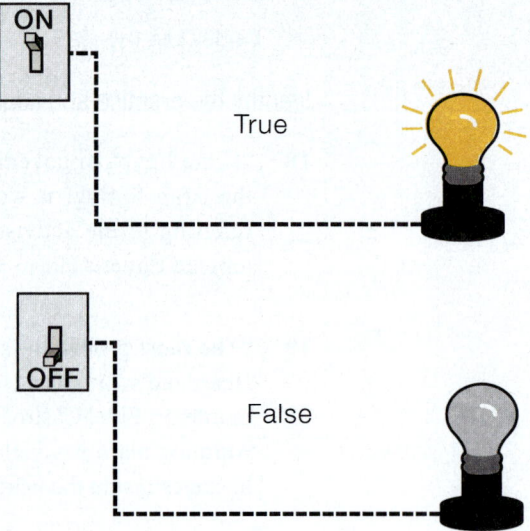

Just as in truth tables, digital circuits can be constructed using logic. The logical connectives for digital circuits are called logic gates. We can add NOT gates, OR gates, or AND gates to any circuit to manipulate the output. As you might imagine, the NOT gate produces an opposite value of the switch as its output. In fact, all the logic gates behave in exactly the same manner as logic symbols in truth tables. For example, suppose we'd like the light to be on when the switch is actually off (or false). We can add a NOT gate to our circuit so the output means the light is on when the switch is off, as shown below.

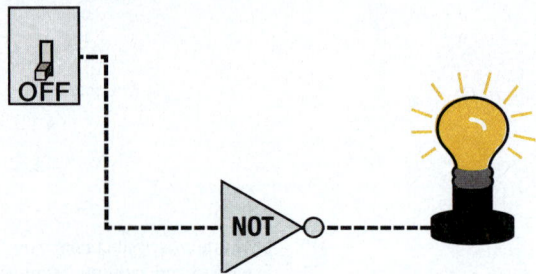

1. What would happen to the light bulb in this circuit if the switch is on?

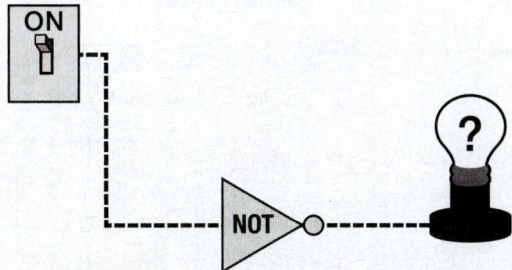

When we have two switches, we can introduce the OR gate so that the light bulb is on if at least one switch is on.

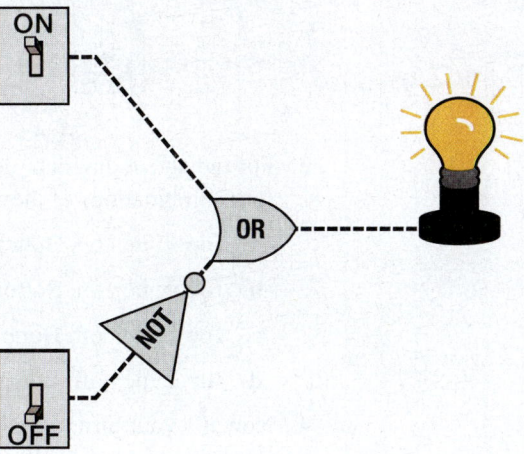

Notice that if we reverse the position of each switch, the light bulb would still remain on. The same is true if both switches are on. However, if both switches were in the off position in this circuit, the light bulb would then turn off.

2. Now think about what would happen to the light bulb with a circuit where there is an AND gate in place. Fill in each blank below.

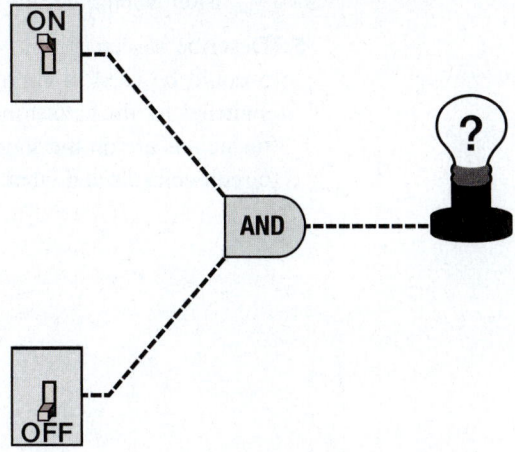

   a. If one switch is on and one is off, the light bulb is _____.

   b. If both switches are on, the light bulb is _____.

   c. If both switches are off, the light bulb is _____.

Of course digital circuits can have multiple gates as well as switches, like the one shown below.

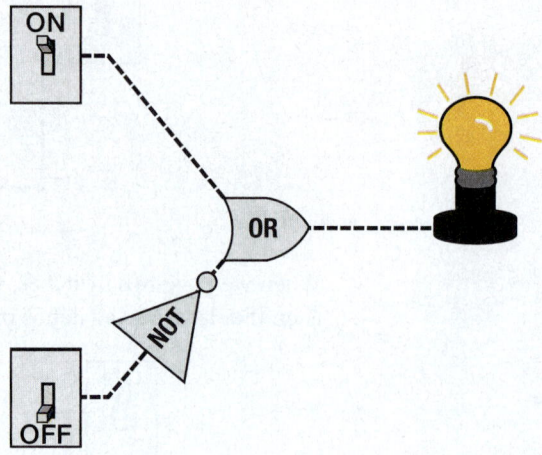

3. For the above diagram, determine the output of the light bulb for each of the four combinations of the switch positions.

   a. Top switch on, Bottom switch off: _____

   b. Top switch on, Bottom switch on: _____

   c. Top switch off, Bottom switch on: _____

   d. Top switch off, Bottom switch off: _____

4. Now it's your turn to design some circuits. (Part c. will require some thought, maybe even some trial and error on your part.)

   a. Design a circuit with two switches and at least two gates in which the light is on when at least one switch is off.

   b. Design a tautology circuit with two switches and at least two gates (that is, the light bulb is always on).

   c. Design a circuit with two switches and at least two gates in which the light is on when precisely one switch is on.

5. Describe a scenario in which the circuits you designed in Questions 4 and 5 could be used to control something. For example, they could control the switches for the buzzer in a quiz match or spotlights for a concert when certain musicians are on the stage. Explain your scenario in detail, noting what each circuit controls and when it will and won't be activated.

CHAPTER 3 REVIEW

Section 3.1 Logic Statements and Their Negations

Definitions

Statement: A mathematical **statement** is a complete sentence that asserts a claim which is either true or false, but not both at the same time.

Paradox: A **paradox** is a sentence that contradicts itself and therefore has no single truth value. A paradox cannot be a mathematical statement.

Negation: The **negation** of a statement is the logical opposite of that statement, or its denial. Negations always have the opposite truth value of the original statement. Negations are denoted by the symbol ~.

Compound Statements: A **compound statement** is composed of two or more statements joined together by connective words such as *and*, *or*, or *implies*.

Conjunction: If a and b are statements, then "a and b" is a compound statement called a **conjunction**. The symbol $\wedge$ is used to represent a conjunction.

Disjunction: If a and b are statements, then "a or b" is a compound statement called a **disjunction**. The symbol $\vee$ is used to represent a disjunction.

Conditional: If a and b are statements, then "if a, then b" is a compound statement called a **conditional**. The symbol $\Rightarrow$ is used to represent a conditional statement.

Biconditional: If a and b are statements, then "a if and only if b" is a compound statement called a **biconditional**. The symbol $\Leftrightarrow$ is used to represent a biconditional statement.

Section 3.2 Truth Tables

Definitions

Conjunction Truth Value: If a and b are statements, then the **conjunction** "a and b" is true only when both a and b are true; otherwise, the conjunction is false.

Disjunction Truth Value: If a and b are statements, then the **disjunction** "a or b" is always true unless a and b are both false.

Conditional Truth Value: If a and b are statements, then the **conditional** "if a, then b" is always true unless a is true and b is false.

Biconditional Truth Value: If a and b are statements, then the **biconditional** "a if and only if b" is true only if a and b have the same truth value; that is, either both are true or both are false.

Tautology: A **tautology** is a statement that is true in all possible circumstances.

Section 3.3 Logical Equivalence and De Morgan's Laws

Definitions

Logically Equivalent Statements: Logically equivalent statements are statements that have exactly the same truth values in all corresponding circumstances. Equivalence is denoted with the symbol ≡.

De Morgan's Laws

1. $\sim(p \wedge q) \equiv \sim p \vee \sim q$
2. $\sim(p \vee q) \equiv \sim p \wedge \sim q$

Negation of Conditional Statements

$\sim(p \Rightarrow q) \equiv p \wedge \sim q$

Section 3.4 Valid Arguments and Fallacies

Definitions

Argument: A logical **argument** consists of a series of logical statements.

Premise: A **premise** is a statement at the beginning of an argument used to support a particular conclusion.

Conclusion: A **conclusion** is the ending statement of an argument.

Valid Argument: A **valid argument** is a deductive argument where the conclusion is always guaranteed from the premises.

Invalid Argument: An **invalid argument** is a deductive argument where the conclusion is not always guaranteed from the premises.

Sound Argument: A **sound argument** is a valid argument that uses true premises.

Fallacy: A **fallacy** is an error in reasoning that leads to an invalid argument.

Procedures

Using a Truth Table to Determine the Validity of an Argument

1. Express the argument in symbolic form as an *if-then* statement. The *if* portion will consist of the premises joined together with conjunctions. The *then* portion is the conclusion. For example, if there are *n* premises, the statement will be of the form

    $$[(\text{premise 1}) \wedge (\text{premise 2}) \wedge \ldots \wedge (\text{premise } n)] \Rightarrow \text{conclusion.}$$

2. Construct a truth table for the *if-then* statement.

3. If the implication is a tautology, then the argument is valid, otherwise the argument is invalid.

CHAPTER 4

Ratios, Percentages, Rates, and Proportionality

🔖 SECTIONS

Introduction

In 1984, the Task Force for Child Survival was founded with the help of five international organizations: UNICEF, Rockefeller Foundation, World Health Organization, The World Bank, and the United Nations Development Programme. Convening at the Bellagio conference in Italy, they set out to raise childhood immunization rates in developing countries around the globe. Before long, they were also helping to eliminate or reduce tropical diseases, polio, intestinal worms, river blindness, as well as childhood and maternal mortality. In 2009, their work stretched so far that the name was changed to the Task Force for Global Health. Much of what they do is focused on the field of epidemiology by investigating the complex root causes of public-health issues and working to develop effective strategies to prevent them. As a result of their efforts, the Task Force for Global Health has helped 4 countries eliminate river blindness and has recorded a 99% decrease in polio cases!

In epidemiology, ratios, rates, and proportions are all numerical ways to paint a picture of the way that diseases influence and affect our human lives. According to the CDC, the *death-to-case ratio* is "the number of deaths attributed to a particular disease during a specified period divided by the number of new cases of that disease identified during the same period." The fraction of clinic patients who test positive for the flu on a yearly basis can be given by percentages, and how quickly new cases of breast cancer occur in a population is depicted using rates of occurrence.

Throughout the chapter we will learn how to calculate each of these descriptors and consider their uses. We'll also examine some common percentage faux pas and possibly rethink the saying, "The numbers don't lie."

4.1 PROPORTIONS, PERCENTAGES, AND RATIOS

OBJECTIVES

1. Write and interpret proportions as fractions.

2. Write and interpret proportions as percentages.

3. Calculate unknown values using proportions as fractions and percentages.

4. Write and interpret ratios.

Fun Fact

Have you been "ratioed" on Twitter? If the number of comments your tweet receives is larger than the number of likes or retweets, indicating that the tweet is more disliked than liked, you've been ratioed!

Proportional relationships—how quantities relate to one another—are one of the most general, yet foundational ideas in mathematics and its applications. For instance, in physics, Hooke's Law states that the length of a spring is proportional to the weight that it's bearing. In geometry, the circumference of a circle is proportional to its radius. And statisticians study sectors of society by looking at what proportion of the population has a specific trait or opinion. On some scale, all of us understand from a very early age the idea that things can be divided up. Take, for instance, a child who says "Mine!" when asked to share a bite of their ice cream cone or three friends haggling over how to fairly slice a pie. We talk about who gets the "lion's share" of a budget, and so on. These are ideas about proportional relationships.

Think for a moment about the following questions:

What proportion of students graduate in 4 years?

What percentage of students take classes fully online?

What is the ratio of students to faculty for in-person classes?

Each of these questions is seeking a picture of how students are divided up in different scenarios. Proportions and ratios help us formally communicate these relationships.

We'll begin with the concept of proportions.

OBJECTIVE 1 Proportions

First of all, let's agree that the English language has a beautiful and complex structure. A single word can literally have ten different meanings, even to the point that a word can have two definitions that are the exact opposite of each other! The word *proportion* is one that has multiple meanings. You might recall hearing phrases such as "solve the proportion" or "is proportional to" or "what proportion is that?" The word *proportion* means something slightly different in each of these situations. For now, however, what we will be focusing on is the statistical meaning of the word *proportion*, which is that a proportion is a fraction that compares a portion to the whole. For instance, suppose we have a sample of four pets: a cat, a hamster, a fish, and a bird. We might ask what proportion of the pets have four legs. Only two pets (the cat and the hamster) have four legs.

Therefore, the proportion of pets with four legs is $\frac{2}{4}$ or 0.50.

Think Back

Recall that fractions are of the form $\frac{\text{Numerator}}{\text{Denominator}}$.

PROPORTION

A **proportion** is a fraction of a whole.

Consider Table 4.1.1, which provides the numbers of vehicles sold for select automotive brands in the United States in May 2020.

Table 4.1.1: Sales of Select US Automotive Brands May 2020

Brand	Number of Sales May 2020
Acura	10,341
Genesis	1350
Honda	110,636
Hyundai	57,619
Kia	45,727
Mazda	24,933
Subaru	51,988
Volvo	9519
Total	**312,113**

Proportions give us a way to communicate exactly what car sales looked like among the different brands that month. We can communicate how the car sales were distributed among the brands by talking about each brand's proportion of cars sold. For instance, the proportion of Hondas sold in the United States in May 2020 compares the number of Hondas sold to the total number of cars sold by the select automotive brands during May.

$$\text{Proportion of Hondas Sold} = \frac{110,636}{312,113}$$

Similarly, the proportion of Volvos sold is $\dfrac{9519}{312,113}$. These proportions convey what market share each brand secured in May 2020. Note that if we were to find the proportion of cars sold for each car brand in the table, and then add them all together, the total would be 1.

Commonly, when we express proportions as fractions, we write them in lowest terms; that is, as a reduced fraction. This means that we divide both the numerator and denominator by the greatest common divisor (GCD). In this case, the Honda proportion is in lowest terms, but the Volvo proportion can be reduced by a factor of 19.

Reducing Fractions to Lowest Terms

Proportion of Hondas Sold: $\dfrac{110,636}{312,113}$

Proportion of Volvos Sold: $\dfrac{9519 \div 19}{312,113 \div 19} = \dfrac{501}{16,427}$

> **💬 Think Back**
>
> The greatest common divisor (GCD)—also called the greatest common factor (GCF)—is the largest integer that divides two numbers without a remainder.

Example 4.1.1

Finding Proportions

The pie chart in Figure 4.1.1 shows the number of years it takes for students to graduate with a bachelor's degree for a certain state college system. Use the pie chart to determine the proportions of students earning a bachelors who graduate within

a. 4 years,

b. 5 years, and

c. 6 years.

Number of Years to Graduation

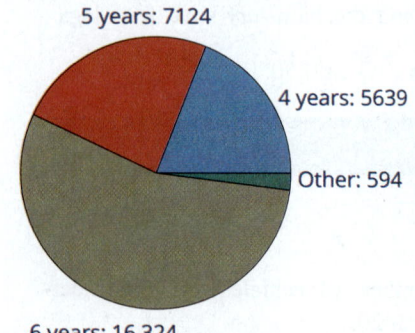

5 years: 7124

4 years: 5639

Other: 594

6 years: 16,324

Figure 4.1.1

Write each proportion as a fraction of the total number of students.

Solution

Since a proportion is a fraction of the whole, first we need to know the total number of students by adding all of the categories in the pie chart.

Total Number of Students: $5639 + 7124 + 16{,}324 + 594 = 29{,}681$

Each proportion is found by creating a fraction with the respective numbers for each category. We are only asked to find the proportions of 4-, 5-, and 6-year graduates. So we will not need to determine the proportion of "other" students.

a. Proportion of Students Who Graduate in 4 Years $= \dfrac{5639}{29{,}681}$

b. Proportion of Students Who Graduate in 5 Years $= \dfrac{7124}{29{,}681}$

c. Proportion of Students Who Graduate in 6 Years $= \dfrac{16{,}324}{29{,}681}$

Note that we will not reduce the fractional answers here since we were asked to give each proportion as a fraction of the total number of students.

OBEJCTIVE 2 Percentages

Let's go back to our car sales example. We can see by the number of Hondas sold (110,636) compared to the number of Volvos sold (9519) that Hondas were more popular in May 2020. But do the proportions $\dfrac{110{,}636}{312{,}113}$ and $\dfrac{501}{16{,}427}$ give you a sense of how much more popular the Hondas were? Or what share of the sales each received relative to all the other brands?

Sometimes, at a glance, proportions written as a fraction are not easy to interpret. As a result, proportions are often expressed as percentages so that they can more easily be used for comparisons. Recall that a *percentage* is a fraction where the denominator is 100, or a rate per 100, and is most commonly denoted using a percent sign, %.

> 📖 **PERCENTAGE**
>
> A **percentage** is an amount per 100, expressed using a percent sign, %.

Think Back

To convert a fraction to a percentage, divide the numerator by the denominator and multiply by 100%.

$$\frac{3}{8} = 0.375 = 37.5\%$$

fraction to decimal to percent

With the example of car sales, by changing the fractions to percentages, we see the relative sizes of the car sales for Hondas and Volvos.

Percentage of Hondas Sold: $\dfrac{110{,}636}{312{,}113} \approx 0.354 = 35.4\%$

Percentage of Volvos Sold: $\dfrac{501}{16{,}427} \approx 0.030 = 3\%$

Notice that we now have a sense of relative scale. We can easily see that Hondas, at 35.4% of sales, were about 10 times more popular than Volvos, at 3% of the total sales.

In general, calculating a proportion and expressing it as both a nonreduced fraction and a percentage provides added insight into the way a group of things are divided up. Table 4.1.2 contains new columns showing the proportions as both fractions and percentages for each car brand, making it easier to see how the sales compare among the brands. Notice that in the bottom row of the table, the totals confirm that the fractions add to one and the percentages add to 100%.

Table 4.1.2: US Best-Selling Automotive Brands May 2020

Brand	Number of Sales May 2020	Proportion as a Fraction	Proportion as a Percentage
Acura	10,341	$\dfrac{10{,}341}{312{,}113}$	3.3%
Genesis	1350	$\dfrac{1350}{312{,}113}$	0.4%
Honda	110,636	$\dfrac{110{,}636}{312{,}113}$	35.4%
Hyundai	57,619	$\dfrac{57{,}619}{312{,}113}$	18.5%
Kia	45,727	$\dfrac{45{,}727}{312{,}113}$	14.7%
Mazda	24,933	$\dfrac{24{,}933}{312{,}113}$	8.0%
Subaru	51,988	$\dfrac{51{,}988}{312{,}113}$	16.7%
Volvo	9519	$\dfrac{9519}{312{,}113}$	3.0%
Total	312,113	$\dfrac{312{,}113}{312{,}113}$	100%

Example 4.1.2

Finding Proportions

A company is looking to open a new manufacturing plant in one of three possible communities. They know that to be a viable location for a new plant, the community needs to consist of at least 35% of residents who hold a college degree. Table 4.1.3 shows the survey results of a sample of adults in each of the three communities. Determine which communities would meet the minimum requirement for degrees.

Table 4.1.3: Random Sample of Adults

	Community A	Community B	Community C
Have a Degree	3720	3996	4400
Do Not Have a Degree	6290	8806	7170
Total	10,020	12,802	11,570

Solution

For each community, calculate the proportion of adults who have a degree and then write it as a percentage to determine if it meets the 35% cutoff.

Community A: $\dfrac{3730}{10,020} \approx 37.23\%$

Community B: $\dfrac{3996}{12,802} \approx 31.21\%$

Community C: $\dfrac{4400}{11,570} \approx 38.03\%$

So both communities A and C meet the required minimum percentage.

✔ Skill Check 4.1.1

In baseball and softball, a player's batting average is the proportion of the number of hits to the number of "at bats." What is Jessica's batting average if she had 5 hits in a game with 12 at bats? Express the answer as both a fraction and a percentage rounded to the nearest hundredth.

Whether as fractions or percentages, proportions help to paint a picture of how one component of a group compares to the whole. So far, we have considered one group of things (either cars sold or adults with degrees) and compared different proportions within that singular group. However, what happens when we compare proportions that come from different groups? Consider the following situation.

One definition of *voter turnout* is the proportion of the voting-age population that votes in an election. Records dating back to 1828 show that the largest voter turnout for a US presidential election happened in 1876, when a whopping 81.8% of eligible voters voted. The smallest voter turnout to date occurred in 1924, when less than half of the voting-age public, 48.9%, decided to cast a ballot.[1]

Stop for a moment and consider the following question:

> Does "81.8% of eligible voters" imply that more people voted in 1876 than in 1924, when the turnout was "48.9% of eligible voters"?

The simple answer is no, not necessarily.

The proportion in 1876 is certainly higher. But, before we go any further, we want to begin to etch in your thoughts a fundamental question that you should always ask when dealing with proportions: "*Out of how many?*" In our case, you should be thinking, "81.8% out of how many? 48.9% out of how many?"

You don't have to be a historian to realize that the number of eligible voters changed between 1828 and 1924. For starters, the landscape for who was given a vote in the United States changed drastically; for instance, women were given the right to vote in 1920. In addition, the population of the nation increased about sixfold. The number of eligible voters increased from 10,291,759 in 1876 to 59,503,286 in 1924.[2]

With this new information, we are able to not only compare the proportions of voters, but also the number of eligible persons who voted in each election. We do this by multiplying the percentages by their respective populations.

---

1   "Voter Turnout in Presidential Elections," The American Presidency Project, University of California, Santa Barbara, last modified January 22, 2021, https://www.presidency.ucsb.edu/statistics/data/voter-turnout-in-presidential-elections.

2   Inc. Congressional Quarterly. *Congressional Quarterly's Guide to U.S. Elections.* Edited by John L. Moore, Jon P. Preimesberger, and David R. Tarr. Washington, D.C.: CQ Press, 2001.

$$1876 \text{ US Presidential Voter Turnout: } 81.8\% \times 10{,}291{,}759 = 8{,}418{,}659$$

$$1924 \text{ US Presidential Voter Turnout: } 48.9\% \times 59{,}494{,}119 = 29{,}092{,}624$$

Notice that the percentage in 1876 (81.8%) was almost double the percentage in 1924 (48.9%). But now we can see that in 1924 the actual number of people who voted was more than three times as many as in 1876.

The bottom line is this: when we want to use proportions to draw conclusions, *the sizes of the groups matter.*

Example 4.1.3

Using Percentages

In the United States, 13.7% of all students aged 3 through 21 received special education services in 2017–2018. The percentage per state varied from 9.2% in Texas to 19.2% in New York. Table 4.1.4 shows a sample of data from six states. The proportions of students in special education in the states as well as the overall numbers of students between ages 3 and 21 are given. The states are ordered, from lowest to highest, according to the proportion of students in special education in each state.

Table 4.1.4: Sample of States by Proportion of Students Receiving Special Education Services

State	Proportion of Students in Special Education	Students Aged 3–21
Texas	9.2%	7,769,439
California	12.2%	9,596,953
Florida	13.7%	4,512,312
Wyoming	16.6%	143,218
Pennsylvania	18.6%	2,907,692
New York	19.2%	4,384,186

Source: "Digest of Education Statistics: 2018," National Center for Education Statistics, last modified December 2019, https://nces.ed.gov/programs/digest/d18/index.asp.

a. Determine the number of students in special education in each of the states listed, and then use these numbers to order the states from highest to lowest. (Round your answer to the nearest whole number of students.)

b. Suppose that a new federal initiative to provide money to education departments across the states is in the planning stages. A committee is considering how to disperse federal funds to states based on their populations of students in special education. Discuss whether the committee should base its decision on the proportions of students or the number of students in special education from each state.

Solution

a. To determine the number of students in special education in each state, multiply the percentage by the respective number of students aged 3–21 as shown in Table 4.1.5, ordering the states by number of students in special education, from lowest to highest.

Table 4.1.5: Ranking of States by Number of Students in Special Education

	State	Proportion of Students in Special Education	Students Aged 3–21	Number of Students in Special Education
1	Wyoming	16.6%	143,218	$0.166 \cdot 143,218 = 23,774$
2	Pennsylvania	18.6%	2,907,692	$0.186 \cdot 2,907,692 = 540,831$
3	Florida	13.7%	4,512,312	$0.137 \cdot 4,512,312 = 618,187$
4	Texas	9.2%	7,769,439	$0.092 \cdot 7,769,439 = 714,788$
5	New York	19.2%	4,384,186	$0.192 \cdot 4,384,186 = 841,764$
6	California	12.2%	9,596,953	$0.122 \cdot 9,596,953 = 1,170,828$

b. Although the proportions give a sense of relative scale for each state, the initiative funding should be scaled according to the number of students in special education. Notice that although Wyoming is in the middle of the rankings proportionally, the actual number of students in special education is much, much smaller simply because Wyoming's population is smaller. On the other hand, Texas ranks lowest proportionally, but is in the top half for number of students in special education.

OBEJCTIVE 3 Using Proportions

In the last example, we found the number of students in special education in each state by using the percentage of students in special education and the total number of students. We can also use proportions in fraction form to calculate the size of a particular category. The next two examples do just that.

Example 4.1.4

Finding Quantities from Proportions as Fractions

Suppose that the proportion of people who are left-handed is approximately $\frac{1}{11}$.

If there are 45 people in Room A and 74 people in Room B, estimate how many left-handed people are in each room.

Solution

To find the number of left-handed people in each room, multiply the proportion $\frac{1}{11}$ by the total number of people in each room.

> **Think Back**
>
> A vinculum is a horizontal line placed above a number or group of numbers in a decimal and is used to indicate unending repetition.

Estimated Number of Left-Handed People in Room A: $\frac{1}{11} \cdot 45 = 4.\overline{09}$

Therefore, we would expect there to be about 4 left-handed people in a room of 45.

Estimated Number of Left-Handed People in Room B: $\frac{1}{11} \cdot 74 = 6.\overline{72}$

Therefore, we would expect there to be about 7 left-handed people in a room of 74.

💬 **Think Back**

For very large or very small numbers, an alternative method to express decimals, called scientific notation, is often used. Scientific notation requires that we rewrite a number to be between 1 and 10 and then multiply the number by a power of 10 equal to the number of decimal places represented. So, 3,658,000,000,000,000 can be rewritten in scientific notation as 3.658×10^{15}. Small numbers less than 1 will have negative powers of 10. For example, 0.0000223 is written 2.23×10^{-5}.

Numbers in scientific notation can be multiplied by multiplying the respective parts of the notation together and using the product rule for exponents.

For example, the two numbers above are multiplied together as follows.

$$\left(3.658 \times 10^{15}\right)\left(2.23 \times 10^{-5}\right)$$
$$= \left(3.658 \cdot 2.23\right) \times \left(10^{15} \cdot 10^{-5}\right)$$
$$= 8.15734 \times 10^{10}$$

Example 4.1.5

Using Proportions and Percentages

Earth consists of three parts: crust, mantle and core. Earth's core makes up 31.5% of Earth's mass. The mantle makes up 68.1% of its mass. Although 20 to 30 miles thick, the crust only makes up a small proportion of its mass—the remaining 0.4%. If the total mass of Earth is 5.9742×10^{24} kg, what is the mass of Earth's crust?

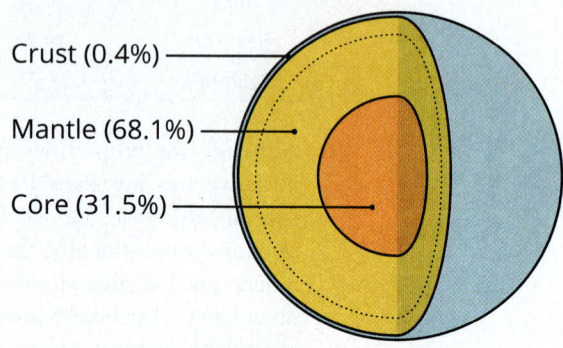

Crust (0.4%)

Mantle (68.1%)

Core (31.5%)

Figure 4.1.2

Solution

We are given the fact that Earth has a total mass of 5.9742×10^{24} kg, and its crust is 0.4% of the overall mass. We simply need to multiply the two quantities together to find the amount of mass that is crust. Because we are given the proportion of crust as a percentage, we need to first write it as a decimal before multiplying the two together.

$$\left(5.9742 \times 10^{24}\,\text{kg}\right)\left(0.4\%\right) = \left(5.9742 \times 10^{24}\,\text{kg}\right)\left(0.004\right) \quad \text{Change the percent to a decimal.}$$

$$= \left(5.9742 \times 10^{24}\,\text{kg}\right)\left(4 \times 10^{-3}\right) \quad \text{Change the decimal to scientific notation.}$$

$$= \left(5.9742 \cdot 4\right) \times \left(10^{24} \cdot 10^{-3}\right) \quad \text{Regroup the products.}$$

$$= 23.8968 \times 10^{21}\,\text{kg} \quad \text{Multiply.}$$

$$= 2.38968 \times 10^{22}\,\text{kg} \quad \text{Adjust the decimal for scientific notation.}$$

So the mass of Earth's crust is 2.38968×10^{22} kg.

OBJECTIVE 4   Ratios

Another way to communicate the shared relationships among parts of a whole is to use the concept of ratio. A *ratio* provides a comparison of measured quantities of the same type—people, distance, objects, animals—and uses units of identical dimension. In other words, it describes the relationships among similar parts of a whole.

📖 RATIO

A **ratio** is a comparison of measured quantities whose units are of the same type. It is most often written with a colon between quantities; for example, $a : b$.

Suppose a flag is colored red and blue in the ratio 1 : 2 (read "1 to 2"). This one simple statement tells you a lot of information about the way the colors are proportioned, or divided up, on the flag. First, it tells you that the flag can be cut up into 3 equal parts where 1 part is red and the remaining 2 parts are blue. This flag ratio also tells you that there is twice as much blue as there is red and that the proportion of each color on the flag is $\frac{1}{3}$ red and $\frac{2}{3}$ blue.

Figure 4.1.3: Examples of Flags Colored Red and Blue in the Ratios 1 : 2

Let's take note of a few foundational points about ratios that we've just seen:

1. The sum of all of the components of a ratio determines the total number of equal parts in the whole.

2. Each individual component of a ratio can be compared to the whole by creating a fraction whose numerator is the number of parts in that component and whose denominator is the total number of parts in the whole.

3. The individual components of the ratio can also be compared to each other using fractions comprised of their respective numbers of parts.

Using fractions, we can tease out all of these individual relationships given in the single flag ratio

Fractions Related to the Flag Ratio 1 : 2

$\frac{1}{3}$ represents the red coloring (1) compared to the whole (2 + 1) of the flag

$\frac{2}{3}$ represents the blue coloring (2) compare to the whole (2 + 1) of the flag

$\frac{1}{2}$ represents the red coloring (1) compared to the blue coloring (2)

$\frac{2}{1}$ represents blue coloring (2) compared to the red coloring (1)

That's a lot of information about the interconnections between the colors that we captured from the one simple ratio given! Notice, however, one thing you don't know about the flag from the ratio given is its size. One of the principles of a ratio is that the relationship it represents holds true no matter what units of measure are used. In other words, whatever size the flag is—3 ft by 5 ft, 2 cm by 8 cm, or 8 yd by 15 yd—the ratio of red to blue coloring will be the same: twice as much blue as red.

With our flag example, the ratio tells us a lot about the way the two colors compare to one another. But a ratio is not limited to describing only two components. In theory, there can be an unlimited number of components of a ratio. For example, a cocktail drink might call for a ratio of 2 : 1 : 1—alcohol : sour : sweet, meaning the liquid in the drink consists of 4 equal parts—2 parts alcohol : 1 part sour : 1 part sweet. Notice that the order of the ratio descriptors is important. The first number given refers to the first item in the list, the second number corresponds to the second descriptor, and so on. Think how differently a drink would taste if you didn't know which number went with which ingredient. Again, observe that the size of the drink is not indicated by the ratio, just the relationship of its 4 equal parts.

Example 4.1.6

Interpreting Ratios

One cooking website claims that the essential ratio for the weights of the main ingredients for the ultimate sugar cookie is 3 : 2 : 1—flour : fat : sugar.

a. Write the fractions to represent the following relationships between flour, fat, and sugar as indicated by the ratio.

> flour to fat   flour to sugar   flour to whole
> fat to whole   fat to sugar   sugar to whole

b. If a recipe calls for 226 grams of fat, how much sugar should be used to make the ultimate sugar cookie?

Solution

a. First note that there are 6 parts in the whole. The fractions associated with the recipe ratio 3 : 2 : 1 are as follows.

$$\text{Flour to Fat} = \frac{3}{2} \quad \text{Flour to Sugar} = \frac{3}{1} \quad \text{Flour to Whole} = \frac{3}{6}$$

$$\text{Fat to Sugar} = \frac{2}{1} \quad \text{Fat to Whole} = \frac{2}{6} \quad \text{Sugar to Whole} = \frac{1}{6}$$

b. The fractions from part a. tell us that the fat to sugar ratio is 2 to 1, so there will be twice as much fat as sugar. If there are 226 grams of fat, we need 113 grams of sugar.

Example 4.1.7

Writing Ratios

On a university campus, the fall enrollment data shows that the new freshman class consists of 214 out-of-state residents and 321 in-state residents. Write the ratio of in-state residents to out-of-state residents and determine the number of students in the freshman class.

Solution

Since the ratio we are interested in is *in-state residents : out-of-state residents*, the number of in-state residents must come first. So, the ratio of in-state to out-of-state is as follows.

$$321 : 214$$

The number of students in the freshman class will consist of both the in-state and out-of-state residents. Adding these together we have the following equation.

Number of Students in the Freshman Class = 214 + 321 = 535

Since ratios encapsulate how equal parts of the whole are allocated, we can often express them in more than one way. In Example 4.1.7, the ratio 321 : 214 can be reduced to the ratio 3 : 2 by dividing both numbers by their greatest common factor, 107. Therefore, 321 : 214 and 3 : 2 are considered to be equivalent ratios, because they express the same relationship between the parts.

We can also find equivalent ratios by multiplying or scaling up both parts. For instance, take the ratio 4 : 1. Multiplying each part by 2 gives the equivalent ratio 8 : 2, or multiplying both parts by 3 gives the equivalent ratio 12 : 3. In fact, every ratio has an unlimited number of equivalent ratios that can be found by using either multiplication or division.

📖 EQUIVALENT RATIOS

Equivalent ratios are ratios that express the same relationship.

Example 4.1.8

Writing Equivalent Ratios

For each ratio, write two equivalent ratios, one of which is in the lowest terms possible.

a. 90 : 120

b. 5 : 15 : 10

Solution

a. The ratio in lowest terms is found by dividing both components of the ratio by the greatest common factor, 30.

90 : 120 is equivalent to 3 : 4

We can scale up a ratio by multiplying by any number. In this case, we will use 2.

90 : 120 is equivalent to 180 : 240

b. The ratio in lowest terms is found by dividing all three components of the ratio by the greatest common factor, 5.

$$5 : 15 : 10 \text{ is equivalent to } 1 : 3 : 2$$

We can scale up a ratio by multiplying by any number. In this case, we will use 3.

$$5 : 15 : 10 \text{ is equivalent to } 15 : 45 : 30$$

> ✔ **Skill Check 4.1.2**
>
> State law requires that a day care have a maximum class size of 30 children for every 2 providers. Write the ratio of children to providers in lowest terms.

Another way to find equivalent ratios is to set up an equation and solve for x. This equation is called a *proportional equation*. More on proportionality will be covered in Section 4.5.

Example 4.1.9

Finding Equivalent Ratios with Proportional Equations

The local elementary school is taking its first graders on a field trip to an aquarium. Aquarium guidelines state that a ratio of 3 adult chaperones for every 10 students must be maintained.

a. If 90 students plan to go to the aquarium, how many adult chaperones are needed?

b. If 143 people go on the field trip, how many of the attendees are adults?

Solution

Since the ratio of adults to children is 3 : 10, we know that each group of 13 attendees should consist of, on average, 3 adults and 10 children. This means that the fraction of adults to attendees is $\dfrac{3}{13}$ and the fraction of children to attendees is $\dfrac{10}{13}$.

a. We've been asked to find the number of adults needed if 90 students attend. This ratio can be described by $x : 90$ and is equivalent to 3 : 10.

We can use these ratios to set up the proportional equation $\dfrac{x}{90} = \dfrac{3}{10}$, and then solve for x.

$$\frac{x}{90} = \frac{3}{10}$$

$$\frac{x}{90} \cdot 90 = \frac{3}{10} \cdot 90^{9}$$

$$x = 27$$

This gives the equivalent ratio of 27 : 90, which means that 27 adults would need to chaperone the field trip if 90 students attend.

b. Here we are asked to find the number of adults in a group of 143 attendees. We need to use the fractions that we derived from the ratio. This fraction can be described by $\dfrac{x}{143}$ and is equivalent to $\dfrac{3}{13}$.

We can use these fractions to set up the proportional equation $\frac{x}{143} = \frac{3}{13}$, and then solve for x.

$$\frac{x}{143} = \frac{3}{13}$$

$$\frac{x}{\cancel{143}} \cdot \cancel{143} = \frac{3}{\cancel{13}} \cdot \cancel{143}^{11}$$

$$x = 33$$

This gives the equivalent fraction of $\frac{33}{143}$, which means 33 of the attendees are adults if the group consists of 143 people.

Example 4.1.10

Writing and Interpreting Ratios

A study by an online health journal claims that although fish has long been a food of choice for helping to control heart disease because it is high in omega fatty acids, farm-raised tilapia might actually be dangerous to the heart. That is to say, not all omega fatty acids are beneficial. The omega fatty acid ratios in farm-raised tilapia actually give undesirable amounts of omega-6 acids. In farm-raised tilapia, the ratio of these potentially detrimental long-chain omega-6 fatty acids (ω-6) to the beneficial long-chain omega-3 fatty acids (ω-3) averaged about 11 : 1. Table 4.1.6 shows a sample of the amounts of ω-6 and ω-3 fatty acids in a tablespoon of different types of fish oil.

Table 4.1.6: Omega Fatty Acids in a Tablespoon of Various Fish Oils

Fish Oil	Long-Chain Omega-6 Fatty Acids (ω-6) (mg)	Long-Chain Omega-3 Fatty Acids (ω-3) (mg)
Herring	39	1509
Salmon	92	4657
Sardine	239	3096
Cod Liver	127	2557
Menhaden	159	3624

Source: FoodData Central, U.S. Department of Agriculture, accessed February 16, 2021, https://fdc.nal.usda.gov/index.html.

a. Write a ratio for each fish oil so that it can be compared to the given ratio for farm-raised tilapia.

b. Write an equivalent ratio for the fatty acids in each fish oil so that you can more easily compare it to the tilapia. Round to the nearest ten-thousandth, if necessary. Explain how the ratios of fatty acids in the fish oil compare to that of the tilapia.

Solution

a. In the information given, we were told that the ratio 11 : 1 for farm-raised tilapia was the ratio of ω-6 to ω-3. So, for comparison we need to list the ω-6 fatty acids first in each of the new ratios. For each ratio, write the numbers in the order they appear in the table, since the ω-6 values are listed first.

Table 4.1.7: Ratio of ω-6 to ω-3 in Various Fish Oils

Fish Oil	Long-Chain Omega-6 Fatty Acids (ω-6) (mg)	Long-Chain Omega-3 Fatty Acids (ω-3) (mg)	Ratio
Herring	39	1509	39 : 1509
Salmon	92	4657	92 : 4657
Sardine	239	3096	239 : 3096
Cod Liver	127	2557	127 : 2557
Menhaden	159	3624	159 : 3624

b. In order to compare the ratios of fatty acids, we can create equivalent ratios so that they all have a common form. Since the farm-raised tilapia has a 1 on the right side of the ratio, we'll divide each ratio by the right-hand number so that it also will have a 1.

Table 4.1.8: Ratio of ω-6 to ω-3 in Various Fish Oils

Fish Oil	Ratio	Division	Equivalent Ratio
Herring	39 : 1509	$\frac{39}{1509} : \frac{1509}{1509}$	0.0258 : 1
Salmon	92 : 4657	$\frac{92}{4657} : \frac{4657}{4657}$	0.0198 : 1
Sardine	239 : 3096	$\frac{239}{3096} : \frac{3096}{3096}$	0.0772 : 1
Cod Liver	127 : 2557	$\frac{127}{2557} : \frac{2557}{2557}$	0.0497 : 1
Menhaden	159 : 3624	$\frac{159}{3624} : \frac{3624}{3624}$	0.0439 : 1
Tilapia			11 : 1

Now that all the ratios are in a common form, we can see that the other fish have much lower ratios of omega-6 fatty acid (ω-6) to omega-3 fatty acid (ω-6) than tilapia. The closest ratio is that of sardines, and it has more than a hundredfold difference!

Example 4.1.11

Using Proportions to Express a Ratio

A mixing ratio for automotive paint is given as 4 : 2 : 1—base product : thinner : hardener. The thinner is added to aid in the flow of the paint through the spray gun and incorrect amounts can result in runs or dry patches in the paintwork. The hardener is to help with faster drying times, but too much or too little of it can alter suggested curing periods.

a. Express the proportion of each component of the mixture as a fraction compared to the whole.

b. Determine how much of each component is needed if you want to make 1.5 gallons of paint. Round to the nearest thousandth.

c. Determine the best way for a painter to accurately create 1.5 gallons of mixed paint.

Solution

a. The proportion of each component will be a fraction of the total amount of product. First, notice that there are seven equal parts in this ratio: 4 base product + 2 thinner + 1 hardener = 7 total. Therefore, the proportions are as follows.

$$\text{Base Product: } \frac{4}{7}$$

$$\text{Thinner: } \frac{2}{7}$$

$$\text{Hardener: } \frac{1}{7}$$

b. Altogether, we want to have 1.5 gallons of paint, so we need to know how much of each component we should mix in. In order to determine the amount of each component needed for 1.5 gallons of paint, we multiply each proportion by the total amount of paint needed.

$$\text{Amount of Base Product: } \frac{4}{7} \cdot 1.5 \text{ gallons} \approx 0.857 \text{ gallons}$$

$$\text{Amount of Thinner: } \frac{2}{7} \cdot 1.5 \text{ gallons} \approx 0.429 \text{ gallons}$$

$$\text{Amount of Hardener: } \frac{1}{7} \cdot 1.5 \text{ gallons} \approx 0.214 \text{ gallons}$$

c. Let's take a moment to notice that the decimal amounts found in part b. of each component required for 1.5 gallons of paint will need to be measured with great precision to ensure that there are no problems painting the car. In reality, that would be very difficult to achieve. To avoid this situation, a painter could choose to mix the paint using the ratios rather than the estimated decimals so that measuring is easier.

If we start with 1 gallon of base product, we know that the amount of thinner will need to be half of that, or 0.5 gallons, and the amount of hardener will be half of the thinner amount, or 0.25 gallons. Take a moment and verify that these amounts are still in the correct ratio given originally. Using this method, the painter will end up with 1.75 gallons of paint that is accurately mixed, from which it's easy to measure out the 1.5 gallons needed.

Example 4.1.12

Finding Proportions and Ratios

Suppose you earned a B on your last History 102 test by getting 80% of the questions correct.

a. If there were 60 equally weighted questions on the test, how many did you answer correctly?

b. What was your ratio of correct to incorrect answers on the test?

Solution

a. We know that you correctly answered 80%, or $\dfrac{80}{100}$, of the questions. We can find 80% of the 60 questions by multiplying the two numbers together as shown.

$$\frac{80}{100} \cdot 60 = 48$$

Therefore, you correctly answered 48 questions on the test.

b. Since we know from part a. that you correctly answered 48 questions, we can subtract this from the total number of questions to find the number of questions that you answered incorrectly

$$60 - 48 = 12 \text{ incorrect answers}$$

Therefore, the ratio of correct to incorrect answers would be written as 48 : 12, or 4 : 1.

Skill Check Answers

1. $\dfrac{5}{12} \approx 41.67\%$ **2.** 15 : 1 children to providers

4.1   Exercises

✔ CONCEPT CHECK

1. A proportion is a _____ of a whole.

2. To write a proportion as a fraction in lowest terms, divide both the numerator and the denominator by the _____.

3. A _____ is an amount per 100.

4. True or False: If a flag has the colors green and purple in the ratio 1 : 2, there is twice as much purple as there is green.

5. True or False: The ratio 3 : 1 is equivalent to the ratio 1 : 3.

◊ PRACTICE

Change each fraction to a percentage. Round your answer to the nearest hundredth when necessary.

6. $\dfrac{20}{45}$

7. $\dfrac{1}{5}$

8. $\dfrac{2}{3}$

9. $\dfrac{61}{212}$

10. Which is greater, half of 22 or one-fifth of 95?

11. Which is greater, a third of 36 or one-sixth of 138?

12. Which is greater, half of 42 or twenty-five percent of 120?

13. Which is greater, one-third of 93 or thirty percent of 110?

14. The ratio of 9.2 to 24 is the same as the ratio of _____ to 6.

15. The ratio of 1.2 to 32 is the same as the ratio of 3.6 to _____.

Find the equivalent ratio.

16. **a.** $1:7 = \_:14$

    b. $1:7 = \_:21$

    c. $1:7 = 4:\_\_\_$

    d. $1:7 = 5:\_\_\_$

17. **a.** $2:5 = 4:\_\_\_$

    b. $2:5 = \_\_\_:35$

18. **a.** $10:3 = \_\_\_:9$

    b. $10:3 = 5:\_\_\_$

19. **a.** $5:9 = 2.5:\_\_\_$

    b. $5:9 = \_\_\_:13.5$

For each ratio, write two equivalent ratios, one of which is in the lowest terms possible.

20. $25:45$

21. $24:60$

22. $12:36:72$

23. $25:75:45$

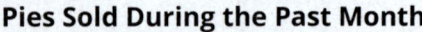

24. For their social media account, a local bakery created a pie chart indicating the popularity of the pies they sold over the past months. Use the pie chart to determine the proportion of pies sold that were **a.** apple and **b.** French silk. Write each proportion in lowest terms.

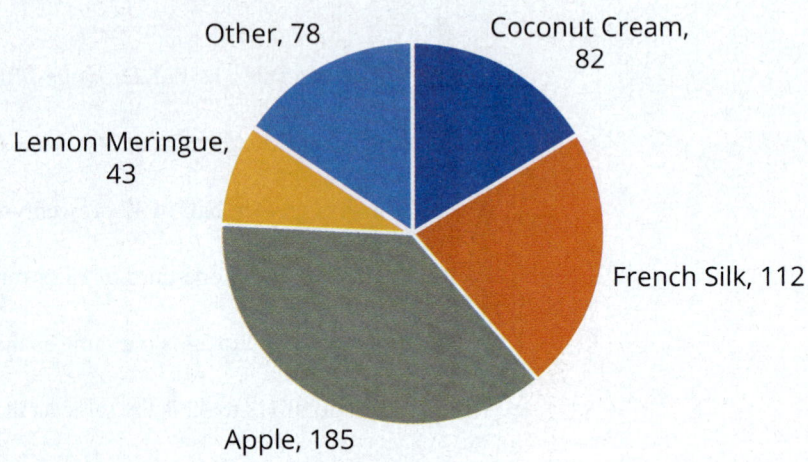

Pies Sold During the Past Month

Other, 78

Coconut Cream, 82

Lemon Meringue, 43

French Silk, 112

Apple, 185

25. A Richardson's family monthly budget for necessary expenses is shown in the table. Use the table to determine the proportion of the budget assigned to **a.** childcare and **b.** groceries. Write each proportion in lowest terms.

Richardson Family Monthly Budget	
Childcare	$750
Groceries	$800
Utilities	$300
Vehicle Loan Payment and Expenses	$525
Other	$425

26. Alexander expects to spend $10,500 on college tuition and expenses for the next school year. Of that total amount, he expects to spend $1200 on textbooks and course supplies. Write the proportion of textbooks and course supplies to college tuition and expenses in lowest terms.

27. A 950 calorie meal from a fast food restaurant is made up of a cheeseburger (520 calories), fries (220 calories), and a fountain drink (210 calories). Write the calories in the cheeseburger and the calories in the entire meal as a proportion in lowest terms.

28. If there are 48 students in a class and 32 are seniors, what proportion of students are *not* seniors?

29. In a survey, 99 out of 100 people are *not* left-handed. What proportion of people are left-handed according to this survey?

30. Nearly two-in-ten American women end their childbearing years without having borne a child. What proportion of American women end their childbearing years with having borne a child?

31. A survey indicated that 11 out of 100 students studying at a university reported having a disability. What proportion of students did *not* report having a disability?

32. If the drinks portion of a bill costs $18 and the total bill was $30, what percentage of the total was the cost of the drinks?

33. College University sent out 2000 requests to alumni for scholarship fund donations. In response, the university received 325 donations. What percent of alumni donated to the scholarship fund?

34. A student at a state college is expected to spend $7400 per semester on tuition, fees, and other expenses. If a student at the same college is expected to spend $700 on textbooks and course supplies, what percentage of the total is the cost of textbooks and course supplies?

35. If the monthly expense for a car is $600 and the cost of fuel for a month is $180, what percentage of the total expenses is the cost of fuel?

36. A city parks and recreation department is planning to build a new community park with a playground in one of three possible locations. The park will be built only in an area where at least 60% of the residents approve of the location. The table shows the results of a survey of city residents. Determine which locations meet the minimum approval requirement.

Citizen Approval of Park Location

	Location A	Location B	Location C
Approve of Location	17,739	13,298	15,265
Disapprove of Location	5628	9766	7947
No Opinion	1416	1719	1571
Total	24,783	24,783	24,783

37. A brand of cookie is planning to release a new special edition flavor selected by fans, but will only release a flavor if at least 75% of fans think the potential flavor is delicious and not gross. The table shows the results of an online poll. Determine which flavors meet the minimum approval requirement.

Fan Flavor Poll Results

	Green Tea	Toffee	Peach Mango
Delicious	6341	8826	9402
Gross	4227	2753	3283
Total	10,568	11,579	12,685

38. From 2006 to 2016, the number of people in the US age 65 and older increased by 33%. Six of the states that experienced a large increase are given below along with the proportion of their state population that is age 65 and older and the number of residents in the state. Determine the number of residents age 65 and older in each of the states listed, and then use these numbers to order the states from largest to smallest. (Round your answer to the nearest whole number of residents.)

State	Proportion of Residents Age 65 and Older	# of Residents
Utah	10.5%	3,058,705
Colorado	13%	5,719,415
Nevada	15%	2,940,947
Idaho	15%	1,699,927
South Carolina	16.7%	4,971,449
Delaware	17.5%	954,000

39. In the 2020 election, 66.7% of the Eligible Voting Population in the country voted. The proportion of eligible voters who voted is given for five states, as well as the number of eligible voters in the state. Determine the number of votes cast in each of the states listed, and then use these numbers to order the states from largest to smallest. (Round your answer to the nearest whole number of votes.)

State	Proportion of Eligible Voters who Voted	# of Eligible Voters
Tennessee	59.81%	5,124,867
Louisiana	64.61%	3,373,932
Illinois	67.02%	9,027,082
Iowa	73.25%	2,321,131
Minnesota	79.96%	4,118,462

40. A car's weight depends on all of its component materials, with steels, aluminum, plastics, and rubber being some of the biggest contributors. Approximately 53.41% of the weight comes from steels, 9.96% from aluminum, 8.24% from plastics, 4.93% from rubber, and 23.46% from other materials. If a car weighs 3994 lb, what is the weight of the plastics used in the car?

41. A website reports that 14% of the world's population is Hindi and 33% are Christian. If there are approximately 6,840,507,000 people in the world, how many are Hindi and how many are Christian according to the website?

42. There is a herd of cattle out on the range. Fifty-two percent of the cattle are male. If the herd consists of 1175 animals, how many are female?

43. Seventy percent of the animals in a zoo are herbivores. If there are 821 animals in the zoo, how many are *not* herbivores? Round your answer to the nearest whole number.

44. According to a survey by Airlines for America taken in 2017, 88% of Americans had flown on a commercial airline in their lifetime. If there are 75 people in Room A and 50 people in Room B, estimate how many people in each room had flown on a commercial airline in their lifetime.

45. The US Census Bureau's Educational Attainment report released in 2020 stated that approximately 23% of people 25 and older had earned a bachelor's degree. If City A has 10,726 citizens aged 25 and older and City B has 8253 citizens aged 25 and older, approximately how many citizens aged 25 and older from each city have a bachelor's degree?

46. Approximately 1.5% of the population have a peanut allergy. If East Central Middle School has 1180 students and West Chester Middle School has 1373 students, approximately how many students at each school have a peanut allergy?

47. Approximately 2% of the population has naturally red hair. The population of Utah is approximately 3.2 million and the population of South Dakota is approximately 885,000. Approximately how many people with naturally red hair live in each state?

48. Three-fourths of the students at the local community college live in the county. If there are 2004 students enrolled, how many do *not* live in the county?

49. At a recent education conference, one-fourth of the total participants came to learn about online courses. The rest came to learn about other educational practices. If 92 participants showed up, how many were *not* there to learn about online courses?

50. A general guideline for making cocktails at home is to follow the ratio 2 : 1 : 1—spirit : sweet : tart.

   a. Write a fraction to represent the relationship of spirit to tart as indicated by the ratio.

   b. If 4 ounces of spirit was used to mix a drink, how much of the tart ingredient would be required?

51. The Wilton icing coloring color chart states that to create burgundy colored icing, you need to add food coloring in the ratios 3 : 2 : 2—rose : red-red : black.

   a. Write a fraction to represent the relationship of rose to red-red as indicated by the ratio.

   b. If 12 drops of rose food coloring were added to the icing, how many drops of black food coloring should be added to make burgundy icing?

52. A website claims that the ideal fertilizer for flowering shrubbery is created by combining two different fertilizers with water at a ratio of 5 : 3 : 2—water : Sunny Shrub fertilizer : Flower Friends fertilizer.

    a. Write a fraction to represent the relationship of Flower Friends fertilizer to water as indicated by the ratios.

    b. If a gardener fills a container with 40 fluid ounces of water, how many fluid ounces of Sunny Shrub fertilizer should be added to create this ideal fertilizer for flowering shrubbery?

53. A recipe for hard candy indicates that the ingredients should be mixed at a ratio of 8 : 6 : 2—granulated sugar : corn syrup : water.

    a. Write a fraction to represent the relationship of corn syrup to the whole as indicated by the ratio.

    b. If someone following the recipe only had 4 cups of sugar, how many cups of corn syrup would be needed to make the hard candy?

54. Each week, a city-run afterschool tutoring program has 72 tutor volunteers and helps 243 students. Write the ratio of tutor volunteers to students and determine the number of people who participate in the afterschool tutoring program.

55. A recent population study of elephants revealed that due to population declines there are only 50,000 wild Asian elephants and 415,000 wild African elephants in the world. Write the ratio of Asian elephants to African elephants and calculate the total number of elephants living in the wild according to this study.

56. As part of a weeklong fundraising campaign, club members sold 476 brownies and 392 cupcakes during bake sales. Write the ratio of cupcakes to brownies and determine the total number of baked goods sold during the bake sales.

57. At a state college, a recent survey performed by the campus technology department revealed that 10,473 students had a phone with a Google Android operating system and 12,748 students had a phone with an Apple iOS operating system. Write a ratio of students with Apple iOS to students with Google Android and determine the total amount of students surveyed.

58. If 9 of the 21 members of a tour group are children, what is the ratio of children to adults on the trip?

59. Triangles A and B are *similar* triangles. In other words, triangle B has exactly the same shape as Triangle A, just a different scale. Triangle A has sides 45 cm, 54 cm, and 99 cm. If Triangle B has sides of lengths 90 cm, 108 cm, and 198 cm, what is the ratio of the lengths of the sides of Triangle A to the lengths of the sides of Triangle B?

60. A bag contains the following marbles: 8 black marbles, 17 blue marbles, 7 brown marbles, and 14 green marbles.

   a. What is the ratio of brown marbles to black marbles?

   b. What is the ratio of brown marbles to all the marbles in the bag?

61. A recent study showed that in American universities there were 18 million college students and 1.3 million faculty. Write the overall student to faculty ratio for universities across America in lowest terms.

62. Christopher had acquired 12 of the 20 items requested by his supervisor. Write the ratio of acquired items to nonacquired items in lowest terms.

63. Almost 30% of the world's billionaires live in 15 major cities. The city with the most billionaires is New York City with 113 billionaires among its population of approximately 8.4 million. This equates to 1 billionaire for each 74,336 residents, which we can write as 1 : 74,336. The data for five other cities is given below.

City	# of Billionaires	Population
Mumbai, India	38	20,000,000
Los Angeles, USA	44	4,000,000
Beijing, China	57	20,400,000
London, UK	66	9,300,000
Moscow, Russia	73	12,500,000

   a. Write a unit ratio for each of the given cities, rounding to the nearest whole person.

   b. Compare the ratio of billionaires for each city to New York City.

64. An 8 oz cup of brewed coffee contains 96 mg of caffeine. The caffeine content of four other beverages is given below.

Beverage	Size in oz	Caffeine in mg
Espresso	1	64
Black tea	16	94
Green tea	12	42
Cola	24	66

   a. Write an equivalent ratio for the caffeine content of each beverage.

   b. Compare the caffeine content of each beverage to that of brewed coffee.

65. If the ratio of tourists to locals is 2 : 3 and there are 40 tourists at the opening of a new museum, how many locals are at the opening?

66. A powdered drink mix calls for 3 scoops powder to 8 ounces water. How much powder do you need if you have 32 ounces of water?

67. It was recently estimated that uninsured vehicles outnumber insured ones by about six to five. If there are 2002 vehicles in a county, how many of them are uninsured?

68. One group (A) contains 125 people. One-fifth of the people in group A will be selected to win $15.00 grocery cards. There is another group (B) in a nearby town that will receive the same number of grocery cards, but there are 585 people in that group.

   a. What will be the ratio of nonwinners in group A to nonwinners in group B after the selections are made?

   b. What is the ratio of winners to people for group B?

4.1 PROJECT

PROPORTIONS, PERCENTAGES, AND ELECTIONS

Our political system is based on a voting method called the plurality method. In the plurality method, voters are presented with a list of candidates and each voter selects their first-choice candidate. The choice with the greatest number of first place votes is declared the winner.

A small town decided to ask voters to do something slightly different during an election to fill a town council seat: they were asked to rank the four candidates in order of preference, from first to fourth. The following table, called a preference table, summarizes all of the votes that were cast during the election. For example, the first column shows that 540 voters picked Smith as their first choice, Thomas as their second choice, Orlando as their third choice, and Patel as their last choice.

Preference Table for the Town Council Election

	540	250	160	140
1st choice	Smith	Thomas	Patel	Orlando
2nd choice	Thomas	Patel	Thomas	Thomas
3rd choice	Orlando	Orlando	Orlando	Patel
4th choice	Patel	Smith	Smith	Smith

1. Determine the total number of voters who participated in this election.

2. How many voters chose Smith as their first choice? How many chose Smith as their last choice?

3. What proportion of voters chose Smith as their first choice? What proportion of voters chose Smith as their last choice? Write each proportion as a fraction of the total number of voters, reduced to lowest terms.

4. What percentage of voters chose Smith as their first choice? What percentage of voters chose Smith as their last choice? Round your answer to the nearest hundredth, if necessary.

5. According to the plurality method, which candidate should be declared the winner?

6. Considering the entire preference table, is there anything that strikes you as odd about the chosen winner?

7. Is there a candidate that would be less polarizing?

4.2 USING PERCENTAGES

OBJECTIVES

1. Calculate tip amounts.

2. Calculate sales price amounts.

3. Calculate sales tax.

4. Calculate percent increase and percent decrease.

5. Understand percentage faux pas.

Think Back

Remember, to find 10% of an amount, move the decimal point one place to the left.

Helpful Hint

Find 10% of the pretax bill by moving the decimal one place to the left. Use multiples of 10% to calculate the desired tip amount.

As we saw in the previous section, we can describe proportions or compare values using percentages. Percentages are a useful and simplified means of conveying size or scale or value. In this section, the primary focus will be to improve your skills with percentages, pointing out some useful tips along the way. We'll start with the uber practical—tipping

OBJECTIVE 1 Service Tipping

The "Worldwide Tipping Guide" published by World Travelers of America states that not every country in the world considers it a good thing to tip servers at restaurants. In fact, tipping is seen as an insult in some countries. However, according to the guide, the United States has the highest expectations for tipping in restaurants of any country at 15–20%. Even though there are phone applications that allow you to calculate a tip quickly, it is smart to have a general idea of what the amount should be in case you inadvertently type in the wrong amount.

Suppose you want to tip 20%. The easiest way to quickly calculate this amount is to start with 10% of the bill and then double it. To find 10% of a number, you simply move the decimal point one place to the left.

For instance, consider the following calculations.

$$10\% \text{ of } \$25.00 = \$2.50$$

$$10\% \text{ of } \$349.00 = \$34.90$$

$$10\% \text{ of } \$1275.00 = \$127.50$$

Once you know 10% of the total, you can find 20% by doubling that amount.

$$20\% \text{ of } \$25.00 = 2 \cdot (\$2.50) = \$5.00$$

$$20\% \text{ of } \$349.00 = 2 \cdot (\$34.90) = \$69.80$$

$$20\% \text{ of } \$1275.00 = 2 \cdot (\$127.50) = \$255.00$$

The last bill is quite large, so 20% seems rather much for a tip. Instead, you decide for this party to tip only 15%. No problem. Start with the 10% you found earlier. To find 15%, simply add on half of that amount (that is, 5%).

$$\begin{array}{c} (10\%) \quad (5\%) \\ 15\% \text{ of } \$1275.00 = \$127.50 + \dfrac{\$127.50}{2} \\ = \$127.50 + \$63.75 \\ = \$191.25 \end{array}$$

A similar idea works for 5% tips as well. Start with 10% of the amount and then halve it to find 5% of an amount quickly. You're now well equipped to find any multiple of 5: 10%, 15%, 20%, or even 25%.

Note that tips are normally calculated on the pretax portion of the bill only. Since tipping is a voluntary expression of thanks, it is perfectly acceptable to use rounding to make the calculations easier. Rounding up a tip would likely never be frowned upon by the recipient.

Example 4.2.1

Finding Tip Amounts

The following are suggested tipping percentages from the "Worldwide Tipping Guide." Find each suggested tip amount based on the pretax bill total without using a calculator.

Table 4.2.1: Suggested Tipping Percentages

Service	Suggested Tipping %	Pretax Bill Amount
a. Buffet Wait Service	10%	$16.87
b. Hair Salon	15%	$43.00
c. Tattoo Artist	20%	$125.00

Solution

a. Buffet Wait Service

To find a 10% tip on the service, simply move the decimal one place to the left.

$$10\% \text{ of } \$16.87 = \$1.687 \approx \$1.69$$

b. Hair Salon

To find a 15% tip on the service, begin by finding 10% in the same way as in part a.

$$10\% \text{ of } \$43.00 = \$4.300 = \$4.30$$

Then add on half of the 10%.

$$15\% \text{ of } \$43.00 = \$4.30 + \$2.15 = \$6.45$$

c. Tattoo Artist

To find 20% tip on the service, begin by finding 10%, and then double it.

$$10\% \text{ of } \$125.00 = \$12.50$$

$$20\% \text{ of } \$125.00 = \$12.50 \cdot 2 = \$25.00$$

✔ **Skill Check 4.2.1**

Suppose your lunch bill is $9.84. Approximately how much would a 15% tip be? 20%?

OBJECTIVE 2 Sale Prices

Let's turn our attention to percentages concerning consumer purchases—taxes, sale prices, and discounts. When determining taxes or discounts on items, we often refer to the *list price* of the item. This is the amount at which an item is listed for public sale. The *discount* is the amount of money the price is reduced by and the *sale price*, or *net price*, is the cost of the item after the discount.

📖 LIST PRICE, DISCOUNT, AND SALE PRICE

List Price

The price of an item as it is listed for public sale is the **list price**.

Discount

A **discount** is a reduction made from the list price. This is often given as a percentage of the list price.

Sale Price

The **sale price**, also called the **net price**, is the actual cost of an item after any discounts are subtracted.

Sale Price = List Price − Discount

Practically speaking, we all love to see a "sale" sign in the store. Sometimes the reduced prices are indicated on the price tag, but often we are left calculating the discount on our own in order to determine the sale price. This is where estimation becomes useful and convenient. Since sale discounts usually occur in increments of 5% or 10%, we like to think of it in the same manner as the tipping calculation —just subtracting the amount instead of adding as the final step.

For example, suppose you found the perfect sofa for your new apartment. The store is having its annual End of Season Sale and is offering 40% off on its entire stock. Assuming you are in the store without a calculator, estimate the discounted price of your dream sofa if the list price is $496.00.

Remember that finding 10% of an amount allows us to easily find discounts that are a multiple of 10%. We start just as we did with tipping and find 10% of the list price by moving the decimal point one place to the left.

Finding 10% of the List Price: 10% of $496.00 = $49.60

Since this is an in-store estimate for yourself, we can simplify the calculation by rounding this discount to $50. If 10% is approximately $50 off, then 40% would equate to four $50 discounts, giving you $200 off! Your dream sofa now only costs approximately $296.00—a bargain you can't afford to miss. Another way to estimate the discount is to first round the cost of the sofa from $496.00 to $500.00. Then, 10% of $500.00 would be $50.00. A discount of 40% would then equate to four $50.00 discounts, or $200.00 off. Either way you calculate your estimate, the discount is approximately $200.

Suppose the sales assistant offers you an additional 10% off if you pay cash in the store that day. What would your final cost be then?

What you need to keep in mind here is that the additional discount is 10% off the sale price, not 10% off the list price. Let's calculate 10% of the estimated sale price by moving the decimal point.

Finding 10% of the Sale Price: 10% of $296.00 = $29.60, or almost $30

Alternatively, we can round $296.00 to $300, and then calculate 10% to estimate the discount.

Rounding the Sale Price First and Then Finding 10%: 10% of $300 = $30

So the additional discount means that the price of your dream sofa would be calculated as follows if you pay cash.

$$\$296.00 - \$30 = \$266.00, \text{ or about}$$

$$\$300.00 - \$30 = \$270.00$$

Be careful not to be misled into thinking that 40% off the list price, plus an additional 10% off is the same as receiving 50% off the list price. If the sofa was 50% off to begin with, you could just halve the list price to find the sale price.

$$50\% \text{ Off the List Price: } \frac{\$496.00}{2} = \$248.00$$

As we've just seen, 40% off plus an additional 10% off results in a sale price of approximately $266.00, which is more than 50% of the list price. It is important to remember that in this situation you cannot just add the percentages together since they do not apply to the same quantity.

Example 4.2.2

Calculating Sale Price

A clothing store is having a 50% off sale and then offered you an additional 50% off any one sale item. Suppose you are interested in an item that has a listed price of $42.99. Show that the additional 50% off does not mean that you get that item for free.

Solution

This scenario points to a common error with percentages. In order to add percentages together, they must apply to the same quantity. In this case, the first 50% is taken off the list price, but the second 50% is taken off the already reduced price. So we cannot add them together. The calculation requires two steps.

Begin by calculating the first discount. Convert the percentage to a decimal and multiply by the list price. We will first round the cost of the item from $42.99 to $43.00.

$$\$43.00 \cdot 0.50 = \$21.50$$

Because the percentage is exactly 50%, there is no need to subtract the discount from the list price.

Now we apply the additional 50% discount in the same manner. Convert the percentage to a decimal and multiply by the discounted price.

$$\$21.50 \cdot 0.50 = \$10.75$$

Although the item is not free, it will cost only $10.75—a considerable reduction in price!

✔ **Skill Check 4.2.2**

Find the sale price of a new sweater that is 40% off the list price of $88.00.

OBJECTIVE 3 Sales Tax

Sales tax is a charge collected by the government for the sale of certain goods. It is a percentage of the price of the item being sold. The amount of the sales tax, as well as what goods are taxed, varies by state as well as country. Unlike in Europe, the sales tax on goods in the United States is not typically included in the advertised sticker price. Instead, it is added at checkout. In 2020, state-enacted sales tax rates ranged from 2.9% in Colorado to 7% in several states.

📖 SALES TAX

A charge collected by the government for the sale of goods is called a **sales tax**.

Sales Tax = (Selling Price) · (Sales Tax Percentage as a Decimal)

One way to determine the final cost of an item including tax is to calculate the tax amount and then add it onto the selling price.

$$\begin{array}{r} \text{Selling Price} \\ + \qquad \text{Tax} \\ \hline \text{Final Cost} \end{array}$$

A second method is to combine the two-step process into one step by using a **multiplier**, or a number you can multiply by that represents the increase or decrease. For instance, a 3% tax on an item means you will actually pay 103% for that item. This is the same as multiplying the price by 1.03. To find a multiplier, first add the percentage to 100% for an increase (or subtract the percentage from 100% for a decrease), and then convert it to a decimal.

$$\text{Sales Tax Multiplier: } \frac{100\% + \text{Sales Tax}\%}{100\%}$$

For instance, if a sales tax rate is 4.2%, the multiplier is determined as follows.

$$\frac{100\% + 4.2\%}{100\%} = \frac{104.2\%}{100}$$
$$= 1.042\%$$

Both methods are shown using an item that costs $25.00 and a sales tax rate of 4.2%; notice that both arrive at a total cost of $26.05.

Calculating Sales Tax and Then Adding It On:

($25.00)(0.042) = $1.05

$25.00 + $1.05 = $26.05

Using the Multiplier Method:

($25.00)(1.042) = $26.05

Example 4.2.3

Calculating Cost with Sales Tax

A car is priced at $10,988 and the state sales tax rate is 7.2%. The county also collects an additional tax at a rate of 1.6%.

a. Determine the total price of the car including both taxes using the first method of calculating the sales tax and then adding it onto the list price.

b. Use a multiplier to find the cost of the car with both taxes included.

c. The dealership offers to either pay your sales taxes or give you a 10% discount on the list price of the car. Which is the better deal for you?

Solution

a. To find the total price of the car we will calculate the sales tax for each of the tax rates and then add the taxes to the list price.

$$\text{Tax at 7.2\%: } \$10,988 \cdot 0.072 \approx \$791.14$$

$$\text{Tax at 1.6\%: } \$10,988 \cdot 0.016 \approx \$175.81$$

$$\text{Price of the Car with Taxes: } \$10,988 + \$791.14 + \$175.81 = \$11,954.95$$

b. Because both tax percentages are being applied to the same quantity (i.e., the list price of the car), we can find the multiplier by adding both of the sales tax percentages to 100% and then converting to a decimal.

$$\frac{100\% + 7.2\% + 1.6\%}{100\%} = \frac{108.8\%}{100\%} = 1.088$$

Our multiplier is 1.088.

Multiplying by the list price determines the total cost including both sales taxes.

$$\text{Price of the Car with Taxes: } \$10,988 \cdot \$1.088 \approx \$11,954.94$$

c. First, if the dealership pays the taxes owed on the car, you end up buying the car for the list price only, which is $10,988.

The second option is a 10% discount. We will calculate the discounted price by using a multiplier, which is $\dfrac{100\% - 10\%}{100\%} = 0.90$.

$$\text{Discounted Price: } \$10,988 \cdot 0.90 = \$9,889.20$$

We now need to calculate the cost with taxes on this discounted price using the multiplier we calculated in part b.

$$\text{Discounted Price with Taxes: } \$9,889.20 \cdot 1.088 \approx \$10,759.45$$

Therefore, the 10% discount is the better deal for the car.

> ✔ **Skill Check 4.2.3**
>
> Find the total price of a new 32″ TV that sells for $219.99 if purchased in New York, where the sales tax is 4%, and if it is purchased in South Carolina, where the sales tax is 6%.

OBJECTIVE 4 Percentage Increase and Percentage Decrease

The three topics we just covered (tipping, sales prices, and sales tax) are all examples of *percentage increases* or *percentage decreases*. In other words, we actually changed the original amount by either adding or subtracting a certain percentage of that amount. In each case, we were told by what percentage we should either increase or decrease the amount. However, sometimes we might be given two amounts, a list price and a sale price, and want to calculate the percentage change. You can use the following formulas to find any percentage increase or decrease.

ABSOLUTE CHANGE AND PERCENTAGE CHANGE

Absolute Change

The **absolute change** between two amounts is the absolute value of the difference between the two numbers.

$$\text{Absolute change} = |\text{New Amount} - \text{Original Amount}|$$

Percentage Change

The **percentage change** between two amounts is the absolute change expressed as a percentage of the original value.

$$\text{Percentage Change} = \frac{\text{Absolute Change}}{\text{Original Amount}} \cdot 100\%$$

> **Think Back**
>
> The absolute value of any number is the distance that number lies from zero without regard to sign. For example, the absolute value of 8 is 8, and the absolute value of –8 is also 8. To indicate absolute value symbolically, we use a vertical bar on both sides of the number; that is, $|8| = 8$ and $|-8| = 8$.

Example 4.2.4

Determining a Discount

Andrea bought a stereo on sale for $198.45. The list price of the stereo was $330.75. What percentage discount did Andrea receive?

Solution

Before we can determine the percentage of the discount, or the percentage change, we need to know the actual dollar amount of the discount, or the absolute change. Since absolute change is an absolute value, the order of the subtraction does not make a difference.

$$\text{Absolute Change} = |\text{New Amount} - \text{Original Amount}|$$
$$= |\$198.45 - \$330.75|$$
$$= \$132.30$$

To determine what percentage change this is, divide the absolute change (or discount amount) by the list price, or original amount, as follows.

$$\text{Percentage Change} = \frac{\text{Absolute Change}}{\text{Original Amount}} \cdot 100\%$$
$$= \frac{\$132.30}{\$330.75} \cdot 100\%$$
$$= 40\%$$

So the list price of the stereo was discounted by 40%.

> **Skill Check 4.2.4**
>
> Max bought a computer on sale for $432. The list price of the computer was $1200. What percentage discount did Max receive?

Example 4.2.5

Computing Percentage Decrease

During the pandemic of 2020, the stock market recorded some of its biggest losses since the Great Depression. Shares for one Ohio-based bank hit a price of $14.50 in the middle of July. If the shares started the year off at a price of $19.33, find the overall percentage decrease from January to July. Round your answer to the nearest whole percent.

Solution

Just as in the previous example, to find the percentage decrease, or percentage change, we must calculate the absolute change and then divide that by the original amount. We can write this as one step instead of two as follows.

$$\text{Percentage Change} = \frac{\left|\text{New Amount} - \text{Original Amount}\right|}{\left|\text{Original Amount}\right|} \cdot 100\%$$

$$= \frac{\left|\$14.50 - \$19.33\right|}{\left|\$19.33\right|} \cdot 100\%$$

$$\approx 25\%$$

This indicates that the percentage decrease of the bank's share price was approximately 25% over the time period.

Example 4.2.6

Finding Percentage Change

A student group is trying to increase its monthly attendance.

a. If after a big recruiting push, the meeting attendance increased from 25 to 38 people, what was the percentage change in attendance?

b. During the third meeting, only 31 people attended. What was the percentage change from the second meeting to the third? Round your answer to the nearest whole percent.

c. For advertising purposes, the organization decides to promote the overall increase in attendance for the two-month span since it has increased from the original number of attendees. Calculate the percentage change from Meeting 1 to Meeting 3.

Solution

a. To find the percentage change, we need to know the absolute change in attendance between the first meeting and the second meeting. Use subtraction to find this value.

$$|38 - 25| = 13 \text{ more people came to the second meeting}$$

Next, substitute the appropriate values into the equation to find the percentage change.

$$\frac{\text{Absolute Change}}{\text{Original Amount}} \cdot 100\% = \frac{13}{25} \cdot 100\% = 52\%$$

Since the attendance had an increase from the original amount, we say there was a 52% increase in attendance.

b. Now, if at the next meeting only 31 people attend, what is the percentage change from Meeting 2 to Meeting 3?

This time, the absolute change in attendance is as follows.

$$\text{Absolute Change} = |\text{Meeting 3} - \text{Meeting 2}| = |31 - 38| = 7$$

Substituting this into our formula, we calculate the percentage change as follows.

$$\frac{7}{38} \cdot 100\% \approx 18.42\%.$$

Since this was a decline in attendance, we say there was approximately an 18.42% decrease from Meeting 2 to Meeting 3.

c. Recall that, in Meeting 1, there were 25 people in attendance and in Meeting 3, there were 31 people. We need to subtract these numbers to find the absolute change in attendance.

$$\text{Absolute Change} = |\text{Meeting 3} - \text{Meeting 1}| = |31 - 25| = 6$$

So our percentage change is as follows.

$$\frac{6}{25} \cdot 100\% = 24\%$$

Since this was an increase in attendance from Meeting 1 to Meeting 3, we say there was a 24% increase overall.

Once again, let's note something that cannot be done in the situation from Example 4.2.6. It might be tempting to rationalize that since there was a 52% increase from the first meeting followed by an 18.42% decrease the second meeting, we could just subtract the two percentages to find the overall percentage change. However, we have just shown that not to be the case. (Check this for yourself.) Remember that you cannot manipulate percentages by adding or subtracting them unless they reference the same quantity. It is such a misinterpretation of percentage change that we are not even going to write down the subtraction, lest someone should mistakenly see it written as an accepted form of mathematics!

OBJECTIVE 5 Percentage Faux Pas

Unintentionally or not, very often percentages are used incorrectly and lead people down the wrong path. You should be careful to avoid making these mistakes yourself and circumvent being fooled by abuses of percentages. The following is a look at some of the more common misuses of percentages.

Avoid Comparing Percentages of Groups of Very Different Sizes

When using percentages to compare things, it is easy to make errors if we are not careful and we only look at the percentages while ignoring the portion they represent, as we saw in Section 4.1 with voting percentages in the United States from 1876 and 1924. Remember, when percentages are involved, ask yourself "out of how many?"

Example 4.2.7

Avoiding Percentage Misuse—Different-Sized Populations

A recent headline claimed that 75% of people living in Alaska favor free skis for schoolchildren while only 40% of Californians do. If the population of Alaska is around 736,000 and the population of California is roughly 39,000,000, is it fair to say that the number of Alaskans who are in favor of the free skis is almost twice that of the number of Californians?

Solution

Although the percentage of Alaskans (75%) who favor free skis is almost double that of the percentage of Californians (40%), we need to look at the sizes of the populations to determine if the number of people in favor is actually double.

Alaska's population is around 736,000, while California's population is roughly 39,000,000. Given that information, we can find the actual number of people in favor of free skis in each state.

$$\text{Alaska: } 75\% \text{ of } 736{,}000 = (0.75)(736{,}000) = 552{,}000$$

$$\text{California: } 40\% \text{ of } 39{,}000{,}000 = (0.40)(39{,}000{,}000) = 15{,}600{,}000$$

We can now see that 75% of Alaskans is a much, much smaller number than 40% of Californians. Comparing 40% to 75% when the populations are not even in the same ballpark is misleading, to say the least. In fact, just 2% of California's population is greater than the entire population of Alaska.

> **Helpful Hint**
>
> The ski example is another situation where we should avoid adding percentages together. If we add the percentages of people in both states who favor free skis it would appear that 115% of people favor the idea!

Comparing percentages of different groups when the sizes of the groups are very different can often result in misleading conclusions. However, when the groups are roughly the same size, these issues disappear. For instance, it would be reasonable to compare percentages from the populations of South Carolina (5,200,000 in 2020) and Alabama (4,900,000 in 2020) since their population sizes are more similar.

Avoid Averaging Percentages

Another no-no when manipulating percentages is trying to average them. In general, averaging percentages almost always yields incorrect answers. There are exceptions, but to avoid confusion or misuse, stay away from averaging percentages.

Example 4.2.8

Avoiding Percentage Misuse—Averaging Percentages

Consider our previous example about approval for free skis to schoolchildren in Alaska and California. Calculate the percentage of people in both states who favor the idea. (Round your answers to the nearest tenth of a percent, if necessary.) Then, compare that to the incorrect method of adding the two percentages together and dividing by 2. We've listed the statistics once again for your reference.

- Alaska has a population of 736,000 and 75%, or 552,000 people, favor free skis for schoolchildren.

- California has a population of 39,000,000 and 40%, or 15,600,000 people, favor free skis for schoolchildren.

Solution

To calculate the percentage of people who are in favor of the idea from both states, add the number of votes in favor in the numerator and add the total populations in the denominator.

Correct Method to Determine Percentage of Votes in Favor:

$$\frac{\text{AK Population in Favor} + \text{CA Population in Favor}}{\text{AK Population} + \text{CA Population}}$$

The combined percentage of people in both states in favor of free skis is calculated as shown.

$$\frac{552,000+15,600,000}{736,000+39,000,000} = \frac{16,152,000}{39,736,000} \approx 0.40648 \approx 0.406 \text{ or } 40.6\%$$

The following is an *incorrect* method to find the average of the percentages.

Incorrect Averaging of Percentages: $\frac{75\%+40\%}{2} = \frac{115\%}{2} \approx 57.5\%$

Because the sizes of the populations are not the same, we cannot simply average the two percentages as shown.

> **✔ Skill Check 4.2.5**
>
> A survey of 50 adults aged 21–30 indicated that 62% preferred the beach over the mountains. An additional 500 adults aged 31–40 were surveyed and 42% said they also prefer the beach over the mountains. What is the combined percentage of 21- to 40-year-olds surveyed who prefer the beach over the mountains?

Avoid Shifting Reference Points

Recall from our discussion in Section 4.1 that we can use percentages, as well as proportions and ratios, to communicate how quantities are distributed within a group. If we're not careful though, we can unintentionally paint a misleading picture even when using factually accurate numbers. Consider the following example.

> **Example 4.2.9**

Avoiding Percentage Misuse—Shifting Reference Points

Campus police are proud to report that car break-ins on campus decreased by 25% this year compared to only a 20% decline the previous year. They promoted the fact that their law enforcement efforts reduced the number of break-ins more this year than in the previous year. Use Table 4.2.2 to determine if their claim is justified.

Table 4.2.2: Campus Car Break-In Report

Report Year	# of Car Break-Ins	Year-to-Year Percentage Change
2020	10	
2021	8	20% decrease
2022	6	25% decrease

Solution

In the last column of Table 4.2.2, we can see that, just as the police department claims, car break ins are decreasing by a larger percentage each year—20% the first year, 25% the second year. However, notice that the absolute reduction in the number of car break-ins remains exactly 2 fewer each year. In 2021, a reduction of 2 break-ins equates to a 20% decrease, but in 2022 a reduction of 2 break-ins is a 25% decrease. Even though the number of break-ins reduces by the same amount from year to year, the absolute difference of 2 becomes a larger percentage of the shrinking overall number of break-ins. In other words, the reference point changes each year, causing the percentage decrease from year to year to become larger. You'll need to decide for yourself the validity of the department's claims since they are correct in saying that break-ins decreased by a larger *percentage* in 2022, although the *number* of break-ins decreased by the same amount for both years

As Example 4.2.9 illustrates, it is always wise to have an eye on the underlying numbers before comparing percentages so that you do not fall victim to shifting reference points. We like to assume that these types of percentage faux pas examples are done unintentionally in society. But one does wonder if the cynical phrase "you can make numbers say anything you want" originates from exactly these types of occurrences.

Skill Check Answers
1. $1.50, $2.00 **2.** $52.80 **3.** $228.79, $233.19 **4.** 64% **5.** 43.8%

4.2   Exercises

✔ CONCEPT CHECK

1. The price of an item as it is listed for public sale is the _____.

2. The _____ is the actual cost of an item after any discounts are subtracted.

3. One way to apply a 4% tax to an amount is to multiply the amount by _____.

4. The _____ between two amounts is the absolute value of the difference between the two numbers.

5. The percentage change between two amounts is equal to the absolute change divided by the _____.

◉ PRACTICE

Find the absolute change and the percentage change for each situation.

6. 270 is increased to 1134

7. 14 is increased to 70

8. 150 is decreased to 39

9. 475 is decreased to 152

10. 39 is increased to 69.42

11. 62 is decreased to 42.78

🚀 APPLICATIONS

Solve. Round your answer to the nearest hundredth, if necessary.

12. After taking an Uber where the bill came to $27.50, you decide to tip the driver 10%. How much will the tip be?

13. You and a group of 5 friends dined at a restaurant. When bringing the bill, the server tells you that a 15% tip is included automatically for groups of 6 or more. If your pretax bill was $112.40, how much was the tip?

14. Sofia's bill at the hair salon was $85. If she wants to tip 20%, how much tip will she leave?

15. Jaylon and his friend had a pizza delivered, and the bill was $24.60. Jaylon says they should tip the driver 15%; his friend says 10% is sufficient. How much more would the driver receive if they tip 15%?

16. Kyle is getting a tattoo tomorrow and the artist told him the total will be $225. If Kyle decides to tip 20%, what amount will he tip?

17. Cheryl ordered grocery delivery after her hip surgery. If her grocery bill was $132.70 and she wants to tip the delivery driver 10%, how much will the tip be?

18. Your dinner bill was $18.00. If you leave a 20% tip, how much will the tip be?

19. If you wish to leave a 15% tip and your bill is $9.50, how much tip will you leave?

20. A furniture store advertises 30% off all couches for their grand opening sale. What would be the sale price of a couch originally priced at $849.00?

21. A local coffee shop offers a 5% discount if you bring your own reusable mug. What is the discounted price of a cup of coffee that is normally $2.80?

22. While shopping for a used car, Cait receives competing offers from two dealerships. Dealer A offers a 10% discount on a $11,900 car; Dealer B offers a 15% discount on a $12,400 car. After the discount, which dealer's car costs less?

23. An online company is advertising a Presto Pizza Oven on sale for 45% off the original price of $69.99. What is the sale price for the oven?

24. A furniture store is having an annual sale where everything is 20% off and is offering an additional 15% off to anyone who finances through a store credit card. If the couch you are interested in originally costs $950, how much would you need to finance if you used the store credit card?

25. A power washing company has a package to power wash your house, patio, and driveway for $780. The company placed an ad in the local newspaper with a 15% off coupon and will take an additional 5% off if you place a sign advertising the company in your front lawn for a week. How much would you pay for the power washing service if you use the coupon and place the sign in your front lawn?

26. A catering company charges $1200 to cater a small company lunch. The catering company offers at 10% discount if the lunch is scheduled more than six weeks ahead of time and an additional 10% discount if the payment is paid in full at the time of scheduling. How much would the catered lunch cost if it was scheduled 8 weeks ahead of time and paid in full at the time of booking?

27. At a local optometrist, the regular price of a new pair of prescription sunglasses, including lenses and frames, is $380. The optometrist is offering a 20% discount on sunglasses to new patients. If a new patient has vision insurance that covers 20% of the cost of eyewear after any discounts, how much will the patient pay for a pair of prescription sunglasses?

28. An artist is selling limited edition prints of select paintings from their latest gallery show. If you purchase a set of three prints, you can save 10% on the total price. People who purchased prints from the previous print release receive a discount code for an additional 15% off during a presale. If you plan to purchase a set of three prints for a total of $350, how much would you pay if you made the purchase during the presale?

29. A shirt is on sale for 40% off, and you have an additional 20% off coupon. True or false: The shirt will ultimately be 60% off the original price?

30. Get Lucky is having a clearance sale of their winter items. All sale items are marked an additional 10% off the sale price. If a coat that originally costs $110 is on sale for 25% off, what is the new sale price? Is this a reduction of 35% off the original price?

31. A ticket to see Pink in concert costs $117.80 at regular price. An online ticket company is offering tickets at a 25% discount. Your friend bought a ticket at the discounted price but found out he can't go. He's offering the ticket to you at an additional 10% off what he paid. What price could you get the concert ticket for if you bought it from your friend?

32. The wholesale cost of a new sofa is $240. The price of the sofa is marked up 40% before putting it on the showroom floor. During a showroom sale, the sofa was marked down 40%. What is the price of the sofa now?

33. The stock price for KTD was $54.56 in January 2014. In February the stock had increased by 6% but in March the price fell by 6%. What was the price of KTD stock in March 2014?

34. Liza stays overnight in a hotel that charges 6% sales tax plus a 2% lodging tax in addition to the price of the room. If the room price was $109, what was her total bill?

35. A local restaurant is required to add a 5.5% sales tax plus a 1.5% hospitality tax to each patron's bill. Ignoring tip, if a customer's pre-tax bill is $32.62, what would be the total bill?

36. A new wireless printer costs $99.99 in the store. What would your total cost be if sales tax is 9.5%?

37. You purchase a pair of jeans with a list price of $175. If the sales tax is 8.5%, what is the total cost of the jeans? Round your answer to the nearest cent.

38. Jamie found a receipt for an MP3 player for $127.59, tax included. If the sales tax rate was 9%, what was the selling price of the MP3 player? Round your answer to the nearest cent.

39. The local sales tax is 8%. If a pair of shoes sells for $115, what is the total cost after tax?

40. Your total purchase at a department store is $173.56. The local sales tax is 8.75%. How much will you pay if you have a 30% store coupon?

41. A local store is having a sale. All waterproof shoes are 40% off. If a pair of rain boots normally costs $13.72 plus 7% sales tax, how much will it cost to buy the rain boots during the sale? (**Hint:** Discounts are applied before sales tax is added.)

42. A coffee and a Danish from a local café cost $6.15 plus a 6.25% tax. How much would a person spend on a breakfast of coffee and Danishes if they purchased this meal twice a week for a year (52 weeks)?

43. Two movie tickets, a medium popcorn, and two drinks costs $40.50 plus a 7.75% sales tax. If a couple attends a movie once a month for a year, how much would they spend on tickets and concessions?

44. To complete an outfit for an upcoming event, you purchase shoes online for $119.95 and decide to also buy a pack of socks for $15. The sales tax applied to the full purchase price is 6.2%. What is the total purchase price of the shoes and socks?

45. Four pavilion seating tickets to see the Cleveland Orchestra during a summer concert costs a total of $343. During the check out process, you are asked whether you'd like to make a donation to the orchestra and decide to donate $50. If the sales tax is 6.75% and the donation is not taxed, what is the total price for the four tickets and the donation?

46. In Louisiana, a pack of cigarettes costs about $4.82 plus 4% sales tax. How much would a person spend on cigarettes in a year in Louisiana if he smoked one pack of cigarettes on each of the 365 days of the year?

47. You owe $257.43 on your credit card. Because you have not paid your bill on time, you have to add a past due charge of $25, an over limit fee of $40, and a service charge of 9% of the amount owed on the credit card (without the past due charge and over limit fees). If you wish to send a check for the full amount immediately, how much should you write the check for?

Solve each problem. Round your answer to the nearest hundredth, if necessary. (**Hint:** Commission is a percent of the house's selling price that is added to the realtor's income.)

48. You are selling your house with a realtor who earns a 7% commission fee. If your house sells for $185,000, how much commission will the realtor earn?

49. If you are selling your house with a local realtor who requires a 6% commission fee, what can you expect to pay the realtor if your house sells for $139,000?

50. Suppose you sell your house for $215,000 but still owe $175,000 on the mortgage. If your real estate agent gets a 6% commission, what is the amount of money you'll make on this sale?

51. You are about to list your house as For Sale By Owner. Suppose you need to allow for a 3% commission for the buyer's realtor, and you need to receive a minimum of $92,500 on the sale. What is the minimum price you can list the house at to accommodate both of these requirements? Round to the nearest hundred dollars.

52. If a laptop computer regularly costs $879 and is on sale for $599, what percentage discount is being given?

53. A poll in Russia looked at the support of Vladimir Putin's United Russia party. At that time, United Russia held 315 of the 450 seats in the State Duma, but the poll predicted this would fall to 252. If this was the case when the elections happened, calculate the percentage change in the number of seats held by the United Russia party.

54. Suppose that a local newspaper reported that Cityville and Beautytown both had 50 murders last year. However, 5 years ago, Cityville had 42 murders while Beautytown had just 29. Calculate the absolute change and the percentage change for each city over the five years. Could the newspaper make the claim that crime had gone up by the same amounts in each city? Explain.

55. The United States population was estimated to be 312,681,446 in 2011, estimated to be 285,102,075 in 2001, and estimated to be 77,584,000 in 1901. Calculate the absolute change and percentage change between each of the figures given (that is, 1901 to 2001, 1901 to 2011, and 2001 to 2011).

56. In 2020, a total of 23.6 million people tuned in to watch the Oscars. In 2021, the viewership of the rewards show fell by 56%. How many people watched the Oscars in 2021? Round your answer to the nearest tenth of a million.

57. For the 1998–1999 academic year, the average cost of attending a public in-state college was $12,000. By the 2018–2019 academic year, the average cost of attending a public in-state college increased by approximately 78%. What was the average cost of attending a public in-state college during the 2018–2019 academic year?

58. If the Chevron Corp stock increased 2.44% from $102.73, what was the absolute change in stock price? Round your answer to the nearest cent.

59. Over a week, the average price of gas decreased by 18% from $2.65 per gallon. What was the absolute change in gas price? Round your answer to the nearest cent.

60. Two companies decreased overhead costs by 12% during the last year. Suppose Company A originally had an overhead cost of $85,000 and Company B originally had an overhead cost of $1,250,000. Do you believe the decrease in overhead costs between the two companies was the same? Justify your answer.

61. A news outlet reported that fifty million Americans visited online retail sites on Black Friday, representing an increase of 35 percent versus a year ago. How many Americans visited online retail sites on Black Friday last year?

62. If Apple stock fell 0.93% from $363.57 per share, what was the absolute change in stock price?

63. Two companies report a 15% gain in online sales over the last year. Suppose Company Q had 25 online sales and Company Z had 150 online sales last year. Do you believe their growth was the same? Justify your answer.

✎ WRITING & THINKING

64. You decide to interview 9th graders from two different schools on whether they agreed with school dress codes or not. Your interview revealed that 18% of the 600 9th graders from Riverside High agree with the school dress code, and 30% of the 50 9th graders from Lakeview High agree. Is it fair to say that the number of 9th graders from Lakeview High who agree with the dress code is almost twice that of the number of 9th graders from Riverside High? Explain your reasoning.

4.2 PROJECT

COUPONS, DISCOUNTS, AND SALES TAX

Many retail stores offer shoppers incentives in the form of coupons. Sometimes these coupons are advertised as having the same value as cash, with the disclaimer that they have to be used before taxes. In this activity we will explore the use of such store-cash promotions.

Suppose you have received a coupon for your favorite department store Tohl's. The coupon states that you have 10 dollars' worth of Tohl's bucks that can be used when you spend at least 50 dollars in the store.

You decide to go to Tohl's and purchase four items: two T-shirts, a pair of socks, and a pair of jeans. The prices for these items are recorded in the following table.

Math Rocks T-shirt	$13.99
Nickelback T-shirt	$12.99
Pokémon Socks	$8.99
Levi's 550 Jeans	$27.99
Total	$63.96

1. If you do not use your 10-dollar Tohl's bucks coupon, what would be your total, considering that the county where the store is located has a 6.75% sales tax? Round your answer to the nearest cent.

2. Now assume that there is a sale going on and everything in the store is 15% off. What would be your total without using the $10 of Tohl's bucks and with the same 6.75% sales tax? Round your answer to the nearest cent.

3. Now, let's investigate what happens if you use the Tohl's bucks coupon. There are two options: the coupon is applied after the 15% discount or the coupon is applied before the 15% discount. Fill in the following table to compare the amount due with the two options. Round each calculation to the nearest cent, if necessary.

Option 1		Option 2	
Total	$63.96	Total	$63.96
15% Discount		Tohl's Bucks	−$10.00
Tohl's Bucks	−$10.00	15% Discount	
Total before Tax		Total before Tax	
Sales Tax		Sales Tax	
Amount Due		Amount Due	

Which option do you believe is the one actually used by stores? Explain your reasoning.

4.3 RATES, UNIT RATES, AND RATES OF CHANGE

OBJECTIVES

1. Identify and use rates.

2. Calculate and compare unit rates.

3. Calculate rates of change.

Helpful Hint

Notice that 8 cents per dollar is $\dfrac{8}{100}$, which is 8%.

OBJECTIVE 1 Rates

In an earlier section, we looked at how we can use proportions, percentages, and ratios to communicate the relationship among quantities and how they compare to one another. Now, we turn our attention to another method used to describe how quantities relate to each other—rates. A *rate* is a fractional comparison between two quantities that are not necessarily in the same units. For instance, the following examples are all rates.

Fuel Consumption:	$\dfrac{23 \text{ miles}}{1 \text{ gallon}}$	read "23 miles per gallon"
Price of Rice:	$\dfrac{\$7.99}{5 \text{ pounds}}$	read "$7.99 per 5 pounds"
Heart Rate:	$\dfrac{73 \text{ beats}}{1 \text{ minute}}$	read "73 beats per minute"
Sales Tax Rate:	$\dfrac{8 \text{ cents}}{1 \text{ dollar}}$	read "8 cents per dollar"

Notice that in each of these rates, the differing units are written as part of the rate for clarity. The word "per" is used to denote the division bar when reading rates aloud.

📖 RATE

A **rate** is a fraction used to compare two quantities that are not necessarily in the same units.

The last rate mentioned above is one very familiar to us all, a sales tax rate. You've more commonly seen a sales tax rate written not as a fraction but as a percentage. When the two quantities in a rate are in the same unit of measurement, we can write it as a percentage just as we did in Section 4.2. The sales tax rate written above is 8 cents out of 100 cents, or 8%.

$$\frac{8 \text{ cents}}{1 \text{ dollar}} = \frac{8 \text{ cents}}{100 \text{ cents}} = 8\%$$

Consider the other rates: fuel consumption, price of rice, and heart rate. These rates are in differing units of measurement, and hence it would not make sense to talk about them as percentages.

Example 4.3.1

Writing Rates

The click-through rate (CTR) in internet marketing is the rate at which ads are being viewed by potential customers. The rate is calculated by dividing the total number of clicks on the ad by the total number of people who view the ad (impressions). Table 4.3.1 gives the number of impressions and clicks for three different versions of an ad for a start-up company. Write the click-through rate for each version and then change it to a percentage by dividing.

Table 4.3.1: Impressions vs. Clicks

	Impressions (Views)	Clicks
Ad A	34,271	592
Ad B	36,859	471
Ad C	31,022	519

Solution

We are told that the click-through rate is the number of clicks divided by the number of views: $\text{CTR} = \dfrac{\text{clicks}}{\text{views}}$. Once found, divide the numerator by the denominator and multiply by 100% to get the rate as a percentage.

Table 4.3.2: Click-Through Rates

	Impressions (Views)	Clicks	CTR	CTR as a Percentage
Ad A	34,271	592	$\dfrac{592}{34,271}$	1.73%
Ad B	36,859	471	$\dfrac{471}{36,859}$	1.28%
Ad C	31,022	519	$\dfrac{519}{31,022}$	1.67%

While defining what makes a good click-through rate is still a highly debated topic, it is generally considered that a 2% rate is successful.

Example 4.3.2

Identifying Rates

BuyinBulk offers 30 pounds of candy for $64.49 on its website. Which of the following fractions represents the rate given by their website?

a. $\dfrac{1}{\$64.49}$

b. $\dfrac{30\,\text{lb}}{\$64.49}$

c. $\dfrac{30\,\text{lb}}{1}$ **d.** $\dfrac{\$64.49}{30\,\text{lb}}$

Solution

There are actually two correct ways to write the rate from the website: answers b. and d.

Choice b. provides pounds per dollars.

$$\frac{30\,\text{lb}}{\$64.49}$$

Choice d. provides price per 30 pounds.

$$\frac{\$64.49}{30\,\text{lb}}$$

Both a. and c. are incorrect because neither fraction compares the amount of candy to the price of the candy.

Example 4.3.3

Using Rates

A patient in a hospital is prescribed IV fluids with a rate of 5 grams (g) of sugar per 100 milliliters (mL) of water.

a. If the patient is prescribed 1500 mL of fluids, how many grams of sugar will be delivered?

b. If the patient receives the fluids 4 times a day at the dosage in part a., how much sugar is delivered per day from the fluids?

c. If each gram of sugar contains 4 calories, how many calories are being delivered each day from the fluids the patient receives at the dosage in part b.?

Solution

a. Begin by writing down the rate of sugar delivered in the IV fluids.

$$\text{Sugar to Water Rate of IV Solution} = \frac{5\,\text{g}}{100\,\text{mL}}$$

Since we want to know the amount of sugar contained in 1500 mL of fluids, we need to scale up the rate so that the denominator will have the prescribed amount of fluids. We can do this by multiplying the numerator and denominator by 15.

Sugar to Water Rate of IV Solution with 1500 mL of Water

$$= \frac{5\,\text{g}\cdot 15}{100\,\text{mL}\cdot 15} = \frac{75\,\text{g}}{1500\,\text{mL}}$$

Therefore, 75 grams of sugar is delivered.

b. If the patient receives the solution 4 times a day, multiply the sugar content in one dose of fluids by 4 to find how much sugar is delivered per day.

$$\text{Sugar Content in 4 Doses} = 75\,\text{g}\cdot 4 = 300\,\text{g}$$

Thus, 300 grams of sugar are delivered per day.

c. The calories delivered in 4 dosages can be calculated as follows.

$$\text{Calories in 4 Doses} = 300 \text{ g} \cdot \frac{4 \text{ calories}}{1 \text{g}} = 1200 \text{ calories}$$

Therefore, 1200 calories are being delivered.

> **💬 Helpful Hint**
>
> Density is a rate of mass to volume and is represented by the Greek letter ρ, pronounced "rho."
>
> $$\rho = \frac{\text{Mass}}{\text{Volume}}$$

Example 4.3.4

Using Rates

Gold is relatively heavy when compared to other metals. It has a density of 19.3 g/cm³, whereas lead has a density of 11.4 g/cm³, and aluminum's density is 2.7 g/cm³. If a gold bar has a volume of 51.15 cm³, what is its mass?

Solution

Density is a rate of mass to volume, $\dfrac{\text{Mass}}{\text{Volume}}$, and we are given the density of gold.

$$\text{Density of Gold} = \frac{\text{Mass}}{\text{Volume}} = \frac{19.3 \text{ g}}{1 \text{ cm}^3}$$

Now, we need to multiply the numerator and denominator by the required volume.

$$\text{Density of a Gold Bar with Volume } 51.15 \text{ cm}^3 = \frac{19.3 \text{ g} \cdot 51.15}{1 \text{ cm}^3 \cdot 51.15} \approx \frac{987.20 \text{ g}}{51.15 \text{ cm}^3}$$

This tells us the mass of the gold bar is 987.20 grams.

> **💬 Helpful Hint**
>
> When written out, the word "per" indicates that the denominator amount of a rate is about to follow.

Let's consider another familiar rate example, miles per gallon (mpg). The "per gallon" phrase implies that the comparison is between miles traveled and one gallon of gasoline. For instance, 36 mpg for highway driving means that for every gallon of fuel, the car can be driven 36 miles on the highway. Traditionally, society has moved away from writing this rate as a fraction, such as $\dfrac{36 \text{ miles}}{1 \text{ gallon}}$, and has instead moved to using the mpg notation, such as 36 mpg.

Example 4.3.5

Working with Miles per Gallon

As a college graduation present, Katie's parents helped her buy her first new vehicle. She chose one that claims a fuel efficiency of 29 miles per gallon (mpg) on the highway.

a. Check how close the fuel consumption on Katie's new car is to the manufacturer's claim if she drove 309 miles on 11 gallons of gas. Round your answer to the nearest hundredth.

b. What might explain any discrepancies between the manufacturer's rate and Katie's car performance? Should she return the car to the dealer?

Solution

a. Begin by setting up the rate of miles per gallons, $\dfrac{\text{miles}}{\text{gallons}}$. Katie drove 309 miles on 11 gallons of gas.

Notice that the number of miles will be in the numerator and the number of gallons will be in the denominator of the rate.

$$\text{Katie's Recorded Rate of Fuel Consumption} = \frac{309 \text{ miles}}{11 \text{ gallons}}$$

To compare Katie's fuel consumption rate to that of the manufacturer's claim, we need to divide the numerator and denominator by 11, so that the new denominator is 1 gallon.

$$\text{Katie's Recorded Rate of Fuel Consumption} = \frac{309 \text{ miles} \div 11}{11 \text{ gallons} \div 11} \approx \frac{28.09 \text{ miles}}{1 \text{ gallon}}$$
$$= 28.09 \text{ mpg}$$

Therefore, Katie registered approximately 1 mile per gallon less than was claimed.

b. One possible explanation is that Katie did not drive all 309 miles on the highway. We are not told the specifics of her journey.

It also might be a case of accuracy on Katie's end. Suppose she actually used 10.6 gallons to drive for 309.2 miles. Then the mpg is $\dfrac{309.2 \text{ miles} \div 10.6}{10.6 \text{ gallons} \div 10.6} \approx \dfrac{29.17 \text{ miles}}{1 \text{ gallon}} = 29.17 \text{ mpg}$.

This does not seem to warrant Katie returning the new car.

> ✔ **Skill Check 4.3.1**
>
> If rice costs $7.99 for a 5-pound bag, determine the cost per pound of the rice. Round your answer to the nearest cent.

OBJECTIVE 2 Unit Rates

When we compare things to a single unit, such as miles driven per one gallon of fuel or cost per one pound of rice, it is referred to as a unit rate. A unit rate, by definition, will have a denominator of one unit. Unit rates are often expressed using the word "per" followed by a specific unit—things such as *population per county*, *price per item*, *calories per granola bar*, and *per diem food allowance*.

> 📖 **UNIT RATE**
>
> A **unit rate** is a rate comparing two measured quantities, one of which is a single unit written in the denominator of the fraction.

Unit rates can be used to compare prices when you're grocery shopping. If you want to know which is the better deal between a smaller box of cereal at $2.99 for 9.8 ounces or a 14-ounce box for $4.49, write each rate as a unit rate. There are two unit rate options to choose from: we can find either the price per ounce or the number of ounces per dollar.

Possible Unit Rates for Cereal Boxes:

$$\frac{\text{Price}}{1 \text{ Ounce}} \quad \text{or} \quad \frac{\text{Ounces}}{1 \text{ Dollar}}$$

Either rate will provide us with a comparison of the cereal boxes; we will use the unit rate of price per ounce.

Rates of Dollars to Ounces for the Cereal Boxes:

$$\text{Smaller Cereal Box} = \frac{\$2.99}{9.8 \text{ oz}} \quad \text{Larger Cereal Box} = \frac{\$4.49}{9.8 \text{ oz}}$$

In order for each rate to become a unit rate, the denominator needs to be 1 oz. To achieve this, we divide the numerator and denominator of each fraction by its respective denominator.

$$\text{Unit Rate for the Smaller Box of Cereal: } \frac{\$2.99 \div 9.8}{9.8 \text{ oz} \div 9.8} \approx \frac{\$0.305}{1 \text{ oz}}$$

or approximately 30.5¢ per ounce

$$\text{Unit Rate for the Larger Box of Cereal: } \frac{\$4.49 \div 14}{14 \text{ oz} \div 14} \approx \frac{\$0.321}{1 \text{ oz}}$$

or approximately 32.1¢ per ounce

We can now see that the larger box is actually more expensive per ounce than the smaller one. Buyer beware! Bulk size isn't always the best deal.

In reality, you can easily compare the unit prices for items in the grocery store without having a calculator on hand. The shelving stickers will list the unit price underneath the container price in most stores. Take the extra time to check for yourself and make the consumer-savvy purchase.

> **Helpful Hint**
>
> To convert rate to a unit rate, divide the numerator of the fraction by the denominator and keep the units.

Figure 4.3.1

Example 4.3.6

Finding Unit Rates

Determine which type of wood mulch is the least expensive per cubic foot.

    Option A: Pine bark mulch at $3.77 for 3 cubic feet

    Option B: Aromatic cedar mulch at $3.98 for 2 cubic feet

    Option C: Evergreen red mulch at $3.33 for 2 cubic feet

Solution

Let's begin by noticing that we can immediately tell that Option B is more expensive than Option C, so we really only need to compare Options A and C. To do this, we can use the unit rates of each type of mulch to determine which is the least expensive. To find the unit rates for each mulch, begin by writing each rate as a fraction; that is, write each as $\dfrac{\text{Price}}{\text{Volume}}$. Then convert each to a unit rate using division.

$$\text{Option A: } \frac{\$3.77}{3 \text{ cubic feet}} = \frac{\$3.77 \div 3}{3 \text{ cubic feet} \div 3} \approx \frac{\$1.26}{1 \text{ cubic foot}} \text{ or } \$1.26 \text{ per cubic foot}$$

$$\text{Option C: } \frac{\$3.33}{2 \text{ cubic feet}} = \frac{\$3.33 \div 2}{2 \text{ cubic feet} \div 2} \approx \frac{\$1.67}{1 \text{ cubic foot}} \text{ or } \$1.67 \text{ per cubic foot}$$

Once converted to unit rates, we can see that Option A, the pine bark mulch, is the least expensive mulch in our list at $1.26 per cubic foot.

> **✔ Skill Check 4.3.2**
>
> Helium gas (He) was placed in a container fitted with a porous membrane. The helium effused, or passed, through the membrane at the rate of $\dfrac{1.5 \text{ L}}{24 \text{ hr}}$. Find the unit rate of effusion per hour for the helium.

Example 4.3.7

Using Unit Rates

Emma is traveling to Italy for spring break and is checking the exchange rate between dollars and euros (€). Table 4.3.3 shows a certain section of the exchange rates that Emma found listed on the internet

Table 4.3.3: Euro Exchange Rates

Currency Name	Units per EUR	EUR per Unit
US Dollar (USD)	1.139700	0.877448
Argentine Peso	81.189972	0.012317
Brazilian Real	6.119143	0.163422
British Pound	0.907733	1.101645
Canadian Dollar	1.551809	0.644409

Fun Fact

Exchange rates are fluid; that is, they are subject to change at any time. Because financial markets around the world are functioning 24-7, rates are constantly being updated.

a. What is the unit exchange rate of US dollars to euros?

b. What is the unit exchange rate of euros to US dollars?

c. Which is worth more, a US dollar or a euro?

d. If Emma is considering taking $200 for spending money, approximately how many euros will she have to spend?

e. Once in Italy, Emma falls in love with a leather bag priced at €150. Use the exchange rate to find out approximately how much the bag would cost in US dollars, rounded to the nearest dollar.

Solution

a. The format of this table is very similar to many others you will find online. Reading it correctly is half the battle. The column labeled "Units per EUR" shows the amount of corresponding currency that is equivalent to one euro. For example, approximately 81 Argentine pesos equals 1 euro. If we want a rate of US dollars to euros, the unit rate for euros should be in the denominator. Reading across the first row, we see that there are 1.1397 US dollars to each euro.

$$\text{US Dollars to Euros Rate} = \frac{\text{USD}}{\text{EUR}} = \frac{\$1.139700}{€1}$$

b. This time we need the rate for euros to US dollars, so the dollar amount will be in the denominator. The column labeled "EUR per Unit" provides the number of euros equivalent to one unit of the currency listed.

$$\text{Euros to US Dollars Rate} = \frac{\text{EUR}}{\text{USD}} = \frac{€0.877448}{\$1}$$

c. Notice that both of the rates we found in parts a. and b. express the relationship between US dollars and euros. We just need to think carefully about what they are saying. The US dollars to euro rate in part a. says that each euro is worth 1.14 US dollars. Therefore, because it costs more than one dollar to buy one euro, we say the euro is worth more than the dollar. Similarly, the euros to dollars rate in part b. says that each US dollar is worth 0.88 euros; that is, less than 1 euro.

d. To change US dollars to euros, we can multiply by the rate we just found in part b. so that dollars will cancel, and we'll be left with euros.

$$\text{Changing US Dollars to Euros: } \cancel{\$}200 \cdot \frac{\text{€}0.877448}{\cancel{\$}1} = \text{€}175.4896$$

Therefore, Emma's $200 USD will give her approximately €175 to spend.

e. To change euros to US dollars, we will use the rate we found in part a. so that euros will cancel and leave dollars in the numerator.

$$\text{Changing Euros to US Dollars: } \cancel{\text{€}}150 \cdot \frac{\$1.1397}{\cancel{\text{€}}1} = \$170.955$$

This means that the leather bag will cost approximately $171.

Example 4.3.8

Using Rates for Comparison

You and your friends are deciding what you should do together Friday night. Here are the options that you've come up with:

1. Laser tag: $13.00 per person for 2 games

2. Laser tag: $18.00 per person for 3 games

3. County Fair: $20.00 wristbands allowing all-you-can-ride access for the night.

To make the decision easier, present your friends with rates for the evening's options based on the amount of time each option would take.

a. Write down a rate for each option if it takes 30 minutes to play a game of laser tag and you would stay at the fair from 8 p.m. until 10 p.m.

b. Discuss whether it would be useful to convert the rates from part a. into unit rates

Solution

a. Since we are asked to write rates based on cost and time, we will write each option as a fraction with cost in the numerator of the fraction and time in the denominator.

We are given the costs for 2 options of laser tag. Since it takes 30 minutes to play one game of laser tag, the denominator is calculated by multiplying the number of games played by 30 minutes.

$$\text{Option 1—Laser Tag: } \frac{\$13.00}{2 \cdot 30 \text{ min}} = \frac{\$13.00}{60 \text{ min}}$$

$$\text{Option 2—Laser Tag: } \frac{\$18.00}{3 \cdot 30 \text{ min}} = \frac{\$18.00}{90 \text{ min}}$$

For the fair option, the rate is based on staying at the fair from 8 p.m. until 10 p.m., which is 2 hours of unlimited rides.

$$\text{Option 3—Fair Rides: } \frac{\$20.00}{2 \text{ hr}}$$

b. Technically, we can find the rate per hour for each of the three options. However, it would be misleading to present them to your friends because each option is not available at the unit rate. For instance, although the fair costs $10 per hour of rides, an hour of unlimited rides is not offered as an option. So the mixed rates are the best to use to decide which entertainment to choose for Friday night.

OBJECTIVE 3 Rates of Change

Whether we use miles per gallon, drops per minute of an IV, calories per dosage, or population density per county, we can communicate all types of everyday comparisons using rates. Often though, considering how certain rates change is also helpful. For instance, consider the following scenarios.

- A medical researcher might be interested in the rate at which the radius of an artery changes with temperature alterations.

- An entrepreneur might be interested in the rate at which gym membership changes over time.

- An engineer might be interested in the rate at which the length of a steel rod changes with heat over time.

- A microbiologist might be interested in the rate at which the number of bacteria in a culture changes over time.

- A city planner might be interested in the rate at which a population grew over the past ten years.

All of these examples are considering a specific type of rate called *rate of change*. You can think of *rate of change* as "the pace at which one quantity changes in relation to another quantity that also changes."

> ### 📖 RATE OF CHANGE
>
> The **rate of change** of one quantity with respect to another is how much the first quantity changes for each unit change of the second quantity.
>
> $$\text{Rate of Change} = \frac{\text{Change in 1st Quantity}}{\text{Change in 2nd Quantity}}$$

> ### 💬 Helpful Hint
>
> $$\text{Acceleration} = \frac{\text{Change in Velocity}}{\text{Change in Time}}$$

Let's talk about a common rate of change we are all familiar with—acceleration. When you are in a car and feel the force of being either pushed backwards or pulled forwards, you aren't feeling the speed of the car at that moment, you're feeling the rate at which the speed of the car is changing, or the *acceleration*. Acceleration is the rate of change of velocity with respect to time.

Example 4.3.9

Calculating Rate of Change

Table 4.3.4 shows the speed of racers at various times during a game of Mario Kart.

Table 4.3.4: Speed of Racers at Various Times

		Velocity at Time Mark		
	Time	Mario	Luigi	Peach
Start	00:00:00.00	0 m/s	0 m/s	0 m/s
Time 1	00:00:25.00	35 m/s	30 m/s	25 m/s
Time 2	00:00:50.25	12 m/s	15 m/s	11 m/s
Time 3	00:01:01.02	55 m/s	58 m/s	785 m/s
Time 4	00:01:01.20	7 m/s	42 m/s	50 m/s

a. What is Mario's acceleration between the start and Time 1

b. Does Luigi or Peach decelerate more between Times 1 and 2?

c. One of the characters gets shrunk, and consequently is not able to travel as fast, right before Time 4. Which one do you think it is?

d. The human body reacts adversely to high levels of acceleration, with loss of consciousness for accelerations over 60 m/s^2. Is it feasible for human drivers to safely experience Peach's acceleration between Times 2 and 3 when she had a speed boost?

Solution

a. Acceleration is the rate of change between velocity and time. To calculate acceleration, the change in velocity will be in the numerator and the change in time will be in the denominator.

Mario's acceleration between the start and Time 1 is calculated as follows.

$$\text{Mario's Acceleration: } \frac{35 \text{ m/s} - 0 \text{ m/s}}{25 \text{ s} - 0 \text{ s}} = \frac{35 \text{ m/s}}{25 \text{ s}} = 1.4 \text{ m/s}^2$$

Notice that acceleration is measured in meters per second per second, which we refer to as "meters per second squared."

b. In order to find out who decelerates more, we need to calculate the change in velocity between Times 1 and 2 for both Luigi and Peach.

$$\text{Luigi's Acceleration: } \frac{15 \text{ m/s} - 30 \text{ m/s}}{50.25 \text{ s} - 25 \text{ s}} = \frac{-15 \text{ m/s}}{25.25 \text{ s}} = -0.59 \text{ m/s}^2$$

$$\text{Peach's Acceleration: } \frac{11 \text{ m/s} - 25 \text{ m/s}}{50.25 \text{ s} - 25 \text{ s}} = \frac{-14 \text{ m/s}}{25.25 \text{ s}} = -0.55 \text{ m/s}^2$$

So Luigi decelerates at a faster rate than Peach between Times 1 and 2. Notice that even though Peach is going slower at Time 2, Luigi had the bigger change in velocity.

c. If we consider the velocity of each racer at Time 4, we can see that Mario is going the slowest at 7 m/s, so he is most likely to have been shrunk. We do not have to consider acceleration here.

d. Peach's acceleration between Times 2 and 3 is calculated as follows.

$$\text{Peach's Acceleration: } \frac{785 \text{ m/s} - 11 \text{m/s}}{61.02 \text{ s} - 50.25 \text{ s}} = \frac{774 \text{ m/s}}{10.77 \text{ s}} = 71.87 \text{ m/s}^2$$

Therefore, human drivers could not safely experience this acceleration.

Example 4.3.10

Calculating Rate of Change

The Rock 'n' Roll Marathon in Nashville, Tennessee, is known to be a hilly racecourse. Figure 4.3.2 shows the average change in elevation along the 42.195-kilometer route. Use the chart to answer the following questions.

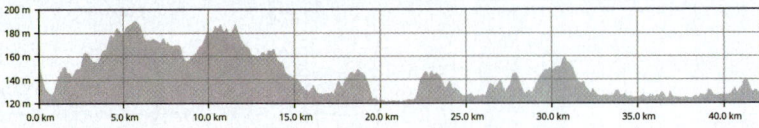

Figure 4.3.2: 2019 Rock 'n' Roll Nashville Marathon Elevation Chart

a. At what point in the race are runners at the highest elevation on the course?

b. What is rate of change of elevation over the first 5 kilometers?

c. Explain if there is a level section of the course where runners are neither going up a hill or down one.

d. How would you determine where the steepest incline and steepest decline are likely to be along the course based on the chart?

Solution

a. The runners will be at the course's highest elevation when the line graph is at the highest point on the chart. This appears to be around the 6-kilometer marker.

b. To find the rate of change we find the difference in elevation from the starting point to the 5-kilometer marker in the numerator and the distance covered in the denominator.

Rate of Change of Elevation over First 5 Kilometers:

$$\frac{180 \text{ m} - 150 \text{ m}}{5 \text{ km} - 0 \text{ km}} = \frac{30 \text{ m}}{5 \text{ km}} = 6 \text{ m/km}$$

c. When the course is level, the elevation is constant as the distance changes. We can see that from about the 20-kilometer marker to approximately the 22.5-kilometer marker the elevation remains at 123 meters and therefore the course is flat.

d. Steep inclines and declines occur when the change in elevation happens quickly, not necessarily at the highest or lowest elevations along the route. Therefore, steep inclines or declines will occur at the places along the chart where the rate of change is the greatest; that is, the elevation has a large jump over a short distance.

Example 4.3.11

Drawing Conclusions from Rate of Change

The following chart shows the United States population growth rates from 1960 to 2019.[3]

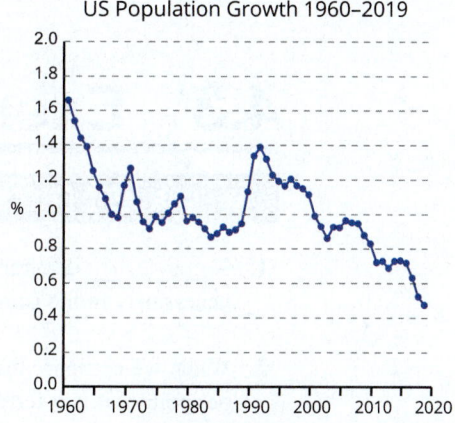

US Population Growth 1960–2019

Figure 4.3.3

3   Source: "Population growth (annual%) – United States", The World Bank, accessed February 17, 2021, https://data.worldbank.org/indicator/SP.POP.GROW?locations=US.

The population of the United States changes continuously, with people being born, immigrating, emigrating, and dying every second. The chart above shows not the size of the population over time, but the rates at which the population was changing. Determine if the following statements are true or false. Explain.

a. The US population had an annual increase of approximately 0.8% in 2010.

b. The population of the US in 1970 was a little more than 120 million.

c. The rate of change in the population of the US between 1990 and 1992 was $\frac{1.39-1.13}{2}$.

d. The US population has been declining since the 1990s.

Solution

a. True. The chart gives the percentage increase for the US population for each year over the previous year. In 2010, the population grew by about 0.8% over 2009.

b. False. The chart does not provide any information about the actual size of the US population at any point in history. However, the population growth was about 1.2% in 1970.

c. False. The calculation shows the rate of change in the US population growth rate percentage, not the rate of change of the actual population

d. False. The US population increased in every year on the chart. It has just been increasing at a slower rate since 1990. For the population to be decreasing, the growth rate would need to drop below 0%.

Rate of change is a cornerstone of the branch of mathematics called differential calculus. It is the study of how quantities change from moment to moment and is used in disciplines such as physics, engineering, kinesiology, and economics, among many others.

Skill Check Answers

1. \$1.60 per pound **2.** $\frac{1.5\,L \div 24}{24\,hr \div 24} = \frac{0.0625\,L}{1\,hr}$

4.3   Exercises

✔ CONCEPT CHECK

1. A _____ is a fraction used to compare two quantities that are not necessarily in the same unit.

2. When we compare things to a single unit, such as miles per gallon or cost per pound, it is referred to as a _____.

3. The _____ of one quantity with respect to another is how much the first quantity changes for each unit change of the second quantity.

4. True or False: The ratios $\dfrac{5 \text{ tacos}}{\$7}$ and $\dfrac{\$7}{5 \text{ tacos}}$ both represent a local restaurant's advertisement of 5 tacos for $7.

5. True or False: If a car has an advertised mileage rate of 20 mpg, that means it uses 20 gallons for ever mile driven.

💡 PRACTICE

Write each rate described.

6. An ad for a local vitamin and nutrition shop advertises that a certain brand of protein bar is on sale for $38 per box of 18.

7. The price of tuition for a semester at an in-state university is $4800 for 15 credit hours.

8. The cost of tuition at a local daycare center is $1150 for 20 days of care per month.

9. Google Voice advertises its competitive price for international calling on mobile phones as 15 cents per minute from Mexico.

10. A freelance online editorial company offers document indexing for $3.50 for 12 indexable printed pages.

11. The Department of Veterans Affairs offers tuition assistance for active military personnel under a new GI Bill that pays up to $1473.00 per month for full-time status at a higher education institution.

12. The local grocery store offers 10 cans of tomato sauce for $10 in the weekly sales ad.

13. Grayson needs to work 65 hours every three weeks to meet the quota for his school internship.

Write each rate described and then convert to a percentage. Round to the nearest tenth, if necessary.

14. In Parma, Ohio, for the May 4, 2021, primary elections, a total of 10,420 people voted out of 51,697 eligible voters in the city.

15. During the 2020 tax year, a tax return form indicates that a filer who earned $32,750 paid $3732 in federal taxes.

16. During the 2020 tax year, a tax return form indicates that the filer paid $23,960 in federal taxes and earned a total of $124,500.

17. For the 2019 enrollment period at Clemson University, a total of 3932 new freshman enrolled out of the 29,070 people who applied to the school.

Choose the correct from of the rate given.

18. Packs of noodle side dishes are on sale for $5 when you purchase 4 packs.

  a. $\dfrac{\$4}{5 \text{ packs}}$

  b. $\dfrac{4 \text{ packs}}{\$5}$

  c. $\dfrac{\$5}{4 \text{ packs}}$

  d. $\dfrac{5 \text{ packs}}{\$4}$

19. Cooking instructions recommend roasting a 14-pound turkey for approximately 3 hours.

  a. $\dfrac{3 \text{ hours}}{14 \text{ pounds}}$

  b. $\dfrac{14 \text{ pounds}}{3 \text{ hours}}$

  c. $\dfrac{1 \text{ hour}}{14 \text{ pounds}}$

  d. $\dfrac{3 \text{ hours}}{1 \text{ pound}}$

20. To advertise their weekly special, Carol needs a sign made. One pen costs $3.18, but she is going to offer 3 pens for the price of one. Which is the correct sign?

  a. $3.18/pen

  b. 3 pens/$3.18

  c. $9.54/3 pens

21. Tyson can run 4 km in 30 minutes.

  a. 4 km/30 min

  b. 7.5 km/min

  c. 2 km/hour

  d. 1 hour/2 km

🚀 APPLICATIONS

Solve. Round your answer to the nearest hundredth, if necessary.

22. A patient in a hospital is prescribed IV fluids with a rate of 5 grams of sugar per 100 milliliters of water. If a patient is prescribed 1000 milliliters of fluids, how many grams of sugar will be delivered?

23. A patient in a hospital is prescribed IV fluids with a rate of 9 grams of sodium chloride per 1000 milliliters of water. If a patient is prescribed 1500 milliliters of fluids, how many grams of sodium chloride will be delivered?

24. A patient in a hospital is prescribed IV fluids with a rate of 4.5 grams of sodium chloride per 1000 milliliters of water. If a patient is prescribed 500 milliliters of fluids, how many grams of sodium chloride will be delivered?

25. To treat the water in a swimming pool, it is recommended to add 10 milliliters of chlorine per 1000 liters of water. How much chlorine is needed to treat the water in a pool that holds 34,000 liters?

26. Lucinda earns $16.32 per hour working as a pharmacy technician. If Lucinda worked 171.5 hours last month, what was her gross pay for the month?

27. The average rainfall in Phoenix is 8.04 inches per year. If this average is maintained, how many inches of rain would Phoenix receive over 7 years?

28. The density of a cooking oil is 0.93 g/mL. By comparison, the density of water is approximately 1 g/mL. What is the mass of 1 cup of the oil? (Note that 1 cup is approximately 236.59 mL.)

29. The fastest speed recorded for English language typing is 216 words per minute. If the record holder maintains that rate for 11.5 minutes, how many words would be typed?

30. An SUV is advertised to have fuel tank capacity of 23 gallons and can travel 330 miles on a full tank of gas. Find the miles per gallon the SUV gets.

31. If your car can go 433 miles on a full tank of gas, find the miles per gallon your car gets if the tank holds 17.2 gallons.

32. You are planning to buy a mid-sized car and plan to factor fuel efficiency into your decision. Using the given information, decide which mid-sized car gets the best gas mileage.

    a. 2021 Honda Civic: 397 miles on 12.4 gallons

    b. 2021 Mazda Mazda6: 476 miles on 16.4 gallons

    c. 2021 Toyota Camry: 459 miles on 14.4 gallons

33. You want to buy a used truck and are interested in the fuel consumption rate for each of the three trucks you are considering. Using the information given, decide which of the used trucks gets the best gas mileage.

    a. 2008 Chevy Silverado: 500 miles on 29.4 gallons

    b. 2006 Toyota Tundra: 428 miles on 26.4 gallons

    c. 2007 Ford F-150: 250 miles on 15.6 gallons

34. Two competing stores sent out fliers for back-to-school sales and listed similar items for sale. Use unit prices to decide which store to buy each item from if you need to purchase all of the items listed.

    Price Comparison

Item	Closeout Closet	Discount Dollar
Mechanical Pencils	8 for $2.39	10 for $2.64
Pens	10 for $1.49	10 for $1.57
Notebooks	3 for $4.99	5 for $9.99
Wide-Ruled Paper	3 for $2.50	4 for $3.00

35. It costs $130.11 per week to rent a standard-size car in Nashville, TN. Find the price per day it costs to rent this car.

36. It costs $2.01 to drive 25 miles in a new 2012 Hybrid car. How much does it cost per mile to drive 25 miles?

37. It costs $4.78 to drive 25 miles in a 2012 sedan. How much does it cost per mile to drive 25 miles?

38. Ron gets paid $51,630 per year. If he works 2087 hours in a year, find his hourly rate of pay.

39. If a 4-pound bag of sugar costs $2.99, what is the unit cost for a pound of sugar?

40. The advertised special is $9.99 for 16 ounces of frozen shrimp. Find the unit cost of the shrimp.

41. Eggplants are 4 for $5.00. What is the cost per eggplant? How much eggplant do you get per dollar?

Use the following table of exchange rates for Exercises 42–43.

US Dollar Exchange Rates		
US Dollar (USD)	**Units per USD**	**USD per Unit**
British Pound (GBP)	0.711576	1.405332
Indian Rupee (INR)	73.669595	0.013574
Australian Dollar (AUD)	1.294470	0.772517
Canadian Dollar (CAD)	1.212872	0.824489
Japanese Yen (JPY)	109.642453	0.009121

42. Use the table to answer the following questions.

   a. What is the unit exchange rate of US dollars to British Pounds?

   b. What is the unit exchange rate of British Pounds to US dollars?

   c. If a pair of shoes costs 75 US dollars, approximately how many British Pounds do the shoes cost?

43. Use the table to answer the following questions.

   a. What is the unit exchange rate of US dollars to Japanese Yen?

   b. What is the unit exchange rate of Japanese Yen to US dollars?

   c. If a comic books costs 500 Japanese Yen, approximately how many US Dollars does the comic book cost?

44. The projected velocity of a space shuttle during its first three minutes in the air is given below.

Time (s)	Altitude (m)	Velocity (m/s)
0	-8	0
20	1244	139
40	5377	298
60	11,617	433
80	19,872	685
100	31,412	1026
120	44,726	1279
140	57,396	1373
160	67,893	1490
180	77,485	1634

a. What is the space shuttle's acceleration during the first 20 seconds?

b. Does the space shuttle accelerate more between 20–40 seconds or between 120–140 seconds?

45. The civilian unemployment rate as determined by the US Bureau of Labor Statistics[4] is shown for 2002–2021.

Civilian unemployment rate, seasonally adjusted

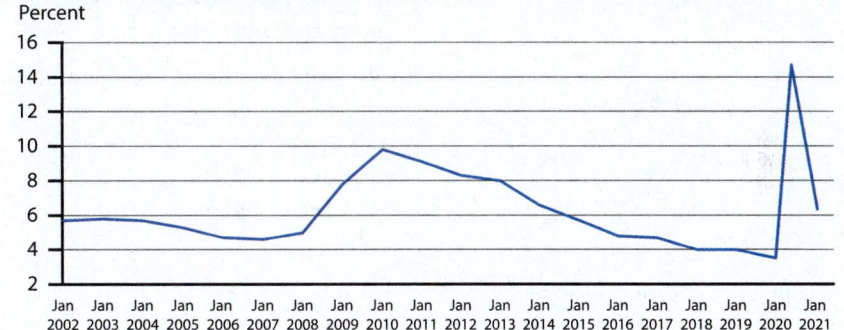

a. When was the unemployment rate the highest?

b. What is the average rate of change from January 2012 to January 2016?

c. Are there any intervals where the unemployment rate is not changing?

d. What interval has the steepest decline?

---

4   Source: "Civilian unemployment rate," Bureau of Labor Statistics, accessed December 8, 2021, https://www.bls.gov/charts/employment-situation/civilian-unemployment-rate.htm.

46. The Consumer Price Index,[5] which is "a measure of the average change over time in the prices paid by urban consumers for a market basket of consumer goods and services," is shown for May 2007 to May 2021.

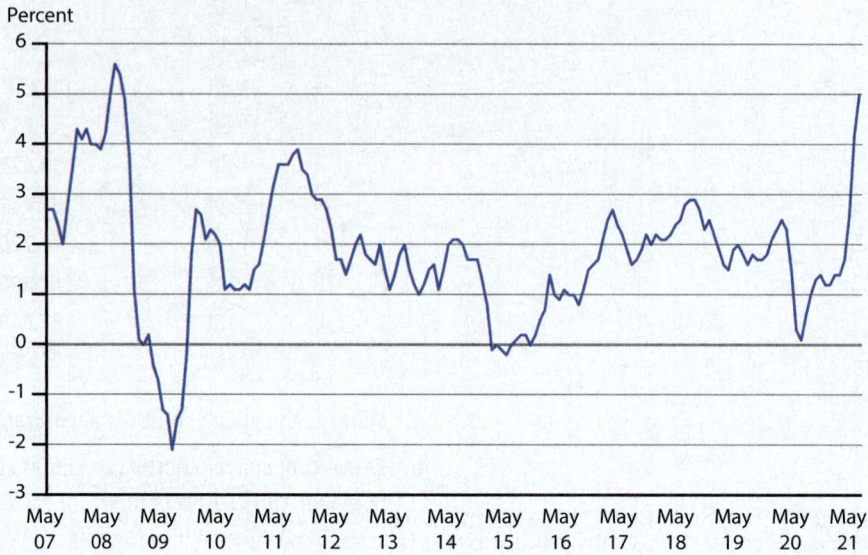

Consumer Price Index, May 2007-May 2021

a. When was the Consumer Price Index the lowest?

b. What is the average rate of change from May 2015 to May 2020?

c. Are there any intervals where the Consumer Price Index appears to have no change over time?

d. What interval has the steepest decline?

5   Source: "Consumer prices increase 5.0 percent for the year ended May 2021," Bureau of Labor Statistics, accessed December 8, 2021, https://www.bls.gov/opub/ted/2021/consumer-prices-increase-5-0-percent-for-the-year-ended-may-2021.htm.

4.3 PROJECT

WHICH TIRE IS THE BEST OPTION?

Calculating unit rates can help you make better decisions when it comes time to spend your hard-earned money. For example, dealing with car maintenance can be a frustrating experience. How can a person make an informed decision when replacing tires?

In this project, you will explore the use of unit rates to help you choose a new set of tires for your car.

You have budgeted a maximum of $370 for a set of new tires. The following table shows two different brands of tires that would fit your vehicle.

Name	Raptor VR	Solus TA31
Price per Tire	$83.26	$90.00
Set of 4 Tires	$333.04	$360.00
Mileage Warranty	55,000 miles	65,000 miles

1. If total cost is your only consideration, which tire is the "best" choice?

2. The Raptor VR has a 55,000-mile warranty. Find the rate of miles to dollars for one Raptor VR tire. Write your answer as a unit rate.

3. The Solus TA31 has a 65,000-mile warranty. Find the rate of miles to dollars for one Solus TA31 tire. Write your answer as a unit rate.

4. What do the rates in parts 2 and 3 measure?

5. If getting the most for your money and staying within your budget is your goal, which tire seems like a better buy? Explain your reasoning. (**Hint:** Compare the percentage change between the two prices and between the two rates of miles per dollar.)

4.4 USING RATES FOR DIMENSIONAL ANALYSIS

OBJECTIVES

1. Convert measurements using dimensional analysis.

In the late 1990s, after almost 10 months of travel to Mars, the NASA Mars Climate Orbiter unexpectedly burned and broke into pieces in space instead of entering Mars's orbit as anticipated. Why? The simple answer is conversions, or lack of conversions more specifically. The Mars Climate Orbiter was a collaboration between the navigation team at NASA's Jet Propulsion Laboratory (JPL) and the design team from Lockheed Martin Astronautics. The Lockheed Martin team provided crucial acceleration data in the US Customary System of measurement unit *pounds*, while the software used by the engineers at JPL read the data as if it were using the metric system unit *newtons*. Somewhere, the units failed to be converted, resulting in the loss of the $125 million space probe—a very expensive mistake!

Had scientists converted the measurements from the US Customary System into the metric system using dimensional analysis, the whole fiasco could have been avoided.

OBJECTIVE 1 Dimensional Analysis

Dimensional analysis is a method used to convert one unit of measurement to another. It is often discussed when converting between the US Customary System of measurement and the metric system. In fact, Section 7.5 in this book discusses just that. Here we are going to apply the same principles to specific situations involving various rates.

Formally, *dimensional analysis* is a problem-solving method that uses the fact that any number or expression can be multiplied by the number 1 without changing the value. *Conversion factors* are used to change from one particular unit of measurement to another. When needed, we will provide equivalences to use as conversion factors. The internet is a quick and easy resource for a more exhaustive list.

📖 CONVERSION FACTOR AND DIMENSIONAL ANALYSIS

Conversion Factor

A **conversion factor** is a fraction that is equal to 1 but contains different units of measurement in the numerator and denominator.

Dimensional Analysis

Dimensional analysis is a method of using conversion factors to change from one unit of measurement to another.

The following is a small sample of conversion factors. Take a moment to confirm that they all equal 1 because the numerator and denominator of each factor represent the same quantity.

Table 4.4.1: Examples of Conversion Factors

$\dfrac{24\text{ hours (hr)}}{1\text{ day}}$	$\dfrac{16\text{ ounces (oz)}}{1\text{ pound (lb)}}$	$\dfrac{1\text{ meter (m)}}{100\text{ centimeters (cm)}}$
$\dfrac{1\text{ pound (lb)}}{0.454\text{ kilogram (kg)}}$	$\dfrac{3\text{ feet (ft)}}{1\text{ yard (yd)}}$	$\dfrac{2.54\text{ centimeters (cm)}}{1\text{ inch (in.)}}$
$\dfrac{1\text{ mile (mi)}}{1.61\text{ kilometer (km)}}$	$\dfrac{1\text{ gallon (gal)}}{3.785\text{ liters (L)}}$	$\dfrac{1\text{ cup (c)}}{8\text{ fluid ounces (fl oz)}}$

Note: The conversion factors for pound to kilogram, mile to kilometer, and gallon to liter are not exact.

By definition, a conversion factor contains different units of measurement in its numerator and denominator. Since the fraction is equal to 1, there is no inherent restriction on which measurement is placed in the numerator. For instance, since there are 3 feet in a yard, both $\dfrac{3\text{ feet}}{1\text{ yard}}$ and $\dfrac{1\text{ yard}}{3\text{ feet}}$ are viable conversion factors.

Each particular situation will dictate which orientation to use.

> **Helpful Hint**
>
> Recall that mpg means "miles per gallon."

Example 4.4.1

Using Dimensional Analysis with One Conversion Factor

a. A new hybrid car sold in the United States reports a fuel usage of 51 mpg. Determine how many kilometers per gallon this is equivalent to.

b. If the fuel consumption standard for the European Union is approximately 24.39 km/L, does this car meet the standard?

Solution

a. Let's begin by writing the rate we were given as a fraction.

$$\text{Fuel Usage Rate: } \frac{51\text{ mi}}{1\text{ gal}}$$

Next, we need a conversion factor that relates miles and kilometers. Table 4.4.1 shows that 1 mile is approximately 1.61 kilometers, so the following fractions are both conversion factors.

$$\text{Conversion Factors Using Miles and Kilometers: } \frac{1\text{ mi}}{1.61\text{ km}} \text{ and } \frac{1.61\text{ km}}{1\text{ mi}}$$

The conversion factor is shown in both formats to emphasize the point that both are appropriate factors, but we want to choose the most useful one in this situation. We wish to know kilometers per gallon, or $\dfrac{\text{km}}{\text{gal}}$. Notice that we need to end up having kilometers in the numerator, so we will use the second form of the conversion factor in our multiplication.

$$\text{Miles to Kilometers Conversion: } \frac{51\,\cancel{\text{mi}}}{1\text{ gal}} \cdot \frac{1.61\text{ km}}{1\,\cancel{\text{mi}}} = \frac{82.11\text{ km}}{1\text{ gal}}$$

Since the units are now kilometers per gallon, we are finished. This means that 51 miles per gallon are approximately 82.11 kilometers per gallon.

b. To determine if the car meets the European Union standards, we need to change the rate we found in part a. from gallons to liters. Table 4.4.1 shows that 1 gallon is approximately 3.785 liters. Since we want to ultimately end up with liters in the denominator, the conversion factor should be written in the form $\dfrac{1 \text{ gal}}{3.785 \text{ L}}$. Using multiplication, we have the following conversion.

Miles to Kilometers Conversion: $\dfrac{82.11 \text{ km}}{1 \text{ gal}} \cdot \dfrac{1 \text{ gal}}{3.785 \text{ L}} \approx \dfrac{21.69 \text{ km}}{1 \text{ L}}$

Since the European Union standard is 24.39 km/L, the new hybrid does not quite meet the standard.

✔ **Skill Check 4.4.1**

Convert 180 centimeters to inches.

Example 4.4.2

Using Dimensional Analysis with One Conversion Factor

A pesticide label recommends using its product at a rate of 10 milliliters concentrate per 5 liters of water to cover an area of 40 square meters. A 5-gallon sprayer is being used for the mixture. How many milliliters of concentrate should be added if the sprayer's container is filled with water to the 5-gallon mark? Recall that 1 gallon is approximately 3.785 liters.

Solution

Begin by writing both the recommended rate of concentrate to water (which is given), as well as the conversion factor for gallons to liters.

Recommended Pesticide Rate: $\dfrac{10 \text{ mL concentrate}}{5 \text{ L water}}$

Conversion Factors Using Gallons and Liters: $\dfrac{1 \text{ gal}}{3.785 \text{ L}}$ or $\dfrac{3.785 \text{ L}}{1 \text{ gal}}$

We wish to find the rate of milliliters to gallons, or $\dfrac{\text{mL}}{\text{gallons}}$. So we will use the second form of the conversion factor, which has gallons in the denominator, in our multiplication.

Helpful Hint

When setting up a conversion factor, diagonal units should cancel so that you are left with the desired units in your answer.

Pesticide Conversion to Gallons:

$\dfrac{10 \text{ mL concentrate}}{5 \text{ L water}} \cdot \underbrace{\dfrac{3.785 \text{ L}}{1 \text{ gal}}}_{\text{Converts L to gal}} = \dfrac{37.85 \text{ mL concentrate}}{5 \text{ L water}}$

Since the units are now milliliters per 5 gallons, we are finished. Thus, 37.85 mL of concentrate should be used in the 5-gallon sprayer.

Example 4.4.3

Using Dimensional Analysis with More Than One Conversion Factor

The density of mercury is given as a rate of 13.6 g/mL. What is the density of mercury in pounds per liter if 1 pound is equal to 0.454 kilograms?

Solution

Begin by writing the rates that were given, the density of mercury as well as the conversion factor for pounds to kilograms.

$$\text{Density of Mercury: } \frac{13.6 \text{ g}}{1 \text{ mL}}$$

Note that we will need to change grams to pounds and milliliters to liters. The given density is in grams, which will need to change to kilograms before we can use the one conversion factor.

$$\text{Conversion Factors Using Pounds and Kilograms: } \frac{1 \text{ lb}}{0.454 \text{ kg}} \text{ or } \frac{0.454 \text{ kg}}{1 \text{ lb}}$$

Notice that the first form of the conversion factor has pounds in the numerator, so we will use that one. The final form should have liters in the denominator, so we will need to use a second conversion factor to transpose milliliters into liters. There are 1000 milliliters in 1 liter, so the conversion factors are $\frac{1000 \text{ mL}}{1 \text{ L}}$ or $\frac{1 \text{ L}}{1000 \text{ mL}}$. Again, we will use the first form since we need to end up with liters in the denominator.

Multiplying each conversion factor by the original rate, remembering that the order of multiplication is flexible, we have the following.

$$\text{Density Conversion to Pounds per Liter:}$$

$$\frac{13.6 \text{ g}}{1 \text{ mL}} \cdot \underbrace{\frac{1000 \text{ mL}}{1 \text{ L}}}_{\text{Converts to L to mL}} \cdot \underbrace{\frac{1 \text{ kg}}{1000 \text{ g}}}_{\text{Converts g to kg}} \cdot \underbrace{\frac{1 \text{ lb}}{0.454 \text{ kg}}}_{\text{Converts kg to lb}} = \frac{13,600 \text{ lb}}{454 \text{ L}}$$

So the density of mercury is $\frac{13,600 \text{ lb}}{454 \text{ L}}$. We can reduce this to lowest terms by dividing both the numerator and denominator by the greatest common factor 2.

$$\text{Reduced to Lowest Terms: } \frac{13,600 \text{ lb} \div 2}{454 \text{ L} \div 2} = \frac{6800 \text{ lb}}{227 \text{ L}}$$

✔ **Skill Check 4.4.2**

Determine how many minutes are in three weeks.

Example 4.4.4

Using Dimensional Analysis with More Than One Conversion Factor

In medical fields, the drip factor for a liquid administered through an IV is given in gtt/mL. The flow rate of an IV is measured in gtt/min. For each of the following situations, use the doctor's orders to determine how many drops per minute (in whole numbers) the doctor is requesting. (Note: The abbreviation gtt comes from the Latin word guttae, which means "drops.")

a. A doctor's order calls for an infusion of 1500 mL of a solution over 12 hours with a drip factor of 15 gtt/mL.

b. A doctor's order instructs the infusion of 300 mL of ampicillin sodium over 40 minutes with a drip factor of 20 gtt/mL.

c. A doctor's order instructs the infusion of 250 mL of red blood cells, twice over 4 hours with a drip factor of 15 gtt/mL.

Solution

In each case, the doctor's order is given in milliliters over a period of time. Since we are asked to find drops per minute, we need to use conversion factors to make sure the time unit is in minutes (if it's not already in minutes) and the amount of liquid is in number of drops instead of milliliters. We begin by writing the doctor's order as a rate and determining what conversion factors we need to multiply by.

a. The rate in this doctor's order is given as $\dfrac{1500\,\text{mL}}{12\,\text{hr}}$. Notice that we will need

to change hours to minutes and milliliters to drops. The conversion factor comparing hours to minutes is either $\dfrac{60\,\text{min}}{1\,\text{hr}}$ or $\dfrac{1\,\text{hr}}{60\,\text{min}}$. Because we need to cancel the hours out of the denominator of the rate, we need to use the second conversion factor, which has hours in the numerator. We will also use the drip factor, which was given as $\dfrac{15\,\text{gtt}}{1\,\text{mL}}$.

Multiplying the original rate by both conversion factors, we have the following equation.

$$\frac{1500\ \cancel{\text{mL}}}{12\ \cancel{\text{hr}}}\cdot\frac{1\ \cancel{\text{hr}}}{60\ \text{min}}\cdot\frac{15\ \text{gtt}}{1\ \cancel{\text{mL}}}=\frac{22{,}500\ \text{gtt}}{720\ \text{min}}$$

After multiplying, we have a rate showing the number of drops over 720 minutes. Since we want the unit rate per minute, we need to divide both the numerator and denominator by 720.

$$\frac{22{,}500\ \text{gtt}\div720}{720\ \text{min}\div720}=\frac{31.25\ \text{gtt}}{1\ \text{min}}$$

Therefore, the doctor is ordering approximately 31 drops per minute (gtt/min), rounded to the nearest whole number of drops.

b. We will follow the same procedure as we did in part a. The doctor's order gives an overall rate of $\dfrac{300\,\text{mL}}{40\,\text{min}}$. Notice that the order is already in minutes, so there is no need for a time conversion factor. However, we still need a drip factor in order to change milliliters to drops. It is given as $\dfrac{20\,\text{gtt}}{1\,\text{mL}}$. Multiplying, we get the following rate in drops over minutes.

$$\frac{300\ \cancel{\text{mL}}}{40\ \text{min}}\cdot\frac{20\ \text{gtt}}{1\ \cancel{\text{mL}}}=\frac{6000\ \text{gtt}}{40\ \text{min}}$$

Finally, we change to a unit rate per minute by dividing the numerator and denominator by 40.

$$\frac{6000\ \text{gtt}\div40}{40\ \text{min}\div40}=\frac{150\ \text{gtt}}{1\ \text{min}}$$

Therefore, the doctor is ordering 150 drops per min (gtt/min).

c. In the last scenario, the doctor's instructions ask for 250 mL to be given twice over 4 hours. Thus, the overall rate is $\dfrac{2\cdot250\ \text{mL}}{2\cdot1\ \text{mL}}=\dfrac{500\ \text{mL}}{4\ \text{hr}}$. Again, we will

need to change hours to minutes using $\dfrac{1\,\text{hr}}{60\,\text{min}}$ and change milliliters to drops using the drip factor given as $\dfrac{15 \text{ gtt}}{1 \text{ mL}}$. We will also condense steps and divide at

the end so that we have a unit rate per minute.

$$\frac{500 \ \cancel{\text{mL}}}{4 \ \cancel{\text{hr}}} \cdot \frac{1 \ \cancel{\text{hr}}}{60 \ \text{min}} \cdot \frac{15 \ \text{gtt}}{1 \ \cancel{\text{mL}}} = \frac{7500 \ \text{gtt}}{240 \ \text{min}}$$

$$= \frac{7500 \ \text{gtt} \div 240}{240 \ \text{min} \div 240}$$

$$= \frac{31.25 \ \text{gtt}}{1 \ \text{min}}$$

Therefore, the doctor is ordering approximately 31 drops per min (gtt/min), rounded to the nearest whole number of drops.

Example 4.4.5

Investigating a Conversion Error

In 1983, Air Canada's Boeing 767 jet famously ran out of fuel midflight in what became known as the "Gimli Glider" incident. Instead of calculating the number of liters needed to hit a required 22,300 kilograms of fuel, the crew calculated how many liters were needed to hit 22,300 pounds—just about half of the fuel required. In the end, the skilled pilots were able to land the plane by gliding the last 100 kilometers, bringing all on board to safety

Follow the given steps to correctly calculate how much fuel should have been loaded onto the plane. The fuel requirement for the plane to make the trip was 22,300 kilograms. A float stick check indicated that the tanks contained 7682 liters of fuel already.

a. Calculate the additional amount of fuel in kilograms that needed to be added to the fuel tanks in order to meet the requirement if the mass of 1 liter of fuel in this situation was 0.803 kilograms.

b. Convert the fuel amount found in part a. from kilograms to liters.

Solution

a. The plane needs 22,300 kilograms of fuel to make the trip. We are told there are 7682 liters of fuel already on board. So we will need to subtract the amount of fuel that is already in the plane's tanks from the amount required. Before we can subtract, we need to convert the fuel amount in the tanks from liters to kilograms. The volume of jet fuel varies with temperature. In this case, the mass of 1 liter of fuel was 0.803 kilograms. Therefore, we will use the conversion factor $\dfrac{0.803 \text{ kg}}{1 \text{ L}}$.

Mass of Fuel Already in the Plane's Tanks: $7682 \text{ L} \cdot \dfrac{0.803 \text{ kg}}{1 \text{ L}} \approx 6169 \text{ kg}$

Now we can subtract since both measurements are in kilograms.

Mass of Additional Fuel Required: $22{,}300 \text{ kg} - 6169 \text{ kg} = 16{,}131 \text{ kg}$

Therefore, 16,131kg of fuel needed to be added to the fuel tanks.

b. Now we need to convert the required fuel amount, which is in kilograms, into a volume using liters. The conversion factor is the same, but with the numerator and denominator flipped.

Volume of Additional Fuel Required: $16{,}131 \text{ kg} \cdot \dfrac{1 \text{ L}}{0.803 \text{ kg}} \approx 20{,}088 \text{ L}$

In comparison, using an incorrect conversion factor, the crew calculated that only 4917 liters of fuel needed to be added to the plane's tanks, which is why they ran out of gas!

Skill Check Answers
1. 70.87 **2.** 30,240

4.4 Exercises

✔ CONCEPT CHECK

1. A conversion factor is a fraction that is equal to ___ but contains different units of measurements in the numerator and the denominator.

2. _____ is a method of using conversion factors to change from one unit of measurement to another.

3. When setting up a conversion factor, _____ units should cancel so that you are left with the desired units in your answer.

4. True or False: Since there are 12 inches in a foot, $\dfrac{12 \text{ inches}}{1 \text{ foot}}$ and $\dfrac{1 \text{ foot}}{12 \text{ inches}}$ are both viable conversion factors.

5. True or False: Only one conversion factor can be used in in a dimensional analysis problem.

🚀 APPLICATIONS

Solve. Round your answers to the nearest thousandth, if necessary.

6. A regular Hershey's bar weighs approximately 0.29 ounces per inch. Determine how many ounces per foot this is. (**Hint:** There are 12 inches per foot.)

7. Sterling silver in the form of #2-gauge wire weighs 3.4 ounces per foot. Determine how many ounces per inch this is. (**Hint:** There are 12 inches per foot.)

8. A cyclist travels 18 miles per hour. Determine how many miles per minute the cyclist travels.

9. A car travels at a speed of 2 kilometers per minute. Determine the speed of the car in kilometers per hour.

10. Diet Coke contains approximately 31 milligrams of caffeine per cup. Determine how many milligrams per fluid ounce this is.

11. Barbecue potato chips contain 150 calories per ounce. Determine how many calories per pound this is.

12. All-purpose flour costs $0.99 per pound. What is the price of flour per kilogram?

13. A hardware store loans out toolsets at a rate of $5 per hour. Determine how much tool rental costs per day.

14. A water faucet left running can use 900 liters of water per hour. Determine how many gallons per hour this is.

15. A machine uses 0.5 gallons of fuel per hour. Determine how many liters per hour this is.

16. A road sign indicates that the speed is 50 miles per hour. What is the speed in kilometers per hour?

17. A road sign indicates that the speed is 100 kilometers per hour. What is the speed in miles per hour?

18. Hair grows at an average rate of 0.5 inches per month. What is the average growth rate of hair in centimeters per month?

19. For the first 30 years of its lifespan, a dawn redwood tree grows an average of 1.22 meters per year. Determine the growth rate of the tree in centimeters per year.

20. Guidelines recommend that paper lawn bags contain no more than 50 pounds of material per bag. Determine the recommended capacity in kilograms per bag.

21. According to an EPA study, the average American produces 4.5 pounds of trash per day. How many ounces of trash per day does the average American produce?

22. A motorcycle gets an average fuel efficiency of 25 km/L. Determine the fuel efficiency in miles per gallon.

23. An SUV gets an average fuel efficiency of 14 mi/gal. Determine the fuel efficiency in kilometers per liter.

24. The maximum speed of a housefly is 5 miles per hour. Determine the speed in kilometers per minute.

25. The average speed of a tree swallow is 20 miles per hour. Determine the speed in kilometers per minute.

26. Lumber varies in weight depending on the type of tree it is made from. A 2 x 6 Sitka Spruce lumber has a weight of 1.6 pounds per linear foot. Determine the weight in ounces per yard.

27. A pool is losing water at a rate of 3 fluid ounces per hour. Determine the rate of water loss in cups per day.

28. A small mill can process wood chips into sawdust at a rate of 75 kilograms per hour. Determine the rate in pounds per day.

29. A water faucet left running can use 240 gallons of water per hour. Determine this rate in liters per minute.

30. A pedometer is a device that records the number of steps taken, and many watches can measure this as well. Assuming a consistent walking pace, convert the following into steps per minute.

    a. Walking 4 miles per hour with a stride length of 2100 steps in a mile

    b. Walking 2.75 miles per hour with a stride length of 1920 steps in a mile

    c. Walking 3.25 miles per hour with a stride length of 2005 steps in a mile

31. Tenisha makes jewelry and sells it via an online craft site. For each of the following scenarios, express Tenisha's earnings in dollars per hour.

    a. Making 3 bracelets in 50 minutes and selling them for $15 each.

    b. Making 2 necklaces in 105 minutes and selling them for $22 each.

    c. Making 1 pair of earrings in 17 minutes and selling them for $13 each.

32. Convert the following fuel consumption measurements from miles per gallon to kilometers per liter. Use 1 mile is approximately 1.609 kilometers and 1 gallon is approximately 3.785 liters.

    a. 28 miles per gallon

    b. 21 miles per gallon

    c. 32.5 miles per gallon

33. Recall from Example 4.4.4 that the drip factor for a liquid administered through an IV is given in gtt/mL. The flow rate of an IV is measured in gtt/min. For each of the following situations, determine how many drops per minute (in whole numbers) should be given.

    a. 300 mL given over 2 hours with a drip factor of 20 gtt/mL

    b. 90 mL given over 0.5 hours with a drip factor of 15 gtt/mL

    c. 1200 mL given over 10 hours with a drip factor of 20 gtt/mL

34. In 1628 the Swedish warship Vasa sank less than a mile into its first voyage. One factor may have been the use of different rulers during construction, with some calibrated to Swedish feet which had 12 inches and some calibrated to Amsterdam feet which had 11 inches. If the Vasa was 69 meters long, find the ship's length in Swedish feet and Amsterdam feet. Let 1 meter equal 39.37 inches.

35. After NASA's Hubble telescope first launched in 1990, the pictures it took were fuzzier than expected. The cause was an error in the shape of the telescope's main mirror: it was 2.2 microns too flat. If a micron is 1/1000 of a millimeter and 1 millimeter is approximately 0.03937 inches, how large was the error in inches?

36. The day before the start of the 2014 Sochi Winter Olympics, an error was discovered in the biathlon track. The loop was supposed to be 2.5 kilometers but was 40 meters too short. If an athlete completed 3 loops on the erroneous track, how many feet short was their distance? Use 1 meter is approximately 3.281 feet.

4.4 PROJECT

LITERS, GALLONS, AND FUEL EFFICIENCY

According to a 2019 report produced by the United Nations, fossil fuel emissions from energy use grew 2.0% in 2018. These emissions reached a record high of 37.5 gigatons of equivalent CO_2 per year. In order to try to alleviate the issue, many countries are working on requiring that cars become more fuel efficient.

In this activity, you will investigate a few different units used to measure fuel efficiency and how to compare such units.

In the US, we define fuel efficiency as the number of miles a car can travel using one gallon of fuel. The unit of choice for this measurement is miles per gallon (mpg). This number is the rate of miles traveled per gallons used. European countries commonly define fuel efficiency as the amount of fuel a car requires to travel a fixed distance. The unit of choice for this measurement is liters per 100 kilometers $\left(\dfrac{L}{100 \text{ km}} \right)$. This number is the rate of liters of fuel used per 100 kilometers traveled. Notice that mpg is a rate of distance per volume of fuel while $\dfrac{L}{100 \text{ km}}$ is a rate of volume of fuel per distance.

The 2021 Honda Accord has a reported highway fuel efficiency of 38 mpg. Complete the following steps to find the Accord's equivalent highway fuel efficiency in the European standard of $\dfrac{L}{100 \text{ km}}$.

Recall that 1 US gallon is equal to 3.785 liters and that 1 mile is approximately equal to 1.61 kilometers.

1. Determine how many 100 kilometers are in 38 miles. Round your answer to the nearest hundredth, if necessary. (**Hint:** Find the number of kilometers in 38 miles and divide your answer by 100.)

2. Organize this information using the table below.

Distance	Miles	38
	100 Kilometers	
Fuel Volume	Gallons	1
	Liters	

3. Now, divide the number of liters per number of 100 kilometers to find the Accord's equivalent European standard fuel efficiency. Round your answer to the nearest hundredth, if necessary.

4. You may have noticed that the two rates reflect the fuel efficiency of a vehicle, but they are not really measuring the same quantity. As more governments require higher mpg from cars, would the corresponding $\dfrac{\text{L}}{100 \text{ km}}$ also increase as the mpg increase? Why or why not?

4.5 PROPORTIONALITY

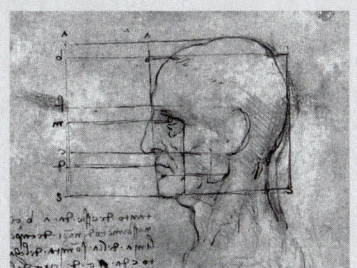

Figure 4.5.1

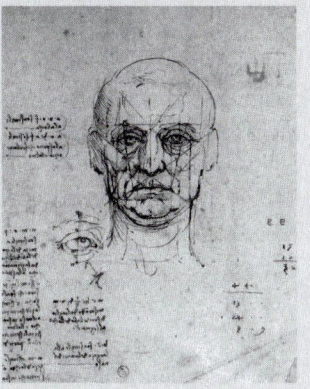

Figure 4.5.2

OBJECTIVE 1 Constant of Proportionality

Let's turn our attention to *proportionality*—what do we mean when we say things are "proportional to one another"? Any of the following phrases might sound familiar to you when proportionality is discussed.

- "as one increases, so does the other"

- "a comparison relating two ratios"

- "if one value is multiplied by a number, so is the other"

- "when one thing is proportional to another, we can set up two fractions that are equivalent"

These explanations are all dancing around the idea of proportionality, but they don't fully capture all that is involved. Let's think back to our previous talk about ratios. In the example about the colors in a flag, we saw that the colors of the flag always obeyed several proportional relationships based on the ratio given. Recall that the flag was colored red and blue in the ratio 1 : 2, and from that we are able to know that all of the following fractions are true.

Fractions Related to the Flag Ratio 1 : 2

$\frac{1}{3}$ represents the red coloring (1) compared to the whole (2 + 1) of the flag

$\frac{2}{3}$ represents the blue coloring (2) compared to the whole (2 + 1) of the flag

$\frac{1}{2}$ represents the red coloring (1) compared to the blue coloring (2)

$\frac{2}{1}$ represents the blue coloring (2) compared to the red coloring (1)

Up until now we have not brought algebra into play. Introducing variables that represent the areas of the blue and red parts allows us to use the ratio to create an equation that expresses that proportionality. For instance, if we let the variable B represent the blue area on the flag and the variable R represent the red on the flag, we can write the proportions algebraically as follows.

Proportion of Blue to Red on the Flag:

$$B = 2R$$

Notice that expressing the ratio in this way is actually saying something more powerful than "as the blue increases by 1, the red increases by 2." The blue can increase by any amount, say 0.84, and the red will increase by just the right amount (in this case, $2 \cdot 0.84 = 1.68$) so that the two colors always have the same proportional relationship; that is, the relationship between the two variable quantities are algebraically tied together in this way.

Helpful Hint

If *A* is proportional to *B*, then *B* is also proportional to *A*.

 Helpful Hint

The term **invariant** literally means "does not vary;" that is, it is a value that remains unchanged.

📖 **PROPORTIONAL TO**

A variable quantity *A* is **proportional to** a variable quantity *B* when there exists an invariant *k* (the constant of proportionality) such that $A = kB$.

This definition formalizes the idea that a proportional relationship is a relationship between three quantities, two of which can change at will and a third (*k*) that is constant. Specifically, the fixed constant is an *invariant* called the *constant of proportionality*.

In the flag equation $B = 2R$, the constant of proportionality *k* is 2, which tells us that there is twice as much blue as red. Similarly, we might instead write $R = \dfrac{1}{2}B$, with a constant of proportionality *k* of $\dfrac{1}{2}$. Notice that both of these constants are two of the ratio fractions that we found earlier. In general, we can think of the constant of proportionality as a rate or a conversion factor that connects the two variable quantities that are proportional to one another.

Example 4.5.1

Finding the Constant of Proportionality

The scale of a model car is often given in a ratio format that indicates the relative size of the model to the real object. One particular size is built on a one eighteenth scale to the original full-size car. This indicates that for every 1-inch length on the model, the corresponding part of the actual car is 18 inches long. Find the constant of proportionality for the length of the scale model to the length of the actual car.

Solution

Since the length of the model car is proportional to that of the full-size car, we know that a constant of proportionality exists.

We can use the formula $M = kR$ to set up an equation for the proportionality of the car, where *M* represents the length of the model car and *R* represents the length of the real car. Since the model car is built on a one eighteenth scale to the real car, we know that $M = 1$ and $R = 18$. Substituting these values into the equation and solving for *k* gives the following.

$$M = kR$$
$$1 = k18$$
$$\frac{1}{18} = k$$

Thus, the constant of proportionality is $\dfrac{1}{18}$. Notice that this tells us the model has a length that is $\dfrac{1}{18}$ the length of the real car. We might also write $R = 18M$, which implies that the real car has a length that is 18 times the length of the model car.

✔ **Skill Check 4.5.1**

Which of the following have a constant of proportionality equal to 3?

a. $3y = 3x$

b. $3y = 2x$

c. $5y = 15x$

d. $4y = 12x$

Once we know the constant of proportionality for a proportional relationship *and* one of the quantities in the relationship, we find the unknown quantity in the relationship. By definition, we know that the two quantities are related in a specific way defined by the constant of proportionality.

Example 4.5.2

Using the Constant of Proportionality

Minimundus in Klagenfurt, Austria, is an outdoor museum that is home to over 150 miniature models of prominent architecture from all over the globe. The museum has everything from the Eiffel Tower to St. Peter's Basilica to the US White House. Whenever possible, the same building materials used for the actual building are used in all of the models. Each replica is a built on a one twenty-fifth scale of the original building.

a. Find the constant of proportionality that connects the heights of the real buildings R to the heights of the model buildings M.

b. The model of the Sagrada Familia Basilica at Minimundus stands 22.4 ft. Find the height of the actual basilica in Barcelona, Spain.

c. If the Eiffel Tower in Paris is 1063 ft, how tall is the model tower in Minimundus?

d. The State Dining Room in the US White House measures approximately 48 feet by 36 feet and can seat 140 people. If miniature people are scaled at the same ratio as the model buildings, how many could be seated in the State Dining Room of the Minimundus model?

Solution

a. As we saw in the previous example, we can use the formula $M = kR$ to set up an equation for the proportionality of the model, where M is the height of the model and R is the height of the actual building. Since the model is built on a one twenty-fifth scale of the original building, we know that $M = 1$ and $R = 25$. Substituting these values into the equation and solving for k gives the following.

$$M = kR$$
$$1 = k25$$
$$\frac{1}{25} = k$$

Therefore, the constant of proportionality for the heights of the actual buildings to the heights of the model is 25. This means that a real building's height is 25 times the height of its model. And the models are $\frac{1}{25}$ the size of the real buildings. So we can use the equivalent equations $R = 25M$ and $M = \frac{1}{25}R$.

b. Now that we know the constant of proportionality, we can easily find the height of the Sagrada Familia Basilica in Barcelona.

We know the real height is 25 times the height of the Minimundus model.

$$R = 25M$$
$$= 25(22.4 \text{ ft})$$
$$= 560 \text{ ft}$$

The height of the actual basilica in Barcelona, Spain is 560 feet.

c. If the Eiffel Tower in Paris is 1063 ft tall, we know from the constant of proportionality that the model is $\frac{1}{25}$ of that height. So we have the following equation.

$$M = \frac{1}{25}R$$
$$= \frac{1}{25}(1063 \text{ ft})$$
$$= 42.52 \text{ ft}$$

Therefore, the model tower is approximately 42.5 feet tall.

d. Because the people are scaled at the same ratio as the model buildings, the miniature State Dining Room would be able to seat the same number of people as the real White House State Dining Room, 140.

So far, we've only found the constants of proportionality that arise from ratios, but they can occur in many other ways. Often, they arise from a rate or a conversion factor connecting two quantities.

One such connection dates back to the 17th century when Robert Hooke discovered a principle of physics connecting the force applied to a spring and its elasticity. Hooke's Law states that the length of a spring will always be proportional to the weight on the spring at a constant rate, no matter which units are used. Let's look at that algebraically.

If we let the length of the spring be represented by x, and the weight (or force) on the spring be represented by F, then we can create a rate $\frac{F}{x}$. Since this is a proportional relationship, it must always equal some constant; that is, the constant of proportionality, which in this case is k. This gives us the following formula.

$$\text{Hook's Law: } \frac{F}{x} = k$$

The formal definition of Hooke's Law is usually written in the form shown in the definition box.

📖 HOOKE'S LAW

The proportional relationship between the force applied to a spring and the length of the spring is given by

$$F = kx,$$

where F is the force applied to a spring, x is the displacement of the spring (either stretched or compressed), and k is the spring constant.

Notice that the formal definition of Hooke's Law is in the form of a proportional equation where k is the constant of proportionality.

Example 4.5.3

Using Hooke's Law

A force of 200 newtons (N) stretches a spring by 4 meters (m).

a. What is the value of the constant of proportionality (the spring constant)?

b. How much force would be needed to stretch the spring 4.5 meters?

Solution

a. We know the spring constant is the ratio between the force and the distance the spring is stretched. We can use the formula $F = kx$ to set up an equation for the proportionality of the spring, where F is the force applied to the spring and x is the length of the spring.

$$F = kx$$
$$200 \text{ N} = k(4 \text{ m})$$
$$\frac{200 \text{ N}}{4 \text{ m}} = k$$
$$50 \frac{\text{N}}{\text{m}} = k$$

Therefore, the constant of proportionality is $50 \frac{\text{N}}{\text{m}}$. This tells us is that in order to stretch this spring a distance of 1 meter, a weight or force of 50 newtons is required.

b. In order to stretch the spring 4.5 meters, we can multiply the constant of proportionality that we found in part a. by the length of 4.5 meters.

$$F = 50 \frac{\text{N}}{\cancel{\text{m}}} \cdot 4.5 \cancel{\text{m}} = 225 \text{ N}$$

Therefore, a force of 225 newtons is needed to stretch the spring 4.5 meters.

Example 4.5.4

Exploring the Relationship between the Constant of Proportionality and *Solving a Proportion*

Suppose you want to know how much money you've managed to save in your penny jar, but you don't want to count out the pennies directly. You have the following information.

• 25 pennies weigh 2.20 ounces.

• The weight of all the pennies in your jar is 148 ounces.

Use the fact that the number of pennies in your jar is proportional to the weight of those pennies and find out how much you have saved in pennies.

Solution

We're going to use two different methods to find the amount of money in the jar—one using the constant of proportionality and one by *solving a proportion*.

Finding the Constant of Proportionality

In this first method, will use the equation $n = kw$, where n is the number of pennies and w is the weight of the pennies. Since we know that 25 pennies weigh 2.20 ounces, we can substitute $n = 25$ and $w = 2.20$ into the equation and solve for k.

$$n = kw$$
$$25 \text{ pennies} = k(2.20 \text{ oz})$$
$$\frac{25 \text{ pennies}}{2.20 \text{ oz}} = k$$
$$11.364 \, \frac{\text{pennies}}{\text{oz}} \approx k$$

In other words, there are approximately 11.364 pennies for every ounce. Notice we can think of this as a rate of pennies per ounce.

Using this constant of proportionality, we can use the equation $n = \left(11.364 \, \dfrac{\text{pennies}}{\text{oz}}\right)w$. We are now ready to substitute the weight of the pennies for the variable w.

$$n = 11.364 \, \frac{\text{pennies}}{\cancel{\text{oz}}} \cdot 148 \, \cancel{\text{oz}}$$
$$= 1681.872 \text{ pennies}$$

This tells us that there are approximately 1682 pennies, or \$16.82, in the jar.

Solving a Proportion

Another way to determine the amount of money in the jar is to set up two proportions that we know to be equal to one another. Since the number of pennies is proportional to the weight of the pennies, the fraction $\dfrac{\text{\# of pennies}}{\text{weight}}$ is a constant of proportionality. Thus, it will always be true that corresponding fractions will be equivalent.Once again, we will begin by letting n be the unknown number of pennies in the jar and use the proportion of the number of pennies to the weight of the pennies. We know that 25 pennies weigh 2.20 ounces and that n pennies weigh 148 ounces. Setting these two equivalent fractions equal gives us the following equation.

$$\frac{25 \text{ pennies}}{2.20 \text{ oz}} = \frac{n \text{ pennies}}{148 \text{ oz}}$$

We want to find the value of n. We do this by multiplying each side of the equation by 148 to find our answer.

$$148 \, \cancel{\text{oz}} \cdot \frac{25 \text{ pennies}}{2.20 \, \cancel{\text{oz}}} = \frac{n \text{ pennies}}{\cancel{148 \text{ oz}}} \cdot \cancel{148 \text{ oz}}$$
$$1681.8 \text{ pennies} \approx n$$

This method also tells us there are approximately 1682 pennies, or \$16.82, in the jar.

Either method is correct. This second method of setting two equivalent fractions equal to one another and solving for an unknown quantity in one of the fractions is often referred to as solving a proportion.

Notice that although either method allows us to find the amount of money in the jar, there is a difference in what the two methods provide in the solution. In method b., we get a solution to this particular problem; that is, how many pennies are in *this* jar. However, in method a., by finding the constant of proportionality first, we could determine the number of pennies that are in any given amount of weight, not just the ones in this particular 148 oz jar. In the big picture, this is far more useful information.

✔ Skill Check 4.5.3

Use the information in Example 4.5.4 to find the monetary value of a stack of pennies that weighs 53 ounces.

So far, we have explored several instances of finding and using the constant of proportionality. It is important to realize that when two quantities are proportional, it means there is more to the relationship than the existence of a conversion factor that relates one to the other. Just as we have stated before, it means that the two are connected by an algebraic formula. Sometimes we need to combine that algebraic formula with other algebraic techniques to reach the desired result. The next example illustrates just such a situation.

Example 4.5.5

Using Proportionality in Business

Raw mixed nuts make a healthy and delicious snack. Market research suggests that consumers will impulse buy a snack bag for 99¢ while standing in the checkout line. It also found that consumers prefer that the weight ratio of cashews, Brazil nuts, and almonds to be 2 : 3 : 5.

On the financial side, a company can purchase, repackage, and distribute cashews for 81¢ per ounce, Brazil nuts for 75¢ per ounce, and almonds for 63¢ per ounce. Suppose a company wants to make a snack bag to display at the checkout line. Follow the steps to determine what size of bag would be needed, in ounces, to maintain the nut ratio and provide a 10¢ margin of profit on each snack bag.

Each step in the following list will walk through the steps using algebra techniques to write and solve equations based on the information given. As always, the first step is to define the variables that are needed. Then we will construct an equation for the price of the snack bag and a separate equation connecting the ratio given and the cost of each nut. Lastly, we will combine these equations to determine the size of the bag.

a. Define the variables involved.

b. Write an equation that connects the sales price of the bag to the company's cost for the nuts.

c. Because the cost of each type of nut varies, we need an expression for the proportion of the total weight of the snack bag for each kind of nut. Use the ratio given to write expressions for the weight of each kind of nut.

d. Write an equation that connects the weight of nuts in a bag to the company's cost for the nuts.

e. Use the equations from parts b. and d. to determine what the weight of a 99¢ bag should be.

Solution

a. The scenario describes two variables: weight of the snack bag and the company's cost for the snack bag. We will define the variables as follows.

w = weight of the snack bag in ounces

c = company's cost for the snack bag in cents

b. We are told that the company wants to make 10¢ profit per bag. Therefore, the price of the snack bag, 99¢, will need to be 10¢ more than the company's cost (c).

Equation Connecting Sales Price to Company's Cost:

$$99¢ = c + 10¢$$

c. First let's consider what we know from the ratio of weights of the nuts that we were given.

$$2 : 3 : 5\text{—cashews : Brazil nuts : almonds}$$

Since a ratio divides a whole into equal parts, we know the weight of each bag (w) in this case is divided into $2 + 3 + 5 = 10$ equal parts. Therefore, each part will have weight of $\dfrac{w}{10}$.

From there, we can represent the weight of each type of nut in a bag by the number of its parts. We will use Table 4.5.1 to help us keep track of the details of each expression.

Table 4.5.1: Mixed Nut Bag

Type of Nut	Number of Parts in the Ratio	Portion of the Total Weight
Cashews	2	$\dfrac{2w}{10}$
Brazil nuts	3	$\dfrac{3w}{10}$
Almonds	5	$\dfrac{5w}{10}$

d. For each nut, the company's cost is computed by multiplying the weight of the nut in the bag, which we found in part c., by its respective cost, which we were given. Once again, we'll use a table to keep track of the details.

Table 4.5.2: Mixed Nut Bag

Portion of the Total Weight	Company's Cost per Type of Nut (in Cents per Ounce)	Company's Cost per Bag (in Cents per Ounce)
Cashews: $\dfrac{2w}{10}$	81	$\dfrac{2w}{10} \cdot 81 = \dfrac{162w}{10}$
Brazil nuts: $\dfrac{3w}{10}$	75	$\dfrac{3w}{10} \cdot 75 = \dfrac{225w}{10}$
Almonds: $\dfrac{5w}{10}$	63	$\dfrac{5w}{10} \cdot 63 = \dfrac{315w}{10}$

The company's overall cost (c) for nuts in every bag that they produce, no matter what size bag, is the sum of the expressions in the last column of Table 4.5.2. So the company's total cost (in ¢) for the nuts in any size bag where w represents the weight in ounces is calculated using the following formula.

$$c = \frac{162w}{10} + \frac{225w}{10} + \frac{315w}{10}$$

$$= \frac{702w}{10}$$

$$= 70.2w$$

Note that we have just determined the constant of proportionality for the cost in cents compared to the weight in ounces for each bag. It is 70.2¢ per oz.

e. The two equations found in parts b. and d. are $99 = c + 10$ and $c = 70.2w$. Notice that because both equations include the variable c, we can use substitution to eliminate the variable c. We will take the second equation and substitute it into the first one and solve for w.

$$99 = 70.2w + 10$$

$$89 = 70.2w$$

$$1.27 \approx w$$

Therefore, the 99¢ snack bag should contain approximately 1.27 ounces of nuts for the company to make 10¢ profit.

OBJECTIVE 2 Determining Proportionality

How can you determine if one quantity is proportional to another? There are three methods: numerically, graphically, or algebraically. The first method allows you to establish the possibility of proportionality by checking whether the corresponding pairs of values in the relationship increase or decrease at the same pace.

For instance, consider the pairs of corresponding values between the number of vials that are filled at a plant and the time it takes in seconds to fill them, as shown in Table 4.5.3. Confirm for yourself that the values increase at the same rate; that is, as the value of x increases by 1, the value of y increases by 0.5.

Table 4.5.3: Number of Vials Filled per Second

The Number of Vials Filled, x	1	2	3	4	5	6
The Time in Seconds, y	0.5	1	1.5	2	2.5	3

Because the values increase at the same rate each time, we can conclude from this snapshot of the data that there is a proportional relationship between the two quantities. This method is the least reliable in that it's impossible to check all possible values. However, we can use this method to show that two quantities are *not* proportional.

The second method for confirming proportionality is to consider a graph of the pairs of values. **If a relationship is proportional, the graph of the corresponding pairs of values will be a line that passes through the origin, and the slope of that line is the constant of proportionality.**

Consider the following graph of the value of a US dollar to that of a Mexican peso. The exchange rate illustrated by this graph, which is the constant of proportionality, is 22.44 pesos per dollar.

US Dollars to Mexican Pesos

Figure 4.5.3

Recall that the slope of a line tells you the rate at which the y-values change as the x-values change. You might remember learning to calculate slope as "the change in y over the change in x." This is a rate of change, as we discussed in Section 4.3, and it is also the constant of proportionality if x and y are proportional. If the line passes through the origin, the variables are proportional.

The last method is to use algebra. From the definition, we know values that have a proportional relationship can be written in the form $A = kB$. So any such equation is representing a proportional relationship.

Consider the volume of a cylindrical container. The volume is defined to be the area of the base times the height of the container. If we let A be the area of the base and h be the height of the container, then the volume can be written as follows.

$$\text{Volume} = \text{Area of Base} \cdot \text{Height}$$
$$v = Ah$$

Notice that this equation is in the required format for a proportion, and A (the area of the base) is the constant of proportionality where the base is a fixed size and shape and the height can vary.

So for a cylindrical container with a base of a given size, the volume it contains is proportional to its height.

Example 4.5.6

Determining Whether Quantities Are Proportional

Consider a stack of traffic caution cones. Is the height of the cones in the stack proportional to the number of cones in the stack?

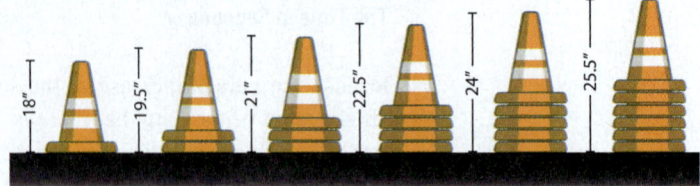

Figure 4.5.4

Solution

In order to confirm whether these quantities are proportional, we can use any one of the three methods discussed. However, let's examine all of the methods—pairs of values, a graph, and an equation—so that we can see how each technique could be applied.

Without any values, intuitively we might assume that they are proportional because the height of the stack increases by the same amount each time a cone is added since the cones are all the same height. Consider the following table that gives the varying heights of the stack as more cones are added.

Table 4.5.4: Traffic Cone Stacks

Number of Cones in the Stack	0	1	2	3	4	5	6
Height of the Stack (inches)	0	18	19.5	21	22.5	24	25.5

Notice that the number of cones in the stack increases by 1 each time. Similarly, the increase in height is 1.5 inches with each addition, except for the first increase. The first increase is 18 inches. So although the increase is same for all the others, to be proportional, the increase must be the same at all times. From this, we can conclude that the number of cones in a stack and its height are *not* proportional.

If we graph the paired values, we can also confirm that they are not proportional.

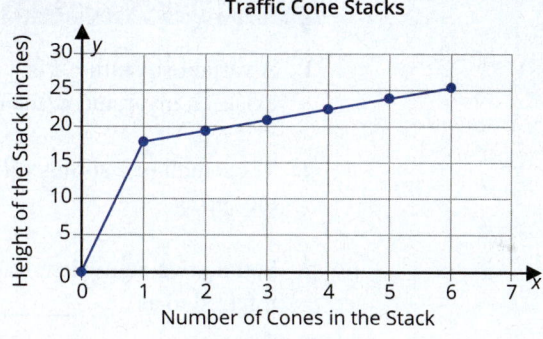

Figure 4.5.5

Notice that the graph of the paired values is not a straight line, although it does go through the origin.

You might notice that the problem lies with a stack of 0 cones. Suppose we discard the stack with no cones and consider only stacks that have at least one cone in them. Here is the graph without the 0-cone stack along with its equation.

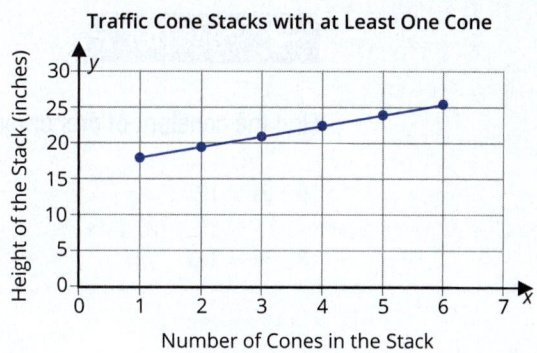

Figure 4.5.6

Heights of Traffic Cone Stacks:

$$h = c(1.5) + 16.5$$

In this equation, h = height in inches of the stack and c = number of cones in the stack.

Two things to note here. First, looking at the graph, if we extend the line, it no longer intersects the origin as it should for proportional relationships. Second, the equation does not adhere to the correct form. The extra 16.5 shatters the form of $y = kx$ necessary for a proportional equation.

Skill Check Answers
1. c. and d. **2.** a. and b. **3.** $6.02

4.5   Exercises

✔ CONCEPT CHECK

1. A variable quantity A is _____ to a variable quantity b when there exists an invariant, k (the constant of proportionality), such that $A = kb$.

2. The length of a spring will always be proportional to the _____ on the spring.

3. Setting two equivalent fractions equal to one another and solving is often referred to as _____.

4. When two quantities are proportional, they are connected by an _____ formula.

5. If a relationship is proportional, the graph of the corresponding values will be a line that passes through the _____, and the slope of that line is the constant of proportionality.

💡 PRACTICE

Find the constant of proportionality.

6. $3y = 12x$

7. $7a = 35b$

8. $4r = 36s$

9. $5h = 55w$

Solve.

10. A force of 500 newtons stretches a spring by 2 meters.

    a. What is the constant of proportionality?

    b. How much force would be needed to stretch the spring 5 meters?

11. A force of 70 newtons stretches a spring by 5 meters.

    a. What is the constant of proportionality?

    b. How long would the spring stretch if a force of 42 newtons was used?

12. A force of 240 newtons stretches a spring by 4 meters.

    a. What is the constant of proportionality?

    b. How much force would be needed to stretch the spring 7.5 meters?

13. A force of 125 newtons stretches a spring 5 meters.

    a. What is the constant of proportionality?

    b. How long would the spring stretch if a force of 225 newtons was used?

🚀 APPLICATIONS

14. If a model car is an S scale model car, then for every 1-inch length on the model, the corresponding part of the actual car is 64 inches long. Find the constant of proportionality for the length of the scale model to the length of the actual car.

15. On a map for a local community park, the scale indicates that every inch of length on the map corresponds to 48 inches of length of the park. Find the constant of proportionality for the scale of the map to the actual park.

16. The scale on a map of a town center indicates that one inch on the map is equivalent to 60 inches in the town center. Find the constant of proportionality for the length on the map to the length in the town center.

17. For an HO scale model train, each inch of length of the on the model train corresponds to 87 inches on the actual train. Find the constant of proportionality for the length on the model train to the length on the actual train.

18. The scale given on a blueprint of a city park indicates that each inch on the blueprint is equivalent to 8 feet of the actual park. If a path in the park is 160 feet long, how long is the path on the blueprint?

19. For a G scale model train, each inch of length of the model train is equivalent to 25 inches of length on the actual train. If a real boxcar is 665 inches long, how long is the G scale model of the boxcar?

20. At Minimumdus, an outdoor museum in Klagenfurt, Austria, each building replica is a scaled version of the original with a ratio of 1 : 25. If the actual Taj Mahal is 73 meters tall, how tall is the replica at Minimundus?

21. A Cuda Dragster car model has a scale of 1 : 25. If the tire height of the model is 1.06 inches, what is the height of the life-size tire?

22. The Clarksville football team outscored its opponents 5 : 3 last season. If their opponents scored 78 points, how many points did Clarksville score?

23. The scale of a map is 1 in. = 11.9 miles. How many inches will be drawn on the map to represent 30.9 miles?

24. Find the actual width of a building if the model of the building is 5 cm wide by 58.7 cm long and the actual length of the building is 140.9 m.

25. The ratio of outdoor swimmers to indoor swimmers at the recreation center is 3 : 2. If the center has 55 swimmers altogether, how many of them are indoor swimmers?

26. On a globe model of the Earth, one centimeter is equivalent to 650 kilometers on the actual Earth. The straight-line distance from Juneau, AK, to Orlando, FL, is approximately 5215 kilometers. What would the distance between the two cities be on the globe model?

27. You are packing to move and need to pack your collection of paperback novels. The moving company you hired requires that small boxes weigh no more than 45 pounds. You weigh a few of your books and determine that 4 novels weigh approximately 3.6 pounds. If you have 125 novels, approximately how many boxes will you fill with books?

28. A contest is being held to guess the number of candies in a jar where the winner will receive a $50 gift card. You make an educated guess that the jar holds 17 ounces of candy and experiment with a pack of candy you purchased yourself to determine that 20 individual candies weigh 0.15 ounces. Using this information, how many candies would you guess are in the jar?

29. A wedding planner is assembling small gift bags for an upcoming wedding reception. The couple set aside $240 of their wedding budget for the gifts, $120 of which is specifically for chocolates. The planner found chocolate hearts that are sold in packs of 8 for $2.50 per pack. How many chocolate hearts can the wedding planner purchase for the gift bags?

30. A general goods store at a campsite needs to restock their inventory of fire starters. The store's distributor sells boxes of 12 fire starter kits for $16. If the store's budget for ordering fire starters is $260, how many fire starter kits can be purchased? (**Hint:** You can only purchase whole boxes.)

31. While packing up donations for distribution, a worker at a donation center forgot to write the number of T-shirts that were packed into a box before sealing it with tape. Instead of unpacking the box, the worker weighed a few T-shirts and discovered that a stack of four T-shirts weighs an average of 550 grams. The worker also estimates that the weight of the contents of the box is 16 550 grams. Approximately how many T-shirts are in the box?

32. A local food bank received a shipment of 3 large boxes each containing 12-count individual mac and cheese dinners. Employees determine that each of the 12-count dinner boxes weighs 25 pounds and each of the large boxes weighs 100 pounds. How many individual dinners were received?

33. A coffee roasting company offers a blend made of coffee beans from Brazil, Colombia, and Peru in a ratio of 4 : 5 : 2. The company pays $3.50/lb for the beans from Brazil, $4.00/lb for the beans from Colombia, and $4.50/lb for the beans from Peru. If the production steps (such as roasting, packaging, and shipping), cost the company an additional $3.00 per pound, what is the total cost of producing one pound of the coffee blend?

34. Instructions for a chemical procedure state to mix salt, baking soda, and water in a 4 : 5 : 21 ratio by mass. How many grams of water would be required to make a mixture that contains 60 grams of salt?

35. A screen printing company finds that for each T-shirt it produces, 40% of the cost is for materials, 50% is for labor, and 10% is for advertising. If a batch of 100 shirts required $180 in materials, what was the total cost of producing one shirt in the batch?

36. You want to estimate the number of sheets of office paper that you have on your desk. You know the following.

    1. The standard pack of paper contains 500 sheets and is 52 millimeters high.

    2. The height of the stack you have is 20 millimeters.

    Use the fact that the height of a stack is proportional to the number of sheets of paper in it to find the number of sheets in your stack.

37. You need to determine how many cans a machine can produce in half an hour. You know that the average amount of cans produced in a 6-hour shift is 3000. Use the fact that the number of cans produced is proportional to time to find the number of cans that is produced in half an hour.

38. The following table contains the amount of simple interest paid on a loan in relation to the length of the loan.

Length of the Loan (years)	1	2	3	4	5	6
Interest ($)	70	140	210	280	350	420

Is the interest proportional to the length of the loan? If it is, find the constant of proportionality. If not, explain why.

39. The following table contains the amount of simple interest paid on a loan in relation to the length of the loan.

Length of the Loan (years)	1	2	3	4	5	6
Interest ($)	308.39	635.80	983.40	1352.45	1744.25	2160.22

Is the interest proportional to the length of the loan? If it is, find the constant of proportionality. If not, explain why.

40. Consider the distance covered by a car moving at a constant speed.

Time (minutes)	3	5	10	15	20
Distance (miles)	2.4	4	8	12	16

Is the distance proportional to the time spent? If it is, find the constant of proportionality. If not, explain why.

41. Consider the weight of a bottle of milk in relation on the amount of milk in the bottle.

Amount of Milk (milliliters)	300	400	550	750
Weight (grams)	800	900	1050	1250

Is the amount of milk proportional to the weight of the bottle? If it is, find the constant of proportionality. If not, explain why.

42. The air temperature in Anchorage, Alaska, during a day in 2022 is shown in the table.

Hours after Midnight	3	6	9	12	15
Temperature (°C)	1	2	3	5	6

Is the temperature proportional to the number of hours passed after midnight? If it is, find the constant of proportionality. If not, explain why.

43. The table shows the amount of flour needed to make a loaf of bread.

Flour (grams)	400	1200	1600	2800
Bread (loafs)	1	3	4	7

Is the amount of flour proportional to the number of loafs of bread baked? If it is, find the constant of proportionality. If not, explain why.

44. The growth of a child during the first 6 months of life is shown in the following table.

Age (months)	1	2	3	4	5	6
Height (centimeters)	55	58	61	64	67	70

Is the height of the child proportional to the age of the child? If it is, find the constant of proportionality. If not, explain why.

45. The temperature of water heated over a span of 8 minutes is shown in the table.

Time (minutes)	1	2	4	6	8
Temperature (°C)	30	40	60	80	100

Is the temperature of the water proportional to the time it was heated? If it is, find the constant of proportionality. If not, explain why.

46. The following table shows the amount of water needed to fill a cylindrical tank depending on the tank's diameter.

Diameter (meters)	1	2	3	4	5
Volume of Water (liters)	1570	6280	14,140	25,130	39,270

Is the volume proportional to the diameter of the tank? If it is, find the constant of proportionality. If not, explain why.

4.5 PROJECT

WHY ARE CELLS SO SMALL?

The germ theory of disease is the fundamental theory of medicine that says infectious diseases are caused by microscopic organisms called pathogens. The theory wasn't accepted until the end of the 19th century, in part because pathogens are invisible to the naked eye.

A main type of pathogen is bacteria (singular, bacterium). These are one-celled (or unicellular) living organisms present almost everywhere on Earth.

In this project, we will explore one possible reason that bacteria are so small.

In order to simplify matters, let's assume we are going to study a bacterium that is perfectly spherical.

Biologists believe that the ability of a bacterium to obtain resources (such as food) is proportional to its radius. They also believe that the bacterium's need (or demand) for resources is proportional to the square of its radius. If the demand for resources is less than or equal to the ability to obtain resources, then the bacterium is able to sustain itself.

The table below contains the values for the ability to obtain resources (A) and the demand for resources (D) for a hypothetical bacterium. Units are omitted for clarity.

Radius (R)	Ability (A)	Radius Squared (R^2)	Demand (D)
1.5	0.75	2.25	0.45
2	1	4	0.8
2.5	1.25	6.25	1.25

1. Assume that A is proportional to R; that is, $A = k_1 R$ for a constant of proportionality k_1. Determine the value of k_1.

2. Assume that D is proportional to R^2; that is, $D = k_2 R^2$ for a constant of proportionality k_2. Determine the value of k_2.

3. When the radius doubles from 1.5 to 3.0, how many times does the ability to obtain resources increase by?

4. Determine the demand (D) from a radius of $r = 3$. How many times larger is that than the demand (D) when the radius is equal to 1.5?

5. How can these numbers help explain why most unicellular organisms are usually small?

Chapter 4 Exercises

Answer each question. Round your answer to the nearest hundredth, if necessary.

1. If six out of every ten individuals in a population carry a gene for a defective enzyme, how many individuals carry the normal gene in a population of 900?

2. As part of a house renovation, the total cost to remodel the bathrooms was $4750. The total cost of bathroom fixtures was $3750. Write the cost of the fixtures and the total cost of the bathroom renovation as a proportion in lowest terms.

3. The ratio of apples to oranges in a basket is 3 to 4. What fraction of fruit in the basket consists of apples?

4. Suppose that the ratio of males to females in the sales field of a large company is $4:5$. If there are 81 sales representatives in the company, how many are male and how many are female?

5. If Whitt wins 3 out of every 4 games of Solitaire that he plays on his computer, how many games would he need to play so that he wins 51 games?

6. Your dinner bill was $30.00. If you leave a 15% tip, how much will this be?

7. If you wish to leave a 20% tip and your bill is $58.50, how much tip will you leave?

8. A garden center is having a Memorial Day sale on flowers. All flowers are 15% off, but if you buy 3 flats of flowers, you save an additional 10%. How much would you pay if you purchase 3 flats of flowers that are priced at $12 per flat?

9. The wholesale price of a laptop is $525. The price of the laptop is marked up by 45% before it is listed as for sale on the store website. During a back-to-school sale, the laptop is marked down by 15%. What is the price of the laptop during the back-to-school sale?

10. Nate just moved from a state with an 8% sales tax to a state with a 6% sales tax. If he buys a shirt for $35.00, how much less tax does he pay in his new state?

11. Last year, a new computer sold for $1600. This year, the same computer can be purchased for $1200. What percentage discount is being given from the previous year?

12. Suzanne's salary grew from $40,000 in 2000 to $56,000 in 2012. During the same time period, Mike's salary grew from $38,000 to $53,000. Whose salary grew more in absolute change? In percentage change?

Write each rate described and then convert it to a unit rate.

13. The university reported that there were 42 freshmen in the entering class for every 3 advisors.

14. Coach Burge wanted the entire class to be able to do 90 sit-ups in 2 minutes by the end of the semester.

Solve each problem. Round your answer to the nearest hundredth, if necessary.

15. Which is the better value for the money, $5.99 for a package of 8 cold remedy capsules or $2.85 for 2 capsules? Why?

16. Suppose at bulk rate, instant potato flakes cost 12.04 cents, or $0.1204, per ounce.

   a. If a box costs $77.05, how many pounds of potato flakes are in the box? (**Hint:** There are 16 ounces in a pound.)

   b. What would a 2-pound box of potato flakes cost?

17. To treat the water in a swimming pool, it is recommended to add 6.4 ounces of chlorine per 5000 gallons of water. How much chlorine is needed to treat the water in a pool that holds 24,000 gallons?

18. An advertisement for a hybrid sedan claims that the car can travel 600 miles on a full tank of gas. IF the fuel tank capacity for the sedan is 11.4 gallons, what is the miles per gallon that the sedan gets?

19. Convert the following rainfall measurements from inches per hour to millimeters per minute. Note that 1 inch is equivalent to 25.4 mm.

   a. 0.5 inches per hour

   b. 2 inches per hour

   c. 0.01 inches per hour

20. A vehicle requires 0.25 gallons of fuel to travel 1 kilometer. Determine how many gallons per mile are required.

21. A restaurant sells soup in a quantity of 20 fluid ounces per container. Determine this amount as cups per container.

22. A garden snail can travel 6 centimeters per minute. Determine the speed of the snail in inches per hour.

23. A sculpture for the lawn in front of the new art gallery is 38.5 feet long. A model is being made for a display inside the gallery. The model is 14.2 inches long. Based on this scale, how long would an actual bench be on the lawn if the bench in the scale model is 2 inches?

24. A 180-pound student burns 120 calories every half hour of playing ultimate Frisbee. How many calories would he burn in

   a. 10 minutes?

   b. One hour and 45 minutes?

25. On a G scale model train, each inch of length corresponds to 25 inches of length on the actual model train. Find the constant of proportionality for the length on the model train to the length on the actual train.

26. A force of 144 newtons stretches a spring 4 meters.

   a. What is the constant of proportionality?

   b. How much force would be needed to stretch the spring 7.5 meters?

27. A model car has a length of 5.75 inches and a width of 2.25 inches. Find the length of the actual car if the actual car has a width of 72 inches.

CHAPTER 4 PROJECT

The Federal Budget—Where Do My Taxes Go?

When it comes to spending money, the United Stated government is similar to a family. There are bills to pay, and in order to pay these bills, the government must generate revenue and create a budget to allocate those funds. The main source of revenue for the government comes from collecting taxes. When the government spends exactly as much as it generates in revenue, we say that the government has a balanced budget. In this project, you will examine the 2019 US federal budget and compare it to a family budget and the 1999 US federal budget.

The following table[6] lists the main expenditure categories for the fiscal year 2019.

US Federal Government 2019 Budget	
Expenditure Category	Approximate Amount Spent (US dollars)
Social Security	1 trillion
Medicare and Medicaid	1.1 trillion
Defense	697 billion
Nondefense	661 billion
Interest on Debt	375 billion
Other	567 billion
Total	4.40 trillion

1. The entire revenue from federal taxes in fiscal year 2019 was approximately equal to 3.5 trillion dollars. Determine the difference between the total expenditures and the revenues in 2019.

2. Since the government spent more than the revenue brought in, we call the difference you found in part a. the *fiscal deficit* for 2019. The fiscal deficit is typically given as a positive value. Economists often compare the fiscal deficit to the entire economic output of the nation. The economic output is the total amount of money produced by the economy, which is known as the Gross Domestic Product (GDP). Perform an internet search and find the United States 2019 GDP.

3. What percentage of the 2019 GDP was the federal deficit in 2019? (Round your answer to the nearest tenth.) This number is known as the deficit-to-GDP rate. Would it be better for the country to have a higher or lower deficit-to-GDP rate? Explain your reasoning.

4. There is a lot of discussion among economists about what is considered an acceptable deficit-to-GDP rate. The general consensus is that a number under 2% is acceptable as a rate that isn't too concerning. If all the US expenditures remained the same in 2019, what would the revenue (in dollars) have to be to attain a 2% deficit-to-GDP rate? Without increasing economic output (that is, keeping the GDP the same), how could the government generate more revenue?

5. Some amounts in the budget cannot be changed. For example, Social Security must be paid to everyone that is entitled to receive it. On the other hand, the amounts for other certain categories are set by Congress and can be negotiated. Perform an internet search on discretionary expenses versus non-discretionary expenses. Write down the definition of each expense and provide an example of each expense in the federal budget.

6. If the total revenue and GDP remain the same, what change to the budget would be necessary to attain a 2% deficit-to-GDP rate? What specific area(s) would you target for making the change? Explain your reasoning.

7. Let's compare the deficit to the actual revenues. What percentage of the 2019 total revenue is the 2019 deficit?

---

6   "Policy Basics: Where Do Our Federal Tax Dollars Go?" Center on Budget and Policy Priorities, Updated April 9, 2020, https://www.cbpp.org/research/federal-budget/where-do-our-federal-tax-dollars-go.

8. Now, suppose that a family of four has an income of $63,000 per year. If this family had the same deficit to revenue rate that the federal government had in 2019, how much would they have to borrow every year to pay their bills?

9. List some methods that the family (and the federal government) could use to balance its budget. In both cases, list some discretionary and non-discretionary expenses that could be adjusted.

In 1999, things were a bit different for the US economy. The 1999 budget had expenditures totaling 1.7 trillion dollars and revenues of 1.826 trillion. Back then, the government was bringing in more money than it spent. In cases like this, we call the difference between revenue and expenditure a *surplus*.

10. Calculate the 1999 federal surplus in billions of dollars by finding the difference between the expenditures total and the revenues.

11. Perform an internet search to find the United States 1999 GDP.

12. Determine the surplus-to-GDP ratio in 1999.

13. What was the percentage change in government spending between 1999 and 2019?

14. Now, determine the percentage change in GDP between 1999 and 2019.

15. Argue how the difference between the percentage change of government spending and the percentage change in GDP can be used to partially explain the emergence of large deficits in the federal budget.

16. Calculate the percentage of the revenue that was the surplus in 1999.

17. Suppose that our family of four had the surplus-to-revenue ratio that the federal government had in 1999. How much would they have saved at the end of the year?

CHAPTER 4 REVIEW

Section 4.1 Proportions, Percentages, and Ratios

Definitions

Proportion: A **proportion** is a fraction of a whole.

Percentage: A **percentage** is an amount per 100, expressed using a percent sign, %.

Ratio: A **ratio** is a comparison of measured quantities whose units are of the same type. It is most often written with a colon between quantities; for example, $a:b$.

Equivalent Ratios: **Equivalent ratios** are ratios that express the same relationship.

Section 4.2 Using Percentages

Definitions

List Price: The price of an item as it is listed for public sale is the **list price**.

Discount: A **discount** is a reduction made from the list price. This is often given as a percentage of the list price.

Sale Price: The **sale price**, also called the **net price**, is the actual cost of an item after any discounts are subtracted.

$$\text{Sale Price} = \text{List Price} - \text{Discount}$$

Sales Tax: A charge collected by the government for the sale of goods is called a **sales tax**.

$$\text{Sales Tax} = (\text{Selling Price}) \cdot (\text{Sales Tax Percentage as a Decimal})$$

Absolute Change: The **absolute change** between two amounts is the absolute value of the difference between the two numbers.

$$\text{Absolute change} = |\text{New Amount} - \text{Original Amount}|$$

Percentage Change: The **percentage change** between two amounts is the absolute change expressed as a percentage of the original value.

$$\text{Percentage Change} = \frac{\text{Absolute Change}}{\text{Original Amount}} \cdot 100\%$$

Section 4.3 Rates, Unit Rates, and Rates of Change

Definitions

Rate: A **rate** is a fraction used to compare two quantities that are not necessarily in the same units.

Unit Rate: A **unit rate** is a rate comparing two measured quantities, one of which is a single unit written in the denominator of the fraction.

Rate of Change: The **rate of change** of one quantity with respect to another is how much the first quantity changes for each unit change of the second quantity.

$$\text{Rate of Change} = \frac{\text{Change in 1st Quantity}}{\text{Change in 2nd Quantity}}$$

Section 4.4 Using Rates for Dimensional Analysis

Definitions

Conversion Factor: A **conversion factor** is a fraction that is equal to 1 but contains different units of measurement in the numerator and denominator.

Dimensional Analysis: Dimensional analysis is a method of using conversion factors to change from one unit of measurement to another.

Section 4.5 Proportionality

Definition

Proportional to: A variable quantity A is **proportional to** a variable quantity B when there exists an invariant k (the constant of proportionality) such that $A = kB$.

Formula

Hooke's Law: The proportional relationship between the force applied to a spring and the length of the spring is given by

$$F = kx,$$

where F is the force applied to a spring, x is the displacement of the spring (either stretched or compressed), and k is the spring constant.

CHAPTER 5

Algebra: Equations, Inequalities, and Functions

🔖 SECTIONS

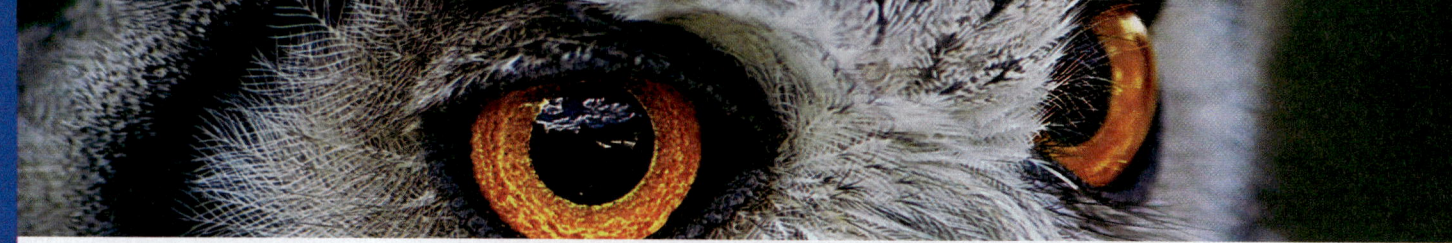

Introduction

Just as visual art represents our world with pictures, and music represents our world with lyrics and melodies, mathematics provides a means of representing the world numerically. Linear, quadratic, exponential, and logarithmic equations can be used to model a wide variety of different scenarios, from paying down a loan to tracking the growth rate of bacteria. Once created, these models can be used to inform everything from the choices we make that affect our everyday lives (such as how to proceed with personal finance decisions) to decisions that affect the world (such as whether to declare a global pandemic).

The Center for Disease Control and Prevention looks at mathematical models to determine how quickly a virus is spreading and also looks at the clinical severity, which is the seriousness of the illness associated with the infection. The CDC has determined that a virus such as influenza (a contagious respiratory illness that infects the nose, throat, and lungs) can spread when an infected person comes within 6 feet of someone not infected. Assuming that a given person who is infected with the virus could come in contact with 10 uninfected people each day, the spread of the virus would occur very rapidly if all 10 uninfected people actually become infected with the virus. If this pattern were to continue, those 10 people could, in turn, infect 10 more each. After only one week, there would be 1,000,000 people infected. After two weeks, the number infected would be greater than the entire population of the world. Thankfully, not everyone is susceptible to the virus on first contact, while others take a vaccination to help prevent them from contracting the virus. Situations such as this are carefully studied by biologists, in conjunction with mathematicians, to model the growth of diseases in an effort to best address the needs of a population.

In this chapter, we will look at several different types of equations and real-world examples to illustrate the usefulness of mathematical modeling.

OBJECTIVES

1. Solve linear equations.

2. Solve formulas for a specified variable.

3. Evaluate functions.

> **Think Back**
>
> A variable is a symbol (often a letter) used to represent an unknown quantity.

OBJECTIVE 1 Linear Equations

A friend of mine once had an Uber driver engage her in conversation about what she did for a living. The driver, upon hearing that she was a mathematician, responded with the ever typical, "Not sure why people need to learn that stuff. I mean, I've made it just fine my entire life without using any of it."

To which she replied, "Not that you're aware of."

The driver continued, "I mean why do we need that sign that tells you which number is bigger? Can't you just look at two numbers and see which one is bigger?"

"Yes, you're right," my friend replied. "As long as you know exactly what both numbers are. What happens when you don't know one of the numbers?"

Enter algebra. A handy means by which we can manipulate and calculate quantities and measurements when we don't have all the pieces; that is, when some numbers are unknown quantities, or *variables*.

At one point, you too might have thought or even said aloud, "When will I ever use algebra once I leave school?" What we don't realize is that when we are confronted with a scenario in our lives that could feasibly be solved by using a linear *equation* or *formula*, most often the calculation (or a rough estimate) is done in our heads—especially when the setup is not too involved. Situations that involve a *variable* amount of something are commonplace in our lives.

- How much tax will you pay on your meal? *Well, that depends on the cost of the meal.*

- What will the total cost of a shared car ride be? *That depends on how far you ride and how many people are sharing the cost.*

- Should you take the sales job with a flat salary or one with a per sale commission? *That may depend on how much risk you are comfortable with.*

This section looks at using algebra to solve equations (specifically, linear equations) that represent situations such as the ones just mentioned. Recall that when we use the term **equation**, we mean a mathematical statement that consists of two algebraic expressions joined by an equal sign that indicates the two expressions are equal in value. When we specifically talk about **linear equations**, we are only considering those equations where the variables have an exponent of one. The rate of change between the variables is a constant rate, and when we graph a linear equation, the graph is a line. We'll look more at the characteristics of linear equations throughout this chapter.

Despite the Uber driver's reservations, linear equations really do have a wide variety of applications. Let's consider a real-life use.

Everett is planning an after-party for a theatrical production. One venue he contacted charges a set fee of \$1300 to rent the space for 6 hours, plus \$12.45 per person for food and nonalcoholic drinks. Everett doesn't have a firm count on exactly how many people will attend, but he would like to have an idea of how much he should expect to pay for the venue. To model this scenario, we can represent Everett's cost with an equation that captures all of the venue information.

To express this equation algebraically, we need to introduce a variable to represent what Everett doesn't know—the number of guests. Customarily, algebraic variables are represented by lowercase letters of the alphabet. Here we will use x to represent the number of guests and use y to represent the total cost of the venue rental. The guest count x is multiplied by the cost per guest, $12.45, and then added to the set fee of $1300. The relationship between the number of guests and the total rental cost of the venue is then expressed with the following equation.

Linear Equation Representing the Total Cost to Rent the Venue

$$y = \$12.45x + \$1300$$

Notice that each time the number of guests increases (or decreases) the price will change by a constant amount of $12.45, which is a sign that the equation is linear. The x-value for the number of guests is referred to as the **independent variable**, and the y-value for the price is the **dependent variable** because its value *depends* on the other variable.

📖 DEPENDENT AND INDEPENDENT VARIABLES

Dependent Variable

The value of the **dependent variable** changes with respect to the value of the independent variable.

Independent Variable

The value of an **independent variable** does not rely on the values of the other variables in an expression or function, and its value determines the values of the other variables.

Everett has modeled his scenario by creating a linear **formula** for the cost of the venue. In other words, a rule set up to quickly establish the cost of the venue as it relates to the number of guests and the rental fee. It's important to note that the equations and formulas don't necessarily have to be linear. In fact, we will look at quadradic, exponential, and logarithmic equations later in the chapter.

Using this formula, Everett can easily calculate the change in venue cost as the number of guests fluctuates. He can also answer other questions that might arise, such as "If the budget mandates that he cannot exceed $4000 on the event, what is the maximum number of guests that could attend?" or "How does this venue compare with another venue that charges less per guest but more for the space?"

Let's consider the budget question. If Everett has $4000 to work with, how many guests can attend? By substituting $4000 in for the y-value, we have the following equation that can be solved for x, which is the number of guests.

$$\$4000 = \$12.45x + \$1300$$

You might recall from previous math classes that when you are given an algebraic equation and instructed to "solve it," what you are solving for is the collection of numbers that make the statement true. This number, or set of numbers, is called the *solution* or the *root* of the equation. We can verify that those particular numbers are solutions by substituting them back into the original equation. A solution will make the equation true.

 Helpful Hint

Simplify means to make easier to understand or reduce the number of parts.

Solve means to find a solution to a problem or question.

There are 3 types of linear equations:

1. Those with *one solution* (that is, only one number will make the equation true).

2. Those with *infinitely many solutions* (that is, any number will make the equation true).

3. Those with *no solutions* to the equation (that is, no number or set of numbers exist that will make the equation true).

📖 **POSSIBLE SOLUTIONS TO LINEAR EQUATIONS**

One solution—there exists only one number that makes the equation true.

Infinitely many solutions—any number makes the equation true.

No solution—no number exists that makes the equation true.

An equation with one solution seems easy to understand, but what about the other two categories? When we say an equation has no solutions, it means there does *not* exist a number that will make the equation true.

For instance, take the equation $x = x - 1$. It doesn't matter what number you substitute in for x, it will never be the case that a number is equal to itself minus 1. Thus, this equation has *no solution*.

Now, consider the equation $0x + 50 = 50$. Similarly, it doesn't matter what number you substitute in for the variable x, on the left-hand side you will always get 50. (Try it for yourself with any number!) Since both sides of the equation will always equal 50—thus making the equation a true statement at all times—we say the equation has *infinitely many solutions*.

Regardless of the number of solutions to a particular equation, when you begin the process of solving an equation through algebraic manipulations, your end goal is the same: isolate the variable on one side of the equation with a coefficient of 1. Through this process, you discover how many solutions there are to any particular equation.

Part of the beauty of mathematics is that it allows us to solve equations in different manners, some that might appear to be following different routes at first glance. However, we often ask "How can I do this most efficiently?" With this question in mind, it's helpful to have a systematic approach you can consistently follow to arrive at the solution of an equation. For simplicity, only one approach is covered here.

💬 **Think Back**

A **coefficient** is the number (or constant) in front of a variable. Variables without a number written before them are understood to have a coefficient of one.

💬 **Think Back**

The **least common denominator** (LCD) is the **least common multiple** (LCM) of all of the denominators, or the smallest number that is divisible by all denominators.

📋 **STEPS FOR SOLVING LINEAR EQUATIONS IN ONE VARIABLE**

Step 1: Clear the equation of fractions and decimals by multiplying both sides of the equation by the least common denominator (LCD).

Step 2: Simplify the expressions on both sides by clearing the equation of parentheses. Repeat Step 1 if fractions or decimals remain; otherwise move to Step 3.

Step 3: the expressions on both sides by combining like terms.

Step 4: Isolate the variable on one side of the equation.

Step 5: Make the variable coefficient equal to 1

Let's begin by solving Everett's equation to determine how many guests can attend his function with a budget of $4000.

Example 5.1.1

Solving a Linear Equation

Determine the maximum number of guests that can attend the after-party Everett is planning by solving the following equation.

$$\$4000 = \$12.45x + \$1300$$

Solution

Notice that the equation is already simplified and there are no fractions, decimals, or parentheses that need to be cleared. So according to the steps for solving linear equations in one variable, we can start with step 4: isolate the variable on one side of the equation.

$$\$4000 = \$12.45x + \$1300$$
$$\$2700 = \$12.45x \qquad \text{Step 4: Subtract \$1300 from both sides.}$$
$$216.87 \approx x \qquad \text{Step 5: Divide both sides by \$12.45 to}$$
$$\text{make the coefficient equal to 1.}$$

This tells us that Everett can have a maximum of 216 people attend the party and still not exceed his budget.

Not all linear equations involve whole positive integers. In the next examples, we will see how the same sequence of steps works no matter the type of numbers that are involved.

💬 Think Back

PEMDAS is a helpful way to remember the correct sequence of calculations when simplifying.

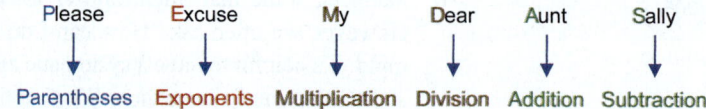

Please	Excuse	My	Dear	Aunt	Sally
↓	↓	↓	↓	↓	↓
Parentheses	Exponents	Multiplication	Division	Addition	Subtraction

Perform the multiplication and division as it appears from left to right in the expression. Similarly, do the same with addition and subtraction.

Example 5.1.2

Solving Linear Equations

For each equation, use the steps for solving linear equations to solve the given linear equation for y. State if there is one solution, no solution, or infinitely many solutions.

a. $4.1y + 2.6(y + 4) = 64$

b. $\dfrac{5w - 4}{4} + \dfrac{10}{8} = \dfrac{10w + 2}{8}$

c. $0.28t + 0.56(t - 3) = 0.84t - 1.86$

Solution

a. The variable is y this time rather than x as we have seen before, but the variable y only ever appears with an exponent of 1. So we know that $4.1y + 2.6(y + 4) = 64$ is a linear equation. Notice that the numbers 4.1 and 2.6 contain tenths. We can multiply both sides of the equation by 10 to eliminate the decimals. Afterward, we need to clear the equation of parentheses. That is, we need to distribute the number on the outside of the parentheses, multiplying it by everything on the inside.

$$4.1y + 2.6(y + 4) = 64$$

$$41y + 26(y + 4) = 640 \quad \text{Step 1: Multiply both sides by 10 to eliminate the decimals.}$$

$$41y + 26y + 104 = 640 \quad \text{Step 2: Remove the parentheses by distributing 26.}$$

$$67y + 104 = 640 \quad \text{Step 3: Combine like terms on the left-hand side of the equation.}$$

$$67y = 536 \quad \text{Step 4: Subtract 104 from both sides to isolate the variable } y.$$

$$y = 8 \quad \text{Step 5: Divide both sides by 67 to make the coefficient equal to 1.}$$

From this we know that only a single solution is possible. We'll take a moment to check that the solution does indeed solve the equation. To do this, we substitute the answer for the variable back into the original equation.

$$4.1y + 2.6(y + 4) = 64$$
$$4.1(8) + 2.6(8 + 4) = 64$$
$$4.1(8) + 2.6(12) = 64$$
$$32.8 + 31.2 = 64$$
$$64 = 64$$

Since the result is a true statement when we substitute the solution back into the original equation, we know that this equation has only one solution, $y = 8$.

Note that it is never a bad idea to check your answers when solving an equation. We will leave that for you to practice on your own after each example for the remainder of the section.

b. The equation $\dfrac{5w - 4}{4} + \dfrac{10}{8} = \dfrac{10w + 2}{8}$ is a linear equation in w, so we can follow the steps for solving a linear equation.

The first step in our guidelines is to clear the equation of fractions or decimals. To do this, multiply both sides of the equation by the LCD, which is 8 in this case.

$$\frac{5w - 4}{4} + \frac{10}{8} = \frac{10w + 2}{8}$$

$$8\left(\frac{5w - 4}{4} + \frac{10}{8}\right) = 8\left(\frac{10w + 2}{8}\right) \quad \text{Step 1: Remove the fractions by multiplying both sides by the LCD, 8.}$$

$$10w - 8 + 10 = 10w + 2 \quad \text{Step 2: Remove the parentheses by distributing 8.}$$

$$10w + 2 = 10w + 2 \quad \text{Step 3: Combine the like terms on each side.}$$

When we reach a place where both sides of the equation are the same, we can stop. This signifies that the equation is always true. That is, any number we substitute in for x will make the equation true.

Therefore, this equation has *infinitely many solutions*; that is, the solution set is all real numbers.

c. Again, notice that $0.28t + 0.56(t - 3) = 0.84t - 1.86$ is a linear equation in t, so we can follow the steps for solving a linear equation.

Step 1 says to eliminate any fractions or decimals. The decimal numbers all contain hundredths, so multiplying both sides of the equation by 100 will remove the decimals.

$$0.28t + 0.56(t - 3) = 0.84t - 1.86$$

Step 1: Remove the decimals by multiplying both sides by 100.

$$28t + 56(t - 3) = 84t - 186$$

Step 2: Clear the equation of parentheses by distributing 56.

$$28t + 56t - 168 = 84t - 186$$

Step 3: Combine like terms on each side of the equation.

$$84t - 168 = 84t - 186$$

Step 4: Subtract $84t$ from both sides.

$$-168 = -186$$

Notice that once we eliminated the variable t, we are left with a false statement. This indicates that our original equation does not have any solutions.

Therefore, we say that there is *no solution* to this equation.

> ✔ **Skill Check 5.1.1**
>
> Solve the following linear equation.
>
> $$\frac{4}{13}(x - 3) - \frac{15}{13} = -5$$

> 💬 **Helpful Hint**
>
> A **rate** is a fraction used to compare two quantities that are not necessarily in the same units.

As we saw in Chapter 4, a *proportional equation* is a special type of equation stating that two rates are equal. Solving a proportional equation for the unknown amount is often referred to as "solving a proportion." Our next example involves just that—solving a proportion. Notice that this is simply an equation involving two fractions, and we will use the same steps to solve the equation.

Example 5.1.3

Solving a Proportional Equation

In chemistry, the law of definite proportions states that every chemical compound contains fixed and constant proportions (by mass) of its component elements. Specifically, any sample of pure water contains 11.19% hydrogen and 88.81% oxygen by mass. Consider a sample of pure water that contains 2.5 grams of hydrogen. Find x, the amount of oxygen in the sample, given the following proportional equation.

$$\frac{11.19}{88.81} = \frac{2.5}{x}$$

Solution

We'll continue to use the same steps as before to solve the equation. Notice that the left-hand side of the equation is simply a number, so we can go ahead and multiply both sides of the equation by the LCD.

✔ **Skill Check 5.1.2**

In October 2020, the following proportional equation represented the monetary value between Euros and American dollars. Solve the equation to determine the value of 200€ in American dollars at that time.

$$\frac{1€}{\$1.18} = \frac{200€}{x}$$

$$\frac{11.19}{88.81} = \frac{2.5}{x}$$

$$88.81x\left(\frac{11.19}{88.81}\right) = 88.81x\left(\frac{2.5}{x}\right)$$

$$11.19x = 222.025$$

$$x \approx 19.84$$

Step 1: Clear the equation of fractions by multiplying both sides by the LCD, which is $88.81x$.

Step 2: Remove the parentheses by distributing the $88.81x$.

Step 5: Make the coefficient equal to 1 by dividing both sides by 11.19.

So there are approximately 19.84 grams of oxygen in the sample of water.

OBJECTIVE 2 Formulas

Sometimes we are given an equation or formula that has more than one variable and we are asked to solve it in terms of a specific variable; that is, identify a new equation that is equivalent to the original equation. We'll use the same steps as we have used previously, and treat the variables as if they were numbers, applying any operations we would normally use in order to move the variables around in the equation.

Example 5.1.4

Solving for a Variable

The formula for the perimeter of a rectangle is $P = 2l + 2w$, where P is the perimeter, l is the length, and w is the width. Solve the formula for w.

Solution

Notice that there are three variables in this linear equation: P, l, and w. Since we want to solve for w, we will treat the other two variables as if they were numbers.

✔ **Skill Check 5.1.3**

The volume of a cube V is given by the formula $V = lwh$, where l is the length, w is the width, and h is the height. Solve the equation for the width.

$$P = 2l + 2w$$

$$P - 2l = 2w$$

$$\frac{P - 2l}{2} = w$$

Step 4: Subtract the number $2l$ from both sides to isolate the variable w.

Step 5: Divide both sides by 2 to make the coefficient equal to 1.

OBJECTIVE 3 Functions

A function is similar to an equation or formula in that it describes the relationship between variables for a given situation, but we might not be able to express the relationship between the variables with an equation. For instance, I might describe a function of a variable T to be the value of the New York Stock Exchange at particular times. However, we cannot express that function as an algebraic expression. More precisely, a **function** is a set of ordered pairs of the form (x, y) in which no two coordinates have the same x-value. In other words, for each x-coordinate, there is exactly one y-coordinate.

> ### 📖 FUNCTION
>
> A **function** is a set of ordered pairs in the form (x, y) in which each value of the dependent variable is associated to a unique independent variable.

Sometimes the mechanism to establish the value of y for any given value of x can be described by an algebraic formula or rule; sometimes it cannot. Either way, the notation used to represent y as a function of x is $y = f(x)$ and is read as "y equals f of x." This notation indicates that for a given value of x, we substitute that value into the function f to find a value for the function that we call y. In this way, we determine a set of coordinate pairs (x, y) based on the input (independent variable x) and its output (dependent variable y).

For our purposes, we will only consider functions that can be described by an algebraic expression. Equations, formulas, and functions don't necessarily have to be linear; they can be quadradic, cubic, exponential, logarithmic, and lots and lots more.

Regardless of the type of function, the process of evaluating a function means substituting a given value for the independent variable into the function to determine the value of the dependent variable. The variables don't always have to be called x and y, but whatever their names, it will be important to know which is the independent variable and which is the dependent variable. Example 5.1.5 gets us started with evaluating functions.

Example 5.1.5

Evaluating Functions

Evaluate the following functions for the given values.

a. Find $f(5)$ for $f(x) = x^3 - 4$.

b. Find $f(-3)$ for $f(t) = t^2 - 2t - 5$.

c. Find $f(15)$ for $f(x) = \begin{cases} -x, & x < 0 \\ x - 1, & x \geq 0 \end{cases}$

d. Find $f(2)$ for $f(x) = \log 2x$.

Solution

To find the solution, we need to evaluate each function at the given value; that is, substitute the appropriate number in for x in each function.

By Hand

a. The notation $f(5)$ indicates that $x = 5$, so we substitute 5 in for the variable x in the function $f(x) = x^3 - 4$.

$$f(5) = 5^3 - 4 = 125 - 4 = 121$$

b. This time, the independent variable is t, but we still follow the same procedure. We substitute $t = -3$ into the function $f(t) = t^2 - 2t - 5$. Be careful to correctly include the negative sign when substituting -3 into the function.

$$f(-3) = (-3)^2 - 2(-3) - 5 = 9 + 6 - 5 = 10$$

c. This function is called a piecewise function because it has different rules for different values of *x*. In this case, there are two pieces to the function: one for input values less than 0 and one for those greater than or equal to 0. Since we are evaluating a variable that is larger than 0 (that is, *x* = 15), we need to use the second piece of the function *f*(*x*) = *x* − 1. We can ignore the other piece of the function. Thus, we have the following.

$$f(15) = 15 - 1 = 14$$

d. Even though the function may be less familiar to you, the same process applies. Since we are looking for *f*(2), we substitute the value 2 in for *x* in the function.

$$f(2) = \log[2(2)] = \log 4 \approx 0.602$$

TI-83/84 Plus

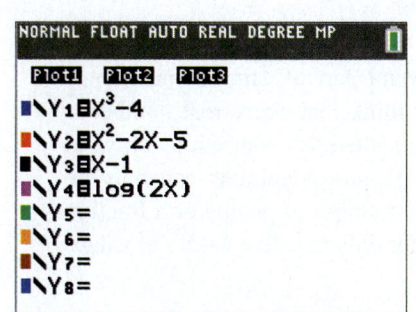

Figure 5.1.1

To evaluate functions in a calculator, we first store the functions by pressing [y=] and typing in the right-hand side of the function using [X,T,θ,n] for the variable, *x*. Let's type the function in part a. for Y1 and the function in part b. for Y2.

The function in part c. is called a piecewise function because it has different rules for different values of *x*. In this case, there are two pieces to the function: one for input values less than 0 and one for those greater than or equal to 0. Since we are evaluating a variable that is larger than 0 (that is, *x* = 15), we need to use the second piece of the function *f*(*x*) = *x* − 1. We can ignore the other piece of the function and type X-1 for Y3.

Finally, type the function in part d. for Y4.

Press [2nd] [mode] to return to the home screen. Press [vars], highlight the Y-VARS menu, and select Function.... To evaluate the first function, which we stored as Y1, select Y1. To evaluate this function at *x* = 5, type (5) and press [enter]. The calculator returns 121.

For part b., we'll press [vars], highlight the Y-VARS menu, and select Function... again, but this time select Y2 and type (-3) to evaluate this function at *x* = −3. The calculator returns 10.

For part c., press [vars], highlight the Y-VARS menu, and select Function.... Then select Y3, type (15), and press [enter]. The calculator returns 14.

Finally, we'll press [vars], highlight the Y-VARS menu, select Function..., and select Y4. We want to evaluate this function at *x* = 2, so type (2) and press [enter]. The calculator returns 0.6020599913.

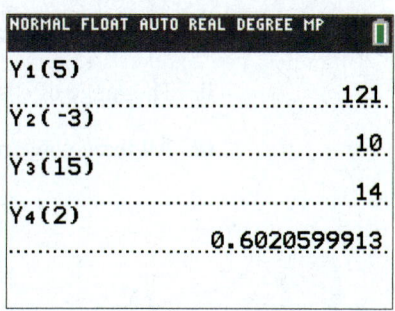

Figure 5.1.2

✔ **Skill Check 5.1.4**

Evaluate the following functions for the given values, round to the nearest thousandth if necessary.

a. Find *f*(−2) for *f*(*x*) = 2*x*² + 5.

b. Find *f*(−1) for $f(w) = \dfrac{(w-3)^2}{w}$.

c. Find *f*(3) for *f*(*t*) = 22 − (*t* + 2).

We call the set of input values for a function the **domain** of the function and we call the set of output values the **range**. A function assigns each element of its domain to exactly one element in the range.

> ### 📖 DOMAIN AND RANGE
>
> **Domain**
>
> The **domain** of a function is the set of all input values (independent variable) that result in a real number value for the output values (dependent variable).
>
> **Range**
>
> The **range** of a function is the set of all possible output values (dependent variable) that correspond to the domain values.

> **💬 Helpful Hint**
>
> Each element of the domain is assigned to exactly one element in the range but not vice versa. For instance, in the function $f(x) = x^2$, both $f(2)$ and $f(-2)$ equal 4.

Establishing the domain and range is an important part of using a function to model a scenario. Although you might initially think that every real number is a member of the domain or the range, there are sometimes restrictions on what numbers we can use. For instance, counts of people or animals can only be nonnegative integers—you can't have a negative number of people or a fraction of a person. At other times, situations might call for only negative values or values between two extremes.

When considering the range, it's helpful to think about the maximum and minimum values that could be obtained by the function and whether all values in between the extremes are possible. Just as with the input values, there are situations where some output values are not possible. Without clearly establishing the domain and range, you run the risk of producing solutions to a problem that make algebraic sense but are not realistically applicable.

Example 5.1.6

Determining Domain and Range

Answer the following questions about the domain and the range.

a. Use your knowledge of real numbers to determine how their properties limit the domain and range of the function $f(x) = \sqrt{x-2}$.

b. The graph in Figure 5.1.3 is a plot of a radial wave function describing the energy states of a hydrogen atom.[1] Determine if the following statements about the graph are true or false. If false, explain why.

  i. The domain of a looks to be between 0 and 0.11, inclusive.

  ii. The range of c is a subset of the range of b.

  iii. All three wave functions have the same domain.

---

1   Benjamin Obi Tayo, "Hydrogen Atom," Medium, February 11, 2019, https://medium.com/modern-physics/hydrogen-atom-4ca1599e94a8

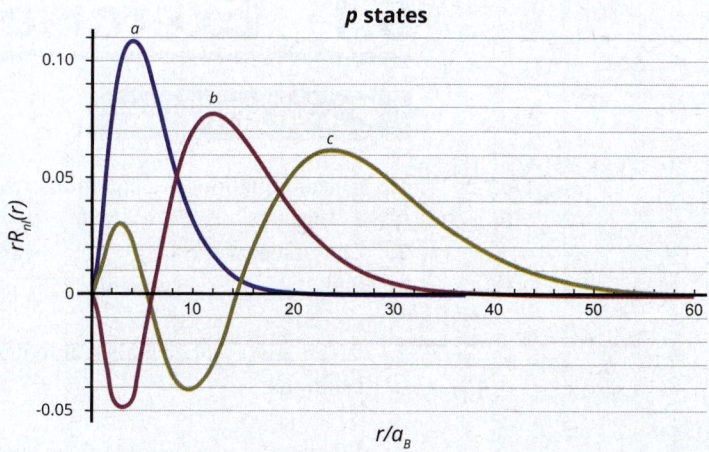

Figure 5.1.3

c. The function P represents the number of penguin eggs laid each season in the Antarctic Peninsula based on the number of adult penguins and climate conditions. What can you say about the domain and range of the function?

Solution

a. We know that we can only take the square root of nonnegative real numbers. Therefore, the expression under the radical sign must be a nonnegative number; that is, $x - 2 \geq 0$. We can solve this inequality to identify the restrictions on x.

$$x - 2 \geq 0$$
$$x \geq 2$$

Therefore, the domain only includes numbers that are greater than or equal to 2.

Since $x - 2$ can take on any nonnegative real value when x is in the domain, the range (which is all possible square roots of these values) consists of all nonnegative real numbers.

b. i. The domain of a looks to be between 0 and 0.11, inclusive.

The values of the domain lie along the x-axis and the values of the range lie along the y-axis. Therefore, this first statement is false. The domain is not between the two numbers given; however, the range, or the output numbers, appears to fall between 0 and 0.11 inclusive.

ii. The range of c is a subset of the range of b.

We can see that the values for the range (y-axis) of c fall within the range values of b. Therefore, this statement is true.

iii. All three wave functions have the same domain.

All three wave functions are defined for all values of the independent variable between 0 and 60. Therefore, this statement is true.

c. Because the function is based on counting the number of live animals, both the domain and range have to be at least 0, and integers; that is, they cannot be negative numbers, fractions, or decimals. This will be true for all living things.

5.1 Exercises

✔ CONCEPT CHECK

1. Linear equations are equations where the variable has an exponent of ____.

2. The value of the _____ variable changes with respect to the value of the _____ variable.

3. When any number makes an equation true, the equation has _____ solutions.

4. When solving a formula for a specific variable, treat the variables as if they are _____.

5. A function is a set of _____ where each value of the dependent variable is associated with a unique independent variable.

💡 PRACTICE

Solve each equation for the variable.

6. $4 = -6w$

7. $\dfrac{-7}{4}x = -2$

8. $-2(4v + 4) = -8v + 7$

9. $z + 0.2z - 3.6 = 5.8(z - 3)$

10. $2(x + 5) - 2 = 4(x + 2) - 2x$

11. $3.2x - 1 = 2(1.6x + 1)$

12. $3(x + 2) - 1.2 = 1.2x + 4.8 + 1.8x$

13. $10.3z - 9.3z - 5.4 = 5.5$

14. $\dfrac{4}{y} = \dfrac{4}{y - 1}$

15. $\dfrac{y - 1}{6} = \dfrac{y + 1}{3}$

16. $\dfrac{x - 10}{10} = \dfrac{x - 7}{8}$

17. $\dfrac{1}{2} = \dfrac{x - 1}{2x - 2}$

Solve each formula for the indicated variable.

18. Solve $P = R - C$ for C.

19. Solve $v = k + gt$ for k.

20. Solve $L = 2\pi rh$ for r.

21. Solve $I = Prt$ for r.

22. Solve $A = \dfrac{1}{2}bh$ for h.

23. Solve $V = \pi r^2 h$ for h.

Evaluate each function for the given value. Round your answer to the nearest hundredth, if necessary.

24. Find $f(7)$ for $f(x) = x^3 - 12$.

25. Find $f(-2)$ for $f(t) = t^3 + 5$.

26. Find $f(10)$ for $f(x) = x^2 - 2x + 3$.

27. Find $f(9)$ for $f(t) = t^2 + 100t - 125$.

28. Find $f(5)$ for $f(x) = \begin{cases} 2x, & x < 0 \\ x + 7, & x \geq 0 \end{cases}$.

29. Find $f(1.5)$ for $f(t) = \begin{cases} 15 - 2t, & t \leq 5 \\ 8t - 5, & t > 5 \end{cases}$.

30. Find $f\left(\dfrac{1}{3}\right)$ for $f(x) = \log 3x$.

31. Find $f(4)$ for $f(x) = \log 3x$.

32. Use your knowledge of real numbers to determine how their properties limit the domain and range of the function $f(x) = \sqrt{x + 7}$.

33. Use your knowledge of real numbers to determine how their properties limit the domain and range of the function $f(x) = 2x^2 + 5$.

🚀 **APPLICATIONS**

Solve. Round your answer to the nearest hundredth, if necessary.

34. When setting up a store front in an online marketplace, the seller needs to pay a base fee of $15 for the website space and $0.50 per item listed. A seller wants to spend no more than $40 to list items in the online store. Solve the equation $40 = 15 + \$0.50x$ for x to determine the maximum number of items the seller can list.

35. A carnival costs $20 to enter and each carnival game costs $1.50 to play. Sam wants to win a stuffed animal for his date, but only wants to spend a total of $50 on non-food purchases. Solve the equation $50 = $20 + $1.50x$ for x to determine how many attempts he can make at winning his date a stuffed animal.

36. The Reynold's want to build a decorative fence around their vegetable garden. A garden fence needs to be 72 inches tall to keep out deer. They select a solid base frame that is 24 inches tall and a mesh fence with 2-inch squares for the top of the fence. Determine how many squares high the mesh needs to be by solving the equation $72 = 24 + 2x$ for x.

37. The Geology department is raising money for scholarships by hosting the annual Sedimental Evening Dinner. The music department provides free entertainment, and the hospitality management department hosts as part of student course requirements. The department would like to raise $3000 for scholarship and will spend $760 on food and decorations. The dining hall can accommodate 80 guests. Solve the equation $3000 = 80x − 760 to determine the price the department should charge per guest.

38. Suppose corn chips cost 22.5 cents per ounce. Use the given equation to determine how many ounces are in the bag of chips if a bag costs $3.03.

$$\frac{\$0.225}{1 \text{ oz}} = \frac{\$3.03}{x \text{ oz}}$$

39. Suppose cereal cost 24.5 cents per ounce. Use the given equation to determine the cost of a box of cereal if a bag contains 20 oz.

$$\frac{\$0.245}{1 \text{ oz}} = \frac{\$x}{20 \text{ oz}}$$

40. An entry level civil engineer's annual pay is $62,348 based on 52 weeks per year. Due to the economy, his company is having to cut back on the number of weeks that it employs its civil engineers. The firm plans on cutting the work year to 49 weeks and keep the same rate of pay. Use the given equation to determine how much should the civil engineer expect his annual pay to decrease. Round your answer to the nearest dollar.

$$\frac{\$62,348}{52 \text{ weeks}} = \frac{\$x}{49 \text{ weeks}}$$

41. You're making dessert, but your recipe needs adjustment. Your peanut butter cookie recipe makes 4 dozen cookies, but you need 1 dozen cookies. The recipe requires $1\frac{1}{2}$ cups of butter, 3 teaspoons of almond extract, and $2\frac{1}{2}$ cups of peanuts. Use the given equations to determine how much of each of these ingredients are necessary for 1 dozen cookies. Simplify your answer.

    a. butter: $\dfrac{1\frac{1}{2}\text{ cups}}{4\text{ dozen}} = \dfrac{x}{1\text{ dozen}}$

    b. almond extract: $\dfrac{3\text{ teaspoons}}{4\text{ dozen}} = \dfrac{x}{1\text{ dozen}}$

    c. peanuts: $\dfrac{2\frac{1}{2}\text{ cups}}{4\text{ dozen}} = \dfrac{x}{1\text{ dozen}}$

42. In business, profit for producing and selling an item is modeled by the equation $P = R - C$, where P is the profit earned, R is the revenue earned from sales, and C is the cost to produce the item. Find the revenue earned by the company if they made a profit of \$3.2 million and incurred a cost of \$1.8 million.

43. The distance traveled by an object can be described by $d = rt$, where r is the objects rate (or speed) and t is the amount of time the object traveled. If you need to drive a distance of 105 miles and can travel at an average speed of 60 miles per hour, how many hours will the trip take?

44. The following graph displays the change in air temperature over the last three days. Determine if the following statements about the graph are true or false. If false, explain why.

    a. The graph of Day 2 has a domain between −4 and 11, inclusive.

    b. The graph of Day 1 and the graph of Day 3 have the same range.

    c. The range of Day 3's graph is a subset of the combined range of Day 1's graph and Day 2's graph.

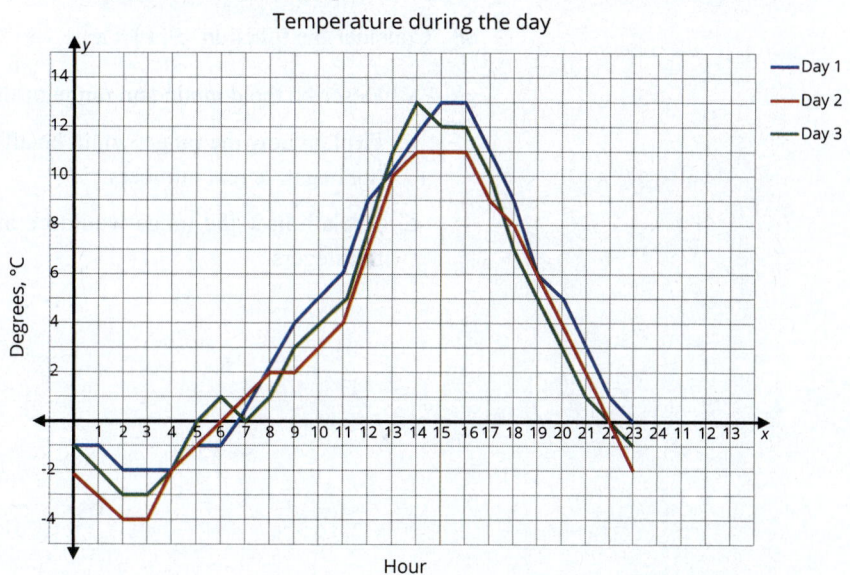

45. The following graph shows the number of items (in thousands) produced by a factory each year from 2018 to 2020. Determine if the following statements about the graph are true or false. If false, explain why.

a. The ranges of all three graphs are the same.

b. The domains of all three graphs are the same.

c. The range of the 2019 graph is a subset of the range of the 2020 graph.

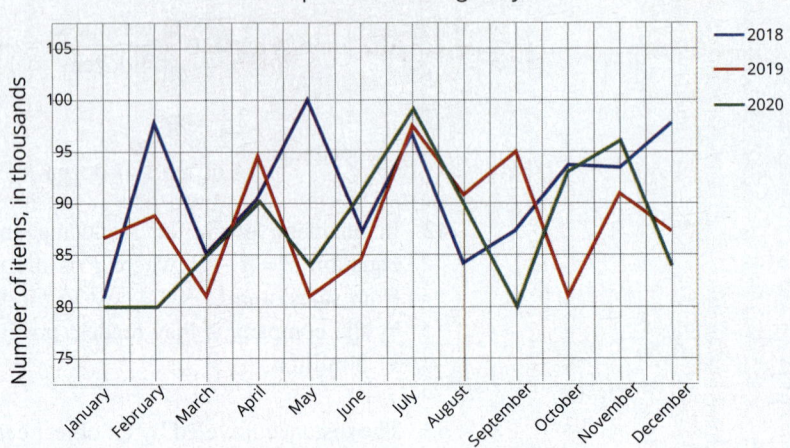

Items produced during the year

46. The function M represents the number of a specific model of car manufactured by an automotive company in the year 2022 based on sales of that model car in the year 2021. What can you say about the domain and range of the function M?

47. The function C represents the cost of a wedding reception where the price to rent the venue is $7000 and the price of food and decorations depends on the number of people attending. The venue has a maximum capacity of 150 guests. What can you say about the domain and range of the function C?

✏️ **WRITING & THINKING**

48. Consider the function $f(x) = x^2$.

a. Describe the domain and range of the function f.

b. Explain how the range would be affected if the domain was restricted to nonnegative real numbers.

c. Explain how the range would be affected if the domain was restricted to integers

5.1 PROJECT

MAKING ELECTRIC DECISIONS

A door-to-door salesperson cold calls at your house (that is, shows up at your house without prior contact) and tells you about the great deal you can get if you switch your electric provider to a clean, renewable electric supplier they represent. You don't have a current electric bill readily available, so you take down information to compare later. The salesperson tells you that the electric supplier uses 35% renewable energy and that you will be charged an introductory fixed rate of $0.0639 per kilowatt-hour (kWh) with a base-service fee of $73 per month, which includes distribution and fees.

1. Use the information provided by the salesperson to create a function to represent the monthly cost of electric service through the supplier, where x represents the kilowatt-hours used per month.

   You find a copy of your electric bill and discover that you currently pay a fixed rate of $0.0724 per kilowatt-hour and a base-service fee of $65 per month. You also notice that your current supplier uses 15% renewable energy.

2. Use this information to create a function to represent the monthly cost of electric service through your current supplier, where x represents the kilowatt-hours used per month.

3. Your electric usage for the past 6 months are shown in the second column of the following table. In the third column, use the function from part 1 to calculate the monthly cost of service with the new electric supplier. In the fourth column, use the function from part 2 to calculate the monthly cost of service with the supplier you are currently using. Round your answers to the nearest cent.

Electric Service Cost Comparison

Month	Electric Usage (in kWh)	New Supplier Cost (in $)	Current Supplier Cost (in $)
December	542		
January	952		
February	638		
March	626		
April	587		
May	597		

4. Based on these calculations, would you spend less money on electric service by switching to the new supplier or keeping your current supplier?

5. The information from the functions and the table might not provide you with all the information you need to make a decision. Describe two additional types of information you should consider before making a decision and how that information would be useful.

6. Based on the current information and any addition situations you've considered, would you switch to the new electric provider or stay with your current one?

5.2 LINEAR MODELING

OBJECTIVE 1 Linear Models

Buying a home is one of the largest purchases most Americans make in their lives. Tiffany, who is a first-time home buyer, is a little nervous about making such a big purchase, and rightly so. She found a house that she likes and wants to know if the asking price is a fair price for the particular neighborhood. Take a look at the graph in Figure 5.2.1 and see what you think.

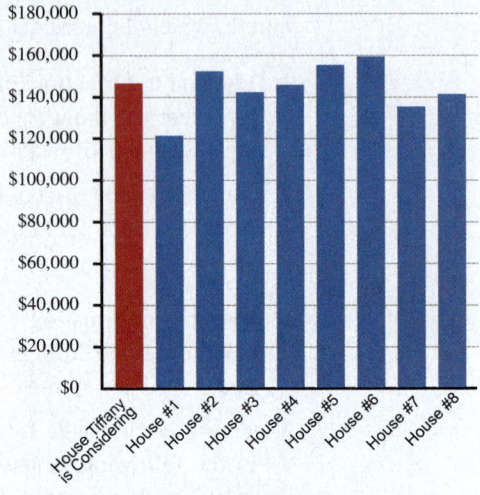

Figure 5.2.1: Houses in the Neighborhood that Sold in the Last 18 Months

If you deduced that the price of the house Tiffany is interested in (the first bar in the graph) is somewhere in the middle of the pack of house prices in that neighborhood, you'd be right based on the information given. Perhaps this information would put Tiffany a little more at ease with the asking price of the house she likes. However, there is more to consider than just price to make sure that Tiffany makes a wise investment. Along with price, let's consider another piece of information—the size of each of the houses. Have a look at Table 5.2.1, which shows both the square footage and sales price for each house sold in that neighborhood over the past 18 months.

Table 5.2.1: Sale Price vs. Square Footage

Houses	Square Footage	Price
House Tiffany is Considering	1200	$146,300
House #1	1220	$120,995
House #2	2060	$152,000
House #3	1750	$142,000
House #4	1970	$145,500
House #5	2250	$155,000
House #6	2200	$159,000
House #7	1430	$135,000
House #8	1700	$141,000

Seeing this information in a chart is informative but makes it hard to interpret how square footage relates to price in that neighborhood. From the table, we can see that Tiffany's potential house has the smallest square footage out of all the houses, but not the smallest price. To better understand the connection between price and square footage, it would be helpful to see both pieces of information together. We can visually depict these two-dimensional data by graphing them on the Cartesian coordinate system. Each square footage and its respective house price can be represented as a point on the graph using an *ordered pair*, (*x*, *y*).

We'll let the square footages of the houses be the *x*-coordinate and the prices be the *y*-coordinate, as shown in Table 5.2.2.

Table 5.2.2: Ordered Pairs

x (Square Footage)	*y* (Sale Price, $)
1200	146,300
1220	120,995
2060	152,000
1750	142,000
1970	145,500
2250	155,000
2200	159,000
1430	135,000
1700	141,000

Figure 5.2.2 shows each coordinate pair as a point on the Cartesian plane. Remember when plotting points on a graph to always move left or right along the *x*-axis first and then up or down along the *y*-axis.

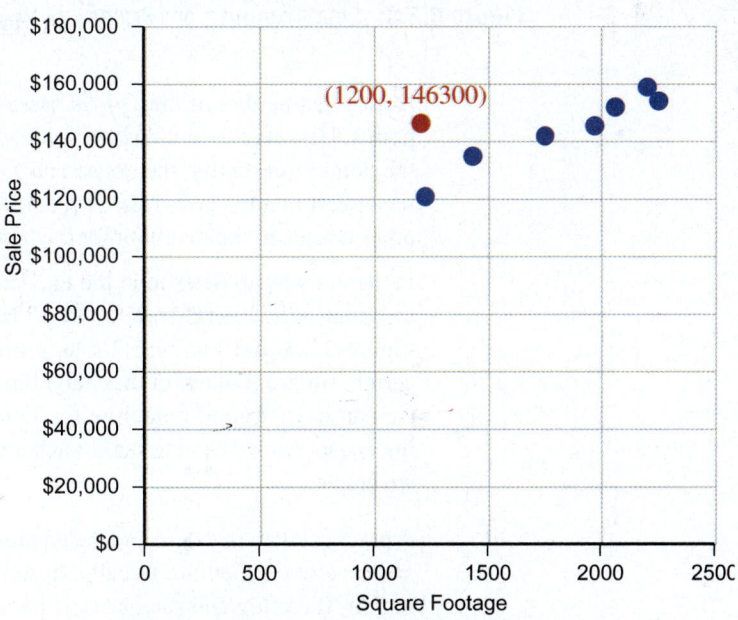

Figure 5.2.2: Square Footage vs. Sale Price

Looking at the ordered pairs of data, we can now better compare Tiffany's potential house to the other houses that have recently sold in the neighborhood. The house Tiffany is interested in appears to no longer be in the middle of the pack. As you can see, there was a house with essentially the same square footage as Tiffany's potential house that sold for $25,000 less. It seems that she should be getting more square feet for the asking price of her house, or at least be competitive in making a lower offer to the seller.

Notice that the points representing the houses that were previously sold almost form a line, and the point representing Tiffany's potential house is way off that line and is more of an outlier. To help Tiffany calculate an offer that is in keeping with the neighborhood, we can create a mathematical model based on the line that the previously sold houses closely follow. To specify that line, we can join together the points representing the lowest-priced house and the highest-priced house, as shown in Figure 5.2.3. Similarly, we can form a line with the points of the houses with the smallest and largest square footages as shown in Figure 5.2.4.

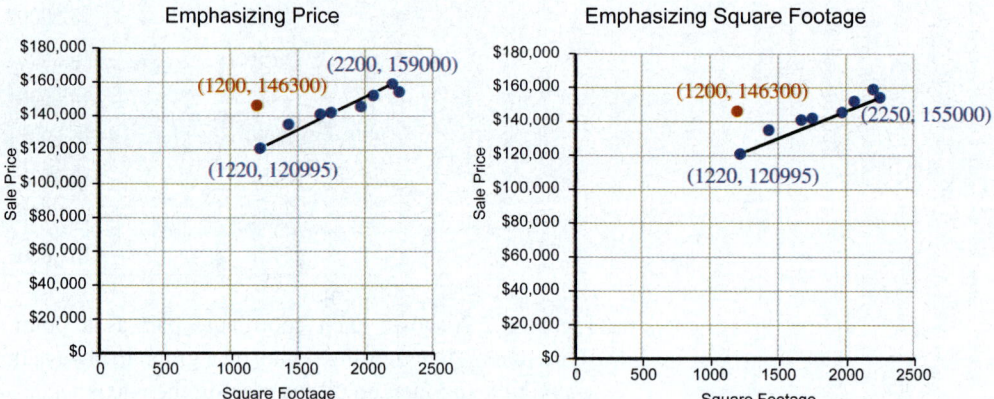

Figure 5.2.3: Line Created from Prices   **Figure 5.2.4:** Lines Created from Square Footage

Notice that neither of these lines passes perfectly through all of the housing points. However, both lines are somewhat close to all of the points—except for the point representing the house that Tiffany is interested in. The graphs seem to suggest that the house she is considering is not priced in accordance with the other houses in the neighborhood when comparing square footage.

Is there a way to determine the expected price of the house Tiffany is interested in based on the neighborhood data? Yes. We can use the lines we've drawn in Figure 5.2.3 and Figure 5.2.4 as a model to estimate how much a house of a certain square footage in that neighborhood should be priced. And consequently, we can have a good guideline for Tiffany to consider when making an offer on the house. All we need to make such an estimation is the equation of the lines that we created.

A practical way to determine the equation of any line is to use the slope of the line and a point on the line. Recall that the slope of a line is a number that describes both its direction and steepness. If the slope is positive, then the line is *rising* left to right. If the slope is negative, the line *is falling* left to right. Additionally, the larger the absolute value of the slope is, the *steeper* the line is. A gentle incline such as a wheelchair ramp will have a small slope, but a ladder leaning against a wall will have a larger slope. If a line does not lean either left or right, then it is either a horizontal line and has a slope of zero, or it is a vertical line and has a slope that is undefined. By definition, slope is the ratio of the change in *y*-values to the change in x-values. One of the ways to find the slope of a line is to use any two points on the line in the slope formula.

Helpful Hint

It does not matter which of the points we choose to label (x_1, y_1) for the slope formula. What is important is that the subtraction of coordinates occurs in the same order in both the numerator and denominator. That means that

$$m = \frac{y_2 - y_1}{x_2 - x_1} = \frac{y_1 - y_2}{x_1 - x_2}.$$

$f(x)$ SLOPE FORMULA

The slope of a line passing through the points (x_1, y_1) and (x_2, y_2) is given by the following.

$$m = \frac{y_2 - y_1}{x_2 - x_1}$$

Helpful Hint

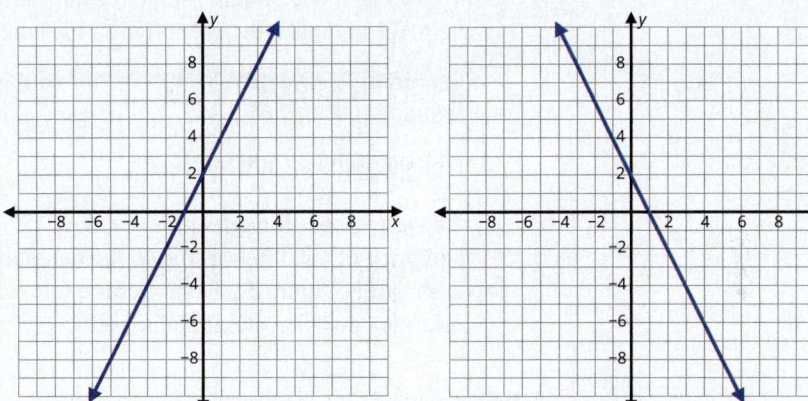

Figure 5.2.5: Positive Slope   Figure 5.2.6: Negative Slope

✔ Skill Check 5.2.1

Find the slope of the line passing through each pair of points.

a. $(-3, 12)$ and $(15, 12)$

b. $(2, 4)$ and $(5, 1)$

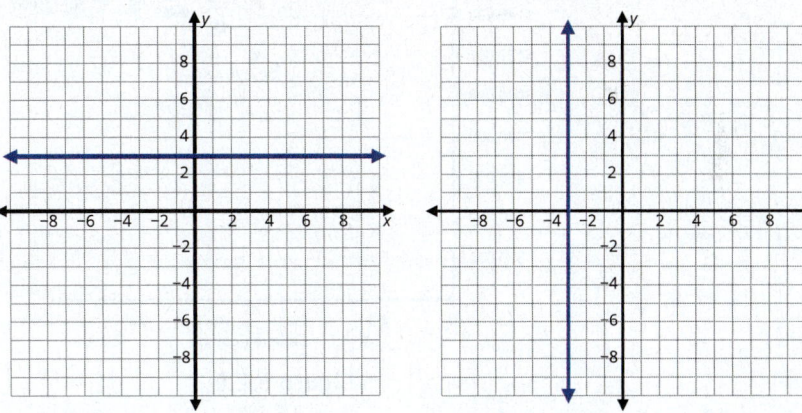

Figure 5.2.7: Zero Slope   Figure 5.2.8: Undefined Slope

Knowing the slope of a particular line, along with any point on the line, we can write the equation of that line using the **point-slope form** of an equation and then arrange the equations in **slope-intercept form**. Although these equations have different names, they are simply two versions of the same line. In fact, a third version of the same line is the standard form, which is when the equation of a line is in the form $Ax + By = C$.

f(x) FORMULAS FOR THE EQUATION OF A LINE

The **point-slope form** of the equation for a line, where m is the slope of the line and (x_1, y_1) is a point on the line, is

$$y - y_1 = m(x - x_1).$$

The **slope-intercept form** of the equation for a line, where m is the slope and b is the y-intercept, is

$$y = mx + b.$$

We can now use both the slope formula and formulas for the equation of a line to help Tiffany make the best offer on the house she found.

Example 5.2.1

Modeling with a Linear Equation

As we've seen throughout the section, Tiffany has data about the square footage and price of each house sold in the neighborhood. Using this information, which is shown in Figure 5.2.9, answer the following questions to help Tiffany make an informed decision about the house.

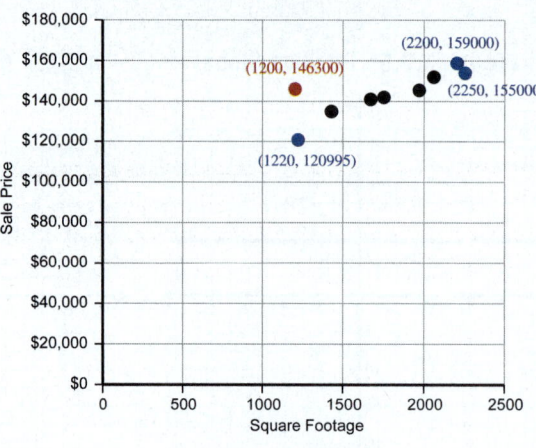

Table 5.2.3: Ordered Pairs

x (Square Footage)	y (Sale Price, $)
1200	146,300
1220	120,995
2060	152,000
1750	142,000
1970	145,500
2250	155,000
2200	159,000
1430	135,000
1700	141,000

Figure 5.2.9

a. First consider the houses based on **price**. Find the slope-intercept form of the equation for the line that passes through the points representing the lowest-priced house, (1220, 120995), and the highest-priced house, (2200, 159000). Round the slope to the nearest hundredth.

b. Next consider the houses based on **square footage**. Find the slope-intercept form of the equation for the line passing through the points representing the house with the smallest square footage, (1220, 120995), and the house with the largest square footage, (2250, 155000). Round the slope to the nearest hundredth.

c. Use the equations found in parts a. and b. to find two estimates of the price Tiffany should pay for the 1200 square foot house she is interested in.

Solution

a. We will let the lowest-priced house be Point 1 and the highest-priced house be Point 2. Therefore, $(x_1, y_1) = (1220, 120995)$ and $(x_2, y_2) = (2200, 159000)$.

By Hand

We can write the equation of the line using point-slope form since we are given two points. Begin by finding the slope of the line. Substitute the coordinate values into the slope formula and simplify, as shown.

$$m = \frac{y_2 - y_1}{x_2 - x_1}$$
$$= \frac{159{,}000 - 120{,}995}{2200 - 1220}$$
$$\approx 38.78$$

So the slope of the line is approximately 38.78.

Next, substitute one of the points and the slope into the point-slope formula and simplify. Since it doesn't make a difference which point we choose, we will use the point representing the lowest-priced house, (1220, 120995).

$$y - y_1 = m(x - x_1)$$

$y - 120{,}995 = 38.78(x - 1220)$ **Substitute known values.**

$y - 120{,}995 = 38.78x - 47{,}311.6$ **Simplify.**

$y = 38.78x + 73{,}683.4$ **Isolate y by adding 120,995 to both sides.**

> **💬 Helpful Hint**
>
> When using the by hand method, a more precise equation can be found by keeping the slope in fraction form until the final step of finding the equation.

TI-83/84 Plus

To find the slope-intercept form of the equation using a TI-83/84 Plus calculator, begin by entering the points in the list function of the calculator. Press [stat], select Edit..., and enter the x-values for L1 and the y-values for L2. Press [stat] again, highlight CALC, and select LinReg(ax+b). Move your cursor down to highlight Calculate and press [enter].

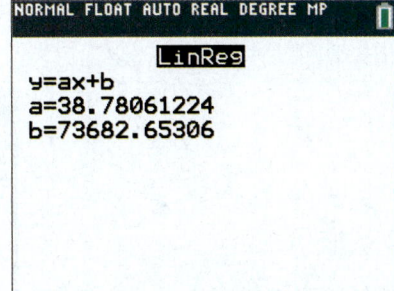

Figure 5.2.10   Figure 5.2.11

The output screen tells us that the slope of the line is $a = m \approx 38.78$ and the y-intercept is $b \approx 73{,}682.65$. Thus, the equation of the line in the form $y = mx + b$ is as follows.

$$y = 38.78x + 73{,}682.65$$

So the slope-intercept form of the equation for the line that passes through the points representing the lowest-priced house and the highest-priced house is $y = 38.78x + 73{,}682.65$. (Note that the equation found by hand differs slightly from the one calculated with technology. The difference is due to rounding. The calculator provides a more exact answer because it does not round before any multiplication.)

b. We will let the house with the smallest square footage be Point 1 and the house with the largest square footage be Point 2. Therefore, $(x_1, y_1) = (1220, 120995)$ and $(x_2, y_2) = (2250, 155000)$.

By Hand

Once again, we begin by finding the slope of the line. Substitute the coordinate values into the slope formula.

$$m = \frac{y_2 - y_1}{x_2 - x_1}$$
$$= \frac{155{,}000 - 120{,}995}{2250 - 1220}$$
$$\approx 33.01$$

So the slope of the line is approximately 33.01.

We will use the point representing the largest house, $(2250, 155000)$, and the slope in the point-slope formula to determine the equation.

$$y - y_1 = m(x - x_1)$$
$$y - 155{,}000 = 33.01(x - 2250)$$
$$y - 155{,}000 = 33.01x - 74{,}272.5$$
$$y = 33.01x + 80{,}727.5$$

TI-83/84 Plus

Begin by entering the points in the list function of the calculator. Press [stat], select `Edit...`, and enter the x-values for L1 and the y-values for L2. Press [stat] again, highlight `CALC`, and select `LinReg(ax+b)`. Move your cursor down to highlight `Calculate` and press [enter].

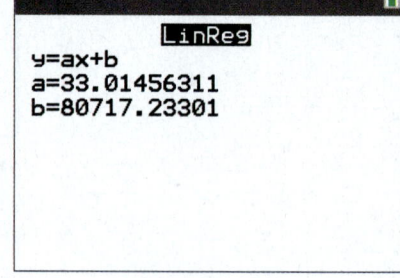

Figure 5.2.12   **Figure 5.2.13**

The output screen tells us that the slope of the line is $a = m \approx 33.01$ and the y-intercept is $b \approx 80{,}717.23$. Thus, the equation of the line in the form $y = mx + b$ is as follows.

$$y = 33.01x + 80{,}717.23$$

💬 Helpful Hint

Although we estimated the line that looked to be closest to most of the houses in the graph by selecting two points, there is a mathematical process to find the **line of best fit**, which we will cover more thoroughly in Section 12.3.

Again, the calculator provides the more precise answer. So the slope-intercept form of the equation for the line that passes through the points representing the smallest house and the largest house is as follows.

$$y = 33.01x + 80,717.23$$

c. We are asked to use the equations found in parts a. and b. to determine the price for a house a house that is 1200 square feet. Since x represents the square footage, we substitute 1200 in for x in both equations as shown below. We will use the more precise equations given by the calculator.

Equation based on lowest price (**a**)   Equation based on smallest square footage (**b**)

$$y = 38.78x + 73,682.65 \qquad\qquad y = 33.01x + 80,717.23$$
$$= 38.78(1200) + 73,682.65 \qquad = 33.01(1200) + 80,717.23$$
$$= 120,218.65 \qquad\qquad\qquad = 120,329.23$$

Notice that by using both equations we have created a price range for a 1200 square foot house that would be in line with the other houses in the neighborhood. Based on the data known, we can say that a house with 1200 square feet should cost between \$120,219 and \$120,329. Since the house Tiffany is interested in is actually listed at \$146,300, based on our model it seems that it is overpriced for the neighborhood.

> ✔ **Skill Check 5.2.2**
>
> Find the slope-intercept form of the equation for line passing through the points (31, 85) and (27, 98).

OBJECTIVE 2 Graphing Lines

The different configurations of an equation each have their own advantages. By simply looking at an equation in slope-intercept form or point-slope form we can easily identify the slope, m, as shown.

Slope

$$y = mx + b \quad y - y_1 = m(x - x_1)$$

Both forms also allow us to identify a point on the line. In the slope-intercept form, the point we are given is the y-intercept (as the name suggests) with coordinates $(0, b)$. And in the point-slope form, the point is (x_1, y_1).

> 💬 **Think Back**
>
> Recall that the coordinates of the y-intercept are (0, b).

Point on the Line
$$\underset{(0,b)}{y = mx + b} \quad \underset{(x_1, y_1)}{y - y_1 = m(x - x_1)}$$

For instance, consider the line $2x - 3y = 12$. Its slope-intercept form is $y = \frac{2}{3}x - 4$, and the point-slope form is $y + 2 = \frac{2}{3}(x - 3)$. Given these different forms of the equation, it's easy to identify the slope of the line as $m = \frac{2}{3}$, the y-intercept as $(0, -4)$, and $(3, -2)$ as a point on the line.

> ✔ **Skill Check 5.2.3**
>
> Identify the slope and y-intercept, along with another point on the line $14x - 2y = 20$, given that $y = 7x - 10$ and $y - 11 = 7(x - 3)$ are equivalent equations.

When we model linear situations, it is often useful to transition between the different forms of the equation and its visual representation on the graph. Being able to identify these properties from a given equation, helps us in this transition. Example 5.2.2 illustrates this for us.

Example 5.2.2

Graphing a Linear Equation from Slope-Intercept Form and Point-Slope Form

Ethan is giving a PowerPoint presentation in his international business class about a start-up business idea. For visual effect, he intends to have a slide showing the equations of the lines used in his analysis to find the break-even point where the venture will actually be making money. The equations of the lines to be shown on the slide are given here. Graph each of the lines in parts a. and b. on the same graph.

a. $y = 2x + 3$

b. $y - 9 = \dfrac{10}{3}(x - 3)$

c. Using the graph of the lines from parts a. and b., estimate the break-even point for Ethan's venture.

Solution

a. First, we graph the equation $y = 2x + 3$ on the coordinate plane.

> **Helpful Hint**
>
> In order to graph a linear equation, regardless of the form it's presented in, we can simply rearrange the equation so that it is in slope-intercept form.

By Hand

The first equation, $y = 2x + 3$, is given in slope-intercept form, $y = mx + b$. We can see the slope is $m = 2$ and the y-intercept is $b = 3$ or $(0, 3)$.

When creating a graph using a point and the slope, always plot the point on the graph first and then, starting from that point, count the slope as $\dfrac{\text{rise}}{\text{run}}$ to identify a second point on the line. Since the slope $m = 2$ is a whole number, we can make the slope a fraction by putting it over 1. Thinking of the slope as $m = \dfrac{2}{1}$ is more convenient for graphing purposes since the rise and run are both clearly stated. We can now draw the line passing through the two points.

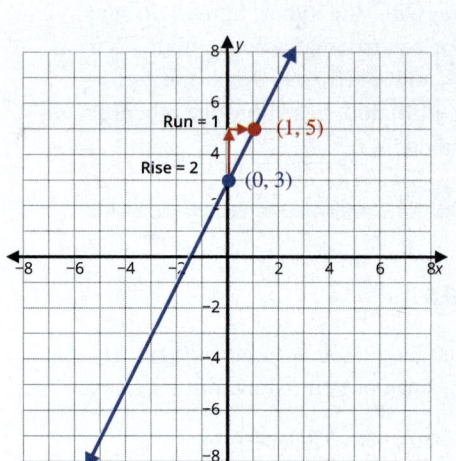

Figure 5.2.14

> **TECH TIP**
>
> Pressing [window] or [zoom] will allow you to adjust the display window on the calculator.

TI-83/84 Plus

In order to graph equations on the TI-83/84 Plus, the equations must be written in slope-intercept form so that the y-value is by itself on one side of the equal sign. This equation is already in the correct form, so we can input it into the calculator as it is. Press [y=] and clear any equations currently in the calculator. Input the right-hand side of the equation, 2x+3, for Y1, pressing [x,t,θ,n] for the variable, x. Press [graph].

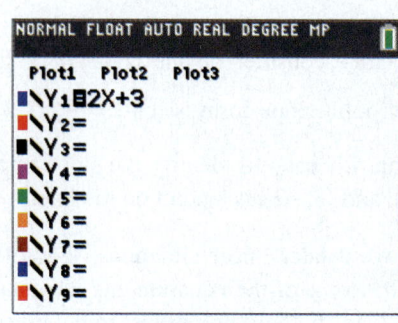

Figure 5.2.15

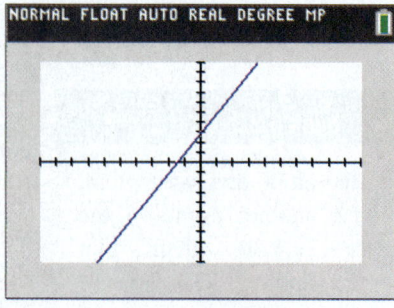

Figure 5.2.16

b. Next, we graph the equation $y - 9 = \frac{10}{3}(x-3)$ on the same coordinate plane.

By Hand

The second equation $y - 9 = \frac{10}{3}(x-3)$ is given in point-slope form: $y - y_1 = m(x - x_1)$. Thus, the slope m is $\frac{10}{3}$ and a point on the line is $(3, 9)$.

We plot the point $(3, 9)$ first and then count the slope (up 10 and over 3) to find a second point on the line.

TI-83/84 Plus

Before we can enter this equation into the calculator, the equation must be written in slope-intercept form with the y-value by itself on one side of the equal sign. To do this, we simplify and rearrange the equation by clearing it of the parentheses and isolating the variable y.

$$y - 9 = \frac{10}{3}(x-3)$$

$$y - 9 = \frac{10}{3}x - 10 \qquad \text{Distribute.}$$

$$y = \frac{10}{3}x - 1 \qquad \text{Add 9 to each side.}$$

Now that we have the equation in slope-intercept form, press $\boxed{\text{y=}}$ and input the right-hand side of the equation for Y2. Press $\boxed{\text{graph}}$.

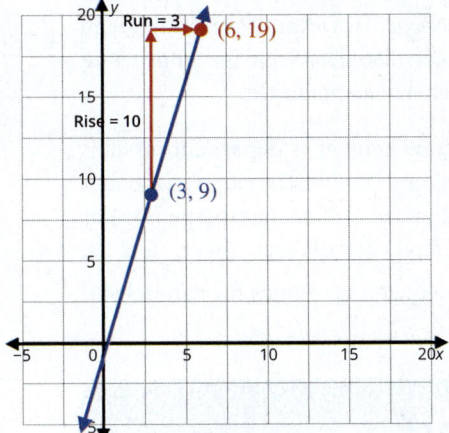

Figure 5.2.17

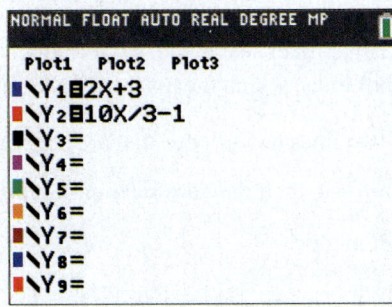

Figure 5.2.18

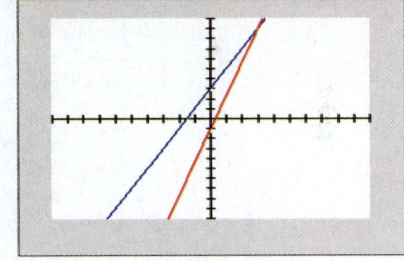

Figure 5.2.19

So Ethan's PowerPoint slide would have the following graph showing both lines.

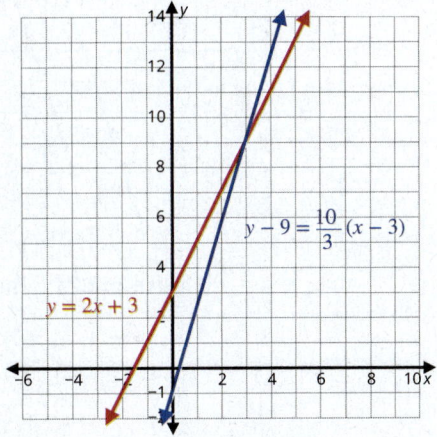

Figure 5.2.20

c. The break-even point for Ethan's venture occurs where the lines intersect. We can estimate that this happens around the point (3, 9). We will cover break-even points and how to find them precisely in a later section.

OBJECTIVE 3 Parallel and Perpendicular Lines

Through our example about Tiffany buying a house, we observed that in order to find the equation of a line, we need to know two things: the slope of the line and a point on the line. Sometimes we will be given the slope of the line, but other times we will need to calculate the slope. As we have seen, if we know two points on the line, we can use the formula to calculate the slope. However, if we aren't given enough information to identify two points, we can also determine the slope using the properties of parallel and perpendicular lines when applicable.

Whether lines are parallel or perpendicular to one another is dependent on their slopes. Parallel lines are lines that have equal slopes; that is, the slope of the first line (m_1) is equal to the slope of the second line (m_2). Lines are perpendicular when they have slopes that are negative reciprocals of each other; that is, $m_1 = -\dfrac{1}{m_2}$. Example 5.2.3 looks at finding the equations of lines for parallel and perpendicular lines.

📖 PARALLEL AND PERPENDICULAR LINES

Consider two lines $y = m_1x + b_1$ and $y = m_2x + b_2$.

If the two lines have the same slope (that is, $m_1 = m_2$), then they are **parallel lines**, which means they never intersect.

If the two lines have slopes that are negative reciprocals of each other (that is, $m_1 = -\dfrac{1}{m_2}$), then they are **perpendicular lines**, which means they intersect at right angles.

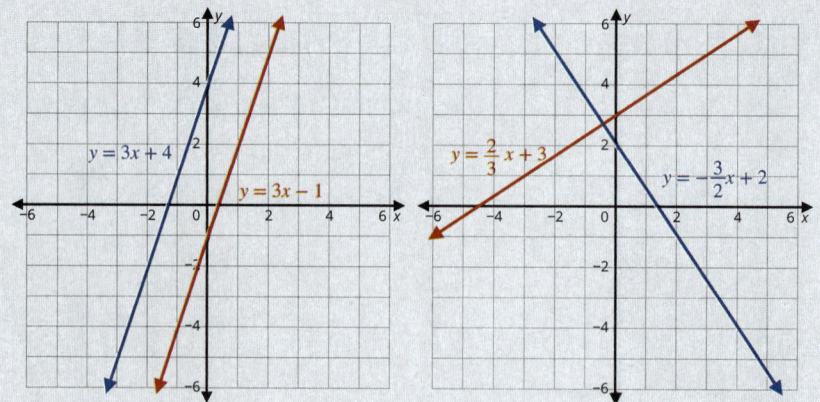

Figure 5.2.21: Parallel Lines   **Figure 5.2.22:** Perpendicular Lines

Example 5.2.3

Finding the Equations of Parallel and Perpendicular Lines

AutoCAD, a computer-aided design and drafting software, has been widely used for almost 40 years. As part of an introductory AutoCAD class, students need to be able to draw lines that are both parallel and perpendicular to other lines. To understand the calculations that the software automates, they are given criteria and asked to find the equation of each line based on the information given. Use the criteria to write the equation of the desired line in slope-intercept form.

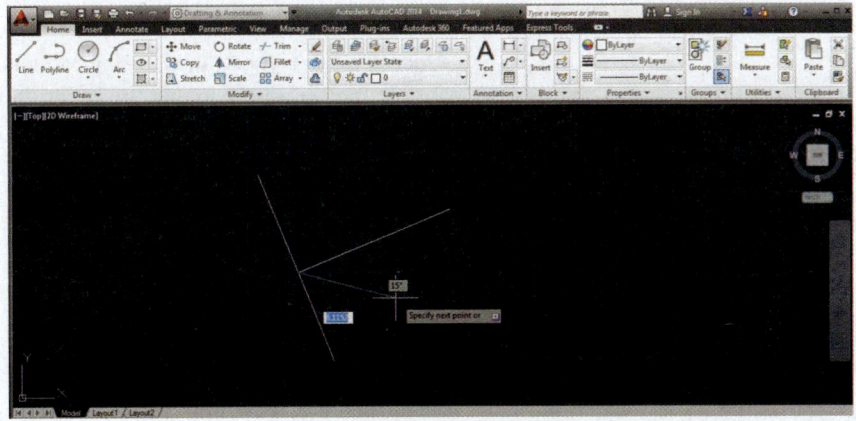

Figure 5.2.23

a. The line passes through the point $(-2, 3)$ and is parallel to the line with the equation

$$y - 7 = -4\left(x - \frac{2}{9}\right).$$

b. The line passes through the point $(8, -11)$ and is perpendicular to the line with the equation

$$y - 18 = -\frac{2}{3}(x - 21).$$

Solution

a. In order to find the equation of the line using the point-slope formula, we need the value of the slope along with the coordinates of a point on the line. We are given a point that lies on the line, $(-2, 3)$. To determine the slope, we can use fact that parallel lines have equal slopes. The line we are given is written in point-slope form: $y - 7 = -4\left(x - \frac{2}{9}\right)$. Therefore, we can simply identify its slope as $m = -4$. So the new line will also have a slope of -4.

Once we have substituted in the values, we will simplify the equation and isolate the variable y so that the equation is in slope-intercept form.

$$y - y_1 = m(x - x_1)$$
$$y - 3 = -4(x - (-2)) \qquad \text{Substitute known values.}$$
$$y - 3 = -4x - 8 \qquad \text{Simplify.}$$
$$y = -4x - 5 \qquad \text{Isolate } y \text{ by adding 3 to both sides.}$$

So the equation of the line, in slope-intercept form, that passes through the point $(-2, 3)$ and is parallel to the line $y - 7 = -4\left(x - \dfrac{2}{9}\right)$ is $y = -4x - 5$.

b. Once again, we need a point and the slope in order to write the equation. We are given a point $(8, -11)$ and the fact that the line is perpendicular to a particular line. The line we are given is written in slope-intercept form: $y = -\dfrac{2}{3}(x - 21)$. We can identify its slope as $m = -\dfrac{2}{3}$. The slope of the line we are interested in will be the negative reciprocal of this slope. Therefore, the slope of the new line is $m = \dfrac{3}{2}$.

Once we have substituted in the values, we will simplify the equation and isolate the variable y so that the equation is in slope-intercept form.

$$y - y_1 = m\left(x - x_1\right)$$

$$y - (-11) = \frac{3}{2}(x - 8) \qquad \text{Substitute known values.}$$

$$y + 11 = \frac{3}{2}x - 12 \qquad \text{Simplify.}$$

$$y = \frac{3}{2}x - 23 \qquad \text{Isolate } y \text{ by subtracting 11 from both sides.}$$

So the equation of the line in slope-intercept form that passes through the point $(8, -11)$ and is parallel to the line $y = -\dfrac{2}{3}(x - 21)$ is $y = \dfrac{3}{2}x - 23$.

> ✔ **Skill Check 5.2.4**
>
> What is the slope of the line which is perpendicular to the line $y = -\dfrac{3}{4}x + 8$?

As mentioned earlier, the slope of a horizontal line is zero and the slope of a vertical line is undefined. Consequently, the equations of horizontal and vertical lines equations appear a bit differently than the standard forms we have mentioned so far.

For a horizontal line, where $m = 0$ the equation $y = mx + b$ becomes $y = 0x + b$, which simplifies to $y = b$, where b still represents the y-intercept. To graph the equation, draw a horizontal line through the y-intercept, $(0, b)$, as illustrated in Figure 5.2.24.

Just as all horizontal lines have the form $y = b$, vertical lines are also all of similar form. Because their slope is undefined, m is no longer in the equation. The equations of vertical lines are of the form $x = a$, where a is the x-intercept. Again, we simply draw a vertical line through the x-intercept, $(a, 0)$, as illustrated in Figure 5.2.25.

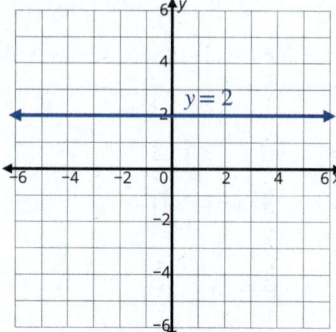

Figure 5.2.24: Graph of $y = 2$

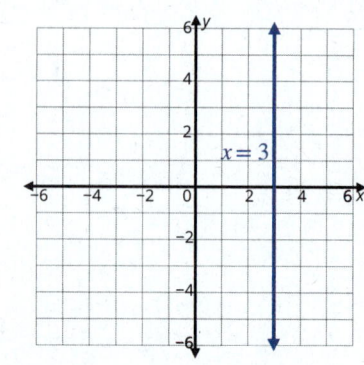

Figure 5.2.25: Graph of $x = 3$

📖 HORIZONTAL AND VERTICAL LINES

The graph of a horizontal line $y = b$ will pass through the point $(0, b)$.

The graph of a vertical line $x = a$ will pass through the point $(a, 0)$.

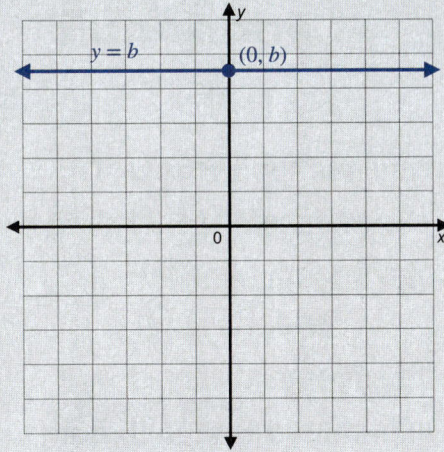

Figure 5.2.26: Horizontal Line

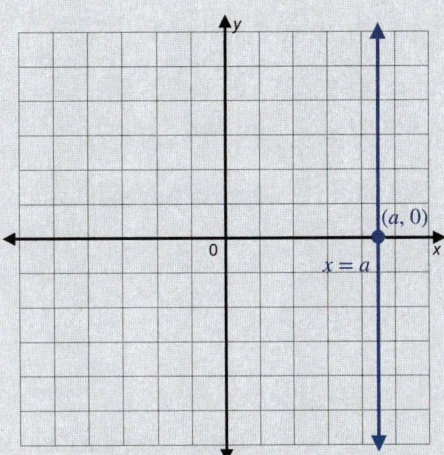

Figure 5.2.27: Vertical Line

We will investigate horizontal and vertical lines further in later sections when we consider modeling with linear inequalities.

Skill Check Answers

1. a. $m = 0$, the line is horizontal; **b.** $m = -1$, the line falls from left to right

2. $y = -3.25x + 185.75$ **3.** $m = 7$, $(0, -10)$, $(3, 11)$ **4.** $\dfrac{4}{3}$

5.2   Exercises

✔ CONCEPT CHECK

1. The _____ of a line is the ratio of the change in y-values to the change in x-values.

2. The _____ form of an equation is $y - y_1 = m(x - x_1)$.

3. The _____ form of an equation is $y = mx + b$.

4. _____ lines have slopes that are negative reciprocals of each other.

5. _____ lines have slopes that are equal to each other.

💡 PRACTICE

Find the slope of the line determined by each pair of points.

6. $(8, -3), (5, 9)$

7. $(9, -2), (11, -2)$

8. $(8, 3), (8, -3)$

9. $(1, 1), \left(2, \dfrac{3}{2}\right)$

Find the slope and y-intercept of each linear equation.

10. $y = 2x - 7$

11. $y = -4x$

12. $2x + 4y = -5$

13. $-x + 2y = 3$

14. $y = 7$

15. $x = 10$

Find the slope-intercept form of the equation for the line passing through each pair of points.

16. $(12, 83)$ and $(9, 59)$

17. $(-8, 76)$ and $(17, -74)$.

18. $\left(-\dfrac{5}{4}, -\dfrac{13}{4}\right)$ and $\left(\dfrac{5}{3}, -\dfrac{23}{6}\right)$

19. $(5.5, 45)$ and $(-10, -85.2)$

20. Identify the slope and y-intercept, along with another point on the line $14x - 2y = 64$, given that $y = 7x - 32$ and $y - 24 = 7(x - 8)$ are equivalent equations.

21. Identify the slope and y-intercept, along with another point on the line $5x + 4y = 18$, given that $y = -\dfrac{5x}{4} + \dfrac{9}{2}$ and $y - \dfrac{5}{4} = -\dfrac{5}{4}\left(x - \dfrac{13}{5}\right)$ are equivalent equations.

22. Identify the slope and y-intercept, along with another point on the line $-22x + 6y = -74$, given that $y = \dfrac{11x}{3} - \dfrac{37}{3}$ and $y + \dfrac{49}{12} = \dfrac{11}{3}\left(x - \dfrac{9}{4}\right)$ are equivalent equations.

23. Identify the slope and y-intercept, along with another point on the line $19x + 5y = 7$, given that $y = -3.8x + 1.4$ and $y + 48 = -3.8(x - 13)$ are equivalent equations.

Find the equation for each line described. Write your answer in slope-intercept form.

24. A line passes through the point $(1, 7)$ and is parallel to a line with equation $y = 3x + 12$.

25. A line passes through the point $(-3, 5)$ and is parallel to a line with equation $y = -\dfrac{2}{3}x - 10$.

26. A line passes through the point $(0, -2)$ and is parallel to a line with equation $y + 5 = \dfrac{1}{2}(x + 5)$.

27. A line passes through the point $(6, 0)$ and is parallel to a line with equation $y - \dfrac{1}{2} = -\dfrac{1}{4}(x + 1)$.

28. A line passes through the point $(5, 3)$ and is perpendicular to a line with equation $y = 5x + 11$.

29. A line passes through the point $(-4, 8)$ and is perpendicular to a line with equation $y = -8x + 9$.

30. A line passes through the point $(0, 9)$ and is perpendicular to a line with equation $y - 3 = 2\left(x + \dfrac{1}{2}\right)$.

31. A line passes through the point $(-3, 0)$ and is perpendicular to a line with equation $y + 6 = \dfrac{3}{5}(x + 15)$.

Graph each line.

32. $y = x + 8$

33. $-2x = y - 6$

34. $y - 3 = \dfrac{1}{2}(x + 6)$

35. $y + 2 = -\dfrac{1}{3}(x + 12)$

36. $x = -5$

37. $y = 2.5$

APPLICATIONS

38. The following table provides the mileage and list price of pre-owned 2018 Hyundai Elantras being sold within a 50-mile radius. Round your answers to the thousdandth, if necessary.

Pre-Owned 2018 Hyundai Elantra	
Mileage (miles)	**Price (dollars)**
14,028	19,431
26,477	18,877
16,838	19,947
32,701	18,455
29,359	20,157
17,754	19,891
31,484	18,995

a. Use the car with the lowest mileage and the car with the highest mileage to create an equation to estimate the price the car should be. Let x be the mileage and y be the price.

b. Use this equation to estimate the price a dealership should use to sell a 2018 Hyundai Elantra with a mileage of 20,000 miles.

c. Explain some of the limitations of the equation created in part a.

39. The following table shows the mileage and list price of pre-owned 2018 Ford Focuses being sold within a 50-mile radius.

Pre-Owned 2018 Ford Focus	
Mileage (miles)	**Price (dollars)**
50,868	14,717
72,208	13,907
37,385	14,839
90,696	12,889
25,344	16,250
45,069	13,863
28,908	16,895

a. Use the car with the lowest mileage and the car with the highest mileage to create an equation to estimate the price the car should be based on mileage. Let x be the mileage and y be the price.

b. Use this equation to estimate the price a dealership should use to sell a 2018 Ford Focus with a mileage of 50,000 miles.

c. How does the price obtained from the formula compare to the Ford Focus that has 50,868 miles? Explain why the prices might be similar or different.

40. Refer to the table from Exercise 38.

    a. Use the car with the lowest price and the car with the highest price to create an equation to estimate the price the car should be based on mileage. Let x be the mileage and y be the price.

    b. Use this equation to estimate the price a dealership should use to sell a 2018 Hyundai Elantra with a mileage of 20,000 miles.

    c. Compare the answer from part b. to the answer from Exercise 38 part b. Explain why the values are similar or different.

41. Refer to the table from Exercise 39.

    a. Use the car with the lowest price and the car with the highest price to create an equation to estimate the price the car should be based on mileage. Let x be the mileage and y be the price.

    b. Use this equation to estimate the price a dealership should use to sell a 2018 Ford Focus with a mileage of 50,000 miles.

    c. Compare the answer from part b. to the answer from Exercise 39 part b. Explain why the values are similar or different.

42. The following table shows the area of apartments along with the monthly rent.

Apartment Rent

Area (square feet)	Price (dollars)
504	1450
723	2180
465	1040
812	1970
389	930
637	1320
558	1290

    a. Use the apartment with the smallest area and the apartment with the largest area to create an equation that can be used to estimate the price of rent based on the area. Let x be the area and y be the price.

    b. Use this equation to estimate the price a landlord should use to rent out an apartment with an area of 500 square feet. Round your answer to the nearest cent.

    c. How does the price obtained from the formula compare to the apartment that is 504 square feet large? Explain why the prices might be similar or different.

43. The local candy store has a bin of candies where customers can fill their own bag with whatever amount they'd like, and they are charged by depending on how much the bag weighs. The following table shows the size and price of several different bags.

Bag of Candies

Size (grams)	Price (dollars)
510	16.50
600	19.50
340	13.00
420	14.00
280	12.50
630	18.00
550	13.50

a. Use the smallest bag and the largest bag to create an equation that can be used to estimate the price of a bag based on size. Let x be the size and y be the price.

b. Use this equation to estimate how much a bag filled with 500 grams of candies would cost.

c. How does the price obtained from the formula compare to the 510-gram bag of candies? Explain why the prices might be similar or different.

44. Refer to the table from Exercise 42.

a. Use the apartment with the lowest price and the apartment with the highest price to create an equation that can be used to estimate the price of rent based on the area. Let x be the area and y be the price.

b. Use this equation to estimate the price a landlord should use to rent out an apartment with an area of 500 square feet.

c. Compare the answer from part b. to the answer from Exercise 42, part b. Explain why the values are similar or different.

45. Refer to the table from Exercise 43.

a. Use the bag with the lowest price and the bag with the highest price to create an equation that can be used to estimate the price of a bag based on size. Let x be the mass and y be the price.

b. Use this equation to estimate how much a bag filled with 500 grams of candies would cost.

c. Compare the answer from part b. to the answer from Exercise 43, part b. Explain why the values are similar or different.

46. A biologist does research on two populations of bacteria, one pathogenic and the other nonpathogenic. She knows that the nonpathogenic bacteria suppress the growth of the pathogenic bacteria. The biologist estimates the population growth for the two types of bacteria with the following equations.

    a. Pathogenic: $y = 5 + \dfrac{t}{4}$

    b. Nonpathogenic: $y = \dfrac{3t}{2}$

Graph each of the lines on the same graph and use the graph to determine the number of days t required for the nonpathogenic population to suppress the pathogenic population, if this happens at all.

47. Sandy started a flower-delivery business. At first, the expenses exceeded her income, so she hired an expert to estimate when the business would start to earn profit. Based on the provided data, the expert obtained two models describing the expenses and the income of the business.

    a. Expenses: $y = \dfrac{3t}{2} + 1$

    b. Income: $y = 2t - 3$

Plot the two functions on the same graph. Use the graph to determine the number of months t required for the flower-delivery business to begin to earn a profit.

48. The following equations represent the amount of a product that consumers want to buy (quantity demanded) and the amount that manufacturers are ready to sell (quantity supplied) depending on the price of the product.

    a. Quantity demanded: $Q_d = 20 - 2P$

    b. Quantity supplied: $Q_s = 3P + 10$

Plot the two functions on the same graph. Use the graph to determine the equilibrium price P (given in hundreds of dollars) where supply and demand meet.

49. Suppose that the addition of some element to an alloy increases the strength and decreases the thermal conductivity of the alloy. Expressed in the same units, these properties depend on the mass of the element x (in grams per 100 g of alloy) as follows.

    a. Thermal conductivity: $y_t = 15 - \dfrac{x}{2}$

    b. Strength: $y_s = \dfrac{1}{3}(x + 5)$

Graph each of the lines on the same graph and use the graph to determine the mass of the element needed to get an alloy of optimal properties.

5.2 PROJECT

THE THEORY OF THE STORK

It has been humorously stated that there are two theories concerning the origin of children: the Theory of Sexual Reproduction (ThoSR) and the Theory of the Stork (ThoS)[2]. Noticeably absent is the Theory of the Cabbage Patch, which, for whatever reason, has yet to gain traction. With respect to ThoS, it was observed that in the German state of Lower Saxony, between 1970 and 1985, the number of out-of-hospital births decreased while there was also a decline in recorded pairs of storks. Between 1985 and 1995, both of those numbers remained relatively unchanged. Meanwhile, in Brandenburg, the countryside around Berlin, the stork population showed a decline between 1990 and 1991, followed by a general increasing trend from 1991 to 1999. In the city of Berlin, out-of-hospital births showed the same trend. Below is a table with these numbers recorded, rounded to the nearest 10.

Year	Pairs of Storks (x)	Out-of-Hospital Births (y)
1990	970	900
1991	850	790
1992	980	780
1993	1210	890
1994	1280	960
1995	1270	1080
1996	1360	1070
1997	1120	1250
1998	1320	1130
1999	1370	1200

1. Plot the points on a graph with pairs of storks as the x-coordinate and out-of-hospital births as the y-coordinate.

2. Find the slope-intercept form of the equation for the line passing through the data points representing the lowest number of pairs of storks and the highest number of pairs of storks.

3. Find the slope-intercept form of the equation for the line passing through the data points representing the lowest number of out-of-hospital births and the highest number of out-of-hospital births.

4. Find a third slope-intercept form of a line passing through two data points of your own choosing that you think may yield a line that better "fits" the plotted points.

5. Using the equations for the lines you obtained in parts 2, 3, and 4, choose any x-value from the table and calculate the corresponding y-values. How does each y-value compare to the actual observed number of out-of-hospital births for that number of pairs of storks?

6. Sketch the three lines on the graph with the plotted data points. Which of the three lines would you consider the "best fit"?

---

2   Thomas Höfer, Hildegard Przyrembel, and Silvia Verleger, "New Evidence for the Theory of the Stork," Paediatric and Perinatal Epidemiology, Volume 18 (January 2004): 88–92.

OBJECTIVES

1. Solve different types of systems of linear equations.

2. Model linear equations with cost and revenue functions.

OBJECTIVE 1 Systems of Linear Equations

In the last section, we developed techniques to help us model situations with linear equations and visualize them as straight-line graphs. In this section, we will turn our attention to focus on two linear equations at the same time, or what is called a *system of two equations in two variables*.

If we draw a system of two lines on a Cartesian coordinate plane, one of three things will occur:

1. The lines will *never* intersect.

2. The lines will be the *same* line and intersect in infinitely many points.

3. The lines will intersect in exactly *one* point.

Table 5.3.1: Types of Systems of Linear Equations

System

$$\begin{cases} y = -x + 5 \\ x + y = 3 \end{cases} \qquad \begin{cases} y = -x + 5 \\ 2x + 2y = 10 \end{cases} \qquad \begin{cases} y = -x + 5 \\ y = 2x + 2 \end{cases}$$

Graph

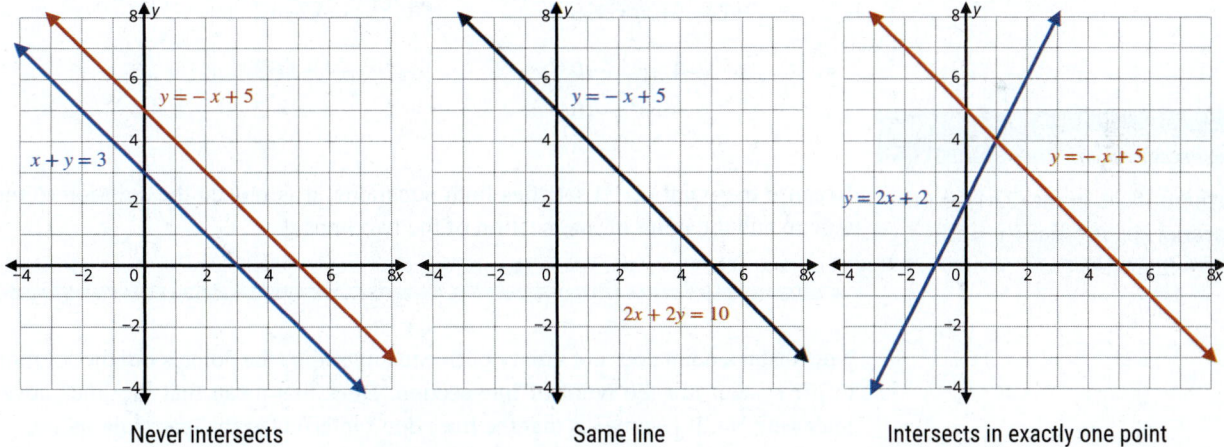

Never intersects   Same line   Intersects in exactly one point

How does the first scenario happen? The only way that two lines will never intersect, no matter how long you follow the lines, is if their slopes are equal to one another and they don't have the same *y*-intercept; in other words, when the lines are *parallel* to one another.

However, if the lines have the same slope *and* the same *y*-intercept, the second scenario occurs. Even though the equations might appear in different forms, equations that have the same slope and *y*-intercept actually represent the same line and have infinitely many points of intersection.

The third scenario listed arises when two lines have different slopes. When this is the case, the lines must intersect one another at *exactly one* point. This is the scenario where we will focus our attention, and we will look at ways to find that point of intersection.

> ### 💬 Helpful Hint
>
> Systems of equations are considered **consistent** if they have either one or infinitely many solutions in common. They are considered **inconsistent** if they have no solution in common.

Before we discuss methods of finding the intersection of two lines (and hence solving a system of linear equations), it's important to think about what we are searching for. When two lines cross, or intersect one another, they do so at a single point. When we speak of their intersection, we are speaking of the singular point that lies on both lines. That is to say, the coordinates of the point satisfy both equations. The first example is simply that: checking to see if a point is the intersection of two given lines; in other words, verifying the solution to a given system of equations.

Example 5.3.1

Verifying the Point of Intersection of Two Lines

Determine if the point (2, 3) is the point of intersection for the given lines. In other words, is (2, 3) the solution to the system of equations?

$$\begin{cases} -2x - 2y = -10 \\ -0.5x + y = 2 \end{cases}$$

Solution

In order to determine if a point is a solution to a system of equations, we must substitute the x- and y-coordinates into each equation and verify that both are satisfied.

Verifying the 1st Equation:	Verifying the 2nd Equation:
$-2x - 2y = -10$	$-0.5x + y = 2$
$-2(2) - 2(3) \overset{?}{=} -10$	$-0.5(2) + (3) \overset{?}{=} 2$
$-4 - 6 \overset{?}{=} -10$	$-1 + 3 \overset{?}{=} 2$
$-10 = -10$	$2 = 2$
☑	☑

✔ **Skill Check 5.3.1**

Determine if the point (2, 4) is the intersection of the lines $y = 5x - 6$ and $3x - 0.1y = 1$.

Because the point (2, 3) satisfies both equations, it is indeed the solution to the system and the point of intersection of the two lines.

Note that if a point does not satisfy **both** equations, then the point is **not** the solution to the system nor the point of intersection. Does this mean that the lines never intersect? No. It just means that the lines don't intersect at that particular point.

Now let's turn our attention to two methods of finding the point of intersection for a system of equations: an algebraic method and a graphing method. Graphically, we can plot both lines on the same graph and find their point of intersection by observation. This gives us a visual representation of the whole scenario, as well as the point of intersection itself. However, as you might imagine, this method is sometimes cumbersome, as it has the potential to necessitate a very large graph —say the point of intersection is (−2000, 11987)—or to create ambiguity if the point of intersection has decimal coordinates—say $\left(1.35, \dfrac{11}{98}\right)$. It is for this reason that technology is the best way to find the point of intersection using the graphing method. However, using algebra to find the intersection is always precise.

The algebraic method we will use is often referred to as the "substitution method." The idea is that if we know two things are equal to each other, we can *substitute* one for the other. The process involves solving one of the equations for a particular variable, and then using that expression as the substitute for a variable in the other equation. The steps are summarized as follows.

> ### ☰ SOLVING A SYSTEM OF LINEAR EQUATIONS BY SUBSTITUTION
>
> 1. Solve one of the equations for one of the variables.
> 2. Substitute the expression found in Step 1 into the other equation and solve for the remaining variable.
> 3. Re-substitute the value found in Step 2 into the original equation from Step 1 to find the corresponding coordinate.

Because this algebraic method is more precise than graphing, we will solve systems algebraically to find the point of intersection and then use a graphing calculator to check our answer.

Example 5.3.2

Solving a System of Linear Equations

Solve the following linear system of equations algebraically. Check your answer using a graphing calculator.

$$\begin{cases} y = 2x + 4 \\ -x + y = 1 \end{cases}$$

Solution

Step 1: Since the first equation given is already solved for y, we can move on to step 2.

Step 2: the expression $2x + 4$ as a substitution for y in the second equation given.

$$-x + (2x + 4) = 1$$

Now we have a single equation in one variable that we can solve. Solving for x we get the following.

$$x + 4 = 1$$
$$x = -3$$

Step 3: Using the original equation from step 1, find the y-coordinate by substituting in $x = -3$.

$$y = 2x + 4$$
$$y = 2(-3) + 4$$
$$y = -6 + 4$$
$$y = -2$$

Thus, our point of intersection for the two lines is $(-3, -2)$.

To confirm that our point is correct, we will graph both lines on the same set of axes. In theory we could graph both lines by hand, as we saw in Section 5.2, but we will use technology as a more precise method of locating the intersection.

Check Graphically

To check the solution on a TI-83/84 Plus calculator, press [y=] and clear any equations currently in the calculator. The first equation is already in the required form, with the y-variable isolated on one side, but the second equation will need to be rewritten as $y = x + 1$. Now we can input the equations for Y1 and Y2 using the slope-intercept forms found and press [graph].

> **⌁ TECH TIP**
>
> Pressing [window] or [zoom] will allow you to adjust the display window on the calculator.

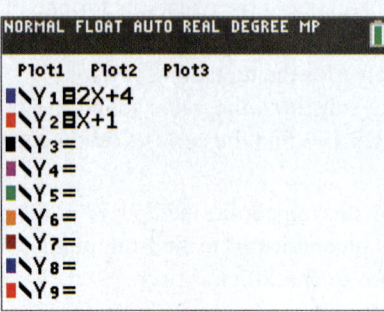

Figure 5.3.1

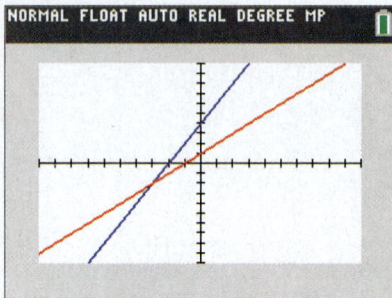

Figure 5.3.2

Once the lines are graphed, to find the intersection of the lines, press [2nd] [trace] and select intersect. You'll see a blinking cursor along the first line. Press [enter], and there should now be a blinking cursor on the second line. Press [enter] again, and you'll see Guess? at the bottom of the screen. Press [enter] a third time, and at the bottom of the screen the calculator will display its best approximation of the intersection of the two lines.

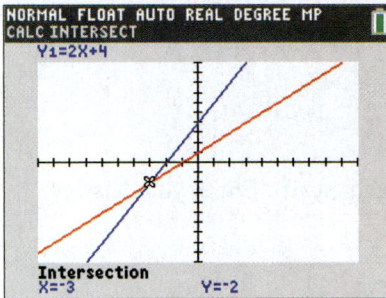

Figure 5.3.3

The calculator confirms that the point of intersection is $(-3, -2)$.

Example 5.3.3

Solving a System of Linear Equations

Find the point of intersection between the lines $x + y = 4$ and $2x - 3y = 6$ using algebra. Check your answer using a graphing calculator.

Solution

Step 1: Notice that this time neither of the equations is in slope-intercept form; that is, neither of them is solved for one variable. Thus, we need to solve one of the equations for one of the variables. We will use the first equation and solve for y. Note that it does not make a difference which equation or which variable is chosen.

Solve the First Equation for y

$$x + y = 4$$

$$y = -x + 4$$

Step 2: Use the expression $-x + 4$ as a substitution for y in the second equation given and solve.

$$2x - 3y = 6$$
$$2x - 3(-x + 4) = 6$$
$$5x - 12 = 6$$
$$5x = 18$$
$$x = \frac{18}{5}$$

Step 3: Using the original equation from step 1, find the y-coordinate by substituting in $x = \frac{18}{5}$.

$$x + y = 4$$
$$\frac{18}{5} + y = 4$$
$$y = \frac{2}{5}$$

Thus, the point of intersection is $\left(\frac{18}{5}, \frac{2}{5} \right)$. We can now use the calculator to check our answer.

Check Graphically

Press [y=] and clear any equations currently in the calculator. Using the slope-intercept form of each equation ($y = -x + 4$ and $y = \frac{2}{3}x - 2$), input the equations for Y1 and Y2 and press [graph]. Press [2nd] [trace] and select **interact** and press [enter] three times. At the bottom of the screen, the calculator gives the point of the intersection in decimal form. This confirms our findings, as $\frac{18}{5} = 3.6$ and $\frac{2}{5} = 0.4$.

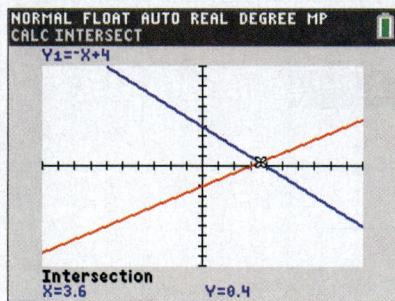

Figure 5.3.4

So far, the systems of linear equations that we have solved all have intersected in one point. At the beginning of this section, we learned that two lines can either intersect in one point, never intersect (be parallel), or always intersect (be the same line). Let's consider what the solution looks like algebraically for a system of two lines that are parallel or that are the same line.

Recall from Section 5.2 that if lines are parallel or are the same line, they have the same slope. By arranging both equations into slope-intercept form, we can easily determine if the slopes of the lines are the same. If the slopes are equal and the y-intercepts are also equal, then the lines are identical. If the slopes are equal and the y-intercepts are different, then the lines are parallel. Example 5.3.4 illustrates this last scenario.

Example 5.3.4

Solving a System of Linear Equations

Find the point of intersection for the lines $y = -4x + 3$ and $8x + 2y = 10$. Check your answer using a graphing calculator.

Solution

Step 1: Since the first equation given is already solved for y, we can move on to step 2.

Step 2: Use the expression $-4x + 3$ as a substitution for y in the second equation given and solve.

$$8x + 2y = 10$$
$$8x + 2(-4x + 3) = 10$$
$$8x - 8x + 6 = 10$$
$$6 = 10$$
$$\boxed{\times}$$

Because we have a false statement, we can stop. A false statement means that there is not a point of intersection between the lines. Thus, the lines are parallel.

Check Graphically

Press $\boxed{y=}$ and clear any equations currently in the calculator. Input the equations for Y1 and Y2 using the slope-intercept forms found and press $\boxed{\text{graph}}$.

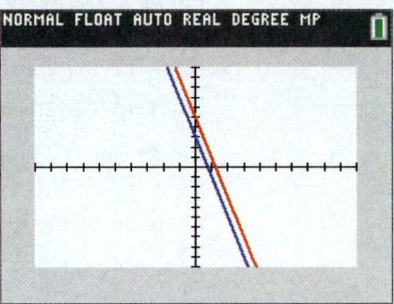

Figure 5.3.5

TECH TIP

When you graph equations that are the same line on a TI-83/84 Plus calculator, it draws the first line, then draws the second line over the first. If you're not careful, you might see what looks like one line even though both have been graphed.

The two lines appear to be parallel; that is, no point of intersection is visible. To confirm this, press $\boxed{\text{2nd}}\boxed{\text{trace}}$ and select `intersect`. As in the previous examples, press $\boxed{\text{enter}}$ three times. Instead of returning a point of intersection, the calculator returns an error. This is because the lines never intersect. So, we've confirmed they are parallel.

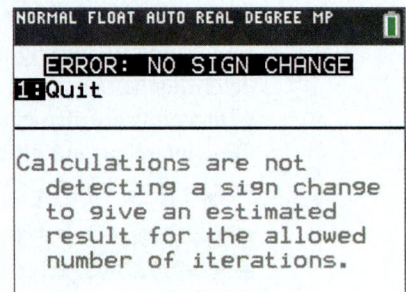

Figure 5.3.6

✔ Skill Check 5.3.2

Find the point of intersection between the lines $2y = -6x - 20$ and $3x - y = -10$ using algebra. Check your answer using a graphing calculator.

OBJECTIVE 2 Modeling Cost and Revenue with Systems of Linear Equations

Example 5.3.5

Modeling with Linear Systems

As a recruitment officer, Asher is responsible for planning outings for his university organization. For the next outing, he's considering The Escape Station, which charges $135 plus $15 per person to rent out the space. The organization guidelines indicate that he can charge at most $20 per person for the trip. Asher needs to determine how many people would need to go on the trip to make it viable for him to rent The Escape Station if he charges $20 per person.

Write two linear functions, one for the cost of renting The Escape Station, $C(x)$, and one for the amount of money Asher will collect, $R(x)$. Use the functions to find out how many people need to go on the trip in order for Asher to not lose any money.

Solution

We can begin to help Asher with his question by setting up two linear functions to model the scenario. As we did back in Section 5.1, we will begin by introducing variables to represent the unknowns. We will let x be the number of people attending the outing. The function $C(x)$ represents the cost of renting The Escape Station and the function $R(x)$ represents the amount of money that Asher collects.

We know that Asher will collect $20 for each person that attends. So the function for the money he receives is $R(x) = 20x$. This is often referred to as the *revenue function*.

At the same time, we know the cost of the rental space will be $15 per person plus $135 for the room. The function that connects Asher's total cost with the number of attendees is then $C(x) = 15x + 135$. This is often referred to as the *cost function*.

Now that we have both functions, we can find their point of intersection, which is called the *break-even point*. This point represents when the cost of the trip and the revenue Asher will collect are equal to one another, and hence will tell us the number of people he needs to make the trip viable.

We set the functions equal to each other and solve for x.

$$R(x) = C(x)$$
$$20x = 15x + 135$$
$$5x = 135$$
$$x = 27$$

Using substitution and either of the functions, we can find the y-coordinate of the point of intersection. We'll use the revenue function $R(x)$.

$$R(x) = 20x$$
$$R(27) = 20(27)$$
$$R(27) = 540$$

Hence, the point of intersection is at (27, 540). This means that when there are 27 people on the trip, Asher will collect $540 and then owe The Escape Station $540, breaking even on the trip. Fewer people attending means Asher will owe more money than he collects, and any extra attendees over 27 means he will have money left over to spend on other things.

As we alluded to in Example 5.3.5, equations that describe cost and revenue are often used in business models. Along with profit functions, the cost and revenue functions are foundational elements for any business model. The *cost* function calculates the total amount of money required to make a certain quantity of product. This might include things such as start-up costs, running costs like rent and supplies, or monthly salaries. The *revenue* function is the amount of money collected by selling a certain quantity of product. And the *profit* function calculates the amount of profit the business can expect from producing and selling a certain quantity of the items.

> **💬 Helpful Hint**
>
> The break-even point occurs when $R = C$, which is also when $P = 0$.

Often businesses, especially start-up companies, are interested in what is called the *break-even* point. The **break-even point** is the quantity of units that needs to be produced and sold so that the amount of money spent by the business is equal to the amount of money being collected. In other words, the business is neither making any profit nor losing any money—it is simply "breaking even." This is an important place to identify because it means that the sales of the company are enough to cover the cost of running the company. We can determine the break-even point by solving the system of equations consisting of the cost and revenue equations.

📖 BUSINESS MODELING FUNCTIONS

Cost Function (C)

The **cost function** calculates the total cost C of producing x units; it is equal to the original fixed cost plus the production cost for each unit multiplied by the number of units x.

$$C = \text{Fixed Cost} + (\text{Production Cost per Unit})x$$

Revenue Function (R)

The **revenue function** calculates the revenue received R; it is equal to the price per unit multiplied by the number of units sold x.

$$R = (\text{Price per Unit})x$$

Profit Function (P)

The **profit function** calculates the amount of profit P created for x number of units made and sold and is found by subtracting the cost from the revenue.

$$P = R - C$$

Break-Even Point

The quantity of units to be produced and sold so that the cost of production equals the revenues for a product is called the **break-even point**.

Example 5.3.6

Determining the Break-Even Point

Harley is drawing up a business plan to produce and sell a new in-store price scanner. She has determined that the initial start-up cost will be $18,000, and it will cost her $14 to produce each scanner. Harley plans to sell each scanner to stores for $30. Let x be the number of scanners produced.

a. Write the cost function for producing the scanners.

b. Write the revenue function for selling the scanners.

c. Write the profit function for Harley's business.

d. Determine the break-even point for Harley's new scanner business.

e. If Harley sells 2500 scanners, what profit can Harley expect to make?

Solution

a. The cost function includes the fixed cost, which is $18,000, plus the cost for each scanner to be produced. The production cost per scanner is $14 multiplied by the number of scanners produced, x.

Cost Function

$$C(x) = 18{,}000 + 14x$$

b. The revenue function consists of the sale price of each scanner, $30, multiplied by the number of scanners sold, x.

Revenue Function

$$R(x) = 30x$$

c. The profit function is determined by subtracting the cost function from the revenue function, $P(x) = R(x) - C(x)$.

Profit Function

$$P(x) = 30x - (18{,}000 + 14x)$$
$$P(x) = 16x - 18{,}000$$

d. The break-even point is found by solving the system of equations consisting of the cost and revenue functions. The intersection will happen when the cost is equal to the revenue. Therefore, we need to set the two equations equal to one another and solve for x.

$$30x = 18{,}000 + 14x$$
$$16x = 18{,}000$$
$$x = 1125$$

Therefore, the break-even point for Harley is when she produces and sells 1125 scanners.

e. To determine the profit Harley will make if she sells 2500 scanners, we need to substitute 2500 into the profit function found in part c.

✔ **Skill Check 5.3.3**

Suppose you are planning to manufacture a new board game that requires an initial cost of $5000 and an additional $11 for each game produced. You plan to then sell each board game for $28. Write the cost function and the revenue function for this venture. Then, find the break-even point for your board game business.

$$P(x) = 16x - 18,000$$
$$P(x) = 16(2500) - 18,000$$
$$P(x) = 22,000$$

Therefore, Harley will make a profit of $22,000 when she sells 2500 scanners.

The break-even point might not always be an integer. Note that for many businesses, it is only possible to sell in whole units, and not in partial units. This means that if the break-even point is a fraction, you should always round up to the next whole number so that the company is actually covering *all* costs and not making a loss of even a small fraction.

Skill Check Answers

1. $x = -\dfrac{19}{2}$ 2. $(2, 4)$ is not the point of intersection   3. $C(x) = 5,000 + 11x$;

$R(x) = 28x$; The break-even point is approximately 294.12, which rounds up to 295.

5.3   Exercises

✔ **CONCEPT CHECK**

1. When drawing a system of linear equations, one of three things will occur:

   a. The lines will _____.

   b. The lines will _____.

   c. The lines will _____.

2. The first step in solving a system of linear equations is to _____ one of the equations for one of the _____.

3. The break-even point is the amount of units that need to be produced and sold so that the _____ by the business is equal to the amount of _____.

4. The _____ function is equal to the price per unit multiplied by the number of units sold.

5. The _____ function is equal to the original fixed cost plus the production cost for each unit multiplied by the number of units.

💡 **PRACTICE**

Determine if the given point is the point of intersection for the pair of lines.

6. $(3, -8)$; $\begin{cases} y = -5x + 7 \\ 3x + y = 1 \end{cases}$ 7. $(6, -3)$; $\begin{cases} 0.5x - 2y = 9 \\ x + 3y = 13 \end{cases}$

8. $(-4, -5)$; $\begin{cases} 3x + 4y = -32 \\ 2x + y = -13 \end{cases}$ 9. $\left(\dfrac{1}{4}, 2\right)$; $\begin{cases} -8x = -2y \\ 6x - 3y + 6 = 0 \end{cases}$

Solve the system of equations.

10. $\begin{cases} 4x + 3y = 1 \\ 2x - 2y = 4 \end{cases}$ 11. $\begin{cases} 2x + 4y = 8 \\ 6x - 8y = 4 \end{cases}$

12. $\begin{cases} y = 3x - 1 \\ y = 3x + 2 \end{cases}$ 13. $\begin{cases} -\dfrac{2}{3}x + \dfrac{1}{2}y = \dfrac{4}{7} \\ \dfrac{2}{12}x - \dfrac{1}{8}y = -\dfrac{1}{7} \end{cases}$

14. $\begin{cases} \dfrac{3}{4}x + 3y = \dfrac{4}{3} \\ -\dfrac{1}{4}x - y = \dfrac{7}{3} \end{cases}$ 15. $\begin{cases} -4x + y = 1 \\ 4x + 4y = -16 \end{cases}$

16. $\begin{cases} 6x + y = 30 \\ 6x + y = -19 \end{cases}$ 17. $\begin{cases} x + y = -1 \\ -3x + y = -1 \end{cases}$

18. $\begin{cases} 9x + 3y = 15 \\ 3x + y = 5 \end{cases}$ 19. $\begin{cases} y = 4x + 12 \\ 2y = 8x + 24 \end{cases}$

20. $\begin{cases} -5x + y = -12 \\ 5y = 10x - 15 \end{cases}$ 21. $\begin{cases} 6x - y = 11 \\ -6y = -8x + 10 \end{cases}$

22. The profit function for a company selling x teapots is given by $P(x) = 2.4x - 3230.50$. Find the company's profit if 17,000 teapots are sold.

23. If a company sells 3000 units of their product, what profit can they expect to make if their profit function is given by $P(x) = 0.54x - 841.75$?

🚀 APPLICATIONS

24. Twelve gallons of a salt solution consists of 30% salt. This solution was created by mixing a 40% solution with a 25% solution. How many gallons of each of the solutions was used to create the 30% solution? Let $x =$ the number of gallons of the 40% solution and $y =$ the number of gallons of the 25% solution. The corresponding model is the following system of equations.

$$\begin{cases} x + y = 12 \\ 0.4x + 0.25y = 0.3(12) \end{cases}$$

25. A student bought a calculator and a textbook for an algebra course. He told his friend that the total cost was $170 (without tax) and that the calculator cost $20 more than four times the cost of the textbook. What was the cost of each item? Let $x =$ the cost of a calculator and $y =$ the cost of the textbook. The corresponding modeling system is $\begin{cases} x + y = 170 \\ x = 4y + 20 \end{cases}$. Solve the system by using the method of substitution.

26. On a round-trip to the south of the state, the pilot of a small personal plane recorded the average speed of the plane at 159 mph when flying with a tailwind. On the return trip, the plane averaged 133 mph flying back into the same wind. What was the speed of the plane in still air and the speed of the wind? Let x = the speed of the plane, and y = the speed of the wind. The corresponding model is the following system of equations. Solve the system to find the speed of the plane and the wind.

$$\begin{cases} x + y = 159 \\ x - y = 133 \end{cases}$$

27. The theater department brought in $3072 from the sale of 723 tickets to the spring play. Unfortunately, the record of how many types of tickets were sold was lost. The department would like to know how many student tickets were bought for $2 and how many nonstudent tickets were bought at $5. Let x = the number of non-student tickets and y = the number of student tickets. The corresponding model is the following system of equations. Determine how many of each type of ticket was sold.

$$\begin{cases} 5x + 2y = 3072 \\ x + y = 723 \end{cases}$$

28. Ten liters of an acid solution contains 42% sulfuric acid. This mixture was obtained by combining a 30% solution with a 50% solution. How many liters of each solution was used? Let x = the number of liters of the 30% solution and y = the number of liters of the 50% solution. The corresponding model is the following system of equations. Solve the system to find the number liters used for each type of solution.

$$\begin{cases} 0.3x + 0.5y = 0.42(10) \\ x + y = 10 \end{cases}$$

29. Suppose that $10,000 was invested in two accounts. In one year, the 7.5% mutual fund account earned $585.45 more interest than the 3.47% CD. How much was originally invested in each account? Let x = the amount deposited in the mutual fund and y = the amount deposited in the CD. The corresponding model is the following system of equations. Determine the amount deposited in each account.

$$\begin{cases} x + y = 10,000 \\ 0.075x = 585.45 + 0.0347y \end{cases}$$

30. A farmer plans to sell harvested tomatoes at a nearby market. The cost of growing tomatoes is $165 plus $0.50 for each pound of the product. The farmer plans to sell the tomatoes at $2 per pound. He wants to know how many pounds he'll need to sell in order to recoup the costs of growing tomatoes.

Write two linear functions, one for the cost of growing the tomatoes, $C(x)$, and one for the amount of money the farmer will earn, $R(x)$. Use the functions to find out how many pounds of tomatoes need to be sold in order for the farmer to not lose any money.

31. In some countries, there is a limit on water consumption per person per day. If this rate is exceeded, water supply is paid at increased rates. Suppose that a family spends 270 liters of water per day for general household needs, and each member of the family also spends 50 liters of water individually. The water consumption rate per person in this region is 140 liters per day, and this family does not exceed the allowed rate.

Write two linear functions, one for the amount of water the family spends, $C(x)$, and one for the amount of water specified by the consumption rate, $R(x)$. Use the functions to find the maximum number of people in this family given that they do not overpay for the water supply services.

32. You are investing in a new outdoor game similar to a corn-toss game, but for golf. The start-up cost includes an initial lump sum of \$9200. The production cost for each game is \$17.43 and they will be sold for \$84.99. Let x represent the number of games produced.

   a. Write the cost function, C.

   b. Write the revenue function, R.

   c. Write the profit function, P.

   d. Determine the break-even point and describe what this point means.

33. A shoe company that manufactures waterproof running shoes has a fixed cost of \$100,900. Each pair of shoes sells for \$37.50 wholesale and costs \$27.60 to produce, \$1.00 to ship, \$0.30 for shipping insurance, and \$3.00 in custom taxes. Let x = the number of pairs of running shoes produced.

   a. Write the cost function, C.

   b. Write the revenue function, R.

   c. Write the profit function, P.

   d. Determine the break-even point and describe what this point means.

34. A technology company is developing a new data scanner for use in hospitals. Overhead costs for manufacturing the data scanners requires a fixed cost of \$328,600. Each scanner will sell for \$10,000 and cost \$4700 to produce. Let x represent the number of scanners produced.

   a. Write the cost function, C.

   b. Write the revenue function, R.

   c. Write the profit function, P.

   d. Determine the break-even point and describe what this point means.

35. A ride share company is recruiting drivers. They advertise that each driver will earn on average, $18.26 per hour. The company charges each driver a fee of 25% of each fare. Before deciding if you want to become a driver, you calculate the hidden costs for the car you already own. Insurance, gas, maintenance, and depreciation will cost you $8.16 per hour. Let $x =$ the number of hours you drive for the ride share company. Answer the following questions based on the average earnings per hour as advertised by the company.

  a. Determine the cost function if you decide to drive for the company.

  b. Determine your revenue function.

  c. Determine your profit function.

  d. Suppose you drive 5 hours a day for 6 days the first week and 6.5 hours a day for 4 days the second week. How much did you make your first two weeks as a driver?

  e. Some drivers say that the average profit that they make per hour is actually higher than you calculated because they would be paying insurance, maintenance and have depreciation on their car anyway. Do you agree? Explain your reaction.

36. Maggie decides to become a make-up artist. To do this, she is taking a course for $3300 and purchases the necessary materials for $1500. She also estimates having to spend $20 per client on supplies. Maggie expects to receive an average of $80 per client. Let x be the number of clients.

  a. Write the cost function for Maggie's business.

  b. Write the revenue function.

  c. Write the profit function for the business.

  d. Determine the x-value of the break-even point.

  e. If Maggie serves 100 clients, what profit can she expect to make?

37. A company sells its product for $40 each. Variable costs are $12 per item, and the total fixed costs are $56,000. Let x be the number of products sold.

  a. Write the cost function for the company.

  b. Write the revenue function.

  c. Write the profit function for the company.

  d. Determine the x-value of the break-even point for the company.

  e. If the company sells 3000 items, what profit can it expect to make?

5.3 PROJECT

THE NBA AND SHOT DISTRIBUTION

In this project, you will use systems of linear equations and basketball statistics to take a partial set of information to find additional specific details.

During the 2020–21 season[3], the teams in the National Basketball Association (NBA) averaged taking 88.4 shots per game (the 3-point shots attempted plus the 2-point shots attempted). The teams scored an average of 112.1 points per game with 17 of those points coming from free throws. The average shooting percentages were 36.7% for 3-point shots and 53% for 2-point shots.

1. Letting x be the average number of 3-point attempts per game, write an expression to describe the total points made from 3-point shots.

2. Letting y be the average number of 2-point attempts per game, write an expression to describe the total points made from 2-point shots.

3. Use the expressions from parts 1 and 2 to set up a system of linear equations: one equation to describe the number of shots taken and one equation to describe the average number of points scored. (**Note:** Free throws should not be included in the total number of points since we are only considering 3-point shots and 2-point shots.)

4. Solve the system of equations from part 3 to find the average number of 3-point shots attempted and the average number of 2-point shots attempted.

The highest scoring team in the league during the 2020–21 season averaged making 44.7 shots per game and scoring 120.1 points per game with 16.2 of those points coming from free throws. This team averaged taking 37.1 3-point shots and 54.7 2-point shots.

5. Letting a be the percentage of 3-point attempts made per game (as a decimal), write an expression to describe the total number of points scored from 3-point shots.

6. Letting b be the percentage of 2-point attempts made per game (as a decimal), write an expression to describe the total number of points scored from 2-point shots.

7. Use the expressions from parts 5 and 6 to set up a system of linear equations: one equation for the total number of shots made and one equation for the total number of points scored. (Again, note that free throws should not be included in the total number of points since we are only considering 3-point and 2-point shots.)

8. Solve the system of equations from part 7 to find the percent of 3-point shots made per game and the percent of 2-point shots made per game.

---

3   Source: Basketball Reference, https://www.basketball-reference.com/.

5.4 LINEAR INEQUALITIES IN TWO VARIABLES

OBJECTIVES

1. Graph linear inequalities.

2. Graph systems of linear inequalities.

OBJECTIVE 1 Linear Inequalities

The "Diet Problem," studied in the 1930s and 40s, sought to help the US Army minimize the cost of feeding a nutritional diet to soldiers who were in the field. The idea was to regulate the number of calories, vitamins, fats, and other nutrients while aiming for palatable food below a certain cost. The researchers were interested in finding a range of quantities rather than a strict equality, like we've looked at so far. These types of investigations involve **inequalities** rather than equations. Inequalities involve finding "no more than" or "at least" certain amounts.

Helpful Hint

Common Inequality Phrases

>	<	≥	≤
More than	Fewer than	At least	At most
Above	Below	No fewer than	No more than
Exceeds	Up to		

INEQUALITIES

Inequality Symbol

An **inequality symbol** is a relation used between two expressions that are not strictly equal.

$<$ less than $\leq$ less than or equal to

$>$ greater than $\geq$ greater than or equal to

Linear Inequality

A **linear inequality** in the variable x can be written in one of the following forms

$$ax + b < c \quad ax + b > c$$

$$ax + b \leq c \quad ax + b \geq c$$

where a, b, and c are numbers and $a \neq 0$.

When finding the solution to an inequality, we are looking for the range of points that make the statement true rather than the single point, as with an equation. For instance, if we take the simple inequality $x \geq 15$, there are an infinite number of x-values that makes the statement true. In fact, as Figure 5.4.1 shows, any point that has an x-coordinate of at least 15 is a solution to the inequality; the y-coordinate can take any possible value. The points $(18, 2)$, $(15, -1)$, and $(31, -16)$ are all solutions and all fall within the shaded region. The vertical line $x = 15$ acts as a **boundary line** between the points that are solutions to the inequality and those that are not.

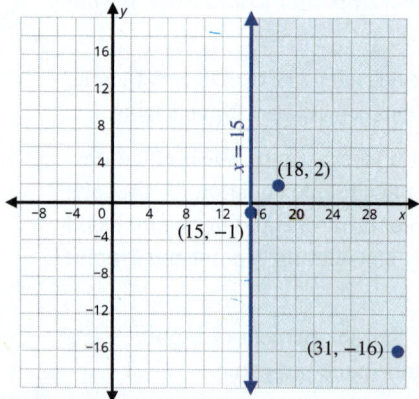

Figure 5.4.1: Graph of $x \geq 15$

BOUNDARY LINE

A **boundary line** is a line that separates those points that satisfy an inequality from those that do not. It is a representation of the equation created when the inequality sign is replaced with an equal sign.

Helpful Hint

Recall that vertical lines are of the form $x = a$ and horizontal lines are of the form $y = b$.

Notice in Figure 5.4.1 that the graph of the solution to the inequality involves shading the area that contains the solution set. Also notice that the boundary line $x = 15$ is solid. If an inequality contains equality (that is, the inequality is either $\geq$ or $\leq$), points along the line *are included* in the solution set. We represent this by drawing the boundary line as a solid line. However, if it is a strict inequality (that is, the inequality is either $>$ or $<$), the solution set *does not include* the points on the boundary line and the boundary line is drawn as a dashed line. When the boundary line is drawn on a graph, it divides the graph into **half-planes**, where a half-plane is the set of all points on one side of the line, but not including the line itself.

> 💬 **Think Back**
>
> The slope-intercept form of an equation is $y = mx + b$, where m is the slope and b is the y-intercept.

📋 STEPS TO GRAPHING LINEAR INEQUALITIES

1. Graph the boundary line.
   - The line will be solid if the inequality is $\geq$ or $\leq$.
   - The line will be dashed if the inequality is $>$ or $<$.
2. Determine which side of the line to shade by using a test point.
3. Shade the half-plane that contains the solution set.

Let's take a moment to elaborate on Step 2 in the process for graphing linear inequalities. The half-plane that contains the solution set is determined by using a **test point**. A test point is any point not on the boundary line that is substituted into the inequality to *test* whether the inequality is satisfied. If the test point chosen is a solution to the inequality, we know the half-plane containing the test point is the side to be shaded. If the test point doesn't satisfy the inequality, then we shade the half-plane that does not contain the test point.

Example 5.4.1

Graphing a Linear Inequality

Graph the linear inequality.

$$4x - 3y > 12$$

Solution

We begin by graphing the boundary line of the inequality. In order to graph the boundary line by hand or with a graphing calculator, we need to first rearrange the inequality into slope-intercept form.

> 💬 **Helpful Hint**
>
> When solving inequalities, dividing or multiplying both sides by a negative number requires that the inequality sign change direction.

$4x - 3y > 12$	
$-3y > -4x + 12$	Subtract 4x from both sides.
$y < \dfrac{4}{3}x - 4$	Divide both sides by −3, and remember to flip the inequality sign.

By Hand

Replacing the inequality sign with an equal sign gives us the boundary line for the graph.

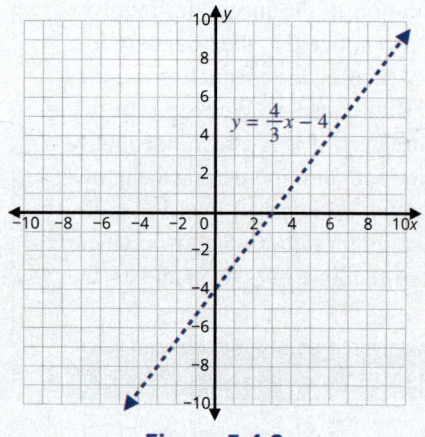

Figure 5.4.2

Boundary Line Equation

$$y = \frac{4}{3}x - 4$$

We can now graph the boundary line by plotting the y-intercept $(0, -4)$ and then counting the slope $m = \frac{4}{3}$. Because the original inequality does not contain equality, we will use a dashed line for the boundary line. This indicates 3inequality.

Now, use a convenient test point to determine which side of the line needs to be shaded. When choosing a test point, we want to choose a point that is obviously not on the boundary line. Since $(0, 0)$ is not on the line, we can use it as our test point and substitute it back into the original equation.

Testing the Point $(0, 0)$

$$4x - 3y > 12$$

$$4(0) - 3(0) \overset{?}{>} 12$$

$$0 > 12$$

☒

Since the point $(0, 0)$ does not make the inequality true, we know that the origin is not in the solution set and hence should not be in the shaded area on the graph. So, we shade the other half-plane.

> **💬 Helpful Hint**
>
> Choosing the origin, $(0, 0)$, as the test point makes substitution easier. However, if the boundary line contains the origin, you will need to choose a different test point.

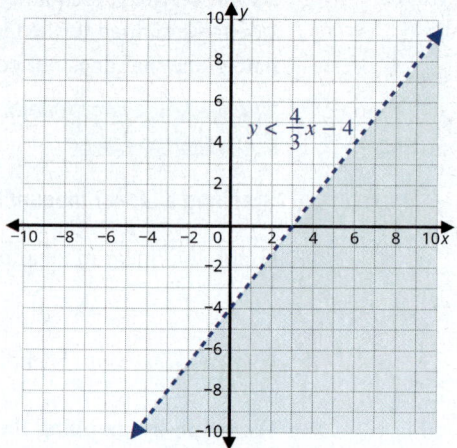

Figure 5.4.3

TI-83/84 Plus

The TI-84 Plus CE has a built-in application that can be used to graph inequalities. To run this app, press [apps] and select `Inequalz`. When prompted, press any key. Enter the right-hand side of the inequality $y < \frac{4}{3}x - 4$ for `Y1`. Then press the left-arrow key until a blinking rectangular box appears around the `Y1=`. Press [enter] and move the cursor down next to the equal sign. Use the right- and left-arrow keys to change the symbol from = to <. Highlight `OK` and press [enter]. Then press [graph].

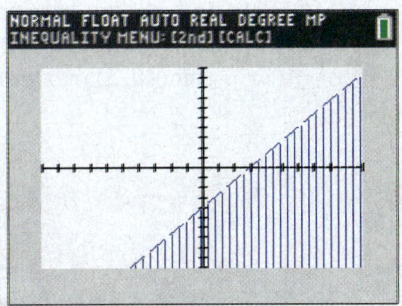

Figure 5.4.4

The calculator shows a graph with shading underneath a dotted boundary line.

Therefore, the solution set consists of all points below the line, but not actually on the line.

✓ **Skill Check 5.4.1**

Graph the linear inequality $2x + y > 0$.

Example 5.4.2

Modeling Using a Linear Inequality

Ivy walks dogs and house-sits pets for the pet sitting service Happy Petz. She earns $20 per 30-minute dog walk and $40 per night for house sitting. Ivy needs to decide if she wants to do this full time over the summer when she'll need to make $1200 per month.

a. Model this situation using a linear inequality. Let x represent the earnings for dog walking and y represent the earnings for house sitting.

b. Can Ivy meet her monthly budget needs if she has 40 dog walking appointments and 5 nights of house sitting during a month?

c. The table below shows Ivy's bookings over the last six months. Do you think Ivy should feel confident that she will be able to meet her budget with the income from Happy Petz?

Table 5.4.1 Ivy's Happy Petz Bookings

	Number of Walking Jobs	Number of Nights House Sitting
Month 1	53	7
Month 2	45	10
Month 3	40	6
Month 4	35	8
Month 5	50	8
Month 6	25	6

Solution

a. To model Ivy's earnings, multiply each job's variable by the amount she earns when doing that job: $20x + 40y$. We know that these earning need to be *at least* $1200, so we use the ≥ symbol. Thus, we have the following inequality model for Ivy's needed minimum monthly income.

$$20x + 40y \geq 1200$$

b. In order to determine if 40 dog walking appointments and 5 house sitting appointments will meet the monthly budget, we can substitute the coordinate pair (40, 5) into the inequality and see if they make it true.

$$20x + 40y \geq 1200$$

$$20(40) + 40(5) \overset{?}{\geq} 1200$$

$$1000 \geq 1200 \; \boxtimes$$

Because the point does not make the inequality true, we know that dog walking 40 times and house sitting 5 times is not enough to meet Ivy's budgetary needs for a month.

c. One way to determine if the last six months are a positive indicator for Ivy to meet her monthly budget is to substitute each data point into the inequality as we did in part b. However, if we graph the linear inequality given, we can see all the possible combinations of house sitting and dog walking that will provide Ivy with the income that she needs per month. This will give her a better picture of her possibilities.

The first step in graphing the inequality is to graph the boundary line. To do so, solve the inequality for y, arranging it into y-intercept form.

$$20x + 40y \geq 1200$$

$$40y \geq -20x + 1200$$

$$y \geq -\frac{1}{2}x + 30$$

By Hand

Replacing the inequality sign with an equal sign gives us the boundary line for the graph.

Boundary Line Equation

$$y = -\frac{1}{2}x + 30$$

We can now graph the boundary line by plotting the y-intercept, $(0, 30)$ and then counting the slope, $m = -\frac{1}{2}$. Because the original equation contains equality, we will use a solid line for the boundary line. This indicates that points on the boundary line *are* included in the solution set for the inequality.

Since $(0, 0)$ is not on the line, we can use it as our test point and substitute the values back into the original inequality.

Testing the Point $(0, 0)$

$$20x + 40y \geq 1200$$

$$20(0) + 40(0) \geq 1200$$

$$0 \geq 1200$$

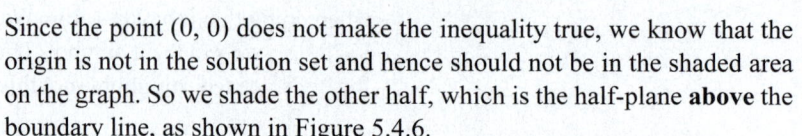

Since the point $(0, 0)$ does not make the inequality true, we know that the origin is not in the solution set and hence should not be in the shaded area on the graph. So we shade the other half, which is the half-plane **above** the boundary line, as shown in Figure 5.4.6.

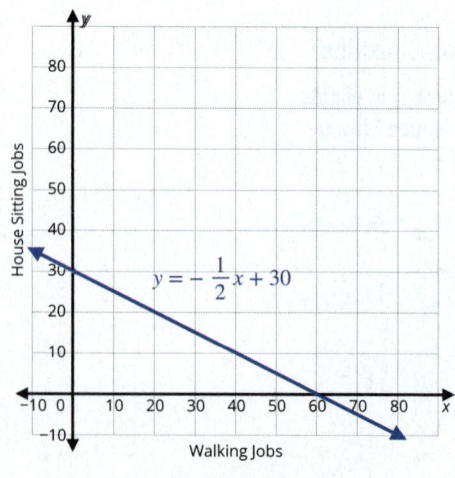

Figure 5.4.5

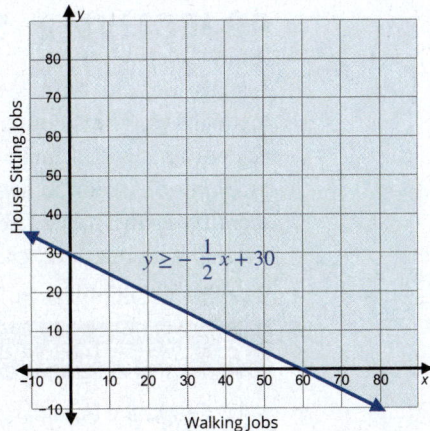

Figure 5.4.6

TI-83/84 Plus

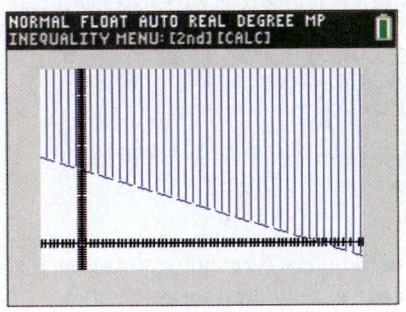

Figure 5.4.7

If the Inequalz app is not running, press [apps] and select `Inequalz`. When prompted, press any key. Press [y=] and enter the right-hand side of the inequality $y > -\dfrac{1}{2}x + 30$ for `Y1`. Then press the left-arrow key until a blinking rectangular box appears around the `Y1=`. Press [enter] and move the cursor down next to the equal sign. Use the right- and left-arrow keys to change the symbol from = to ≥. Move the cursor down to highlight `OK` and press [enter]. Then press [graph].

The calculator shows a graph with shading above a solid boundary line. Therefore, the solution set consists of all points above the line and on the line.

Now that we have graphed the inequality, we can plot each of the last six months on our graph.

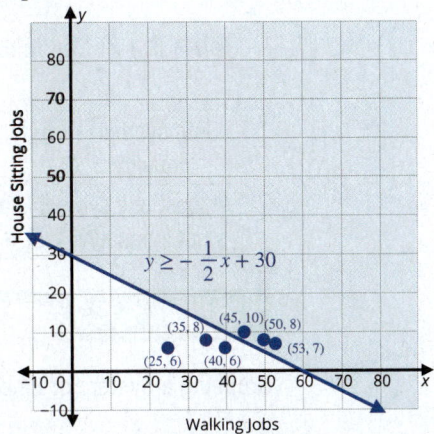

Figure 5.4.8

As we can now see, Ivy has only surpassed her budget goal in 3 out of the last 6 months, or half of the time. Based on this data, Ivy should not depend on meeting her budget goal every month only working for Happy Petz. However, she can use this graph to help her determine the various ways she can meet her goal each month. For instance, if Ivy has 51 walking jobs one month, she only needs 5 house sitting jobs that month to meet her goal.

OBJECTIVE 2 Systems of Linear Inequalities

Just as we did with systems of equations, we can graph systems of linear inequalities. The solution set to a system of inequalities includes all of the pairs of coordinates that make both inequalities true at the same time. The process of graphing the solution set involves graphing each individual inequality and then identifying the part of the graph where the shaded regions for each inequality overlap with one another. This area of intersection is the solution set to the system of linear inequalities.

⁝☰ STEPS TO GRAPHING SYSTEMS OF LINEAR INEQUALITIES

1. Graph the boundary line for each inequality.

   • The line will be solid if the inequality is ≥ or ≤,

   • The line will be dashed if the inequality is > or <.

2. Determine which side of the line to shade for each inequality by using a test point.

3. Lightly shade the half-plane that contains the solution set for each individual inequality.

4. The solution region for the system of inequalities is the intersection of the shaded half-planes from Step 3.

One last thing to note is the point of intersection of the lines. Remember the point of intersection lies on both lines. In order for it to be a solution to the system, it must satisfy both inequalities. The only way for that to happen is for both inequalities to include equality. If this is the case, we use a closed (or solid) circle for the point of intersection. But, if either of the inequalities does not include the equal sign, we must draw an open circle to indicate that the point of intersection is not included in the solution set.

Table 5.4.2: Guidelines for Graphing the Point of Intersection of Linear Inequalities

Both symbols include equality: ≤ or ≥	Closed Circle ●	Indicates the intersection point **is** included in the solution set.
At least one symbol is a strict inequality: < or >	Open Circle ○	Indicates the intersection point **is not** included in the solution set.

Example 5.4.3

Graphing a System of Linear Inequalities

Graph the solution set of the following system of linear inequalities.

$$\begin{cases} -3x + y > 3 \\ \dfrac{2}{5}x + y \le 4 \end{cases}$$

Solution

First, we need to have the equations in slope-intercept form so that we can graph them. Begin by rearranging each inequality into the correct form to obtain the boundary lines.

First inequality	Second inequality
$-3x + y > 3$	$\dfrac{2}{5}x + y \le 4$
$y > 3x + 3$	$y \le -\dfrac{2}{5}x + 4$

By Hand

We will go ahead and determine which side will need to be shaded by testing a point before we draw the lines on the graph. Since both lines have a y-intercept other than 0 and a slope that is not 0, we know that the origin $(0, 0)$ is not on either line. Thus, we can use it as a test point for both inequalities.

Table 5.4.3: Testing the Point $(0,0)$

First Inequality	Second Inequality
$-3x + y > 3$	$\dfrac{2}{5}x + y \le 4$
$-3(0) + (0) > 3$	$\dfrac{2}{5}(0) + (0) \le 4$
$0 > 3$	$0 \le 4$
☒ False statement	☑ True statement

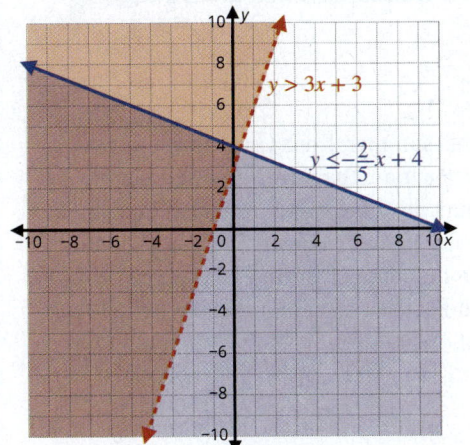

Figure 5.4.9

This indicates that shading will be above the boundary line for the first inequality and below the boundary line for the second inequality.

Now, graph the boundary line for each inequality. Note that the first boundary line should be dashed and the second one should be solid since the inequality contains the equal sign.

$$y = 3x + 3 \qquad \text{Boundary line equation (use dashed line, shade above)}$$

$$y = -\frac{2}{5}x + 4 \qquad \text{Boundary line equation (use solid line, shade below)}$$

TI-83/84 Plus

If the Inequalz app is not running, press $\boxed{\text{apps}}$ and select `Inequalz`. When prompted, press any key. Press $\boxed{\text{y=}}$ and enter the right-hand side of the first inequality for Y1 and the right-hand side of the second inequality for Y2. Change the inequality for Y1 to > and change the inequality for Y2 to ≤. Then press $\boxed{\text{graph}}$.

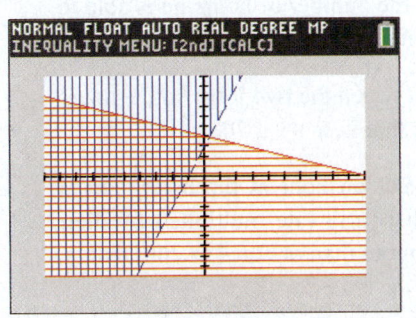

Figure 5.4.10

The solution set for the system of inequalities lies where the shading overlaps. So our solution set is the region of the graph that is shaded in Figure 5.4.11. Since the first inequality does not include the equal sign, we must draw an open circle to indicate that the point of intersection is not included in the solution set.

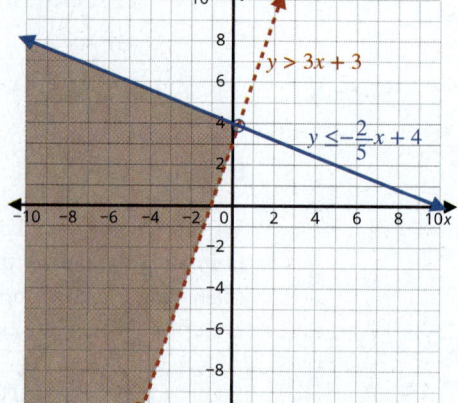

Figure 5.4.11: Solution Area

Determine if the point of intersection for the given system of linear inequalities should be an open circle or a closed circle.

$$\begin{cases} 3x - y \geq 20 \\ x + \dfrac{5}{2}y \leq 4 \end{cases}$$

Example 5.4.4

Modeling using a System of Linear Inequalities

In order to make as much money as possible this summer, Elisha has taken on two summer jobs. He works at a local fast-food restaurant chain and as a bellman at a hotel. He knows he can work at most 70 hours next week and needs to make at least $260. Elisha earns $9.11 per hour at the fast-food chain and $11 per hour at the hotel. Graph the solution set to determine the number of hours he'd need to work at each job to earn $260 in one week. Find at least one combination of hours Elisha could work at each job to meet his goal.

Solution

Begin by defining the variables that will be needed.

x = number of hours Elisha works at the fast-food restaurant

y = number of hours Elisha works at the hotel

We can then write an inequality that represents the number of hours he is able to work and an inequality representing the pay he needs to receive.

Elisha can work at most 70 hours in the week between the two jobs; so the hours added together must be less than or equal to 70; that is, $x + y \leq 70$.

Earning $9.11 an hour at the restaurant plus $11 an hour at the hotel, Elisha needs to earn at least $260 in the week. Each hourly rate multiplied by their respective variable and added together must be more than or equal to 260; that is, $9.11x + 11y \geq 260$.

Graph each inequality by rewriting it in slope-intercept form and determine the boundary equations. Round to the nearest thousandths.

Hours	Earnings
$x + y \leq 70$	$9.11x + 11y \geq 260$
$y \leq -x + 70$	$11y \geq -9.11x + 260$
$y = -x + 70$ Boundary line	$y \geq 0.83x + 23.64$
equation (use solid line)	$y = -0.83x + 23.64$ Boundary line
	equation
	(use solid line)

By Hand

Once again, we will determine which side will need to be shaded by testing a point before we draw the lines on the graph. Since both lines have a y-intercept other than 0, and a slope that is not 0, we know that the origin $(0, 0)$ is not on either line. Thus, we can use it as a test point for both inequalities.

Table 5.4.4: Testing the Point (0,0)

Hours Inequality	Earnings Inequality
$x + y \leq 70$	$9.11x + 11y \geq 260$
$(0) + (0) \leq 70$	$9.11(0) + 11(0) \geq 260$
$0 \leq 70$	$0 \geq 260$
☑ True statement	☒ False statement

This indicates that we shade below the boundary line representing the hours inequality and above the boundary line representing the earnings inequality. Since the number of hours Elisha needs to work can only be a positive number, we only need to focus on the portion of the graph where both x and y are positive; that is, the first quadrant.

TI-83/84 Plus

If the Inequalz app is not running, press apps and select `Inequalz`. When prompted, press any key. Press y= and enter the right-hand side of the first inequality for `Y1` and the right-hand side of the second inequality for `Y2`. Change the inequality for `Y1` to ≤ and change the inequality for `Y2` to ≥. Then press graph.

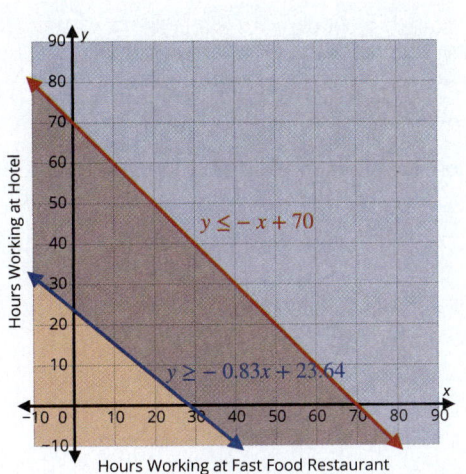

Hours Working at Fast Food Restaurant

Figure 5.4.12

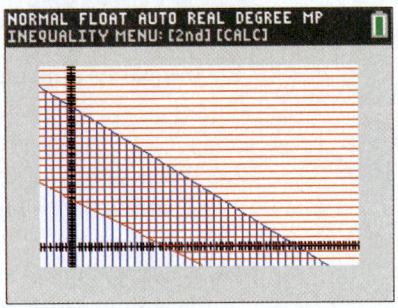

Figure 5.4.13

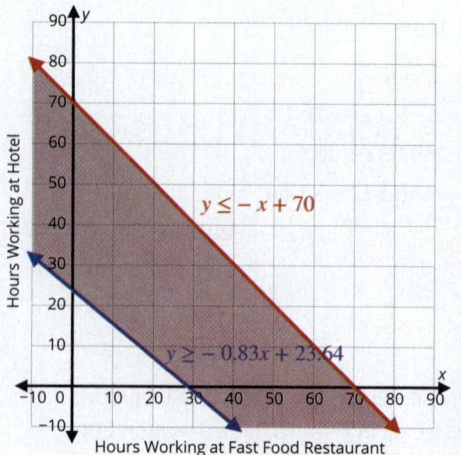

Figure 5.4.14: Solution Area

Notice that the area where the shaded regions overlap is the solution to the number of hour combinations that Elisha could work at the two jobs. We will only look at the first quadrant of the graph since, practically speaking, the number of hours that Elisha can work must be either 0 or positive.

Note that the point of intersection is included in the solution set and is graphed as a closed circle since both lines are solid. However, the point of intersection will not be shown in Figure 5.4.12, Figure 5.4.13, or Figure 5.4.14 since we are only focusing on the first quadrant.

Any point in the shaded area is a combination of hours that Elisha could work and meet his goal of earning $260 for the week. One such combination is the point (20, 50). This means that Elisha could work 20 hours at the fast-food restaurant and 50 hours at the hotel and meet his goal. In fact, 20 hours at the restaurant earns $182.20 and 50 hours at the hotel earns $550. This would put Elisha well past his goal at $732.20 for the week.

Skill Check Answers

1.

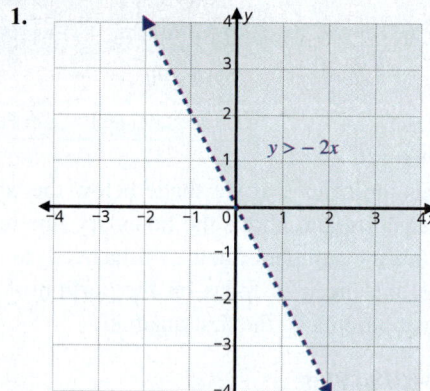

2. Closed circle at the point of intersection

5.4 Exercises

✔ CONCEPT CHECK

1. Inequality symbols are used between two expressions that are not strictly _____.

2. When graphing the solution of an inequality equation, the _____ separates the points that satisfy the inequality from those that do not.

3. When graphing an inequality equation, a _____ can be used to determine which half-plane to shade.

4. When solving a system of linear equations, the solution region for the system is the _____ of the shaded half-planes.

5. When graphing a system of linear equations, an open circle indicates that the intersection point is _____ in the solution set.

💡 **PRACTICE**

Determine if the points given are part of the solution set for the linear inequality shown in the graph.

6.   **a.** $(1, 4)$

      b. $(3, 8)$

      c. the origin

7.   **a.** $(-5, 5)$

      b. $(5, 6)$

      c. the origin

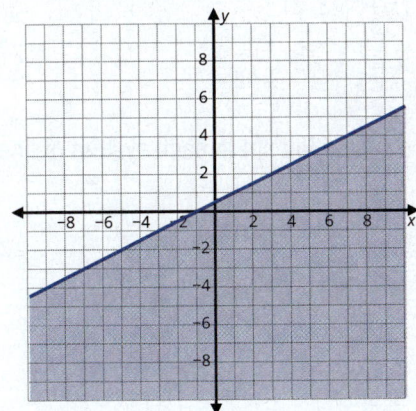

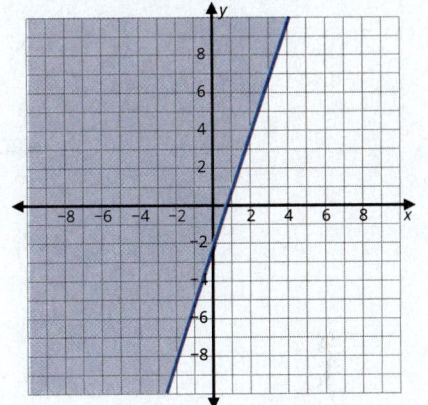

8.   **a.** $(-3, 0)$

      b. $(1, 8)$

      c. the origin

9.   **a.** $(-1, 5)$

      b. $(1, 0)$

      c. the origin

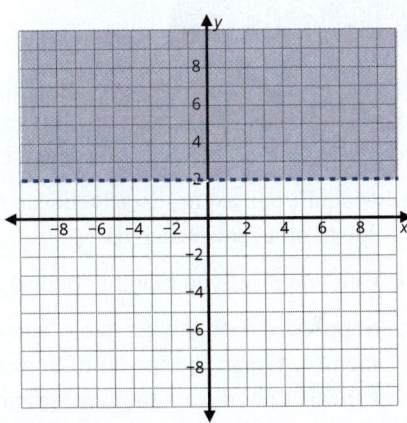

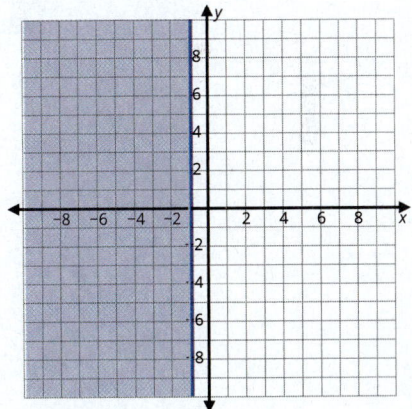

Graph the solution set for each linear inequality given.

10. $6x + 5y < -30$

11. $4y - 6x > 12$

12. $-y - 2 \leq 2x$

13. $-y + 4 \geq \dfrac{1}{2}x$

14. $8y + 6x \geq 8y + 12$

15. $y - 1 \leq 0$

16. $6x + 8 < -6y + 8$

17. $4 - 6y > 22 - 6x$

18. $4x \geq -5y + 40$

19. $-3y \leq -5x - 15$

20. $y + 4 < -x$

21. $\dfrac{4x + 5y}{20} > 1$

22. $\dfrac{2x-3y}{11} \le -1$

23. $\dfrac{2}{3}x - \dfrac{1}{6}y > 4$

Determine if the point of intersection for the given system of linear inequalities should be an open circle or a closed circle on the graph.

24. $\begin{cases} x - 4y > 0 \\ \dfrac{1}{3}x - \dfrac{3}{2}y \le -1 \end{cases}$

25. $\begin{cases} y < 5 \\ \dfrac{5}{6}x - 7x > 13 \end{cases}$

26. $\begin{cases} 7x + 5y \le 21 \\ x + \dfrac{1}{5}y \ge 2 \end{cases}$

27. $\begin{cases} x \ge -17 \\ \dfrac{3x+y}{2} \le 8 \end{cases}$

Graph the solution set to each system of linear inequalities.

28. $\begin{cases} \dfrac{1}{4}x + y > 10 \\ 7x + y \le -1 \end{cases}$

29. $\begin{cases} -x + 3y > 2 \\ 5x + \dfrac{1}{2}y > 3 \end{cases}$

30. $\begin{cases} 2x > -1 \\ y \le 6 \end{cases}$

31. $\begin{cases} x - 0.5y \le 1.3 \\ 3x + 8y \le 24 \end{cases}$

32. $\begin{cases} 4x + 6y \ge 12 \\ 2.8x + 1.1y \ge 6.2 \end{cases}$

33. $\begin{cases} 7x - y \le 14 \\ \dfrac{x+y}{-3} \ge -5 \end{cases}$

🚀 APPLICATIONS

34. Suppose you make $2400 per month. Your obligatory payments, such as rent, your cell phone bill, etc., are $1140 each month. You spend the rest of the money on your daily needs. Assume there are 30 days in a month.

   a. Model your expenses as a linear inequality. Let x be the amount you spend on average each day.

   b. Can you meet you monthly budget if you spend $45 per day?

   c. Suppose the following table shows your expenses over the last six months. Graph the model of expenses created in part a along with the expenses per day on the same graph. (**Hint:** Let y represent the month.)

	Expenses per Day
Month 1	$46
Month 2	$42
Month 3	$44
Month 4	$34
Month 5	$38
Month 6	$39

   d. What daily spending amount should you aim for to meet your monthly budget? Do you think you should feel confident that you will be able to meet your budget?

35. Shelly's business sells handmade chocolate. She rents a small shop that costs her $1200 per month, including taxes. She sells two sets of sweets, a standard set for $14, and a premium set for $22. The cost of making a standard set is $6, and the cost of making a premium set is $10.

    a. Model Shelly's monthly profit as a linear inequality. Let x be the number of standard sets sold and y be the number of premium sets sold.

    b. Can Shelly make any profit if she sells 94 standard sets and 26 premium sets in a month?

    c. Suppose the following table shows Shelly's sales over the last six months. Graph the model of monthly profit created in part a along with the sets sold day on the same graph.

    d. Do you think she should feel confident that her business is profitable?

	Standard Sets Sold	Premium Sets Sold
Months 1	87	42
Months 2	94	35
Months 3	102	32
Months 4	76	44
Months 5	98	36
Months 6	83	36

36. Suppose you have a budget of $500 for the flight and hotel for your vacation. The hotel you found is $60 a night and your flight is $150. Let x be the number of nights you stay.

    a. Write the modeling inequality for your accommodation and flight vacation budget.

    b. Graph the inequality.

    c. What is the maximum number of nights you can afford to stay in the accommodation?

    d. How many points on the graph are solutions to the inequality and an appropriate number of nights to stay for your vacation. Explain your reasoning.

37. The city charges $250 for a permit to use a city park for a function. To pay for the permit, a club is charging a fee of $0.75 for each member and $1.25 for each guest who is not a member. Let x be the number of members attending and y be the number of nonmembers attending.

    a. Write the modeling inequality for the number of guests needed so that the park permit is paid for.

    b. Graph the inequality.

    c. Would the permit be covered if only 300 members and no non-members attended?

    d. how many nonmembers must attend if no members attend.

    e. Identify one possible combination of member and nonmember guest counts that would cover the cost of the permit.

38. Oliver supplements his income by driving for a shared-ride company and driving for a food delivery app. On average, he makes $17.50 an hour driving for the shared-ride and $15 an hour for the delivery app. Oliver has set a goal of making a minimum of $50 a day by driving for these companies. Let x be the number of hours driving for the shared-ride company and y be the number of hours driving for the food delivery app.

   a. Write the modeling inequality for the number of hours Oliver needs to drive for each company so that he reaches his daily goal.

   b. Graph the inequality.

   c. If Oliver drove 2 hours for each company in one day, would he meet his goal for the day?

   d. Identify one way Oliver could meet his goal of supplementing his income by at least $50 by driving for the two companies.

   e. Is the point (15, 10) a solution to the inequality? If so, is it possible for Oliver to meet his supplement goal using this solution? Explain your reasoning.

39. A shipping company delivers glass vases. The company is paid $0.80 for each safely delivered vase and is fined $4.50 for each broken vase. The company expects to earn at least $1500 in a delivery. Let x be the number of unbroken vases delivered and y be the number of broken vases.

   a. Write the modeling inequality for the shipping company's delivery earnings.

   b. Graph the inequality.

   c. Use the graph in part a to determine the minimum number of unbroken vases that must be delivered in order to earn $1500.

   d. Identify one possible combination of safely delivered vases and broken vases that allows the company to earn $1500 given that at least one vase gets broken.

   e. Explain why the point (1500, −100) is in the solution set for the inequality but is not a plausible solution for the shipping company.

40. Kashton is opening a fitness center for an apartment complex. As manager, he would like a full schedule of hot yoga classes as well as aerobic exercise classes. Each yoga class requires a time slot of $1\frac{1}{2}$ hours while aerobic exercise classes require 1 hour. Kashton would like to have at least 20 hours of classes scheduled each week. Yoga instructors are paid $35 per class and instructors for the aerobic exercise classes earn $30 per class. Kashton has a budget of $1000 per week to spend on instructors for the fitness classes. Let x be the number of yoga classes and y be the number of aerobic classes.

    a. Write two inequalities—one that represents the number of hours of classes the manager wants to have and one to represent the instructor pay required for the classes.

    b. Graph the solution set for the system of inequalities.

    c. Identify at least one combination of classes that meets Kashton's requirements.

    d. If Kashton wants to offer the same amount of yoga and aerobic classes each week, what is the maximum number of each that he is able to offer and stay in his budget?

41. Because of her class schedule, Fern can work at most 13 hours per week. She would like to make at least $60 a week. She makes $15 an hour tutoring and $8.50 an hour proofreading essays. She tutors in whole hours only but does proofreading by the half-hour. Let x be the number of hours she tutors per week and y be the number of hours she does proof reading.

    a. Write two inequalities—one that represents the number of hours Fern tutors and one to represent her pay.

    b. Graph the solution set for the system of inequalities.

    c. What is the minimum number of hours Fern can tutor without any proof-reading work and make her goal?

    d. What is the minimum amount of proofreading that Fern can do and make her goal if she doesn't tutor at all in the week?

    e. Identify at least one combination that Fern can work and meet her goal if she does some of each type of work.

42. A nutrition plan requires that at lunch you eat a minimum of 35 gram of protein and have at most 21 grams of fat. A suggested meal includes steak and eggs. Strip steak contains 7 grams of protein and 1.9 grams of fat per ounce and eggs contain 6 grams of protein and have 4.5 grams of fat each. Let x be number of ounces of strip steak and y be number of eggs.

    a. Write two inequalities—one that represents the required amounts of protein and one to represent the fat.

    b. Graph the solution set for the system of inequalities.

    c. What is the maximum number of eggs allowed at lunch on the nutrition plan?

    d. Identify an amount of steak and eggs that is allowed if the most precise you can be with measuring the steak is to the nearest half an ounce?

43. The last page in the newspaper is reserved for ads that can be purchased. There are two types of ads. The first type is a 215-character short ad, and the second type is a 380-character long ad. The page can contain at most 4000 characters. The cost of placing a short ad is $110, and the cost of placing a long one is $200. The newspaper wants to earn at least $1400 on the ads. Graph the solution set to the corresponding inequality and determine the number of ads the newspaper needs to place to fit the page and earn the required amount. Find at least one combination of how many of each type of ad would meet the goal.

44. To lose weight more effectively, a trainer recommended Charlie add cardio trainings to his weight loss plan. He should burn at least 3500 additional calories per week. Charlie chose two types of cardio exercises, running and cycling. When running he will burn about 670 calories per hour, and when cycling he will burn 530 calories per hour. He also figured that he could not devote more than seven hours a week to this. Graph the solution set to determine the time Charlie needs to spend on each type of exercise to fit the trainer's expectations and the time constraints. Find at least one combination of hours Charlie should spend on each type of exercise to meet the goal.

45. A factory produces two types of glue, regular and increased adhesiveness. Only one type of glue can be produced at a time, and the factory operates 24 hours per day. The factory can produce 500 liters of regular glue or 420 liters of advanced glue in an hour, and they need to make at least 9000 liters each day. Graph the solution set to determine the time the factory should work producing each type of glue to implement the production plan. Find at least one combination of hours the factory should spend making each type of glue to meet the goal.

46. Andy is traveling to another city by bus and will not be able to charge his phone. He will have to spend at least fourteen hours on the road. To pass the time, he plans to listen to music and watch videos. The battery capacity of his phone is 2800 mAh (milliamperes/hour). Listening to music uses 43 mA per hour and watching videos uses 165 mA per hour. Graph the solution set to determine how long Andy should use his phone in each mode so that it doesn't run out of power along the way. Find at least one combination of hours spent on each activity that meets the goal.

5.5 LINEAR PROGRAMMING

OBJECTIVE 1 Objective Functions

In Section 5.4, we began to analyze how someone could share their work hours between two summer jobs. In the scenario we explored, Elisha had to decide how to best share out his time between the fast-food restaurant and the hotel. In reality, there are more details than we were initially able to take into account. Certainly, Elisha would have to work a minimum number of hours at each job just to keep his position. Also, we may need to explore more than one idea of "best." The solution that allows him to meet his budget while working the fewest hours may be very different than the solution that allows him to make the most money.

Finding the "best" option possible that satisfies particular criteria is called *optimization*: a process used in mathematics, computer science, operations research, and many other fields to find the most effective, useful, or cost-effective solution. In mathematics, one such optimization process is called **linear programming**. It uses systems of linear inequalities, like we saw in Section 5.4, to find the maximum or minimum of a particular function. Although optimization can be done with many inequalities in many variables, we will look at situations that involve two-variable inequalities that we can graph in the Cartesian coordinate system.

The **objective function**, also called the optimization function, is a function that expresses the relationship between two variables, x and y, in the form of a third variable, z, which is to be maximized or minimized. The linear inequalities that express the limitations of each variable are called the **constraints**.

📖 LINEAR PROGRAMMING

Linear Programming

The mathematical process of **linear programming** is a method to optimize a particular quantity in a mathematical model constrained by linear inequalities.

Objective Function

An **objective function**, or *optimization function*, is a function that expresses the quantity that is to be optimized.

Constraints

Constraints in linear programming are the linear inequalities that express the limitation of each variable or combination of variables.

In Elisha's scenario, the constraints are the linear inequalities that express the work requirements of his two jobs along with his overall restriction of working no more than 70 hours during the week. For instance, since he could work at most 70 hours, one constraint is the inequality $x + y \leq 70$, where x is the number of hours worked at the fast-food restaurant and y is the number of hours worked at the hotel.

The idea of *best* for Elisha would be introduced by choosing the objective function to measure either the number of hours he works or the amount of salary he earns. Example 5.5.1 looks at how to write an objective function.

Example 5.5.1

Writing an Objective Function

A nutrition specialist is designing a meal plan for a hospital patient where protein, carbohydrates, and fats need to be monitored. The patient needs a fat-free diet where protein counts for at most 20% of their daily calories while the remaining 80% is from carbohydrates. To maintain the patient's body weight, the nutritionist needs to maximize their overall daily calorie intake.

Let x represent the weight in grams of protein and y represent the weight in grams of carbohydrates. If a gram of protein and a gram of carbohydrate each provide four calories, write the objective function that describes the patient's daily calorie intake.

Solution

We're told some of the constraint information that guides the makeup of the patient's diet. To construct the objective function, we need to concentrate our attention on what values need to be maximized (or in some cases minimized). In this case, the total daily calorie intake needs to be maximized. Each gram of protein and each gram of carbohydrate that the patient eats will provide 4 calories of energy. Hence, the objective function that models their total calorie intake can be found by multiplying the calories by each variable. We introduce a new variable z to represent the calories that will be optimized.

Caloric Objective Function

$$z = 4x + 4y$$

Helpful Hint

A convex polygon is a polygon in which all vertices point away from the interior of the shape. A concave polygon is a polygon that is not convex.

Table 5.5.1: Convex and Concave Polygons

Convex Polygon	Concave Polygon

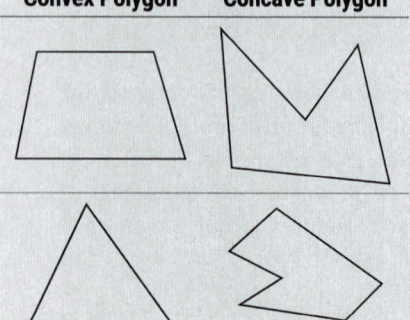

OBJECTIVE 2 Linear Programming

Once we have an objective function and constraints, all we need is to find suitable values for each of the variables; that is, to find the points that satisfy all of the requirements. Once we have points, we then need to determine which one maximizes (or minimizes) the objective function. Finding the values of x and y that make z the largest or smallest value possible would require us to check an infinite number of points if it weren't for the Fundamental Theorem of Linear Programming. This theorem helps us narrow down the places where a maximum or minimum could occur in an objective function to just a few possibilities. Loosely speaking, it states that if a maximum or minimum of an objective function exists, it will happen at one of the vertices of the polygon formed by the graphs of the constraint inequalities.

🔑 FUNDAMENTAL THEOREM OF LINEAR PROGRAMMING

The maximal and minimal values of a linear function over a convex polygonal region are found at the corners of the region.

So what does that mean for us? It means that if we are given constraints in the form of linear inequalities, we can graph them. Their intersections form a polygon which encloses the set of points that satisfy all of the inequalities simultaneously. This set of points is called the **feasible region**.

> 📖 **FEASIBLE REGION**
>
> The **feasible region** is the set of all possible points that satisfy the constraints of an optimization problem.

> 💬 **Helpful Hint**
>
> *Maxima* and *minima* are the respective plurals of *maximum* and *minimum*. The minimum and maximum values are referred to as *extrema values*.

The Fundamental Theorem of Linear Programming tells us that although all points in the feasible region are possible maxima or minima, we only need to look at the corners of the feasible region to find the extrema values. This helps us drastically reduce the number of points we need to consider in order to find the maxima or minima.

The images in Figure 5.5.1 show examples of polygons formed by the constraints in different linear programing scenarios.

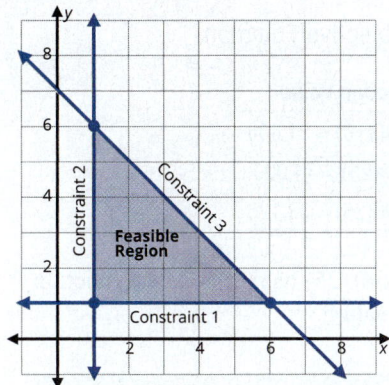 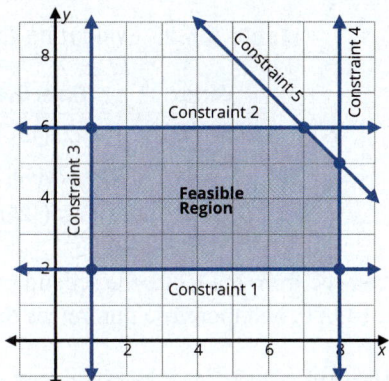 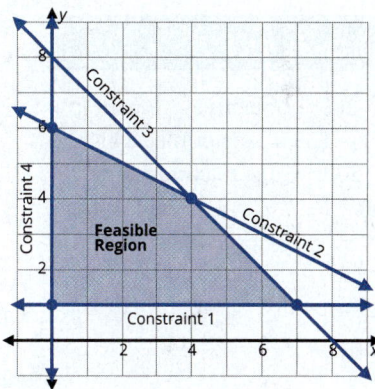

Figure 5.5.1: Examples of Linear Programming Scenarios

When we use linear programming, our goal is to graph the inequalities of the constraints and find the intersection points, which are the vertices of the polygon formed. Then, we check each vertex to determine which points give the maximum and minimum values for the objective function.

> ☰ **STEPS FOR SOLVING AN OPTIMIZATION OR LINEAR PROGRAMMING PROBLEM**
>
> 1. Define the objective function.
>
> 2. Define the constraints using linear inequalities.
>
> 3. Graph the constraint inequalities.
>
> 4. Identify the vertices of the feasible region; that is, the polygon formed by intersection of the lines.
>
> 5. Identify any maximum or minimum values of the objective function by finding the values of the objective function at each of the vertices.

Although these steps might seem like a lot, you have already conquered all but the final step. Once we know the vertex, we can substitute these into the objective function to help us identify a maximum or minimum point. Example 5.5.2 takes a look at this last step.

Example 5.5.2

Determining Maximum or Minimum Points

A student is determining the minimum point of the objective function $z = 225x + 200y$. After graphing the constraints, the student finds the following three vertices of the feasible region polygon.

1. $(40, 110)$

2. $(40, 25)$

3. $(125, 25)$

Which of the points minimizes the objective function?

Solution

Since the coordinates of the vertices have been found, we need to only evaluate the objective function at each of the points, as Table 5.5.2 shows.

✔ Skill Check 5.5.1

Given the objective function $13x + 5y - 125 = z$, determine which of the following maximizes the function.

a. $(36, 0)$

b. $(11, 14)$

c. $(0, 36)$

Table 5.5.2: Evaluating the Objective Function

Vertex	Objective Function Value
$(40, 110)$	$z = 225(40) + 200(110) = 31{,}000$
$(40, 25)$	$z = 225(40) + 200(25) = 14{,}000$
$(125, 25)$	$z = 225(125) + 200(25) = 33{,}125$

As we can see from the table, the second point $(40, 25)$ has the smallest function value at 14,000, and therefore minimizes the equation.

Example 5.5.3

Finding the Maximum Value of an Objective Function

The objective function $z = 3x + 4y$ is subject to the following constraints.

$$x - y \leq 2$$
$$x + 2y \leq 14$$
$$x \geq 0$$
$$y \geq 0$$

Find the maximum value of z.

Solution

Since we are given the objective function as well as the constraints in the form of linear inequalities, we will start the process of finding the maximum by graphing the inequalities. As we know, graphing these inequalities by hand or by the calculator requires that the inequalities be written in slope-intercept form. So the first step is to rearrange each of the constraints into this form. They are listed for you here. (We leave this for you to check on your own.)

Original Constraints	Slope-Intercept Form
$x - y \le 2$	$y \ge x - 2$
$x + 2y \le 14$	$y \le -\dfrac{1}{2}x + 7$
$x \ge 0$	$x \ge 0$
$y \ge 0$	$y \ge 0$

By Hand

In Table 5.5.3, we identified each of the necessary pieces of information needed to graph each of the inequalities by hand. Remember to use a test point to determine which half plane to shade. We list each shaded region as "above" or "below" or "to the right" of the line.

Table 5.5.3

Original Constraint	Slope-Intercept Form	Boundary Line Equation	Slope (*m*) & *y*-intercept (*b*)	Solid or Dashed line	Shading
$x - y \le 2$	$y \ge x - 2$	$y = x - 2$	$m = 1, b = -2$	Solid	Above
$x + 2y \le 14$	$y \le -\dfrac{1}{2}x + 7$	$y = -\dfrac{1}{2}x + 7$	$m = -\dfrac{1}{2}, b = 7$	Solid	Below
$x \ge 0$	$x \ge 0$	$x = 0$	Vertical line; *y*-axis	Solid	To the right
$y \ge 0$	$y \ge 0$	$y = 0$	Horizontal line; *x*-axis	Solid	Above

Figure 5.5.2 shows each constraint inequality with its appropriate shading.

The intersection of all of the shaded areas forms the feasible region, as shown in Figure 5.5.3.

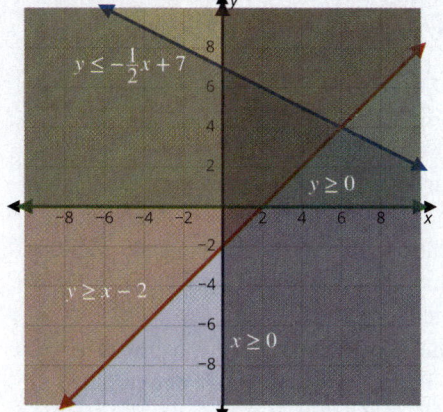

Figure 5.5.2: Graphed Constraints

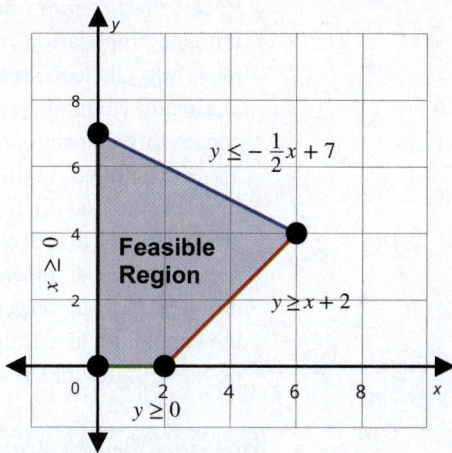

Figure 5.5.3: Feasible Region

We know that the maximum of the objective function will happen at one of the vertices of the polygon. To find the coordinates of each vertex, we need to solve a system of equations for each pair of constraints by setting them equal to one another. In some instances you may be able to "eyeball" the coordinates for each of the vertices; however, using algebra will always produce a precise answer even when you don't have a precise graph. The first vertex we will find is between the first two lines.

$$y = x - 2$$

$$y = -\frac{1}{2}x + 7$$ System of equations

$$x - 2 = -\frac{1}{2}x + 7$$ Set the equations equal to one another to eliminate a variable.

$$x + \frac{1}{2}x = 7 + 2$$ Isolate x on one side of the equation.

$$\frac{3}{2}x = 9$$ Combine like terms.

$$x = 9 \cdot \frac{2}{3}$$ Solve for x.

$$x = 6$$

$$y = (6) - 2 = 4$$ Substitute the x-value to solve for y.

$$(6, 4)$$ Vertex of the polygon

We can do the same process for each of the other two pairs of equations.

$$y = x - 2$$
$$y = 0$$ System of equations

$$0 = x - 2$$ Substitute to find x.

$$x = 2$$

$$(2, 0)$$ Vertex of the polygon

$$y = -\frac{1}{2}x + 7$$ System of equations
$$x = 0$$

$$y = -\frac{1}{2}(0) + 7$$ Substitute to find y.

$$y = 7$$

$$(0, 7)$$ Vertex of the polygon

Notice that the last vertex is at the origin (0, 0) where the x-axis and y-axis meet. We will not need to solve a system of equations to find this point.

TI-83/84 Plus

Recall that to graph inequalities on the TI-84 Plus CE, we need to have the Inequalz application running. If it is not already running on your calculator, press [apps], select `Inequalz`, and press any key. If the Inequalz application is already running, press [y=]. Now use the slope-intercept form of the three constraints that involve y and enter the right-hand side of each inequality as Y1, Y2, and Y3. (While the calculator cannot graph a vertical line for $x \geq 0$, it is understood that we are only interested in the region where x-values are positive.) For each inequality, press the left-arrow key until a blinking rectangular box appears around the Y1=. Press [enter] and move the cursor down next to the equal sign. Press the right- and left-arrow keys to change the equal sign to a $\geq$ sign for Y1 and Y3 and a $\leq$ sign for Y2. Move the cursor down to highlight OK and press [enter]. Then press [graph].

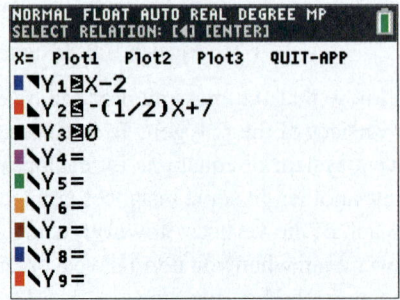

Figure 5.5.4

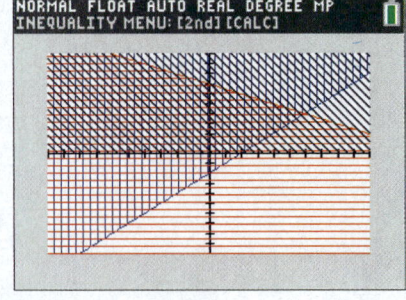

Figure 5.5.5

We know that maximum of the objective function will occur at one of the vertices of the polygon that forms the feasible region. To find the points of intersection, press [2nd] [trace] and select `intersect`. A blinking cursor will appear along the first line and the inequality Y1 ≥ X-2 will appear above the graph. Use the left- and right-arrow keys to move the cursor close to where Y1 intersects Y2. Press [enter]. A blinking cursor will be on the second line (and the inequality Y2 ≤ -(1/2)X+7 will appear above the graph). Press [enter] again, and `Guess?` appears at the bottom of the screen. Press [enter] a third time, and at the bottom of the screen the calculator will display its best approximation of the intersection of the two lines, X=6, Y=4 (Figure 5.5.6).

We'll use the process again to find the intersection between Y1 and Y3. Press [2nd] [trace] and select `intersect`. Use the left- and right-arrow keys to move the cursor close to where Y1 intersects Y3 (which is where Y1 intersects the *x*-axis) and press [enter]. The cursor will jump to Y2; press the down-arrow to move the cursor to Y3 and press [enter] two more times. At the bottom of the screen, the calculator will display its best approximation of the intersection of the two lines, X=2, Y=Ø (Figure 5.5.7).

We didn't graph the line *x* = 0, but we do need to know the point where Y2 intersects the *y*-axis. To find this value, press [2nd] [trace] and select `value`. At the bottom of the screen, you'll be prompted to enter an *x*-value. Since we want to know the value of *y* when *x* is 0, enter Ø. The cursor will first appear on Y1; press the up-arrow until the cursor appears on Y2. At the bottom of the screen, the calculator will return X=Ø, Y=7 (Figure 5.5.8).

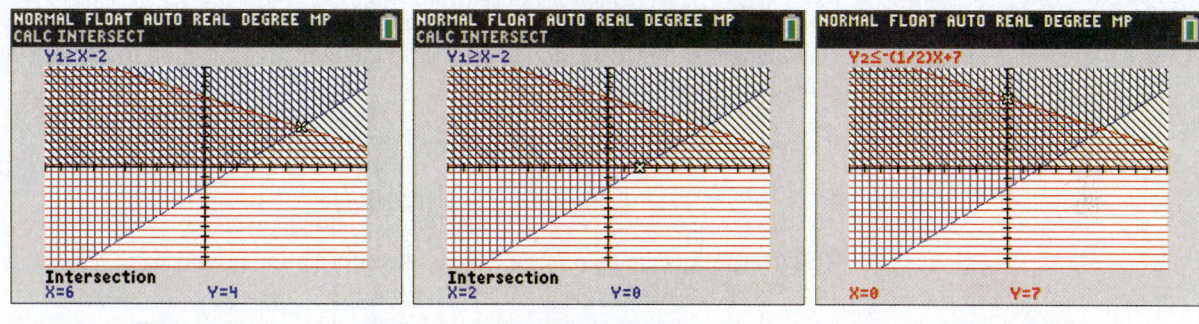

Figure 5.5.6   **Figure 5.5.7**   **Figure 5.5.8**

Therefore, the vertices of the polygon—which is the feasible region—are (0, 0), (0, 7), (6, 4), and (2, 0).

We now substitute these points into the objective function $z = 3x + 4y$ to find the maximum.

$$z = 3x + 4y$$
$$(0,0): \quad z = 3(0) + 4(0) = 0$$
$$(0,7): \quad z = 3(0) + 4(7) = 28$$
$$(6,4): \quad z = 3(6) + 4(4) = 34$$
$$(2,0): \quad z = 3(2) + 4(0) = 6$$

The third substitution shows the maximum of *z* is 34 and occurs at the point (6, 4).

✔ **Skill Check 5.5.2**

Determine the three vertices of the polygon formed by the following constraints.

a. $x - 2y \le 100$

b. $y \le -\dfrac{1}{2}x + 35$

c. $x \ge 0$

d. $y \ge 0$

Optimization is often used to maximize resources, optimize personnel, or minimize costs, which are all goals that businesses aim to achieve. The following example follows the steps of linear programming through one such scenario for a business entrepreneur.

Example 5.5.4

Modeling with Linear Programming

An entrepreneur is opening an online business selling original handcrafted items. To start, they will focus on two types of art: original paintings framed in reclaimed wood and a set of wooden placemats and coasters. Each framed piece is priced so that the profit is $105. For the artwork, it takes 4 hours to cut and sand the wood for one frame, 2.5 hours to assemble the artwork on the frame, and 1.5 hours to put the finishing touches on it. Each set of placemats and coasters is priced for a profit of $57. However, the placemat and coaster sets are slightly faster to make than the artwork. Each set requires 2 hours to cut the wood, 1 hour to assemble, and 2 hours for finishing touches. With friends who have committed to help get the business off the ground, the entrepreneur has manpower equivalent to 165 hours a week to devote to making the pieces: 70 hours for cutting the wood, 40 hours for assembly work, and 55 hours for finishing touches. How many of each piece should be made in order to maximize the monthly profit?

Solution

To maximize monthly profit, the objective function will be based on the profit z from each item. If we let x be the number of framed artwork pieces and y be the number of placemat and coaster sets, the objective function to maximize the amount of profit is as follows.

Objective Function for Maximum Profit

$$z = 105x + 57y$$

The constraints are then formed using the information provided.

| Constraint inequality for cutting wood (4 hours for the frame, 2 hours for the set): | $4x + 2y \leq 70$ |

| Constraint inequality for assembling (2.5 hours for the frame, 1 hour for the set): | $2.5x + y \leq 40$ |

| Constraint inequality for finishing touches (1.5 hours for the frame, 2 hours for the set): | $1.5x + 2y \leq 55$ |

Notice that since only a positive number of items can be made, two more constraints are needed.

Other constraints for profit: $x \geq 0$ and $y \geq 0$

In Table 5.5.4, we have put each constraint into slope-intercept form and identified the necessary information needed for graphing each by hand.

Table 5.5.4

Original Constraint	y-Intercept Form	Boundary Line Equation	Slope (m) and y-Intercept (b)	Solid or Dashed line	Shading
$4x + 2y \leq 70$	$y \leq -2x + 35$	$y = -2x + 35$	$m = -2$, $b = 35$	Solid	Below
$2.5x + y \leq 40$	$y \leq -2.5x + 40$	$y = -2.5x + 40$	$m = -2.5$, $b = 40$	Solid	Below
$1.5x + 2y \leq 55$	$y \leq -0.75x + 27.5$	$y = -0.75x + 27.5$	$m = -0.75$, $b = 27.5$	Solid	Below
$x \geq 0$	$x \geq 0$	$x = 0$	Vertical line; y-axis	Solid	To the right
$y \geq 0$	$y \geq 0$	$y = 0$	Horizontal line; x-axis	Solid	Above

Figure 5.5.9 shows each constraint inequality with its appropriate shading, whether done by hand or with technology.

Figure 5.5.10 shows only the intersection of all of the shaded areas, or the feasible region. We can now see the vertices that need to be found in order to identify a maximum profit.

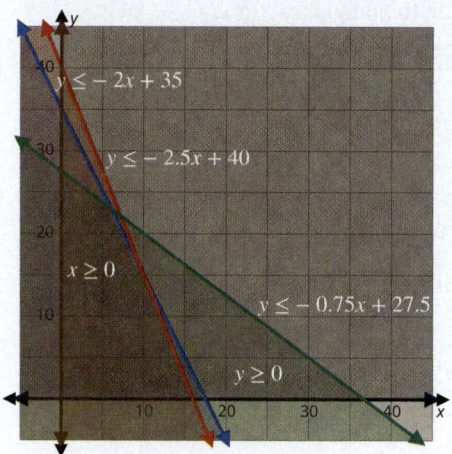

Figure 5.5.9

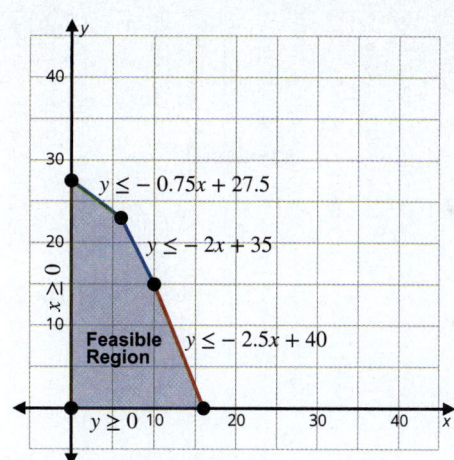

Figure 5.5.10: Feasible Region

Once again, we need to solve the systems of equations for each pair of lines in order to find the vertices of the polygon. Notice that the lines $y = -2.5x + 40$ and $y = -0.75x + 27.5$ don't intersect at one of our corners, so we do not need to find their intersection. In fact, not all of the intersections of the constraint lines are the corners of the area we are interested in. It's important to notice which lines form the vertices of the feasible region; the feasible region is the intersection of all of the overlapping shaded regions.

System 1

$$y = -2x + 35$$
$$y = -2.5x + 40$$

System of equations

$$-2x + 35 = -2.5x + 40$$

Substitute.

$$-2x + 2.5x = 40 - 35$$

Solve for x.

$$0.5x = 5$$
$$x = 10$$
$$y = -2(10) + 35 = 15$$

Substitute to find y.

$$(10, 15)$$

Vertex of the polygon

System 2

$$y = -2x + 35$$
$$y = -0.75x + 27.5$$

System of equations

$$-2x + 35 = -0.75x + 27.5$$

Substitute.

$$-2x + 0.75x = 27.5 - 35$$

Solve for x.

$$-1.25x = -7.5$$
$$x = 6$$
$$y = -2(6) + 35 = 23$$

Substitute to find y.

$$(6, 23)$$

Vertex of the polygon

System 3

$$y = -0.75x + 27.5$$
$$x = 0$$

System of equations

$$y = -0.75(0) + 27.5$$

Since x is given, substitute to find y.

$$y = 27.5$$
$$(0, 27.5)$$

Vertex of the polygon

System 4

$$y = -2.5x + 40$$
$$y = 0$$

System of equations

$$-2.5x + 40 = 0$$

Since y is given, substitute to find x.

$$-2.5x = -40$$
$$x = 16$$
$$(16, 0)$$

Vertex of the polygon

Once again, the origin is one of the corners because of the two constraints $x \geq 0$ and $y \geq 0$.

So the vertices of the polygon for the feasible region are (10, 15), (6, 23), (0, 27.5), (16, 0), and (0, 0).

We now substitute these points into the objective function for profit to find any maximum that may occur.

$$z = 105x + 57$$

$$(10, 15): \quad z = 105(10) + 57(15) = 1905$$

$$(6, 23): \quad z = 105(6) + 57(23) = 1941$$

$$(0, 27.5): \quad z = 105(0) + 57(27.5) = 1567.5$$

$$(16, 0): \quad z = 105(16) + 57(0) = 1680$$

$$(0, 0): \quad z = 105(0) + 57(0) = 0$$

Therefore, the maximum of $z = 1941$ occurs at the point (6, 23). This means that the entrepreneur can realize their maximum profit of $1941 when they make and sell 6 framed art pieces and 23 sets of placemats.

Skill Check Answers

1. a.   2. (0, 0), (70, 0), (0, 35)

5.5   Exercises

✔ CONCEPT CHECK

1. Linear programming is a process that optimize certain quantities in a mathematical model constrained by _____.

2. The _____ expresses the quantity that is to be optimized.

3. Constraints express the _____ of each variable or combination of variables.

4. The Fundamental Theorem of Linear Programming states that the maximal and minimal values are found at the _____ of the region.

5. The _____ is the set of all points that satisfy the constraints of an optimization problem.

💡 PRACTICE

6. Suppose you plan to build a rectangular pool on your property. You want the pool to have a volume of 350 square feet with the depth of 5 feet. To save on materials for the pool walls and pool bottom, you need to minimize the corresponding surface area. Let x represent the length of the pool and y represent the width of the pool. Write the objective function that describes the surface area of the open pool.

7. A construction company took a job for a fee of $80,000. If the project is finished ahead of schedule, the customer will pay an additional $2000 for each day saved. There are a limited number of employees in the company who can work on this project, so they are offered overtime. For each overtime day, the employees will receive a bonus of $200. The company wants to maximize the profit on the project. Let x represent the number of days saved and y represent the number of days of overtime work. Write the objective function that describes the revenue for the project.

8. A florist has 200 roses a day and uses them to make two types of bouquets. The small bouquets consist of 10 roses and sell for $35. The large bouquets consist of 25 roses and sell for $80. The florist wants to maximize his revenue. Let x represent the number of small bouquets and y represent the number of large bouquets. Write the objective function that describes the revenue for selling the bouquets.

9. Suppose you need a task to be done as soon as possible. You can hire two types of workers, the full-time ones working 10 hours a day and the part-time ones working 6 hours a day. You have a limited budget for the task, and the full-time workers will need to be paid more per hour than the part-time ones. You want to maximize the number of working hours per day depending on the number of workers of each type. Let x represent the number of full-time workers and y represent the number of part-time workers. Write the objective function that describes the number of working hours per day.

10. Daniel has a job in cargo transportation and wants to maximize his earnings. He needs to transport several 200-pound and 600-pound loads. The carrying capacity of his truck is 45,000 pounds. He is paid $50 for transporting a lighter load and $100 for a heavier one. The number of lighter loads should be no more than one-third of the total number transported. Let x represent the number of lighter loads and y represent the number of heavier loads. Write the objective function that describes the amount Daniel earns for transporting the cargo.

11. A factory produces three types of cardboard boxes. The amount of cardboard needed to make each type of box is 6 ft², 4 ft², and 9 ft², respectively. The first type of box sells for $1.50, the second for $1.25, and the third for $2.10. You need to maximize the revenue for selling these types of boxes. Let a represent the number of boxes of the first type, let b represent the number of boxes of the second type, and let c represent the number of boxes of the third type. Write the objective function that describes the revenue.

12. Suppose you need to determine the minimum point of the objective function $z = 380x - 415y$. After graphing the constraints, you find the following three vertices of the feasible region polygon.

   a. $(100, -45)$

   b. $(-215, 310)$

   c. $(0, 70)$

   Which of the points minimizes the objective function?

13. Consider the objective function $z = -220x + 185y$. The following three vertices of the feasible region polygon were found after graphing the constraints.

    a. (365, 450)

    b. (−160, 95)

    c. (−5, −270)

    Which of the points minimizes the objective function?

14. The following three points are the vertices of the feasible region polygon for the objective function $z = -80x - 325y$.

    a. (130, 90)

    b. (320, 0)

    c. (−170, 120)

    Which of the points minimizes the objective function?

15. Laura needs to find the maximum point of the objective function $z = 135x + 295y$. After graphing the constraints, she identifies the following three vertices of the feasible region polygon.

    a. (0, 400)

    b. (400, 0)

    c. (250, 250)

    Which of the points maximizes the objective function?

16. Consider the objective function $z = 340x - 230y$. The following three vertices of the feasible region polygon were found after graphing the constraints.

    a. (60, 10)

    b. (100, 100)

    c. (0, −100)

    Which of the points maximizes the objective function?

17. The following three points are the vertices of the feasible region polygon for the objective function $z = -55x + 410y$.

    a. (−125, 300)

    b. (0, 405)

    c. (−15, 395)

    Which of the points maximizes the objective function?

18. The objective function $z = -5x - 7y$ is subject to the following constraints.

$$y - x \geq 6$$
$$3x + 2y \leq 12$$
$$x \geq 0$$
$$y \geq 0$$

Find the maximum value of z.

19. The objective function $z = -12x + 8y$ is subject to the following constraints.

$$x + y \leq 20$$
$$5x - y \geq 10$$
$$x \geq 0$$
$$y \geq 0$$

Find the maximum value of z.

20. The objective function $z = 3x - 19y$ is subject to the following constraints.

$$x + y \leq 2$$
$$8y - 4x \leq 10$$
$$x \geq 0$$
$$y \geq 0$$

Find the maximum value of z.

21. The objective function $z = 9x + 4y$ is subject to the following constraints.

$$x + y \geq 12$$
$$15y - 3x \leq 18$$
$$x \leq 24$$
$$y \geq 0$$

Find the minimum value of z.

22. The objective function $z = -x - 6y$ is subject to the following constraints.

$$x + y \leq -15$$
$$12x - 8y \geq 20$$
$$y \geq -13$$

Find the minimum value of z.

23. The objective function $z = 13x - y$ is subject to the following constraints.

$$2x + 5y \leq 20$$
$$-x + 5y \geq 5$$
$$x \geq 0$$
$$y \geq 0$$

Find the minimum value of z.

🚀 **APPLICATIONS**

24. A plane carrying food and water to a resort island can carry a maximum of 30,000 pounds and is limited in space to carrying no more than 575 cubic feet. Each container of food weighs 175 pounds and takes up 5 cubic feet. Each container of water weighs 60 pounds and takes up 1 cubic foot of cargo space. Hotels on the island will buy the food for 14 dollars a pound and the water for 3 dollars per pound. Let x represent the number of food containers and y represent the number of water containers.

    a. Determine the objective function to represent the revenue generated by the plane.

    b. Write the constraint inequality that describes the amount of weight the plan can carry.

    c. Write the constraint inequality that describes the amount of cargo the plane can hold.

    d. List any other constraints.

    e. What is the optimum number of containers of each item that will maximize the revenue generated by the plane?

    f. What is the maximum revenue?

25. To prevent pests, an orchard can have no more than 6 times as many apple trees as peach trees. Also, 2 times the number of apple trees plus 3 times the number of peach trees must not exceed 180. The revenue from a single apple tree is $157 and the revenue from a single peach tree is $423. Let x represent the number of apple trees and y represent the number of peach trees.

    a. Determine the objective function to represent the revenue earned.

    b. Write the constraint inequality that describes that an orchard can have no more than 6 times as many apple trees as peach trees.

    c. Write the constraint inequality that describes that the number of apple trees plus 3 times the number of peach trees must not exceed 190.

    d. List any other constraints.

    e. Determine the number of each type of tree that will maximize revenue.

    f. What is the maximum revenue?

26. On your birthday, your great aunt gave you $19,000. You would like to invest at least $4000 of the money in municipal bonds yielding 4% and no more than $2000 in Treasury bills yielding 6%. How much should be placed in each investment in order to maximize the interest earned in one year? Assume simple interest applies. Let x represent the amount of money in municipal bonds and y represent the amount of money in Treasury bills.

   a. Write the objective function to represent the interest earned in one year.

   b. Write a constraint that describes the money being invested in municipal bonds.

   c. Write a constraint that describes the money being invested in Treasury bills.

   d. Write any additional constraints.

   e. How much should be placed in each investment in order to maximize the interest earned in one year?

27. A volunteer has been asked to drop off some supplies at a facility housing victims of a hurricane evacuation. The volunteer would like to bring at least 132 bottles of water, 82 first aid kits, and 45 security blankets on his visit. The relief organization has a standing agreement with two companies that provide victim packages. Company A can provide packages of 6 water bottles, 4 first aid kits, and 3 security blankets at a cost of $2.50. Company B can provide packages of 9 water bottles, 5 first aid kits, and 1 security blanket at a cost of $2.00. Let a represent the number of packages provided by Company A and b represent the number of packages provided by Company B. (**Note:** The companies cannot provide partial packages.)

   a. Write the objective function to represent the relief organization's cost of supplies.

   b. Write the constraint inequality that describes the number of bottles of water.

   c. Write the constraint inequality that describes the number of first aid kits.

   d. Write the constraint inequality that describes the number of security blankets.

   e. List any other constraints.

   f. How many of each package should the volunteer pick up to minimize the cost?

   g. What amount does the relief organization pay?

28. An outdoor recreation supply company's two best-selling hammocks are made from cotton and nylon rope. The best-selling hammock requires 2.5 yards of cotton and 100 feet of rope. The second best-selling hammock needs 4 yards of cotton and 150 feet of rope. The company has 450 yards of cotton and 3000 feet of rope in stock. The best-selling hammock makes a profit of $21 for the company and the second best-selling hammock makes a profit of $33. Let x be the number of best-selling hammocks and y be the number of second best-selling hammocks.

    a. Write the objective function to represent the profit from the hammocks.

    b. Write the constraint inequality that describes the number of yards of cotton.

    c. Write the constraint inequality that describes the number of feet of rope.

    d. List any other constraints.

    e. Based on the stock the company has, how many of each hammock should the company make to maximize their profit?

29. A transport company can purchase at most 9 buses with 30 seats and at most 7 buses with 50 seats. The selling price of a 30-seat bus is $45,000 and the selling price of a 50-seat bus is $60,000. The company can allocate $780,000 for the purchase of the buses. How many buses of each type should the company buy to maximize the total number of bus seats available?

30. To produce tables and cabinets, a furniture factory uses two types of wood: oak and pine. A minimum of 6 cubic feet of oak and 5 cubic feet of pine are required to produce one table. A minimum of 4 cubic feet of oak and 10 cubic feet of pine are required to produce one cabinet. Due to limited stock, the factory can use no more than 1400 ft^3 of oak and 2300 ft^3 of pine. Determine how many tables and cabinets the factory should produce to maximize the revenue if a table sells for $1300 and a cabinet sells for $1900.

31. For optimum prenatal care, an ob-gyn doctor recommends that expecting mothers taking a combination of supplements which included folic acid, calcium, and iron. The following table shows the recommended daily dosage for each essential nutrient along with two possible brands of multi-vitamins that contain various amounts of each. Each Baby Bliss Prenatal pill costs $0.40 and each Natural Gift Prenatal pill costs $0.34. How many pills of each brand should an expecting mother take each day to meet the recommended dosage and minimize the cost of supplements? Let x be the number of Baby Bliss Prenatal pills and y be the number of Natural Gift Prenatal pills.

	Recommended daily dosage	Baby Bliss Prenatal ($0.40)	Natural Gift Prenatal ($0.34)
Folic Acid	0.4 mg	0.1 mg	0.1 mg
Calcium	1000 mg	200 mg	400 mg
Iron	27 mg	9 mg	6 mg

5.5 PROJECT

MAXIMIZING PROFITS ON JEWELRY SALES

Eve likes to create jewelry as a hobby and has decided to open an Etsy shop to sell some of her creations in hopes of making some extra money. On average, it costs Eve $9 to make a necklace and $6 to make a pair of earrings. She wants to earn a profit (before taxes and fees) of $11 on each necklace sold and $8 on each pair of earrings sold. Eve has a starting budget of $180 and doesn't want to open her Etsy shop with more than 15 of either type of jewelry. Eve's goal with the Etsy shop is to maximize her potential profit. In this project, you will help Eve determine how many necklaces and how many pairs of earrings she should create to set up her Etsy shop.

Let x represent the number of necklaces and y represent the number of pairs of earrings.

1. What is the objective function for this situation? Explain what the function represents.

2. Describe the constraints of the situation and write a linear inequality to represent each one.

3. Graph the constraints from part 2.

4. What are the vertices of the feasible region?

5. Find the maximum of the objective function. Explain what this result means. Does the value make sense? Explain why or why not.

6. Suppose Eve decided to not limit herself to 15 of each type of jewelry. Would this change the initial setup of her Etsy shop? Explain why or why not.

OBJECTIVES

1. Model quadratic functions.

2. Write systems of quadratic equations.

OBJECTIVE 1 Quadratic Equations

In days gone by, baseball was a game that was judged and coached by eye. Swarms of baseball scouts scoured high school and college games for players who looked like they had big-league potential. But that all changed in the 1990s when the Oakland Athletics pioneered using a more data-based approach, and the era of sabermetrics began. Sabermetricians collect and analyze and model every aspect of the game of baseball to get an advantage over their opponents. So far in this chapter, we have looked at equations and inequalities that have all been linear. But not every situation can be modeled using a linear equation, and that is certainly true of sabermetrics.

Consider, for instance, how the exit velocity of a baseball that's just been hit can affect the ball's trajectory. If you've seen the path of a homerun baseball, you probably know that it is not a straight line from the bat to where the baseball lands, and so the path cannot be modeled by a linear equation.

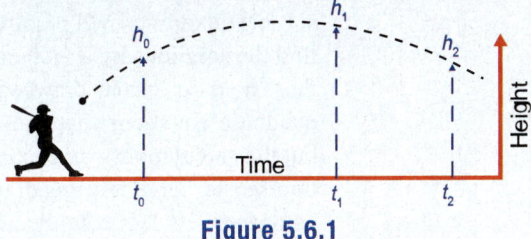

Figure 5.6.1

In fact, the trajectories of baseballs, soccer balls, footballs, and basketballs can all be modeled by another class of polynomials called *quadratics*—expressions that have an exponent of 2. Figure 5.6.2 shows the difference between the graph of a linear expression on the left and that of a quadratic expression on the right. Notice that the quadratic does not change at a constant rate like the linear expression does.

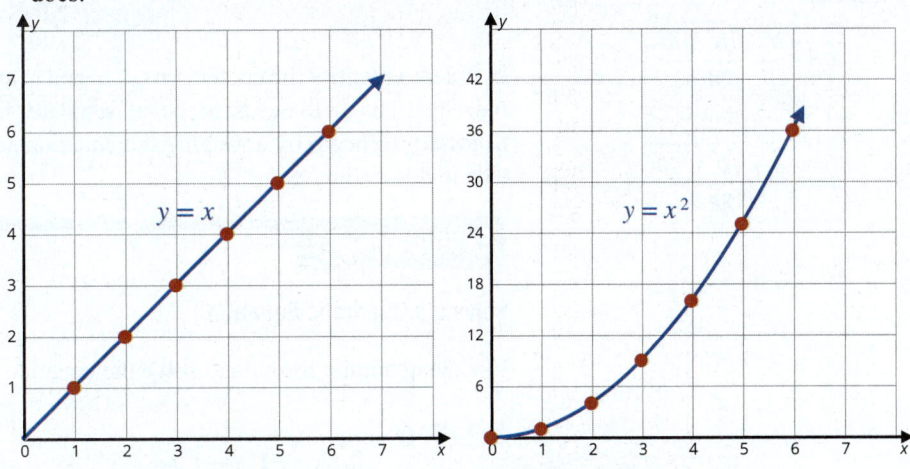

Figure 5.6.2: Graphs of Linear and Quadratic Equations

In this section, we will develop the tools to analyze and manipulate this new algebraic class. First, let's look more into how a quadratic equation differs from a linear equation.

📖 **QUADRATIC EQUATION**

A quadratic equation in x is written in standard form as

$$ax^2 + bx + c = 0,$$

where a, b, and c are numbers and $a \neq 0$.

Notice that in a quadratic equation the largest exponent is 2, not 1 like in a linear equation. This is what causes a quadratic function to grow differently than a linear function.

As we have seen throughout this chapter, there are numerous scenarios where manipulating and solving linear equations provides insights we would not otherwise be able to realize. This is no less the case for quadratic equations than for linear. Quadratic equations arise when we are modeling the trajectory of any projectile from a baseball to a rocket, but they can also be used to model situations such as gravitational forces, air resistance, and the concentration of a drug in a patient's bloodstream over time. In Section 5.1, we developed the algebra that we needed to solve linear equations, but those steps won't work to solve quadratics; solving quadratics will require a different technique. In some cases, we can easily find the solutions by a method of *factoring* or *completing the square*, but each of these methods has its drawbacks. Factoring is quick, but it only works on specific quadratic equations that can be factored. Completing the square always works, but the calculations can become tedious. The **Quadratic Formula**, however, can be used to solve any quadratic equation regardless of its factors and provides comparatively less complicated calculations.

💬 **Helpful Hint**

The ± sign in the quadratic formula indicates that there are two solutions to the equation: one using the plus sign and one using the minus sign in the formula.

$$x = \frac{-b + \sqrt{b^2 - 4ac}}{2a}$$

and

$$x = \frac{-b - \sqrt{b^2 - 4ac}}{2a}$$

$f(x)$ **QUADRATIC FORMULA**

A quadratic equation written in standard form, $ax^2 + bx + c = 0$, where a is nonzero, can be found using the quadratic formula.

$$x = \frac{-b \pm \sqrt{b^2 - 4ac}}{2a}$$

In order to identify the coefficients a, b, and c, and then use the quadratic formula, the equation itself needs to be in standard form. In other words, it may be necessary to begin by arranging the equation so that the variables are all on one side in descending order.

Example 5.6.1

Solving a Quadratic Equation

Use the quadratic formula to solve the equation $5x^2 - 31x + 6 = 0$.

Solution

To use the quadratic formula, the equation must be in the form $ax^2 + bx + c = 0$. Since we are given the equation in standard form, we can identify a, b, and c.

$$5x^2 - 31x + 6 = 0$$

$$a = 5 \quad b = -31 \quad c = 6$$

Now, substitute these values into the quadratic formula.

$$x = \frac{-b \pm \sqrt{b^2 - 4ac}}{2a}$$

$$x = \frac{-(-31) \pm \sqrt{(-31)^2 - 4(5)(6)}}{2(5)}$$ Substitute appropriate values.

$$x = \frac{31 \pm 29}{10}$$ Simplify.

$$x = \frac{31 + 29}{10} = 6 \text{ and } x = \frac{31 - 29}{10} = \frac{1}{5}$$ Simplify using both operations.

So $x = 6$ and $x = \frac{1}{5}$.

We need to determine if either (or both) of the solutions we found to the linear equations satisfies the original quadratic equation. Check each by substituting back into the original equation.

Check:

$x = 6$ $x = \frac{1}{5}$

$$5(6)^2 - 31(6) + 6 \overset{?}{=} 0 \qquad\qquad 5\left(\frac{1}{5}\right)^2 - 31\left(\frac{1}{5}\right) + 6 \overset{?}{=} 0$$

$$180 - 186 + 6 \overset{?}{=} 0 \qquad\qquad\qquad \frac{1}{5} - \frac{31}{5} + 6 \overset{?}{=} 0$$

☑   ☑

Thus, we have verified that both of these numbers are solutions to the quadratic equation.

> **✔ Skill Check 5.6.1**
>
> Identify a, b, and c in the quadratic equation $3x^2 = 5x - 2$.

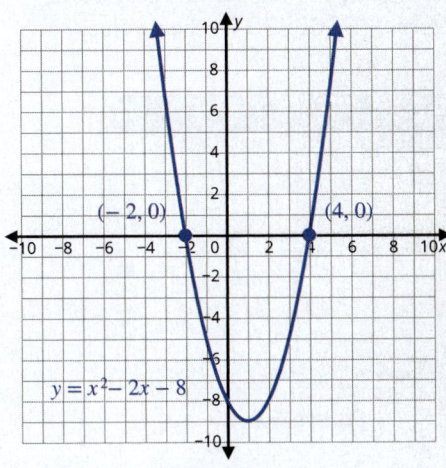

Figure 5.6.3

As we saw when we analyzed linear equations, the roots (or zeros) of an expression are the values of the independent variable that make the expression evaluate to 0. As Figure 5.6.3 shows, the roots of a quadratic equation are the places where the graph crosses the x-axis. A quadratic may have two, one, or no real roots depending on the various coefficients.

Example 5.6.2

Solving a Quadratic Equation

Find the roots of the expression $2x^2 - 7x - 3$.

Solution

The graph of a quadratic will cross the x-axis when x is one of the roots. To find the roots we set the expression equal to zero and solve.

$$2x^2 - 7x - 3 = 0$$

By Hand

Since the equation is in standard form, we can identify a, b, and c.

$$a = 2 \quad b = -7 \quad c = -3$$

Substitute these values into the quadratic formula and simplify.

$$x = \frac{-b \pm \sqrt{b^2 - 4ac}}{2a}$$

$$x = \frac{-(-7) \pm \sqrt{(-7)^2 - 4(2)(-3)}}{2(2)} \qquad \text{Substitute the appropriate values.}$$

$$x = \frac{7 \pm \sqrt{49 + 24}}{4} \qquad \text{Simplify.}$$

$$x = \frac{7 \pm \sqrt{73}}{4}$$

$$x = \frac{7 + \sqrt{73}}{4} \approx 3.886 \quad \text{and} \quad x = \frac{7 - \sqrt{73}}{4} \approx -0.386$$

TI-83/84 Plus

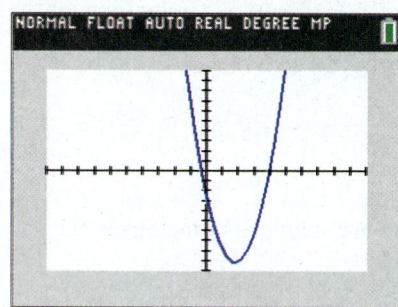

Figure 5.6.4

To find the roots by graphing on a TI-83/84 Plus, press $\boxed{y=}$ and clear any equations currently in the calculator. Type $2x^2 - 7x - 3$ for **Y1** and press $\boxed{\text{graph}}$.

We can now press $\boxed{\text{2nd}}$ $\boxed{\text{trace}}$ and select **zero** to find the roots. You'll see a blinking cursor along the curve and **LeftBound?** at the bottom of the screen. Move the cursor to the left of one of the locations where the curve crosses the x-axis and press $\boxed{\text{enter}}$. The bottom of the screen should now show **RightBound?**. Move the cursor to the right of that same root and press $\boxed{\text{enter}}$ again; the bottom of the screen will now show **Guess?**. Press $\boxed{\text{enter}}$ a third time, and the bottom of the screen the calculator will display its best approximation of the x-value of that root, **-0.386001**. Repeat this process with the other root, and the calculator will return **3.8860009** for the x-value.

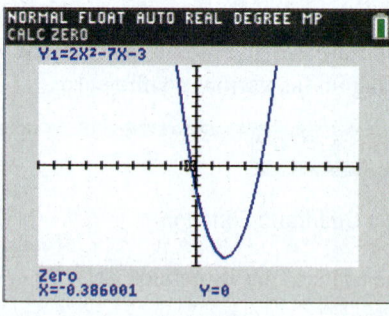

Figure 5.6.5

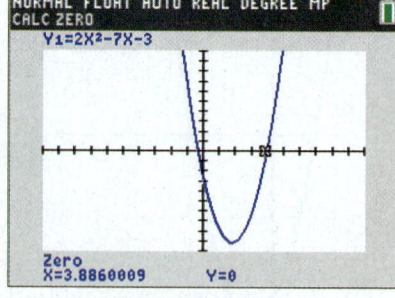

Figure 5.6.6

✔ **Skill Check 5.6.2**

Find the roots of the equation.

$$4x^2 - 12x = 27$$

Therefore, the roots of the expression $2x^2 - 7x - 3$ are $x \approx 3.886$ and $x \approx -0.386$.

Example 5.6.3

Modeling with a Parabola

An architect firm has been hired to design a new suspension bridge to be included in a town's overall plan to revive the downtown area. The outermost support towers extend 80 feet above the deck (or road) of the bridge. Two inner decorative pillars are 20 feet tall and reach to the main cable. A simplified rendering of the bridge is shown in Figure 5.6.7.

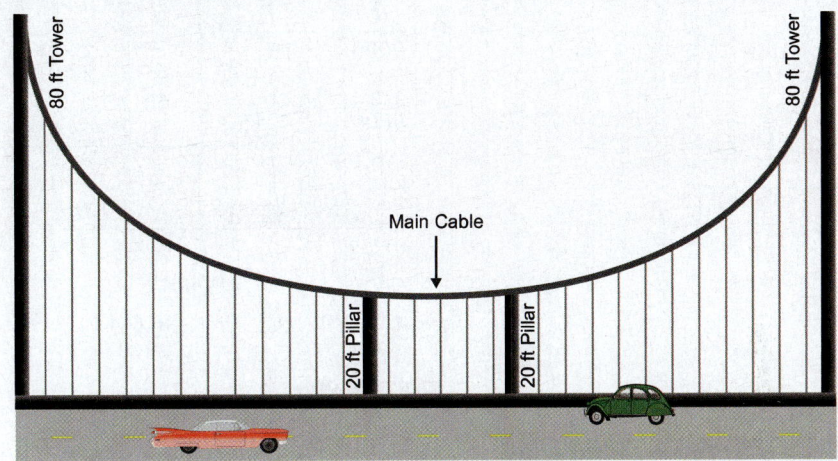

Figure 5.6.7: Suspension Bridge with Two Decorative Pillars

The shape of the main cable can be modeled by the quadratic equation $y = \frac{7}{9000}(x)(x - 600) + 80$, where x is the horizontal distance from the leftmost tower and y is the vertical height from the deck to the cable. If the leftmost tower is placed at 0 on the x-axis, how far away from the leftmost tower do each of the two 20-foot pillars need to be? Round your answer to the nearest tenth of a foot.

Solution

We are told that the shape of the main cable can be modeled by the quadratic $y = \frac{7}{9000}(x)(x - 600) + 80$. Every point on this quadratic has an x-value that is the horizontal distance from the leftmost tower and a y-value that is the height of the main cable above the deck. We are told that the height of each decorative pillar is 20 feet; in other words, $y = 20$.

By Hand

In order to use the quadratic formula, we need to substitute this into the equation and rearrange it into standard form.

Fun Fact

The Clifton Suspension Bridge in Bristol, UK, is similar to the bridge modeled in Example 5.6.3. The bridge has a length of 1352 feet and is 331 feet tall.

Figure 5.6.8

Photo by Gothick, Creative Commons Attribution—Share Alike 3.0 Unported license.

$$y = \frac{7}{9000}(x)(x-600)+80$$

$$20 = \frac{7}{9000}(x)(x-600)+80 \qquad \text{Substitute } y = 20 \text{ into the equation.}$$

$$20 = \frac{7}{9000}\left(x^2 - 600x\right)+80 \qquad \text{Distribute } x.$$

$$20 = \frac{7}{9000}x^2 - \frac{7}{15}x + 80 \qquad \text{Write in standard form.}$$

$$0 = \frac{7}{9000}x^2 - \frac{7}{15}x + 60 \qquad \text{Substitute values into the quadratic formula.}$$

$$a = \frac{7}{9000} \quad b = -\frac{7}{15} \quad c = 60$$

$$x = \frac{-\left(-\dfrac{7}{15}\right) \pm \sqrt{\left(-\dfrac{7}{15}\right)^2 - 4\left(\dfrac{7}{9000}\right)(60)}}{2\left(\dfrac{7}{9000}\right)}$$

$$x \approx 413.389 \quad \text{and} \quad x \approx 186.611$$

TI-83/84 Plus

Press $\boxed{y=}$ and clear any equations currently in the calculator. Type $\frac{7}{9000}(x)(x-600)+80$ for **Y1**. Now press $\boxed{\text{graph}}$.

TECH TIP

If you do not see a parabola on your calculator screen after pressing $\boxed{\text{graph}}$, you'll need to adjust the minimum and maximum of your x- and y-axes. Press $\boxed{\text{window}}$ and set **Xmin** to –50, **Xmax** to 500, **Ymin** to –10, and **Ymax** to 50.

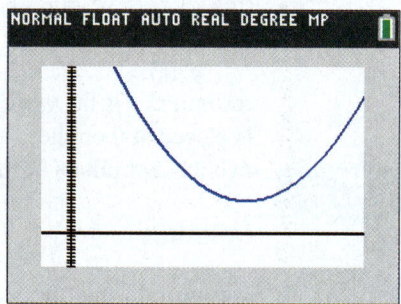

Figure 5.6.9

Press $\boxed{\text{trace}}$ and move the cursor until the y-value at the bottom of the screen is close to 20. The calculator will show the corresponding x-value. Since the cursor will not stop at exactly $y = 20$, you might need to check the x-value in one location where y is slightly less than 20 and another location where y is slightly greater than 20.

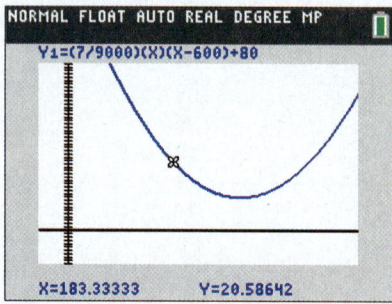

Figure 5.6.10

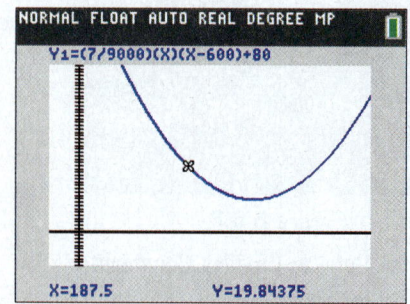

Figure 5.6.11

The corresponding *x*-value is somewhere between 183.33333 and 187.5. Since we don't know the exact *x*-value, we can average 183.33333 and 187.5, which gives us 186.6.

Repeat this process for the other location on the graph where *y* is 20.

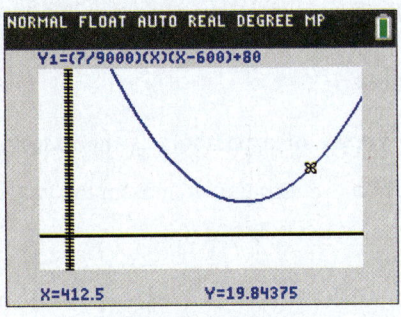

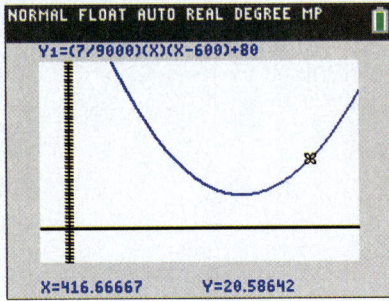

Figure 5.6.12   **Figure 5.6.13**

The corresponding *x*-value is somewhere between 412.5 and 416.66667. We again average these numbers and get 413.4.

Thus, the decorative pillars should be placed at approximately 186.6 feet and 413.4 feet from the leftmost tower.

As we saw in the previous example, the graph of a quadratic equation is a U-shaped curve, which is called a parabola. The trajectories of fireworks, water fountain streams, and baseballs are all examples of this same curvature. A **parabola** is a symmetrical U-shaped curve around a vertical or horizontal line. Some parabolas open upward like a smile, some open downward like a frown, while others open to the left or right. The curve has an extreme point, called a **vertex**, that is either the parabola's minimum point or maximum point, depending on whether the curve opens upward or downward.

📖 PARABOLA AND VERTEX

Parabola

The graph of a quadratic equation is called a **parabola**. It is a symmetrical U-shaped curve around a vertical or horizontal line.

Vertex

The **vertex** of a parabola is the extreme point representing the quadratic equation's minimum or maximum value. It occurs at the point $\left(\dfrac{-b}{2a}, y\right)$, given that $ax^2 + bx + c = y$.

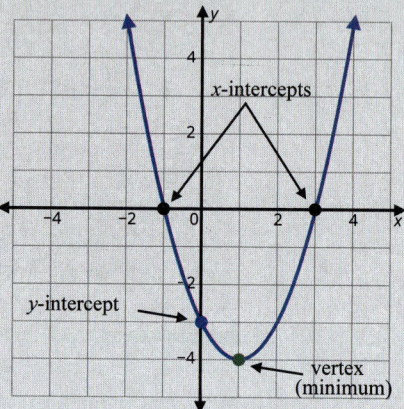

Figure 5.6.14

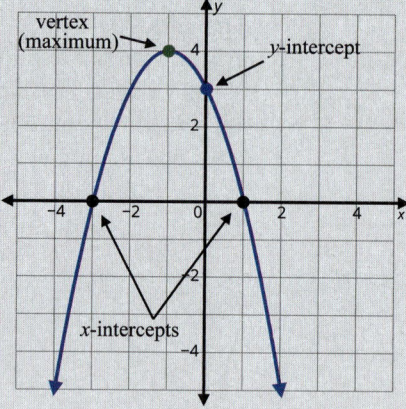

Figure 5.6.15

Example 5.6.4

Modeling with a Parabola

As part of new landscaping for an art museum, an outdoor water fountain is being designed. The water jets will create a parabolic stream of water modeled by the equation $y = -2x^2 + 6x + 5$, where y is the height of the jet of water and x is the horizontal distance of the jet of water from the water nozzle, both in meters.

Figure 5.6.16

a. The design includes lights placed so that the water stream ends on them. Determine how far from the nozzle each light needs to be placed so that the jet of water lands on it.

b. What is the maximum height the jet of water will reach?

c. If the designer would like to have varying heights for the jets of water but have the streams land at the same distance, can the same equation be used? Explain.

d. How would you change the equation to produce a water jet that goes half as high but still lands in the same spot?

Solution

a. The lights will need to be placed where the water hits the ground, which is where $y = 0$. This point is the x-intercept of the parabola. To find this distance of the nozzle to the light, we need to solve the equation $0 = -2x^2 + 6x + 5$.

The equation is in standard form, so we can identify a, b, and c and use the quadratic formula to find the solutions.

$$a = -2 \quad b = 6 \quad c = 5$$

Substitute these values into the quadratic formula and simplify.

$$x = \frac{-(6) \pm \sqrt{(6)^2 - 4(-2)(5)}}{2(-2)}$$

$$x = \frac{-6 \pm \sqrt{76}}{-4}$$

$$x = \frac{-6 + \sqrt{76}}{-4} \approx -0.68 \qquad x = \frac{-6 - \sqrt{76}}{-4} \approx 3.68$$

Because we are interested in distance, we need the positive value of x. So the lights should be placed 3.68 meters from the nozzle.

b. The water will reach its highest point at the vertex of the parabola.

By Hand

To find this height (that is, the y-coordinate of the vertex), we first find the x-coordinate using $x = \dfrac{-b}{2a}$. From the quadratic we have that $a = -2$ and $b = 6$. Therefore, we have the following.

$$x = \frac{-b}{2a} = \frac{-(6)}{2(-2)} = \frac{-6}{-4} = 1.5$$

Now that we have the x-coordinate, substitute this value back into the equation to find the maximum height of the water; that is, the y-coordinate.

$$y = -2(1.5)^2 + 6(1.5) + 5$$
$$y = 9.5$$

Thus, the vertex occurs at the point (1.5, 9.5).

TI-83/84 Plus

To find the vertex, first graph the parabola by pressing $\boxed{\text{y=}}$. Type the right-hand side of the equation $-2x^2 + 6x + 5$ for **Y1**, using $\boxed{\text{X,T,}\theta\text{,n}}$ for the variable, x, and press $\boxed{\text{graph}}$.

TECH TIP

If you do not see a parabola on your calculator screen after pressing $\boxed{\text{graph}}$, you'll need to adjust the minimum and maximum of your x- and y-axes. Press $\boxed{\text{window}}$ and change **Xmin** and **Ymin** to −10, and change **Xmax** and **Ymax** to 10.

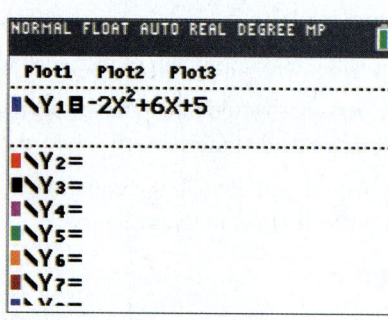

Figure 5.6.17

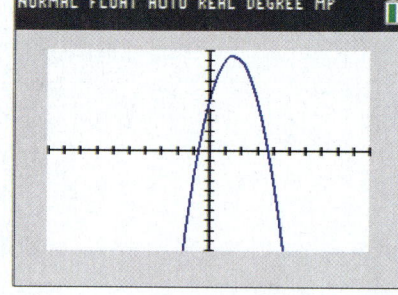

Figure 5.6.18

Press $\boxed{\text{2nd}}\boxed{\text{trace}}$ and select **maximum**. A blinking cursor will appear on the curve. Use the right- and left-arrow keys to move the cursor somewhere to the left of the maximum and press $\boxed{\text{enter}}$, then move the cursor to the right side of the maximum and press $\boxed{\text{enter}}$. The word **Guess?** will appear at the bottom of the screen. Press $\boxed{\text{enter}}$ again, and the calculator will return both coordinates of the maximum: x=1.4999985 and y=9.5.

This means that the water reaches its maximum height of 9.5 meters when it is 1.5 meters away from the nozzle.

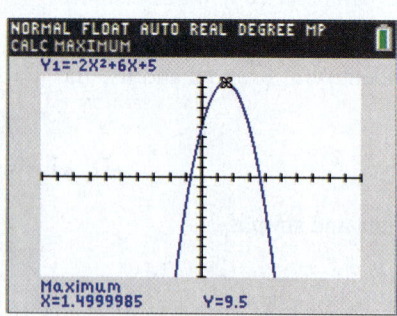

Figure 5.6.19

c. No; the same equation cannot be used for different heights. However, the equation can be used as a basis for new equations that will have different maximums and the same starting and ending points for the jets. Because we know that the arc of the fountain must begin and end at the same point no matter the height that it reaches, the equation must be a quadratic equation that has the same solutions as the original equation. The only such equations are multiples of the original equation. In other words, if the vertex needs to be three times as high, we can multiply the entire original equation by 3 to construct the new equation.

d. To produce a water jet that reaches half as high, multiply the entire original equation by $\frac{1}{2}$ to construct the new equation. The new equation would be $y = x^2 + 3x + 2.5$.

✔ **Skill Check 5.6.3**

Determine the coordinates of the vertex of the equation $y = x^2 - 6x - 7$.

Example 5.6.5

Modeling with Quadratic Functions

According to the National Highway Traffic Safety Administration, the stopping distances of a vehicle under various conditions are modeled with the following quadratic functions, where x is the speed of the car in miles per hour, and the value of the function is the distance, in feet, required to stop under those conditions.

Stopping Distance under Normal Conditions: $\quad f(x) = 0.0086x^2 + 0.671x - 1.97$

Stopping Distance While Using a Cell Phone: $\quad g(x) = 0.0086x^2 + 1.11x - 1.37$

Stopping Distance on a Wet Road: $\quad\quad\quad\quad h(x) = 0.02x^2 + 0.5x$

Each of the curves are depicted in Figure 5.6.20. Note, that although these formulas are defined for any value of x, for this application, we are only interested in the positive values shown in the graph. In the next section, we will further explore why this is the case.

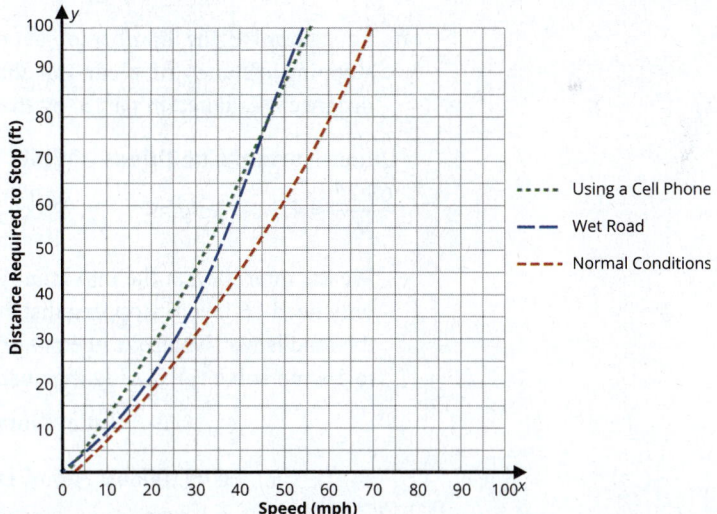

Figure 5.6.20: Speed and Feet Required to Stop Under Various Conditions

a. Determine the stopping distance required for each condition when a car is traveling at 60 mph.

b. If the average length of a car is 16 feet, how many car lengths are required for the stopping distances found in part a.?

c. Find the fastest speed a car could be traveling (to the nearest mile per hour) and still be able to stop within one average car length under both normal conditions and when the road is wet.

d. Use the graphs in Figure 5.6.20 to compare the stopping distances required for each driving condition.

Solution

a. In order to determine the stopping distance for a car traveling at 60 mph in each of the conditions, we need to evaluate each function at $x = 60$.

Under Normal Conditions

$$f(x) = 0.0086x^2 + 0.671x - 1.97$$

$$f(60) = 0.0086(60)^2 + 0.671(60) - 1.97$$

$$f(60) = 69.25$$

Stopping Distance Required: 69.25 feet

While Using a Cell Phone

$$g(x) = 0.0086x^2 + 1.11x - 1.37$$

$$g(60) = 0.0086(60)^2 + 1.11(60) - 1.37$$

$$g(60) = 96.19$$

Stopping Distance Required: 96.19 feet

On a Wet Road

$$h(x) = 0.02x^2 + 0.5x$$

$$h(60) = 0.02(60)^2 + 0.5(60)$$

$$h(60) = 102$$

Stopping Distance Required: 102 feet

b. To determine the number of average car lengths that are required for the stopping distance of a car traveling 60 mph, we can divide each stopping distance calculated in part a. by the average car length of 16 feet.

Under Normal Conditions	While Using a Cell Phone	On a Wet Road
$\dfrac{69.25}{16} \approx 4.3$ car lengths	$\dfrac{96.19}{16} \approx 6.0$ car lengths	$\dfrac{102}{16} \approx 6.4$ car lengths

c. We are interested in the maximum speed that a car could be traveling and still only need 16 feet of stopping distance under both normal conditions and when the road is wet. Thus, we need to set each of the two respective functions equal to 16 and solve for x using the quadratic formula.

Under Normal Conditions

$$16 = 0.0086x^2 + 0.671x - 1.97$$

$$0 = 0.0086x^2 + 0.671x - 17.97$$

$$a = 0.0086 \quad b = 0.671 \quad c = -17.97$$

$$x = \frac{-(0.671) \pm \sqrt{(0.671)^2 - 4(0.0086)(-17.97)}}{2(0.0086)}$$

$$x \approx 21.084, -99.107$$

Since the car can only travel at a positive speed, we discard the negative value of x. Thus, under normal conditions, a car could be traveling at most 21 mph and still stop within 16 feet.

On a Wet Road

$$16 = 0.02x^2 + 0.5x$$

$$0 = 0.02x^2 + 0.5x - 16$$

$$a = 0.02 \quad b = 0.5 \quad c = -16$$

$$x = \frac{-(0.5) \pm \sqrt{(0.5)^2 - 4(0.02)(-16)}}{2(0.02)}$$

$$x \approx 18.423, -43.4232$$

Thus, on a wet road, a car could be traveling at most 18 mph and still stop within 16 feet.

d. Using the graphs in Figure 5.6.20, we can see that the distance required to stop while talking on a cell phone versus normal driving conditions is greater at all speeds, thus making it not safe to talk on a cell phone while driving. The stopping distances for driving on a wet road is similar to that of talking on a cell phone.

OBJECTIVE 2 Multiple Functions

Sometimes modeling a situation requires more than one type of function. Let's return to our discussion of sabermetrics. Example 5.6.6 looks at using both a linear function as well as a quadratic function to model the flight of a baseball.

Example 5.6.6

Using Multiple Functions

In 2020, the longest homerun hit in Major League Baseball (MLB) was made by the 23-year-old Atlanta Braves right fielder, Ronald Acuña Jr. He hit the ball 495 feet, with an exit velocity of 112.9 mph.

Physics tells us that an object projected at time $t = 0$ from a point y_0 feet above ground with an initial vertical velocity of v_0 feet per second has a height above the ground after t seconds that is modeled by the function $h(t) = -16t^2 + v_0 t + y_0$. At that same time t, the distance the object has traveled is modeled by the function $d(t) = v_1 t$ where v_1 is the initial horizontal velocity.

> **Helpful Hint**
>
> We use subscripts of 0 on variables to denote the beginning, or initial, values. For instance, the initial height of a falling object is denoted as y_0 and the initial velocity is denoted as v_0.

The ball was 4 feet off of the ground when Ronald Acuña hit it and had an initial vertical velocity of 87.85 ft/sec and a horizontal velocity of 140.42 ft/sec. Answer the following questions.

a. Write the quadratic function for the height of the ball above the ground after t seconds, $h(t)$, in feet.

b. Write the linear function for the distance the ball will travel after t seconds, $d(t)$, in feet.

c. How high was Acuña's homerun ball after 2 seconds?

d. During what time interval was the homerun ball more than 100 feet above the ground?

e. How long was the ball in flight before it became a homerun because it passed over the outfield fence at a distance of approximately 395 feet?

f. The outfield fence at Truist park where Acuña hit the homeroom is 8 feet 8 inches tall. How high above the fence was the ball when it went over the fence?

Solution

a. We can write the quadratic function for the height $h(t)$ by substituting the values for v_0 and y_0 into the function model. We were given that the starting point was 4 feet above ground, so $y_0 = 4$. We were also told that the initial vertical velocity is 87.75 ft/sec, so $v_0 = 87.75$. Therefore, the function for the height above the ground after t seconds is as follows.

$$h(t) = -16t^2 + 87.75t + 4$$

In this function, time t is the independent variable and the height of the projectile $h(t)$ is the dependent variable.

b. We can write the linear function for the distance traveled $d(t)$ by substituting the value for the initial horizontal velocity v_1. We're told the initial horizontal velocity is 140.42 ft/sec. Therefore, the function for the distance traveled after t seconds is as follows.

$$d(t) = 140.42t$$

In this function, time t is the independent variable and the distance the ball traveled $d(t)$ is the dependent variable.

c. We can use the function we found in part a. to determine how high the ball will be after a time of 2 seconds. In other words, we need to find $h(2)$. Using the formula, we have the following.

$$h(t) = -16t^2 + 87.75t + 4$$
$$h(2) = -16(2^2) + 87.75(2) + 4$$
$$h(2) = 115.5$$

Thus, after 2 seconds the ball was 115.5 feet in the air.

d. The height of the ball is the dependent variable $h(t)$, and we are asked to find the values of t that make the function more than 100. In other words, we need to find the values of t such that $h(t) > 100$. Algebraically, we have the following.

$$-16t^2 + 87.75t + 4 > 100$$

Figure 5.6.21 is graph of the function along with a line indicating the cutoff height of 100 ft.

By Hand

As you can see in Figure 5.6.21, the height of the curve is more than 100 feet between the two points of intersection involving the function and the line. Therefore, we need to find the two intersection points that occur when the function is equal to 100. We can do this by changing the inequality sign to an equal sign, and then solving for t by using the quadratic formula.

$$-16t^2 + 87.75t + 4 = 100$$
$$-16t^2 + 87.75t - 96 = 0 \qquad \text{Write in standard form.}$$

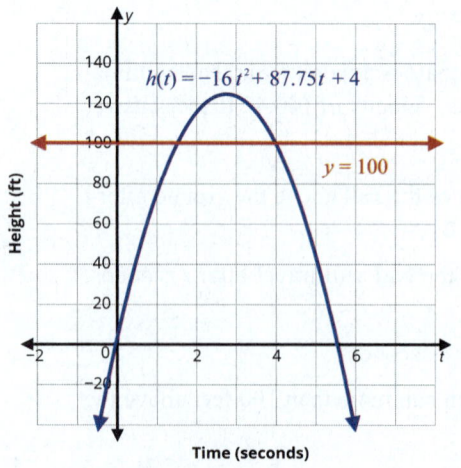

Figure 5.6.21

So we have $a = -16$, $b = 87.75$, and $c = -96$. Substitute these values into the quadratic formula and simplify.

$$t = \frac{-87.75 \pm \sqrt{87.75^2 - 4(-16)(-96)}}{2(-16)}$$

$$t = \frac{-87.75 + \sqrt{1556.0625}}{-32} \quad \text{and} \quad t = \frac{-87.75 - \sqrt{1556.0625}}{-32}$$

$$t \approx 1.51 \quad \text{and} \quad t \approx 3.97$$

TI-83/84 Plus

Using a graphing utility, we can graph both $h(t)$ and the horizontal line $y = 100$. (When graphing $h(t)$, we will use $\boxed{\text{X,T,}\theta\text{,n}}$ for the variable t, but it will appear as x in the calculator.) Press $\boxed{\text{y=}}$ and enter $-16x^2 + 87.75x + 4$ for Y1 and 100 for Y2. If needed, adjust your viewing window by pressing $\boxed{\text{window}}$ and set Xmin to -2, Xmax to 10, Ymin to -10, and Ymax to 150. Then press $\boxed{\text{graph}}$.

Notice that the two places where the function and the horizontal line intersect are the endpoints of the interval when the height is above 100. To find the two points of intersection, press $\boxed{\text{2nd}}\boxed{\text{trace}}$ and select **intersect**. Move the cursor close to one of the points of intersection and press $\boxed{\text{enter}}$ three times; at the bottom of the screen the calculator will display its best approximation of the intersection: x≈1.5094701 (Figure 5.6.22). Repeat this process for the other point of intersection. The calculator returns a value of x≈3.9749049 (Figure 5.6.23).

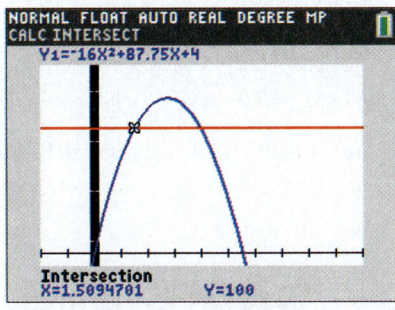

| Figure 5.6.22 | Figure 5.6.23 |

Therefore, the ball is above 100 feet when $1.51 < t < 3.97$.

e. The ball became a homerun when it went over the fence, which was 395 feet away. Thus, we need to set our distance function equal to 395 and solve for t.

Substituting in the distance function we have the following.

$$395 = 140.42t$$

$$\frac{395}{140.42} = t$$

$$t \approx 2.81 \text{ sec}$$

Therefore, we know that the ball was in the air for 2.81 seconds before going over the fence to score the homerun.

f. To determine how high above the fence the ball was when it went over the fence, we first need to determine the height of the ball as it went over the fence. Then, we need to subtract the height of the fence from the height of the ball.

From part e., we know that the ball went over the fence after 2.81 seconds. We can use that value for t in the height function to calculate how high the ball was when it went over the fence. Using the formula, we have the following.

$$h(t) = -16t^2 + 87.75t + 4$$
$$h(2.81) = -16(2.81)^2 + 87.75(2.81) + 4$$
$$h(2.81) = 124.2399$$

Thus, Acuña's homerun ball was approximately 124.24 feet off of the ground when it went over the fence.

Now we subtract the height of the fence, 8 feet 8 inches, from the ball's height. Before we can subtract, both measurements need to be in the same format. We'll convert the fence height to a decimal form of feet: $8 \text{ in.} \left(\dfrac{1 \text{ ft}}{12 \text{ in.}} \right) \approx 0.67 \text{ ft.}$

This gives that the fence is 8.67 feet tall. Now we can subtract.

$$124.24 - 8.67 = 115.57$$

Thus, the homerun ball was 115.57 feet above the fence when it went over it for a homerun.

✔ **Skill Check 5.6.4**

Suppose a rocket is fired vertically upward from 1200 feet above the ground, with an initial velocity of 620 ft/sec. Write the quadratic model for its height $h(t)$ in feet above the ground after t seconds and determine how long the projectile will be in flight.

Skill Check Answers

1. $a = 3, b = -5, c = 2$ **2.** $x = -\dfrac{3}{2}$ and $x = \dfrac{9}{2}$ **3.** $(3, -16)$

4. $h(t) = -16t^2 + 620t + 1200$; approximately 40.6 seconds

5.6 Exercises

1. A quadratic equation in x is written in standard form as _____.

2. The _____ is used to determine the solutions of a quadratic equations.

3. To use the quadratic formula, the quadratic equation needs to be written in _____.

4. A parabola is a symmetrical U-shaped curve around a _____ line.

5. The _____ of a parabola is the extreme point that represents the quadratic equations minimum or maximum.

> 💡 **PRACTICE**

Use the quadratic formula to solve each equation.

6. $x^2 + 2x - 3 = 0$

7. $x^2 - 6x - 7 = 0$

8. $x^2 = -14x - 49$

9. $2x^2 + 5x - 3 = 0$

10. $2x^2 = 2x + 40$

11. $2x^2 - 3x - 1 = 0$

12. $3x^2 + 2x - 2 = 0$

13. $16x^2 + 8x - 1 = 0$

14. $(x + 5)(x - 1) = -3$

15. $(x + 4)(x - 2) = -4$

> 🚀 **APPLICATIONS**

16. The function $C(x) = 0.0086x^2 + 1.11x - 1.37$ represents the stopping distance in feet while talking on a cell phone and driving at a speed of x mph. What distance will it take you to stop while talking on a cell phone if you are driving 65 mph? 75 mph?

17. The function $W(x) = 0.02x^2 + 0.5x$ represents the stopping distance in feet on a wet road when driving at a speed of x. What distance will it take you to stop while driving on a wet road if you are driving 65 mph? 35 mph?

Use the fact that a projectile can be modeled by the function
$h(t) = -16t^2 + v_0t + y_0$ to solve the following problems.

18. A projectile is fired vertically upward from a height of 200 feet above the ground, with an initial velocity of 1100 ft/sec.

   a. Write a quadratic equation to model the projectile's height in feet above the ground after t seconds.

   b. During what time interval will the projectile be more than 8000 feet above the ground? Round your answer to the nearest hundredth.

   c. What is the total flight time of the projectile? Round your answer to the nearest hundredth.

19. A ball is thrown straight up, from ground zero, with an initial velocity of 55 feet per second. Find the maximum height attained by the ball and the time it takes for the ball to return to ground zero.

20. From the top of a 250-foot-tall building, a ball is thrown straight up with an initial velocity of 25 feet per second. Find the maximum height attained by the ball and the time it takes for the ball to hit the ground.

21. A ball is dropped from the top of a 1250-foot-tall building. How long does it take the ball to hit the ground? (Note that $v_0 = 0$.)

22. Grayson drops a rock into a well in which the water surface is 275 feet below ground level. How long does it take the rock to hit the water surface? (Note that $v_0 = 0$.)

23. A parabola-shaped floodlight is attached to the ceiling with four cables. The outermost support cables are 3 feet long and the inner cables are 1.5 feet long. The shape of the floodlight can be modeled by the quadratic equation $y = -\dfrac{x^2}{8} - 1$, where x is the horizontal distance from the center of the floodlight and y is the vertical distance from the ceiling to the floodlight. Let the center of the floodlight be placed at 0 on the x-axis. What does the distance between the inner cables and the outer cables need to be? A simplified rendering of the floodlight is shown in the following figure.

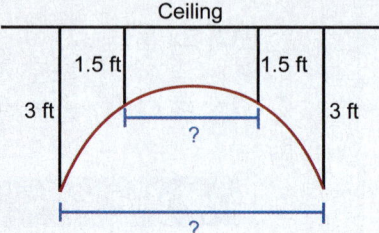

24. A domed roof is mounted on two types of columns. The outermost support columns are 15 feet high, and the interior decorative columns are 18 feet high. A simplified rendering of the construction is shown in the following figure. The shape of the roof can be modeled by the quadratic equation $y = -\dfrac{(x-10)^2}{25} + 19$, where x is the horizontal distance from the leftmost column and y is the vertical height from the floor to the roof. Let the leftmost column base be placed at the origin. What is the distance between the two inner columns? What is the distance between the two outer columns?

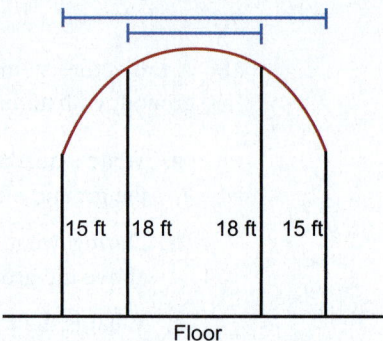

25. The shape of the bottom of a pool can be modeled by the quadratic equation $y = \dfrac{(x-4)^2}{8} - 2$, where x is the horizontal distance from the leftmost pool edge and y is the depth of the pool. Let the leftmost pool edge be placed at the origin. What is the distance from the edge to the place where the depth is 1.5 meters? What is the length of the pool? A simplified rendering of the pool is shown in the following figure.

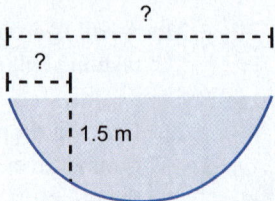

26. Due to their reflectivity, parabolas are used in the construction of amphitheaters so that the audience can clearly hear the actors. Suppose the amphitheater is built based on the parabola where x is the horizontal distance from the center of the amphitheater and y is the vertical height from the scene to the seats on the specific row. Let the center of the theater be placed at 0 on the x-axis. What is the distance between the center and the seats at the 2.5-meter height? What is the distance between the center and the seats at the 10-meter height? A simplified rendering of the amphitheater is shown in the following figure.

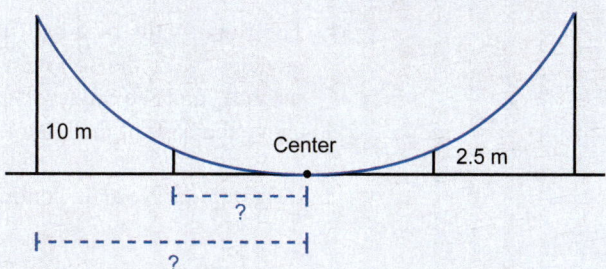

27. After a golfer hits a ball that is lying on the ground, it flies along the trajectory given by the quadratic equation $y = -\dfrac{x^2}{4} + \dfrac{3x}{2} + \dfrac{7}{4}$, where y is the height of the ball above the ground and x is the horizontal coordinate of the ball, both in meters. Assume the land from the initial location of the ball to the hole is flat.

   a. Determine how far the hole must be from the golfer for the ball to land in it.

   b. What is the maximum height the ball will reach?

28. A suspension bridge has the shape of a parabola that can be described by the quadratic equation $y = \dfrac{1}{20}\left(x^2 - 4x - 21\right)$, where x and y are the horizontal and vertical coordinates of the points on the bridge, both in meters, and the bridge connects to its supports at the level of $y = 0$.

   a. Determine the distance between the bridge supports.

   b. What is the maximum "depth" of the bridge?

29. The cross section of a glass is a parabola described by the quadratic equation $y = x^2 + x - \dfrac{35}{4}$, where x and y are the horizontal and vertical coordinates of the points on the glass, both in centimeters, and the edge of the glass is on the level of $y = 0$.

   a. Determine the diameter of the glass.

   b. What is the maximum depth of the glass?

30. An arch in the form of a parabola is installed at the entrance to the city. It can be described by the quadratic equation $y = -\dfrac{x^2}{4} + \dfrac{7x}{2}$, where y is the height of the arch and x is the horizontal distance from one of the arch bases, both in meters.

    a. Determine the distance between the arch bases.

    b. What is the maximum height of the arch?

31. The flight paths of a ball thrown from a hill under various conditions are modeled with the following quadratic functions, where x is the time in seconds, and the value of the function is the height, in meters, of the ball above the base of the hill.

Height under Normal Conditions: $f(x) = -\dfrac{x^2}{40} - \dfrac{x}{10} + 8$

Height When It's Windy: $g(x) = -\dfrac{2x^2}{49} + 8$

Height When It's Rainy: $h(x) = -\dfrac{x^2}{20} - \dfrac{x}{15} + 8$

Each of the curves are depicted in the following figure.

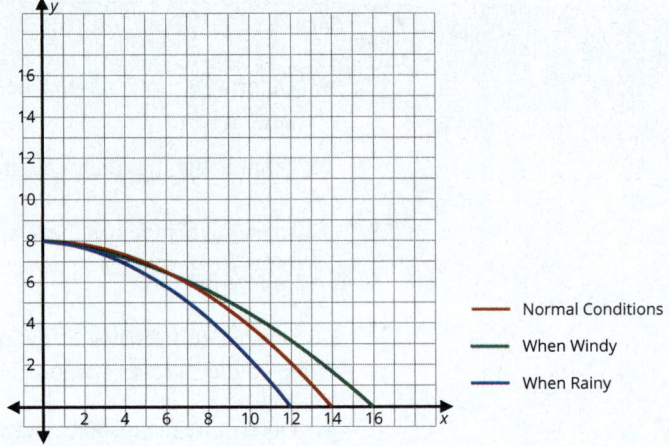

    a. Determine the height of the ball for each condition after 10 seconds.

    b. If a person who is 1.7 meters tall stands at the base of the hill, how far above the person will the ball be after 10 seconds?

    c. If a person can catch the ball when it is 1.7 meters above the ground, how long will the ball fly before being caughtunder normal conditions and when it's rainy.

    d. Use the graphs in the figure to compare the heights of the ball for each of the weather conditions.

32. A biologist studied the growth rate of geraniums under various conditions for 60 days. His results are modeled with the following quadratic functions, where x is the time in days, and the value of the function is the height of the plant, in centimeters, corresponding to that time.

No Specific Conditions: $f(x) = 0.006x^2 + 0.021x + 12$

With an Additional Light Source: $g(x) = 0.01x^2 + 0.013x + 12$

With Fertilizers: $h(x) = 0.009x^2 - 0.03x + 12$

Each of the curves are depicted in the following figure.

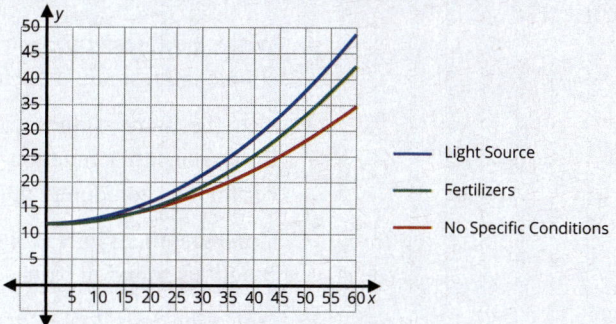

   a. Determine the height of the plant for each condition after 30 days.

   b. At the beginning of the experiment, all three plants had the same height, 12 centimeters. By how many times did the plants grow from their original size in 30 days?

   c. Find the time required for the plant to grow twice its original size when there is an additional light source and when there are fertilizers.

   d. Use the graphs in the figure to compare the heights of the plant for each growth condition.

33. You may have noticed that sometimes the water in ponds or aquariums can turn blue-green. This is due to the increased reproduction of cyanobacteria. The reproduction rate of cyanobacteria is described by an equation of the form $N(t) = N_0 e^{rt}$, where N_0 is the initial population size (given in millions of cells per liter), r is the growth rate, and t is the time in hours.

Among other factors, the growth rate is dependent on the concentration of phosphates in the water. This dependency is modeled by the linear function $r = 2p$, where p is the concentration of phosphates given in milligrams per liter.

   a. Write the exponential function of the population size $N(t)$ if the concentration of phosphates is normal; that is, equals 1 mg/L.

   b. Write the exponential function of the population size $N(t)$ if the concentration of phosphates is twice the norm and the initial population size is 3.3×10^6 cells/L.

   c. Use the function from part b. to calculate the population size after 2 hours.

   d. Use the function from part a. to calculate the time required for the population to double.

34. When you put your money into a savings account, its future value depends, among other things, on the way the interest is calculated. Let's consider two options.

    Simple Interest: The future value of an account with a simple interest is calculated by the formula $A = P(1+rt)$, where P is the initial deposit, r is the annual percentage rate (APR) written as a decimal, and t is the length of the deposit in years.

    Continuous Compound Interest: The future value of an account with a continuous compound interest is calculated by $A = Pe^{rt}$, with the variables having the same meanings.

    a. Write the linear function for the future value of an account using the simple interest rate if the initial deposit is $10,000 and the APR is 2.43%.

    b. Use the same values for the initial deposit and the APR to write the exponential function for the future value of an account using the continuous compound interest.

    c. Compare the future values of the accounts after 5 years using the functions found in parts a. and b.

    d. How long does it take for the accounts to earn the first $1000 of interest? Use the functions obtained in parts a. and b. to determine how long it takes the accounts to earn the first $1000 of interest.

35. Consider a diver jumping from a platform into the pool. His height above the water can be modeled by the quadratic equation $h(t) = -5t^2 + v_0t + y_0$, where t is the number of seconds after the jump, v_0 is the initial vertical velocity of the diver, and y_0 is the height of the platform in meters. At the same time, the horizontal distance from the diver to the platform is described by the function $d(t) = v_1t$, where v_1 is the initial horizontal velocity.

    Assume that the height of the platform is 6 meters, the initial vertical velocity is 3.9 m/s, and the initial horizontal velocity is 2.32 m/s. Answer the following questions.

    a. Write the quadratic function for the height of the diver above the water after t seconds.

    b. Write the linear function for the horizontal distance the diver will travel after t seconds.

    c. How high above the water will the diver be after 0.8 seconds?

    d. How long will it take for the diver to hit the water?

    e. What is the maximum height above the water the diver can reach?

    f. At what time t will the height in part e. be reached?

36. Suppose the revenue of a company is described by the linear function $R(p,Q) = pQ$, where p is the price of an item produced and Q is the number of items sold. The quantity Q is linearly dependent on the price and can be modeled by $Q(p) = -2000p + 134{,}000$. Suppose 15,000 items were sold last month. Answer the following questions.

   a. Write the linear function for last month's revenue using the given number of items sold.

   b. Write the quadratic function for the revenue, assuming that the quantity is a function of price.

   c. What is the revenue if the price of the item is $23? Use the function obtained in part b.

   d. What is the maximum revenue the company can achieve based on the function obtained in part b?

   e. What price corresponds to the maximum revenue found in part d?

5.6 PROJECT

THE WEIGHTLESSNESS OF PARABOLIC ARCS

A reduced-gravity aircraft is an aircraft that can simulate weightlessness of its passengers and contents by following a parabolic flight path. While zero gravity (zero-g) is not perfectly attained, the simulation is close enough to zero-g to train astronauts and film movie scenes. This type of aircraft has lovingly been given the nickname "vomit comet" due to two-thirds of all passengers experiencing airsickness during the 40 to 60 parabolic maneuvers of the flight.

According to NASA, the function $f(t) = -4.9t^2 + 87.21t + 9144$ can be used to describe the altitude in meters of a certain reduced-gravity aircraft t seconds after the start of the parabolic maneuver. Reduced gravity occurs during the entire parabolic arc of the maneuver.

1. Determine the altitude when reduced gravity starts and ends.

2. How long does the reduced-gravity period last? Round your answer to the nearest tenth of a second.

3. What is the maximum height attained by the aircraft during the parabolic maneuver? At what time into the parabolic maneuver is this height attained? Round your answer to the nearest tenth.

Suppose an 80-second movie scene takes place in zero-g. The production crew needs to plan the film sequence to minimize the cost of renting a reduced-gravity aircraft.

4. The 80-second scene would need to be split up across multiple periods of weightlessness and then stitched together in editing. What is the minimum number of parabolic arcs the movie crew would need to film the entire scene once?

5. If it takes the aircraft approximately 5 minutes from the end of one parabolic arc to set up to start another parabolic arc, how long would it take to film the 80-second scene one time?

The production crew learns that another company with a reduced-gravity aircraft can follow a parabolic arc of $f(t) = -4.9t^2 + 98.2t + 8930$ to increase the time spent in weightlessness. The cost is 15% more than the initial company.

6. Determine the length of each reduced-gravity period in seconds with the second company. Round your answer to the nearest tenth.

7. Assuming the aircraft needs the same 5 minutes from the end of a parabolic arc before starting another, determine the total time it would take to film the 80-second scene. (**Hint:** Determine the number of arcs needed to film the entire scene.)

8. Discuss the pros and cons of choosing each reduced-gravity aircraft to film the 80-second scene.

5.7 EXPONENTIAL AND LOGARITHMIC FUNCTIONS

OBJECTIVES

1. Model with exponential functions.

2. Model the frequency of music notes with exponential functions.

3. Compare linear growth and exponential growth.

4. Model with logarithmic functions.

Just as we have seen that there are many real-life phenomena that are best described by quadratic rather than linear functions, there are still others that are not described by polynomials at all. Scenarios that describe how populations vary over time (whether they are city, state, or national populations), the growth of bacteria, or the spread of a virus or a meme are best described using **exponential functions**. On the other hand, scenarios that describe the dispersion of energy phenomenon such as the strength of earthquakes, how loud music is at a distance, or even the intensity of the light from different stars are modeled by **logarithmic functions**. This section looks at both of these types of functions and how they differ from linear and quadratic growth.

OBJECTIVE 1 Exponential Functions

Exponential functions model scenarios where growth is quick and proportional to the previous quantity. For instance, when we earn money from investments, or pay interest on loans and mortgage payments, the calculations are done with exponential functions. While Chapter 6 explores these financial applications in more detail, in this section we'll consider other situations where growth is rapid because it builds on itself. Regardless of the application, what separates exponential functions from polynomials is that the exponent is a variable.

EXPONENTIAL FUNCTION

An **exponential function** is a function of the form $f(x) = ab^x$, where $b > 0$, $b \neq 1$, a is the initial value, and x is a real number.

Remember that the graphs of linear functions are lines, while quadratic functions follow a U-shaped path. Exponential functions, as we will see, start with a slow growth and then accelerate quickly. In fact, they increase much more quickly than any polynomial can.

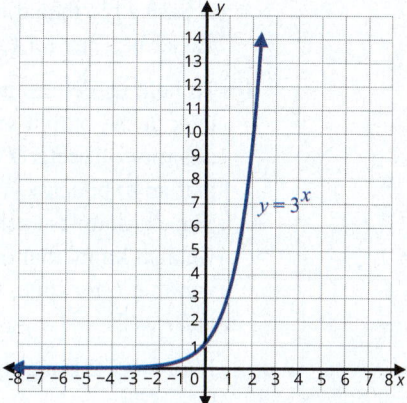

$y = 3^x$

Figure 5.7.1: Graph of an Exponential Function

The following example is adapted from a famous problem developed by Professor Bartlett to illustrate the growth power of exponential functions.

Example 5.7.1

Using an Exponential Function

A certain type of bacteria reproduces in such a way that every minute, the number of bacteria doubles. At 8:00 a.m., or time $t = 0$, a single bacterium is placed in a jar. At 8:01 a.m., there are two bacteria in the jar, and at 8:02 a.m., there are four bacteria. Assume the bacteria continues to reproduce in this manner, with no loss in the number of bacteria, and that the jar is full of bacteria at 8:30 a.m. Table 5.7.1 displays the pattern of growth in the bacteria. Note that time $t = 0$ represents the starting time of 8:00 a.m.

Table 5.7.1: Bacteria Growth

Time (minutes)	Number of Bacteria
0	1
1	2
2	4
3	8
4	16
5	32
⋮	⋮

a. At what time will the jar be half full of bacteria?

b. Using what you already know, develop a function to model this data.

Solution

a. At first glance, you might conclude that the jar will be half full halfway through the time period; that is, at 8:15 a.m. This would be incorrect because this situation cannot be represented by a linear function—the increase between each y-value is not consistent. Since the number of bacteria doubles every minute, the jar will be half full one minute prior to the end time. In other words, at 8:29 a.m. the jar is half full, and when the number of bacteria doubles at 8:30 a.m., the jar is full. We can work backward and know that at 8:28 a.m., the jar is a quarter full.

b. To help us determine what type of equation will model the data, we need to look at the output values in the table (1, 2, 4, 8, 16, 32, …). Notice that the quantity of bacteria doubles during each unit of time increase. This indicates that an exponential function of the form $f(x) = ab^x$ will model this behavior. Since it doubles each time, the base b of the function is 2. From the definition, we also know that a will be the initial value. So in this case $a = 1$. Using the variable t, for time, the exponential function that models this bacteria growth is $f(t) = 2^t$.

Figure 5.7.2 is a graph of the bacteria function.

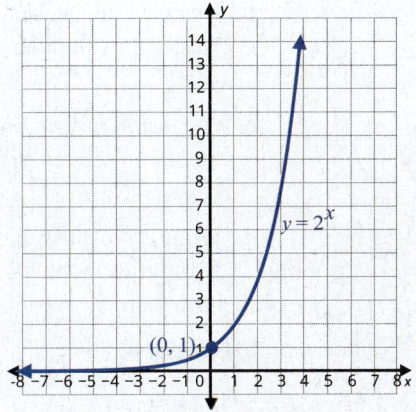

Figure 5.7.2: Bacteria Growth

✔ **Skill Check 5.7.1**

Use Example 5.7.1 to answer the following questions

a. At what time will the jar of bacteria be $\frac{1}{8}$ full?

b. How full is the jar at 8:26 a.m.?

Table 5.7.2: 5% Population Growth

Time (Years)	Population
0	100
1	105
2	110.25
3	115.7625
4	121.550625

Population growth is another common illustration of exponential growth. Consider the population growth of the American bison after wildlife conservation regulations were put in place to keep the bison from becoming extinct. Although herds started small, with protections and resources in place, their numbers could grow quickly. For instance, suppose a herd initially consisted of 100 bison and grew at a rate of 5% each year. After the first year, the new population would be the initial population of 100 plus an additional 5% of 100, or 5 new bison. This means the new population would be 105 after the first year. Table 5.7.2 shows the growth of this population over four years.

We can use the growth rate of a population to create a function that models the growth. When the rate of change is a fractional amount, as in this population example, we can determine b in the general exponential function $f(x) = ab^x$ by adding the rate of change to 1. Thus, b becomes $b = (1 + r)$, where r is the rate of change as a decimal. So the expanded version of the exponential function becomes $f(x) = a(1 + r)^x$. Note that if the rate of change is decreasing, then r is negative.

📖 **EXPANDED FORM OF AN EXPONENTIAL FUNCTION**

The **expanded form of an exponential function** is a function of the form $f(x) = a(1 + r)^x$, where a is the initial value, r is the rate of change, and x is any real number.

In our bison example, the rate of change for the population growth is 5%, so $b = 1 + 0.05 = 1.05$. Thus, the bison population growth can be modeled by the function $f(x) = 100(1.05)^x$, where x is the number of years since the initial population was established.

💬 **Helpful Hint**

When using a function to estimate the amount of living things (such as bacteria, animals, and people), if the function results in a fractional number, you should round up to the next whole number.

Example 5.7.2

Predicting Population Growth

In 2020, the metro area population of Nashville, Tennessee, was 1,249,000. This was a 2.04% increase from 2019.

a. If the population is expected to continue to increase at the same rate per year, what will the population be in 2040?

b. Using your answer from part a., determine what the percentage change in Nashville's population would be after 20 years with a 2.04% increase each year.

c. A consulting firm predicts that with current trends, Nashville could grow at a rate of 3% per year. Calculate the difference that this larger growth rate would have on Nashville's population by 2040.

d. Using your answer from part c., determine what the percentage change in Nashville's population would be after 20 years with the predicted 3% increase each year.

Solution

a. We are told that the initial population for Nashville is 1,249,000; that is, $a = 1,249,000$. The percent change each year is predicted to stay the same at 2.04%; therefore, $r = 0.0204$. Since $2040 - 2020 = 20$, we are interested in the population after 20 years, so $x = 20$. Substituting these into the expanded exponential formula, we can find the population.

$$f(x) = a(1+r)^x$$
$$f(20) = 1,249,000(1+0.0204)^{20}$$
$$f(20) \approx 1,870,559.11$$

This means that the population would be approximately 1,870,560 people after 20 years assuming a growth rate of 2.04%.

b. Recall from an earlier section that percentage change is given by the formula

$$\text{Percentage Change} = \frac{\text{Absolute Change}}{\text{Original Amount}} \cdot 100\%. \text{ This gives us the following.}$$

Percentage Change with a 2.04% Yearly Growth
$$= \frac{1,870,560 - 1,249,000}{1,249,000} \cdot 100\% \approx 49.8\%$$

This means that after 20 years, Nashville's population will have grown by nearly 50%.

c. We will use the same process as we did in part a., but the growth rate is now 3% instead of 2.04%. The values for the initial population and the number of years remain the same; that is, $a = 1,249,000$ and $x = 20$.

$$f(x) = a(1+r)^x$$
$$f(20) = 1,249,000(1+0.03)^{20}$$
$$f(20) \approx 2,255,832.93$$

This means that Nashville's population would be about 2,255,833 people after 20 years with a growth rate of 3%.

d. The percentage change in the population with a 3% growth each year would be

Percentage Change with a 3% Yearly Growth
$$= \frac{2,255,833 - 1,249,000}{1,249,000} \cdot 100\% \approx 80.6\%.$$

This means that after 20 years of a 3% increase to the population each year,

Nashville will experience a population growth of almost 81%!

Notice that the predicted increase in growth rate from 2.04% to 3% is only a 0.96% increase each year, but over time would result in around 20% more people. This shows the long-term impact of exponential functions.

OBJECTIVE 2 Music and Exponential Functions

Let's turn our attention to performance art, specifically that of music. Musical sound can evoke a wide range of emotions, whether it is created vocally, with a stringed instrument such as a violin or harp, with drums, or any other instruments. The sounds that instruments make all travel in waves of energy much like the waves of an ocean.

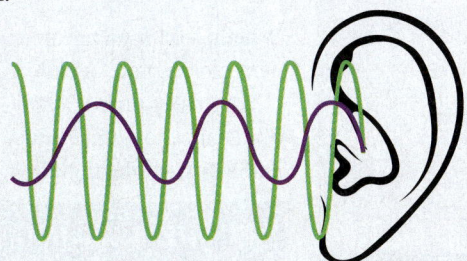

Figure 5.7.3

The **frequency** of a sound is the number of times a wave of sound energy completes a cycle in one second. We measure frequency in terms of cycles per seconds, or hertz (Hz). Frequency has an audible range of 16 Hz to 16,000 Hz. Sounds with a higher *pitch* have higher frequencies and sounds with a lower pitch have lower frequencies. **Pitch** is the tonal quality of sound that makes an instrument (or voice) sound high or low to our ear. Although frequency is objective and measurable, pitch is more subjective and might vary based on the individual listening. That is, two individuals might differ in the perceived pitch of a note, but the frequency will be exactly the same.

📖 FREQUENCY AND PITCH

Frequency

The **frequency** of a sound is the number of energy wave cycles completed in one second. Frequency is measured in hertz, Hz.

Pitch

Pitch is the tonal quality of sound that describes how low or high an instrument (or voice) sounds to our ears.

💬 Fun Fact

The strings on a guitar are tuned to increase in pitch from one side to the other. However, strings on a ukulele are not in order of pitch.

On a piano, each key (or note) has a standard frequency. When a key is pressed on the piano, the sound that plays is the result of a small felt-covered hammer striking a string. The strings of the piano are arranged in such a way that the notes progress from lower to higher pitches as you move up the keyboard from left to right. The notes on the left end of a piano keyboard have a very low pitch while the notes on the right end have a much higher pitch.

Figure 5.7.4 shows a portion of a piano keyboard that will give us a basis for how musical notes are represented. The white keys represent the notes A, B, C, D, E, F, and G. These notes are repeated over the entire keyboard.

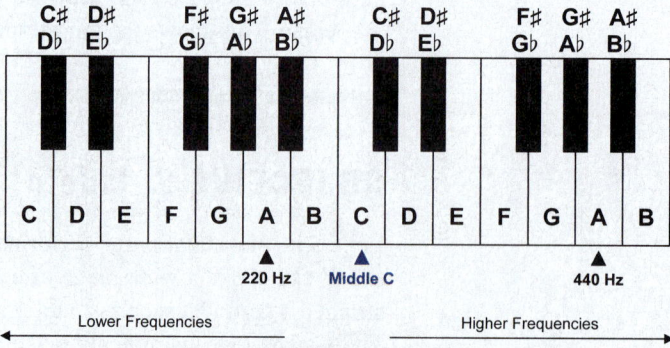

Figure 5.7.4

Helpful Hint

As Figure 5.7.4 indicates, a label is made for the note called middle C. This note gets its name from being the 4th C of eight Cs on the piano; that is, it is the C located in the middle of the piano.

Whether on a piano or on another instrument, the distance from one note to the next closest note is called a **half step**. Using Figure 5.7.4, if we start with middle C, a half step up is C♯ and a half step down is B. We use this smallest measurement of distance to define larger distances on the piano. A **whole step** is defined as two half steps, while an **octave** covers 12 half steps.

STEPS AND OCTAVES

Half Step

A **half step** is the distance between one note and the next nearest note on a piano.

Whole Step

A **whole step** is defined as the interval between two half steps on a piano.

Octave

An **octave** is the interval of notes between 12 half steps on a piano.

Skill Check 5.7.2

Starting at F in Figure 5.7.4, name the notes that are one half step up on a keyboard and one half step down on a keyboard.

Helpful Hint

If you don't have access to a piano, you can use the virtual piano on the website http://www.virtualpiano.net/.

Keys that are an octave apart have the same name, as you can see in Figure 5.7.4. The notes they produce sound alike, too. Try listening for yourself on a piano. Take a moment and listen to notes that are an octave apart. For example, play different Cs on the piano and notice how they sound similar. In an octave, the highest note always has twice the frequency of the lowest note. So as you play Cs up the piano, the frequency of the note is doubling each time. This is where the idea of pitch comes in. No matter where you play on the piano, an A is an A is an A. This is true for every instrument. Even a person with a very low voice can sing in pitch with a person who has a high voice by singing in different octaves, but the notes they sing always have frequencies that are powers of two apart.

Fixed frequencies are used when tuning musical instruments. For instance, on a piano, the note of A above middle C (we will use the notation A_4) has a frequency of 440 Hz, the note of A below middle C (referred to as A_3) has a frequency of $440 \div 2 = 220$ Hz, and A_5 (the note an octave above A_4) has a frequency of $440 \cdot 2 = 880$ Hz. In fact, when an orchestra begins to tune up before a concert, they traditionally tune based on a note played by the oboe, which is 440 Hz. When the orchestra is perfectly in tune, every instrument can match that frequency. Higher pitched instruments play an octave above and tune to an 880 Hz frequency, and lower pitched instruments play an octave below and tune to a 220 Hz frequency.

We know that 12 equal half steps on a piano make up a whole octave. Each step increases the frequency by the same factor so that all 12 together doubles the frequency. The transition between consecutive notes on a piano differs by a factor of the 12th root of 2, or approximately 1.059463. Thus, the frequencies between one note and another can be modeled by an exponential function.

📖 FREQUENCY OF MUSICAL NOTES

The **frequency of a music note** in relation to a specific note is modeled by the following function.

$$F(x) = F_0(1.059463)^x$$

Here, F_0 is the reference frequency and x is the number of half steps up from F_0.

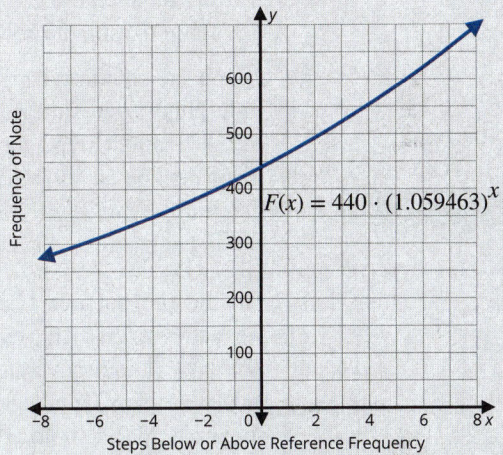

Figure 5.7.5: Frequency of Notes in Reference to A above Middle C

Example 5.7.3

Finding the Frequency of Musical Notes

The frequency of the note A_4 is 440 Hz.

a. Find the frequency of the note D, which is five half steps above A_4.

b. Find the frequency of the note F♯, which is three half steps below A_4.

c. A wine glass can be shattered by singing a note that is the resonant frequency of the glass at a very loud volume. In a demonstration, an opera singer is trying to shatter a glass that has a resonant frequency of 556 Hz. Use the graph of the function $F(x) = (440)1.059463^x$ to estimate how many half steps above A_4 the note is.

Solution

a. We will use the formula for the frequency notes on a piano. To determine the frequency of the note five half steps above A_4, we let x equal 5. A_4 is our reference frequency, so F_0 is 440 Hz. Substituting in the formula, we have the following.

$$F(x) = F_0 \cdot 1.059463^x$$
$$F(5) = (440)1.059463^5$$
$$F(5) \approx 587.56 \text{Hz}$$

Therefore, the note D above A_4 has a frequency of approximately 588 Hz.

b. We can use the same formula as we did in part a., where F_0 remains the same. This time the note we are interested in is 3 half steps below A_4, so x will equal −3.

$$F(x) = F_0 \cdot 1.059463^x$$
$$F(-3) = (440)1.059463^{-3}$$
$$F(-3) \approx 369.91 \text{Hz}$$

Therefore, the note F♯ below A_4 has a frequency of approximately 370 Hz.

c. Instead of attempting to solve the function for x, we will use a graphing calculator to estimate how many half steps above A_4 the glass-shattering note is.

Graph the equation $F(x) = (440)1.059463^x$ by pressing $\boxed{y=}$ and typing in the right-hand side of the equation. We know from part a. that the frequency of a note five half steps above A_4 is 587.56. This means that the y-value on our graph will be 587.56 when x is 5. This tells us that in order to see what we're graphing; we need the y-axis to go to a number larger than the default 10. Press $\boxed{\text{window}}$ and change **Ymax** to 1000. Press $\boxed{\text{graph}}$.

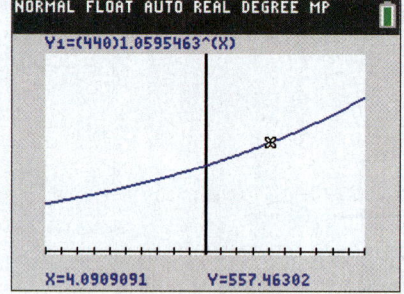

Figure 5.7.6

To find the number of half-steps above A_4 that a note must be to have a frequency of 556 Hz, we need to find what the value of x is when y is 556. Press $\boxed{\text{trace}}$ and move the cursor until the y-value at the bottom of the screen is close to 556. The x and y-values are shown at the bottom of the calculator screen. At y=557.46302, we see that x=4.0909091.

So for a note to be capable of shattering the wine glass, it would need to be approximately 4 half steps above A_4.

OBJECTIVE 3 Comparing Linear and Exponential Growth

We have mentioned several times that exponential growth is different from the linear growth that we have seen previously. The next example will illustrate how different those growth patterns are.

Example 5.7.4

Comparing Linear and Exponential Models

Helpful Hint

Net income is the amount of revenue a business has left after subtracting all expenses, taxes, and costs.

Let's compare two relatively new companies who began operations at the same time, Portrait Inc and SO Interiors. In their third year of business, Portrait Inc had a net income of $300,000 while SO Interiors had a net income of $190,000. Both companies built a business plan for future growth beyond the third year. Portrait Inc projected an increase in net income of $25,000 per year, while SO Interiors' plan aimed to increase its net income by 15% each year.

a. Create a function model for the projected net income over time for both companies.

b. How much is each company predicted to be making in 10 years?

c. Approximately how many years would it take for the two companies have the same net income? Based on their business plans, will the two companies ever have equal net incomes again?

d. If both companies were able to meet their net income growth goals, which company would you choose to invest in? Why?

e. In 2020, Apple was the world's most profitable company reporting a net income of approximately $59 billion, overtaking Saudi Aramco for the title. How many years would it take SO Interiors to reach the same level of income as Apple did in 2020?

Solution

a. To create a function for Portrait Inc's net income, let's begin by noticing what type of growth Portrait Inc is projecting. Since they project a constant increase of $25,000 each year, we know that a linear function will model this type of growth, and that the function's slope is equal to 25,000. We'll let x represent the number of years of growth and $f(x)$ represent the net income. The function is then the starting amount of $300,000 plus the increase $25,000 times the number of years of growth.

$$f(x) = 300{,}000 + 25{,}000x$$

For SO Interiors, notice that they project a percentage increase each year. Thus, this growth is exponential in nature, and we can use the exponential formula. We are told that the initial net income is $190,000; that is, $a = 190{,}000$. The percent change each year is predicted to be 15%; therefore, $r = 0.15$. Substituting these into the exponential formula we have the following.

$$g(x) = a(1+r)^x$$
$$g(x) = 190{,}000(1+0.15)^x$$

b. We can use the functions from part a. to determine the amount of net income each company is predicting after 10 years.

Portrait Inc

$f(x) = 300{,}000 + 25{,}000x$

$f(10) = 300{,}000 + 25{,}000(10)$

$f(10) = 550{,}000$

SO Interiors

$g(x) = 190{,}000(1+0.15)^x$

$g(10) = 190{,}000(1+0.15)^{10}$

$g(10) \approx 768{,}655.97$

Thus, Portrait Inc is predicted to have a net income of $550,000 after 10 years, while SO Interiors predicts their net income to be approximately $768,655.97.

c. In order to determine when the two companies have the same net income, we can graph the functions on the same graph and determine where they intersect. We will use a TI-83/84 Plus calculator to graph the functions.

Press $\boxed{\text{y=}}$ and clear any equations currently in the calculator. Input the equation $300{,}000 + 25{,}000x$ for Y1 and $190{,}000(1 + 0.15)^x$ for Y2. In order to see the point of intersection between these two functions, we'll need to adjust our window. We know from part b. that $f(10) = 550{,}000$ and $g(10) = 768{,}655.97$. We'll start with a window that will show those values by setting Xmin to -2

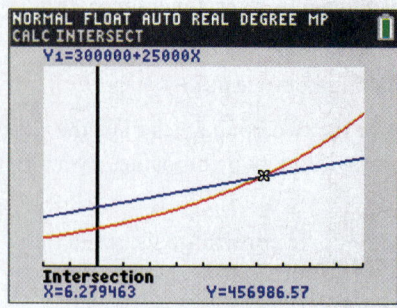

Figure 5.7.7

and Xmax to 10 and setting Ymin to −2 and Ymax to 1,000,000. Press [graph] and then press [2nd][trace] and select intersect. Move the cursor close to where the two functions appear to intersect and press [enter] three times.

At the bottom of the screen, the calculator gives the point of intersection as x=6.279463, y=456,986.57.

Thus, the two companies are projected to have the same net income at approximately 6.3 years.

Since Portrait Inc's projected net income is increasing linearly, and SO Interiors project net income is increasing exponentially, the net incomes will never be equal again. SO Interiors will continue to grow at a faster rate and Portrait Inc will not catch up if it only has linear growth.

d. If you are able to invest money for more than 6.3 years, then SO Interiors is the better long-term investment because their income will continue to grow at a faster rate. However, you might choose Portrait Inc if you prefer a quick return of your investment.

e. We are told the future value we want to reach is 59,000,000,000, so $g(x) = 59{,}000{,}000{,}000$. We are asked to find the time x if the rate stays at 15%. Substituting this into in our function for SO Interiors gives the following.

$$59{,}000{,}000{,}000 = 190{,}000(1 + 0.15)^x$$

Because the variable x is in the exponent, we need to use logarithms to solve the equation.

$$59{,}000{,}000{,}000 = 190{,}000\left(1 + 0.15\right)^x$$

$$\frac{59{,}000{,}000{,}000}{190{,}000} = \frac{\cancel{190{,}000}\,(1.15)^x}{\cancel{190{,}000}}$$ Take the log of both sides.

$$310{,}526.315789 \approx 1.15^x$$ Recall that $\log 1.15^x = x\log 1.15$.

$$\log 310{,}526.315789 = \log 1.15^x$$

$$\frac{\log 310{,}526.315789}{\log 1.15} = \frac{x\,\cancel{\log 1.15}}{\cancel{\log 1.15}}$$

$$90.48 \approx x$$

 Helpful Hint

When a variable is in the exponent of an equation, we can use common logs to solve.

So if they were able to keep increasing their net income at 15% each year, it would take SO interiors approximately $90\frac{1}{2}$ years to reach 59 billion in net income.

OBJECTIVE 4 Logarithmic Functions

 Fun Fact

The word *logarithm* implies exponent or power. It originates from the Greek words *logos* (reckoning) and *arithmos* (number).

As we just saw in the previous example, logarithms can be used to help us solve equations where the variable is in the exponent. They help us answer questions that exponential functions raise. This is because logarithms are inverse operations for exponents. In a sense, logarithmic functions "undo" the work done by the exponential function. Figure 5.7.8 shows the graphs of both the exponential function $f(x) = 2^x$ and its corresponding logarithm $g(x) = \log_2 x$. Notice how the growths compare to one another. Where the exponential function $f(x)$ starts slowly growing and then increases very rapidly, the logarithmic function $g(x)$ has an initial period of rapid growth, but then the growth begins to slow and continues to slow down as the value of the function continues to increase without bound. The two functions are mirror images of each other.

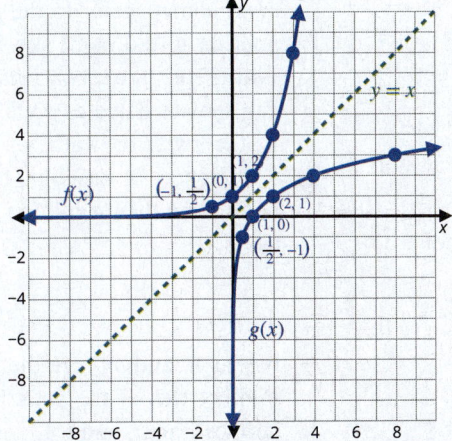

Figure 5.7.8: Exponential Function $f(x)$ vs. Logarithmic Function $g(x)$

📖 LOGARITHM

If $b^x = a$, then the **logarithm** with base b of a is x. Symbolically, we can express this exponential as an equivalent logarithm

$$\log_b a = x,$$

where $b > 0$, $b \neq 1$, and $a > 0$.

Notice that the logarithmic function is an example of a function that isn't evaluated as an algebraic expression like we have seen so far. To evaluate the function, we have to use a type of procedure. For instance, suppose we have two functions $f(x) = 3x + 1$ and $g(x) = \log_3 x + 1$. Evaluating the first linear function for $x = 9$ is easy. We multiply by 3 and add one. Thus, $f(9) = 3(9) + 1 = 28$. However, substituting in the second logarithmic function gives us $g(9) = \log_3 9 + 1$. We must first determine what the "log of 9 to base 3" is before we can add 1. The definition of a logarithm tells us the procedure for finding this. We know that $\log_3 9$ is the number you raise 3 to in order to get 9; that is, $3^x = 9$. We know that you square 3 to get nine, so $\log_3 9 = 2$. Now we can evaluate the function: $g(9) = \log_3 9 + 1 = 2 + 1 = 3$.

Carrying out this procedure can sometimes be straightforward, but most often it is much more involved. For instance, calculating $g(10)$ would involve establishing $3^x = 10$. Conveniently, today's technology can calculate logarithms with the push of a button; but you have to push the right button. When a logarithm has a base of 10 (that is, $b = 10$), we refer to this as the **common logarithm**, which is usually expressed without writing the 10. In other words, we simply write $\log a = x$ instead of $\log_{10} a = x$. This means that the log button on a calculator calculates logs with base 10. The natural logarithm, represented by ln, calculates logs with the base e, which is approximately equal to 2.718281828.

Example 5.7.5

Evaluating Logarithms

Astronomers refer to the magnitude of a star or planet by how bright an object appears in the night sky. The brightest stars are called *first magnitude*, slightly fainter stars are *second magnitude*, and so on until the faintest stars the naked eye can see are listed as *sixth magnitude*. This system means that the brighter a star is, the lower the value of its magnitude. Apparent magnitude uses one object as a reference, or baseline, and measures other objects against it. The formula for comparing the magnitude of two objects, m_1 and m_{ref}, is given by

$$m_1 - m_{ref} = -2.5 \log_{10}\left(\frac{l_1}{l_{ref}}\right),$$

where m_1 and m_{ref} are the magnitudes for a star m_1 and a reference star m_{ref}. The luminosities of the stars, which is the amounts of energy they emit from their surfaces, are l_1 and l_{ref}.

The luminosity of the sun is approximately 400,000 times more than that of the full moon. That means $\left(\frac{l_{sun}}{l_{moon}}\right) = 400,000$. Find the difference between the apparent magnitudes of the sun and the full moon, $m_{sun} - m_{moon}$.

Solution

To find the difference between the apparent magnitudes of the sun and moon, we substitute the given value of $\left(\frac{l_{sun}}{l_{moon}}\right)$ into the formula.

$$m_{sun} - m_{moon} = -2.5 \log_{10}\left(\frac{l_{sun}}{l_{moon}}\right)$$

$$m_{sun} - m_{moon} = -2.5 \log_{10}\left(400,000\right)$$

$$m_{sun} - m_{moon} \approx -14.01$$

Thus, the difference in apparent magnitude between the sun and the moon is about −14. That is, the apparent magnitude of the sun is 14 less than that of the moon.

🗠 TECH TIP

The log button on a calculator is a logarithm base 10, so to calculate $-2.5 \log_{10}(400,000)$, type -2.5, then press [log], and then type 400000 and), and press [enter].

✔ Skill Check 5.7.3

Find $f(23)$, if $f(x) = \dfrac{\log x}{\log 4}$.

As we saw in Example 5.7.5 with the stars, logarithms help us model patterns that deal with the intensity of natural phenomena. This is also the case with earthquakes. The intensity of earthquakes can vary widely, and the scale of their intensity is sometimes hard to wrap our minds around. Take for example an earthquake in January 2020 with a magnitude of 7.7 on the Richter scale. It struck in the Caribbean between Jamaica, the Cayman Islands, and Cuba and prompted tsunami warnings and evacuations as far away as Florida. However, earthquakes that measure in the 2 or 3 range happen on a daily basis across the United States and no such warnings are issued. What prompts a warning to evacuate in the first case, but not in the latter?

Figure 5.7.9

A logarithmic scale called the Richter scale allows us to compare the earthquake numbers in a meaningful way. Since it is a scale based on a logarithm in base 10, we can say that for each increase in one unit on the scale, there is an increase of 10 times in the intensity of the earthquake. So a jump from a level 3 earthquake to a level 7 earthquake is somewhere in the range of $10 \cdot 10 \cdot 10 \cdot 10 = 10{,}000$ times the intensity!

The formula for the magnitude of the earthquake on the Richter scale R can be simplified to the following equation.

Magnitude on the Richter Scale

$$R = \log\left(\frac{I}{I_0}\right)$$

In the equation, I_0 is the intensity of an earthquake that is barely felt (a zero-level earthquake) and has the value $I_0 = 1$. I is the intensity of the earthquake measured relative to the smallest seismic activity that can be measured.

Example 5.7.6

Using Logarithmic Functions

Every increase of 1 in the Richter scale means the magnitude of the earthquake is 10 times more powerful. Use the function for the Richter scale to answer the following questions.

a. What is the magnitude of an earthquake that is 1000 times stronger than a zero-level earthquake?

b. If $I_0 = 1$, what was the intensity of level I of the Caribbean earthquake in 2020 that measured 7.7?

c. The strongest earthquake ever recorded hit Chile in 1960 and was estimated to have been about 9.5 on the Richter scale. Find the intensity I of this earthquake if $I_0 = 1$.

Solution

a. Since $R = \log\left(\dfrac{I}{I_0}\right)$, an intensity of 1000 times that of a zero-level earthquake would mean the following.

$$R = \log\left(\frac{I}{I_0}\right) = \log\left(\frac{1000}{1}\right) = \log 1000 = 3$$

Thus, an earthquake that is 1000 times as intense as a zero-level earthquake has a Richter scale value of $R = 3$.

b. We are told that $I_0 = 1$ and the Caribbean earthquake in 2020 measured 7.7; that is, $R = 7.7$. We can substitute these values into the function and solve to find out the intensity level.

$$R = \log\left(\frac{I}{I_0}\right)$$

$$7.7 = \log\left(\frac{I}{1}\right)$$

$$7.7 = \log I$$

Just as logarithms helped us solve equations involving exponents, exponents can help us solve logarithmic equations. We can use the definition of a logarithm to determine what I is now. Recall that $b^x = a$ when $\log_b a = x$. From this, we know that $x = 7.7$, $a = I$, and $b = 10$. Substituting these into the equation, we have the following.

$$10^{7.7} = I$$
$$50,118,723.36 \approx I$$

So the intensity level is approximately $I \approx 50,118,723$ which means the Caribbean earthquake was over 50 million times stronger than a zero-level earthquake.

c. We can substitute the values for the scale number, $R = 9.5$ and $I_0 = 1$, into the function to obtain the following.

$$R = \log\left(\frac{I}{I_0}\right)$$

$$9.5 = \log\left(\frac{I}{1}\right)$$

$$9.5 = \log I$$

Once again, we know that $x = 9.5$, $a = I$, and $b = 10$.

$$10^{9.5} = I$$
$$3,162,277,660.17 \approx I$$

This means that the Chilean earthquake was over 3.16 billion times stronger than a zero-level earthquake.

✔ **Skill Check 5.7.4**

Determine how much stronger the Chilean earthquake was than the Caribbean earthquake.

Summary

Table 5.7.3 gives a summary of the four different functions we've covered in the chapter and ways to identify them along with the shapes of their graphs.

Table 5.7.3: Function Characteristics and Graph Shapes

Function Type	Characteristics	Shape of Graph
Linear	Largest exponent is 1. Change is constant.	
Quadratic	Largest exponent is 2. Curve is U-shaped.	
Exponential	The exponent is a variable. Growth becomes more rapid.	
Logarithmic	Involves logarithms. Growth becomes slower and slower.	

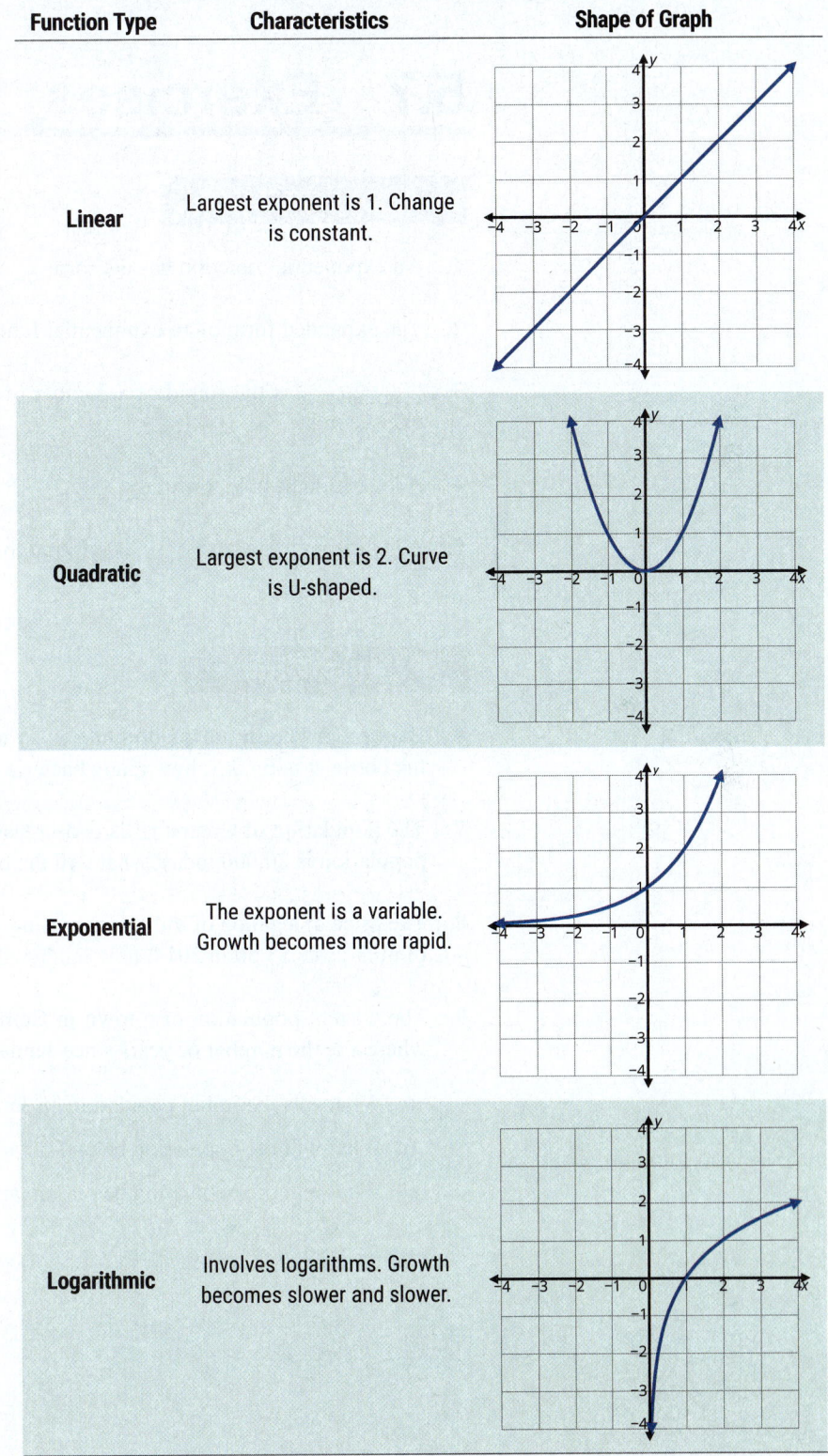

Skill Check Answers

1. a. 8:27 a.m. **b.** $\dfrac{1}{16}$ full   **2.** One half step up is F♯ and one half step down is E.

3. $f(23) \approx 2.262$

4. About 63 times stronger.

5.7   Exercises

✔ CONCEPT CHECK

1. An exponential function has the form _____.

2. The expanded form of an exponential function has the form _____.

3. _____ is the sound of a number of energy wave cycles completed in one second.

4. A logarithm has the form _____.

5. A logarithm has a base of __ is referred to as a common logarithm.

🚀 APPLICATIONS

6. Bacteria in a bottle are quadrupling every minute. If the number of bacteria in the bottle at noon is 1, how many bacteria are in the bottle at 12:10 p.m.?

7. The population of Greene Hills is decreasing at a rate of 2% per year. If the population is 20,000 today, what will the population be in 10 years?

8. The price of a gallon of milk is increasing at a rate of 1% per year. If a gallon of milk costs $3.50 in 2014, how much will a gallon of milk cost in 2017?

9. The current population of a town in California is modeled by the function where x is the number of years since January 1, 2022 .

   a. What was the initial population on January 1, 2022?

   b. What will the population be on January 1, 2025?

   c. What will the population be on January 1, 2030?

10. A checkerboard has 64 squares that alternate black and white. A grain of rice weighs 0.03 grams. Assume 1 grain of rice is placed on the first square, 2 grains are placed on the second square, 4 grains are placed on the third square, and so on.

    a. How many grains will be placed on the 14th square?

    b. How many grains will be placed on the 25th square?

    c. How much does the rice on the 20th square weigh?

    d. How much does the rice on the 30th square weigh?

Use the given information and logarithmic equation to answer each question.

pH = –log[H+] where [H+] is the hydrogen ion concentration, measured in moles per liter. Solutions with a pH-value of less than 7 are acidic; solutions with a pH-value of greater than 7 are basic; solutions with a pH-value of 7 (such as pure water) are neutral.

11. Suppose that you test apple juice and find that the hydrogen ion concentration is [H+] = 0.0003. Find the pH value and determine whether the juice is basic or acidic. Round your answer to the nearest hundredth.

12. You test some ammonia and determine the hydrogen ion concentration to be [H+] = 1.3×10^{-9}. Find the pH value and determine whether the ammonia is basic or acidic. Round your answer to the nearest hundredth.

13. An orange has a pH of about 3. Find the concentration of hydrogen ions in an orange.

14. An antacid tablet has a pH of 11. Find the concentration of hydrogen ions in the tablet.

15. Acid rain has a pH of 4 and normal rain has a pH of 6. Compare the pH of each and determine how much more acidic acid rain is than normal rain.

Solve each problem. Round your answer to the nearest thousandth, if necessary. Assume middle C has a frequency of 261 Hz.

16. If a note has a frequency of 27.5 Hz, find the frequencies of the notes one, two, three, four, and five octaves higher.

17. If a note has a frequency of 220 Hz, find the frequencies of the notes one, two, and three octaves higher.

18. If a note has a frequency of 4176 Hz, how many octaves higher is the note than middle C?

19. If a note has a frequency of 1044 Hz, how many octaves higher is the note than middle C?

20. What is the frequency of the note that is nine half steps below middle C?

21. What is the frequency of the note that is nine half steps above middle C?

22. Two podcasts, one on psychology and one on medicine, were both launched about a year ago. By now, the psychological podcast has 23,000 subscribers, and the medical one has 34,000 subscribers. Both podcasts estimate their future growth. The psychology podcast projects an increase of 45% in the number of subscribers each year, while the medical podcast plans to increase its coverage by 28,000 subscribers per year.

    a. Create a function model for the projected growth over time for both podcasts.

    b. How many subscribers is each podcast predicted to have in 5 years?

    c. Approximately how many years would it take for the two podcasts to have the same number of subscribers?

    d. How many years would it take the psychology podcast to reach 1 million of subscribers?

23. Suppose you want to borrow $50 from either your brother or your sister for a week. They each offer you a different condition for borrowing money. Your brother asks for an additional $5 for each day that you are late paying back the money. Your sister says she will increase the total amount you owe her by 10% per day for each day you are late returning the money. Your brother and sister both want the full amount paid back all at once.

    a. Create function models for the money you need to return to both siblings based on the number of days you are late paying back the money.

    b. How much money do you need to return in both cases if are 3 days late in paying back the money?

    c. Assuming you are unable to pay the money back on time, after how many days would you owe equal amounts to your siblings?

    d. If you knew that you would be at least 2 days late paying back the money, who would you borrow money from?

Use the given information and logarithmic equation to answer each question. Round your answer to the nearest whole number, if necessary.

Sound intensity, given in decibels, is measured by the formula

$db = 10\log\left(\dfrac{I}{I_0}\right)$. I is the intensity of the sound in question relative to the

reference value I_0 which has the value $I_0 = 1$.

24. A dog bark is 400 times as intense as the reference value I_0. How many decibels is a dog bark? Round your answer to the nearest whole decibel.

25. A jet engine is about 155 decibels. What is the intensity level of the jet engine? Round your answer to the nearest whole number.

26. The threshold of hearing loss is about 125 decibels. If a gunshot has an intensity of 5.012×10^{12}, should you wear ear protection to prevent hearing loss?

Use the given information and logarithmic equation to answer each question. Round your answer to the nearest hundredth, if necessary.

The magnitude of an earthquake on the Richter scale is measured by

$R = \log\left(\dfrac{I}{I_0}\right)$

27. What is the magnitude of an earthquake that is 5×10^8 times as intense as a zero-level quake?

28. A dump truck has an intensity of about 975. If a seismograph will register a quake of anything larger than a magnitude of 3, will the dump truck cause a small earthquake on the seismograph?

✏ WRITING & THINKING

29. In a given 30-day month, would you rather get a penny on day 1, two pennies on day 2, and continue to double that amount each day from the previous day, or start with a quarter on day 1, 2 quarters on day 4, 4 quarters on day 7 and continue to double the amount from the previous amount every 3rd day? Explain your reasoning.

5.7 PROJECT

USING NEWTON'S LAW OF COOLING TO AVOID COFFEE BURNS

Newton's Law of Cooling gives the temperature of an object (such as a pan or a cup of liquid) as it cools over time to the temperature of its surrounding environment (such as the air in a room). The law can be described with the following function.

$$T(t) = T_e + (T_0 - T_e)e^{-kt}$$

In this function, $T(t)$ is the temperature of the object at time t, in minutes. The value T_e represents the temperature of the surrounding environment, T_0 represents the initial temperature of the object, and k is a constant of variation, which depends on the object.

When brewing coffee, the ideal water temperature is between 195 and 205 degrees Fahrenheit. Suppose some freshly brewed 205 °F coffee was poured into a ceramic mug with no lid and left in a 73 °F room. Five minutes after being poured, the coffee is 161.9 °F. The constant of variation can be calculated as $k = 0.079$.

1. Use the given information to create a function for the temperature of the coffee as it cools over time.

2. Use the function from part 1 to calculate the temperature of the coffee at the time intervals stated in the following table. Round the temperatures of the coffee to the nearest tenth of a degree.

Time (minutes)	Temperature of Coffee (°F)	Time (minutes)	Temperature of Coffee (°F)
0		5	
1		6	
2		7	
3		8	
4		9	

3. Does the calculated value match the recorded value after 5 minutes? Explain why it does or does not match.

4. What is the lowest temperature that the coffee can reach in these conditions? Explain your reasoning.

According to the University of Wisconsin–Madison, hot liquids can cause third-degree burns at the following exposure periods and temperatures: 5 seconds at 140 °F, 2 seconds at 149 °F, and 1 second at 156 °F.

5. After how many minutes is the coffee safe to drink? Explain your reasoning.

6. A study of 300 participants showed that the preferred drinking temperature of coffee is between 125 °F and 165 °F. According to the table in part 2, when does the temperature of the coffee enter this preferred interval?

7. The same study determined that the optimal drinking temperature is approximately 136 °F. When is the coffee at that optimal temperature?

8. If a lid were added to the mug, would that change the cooling rate of the liquid? Explain your reasoning.

Chapter 5 Exercises

1. Evaluate the function $f(x) = x^2 + 2x - 3$ for the given values.

x	$f(x) = x^2 + 2x - 3$
−3	
−2	
0	
1	
2	
3	

Determine the dependent and independent variables. Then write a function to represent each situation and answer each question.

2. A car purchased in 2010 had an initial value of $19,500. If the car depreciates at $1200 per year, write an equation that represents the depreciation. What is the value of the car after 5 years?

3. The cost to rent a car for a day is an initial cost of $35 and $0.25 per mile driven. Write an equation that represents to the cost of the car rental. What is the total cost if you drove 250 miles?

4. Jeff delivers pizza for Papa Jim's Pizza. Papa Jim's pays Jeff an hourly wage of $6.25. In addition to his hourly wage, Jeff earns $0.25 for every pizza he delivers. Write an equation that represents the wages earned by Jeff, where x represents the number of pizzas delivered when Jeff works a five-hour shift. If Jeff delivers 21 pizzas during his shift, how much money will he earn?

Graph each equation.

5. $y = -2x$

6. $y = 3x + 1$

7. $y = \dfrac{2}{3}x - 1$

Find the slope and y-intercept of each linear equation.

8. $y = -3x + 4$

9. $y = \dfrac{3}{4}x - 3$

10. A medical conference has two phases. The first phase consists of a general opening speech lasting 40 minutes and several presentations for 50 minutes each. The second phase consists of the participants' reports, each lasting 55 minutes, and a fifteen-minute conclusion. The conference organizer wants to know how many speakers should participate in the conference in order for the duration of the two phases to be the same.

Write two linear functions, one for the duration of the first phase, $C(x)$, and one for the duration of the second phase, $R(x)$. Use the functions to find out how many people need to participate in the conference in order for both stages to have the same duration.

11. Mark is a beginner illusionist. To start performing, he needs to invest $5400 on props. Also, renting a room for each performance will cost him $1400. He plans to sell tickets for $20, and the capacity of the room is 150 people. Suppose that, on average, the hall will be two-thirds full at each performance. Let x be the number of performances.

   a. Write the cost function for holding performances.

   b. Write the revenue function.

   c. Write the profit function for Mark.

   d. Determine the x-value of the break-even point for his business.

   e. If Mark holds 20 performances, what profit can he expect to make?

12. Graph the linear inequality $3x - 5y > 22$.

13. Graph the system of linear inequalities: $\begin{cases} 3x - 4y > 0 \\ \quad\ y \geq 2x - 5 \end{cases}$

14. A cereal manufacturer is creating a new cereal that is a mix of two previously established cereals: cinnamon squares and chocolate puffs. Cinnamon squares cost $0.15 per ounce to make and chocolate puffs cost $0.19 per ounce to make. Standard packaging costs $0.45 per box. The company wants to minimize the cost of creating this cereal. Let x represent the number of ounces of cinnamon squares and let y represent the number of ounces of chocolate puffs that are added to each box of cereal. Write the objective function that describes the cost to create each box of cereal.

15. The objective function $z = 2x + 5y$ is subject to the following constraints.

   $x + 2y \leq 40$

   $y \leq 15$

   $x \geq 0$

   $y \geq 0$

   Find the maximum value of z.

16. A basketball's height (in feet) is a function of time in flight (in seconds), modeled by an equation such as $h(t) = -16t^2 + 40t + 6$.

   a. Describe the graph of the function in terms of time in flight and height.

   b. Describe how you could use the given function relating time in flight to height to find when the shot will reach the height of the basket (10 feet)?

   c. Describe how you could find the time when the ball would hit the floor if it missed the basket entirely?

17. The figure contains the graph of a function, $f(x) = -0.00875x^2 + 0.775x + 13$, that measures the average fuel economy for a certain model of car based on the speed x of the car (in miles per hour).

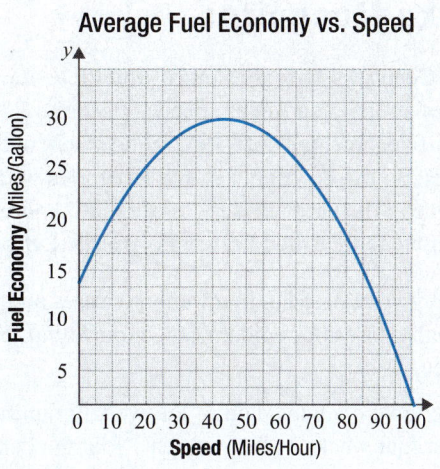

Average Fuel Economy vs. Speed

a. What is the miles per gallon average if the vehicle is traveling 80 miles per hour?

b. What is the miles per gallon average if the vehicle is traveling 40 miles per hour?

c. At approximately what speed will the car have the greatest fuel economy? Round your answer to the nearest whole mile per gallon.

18. The function $C(x) = 0.0086x^2 + 1.11x - 1.37$ represents the stopping distance in feet while talking on a cell phone and driving at a speed of x mph. What distance will it take you to stop while talking on a cell phone if you are driving 50 mph? 80 mph?

19. Whispertown has a population growth model of $P(t) = at^2 + bt + P_0$ where P_0 is the initial population. Suppose that the projected future population of Whispertown t years after January 1, 2012, is described by the quadratic model $P(t) = 0.8t^2 + 6t + 24{,}000$.

a. What is the projected population of Whispertown on January 1, 2030?

b. In what month and year will the population reach 60,000?

20. A rifle fires a projectile vertically upward from a height of 150 feet above the ground, with an initial velocity of 1250 ft/sec. Recall that projectiles are modeled by the function $h(t) = -16t^2 + v_0 t + y_0$.

a. Write a quadratic equation to model the height in feet above the ground after t seconds.

b. During what time interval will the projectile be more than 10,000 feet above the ground?

c. After how many seconds will the projectile hit the ground?

21. A ball is dropped from the top of a 750-foot tall building. How long does it take the ball to hit the ground? Recall that projectiles are modeled by the function $h(t) = -16t^2 + v_0 t + y_0$. (Note that $v_0 = 0$.)

CHAPTER 5 PROJECT

Those Pesky Mosquitos

According to the Center for Disease Control and Prevention, the Zika virus is a disease "spread to people primarily through the bite of an infected Aedes species mosquito." Although the virus was first discovered in 1947, the first human case wasn't documented until 1952. Since then, outbreaks have been reported in Africa, Southeast Asia, and the Pacific Islands. So far, 86 countries and territories worldwide have reported evidence of mosquito-transmitted Zika infection. One of the latest epidemics of the Zika virus occurred between 2013 and early 2014, on a cluster of islands in the South Pacific called French Polynesia. Let's have a look at how fast it spread.

1. Based on what you've heard or know about viruses and your knowledge about functions, what type of growth do you think usually describes epidemics—exponential, polynomial, logarithmic, or linear? Explain your reasons for your choice.

2. The following table contains data of the weekly number of suspected Zika cases in French Polynesia in 2013 during the first four weeks of the outbreak. Plot the data on a graph.

Week #	New Cases	Cumulative Cases
1	49	49
2	191	240
3	369	609
4	331	940
5	333	1273

3. Here are five different functional models that might represent the growth of the number of Zika cases, where x represents the week number and y represents the number of cumulative cases.

    1. Linear $y = 258.74x$

    2. Logarithmic $y = 937.37\ln(x) - 202.03$

    3. Quadratic $y = 31.357x^2 + 134.93x - 122$

    4. Power $y = 55.278x^{2.0101}$

    5. Exponential $y = 47.399e^{(0.6737x)}$

For each function model listed, create a graph for $1 \le x \le 5$, along with the Zika case data from part 2. Be sure your graphs clearly label the function and the actual data.

4. Which of the graphs in part 3 do you think best models the Zika data? Why?

5. On the graphs from part 3, extend the graphs by plotting the functions for $6 \le x \le 10$.

6. Which of these functions do you think will best model the growth of the number of Zika cases over weeks 6 through 10? Is it the same function as you choose in part 4? If not, what caused you to change your decision?

7. The following table contains the actual data for the spread of the Zika virus during weeks 6 through 10 of the epidemic.

Week #	New Cases	Cumulative Cases
6	571	1844
7	742	2586
8	955	3541
9	1029	4570
10	883	5453

Plot the actual data for weeks 6 through 10 on each of the function graphs. Is the function you chose in Step 6 still the best model for growth over weeks 1 through 10? Why or why not?

8. The following table contains the data for weeks 11 through 20. Plot the Zika data for weeks 1 through 20 on each of the function graphs. Discuss when each function ceases to be a good model for the data and why that might be.

Week #	New Cases	Cumulative Cases
11	682	6135
12	512	6647
13	412	7059
14	381	7440
15	343	7783
16	256	8039
17	247	8286
18	142	8428
19	82	8510
20	17	8581

Ioos S, Mallet HP, Leparc Goffart I, Gauthier V, Cardoso T, Herida M. "Current Zika virus epidemiology and recent epidemics."

Med Mal Infect (July 2014): 44(7):302-7 doi: 10.1016/j. medmal.2014.04.008. Epub 2014 Jul 4.

9. The population of French Polynesia at the time of the outbreak was about 270,000. The population of the United States in 2016 is approximately 322,762,000. How could you modify your function to model a potential spread of the Zika virus over the United States?

10. Discuss if it is reasonable to use your modified Zika function model for French Polynesia as a model for the United States.

▤ CHAPTER 5 REVIEW

Section 5.1 Linear Equations and Functions

Definitions

Dependent Variable: The value of the **dependent variable** changes with respect to the value of the independent variable.

Independent Variable: The value of an **independent variable** does not rely on the values of the other variables in an expression or function, and its value determines the values of the other variables.

Possible Solutions to Linear Equations

   One solution—there exists only one number that makes the equation true.

   Infinitely many solutions—any number makes the equation true.

   No solution—no number exists that makes the equation true.

Function: A **function** is a set of ordered pairs in the form (x, y) in which each value of the dependent variable is associated to a unique independent variable.

Domain: The **domain** of a function is the set of all input values (independent variable) that result in a real number value for the output values (dependent variable).

Range: The **range** of a function is the set of all possible output values (dependent variable) that correspond to the domain values.

Procedure

Steps for Solving Linear Equations in One Variable

Step 1: Clear the equation of fractions and decimals by multiplying both sides of the equation by the least common denominator (LCD).

Step 2: Simplify the expressions on both sides by clearing the equation of parentheses. Repeat Step 1 if fractions or decimals remain; otherwise, move to Step 3.

Step 3: Simplify the expressions on both sides by combining like terms.

Step 4: Isolate the variable on one side of the equation.

Step 5: Make the variable coefficient equal to 1.

Section 5.2 Linear Modeling

Definitions

Parallel and Perpendicular Lines

Consider two lines $y = m_1 x + b_1$ and $y = m_2 x + b_2$.

If the two lines have the same slope (that is, $m_1 = m_2$), then they are **parallel lines**, which means they never intersect.

If the two lines have slopes that are negative reciprocals of each other (that is, $m_1 = -\dfrac{1}{m_2}$), then they are **perpendicular lines**, which means they intersect at right angles.

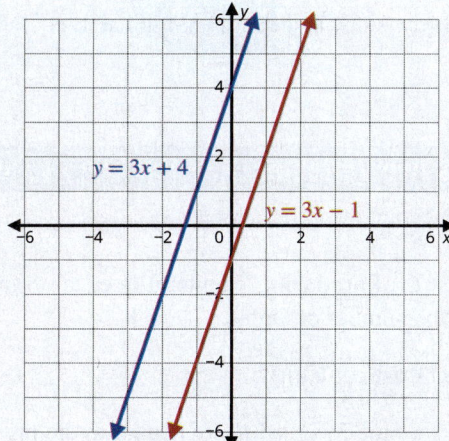

Figure 5.2.21: Parallel Lines

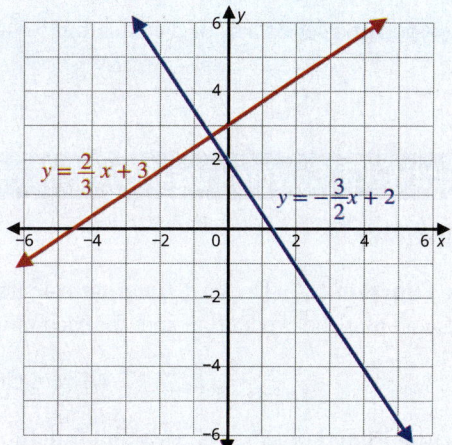

Figure 5.2.22: Perpendicular Lines

Horizontal and Vertical Lines

The graph of a **horizontal line** $y = b$ will pass through the point $(0, b)$.

The graph of a **vertical line** $x = a$ will pass through the point $(a, 0)$.

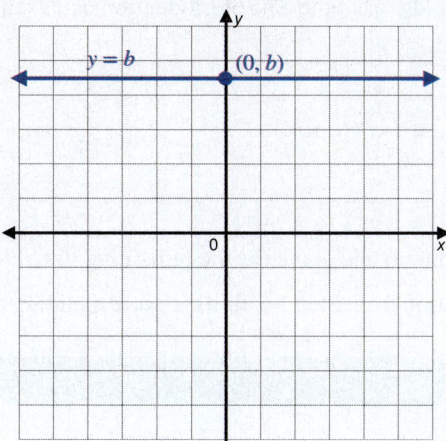

Figure 5.2.26: Horizontal Line

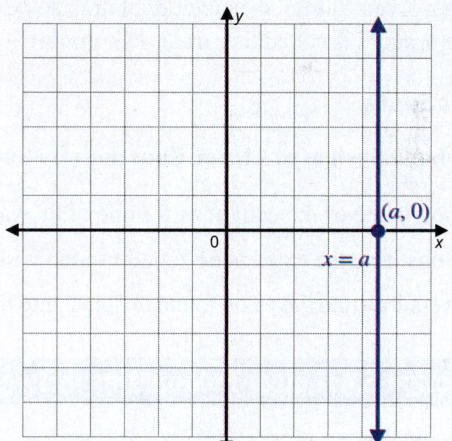

Figure 5.2.27: Vertical Line

Formulas

Slope Formula

The slope of a line passing through the points (x_1, y_1) and (x_2, y_2) is given by the following.

$$m = \frac{y_2 - y_1}{x_2 - x_1}$$

Formulas for the Equation of a Line

The **point-slope form** of the equation for a line, where m is the slope of the line and (x_1, y_1) is a point on the line, is

$$y - y_1 = m(x - x_1).$$

The **slope-intercept form** of the equation for a line, where m is the slope and b is the y-intercept, is

$$y = mx + b.$$

Section 5.3 Solving Linear Systems of Equations in Two Variables

Definitions

Cost Function (C): The **cost function** calculates the total cost C of producing x units; it is equal to the original fixed cost plus the production cost for each unit multiplied by the number of units x.

$$C = \text{Fixed Cost} + (\text{Production Cost per Unit})x$$

Revenue Function (R): The **revenue function** calculates the revenue received R; it is equal to the price per unit multiplied by the number of units sold x.

$$R = (\text{Price per Unit})x$$

Profit Function (P): The **profit function** calculates the amount of profit P created for x number of units made and sold and is found by subtracting the cost from the revenue.

$$P = R - C$$

Break-Even Point: The quantity of units to be produced and sold so that the cost of production equals the revenues for a product is called the **break-even point**.

Procedure

Solving a System of Linear Equations by Substitution

1. Solve one of the equations for one of the variables.

2. Substitute the expression found in Step 1 into the other equation and solve for the remaining variable.

3. Re-substitute the value found in Step 2 into the original equation from Step 1 to find the corresponding coordinate.

Section 5.4 Linear Inequalities in Two Variables

Definitions

Inequality Symbol: An **inequality symbol** is a relation used between two expressions that are not strictly equal.

$$< \text{ less than} \qquad \leq \text{ less than or equal to}$$

$$> \text{ greater than} \qquad \geq \text{ greater than or equal to}$$

Linear Inequality: A **linear inequality** in the variable x can be written in one of the following forms

$$ax + b < c \quad ax + b > c$$

$$ax + b \leq c \quad ax + b \geq c$$

where a, b, and c are numbers and $a \neq 0$.

Boundary Line: A **boundary line** is a line that separates those points that satisfy an inequality from those that do not. It is a representation of the equation created when the inequality sign is replaced with an equal sign.

Procedures

Steps to Graphing Linear Inequalities

1. Graph the boundary line.
   - The line will be solid if the inequality is $\geq$ or $\leq$.
   - The line will be dashed if the inequality is $>$ or $<$.

2. Determine which side of the line to shade by using a test point.

3. Shade the half-plane that contains the solution set.

Steps to Graphing Systems of Linear Inequalities

1. Graph the boundary line for each inequality.
   - The line will be solid if the inequality is $\geq$ or $\leq$.
   - The line will be dashed if the inequality is $>$ or $<$.

2. Determine which side of the line to shade for each inequality by using a test point.

3. Lightly shade the half-plane that contains the solution set for each individual inequality.

4. The solution region for the system of inequalities is the intersection of the shaded half-planes from Step 3.

Section 5.5 Linear Programming

Definitions

Linear Programming: The mathematical process of **linear programming** is a method to optimize a particular quantity in a mathematical model constrained by linear inequalities.

Objective Function: An **objective function**, or optimization function, is a function that expresses the quantity that is to be optimized.

Constraints: Constraints in linear programming are the linear inequalities that express the limitation of each variable or combination of variables.

Feasible Region: The **feasible region** is the set of all possible points that satisfy the constraints of an optimization problem.

Procedure

Steps for Solving an Optimization or Linear Programming Problem

1. Define the objective function.
2. Define the constraints using linear inequalities.
3. Graph the constraint inequalities.
4. Identify the vertices of the feasible region; that is, the polygon formed by intersection of the lines.
5. Identify any maximum or minimum values of the objective function by finding the values of the objective function at each of the vertices.

Theorem

Fundamental Theorem of Linear Programming: The maximal and minimal values of a linear function over a convex polygonal region are found at the corners of the region.

Section 5.6 Modeling with Quadratics

Definitions

Quadratic Equation: A quadratic equation in x is written in standard form as

$$ax^2 + bx + c = 0,$$

where a, b, and c are numbers and $a \neq 0$.

Parabola: The graph of a quadradic equation is called a **parabola**. It is a symmetrical U-shaped curve around a vertical or horizontal line.

Vertex: The **vertex** of a parabola is the extreme point representing the quadratic equation's minimum or maximum value. It occurs at the point $\left(\dfrac{-b}{2a}, y \right)$, given that $ax^2 + bx + c = y$.

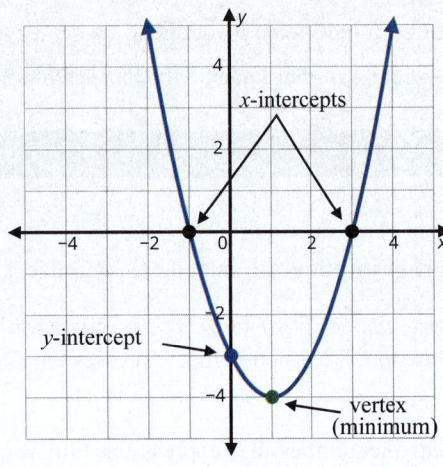

Figure 5.6.14

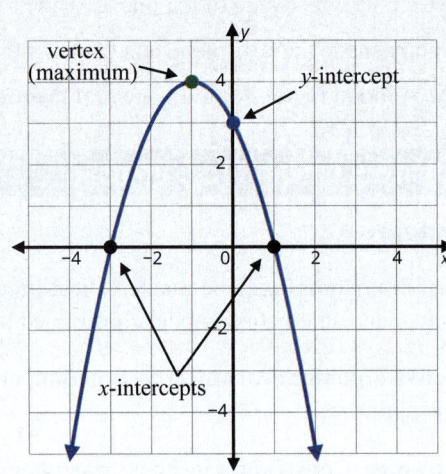

Figure 5.6.15

Formula

Quadratic Formula: A quadratic equation written in standard form, $ax^2 + bx + c = 0$, where a is nonzero, can be found using the quadratic formula.

$$x = \frac{-b \pm \sqrt{b^2 - 4ac}}{2a}$$

Section 5.7 Exponential and Logarithmic Functions

Definitions

Exponential Function: An **exponential function** is a function of the form $f(x) = ab^x$, where $b > 0$, $b \neq 1$, a is the initial value, and x is a real number.

Expanded Form of an Exponential Function: The **expanded form of an exponential function** is a function of the form $f(x) = a(1 + r)^x$, where a is the initial value, r is the rate of change, and x is any real number.

Frequency: The **frequency** of a sound is the number of energy wave cycles completed in one second. Frequency is measured in hertz, Hz.

Pitch: **Pitch** is the tonal quality of sound that describes how low or high an instrument (or voice) sounds to our ears.

Half Step: A **half step** is the distance between one note and the next nearest note on a piano.

Whole Step: A **whole step** is defined as the interval between two half steps on a piano.

Octave: An **octave** is the interval of notes between 12 half steps on a piano.

Frequency of Musical Notes: The **frequency of a music note** in relation to a specific note is modeled by the following function.

$$F(x) = F_0(1.059463)^x$$

Here, F_0 is the reference frequency and x is the number of half steps up from F_0.

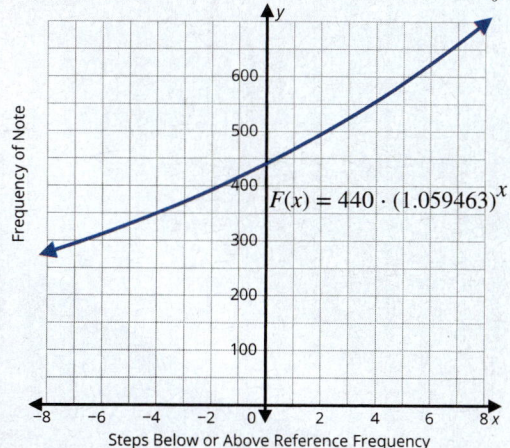

Figure 5.7.5: Frequency of Notes in Reference to A above Middle C

Logarithm: If $b^x = a$, then the **logarithm** with base b of a is x. Symbolically, we can express this exponential as an equivalent logarithm

$$\log_b a = x,$$

where $b > 0$, $b \neq 1$, and $a > 0$.

CHAPTER 6

Finance

SECTIONS

Introduction

The dictionary defines *savvy* as "shrewd and knowledgeable; having common sense and good judgment." It's fair to say that financially, we as consumers want and need to be savvy. Fortunately, the goal of this chapter is to help us do just that!

In 2020, the credit card industry made over $100 billion off of consumers, charging astronomical amounts in interest and fees. The average interest rate for a new credit card in 2020 was 18.04%, and the average minimum payment was 2%. The following graph shows the payoff schedule for borrowing $5000 at 18.04% and only paying the minimum amount due every month for 30 years. Notice that after 30 years, the balance is still $833! Given the low likelihood that we still own or use an item that we purchased 30 years ago, nobody wants to still be making payments after that long.

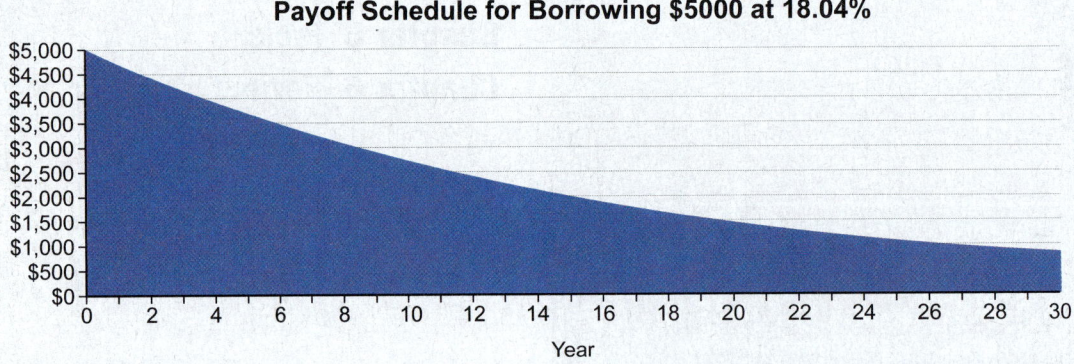

Payoff Schedule for Borrowing $5000 at 18.04%

While credit cards make purchasing items easier, that convenience can be both a good thing and a bad thing. A savvy consumer reaps the benefits and only uses a credit card for convenience by paying the balance in full each month so that no interest is accrued. But that convenience becomes a burden when debt begins to accumulate. In fact, people almost always spend more if they pay with a credit card than if they hand over cash instead. So unless we are mindful of our spending, we often do not realize how much we have spent until the bill comes due.

Being a savvy consumer can also mean forming and sticking to a budget. Knowing how we will spend each dollar helps us determine where our hard-earned money goes. It also helps us plan for the future. Budgeting can help minimize the amount of hard-earned money that is consumed by interest and instead let interest work to grow future wealth for us. Did you know that the average annual return on investments from 1928 to 2020 was 11.24%? If we invest $100 each month for 30 years at just 10% interest (which is below that average), we will end up wtih $229,916. That's $193,816 in interest that our investment earned. Now that's savvy—getting our money to work for us!

OBJECTIVES

1. Calculate simple interest.

2. Calculate future value for simple interest.

3. Calculate future value for compound interest.

4. Calculate future value for continuous compound interest.

5. Calculate annual percentage yield.

Albert Einstein is reported to have once said, "the most powerful force in the universe is compound interest." Why? Because compound interest is an exponential function; in short, the more of something there is, the faster it grows. This can be a good thing if you are the one investing money or are the lender. For instance, suppose that when you turn 18, you make a one-time investment of $5000 with a 6% interest rate. If left untouched, you will have over $77,000 by the time you retire at age 65. However, if you are the one needing to borrow money, interest (that is, the cost of borrowing) is not as gratifying. If you place $1000 on a credit card with a 15% interest rate and pay only the minimum payment required each month, it will take you almost nine years to pay off the debt and cost you an extra $730 in additional interest payments over that time period.

This section looks at the two main ways interest is calculated—simple interest and compound interest. Formally, *interest* is the amount of money that is charged by a lender for borrowing money. The *initial value* or *principal* is the sum of money on which the interest is charged. We use the same term whether we are investing or borrowing. The principal amounts in our examples above are $5000 and $1000.

📖 INTEREST AND PRINCIPAL

Interest

The amount charged by a lender for borrowing money is the **interest**.

Principal

The **principal** is the sum of money on which interest is charged.

When lending money, lenders must decide how much to charge. Usually the amount charged is not a single fixed price but is a certain percentage of the principal. The percentage charged for borrowing money is referred to as the *interest rate*. If the rate is calculated on an annual basis, it is called the *annual percentage rate (APR)*. We will discover later that lending businesses often advertise a certain APR in order to appeal to potential customers, but other factors play just as important a role in determining how much money is involved. Things like whether the interest is simple or compound and how long you accrue interest can have large impacts.

📖 INTEREST RATE AND ANNUAL PERCENTAGE RATE (APR)

Interest Rate

The **interest rate** is the percentage of the principal that is charged to the borrower.

Annual Percentage Rate (APR)

The **annual percentage rate (APR)** is the yearly interest rate that is charged for borrowing money, including fees. APR is normally given as a percentage per year.

As we mentioned, there are two basic types of interest: simple interest and compound interest. The difference between the two is how often the interest is calculated throughout the loan period. Simple interest is charged on the original principal amount only, whereas compound interest is calculated more than once and is computed on the original amount plus any previously earned interest. You can think of it as "interest on top of interest." Compound interest builds on itself and usually has a smaller percentage rate compared to simple interest, which is slower to accumulate and hence often charged at a bigger percentage rate.

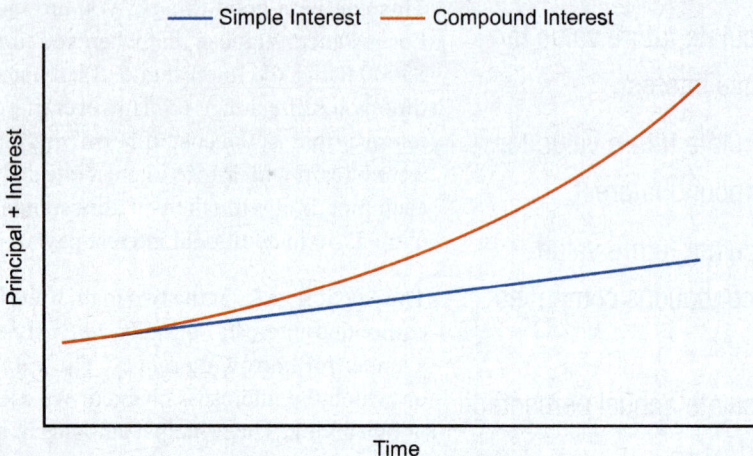

FIGURE 6.1.1: Simple Interest vs Compound Interest at the Same Interest Rate

OBJECTIVE 1   Simple Interest

Simple interest is the most straightforward interest to calculate. This interest is calculated by multiplying the annual percentage rate by the principal amount. If the loan is for more than one year, we need to multiply by the number of years as well.

📖 **SIMPLE INTEREST**

The amount of **simple interest** earned on a loan is calculated with the following formula.

$$I = Prt$$

Here, P is the principal, r is the APR written as a decimal, and t is the length of the loan in years.

Helpful Hint

A credit score is a number that ranks your risk as a borrower.

Example 6.1.1

Calculating Simple Interest

An online lending company is offering simple-interest personal loans based on consumer credit scores, as shown in Table 6.1.1. The better your credit score, the lower the interest rate.

Table 6.1.1: Simple Interest

Principal	Minimum Credit Score Required	Interest Rate (APR)	Length of Loan	Interest Accrued
$3500	580	21.99%	4 years	
$3500	680	9.99%	4 years	
$3500	780	6.95%	4 years	

a. Determine the amount of interest that would be accrued for each tier with a $3500 principal.

b. How much money will be paid to the lending company at the end of four years? Note that while sometimes companies charge one-time loan fees, this company does not.

c. Suppose you have worked hard to improve your credit score and it moves up a tier. Which provides the most significant savings in interest: to move from the bottom tier (580) to the middle one (680), or to move from the middle tier (680) to the top one (780)?

Solution

a. Although we are calculating simple interest for three cases in this scenario, the principal P and the time t are the same for each: $P = \$3500$ and $t = 4$. The interest rate is the only value that changes.

By Hand

The multiplication using the simple interest formula is shown in Table 6.1.2 for each credit-score tier. Note that since the interest rate is given as a percentage, it will need to be changed to a decimal before substituting it in the formula.

Table 6.1.2: Simple Interest

Principal	Minimum Credit Score Required	Interest Rate (APR)	Length of Loan	Interest Accrued
$3500	580	21.99%	4 years	$I = (\$3500)(0.2199)(4) = \$3078.60$
$3500	680	9.99%	4 years	$I = (\$3500)(0.0999)(4) = \$1398.60$
$3500	780	6.95%	4 years	$I = (\$3500)(0.0695)(4) = \$973.00$

Microsoft Excel

Excel can be used to calculate all three interest amounts. For clarity, we'll begin by labeling the columns "Principal", "Interest rate", "Length of loan", and "Interest accrued", in A1 through D1, respectively. Next enter the necessary data into the first 3 empty columns; that is, enter the values for principal in column A, interest rate in column B, and length of loan in column C. In cell D2, input the formula for interest "=A2*B2*C2" and press Enter. The value will appear as $3,078.60.

Drag the bottom right-hand corner of cell D2 down through cell D4. This will copy the formula into cells D3 and D4 while updating the referenced cells. The values $1,398.60 and $973.00 will appear in cells D3 and D4, respectively.

D2	▼ ⋮	× ✓ *fx*	=A2*B2*C2

◢	A	B	C	D
1	Principal	Interest rate	Length of loan	Interest accrued
2	$3,500	21.99%	4	$3,078.60
3	$3,500	9.99%	4	$1,398.60
4	$3,500	6.95%	4	$973.00

FIGURE 6.1.2

Therefore, the interest accrued on a $3500 loan would be $3078.60 for the lowest credit-score tier, $1398.60 for the middle tier, and $973.00 for the highest tier.

b. To determine the total amount of money that will be paid to the lender, we need to add the calculated interest from part a. to the original principal of $3500.

By Hand

Table 6.1.3 shows the addition in the last column.

Table 6.1.3: Simple Interest

Principal	Minimum Credit Score Required	Interest Rate (APR)	Length of Loan	Interest Accrued	Total Amount Paid to Lender
$3500	580	21.99%	4 years	$3078.60	$3078.60 + $3500 = $6578.60
$3500	680	9.99%	4 years	$1398.60	$1398.60 + $3500 = $4898.60
$3500	780	6.95%	4 years	$973.00	$973.00 + $3500 = $4473.00

Microsoft Excel

In an empty column, we need to add the interest found in column D to the principal found in column A. Enter the label "Amount Paid" in cell E1. Next, input "=D2+A2" into cell E2 and press Enter. The value will appear as $6,578.60. Again, drag the lower right-hand corner of E2 down to copy the formula into the other two cells while updating the referenced cells. The values $4,898.60 and $4,473.00 will appear in E3 and E4, respectively.

E2	▼	⋮	× ✓	*fx*	=D2+A2	

	A	B	C	D	E
1	Principal	Interest rate	Length of loan	Interest accrued	Amount Paid
2	$3,500	21.99%	4	$3,078.60	$6,578.60
3	$3,500	9.99%	4	$1,398.60	$4,898.60
4	$3,500	6.95%	4	$973.00	$4,473.00

FIGURE 6.1.3

Thus, at the end of four years, borrowing $3500 at the lowest credit-score tier would mean paying the lender $6578.60. At the middle tier, a borrower would pay a total of $4898.60. And at the highest tier, $4473.00 would be paid.

c. The savings in interest between each tier can be found by subtracting.

Difference between Bottom Tier (580) and Middle Tier (680):

$$3078.60 - \$1398.60 = \$1680$$

Difference between Middle Tier (680) and Top Tier (780):

$$1398.60 - \$973.00 = \$425.60$$

While both credit scores increased by 100 points, moving from the bottom tier to the middle tier can save you almost four times as much as moving from the middle tier to the top tier.

OBJECTIVE 2 Future Value for Simple Interest

In Example 6.1.1, we saw that when the loan period ended, the amount repaid to the lender was the principal amount plus any interest that had accrued. This amount due is called the *future value*, A, of the loan. We can actually calculate this amount all in one step using a bit of algebra.

$$A = \text{Principal} + \text{Interest}$$
$$= P + I$$
$$= P + Prt$$
$$= P(1 + rt) \qquad \text{Distribution property}$$

Let's take a moment to verify this formula for the bottom tier of Example 6.1.1. The principal was $P = \$3500$ and the rate of interest was $r = 0.2199$ over 4 years. Substituting these values into the formula for future value, we have the following equation.

Future Value of Bottom-Tier Loan in Example 6.1.1:

$$A = \$3500(1 + 0.2199 \cdot 4) = \$6578.60$$

Notice that this matches what we previously found.

📖 FUTURE VALUE FOR SIMPLE INTEREST

The **future value**, A, of a loan is the amount to be repaid at the end of the loan period. It is the principal amount plus any interest accrued. Using **simple interest**, future value is calculated with the following formula.

$$A = P(1 + rt)$$

Here, P is the principal, r is the APR written as a decimal, and t is the length of the loan in years.

Retail stores often offer special in-store financing for purchases over a certain amount. They entice consumers by advertising a "no-interest, same-as-cash" loan. But beware, these interest-free loans are often short-term and have very high interest rates after the free period is over. Example 6.1.2 examines such an offer.

Example 6.1.2

Calculating Future Value for Simple Interest with a Partial Year

Ian is purchasing a new TV by using an in-store offer. The store is offering a 90-day, same-as-cash loan to make the purchase. This means that at the end of 90 days, if Ian has paid off the cost of the television, he owes no added interest charge. However, if Ian does not pay off the entire cost of the TV, he owes simple interest on the original purchase amount, calculated over the entire 90 days. The price of the television Ian is purchasing is $2650 (tax included) with an annual interest rate of 27.52%.

a. How much would Ian owe on the 91st day if he made no payments during the first 90 days?

b. How much would Ian owe *in interest* on the 91st day if he paid off $2000 of the cost of the TV by the 90th day?

Solution

a. Since Ian did not pay off the loan during the 90 days, he will owe the future value of the original loan amount. This includes the simple interest calculated over the 90 days plus the original purchase price P of $2650.

In order to use the formula for future value, we need time t to be in years. The period of time for this loan was 90 days. As a fraction of a year that would be $t = \dfrac{90}{365}$. The interest rate must also be in decimal form, so $r = 0.2752$.

$$A = P(1 + rt)$$

$$= \$2650\left(1 + 0.2752\left(\frac{90}{365}\right)\right)$$

$$\approx \$2829.82$$

Therefore, the total amount Ian must pay for the television on the 91st day is $2829.82, the sum of the principal and the interest incurred.

b. Because the conditions of the loan say that the entire price of the TV must be paid by the end of the 90 days, Ian does not gain any reduction in interest for paying a portion of the loan back. Therefore, Ian will owe the same amount of interest that must be paid in part a. Remember, the future value we found includes the original loan amount. To determine what part of that is interest, subtract the original price of the TV from the future value.

Interest Owed: $2829.82 − $2650 = $179.82

So if Ian does not pay off the entire purchase price in 90 days, he will owe $179.82 in interest.

✔ **Skill Check 6.1.1**

Find the total cost of a loan for $4320 that has a simple interest rate of 15% for 18 months.

💬 **Fun Fact**

Benjamin Franklin was keenly aware of the power of compound interest. Upon his death in 1970, he left a sum of about $4500 each to the cities of Philadelphia and Boston to be invested in a trust for 200 years. After the first 100 years, the cities could withdraw up to 75% of the amount and after the second 100 years, they could withdraw the rest. Had the cities not withdrawn any money after the first 100 years, each city would have had a balance of about $77,816,613.67 after 200 years. Visit http://mathcentral.uregina. ca/beyond/articles/CompoundInterest/ Money.html to learn more.

OBJECTIVE 3 Compound Interest

As we mentioned earlier, you can think of compound interest as "interest on top of interest." In other words, the interest charged is calculated more than once and is applied not only to the principal, but also to any interest that has previously accrued. This is the reason that compound interest grows more rapidly over time.

For example, suppose we have a $500 loan with an interest rate of 4% per year that is compounded once per year. The interest after the first year would be 4% of $500, or $0.04 \cdot 500 = \$20$. This amount is added to the principal of $500, and the new adjusted principal becomes $520. The second year's interest is 4% of $520, or $20.80. In the same manner as before, the interest is added to the principal and the adjusted principal is $540.80. This means that after two years, we have accrued $40.80 in interest. Compare that to simple interest, where we would have accrued $40 over two years.

You may have noticed that compound interest, when compared with simple interest, needs one more factor in order to be calculated—the number of times the interest will be calculated, or *compounded*. The variable n represents the number of compounding intervals in a year. For instance, if interest is compounded quarterly, n is 4. If it is compounded monthly, n is 12. Table 6.1.4 gives some common compound frequencies along with their terminology for reference.

Table 6.1.4: Common Compounding Intervals

Compound Frequencies	Number of Compounds per Year, n
Annually	1
Semiannually/Biannually	2
Quarterly	4
Monthly	12
Weekly	52
Daily	365

> **Helpful Hint**
>
> Linear functions grow at a steady rate. Exponential functions grow in relation to how big they are. **The bigger it gets, the faster it grows.**

> **FUTURE VALUE FOR COMPOUND INTEREST**
>
> The **future value**, A, of a loan is the amount to be repaid at the end of the loan period. It is the principal amount plus any interest accrued. Using **compound interest**, future value is calculated with the following formula.
>
> $$A = P\left(1 + \frac{r}{n}\right)^{nt}$$
>
> Here, P is the principal, r is the APR written as a decimal, t is the length of the loan in years, and n is the number of compounding intervals per year.

Notice that the formula for the future value for compound interest differs from the formula to find the future value for simple interest. The compound interest formula will grow exponentially while the simple interest formula grows linearly.

Example 6.1.3

Calculating Future Value for Compound Interest

Lilly deposits $12,000 into an account with an annual interest rate of 1.05% compounded monthly. If she leaves the money in the account for 10 years, what will the future value be at the end of this time period? Round to the nearest cent.

Solution

We can use the compound interest formula to compute the future value for Lilly after 10 years.

By Hand

In this case, $P = \$12,000$, $r = 0.0105$, and $t = 10$. Since interest is compounded monthly, $n = 12$.

$$A = P\left(1 + \frac{r}{n}\right)^{nt}$$

$$= \$12,000\left(1 + \frac{0.0105}{12}\right)^{(12)(10)}$$

$$= \$12,000(1.000875)^{120}$$

$$\approx \$13,327.92$$

TI-83/84 Plus

- The TI-83/84 Plus has a built-in financial app called TVM (which stands for Time, Value, Money) that can be used to find compound interest. To access this, press [apps], select Finance..., and then select TVM Solver. You will be prompted to enter values for the following variables.

N is the total number of times the account is compounded;

I% is the interest rate as a percentage;

PV is the present value;

PMT is the monthly payment, which is 0 if no monthly payments are being made;

FV is the future value, which is what we'll be solving for;

P/V = C/Y is the number of compounding periods per year.

- In this case, the interest is compounded monthly for 10 years. So N = 12 · 10 = 120.

- For I%, enter 1.05. Remember that when using the TVM Solver, we use the percentage form of the rate, not the decimal form.

- The present value is the amount that is initially deposited into the account, which is $12,000. Since we want the future value (the value paid out) to be positive, we need the present value (the value paid into the account) to be negative. So PV = –12000.

- There are no monthly payments being made, so PMT = 0.

- We are trying to find the future value, so we will come back to the FV entry after filling out the other values.

- Because the interest is being compounded monthly, P/Y = C/Y = 12. (Note that once you enter a value for P/Y, the value of C/Y is automatically updated to that same value.)

- The final option for PMT: is either END or BEGIN. This refers to whether the payments are being made at the beginning of each period or at the end. Typically, payments are made at the end of each period, so be sure that END is highlighted.

- Once these values are filled out, return your cursor to the FV entry and press [clear]. To solve for this value, press [alpha] [enter]. The value is calculated as 13327.91542.

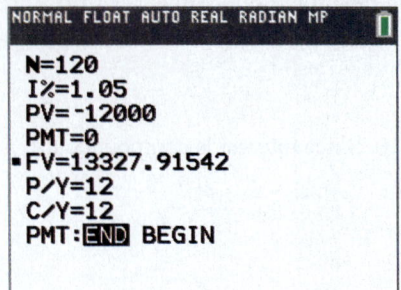

Figure 6.1.4

Microsoft Excel

Excel has a built-in financial function **FV(rate, nper, pmt, [pv], [type])** that calculates the future value of an account given certain information about the account, such as the interest rate and the initial value. The specific arguments for this function are defined as follows:

rate is the interest rate per period. This is the interest rate divided by the number of compounding periods per year.

nper is the total number of compounding periods. This is the number of periods per year multiplied by the length of the loan in years.

pmt is the payment made each period. This can be left blank if no payments are made.

pv is the present value, or principal amount. Default is 0.

type refers to whether payments are being made at the beginning of each period or at the end; 0 = end of the period, 1 = beginning of the period. Default is 0.

- In this example, we have an interest rate of 0.0105 that is being compounded 12 times a year, so for the first argument in this Excel function, we have rate = 0.0105/12. Be careful to not use just the interest rate here.

- Because the interest is being compounded monthly for 10 years, we will use $12 \cdot 10 = 120$ for nper.

- There are no payments being made, so we will use pmt = 0, or leave that argument empty.

- The present value is the amount that is initially deposited into the account, which is $12,000. Since we want the future value (the value paid out) to be positive, we need the present value (the value paid into the account) to be negative. So we'll use −12000 for pv.

Typically, payments are made at the end of each period, so we will use type = 0, or leave that argument empty. In an empty cell, type "=FV(0.0105/12,12*10,,-12000,0)" and press Enter. The future value is given as 13,327.92.

A1	▼	:	×	✓	fx	=FV(0.0105/12,120,,-12000,0)		
	A	B	C	D	E	F	G
1	$13,327.92						

FIGURE 6.1.5

✔ **Skill Check 6.1.2**

Use the information in Example 6.1.3 to find the future value of Lilly's account after 20 years.

Therefore, at the end of 10 years Lilly will have a total of $13,327.92 in her account.

The power of compound interest lies not only in the interest rate, but also in the number of times the interest is compounded per year. The following example compares the growth of an investment with different numbers of yearly compounding intervals.

Example 6.1.4

Comparing Compound Interest Rates and Intervals

In 2010, Anthony invested money in an IRA (Individual Retirement Account) that earns 2.01% APR and is compounded quarterly. Currently, a bank is offering new customers an IRA that earns 1.49% APR, compounded daily. If Anthony has $21,045 at the moment, where is the most profitable place for his money to be over the next 12 years?

Solution

To determine the most profitable place for Anthony to keep his money, we need to compare the future values for each IRA option. In both cases, the principal amount is $21,045. The rates and compounding intervals for each account are restated before each calculation.

By Hand

Current IRA:

$P = \$21,045$

$r = 0.0201$

$t = 12$

$n = 4$

$$A = P\left(1 + \frac{r}{n}\right)^{nt}$$

$$= \$21,045\left(1 + \frac{0.0201}{4}\right)^{(4)(12)}$$

$$\approx \$26,769.39$$

New IRA:

$P = \$21,045$

$r = 0.0149$

$t = 12$

$n = 365$

$$A = P\left(1 + \frac{r}{n}\right)^{nt}$$

$$= \$21,045\left(1 + \frac{0.0149}{365}\right)^{(365)(12)}$$

$$\approx \$25,165.13$$

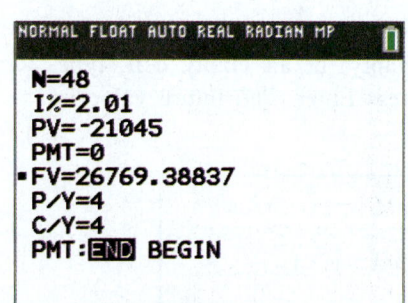

Figure 6.1.6

TI-83/84 Plus

As in the previous example, we can use the TVM Solver to find the future value in both scenarios.

The present value of $21,045 will be the same for both options: PV = −21045. Also, PMT = 0 for both options because there are no monthly payments.

The interest in the current IRA is compounded quarterly over the 12 years, so we'll use N = 4 . 12 = 48 and P/Y = C/Y = 4.

The interest in the new IRA is compounded daily over the 12 years, so we'll use N = 365 · 12 = 4380 and P/Y = C/Y = 365.

The calculator returns 26769.38837 for the current IRA and 25165.13118 for the new one.

Microsoft Excel

We will use the **FV(rate, nper, pmt, [pv], [type])** function to determine the current IRA amount and the new IRA amount.

- The original IRA has an IRA of 2.01% compounded quarterly, so we will use rate = 0.0201/4.

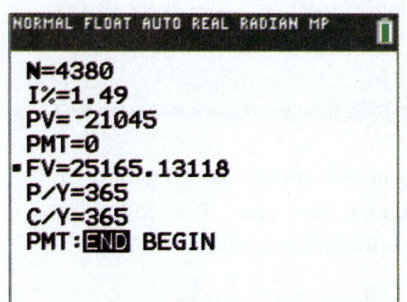

Figure 6.1.7

- Because the interest is being compounded quarterly for 12 years, we'll use nper = 4 · 12.

- The new IRA has an IRA of 1.49% compounded daily, so we will use rate = 0.0149/365.

- Because the interest is being compounded daily for 12 years, we'll use nper = 365 · 12.

For both accounts, we will use 0 for pmt and –21045 for pv.

To keep track of the two accounts, enter "Current IRA" in cell A1 and "New IRA" in cell B1.

Calculate the future value of the current IRA in cell A2 by typing "=FV(0.0201/4,4*12,,-21045,0)" and pressing Enter. This gives a value of 26,769.39.

Calculate the future value of the new IRA in cell B2 by typing "=FV(0.0149/365,365*12,,-21045,0)" and pressing Enter. This gives a value of 25,165.13.

B2	▼	⋮	× ✓ _fx_	=FV(0.0149/365,365*12,,-21045,0)		
	A	B	C	D	E	F	G
1	Current IRA	New IRA					
2	$26,769.39	$25,165.13					

FIGURE 6.1.8

Therefore, Anthony's current IRA will earn more than the new offer. Notice that the increase in number of compounding intervals was not enough to outweigh the difference in interest rate. So Anthony would do better to leave his money in the current IRA account.

OBJECTIVE 4 Continuous Compound Interest

In theory, instead of compounding interest a certain number of times per year, an account balance could be constantly earning money if it were compounding literally continuously—every moment of every day of every week of every year. In other words, it would have an infinite number of compounding intervals. We call this *continuous compound interest*. Practically speaking, an infinite number of intervals is impossible; however, continuous compounding is an important concept in the financial world. It is a way to determine the largest amount of interest that could possibly accrue during a given time frame.

Interest that is continuously compounded uses a formula similar to that of interest compounded n times per year, except that the variation among the number of compounding intervals per year is accounted for using the irrational number e, which is approximately equal to 2.718281828459.

FUTURE VALUE FOR CONTINUOUS COMPOUND INTEREST

The **future value**, A, of a loan is the amount to be repaid at the end of the loan period. It is the principal amount plus any interest accrued. Using **continuous compound interest**, future value is calculated with the following formula.

$$A = Pe^{rt}$$

Here, P is the principal, r is the APR written as a decimal, and t is the length of the loan in years.

Example 6.1.5

Calculating Continuous Compound Interest

Suppose an investment earns 1.8% interest compounded continuously.

a. Find the future value of a $5000 investment after 20 years.

b. How long will it take the investment in part a. to be worth $6000?

c. In general, how long will it take the value of this type of investment to triple?

Solution

a. We are given the following quantities to use in the formula for future value when the interest is compounded continuously.

$$P = \$5000 \quad r = 0.018 \quad t = 20$$

So the future value is calculated as follows.

$$\begin{aligned} A &= Pe^{rt} \\ &= \$5000e^{(0.018)(20)} \\ &\approx \$7166.65 \end{aligned}$$

After 20 years, the investment will be worth $7166.65.

b. In this question, we are told the future value we want to reach is $6000, so $A = \$6000$. We are asked to find the time t if the rate stays at $r = 0.018$. Let's fill in what we know in our formula for future value.

$$A = Pe^{rt}$$

$$\$6000 = \$5000e^{(0.018)t}$$

Notice that the variable t is in the exponent. We can solve the equation for t by dividing and using natural logarithms.

$$6000 = 5000e^{(0.018)t}$$

$$\frac{6000}{5000} = \frac{\cancel{5000}e^{(0.018)t}}{\cancel{5000}}$$

$$1.2 = e^{(0.018)t}$$

$$\ln(1.2) = \ln\left(e^{(0.018)t}\right) \qquad \text{Take the natural log of both sides.}$$

$$\ln(1.2) = 0.018t \qquad \text{Recall that } \ln(e^x) = x.$$

$$\frac{\ln(1.2)}{0.018} = \frac{\cancel{0.018}t}{\cancel{0.018}}$$

$$10.13 \approx t$$

Therefore, it will take 10.13 years to reach $6000. (Note that 10.13 years is approximately 10 years and one and half months.)

c. In order for the value to triple, the future value needs to be three times the amount of the principal. So if we let P represent the amount of the principal, then that amount tripled would be $3P$. We can substitute this information along with the rate of 0.018 in the equation and solve for time once again.

$$A = Pe^{rt}$$

$$3P = Pe^{(0.018)t}$$

$$\frac{3\cancel{P}}{\cancel{P}} = \frac{\cancel{P}e^{(0.018)t}}{\cancel{P}}$$

$$3 = e^{(0.018)t} \qquad \text{Take the natural log of both sides.}$$

$$\ln(3) = \ln\left(e^{(0.018)t}\right) \qquad \text{Recall that } \ln(e^x) = x.$$

$$\ln(3) = 0.018t$$

$$\frac{\ln(3)}{0.018} = \frac{\cancel{0.018}t}{\cancel{0.018}}$$

$$61.03 \approx t$$

Therefore, it will take approximately 61 years for an investment to triple if it is compounded continuously at a rate of 1.8%. Notice that the time it takes to triple the value is the same regardless of the amount of the initial investment.

✔ **Skill Check 6.1.3**

Find the future value of a $1000 nvestment after 15 years if the investment earns 2.5% interest compounded continuously.

OBJECTIVE 5 Annual Percentage Yield

Consider for a moment the total amount of interest that accumulates during the year when interest is compounded. The following spreadsheet is an example of how the interest grows on a savings account throughout the year with a $100 initial principal and a 5% APR, compounded quarterly.

Table 6.1.5: Savings Account with $P = \$100$, APR = 5%, and Compounded Quarterly

Compound Period	Amount of Interest	Amount in Savings after Compounding
1st	$1.25	$101.25
2nd	$1.27	$102.52
3rd	$1.28	$103.80
4th	$1.30	$105.10

Although the Annual Percentage Rate (APR) of 5% is used in the compounding calculations along the way, when we get to the end, we can see that absolute change in the principal amount from $100 to $105.10 is more than 5% overall. (Notice that the total would be $105 if it were 5%.)

When we take compounding into account, the practical yearly interest rate is no longer 5%, it is effectively 5.1%. It makes sense that this rate is higher, since at each compounding period, we earn interest on the original principal plus the interest earned in the previous periods. This practical interest rate, which reflects the compounding intervals, is called the *annual percentage yield (APY)* or the *effective interest rate* for a given period.

📖 ANNUAL PERCENTAGE YIELD (APY)

The **annual percentage yield (APY)** is the simple annual interest rate earned in a given year that accounts for the effects of compounding. **APY** is calculated using the following formula.

$$APY = \left(1 + \frac{r}{n}\right)^n - 1$$

Here, r is the APR written as a decimal and n is the number of compounding intervals per year.

Note that you do not need to know the principal amount being invested or borrowed in order to calculate the annual percentage yield. This is a rate, and as such is independent of the principal.

Example 6.1.6

Calculating Annual Percentage Yield (APY)

Determine the annual percentage yield, or the effective interest rate, for $500 invested at 2.05% over 10 years compounded daily. Write the APY as a percentage, rounded to the nearest hundredth.

Solution

First notice that as we mentioned before, the annual percentage yield is independent of the principal or the number of years of an investment. All we actually need to know is the annual interest rate and the number of compounding intervals.

By Hand

For this example, $r = 0.0205$ and $n = 365$. Substituting these values into the formula for **APY**, we have the following calculation.

$$
\begin{aligned}
APY &= \left(1 + \frac{r}{n}\right)^n - 1 \\
&= \left(1 + \frac{0.0205}{365}\right)^{365} - 1 \\
&\approx 0.020711 \\
&= 0.0207
\end{aligned}
$$

TI-83/84 Plus

The TI-83/84 Plus does not have a function that computes the APY directly. Instead, we first need to use the TVM Solver to find the future value of this account after the 10 years. We will then use the TVM Solver a second time with this future value to find the APY.

1st Step

We will use 2.05 for **I%** and −500 for **PV**. **PMT** = Ø because there are no monthly payments. The interest rate in this account is compounded daily for 10 years, so we'll use **N** = 365.10 = 3650 and **P/Y** = **C/Y** = 365. Solving for **FV**, we get 613.7589993.

2nd Step

We now use the solver again with the FV we just found. This time we are trying to find the simple annual interest rate necessary for a $500 investment to grow to the future value we just found (613.7589993) in 12 years, *without* the interest being compounded daily.

We need to change **P/Y** = **C/Y** to 1 and update the value of **N** to 1 · 1Ø = 1Ø. To find this new interest rate, clear the existing value for **I%** and solve by pressing [alpha] [enter].

The calculator returns 2.071098066.

This means the annual percentage yield for $500 invested at 2.05% over 10 years compounded daily is approximately 2.07%

NORMAL FLOAT AUTO REAL RADIAN MP

```
 N=3650
 I%=2.05
 PV=-500
 PMT=0
■FV=613.7589993
 P/Y=365
 C/Y=365
 PMT:END BEGIN
```

Figure 6.1.9

NORMAL FLOAT AUTO REAL RADIAN MP

```
 N=10
■I%=2.071098066
 PV=-500
 PMT=0
 FV=613.7589993
 P/Y=1
 C/Y=1
 PMT:END BEGIN
```

Figure 6.1.10

✔ **Skill Check 6.1.4**

Determine the APY for a 4.2% interest rate compounded monthly. Round to the nearest hundredth.

Both APR and APY are used by banks and financial institutions to describe your interest rate, whether for loans or investments, but they mean different things. The difference is compounding, or the ability to earn interest on interest. Since the APY is a larger percentage than the annual percentage rate (APR) when compounding occurs, banks will often publicize the APY for investment accounts to entice investors to deposit money with their institution. Conversely, when advertising interest rates for loans, lenders will use the APR to entice borrowers. So when researching for the best option it's important to be sure you're comparing apples to apples; that is, comparing APYs to APYs offered by the financial institutions you're evaluating, not an APR versus an APY.

Example 6.1.7

Comparing APR versus APY

Suppose that the APD Bank of the South advertises the following rates for their personal loans. Determine the APY for each of the loan categories to complete Table 6.1.6. Write each APY as a percentage, rounded to the nearest hundredth.

Table 6.1.6: APD Bank of the South Personal Loan Rates

Loan Amount	APR*	APY
< $20,000	9.95%	
$20,000 – $99,999	6.49%	
≥ $100,000	5.95%	

*Interest is compounded daily, and accrued interest is posted to the account monthly.

Solution

To find the APY for each loan category, we need the published APR as well as the number of compounding intervals per year. Notice that the footnote tells us that interest is compounded daily, so $n = 365$.

By Hand

Loans < $20,000   Loans $20,000 – $99,999   Loans ≥ $100,000

$r = 0.0995$ and $n = 365$ $r = 0.0649$ and $n = 365$ $r = 0.0595$ and $n = 365$

$$\text{APY} = \left(1 + \frac{r}{n}\right)^n - 1 \qquad \text{APY} = \left(1 + \frac{r}{n}\right)^n - 1 \qquad \text{APY} = \left(1 + \frac{r}{n}\right)^n - 1$$

$$= \left(1 + \frac{0.0995}{365}\right)^{365} - 1 \qquad = \left(1 + \frac{0.0649}{365}\right)^{365} - 1 \qquad = \left(1 + \frac{0.0595}{365}\right)^{365} - 1$$

$$\approx 0.1046 \qquad\qquad \approx 0.0670 \qquad\qquad \approx 0.613$$

$$= 10.46\% \qquad\qquad = 6.70\% \qquad\qquad = 6.13\%$$

Microsoft Excel

For clarity, we'll label the columns "APR" and "APY" in A1 and B1, respectively. Enter the three different rates given as percentages in column A.

There is not a built-in function specifically for finding the APY, but we can type in the formula by using an equal sign followed by "(1+r/n)^n-1", where $n = 365$ and r is the value in column A. In cell B2, type "=(1+A2/365)^365-1" and press Enter. (Notice that we reference cell A2 for the rate.) The value in cell B2 appears as 0.1046. (Note that the number of decimal places will vary depending on the width of the cell).

Click on cell B2 and drag the bottom right corner of the cell down through cell B4. This will copy the formula into cells B3 and B4. (Notice that the reference to cell A2 is updated to refer to cells A3 and A4 in the respective rows.) The values 0.06705 and 0.0613 will appear in cells B3 and B4, respectively.

| B2 | | ▼ | : | × | ✓ | fx | =(1+A2/365)^365-1 | | |

	A	B	C	D	E	F
1	APR	APY				
2	9.95%	0.104603				
3	6.49%	0.067046				
4	5.95%	0.061301				

FIGURE 6.1.11

Converting these APYs to percentages and rounding to the nearest hundredth we have 10.46%, 6.70%, and 6.13%.

Now, we are ready to complete the table. Notice that the difference between the APY and the corresponding APR is not the same for all levels, even though the compounding periods are the same.

Table 6.1.7: APD Bank of the South Personal Loan Rates

Loan Amount	APR*	APY
< $20,000	9.95%	10.46%
$20,000 – $99,999	6.49%	6.70%
≥ $100,000	5.95%	6.13%

*Interest is compounded daily, and accrued interest is posted to the account monthly.

Skill Check Answers
1. $5292   **2.** $14,802.78   **3.** $1454.99   **4.** 4.28%

6.1 Exercises

✔ CONCEPT CHECK

1. The amount of money that you are charged for _____ money is called interest.

2. The two types of interest are _____ interest and _____ interest.

3. With compound interest, the interest charged applies not only to the _____, but also to any interest that has already been accrued.

4. The largest amount of interest that could possibly accrue over a given time period is _____.

5. True or False: Simple interest grows exponentially.

> ○ **PRACTICE**

Calculate the simple interest for each situation. Round your answer to the nearest cent, if necessary.

6. Determine the interest owed on $800 for 5 years at a rate of 8.5%.

7. Determine the interest owed on $5000 for 2 years at a rate of 10%.

8. Determine the interest owed on $1200 for 30 months at a rate of 19.5%.

9. Determine the interest owed on $550 for 5 years at a rate of 8.5%.

10. Determine the simple interest owed on $2700 for 3 years at a rate of 15%.

11. Determine the simple interest owed on $1050 for 18 months at a rate of 8%.

12. Determine the future value owed on $1300 after 5 years at a simple interest rate of 3.2%.

13. Determine the future value owed on $3200 after 4 years at a simple interest rate of 4%.

Use the compound interest formula to find the following for each situation.

a. Calculate the total amount in the account after the given time period.

b. Determine the amount of interest earned for each.

14. $P = \$2500$, $r = 6.5\%$ compounded weekly, $t = 10$ years

15. $P = \$3500$, $r = 4.5\%$ compounded monthly, $t = 10$ years

16. $P = \$2500$, $r = 6.5\%$ compounded daily, $t = 10$ years

17. $P = \$5650$, $r = 8\%$ compounded biannually, $t = 15$ years

18. $P = \$2500$, $r = 6.5\%$ compounded yearly, $t = 10$ years

19. $P = \$15,000$, $r = 6\%$ compounded semiannually, $t = 25$ years

20. $P = \$12,500$, $r = 8\%$ compounded biweekly, $t = 15$ years

21. $P = \$7300$, $r = 19.9\%$ compounded weekly, $t = 20$ years

22. Angela deposits $2500 into an account with an APR of 5.5% for 10 years. Find each of the following.

    a. Amount of interest earned if the account is compounded annually

    b. Amount of interest earned if the account is compounded monthly

    c. Amount of interest earned if the account is compounded weekly

    d. Amount of interest earned if the account is compounded daily

    e. Amount of interest earned if the account is compounded continuously

23. David deposits $4000. Determine the APY for each of the following.

    a. APR of 5% compounded monthly

    b. APR of 5% compounded weekly

    c. APR of 5% compounded daily

    d. APR of 7.5% compounded monthly

    e. APR of 7.5% compounded weekly

    f. APR of 7.5% compounded daily

24. Determine the simple interest earned on $10,000 after 10 years if the APR is each of the following rates.

    a. 3%

    b. 6%

    c. 12%

    d. 24%

🚀 APPLICATIONS

25. Julia takes out a simple-interest loan to buy a new TV set. If she borrows $1500 for 2 years at a rate of 12%, how much interest will she have to pay back?

26. Suppose a friend asked you to lend him $3500 and promised to return the money several years later with a 10% annual interest. How much interest will he pay you if he decides to return all the money in two and a half years?

27. Suppose your bank gives you a special offer for a simple interest loan of $10,000 for 5 years at an annual interest rate of 5%. How much will you owe the bank at the end of the loan term?

28. Kamil deposits $3000 in an account with a simple interest rate of 2.2%. If he decides to take all the money from the account after 8 years, how much will he withdraw?

29. A lending company offers small personal loans with a simple interest rate. If the loan is paid off in full within 30 days, the annual interest rate is 14%. If you take more than 30 days to return the full loan amount, the interest is calculated at an annual rate of 31% on the original loan amount. How much would you need to pay back to the company if you borrowed $400 and returned the full loan amount after 15 days?

30. Lina is purchasing a new washing machine on credit at a simple annual interest rate of 24.99%. How much will she owe the store in 60 days if the price of the washer is $650?

31. Emiel deposits $700 on a savings account with a simple annual interest rate of 6.5%. What will the future value of his account be after 180 days?

32. Bart deposits $6000 into an account with an annual percentage rate of 3.1% compounded daily. What will the future value of his account be after 5 years?

33. A bank offers Eleanor a loan of $9000 for 8 years with an annual interest rate of 5.5% compounded monthly. What will the future value of the loan be at the end of the loan period?

34. Barbara added $2100 to her bank account that already had $6200. How much will she have in the account after 2 years if the interest is compounded quarterly at an annual rate of 2.4%?

35. To buy a car, Cornelis takes out a loan of $13,500 for 15 years with an annual interest rate of 1.35% compounded semiannually. What is the total cost of the car, including interest?

36. Laura needs $3600 to renovate her condo and plans to take out a two-year loan for this purpose. After searching through available bank offers, she narrowed the choice to two banks. The first bank offered an APR of 4.4% compounded semi-annually, and the second bank offered an APR of 3.9% compounded monthly. Which of the offers should Laura choose?

37. Ralph ponders which of two banks he should place his money in. The first bank offers an APR of 2.15% compounded monthly. The second offers an APR of 2.29% compounded quarterly. Ralph has $12,000 and plans to leave it in the bank for at least 7 years. Which of the banks should he choose?

38. Donald invests $7000 into an account that pays 2.3% interest compounded continuously.

    a. Find the future value of his account in 6 years.

    b. How long will it take for his account to be worth $10,000?

39. In order to get an implant in place of her extracted tooth, Rachel takes out a five-year loan for $4000 with a 5.58% interest rate compounded continuously.

    a. How much will she need to repay at the end of the loan term?

    b. If she wants her total repayment to be no more than $5000, for how long can she borrow the money?

40. If you deposit $4500 into an account with an interest rate of 2.19% compounded continuously, how much will you have in the account in 10 years?

41. Suppose that in 2020 Amanda deposited $14,000 into an account that earns 1.89% APR and is compounded continuously. What will be the value of her investment in 2030?

42. After getting a part-time job, Evelyn decided to move from the student dorm to her own apartment. She borrowed $1600 to pay the deposits to rent the apartment and decided to pay it back over the course of two years. How much will she need to pay in total if the interest rate is 4.4% compounded continuously?

43. Samuel buys a new refrigerator using an in-store offer for a loan with an APR of 22.5% compounded continuously. If the refrigerator costs $1250, how much will Samuel owe the store after half a year?

44. After selling an old car she got from her parents for $11,000, Leah decided to place the money in a savings account with an APR of 3.07% compounded continuously. How long will it take for the money in the account to increase by $1000?

45. How long would it take for $26,000 invested at a rate of 4.09% compounded continuously to reach the value of $30,000?

46. Joshua takes a loan of $2000 with an APR of 8.01% compounded continuously. How long will it take for the amount owed to increase by $700?

47. How long would it take for an amount owed to double if the interest rate is 10.39% compounded continuously?

48. A bank is offering new customers an IRA that earns 1.75% APR compounded monthly. Adam uses this offer to deposit $15,000 for 10 years. What is the annual percentage yield for this deposit?

49. Peter invests $34,000 for six years into an account with an APR of 2.12% compounded daily. Find the annual percentage yield for this investment.

50. If $9000 is invested for 4 years at an interest rate of 7.19% compounded semi-annually, what will be the annual percentage yield?

51. Megan places $6000 into a savings account for one year. The APR for this account is 1.65% and is compounded quarterly. What is the annual percentage yield for her deposit?

52. Assume you wish to borrow $500 for two weeks and the amount of interest you must pay is $20 per $100 borrowed. What is the APR at which you are borrowing money? Round your answer to the nearest hundredth.

53. Suppose your salary in 2022 is $65,000. If the annual inflation rate is 4%, what salary do you need to make in 2030 in order for it to keep up with inflation?

54. Suppose the First Bank of Lending offers a CD (Certificate of Deposit) that has a 6.45% interest rate and is compounded quarterly for three years. You decide to invest $5500 into this CD.

    a. Determine how much money you will have at the end of the three years?

    b. Find the APY.

55. The First Bank of Lending lists the following APR for loans. Determine the APY, or effective interest rate for each category.

First Bank of Lending Loan APR	
Loan Amount	**APR**
< $20,000	11.25%
$20,000–$99,999	8.99%
≥ $100,000	5.75%
*Interest rates are compounded quarterly	

✎ WRITING & THINKING

56. You are purchasing a new computer using the store's "90 days same as cash" deal. If the cost of the computer is $1575 (tax included) with an annual interest rate of 19.99%, how much would you owe on the 91st day if you make no payments during the first 90 days?

6.1 PROJECT

INFLATION AND TIME TRAVEL: THE VALUE OF MONEY OVER TIME

In 1979, the federal minimum wage in the United States was $2.90 per hour. By 2019, that number increased to $7.25 per hour. While there is a clear increase in the minimum wage in dollars, it is hard to compare the two figures without knowing more about how the economy changed between 1979 and 2019.

In this activity, you will explore different ways to compare the value of money across different time periods.

Economists define inflation as the rise of price levels in the economy over time. The average yearly rate of inflation between 1979 and 2019 was 3.2%. That means prices of goods and services went up by an average of 3.2% every year between 1979 and 2019.

1. Suppose that an item cost $1.00 in 1979. What was the price of that item in 2019 when adjusted for inflation? (**Hint:** You can think about this in terms of an investment that compounds interest—if you invested $1.00 in 1979 at 3.2% compound annually, how much would you have in 2019?)

The answer to part 1 represents the value of $1.00 of 1979 currency in 2019 when adjusted by inflation.

2. What is the 2019 value of the 1979 minimum wage adjusted by inflation? Is this value more or less than the actual 2019 minimum wage?

3. Purchasing power is considered the amount of goods and services that can be bought with a single unit of money. Is the purchasing power of workers in 2019 lower or greater than the purchasing power of workers in 1979?

4. Perform an internet search to compare the average yearly tuition at a 4-year public university between 1979 and 2019. Did tuition increase faster or slower than the inflation rate over the same time period?

During the same time period, the average compensation for company CEOs in the US has increased by 940%.

5. Determine the 2019 value of the 1979 minimum wage if it had grown at the same rate as the CEO compensation.

6. Discuss some ways to adjust the disparity between the highest and lowest paid employees in our economy.

6.2 SAVING AND INVESTING

OBJECTIVES

1. Calculate present value and annuity payments.

2. Calculate future value of an annuity.

3. Calculate future value of retirement plans.

4. Read a stock market chart.

Fun Fact

Born and raised near Atlanta, Georgia, Bill Copeland (1946–2010) was an American poet, writer and historian whose works focused on the Holocaust, World War II, and Native Americans.

In 2003, a small city on the southwest coast of Alabama was warned by a pension expert that if they continued on with their current spending without any changes, the city would run out of money for their retirement fund by mid-2009. In September 2009, the retirement fund did just that—it ran out. As a result, even though it was illegal, the city stopped paying monthly pensions to its 150 retired employees: firemen, police officers, and city workers. This historic first was a warning not only to other cities, but to individuals as well. We all need to responsibly plan for our futures.

Saving for the future is often referred to as "building a nest egg"—a phrase you've heard before; but what is a nest egg and how does someone build one? A *nest egg* is a sizeable amount of money that has been saved or invested for later use. Most commonly, nest eggs are earmarked for long-term uses like retirement, buying a house, or paying for college. Whatever the end objective, there are a few standard principles to help you build your nest egg:

- Set a Goal.

- Start now.

- Pay fewer taxes on earnings.

- Diversify.

OBJECTIVE 1 Set a Goal: Take Advantage of Compound Interest

> The trouble with not having a goal
> is that you can spend your life
> running up and down the field and never score
> —Bill Copeland

How much money do you need to save and when should you start saving? Knowing what you are aiming for is half the battle. Setting specific financial goals helps you keep track of your progress and hold yourself accountable. Once you have a monetary goal, there are ways to use interest-bearing accounts to reach that goal—you can either deposit a lump sum of money and begin to earn interest, or you can make regular deposits along the way while also earning interest over time. Either way, you benefit more by having time work for you, as we saw in the last section.

For example, many states have college savings plans that allow parents to deposit a lump sum amount of money into their child's college fund based on current college costs. The amount deposited will then earn enough interest to cover the predicted cost of college 18 years in the future. The only catch is you must have a lump sum payment to invest now.

Using algebra, we can rearrange the future value formula for compound interest, which was introduced in Section 6.1, to determine how much money needs to be invested *now* in order to meet our goal in the *future*. This amount needed to invest now is called the *present value*.

Helpful Hint

The terms *present value*, used here, and *principal*, used earlier, both refer to an initial amount of money and are often used interchangeably.

PRESENT VALUE WITH COMPOUND INTEREST

The amount of principal needed now in order to reach a future value amount is called the present value (PV). The **present value with compound interest** is calculated with the following formula.

$$PV = \frac{A}{\left(1 + \dfrac{r}{n}\right)^{nt}}$$

Here A is the future value, r is the APR written as a decimal, n is the number of compound intervals per year, and t is time in years.

Fun Fact

The online cost predictor in Example 6.2.1 uses a national average to determine the annual percentage increase of higher education costs. However, individual states may have lower or higher percentage increases. For instance, Georgia's annual cost of higher education has only increased 1.9% a year during the years 2016 through 2020.

Example 6.2.1

Calculating Present Value

An online college cost predictor indicates that the average cost for four years of college at a public institution with in-state tuition is currently $87,800. With a 5% annual cost increase, it predicts that in 18 years students will need $227,000 on average to attend the same type of school. What amount of money should be invested now so that an account with an APR of 3% compounded monthly will accrue enough money to cover the estimated cost of college in 18 years? Round your answer to the nearest cent.

Solution

Begin by identifying the information needed to find present value: future value, interest rate, number of compounding periods, and the number of years to invest. Notice that there are two percentages given. We will need to use the APR in the formula, not the annual cost increase.

By Hand

From the information given we know the following quantities.

$$A = \$227{,}000 \quad r = 3\% = 0.03 \quad n = 12 \quad t = 18$$

Substituting these values into the present value formula, we have the following equation.

$$PV = \frac{A}{\left(1 + \dfrac{r}{n}\right)^{nt}}$$

$$= \frac{\$227{,}000}{\left(1 + \dfrac{0.03}{12}\right)^{(12)(18)}}$$

$$\approx \$132{,}373.03$$

TI-83/84 Plus

As in the previous lesson, we can press $\boxed{\text{apps}}$, then select `Finance...` and use the TVM Solver.

- In this case, the interest is compounded monthly for 18 years, so $N = 12 \cdot 18 = 216$.

- For I%, enter 3. Remember that when using the TVM Solver, we use the percentage form of the rate, not the decimal form.

- The present value is what we need to solve for, so we will come back to the PV entry after filling out the other values.

- There are no monthly payments being made, so PMT = Ø.

- The future value is the amount we need to have in 18 years, which is $227,000. The future value and the present value will always have opposite signs; in this situation, we will use a negative value for FV so that the calculator will return a positive number for the present value that we are solving for, so FV = −227ØØØ.

- Because the interest is being compounded monthly, P/Y = C/Y = 12.

- We'll assume that payments are made at the end of each period, so be sure that END is highlighted in the final row.

Once these values are filled out, return your cursor to the PV entry and press $\boxed{\text{clear}}$. To solve for this value, press $\boxed{\text{alpha}}$ $\boxed{\text{enter}}$. The calculator returns 132373.Ø264.

Microsoft Excel

Excel has a built-in financial function **PV(rate, nper, pmt, [fv], [type])** that can be used to calculate the present value needed to reach a certain value in the future. The specific arguments for this function are similar to the arguments for the future value function that we used in the previous lesson:

rate is the interest rate per period. This is the interest rate divided by the number of compounding periods per year.

nper is the total number of compounding periods. This is the number of periods per year multiplied by the length of the loan in years.

pmt is the payment made each period. This can be left blank if no payments are made.

fv is the future value. Default is 0.

type refers to whether payments are being made at the beginning of each period or at the end; 0 = end of the period, 1 = beginning of the period. Default is 0.

- Since the interest rate is 0.03 and it is being compounded monthly, which is 12 times a year, we use rate = 0.03/12. Be careful to not use only the interest rate here.

- Because the interest is being compounded monthly for 18 years, we will use $12 \cdot 18$ for nper.

- There are no payments being made, so we will use pmt = 0, or leave that argument empty.

- The future value is the amount we need to reach at the end of the 18 years, which is $227,000. The future value and the present value will always have opposite signs; in this situation we will use a negative value for fv so that Excel will return a positive number for the present value that we are solving for, so fv = −27,000.

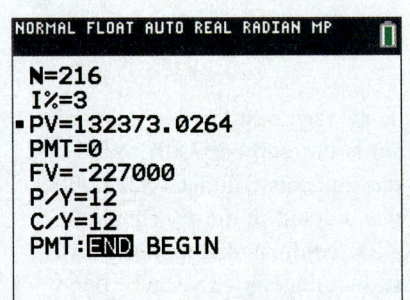

Figure 6.2.1

NORMAL FLOAT AUTO REAL RADIAN MP

N=216
I%=3
▪PV=132373.0264
PMT=0
FV=-227000
P/Y=12
C/Y=12
PMT:**END** BEGIN

- Typically, payments are made at the end of each period, so we will use type = 0 or leave that argument empty.

In an empty cell, type "=PV(0.03/12,12*18,,-227000,0)", and press Enter. The present value is given as $132,373.03.

A1	▼	⋮	×	✓	fx	=PV(0.03/12,12*18,,-227000,0)

	A	B	C	D	E	F	G
1	$132,373.03						

FIGURE 6.2.2

This means that if you can deposit 132,373.03 into the account now, then after 18 years there will be $227,000 in the account.

✓ Skill Check 6.2.1

How much should you deposit now in an account with an APR of 7% compounded monthly if you wish for the account to have a balance of $200,000 in 25 years?

After working through the previous example, you might have a bit of sticker shock. Don't worry, you are in the majority if your first thought was "That's great. But I don't have $132,000 at my disposal to invest in anything." Instead of depositing a lump sum of money, another way to meet your goal is by making regular payments into an interest-bearing account. These payments are usually far less than a large lump sum payment, but compounding interest and time can still work in your favor. The significant point with this type of savings plan is that consistent payments are ongoing. The fixed, regular payments are referred to as *annuity* payments.

📖 ANNUITY

An **annuity** is a sequence of regular payments made into an account, or taken out of an account, over time.

As it happens, you might already have an established annuity. Common examples of annuities are regular home mortgage payments, monthly car payments, monthly insurance payments, and pension payments. In the same manner that we determined what the initial lump sum would need to be to meet a future goal, we can use a formula to determine how much our annuity payment (the regular payment amount) needs to be to meet that future goal.

📖 ANNUITY PAYMENT AMOUNT

The **annuity payment amount (PMT)**, or regular payment amount, needed to be deposited at the end of each compounding period in order to reach a future goal with an annuity savings account is calculated with the following formula.

$$\text{PMT} = \text{FV} \cdot \frac{\left(\dfrac{r}{n}\right)}{\left[\left(1+\dfrac{r}{n}\right)^{nt} - 1\right]}$$

Here, FV is the future value desired, r is the APR written as a decimal, n is the number of compounding intervals per year, and t is time in years.

When determining the payment amount required to reach a goal for the future value, it is best to round up to the nearest dollar if the formula results in a decimal value. This ensures that we are making the minimum deposit required to either reach or exceed our goal.

Let's return to the first example in which we wished to save enough money to pay for the predicted cost of college 18 years from now. Instead of depositing a lump sum of $132,373.03 into an investment account, let's determine the monthly annuity payment amount needed to reach the college fund goal.

Example 6.2.2

Determining Annuity Payments in Order to Meet a Goal

It is estimated that in 18 years, the average cost for four years of college at a public institution with in-state tuition will be $227,000.

a. Determine what monthly annuity payment amount is needed for an account with an APR of 3% to accrue enough money to cover the estimated cost of college in 18 years. Round your answer up to the nearest dollar, if necessary.

b. How much of the $227,000 will ultimately be deposited into the account and how much is interest earned?

Solution

a. The formula to determine the annuity payment amount requires knowing not only the goal amount and date, but also the annual percentage rate along with the number of compounding intervals per year.

By Hand

We are told that FV = $227,000, $r = 0.03$, $n = 12$, and $t = 18$. Substituting these values into the annuity payment formula, we have the following equation.

$$\text{PMT} = \text{FV} \cdot \frac{\left(\dfrac{r}{n}\right)}{\left[\left(1+\dfrac{r}{n}\right)^{nt} - 1\right]}$$

$$= \$227,000 \cdot \frac{\left(\dfrac{0.03}{12}\right)}{\left[\left(1+\dfrac{0.03}{12}\right)^{(12)(18)} - 1\right]}$$

$$\approx \$793.87$$

TI-83/84 Plus

We can again use the TVM Solver, but this time we'll be solving for the PMT. The account in this example has the same terms as the account in Example 6.2.1, so many of the values will be the same. Instead of trying to find the necessary present value, however, we are starting with a present value of 0 and finding the monthly payment amount needed to meet the goal.

- Again, the interest is compounded monthly for 18 years, so N = 12 · 18 = 216.

- The interest rate is still 3%, so I% = 3.

- In this scenario, we are not starting with any money in the account, so PV = Ø.

- The monthly payment amount is what we need to solve for, so we will come back to the PMT entry after filling out the other values.

- The future value is the amount we need to have in 18 years, which is again $227,000. Since the future value and the PMT amount will always have opposite signs, we will use a negative value for FV so that the calculator will return a positive value for the regular payment amount. This means FV = -227000.

- Because the interest is being compounded monthly, P / Y = C / Y = 12.

- We'll assume that payments are made at the end of each period, so be sure that END is highlighted in the final row.

Once these values have been filled out, return your cursor to the PMT entry and press clear. To solve for this value, press alpha enter. The calculator returns 793.8718699.

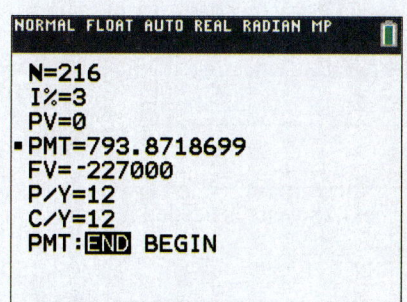

Figure 6.2.3

Microsoft Excel

Excel has a built-in financial function **PMT(rate, nper, pv, [fv], [type])** that can be used to calculate the amount that each monthly payment should be in order to reach a future goal. The specific arguments for this function are similar to the arguments for the future value and present value functions.

rate is the interest rate per period. This is the interest rate divided by the number of compounding periods per year.

nper is the total number of compounding periods. This is the number of periods per year multiplied by the length of the loan in years.

pv is the present value, or principal amount. Default is 0.

fv is the future value. Default is 0.

type refers to whether payments are being made at the beginning of each period or at the end; 0 = end of the period, 1 = beginning of the period. Default is 0.

The account in this example has the same terms as the account in Example 6.2.1, so many of the values will be the same. Instead of trying to find the necessary present value, however, we are starting with a present value of 0 and finding the needed monthly payment amount.

- Again rate = 0.03/12. Be careful to not use just the interest rate here.

- Because the interest is being compounded monthly for 18 years, we will use 12 · 18 for nper.

- We are not starting with any money in this account, so we will use pv = 0, or leave that argument empty.

- The future value is the amount we need to reach at the end of the 18 years, which is $227,000. Since the future value and the PMT amount will always have opposite signs, we will use a negative value for fv so that the calculator will return a positive value for the regular payment amount. So we'll use –227000 for fv.

- Typically, payments are made at the end of each period, so we will use type = 0, or leave that argument empty.

In an empty cell, type "=PMT(0.03/12,12*18,0,-227000,0)" and press Enter. The payment amount is given as $793.87.

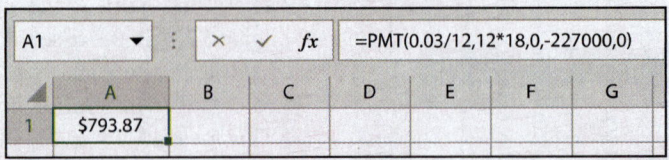

A1	▼	⋮	×	✓	fx	=PMT(0.03/12,12*18,0,-227000,0)

	A	B	C	D	E	F	G
1	$793.87						

FIGURE 6.2.4

Therefore, a monthly deposit of $794 over the next 18 years is needed in order to meet the goal of saving $227,000.

b. We know that 18 years of monthly payments of $794 are required to meet the goal. We can calculate the overall contribution that is deposited by multiplying the monthly payment amount by the number of payments a year and then by the number of years.

$$\text{Payment Amount} \cdot \text{Payments per Year} \cdot \text{Number of Years} = \$794 \cdot 12 \cdot 18$$
$$= \$171,504$$

The interest earned is then the difference between the goal and the contributed amount. Subtracting, we have the following equation.

$$\$227,000 - \$171,504 = \$55,496$$

So the total amount deposited into the annuity account over the 18 years is $171,504, and the total amount of interest earned is $55,496.

While the savings methods from Examples 6.2.1 and 6.2.2 are two possibilities to help cover the cost of higher education, there are many other opportunities to meet that goal. Governmental and private scholarships, grants, and loans are just a few. We'll look more at loans in a later section.

OBJECTIVE 2   Start Now: Take Advantage of Annuities

If you start now,
you'll begin seeing results one day earlier
than if you start tomorrow.
—Author Unknown

As we have seen, time is a large factor in earning interest. Letting funds sit untouched over a long period of time can help us meet our goals. But even without a specific goal in mind, forming good habits of saving and investing is very wise. The main thing is to begin. Start putting money in an account today and make it a regular commitment—it doesn't matter the size of the payment. Even a small amount each month or each week can ultimately make a big difference.

In Section 6.1, we saw how we can calculate the future value of a lump-sum investment by knowing the principal amount along with the annual interest rate and the compounding intervals. We can do the same for annuity accounts. The formula for future value of an annuity takes into account not only the compound interest for each interval, but also the regular payment that is added at the end of each interval.

> ### 📖 FUTURE VALUE OF AN ANNUITY
>
> The **future value (FV) of an annuity** account is calculated with the following formula.
>
> $$FV = PMT \cdot \frac{\left[\left(1+\dfrac{r}{n}\right)^{nt} - 1\right]}{\left(\dfrac{r}{n}\right)}$$
>
> Here, PMT is the payment amount deposited at the end of each compounding interval, r is the APR written as a decimal, n is the number of compounding intervals per year, and t is time in years.

Example 6.2.3

Calculating Future Value of an Annuity

Donovan began placing $75 per month into an annuity savings plan when he received his first official paycheck in August 2021. The savings plan earned 1.5% APR, compounded monthly.

a. Calculate the future value of Donovan's savings in August 2031. Round your answer to the nearest cent, if necessary.

b. Determine the amount of money that Donovan will personally put in the savings plan over the period given and the amount of interest that will be earned.

Solution

a. The future value formula for annuity can be used to calculate how much Donovan will have in his savings by August 2031.

> **By Hand**
>
> We are told that Donovan contributes $75 at the end of each compounding interval, which is monthly. Therefore, PMT = $75 and $n = 12$. The APR given is 1.5%, so $r = 0.015$. The time period is from August 2021 through August 2031, so $t = 10$.
>
> Substituting these values in the annuity formula to find the future value, we have the following equation.

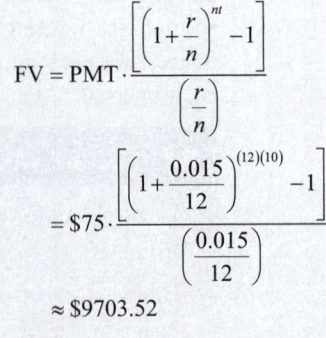

$$FV = PMT \cdot \frac{\left[\left(1 + \dfrac{r}{n}\right)^{nt} - 1\right]}{\left(\dfrac{r}{n}\right)}$$

$$= \$75 \cdot \frac{\left[\left(1 + \dfrac{0.015}{12}\right)^{(12)(10)} - 1\right]}{\left(\dfrac{0.015}{12}\right)}$$

$$\approx \$9703.52$$

TI-83/84 Plus

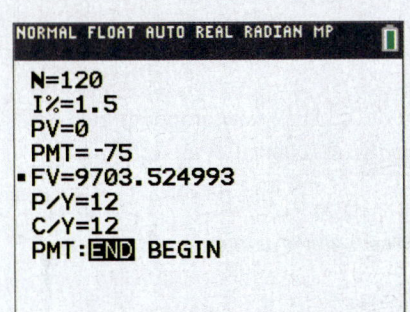

Figure 6.2.5

We can again use the TVM Solver to solve for the future value, except this time PMT will not be 0.

- The interest is compounded monthly for 10 years, so N = 12 · 10 = 120.

- The interest rate is 1.5 %, so I% = 1.5.

- In this scenario, we are not starting with any money in the account, so PV = 0.

- We saw earlier that the future value and the PMT amount will always have opposite signs. This time we will use a negative value for PMT so that the calculator will return a positive number for the future value. Since Donovan is making monthly payments of $75, we have PMT = −75.

- The future value is the amount we need to solve for, so we will come back to the FV entry after filling out the other values.

- Because the interest is being compounded monthly, P / Y = C / Y = 12.

- We'll assume that payments are made at the end of each period, so be sure that END is highlighted in the final row.

Once these values have been filled out, return your cursor to the FV entry and press [clear]. To solve for this value, press [alpha] [enter]. The calculator returns 9703.524993 .

Microsoft Excel

We can use the built-in financial function **FV(rate, nper, pmt, [pv], [type])**, which we used in the previous section.

- Since the interest rate is 0.015 and it is being compounded monthly, we use rate = 0.015/12.

- Because the interest is being compounded monthly for 10 years, we will use 12 · 10 for nper.

- We saw earlier that the future value and the monthly payment amount will always have opposite signs. This time we will use a negative value for pmt so that the calculator will return a positive number for the future value. Monthly payments of $75 are being made, so we use −75 for pmt.

- We are not starting with any money in this account, so we will use pv = 0, or leave that argument empty.

- Typically, payments are made at the end of each period, so we will use type = 0 or leave that argument empty.

In an empty cell, type "=FV(0.015/12,12*10,-75,0,0)", and press Enter. The future value is given as $9,703.52.

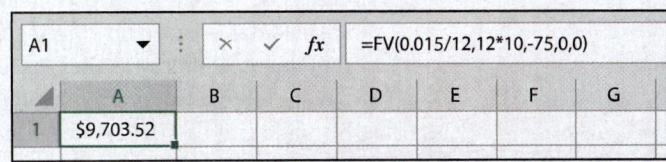

FIGURE 6.2.6

Therefore, after 10 years, Donovan's savings account will contain approximately $9703.52.

b. To find the amount of money that Donovan will personally add to the savings plan, we need to multiply the monthly payment amount by the number of payments each year and then by the number of years he plans to continue his payments. This gives us the following calculation.

$$\text{Payment Amount} \cdot \text{Payments per Year} \cdot \text{Number of Years} = \$75 \cdot 12 \cdot 10$$
$$= \$9000$$

So over 10 years Donovan will contribute $9000 to his savings plan. Since we know that the account will contain $9703.52, the interest earned in the account is $9703.52 − $9000 = $703.52.

✔ Skill Check 6.2.2

Use the annuity formula to calculate the future value after six months of monthly $50 payments for an annuity savings plan that compounds monthly with an APR of 0.8%.

OBJECTIVE 3 Pay Fewer Taxes on Investment Earnings

A person doesn't know how much he has to be thankful for until he has to pay taxes on it. —Ann Landers

Retirement savings are what most people think of when they think of nest eggs and their goals. Because it is how you will care for yourself later on, planning for retirement should be a top priority. But if you wish to maximize your future spending, you can benefit from tax advantages that retirement savings accounts offer. These accounts are different from other investment accounts because they have different tax stipulations and limitations.

Unlike traditional savings accounts, taxes collected on earnings in retirement accounts can be delayed or paid up front in order to save more later. However, most retirement funds require you, by federal law, to leave your investment in the account until you are $59\frac{1}{2}$ years old to avoid paying penalties. Let's take a quick look at three types of retirement savings accounts and some of their advantages: Traditional IRAs, Roth IRAs, and 401(k)s.

💬 Helpful Hint

Similar to 401(k), educators and nonprofit workers are offered 403(b) and government employees are offered 457(b) plans.

- A **401(k)** is an employer-initiated retirement account that's offered as an employee benefit. As an employee, you are allowed to contribute a portion of your paycheck into this account before you pay taxes on it while deferring the taxes until later. This has an immediate effect of reducing the overall amount of taxes that you pay in that particular year. For example, if you earn $60,000 one year but contributed $5000 to your 401(k), you only pay taxes on $55,000. As an added benefit, some employers match their employee contributions to a 401(k) up to a certain amount; that is, they contribute the same amount of money to your account as you do each month. This is often a bargaining tool when negotiating a new contract with an employer. However, a potential downside is that when you make a withdrawal from a 401(k) later on, taxes are calculated based on your income at that time. For some people, their yearly retirement withdrawal may exceed their yearly salary, so they may end up paying a higher rate. The federal government requires minimum withdrawals be made from a 401(k) once you reach the age of 72.

- **Traditional Individual Retirement Accounts (IRAs)** are not employer initiated; as the name suggests, they are opened by individuals. Unlike bank accounts, you pay no taxes on the deposits contributed or the investment earnings until you withdraw the money, which helps the account to grow faster. As of 2020, you are allowed IRA investments up to $6000 a year, or $7000 if you are age 50 or older. The limits allowed for IRAs are adjusted periodically by the Internal Revenue Service and are updated on their website, irs.gov. After the age of $59\frac{1}{2}$, an IRA withdrawal is taxed as current income, and IRAs have mandatory withdrawals after the age of 72.

- **Roth IRAs** differ from traditional IRAs in that contributions are taxed before they are deposited. Although this leads to more taxes in the present, neither the deposit nor any interest generated within the Roth IRA are taxed when you withdraw them in retirement. This is a positive if your tax bracket is higher when you retire; that is, the percentage of taxes you owe gets larger as your yearly income increases over your lifetime. With Roth IRAs, after 5 years (and age $59\frac{1}{2}$) you may also withdraw contributions you've made without penalty before retirement age. However, withdrawing earnings early will likely incur a penalty. Additionally, a Roth IRA does not require the owner to begin taking withdrawals at the age of 72 as is the case with other retirement savings plans. In other words, you can continue to earn interest on the entire account.

Despite the special names for retirement accounts, they are all fundamentally annuity accounts. The same formulas we have been using to find future values and payment amounts are still appropriate.

Example 6.2.4

Calculating Future Value of a Retirement Account

From June 2020 to June 2031, an employee makes monthly contributions of $123.25 to a 100% employer-matched 401(k), which is invested at an average APR of 6.3% compounded monthly. After June 2031, the employee leaves the company but keeps the retirement account.

a. Calculate the future value of the account when the employee leaves the company in June 2031.

b. Calculate the future value of the account in June 2045, assuming the employee makes no additional contributions or withdrawals after leaving?

c. How much interest is earned in the 401(k) from June 2020 to June 2045?

d. *The four percent rule of thumb* advises that your savings should last through 30 years of retirement if you withdraw only 4% of your nest egg during the first year of retirement. Based on the 4% rule of thumb, how much could the employee start withdrawing per month in June 2045, assuming they are old enough at that point to not incur a penalty? Round your answer to the nearest cent.

Solution

a. The future value formula for an annuity can be used to calculate how much the employee will have in the 401(k) by June 2031.

By Hand

We are told that the employee contributes $123.25 each month. But since this amount is matched by the employer, the actual PMT is $123.25 \cdot 2 =$ $246.50. The number of compounding intervals per year is 12. The APR given is 6.3%, so $r = 0.063$. The time period is from June 2020 through June 2031, so $t = 11$ years.

Substituting these values in the annuity formula to find the future value, we have the following equation.

$$FV = PMT \cdot \frac{\left[\left(1+\dfrac{r}{n}\right)^{nt} - 1\right]}{\left(\dfrac{r}{n}\right)}$$

$$= \$246.50 \cdot \frac{\left[\left(1+\dfrac{0.063}{12}\right)^{(12)(11)} - 1\right]}{\left(\dfrac{0.063}{12}\right)}$$

$$\approx \$46,768.51$$

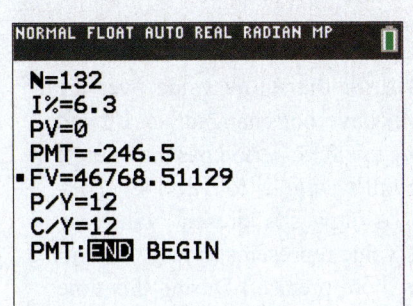

Figure 6.2.7

TI-83/84 Plus

This is another scenario where we can use the TVM Solver to solve for the future value.

- The interest is compounded monthly for 10 years, so $N = 12 \cdot 11 = 132$.

- The interest rate is 6.3%, so $I\% = 6.3$.

- In this scenario, we are not starting with any money in the account, so $PV = \emptyset$.

- Since the employee's monthly contributions of $123.25 are being matched by her employer, the monthly payments into the account are $2 \cdot \$123.25$, so we have $PMT = -246.5$.

- The future value is the amount we need to solve for, so we will come back to the FV entry after filling out the other values.

- Because the interest is being compounded monthly, $P/Y = C/Y = 12$.

- We'll assume that payments are made at the end of each period, so be sure that END is highlighted in the final row.

Once these values have been filled out, return your cursor to the FV entry and press clear. Press alpha enter to solve for this value. The calculator returns 46768.51129.

Microsoft Excel

We can again use the built-in financial function **FV(rate, nper, pmt, [pv], [type])** with rate = 0.063/12, nper = 12 · 11, and pmt = −246.5.

In an empty cell, type "=FV(0.063/12,12*11,-246.50,0,0)" and press Enter. The future value is given as $46,768.51.

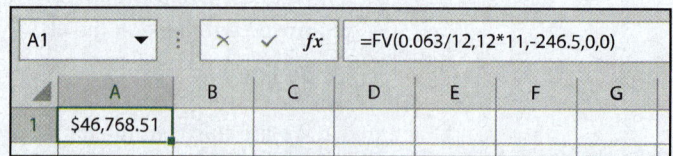

FIGURE 6.2.8

Therefore, after 11 years, the 401(k) account will contain approximately $46,768.51.

b. From 2031 to 2045, the account will still accrue interest, but monthly payments will no longer be added to the account. So now we need to use the future value formula for compound interest with a principal of $46,768.51. There are 14 years between 2031 and 2045 where interest continues being compounded monthly at a rate of 6.3%.

By Hand

$$A = P\left(1 + \frac{r}{n}\right)^{nt}$$

$$= \$46,768.51\left(1 + \frac{0.063}{12}\right)^{(14)(12)}$$

$$\approx \$112,719.56$$

TI-83/84 Plus

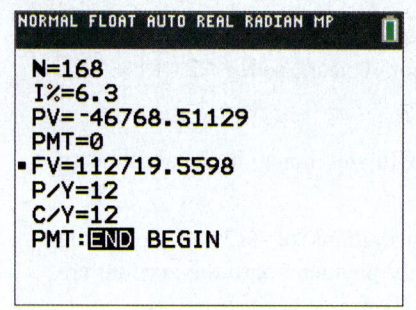

Figure 6.2.9

We'll use the TVM Solver again to solve for the future value over the next time period. The terms of the 401(k) have not changed, so I% and P/Y = C/Y are unchanged. However, since this time period has the interest compounding monthly for 14 years, we will change N to $12 \cdot 14 = 168$. The future value we found in part a. is now the present value, so PV = -46768.51129. (Although this is value represents money, we will not round here so that the final answer is more precise.) During this time period, no monthly payments are being made, so PMT = 0. Once all of these values have been filled out, return the cursor to the FV entry and press clear. Solve for this value by pressing alpha enter. The calculator returns 112719.5598.

Microsoft Excel

We can again use the built-in financial function, **FV(rate, nper, pmt, [pv], [type])**, with rate = 0.063/12, nper = 12 · 14, pmt = 0, and pv = -46768.51.

In an empty cell, type "=FV(0.063/12,12*14,0,-46768.51,0)" and press Enter. The future value is given as $112,719.56.

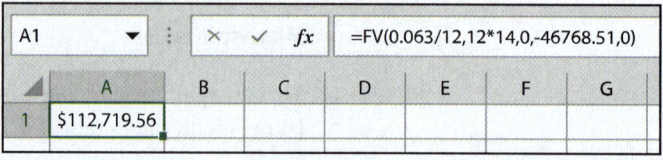

FIGURE 6.2.10

Therefore, after 25 years the 401(k) will have approximately $112,719.56 in the account.

c. The interest on the account can be found by subtracting the amount of the contributions from the future value of the account after 25 years.

Remember that there were 12 payments of $246.50 each year for 11 years.

$$12 \cdot \$246.50 \cdot 11 = \$32,538$$

So over the 25 years $32,538 was contributed to the 401(k). The interest the account earned is then $112,719.56 − $32,538 = $80,181.56.

d. The 4% rule of thumb suggests that withdrawing only 4 % out of a retirement fund in the first year will enable the savings to last through 30 years of retirement. After calculating 4 %, we need to divide by 12 to determine the monthly withdrawal amount. Since the account will have $112,719.56 in June 2045, the calculation is as follows.

Suggested Maximum Monthly Withdrawal:

$$\frac{\$112,719.56 \cdot 0.04}{12} \approx \$375.73$$

Therefore, the employee should only withdraw $375.73 per month from the retirement account beginning in June 2045.

OBJECTIVE 4 Diversify with Stocks and Bonds

> How many millionaires do you know
> who have become wealthy by investing in savings accounts?
> I rest my case
> —Robert G. Allen

So far we've only discussed cash investments as a way to save money; that is, depositing money into a bank account or some other interest-bearing account. These are relatively safe, or low risk, ways to grow your nest egg. However, along with low risk often comes low interest. Often the rates on interest-bearing accounts do not even keep pace with rising inflation rates. In other words, relying only on savings accounts to grow your retirement fund requires that you start saving very early and wait a very long time to begin spending it. To balance the slower growth of bank savings accounts, many financial advisors recommend building your nest egg by having a variety of investments, some of which are considered higher risk like *stocks*. Along with higher risk comes the chance of higher gain

Stocks

Stocks are investments people make in a corporation. When you buy the stock of a company, you are effectively buying a share in the ownership of that company and are now a **shareholder**. The number of shares you purchase determines how much of the company you own. The more shares you buy, the bigger the slice of the company is yours. As a shareholder, if the company is successful, then so are you! The value of your shares grows just like the value of the company. Of course, if the company does poorly, you also suffer.

> **Helpful Hint**
>
> A stock exchange is an entity that facilitates the buying and selling of stocks, shares, and bonds. There are 60 exchanges globally. The top 16 are referred to as the "$1 Trillion Club" and comprise 87% of the global market capitalization. The New York Stock Exchange and NASDAQ are the two largest exchanges.

One of the ways to make money from investing in stocks is to receive a portion of the profit. This money, divided among shareholders, is called a **dividend**. The larger number of shares you own, the bigger the dividend you receive. However, not all companies issue dividends. Instead, they sometimes choose to reinvest profit back into the company to help with its growth.

Another way to earn money with stocks is to sell your shares for more than you purchased them. For instance, let's say you invest $2000 in a company whose stock price is $50 a share. You then own 40 shares. As the company does well, the value of the stock increases to $75 a share (a 50 % increase); the value of your investment also increases to $3000. You could then sell those shares to another investor for a profit of $1000. The profit from the sale of an asset such as stock is referred to as a **capital gain**.

Both dividends and capital gains are considered income and, hence, are taxable. However, if you hold on to stocks for more than a year or own stock with qualified dividends, the *long-term capital gains* are generally taxed at a lower rate than other taxable income. For instance, if you earn less than $40,000 per year, the tax rate for long-term capital gains is 0 %.

So how do you know what stocks are available to purchase or which ones to invest in? In the past, in order to get information about stocks, the public relied on printed newspapers or speaking with someone by phone who was physically at the stock market. These days, we can get real-time information on the internet or through smartphone apps and observe prices change right before our eyes. Regardless of the website or app that you choose to use, there is a set of standard information available about all stocks. The following is a look at some of the most important information about the Tesla stock as viewed in the Yahoo Finance app.

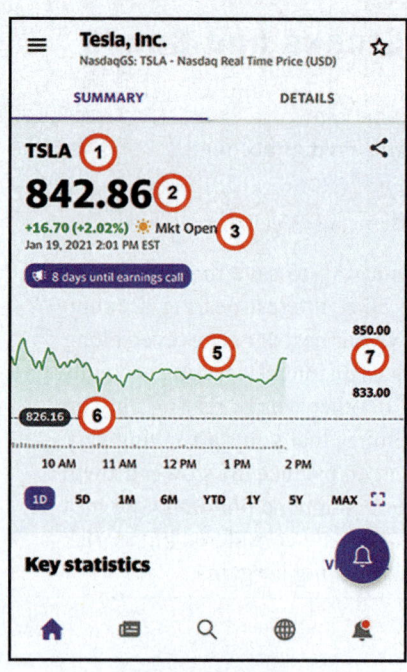

Figure 6.2.11

1. **Ticker Symbol** Every stock that is publicly traded has a symbol consisting of a few characters that identify the company. A three-letter symbol usually means the stock trades on the New York Stock Exchange. A four-letter symbol means it probably trades on NASDAQ.

2. **Trading Price** Current price of the stock for one share.

3. **Gain or Loss** Smaller numbers in either green indicating a gain (noted with a +) or red indicating a loss (noted with a −) for the day. The percentage is the percentage increase or percentage decrease from the previous price to the current price

4. **After-Hour Price** This represents trading that occurred outside of normal business hours. The market hours for US exchanges are 9:30 am to 4:00 pm EST. You can watch this value to predict what the opening price of the stock might be. Because the market is open in Figure 6.2.11, no after-hours price is shown

5. **Graph** The graph shows how the price of the stock has fluctuated over a period of time of your choosing: 1 day, 5 days, 1 month, 6 months, year to date (YTD), 1 year, 5 years, entire stock history (MAX).

6. **Previous Close** The price of the stock when the market closed the previous day. A difference between this price and the opening price reflects any reactions or concerns that investors might have while the US market is closed. World events happen 24 hours a day and don't respect normal business hours.

7. **Day's Range** The lowest and highest prices that the stock has traded for the day during business hours. This does not specify how long either the low or high price held. It could have only held for a few minutes or several days.

8. **Market Cap** Short for *market capitalization*, this is the total value of the company; it's found by multiplying the number of shares by the price of each share. By dividing this number by the share price, we can determine how many shares the company is currently offering. Tesla's current worth of $798,780,000,000 with a share cost of $842.86 tells us there are approximately $\frac{\$798,780,000,000}{\$842.86} \approx 947,701,872$ shares offered at the moment.

9. **Volume** The number of shares that have been bought and sold for the day. As a number on its own, 17,540,000 may not have much value unless you've been watching the stock over some time and have a reference point for how many shares are usually traded a day. That's where Avg Volume (3m) comes in. This is a three-month average of shares traded per day.

10. **1y Target Est** The one-year target estimate is the price that analysts have predicted the stock will be one year from now. A price of $500.27 means that analysts believe Tesla is headed for a decrease in value over the next year. The bottom line is that it is an estimate but it can help you evaluate potential risk or reward.

11. **P/E** Price-to-earnings ratio is calculated by dividing a stock's share price by its earnings per share; that is, $\text{PE} = \dfrac{\text{Price per Share}}{\text{Earning per Share}}$. The PE ratio provides you a guide as to how much you can expect to invest in a company in order to receive one dollar of that company's earnings. For instance, you can expect to pay $1611.26 to get a $1 in return for investing in Tesla. In general, if you want to "buy low and sell high," you would want to buy stocks when PE ratios are low and sell when PE ratios are high.

12. **EPS (TTM)** Earnings per share over the past 12 months. As a shareholder, the higher this number, the better!

13. **Beta** This measures how volatile a stock's price has been in recent history compared to the overall market. A beta score of 1 means the stock's price activity is consistent with the movement of the overall market. Greater than 1 indicates more unpredictability than the average stock while less than 1 means the stock price has been more stable than the overall market. Generally, investors with very low tolerance to risk avoid stocks with a high beta. While stocks with a higher beta value add risk of loss to a portfolio, they also have the possibility of increasing return. Most stocks have beta values between 0 and 3. In relation to the rest of the stock market, Tesla is twice as volatile as the rest of the stock market with a beta score of 2.19.

14. **Dividend** This value indicates the annual dividend each share receives. If a company does not give out dividends, N/A is displayed. While large dividends can be attractive to investors, it should not be the single decision maker when deciding to buy a stock. Companies that are facing financial trouble often raise their dividends to attract new investors.

Although this is a lot of information, we've only identified the key statistics displayed on the first two screens of the app. Most stock chart apps also provide information about the company, any related news articles, and sometimes even expert advice about whether the stock is a good purchase or not.

Figure 6.2.12

Example 6.2.5

Reading a Stock Chart

Use the Apple stock listing from the Yahoo Finance app to answer each question.

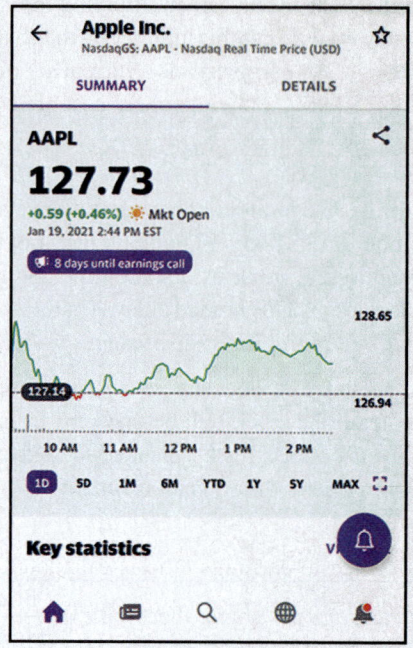

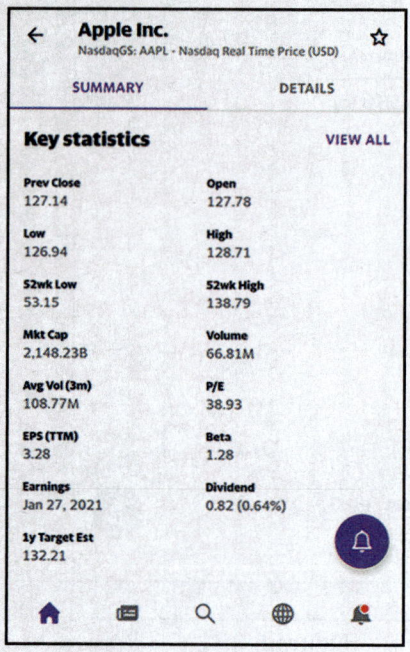

Figure 6.2.13   **Figure 6.2.14**

a. What is the ticker symbol for Apple?

b. What is the current price of the stock?

c. What do analysts predict the price of Apple stock will be in one year?

d. Is this stock more or less volatile than the market as a whole? How do you know?

e. Does Apple pay dividends to its shareholders? If so, how much would you receive if you owned 50 shares?

f. How many shares has Apple issued?

Solution

a. The ticker symbol for Apple is AAPL, found at the top of the summary screen.

b. Apple currently has a stock price of $127.73.

c. Using the 1y Target Est we can see that analysts predict Apple stock will have a value of $132.21 in one year.

d. Because the beta number of 1.28 is more than 1, we know that Apple's stock price is slightly more volatile than the market in general.

e. Yes, Apple does pay dividends to its shareholders. At the moment, if you owned 50 shares, you would receive $0.82 · 50 = $41.00.

f. The market cap number tells us the stock value of the company. We can divide this number by the price of the stock to find approximately how many shares Apple has issued. Apple's market cap is 2,148.23B, or 2,148,230,000,000. The current stock price is $127.73.

$$\frac{2,148,230,000,000}{127.73} \approx 16,818,523,448$$

Therefore, Apple has issued approximately 16.8 billion shares.

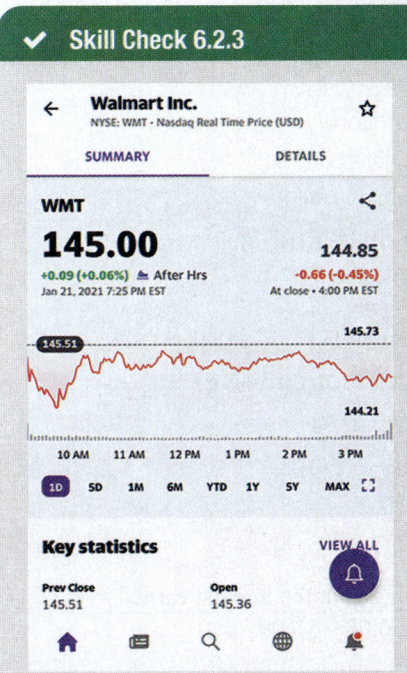
Bonds

Bonds are another way to invest in companies. But unlike stocks, when you purchase a bond, you do not have part ownership in a company. Instead, you are acting more like a bank. Bonds are more of a loan from you to the company or government that offers them. When governments and companies find themselves in need of cash, either for debt purposes or to fund new projects, they sell bonds to investors. In return, they promise to pay back the full amount of the bond plus interest after a set time; this is called the **maturity date** of the bond. Bonds are less risky than stocks in that you know the terms of the interest rate and the length of time. However, if a company goes bankrupt during the bond period, you may not get your investment back. But you are at the front of the line, ahead of stockholders, to claim your funds. Example 6.2.6 looks at a bond investment.

Example 6.2.6

Calculating Interest on US Treasury Bonds

A Series EE bond is a US government savings bond. It is sold at its face value and pays a fixed interest rate until it reaches 30 years or you cash it in, whichever comes first. To calculate the interest on a Series EE bond, a combination of simple interest and a form of compound interest is used. The bond earns simple interest on its face value every month. However, the interest is compounded once every six month, and simple interest is then calculated on the new compounded amount. This process is repeated throughout the life of the bond.

The minimum length of ownership is one year. If you redeem the bond before five years, you forfeit the interest earned from the previous three months. There is no penalty for redemption after five years. Regardless of the interest rate, the bond is guaranteed by the US government to be worth twice its face value 20 years after purchase.

Suppose you purchase an EE bond with a face value of $500 and a 0.10 % APR.

a. Determine the profit earned on the bond if you redeem it after 74 months.

b. What is the value of the bond 20 years after it was purchased?

Solution

a. The profit earned on the bond is the value of the bond at 74 months minus the original cost of the bond. Note that 74 months is equal to 6 years and 2 months. Since the bond earns interest monthly but the earned interest only compounds every 6 months, we can use the future value for compound interest formula, with $n = 2$, to calculate the value of the bond after 6 years. Then, we can use the simple interest formula to calculate the interest earned over the final 2 months.

To find the value after 6 years, we know that $P = \$500$, $r = 0.001$, $n = 2$, and $t = 6$. Substitute these values into the formula and simplify.

$$A = P\left(1+\frac{r}{n}\right)^{nt}$$

$$= \$500\left(1+\frac{0.001}{2}\right)^{(2)(6)}$$

$$\approx \$503.01$$

So after 6 years, the value of the bond is $503.01. Over the first 6 years of the bond, $503.01 − $500 = $3.01 interest was earned.

To find the interest earned over the final two months of the bond, we use the simple interest formula $I = Prt$ with $P = \$503.01$, $r = 0.001$, and $t = \frac{2}{12} = \frac{1}{6}$.

$$I = Prt$$

$$= \$503.01(0.001)\left(\frac{1}{6}\right)$$

$$\approx \$0.08$$

After 2 months, the bond earned $0.08. Adding this to the interest earned over the first 6 years gives a total profit of $3.01 + $0.08 = $3.09.

b. A Series EE bond is guaranteed to double in face value 20 years after purchase. This means that the government makes a one-time adjustment to bring the value to the guaranteed amount, if necessary. In this case, if the value of the bond, which was bought for $500, has not exceeded the $1000 mark after 20 years, the government will adjust its value to be $1000.

To find the value of the bond 20 years after purchase, we use $P = \$500$, $r = 0.001$, $n = 2$, and $t = 20$. Substituting these values into the future value for compound interest formula and simplifying gives the following.

$$A = P\left(1+\frac{r}{n}\right)^{nt}$$

$$= \$500\left(1+\frac{0.001}{2}\right)^{(2)(20)}$$

$$\approx \$510.10$$

Since the future value after 20 years does not exceed the $1000 mark, the government will adjust the value of this bond to make it worth $1000. Therefore, after 20 years, this bond would be worth $1000.

Now that you know about different investment options, the question becomes "Where should you invest your money?" Many online tools exist to help you decide what is best for your situation. One thing to consider is how bonds and stocks have performed historically. The Bloomberg Barclays US Aggregate Bond Index is a broad benchmark for bonds. It looks at the value of approximately 43% of the total US bond market to gauge the value of bonds across the market. Similarly, the S&P 500 Index measures the performance of the 500 largest companies in the US stock market. Table 6.2.1 compares the performance of bonds and stocks by looking at these two indices.

Notice that over the past 20 years, bonds tend to do their best when stocks are at their worst and vice versa. An exception to this was during the 2008, which shows the result of the United States housing bubble.

A general overview of different types of investments like this might have you wondering if managing your investments yourself is plausible. The short answer is yes! But for those who would prefer a professional investor to manage their investments, one option is a *mutual fund*. In short, a **mutual fund** is an investment where a professional fund manager pools together the resources of many individual investors and purchases a combination of stocks and bonds in order to maximize profit. When you invest in a mutual fund you might own a portion of lots of different companies or bonds. It is a means of attaining relatively large diversity in your investments with relatively low cost or risk to the investor. Whether you choose to decide on the particular ways to invest your money, or rely on others' expertise, there are certain fees associated with buying and selling assets that need to be factored in the overall cost of investment.

Many options exist in which you can invest money and prepare for your future. For instance, you're doing just that by earning a higher education degree. If you have started on the path to home ownership, your real estate investment is also adding to your assets. Pat yourself on the back for those smart decisions and then keep the momentum going. It's a satisfying feeling to watch your investments work for you.

Table 6.2.1: Return % of Bonds and Stocks

	Bloomberg Barclays US Agg Index (Bonds)	S&P 500 Index(Stocks)
2000	11.63%	− 9.11%
2001	8.43%	− 11.89%
2002	10.26%	− 22.10%
2003	4.10%	28.68%
2004	4.34%	− 10.88%
2005	2.43%	4.91%
2006	4.33%	15.79%
2007	6.97%	5.49%
2008	5.24%	− 37.00%
2009	5.93%	26.46%
2010	6.54%	15.06%
2011	7.84%	2.11%
2012	4.22%	16.00%
2013	− 2.02%	32.39%
2014	5.97%	13.46%
2015	0.55%	1.25%
2016	2.65%	12.00%
2017	3.54%	21.70%
2018	− 0.05%	− 4.38%
2019	8.72%	31.49%

Source: Thomas Kenny, "Aggregate Bond Index Returns vs. Stocks '80-'18," The Balance, last modified November 18, 2020, https://www.thebalance.com/stocks-and-bonds-calendar-year-performance-417028

Skill Check Answers
1. $34,931.95 **2.** $300.50 **3. a.** Walmart, Inc. **b.** $145 **c.** Lower, because the after-hours price is showing a small loss.

6.2   Exercises

✔ CONCEPT CHECK

1. The _____ is the amount you need to invest now in order to meet a future goal.

2. A sequence of _____ made into or taken out of an account is called an annuity.

3. If you buy stock in a company, you are buying a share in the ownership of that company, and you become a _____.

4. A dividend is the money that is divided amongst shareholders when a _____ earns a profit.

5. True or False: Contributions to a Roth IRA are taxed before they are deposited, but withdrawals are not taxed.

💡 PRACTICE

For each situation, use the formula for present value of money to calculate the amount you need to invest now in one lump sum. Round your answer to the nearest cent, if necessary.

6. $25,000 after 10 years with an APR of 8% compounded monthly

7. $25,000 after 10 years with an APR of 12% compounded monthly

8. $100,000 after 18 years with an APR of 6% compounded quarterly

9. $1,000,000 after 40 years with an APR of 10% compounded monthly

10. $22,000 after 12 years with an APR of 7% compounded quarterly

11. $54,000 after 8 years with an APR of 9% compounded semiannually

12. $115,000 after 15 years with an APR of 11% compounded monthly

13. $800,000 after 30 years with an APR of 13% compounded quarterly

A CD, or "certificate of deposit," is a type of savings account with a fixed rate and term, meaning you can only redeem it when the term is over. Answer the following questions about CDs. Round your answers to the nearest cent, if necessary.

14. What is the profit earned on a 6-month CD with a face value of $600 and a 0.45% APR compounded monthly?

15. What is the profit earned on a 1-year CD with a face value of $2000 and a 0.51% APR compounded monthly?

16. What is the profit earned on a 3-year CD with a face value of $1500 and a 0.71% APR compounded monthly?

17. What is the profit earned on a 5-year CD with a face value of $800 and a 0.77% APR compounded monthly?

🚀 APPLICATIONS

18. Revere College predicts that in 18 years it will take $200,000 to attend the college for four years. Debbie wishes to save money for her child's college fund. How much should Debbie put aside in an account with an APR of 9% compounded monthly in order to have $200,000 in the account in 18 years?

19. Richard just entered a university, and his parents are planning to give him a car upon graduation. How much should they invest into an account with an APR of 10% compounded monthly if they estimate that in four years the car will cost $15,000? Round your answer to the nearest cent.

20. Bella plans to buy a new laptop that should be released in a year. She predicts that the laptop will cost about $2500. How much should she deposit now into an account with an APR of 5% compounded monthly so that the account accrues enough money to cover the cost of the laptop when it is released? Round your answer to the nearest cent.

21. Anita sets a goal to have $100,000 in her account when she turns 40. What amount of money should she invest now at an APR of 4% compounded monthly to reach her goal if she is 22 this year? Round your answer to the nearest cent.

22. Raul's vision started to deteriorate while he was in middle school, and by the time he entered high school, he had to wear glasses. Together with his parents, he plans to undergo a laser vision correction in 6 years when he turns 21. They estimate the cost of the surgery at that time will be $5500. How much should they deposit into an account with an APR of 8% compounded quarterly to obtain the desired sum in 6 years? Round your answer to the nearest cent.

23. Kevin learns that the game he has long awaited will be released in two years. He is not sure that his computer will be able to handle this game, so he decides to buy a new one. He estimates that the cost of the computer will be $3200. If he wants to have enough money to buy it by the time the game is released, how much should he invest into an account with an APR of 4.5% compounded monthly? Round your answer to the nearest cent.

24. Carol wants to build a pool in her backyard in 5 years, so she decides to set aside a certain amount of money for this purpose. If the average cost of building a pool is $50,000, how much should she invest into an account with an APR of 8.2% compounded quarterly to reach her goal? Round your answer to the nearest cent.

25. Alexis and Will are purchasing a home. They wish to save money for 10 years and purchase a house that has a value of $180,000 with cash. If they deposit money into an account paying 12% interest, how much do they need to deposit each month in order to make the purchase?

26. Marilyn wishes to retire at age 65 with $2,000,000 in the bank. At the age of 21, she decides to begin depositing money into an account with an APR of 11%. What is the monthly payment Marilyn must make in order to make this happen?

27. Suppose you wish to retire at the age of 65 with $80,000 in savings. Determine your monthly payment into an IRA if the APR is 7.5% and you begin making payments at

    a. 20 years old.

    b. 30 years old.

    c. 40 years old.

28. Suppose you'd like to save enough money to pay cash for your next car. The goal is to save an extra $26,000 over the next 6 years. What amount of quarterly payments must you make into an account that earns 5.5% interest in order to reach your goal?

29. Willie deposits a fixed monthly amount into an annuity account for his child's college fund. He wishes to accumulate a future value of $75,000 in 15 years.

    a. Assuming an APR of 3.5%, how much money should Willie deposit monthly in order to reach his goal?

    b. How much of the $75,000 will Willie ultimately deposit in the account, and how much is interest earned?

30. Francis hopes that in 20 years, he will be able to purchase his own house. He estimates the amount of money needed for this will be $450,000. How much should he deposit each month into an account with an APR of 4.4% compounded monthly to reach his goal? Round your answer to the nearest dollar.

31. Gabriel plans to retire when he has $1,500,000 in his bank account, and he does not want to work more than 30 years. If his account has an APR of 5.4%, determine the minimum monthly annuity payment he would need to make. Round your answer to the nearest dollar.

32. Sophia has decided to undergo rhinoplasty when she turns 25. She estimates that the procedure will cost about $8000. Determine the annuity payments she should deposit quarterly into an account with an APR of 3.2% to reach her goal if she is 21. Round your answer to the nearest dollar.

33. Samantha wants to open a clothing store using her own money. She estimates that she will need about $80,000. How much should she deposit quarterly into an account with an APR of 2.4% in order to accrue enough money in ten years? Round your answer to the nearest dollar.

34. Hasan needs $1200 to buy a new graphics card for his computer.

    a. How much should he deposit monthly into an account with an APR of 2.2% to accrue the desired amount in a year? Round your answer to the nearest dollar.

    b. How much of the graphics card cost will be deposited by Hasan, and how much will be the interest?

35. Noah has always dreamed of having a yacht. When he was 18, he decided to start saving money to buy one. He learned that the cost of a yacht starts at $500,000.

    a. If his goal is to own a yacht by age 30, how much should he deposit each quarter into an account with an APR of 8.7% in order to have enough money to purchase a $500,000 yacht? Round your answer to the nearest dollar.

    b. If he reaches his goal, how much of the $500,000, will come from deposits, and how much will come from interest?

36. Blake starts an IRA (Individual Retirement Account) to save for retirement at the age of 22. He deposits $450 each month. The IRA has an average annual interest rate of 7%.

    a. How much money will he have saved upon retirement at the age of 65?

    b. Determine the amount of money Blake deposited over the length of the investment and how much he made in interest.

37. Jimmie has a job at an advertising agency earning $54,000 per year. Jimmie is currently 26 years old and wishes to retire at age 67 with a retirement income of $75,000. How much money would Jimmie need to invest each month into a growth stock mutual fund with an interest rate of 6.5% in order to withdraw $75,000 per year without reducing the principal?

38. After getting married, Liam and Olivia decide to start a shared annuity savings plan. They plan to deposit $1500 every quarter into an account that earns 2.1% APR compounded quarterly. How much will they have in the account on the day of their Pearl Anniversary, which is the 30th wedding anniversary? Round your answer to the nearest cent.

39. After graduating from college, Ela decides that she wants to start saving so that she can move out of her parents' home and start a new chapter in her life. She starts placing $90 each month into an account with an APR of 3.4% compounded monthly. How much will she have in her account if she does this faithfully for five years? Round your answer to the nearest cent.

40. When Celine was ten, she asked her parents to start placing $10 from the money they gave her each month into a savings account. If the account earns 4.1% APR compounded monthly, how much will Celine have when she turns 18? Round your answer to the nearest cent.

41. Alexander had difficulty finding a new job, so he decided to create an emergency fund in case this happened again. He plans to deposit $2000 every quarter into an account with an APR of 5.8% compounded quarterly. How much will he have in his emergency fund after 10 years? Round your answer to the nearest cent.

42. Elias decided he would go on a trip to Australia when he saved enough money. He started to deposit $150 each month in an account with an APR of 4.8% compounded monthly.

    a. How much will he save in three years? Round your answer to the nearest cent.

    b. How much of the saved amount will be deposited by Elias, and how much will be the interest?

43. When Mia was 21, she started depositing $500 each quarter into an account that earns 7.2% APR compounded quarterly.

    a. Find the future value of her account in 20 years. Round your answer to the nearest cent.

    b. Determine how much of the future value Mia will personally put into the account and how much will she earn in interest.

44. When she was 35, Emma started an IRA account with an APR of 7.3% compounded quarterly. Until she turned 50, she contributed $1400 every quarter. Then, she stopped making deposits and left the account to accrue interest for the next ten years.

    a. Find the value of the account when Emma was 50.

    b. Find the value of the account when she turned 60.

    c. How much interest did the account earn until the moment Emma turned 60?

    d. According to the four percent rule of thumb, how much could Emma start to withdraw each quarter after she turned 60? Round your answer to the nearest cent.

45. In the first twelve years after opening an IRA account, Hugo deposited $280 into it every month. After that, he was promoted to a higher position and from then on decided to deposit $420 each month. The APR for the account is 5.4% compounded monthly.

    a. Calculate the future value of the account after the first twelve years.

    b. Calculate the future value of the account after ten years of increased deposits.

    c. How much interest is earned over the twenty-two years?

    d. If after twenty-two years Hugo is old enough to not incur any penalties, how much can he start withdrawing each month according to the four percent rule of thumb? Round your answer to the nearest cent.

46. Use the Facebook stock listing in the following table to answer each question.

Facebook Inc. (FB) ↑ 333.65 +0.69 (+0.21%)	
Prev. Close	Open
332.96	329.19
Low	High
325.80	334.38
52wk Low	52wk High
244.69	384.33
Mkt Cap	Volume
938.76B	21,767,335
Avg Vol (3m)	P/E
14,696,273	24.69
EPS (TTM)	Beta
13.49	1.34
1y Target Est	Dividend
419.77	N/A

a. What is the ticker symbol for Facebook?

b. What is the percentage change in the stock price for the day?

c. Does Facebook pay dividends to its shareholders?

d. Has the stock price increased or decreased while the market was closed, and by how much?

e. Is the number of shares bought and sold for the day greater or less than the average?

f. How many shares has Facebook issued?

47. Use the Nike stock listing in the following table to answer each question.

Nike Inc. (NKE) ↓ 148.72 -0.74 (-0.50%)	
Prev. Close	Open
149.46	148.00
Low	High
147.52	149.46
52wk Low	52wk High
118.80	174.38
Mkt Cap	Volume
236.61B	5,512,446
Avg Vol (3m)	P/E
5,910,602	39.70
EPS (TTM)	Beta
3.76	0.89
1y Target Est	Dividend
173.78	1.10 (0.75%)

    a. What is the ticker symbol for Nike?

    b. What is the change in the stock price for the day?

    c. What do analysts predict the price of Nike stock will be in a year?

    d. Has the stock price increased or decreased while the market was closed, and by how much?

    e. How many shares has Nike issued?

    f. How much would you receive in dividends if you owned 70 shares?

✏ WRITING & THINKING

48. Why is it important for the payments in the formula for the future value of an annuity to be made at the end of each compounding interval and not in the middle or any other part of it?

49. If you follow the four percent rule of thumb in the future, will it guarantee that your retirement funds will last through 30 years of retirement?

6.2 PROJECT

PAYOUT ANNUITIES: WHAT HAPPENS AFTER YOU RETIRE?

According to Northwestern Mutual, 56% of adults in the US don't know how much money they will need to save to retire. This can be problematic when setting up your long-term financial goals. In this activity, you will explore a financial instrument, called a payout annuity, that can be used to invest after retirement and maintain a steady income.

Suppose you want to have an after-retirement annual income of $50,000 for 20 years.

1. Suppose you plan to place your retirement fund into an account that does not earn interest. How much money would you need in the account by the time you retire?

Without further investing, your retirement fund will sit idle when it could be earning interest. A payout annuity provides regular withdrawals while allowing your balance to earn interest. The following formula is used to calculate the value of a payout annuity that compounds annually and has annual withdrawals.

$$P = \frac{d\left(1-\left(1+r\right)^{-N}\right)}{r}$$

Here, P is the starting balance of the account (that is, the size of your retirement fund), d is the regular annual withdrawal, r is the annual interest rate as a decimal, and N is the number of years you plan to take withdrawals.

2. Suppose that you will invest your retirement fund (the value you found in part 1) for 20 years at an interest rate of 7% per year. Up to how much could you withdraw yearly in this case and still meet your goal? In other words, what is the value of d in the annuity formula in this case?

3. Why is the value you found in part 2 larger than $50,000? Where is the extra money coming from?

4. Suppose you want to keep your withdrawal at $50,000 per year. At the same 7% interest per year, what starting principal would you need if you want to run out of money in the account after 20 years?

5. Discuss the reasons you might want to start your retirement with a higher principal than the one found in part 4.

OBJECTIVES

1. Compare sources for borrowing money.

2. Calculate values for fixed installment loans.

3. Calculate the monthly payment for a mortgage.

4. Create an amortization table.

5. Calculate values related to credit card debt.

6. Read credit reports.

Sometimes we need to borrow money in order to make purchases, bridge the gap between paychecks, or cover unexpected financial emergencies. Beyond asking a friend or family member, many options exist in the financial world to help you meet those needs. **Creditors**, or **lenders**, come in all shapes and sizes. These are organizations, institutions, or even private entities that lend you money with the idea that when you pay the money back, you will also pay any fees and interest. We're going to begin by giving a brief rundown of some of the most widely used lenders that are available, along with some of the pros and cons of each.

OBJECTIVE 1 Sources for Borrowing Money

Banks The most traditional source for borrowing money is a brick-and-mortar bank. Banks' long-standing methods of loaning money have allowed them to develop a seemingly unending list of loan options with nuanced requirements and fees. They typically charge relatively lower interest rates, have the ability to offer larger loan amounts, and often give perks to existing customers such as reduced or waived fees. Even though banks have the benefit of personal customer service, on the downside they often charge high application fees or service fees.

Credit Unions The difference between credit unions and banks is that credit unions are nonprofit organizations. Credit unions often offer better interest rates and lower fees than banks. In order to qualify for a loan from a credit union, you must be a member. This allows you the added benefit that loan officers are willing to consider other factors beyond your single credit score so that a low credit score doesn't automatically disqualify you. A downside is that they usually have fewer loan options to choose from.

Financing Companies An alternative to traditional banks or credit unions, financing companies deal strictly in providing loans and make their money on the interest they charge to clients. Today, an increasing number of financing companies are online-only lenders. Because they don't have the expenses of maintaining physical branches, their overhead costs are lower. They often offer smaller loans for cars, appliances, or bridging gaps between paychecks.

Peer-to-Peer Lending (P2P) Since 2005, *social lending*, or *crowd lending*, has become a popular mode of borrowing. Eliminating a financial institution as a middleman, this type of lending allows individuals to be either the borrower or the lender. Websites match investors and borrowers based on profiles and need. Lenders determine how much of a loan they are willing to make based on an agreed interest rate. A lender can also choose to offer a portion of the amount needed and not the entire loan. The remaining amount of the loan can then be funded by others on the site. Similar to banks, fees are often incurred along with the interest rate.

Public Agencies The US government, or a sponsored agency, offers loans for extended periods of time at comparatively lower rates. For example, the Department of Education offers loans for undergraduates, graduate students, and parents of undergraduates, as well as ways to consolidate federal student loans. The Department of Housing and Urban Development (HUD) offers many federal programs to help with loans. Different arms of HUD offer specific financial help to different groups of the population: first time home buyers, military, Native Americans, and rural residents to name a few. Small business loans are available through several

Fun Fact

In March 2020, at the outset of the Coronavirus pandemic, the CARES Act offered $2 trillion in government funding to individuals, small businesses, and medical providers, among others, as emergency relief during the global economic disruption.

government loan programs. Quite often government loans have strict criteria for approval and take much longer to receive than other types of loans.

Retirement Funds Most 401(k)s or similar accounts allow you to borrow up to 50% of the money you've contributed to the fund with the agreement that you pay it back by a certain time with interest. Since this is your money to begin with, you are essentially borrowing from yourself. Any interest collected is deposited back into your account. As long as you pay it back, there is no penalty for taking money out before you reach the age of $59\frac{1}{2}$. The downside to borrowing from your retirement plan is that you are losing out on the compound interest it could be earning, which hurts the growth of your nest egg.

Credit Cards Credit cards should be thought of as a short-term small loan. Because of their high interest rates, they are not good for long-term debt. On one hand, there are no application fees once you have a card and money is immediately available. When you pay off the balance at the end of each billing cycle, no interest is charged. In addition, some cards offer incentives for using the card often—reward points to use for cash back, gift cards, travel, or shopping. However, any balance not paid off each month incurs relatively high interest rates. Having too many credit cards can also reduce your credit score and chance of getting loans from other institutions.

Pawn Shops A means to use physical assets you already own, these types of loans use an object of value that you own as collateral. The pawn shop keeps the object until you pay back the money with interest. If you fail to pay back the money, then they own the object and can sell it to reclaim their money. Because you are leaving collateral with the pawn shop, there are no credit checks or other requirements. Aside from possibly losing your possessions, the biggest downside to these types of loans is the high interest rates that are charged to reclaim your possession.

> **💬 Helpful Hint**
>
> Collateral is something used as security for the repayment of a loan. It is forfeited if the loan is not paid back.

This is certainly not an exhaustive list of sources that will loan money, and each one has its own benefits and downsides. The takeaway is that there is not a "one size fits all" to loans, and it is wise to explore different options based on your needs and particular situation. Regardless of the lender you choose, there are several standard ways money is lent and interest is repaid. We will take a look at two of these: *fixed installment loans* and *open-ended loans (revolving credit)*.

OBJECTIVE 2 Fixed Installment Loans

For a *fixed installment loan*, the terms for the amount of the loan and the length of time it will take to pay off the loan are agreed upon up front, so that both the lender and the borrower know exactly how much the loan will cost. Because everything is decided at the beginning of the loan period, fixed installment loans usually use fixed interest rates—rates where the APR never changes over the life of the loan.

Sometimes a down payment is required for a fixed installment loan. A *down payment* is a cash payment made up front towards the total purchase price. Once the down payment is subtracted from the original price, the amount remaining is the principal that needs to be *financed*, or borrowed with interest.

FIXED INSTALLMENT LOAN AND DOWN PAYMENT

Fixed Installment Loan

A **fixed installment loan** is a loan with a fixed interest rate and loan period. Repayments are made in equal installments throughout the life of the loan.

Down Payment

A **down payment** is a cash payment made up front towards the total purchase price of the goods or service.

When a fixed installment loan is taken out, several factors go into determining the monthly repayment amount. The initial consideration is the size of the down payment. Down payment requirements usually range between 3% and 35% depending on the amount of the loan and the borrower's credit score. For instance, lenders for car loans often want you to show your commitment to paying back the loan and ask for at least 10% down for used cars and at least 20% down for new cars. When purchasing a house, the minimum down payment can be as low as 3% for first-time homeowners or buyers who qualify for government-backed loans, or as high as 35% for "credit-challenged" borrowers. The general rule of thumb has been that you should put 20% down on a home mortgage to get the best interest rate and avoid having to buy mortgage insurance. In the end, a larger down payment can often mean a better interest rate, lower monthly payments, and lower fees.

The other consideration is the length of time over which the loan will be paid off. For car loans, repayment times are usually between three and six years. For federal student loans, generally you have 10 years after graduating to repay the loan. Home mortgages can range from 15 to 30 years. Although a longer time period means lower payments in all of these cases, it also can mean higher interest rates. We already know from the earlier sections that higher interest rates over longer periods of time mean more interest accrues and hence you pay more. However, a bigger down payment and shorter amount of time can mean a lower interest rate in most cases.

The following formula is used to calculate the regular payment amount for fixed installment loans.

Helpful Hint

Notice that the formula contains a negative exponent in the denominator. Be careful not to drop the negative sign during calculations.

REGULAR PAYMENT FOR FIXED INSTALLMENT LOANS

The amount of a regular payment (PMT) on a fixed installment loan is calculated with the following formula.

$$\text{PMT} = \frac{\left(P \cdot \dfrac{r}{n}\right)}{\left[1 - \left(1 + \dfrac{r}{n}\right)^{-nt}\right]}$$

Here, P is the principal amount of money borrowed, r is the APR in decimal form, n is the number of payments per year, and t is time in years.

Fun Fact

The US Department of Education pays the interest on a Direct Subsidized Loan while you are in school at least half-time. However, you are responsible for paying the interest on a Direct Unsubsidized Loan during all periods.

Example 6.3.1

Determining the Monthly Payment on a Fixed Installment Loan

In order to complete their bachelor's degree in business, a student borrowed $22,946 in subsidized federal loans. Once the six-month grace period is over after graduating, they will begin to repay the loan over 10 years. If the APR is 2.75%, calculate the monthly payment amount the student will be responsible for.

Solution

By Hand

We can use the formula for fixed installment loans to calculate the monthly payment amount. We are told the principal P is $22,946. The payments will be monthly, so $n = 12$. The length of the loan repayment t is 10 years. And the interest rate is 2.75%, so $r = 0.0275$.

Substituting into the formula, we have the following equation.

$$\text{PMT} = \frac{\left(P \cdot \dfrac{r}{n}\right)}{\left[1 - \left(1 + \dfrac{r}{n}\right)^{-nt}\right]}$$

$$= \frac{\left(\$22,946 \cdot \dfrac{0.0275}{12}\right)}{\left[1 - \left(1 + \dfrac{0.0275}{12}\right)^{-(12)(10)}\right]}$$

$$\approx \$218.9302$$

TI-83/84 Plus

As in the previous lesson, we can use the TVM Solver to find the monthly payment. Notice that the present value is the amount being financed, the future value is 0, and we are finding the PMT.

- N is the total number of times the account is compounded. Since the loan is being paid back with monthly payments over 10 years, N = 12 *10 = 120.

- The interest rate is 2.75%, so I% = 2.75.

- The present value and the PMT amount will always have opposite signs; we will use a negative value for PV so that the calculator will return a positive value for PMT. Since the student is borrowing $22,946, we will use PV = -22946.

- The monthly payment amount is what we need to solve for, so we will come back to the PMT entry after filling out the other values.

- At the end of the loan, no money will be owed, so FV = Ø.

- Because the interest is being compounded monthly, P/Y = C/Y = 12.

- We'll assume that payments are made at the end of each period, so be sure that END is highlighted in the final row.

```
NORMAL FLOAT AUTO REAL RADIAN MP
 N=120
 I%=2.75
 PV=-22946
■PMT=218.9301509
 FV=0
 P/Y=12
 C/Y=12
 PMT:END BEGIN
```

Figure 6.3.1

Once these values have been filled out, return your cursor to the PMT entry and press clear. To solve for this value, press alpha enter. The calculator returns 218.9101509.

Microsoft Excel

As in the previous lesson, we can use the **PMT(rate, nper, pv, [fv], [type])** function to calculate the amount that each monthly payment should be in order to pay off the loan.

- In this example, we have an interest rate of 0.0275 that is being compounded 12 times a year, so for the first argument in this Excel function, we have rate = 0.0275/12. Be careful to not use just the interest rate here.

- Since the loan is being paid back with monthly payments over 10 years, nper is $12 \cdot 10 = 120$.

- The present value and the PMT amount will always have opposite signs; we will use a negative value for pv so that Excel will return a positive value for PMT. Since the student is borrowing $22,946, we will use pv = –22946.

- We will use FV = 0, or leave that argument empty, because nothing will be owed by the end of the loan.

- Typically, payments are made at the end of each period, so we will use type = 0, or leave that argument empty.

In an empty cell, type "=PMT(0.0278/12,120,-22946,0,0)" and press Enter. The payment amount is given as $218.93.

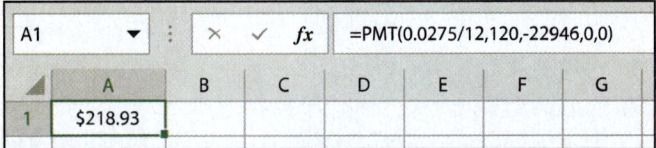

FIGURE 6.3.2

Therefore, after graduating, the student will make monthly payments of $218.93 over 10 years to repay the loans.

Fun Fact

Most auto loans are advertised with the number of months (typically 60 or 72) unlike other loans such as mortgages that are described in terms of years.

Example 6.3.2

Using an Installment Loan to Purchase a New Car

Suppose you are interested in buying a new hybrid car that offers a $500 instant rebate for college graduates. The purchase price is $36,550, including taxes and fees.

a. Assuming you qualify for the instant rebate, determine the monthly payments for a 72-month loan with a down payment of 20% of the price and an APR of 3.99%.

b. At the end of the 72 months, how much will you have paid for the new car?

Solution

a. First, we need to know the amount of money that you will be borrowing. We need to deduct both the instant rebate and the down payment you are going to make.

Begin by finding the down payment amount: 20 % of $36,550

$$\text{Down Payment} = 0.20 \cdot \$36{,}550 = \$7310$$

Subtract both the $500 rebate as well as the down payment to determine the amount you need to finance.

$$\text{Principal to Finance} = \$36{,}550 - \$7310 - \$500 = \$28{,}740$$

By Hand

Now we have $P = \$28{,}740$. A 72-month loan means that $t = \dfrac{72}{12} = 6$ years, and we will make 12 payments a year ($n = 12$) with an APR of 3.99 %, so r is 0.0399. Substituting into the formula for fixed installment loans, we can calculate what your monthly payment will be.

$$\text{PMT} = \frac{\left(P \cdot \dfrac{r}{n} \right)}{\left[1 - \left(1 + \dfrac{r}{n} \right)^{-nt} \right]}$$

$$= \frac{\left(\$28{,}740 \cdot \dfrac{0.0399}{12} \right)}{\left[1 - \left(1 + \dfrac{0.0399}{12} \right)^{-(12)(6)} \right]}$$

$$\approx \$449.5116$$

TI-83/84 Plus

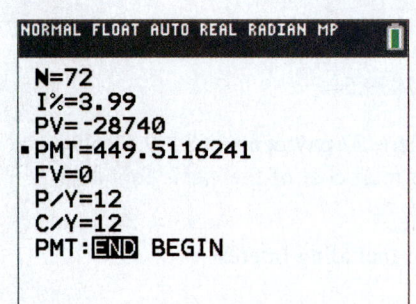

```
NORMAL FLOAT AUTO REAL RADIAN MP    ▯
 N=72
 I%=3.99
 PV=-28740
■PMT=449.5116241
 FV=0
 P/Y=12
 C/Y=12
 PMT:END BEGIN
```

Figure 6.3.3

We can again use the TVM Solver to find the monthly payment. Notice that the present value is the amount being financed, the future value is 0, and we are finding PMT.

- N is the total number of times the account is compounded. Since you are making 72 monthly payments, N = 72.

- The interest rate is 3.99%, so I% = 3.99.

- The present value and the PMT amount will always have opposite signs. We will use a negative value for PV so that the calculator will return a positive value for PMT. Since the amount being financed is PV = −28740.

- The monthly payment amount is what we need to solve for, so we will come back to the PMT entry after filling out the other values.

- At the end of the loan, no money will be owed, so FV = 0.

- Because the interest is being compounded monthly, P/Y = C/Y = 12.

- We'll assume that payments are made at the end of each period, so be sure that END is highlighted in the final row.

Once these values have been filled out, return your cursor to the PMT entry and press clear. To solve for this value, press alpha enter. The calculator returns 449.5116241.

Microsoft Excel

We can again use the **PMT(rate, nper, pv, [fv], [type])** function to calculate the amount that each monthly payment should be in order to pay off the loan.

- In this example, we have an interest rate of 0.0399 that is being compounded 12 times a year, so for the first argument in this Excel function, we have rate = 0.0399/12. Be careful to not use just the interest rate here.

- Since this is a 72-month loan, nper is 72.

- The present value and the PMT amount will always have opposite signs; we will use a negative value for the present value so that Excel will return a positive value for PMT. Since the amount being financed is $28,740, we will use pv = –28740.

- We will use fv = 0, or leave that argument empty, because nothing will be owed by the end of the loan.

- Typically, payments are made at the end of each period, so we will use type = 0, or leave that argument empty.

In an empty cell, type "=PMT(0.03/12,72,-28740,0,0)" and press Enter. The payment amount is given as $499.51.

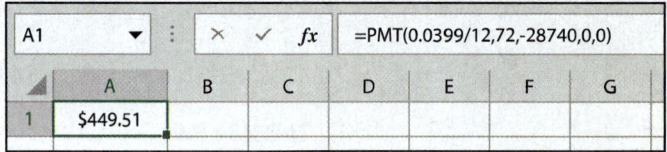

FIGURE 6.3.4

Therefore, the monthly car payment will be $499.51.

b. At the end of the 72 months, you will have made 72 payments of $499.51 plus the down payment of $7310. Therefore, the total cost of the car is found as follows.

Total Cost of the Car to the Buyer, Including Interest

$$72 \cdot \$449.51 + \$7310 = \$39,674.72$$

✔ **Skill Check 6.3.1**

Determine the monthly payment for a new truck that costs $48,000 if the buyer will make a down payment of $10,000. The terms of the loan are 5% APR for 48 months.

Example 6.3.3

Comparing Installment Loan Options

Compare the following new car incentives by determining the total cost of the car for each option and discussing the pros and cons for each.

New Car Price: $27,465

Down Payment from Buyer: $5000

Option A: 0.9% APR financing for 60 months

Option B: $1500 rebate offer from dealership with a 3.75% APR over 48 months

Solution

Option A: 0.9% APR financing for 60 months

For this option, the principal to be financed, or the loan amount, is the price of the car minus the down payment.

$$\text{Loan Amount: } \$27,465 - \$5000 = \$22,465$$

By Hand

In addition to knowing the loan amount, we are told that payments are monthly, so $n = 12$. Also, $r = 0.009$ and $t = \dfrac{60}{12} = 5$ years. The monthly payment is calculated as follows.

$$
\begin{aligned}
\text{PMT} &= \frac{\left(p \cdot \dfrac{r}{n} \right)}{\left[1 - \left(1 + \dfrac{r}{n} \right)^{-nt} \right]} \\[3ex]
&= \frac{\left(\$22,465 \cdot \dfrac{0.009}{12} \right)}{\left[1 - \left(1 + \dfrac{0.009}{12} \right)^{-(12)(5)} \right]} \\[3ex]
&\approx \$383.0446
\end{aligned}
$$

TI-83/84 Plus

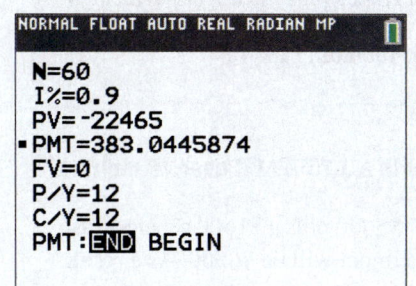

```
NORMAL FLOAT AUTO REAL RADIAN MP

N=60
I%=0.9
PV=-22465
•PMT=383.0445874
FV=0
P/Y=12
C/Y=12
PMT:END BEGIN
```

Figure 6.3.5

Use the TVM Solver to find the monthly payment. Notice that the present value is the amount being financed, the future value is 0, and we are finding PMT.

- N is the total number of times the account is compounded. Since you are making 60 monthly payments, N = 60.

- The interest rate is 0.9%, so I% = 0.9.

- We found that the amount being financed is $22,465, which means that's the initial amount we owe. So PV = -22465.

- The monthly payment amount is what we need to solve for, so we will come back to the PMT entry after filling out the other values.

- At the end of the loan, no money will be owed, so FV = 0.

- Because the interest is being compounded monthly, P/Y = C/Y = 12.

- We'll assume that payments are made at the end of each period, so be sure that END is highlighted in the final row.

Once these values are filled out, return your cursor to the PMT entry and press clear. To solve for this value, press alpha enter. The calculator returns 383.0445874.

Microsoft Excel

Use the **PMT(rate, nper, pmt, [pv], [fv], [type])** function to calculate the amount that each monthly payment should be in order to pay off the loan.

- In this example, we have an interest rate of 0.009 that is being compounded 12 times a year, so for the first argument in this Excel function, we have rate = 0.09/12.

- Since this is a 60-month loan, nper is 60.

- We found that the amount being financed is $22,465, which means that's the initial amount we owe. So pv = –22465.

- We will use fv = 0, or leave that argument empty, because nothing will be owed by the end of the loan.

- Typically, payments are made at the end of each period, so we will use type = 0, or leave that argument empty.

In an empty cell, type "=PMT(0.009/12,60,-22465,0,0)" and press Enter. The payment amount is given as $383.04.

A1	▼	⋮	×	✓	*fx*	=PMT(0.009/12,60,-22465,0,0)		
	A	**B**	**C**	**D**	**E**	**F**	**G**	
1	$383.04							

FIGURE 6.3.6

Therefore, Option A requires a monthly payment of $383.04 over 60 months. The total cost of the car to the buyer is the monthly payments made plus the down payment.

Total Cost of the Car to the Buyer, Including Interest

$$\$383.04 \cdot 60 + \$5000 = \$27{,}982.40$$

Option B: $1500 rebate offer from dealership with a 3.75% APR over 48 months

For this option, the buyer has a down payment of $5000 plus a $1500 rebate to use as part of the down payment; that is, the down payment will be $6500. As a result, the loan amount is calculated as follows.

Loan Amount: $27,465 – $6500 = $20,965.

By Hand

Again, the payments are monthly, so $n = 12$. However, this time the interest rate is 3.75% over 48 months; so $r = 0.0375$ and $t = \dfrac{48}{12} = 4$ years. The payment for each month is calculated as follows.

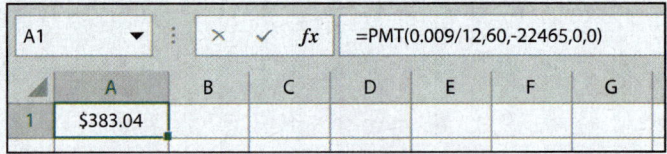

$$PMT = \frac{\left(p \cdot \dfrac{r}{n}\right)}{\left[1 - \left(1 + \dfrac{r}{n}\right)^{-nt}\right]}$$

$$= \frac{\left(\$20{,}965 \cdot \dfrac{0.0375}{12}\right)}{\left[1 - \left(1 + \dfrac{0.0375}{12}\right)^{-(12)(4)}\right]}$$

$$\approx \$471.0281$$

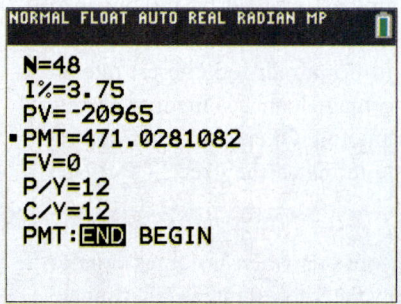

Figure 6.3.7

TI-83/84 Plus

Update the values in the TVM Solver as follows.

- Option B is a 48-month loan, so N = 48.

- The interest rate is 3.75%.

- We found that the amount being financed is $20,965, which means that's the initial amount we owe. So PV = –20965.

- The monthly payment amount is what we need to solve for, so we will come back to the PMT entry after filling out the other values.

- The remaining entries will be the same for Option B as they were for Option A.

Once these values are filled out, return your cursor to the PMT entry and press [clear]. To solve for this value, press [alpha] [enter]. The calculator returns 471.0281082.

Microsoft Excel

Use the PMT function to calculate the amount that each monthly payment should be in order to pay off the loan. For Option B, rate will be 0.0375/12, nper will be 48, and pv will be –20965.

In an empty cell, type "=PMT(0.0375/12,48,-20965,0,0)" and press Enter. The payment amount is given as $471.03.

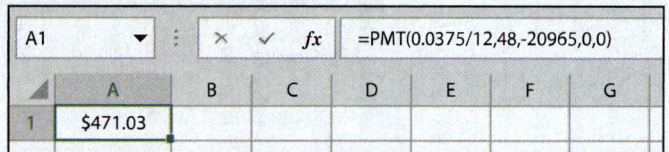

FIGURE 6.3.8

Therefore, Option B requires a monthly payment of $471.03 over 48 months. The total cost of the car to the buyer is the monthly payments made plus the $5000 down payment. We should think of the $1500 rebate as a discount to the buyer, not a cost that they were required to pay. So the buyer's cost is found as follows.

Total Cost of the Car to the Buyer, Including Interest

$$\$471.03 \cdot 48 + \$5000 = \$27,609.44$$

Overall, we can see that Option B is $372.96 cheaper than Option A. If bottom-line savings is the only factor to consider, then Option B is the better way to go. However, the payment each month is lower by $87.99 in Option A. It might be the case that someone cannot afford the higher payment each month, and it would make sense to have the lower payments spread out over an extra year.

OBJECTIVE 3 Mortgage Basics

An installment loan used for the purposes of buying a house or other real estate, where the property is used as collateral, is called a *mortgage*. Because buying a house is one of the biggest purchases most people make in their lifetime, we'll spend some time looking at different aspects of a mortgage. In contrast to a car loan, there are a number of additional costs involved in purchasing a house: inspection costs, surveying work, legal fees, realtor fees, and closing costs paid to the lender for arranging the loan. One fee that some lenders require for setting up the loan is referred to as a *point*. A point is interest prepaid by the buyer to the lender and is equal to 1% of the loan amount (not the purchase price of the house). A mortgage point is typically paid in return for a reduced interest rate on the loan. These costs are usually paid when the mortgage loan is signed, or *closed*, and so are normally not rolled into the borrowed amount. Of course, when setting up a loan, lenders and borrowers can arrange the terms as they see fit for each individual case.

As you might imagine, there are many different options when it comes time to obtain a mortgage. One of the options involves how the interest rate is determined. For *fixed-rate mortgages*, the interest rate is set at the beginning of the loan period and remains the same for the life of the loan, as the name suggests. With *adjustable-rate mortgages* (ARM), however, loans typically start with a low "teaser" interest rate for a period of 3 to 10 years and then the interest rate is allowed to fluctuate based on several things, including the interest rate benchmarks in the market. Of course, there are pros and cons to each option. Historically, fixed-rate mortgages have been seen as the safer option, because the buyer knows the terms of the loan from start to finish. However, although considered more risky, ARMs can benefit buyers who know they will have the loan only a few years—either because they can pay it off quickly or because they will relocate within a few years.

📖 MORTGAGES AND POINTS

Mortgage

An installment loan used for the purposes of buying a house or other real estate where the property is used as collateral is called a **mortgage**.

Fixed-Rate Mortgage

In a **fixed-rate mortgage**, the interest rate remains constant throughout the life of the loan.

Adjustable-Rate Mortgage

In an **adjustable-rate mortgage (ARM)** the interest rate is fixed for a period of time at the beginning of the loan period, and then is allowed to fluctuate over time.

Mortgage Point

A **mortgage point** is interest prepaid by the buyer to the lender and is equal to 1% of the loan amount.

We will focus on fixed-rate mortgages only. Because a mortgage is simply an installment loan, to calculate the monthly payment amount for a fixed-rate mortgage, we use the same formula that we did in the previous examples for buying a car. Once again, we must consider the down payment required, along with any fees or *closing costs* that happen at the time of purchase. The next example looks at comparing two mortgage options.

Example 6.3.4

Calculating Mortgage Payments and Fees

A buyer is purchasing a house for $265,000. Table 6.3.1 shows the particulars a bank is offering to the buyer. Evaluate the mortgage option by answering the following questions.

a. Calculate the required down payment.

b. Calculate the loan amount.

c. Determine the amount the buyer will be expected to pay at closing.

d. Calculate the monthly payment for the loan, not including taxes and insurance.

e. Calculate the buyer's monthly cost, which will include the loan payment, $200 in property taxes, and $83.33 in home insurance.

f. Determine the total cost of the house at the end of the loan period. Include any closing cost required.

g. A second option is offered to the client that has a lower APR but a higher down payment. It requires that the client pay $45,348 at closing and a monthly payment of $1737.03 over 15 years, which includes property taxes and insurance. After 15 years, the client will have paid $307,014 for the house. Discuss the pros and cons of this option compared to the 30-year option.

Table 6.3.1: Bank Mortgage Details

	APR	Down Payment Required	Points Required	Additional Fees
Option: 30-Year Fixed-Rate	3.348%	15%	0.875	$2774

Solution

a. The required down payment is 15% of the price of the house.

Down Payment Required: $0.15 \cdot \$265,000 = \$39,750$

b. The loan amount is the price of the house minus the down payment.

Loan Amount (P): $\$265,000 - \$39,750 = \$225,250$

c. The lender requires 0.875 points, or 0.875% of the loan amount to be paid at closing. This is found by multiplying 0.00875 by the amount of the loan amount that was found in part b.

Fee for Points: $0.00875 \cdot \$225,250 \approx \$1,970.94$

We then add the point amount to the additional fees of $2774 to determine the total fees due to the lender at closing.

Fees Due to Lender: $\$1,970.94 + \$2774 = \$4,744.94$

Therefore, the buyer is responsible for paying $4,744.94 in fees as well as the down payment of $39,750. So, at closing, the buyer will be expected to pay $4,744.94 + $39,750 = $44,494.94.

d. To calculate the monthly payment for the loan, we use the loan amount found in part b. We know that $p = \$225,250$, the interest rate is 3.348 % (so $r = 0.03348$), $t = 30$ years, and $n = 12$ since payments are made monthly.

By Hand

$$PMT = \frac{\left(p \cdot \dfrac{r}{n}\right)}{\left[1 - \left(1 + \dfrac{r}{n}\right)^{-nt}\right]}$$

$$= \frac{\left(\$225{,}250 \cdot \dfrac{0.03348}{12}\right)}{\left[1 - \left(1 + \dfrac{0.03348}{12}\right)^{-(12)(30)}\right]}$$

$$\approx \$992.4583$$

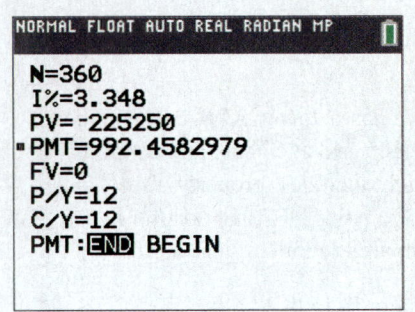

Figure 6.3.9

TI-83/84 Plus

Using the TVM Solver, N is the total number of times the interest is compounded. Since this is a 30-year mortgage with monthly payments, N = 12 · 30 = 360. PV is the amount being financed, which we will enter as –225250. Update the values in the TVM Solver as shown in Figure 6.3.9. Clear the entry for PMT and press alpha enter to solve for that value. The calculator returns 992.4582979.

Microsoft Excel

We will again use the **PMT(rate, nper, pmt, [pv], [fv], [type])** function to calculate the amount that each monthly payment should be in order to pay off the loan. The first argument, rate, is 0.03348/12, and nper is 30 · 12.

In an empty cell, type "=PMT(0.03348/12,30*12,-225250,0,0)" and press Enter. The payment amount is given as $992.46.

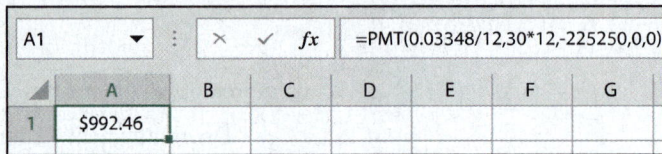

FIGURE 6.3.10

Therefore, the monthly loan payment will be $992.46 over 30 years.

e. The total monthly payment that the buyer will make includes the loan payment, property taxes, and home insurance.

Buyer's Total Monthly Payment Amount:

$$\$992.46 + \$200 + \$83.33 = \$1{,}275.79$$

This is usually referred to as the "mortgage payment," but in reality, it contains all three payments.

f. At the end of the loan period, the total cost of the house to the buyer will be the monthly loan payments (without the taxes and insurance) made over 30 years plus the closing costs that were paid at the beginning of the loan.

Total Cost of the House: ($992.46 · 12 · 30) + $44,494.94 = $401,780.54

g. The big pro for the second option is that over the life of the loan, the client will pay approximately $100,000 less for the house! This is quite an incentive. The client will also finish paying the loan in half the time. However, the second option requires a larger down payment as well as larger monthly payments for 15 years. If this is not financially sustainable, the client should not choose this option no matter how large the potential savings.

OBJECTIVE 4 Amortization Schedules

While you are making payments on a fixed-rate mortgage, the monthly amount stays the same throughout the life of the loan. However, the way the payment is portioned out between interest and principal changes over time. At the beginning of a loan, the majority of the payment is interest, which goes to the lender. But over time, the principal is reduced, and as a result, the interest is also reduced. As a way to show you how your payment is portioned out, lenders provide a breakdown of the payments in a **loan amortization schedule**. Many mortgage financial apps provide amortization schedules. A partial amortization schedule is shown in Table 6.3.2.

Table 6.3.2: Loan Amortization Schedule

Annual Interest Rate	3.50%			
Years	30			
Payments Per Year	12			
Loan Amount	$190,000			
Payment Number	**Payment**	**Principal**	**Interest**	**Balance**
1	($853.18)	($299.02)	($554.17)	$189,700.98
2	($853.18)	($299.89)	($553.29)	$189,401.09
3	($853.18)	($300.77)	($552.42)	$189,100.33
4	($853.18)	($301.64)	($551.54)	$188,798.68
5	($853.18)	($302.52)	($550.66)	$188,496.16
6	($853.18)	($303.40)	($549.78)	$188,192.76
7	($853.18)	($304.29)	($548.90)	$187,888.47
8	($853.18)	($305.18)	($548.01)	$187,583.29
9	($853.18)	($306.07)	($547.12)	$187,277.22
10	($853.18)	($306.96)	($546.23)	$186,970.26
11	($853.18)	($307.85)	($545.33)	$186,662.41
12	($853.18)	($308.75)	($544.43)	$186,353.66

When an amortization schedule is made, the interest is computed using the simple interest formula $I = Prt$, which was covered in Section 6.1, where P is equal to the balance of the loan and t is $\frac{1}{12}$ of a year.

Notice that at the beginning of the mortgage in Table 6.3.2, most of the $853.18 payment that you make is put towards interest; that is, paying the banks fees first. At the end of the first year (the highlighted month 12), the principal balance is reduced by only about $4000 and hence the loan continues accruing a similar amount of interest each month.

If you choose to pay a loan off early (for example, you relocate and sell the house), you have to repay the lender the amount of principal still owed. Although you are ending the interest earning period early, they have still earned profit on lending money because they have collected their interest at the front end of the loan agreement.

As you can imagine, creating an amortization table by hand is a lengthy process. However, spreadsheets like Microsoft Excel make the process much easier. The next example looks at how an amortization table is created using an Excel spreadsheet.

Example 6.3.5

Creating a Loan Amortization Schedule

a. Prepare a loan amortization schedule for a 30-year fixed-rate mortgage for $380,000 with a 3.36% APR.

b. What is the balance of the loan after five years of payments?

c. How many installments (payments) must be made before more than half of the payment is applied to the principal?

Solution

a. Using Microsoft Excel, we can use the finance functions we've seen so far in this chapter to create the amortization schedule. It will be helpful to first make note of the following four values that these functions will refer to.

Annual Interest Rate: 3.36%

Years: 30

Payments Per Year: 12

Amount: $380,000

For the table itself, we'll need five columns. In cell A1, enter "Payment Number"; in cell B1, enter "Payment"; in cell C1, enter "Principal"; in cell D1, enter "Interest"; and in cell E1, enter "Balance".

The **Payment Number** column will start with 1 and continue through 360 (that is, 12 · 30). After entering 2 in the second cell of the column, continue the numbering by highlighting both cells and dragging the lower right-hand corner down the column.

In the **Payment** column, use the **PMT(rate, nper, pmt, pv, [fv], [type])** function to calculate the monthly payment. Remember that rate, the first argument, is the interest rate per period. Since the interest rate here is 3.36 % and there are 12 periods per year, we will use 0.0336/12 for the rate. The next argument, nper, is the total number of compounding periods. Since this loan is a 30-year loan with monthly payments, nper is 12 · 30.

In cell B2, type "=PMT(0.0336/12,30*12,380000,0,0)". Notice that the value ($1,676.81) appears in red text in parentheses. (See Figure 6.3.11.) This indicates that it is a negative amount, or money being paid back to the lender.

In the **Principal** column, use the PPMT function, **PPMT(rate, per, nper, pv, [fv], [type])**, which returns the payment on the principal for a given investment based on periodic, constant payments and a constant interest rate. In other words, this is the portion of the monthly payment that is going toward the principal. The only difference between this function and the PMT function is the second argument, per, which is the period. This value will be 1 for the first payment, 2 for the second, and so on. We will refer to column A for this number.

In cell C2, type "=PMT(0.0336/12,A2,30*12,380000)". The value ($612.81) is returned.

In the **Interest** column, we are showing the amount of each monthly payment that is going toward the interest and not toward the principal. The monthly payment amount is stored in cell B2. The portion of that payment that is going toward the principal is stored in cell C2. If we subtract the value in cell C2 from the value in cell B2, that will tell us the amount going toward the interest.

In cell D2, type "=B2-C2". The value ($1,064.00) is returned.

The **Balance** column will show the total amount of the loan minus the amount that has been paid toward the principal. For the first row, this means the original amount of the loan minus the first principal payment.

In cell E2, type "=380000+C2". (Recall that the value in the Principal column is negative, which is why we are adding instead of subtracting.) The value $379,387.19 is returned.

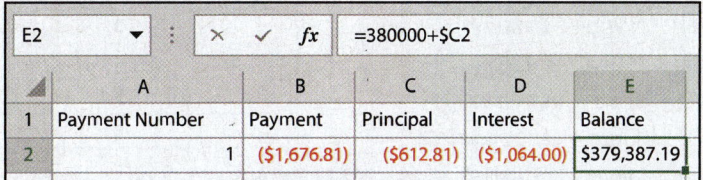

| E2 | ▼ | : | × | ✓ | *fx* | =380000+$C2 |

◢	A	B	C	D	E
1	Payment Number	Payment	Principal	Interest	Balance
2	1	($1,676.81)	($612.81)	($1,064.00)	$379,387.19

FIGURE 6.3.11

We cannot copy these cells for the entire table quite yet. For the second row, select cells B2 through E2 and drag the selection down *one row*. The Payment stays the same, but the amounts in the Principal and Interest columns are updated. We need to adjust the balance column for the second payment since we no longer want to subtract from the original loan amount; instead, we want to subtract from the previous balance in cell E2.

In cell E3, type "=E2+C3". (Recall that the value in the Principal column is negative, which is why we are adding instead of subtracting.) The value $378,722.66 is returned.

We can now select cells B3 through E3 and drag the selection down for the rest of the rows to complete the loan amortization schedule.

A	B	C	D	E
Payment Number	Payment	Principal	Interest	Balance
1	($1,676.81)	($612.81)	($1,064.00)	$379,387.19
2	($1,676.81)	($614.53)	($1,062.28)	$378,772.66
3	($1,676.81)	($616.25)	($1,060.56)	$378,156.41
4	($1,676.81)	($617.97)	($1,058.84)	$377,538.44
5	($1,676.81)	($619.70)	($1,057.11)	$376,918.73
6	($1,676.81)	($621.44)	($1,055.37)	$376,297.29
	↓	↓	↓	↓
58	($1,676.81)	($718.69)	($958.12)	$341,466.43
59	($1,676.81)	($720.71)	($956.11)	$340,745.73
60	($1,676.81)	($722.72)	($954.09)	$340,023.00
61	($1,676.81)	($724.75)	($952.06)	$339,298.26
62	($1,676.81)	($726.78)	($950.04)	$338,571.48
63	($1,676.81)	($728.81)	($948.00)	$337,842.67

FIGURE 6.3.12 (a)

A	B	C	D	E
Payment Number	Payment	Principal	Interest	Balance
1	($1,676.81)	($612.81)	($1,064.00)	$379,387.19
2	($1,676.81)	($614.53)	($1,062.28)	$378,772.66
3	($1,676.81)	($616.25)	($1,060.56)	$378,156.41
4	($1,676.81)	($617.97)	($1,058.84)	$377,538.44
5	($1,676.81)	($619.70)	($1,057.11)	$376,918.73
6	($1,676.81)	($621.44)	($1,055.37)	$376,297.29
	↓	↓	↓	↓
58	($1,676.81)	($718.69)	($958.12)	$341,466.43
59	($1,676.81)	($720.71)	($956.11)	$340,745.73
60	($1,676.81)	($722.72)	($954.09)	$340,023.00
61	($1,676.81)	($724.75)	($952.06)	$339,298.26
62	($1,676.81)	($726.78)	($950.04)	$338,571.48
63	($1,676.81)	($728.81)	($948.00)	$337,842.67

FIGURE 6.3.12 (b)

To make sure that the loan amortization schedule table is correct, you can find the sum of the column containing the principal amounts paid each time. This should add to the original loan amount. To determine how much interest was paid over the life of the loan, you can add the interest column.

b. There will be $5 \cdot 12 = 60$ payments in five years. The table tells us that after the 60th payment is made, a balance of $340,023.00 remains.

c. More than half the payment will be applied to the principal after the 113th payment. The 114th payment shows a flip in the columns so that the principal payment is greater than the interest payment.

✔ **Skill Check 6.3.3**

Based on the loan amortization schedule shown, what is the balance of the $114,000 loan after 2 years of payments?

Amortization

Monthly Payment: (Principal & Interest Only)			$499.89
Month	Principal	Interest	Balance
9	$189.57	$310.32	$112,312.48
10	$190.10	$309.80	$112,122.39
11	$190.62	$309.27	$111,931.77
12	$191.15	$308.75	$111,740.62
13	$191.67	$308.22	$111,548.95
14	$192.20	$307.69	$111,356.75
15	$192.73	$307.16	$111,164.02
16	$193.26	$306.63	$110,970.75
17	$193.80	$306.09	$110,776.95
18	$194.33	$305.56	$110,582.62
19	$194.87	$305.02	$110,387.76
20	$195.41	$304.49	$110,192.35
21	$195.94	$303.95	$109,996.41
22	$196.48	$303.41	$109,799.92
23	$197.03	$302.86	$109,602.90
24	$197.57	$302.32	$109,405.33
25	$198.12	$301.78	$109,207.21
26	$198.66	$301.23	$109,008.55
Total Amount Paid:			$179,964.06
Total Interest Paid:			$65,964.30
Term (Months):			361

Figure 6.3.13

OBJECTIVE 5 Open-Ended Loans (Revolving Credit)

In contrast to fixed installment loans, an *open-ended loan* (or *revolving credit*) is a preapproved loan amount that may be used by the borrower repeatedly; that is, funds can be borrowed, paid back, and borrowed again. The amount that is preapproved is called your *credit limit*. In open-ended loans, the borrower determines when the funds are dispersed and repaid, knowing that if there is a balance owed on the loan, then interest will be charged. Credit cards and home equity lines of credit are the most common forms of open-ended loans.

These types of loans work on the principle that money is lent out for a month. If the money is not repaid in full at the end of the month, and a balance remains, a charge of interest is incurred upon the balance. The month where no interest is accruing is called a *grace period*.

> **Fun Fact**
>
> Usually a derogatory term, Investopedia defines a *deadbeat* as a slang term in the credit world for a credit card user who pays off his or her balance in full and on time every month, thus avoiding the need to pay off the interest that would have accrued on their accounts.

> ▣▣ **OPEN-ENDED LOANS**
>
> **Open-Ended Loans (Revolving Credit)**
>
> An **open-ended loan**, or **revolving credit**, is a preapproved loan amount that may be used by the borrower repeatedly. Interest is applied to any unpaid balance at the end of the payment period.
>
> **Credit Limit**
>
> The maximum amount of an open-ended loan is called your **credit limit**.
>
> **Grace Period**
>
> A **grace period** is a period of time in which no interest accrues on a debt.

Annual Percentage Rates
Your annual interest rate as of August 22:

17.24%	**25.24%**
Purchases Variable Rate	Cash Advances Variable Rate

FIGURE 6.3.14:

Credit Card Interest Rates

Let's turn our attention to credit cards. The interest rate on a credit card is often a *variable interest rate*. This means that the interest rate fluctuates over time based on a particular benchmark, often a country's prime rate of interest. Notice that Figure 6.3.14 shows the APR for a customer's credit card. The APR on the left-hand side is for purchases only and is a variable rate, which means it is subject to change.

The APR on the right-hand side is for cash advances. It also is a variable interest rate, but is quite a bit higher than the rate for purchases. This is standard practice for most credit cards. Unfortunately, most cardholders don't realize that other fees also exist for cash advances. Typically, an up-front fee is charged, usually 2% to 8 % of the amount advanced. Along with the higher interest rate and the fee, there is also no grace period. In other words, the cash advance starts accruing interest the minute it is dispensed from the ATM. Lesson to learn: cash advances on credit cards should only be used in emergencies!

> **Helpful Hint**
>
> Debit cards (or check cards) are not credit cards for borrowing money, they are simply an electronic means by which consumers can access the money already in their bank accounts.

Using credit cards wisely means that you only borrow what you can afford to pay off in one month. If it is the case that you cannot pay off the balance by the end of the month, then you must make a partial payment. The **minimum payment**, which is the minimum amount the lender requires you to pay each month, is usually equal to a percentage of your average balance due, generally 1% to 3%, plus any fees and accrued interest.

For instance, suppose your average daily balance for a certain month is $1000, and your minimum payment for that month is $20. One might assume that it's a good deal to pay $20 per month in order to pay off a debt of $1000. However, remember that the balance is accruing interest every month it goes unpaid. Therefore, this is really not a good deal at all. If the interest rate is as high as 20 % on the balance owed, the $20 payment each month goes almost completely towards interest and has very little impact on reducing the total balance of the credit card. By requiring that you pay only small amounts each month, creditors know that it will take you much longer to pay off your debt, and that you'll end up paying a lot more interest over time. This is how lenders actually make money and stay in business.

Example 6.3.6

Comparing Payoff Times for Credit Cards

The graphs in Figure 6.3.15 show the payment timeline for a credit card with a balance of $1000 and a 14% fixed interest rate where no new purchases are made. Graph A shows the timeline with the 4 % minimum payment made each month. Graph B shows the payment timeline with consistent payments of $40 per month.

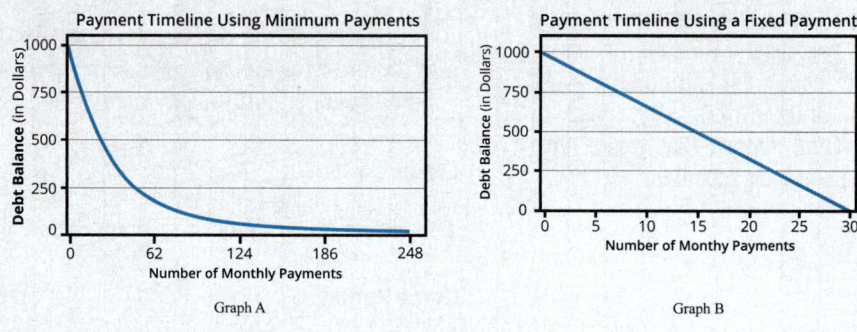

FIGURE 6.3.15

a. Use the graphs to determine approximately how many payments are needed to pay off the $1000 debt in each scenario.

b. How much longer does it take to pay off the debt in Graph A than in Graph B?

Solution

a. In Graph A, the balance is paid off after 248 monthly payments. In Graph B, the balance is paid off in approximately 30 monthly payments.

b. Making minimum payments, it takes 248 months or $\dfrac{248}{12} \approx 20.7$ years. Using payments of $40 a month, it takes $\dfrac{30}{12} = 2.5$ years to pay off the balance. Therefore, it takes approximately $20.7 - 2.5 = 18.2$ years longer to pay off the $1000 debt making only minimum payments.

In May 2009, Congress concluded that certain practices in the credit card industry were neither fair nor transparent to consumers, and so they signed into law the Credit Card Accountability Responsibility and Disclosure Act (Credit CARD Act) with strong support in both the Senate and House of Representatives. Among other things, this act protects consumers from excessive charges and wildly changing APRs. As a result of the Credit CARD Act, many credit card statements now provide information to their customers about paying off the debt.

Figure 6.3.16 shows an example of the information given on a credit card statement. Notice that there is a help line number for consumer debt. Although this number is fictitious as printed in this text, the Credit CARD Act requires that creditors provide a toll-free telephone number for the purpose of providing information about accessing credit counseling and debt management services.

Late Payment Warning:	If we do not receive your minimum payment due by the payment due date listed, you may have to pay a late fee of up to $35.00 and your purchase APR may be increased to the penalty APR of 27.24%.	
Minimum Payment Warning:	If you make only the minimum payment each period, you will pay more in interest and it will take longer to pay off your balance. For example:	
If you make no additional charges and each month you pay...	**You will pay off the balance shown on this statement...**	**You will pay an estimated total of...**
Only the minimum payment due	7 years	$2771
$60	3 years	$2152 (Savings = $619)
If you would like information about credit counseling services, call 1-800-555-0199.		

FIGURE 6.3.16: Credit Card Warnings

This type of warning on a monthly credit card statement is the creditor actually showing you how bad it is for you to only pay them the minimum payment. They even go as far as to tell you how much money you could save! We can use the following formula to calculate the number of fixed payments required to pay down a debt that is accruing interest monthly, just like the credit card company does in these warnings.

📖 **NUMBER OF FIXED PAYMENTS REQUIRED TO PAY OFF CREDIT CARD DEBT**

The **number of fixed payments R required to pay off credit card debt** is calculated with the formula.

$$R = \frac{-\log\left[1 - \frac{r}{n}\left(\frac{A}{\text{PMT}}\right)\right]}{\log\left(1 + \frac{r}{n}\right)},$$

where A is the loan amount, r is the APR in decimal form, n is the number of payments per year, and the payment amount is PMT.

Example 6.3.7

Paying Off Credit Card Debt with a Fixed Payment

Assume you want to buy a new computer that costs $2200 using a credit card that has an APR of 19.99%.

a. Determine how long it will take you to pay off the new computer if you make regular monthly payments of $40.

b. Determine how long it will take you to pay off the computer if you make regular monthly payments of $80.

c. How much is saved by making monthly payments at the higher amount?

Solution

a. We can determine how many $40 payments are needed to pay off the computer by using the fixed payments formula for R.

By Hand

The formula we are using requires that we know the APR, the loan amount, payment amount, and the number of payments. From the information given, these are as follows.

$$r = 0.1999 \quad A = \$2200 \quad \text{PMT} = \$40 \quad n = 12$$

Substituting these values into the formula for the number of fixed payments, we have the following equation.

$$R = \frac{-\log\left[1 - \dfrac{r}{n}\left(\dfrac{A}{\text{PMT}}\right)\right]}{\log\left(1 + \dfrac{r}{n}\right)}$$

$$= \frac{-\log\left[1 - \dfrac{0.1999}{12}\left(\dfrac{\$2200}{\$40}\right)\right]}{\log\left(1 + \dfrac{0.1999}{12}\right)}$$

$$\approx 150.076$$

TI-83/84 Plus

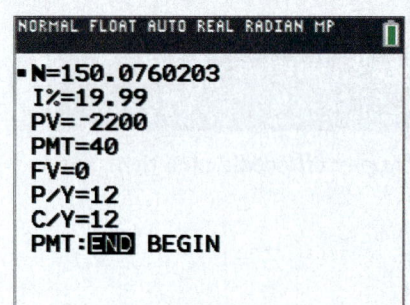

FIGURE 6.3.17

We can use the TVM Solver again, but this time we'll be solving for N.

We've said that N is the total number of times interest is compounded, or the total number of payments. Finding the total number of payments will tell us how long it will take to pay off the computer.

- Since this is the value we are solving for, we will come back to the N entry after filling out the other values.

- The interest rate is 19.99 %, so I% = 19.99.

- We are starting with a balance of $2200, so PV = −2200. (Notice that this value is negative because it is the amount you owe.)

- We are making monthly payments of $40, so PMT = 40.

- At the end of the loan, no money will be owed, so FV = 0.

- Because the interest is being compounded monthly, P/Y = C/Y = 12.

- We'll assume that payments are made at the end of each period, so be sure that END is highlighted in the final row.

Clear the entry for N and press [alpha] [enter] to solve for the number of payments. The calculator returns 150.0760203.

Microsoft Excel

Excel has a built-in financial function **NPER(rate, pmt, pv, [fv], [type])** that can be used to calculate the number of payments necessary to pay off a loan. The specific arguments for this function are similar to the arguments for other financial functions we've discussed in this chapter.

rate rate is the interest rate per period. This is the interest rate divided by the number of compounding periods per year. For this example, rate = 0.1999/12.

pmt is the amount of each monthly payment. For this example, = 40.

pv is the present value, or principal amount. Default is 0. For this example, pv = −2200, because that's the initial balance that is owed.

fv is the future value. For this example, fv = 0, because when the credit card is paid off, nothing will be owed.

type refers to whether payments are being made at the beginning of each period or at the end; 0 = end of the period, 1 = beginning of the period. Default is 0.

In an empty cell, type "=NPER(0.1999/12,40,-2200,0,0)" and press Enter. The number of payments is given as 150.076.

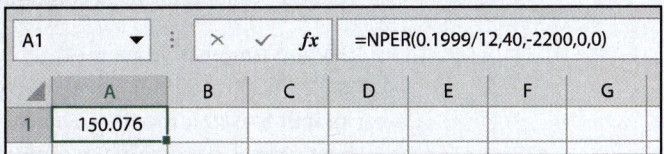

A1	▼	:	×	✓	*fx*	=NPER(0.1999/12,40,-2200,0,0)

	A	B	C	D	E	F	G
1	150.076						

FIGURE 6.3.18

Therefore, it will take approximately 150 monthly payments, or $12\frac{1}{2}$ years, to pay off the debt.

b. For this second option, the only value that changes is the fixed payment amount, PMT = \$80.

By Hand

Substituting into the formula for the number of fixed payments, we have the following.

$$R = \frac{-\log\left[1 - \frac{r}{n}\left(\frac{A}{\text{PMT}}\right)\right]}{\log\left(1 + \frac{r}{n}\right)}$$

$$= \frac{-\log\left[1 - \frac{0.1999}{12}\left(\frac{\$2200}{\$80}\right)\right]}{\log\left(1 + \frac{0.1999}{12}\right)}$$

$$\approx 37.085$$

TI-83/84 Plus

In the TVM Solver, change the **PMT** value to 80 and clear the entry for **N** and press [alpha] [enter] to solve for the new number of payments. The calculator returns 37.084776.

Microsoft Excel

We can again use the **NPER(rate, pmt, pv, [fv], [type])** formula. The only difference this time will be that we need to use pmt = 80 instead of 40.

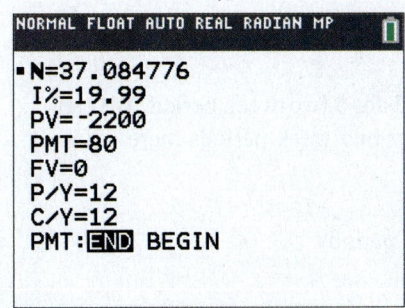

```
NORMAL FLOAT AUTO REAL RADIAN MP    []
▪N=37.084776
 I%=19.99
 PV=-2200
 PMT=80
 FV=0
 P/Y=12
 C/Y=12
 PMT:END BEGIN
```

FIGURE 6.3.19

In an empty cell, type "=NPER(0.1999/12,80,-2200,0,0)" and press Enter. The number of payments is given as 37.08478.

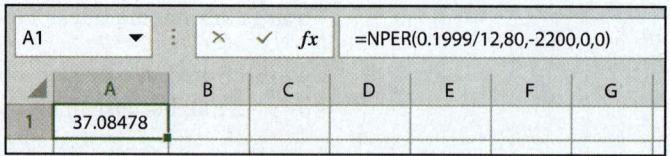

FIGURE 6.3.20

Therefore, it will take approximately 37 monthly payments, or just over three years, to pay off the debt.

c. In part a., we've seen that we could pay off the debt with 150 monthly payments of $40. That would be 150 · $40 = $6000.

In part b., we saw that you need to make 37 monthly payments of $80 to pay off the computer. The second option would be 37 · $80 = $2960.

By the time we've paid off the computer using the second option, we would save $6000 – $2960 = $3040. By doubling the minimum payment from $40 a month to $80 a month, you can save $3040 and be finished with payments in just over 3 years. This is a much wiser financial decision for purchasing the computer with a credit card.

As we discussed earlier, credit cards offer cash advances but at much higher interest rates then purchases. A **payday loan**, sometimes also called a *cash advance*, is similar to this. It is a short-term loan (usually about two weeks) that only requires you be at least 18 years old, have a checking account, and have a history of receiving an income. A typical payday loan is a means to financially make ends meet until your next paycheck. The length of a payday loan can vary from two weeks to 45 days, and lenders advertise their fees rather than interest rates. As a result, when the interest rates are calculated on these short-term loans, they range from 391% to 851%! These loans became popular in the 1990s but have since been made illegal in many states because of the excessive interest rates.

Example 6.3.8

Calculating Interest Rate on a Payday Loan

Assume you wish to borrow $300 for two weeks in the form of a payday loan and the amount of interest you must pay is $25 per $100 borrowed. What is the APR on the payday loan?

Solution

If the interest is $25 per $100, that is a rate of $25 for a two-week period. We can expand that for a year by determining how many two-week periods there are in a year.

$$\frac{52 \text{ weeks}}{2 \text{ weeks}} = 26 \text{ two-week periods.}$$

Now multiply the two-week rate by 26 to see the yearly rate (APR).

$$\text{APR: } 25\% \cdot 26 = 650\%$$

Thus, the APR for the payday loan is 650%!

One might reasonably argue that this APR rate is never actually paid because the loan is for such a short period of time. However, payday loans often become a brutal cycle that is hard to break. The loan along with the interest needs to be paid in two weeks. If a borrower is not able to pay it, they are forced to borrow again and start the cycle over. These high-interest, short-term loans should be procured with caution.

OBJECTIVE 6 Credit Reports

Everyone who has obtained a line of credit has an individual **credit report**. A credit report will provide your bill payment history, current debt (including loans, credit cards, etc.), whether you've ever been sued, arrested, or filed for bankruptcy, and how many new applications for credit you've made. Lenders use your credit report to help make decisions about whether they will approve you for a loan, and if so, what interest rate they will give you. Your landlord, employer, or insurer may also look at your credit report.

National credit reporting agencies gather information and maintain it for your credit report. They in turn give you a **credit score**. A credit score is a number that ranks your risk as a borrower. Depending on the reporting agency, your credit score is a number somewhere between 300 and 850. Having a high score can make it easier to get a loan, rent an apartment, get a lower interest rate on a loan, or lower your insurance rate. In the United States, the Fair Credit Reporting Act requires each of the nationwide credit reporting companies to provide you with a free copy of your credit report, at your request, once every 12 months.

The most commonly used credit score models have a range of 300 to 850. Creditors set their own standards for what constitutes an acceptable score, but Figure 6.3.21 gives a breakdown of general credit score guidelines.

- **Excellent credit:** 720–850
- **Good credit:** 690–719
- **Fair credit:** 630–689
- **Poor credit:** 629 or below

> **Helpful Hint**
>
> The government-approved website https://www.annualcreditreport.com allows you to get a free copy of your credit report every 12 months from each credit of the three reporting companies. You can request all three at once, or as they suggest, spread them out over the course of the year, once every four months, so that you can monitor your credit score.

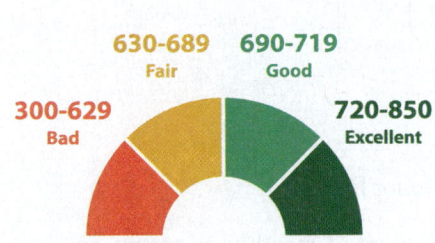

FIGURE 6.3.21

Example 6.3.9

Comparing Effects of Credit Scores on a Loan

Table 6.3.3 displays the interest rates that will be applied to a $25,000 monthly installment loan paid off in 3 years based on a borrower's credit score.

a. Determine the overall costs for a borrower who has a credit score of 660.

b. Determine the overall costs for a borrower who has a credit score of 720.

c. Compare the difference in costs found in parts a. and b.

Table 6.3.3: Interest Rate Tiers

Credit Score	APR
720–850	7.13%
690–719	8.03%
660–689	9.79%
620–659	11.75%
590–619	15.17%
500–589	16.00%

Solution

Begin by determining the monthly payments for each credit score using the payment formula and the appropriate APR. Then multiply the payment amount by 36 months to calculate the total cost of the loan for each level of score. For both scores, the only difference is the interest rate. The other variables needed for the formula remain the same: the $P = \$25,000$, the time $t = 3$, and the payments are monthly so $n = 12$.

a. The table shows that a credit score of 660 would secure an annual interest rate of 9.79%.

By Hand

Substituting the values in the formula we have the following.

$$\text{PMT} = \frac{\left(P \cdot \dfrac{r}{n}\right)}{\left[1-\left(1+\dfrac{r}{n}\right)^{-nt}\right]}$$

$$= \frac{\left(\$25,000 \cdot \dfrac{0.0979}{12}\right)}{\left[1-\left(1+\dfrac{0.0979}{12}\right)^{-(12)(3)}\right]}$$

$$\approx \$804.2171$$

TI-83/84 Plus

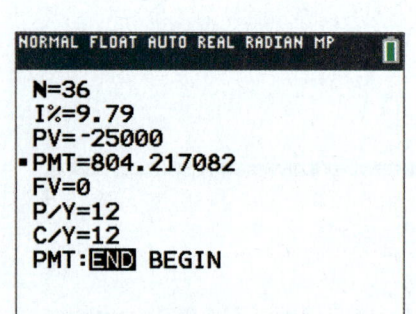

NORMAL FLOAT AUTO REAL RADIAN MP

N=36
I%=9.79
PV=-25000
∎PMT=804.217082
FV=0
P/Y=12
C/Y=12
PMT:**END** BEGIN

FIGURE 6.3.22

We can use the TVM Solver again to solve for **PMT**.

- This loan will be paid off with three years of monthly payments, so N = 12 · 3 = 36.

- The interest rate is 9.79%, so I% = 19.99.

- We are starting with a balance of $25,000, so PV = -25000.

- The monthly payment amount is what we are solving for, so we will come back to the PMT entry after filling out the other values.

- At the end of the loan, no money will be owed, so FV = 0.

- Because the interest is being compounded monthly, P/Y = C/Y = 12.

- We'll assume that payments are made at the end of each period, so be sure that END is highlighted in the final row.

Clear the entry for PMT and press [alpha] [enter] to solve for the number of payments. The calculator returns 804.217082.

Microsoft Excel

We will use the **PMT(rate, nper, pv, [fv], [type])** function with rate = 0.0979/12, nper = 12 · 3, and pv = −25000.

In an empty cell, input "=PMT(0.0979/12,12*3,-25000,0,0)" and press Enter. The payment amount is given as $804.22.

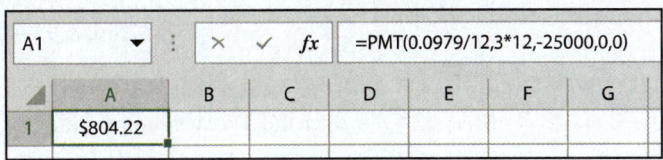

A1	▼	:	×	✓	fx	=PMT(0.0979/12,3*12,-25000,0,0)

	A	B	C	D	E	F	G
1	$804.22						

FIGURE 6.3.23

So after 3 years the total cost of the loan would be as follows.

$$\$804.22 \cdot 36 = \$28{,}951.92$$

b. The table shows that a credit score of 720 would secure an interest rate of 7.13%.

By Hand

Substituting the values in the formula we have the following equation.

$$\mathrm{PMT} = \frac{\left(P \cdot \dfrac{r}{n} \right)}{\left[1 - \left(1 + \dfrac{r}{n} \right)^{-nt} \right]}$$

$$= \frac{\left(\$25{,}000 \cdot \dfrac{0.0713}{12} \right)}{\left[1 - \left(1 + \dfrac{0.0713}{12} \right)^{-(12)(3)} \right]}$$

$$\approx \$773.4142$$

TI-83/84 Plus

In the TVM Solver, change I% to the new rate of 7.13. All other values should stay the same. Clear the value for PMT. Press alpha enter to solve for PMT. The calculator returns 773.4142407.

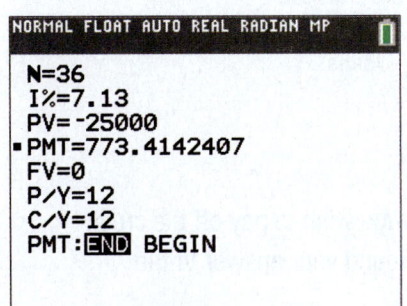

```
NORMAL FLOAT AUTO REAL RADIAN MP

N=36
I%=7.13
PV=-25000
▪PMT=773.4142407
FV=0
P/Y=12
C/Y=12
PMT:END BEGIN
```

FIGURE 6.3.24

Microsoft Excel

We will again use the **PMT(rate, nper, pv, [fv], [type])** function again to calculate the monthly payment with the new credit score.

In an empty cell, type "=PMT(0.0713/12,12*3,-25000,0,0)" and press Enter. The payment amount is given as $773.41 .

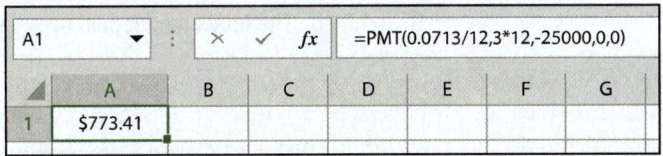

A1	▼	:	×	✓	fx	=PMT(0.0713/12,3*12,-25000,0,0)

	A	B	C	D	E	F	G
1	$773.41						

FIGURE 6.3.25

So after 3 years the total cost of the loan would be as follows.

$$\$773.41 \cdot 36 = \$27{,}842.76$$

c. The difference between the overall cost of the loan for the two credit scores is found by subtracting the results found in parts a. and b.

Difference between Loan Costs over 3 Years:

$$\$28{,}951.92 - \$27{,}842.76 = \$1{,}109.16$$

This means that the better credit score would save the borrower $1,109.16 over the 3-year loan period.

Skill Check Answers
1. $875.11 **2.** $1,031.25 **3.** $109,405.33

6.3   Exercises

✔ CONCEPT CHECK

1. The cash payment made up front towards a total purchase price is called the _____.

2. We say that the amount of a purchase that is being borrowed with interest is the amount that needs to be _____.

3. A mortgage is an installment loan used to buy real estate where the property is used as _____.

4. A loan _____ schedule shows how much of each payment is going toward the principal versus how much is going toward interest.

5. Credit cards are common forms of _____ loans.

💡 PRACTICE

Consider a credit card with a balance of $7000. You wish to pay off the credit card in each scenario. Calculate the following. Round your answer to the nearest cent, if necessary.

   a. The amount of a monthly payment within the time frame given

   b. The total amount paid over the time period

6. APR of 17.99% paid off within 1 year

7. APR of 12.5% paid off within 2 years

8. APR of 24% paid off within 3 years

Consider a credit card with a balance of $5560. You wish to pay off the credit card in each scenario. Calculate the following. Round your answer to the nearest cent, if necessary.

a. The amount of a monthly payment within the time frame given

b. The total amount paid over the time period

9. APR of 14.99% paid off within 1 year

10. APR of 11.99% paid off within 2 years

11. APR of 5.9% paid off within 3 years

12. Consider a 30-year fixed-rate mortgage for $250,000 with a 4.22% APR.

   a. Prepare a loan amortization schedule for the mortgage.

   b. What is the balance of the loan after 7 years of payments?

   c. How many installments (payments) must be made before more than half of the payment is applied to the principal?

13. Consider a 15-year fixed-rate mortgage for $325,000 with a 3.19% APR.

   a. Prepare a loan amortization schedule for the mortgage.

   b. What is the balance of the loan after 7 years of payments?

   c. How many installments (payments) must be made before more than half of the payment is applied to the principal?

🚀 APPLICATIONS

14. Suppose you are interested in buying a new condo. You decide to take out a $200,000 mortgage for 20 years. If the APR is 2.7%, calculate the monthly payment amount you will be responsible for.

15. You want to open your own restaurant, but you don't have enough funds. You take out a small business loan of $75,000 and commit to repaying the loan within 7 years. The APR you were provided is 8%. Calculate the monthly payment amount you will need to pay to the lender.

16. Ben wants to make repairs in his old house. He's decided to try to do all the work at once instead of gradually, so he takes out a consumer loan of $25,000. He believes that he will be able to repay the loan in 3 years. If the APR is 4.7%, what will Ben's monthly payment amount be?

17. Samantha needs surgery. To cover the cost, she takes out a loan of $38,000. The loan was provided for a period of 4 years with an APR of 5.3%. How much does Samantha need to pay monthly for this loan?

18. You need a new laptop and decide to buy it on credit. The price of the laptop is $2300. You are offered a one-year loan with an interest rate of 6.2%.

    a. What will be your monthly payment amount?

    b. At the end of the loan period, how much will you have paid for the laptop?

19. Sheila is buying a new house. The final price of the house after down payment and fees is $140,000. She gets a mortgage for 15 years with an interest rate of 2.9%.

    a. What is her monthly payment amount?

    b. At the end of the loan period, how much will she have paid for the house?

20. Mike bought a new car and financed $25,000 to make the purchase. He financed the car for 60 months with an APR of 6.5%. Determine each of the following.

    a. Mike's monthly payment

    b. Total cost of Mike's car

    c. Total interest Mike pays over the life of the loan

21. Omar wants to purchase three vans for his delivery business. Each van costs $38,000. He wishes to finance the purchase for 48 months and has acquired an APR of 4.5%. Determine each of the following.

    a. Omar's monthly payment

    b. Total cost of Omar's vans

    c. Total interest paid by Omar over the life of the loan

22. Jamal bought a new car for $32,000. He paid a 10% down payment and financed the remaining balance for 36 months with an APR of 4.5%. Determine each of the following.

    a. Jamal's monthly payment

    b. Total cost of Jamal's car

    c. Total interest Jamal pays over the life of the loan

23. Susan wants to buy a new computer from Banana Computers. The company sells a laptop model for $2650. Susan decides to finance the computer for 24 months at an APR of 12.5%. Determine each of the following.

    a. Susan's monthly payment

    b. Total cost of the computer

    c. Total interest paid over the 24 months

24. Calculate the monthly payment for a student loan of $32,000 if the loan is to be repaid over 10 years and the interest rate is 4.2%.

25. Omar borrowed $26,000 in order to complete his higher education. The term of the student loan is 5 years, and the APR is 6.33%. Calculate Omar's monthly payment amount.

26. After getting a bachelor's degree, Lyna decides to continue her education and get a master's degree. She takes out a student loan of $25,000. Calculate the monthly payment Lyna will be responsible for if the term of the loan is 7 years with an APR of 8.14%.

27. A pamphlet handed out during student orientation explains that student loans usually have fixed interest rates and don't have to be repaid until a few months after graduation. The pamphlet gives and example of a student loan of $50,000 that is taken and has to be paid off in 15 years. If the interest rate is 3.67%, what is the monthly payment amount?

28. Tommy gets to choose from one of the new car incentives when he purchases his car next week. He can either choose 0.9% APR financing for 48 months or $1000 cash back with a 4.75% APR over 48 months. Compare the two incentives that Tommy has to choose from if the new car he wishes to buy is $32,457 and he has saved a down payment of $3500.

29. You want to buy a car and finance $20,000 to do so. You can afford a payment of up to $450 per month. The bank offers three choices for the loan: a four-year loan with an APR of 7%, a five-year loan with an APR of 7.5%, and a six-year loan with an APR of 8%. Which option best meets your needs, assuming you want to pay the least amount of interest?

30. Suppose you decide to buy a new house for $270,000. You have saved $30,000 for a down payment, and want to take out a mortgage for the remaining amount. The lender offers you two possible options.

   Option A: 2.9% APR financing for 240 months

   Option B: 3.4% APR financing for 180 months

   Compare the offers by determining the monthly payment and the total cost of the house for each option.

31. You want to upgrade your home appliances. You need $9300 and have saved $2000 so far. For the rest of the amount, you can take out a consumer loan using one of the following options.

   Option A: 4.8% APR financing for 24 months

   Option B: 5.9% APR financing for 18 months

   Compare the offers by determining the monthly payment and the total cost for each option.

32. Linda's parents want to gift her a violin as a birthday present. They choose a model that costs $6250 and make a down payment of $1000. They are offered two loan options for the remaining amount.

   Option A: 7.2% APR financing for 18 months

   Option B: 8.6% APR financing for 12 months

   Compare the offers by determining the monthly payment and the total cost of the violin for each option.

33. Mateo is planning to launch a business and needs an initial capital of $33,000 for his project. He has $8000 saved and plans to take out a loan for the rest of the amount. Assume that he is offered two options for the loan.

    Option A: 5.6% APR financing for 84 months

    Option B: 7.1% APR financing for 60 months

    Compare the offers by determining the monthly payment and the total amount Mateo will pay to the lender for each option.

34. A family is going to buy a new house that costs $195,000. They take out a 30-year fixed-rate mortgage with an APR of 2.857%. The loan requires a down payment of 20% and 1.125 points paid directly to the lender. Additional fees of $1830 will need to be paid at the time of purchase.

    a. Calculate the required down payment.

    b Calculate the loan amount.

35. Joseph plans to purchase a house for $230,000. He is offered a 30-year fixed-rate mortgage with an APR of 3.169% and a down payment of 14%. He will also need to pay 0.915 points to the lender and $2185 in additional fees.

    a. Calculate the required down payment.

    b. Calculate the loan amount.

36. Amir's parents take out a 30-year fixed-rate mortgage with an APR of 2.934% to purchase a new house. The house costs $275,000, and the lender requires a down payment of 12%, 1.275 points, and $2350 in additional fees. Determine the amount Amir's parents can expect to pay at closing.

37. Martha is purchasing a house for $245,000. She takes out a 30-year fixed-rate mortgage with an APR of 2.968% and a down payment of 16%. The additional fees for the loan are $3015 and the lender requires 0.995 points. Determine the amount Martha can expect to pay at closing.

38. Luka wants to buy a new house for $210,000. He plans to take out a 30-year fixed-rate mortgage with an APR of 2.835%. The lender asks him to prepare a down payment of 15% and to pay 1.035 points and $2080 in additional fees.

    a. Calculate the monthly payments for the loan, not including taxes and insurance.

    b. Calculate Luka's monthly cost, which will include the loan payment, $180 in property taxes, and $95.55 in home insurance.

39. Talia is going to purchase a house for $235,000. She is offered a 30-year fixed-rate mortgage with an APR of 2.916%. A down payment of 16% is required and 0.975 points should be paid directly to the lender. Also, she needs to pay $2215 in additional fees at closing.

    a. Calculate the monthly payments for the loan, not including taxes and insurance.

    b. Calculate Talia's monthly cost, which will include the loan payment, $220 in property taxes, and $81.25 in home insurance.

40. Thomas and Ashley are purchasing a house for $215,000. They take out a 30-year fixed rate mortgage with an APR of 2.793%. The lender requires a down payment of 18%, 1.125 points paid directly, and $2350 in additional fees. Determine the total cost of the house at the end of the loan period. Include any closing cost required.

41. Antonio plans to purchase a house that costs $225,000. He is offered a 30-year fixed-rate mortgage with an APR of 2.896%. A down payment of 15% is required, and 0.975 points should be paid directly to the lender. In addition, Antonio needs to pay $2735 in closing fees. Determine the total cost of the house at the end of the loan period. Include any closing cost required.

42. Rachel is purchasing a new camera that costs $3800 for her photography business. Rachel uses a credit card that has an APR of 16.99%.

    a. How long will it take her to pay off the camera if she makes monthly payments of $75?

    b. How much will she pay in the long run for the camera if she makes monthly payments of $75?

    c. How long will it take her to pay off the camera if she makes monthly payments of $150?

    d. How much will she pay in the long run for the camera if she makes monthly payments of $150?

43. Trey is going to buy a new couch for his living room. It costs $1820, and Trey will use a credit card that has an APR of 14.99%.

    a. Determine how long it will take Trey to pay off the debt if he makes regular monthly payments of $30.

    b. Determine how long it will take Trey to pay off the debt if he makes regular monthly payments of $80.

    c. How much can be saved by making monthly payments at the higher amount?

44. In order to purchase a new TV that costs $635, Jessica uses a credit card with an APR of 17.33%.

    a. Determine how long it will take her to pay off the debt if she makes regular monthly payments of $25.

    b. Determine how long it will take her to pay off the debt if she makes regular monthly payments of $55.

    c. How much can be saved by making monthly payments at the higher amount?

45. Assume that Peter needs to borrow $500 for four weeks in the form of a payday loan. He must pay $20 per $100 borrowed as interest. Calculate the APR on the payday loan.

46. Ava is out of money and decides to borrow $400 for two weeks in the form of a payday loan. The amount of interest she will pay is $30 per $100 borrowed. What is the APR on this payday loan?

47. Due to unforeseen circumstances, Mike is going to run out of money before his next paycheck. So, he decides to take out a payday loan of $600 for two weeks. If he pays $35 per $100 borrowed, what is the APR on this loan?

48. Suppose you urgently need money, so you take out a payday loan of $700 for four weeks. The lender says that the amount of interest you will have to pay is $40 per $100 borrowed. Calculate the APR on this payday loan.

49. The following table displays the interest rates for an $8000 monthly installment loan that should be paid off in 2 years based on a borrower's credit score.

Credit Score	APR
780–850	6.1%
740–779	8.4%
720–739	10.0%
680–719	13.2%
620–679	15.9%
520–619	20.0%

    a. Determine the overall cost for a borrower with a credit score of 750.

    b. Determine the overall cost for a borrower with a credit score of 710.

    c. Compare the costs of the loan for these two borrowers.

50. Suppose you want to borrow $20,000 for 5 years. The following table displays the interest rates based on a borrower's credit score.

Credit Score	APR
790–850	6.1%
750–789	8.4%
710–749	10.0%
670–709	13.2%
600–699	15.9%
510–599	20.0%

    a. Determine the overall cost if your credit score is 700.

    b. Determine the overall cost if your credit score is 760.

    c. Compare the costs of the loan for these two credit scores.

✎ **WRITING & THINKING**

51. What if you won the Powerball Lottery with a jackpot of $150 million? Calculate the amount of money you would receive over a 25 year period with each of the following two options. Which option gives you the most money over the 25 years?

    Option 1: Taking all the money at once with a 40% penalty, and pay the income tax of 38% on the lump sum, and investing the remaining amount into an account earning 6% interest for 25 years.

    Option 2: Acquire the money as part of an annuity to be paid out in 25 equal payments over a 25-year period, paying the income tax of 38% on the income from the winnings each year.

52. Think about a situation when it is better to take out a personal loan than to pay with a credit card.

53. Think about a situation when paying off a loan early may be less profitable than repaying it over time.

6.3 PROJECT

CAR LOANS: BRAND NEW OR PRE-OWNED?

According to the Brookings Institution, approximately 76% of working adults in the United States drive to work alone every day. Since owning a car is a big part of our lives, it is important to understand the true cost involved in a car loan. Brand new cars are more expensive but often can be financed at lower interest rates, while pre-owned vehicles cost less but often require a loan at a higher rate. In this activity, you will explore the difference in cost between financing a new vehicle and a pre-owned one.

Consider two options for purchasing a Honda Fit LX in 2020: one was a brand new 2020 model with a manufacturer's suggested retail price (MSRP) of $17,945, and the other was a pre-owned, two-year-old model listed for $15,500. Suppose you have saved $1500 for a down payment and the dealer has already included any applicable fees, including taxes, in the advertised price. You plan on taking 5 years to pay off the loan.

The table below shows the price and interest rate for each option.

	Price	Interest Rate
2020 Honda Fit LX	$17,945	1.9%
2018 Honda Fit LX	$15,000	6.9%

For both the new and the pre-owned Honda Fit LX options, do the following.

1. Compute the amount to be financed considering that you have saved $1500 for a down payment.

2. Use the formula for a regular payment on a fixed installment loan to determine the monthly payment. Round your answer to the nearest dollar.

3. Determine the total amount paid when repaying the car loan.

4. Determine the finance charge for each purchasing option. This is the difference between the total amount paid on the loan and the amount financed.

5. Complete the following table.

	2020 Honda Fit LX	2018 Honda Fit LX
Price	$17,945	$15,000
Interest Rate	1.9%	6.9%
Down Payment		
Amount Financed		
Monthly Payment		
Total Amount Paid		
Finance Charge		

6. The pre-owned car definitely has a lower monthly payment, which might sound appealing when budgeting your expenses. Could you make an argument, using the values in your table, that the money borrowed to purchase the pre-owned vehicle is actually "more expensive" than the money borrowed to purchase the new vehicle? Explain your reasoning.

6.4 FEDERAL REVENUE

OBJECTIVES

1. Determine gross and adjusted gross income.

2. Calculate taxable income.

3. Use an IRS table to calculate federal income tax.

4. Calculate payroll tax items.

The US federal government, just like you and I, has a means of bringing in money (*revenue*) and allotted ways to spend it (*outlay*). We collectively entrust the government to both oversee the inflow of money and then use it to fund different national services and programs. Figure 6.4.1 shows a breakdown of the major sources of revenue for the federal government. The graph represents the $3.462 trillion that was received in 2019.[1]

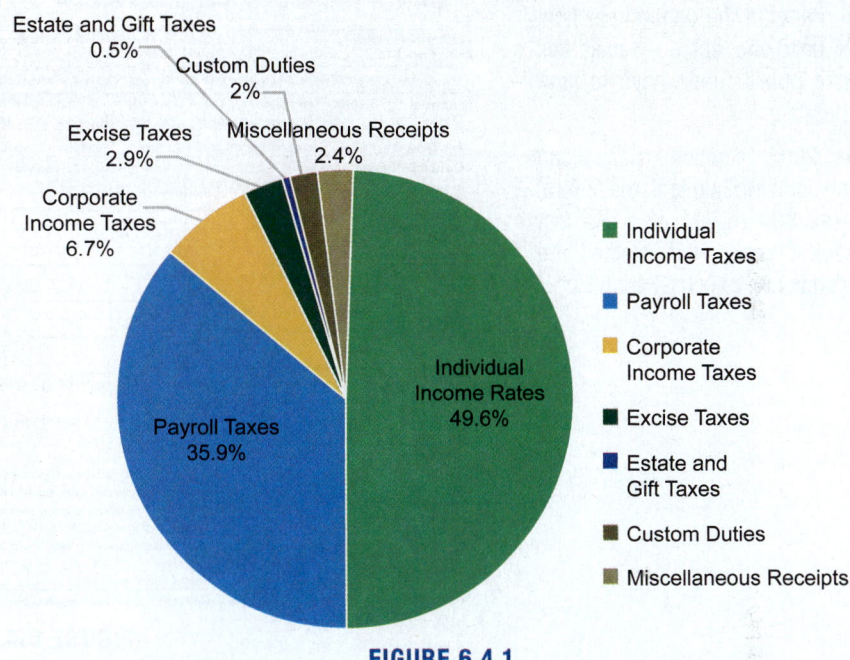

Federal Revenues 2019

- Individual Income Taxes
- Payroll Taxes
- Corporate Income Taxes
- Excise Taxes
- Estate and Gift Taxes
- Custom Duties
- Miscellaneous Receipts

Estate and Gift Taxes 0.5%
Custom Duties 2%
Excise Taxes 2.9%
Miscellaneous Receipts 2.4%
Corporate Income Taxes 6.7%
Individual Income Rates 49.6%
Payroll Taxes 35.9%

FIGURE 6.4.1

As Figure 6.4.1 shows, two of the biggest revenue generators are individual income taxes and payroll taxes. These two areas, which we will take a closer look at, account for almost 90% of all the revenue that the federal government receives. You can see that almost half of the revenue in 2019, approximately $1.718 trillion, came from individual income taxes. In other words, when you pay taxes on money you earn, those taxes are part of the nation's budget. We'll use this section to focus on how your individual contribution to the federal revenue is determined.

OBJECTIVE 1 Gross and Adjusted Gross Income

Income tax is a governmental charge imposed on individuals based on a percentage of earnings they generate in a year. The Internal Revenue Service (IRS) is the division of government that manages the collection of taxes. Although throughout history the laws concerning tax deductions, tax credits, and loopholes seem to have muddied the waters and complicated the tax laws, the basics of paying taxes are not fundamentally complicated. In fact, we will work our way through the major parts on a Form 1040 (Figure 6.4.2), which is what you must fill out each year for the IRS. This form helps you determine how much income tax you must pay. All current federal tax forms, along with instructions about how to fill in each section can be

> **Fun Fact**
>
> Ratified in 1913, the 16th Amendment to the US Constitution established Congress's power to collect income taxes annually in order to fund public services and pay governmental obligations.

1   "Historical Budget Data: Jan 2020," Budget and Economic Data, Congressional Budget Office, last modified January 28, 2020, https://www.cbo.gov/system/files/2020-01/51134-2020-01-historic-albudgetdata.xlsx.

found on the IRS website, www.irs.gov. For our purposes, we will use Form 1040 from 2020.

How much are you taxed?

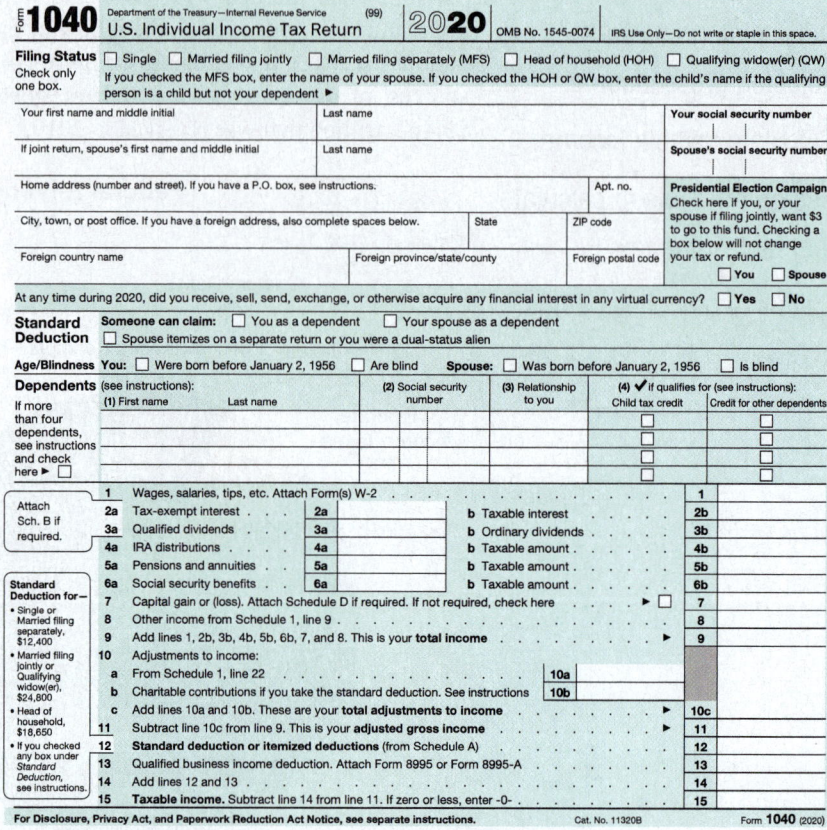

FIGURE 6.4.2

When filling out your tax forms, the top portion of the first page is identification information along with questions that let the government know how you intend to report your earnings; that is, your *filing status* and *number of dependents*. For your filing status, you have five choices:

1. **Single** Normally this status is for taxpayers who aren't married or who are divorced or legally separated under state law.

2. **Married Filing Jointly** If taxpayers are married, they can file a joint tax return.

3. **Married Filing Separately** A married couple can choose to file two separate tax returns. This may benefit them if it results in less tax owed than if they file a joint tax return. Taxpayers may want to prepare their taxes both ways before they choose. They can also use this status if each wants to be responsible only for their own tax.

4. **Head of Household** In most cases, this status applies to a taxpayer who is not married, but there are some special rules. For example, the taxpayer must have paid more than half the cost of keeping up a home for themselves and a qualifying person.

5. **Qualifying Widow(er) with Dependent Child** This status may apply to a taxpayer if their spouse died and they have a dependent child. Other conditions also apply.

The filing status that is chosen affects the amount of tax that an individual owes. It is the first box that you must check on Form 1040, as shown in Figure 6.4.2. Next is your name, address, and social security number.

The sections on *standard deductions* and *dependents* are next on the form. This portion of the form lets you identify any dependents or special filings that you might need to make. For the purposes of this text, we will focus on individuals who file either as single or married filing jointly with no dependents as we walk through Form 1040.

The numbered line portion is where you begin to report your income. Federal taxes are calculated on your earnings throughout the calendar year. These earnings include wages, tips, profits, and even union strike benefits. They can also include money you received from sources other than work such as investment income, rent paid to you by a tenant, alimony, or unemployment benefits awarded to you, to name a few.

Your *total income* or *gross income* for the year is the sum of your gained money, or assets. It is recorded on line 9 of Form 1040. However, the government does not necessarily tax you on all of that income. You are allowed to fine-tune your income to account for different things you paid for during the year, such as tuition and fees for education or interest paid on student loans. Once you've accounted for these items, the total is called your *adjusted gross income*, and it is entered on line 11. The government helps you tally up these adjustments in a worksheet called Schedule 1, Part II. Example 6.4.1 looks at finding adjusted gross income.

📖 GROSS INCOME AND ADJUSTED GROSS INCOME

Gross Income

Gross income is the sum of all forms of earnings, including wages, tips, benefits, investment income, alimony, etc.

Adjusted Gross Income

Adjusted gross income (AGI) is defined as gross income minus any adjustments you are allowed such as education expenses, health savings account contributions, paid alimony, or self-employment expenses.

Example 6.4.1

Determining Gross Income and Adjusted Gross Income

An elementary school speech pathologist is filling in their Form 1040 for the year's taxes. Their annual salary is $54,876, and they earned $27.50 interest from their savings account. They also paid $1398 in student loan interest and contributed $4800 to an IRA retirement account.

a. Calculate the amount of gross income to be entered on line 9 of Form 1040.

b. Use the Schedule 1, Part II worksheet to determine the adjusted gross income to be entered on line 11 of Form 1040.

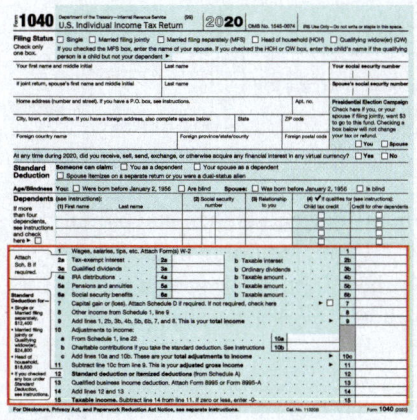

FIGURE 6.4.3: Page 1 of Form 1040

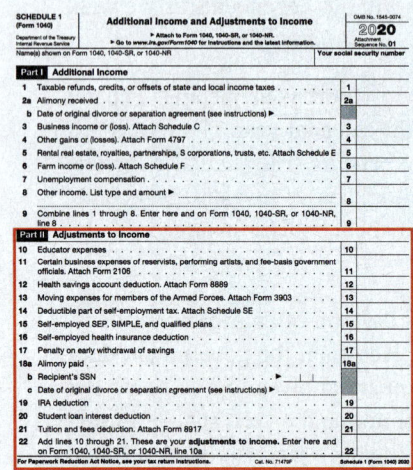

FIGURE 6.4.4: Schedule 1 Form

> 💬 **Helpful Hint**
>
> When entering a line amount, you can round to the nearest whole dollar on your tax return if you choose. However, if you round in one place on the form, you must round everywhere.

Solution

a. Gross income for the speech pathologist includes the wages along with interest from their savings account (taxable interest).

Wages are entered on line 1 and taxable interest is entered on line 2b. We add these together for the gross income.

$$\text{Gross Income: } \$54{,}876 + \$27.50 = \$54{,}903.50$$

Therefore, the gross income (total income) to be entered on line 9 is $54,903.50.

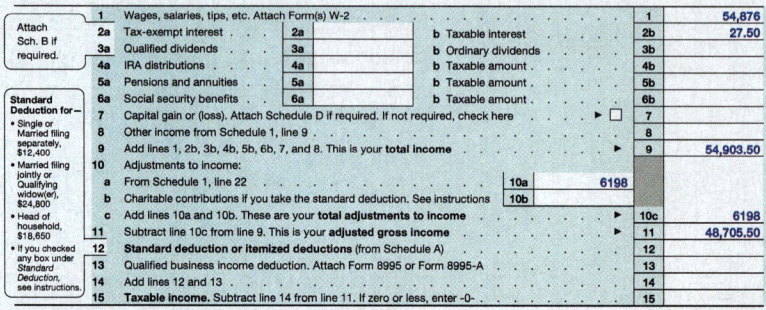

FIGURE 6.4.5

b. The speech pathologist is allowed to adjust the taxable income because of interest paid for student loans and contributions to a retirement account that were made during the year. Using the Schedule 1, Part II worksheet, fill in the appropriate cells. Enter $1398 in student loan interest on line 20 and $4800 on line 19 as an IRA deduction. Line 22 instructs us to add together lines 10 through 21 and enter the total of $6198, as shown in Figure 6.4.6.

$$\text{Adjustments to Income: } \$4800 + \$1398 = \$6198$$

Part II	Adjustments to Income	
10	Educator expenses	10	
11	Certain business expenses of reservists, performing artists, and fee-basis government officials. Attach Form 2106	11	
12	Health savings account deduction. Attach Form 8889	12	
13	Moving expenses for members of the Armed Forces. Attach Form 3903	13	
14	Deductible part of self-employment tax. Attach Schedule SE	14	
15	Self-employed SEP, SIMPLE, and qualified plans	15	
16	Self-employed health insurance deduction	16	
17	Penalty on early withdrawal of savings	17	
18a	Alimony paid	18a	
b	Recipient's SSN . . . . ▶		
c	Date of original divorce or separation agreement (see instructions) ▶		
19	IRA deduction	19	4800
20	Student loan interest deduction	20	1398
21	Tuition and fees deduction. Attach Form 8917	21	
22	Add lines 10 through 21. These are your **adjustments to income.** Enter here and on Form 1040, 1040-SR, or 1040-NR, line 10a	22	6198

For Paperwork Reduction Act Notice, see your tax return instructions. Cat. No. 71479F Schedule 1 (Form 1040) 2020

FIGURE 6.4.6

The adjustments to income total (line 22 of the worksheet) is then entered on line 10a of Form 1040 as shown in Figure 6.4.7. If you are filling in the forms online, some programs will automatically do the addition for you and even transfer the amount from line 22 of the Schedule 1 worksheet to the correct place on Form 1040.

Notice that line 10b on Form 1040 allows for charitable contributions to be counted separately if you choose to take the standard deduction. We will look at this more in the next example. Since the speech pathologist has no charitable contributions to add, we write 6198 on line 10c.

To find the adjusted gross income, we subtract line 10c from line 9 and enter this value on line 11, as instructed.

Adjusted Gross Income: $54,903.50 − $6198 = $48,705.50

So the adjusted gross income is $48,705.50.

Attach Sch. B if required.	**1**	Wages, salaries, tips, etc. Attach Form(s) W-2				**1**	54,876
	2a	Tax-exempt interest . . .	2a		**b** Taxable interest	**2b**	27.50
	3a	Qualified dividends . . .	3a		**b** Ordinary dividends	**3b**	
	4a	IRA distributions . . .	4a		**b** Taxable amount	**4b**	
	5a	Pensions and annuities . .	5a		**b** Taxable amount	**5b**	
Standard Deduction for—	**6a**	Social security benefits . .	6a		**b** Taxable amount	**6b**	
• Single or Married filing separately, $12,400	**7**	Capital gain or (loss). Attach Schedule D if required. If not required, check here . . . . ► ☐				**7**	
	8	Other income from Schedule 1, line 9				**8**	
• Married filing jointly or Qualifying widow(er), $24,800	**9**	Add lines 1, 2b, 3b, 4b, 5b, 6b, 7, and 8. This is your **total income** . . . . ►				**9**	54,903.50
	10	Adjustments to income:					
	a	From Schedule 1, line 22		**10a**	6198		
• Head of household, $18,650	**b**	Charitable contributions if you take the standard deduction. See instructions		**10b**			
	c	Add lines 10a and 10b. These are your **total adjustments to income** . . . . ►				**10c**	6198
• If you checked any box under Standard Deduction, see instructions.	**11**	Subtract line 10c from line 9. This is your **adjusted gross income** . . . . ►				**11**	48,705.50
	12	**Standard deduction or itemized deductions** (from Schedule A)				**12**	
	13	Qualified business income deduction. Attach Form 8995 or Form 8995-A				**13**	
	14	Add lines 12 and 13				**14**	
	15	**Taxable income.** Subtract line 14 from line 11. If zero or less, enter -0-				**15**	

FIGURE 6.4.7

OBJECTIVE 2 Taxable Income

Once you know your adjusted gross income, the government allows for certain *deductions* to be subtracted before calculating your taxes. The easiest way to claim deductions is to take a lump-sum standard deduction based on your filing status, which is set by the IRS and listed on the left of line 12 on Form 1040 each year. If you choose to take the standard deduction allowed, you (or you and your spouse, if filing jointly) are then also allowed to take a deduction for charitable contributions on line 10b. You can claim your contributions up to the following limits.

- $300 if single, head of household, or qualifying widow(er)
- $300 if married filing jointly
- $150 if married filing separately

Alternatively, if you have substantial deductions that add up to more than the standard deductions, you can choose to individually list out the qualifying expenses you've had through the year. This is referred to as *itemizing deductions*. Common deductions might include charitable contributions, interest on mortgages, state and local taxes, investment interest, casualty and theft losses from federally declared disasters, or medical and dental expenses.

Once these deductions have been subtracted, this is now your *taxable income*; in other words, what the government deems subject to taxes. This amount is entered on line 15 of Form 1040.

📖 **DEDUCTION AND TAXABLE INCOME**

Deduction

A tax **deduction** is comprised of expenses incurred by an individual or organization that can be subtracted from their gross income in order to reduce the amount of income that is able to be taxed.

Taxable Income

Taxable income is the amount of income subject to taxes by the government.

Example 6.4.2

Determining Taxable Income

A married taxpayer is filing taxes jointly with his spouse and they do not have any dependents to claim. During the year, they contribute $6000 to a traditional IRA. They also make charitable contributions of $1200 to Feeding America and $500 to Habitat for Humanity. They have no other deductions, and their combined wages are $117,795. Additionally, they earned $327 in interest from their savings accounts, which is taxable.

a. Decide whether the couple should use the 2020 standard deduction amount or itemize their deductions.

b. Determine their taxable income by filling in Form 1040.

Solution

a. To decide whether to use the standard deduction or itemize their deductions, the couple needs to compare their charitable donations to the standard deduction allowed for joint filers. They are allowed a standard deduction for "married filing jointly" of $24,800. This is far more than their charitable contributions of $1700. Therefore, it is better for them to claim the standard deduction and not itemize their deductions. Because of this, they can also claim up to $300 for their donations on line 10b.

b. Based on the information given, the couple will enter their combined wages of $117,795 on line 1 of Form 1040 and enter $327 on line 2b since their earned interest is taxable. Nothing else is entered on lines 3b through 8. On line 9, we enter the sum of these values.

$$\text{Line 9 (Total Income): } \$117{,}795 + \$327 = \$118{,}122$$

Next, we need to fill in the adjustments to income found on line 10. Line 10a is for the total from the Schedule 1, Part II worksheet. Since the only adjustment to income they have is the $6000 they paid into an IRA, we can simply enter this value on line 10a.

As we found in part a., they are taking the standard deduction for "married filing jointly," and hence are allowed an additional deduction of up to $300 for charitable donations. Although their donations were $1700, they can only claim $300. So we fill in 300 on line 10b. Line 10c will be the sum of 10a and 10b, which is $6300.

The couple's adjusted gross income, which is entered on line 11, is found by subtracting line 10c from line 9.

$$\text{Adjusted Gross Income: } \$118,122 - \$6300 = \$111,822$$

On line 12, we enter their standard deduction of 24,800. Subtracting the standard deduction (line 12) from the adjusted gross income (line 11) gives us their taxable income (line 15).

$$\text{Taxable Income} = \text{Adjusted Gross Income} - \text{Deductions}$$

$$\text{Taxable Income} = \$111,822 - \$24,800 = \$87,022$$

Therefore, their taxable income for the year is $87,022.

Attach Sch. B if required.	1	Wages, salaries, tips, etc. Attach Form(s) W-2			1	117,795
	2a	Tax-exempt interest	2a	**b** Taxable interest	2b	327
	3a	Qualified dividends	3a	**b** Ordinary dividends	3b	
	4a	IRA distributions	4a	**b** Taxable amount	4b	
	5a	Pensions and annuities	5a	**b** Taxable amount	5b	
Standard Deduction for—	6a	Social security benefits	6a	**b** Taxable amount	6b	
• Single or Married filing separately, $12,400	7	Capital gain or (loss). Attach Schedule D if required. If not required, check here ▶ ☐			7	
	8	Other income from Schedule 1, line 9			8	
	9	Add lines 1, 2b, 3b, 4b, 5b, 6b, 7, and 8. This is your **total income** ▶			9	118,122
• Married filing jointly or Qualifying widow(er), $24,800	10	Adjustments to income:				
	a	From Schedule 1, line 22	10a	6000		
	b	Charitable contributions if you take the standard deduction. See instructions	10b	300		
• Head of household, $18,650	**c**	Add lines 10a and 10b. These are your **total adjustments to income** ▶			10c	6300
	11	Subtract line 10c from line 9. This is your **adjusted gross income** ▶			11	111,822
• If you checked any box under Standard Deduction, see instructions.	12	**Standard deduction or itemized deductions** (from Schedule A)			12	24,800
	13	Qualified business income deduction. Attach Form 8995 or Form 8995-A			13	
	14	Add lines 12 and 13			14	
	15	**Taxable income.** Subtract line 14 from line 11. If zero or less, enter -0-			15	87,022

FIGURE 6.4.8

✔ **Skill Check 6.4.1**

Determine the standard deduction for a married taxpayer filing separately based on the 2020 Form 1040.

OBJECTIVE 3 Federal Income Tax

Once the taxable income is determined, it's time to calculate the amount of taxes owed. The United States uses a progressive income tax system, which means that the tax rate increases as the taxable amount increases. Table 6.4.1 shows the tax rate for each level of taxable income in 2020. Notice that income up to $9875 is taxed at a rate of 10 %. For the portion of income earned over that amount but less than $40,126, the tax rate increases to 12%, and so on. In other words, income is separated into tiers where each tier is taxed individually.

💬 **Helpful Hint**

You can find up-to-date tax brackets by performing an internet search for "federal income tax brackets." Sometimes including the current year can help narrow down the results.

Taxes for an Income of $12,000

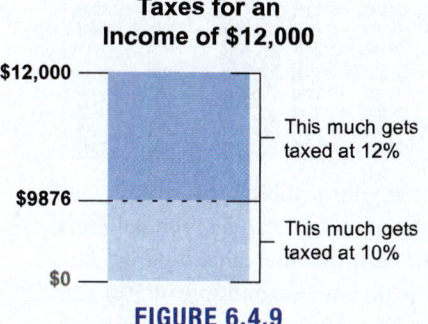

FIGURE 6.4.9

Table 6.4.1: 2020 Federal Tax Rates for Single Individuals

Tax Rate	Taxable Income Bracket
10%	$0 to $9875
12%	$9876 to $40,125
22%	$40,126 to $85,525
24%	$85,526 to $163,300
32%	$163,301 to $207,350
35%	$207,351 to $518,400
37%	$518,401 or more

For instance, suppose your taxable income in 2020 was $12,000. This is how your taxes were calculated.

1st Tier:	10% of $9875	$0.10 \cdot \$9875 = \$987.50$
2nd Tier:	12% of what's left of the taxable income; in this case, $12,000 - $9875 = $2175	$0.12 \cdot \$2125 = \$255.00$
		Total Taxes: $1242.50

Note, it is a common misconception that all income is taxed at the same rate. As we've just shown, only income in the bracketed range is taxed at the given rate. This method continues on up the table as the taxable income increases. Each year, the IRS publishes the tax rates for that year with any adjustments that were made. They also provide tables or worksheets to help you determine the amount of tax you owe where they have already done the calculations for you. If you make less than $100,000 in taxable income, a tax table like the one shown in Table 6.4.2 is published so that you can simply look up the amount of taxes you owe when filling out your Form 1040. As you can imagine, tables like these could go on indefinitely and would take a lot of space to print. Instead, if you make more than $100,000, they provide a worksheet for you to use to help you with the calculation. The next two examples look at each of these methods.

Table 6.4.2: Excerpt from the 2020 Federal Tax Table

21,000

At least	But less than	Single	Married filing jointly *	Married filing separately	Head of a household
21,000	21,050	2,326	2,128	2,326	2,241
21,050	21,100	2,332	2,134	2,332	2,247
21,100	21,150	2,338	2,140	2,338	2,253
21,150	21,200	2,344	2,146	2,344	2,259
21,200	21,250	2,350	2,152	2,350	2,265
21,250	21,300	2,356	2,158	2,356	2,271
21,300	21,350	2,362	2,164	2,362	2,277
21,350	21,400	2,368	2,170	2,368	2,283
21,400	21,450	2,374	2,176	2,374	2,289
21,450	21,500	2,380	2,182	2,380	2,295
21,500	21,550	2,386	2,188	2,386	2,301
21,550	21,600	2,392	2,194	2,392	2,307
21,600	21,650	2,398	2,200	2,398	2,313
21,650	21,700	2,404	2,206	2,404	2,319
21,700	21,750	2,410	2,212	2,410	2,325
21,750	21,800	2,416	2,218	2,416	2,331
21,800	21,850	2,422	2,224	2,422	2,337
21,850	21,900	2,428	2,230	2,428	2,343
21,900	21,950	2,434	2,236	2,434	2,349
21,950	22,000	2,440	2,242	2,440	2,355

24,000

At least	But less than	Single	Married filing jointly *	Married filing separately	Head of a household
24,000	24,050	2,686	2,488	2,686	2,601
24,050	24,100	2,692	2,494	2,692	2,607
24,100	24,150	2,698	2,500	2,698	2,613
24,150	24,200	2,704	2,506	2,704	2,619
24,200	24,250	2,710	2,512	2,710	2,625
24,250	24,300	2,716	2,518	2,716	2,631
24,300	24,350	2,722	2,524	2,722	2,637
24,350	24,400	2,728	2,530	2,728	2,643
24,400	24,450	2,734	2,536	2,734	2,649
24,450	24,500	2,740	2,542	2,740	2,655
24,500	24,550	2,746	2,548	2,746	2,661
24,550	24,600	2,752	2,554	2,752	2,667
24,600	24,650	2,758	2,560	2,758	2,673
24,650	24,700	2,764	2,566	2,764	2,679
24,700	24,750	2,770	2,572	2,770	2,685
24,750	24,800	2,776	2,578	2,776	2,691
24,800	24,850	2,782	2,584	2,782	2,697
24,850	24,900	2,788	2,590	2,788	2,703
24,900	24,950	2,794	2,596	2,794	2,709
24,950	25,000	2,800	2,602	2,800	2,715

Table 6.4.3: Excerpt from the 2020 Federal Tax Table

12,000

At least	But less than	Single	Married filing jointly *	Married filing separately	Head of a household
12,000	12,050	1,246	1,203	1,246	1,203
12,050	12,100	1,252	1,208	1,252	1,208
12,100	12,150	1,258	1,213	1,258	1,213
12,150	12,200	1,264	1,218	1,264	1,218
12,200	12,250	1,270	1,223	1,270	1,223
12,250	12,300	1,276	1,228	1,276	1,228
12,300	12,350	1,282	1,233	1,282	1,233
12,350	12,400	1,288	1,238	1,288	1,238
12,400	12,450	1,294	1,243	1,294	1,243
12,450	12,500	1,300	1,248	1,300	1,248
12,500	12,550	1,306	1,253	1,306	1,253
12,550	12,600	1,312	1,258	1,312	1,258
12,600	12,650	1,318	1,263	1,318	1,263
12,650	12,700	1,324	1,268	1,324	1,268
12,700	12,750	1,330	1,273	1,330	1,273
12,750	12,800	1,336	1,278	1,336	1,278
12,800	12,850	1,342	1,283	1,342	1,283
12,850	12,900	1,348	1,288	1,348	1,288
12,900	12,950	1,354	1,293	1,354	1,293
12,950	13,000	1,360	1,298	1,360	1,298

In Table 6.4.3, notice that you only need to know your taxable income and your filing status. Each block segment contains a range of incomes along with columns for each filing status. (Notice that the range is *at least* the first value *but less than* the second value. This is important if your taxable income is a multiple of 50.) The amount shown where the taxable income line and the filing status column meet is the tax amount owed. Notice that this value is slightly different from the value we just calculated. The table gives the tax amount owed for the middle of the range of taxable income, rounded to the nearest dollar.

Example 6.4.3

Calculating Federal Income Tax Using the IRS Tax Worksheet

Use the partial 2020 IRS tax worksheet shown to calculate the amount of federal income tax charged to a taxpayer who is married filing jointly with a taxable income of $177,839.

2020 Tax Computation Worksheet—Line 16

 See the instructions for line 16 to see if you must use the worksheet below to figure your tax.

Note. If you are required to use this worksheet to figure the tax on an amount from another form or worksheet, such as the Qualified Dividends and Capital Gain Tax Worksheet, the Schedule D Tax Worksheet, Schedule J, Form 8615, or the Foreign Earned Income Tax Worksheet, enter the amount from that form or worksheet in column (a) of the row that applies to the amount you are looking up. Enter the result on the appropriate line of the form or worksheet that you are completing.

Section A—Use if your filing status is **Single.** Complete the row below that applies to you.

Taxable income. If line 15 is—	(a) Enter the amount from line 15	(b) Multiplication amount	(c) Multiply (a) by (b)	(d) Subtraction amount	Tax. Subtract (d) from (c). Enter the result here and on the entry space on line 16.
At least $100,000 but not over $163,300	$	× 24% (0.24)	$	$ 5,920.50	$
Over $163,300 but not over $207,350	$	× 32% (0.32)	$	$ 18,984.50	$
Over $207,350 but not over $518,400	$	× 35% (0.35)	$	$ 25,205.00	$
Over $518,400	$	× 37% (0.37)	$	$ 35,573.00	$

Section B—Use if your filing status is **Married filing jointly** or **Qualifying widow(er).** Complete the row below that applies to you.

Taxable income. If line 15 is—	(a) Enter the amount from line 15	(b) Multiplication amount	(c) Multiply (a) by (b)	(d) Subtraction amount	Tax. Subtract (d) from (c). Enter the result here and on the entry space on line 16.
At least $100,000 but not over $171,050	$	× 22% (0.22)	$	$ 8,420.00	$
Over $171,050 but not over $326,600	$	× 24% (0.24)	$	$ 11,841.00	$
Over $326,600 but not over $414,700	$	× 32% (0.32)	$	$ 37,969.00	$
Over $414,700 but not over $622,050	$	× 35% (0.35)	$	$ 50,410.00	$
Over $622,050	$	× 37% (0.37)	$	$ 62,851.00	$

FIGURE 6.4.10

Solution

This is a partial worksheet from the IRS that is used by taxpayers whose taxable income is more than $100,000. Notice that the worksheet is divided by filing status. The taxpayer needs to use Section B of the worksheet for the filing status of married filing jointly. Begin by entering the taxable income amount in column (a) of the second row (Over $171,050 but not over $326,600.) In column (c) enter the result of multiplying the income by the percentage given, which is 24 %.

$$\text{Column (c): } \$177,839 \cdot 0.24 = \$42,681.36$$

Lastly, subtract the amount given in column (d), which is the adjustment for income that is taxed at the lower tax rate, and enter the result in the column labeled Tax.

$$\text{Column Tax: } \$42,681.36 - \$11,841.00 = \$30,840.36$$

Section B—Use if your filing status is **Married filing jointly** or **Qualifying widow(er).** Complete the row below that applies to you.

Taxable Income. If line 15 is—	(a) Enter the amount from line 15	(b) Multiplication amount	(c) Multiply (a) by (b)	(d) Subtraction amount	Tax. Subtract (d) from (c). Enter the result here and on the entry space on line 16.
At least $100,000 but not over $171,050	$	× 22% (0.22)	$	$ 8,420.00	$
Over $171,050 but not over $326,600	$ **$177,839**	× 24% (0.24)	$ **$42,681.36**	$ 11,841.00	$ **$30,840.36**
Over $326,600 but not over $414,700	$	× 32% (0.32)	$	$ 37,969.00	$
Over $414,700 but not over $622,050	$	× 35% (0.35)	$	$ 50,410.00	$
Over $622,050	$	× 37% (0.37)	$	$ 62,851.00	$

FIGURE 6.4.11

Skill Check 6.4.2

Use the 2020 tax table shown in Table 6.4.2 to calculate the amount of federal income tax charged to a taxpayer who is married filing jointly with a taxable income of $24,400.

Fun Fact

The April 15th deadline began in 1955. Besides moving for holidays and weekends, the deadline has only been extended nationally once. In 2020, due the economic effects of the coronavirus pandemic, the filing deadline was extended to July 15th.

Thus, the amount of federal income tax that should be entered on line 16 of Form 1040 is $30,840.36.

So far in the process, we've been careful to note that what we have worked toward calculating is the amount of money that the government charges you for taxes in the year. It's your responsibility by law to settle up with the IRS on or before April 15th.

Before you pay the tax amount entered on line 16 of Form 1040, you get to account for any taxes you've already paid throughout the year as well as any **tax credits** you might be eligible for.

If you work for someone else, your employer has been helping you chip away at your income taxes throughout the year by automatically withholding a portion from your paycheck. This amount is provided on your W-2 and/or 1099 forms that every employer is required to provide at the beginning of the new year. If you're self-employed, you submit estimated quarterly taxes throughout the year.

A tax credit is not the same as a deduction. A deduction reduces the amount of taxable income; a credit reduces the amount of tax. Think of it as a gift card redeemable to the IRS: it lowers the tax balance you must pay. The government uses credits as a way to encourage certain activities or lessen the tax burden of certain sectors of the public. For instance, the Lifetime Learning Credit is designed to encourage the pursuit of higher education, while the Child Tax Credit is designed to help with the high cost of raising a child.

These tax credits and any taxes withheld by your employer are all added together. That sum is entered as your total payments thus far on line 33.

FIGURE 6.4.12

Now comes the moment of truth—either the IRS owes you a refund because you have overpaid your taxes for the year, or you owe the IRS because you have not yet met your tax obligation. Form 1040 allows for you to compare the appropriate lines and determine your refund or the amount you owe. Finally, you sign and date your form and submit it to the IRS by April 15th, whether by mailing in or dropping off a paper form or sending the form electronically

Example 6.4.4

Determining the Bottom Line

Based on the following Form 1040, did the taxpayer receive a federal tax refund or owe an additional amount to the IRS in 2020?

FIGURE 6.4.13

Solution

On Form 1040, line 24 shows that the total amount of tax that is owed for the year is $3,610. Line 33 shows the sum of the taxes that have already been paid throughout the year as well as any tax credits that were applied, $3,474.50. If line 33 is greater, then the taxpayer receives a refund for the difference. If line 24 is greater, there are more taxes to be paid.

Since line 24 is greater, we need to subtract to determine the amount still owed.

$$\$3,610 - \$3,474.50 = \$135.50$$

Therefore, the taxpayer needed to pay an additional $135.50 to the IRS that year. This value would be entered on line 37.

OBJECTIVE 4 Payroll Tax

As we mentioned earlier, two of the biggest revenue generators for the federal government are individual income taxes and payroll taxes. While individual income taxes made up almost 50% of the revenue in 2019, payroll taxes accounted for another $1.243 trillion, or 36%. Payroll taxes are used to fund three main areas: Social Security, Medicare, and unemployment.

The first two, Social Security tax and Medicare tax, are known as the Federal Insurance Contributions Act (FICA) taxes. Social Security pays benefits to retirees, persons with disabilities, and survivors of deceased workers. Medicare is the national health insurance for people age 65 and older. The cost for these taxes is split between the employee and the employer and, similar to income taxes, your share of the cost as an employee is automatically withheld from your paycheck.

The third tax is mandated by the Federal Unemployment Tax Act (FUTA). It is an employer-paid payroll tax intended to fund state workforce agencies. This tax is not deducted from your paycheck; it is the responsibility of your employer.

FICA taxes are levied as a percentage of a worker's gross income and are decided each year by the IRS. Table 6.4.4 gives the 2020 FICA tax rates. Notice that if you are self-employed, you must pay both the employee's and employer's share of FICA taxes.

Table 6.4.4: 2020 FICA Tax Rates

Tax	Employee Rate	Employer Rate	Self-Employed Rate
Social Security Tax	6.2% of the first $137,700 of gross income	6.2% of the first $137,700 of gross income	12.4% on the first $137,700 of gross income
Medicare Tax	1.45%	1.45%	2.9%
Additional Medicare Tax	0.9% (on earnings over $200,000 for single filers, $250,000 for joint filers, or $125,000 for married filing separately)	N/A	0.9% (on earnings $200,000 for single filers, $250,000 for joint filers, or $125,000 for married filing separately)

Example 6.4.5

Calculating FICA Taxes

A pay stub is shown for an employee who has a gross income of $36,288 annually. Using Table 6.4.4, calculate the missing figures from the pay stub, which is for the first pay period of the year.

a. FICA Medicare tax

b. FICA Social Security tax

c. Current deductions

d. Net pay

Barbara's Bombtastic Bakery
1234 Main Street, Cleveland, OH 44105

EARNINGS STATEMENT

EMPLOYEE NAME			SSN	EMPLOYEE ID	CHECK NO.	PAY PERIOD	PAY DATE
Sample Employee Name			XXX-XX-1234	0000123456	23131	1/1/20 – 1/31/20	2/4/20

INCOME	RATE	HOURS	CURRENT TOTAL	DEDUCTIONS	CURRENT TOTAL	YEAR-TO-DATE
GROSS WAGES	18.90	160	3024.00	FICA MED TAX	a. _____	
				FICA SS TAX	b. _____	
				FED TAX	347.08	347.08

YTD GROSS	YTD DEDUCTIONS	YTD NET PAY	CURRENT TOTAL	CURRENT DEDUCTIONS	NET PAY
3024.00			3024.00	c. _____	d. _____

FIGURE 6.4.14

Solution

a. For someone who is not self-employed, the FICA Medicare tax rate is 1.45% of the gross income. The pay stub shows that the gross income for the current period is $3024. So the Medicare tax withheld in this pay period is calculated as follows.

$$\text{FICA Medicare Tax: } \$3024 \cdot 0.0145 \approx \$43.85$$

b. The FICA Social Security tax rate for employees is 6.2%, for the first $137,700 of gross income. Since this employee's annual gross income is less than that, the 6.2 % tax rate is applied to the entire income earned during the pay period.

$$\text{FICA Social Security Tax: } \$3024 \cdot 0.062 \approx \$187.49$$

c. The current deductions consist of all the taxes withheld for the current pay period. We need to add together the Medicare tax, the Social Security tax, and the federal income tax (FED TAX).

$$\text{Current Deductions: } \$43.85 + \$187.49 + \$347.08 = \$578.42$$

d. To find the net pay, we subtract the current deductions from the current total.

$$\text{Net Pay: } \$3024 - \$578.42 = \$2445.58$$

> ✔ **Skill Check 6.4.3**
>
> Use Table 6.4.4 to calculate the FICA taxes that a self-employed person owed in 2020 if their gross income was $94,538.

This section provided a brief introduction to federal revenue from individual income taxes and payroll taxes. Federal tax laws and the budget for the United States Government are updated annually. For more information on federal taxes, visit the IRS website www.irs.gov. For more information on the United States Government Budget, visit either govinfo at https://www.govinfo.gov/app/collection/budget or the Congressional Budget Office at https://www.cbo.gov/.

Skill Check Answers
1. $12, 400 **2.** $2536 **3.** $14,464.31

6.4 Exercises

✔ CONCEPT CHECK

1. The form you must fill out each year when you file your income taxes is _____.

2. What is an example of an expense you can subtract from your gross income to arrive at your adjusted gross income?

3. Your taxable income is your adjusted gross income once certain _____ have been subtracted.

4. Individual income taxes and _____ taxes are two of the biggest revenue generators for the federal government.

5. True or False: A person's entire taxable income is taxed at the same rate.

PRACTICE

Determine the gross income using the given information.

6. A registered nurse has an annual salary of $63,458, earned $125.72 of interest on their savings account, paid $776 in student loan interest, and contributed $3600 toward a 401(k).

7. A physical therapist assistant has an annual salary of $57,410, earned $723.55 in interest on various investments, paid $276 in student loan interest, and contributed $4200 toward an IRA.

8. A computer programmer has an annual salary of $72,230, earned $1206.39 in interest on various investments, and contributed $3600 toward a 401(k).

9. An epidemiologist has an annual salary of $65,755, earned $158.23 in interest on a savings account, and contributed $4000 toward a 401(k).

10. A dental assistant earns $38,500 per year, paid $700 in student loan interest, and contributed $1200 to a 401(k).

11. A paramedic earns $32,600 per year, earned $266.32 in interest on various investments, and contributed $1200 toward an IRA.

12. A middle-school teacher has an annual salary of $56,250, earned $52.26 in interest on a savings account, paid $654.03 in student loan interest, and contributed $2400 towards an IRA.

   a. Determine the middle-school teacher's gross income.

   b. Determine the middle-school teacher's adjusted gross income.

13. An architect has an annual salary of $78,250, earned $240.88 in interest on a savings account, paid $1250 in student loan interest, and contributed $4800 towards a 401(k).

   a. Determine the architect's gross income.

   b. Determine the architect's adjusted gross income.

Use Table 6.4.1 to calculate the total amount of taxes paid by a single taxpayer for the given income amount.

14. $32,500   15. $24,800   16. $45,650

17. $75,200   18. $175,000   19. $220,700

Use the partial 2020 IRS tax worksheet to calculate the amount of federal income tax charged to each taxpayer.

2020 Tax Computation Worksheet—Line 16

 See the instructions for line 16 to see if you must use the worksheet below to figure your tax.

Note. If you are required to use this worksheet to figure the tax on an amount from another form or worksheet, such as the Qualified Dividends and Capital Gain Tax Worksheet, the Schedule D Tax Worksheet, Schedule J, Form 8615, or the Foreign Earned Income Tax Worksheet, enter the amount from that form or worksheet in column (a) of the row that applies to the amount you are looking up. Enter the result on the appropriate line of the form or worksheet that you are completing.

Section A—Use if your filing status is **Single.** Complete the row below that applies to you.

Taxable income. If line 15 is—	(a) Enter the amount from line 15	(b) Multiplication amount	(c) Multiply (a) by (b)	(d) Subtraction amount	Tax. Subtract (d) from (c). Enter the result here and on the entry space on line 16.
At least $100,000 but not over $163,300	$	× 24% (0.24)	$	$ 5,920.50	$
Over $163,300 but not over $207,350	$	× 32% (0.32)	$	$ 18,984.50	$
Over $207,350 but not over $518,400	$	× 35% (0.35)	$	$ 25,205.00	$
Over $518,400	$	× 37% (0.37)	$	$ 35,573.00	$

Section B—Use if your filing status is **Married filing jointly** or **Qualifying widow(er).** Complete the row below that applies to you.

Taxable income. If line 15 is—	(a) Enter the amount from line 15	(b) Multiplication amount	(c) Multiply (a) by (b)	(d) Subtraction amount	Tax. Subtract (d) from (c). Enter the result here and on the entry space on line 16.
At least $100,000 but not over $171,050	$	× 22% (0.22)	$	$ 8,420.00	$
Over $171,050 but not over $326,600	$	× 24% (0.24)	$	$ 11,841.00	$
Over $326,600 but not over $414,700	$	× 32% (0.32)	$	$ 37,969.00	$
Over $414,700 but not over $622,050	$	× 35% (0.35)	$	$ 50,410.00	$
Over $622,050	$	× 37% (0.37)	$	$ 62,851.00	$

20. A single taxpayer with a taxable income of $110,233

21. A single taxpayer with a taxable income of $165,755

22. A married taxpayer filing jointly with a taxable income of $120,824

23. A married taxpayer filing jointly with a taxable income of $328,100

24. A qualifying widow with a taxable income of $152,070

25. A qualifying widower with a taxable income of $322,085

26. A single taxpayer with a taxable income of $102,780

27. A single taxpayer with a taxable income of $220,490

Based on the given Form 1040, did the taxpayer receive a federal tax refund or owe an additional amount to the IRS in 2020?

28.

Form 1040 (2020)					Page **2**
	16	Tax (see instructions). Check if any from Form(s): 1 ☐ 8814 2 ☐ 4972 3 ☐ _____		16	2,814.53
	17	Amount from Schedule 2, line 3		17	
	18	Add lines 16 and 17		18	2,814.53
	19	Child tax credit or credit for other dependents		19	
	20	Amount from Schedule 3, line 7		20	
	21	Add lines 19 and 20		21	0
	22	Subtract line 21 from line 18. If zero or less, enter -0-		22	2,814.53
	23	Other taxes, including self-employment tax, from Schedule 2, line 10		23	
	24	Add lines 22 and 23. This is your **total tax** ▶		24	2,814.53
	25	Federal income tax withheld from:			
	a	Form(s) W-2	25a	2,596.17	
	b	Form(s) 1099	25b		
	c	Other forms (see instructions)	25c		
	d	Add lines 25a through 25c		25d	2,596.17
• If you have a qualifying child, attach Sch. EIC.	26	2020 estimated tax payments and amount applied from 2019 return		26	
	27	Earned income credit (EIC)	27		
	28	Additional child tax credit. Attach Schedule 8812	28		
• If you have nontaxable combat pay, see instructions.	29	American opportunity credit from Form 8863, line 8	29		
	30	Recovery rebate credit. See instructions	30		
	31	Amount from Schedule 3, line 13	31		
	32	Add lines 27 through 31. These are your **total other payments and refundable credits** ▶		32	0
	33	Add lines 25d, 26, and 32. These are your **total payments** ▶		33	2,596.17
Refund	34	If line 33 is more than line 24, subtract line 24 from line 33. This is the amount you **overpaid**		34	
	35a	Amount of line 34 you want **refunded to you.** If Form 8888 is attached, check here . . . ▶ ☐		35a	
Direct deposit? See instructions.	▶b	Routing number	▶ c Type: ☐ Checking ☐ Savings		
	▶d	Account number			
	36	Amount of line 34 you want **applied to your 2021 estimated tax** . . ▶	36		
Amount You Owe	37	Subtract line 33 from line 24. This is the **amount you owe now** . . . . . ▶		37	
For details on how to pay, see instructions.		Note: Schedule H and Schedule SE filers, line 37 may not represent all of the taxes you owe for 2020. See Schedule 3, line 12e, and its instructions for details.			
	38	Estimated tax penalty (see instructions) . . . . ▶	38		

29.

Form 1040 (2020)					Page **2**
	16	Tax (see instructions). Check if any from Form(s): 1 ☐ 8814 2 ☐ 4972 3 ☐ _____		16	3,765.14
	17	Amount from Schedule 2, line 3		17	
	18	Add lines 16 and 17		18	3,765.14
	19	Child tax credit or credit for other dependents		19	
	20	Amount from Schedule 3, line 7		20	
	21	Add lines 19 and 20		21	0
	22	Subtract line 21 from line 18. If zero or less, enter -0-		22	3,765.14
	23	Other taxes, including self-employment tax, from Schedule 2, line 10 ▶		23	
	24	Add lines 22 and 23. This is your **total tax** ▶		24	3,765.14
	25	Federal income tax withheld from:			
	a	Form(s) W-2	25a	3,893.59	
	b	Form(s) 1099	25b		
	c	Other forms (see instructions)	25c		
	d	Add lines 25a through 25c		25d	3,893.59
• If you have a qualifying child, attach Sch. EIC.	26	2020 estimated tax payments and amount applied from 2019 return		26	
	27	Earned income credit (EIC)	27		
	28	Additional child tax credit. Attach Schedule 8812	28		
• If you have nontaxable combat pay, see instructions.	29	American opportunity credit from Form 8863, line 8	29		
	30	Recovery rebate credit. See instructions	30		
	31	Amount from Schedule 3, line 13	31		
	32	Add lines 27 through 31. These are your **total other payments and refundable credits** ▶		32	0
	33	Add lines 25d, 26, and 32. These are your **total payments** ▶		33	3,893.59
Refund	34	If line 33 is more than line 24, subtract line 24 from line 33. This is the amount you **overpaid**		34	
	35a	Amount of line 34 you want **refunded to you.** If Form 8888 is attached, check here . . . ▶ ☐		35a	
Direct deposit? See instructions.	▶b	Routing number	▶ c Type: ☐ Checking ☐ Savings		
	▶d	Account number			
	36	Amount of line 34 you want **applied to your 2021 estimated tax** . . ▶	36		
Amount You Owe	37	Subtract line 33 from line 24. This is the **amount you owe now** . . . . . ▶		37	
For details on how to pay, see instructions.		Note: Schedule H and Schedule SE filers, line 37 may not represent all of the taxes you owe for 2020. See Schedule 3, line 12e, and its instructions for details.			
	38	Estimated tax penalty (see instructions) . . . . ▶	38		

30. A paystub is shown for an employee who has a gross income of $42,000 annually. Using Table 6.4.4, calculate the missing figures from the paystub, which is for the first pay period of the year.

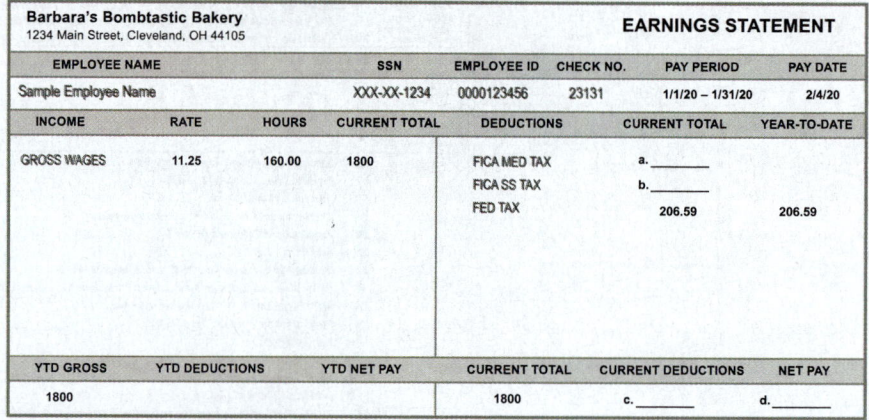

Barbara's Bombtastic Bakery						EARNINGS STATEMENT	
1234 Main Street, Cleveland, OH 44105							
EMPLOYEE NAME				**SSN**	**EMPLOYEE ID**	**CHECK NO.**	**PAY PERIOD**	**PAY DATE**
Sample Employee Name				XXX-XX-1234	0000123456	23131	1/1/20 – 1/31/20	2/4/20
INCOME	**RATE**	**HOURS**	**CURRENT TOTAL**	**DEDUCTIONS**		**CURRENT TOTAL**	**YEAR-TO-DATE**
GROSS WAGES	21.86	160.11	3500	FICA MED TAX		a. _____	
				FICA SS TAX		b. _____	
				FED TAX		401.71	401.71
YTD GROSS	**YTD DEDUCTIONS**		**YTD NET PAY**	**CURRENT TOTAL**	**CURRENT DEDUCTIONS**		**NET PAY**
3500				3500	c. _____		d. _____

31. A paystub is shown for an employee who has a gross income of $21,600 annually. Using Table 6.4.4, calculate the missing figures from the paystub, which is for the first pay period of the year.

Barbara's Bombtastic Bakery						EARNINGS STATEMENT	
1234 Main Street, Cleveland, OH 44105							
EMPLOYEE NAME				**SSN**	**EMPLOYEE ID**	**CHECK NO.**	**PAY PERIOD**	**PAY DATE**
Sample Employee Name				XXX-XX-1234	0000123456	23131	1/1/20 – 1/31/20	2/4/20
INCOME	**RATE**	**HOURS**	**CURRENT TOTAL**	**DEDUCTIONS**		**CURRENT TOTAL**	**YEAR-TO-DATE**
GROSS WAGES	11.25	160.00	1800	FICA MED TAX		a. _____	
				FICA SS TAX		b. _____	
				FED TAX		206.59	206.59
YTD GROSS	**YTD DEDUCTIONS**		**YTD NET PAY**	**CURRENT TOTAL**	**CURRENT DEDUCTIONS**		**NET PAY**
1800				1800	c. _____		d. _____

🚀 **APPLICATIONS**

Determine the taxable income using the given information. See Appendix B for tax forms.

32. A married taxpayer is filing taxes jointly with his spouse, and they do not have any dependents to claim. During the year, they contribute $4800 to a traditional IRA. They also make charitable contributions of $800 to the Wildlife Conservation Society and $500 to Ronald McDonald House Charities. They have no other deductions, and their combined wages are $124,938. Additionally, they earned $374 in interest from their savings accounts, which is taxable.

33. A married taxpayer is filing taxes jointly with her spouse, and they do not have any dependents to claim. During the year, they contribute $3600 to a traditional IRA. They also make charitable contributions of $300 to the Conservation Fund, $250 to the Sheldrick Wildlife Trust, and $450 to the American Lung Association. They have no other deductions, and their combined wages are $110,270. Additionally, they earned $178 in interest from their savings accounts, which is taxable.

34. A single taxpayer is filing taxes and does not have any dependents to claim. During the year, she contributes $1200 to a traditional IRA. She also makes charitable contributions of $250 to the Elephant Sanctuary in Tennessee and $350 to the Parkinson's Foundation. She has no other deductions, and her yearly wage is $45,640. Additionally, she earned $92 in interest from her savings accounts, which is taxable.

35. A single taxpayer is filing taxes and does not have any dependents to claim. During the year, they contribute $1800 to a traditional IRA. They also make charitable contributions of $400 to the American Red Cross and $400 to Feeding America. They have no other deductions, and their yearly wage is $62,700. Additionally, they earned $256 in interest from their savings accounts, which is taxable.

36. A married taxpayer is filing taxes jointly with his spouse and they do not have any dependents to claim. During the year, they contribute $3200 to a traditional IRA. They also make charitable contributions of $125 to Comic Relief, $500 to ProLiteracy Worldwide, and $750 to the Alzheimer's Association. They have no other deductions, and their combined wages are $135,760. Additionally, they earned $730 in interest from their savings accounts, which is taxable.

37. A qualified widow is filing taxes and does not have any dependents to claim. During the year, she contributes $4000 to a traditional IRA. She also makes charitable contributions of $275 to the Trevor Project, $350 to Give Kids the World, and $500 to the Gary Sinise Foundation. She has no other deductions, and her yearly wage is $125,750. Additionally, she earned $220 in interest from her savings accounts, which is taxable.

38. A married taxpayer is filing taxes separately from his spouse and does not have any dependents to claim. During the year, he contributes $2400 to a traditional IRA. He also makes charitable contributions of $175 to the National Council on Aging and $250 to K9s for Warriors. He has no other deductions, and his yearly wage is $56,200. Additionally, he earned $120 in interest from their savings accounts, which is taxable.

39. A single taxpayer is filing taxes and does not have any dependents to claim. During the year, they contribute $1200 to a traditional IRA. They also make charitable contributions of $500 to UNCF and $400 to the Cancer Research Institute. They have no other deductions, and their yearly wage is $73,000. Additionally, they earned $351 in interest from their savings accounts, which is taxable.

✎ WRITING & THINKING

40. Describe how this section has influenced your view on completing taxes.

6.4 PROJECT

VALUE-ADDED TAXES: AN ALTERNATIVE TO INCOME TAXES

The United States government collects income taxes to pay for a portion of government spending. The highest earners, regardless of their consumption habits, are taxed at higher levels. In contrast, the tax rules in many other countries are designed to tax the consumption of citizens rather than the income of citizens with what is called a value-added tax (VAT). In this taxation model, a certain amount of taxes are collected at each stage of the supply chain, from production to final consumer. The amount of VAT that is paid is calculated on the cost of the product minus any of the costs of materials used in the product that have already been taxed.

Let's assume that a country has a VAT rate of 10%. We will follow the supply chain for a loaf of bread, from farmer to baker to supermarket.[2]

a. The farmer grows the wheat and sells it to the baker for $0.30 per unit. The VAT is $0.03 (10% of $0.30) per unit. The baker pays the farmer $0.33 for a unit of wheat, and the farmer sends $0.03 in VAT to the government.

b. The baker makes a loaf of bread and sells it to a store for $0.60 per loaf. The VAT is $0.06 (10% of $0.60) per loaf. Now the store pays the baker $0.66 per loaf, of which $0.06 is VAT. Out of $0.06 VAT collected from the store, the baker only sends to the government $0.03 because he receives a $0.03 credit from the government for the VAT paid to the farmer.

c. The store sells the loaf to a customer for a dollar. The customer pays $1.10. The store sends the government $0.04 total—the $0.10 it collected in VAT on the sale of the bread minus the $0.06 it paid to the baker in VAT, which the store gets back in a credit.

In total, the government received $0.03 from the farmer, $0.03 from baker, and $0.04 from the store. That's $0.10 on a final sale of a dollar loaf of bread for a 10% VAT.

	Selling Price	VAT Collected	VAT Credit	VAT Sent to Government
Farmer	$0.30	$0.03	$0.00	$0.03
Baker	$0.60	$0.06	$0.03	$0.03
Supermarket	$1.00	$0.10	$0.06	$0.04
Total				$0.10

---

2   Example adapted from "How Does a 'Value Added Tax' Work, Anyway?" Derek Thompson, *The Atlantic*, March 1, 2010, https://www.theatlantic.com/business/archive/2010/03/how-does-a-value-added-tax-work-anyway/36834/.

6. Determine the amount of VAT collected, VAT credit, and VAT sent to the government that would be collected at each stage of the supply chain below if the VAT rate were 20%.

	Selling Price	VAT Collected	VAT Credit	VAT Sent to Government
Farmer	$12.00			
Baker	$24.00			
Supermarket	$30.00			
Total				

7. Notice that unlike sales, VAT is charged at all stages of the supply chain instead of only on the final consumer. Discuss why this makes it harder for businesses to avoid paying taxes.

8. VAT rates are usually much higher than sales tax rates; in most of Europe, the VAT rate is around 20%. Discuss why these higher rates might have an impact on consumption.

9. Would you be in favor of replacing the US system of income tax with a VAT? Explain your reasoning.

OBJECTIVES

1. Calculate portions of a monthly budget.

2. Create a personal budget.

Thus far in the chapter, we've examined how to make our finances work for us and what the cost of borrowing is in the long term. A task that many people find challenging is developing a plan for their money in which they are financially robust—taking care of their current needs while planning for the future.

A *budget* is a financial plan for your income over a given period of time. It provides you with a visual map of your money and puts you in control of your funds. Instead of reacting to life's financial ups and down, a budget helps you plan ahead. Contrary to some opinions, it doesn't restrict you like a straitjacket, but instead puts you in the driver's seat of your life.

 BUDGET

A **budget** is a financial plan for your income over a defined period of time.

Maybe you're tired of living paycheck to paycheck, or you'd like to buy a car, or you want to pay off your student loans. Whatever the end goal(s), a budget is a vehicle to get you there. It helps you squeeze every bit of value out of every dollar you earn. We'll first look at the steps to create a budget and then look at some options of how to implement your budget.

OBJECTIVE 1 Steps to Creating a Budget

The big picture of creating a budget is as simple as answering three questions: What is your posttax income (net pay); that is, how much do you have to spend? What are your mandatory expenses? What is left over after you subtract one from the other?

STEPS TO CREATING A BUDGET

1. Calculate monthly income.

2. Calculate monthly expenses.

3. Subtract expenses from income.

4. Allocate any remaining funds or make adjustments.

Step 1: Calculate monthly income.

The first step toward gaining control of your finances is to look at what you have to work with. Write down every source of income you receive. If this is in the form of a paycheck, you'll want to include your net pay; that is, take-home pay after taxes and deductions. Remember to include other types of income like interest on savings accounts, alimony, child support, or any side gigs you might have (such as driving for a rideshare app, making deliveries for Amazon, or even babysitting).

Step 2: Calculate monthly expenses.

Now it's time to account for all the necessities you pay for each month as well as any extra spending. Go back through your bills, your banking account records, or any receipts to write down every penny you spent in the month. It is even better to do this for two or three months in order to get an average of what you spend.

> **Helpful Hint**
>
> Thinking of a budget as a spending plan takes away the negative connotation of the word "budget." You can spend guilt-free knowing your needs are taken care of and you won't run out of money before the month is gone!

You might be apprehensive about facing the cold hard facts about how much you spend each month, but knowledge is power. The following is a sample of some of the common monthly expenses people incur.

- Housing: rent or mortgage
- Transportation: car payments, rental car, public transport, ride sharing, fuel, or car insurance
- Health insurance
- Groceries and household goods
- Mobile phone
- Home internet
- Utilities: water, electricity, and gas
- Debt repayment: student loans or credit cards
- Medical expenses: doctor visits or prescriptions
- Childcare
- Pet care
- Dry cleaning
- Salon services: hair, nails, or massages
- Fitness memberships
- Streaming services or apps
- Gifts
- Clothing
- Dining out or ordering takeout
- Entertainment

If you don't have receipts or a way to account for where your money was spent, then you should make an educated guess to start with while you take the next month to track your spending. Creating a place for you to record your spending (every penny) over a month can be as easy a piece of paper. There are also smartphone apps that keep track of your spending—some even allow you to scan in your receipts and they take care of the rest. Get a head start by assigning expenses to a category as they happen. You don't have to have all the categories figured out from the start—you can always add more.

FIGURE 6.5.1

Example 6.5.1

Calculating Miscellaneous Spending

Determine the approximate monthly cost of each miscellaneous spending activity. Then calculate the overall cost of the three activities in one month.

a. $3.02 on coffee at least 3 times a week

b. Lunch with friends every Friday, averaging $10.65

c. $15 for a friend's birthday (either on a gift or celebrating out with that friend) and 3 friends have birthdays this month

Solution

a. Coffee 3 times a week, for 4 weeks in a month, means we multiply the cost of the coffee by 12.

$$\text{Monthly Coffee Cost: } \$3.02 \cdot 12 = \$36.24$$

b. Eating out with friends each Friday equates to at least 4 times the cost of the meal.

$$\text{Monthly Friday Lunch Cost: } \$10.65 \cdot 4 = \$42.60$$

c. Three friends' birthdays in a month means that you spend $3 \cdot \$15 = \$45$ in gifts that month.

Overall, the miscellaneous expenses for the month add to

$$\$36.24 + \$42.60 + \$45 = \$123.84.$$

✔ **Skill Check 6.5.1**

If a gamer spends 99¢ on average every 3 days to buy extra levels on a game, and also allows himself $5.00 in miscellaneous game spending each month, calculate how much is spent on gaming during the month.

Some expenses don't occur on a monthly basis, but instead happen at other periodic intervals, perhaps semiannually or annually. This doesn't have to mean that there is an unlucky month where you incur this cost all at once. You can **prorate**, or spread out, any expenses you know you will have to pay in the future. For instance, suppose you know your tuition bill comes due at the beginning of each fall and spring semester. You can add together the tuition costs for the two semesters and then divide that total by 12 to determine how much of your monthly budget needs to be set aside for tuition costs.

Example 6.5.2

Prorating Expenses

Imani is self-employed. Based on last year, her projected state and federal taxes for this year will be approximately $11,500.

a. Determine how much Imani should prepay each quarter in order to meet the projected taxes by the end of the year.

b. Determine how much Imani should put aside each month so that she has enough to pay her estimated quarterly taxes.

Solution

a. If Imani is paying her projected taxes quarterly, then the annual estimate should be divided by 4 to find the quarterly amount.

$$\text{Quarterly Taxes Due: } \frac{\$11,500}{4} = \$2875$$

Thus, Imani will need to pay $2875 each quarter in taxes.

b. Although Imani will only pay her taxes quarterly, setting aside funds each month is a smart move. There are two ways to determine the monthly portion: 1. divide the year's approximation by 12 months, or 2. divide the quarterly approximation by 3 months. Both calculations are shown.

Monthly Amount Set Aside For Taxes

Yearly Amount Divided by 12 Months:	Quarterly Amount Divided by 3 Months:
$\dfrac{\$11,500}{12} \approx \$958.33$	$\dfrac{\$2875}{3} \approx \$958.33$

Both methods show that Imani needs to allocate $958.33 each month for her state and federal taxes.

Step 3: Subtract expenses from income.

This step is the epiphany in which the truth comes to light. It might be difficult to acknowledge your financial habits, but it's a powerful and positive step towards controlling your finances.

Example 6.5.3

Determining Total Monthly Expenses

A recent graduate from a liberal arts college acquired a job as a sociologist. Her yearly salary is $34,500. Her employer withholds $5175 in state and federal income taxes and $2639.25 in FICA taxes throughout the year. She has the following monthly costs: transportation is $250, cell phone bill is $70, student loans require $210 in repayment, and rent is $600. She is using the average monthly costs for each of the following in order to gain an idea of other monthly expenses: utilities are $150, home internet is $65, health insurance is $456, and groceries are $270.

a. Determine her monthly net pay amount.

b. Determine how much money is left each month for discretionary spending after all necessities are accounted for.

Solution

a. The monthly net pay amount is the annual salary minus withholdings for all required taxes, divided over 12 months of the year. In this case, we need to subtract $5175 and $2639.25 from the yearly salary before dividing by 12.

Monthly Net Pay: $\dfrac{\$34,500 - (\$5175 + \$2639.25)}{12} = \dfrac{\$26685.75}{12} \approx \$2223.81$

Therefore, the sociologist can expect to collect a monthly paycheck of $2223.81.

b. Let's begin by adding up the monthly expenses, both the known and the estimates from what we are told.

Monthly Expenses:

$250 + $70 + $210 + $600 + $150 + $65 + $456 + $270 = $2071

Now subtract the monthly expenses from the net pay to determine if there is any money remaining for discretionary spending.

$2223.81 - $2071 = $152.81

This means that she is projected to have $152.81 of remaining income each month. On the plus side, there is money remaining. On the downside, there's not too much wiggle room for anything else.

Once you've created a spending plan for your money, you need a way to keep track of your expenses. Doing so helps you stay on budget and make adjustments when necessary. These days, tracking can be done with a paper ledger, any number of budgeting apps, budgeting software, or spreadsheets. Finding the format that works best for you is the most important thing.

One of the biggest advantages to using apps, software, or spreadsheets over a paper ledger is that the arithmetic is done for you. When expenses change month to month or new incomes are introduced, your budget is adjusted accordingly. Example 6.5.4 looks at using a spreadsheet to help create and keep track of a budget.

Example 6.5.4

Creating a Budget with a Spreadsheet

An art gallery curator has an annual income of $63,420. Their net monthly take-home pay is $3435.33, and they have the following monthly expenses.

Table 6.5.1: Monthly Expenses

Rent	$1200.00
Gas	$41.00
Water	$56.00
Electricity	$150.00
Internet	$60.00
Cell Phone	$74.00
Health Insurance	$350.00
Transportation	$420.00
Loan Repayment	$70.00
Food	$440.00
Credit Card	$120.00

a. Additional income includes $30 a week on average from a side consulting job. A total of $45 a month is donated to charities. Use a spreadsheet, such as Microsoft Excel, to create a budget. How much is available each month for expenses not listed in the budget?

b. Identify how the curator could categorize any remaining money at the end of the month. Discuss the possible choices.

Solution

a. To prepare a budget for the curator, we need to add both the incomes and expenses separately. We'll start by creating a column for expenses and one for incomes. Let's start with expenses. In cell A1 of the spreadsheet, type "Expense", and in the adjacent cell, B2, type "Amount". The spreadsheet should look like Figure 6.5.2.

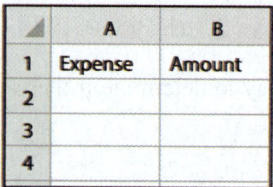

FIGURE 6.5.2

Under the column labeled "Expense" enter the name of each of the curator's expenses and under the column labeled "Amount" enter the amount associated with each expense. Don't forget to include the donations to charity as an expense. The updated spreadsheet should now appear as Figure 6.5.3.

	A	B
1	Expense	Amount
2	Rent	$1,200
3	Gas	$41
4	Water	$56
5	Electricity	$150
6	Internet	$60
7	Cell Phone	$74
8	Health Insurance	$350
9	Transportation	$420
10	Loan Repayment	$70
11	Food	$440
12	Credit Card	$120
13	Donations	$45

FIGURE 6.5.3

To obtain the total amount of expenses, we need a place for the total on row 14. Enter "Expense Total" in A14. In B14, we'll put a formula that will add up all the expenses. Type "=sum(B2:B13)", which instructs Excel to add together the data from column B, row 2 through column B, row 13.

B14		× ✓	fx	=SUM(B2:B13)
	A	B	C	D
1	Expense	Amount		
2	Rent	$1,200		
3	Gas	$41		
4	Water	$56		
5	Electricity	$150		
6	Internet	$60		
7	Cell Phone	$74		
8	Health Insurance	$350		
9	Transportation	$420		
10	Loan Repayment	$70		
11	Food	$440		
12	Credit Card	$120		
13	Donations	$45		
14	Expense Total	$3,026		

FIGURE 6.5.4

Figure 6.5.4 shows that the monthly expenses are $3026.

Next, we'll follow the same process for incomes. Create a set of columns for Income and Amount in columns D and E. (We've intentionally left column C blank for clarity.)

Enter the income labels in column D and the amounts in column E. Note that when we enter the amount for the consulting income, we'll need to write a formula for the monthly income, since we're given the weekly amount. The formula will multiply the weekly amount by 4; that is, "=30*4". (Note that during some months, this will be paid 5 times.) Doing it this way means that if the amount received each week changes, you can simply replace the number 30 with the new weekly income, and the spreadsheet will adjust the monthly amount accordingly. To calculate the total income for the month, enter "=sum(E2:E3)" into cell E14. The budget spreadsheet now looks like Figure 6.5.5.

E14	▼	⋮	×	✓	fx	=SUM(E2:E3)

	A	B	C	D	E
1	Expense	Amount		Income	Amount
2	Rent	$1,200		Salary	$3,435.33
3	Gas	$41		Consulting	$120
4	Water	$56			
5	Electricity	$150			
6	Internet	$60			
7	Cell Phone	$74			
8	Health Insurance	$350			
9	Transportation	$420			
10	Loan Repayment	$70			
11	Food	$440			
12	Credit Card	$120			
13	Donations	$45			
14	Expense Total	$3,026		Income Total	$3,555.33

FIGURE 6.5.5

Finally, we need to subtract the expense total from the income total to find the remaining funds that are available. We'll put this in column G; again, leaving a blank column F for the sake of clarity. We need to tell the spreadsheet to subtract the expense total from the income total by directing it to the cells that contain the total for expenses and incomes. The formula will be in the form Income Total – Expense Total, so type the formula "=E14-B14" into cell G14, as shown in Figure 6.5.6.

G14	▼	⋮	×	✓	fx	=E14-B14

	A	B	C	D	E	F	G
1	Expense	Amount		Income	Amount		Remaining Funds
2	Rent	$1,200		Salary	$3,435.33		
3	Gas	$41		Consulting	$120		
4	Water	$56					
5	Electricity	$150					
6	Internet	$60					
7	Cell Phone	$74					
8	Health Insurance	$350					
9	Transportation	$420					
10	Loan Repayment	$70					
11	Food	$440					
12	Credit Card	$120					
13	Donations	$45					
14	Expense Total	$3,026		Income Total	$3,555.33		$529.33

FIGURE 6.5.6

We can now see that the curator has $529.33 to put towards expenses not in the budget each month. One of the benefits of using a spreadsheet like this is that if any of the expenses or incomes change, the spreadsheet automatically adjusts the final calculation for us.

b. The curator's budget does not currently include categories for entertainment, gifts, monthly subscriptions, savings, or retirement, to name a few. Although it is a personal choice of where to allocate the remaining funds, as we saw in Section 6.2, the earlier savings and retirement accounts begin, the better they perform for you. Regardless of how the curator chooses to allocate the remaining funds, having a plan is always the best option. Otherwise, money seems to find a way of slipping away.

✔ **Skill Check 6.5.2**

Using the spreadsheet from Example 6.5.4, find the amount available for other expenses at the end of the month if the curator's utilities increase by $50.

Step 4: Allocate any remaining funds or make adjustments.

As Step 4 implies, making a budget is not a "one and done" activity. The last two examples both showed an ending monthly balance that was positive. In other words, there was money left unaccounted for at the end of month. Even though there is a sizeable difference between $152.81 and $529.33, having money left over after necessities is still a positive position. But, if your expenses are more than your income, you basically have four choices. You can *earn more*, *spend less*, *dip into your savings*, or *borrow money*.

Section 6.3 looked at the ups and downs of borrowing money, whether that be from a lender or from your own savings account. What about the other two options: earning more money or spending less? Earning more money requires either finding a higher paying job or working side jobs. However, one of the best ways to make ends meet is to spend less money.

So how can you spend less on necessities? One way is to find the best monthly plans for expenses you regularly pay. And for that you should compare, compare, compare! Can you find a better cell phone plan, lower insurance costs for your car, house, or health, or find a low interest loan to consolidate credit card debit? Comparing options takes time, but it has the potential to add to your discretionary budget. Example 6.5.5 looks at comparing cell phone plans.

Example 6.5.5

Comparing Cell Phone Plans

Compare the following cell phone plans. Explain which plan you would choose if you are only interested in a single line?

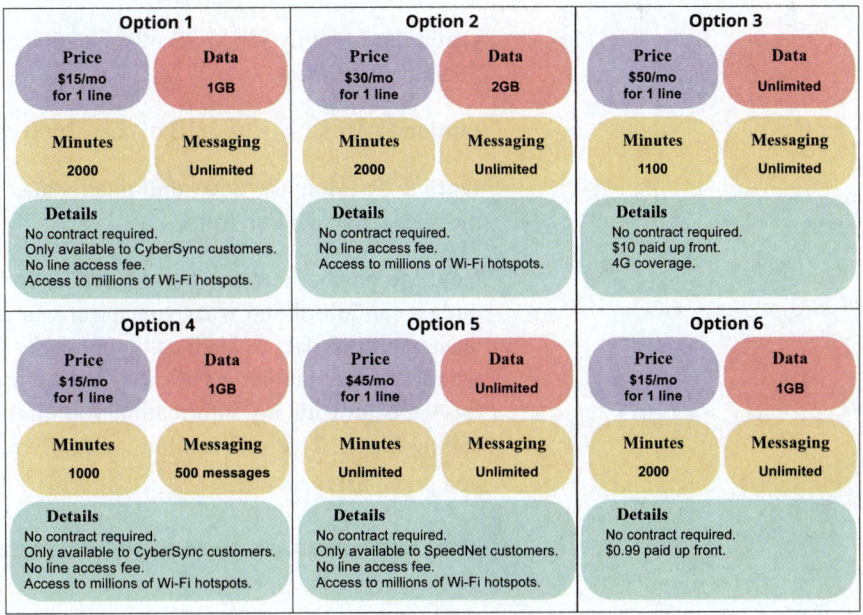

FIGURE 6.5.7

Solution

We are confronted with six options, but it is unlikely that we can easily rank these options best to worst. One option may be cheaper, but also has less data; another option may be more expensive but has unlimited calls; and still another option may look attractive but is only available to certain customers.

Consequently, when you begin to compare any set of options, one of the first things you should do is narrow down your choices by eliminating similar options. In a group of similar things, choose the best one and then discard the others. Continue this process until you have only a small number to choose from. Then you can make a decision based on personal needs or wants.

In this case, notice that Options 1, 4, and 6 are very similar—all are $15 for 1GB of data. The main differences are the unlimited messages and more minutes offered in Options 1 and 6. Therefore we can discard Option 4. Option 1 requires a specific internet provider while Option 6 requires a $0.99 fee. If we are a CyberSync customer, Option 1 is the best choice of the three. If CyberSync is not our provider, we should keep Option 6 in the running instead.

Similarly, Options 3 and 5 differ by $5, but you are required to have a specific internet provider for Option 5. If SpeedNet is your provider, keep Option 5; if not, discard it and keep Option 3.

Option 2 is not similar enough to compare with any other options, so we will keep it as well.

At this point we will assume that options involving particular internet providers are not feasible for this scenario. We are then left with Options 2, 3, and 6. Let's rank them according to price and list the differences.

Option 6: Monthly $15, 1GB of data, 2000 call minutes, unlimited messages, and $0.99 paid up front.

Option 2: Monthly $30, 2GB of data, 2000 call minutes, and unlimited messages.

Option 3: Monthly $50, unlimited data, 1100 call minutes, and unlimited messages. $10 paid up front.

Now the choice between these three options needs to be based on what your budget can afford and what your priorities are for cell phone coverage. Do you *need* unlimited data or 2000 call minutes for work or school? Could you get away with less until your budget allows for the extra? It would be a good idea to inform this decision with any information you might have about your past data use and calling habits. That information is usually available from a cell phone carrier if you previously had one.

OBJECTIVE 2 Budgeting Goals

Ultimately, a sound budget will cover all of your needs, some of your wants, and put you on a path of saving for emergencies and the future. Although everyone's needs and goals are different and nuanced, there are some general budget goals that financial advisors recommend.

One such recommendation is the 50/30/20 rule: 50% of your net income should go to necessities, no more than 30 % for wants, and at least 20% for savings and paying down debt faster (for example, making extra student loan payments).

📖 50/30/20 BUDGETING RULE

The **50/30/20 rule for budgeting** is to allow 50% of your net income for necessities, no more than 30% for wants, and at least 20% for savings and paying down debt.

Example 6.5.6

Preparing a Budget

Starting a new life together, the Garcias have decided to combine their resources and live within their means by creating a budget from the outset. Their combined net income is $3278 a month.

a. Create a monthly budget for the Garcias that follows the 50/30/20 rule. Discuss what you would include in each category.

b. They are considering a new apartment that will raise their housing costs by $175 a month. What factors do the Garcias need to keep in mind if they are trying to stick to their budget?

Solution

a. To follow the 50/30/20 rule, the monthly income should be divided as follows:

Necessities 50%: $3278 \cdot 0.50 = \$1639$	Includes housing, utilities, phone, transportation, insurance, food, and minimum required loan payments
Wants 30%: $3278 \cdot 0.30 = \$983.40$	Includes entertainment, dining out, clothing, exercise costs, and internet/ streaming services
Savings 20%: $3278 \cdot 0.20 = \$655.60$	Includes retirement, savings, and extra payments on debts

b. If a housing move means the Garcias' monthly necessities increase by $175, they should consider if other costs would decrease as a result to help offset the more expensive housing. For instance, their new place could allow for spending less on transportation each month if they are closer to work or able to ride share or use public transportation. They should also compare home insurance and utilities for the new place. Deposits and other one-time costs associated with moving will also need to be taken into consideration.

Example 6.5.7

Comparing Spending Habits

The graph in Figure 6.5.8 gives the side-by-side comparisons by age for average consumer spending in the United States in 2018[3]. Use the graph to answer the following questions.

a. Which age group spent the most on personal insurance and pensions, and approximately how much did they spend annually? Compare this to the age group who spent the least on personal insurance and pensions.

b. Using the graph, estimate how much money consumers under 35 year spent on average for each category in 2018. Based on your estimates, what was the minimum annual income a person in this age group needed to make in order to spend the average amount that year?

c. Based on your estimates you found in part b., does this spending align with a 50/30/20 budget?

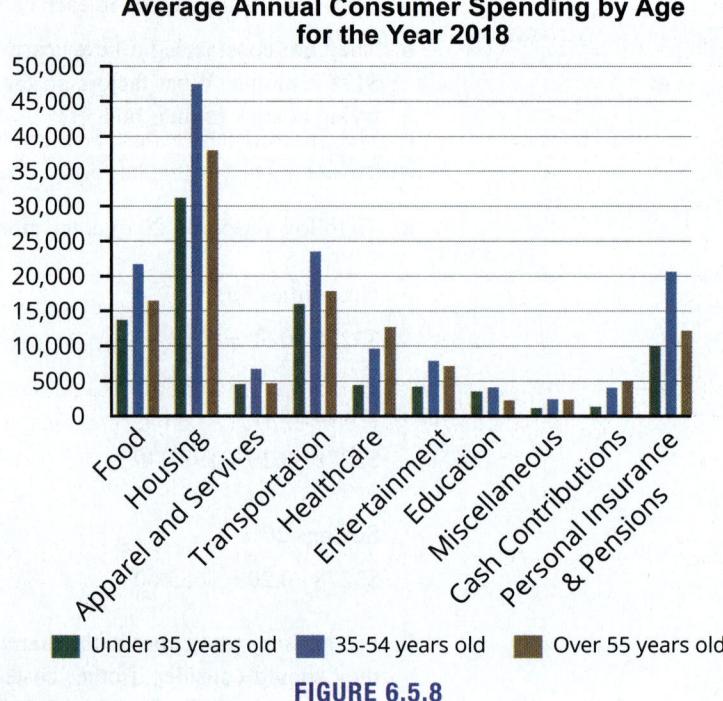

FIGURE 6.5.8

Solution

a. Consumers 35 to 54 years old spent the most on personal insurance and pensions with an average of approximately $20,000 a year. Consumers under 35 to spent half as much on personal insurance and pensions with an approximate average of $10,000 a year.

b. Based on the graph, we can approximate the average spending habits for those under 35 years old in 2018.

---

3   2018 Aggregate Expenditure Shares Tables for Age of Reference Person, U.S. Bureau of Labor Statistics, September 2019 https://www.bls.gov/cex/tables.htm.

Table 6.5.2: Spending Habits for People Under 35 in 2018

Food	$12,550
Housing	$31,000
Apparel and Services	$4000
Transportation	$15,000
Healthcare	$4000
Entertainment	$4000
Education	$3000
Miscellaneous	$1000
Cash Contributions	$1000
Personal Insurance/Pensions	$10,000

The minimum net earnings for a person in this age group would need to cover all of these expenses. Adding them together, we have the following.

$$\$12,500 + \$31,000 + \$4000 + \$15,000 + \$4000 + \$4000 + \$3000 + \$1000 + \$1000 + \$10,000 = \$85,500$$

Therefore, to cover the average spending habits in 2018, a 24- to 34-year-old would have needed an annual income of at least $85,500.

c. Begin by determine the amounts for each category in 50/30/20 budget based on the estimated income of $85,500 found in part b.

Necessities 50%: $85,500 · 0.50 = $42,750

Wants 30%: $85,500 · 0.30 = $25,650

Savings 20%: $85,500 · 0.20 = $17,100

Next, place the different spending types into one of the three categories and find their sum. Note that some categorizing decisions are based on personal needs and reasoning here; as a result, there are many possible allocations. One possible allocation is as follows.

Table 6.5.3: 50/30/20 Budget for People Under 35 in 2018

Necessities	
Food	$12,500
Housing	$31,000
Apparel and Services	$4000
Transportation	$15,000
Healthcare	$4000
Total	**$66,500**

Wants	
Entertainment	$4000
Education	$3000
Miscellaneous	$1000
Cash Contributions	$1000
Total	**$9000**

Savings	
Personal Insurance/Pensions:	**$10,000**
Total	**$10,000**

Based on the way we distributed the spending to the different categories, this does not follow a 50/30/20 budget model. The necessities category is way over budget, while the other two are under budget.

Once you have a spending plan for your money and a way to keep track of it, sticking to it is imperative. Luckily, there are lots of different methods to keep you accountable and help you stay on budget. One such method is the *envelope method* or *cash-only method*. This idea has you divide up your take-home pay into the allotted spending categories and put each amount into an envelope (whether literal or virtual) for the month. You only spend what you have. No credit cards allowed. The amount available in the envelope is what you can spend on that category. The strength of the envelope system is in its ability to help prevent overspending in a certain category. If extra money is needed for a category, it literally has to come out of a different envelope or wait until the next paycheck to replenish it. But a warning, if you overspend in one category then you are liable to be short in another category.

Budgeting, like all new endeavors, requires discipline at the beginning. But the key to budgeting is to start. Start simple, and don't feel like you have to figure it out all at once. If you're not careful, the details of your budget have the potential to get too complicated and then you might be inclined to give up, or to drive yourself crazy, or both! Reevaluating your revenue and expenses on a regular basis will always be important and is a habit worth forming.

Skill Check Answers
1. $14.90 **2.** $479.33

6.5 Exercises

✔ CONCEPT CHECK

1. When making a budget, the first two things you must consider are your monthly _____ and your monthly _____.

2. The last step of making a budget is to allocate any remaining funds or make _____.

3. If you have an expense that doesn't come up every month, but you want to set aside a monthly amount to account for that expense, you can _____ it.

4. According to the 50/30/20 Budgeting Rule, you should allow 50% of your net income for _____, no more than 30% for _____, and at least 20% for savings and paying down debt.

💡 PRACTICE

Determine the approximate monthly cost of each miscellaneous spending activity. Assume there are 4 weeks in a month.

5. $3.18 on coffee and a donut three times a week

6. Lunch five days a week, averaging $11.25 per meal

7. $16.95 on a paperback novel every Tuesday

8. Groceries every Sunday, averaging $65.93

9. $20 on charitable donations, twice a month

10. A tank of gas, averaging $40.95, three times a month

11. A movie ticket for $8.70 along with a popcorn and drink for $11.99 twice a month

12. Dinner with friends, averaging $24.57, followed by drinks, averaging $18.45, every Thursday

For each situation, determine the following. Round your answer to the nearest cent, if necessary.

    a. Determine how much the taxpayer should prepay each quarter to meet the projected taxes by the end of the year.

    b. Determine how much the taxpayer should put aside each month in order to have enough to pay their estimated quarterly taxes.

13. Agustin is self-employed. Based on last year, his projected state and federal taxes for this year will be approximately $18,000.

14. Aya is self-employed. Based on last year, her projected state and federal taxes for this year will be approximately $21,250.

15. Antonio is self-employed. Based on last year, his projected state and federal taxes for this year will be approximately $9,700.

16. Alesia is self-employed. Based on last year, her projected state and federal taxes for this year will be approximately $7,450.

For each situation, prepare a budget using the 50/30/20 rule. Round your answer to the nearest cent, if necessary.

17. After having a child, the MacDonalds have decided to make an effort to live within their means by creating a budget. Their combined net income is $5120 a month.

18. The newly married Youngs are combining their finances and making a budget in order to plan for their future. Their combined net income is $7006 a month.

19. Recent college-graduate Ms. Martin wants to get control of her finances. Her monthly net income is $2472.

20. After a major promotion, Mr. Lee wants to make sure he doesn't suffer from lifestyle inflation, so he creates a budget. His monthly net income is $4114.

🚀 APPLICATIONS

For each situation, determine the following. Round your answer to the nearest cent, if necessary.

    a. Determine the monthly net pay amount.

    b. Determine how much money is left each month for discretionary spending after all necessities are accounted for.

21. A recent college graduate takes an entry-level job as an accountant. His yearly salary is $32,760. His employer withholds $4680 in state and federal income taxes and $2496 in FICA taxes throughout the year. He has the following monthly costs: transportation is $380, cell phone bill is $45, student loan payment is $215, HSA contribution is $250, and rent is $650. He is using the average monthly cost for each of the following to gain an idea of other monthly expenses: utilities are $120, home internet is $55, and groceries are $285.

22. A recent college graduate starts a new job as a middle-school teacher. His yearly salary is $52,500. His employer withholds $7872 in state and federal income taxes and $4008 in FICA taxes throughout the year. He has the following monthly costs: transportation is $575, cell phone bill is $80, student loan payment is $312, retirement fund contribution is $400, and rent is $950. He is using the average monthly cost for each of the following to gain an idea of other monthly expenses: utilities are $165, home internet is $75, and groceries are $350.

23. A recent college graduate acquires a job as a software developer. Her yearly salary is $65,250. Her employer withholds $9312 in state and federal income taxes and $4992 in FICA taxes throughout the year. She has the following monthly costs: transportation is $560, cell phone bill is $100, student loan payment is $280, and rent is $1150. She is using the average monthly cost for each of the following to gain an idea of other monthly expenses: utilities are $195, home internet is $90, health insurance is $420, and groceries are $325.

24. A recent college graduate acquires a job as a civil engineer. Their yearly salary is $68,750. Their employer withholds $9312 in state and federal income taxes and $4992 in FICA taxes throughout the year. They have the following monthly costs: transportation is $475, cell phone bill is $60, student loan payment is $424, HSA contribution is $180, and rent is $900. They are using the average monthly cost for each of the following to gain an idea of other monthly expenses: utilities are $155, home internet is $85, and groceries are $300.

Excel Exercises

For each situation use a spreadsheet, such as Microsoft Excel, to create a budget. Determine how much is available each month for expenses not listed in the budget. Assume there are 4 weeks in month.

25. A respiratory therapist has an annual income of $48,700. Their net monthly take-home pay is $3120, and they have the following monthly expenses. Additional income includes $80 a week on average from a freelance writing job. A total of $30 a month is donated to charities.

Monthly Expenses	
Rent	$950.00
Gas	$70.00
Water	$52.00
Electricity	$112.00
Internet	$65.00
Cell Phone	$58.00
Health Insurance	$320.00
Transportation	$315.00
Loan Repayment	$120.00
Food	$450.00
Credit Card	$168.00

26. An HR specialist has an annual income of $52,000. His net monthly take-home pay is $3302, and he has the following monthly expenses. Additional income includes $100 a week on average from a dog-walking job. A total of $50 a month is donated to charities.

Monthly Expenses	
Rent	$1100.00
Gas	$155.00
Water	$65.00
Electricity	$150.00
Internet	$80.00
Cell Phone	$94.00
HSA Contribution	$250.00
Transportation	$380.00
Loan Repayment	$270.00
Food	$340.00
Credit Card	$85.00

27. An electrician has an annual income of $46,500. Their net monthly take-home pay is $3114, and they have the following monthly expenses. Additional income includes $75 a week on average from a tutoring job. A total of $35 a month is donated to charities.

Monthly Expenses	
Rent	$850.00
Gas	$320.00
Water	$56.00
Electricity	$140.00
Internet	$62.00
Cell Phone	$45.00
Health Insurance	$365.00
Transportation	$415.00
Food	$345.00
Credit Card	$220.00

28. A veterinarian technician has an annual income of $30,200. Her net monthly take-home pay is $1990, and she has the following monthly expenses. Additional income includes $75 a week on average from a dog walking job. A total of $25 a month is donated to charities.

Monthly Expenses	
Rent	$700.00
Gas	$86.00
Water	$56.00
Electricity	$150.00
Internet	$65.00
Cell Phone	$68.00
Health Insurance	$220.00
Transportation	$240.00
Food	$260.00
Credit Card	$90.00

29. A market research analyst has an annual income of $60,300. Her net monthly take-home pay is $3734, and she has the following monthly expenses. Additional income includes $100 a week on average from a tutoring job. A total of $75 a month is donated to charities.

Monthly Expenses	
Mortgage	$1400.00
Gas	$160.00
Water	$86.00
Electricity	$180.00
Internet	$90.00
Cell Phone	$87.00
HSA Contribution	$240.00
Transportation	$380.00
Loan Repayment	$350.00
Food	$360.00
Credit Card	$220.00

30. A insurance agent has an annual income of $44,600 Their net monthly take-home pay is $2912, and they have the following monthly expenses. Additional income includes $80 a week on average from a consulting job. A total of $60 a month is donated to charities.

Monthly Expenses	
Rent	$985.00
Gas	$145.00
Water	$45.00
Electricity	$112.00
Internet	$74.00
Cell Phone	$115.00
Health Insurance	$275.00
Transportation	$380.00
Loan Repayment	$270.00
Food	$310.00
Credit Card	$150.00

✏ WRITING & THINKING

31. Explain the benefits of creating a personal budget.

32. Find four different cell phone plans from at least two different providers. Compare the plans. Pick one plan and explain why it might be best for someone's budget and lifestyle.

6.5 PROJECT

THE COST OF LIVING IN DIFFERENT PARTS OF THE UNITED STATES

The cost of living can vary considerably across the United States. According to the Missouri Economic Research and Information Center, Mississippi offers the lowest average cost of living while Hawaii tops the list as the most expensive state in the nation. In this activity, you will investigate the cost of living in two metropolitan areas.

The figures in the following table represent a reasonable monthly budget for a family of four with two working adults to attain a modest, yet adequate, standard of living in both the Toledo Metro Area in Ohio and in the Providence Metro area in Rhode Island in 2021.

Expense	Toledo, OH	Providence, RI
Housing	$726	$1014
Food	$701	$798
Child Care	$1120	$1659
Transportation	$1164	$1112
Health Care	$895	$793
Other Necessities	$576	$731
Taxes	$643	$901
Monthly Total	$5825	$7008

1. The 2022 minimum wage in Ohio is $9.30 per hour. How many hours must each adult in the family work per month at this pay rate to earn the $5825 dollars needed to live in Toledo?

2. Repeat your calculation for Providence, knowing that the 2022 minimum wage is Rhode Island is $12.25 and the necessary income is $7008?

3. Which city is the one with a true higher cost of living compared to the state's minimum wage.

Now, assume that a regular work week is made of 40 hours and that there are approximately 4.3 weeks in a month.

4. How much money must each adult earn per hour in Toledo to meet their budget assuming they work regular work weeks during a month? How many times more than the minimum wage in Ohio is the hourly rate you calculated? Assume both adults have equal incomes.

5. How much money must each adult earn per hour in Providence in order to meet their budget assuming they work regular work weeks during a month? How many times more than the minimum wage in Rhode Island is the hourly rate you calculated? Assume both adults have equal incomes.

6. Why do you believe there is such a large discrepancy between the minimum wage in each state and the hourly rates you found in parts 4 and 5?

Chapter 6 Exercises

Solve each problem. Round your answer to the nearest cent, if necessary.

1. Determine the interest owed on $1200 for five years at a rate of 6.5%.

2. Assume you are purchasing a new computer with "90 days same as cash" to make the purchase. If the cost of the computer is $1255, tax included, with an annual interest rate of 16.99%, how much would you owe on the 91st day if you make no payments during the first 90 days?

3. Use the compound interest formulas to **a.** calculate the total amount in the account after the given time period and **b.** determine the amount of interest earned.

   $P = \$5500$, $r = 2.5\%$ compounded weekly, $t = 10$ years

   $P = \$4755$, $r = 4.5\%$ compounded monthly, $t = 10$ years

   $P = \$7300$, $r = 19.9\%$ compounded continuously, $t = 20$ years

4. Miguel deposits $2850. Determine the annual percentage yield for each of the following. Round each answer to the nearest hundredth of a percent.

   a. APR of 5% compounded monthly

   b. APR of 5% compounded weekly

   c. APR of 5% compounded daily

5. A couple deposits $12,500 into an account earning 2.75% annual interest for 25 years.

   a. Calculate the future value of the investment if interest is compounded monthly.

   b. Calculate the future value of the investment if interest is compounded weekly.

A savings account is compounded monthly for five years with an APR of 4.99%. For each principal amount, calculate the following. Round your answer to the nearest cent, if necessary.

   a. The future value of the investment

   b. The amount of interest earned

6. $15,000

7. $30,000

8. $60,000

9. $120,000

10. ABC Lending lists the following APR rates for loans. Determine the APY for each category.

First Bank of Lending

Loan Amount	APR
< $20,000	9.75%
$20,000–$99,999	5.99%
≥ $100,000	3.75%

*Interest rates are compounded quarterly

11. Jay and Sybil are purchasing a home. They wish to save money for five years and purchase a house with a value of $195,000 with cash. If they deposit money into an account paying 12% interest, how much do they need to deposit each month in order to make the purchase?

12. Suppose you wish to retire at the age of 65 with $1,000,000 in savings. Determine your monthly payment into an IRA if the APR is 8.5% and you begin making payments at the following ages.

   a. 20 years old

   b. 30 years old

   c. 40 years old

13. Lacy deposits a fixed monthly amount into an annuity account for her child's college fund. She wishes to accumulate a future value of $135,000 in 18 years.

   a. Assuming an APR of 6.5%, how much money should Lacy deposit monthly in order to reach her goal?

   b. How much of the $135,000 will Lacy ultimately deposit in the account, and how much is interest earned?

14. Assume you wish to borrow $750 for two weeks and the amount of interest you must pay is $17 per $100 borrowed. What is the APR at which you are borrowing money? Round your answer to the nearest whole percent.

15. A payday loan is made for eight weeks, where the amount of interest owed per $100 borrowed is $15. Suppose you borrow $1000 for eight weeks.

   a. How much do you owe at the end of eight weeks?

   b. What is the APR for this transaction?

16. Ozzie bought a new car and financed $14,950 of the purchase. He financed the car for 36 months with an APR of 5.75%. Determine each of the following.

   a. Ozzie's monthly payment

   b. Total cost of Ozzie's car

   c. Total interest Ozzie pays over the life of the loan

17. Chelsea and Bill are buying a house on a 30-year mortgage. They can pay $1200 per month for a mortgage. If they have an APR of 4.25%, what is the maximum mortgage that they can take out?

18. Amelia decides to purchase a $215,000 house. She wants to finance the entire balance. She has received an APR of 2.75% for a 15-year mortgage.

   a. What is Amelia's monthly payment?

   b. What is Amelia's total cost if she takes all 15 years to pay off the house?

   c. Over the course of the loan, how much interest will Amelia pay?

   d. If she changed the term to 30 years instead of 15 years, what would her monthly payment be?

   e. With a 30-year mortgage, what is the total cost of the house?

   f. With a 30-year mortgage, how much interest will Amelia pay?

19. You want to buy a car and finance $27,450 to do so. You can afford a payment of up to $600 per month. The bank offers three choices for the loan: a four-year loan with an APR of 5.5%, a five-year loan with an APR of 6.5%, and a six-year loan with an APR of 7%. Which option best meets your needs, assuming you want to pay the least amount of interest?

Consider a credit card with a balance of $4875. You wish to pay off the credit card in each scenario. Calculate the following. Round your answer to the nearest cent, if necessary.

   a. The amount of a monthly payment within the time frame given

   b. The total amount paid over the time period

20. APR of 19.99% paid off within one year

21. APR of 21.5% paid off within two years

22. APR of 29.99% paid off within three years

23. A registered nurse has an annual salary of $57,477, earned $115.27 of interest on their savings account, paid $874 in student loan interest, and contributed $2400 toward a 401(k). What is the nurse's gross income?

24. A dental assistant earns $42,200 per year, paid $860 in student loan interest, and contributed $1200 to a 401(k). What is the dental assistant's adjusted gross income?

25. Keith is self-employed. Based on last year, his projected state and federal taxes for this year will be $19,750.

    a. Determine how much he should prepay each quarter to meet the projected taxes by the end of the year.

    b. Determine how much he should set aside each month in order to have enough to pay the estimated quarterly taxes. Round your answer to the nearest cent, if necessary.

26. William has a job where he has a take-home salary each month of $3375. If William wants to spend no more than 25% of his income on rent, how much rent can William afford?

27. Michael rents an apartment for $750 per month, pays his car payment of $360 per month, has utilities that cost $330 per month, and spends $476 per month on food and entertainment. Determine Jack's monthly expenses.

28. A recent college graduate takes an entry-level job as an accountant. His yearly salary is $30,500. His employer withholds $4296 in state and federal income taxes and $2328 in FICA taxes throughout the year. He has the following monthly costs: transportation is $340, cell phone bill is $65, student loan payment is $179, HSA contribution is $200, and rent is $750. He is using the average monthly cost for each of the following to gain an idea of other monthly expenses: utilities are $100, home internet is $60, and groceries are $250.

    a. Determine the accountant's monthly net pay amount.

    b. Determine how much money is left each month for discretionary spending after all necessities are accounted for.

CHAPTER 6 PROJECT

Defined Benefit Versus Defined Contribution: Two Types of Retirement Plans

In this chapter, you learned about the power of interest and the importance of budgeting in order to plan for your future financial health. In this project, we will explore the two main types of retirement plans available in the United States: defined-benefit plans and defined-contribution plans.

In a *defined-benefit plan* (or a pension plan), an employee is guaranteed life-long income after retirement. The size of the income received after retirement is usually determined by the employee's years of service and salary at the time of retirement. The employer is responsible for all the planning and managing of risk associated with this type of plan.

1. Perform an internet search for "defined-benefit plans". List two advantages and two disadvantages of this type of plan for the employee. List two advantages and two disadvantages for the employer. What kind of employers usually offer a defined-benefit plan?

In a *defined-contribution plan*, the employee (and often the employer) makes contributions to an investment account. Two common plans of this type are 401(k)s and IRAs. The size of income received after retirement is determined by the balance in the account and other market factors. With this type of plan, the employer does not have any responsibility towards the planning or managing the risk of the account.

2. Perform an internet search for "defined-contribution plans". List two advantages and two disadvantages of this type of plan for the employee. List two advantages and two disadvantages for the employer. What kind of employers usually offer a defined-contribution plan?

Suppose that a small town has a pension fund that is expected to make annual payments totaling $500,000 to its local government retirees. The fund must be able to sustain such payments in perpetuity; that is, forever. The easiest way for this to happen is if the town is able to set aside $500,000 every year to pay its retirees.

3. What are some ways the town could secure the funds for the retirees? Think about ways local governments can generate revenue. Do you believe these revenue streams are sustainable in the long run? Explain.

Another possibility to sustain the fund would be to plan ahead. The town could allocate and invest money for its pension fund from the time the first local government employee is hired to the time they retire, 30 years later. In order to simplify our calculations, let's assume that the town will set aside an amount d on year 1 to pay for the first year of retirement for all local government retirees 30 years later. The town council believes it can earn 7% annual interest on the account.

4. Calculate the amount *d* that the city must set aside that first year. Does the 30-year plan work in favor of the town? Explain.

Now, let's assume that the budget committee has allocated the amount found in part 4 for investment each year. Consider the unfortunate event of the market turning sharply downwards, resulting in a 4% rate of return on the account per year instead of the anticipated 7%.

5. Use the answer from part 4 to calculate the amount available in the retirement fund after 30 years? (**Hint:** calculate the future value of a single deposit made on the first year.)

6. What does your answer in part 5 say about the ability of the town to fulfill the pension obligation to its retirees after 30 years?

7. Discuss any changes to the town's predicament if they invested the same amount in a 401(k) without any guarantees of income.

8. If you were the manager of a small town's pension fund, would you prefer to offer a pension fund or a 401(k) to your employees? Explain.

9. If you were an employee of a small town, would you rather be enrolled in a pension plan or a 401(k)? Explain.

CHAPTER 6 REVIEW

Section 6.1 Understanding Interest

Definitions

Interest: The amount charged by a lender for borrowing money is the **interest**.

Principal: The **principal** is the sum of money on which interest is charged.

Interest Rate: The **interest rate** is the percentage of the principal that is charged to the borrower.

Annual Percentage Rate (APR): The **annual percentage rate (APR)** is the yearly interest rate that is charged for borrowing money, including fees. APR is normally given as a percentage per year.

Formulas

Simple Interest: The amount of **simple interest** earned on a loan is calculated with the following formula.

$$I = Prt$$

Here, P is the principal, r is the APR written as a decimal, and t is the length of the loan in years.

Future Value for Simple Interest: The future value, A, of a loan is the amount to be repaid at the end of the loan period. It is the principal amount plus any interest accrued. Using simple interest, future value is calculated with the following formula.

$$A = P(1 + rt)$$

Here, P is the principal, r is the APR written as a decimal, and t is the length of the loan in years.

Future Value for Compound Interest: The **future value**, A, of a loan is the amount to be repaid at the end of the loan period. It is the principal amount plus any interest accrued. Using **compound interest**, future value is calculated with the following formula.

$$A = P\left(1 + \frac{r}{n}\right)^{nt}$$

Here, P is the principal, r is the APR written as a decimal, t is the length of the loan in years, and n is the number of compounding intervals per year.

Future Value for Continuous Compound Interest: The **future value**, A, of a loan is the amount to be repaid at the end of the loan period. It is the principal amount plus any interest accrued. Using **continuous compound interest**, future value is calculated with the following formula.

$$A = Pe^{rt}$$

Here, P is the principal, r is the APR written as a decimal, and t is the length of the loan in years.

Annual Percentage Yield (APY): The **annual percentage yield (APY)** is the simple annual interest rate earned in a given year that accounts for the effects of compounding. **APY** is calculated using the following formula.

$$APY = \left(1 + \frac{r}{n}\right)^{n} - 1$$

Here, r is the APR written as a decimal and n is the number of compounding intervals per year.

Section 6.2 Saving and Investing

Definition

Annuity: An **annuity** is a sequence of regular payments made into an account, or taken out of an account, over time.

Formulas

Present Value with Compound Interest: The amount of principal needed now in order to reach a future value amount is called the present value (PV). The **present value with compound interest** is calculated with the following formula.

$$PV = \frac{A}{\left(1 + \dfrac{r}{n}\right)^{nt}}$$

Here A is the future value, r is the APR written as a decimal, n is the number of compound intervals per year, and t is time in years.

Annuity Payment Amount: The **annuity payment amount (PMT)**, or regular payment amount, needed to be deposited at the end of each compounding period in order to reach a future goal with an annuity savings account is calculated with the following formula.

$$PMT = FV \cdot \frac{\left(\dfrac{r}{n}\right)}{\left[\left(1 + \dfrac{r}{n}\right)^{nt} - 1\right]}$$

Here, FV is the future value desired, r is the APR written as a decimal, n is the number of compounding intervals per year, and t is time in years

Future Value of an Annuity: The **future value (FV) of an annuity** account is calculated with the following formula.

$$FV = PMT \cdot \frac{\left[\left(1 + \dfrac{r}{n}\right)^{nt} - 1\right]}{\left(\dfrac{r}{n}\right)}$$

Here, PMT is the payment amount deposited at the end of each compounding interval, r is the APR written as a decimal, n is the number of compounding intervals per year, and t is time in years.

Section 6.3 Borrowing Money

Definitions

Fixed Installment Loan: A **fixed installment loan** is a loan with a fixed interest rate and loan period. Repayments are made in equal installments throughout the life of the loan.

Down Payment: A **down payment** is a cash payment made up front towards the total purchase price of the goods or service.

Mortgage: An installment loan used for the purposes of buying a house or other real estate where the property is used as collateral is called a **mortgage**.

Fixed-Rate Mortgage: In a **fixed-rate mortgage**, the interest rate remains constant throughout the life of the loan.

Adjustable-Rate Mortgage: In an **adjustable-rate mortgage (ARM)**, the interest rate is fixed for a period of time at the beginning of the loan period, and then is allowed to fluctuate over time.

Mortgage Point: A **mortgage point** is interest prepaid by the buyer to the lender and is equal to 1% of the loan amount.

Open-Ended Loans (Revolving Credit): An **open-ended loan**, or **revolving credit**, is a preapproved loan amount that may be used by the borrower repeatedly. Interest is applied to any unpaid balance at the end of the payment period.

Credit Limit: The maximum amount of an open-ended loan is called your **credit limit**.

Grace Period: A **grace period** is a period of time in which no interest accrues on a debt.

Formulas

Regular Payment for Fixed Installment Loans: The amount of a **regular payment (PMT) on a fixed installment loan** is calculated with the following formula.

$$\text{PMT} = \frac{\left(P \cdot \dfrac{r}{n} \right)}{\left[1 - \left(1 + \dfrac{r}{n} \right)^{-nt} \right]}$$

Here, P is the principal amount of money borrowed, r is the APR in decimal form, n is the number of payments per year, and t is time in years.

Number of Fixed Payments Required to Pay Off Credit Card Debt: The **number of fixed payments R required to pay off credit card debt** is calculated with the formula.

$$R = \frac{-\log\left[1 - \dfrac{r}{n}\left(\dfrac{A}{\text{PMT}} \right) \right]}{\log\left(1 + \dfrac{r}{n} \right)},$$

where A is the loan amount, r is the APR in decimal form, n is the number of payments per year, and the payment amount is PMT.

Section 6.4 Federal Revenue

Definitions

Gross Income: Gross income is the sum of all forms of earnings, including wages, tips, benefits, investment income, alimony, etc.

Adjusted Gross Income: Adjusted gross income (AGI) is defined as gross income minus any adjustments you are allowed such as education expenses, health savings account contributions, paid alimony, or self-employment expenses.

Deduction: A tax **deduction** is comprised of expenses incurred by an individual or organization that can be subtracted from their gross income in order to reduce the amount of income that is able to be taxed.

Taxable Income: Taxable income is the amount of income subject to taxes by the government.

Section 6.5 Budgeting

Definitions

Budget: A **budget** is a financial plan for your income over a defined period of time.

50/30/20 Budgeting Rule: The **50/30/20 rule for budgeting** is to allow 50% of your net income for necessities, no more than 30% for wants, and at least 20% for savings and paying down debt.

Procedure

Steps to Creating a Budget

1. Calculate monthly income.
2. Calculate monthly expenses.
3. Subtract expenses from income.
4. Allocate any remaining funds or make adjustments.

CHAPTER 7

Numeration and Measurement Systems

SECTIONS

Introduction

Throughout history, civilizations have needed ways to measure and compare things. As a result, many different types of number systems were developed over the years. One of the more interesting systems is Oksapmin, a base-27 counting system. The words for the numbers are the names of the body parts. For example, 1 is tip^ na (thumb), 5 is h^th^ta (pinky), and 27 is tan-h^th^ta (pinky on the other side).

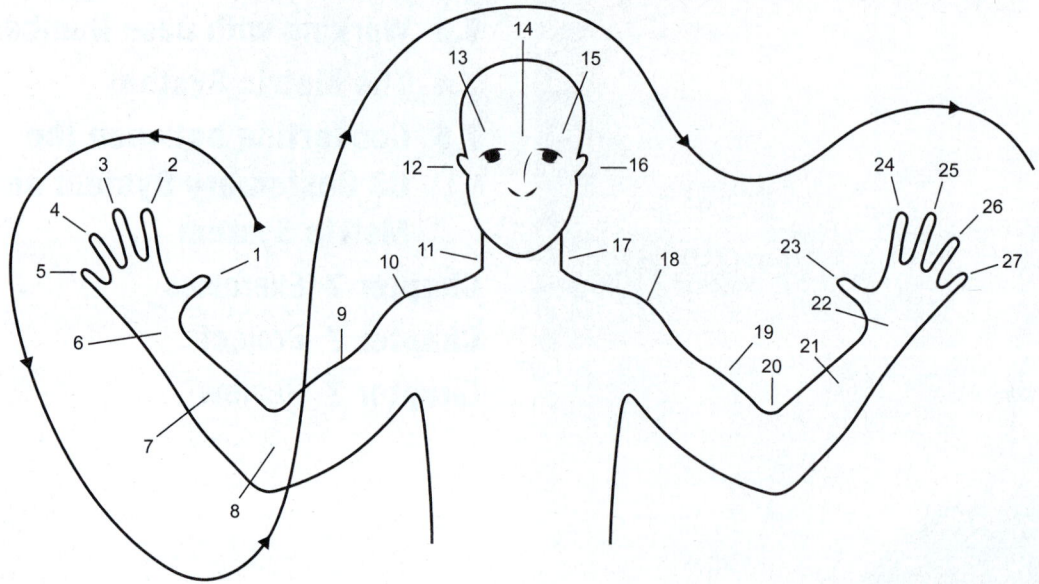

As you can imagine, not everyone adopted the Oksapmin method, and different number systems and units of measurement caused confusion between the different societies as they bartered and traded goods. As the need for international collaboration and a common system of measurement became more apparent, the metric system was created by a group of French scientists in the 18th century. The creators wanted the system to be for all people, for all of time.

The metric system was based on consistent and naturally occurring values, such as the circumference of the earth, rather than on human objects that vary, such as the length of a foot. In 1875, 17 countries, including the United States, signed an international agreement on the maintenance of these new French-created standards of measurement. As of today, 51 nations have signed the agreement known as the International Treaty of the Meter.

Despite signing the treaty, the United States maintains a dual system, which has continued to be problematic over the years. For example, NASA engineers working in the Jet Propulsion Lab used the metric system to calibrate the thrusters for the Mars Climate Orbiter while its partner Lockheed provided crucial acceleration data using the US system. After ten months of travel the Mars Climate Orbiter, which cost $125 million to build, missed its intended orbit due to the different measurement systems used. As a result, the orbiter burned and broke into pieces.

Since the United States continues to have a dual system of measurement, to avoid issues like the ones NASA encountered, we need to be able to work in both the metric and the US customary systems as well as convert between the two. In this chapter, we will begin by exploring several different numeration systems to help us better understand our own systems.

7.1 NUMERAL SYSTEMS BASED ON POSITION

OBJECTIVES

1. Convert decimal numbers to and from expanded form.
2. Convert numbers between the Babylonian system and the Hindu-Arabic system.
3. Convert numbers between the Mayan system and the Hindu-Arabic system.

OBJECTIVE 1 Expanded Form

Numbers are what we use to communicate concepts of *how many* or ideas of *more* or *less*. Fascinatingly, the way we as a society represent numbers in written form has been vastly different across history.

Today's Western world uses the Hindu-Arabic numeral system, also called the *decimal* system, which was widely introduced around AD 800. Most of us probably never think about the fact that our ancestors didn't always use the same numeral system because we've used the same structure for so many generations. There are several qualities about our numeral system that we often take for granted and don't really think about. For instance, think about how many symbols we have to represent all possible numbers. There are infinitely many numbers, yet we can represent all of them with only 10 unique symbols—it is really quite astonishing!

Hindu-Arabic Symbols

0 1 2 3 4 5 6 7 8 9

The Hindu-Arabic numeral system is a positional based system; that is, the position of a digit in the numeral indicates its value. Figure 7.1.1 illustrates this by identifying the positional values for the numeral 5374.32.

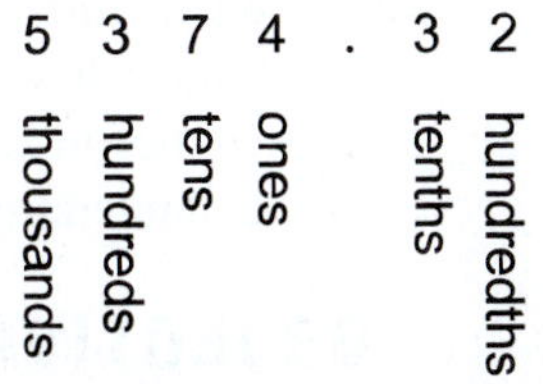

5 3 7 4 . 3 2
thousands
hundreds
tens
ones
tenths
hundredths

Figure 7.1.1: Identifying Positional Values for a Number

Notice that although there are two 3s in the numeral, they represent very different values: the 3 to the left indicates 300, and the 3 to the right indicates $\frac{3}{10}$. We can immediately recognize this simply based on where each 3 is positioned in the numeral.

In addition to relying on positioning, the Hindu-Arabic method of representing numbers also relies on a *base 10* system. This means that each position represents a *power of ten*.

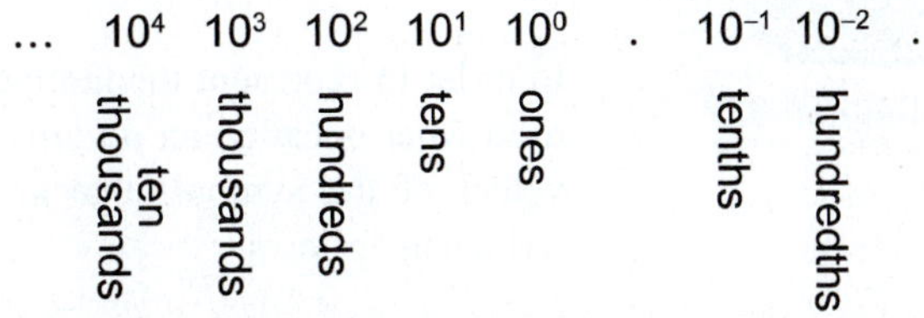

... 10^4 10^3 10^2 10^1 10^0 . 10^{-1} 10^{-2} ...
ten thousands
thousands
hundreds
tens
ones
tenths
hundredths

Figure 7.1.2: Hindu-Arabic Positional Systems

In other words, each time you move a place to the left, the value of the position increases by a multiple of 10. As we will see later, it is not only possible, but sometimes advantageous, to work in numeral systems that have bases other than 10.

Think Back

Recall that exponents are used to indicate repeated multiplication.

Any number raised to the zero power is equal to 1: $x^0 = 1$.

Think Back

Expanded form, or expanded notation, is a way to write out numbers so that one can see the math involved for each digit.

Expanded form clarifies how the Hindu-Arabic numeral system combines both position and multiples of 10. For instance, 5374.32 is equivalent to the following expanded expression.

$$5374.32 = (5 \cdot 10^3) + (3 \cdot 10^2) + (7 \cdot 10^1) + (4 \cdot 10^0) + (3 \cdot 10^{-1}) + (2 \cdot 10^{-2})$$

Example 7.1.1

Converting between Hindu-Arabic Numerals and Expanded Form

a. Write the numeral 10,325 in expanded form.

b. Express the following expanded form as a Hindu-Arabic numeral.

$$(8 \cdot 10^2) + (6 \cdot 10^1)$$

Solution

Helpful Hint

The use of multiplication and then addition, as we saw in Example 7.1.1, shows that the Hindu-Arabic numeral system is both *multiplicative* and *additive*.

a. $10,325 = (1 \cdot 10^4) + (3 \cdot 10^2) + (2 \cdot 10^1) + (5 \cdot 10^0)$

Notice that digit for the place value of 10^3 is a 0, so we do not need to include it in the expanded form, but it is not wrong if you do include it.

b. $\left(8 \cdot 10^2\right) + \left(6 \cdot 10^1\right) = \left(8 \cdot 100\right) + \left(6 \cdot 10\right)$

$$= 800 + 60$$
$$= 860$$

While the Hindu-Arabic numeral system is a positional system based on powers of 10, there have been other positional systems that are based on powers other than 10. For instance, the Babylonian numeral system uses powers of 60 and the Mayan numeral system uses powers of 20. Let's look at each of these systems.

OBJECTIVE 2 Babylonian Numeral System

Fun Fact

The Plimpton 322 is a Babylonian clay tablet circa 1800 BC, which is held in the Columbia University Libraries.

Figure 7.1.4: Plimpton 322

The Babylonian Empire was, by all accounts, a key empire in Mesopotamia from the 18th to 6th centuries BC. Geographically, it covered parts of our modern-day Middle East, what we now call Iraq and Syria. The Babylonian numeral system was developed sometime around 2000 BC. Unlike our Hindu-Arabic numeral system, which uses 10 symbols, the Babylonian system has only 2—one representing the quantity 1 and one representing 10.

Babylonian Symbols

▽   ≺

1   10

In order to represent the numbers 1 through 59, the Babylonian numeral system used what is called an *additive method*. In other words, you simply add up the values of the symbols that are written. For instance, the number 6 would be written as follows.

$$\begin{matrix} ▽\,▽\,▽ \\ ▽\,▽\,▽ \end{matrix} = 6$$

The value 16 would be a ten plus six ones.

$$≺ \begin{matrix} ▽\,▽\,▽ \\ ▽\,▽\,▽ \end{matrix} = 16$$

And the value 46 is similar.

$$\text{≺≺≺≺} \; \text{∇∇∇} \atop \text{∇∇∇} = 46$$

Figure 7.1.3 shows the Babylonian Numeral System for numbers 1 through 59.

Symbol	#	Symbol	#	Symbol	#	Symbol	#	Symbol	#	Symbol	#
∇	1	≺∇	11	≺≺∇	21	≺≺≺∇	31	(40)∇	41	(50)∇	51
∇∇	2	≺∇∇	12	≺≺∇∇	22	≺≺≺∇∇	32	(40)∇∇	42	(50)∇∇	52
∇∇∇	3	≺∇∇∇	13	≺≺∇∇∇	23	≺≺≺∇∇∇	33	(40)∇∇∇	43	(50)∇∇∇	53
∇∇∇ ∇	4	≺∇∇∇ ∇	14	≺≺∇∇∇ ∇	24	≺≺≺∇∇∇ ∇	34	(40)∇∇∇ ∇	44	(50)∇∇∇ ∇	54
∇∇∇ ∇∇	5	≺∇∇∇ ∇∇	15	≺≺∇∇∇ ∇∇	25	≺≺≺∇∇∇ ∇∇	35	(40)∇∇∇ ∇∇	45	(50)∇∇∇ ∇∇	55
∇∇∇ ∇∇∇	6	≺∇∇∇ ∇∇∇	16	≺≺∇∇∇ ∇∇∇	26	≺≺≺∇∇∇ ∇∇∇	36	(40)∇∇∇ ∇∇∇	46	(50)∇∇∇ ∇∇∇	56
∇∇∇ ∇∇∇ ∇	7	≺∇∇∇ ∇∇∇ ∇	17	≺≺∇∇∇ ∇∇∇ ∇	27	≺≺≺∇∇∇ ∇∇∇ ∇	37	(40)∇∇∇ ∇∇∇ ∇	47	(50)∇∇∇ ∇∇∇ ∇	57
∇∇∇ ∇∇∇ ∇∇	8	≺∇∇∇ ∇∇∇ ∇∇	18	≺≺∇∇∇ ∇∇∇ ∇∇	28	≺≺≺∇∇∇ ∇∇∇ ∇∇	38	(40)∇∇∇ ∇∇∇ ∇∇	48	(50)∇∇∇ ∇∇∇ ∇∇	58
∇∇∇ ∇∇∇ ∇∇∇	9	≺∇∇∇ ∇∇∇ ∇∇∇	19	≺≺∇∇∇ ∇∇∇ ∇∇∇	29	≺≺≺∇∇∇ ∇∇∇ ∇∇∇	39	(40)∇∇∇ ∇∇∇ ∇∇∇	49	(50)∇∇∇ ∇∇∇ ∇∇∇	59
≺	10	≺≺	20	≺≺≺	30	≺≺≺≺	40	≺≺≺≺≺	50		

Figure 7.1.3: Babylonian Numeral System

Beyond 59, the Babylonian system used a place value system in a similar manner to the Hindu-Arabic numeral system. The Babylonian numeral system, however, was based on groups of 60, not 10.

$$\ldots \; 60^4 \quad 60^3 \quad 60^2 \quad 60^1 \quad 60^0$$

To indicate a place value shift, a large space was left between the symbols. For instance, the number 5389 would be written as shown here.

∇   ≺≺∇∇∇∇∇∇∇∇∇   ≺≺≺≺∇∇∇∇∇∇∇∇∇

1   29   49

$$5389 = (1 \cdot 60^2) + (29 \cdot 60^1) + (49 \cdot 60^0)$$
$$5389 = 3600 + 1740 + 49$$

Example 7.1.2

Converting Numbers from the Babylonian Numeral System to the Hindu-Arabic Numeral System

Write each Babylonian numeral as a Hindu-Arabic numeral.

a. ≺≺≺ ∇∇∇ ∇∇∇ ∇∇

b. ∇∇∇   ≺≺ ∇

c. ∇∇   ∇∇   ∇∇

Solution

a. The first number does not contain any large spaces between the symbols, so we know that it is a number between 1 and 59. We know the ⟨ symbol represents 10 and the ▽ symbol represents 1. There are three 10s and eight 1s, so this Babylonian numeral can be written as the Hindu-Arabic numeral 38.

b. Because the second numeral contains spaces between the groupings, we will need to multiply by the corresponding place value.

$$\begin{aligned}
▽▽▽ \qquad ⟨⟨\; ▽ \qquad &= (3 \cdot 60^1) + (21 \cdot 60^0) \\
&= 180 + 21 \\
&= 201
\end{aligned}$$

The Babylonian numeral can be written as the Hindu-Arabic numeral 201. Notice that although our system has a 0 as a place holder for the 10s place, the Babylonian numeral system does not contain a symbol for 0.

c. Again, we have spaces between the symbol groupings so we will have to perform multiplication and addition to convert to a Hindu-Arabic numeral.

$$\begin{aligned}
▽▽ \qquad ▽▽ \qquad ▽▽ \quad &= (2 \cdot 60^2) + (2 \cdot 60^1) + (2 \cdot 60^0) \\
&= 7200 + 120 + 2 \\
&= 7322
\end{aligned}$$

Our calculations show that the Babylonian numeral can be written as the Hindu-Arabic numeral 7322.

✔ **Skill Check 7.1.1**

Write the Babylonian numeral as a Hindu-Arabic numeral.

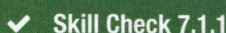

OBJECTIVE 3 Mayan Numeral System

Across the globe from the Babylonians, in what is now referred to as Central America, the Mayan civilization had their own numeral system in base 20. Occurring as early as 1000 BC, it used three symbols to represent the numbers 0, 1, and 5. It is widely believed that their system was the first numeral system to have a symbol for zero.

Mayan Symbols
0   1   5

In order to represent the numbers 0 through 19, the Mayans were similar to the Babylonians in that they used the additive method, adding up the values of the individual symbols drawn. For instance, the number 6 was written as shown.

$$= 6$$

The value 16 would be three lines for 15 (remember, a line has a value of 5) plus one dot for 1.

$$= 16$$

The Mayan numeration system, from 0 through 19, is shown in Figure 7.1.6.

0	1	2	3	4
⬭	●	●●	●●●	●●●●
5	6	7	8	9
▬	● ▬	●● ▬	●●● ▬	●●●● ▬
10	11	12	13	14
▬▬	● ▬▬	●● ▬▬	●●● ▬▬	●●●● ▬▬
15	16	17	18	19
▬▬▬	● ▬▬▬	●● ▬▬▬	●●● ▬▬▬	●●●● ▬▬▬

Figure 7.1.6: Mayan Numeration System

After 19, a place value system was used. Unlike the other two systems we've studied so far, the Mayan's system was vertical instead of horizontal, and each place value was a power of 20. The powers increased from bottom to top.

$$\vdots$$
$$20^4$$
$$20^3$$
$$20^2$$
$$20^1$$
$$20^0$$

To indicate a place value shift, a space was left vertically between the symbols. We'll use the same number as we did when looking at the Babylonian system, 5389, to illustrate the Mayan place value system.

$$5389 = (13 \cdot 20^2) + (9 \cdot 20^1) + (9 \cdot 20^0)$$
$$5389 = 5200 + 180 + 9$$

Example 7.1.3

Converting Numbers from the Mayan System to the Hindu-Arabic Numeral System

Write each Mayan numeral as a Hindu-Arabic numeral.

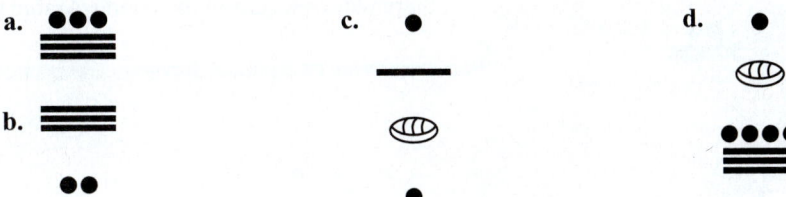

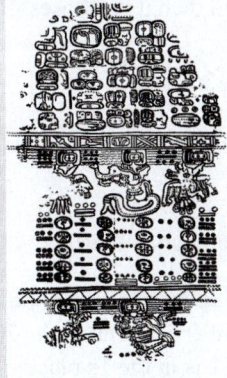
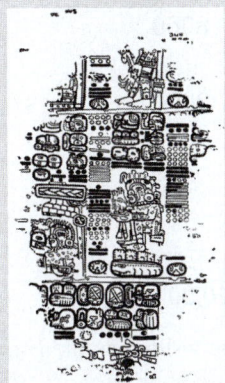

Solution

a. The first number does not contain any large vertical spaces between the symbols, so we know that it is a number between 0 and 19. We know the ● symbol represents 1, and the ▬▬ symbol represents 5. Since there are three 5s and three 1s, the Mayan numeral can be written as the Hindu-Arabic numeral 18.

b. Because the second numeral contains a vertical space between the groupings, we will need to multiply each group by the corresponding place value. Remember that the highest power of 20 is at the top and then it decreases from there.

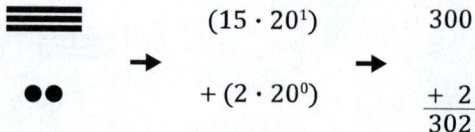

$$(15 \cdot 20^1) \qquad\qquad 300$$
$$+ (2 \cdot 20^0) \qquad\qquad \underline{+\ \ 2}$$
$$\qquad\qquad\qquad\qquad\qquad\quad 302$$

The Mayan numeral can be written as the Hindu-Arabic numeral 302.

c. We have vertical spacing again, so we will need multiplication and addition for the calculation. The top symbol here is for the number 1 and is in the 20^3s place, 5 is in the 20^2s place, 0 is in the 20^1s place, and 1 is in the 1s place.

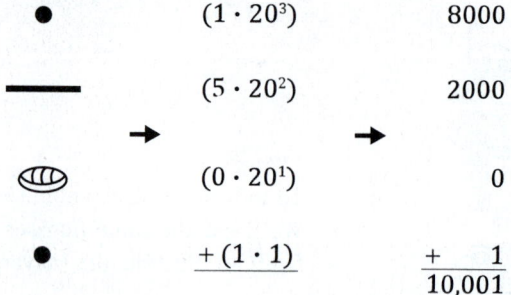

$$(1 \cdot 20^3) \qquad\qquad 8000$$
$$(5 \cdot 20^2) \qquad\qquad 2000$$
$$(0 \cdot 20^1) \qquad\qquad\ \ 0$$
$$+ (1 \cdot 1) \qquad\qquad \underline{+\qquad 1}$$
$$\qquad\qquad\qquad\qquad\qquad 10{,}001$$

Our calculations show that the Mayan numeral can be written as the Hindu-Arabic numeral 10,001.

d. Here the top symbol is in the 20^2s place, the symbol for zero is in the 20^1s place, and the last symbol is in the 1s place.

$$(1 \cdot 20^2) \qquad\qquad 400$$
$$(0 \cdot 20^1) \rightarrow \qquad 0$$
$$+ (19 \cdot 20^0) \qquad \underline{+\ 19}$$
$$\qquad\qquad\qquad\qquad\qquad 419$$

The Mayan numeral can be written as the Hindu-Arabic numeral 419. Notice that a zero character written in the Mayan numeral does not necessarily mean there will be a zero in the Hindu-Arabic numeral.

✔ **Skill Check 7.1.2**

Write the Mayan numeral as a Hindu-Arabic numeral.

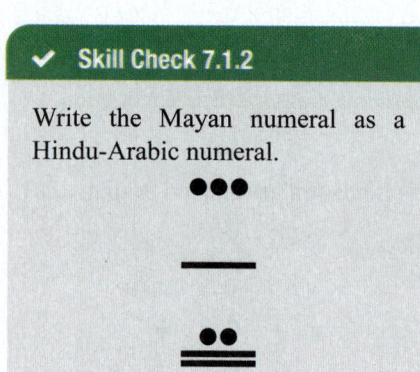

Example 7.1.4

Converting Numbers from the Hindu-Arabic Numeral System to the Babylonian and Mayan Systems

Express each Hindu-Arabic numeral in both the Babylonian and Mayan systems.

a. 8

b. 157

Solution

a. Express 8 as a Babylonian numeral and a Mayan numeral.

Babylonian

Recall that there are only two symbols in the Babylonian system: ∇ represents 1, and $\prec$ represents 10. Since 8 is less than 10, we only need to use the symbol for 1 to write the number.

$$\nabla \nabla \nabla$$
$$\nabla \nabla \nabla$$
$$\nabla \nabla$$

Mayan

To symbolize the number 8 in the Mayan system, we need to use the character for 1 (which is ●) three times stacked on top of the character for 5 (which is ———).

b. Express 157 as a Babylonian numeral and a Mayan numeral.

Babylonian

In order to write 157 using the Babylonian system, begin by determining the largest power of 60 that will divide 157. Since $60^2 = 3600$ is much larger than 157, we know that 60^1 is the largest divisor we need.

Next, divide 60 into 157.

$$157 \div 60 = 2 \text{ with a remainder of } 37$$

This tells us that in expanded Babylonian form, we need $(2 \cdot 60^1) + (37 \cdot 60^0)$ with a large gap in between each set of characters.

$$\nabla \nabla \qquad\qquad \prec \prec \prec \begin{matrix}\nabla \nabla \nabla \\ \nabla \nabla \nabla \\ \nabla\end{matrix}$$

Mayan

In order to write 157 using the Mayan system, begin by determining the largest power of 20 that will divide into 157. Since $20^2 = 400$, we know that 20^1 is the largest divisor we need.

Again, divide 20 into 157.

$$157 \div 20 = 7 \text{ with a remainder of } 17$$

In expanded Mayan form, we need $(7 \cdot 20^1) + (17 \cdot 20^0)$ stacked one on top of the other.

Skill Check Answers
1. 2,199,270 **2.** 1312

7.1   Exercises

✔ CONCEPT CHECK

1. The Hindu-Arabic numeral system uses _____ unique symbols to represent all numbers. In contrast, the Babylonian system used only _____ unique symbols, and the Mayan system used _____ unique symbols.

2. For numbers larger than 59, the Babylonian numeral system uses multiples of _____ when writing numbers.

3. When writing the Mayan numeral system, multiples of _____ are used.

4. It is widely believed that the Mayan system was the first numeral system to have a symbol to represent _____.

💡 PRACTICE

Express the Hindu-Arabic numeral in expanded form.

5. 2305

6. 4072

7. 621.9

8. 58.63

9. 80.02

10. 710.833

Express the expanded form as a Hindu-Arabic numeral.

11. $\left(9\cdot10^2\right)+\left(5\cdot10^1\right)+\left(4\cdot10\right)$

12. $\left(2\cdot10^4\right)+\left(8\cdot10^3\right)+\left(8\cdot10^2\right)+\left(1\cdot10^1\right)+\left(7\cdot1\right)$

13. $\left(3\cdot10^5\right)+\left(6\cdot10^1\right)+\left(2\cdot1\right)$

14. $\left(5\cdot10^3\right)+\left(5\cdot1\right)$

15. $\left(1\cdot10^4\right)+\left(2\cdot10^2\right)+\left(7\cdot10^{-2}\right)$

16. $\left(4\cdot10^2\right)+\left(2\cdot10^1\right)+\left(3\cdot10^{-1}\right)$

Write each Babylonian numeral as a Hindu-Arabic numeral.

17. ⧼⧼⧼ ∇∇∇
　　⧼ ∇∇∇

18. ⧼⧼ ∇∇

19. ⧼ ∇　　⧼⧼⧼ ∇∇∇
　　　　　⧼⧼ ∇∇

20. ∇　　∇∇　　⧼⧼⧼
　　　　　　　⧼

21. ⧼⧼⧼ ∇∇∇　∇∇∇　⧼⧼
　⧼ ∇∇∇　∇∇∇
　　∇∇　　∇∇

22. ⧼⧼⧼ ∇∇∇　　⧼⧼⧼
　⧼⧼ ∇∇∇　　⧼⧼
　　∇∇∇

Write each Hindu-Arabic numeral as a Babylonian numeral.

23. 12   **24.** 37

25. 99   **26.** 175

27. 4270   **28.** 8731

Write each Mayan numeral as a Hindu-Arabic numeral.

29. ●●●   **30.** ●
　　───   ═══

31. ═══   **32.** ≡≡≡
　　●●●●   ───
　　　　　　　　　　　　　　　　⬭

33. ●●●

⬭

●●●

●●●

34. ●●●●
___

●

≡

●●●●
___

Write each Hindu-Arabic numeral as a Mayan numeral.

35. 15   **36.** 19   **37.** 192

38. 261   **39.** 1335   **40.** 4807

41. Write the current calendar year using the Babylonian numeral system and then using the Mayan numeral system.

✎ WRITING & THINKING

42. List any advantages or disadvantages there might be to using the Babylonian numeration system.

43. List any advantages or disadvantages there might be to using the Mayan numeration system.

44. Which numeration system would you prefer to use, the Babylonian system or the Mayan system? Discuss why.

45. Since the original Babylonian system did not use a symbol or place holder for 0, how do you think they were able to represent the number 60?

46. Suppose an ancient manuscript was found that included the Babylonian numeral ∇∇. The text around the numeral suggests that the manuscript documents a convening of a particular country's government officials and that the number is a record of how many officials were in attendance. Based on the text, how many officials were probably in attendance at the convening?

OBJECTIVES

1. Convert numbers between the Egyptian numeral system and the Hindu-Arabic system.

2. Convert numbers between the Ionic Greek numeral system and the Hindu-Arabic system.

3. Convert numbers between the Roman numeral system and the Hindu-Arabic system.

4. Convert numbers between the Chinese numeral system and the Hindu-Arabic system.

In Section 7.1, we looked at three numeral systems that use the idea of bases and positions to express numbers. As you can imagine, across history there have been many, many more ways to express numbers. This section takes a look back at four early numeral systems: the Egyptian numeral system, the Ionic Greek numeral system, Roman Numerals, and the traditional Chinese numeral system. Looking back at historical numeral systems allows us to realize how much evolution has taken place in this fundamental tool. Our modern numeral system provides highly reliable and consistent ways to not only write numbers down, but also more easily carry out complex computations.

OBJECTIVE 1 Egyptian Numeral System

Let's begin with the Egyptian numeral system. The Rosetta Stone—inscribed around 196 BC on behalf of King Ptolemy V to make his claim as pharaoh of Egypt—is our access into the 3000 years of Egyptian history. Written in several languages including Egyptian and Greek, its discovery in 1799 allowed scholars to decipher the Egyptian **hieroglyphics**.

Figure 7.2.1: The Rosetta Stone

As we know from their ability to build the pyramids, the ancient Egyptians had sophisticated mathematical techniques. The discovery of the *Rhind Mathematical Papyrus*, said to be the largest mathematical text we have from the ancient world, shows this sophistication. Containing 84 different calculations, it gives us insight into how the ancient Egyptians thought about mathematics. While we won't learn their mathematical techniques in this section, we will at least learn how they wrote their numerals.

Similar to the Hindu-Arabic numeral system, the numeral system used by the Egyptians is made up of various symbols that represent digits and powers of 10. Table 7.2.1 shows us the hieroglyphic symbols used in the Egyptian numeral system alongside the Hindu-Arabic numerals we use.

> **Fun Fact**
>
> Fractions, with the exception of $\frac{1}{2}$ and $\frac{2}{3}$, were represented as unit fractions with the form $\frac{1}{n}$, where the 1 in the numerator was represented by the symbol ⬯, which means "[one] among." Some of the common fractions used are shown here.
>
> | $\frac{1}{2}$ | $\frac{1}{3}$ | $\frac{2}{3}$ | $\frac{1}{4}$ | $\frac{1}{5}$ |
> |---|---|---|---|---|
> | ═ | ⬯ | | ⬯ | ⬯ |

Table 7.2.1: Hieroglyphics for Egyptian Numerals

Egyptian Hieroglyphic	Hieroglyphic Description	Hindu-Arabic Numeral
│	Vertical line	1
∩	Cattle hobble	10
ℭ	Coiled rope	100

Egyptian Hieroglyphic	Hieroglyphic Description	Hindu-Arabic Numeral
	Lotus plant	1000
	Finger	10,000
	Frog	100,000
	A god with his hands raised above his head	1,000,000

Note: A cattle hobble is a device used to limit an animal's movement by tethering one or more of its legs.

Similar to the Mayan numeral system we looked at in Section 7.1, the Egyptian numeral system uses duplicate characters to represent multiples of numbers. To establish the number being represented, you add together the values of all of the symbols written. For instance, to write the number 7352, you need 7 lotus plants (1000), 3 coiled ropes (100), 5 cattle hobbles (10), and 2 vertical lines (1). That is, the number 7352 would be written as follows.

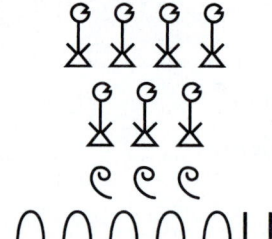

Because the position of the characters does not play a role in the value of the characters, the Egyptians were free to write the characters in any arrangement. However, for ease of reading, we will write them in a straight line.

Example 7.2.1

Converting between the Egyptian Numeral System and the Hindu-Arabic Numeral System

a. Write the following Egyptian numeral as a Hindu-Arabic numeral.

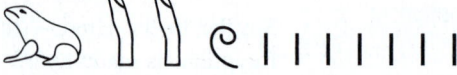

b. Write the following Hindu-Arabic numeral using Egyptian hieroglyphics.

1,325,000

Solution

a. Using Table 7.2.1, we can find the value of each hieroglyphic and then add them together.

= 100,000

= 20,000

= 100

| | | | | | | = 7

So we have the following.

= 100,000 + 20,000 + 100 + 7
= 120,107

b. To begin, we need to write the numeral in expanded form so we can easily see how many symbols of each to write.

$$1{,}325{,}000 = (1 \cdot 1{,}000{,}000) + (3 \cdot 100{,}000) + (2 \cdot 10{,}000) + (5 \cdot 1000)$$

Expanded form conveys how many of each hieroglyphic is needed. Using Table 7.2.1, we can identify the appropriate symbol for each number.

1,000,000 =

100,000 =

10,000 =

1000 =

So the Egyptian hieroglyphic for 1,325,000 is written as follows.

✔ **Skill Check 7.2.1**

Write the Hindu-Arabic numeral 2461 using Egyptian hieroglyphics.

OBJECTIVE 2 Ionic Greek Numeral System

Dating back as early as 7th century BC, the ancient Greeks' numeral systems used letters of their alphabet along with three Phoenician letters to symbolize numbers. Here we'll discuss the system referred to as *Ionic Greek*. Since it's believed that this system was developed in a region composed of several Greek settlements, this system is sometimes also called Milesian numeral system or the Alexandrian numeral system.

Table 7.2.2 shows the symbols used for numbers from 1 to 900 in the Ionic Greek Numeral System.

Table 7.2.2: Ionic Greek Numerals

alpha	beta	gamma	delta	epsilon	vau	zeta	eta	theta
α	β	γ	δ	ε	ϛ	ζ	η	θ
1	2	3	4	5	6	7	8	9

iota	kappa	lambda	mu	nu	xi	omicron	pi	koph
ι	κ	λ	μ	ν	ξ	o	π	ϙ
10	20	30	40	50	60	70	80	90

rho	sigma	tau	upsilon	phi	chi	psi	omega	sampi
ρ	σ	τ	υ	ϕ	χ	ψ	ω	ϡ
100	200	300	400	500	600	700	800	900

Putting a comma (,) before any of the first nine symbols multiplies the value by 1000, allowing the system to have symbols for values up to 9000 using single digits. For instance,

$$, \delta = 4000.$$

A symbol written above M means that value is multiplied by 10,000. For instance,

$$\overset{\delta}{M} = 40,000.$$

As in the Egyptian numeral system, characters written next to one another in the Ionic Greek system are added together. The numbers 68 and 227 would be written as follows.

$$68 = \xi\eta \quad \text{and} \quad 227 = \sigma\kappa\zeta$$

Example 7.2.2

Converting between the Ionic Greek Numeral System and the Hindu-Arabic Numeral System

a. Write the following Greek numeral as a Hindu-Arabic numeral.

$$\pi\beta$$

b. Write the following Hindu-Arabic numeral as a Greek numeral.

$$531$$

Solution

a. Using the Table 7.2.2, we can find the value of each Greek letter and then add them together.

$$\pi = 80$$

$$\beta = 2$$

So $\pi\beta = 80 + 2 = 82$.

b. To begin with, we need to write the numeral in expanded form so that we can easily see which symbols will be needed.

$$531 = 500 + 30 + 1$$

Using Table 7.2.2, we can find the appropriate Greek letter for each value.

$$500 = \phi$$

$$30 = \lambda$$

$$1 = \alpha$$

So $531 = \phi\lambda\alpha$.

> ✔ **Skill Check 7.2.2**
>
> Write the Hindu-Arabic numeral 402 as a Greek numeral.

OBJECTIVE 3 Roman Numeral System

The Romans were much like the ancient Greeks in that they used a selection of letters from their alphabet to represent numerals. However, they only use seven characters and a mixture of addition and subtraction to calculate each written value. Table 7.2.3 shows the seven characters along with their values in Hindu-Arabic numerals.

Table 7.2.3: Roman Numerals

Roman Numeral	Hindu-Arabic Numeral
I	1
V	5
X	10
L	50
C	100
D	500
M	1000

The Roman numeral system is different from the systems that we have seen thus far in that subtraction occasionally takes place based on the position of the characters. There are some rules that we need in order to use the Roman numeral system.

☰ ADDITION AND SUBTRACTION RULES FOR ROMAN NUMERALS

1. A letter may be repeated up to 3 times to indicate multiple values. For example,

$$III = 3$$
$$CC = 200$$

2. If a smaller value is placed on the right of a character, that smaller value is added to the larger one. For example,

$$\overset{5\ 1}{VI} = 5 + 1 = 6$$

$$\overset{100\ 10\ 10}{C\,X\,X} = 100 + 10 + 10 = 120$$

3. If a smaller value is placed on the left of a character, that smaller value is subtracted from the larger one. For example,

$$\overset{1\ 5}{IV} = 5 - 1 = 4$$

$$\overset{100\ 1000}{C\,M} = 1000 - 100 = 900$$

4. A bar is placed above a symbol, or group of symbols, to indicate that the numeral is to be multiplied by 1000. For example,

$$\overline{D} = 500 \cdot 1000 = 500,000$$

$$\overline{VI} = 6 \cdot 1000 = 6000$$

There are a few other rules for subtraction as well.

a. Only subtract *one* number from the left. For instance, the numeral 3 is written as 3 = III, *not* as ~~IIV~~ since we cannot subtract the number 1 from 5 more than once.

b. Only subtract powers of ten (I, X, or C), not powers of 10 multiplied by 5 (V, L, D). For instance, the number 45 is written as XLV, *not* as ~~VL~~.

c. Only subtract a value from a number that is *at most* 10 times greater. For instance, we cannot subtract 1 from 100 (~~IC~~) but must add symbols for 90 and 9 to get to 99. This value is correctly written 99 = XCIX = 90 + 9.

Table 7.2.4: Roman Numerals Up to 1000

Value	Roman Numeral	Value	Roman Numeral	Value	Roman Numeral
One	I	Eleven	XI	Thirty	XXX
Two	II	Twelve	XII	Forty	
Three	III	Thirteen	XIII	Fifty	L
Four	IV	Fourteen		Sixty	LX
Five	V	Fifteen	XV	Seventy	LXX
Six	VI	Sixteen	XVI	Eighty	LXXX
Seven	VII	Seventeen	XVII	Ninety	XC
Eight		Eighteen	XVIII	One hundred	C
Nine	IX	Nineteen		Five hundred	D
Ten	X	Twenty	XX	One thousand	M

(Notice that there are some blanks in Table 7.2.4. Example 7.2.3 shows how these missing numerals are written.)

Table 7.2.4 shows some of the common Roman Numerals still used today. For instance, the numerals 1 through 12 are sometimes seen representing the hours on a clock or watch face. The publishing world uses Roman numerals to indicate volumes of books. In sports, the United States' National Football League (NFL) holds a yearly championship game each February called the Super Bowl. Roman numerals are used to indicate what number Super Bowl it is. The Summer and Winter Olympics, held every four years, also use Roman numerals to record the number of modern games. The numbering of the first modern Olympics started with the Games of the I Olympiad. The games are sequentially numbered, even counting those that were skipped due to war. The XXX Olympiad were held in 2012. The XXXIV Olympiad will be held in 2032.

Example 7.2.3

Converting between the Hindu-Arabic Numeral System and the Roman Numeral System

Complete Table 7.2.4 by writing the following Hindu-Arabic numerals as Roman numerals.

a. Eight

b. Fourteen

c. Nineteen

d. Forty

Solution

To write each of the numbers as Roman numerals, we need to break down each number into quantities represented by the Roman numeral system; that is, 1, 5, 10, etc. (It's a similar concept to making change for someone with a limited pile of coins.)

I	1
V	5
X	10
L	50
C	100
D	500
M	1000

a. $8 = 5 + 3$

$8 = \text{VIII}$

Notice that we did not try to write $10 - 2$ in order to reach 8. This is because you can only subtract one character, not two, from a larger number.

b. $14 = 10 + 4$

$14 = \text{XIV}$

c. $19 = 10 + 9$

$19 = XIX$

Again, notice that we didn't try to write 20 – 1 in order to reach 19. This time it was because 20 is more than 10 times greater than 1, and you can only subtract a value from a number that is *at most* 10 times greater.

d. $40 = 50 - 10$

$40 = XL$

Here we were able to use subtraction because the 50 is not more than 10 times greater than 10, so we can subtract a power of ten from 50 to achieve our value.

> ✔ **Skill Check 7.2.3**
>
> Write the Hindu-Arabic numeral 24 as a Roman numeral.

Example 7.2.4

Converting between the Hindu-Arabic Numeral System and the Roman Numeral System

Write the following Roman numerals as Hindu-Arabic numerals.

a. MDCLX

b. MCMXLII

Solution

In each case, we need to focus on the values of each of the letters so that we can determine where the letters should be grouped for subtraction. Remember that smaller numbers to the left indicate subtraction.

a. 1000   500   100   50   10
     M   D   C   L   X

Since there are no smaller values to the left of any number, we can simply add the values together.

$$MDCLX = 1000 + 500 + 100 + 50 + 10 = 1660$$

b. 1000   100   1000   10   50   1   1
     M   C   M   X   L   I   I

Notice that this time, C (100) is to the left of the larger value M (1000), and X (10) is to the left of the larger value L (50). So we need to group these pairs together for subtraction.

$$M\ CM\ XL\ II = 1000 + (1000 - 100) + (50 - 10) + 2 = 1942$$

OBJECTIVE 4 Chinese Numeral System

The last system that we will look at is the Chinese numeral system. These days, three written numeral systems are used in the Chinese-speaking world: Hindu-Arabic numerals, Chinese characters corresponding to the numerals in the spoken language, and the lesser used Suzhou numerals. We will look at the traditional system of using Chinese characters.

Table 7.2.5 shows the characters for numbers 1 through 10,000 in the Chinese numeral system. Notice that there is a column for the "simple" characters and one for the "formal" characters. Because the simple characters can easily be changed, the formal characters are used for writing money amounts on checks and bank notes, among other important documents, to help prevent fraud. Our focus will be on the simple characters.

> **Fun Fact**
>
> The *Suàn shù shù*, or the *Book on Numbers and Computation*, circa 200 BC, is one of the earliest examples of Chinese mathematics. Consisting of 200 strips of bamboo, it contains 69 mathematical problems with their solutions and methods.

Table 7.2.5: Chinese Numerals

Simple Character	Formal Character	Hindu-Arabic Numeral
一	零	1
二	壹	2
三	貳	3
四	叁	4
五	肆	5
六	伍	6
七	陸	7
八	柒	8
九	捌	9
十	玖	10
百	佰	100
千	仟	1000
万	萬	10,000

In order to form numbers other than the ones shown, addition and multiplication is used based on position in reference to a multiple of ten.

> **▤ ADDITION AND SUBTRACTION RULES FOR THE CHINESE NUMERAL SYSTEM**
>
> 1. A numeral to the right of a multiple of 10 (10, 100, 1000, …) indicates addition. For instance,
>
> $$十六 = 10 + 6 = 16$$
>
> 2. A numeral to the left of a multiple of 10 (10, 100, 1000, …) indicates multiplication. For instance,
>
> $$六十 = 6 \cdot 10 = 60$$
>
> 3. A position holder, 零 is used for when a zero occurs in the middle of a numeral. Only one is needed to show any number of consecutive zeros. For instance,
>
> $$201 = 二百零一$$
> $$2001 = 二千零一$$

Example 7.2.5

Converting between the Hindu-Arabic Numeral System and the Chinese Numeral System

a. Write the following Chinese numeral as a Hindu-Arabic numeral.

$$九十八$$

b. Write the following Hindu-Arabic numeral as a Chinese numeral.

$$407$$

Solution

a. We can use Table 7.2.5 to determine the value for each symbol.

$$\begin{array}{ccc} 9 & 10 & 8 \\ 九 & 十 & 八 \end{array}$$

Because the 9 symbol is to the left of the 10, we will multiply the two together and then add 8, which is to the right of the 10.

$$九 \ 十 \ 八 = (9 \cdot 10) + 8 = 98$$

b. Writing 407 in expanded form we have $(4 \cdot 100) + 7$.

This means that we need the character for 4 to be on the left of the symbol for 100 and the character for 7 to be on the right. However, because there is a 0 in the middle of our numeral, we also need the place holder. The symbols needed are as follows.

$$4 = 四$$

$$100 = 百$$

$$0 = 零$$

$$7 = 七$$

$$407 = 四百零七$$

Skill Check 7.2.4

Write the following Chinese numeral as a Hindu-Arabic numeral.

三十五

Skill Check Answers

1. 2. $\nu\beta$ 3. XXIV 4. 35

7.2   Exercises

✔ CONCEPT CHECK

1. The ancient Egyptian writing system used symbols called _____.

2. In the Egyptian numeral system, _____ symbols are used to represent multiples of numbers.

3. In the Ionic Greek numeral system, a symbol written above ____ means that value is multiplied by 10,000.

4. Unlike previous systems studied, the Roman numeral system involves _____.

5. In the Chinese numeral system, formal characters are used because the simple characters can be easily _____.

💡 PRACTICE

Write each Egyptian numeral as a Hindu-Arabic numeral.

6.

7.

8.

9.

10.

11.

Write each Hindu-Arabic numeral using Egyptian hieroglyphics.

12. 435,072

13. 1128

14. 2,601,011

15. 50,000

16. 71,614

17. 1,003,050

Write each Ionic Greek numeral as a Hindu-Arabic numeral.

18. $\lambda \varepsilon$

19. $\upsilon \mu \delta$

20. $\omega \gamma$

21. $\tau \alpha$

22. $,\zeta \rho \iota$

23. $,\eta \phi \nu \varepsilon$

24. $\gamma \atop M$

25. $\theta \omega \theta \atop M$

Write each Hindu-Arabic numeral as an Ionic Greek numeral.

26. 284

27. 196

28. 2000

29. 1078

30. 50,000

31. 80,424

Write each Roman numeral as a Hindu-Arabic numeral.

32. XXVII

33. XLVIII

34. CCXIX

35. DCLI

36. MCDIV

37. CMLXXV

Write each Hindu-Arabic numeral as a Roman numeral.

38. 35

39. 42

40. 79

41. 163

42. 558

43. 491

Write each Chinese numerals as a Hindu-Arabic numeral.

44. 十六

45. 五十四

46. 七百九十二

47. 一千零八

48. 三千五百二十一

49. 二万五千五百七十五

Write each Hindu-Arabic numeral as a Chinese numeral.

50. 26

51. 103

52. 82

53. 75

54. 649

55. 2002

56. Write today's calendar year using the Egyptian numeral system.

57. Write today's calendar year using the Ionic Greek numeral system.

58. Write today's calendar year using the Roman numeral system.

59. Write today's calendar year using the Chinese numeral system.

✎ **WRITING & THINKING**

60. One of the peculiarities with the Ionic Greek system is that numbers often look like words and vice versa, since the Greek alphabet is used to write numerals. Determine the number that these phonetic spellings also make.

    a. dog = $\delta\omega\gamma$

    b. bird = $\beta\iota\rho\delta$

    c. star = $\sigma\tau\alpha\rho$

    d. rain = $\rho\alpha\iota\nu$

61. Super Bowl LII was played on February 4, 2018, and featured the Philadelphia Eagles against the New England Patriots. The Eagles won the championship by a score of 41 to 33.

    a. What number Super Bowl are the Philadelphia Eagles be able to boast about winning?

    b. Why do you think we use Roman numerals to number the Super Bowl championship games and not our Hindu-Arabic numbers?

    c. Why is an ancient European system used to name the super bowl instead of an ancient American system?

62. The 2024 Summer Olympics will be held in Paris from July 26 through August 11. The games are officially known as the Games of the XXXIII Olympiad.

    a. What number Olympiad will these games be?

    b. Why do you think we use Roman numerals to number the Olympics games and not Ionic Greek numerals since the Greeks started the Olympics?

63. Name at least one weakness of the Egyptian numeral system.

64. Explain why the absence of a zero is not as problematic in the Ionic Greek system as it is in the Babylonian system.

65. Which of the four numeral systems covered in this section do you find the easiest to write numbers in? Why?

7.2 PROJECT

THE ISLAND OF SIX SYMBOLS

In a remote part of the Ocean of Knowledge lies the Island of Six Symbols. The native inhabitants of the island have a nomadic disposition, and directional words and symbols have come to be an important part of their culture. All forms of written communication and mathematical calculation rely on them. In their writing system, there are only six numerals. There are two versions of these six numerals—one for formal documents and one for normal use.

Formal	(fish)	(lightning)	(wave)	(hand)	(bird)	(palm tree)
Normal	→	↓	↖	←	↗	↑

1. From archeological evidence, we have a partial table of the numeral system. Determine the pattern and fill in the blank cells to complete the table. Does this appear to be a positional system?

1		7	↑↑	13		19		25		31		37	
2	↓	8	↑↓	14		20		26		32		38	
3		9		15	↓→	21		27		33		39	
4		10		16	↓←	22		28		34		40	
5		11		17		23	→↗	29		35		41	
6	↖	12		18		24	→↖	30		36		42	

2. Determine the correspondence between these symbols and Hindu-Arabic numerals

Formal	(fish)	(lightning)	(wave)	(hand)	(bird)	(palm tree)
Normal	→	↓	↖	←	↗	↑
Hindu-Arabic						

3. Calculate and express the following in both the Six Symbol system and the decimal system. Write the answer using formal numerals.

   a. (palm tree) + (bird)

   b. (lightning) + (lightning)

   c. (fish) × (wave)

   d. (hand) × (hand)

4. Does the system have a symbol for zero? If not, create a potential symbol for zero and describe the benefits that can be gained by introducing symbols for zero for both formal and normal use.

7.3 WORKING WITH BASE NUMBER SYSTEMS

OBJECTIVES

1. Convert between base number systems.

2. Add, subtract, and multiply in different base number systems.

OBJECTIVE 1 Base Number Systems

Just as we've seen in the first two sections of this chapter, numeration systems throughout history have had varied ways of expressing and manipulating numbers. You might have even decided for yourself that some systems and their functionalities are more appealing or advantageous than others.

The Babylonians had a finite set of symbols (without 0) and used the physical position of the symbols to indicate a multiple of the number 60. However, without a symbol for zero, they lacked an effective way to show with precision how large a number was.

In contrast, the Chinese and Egyptian systems both used powers of 10, but both systems needed a new character to represent each new power of 10. These systems have many symbols with a potentially never-ending need for new symbols as written numbers get larger and larger. However, the Chinese system does have a character for 0 and is consequently able to reduce the amount of space required to write large numbers.

The Hindu-Arabic system is a harmonized approach that captures the ideas that have occurred historically in different number systems. This system uses 10 digits, including 0, and utilizes the zero symbol when necessary to indicate ever increasing powers of ten. With this seemingly simple idea, we can represent any number—no matter how large or small—with precision. We say that the Hindu-Arabic number system is in **base 10.** Using the same principles, we can create a number system with any integer as base. In this section, we will see how this can be done.

Without any other indication, it is assumed that all Hindu-Arabic numerals written are in base 10 and that the powers of ten are written in increasing order as you move right to left.

$$...10^4 \quad 10^3 \quad 10^2 \quad 10^1 \quad 10^0 . 10^{-1} \quad 10^{-2}...$$

However, a base number system doesn't have to be constrained to just base 10. We can use any positive number as the base of a system. To indicate a numeral is written in a base other than 10, a subscript is used on the numeral. For instance, a number in base 5 can be written as 12_5, which is read "one, two, base five."

This is not the number twelve that we are accustomed to. The base number 5 indicates that we will use powers of 5 instead of powers of 10 for each place value. The base also indicates that we will use only 5 digits (0, 1, 2, 3, and 4) to write all numbers in this base system.

We saw in Section 7.1 that writing Hindu-Arabic numerals in expanded form clarifies how a numeral system combines both position and multiples of 10. We can use the same idea to write 12_5 in expanded form and find its equivalent value in base 10.

$$12_5 = (1 \times 5^1) + (2 \times 5^0) = 5 + 2 = 7_{10}$$

Think Back

Any number raised to the zero power is equal to 1.

$$x^0 = 1$$

Think Back

You are already accustomed to interpreting numbers written in different ways. For example, scientific notation is a compact way to write very large and very small numbers with ease; i.e., $35{,}000{,}000{,}000 = 3.5 \times 10^{10}$.

Note: Although we wrote the 10 as a subscript for base 10 here, it is usually omitted. We will write it for clarification purposes when using more than one base.

To illustrate the idea of working in different bases more fully, let's write 10101_2 and 210_3 in expanded forms to identify their equivalent values in base 10.

Let's begin with 10101_2. Base 2 has two digits (0 and 1) and uses powers of 2.

$$...2^4 \quad 2^3 \quad 2^2 \quad 2^1 \quad 2^0$$

Therefore, 10101_2 (read "one, zero, one, zero, one, base 2") is converted to base 10 as shown.

$$10101_2 = \left(1 \times 2^4\right) + \left(0 \times 2^3\right) + \left(1 \times 2^2\right) + \left(0 \times 2^1\right) + \left(1 \times 2^0\right)$$
$$10101_2 = 16 + 0 + 4 + 0 + 1$$
$$10101_2 = 21_{10}$$

Next, we have 210_3. Base 3 has three digits (0, 1, and 2) and uses powers of 3.

$$...3^4 \quad 3^3 \quad 3^2 \quad 3^1 \quad 3^0$$

210_3 (read "two, one, zero, base 3") is converted to base 10 as shown.

$$210_3 = \left(2 \times 3^2\right) + \left(1 \times 3^1\right) + \left(0 \times 3^0\right)$$
$$210_3 = 18 + 3 + 0$$
$$210_3 = 21_{10}$$

Notice that both 10101_2 and 210_3 are equal to 21_{10}.

As you can see, when converting numbers from one base to another, it's helpful to write the numbers in expanded form.

Example 7.3.1

Converting from Other Bases to Base 10

a. Convert the number 1001_2 to base 10.

b. Convert the number 1001_5 to base 10.

Solution

In order to convert to base 10, we need to write the number in expanded form using the appropriate powers.

a. $1001_2 = \left(1 \times 2^3\right) + \left(0 \times 2^2\right) + \left(0 \times 2^1\right) + \left(1 \times 2^0\right)$

$1001_2 = 8 + 0 + 0 + 1$

$1001_2 = 9_{10}$

b. $1001_5 = \left(1 \times 5^3\right) + \left(0 \times 5^2\right) + \left(0 \times 5^1\right) + \left(1 \times 5^0\right)$

$1001_5 = 125 + 0 + 0 + 1$

$1001_5 = 126_{10}$

✔ **Skill Check 7.3.1**

Convert the number 1230_4 to base 10.

There are advantages to having a base number system that suits practical uses. The Babylonians chose a base value of 60, which has a lot of divisors. That property of their system made it possible to easily divide common quantities. However, for computer scientists, using a base that has only 2 digits is more practical. All modern computer science is based on the basic property of semiconductors that have a digital switch for **on** or **off**. Consequently, computer scientists who create microprocessors have developed various techniques based on expressing numbers using base 2 called the binary numeral system. They also developed techniques using numbers based around various powers of 2, such as octal (base 8) and hexadecimal (base 16). Counting in base 16 might sound peculiar to you, but actually it is more familiar than you would think. Recall from unit conversions that there are 16 ounces in a pound, and a weight of 8 pounds 3 ounces seems much more familiar than 131 ounces.

What happens when the base most practical for our use has more than ten digits? Often, we turn to the alphabet to represent the extra characters. For instance, the **hexadecimal system**, which is base sixteen, needs 16 digits. In other words, it needs six more characters other than 0 through 9. Computer programmers have decided to use the uppercase letters A through F (or sometimes lowercase a through f) to represent the digits 10 through 15, as shown in Table 7.3.1.

Table 7.3.1: Characters Used for Base 16 (the Hexadecimal System)

Base 10 Digit	0	1	2	3	4	5	6	7	8	9	10	11	12	13	14	15
Hexadecimal (Base 16) Digit	0	1	2	3	4	5	6	7	8	9	A	B	C	D	E	F

Example 7.3.2

Converting from the Hexadecimal System to Base 10

Convert the number $B2F_{16}$ to base 10.

Solution

We need to write the number in expanded form, using the appropriate Hindu-Arabic numbers in Table 7.3.1 and multiplying by the appropriate powers of 16.

$$B2F_{16} = \left(11 \times 16^2\right) + \left(2 \times 16^1\right) + \left(15 \times 16^0\right)$$
$$B2F_{16} = 2816 + 32 + 15$$
$$B2F_{16} = 2863_{10}$$

Therefore, $B2F_{16}$ is equivalent to 2863 in base 10.

> ✔ **Skill Check 7.3.2**
>
> Use expanded notation, as shown in Example 7.3.2, to show that both 1111001000011000000_2 and $3C8C0_{16}$ are equal to $248,000_{10}$.

An advantage to using a larger base like the hexadecimal system is that it takes less space to represent very large numbers. For instance, look at the difference between writing the number 248,000 in base 2 and base 16.

$$248,000 = 1111001000011000000_2$$
$$248,000 = 3C8C0_{16}$$

So far, we've converted from other bases to base 10. Let's take a look at converting a base 10 number to another base. In order to do the conversion in this direction, we perform a series of divisions using the different powers of the particular base we want to convert to. The following example illustrates this process.

Example 7.3.3

Converting from Base 10 to Other Bases

Convert the given base 10 number to the indicated base.

a. 102 to base 2

b. 138 to base 16

Solution

When converting from base 10 to any other base, it's helpful to begin by examining the powers of the desired base.

a. Converting the base 10 number 102 to base 2 signifies that powers of 2 are desired for each place value. We've listed the powers of two for you, until they exceed 102.

$$\ldots \quad 2^7 \quad 2^6 \quad 2^5 \quad 2^4 \quad 2^3 \quad 2^2 \quad 2^1 \quad 2^0$$
$$\ldots \quad \downarrow \quad \downarrow \quad \downarrow \quad \downarrow \quad \downarrow \quad \downarrow \quad \downarrow \quad \downarrow$$
$$\ldots \quad 128 \quad 64 \quad 32 \quad 16 \quad 8 \quad 4 \quad 2 \quad 1$$

Notice that the eighth position of the base 2 system, which is 2^7 (and equal to 128), is already larger than our numeral 102. Therefore, we need to start the division process with the next smallest number, 2^6, or 64. Using division, we can find out how many multiples of 2^6 we will need.

$$\begin{array}{r} 1 \\ 64\overline{)102} \\ -64 \\ \hline 38 \end{array}$$

The quotient tells us that we will need to use one multiple of 2^6. We then repeat the process by dividing the remainder of 38 by the next smaller power 2^5, or 32. This will show us how many multiples of 2^5 we will need.

$$\begin{array}{r} 1 \\ 32\overline{)38} \\ -32 \\ \hline 6 \end{array}$$

We now know that we will also need to use one multiple of 2^5. Notice that since the remainder here is 6, we can skip dividing by 16 and 8 and move right on to finding how many multiples of 2^2, or 4, are needed.

$$\begin{array}{r} 1 \\ 4\overline{)6} \\ -4 \\ \hline 2 \end{array}$$

Thus, we need one multiple of 2^2. And finally, we divide the last remainder by 2^1.

$$\begin{array}{r} 1 \\ 2\overline{)2} \\ \underline{-2} \\ 0 \end{array}$$

We've found that the base 10 number 102 consists of one multiple of 2^6, one multiple of 2^5, one multiple of 2^2, and one multiple of 2^1. So, in expanded form, we have the following.

$$102 = (1 \times 2^6) + (1 \times 2^5) + (1 \times 2^2) + (1 \times 2^1)$$

When writing the number down in base 2, we need to remember to put 0s in the positions for the multiples of 2 that are not needed.

Thus, the base 10 number 102 written in base 2 is 1100110_2.

b. Converting the base 10 number 138 to base 16 signifies that multiples of 16 are desired for each place value. Again, we list the multiples of 16 until they exceed 138.

$$\begin{array}{ccc} 16^2 & 16^1 & 16^0 \\ \downarrow & \downarrow & \downarrow \\ 256 & 16 & 1 \end{array}$$

Notice that the powers of 16 grow very rapidly. This time we begin the division with 16^1 since $16^2 = 256$ is too large.

$$\begin{array}{r} 8 \\ 16\overline{)138} \\ \underline{-128} \\ 10 \end{array}$$

This division tells us that we need eight multiples of 16^1 in our answer. Since our remainder of 10 is less than 16, no further division is required. In expanded form we have

$$138 = (8 \times 16^1) + (10 \times 16^0)$$

Using Table 7.3.1, the character for 10 in base 16 is A.

Therefore, the base 10 number 138 written in base 16 is $8A_{16}$.

Example 7.3.4

Converting between Base 2 and Base 16

Convert the given number to the indicated base.

a. $B2F_{16}$ to base 2

b. 1101001011001001_2 to base 16

Solution

a. Recall that in Example 7.3.2, we converted $B2F_{16}$ to base 10. One option for converting from base 16 to base 2 is to first convert base 16 to base 10 and then convert base 10 to base 2. That option works, although it requires a significant number of steps. Alternatively, there's a shortcut option for converting directly between base 16 and base 2.

Let's expand Table 7.3.1 to include the first sixteen base 2 equivalents.

Table 7.3.2: Characters Used for Base 16 (the Hexadecimal System)

Base 10 Digit	0	1	2	3	4	5	6	7
Hexadecimal (Base 16) Digit	0	1	2	3	4	5	6	7
Binary (Base 2) Digit	0000	0001	0010	0011	0100	0101	0110	0111

Base 10 Digit	8	9	10	11	12	13	14	15
Hexadecimal (Base 16) Digit	8	9	A	B	C	D	E	F
Binary (Base 2) Digit	1000	1001	1010	1011	1100	1101	1110	1111

We can use this conversion table for values represented in the table. Our given number was $B2F_{16}$. Start by finding B in the hexadecimal column and then locate its equivalent in binary, which is 1011. Doing the same for 2 and F gives us 0010 and 1111, respectively. Thus, we can join them in a string to give the following equivalent in base 2.

This gives us that $B2F_{16}$ is equivalent to 101100101111_2.

b. When converting from base 2 to base 16 using the shortcut method, we work from right to left, splitting the given number into groups of four. This works because the first sixteen binary numbers can each be represented using four characters.

Working from right to left, find the first block of four in our given number.

$$1101001011001\underbrace{1001}_{?}{}_2$$

We then locate 1001 on the conversion table above and find that its base 16 equivalent is 9. Remember that this will be the rightmost digit in our base 16 number.

Continue working from right to left in blocks of four.

$$\underbrace{1101}_{D}\underbrace{0010}_{2}\underbrace{1100}_{C}\underbrace{1001}_{9}{}_2$$

This gives us that 1101001011001001_2 is equivalent to $D2C9_{16}$.

OBJECTIVE 2 Operations in Base Number Systems

Just as with numbers in base 10, we can carry out all of the standard mathematical operations with numbers in other bases. If we want to perform the operations of addition, subtraction, or multiplication, the methods are the same as those we used when working with base 10. The only difference is that after each computation involved, we must make sure the numbers are written appropriately for the base we're computing in. For example, to add 2_3 and 2_3 together, we add the digits $2 + 2$ and get 4. However, this result is 4_{10}, not 4_3. (Note that since the digit 4 is not used in base 3, the number 4_3 does not make any sense.) We need to convert 4_{10} into a base 3 number.

$$4_{10} = (1 \times 3^1) + (1 \times 3^0) = 11_3$$

Thus, $2_3 + 2_3 = 11_3$.

We will write out each example using a long hand approach as it is easier to visualize the computations.

Example 7.3.5

Adding Two Numbers of Like Base

a. Add the numbers 324_5 and 41_5.

b. Add the numbers 102_4 and 33_4.

Solution

a. Remember that when working in base 5, the only digits used are 0, 1, 2, 3, and 4. So if we ever perform an operation and obtain a number bigger than 5, we will need to convert that number back to base 5.

Let's begin by lining up the digits of our numbers in columns for addition.

$$\begin{array}{r} 324_5 \\ +41_5 \\ \hline \end{array}$$

Just as with addition in base 10, add the numbers in the right-hand column first and work your way left, carrying over numbers into the next column, as needed.

$$\begin{array}{r} \overset{1}{3}24_5 \\ +41_5 \\ \hline 0_5 \end{array}$$
$4 + 1 = 5_{10}$, but 5 is not a digit in base 5.
Convert 5_{10} to a base 5 number: $5_{10} = (1 \times 5^1) + (0 \times 5^0) = 10_5$
Write 0 and carry the 1.

$$\begin{array}{r} \overset{1\,1}{3}24_5 \\ +41_5 \\ \hline 20_5 \end{array}$$
$1 + 2 + 4 = 7_{10}$, but 7 is not a digit in base 5.
Convert 7_{10} to a base 5 number: $7_{10} = (1 \times 5^1) + (2 \times 5^0) = 12_5$
Write 2 and carry the 1.

$$\overset{1\,1}{324}_5 \qquad 1 + 3 = 4_5$$
$$+41_5 \qquad \text{No conversion necessary.}$$
$$\overline{420_5}$$

Thus, $324_5 + 41_5 = 420_5$.

b. When working in base 4, the only digits used are 0, 1, 2, and 3. So any number bigger than or equal to 4 will need to be converted back to base 4.

$$\overset{1}{102}_4 \qquad 2 + 3 = 5_{10}, \text{ but 5 is not a digit in base 4.}$$
$$+33_4 \qquad \text{Convert } 5_{10} \text{ to a base 4 number: } 5_{10} = (1 \times 4^1) + (1 \times 4^0) = 11_4$$
$$\overline{1_4} \qquad \text{Write 1 and carry the 1.}$$

$$\overset{1\,1}{102}_4 \qquad 1 + 0 + 3 = 4_{10}, \text{ but 4 is not a digit in base 4.}$$
$$+33_4 \qquad \text{Convert } 4_{10} \text{ to a base 4 number: } 4_{10} = (1 \times 4^1) + (0 \times 4^0) = 10_4$$
$$\overline{01_4} \qquad \text{Write 0 and carry the 1.}$$

$$\overset{1\,1}{102}_4 \qquad 1 + 1 = 2$$
$$+33_4 \qquad \text{No conversion necessary.}$$
$$\overline{201_4}$$

Thus, $102_4 + 33_4 = 201_4$.

✔ **Skill Check 7.3.3**

Add 101_2 and 11_2.

Example 7.3.6

Subtracting Two Numbers of Like Base

a. Subtract the number 41_9 from the number 324_9.
b. Subtract the number 23_6 from the number 212_6.

Solution

a. Begin by lining up the digits appropriately for subtraction and subtract in columns from right to left, borrowing from the column to the left when needed. What we have to remember is that when borrowing for subtraction in any base, we borrow one *multiple of the base*. For instance, we are already accustomed to borrowing 10^1 when we subtract in base 10. Likewise, in base 9 we will borrow 9^1 if necessary. And in base 6, we will borrow 6^1.

$$324_9 \qquad 4 - 1 = 3$$
$$-41_9 \qquad \text{No conversion necessary.}$$
$$\overline{3_9}$$

$$\overset{2}{\cancel{3}}\overset{9+}{2}4_9 \qquad \text{Borrow } 9^1 \text{ from the third column.}$$
$$-\ \ 41_9 \qquad (9+2)-4=7$$
$$\overline{\qquad 73_9} \qquad \text{No conversion necessary.}$$

$$\overset{2}{\cancel{3}}\overset{9+}{2}4_9 \qquad 2-0=2$$
$$-\ \ 41_9 \qquad \text{No conversion necessary.}$$
$$\overline{273_9}$$

Thus, $324_9 - 41_9 = 273_9$.

b. This time when borrowing, we'll borrow a 6^1 since the numbers are in base 6.

$$2\overset{0}{\cancel{1}}\overset{6+}{2}_6 \qquad \text{Borrow } 6^1 \text{ from the second column.}$$
$$-23_6 \qquad (6+2)-3=5$$
$$\overline{\qquad 5_6} \qquad \text{No conversion necessary.}$$

$$\overset{1}{\cancel{2}}\overset{6+0}{\cancel{1}}\overset{6+}{2}_6 \qquad \text{Borrow } 6^1 \text{ from the third column.}$$
$$-2\ 3_6 \qquad (6+0)-2=4$$
$$\overline{\qquad 45_6} \qquad \text{No conversion necessary.}$$

$$\overset{1}{\cancel{2}}\overset{6+0}{\cancel{1}}\overset{6+}{2}_6 \qquad 1-0=1$$
$$-\ 2\ 3_6 \qquad \text{No conversion necessary.}$$
$$\overline{145_6}$$

Thus, $212_6 - 23_6 = 145_6$.

✔ Skill Check 7.3.4

Subtract 121_3 from 211_3.

Example 7.3.7

Multiplying Two Numbers of Like Base

a. Multiply 532_7 and 6_7.

b. Multiply 102_3 by 21_3.

Solution

a. Begin by aligning the numbers for long-hand multiplication.

$$\overset{1}{5}32_7 \qquad 6\times 2 = 12_{10}$$
$$\times\ \ 6_7 \qquad \text{Convert } 12_{10} \text{ to a base 7 number: } 12_{10} = (1\times 7^1)+(5\times 7^0) = 15_7$$
$$\overline{\qquad 5_7} \qquad \text{Write 5 and carry the 1.}$$

$$\overset{2\ 1}{53}2_7 \qquad (6 \times 3) + 1 = 19_{10}$$
$$\underline{\times\ 6_7} \qquad \text{Convert } 19_{10} \text{ to a base 7 number: } 19_{10} = (2 \times 7^1) + (5 \times 7^0) = 25_7$$
$$55_7$$

$$\overset{2\ 1}{53}2_7 \qquad (6 \times 5) + 2 = 32_{10}$$
$$\underline{\times\ 6_7} \qquad \text{Convert } 32_{10} \text{ to a base 7 number: } 32_{10} = (4 \times 7^1) + (4 \times 7^0) = 44_7$$
$$4455_7$$

Thus, 532_7 multiplied by 6_7 is 4455_7.

b. Begin by aligning the numbers for long-hand multiplication.

$$102_3$$
$$\underline{\times 21_3}$$

We multiply here just like long-hand multiplication in base 10 that we are used to. Beginning with the number on the far right-hand side, multiply each digit in the bottom row by each digit in the top row.

$$102_3 \qquad 1 \times 2 = 2_3$$
$$\underline{\times 21_3} \qquad \text{No conversion necessary.}$$
$$2_3$$

$$102_3 \qquad 1 \times 0 = 0_3$$
$$\underline{\times 21_3} \qquad \text{No conversion necessary.}$$
$$02_3$$

$$102_3 \qquad 1 \times 1 = 1_3$$
$$\underline{\times 21_3} \qquad \text{No conversion necessary.}$$
$$102_3$$

Repeat the process for the digit 2 in 213, inserting a place holder 0, just as we would in base 10.

$$102_3 \qquad \text{Insert place holder 0.}$$
$$\underline{\times 21_3}$$
$$102_3$$
$$0_3$$

$$\overset{1}{1}02_3 \qquad 2 \times 2 = 4_{10}$$
$$\underline{\times 21_3} \qquad \text{Convert } 4_{10} \text{ to a base number: } 4_{10} = (1 \times 3^1) + (1 \times 3^0) = 11_3$$
$$102_3$$
$$10_3 \qquad \text{Write 1 and carry the 1.}$$

$$\overset{1}{1}02_3 \qquad (2 \times 0) + 1 = 1_3$$
$$\underline{\times 21_3} \qquad \text{No conversion necessary.}$$
$$102_3$$
$$110_3$$

$$102_3 \quad 2 \times 1 = 2_3$$
$$\underline{\times\ 21_3} \quad \textbf{No conversion necessary.}$$
$$102_3$$
$$2110_3$$

Now we can add the two products together to find the answer, converting any values back into base 3, if necessary.

$$102_3$$
$$\times 21_3$$
$$102_3$$
$$\underline{+2110_3}$$
$$2212_3$$

So 102_3 multiplied by 21_3 is equal to 2212_3.

✔ Skill Check 7.3.5

Multiply $D1_{16}$ and 8_{16}

Skill Check Answers

1. 108

2. $111100100011000000_2 = \left(1 \cdot 2^{17}\right) + \left(1 \cdot 2^{16}\right) + \left(1 \cdot 2^{15}\right) + \left(1 \cdot 2^{14}\right) + \left(1 \cdot 2^{11}\right) + \left(1 \cdot 2^{7}\right) + \left(1 \cdot 2^{6}\right)$
$$= 131,072 + 65,536 + 32,768 + 16,384 + 2048 + 128 + 64$$
$$= 248,000$$

$3C8C0_{16} = \left(3 \cdot 16^4\right) + \left(12 \cdot 16^3\right) + \left(8 \cdot 16^2\right) + \left(12 \cdot 16^1\right)$
$$= 196,608 + 49,152 + 2048 + 192$$
$$= 248,000$$

3. 1000_2 4. 20_3 5. 688_{16}

7.3 Exercises

✔ CONCEPT CHECK

1. The Hindu-Arabic numeration system is in base _____.

2. The number 1348 is written in base _____.

3. The hexadecimal system uses _____ to represent the digits 10 through 15.

4. Numbers in base 2 and base 16 are often used in _____ _____.

5. In the Hindu-Arabic numeration system, the 4th digit represents the _____ place. In the base 2 system, the 4th digit represents the ___ place.

> ## ⚡ PRACTICE

Convert each number to base 10.

6. 201_3

7. 4444_5

8. 11011_2

9. 1010101010100_2

10. 5301_6

11. 1001_8

12. AB_{16}

13. $6D8_{16}$

14. Which is larger, 1111001000110001100_2 or $A9CFE_{16}$?

15. Which is larger, BB_{16} or 60_{32}?

Convert each base 10 number to the indicated base.

16. 52 to base 2

17. 1000 to base 3

18. 2000 to base 5

19. 48 to base 12

20. 66 to base 12

21. 700 to base 2

22. 256 to base 8

23. 542 to base 16

Convert each base 2 number to base 16.

24. 10001_2

25. 111011_2

26. 1010101101_2

27. 11100101100_2

28. 1000100010101100_2

29. 1011000000111000_2

Covert each base 16 number to base 2.

30. 17_{16}

31. $3D_{16}$

32. CAB_{16}

33. FED_{16}

34. $A79B2_{16}$

35. $3D095_{16}$

Perform the indicated operations.

36. Add the numbers 155_8 and 72_8.

37. Add the numbers 63_7 and 51_7.

38. Add the numbers 221_5 and 24_5.

39. Add the numbers 101_2 and 111_2.

40. Add the numbers 1222_3 and 201_3.

41. Add the numbers 45231_6 and 1432_6.

42. Subtract the number 10_8 from 34_8.

43. Subtract the number 41_9 from 324_9.

44. Subtract the number 110011_2 from 110110110_2.

45. Subtract the number $42A_{16}$ from $C0F_{16}$.

46. Subtract the number 506_7 from 2135_7.

47. Subtract the number 2243_5 from 4000_5.

48. Multiply 155_7 and 3_7.

49. Multiply 22_3 and 10_3.

50. Multiply 510_8 and 10_8.

51. Multiply 12_4 and 31_4.

52. Multiply 24_5 and 11_5.

53. Multiply $3A_{16}$ and 4_{16}.

54. An addition table for numbers in base 5 is shown. Fill in the missing numbers. Assume all numbers in the table are in base 5.

+ Base 5	0	1	2	3	4
0	0	1	2	3	4
1	1				
2	2				11
3	3		10		
4	4				

55. An addition table for numbers in base 8 is shown. Fill in the missing numbers. Assume all numbers in the table are in base 8.

+ Base 8	0	1	2	3	4	5	6	7
0	0	1	2	3	4	5	6	7
1			3		5			
2	2							
3		4					11	
4		5		7	10			
5	5							
6		7						
7		10						

56. An addition table for numbers in base 12 is shown. Fill in the missing numbers. Assume all numbers in the table are in base 12, which uses the digits 0, 1, 2, 3, 4, 5, 6, 7, 8, 9, A, B.

+ Base 12	0	1	2	3	4	5	6	7	8	9	10	11
0	0	1	2	3	4	5	6	7	8	9	A	B
1		2										
2			4									
3			5									
4			6				A					
5			7		9							
6			8									15
7		8			B							
8				B								
9									15			
10	A											
11								16				

57. A multiplication table for numbers in base 4 is shown. Fill in the missing numbers. Assume all numbers in the table are in base 4.

× Base 4	0	1	2	3
0	0	0	0	0
1	0		2	
2	0			12
3	0	3		

58. The following table is a multiplication table for numbers in base 5. Fill in the missing numbers. Assume the numbers in the table are in base 5.

× Base 5	0	1	2	3	4
0	0	0	0	0	0
1	0				
2	0		4		13
3	0	3			
4				22	

59. A multiplication table for numbers in base 7 is shown. Fill in the missing numbers. Assume all numbers in the table are in base 7.

× Base 7	0	1	2	3	4	5	6
0	0	0			0	0	0
1			2	3			
2							
3	0				15		
4		4					
5		5				34	
6	0						

✎ WRITING & THINKING

60. Describe one way to convert a number in base 2 to base 16.

61. A piece of an addition table that was in the process of being completed is shown. Based on the information that you have, determine which base is being used for the addition. Assume all numbers in the table are in the necessary base.

	2	3	4
	2	3	4
	3	4	

2	0		4	5	10	
3	0	3	5	10	11	
			10	11	12	

62. A piece of an addition table that was in the process of being completed is shown. Based on the information that you have, determine which base is being used for the addition. Assume all numbers in the table are in the necessary base.

	5	6	7
		6	7
	6	7	8
			10
	8		11
	10		12
		12	

63. A piece of a multiplication table that was in the process of being completed is shown. Based on the information that you have, determine which base is being used for the multiplication. Assume all numbers in the table are in the necessary base.

	4	5	6
		0	0
	4	5	6
	10	12	14
	20		

7.3 PROJECT

THE BAKER'S DOZEN

In the sixteenth century, bakers in the United Kingdom who sold their goods by the dozen (12 items) were obligated to meet specific weight and quality standards. Failing to do so was considered a crime. To avoid punishment, it became a common practice to include an additional item with the dozen purchased to assure the law was properly obeyed. This became known as the baker's dozen.

One particular baker served six customers: Anne, Elinor, Giles, Lancelot, Rose, and Florence. The baker placed their orders on the counter and asked them to verify that their orders were correct and that the orders were, indeed, a baker's dozen each.

Anne nodded her head and confirmed, "Yes. 13." The other five nodded as well. Elinor said, "15 exactly!" Giles smiled, "23 for me. Perfect." Lancelot added, "1101 here. Thank you!" Rose happily counted, "10 for me! Yum!" Finally, Florence, about to sample a morsel, gave a thumbs up and said, "11. I love a baker's dozen!"

The baker looked confused. He knew he had put exactly the same number of items in each customer's box. As the pleased clients left the bakery, the solution to the puzzle dawned on him.

1. Explain the apparent contradiction with the numbers and how each customer arrived at their count. Be sure to support your explanation with useful mathematical expressions and equations.

Later that day, the baker asked his assistant to count the remaining inventory and determine how many items they sold during the day. The assistant counted that there were 22 items left in the inventory. The baker knew they started with 175 items and said, "We've sold 159 items today!" His assistant replied with "No! We sold 243 today."

2. Explain how both the baker and his assistant are correct.

OBJECTIVES

1. Convert units using conversion factors.

2. Convert units within the metric system.

OBJECTIVE 1 Conversion Factors

The United States uses a combination of two measurement systems—the US customary system and SI, the International System of Units (or metric system). The US customary system was a continuation of normality from England at the founding of the United States. It is based on the Imperial system, and it measures length in feet, yards, and miles; mass in ounces and pounds; and temperature in Fahrenheit.

If you grew up with this measurement system, you understand what a gallon looks like thanks to milk jugs and have an idea of how long a foot is thanks to school rulers. Table 7.4.1 displays many of the commonly used measurements and equivalents in the United States.

Table 7.4.1: Measurements Used in the US Customary System

Units of Length	Units of Mass
12 inches (in.) = 1 foot (ft)	16 ounces (oz) = 1 pound (lb)
36 inches = 1 yard (yd)	2000 pounds = 1 ton (T)
3 feet = 1 yard	
5280 feet = 1 mile (mi)	

Units of Capacity	Units of Time
8 fluid ounces (fl oz) = 1 cup (c)	60 seconds (sec) = 1 minute (min)
2 cups = 1 pint (pt) = 16 fl oz	60 minutes = 1 hour (hr)
2 pints = 1 quart (qt)	24 hours = 1 day
4 quarts = 1 gallon (gal)	7 days = 1 week

> **Helpful Hint**
>
> Mass and weight are often referred to interchangeably in our everyday speech. However, scientifically, they are not the same. Weight is actually defined as mass with the effects of gravity pulling on it and it is measured in newtons. If your bathroom scale says you "weigh" 120 pounds, that is really your mass; you actually weigh 533.8 newtons.

Converting between units in the US system is a little cumbersome, but very doable as long as you can remember the facts shown in Table 7.4.1. Using *conversion factors*, we can convert from one unit to another. Recall that a **conversion factor** is a fraction that is equal to 1 but contains different units in the numerator and the denominator. For instance, the following are all conversion factors.

$$\frac{12 \text{ in.}}{1 \text{ ft}} = 1 \qquad \frac{1 \text{ yd}}{3 \text{ ft}} = 1 \qquad \frac{1 \text{ c}}{8 \text{ fl oz}} = 1 \qquad \frac{24 \text{ hr}}{1 \text{ day}} = 1$$

> **Helpful Hint**
>
> Conversion factors are sometimes called *unit fractions* since the numerator and denominator are equivalent values.

> **Think Back**
>
> Recall that fractions are of the form $\frac{\text{numerator}}{\text{denominator}}$.

CONVERSION FACTOR

A **conversion factor** is a fraction equal to 1 that contains different units in the numerator and denominator.

When converting measurements from one unit to another, we can multiply by conversion factors. Because conversion factors are, by definition, always equal to 1, multiplying by them does not change the value of the original expression.

Choosing the most useful conversion factor is the key to easy conversions—units should be placed in the numerator or denominator so that they cancel where needed. For instance, if you want to change 4 yards to feet, you should choose the conversion factor $\frac{3 \text{ ft}}{1 \text{ yd}}$ so that yards cancel out, leaving only feet. However, if you want to change 5 feet into yards, you should use the reciprocal of that conversion factor, which is $\frac{1 \text{ yd}}{3 \text{ ft}}$.

Example 7.4.1

Using Conversion Factors to Convert between US Units of Measure

Convert each measurement to the desired unit of measure using conversion factors.

a. If an extra-large bag of Halloween candy weighs 76.2 ounces, how many pounds of candy are in the bag? Round your answer to the nearest tenth of a pound.

b. A recipe calls for $\frac{1}{3}$ cup of milk. How many fluid ounces would you need?

Solution

a. Begin by choosing a unit fraction that has the desired units in the numerator and the original units in the denominator. For this conversion, we'll need pounds (lb) in the numerator and ounces (oz) in the denominator.

   Using Table 7.4.1, the conversion factor is $\frac{1 \text{ lb}}{16 \text{ oz}} = 1$.

   Convert the value by multiplying the original measurement by the conversion factor.

   $$76.2 \text{ oz} = 76.2 \cancel{\text{ oz}} \cdot \frac{1 \text{ lb}}{16 \cancel{\text{ oz}}} = 4.7625 \text{ lb}$$

   Therefore, there are approximately 4.8 pounds of candy in the bag.

b. We first need to choose an appropriate conversion factor. For this conversion, we'll need fluid ounces in the numerator and cups in the denominator. Using Table 7.4.1, we can write the conversion factor as follows.

   $$\frac{8 \text{ fl oz}}{1 \text{ cup}} = 1$$

   Multiply the original measurement by the conversion factor.

   $$\frac{1}{3} \text{ cup} = \frac{1}{3} \cancel{\text{ cup}} \cdot \frac{8 \text{fl oz}}{1 \cancel{\text{ cup}}} = 2.66\overline{6} \text{ fl oz}$$

   Therefore, approximately 2.7 fluid ounces of milk are needed for the recipe.

✔ Skill Check 7.4.1

Convert 30 inches to feet.

These days, we can employ the assistance of technology by using unit converter apps on phones, typing in a web search, or even just asking aloud for a smart device to convert measurements for us. However, one has to wonder where such an unwieldy system came from. Once upon a time there was a hierarchy to measurements in the Imperial system, but somewhere along the line, we've stopped using many of the intermediate measures. Without them, conversions lack a sense of cohesiveness. We recognize that 5280 feet make a mile, but one might ask, quite reasonably, "Why that amount?"

Well, the intermediate measures that we've lost might tell us.

12 inches in a foot

3 feet in a yard

22 yards in a chain

10 chains in a furlong

8 furlongs in a mile, and

3 miles in a league

Now we can see along the string of measurements that 5280 feet = 1 mile.

$$\frac{3 \text{ feet}}{1 \text{ yard}} \cdot \frac{22 \text{ yard}}{1 \text{ chain}} \cdot \frac{10 \text{ chain}}{1 \text{ furlong}} \cdot \frac{8 \text{ furlong}}{1 \text{ mile}} = \frac{5280 \text{ feet}}{1 \text{ mile}}$$

While this equivalence makes more sense now, it unfortunately does not ease the conversion process.

OBJECTIVE 2 The Metric (or SI) System

In contrast to the US customary system, the International System of Units (SI) bases the transition between all intermediate measurements on multiples of 10. The SI, commonly referred to as the metric system, was created in 1960 by the General Conference on Weights and Measures and is the preferred measurement system around the world. Using the same standard prefixes to indicate multiples of ten, the metric system allows easy conversion between base measurements. Table 7.4.2 displays the most common prefixes used in the metric system.

Table 7.4.2: Common Metric Prefixes for Any Base Unit

Prefix	Multiple	Example (Using Meters)
kilo (k)	Base unit × 1000	1 kilometer (km) = 1000 m
hecto (h)	Base unit × 100	1 hectometer (hm) = 100 m
deka (da)	Base unit × 10	1 dekameter (dam) = 10 m
	Base unit	1 meter (m)
deci (d)	Base unit × 0.1	1 decimeter (dm) = 0.1 m
centi (c)	Base unit × 0.01	1 centimeter (cm) = 0.01 m
milli (m)	Base unit × 0.001	1 millimeter (mm) = 0.001 m

No matter what you're measuring with the metric system (length, mass, or volume, for instance), the prefixes are always the same. Only the base unit of measurement changes. For length, the base unit is meters (m). For mass, it is grams (g). And for capacity, the base unit is liters (L). Table 7.4.3 displays these common base units for the metric system.

Table 7.4.3: Common Units in the Metric System

Prefix	Length	Mass	Capacity
kilo- (1000)	kilometer (km) = 1000 meters	kilogram (kg)	kiloliter (kL)
hecto- (100)	hectometer (hm) = 100 meters	hectogram (hg)	hectoliter (hL)
deka- (10)	dekameter (dam) = 10 meters	dekagram (dag)	dekaliter (daL)
Base unit	meter (m)	gram (g)	liter (L)
deci- $\left(\frac{1}{10}\right)$	decimeter (dm) = 0.1 meters	decigram (dg)	deciliter (dL)
centi- $\left(\frac{1}{100}\right)$	centimeter (cm) = 0.01 meters	centigram (cg)	centiliter (cL)
milli- $\left(\frac{1}{1000}\right)$	millimeter (mm) = 0.001 meters	milligram (mg)	milliliter (mL)

Because the metric system is in base 10, converting between different units in one dimension is as easy as moving decimal places. Think of moving along a number line for the conversions. Figure 7.4.2 shows a conversion line for the metric system.

```
  +------+------+------+------+------+------+------+
 kilo-  hecto-  deka-  Base   deci-  centi-  milli-
  (k)    (h)    (da)   Unit   (d)    (c)     (m)
```

Figure 7.4.2

When you convert from a smaller to a larger unit in the metric system, the decimal moves to the left, and when you convert from a larger to a smaller unit, the decimal point moves to the right.

Example 7.4.2

Converting within the Metric System in One Dimension

Convert each metric unit to the appropriate measurement requested.

a. The men's world record for running 1500 meters was set in July 1998 at 3 min 26 secs by Hicham El Guerrouj. What is this distance in kilometers?

b. A cube of sugar weighs approximately 3 grams. How many centigrams does the cube of sugar weigh?

Solution

a. Notice that we are changing from a smaller unit to a larger unit, so the decimal will move to the left. In order to change 1500 meters into kilometers, we move the decimal 3 places to the left.

$$1500 \text{ m} = 1.5 \text{ km}$$

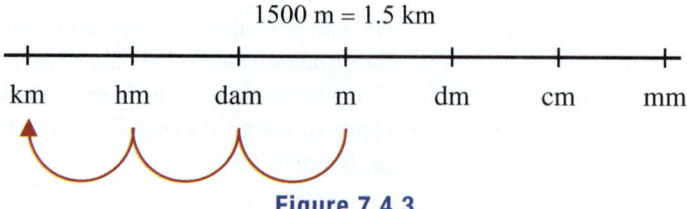

Figure 7.4.3

b. Notice that we are changing from a larger unit to a smaller one, so the decimal will move to the right. In order to change 3 grams into centigrams, we move the decimal 2 places to the right.

$$3 \text{ g} = 300 \text{ cg}$$

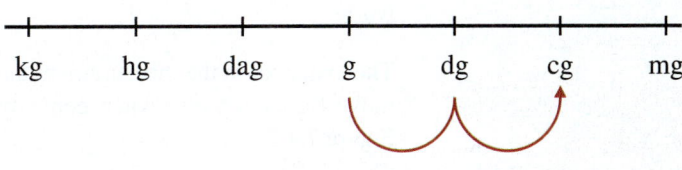

Figure 7.4.4

As we saw with Example 7.4.2, conversions in the metric system (where we simply move the decimal) are much more convenient than conversions within the US customary system (where we need the help of conversion factors). At the same time, we need to be cautious to not just jump to the conclusion that all metric conversions involve moving the decimal one place for each prefix. Let's look at what happens when it comes to converting measurements in more than one dimension, such as area or volume.

Think Back

$Area_{rectangle} = length \cdot width$

$Volume_{rectangular\ box} = length \cdot width \cdot height$

Take for instance, the area of the rectangular space in Figure 7.4.5.

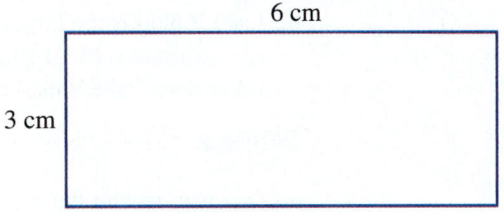

Figure 7.4.5

Area = 3 cm · 6 cm = 18 cm²

Suppose we want to convert to the units of area to mm² instead of cm². One option is to convert the units before the area is calculated. That is, we could choose to change centimeters to millimeters before multiplying the length by the width. Since we know that there are 10mm in each centimeter, we get the measurements shown in Figure 7.4.6.

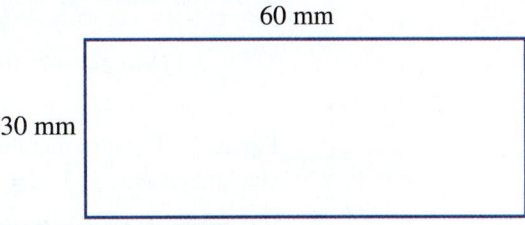

Figure 7.4.6

Area = 30 mm · 60 mm = 1800 mm²

Notice that the area calculation must account for a multiple of 10 in both the length and width when converting. So the decimal actually moves 2 places (not one) when converting cm² to mm².

Here's how that works using conversion factors for 1 cm = 10 mm.

$$18 \text{ cm}^2 = 18 \text{ cm}^2 \cdot \frac{10 \text{ mm}}{1 \text{ cm}} \cdot \frac{10 \text{ mm}}{1 \text{ cm}} = 1800 \text{ mm}^2$$

Note:

The exponent in the measurement indicates the number of places the decimal will move for each prefix when converting within the metric scale, as illustrated in Figure 7.4.7.

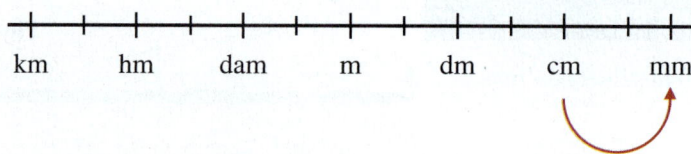

Figure 7.4.7

Example 7.4.3

Converting within the Metric System in Two or Three Dimensions

Convert each metric unit to the appropriate measurement requested.

a. Find the area in km² of a soccer field that measures 105 m by 68 m.

b. Lake Baikal is the largest freshwater lake in the world by volume. It contains approximately 23 615 km³ of fresh water. How many cubic meters of fresh water does Lake Baikal contain?

Solution

a. One way to find the area of the field in km² is to first find the area in m² and then convert to km². To find the area, we multiply the length by the width.

$$\text{Area} = 105 \text{ m} \cdot 68 \text{ m}$$

$$= 7140 \text{ m}^2$$

Converting m² to km² requires a conversion factor comparing kilometers to meters: $\dfrac{1 \text{ km}}{1000 \text{ m}}$. Remember that we need to cancel out meters, so that is the unit to put in the denominator here. Because our unit is m², we need to use two of the conversion factors.

$$7140 \text{ m}^2 = 7140 \text{ m}^2 \cdot \frac{1 \text{ km}}{1000 \text{ m}} \cdot \frac{1 \text{ km}}{1000 \text{ m}} = 0.00714 \text{ km}^2$$

Figure 7.4.8 shows that the decimal shifts two places for each prefix along the conversion line, or $3 \cdot 2 = 6$ places to the left when converting m² to km².

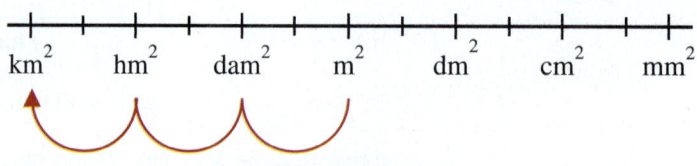

Figure 7.4.8

$$7140 \text{ m}^2 = 0.007140 \text{ km}^2$$

So the area of the soccer field measures 0.00714 km².

b. To determine how many cubic meters of freshwater Lake Baikal contains, we must convert 23 615 km³ to m³. Notice that we need the same conversion factor comparing kilometers to meters that we used in part a. In this case, we want to cancel out kilometers, so we will put that unit in the denominator and use $\dfrac{1000 \text{ m}}{1 \text{ km}}$ as our factor. We will also need to use the conversion factor three times since our units are cubed.

$$23\ 615 \text{ km}^3 = 23\ 615 \text{ km}^3 \cdot \frac{1000 \text{ m}}{1 \text{ km}} \cdot \frac{1000 \text{ m}}{1 \text{ km}} \cdot \frac{1000 \text{ m}}{1 \text{ km}}$$
$$= 23\ 615\ 000\ 000\ 000 \text{ m}^3$$

Figure 7.4.9 shows that the decimal shifts 3 places for each prefix along the conversion line, or $3 \cdot 3 = 9$ places to the right when converting kilometers cubed to meters cubed.

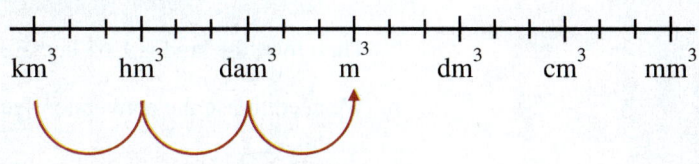

Figure 7.4.9

$$23\ 615 \text{ km}^3 = 23\ 615\ 000\ 000\ 000 \text{ m}^3$$

So Lake Baikal contains approximately 23 615 000 000 000 cubic meters (m^3) of fresh water.

✔ **Skill Check 7.4.3**

Convert 220 cm² to m².

Units of measurement have sometimes arisen as a convenient way to describe commonly used dimensions. For instance, the US customary system has the *acre* to describe area instead of yards squared, and the pint to describe volume instead of cubic inches. Plots of land are commonly sold in acres or parts of acres—not square yards or square miles. Similarly, the metric system has the units *are* for area and *liter* for capacity. Table 7.4.4 shows the equivalence for each of these measurements.

Fun Fact

The term *acre* derives from the Middle English term *aker* and was commonly thought of as the amount of land that could be plowed in one day with oxen pulling a wooden plow.

Table 7.4.4: Measures of Land Area and Capacity

US Customary Measure of Land Area	Metric Measure of Land Area	US Customary Measure of Capacity	Metric Measure of Capacity
1 acre = 4840 yd²	1 are (a) = 100 m²	1 pint = 28.875 in³	1 liter (L) = 1000 cm³
1 acre = 0.0015625 mi²	1 hectare (ha) = 100 a = 10000 m²	1 pint = 0.0167101 ft³	1 mL = 1 cm³

Example 7.4.4

Converting Area and Volume within the Metric System

Convert each metric unit to the appropriate measurement requested.

a. A rectangular piece of land measures 110 m by 150 m. How many hectares is the piece of land?

b. A punch bowl can hold 5500 cm³ of liquid. How many liters of punch can the bowl hold?

Solution

a. To begin, we need to find the area of land by multiplying the lengths together.

$$\text{area} = 110 \text{ m} \cdot 150 \text{ m} = 16\ 500 \text{ m}^2$$

We can now use the conversion factor $\dfrac{1 \text{ ha}}{10\ 000 \text{ m}^2}$ to convert squared meters into hectares.

$$16\ 500 \text{ m}^2 = 16\ 500 \text{ m}^2 \cdot \frac{1 \text{ ha}}{10\ 000 \text{ m}^2} = 1.65 \text{ ha}$$

Therefore, the land is 1.65 hectares.

b. We need to use the conversion factor $\dfrac{1 \text{ L}}{1000 \text{ cm}^3}$ to convert cubic centimeters into liters.

$$5500 \text{ cm}^3 = 5500 \text{ cm}^3 \cdot \frac{1 \text{ L}}{1000 \text{ cm}^3} = 5.5 \text{ L}$$

So, the bowl can hold 5.5 liters of punch.

Skill Check Answers
1. 2.5 feet **2.** 1.5 liters **3.** 0.022 m²

7.4 Exercises

✔ CONCEPT CHECK

1. Conversion factors are always equal to _____.

2. The _____ _____ is the measurement system used in most of the world.

3. The unit for volume in the metric system is the _____.

4. Conversions within the metric system are based on powers of _____.

5. One _____ is equivalent to 100 m².

💡 PRACTICE

Convert each measurement to the desired unit of measure.

6. 2232 in. = _____ yd

7. 13 ft = _____ in.

8. 22 lb = _____ oz

9. 120 oz = _____ lb _____ oz

10. 200 fl oz = _____ cups

11. 5 qt = _____ pt

12. 1008 min = _____ days

13. 4.25 hours = _____ minutes

14. 262 yd = _____ mi

15. 3.5 mi = _____ ft

16. 15.5 gal = _____ pt

17. 4 pt = _____ fl oz

18. 11 acres = _____ yd^2

19. 1000 acres = _____ mi^2

20. How many centimeters are in 364 km?

21. Convert 230 m to kilometers.

22. Convert 0.87 km to centimeters.

23. How many meters are in 123 cm?

24. Convert 21 mm to decimeters.

25. How many kiloliters are in 35 600 cL?

26. Convert 7.22 hL to milliliters.

27. How many grams are in 1.2 kg?

28. Convert 45 422 cg to dekagrams.

29. How many square decimeters are in 726 m^2?

30. Express 40.93 m^2 in square kilometers.

31. How many square centimeters are in 627.2 m^2?

32. How many square millimeters are in 928 cm^2?

33. How many cubic centimeters are in 99 m^3?

34. Express 8503 mm^3 in cubic meters.

35. Convert 345 m^2 to ares.

36. How many square meters are in 1.72 ares?

📈 APPLICATIONS

37. Men ages 19 to 30 are recommended to intake about 13 cups of water a day. Calculate the number of fluid ounces this would be per day. Round your answer to the nearest tenth, if necessary.

38. A Louisiana gumbo recipe calls for 14.5 ounces of stewed tomatoes. How many cups of stewed tomatoes are required? Round your answer to the nearest tenth, if necessary.

39. A king-size mattress is 76 inches wide by 80 inches long. If a queen-size mattress is 5 feet wide and about 6.7 feet long, how much wider (in inches) is the king-size mattress than the queen-size mattress?

40. An online gaming company reports that the average player has spent 832 hours on their sight. Assuming their report is correct, how many days has the average player spent on the gaming sight? Round to the nearest half a day.

41. A local coffee shop is calculating the number of gallons of coffee they sell per day. On average, the shop sells two hundred sixty 16 fl oz cups of coffee a day. How many gallons of coffee per day do they sell?

42. A freight train measures 1.3 miles long when all boxcars are attached. If a standard boxcar is 55 feet 5 inches long, what is the maximum number of boxcars the freight train contains?

43. The outer lane (lane 8) of a running track is 453.66 meters compared to the inside lane (lane 1) which has a length of 400 meters. If a runner makes three complete laps around the track in lane 8, how many kilometers has he run? Round your answer to two decimal places.

44. A paper clip is made from galvanized steel wire. When curled, the zpaperclip measures 4.1 cm in length. Uncurled, it measures 108 mm. What is the maximum number of paper clips that can be made from a straight piece of wire that is 100 m long? Round your answer to the nearest number of whole paperclips.

45. Alex is buying a bag of apples that weighs 720 grams. If the apples are sold for $3.34 per kilogram, how much will Alex pay for the bag of apples? Round your answer to the nearest cent.

46. Forty centigrams of caffeine per day is considered a safe amount. If there are 95 mg of caffeine in one cup of coffee, what is the maximum number of cups of coffee you can drink in a day and stay under the safe amount of caffeine?

47. Trey is reupholstering his dining room chairs. He plans to place brass upholstery tacks every 2 cm along the back edge of the chair. If the back measures 4.45 dm in length, how many tacks will he need for each chair?

48. To help encourage the safe use of acetaminophen, the maximum daily dose has been reduced from 4 g per day to 3 g per day. If an extra strength tablet contains 500 mg of acetaminophen, how many tablets make up the new maximum daily dose?

49. A rectangular plot of land measures 180 m by 50 m. How many hectares is the plot of land?

50. A community garden is being planned. The garden will measure 52 m by 23 m. Calculate the size of the garden in ares.

51. A small fish tank holds 19 liters of water. How many cubic centimeters of water does the tank hold?

52. The Sawyers' water heater broke, and they need to purchase a new one. Their current water heater holds 120 000 cm³ of water. If a new water heater is available in capacities of 30 L, 100 L, and 140 L, which size heater should the Sawyers purchase if they want to buy one that is at least as big as the old one.

✏ WRITING & THINKING

53. What metric unit of length would be best to measure the depth of the Pacific Ocean? Explain your reasoning.

54. What metric unit of length would be best to measure the length of a room? Explain your reasoning.

55. What metric unit of length would be best to measure the thickness of a smartphone? Explain your reasoning.

56. What metric unit would be the best choice to measure the mass of a grain of rice? Explain your reasoning.

57. Rank the following mass measurements from smallest to largest:

100.6 mg   1021 g   0.0000006 kg   0.61 mg

58. Rank the following length measurements from smallest to largest:

3.4 dm   0.341 dam   0.0000034 km   34m

7.5 CONVERTING BETWEEN THE US CUSTOMARY SYSTEM AND THE METRIC SYSTEM

OBJECTIVES

1. Use dimensional analysis to convert units between the US customary system and the metric system.

2. Convert between degrees Celsius and degrees Fahrenheit.

Fun Fact

A large amount of physics is finding laws that connect the seven dimensions together to describe our world. For instance, Einstein's equation $E = mc^2$, describes energy in terms of mass, length, and time.

Think Back

A **conversion factor** is a fraction that is equal to 1 that contains different units in the numerator and denominator.

OBJECTIVE 1 Dimensional Analysis

In geometry, we use *dimensions* to describe objects spatially by the number of measurements of length in independent directions. For instance, a cube is a 3-dimensional object because we measure its length, width, and depth. And a rectangle is 2-dimensional object only having measurements of length and width. However, in the field of physics, dimensions are understood in the physical sense as well as the spatial sense. Dimensions are the metrics that physicists use to describe reality. In the world of physics, there are seven distinct fundamental physical dimensions: length, mass, time, temperature, current (measured in amperes), amount (measured in moles), and luminosity. These are distinct dimensions in the sense that no amount of time can add up to any amount of length. Likewise, no amount of temperature can add up to any amount of mass. These fundamental physical dimensions are what we refer to when we talk about dimensional analysis. In this section we will consider length, mass, and temperature.

When quantities are of the same physical dimension, we can compare them even if they are expressed in different units of measure. For instance, time measured in hours can also be described in minutes or milliseconds, and temperature can be measured in degrees Celsius and then converted to degrees Fahrenheit. We can also apply this **dimensional analysis** to convert between the US customary system and the metric system. By applying the rules of algebra with conversion factors, we are able to change from one unit of measurement to another.

📖 DIMENSIONAL ANALYSIS

Dimensional analysis is a process by which the relationship between different physical quantities of the same dimension can be compared by converting between units of measurement using the rules of algebra.

In order to use dimensional analysis to convert between the US customary system and the metric system, we require conversion factors. Historically, there were customary units for many of the physical dimensions. While the use of some have fallen into disuse, Table 7.5.1 provides a few of the most common equivalences and conversion factors of measures still prevalent today. Remember, when creating conversion factors, it's important to appropriately place units in the numerator and denominator so the units cancel as intended and the answer has the desired units.

Table 7.5.1: Commonly Used Equivalences

	US to Metric	Metric to US
Length	1 in. = 2.54 cm	1 cm ≈ 0.394 in.
	1 ft ≈ 0.305 m	1 m ≈ 3.28 ft
	1 yd ≈ 0.914 m	1 m ≈ 1.09 yd
	1 mi ≈ 1.61 km	1 km ≈ 0.62 mi
Area (Length²)	1 in.² ≈ 6.45 cm²	1 cm² ≈ 0.155 in.²
	1 ft² ≈ 0.093 m²	1 m² ≈ 10.764 ft²
	1 yd² ≈ 0.836 m²	1 m² ≈ 1.196 yd²
	1 mi² ≈ 2.6 km²	
	1 acre ≈ 0.405 ha	1 ha ≈ 2.47 acres
Volume (Length³)	1 tsp ≈ 5 mL	1 L ≈ 61.02 in.³
	1 tbsp ≈ 15 mL	
	1 c ≈ 0.24 L	
	1 pint ≈ 0.47 L	
	1 qt ≈ 0.946 L	1 L ≈ 1.06 qt
	1 gal ≈ 3.785 L	1 L ≈ 0.264 gal
Mass	1 oz ≈ 28.35 g	1 g ≈ 0.035 oz
	1 lb ≈ 0.454 kg	1 kg ≈ 2.205 lb

We'll begin converting between the US customary and metric systems with cases where only one unit of measure needs to be changed. Because these first examples work with a single unit conversion, the conversion factors we construct will be such that the numerator contains the units desired and the denominator contains the units we wish to remove. This will ensure that the units cancel out correctly.

Example 7.5.1

Applying Dimensional Analysis with One Measurement

Convert each measurement using the appropriate conversion factor. Round to the nearest hundredth.

a. 137 pounds to kilograms

b. 420 kilometers to miles

c. 2.8 cubic feet to cubic meters

Solution

a. To convert pounds to kilograms, the conversion factor should contain pounds in the denominator and kilograms in the numerator. Table 7.5.1 gives two equivalences that we could use for a conversion factor: $\dfrac{0.454 \text{ kg}}{1 \text{ lb}}$ or $\dfrac{1 \text{ kg}}{2.205 \text{ lb}}$.

Both are appropriate to use but may produce slightly different answers due to rounding. We will use $\dfrac{0.454 \text{ kg}}{1 \text{ lb}}$.

Notice that since the equivalence we are using is an approximation, our conversion is also an approximation. Hence, our calculations begin with ≈ instead of =.

$$137 \text{ lb} \approx 137 \cancel{\text{lb}} \cdot \frac{0.454 \text{ kg}}{1 \cancel{\text{lb}}} = 62.198 \text{ kg} \approx 62.20 \text{ kg}$$

Therefore, 137 pounds is approximately equal to 62.2 kilograms.

b. To change from kilometers to miles we can use the conversion factor $\frac{0.62 \text{ mi}}{1 \text{ km}}$.

$$420 \text{ km} \approx 420 \cancel{\text{km}} \cdot \frac{0.62 \text{ mi}}{1 \cancel{\text{km}}} = 260.40 \text{ mi}$$

Therefore, 420 kilometers is approximately equal to 260.4 miles.

c. There is not a conversion for cubic feet to cubic meters in Table 7.5.1. However, we can still use the conversion for feet to meters $\frac{0.305 \text{ m}}{1 \text{ ft}}$. Notice that we need to account for the multiple dimensions but are still only changing one unit of measure.

$$2.8 \text{ ft}^3 \approx 2.8 \cancel{\text{ft}^3} \cdot \frac{0.305 \text{ m}}{1 \cancel{\text{ft}}} \cdot \frac{0.305 \text{ m}}{1 \cancel{\text{ft}}} \cdot \frac{0.305 \text{ m}}{1 \cancel{\text{ft}}} \approx 0.079443 \text{ m}^3 \approx 0.08 \text{ m}^3$$

Therefore, 2.8 cubic feet is approximately equal to 0.08 cubic meters.

> ✔ **Skill Check 7.5.1**
>
> Convert 4.3 meters to yards. Round to the nearest hundredth.

Sometimes when we want to convert a single unit of measure to another, the direct conversion quantities are not readily available. Because listing all possible conversions between two systems would create a lengthy table, we are often only given frequently used conversions, such as the ones listed in Table 7.5.1. However, we can use these common conversion factors as stepping-stones to get us from any unit of measure to another. Example 7.5.2 illustrates this.

Example 7.5.2

Converting between Systems When Conversion Factors Are Unknown

Convert 3 cups to milliliters.

Solution

Although Table 7.5.1 does not have an equivalence for cups to milliliters, it does have a conversion for cups to liters. From there, we can use a conversion from liters to milliliters. So we will need to use two conversion factors: $\frac{0.24 \text{ L}}{1 \text{ c}}$ and $\frac{1000 \text{ mL}}{1 \text{ L}}$.

$$3 \text{ c} \approx 3 \cancel{\text{c}} \cdot \frac{0.24 \cancel{\text{L}}}{1 \cancel{\text{c}}} \cdot \frac{1000 \text{ mL}}{1 \cancel{\text{L}}} = 720 \text{ mL}$$

So 3 cups is approximately equal to 720 milliliters.

> ✔ **Skill Check 7.5.2**
>
> Convert 1.5 meters to inches. Round to the nearest hundredth.

Often, more than one unit of measure needs to be converted. For instance, physicists investigating situations involving speed can take measurements in meters per second and might need to convert them to feet per minute; or a chemist measuring the density of objects in pounds per cubic foot might need to convert these values to kilograms per meter cubed. Regardless of the number of units used in a measurement, we employ the same strategy and use multiple conversion factors to switch between systems.

Example 7.5.3

Applying Dimensional Analysis with More than One Measurement

Convert each measurement using the appropriate conversion factors. Round to the nearest hundredth.

a. 32 ft/sec to m/min

b. 2.09 g/mL to lb/gal

Solution

a. To begin converting 32 ft/sec to m/min, notice that the numerator needs to change from feet to meters and the denominator needs to change from seconds to minutes. To change feet to meters, we can use the conversion factor $\dfrac{1 \text{ m}}{3.28 \text{ ft}}$.

(Remember that the measurement you would like to end up with needs to be in the numerator.)

To change seconds to minutes, we can use the conversion factor for 1 min = 60 sec. However, we need to stop and think about how this conversion factor needs to be oriented. Because the unit we want to change (seconds) is in the denominator, we will need to reverse our conventional fraction setup. We will put seconds in the numerator, so the seconds will cancel out, and minutes in the denominator. So the conversion factor to use is $\dfrac{60 \text{ sec}}{1 \text{ min}}$.

Now we can multiply with both conversion factors.

$$32 \text{ ft/sec} \approx \frac{32 \text{ ft}}{1 \text{ sec}} \cdot \frac{1 \text{ m}}{3.28 \text{ ft}} \cdot \frac{60 \text{ sec}}{1 \text{ min}}$$

$$= \frac{1920 \text{ m}}{3.28 \text{ min}}$$

$$\approx 585.3659 \text{ m/min}$$

So 32 ft/sec is approximately 585.37 m/min.

b. 2.09 g/mL to lb/gal

Notice that we need to convert grams to pounds in the numerator and milliliters to gallons in the denominator.

Although we don't have a direct conversion for grams to pounds in Table 7.5.1, we can use two conversions to get us there: grams to ounces and ounces to pounds.

To convert grams to ounces in the numerator, we need grams to be in the denominator of the conversion factor. Therefore, we can use $\dfrac{0.035 \text{ oz}}{1 \text{ g}}$ for grams to ounces. Then, to change ounces to pounds, we need ounces in the denominator of the fraction: $\dfrac{1 \text{ lb}}{16 \text{ oz}}$. These two fractions will take care of the numerator. Now we need to consider the denominator.

Again, we don't have a direct conversion for milliliters to gallons, but we can use milliliters to liters and liters to gallons. Since milliliters is in the denominator to begin with, the conversion factor will need to have milliliters in the numerator: $\dfrac{1000 \text{ ml}}{1 \text{ L}}$. Once we have that first conversion factor, we then know to put liters in the numerator and gallons in the denominator for the other fraction: $\dfrac{1 \text{ L}}{0.264 \text{ gal}}$.

So the conversion requires that we multiply our original measurement by the 4 conversion factors, as shown.

$$2.09 \text{ g}/\text{mL} \approx \frac{2.09 \cancel{g}}{1 \cancel{mL}} \cdot \frac{0.035 \cancel{oz}}{1 \cancel{g}} \cdot \frac{1 \text{ lb}}{16 \cancel{oz}} \cdot \frac{1000 \cancel{mL}}{1 \cancel{L}} \cdot \frac{1 \cancel{L}}{0.264 \text{ gal}}$$

$$\approx \frac{73.15 \text{ lb}}{4.224 \text{ gal}}$$

$$\approx 17.3177 \text{ lb/gal}$$

Therefore, 2.09 g/mL is approximately 17.32 lb/gal.

> ✔ **Skill Check 7.5.3**
>
> Convert 3.42 g/cm to lb/in. Round to the nearest hundredths.

Example 7.5.4

Applying Dimensional Analysis

Two soccer balls are inflated. One ball has a radius of 11 cm and a volume of approximately 5575.28 cm³. The other ball has a radius of 4.3 in., giving it a volume of approximately 333.04 in.³ What is the difference in volume (in cubic centimeters) between the two balls? Round your answer to the nearest hundredth.

Solution

We are asked to find the difference in volume for the two soccer balls in cubic centimeters. Since the volume of the second ball is given in cubic inches, we need to convert its volume to cubic centimeters.

The conversion factor we need to use involves 1 in. = 2.54 cm and places the inches in the denominator so that the inches will cancel out: $\dfrac{2.54 \text{ cm}}{1 \text{ in.}}$. Since we are changing cubic inches, we will need to use the conversion factor 3 times.

$$333.04 \text{ in.}^3 \approx 333.04 \cancel{\text{ in.}} \cdot \frac{2.54 \text{ cm}}{1 \cancel{\text{ in.}}} \cdot \frac{2.54 \text{ cm}}{1 \cancel{\text{ in.}}} \cdot \frac{2.54 \text{ cm}}{1 \cancel{\text{ in.}}} \approx 5457.55 \text{ cm}^3$$

Therefore, the volume of the second soccer ball is approximately 5457.55 cm³.

We can now find the difference in volume between the two soccer balls.

$$5575.28 \text{ cm}^3 - 5457.55 \text{ cm}^3 = 117.73 \text{ cm}^3$$

The first soccer ball has a volume that is 117.73 cm³ more than the volume of the second ball.

OBJECTIVE 2 Temperature

The final dimension we will convert between the US customary system and the metric system is that of temperature. When you travel from the United States to a different country, you might hear that the temperature for the day is going to be 24 degrees. No, you are not going to freeze; it's just that temperature is measured in Celsius, *not* in Fahrenheit as it is in the United States. Since people are accustomed to hearing temperatures in their own systems, the measurement unit is often dropped from speech. In other words, we don't commonly say "The temperature today will reach a high of 24 degrees Celsius."

A temperature of 32 degrees Fahrenheit is equal to a temperature of 0 degrees Celsius. Because zero degrees Celsius is not equal to zero degrees Fahrenheit, we are not able to simply multiply by a conversion factor to change from one measurement to the other. We must also add or subtract 32 to adjust for the different placing of zero on the temperature scales. The conversion formulas help us remember to include the adjustment factor between temperature scales.

> **Fun Fact**
>
> Where do temperature scales set their zero points?
>
> 0° Celsius = water freezes
>
> 0° Kelvin = absence of heat (degrees scaled like Celsius)
>
> 0° Fahrenheit = coldest mixture of ice, salt, and water Daniel Fahrenheit could produce in his lab at the time
>
> 0° Rankine = absence of heat (degrees scaled like Fahrenheit)

𝑓(𝑥) TEMPERATURE FORMULAS

The following formulas are used to convert temperature between degrees Celsius and degrees Fahrenheit, where F = Fahrenheit temperature and C = Celsius temperature.

$$F = \frac{9C}{5} + 32 \qquad C = \frac{5(F-32)}{9}$$

Example 7.5.5

Converting Temperature between Celsius and Fahrenheit

a. Whitley is headed to Brazil and is told that when she gets there, the temperature will be 27 °C. Convert the temperature to Fahrenheit for Whitley so that she can pack appropriately for her trip.

b. Once in Brazil, Whitley would like to make her favorite meal for her host family. The recipe calls for the oven to be at 425 °F. What temperature should Whitley set the oven to if the dial shows only Celsius?

Solution

a. Whitley must use the formula for converting degrees Celsius to Fahrenheit:

$$F = \frac{9C}{5} + 32.$$

$$F = \frac{9 \cdot 27}{5} + 32 \quad \text{Make the appropriate substitution.}$$

$$F = 48.6 + 32 \quad \text{Simplify.}$$

$$F = 80.6$$

So Whitley knows it will be warm when she arrives in Brazil.

b. To convert to Celsius, Whitley must use the formula $C = \dfrac{5(F-32)}{9}$.

$$C = \frac{5(425-32)}{9}$$ Make the appropriate substitution.

$$C = \frac{1965}{9}$$ Simplify.

$$C = 218.3\overline{3}$$

Therefore, Whitley needs to set the oven to approximately 218 °C.

✔ **Skill Check 7.5.4**

Convert 68 °F to °C.

Skill Check Answers
1. 4.70 yards **2.** 59.06 inches **3.** 0.02 lb/in. **4.** 20 °C

7.5   Exercises

✔ CONCEPT CHECK

1. Temperature, time, and length are examples of different physical _____.

2. The relationship between different physical quantities of the same dimension can be compared using _____ _____.

3. Conversion factors are typically _____.

4. The two most common units for measuring temperature are degrees _____ and degrees _____.

5. One inch is equal to exactly _____ centimeters.

💡 PRACTICE

Convert each measurement using the appropriate conversion factor. Round your answer to the nearest thousandth, if necessary. Note that answers may vary depending on the conversion factor used.

6. Convert 52 centimeters to inches.

7. Convert 150 miles to kilometers.

8. What number of yards is equivalent to 800 m?

9. What number of meters is equivalent to 150 ft?

10. Convert 200 ft² to square centimeters.

11. Convert 12 ha to acres.

12. What number of square yards is equivalent to 40 m²?

13. What number of square kilometers is equivalent to 20 mi²?

14. 40 lb = _____ kg

15. 35 g = _____ oz

16. Convert 22 kg to pounds.

17. Convert 42 oz to grams.

18. 16 c = _____ L

19. 6 L = _____ gal

20. Convert 800 in.³ to cubic meters.

21. Convert 100 mL to tablespoons.

22. Convert 198.1 ft to centimeters.

23. What number of inches is equivalent to 760.82 km?

24. 21 ft/sec to m/min

25. 50 km/hr to mi/min

26. 1.35 g/mL to lb/gal

27. 3 oz/cup to g/L

28. 5.22 g/cm to lb/in.

29. 4.6 g/cm to lb/ft

30. 240 ft²/gal to m²/L

31. 25 m²/kg to yd²/lb

32. 45 °C = _____ °F

33. 177 °C = _____ °F

34. −40 °F = _____ °C

35. 275 °F = _____ °C

36. What is 16 °C in degrees Fahrenheit?

37. What is 25 °C in degrees Fahrenheit?

38. Convert 100 °F into degrees Celsius.

39. Convert 12 °F into degrees Celsius.

40. Rank the following mass measurements from smallest to largest:

    2 oz   18 kg   50g   2 lb   35 mg

41. Rank the following length measurements from smallest to largest:

    100 yd   100 m   10 km   10 mi   10 ft   100 cm   100 in.

APPLICATIONS

42. The density of mercury is 13.56 g/mL. What is its density in lb/ft³?

43. The density of gold is 19.32 g/cm³. What is its density in lb/ft³?

44. A toilet flushes with a pressure of 1.6 gal/ft. What is the pressure in SI units (m/L)?

45. A domestic pressure washer sprays water at a rate of 3 GPM (gal/min). What is the spray rate in L/sec?

46. A gas station is charging $2.299 per gallon of gas. What would be the price for a liter of gas?

47. The sign on a freeway in Canada indicates that the speed limit is 100 km/hr. What would the speed limit be in miles per hour?

48. Allison buys embroidery thread for sewing. There are 8.7 yards of thread in each skein. She uses 1 meter. How much thread is left in the skein in meters? Round your answer to the nearest thousandth, if necessary.

49. Suppose it takes John 7 minutes to run 1 mile. How long would it take him to run 3 kilometers? Round your answer to the nearest minute.

50. A website for a new European Low Carb diet provides recipes using metric units of measurement. Convert each ingredient to the indicated measurement. Round to the nearest tenth, if necessary.

    a. 5 mL of vanilla = _____ tsp

    b. 30 g of ground flax = _____ oz

    c. 25 g of almond flour = _____ oz

51. A French recipe for Pain au Chocolat calls for 120 mL of milk and needs the oven preheated to 190 °C. You only have measuring cups in fluid ounces and Fahrenheit on your oven temperature dial. Determine the amount of milk needed in fluid ounces and the temperature for the oven in Fahrenheit. Round your answer to the nearest tenth, if necessary. (**Note:** There are 8 fluid ounces per cup.)

52. The medicine bottle says to take 5 teaspoons every four hours, but your medicine cup only has milliliters. How many milliliters do you need to take?

53. Sandra is on vacation. The hotel clerk at the ski resort tells her that the high temperature for tomorrow is going to be $-7\,°C$. How many degrees Fahrenheit is it going to be?

54. Matt is looking to purchase a new car. The sticker says that the car has an average fuel economy of 18 km/L. How many miles per gallon is that?

55. Logan has made 97 pints of maple syrup. A buyer has offered him $37.28 per gallon. How many gallons does he have made, and how much money will he make if he the buyer purchases all of the maple syrup?

56. The directions for a drink mix say to add 8 teaspoons of sugar for every gallon of liquid. If Sam has 7.6 liters of liquid, how many teaspoons of sugar should he add?

57. Jeff built a model sailboat that can move at a maximum speed of 8 ft/sec. How fast can his sailboat in miles per hour?

58. John's model sailboat can move at a maximum speed of 90 m/min. Whose model sailboat is faster Jeff's or John's, and how much faster is it?

WRITING & THINKING

59. Carlos was trying to make his grandmother's Christmas cinnamon stars called zimtsternes.

    The ingredient conversions calculated by Carlos are shown here.

    350 g unblanched almonds $\rightarrow$ 12 oz unblanched almonds

    400 g icing sugar $\rightarrow$ 13 oz icing sugar

    22.5 mL ground cinnamon $\rightarrow$ 3 tbsp ground cinnamon

    The recipe says to bake them at $140\,°C$ for 20 to 30 minutes. Unfortunately, Carlos did not notice the temperature was given in Celsius and baked them at $140\,°F$. Would you eat these cookies? Explain your answer and include the conversions you based your decision on.

60. When using dimensional analysis with several conversion factors, would it be best to round during intermediate steps or wait until the end of the calculation to round? Explain your reasoning.

7.5 PROJECT

THE IMPORTANCE OF DIMENSIONAL ANALYSIS

As part of a vacation, a group of American friends travel to Toronto, ON, in Canada and rent a car so they can also visit surrounding areas. While in Canada, the group experiences firsthand the usefulness of learning to convert from the metric system to the US customary system. Keep in mind that prices in Canada are in given in Canadian dollars (CAD) and not US dollars (USD). Assume the current exchange rate is 1.00 CAD = 0.80 USD.

1. While renting the car in Toronto, the rental agent tells the group that they can return the car with a full tank of fuel or they can pay 1.999 CAD per liter for the rental company to fill the tank when they return it. Since the current fuel price in the area is 1.279 CAD per liter, the group decides to return the car with a full tank of fuel. The rental car company's policy is to charge the customer for an entire tank of fuel, whether the tank is empty when returned or not. A member of the group looks up the rental car online and learns the fuel tank capacity is 12.5 gallons. How much would the rental company have charged the group to fill the tank with fuel? How much would a full tank of gas have cost the group if they filled it before returning the car?

2. While at a hotel, the group decides to order a pizza from a local pizza parlor. The menu indicates that a small pizza has a diameter of 25 cm, a medium pizza has a diameter of 33 cm, and a large pizza has a diameter of 41 cm. While at home in the states, the group usually orders two 12-inch pizzas. Which pizza size should the group order? Is this an exact match in diameter to the size they usually order?

3. The group plans to drive to Niagara Falls, which is approximately 90 miles from where they are staying in Toronto. The most common speed limit along the route they will take is 100 km/hr. Convert 100 km/hr to mph (miles per hour).

4. Estimate how long it will take to drive from where the group is staying to Niagara Falls (include a time buffer of 15 minutes to account for traffic and any areas with a lower speed limit). Round your answer to the nearest quarter of an hour.

5. The check-in time at their hotel in Niagara Falls is 3:00 p.m. What's the earliest they should leave Toronto if they want to be able to check in upon arrival?

6. Before heading to the falls, the group stops at a grocery store to pick up snacks. They purchase 1.22 kg of grapes that costs 6.08 CAD/kg and 2.39 kg of trail mix that costs 15.67 CAD/kg. (Assume these foods are not taxed.) Determine the cost of the grapes and the trail mix in USD/oz.

7. The grocery store allows payment for the purchase when using a credit card in either USD or CAD. If the credit card used to pay for the purchase is based in the US, a conversion fee of 2.8% is applied if payment is made using CAD. How much would the conversion fee be if the payment is processed in CAD?

8. While in Niagara Falls, the group reads a sign that says the Horseshoe Falls has a width of 670 meters and a height of 51 meters. The sign also indicates that the flow rate is approximately 2,300,000 liters of water per second. Convert the width and height measurements to feet.

9. Determine the flow rate in gallons of water per second.

Chapter 7 Exercises

1. Express the Hindu-Arabic numeral 175.83 in expanded form.

2. Express the expanded form $\left(4 \cdot 10^3\right) + \left(1 \cdot 10^1\right) + \left(7 \cdot 10^0\right)$ as a Hindu-Arabic numeral.

3. Write the Babylonian numeral as a Hindu-Arabic numeral.

4. Write the Hindu-Arabic numeral 101 as a Babylonia numeral.

5. Write the Mayan numeral as a Hindu-Arabic numeral.

6. Write the Hindu-Arabic numeral 4639 as a Mayan numeral.

7. Write the Egyptian numeral ⌒⌒⌒⌒⌒ as a Hindu-Arabic numeral.

8. Write the Hindu-Arabic numeral 1,203,401 as an Egyptian numeral.

9. Write the Ionic Greek numeral $\omega\,\mu\,\zeta$ as a Hindu-Arabic numeral.

10. Write the Hindu-Arabic numeral 3704 as an Ionic Greek numeral.

11. Write the Roman numeral MXLVII as a Hindu-Arabic numeral.

12. Write the Hindu-Arabic numeral 716 as a Roman numeral.

13. Write the Chinese numeral 六千零三十九 as a Hindu-Arabic numeral.

14. Write the Hindu-Arabic numeral 20,075 as a Chinese numeral.

15. Convert 2211_3 to base 10.

16. Convert 125 to base 4.

17. Convert 101011011_2 to base 16.

18. Convert $90D_{16}$ to base 2.

19. Add the numbers 330_5 and 124_5.

20. Subtract the number 65_7 from 320_7.

21. Multiply 122_3 and 11_3.

Convert each measurement to the desired unit of measure. Round to the nearest thousandth, if necessary.

22. 516 in. = _____ ft

23. 13.5 lb = _____ oz

24. 72 fl oz = _____ pt

25. 40 hr = _____ min

26. 35 mg = _____ dag

27. 15.24 hL = _____ cL

28. 2.786 km = _____ dm

29. 5775 ha = _____ m²

30. 27 mL = _____ cm³

31. How many yards are in 52 meters?

32. How many square centimeters are in 100 square inches?

33. How many quarts are in 15 liters?

34. How many grams are in 6.5 ounces?

35. What is 85°F in degrees Celsius?

36. If a rock is thrown at a speed of 82 miles per hour, how fast is the rock moving in kilometers per minute?

37. Propane gas weighs approximately 4.2 lb/gal. How much does it weigh in grams per milliliter?

38. Kamila determined that her emergency generator uses approximately 3 gallons of propane per hour. If her propane tank holds 100 kilograms, how long will it last?

CHAPTER 7 PROJECT

A Meeting of Civilizations

A wandering wormhole picked up a community from each of the following civilizations and placed them in relatively close distance to each other on a new, habitable planet: Mayans, Babylonians, and Greeks. Once the Babylonians adjusted to their new surroundings, they set out exploring and discovered the Mayan civilization. After working through the communication difficulties of having different languages, the two communities decided to set up a trade system. However, since their numeration systems were so different, they first had to figure out how to translate values between the two systems.

1. First, we need to determine the basics of each numeration system. Determine the base number systems used by the Babylonians and the Mayans.

2. To initiate numerical conversions between the two civilizations, the Babylonians decided to create a table of the numerals from 1 through 59. Fill out the table (provided on the next page) of the first 59 values of the Mayan numeration system. (The Babylonian numeration system has been provided.)

3. Describe how place value is represented in each numeration system. What kind of difficulties or confusions might arise from the differences in writing larger numbers between the systems.

4. Mathematical communication between the two civilizations holds some difficulty because each uses a different base number system and has different symbols for numerals. The Mayan system adds another layer of complexity by having a symbol for zero. How would you explain the use of the symbol zero to someone who is not familiar with the idea? What difficulties might arise from one system having a zero symbol and the other not having it? Would you, as a user of the Mayan system, try to convince the users of the Babylonian system to adopt the zero symbol or would you give up the zero symbol for communication purposes?

5. To convert numbers from the Babylonian system to the Mayan system (and vice versa), the two civilizations would not have converted the values first to base 10. This is because neither civilization used a base 10 system. Describe a method the two civilizations might use to convert numbers to and from each system. (**Hint:** Are there any patterns? Would a larger conversion table be necessary? Can you think of any other methods they might use?)

6. After setting up a trade system with the Mayans, the Babylonians continued their exploration of their new world and discovered the Greek civilization. After working through the communication difficulties of having different languages, the two communities decided to set up a trade system. Determine the base number system used by the Greeks. Does the Greek system a positional system?

7. The Babylonians decided to add the Greek numerals to their table of Babylonian and Mayan numerals. Add the first 59 numerals of the Greek system to the table.

8. Which of the two systems of most similar? Explain how the two systems are similar and explain how they are different.

9. To simplify trade, the three civilizations adopted a unified currency: gold coins. The Babylonians are the first to advertise a deal to both civilizations. They are willing to sell ⪡⪡ ∇∇∇ bales of hay for ∇∇∇ gold coins. Convert these values to both the Mayan and the Greek system.

10. The Greeks advertise that they are willing to sell bundles of wool to the other two civilizations. The wool comes in uniformly sized bundles. They are willing to sell ρ bundles of wool for μ ε gold coins. Convert these values to both the Babylonian and the Mayan system.

11. The Mayans consider the offer from the Greeks and respond that they are only willing to pay ⊙⊙ gold coins for ⊙⊙ bundles of wool. The Greeks counteroffer with Ϙε bundles of wool for μ gold coins. The Mayans agree to the new deal as long as the Babylonians can also have the same deal. The Greeks accept. Translate the Mayan offer to Greek numerals and translate the Greek counteroffer into both Mayan numerals and Babylonian numbers.

Babylonian	Mayan	Greek	Babylonian	Mayan	Greek	Babylonian	Mayan	Greek
▽			＜＜ ▽			＜＜＜＜ ▽		
▽▽			＜＜ ▽▽			＜＜＜＜ ▽▽		
▽▽▽			＜＜ ▽▽▽			＜＜＜＜ ▽▽▽		
▽▽▽ ▽			＜＜ ▽▽▽ ▽			＜＜＜＜ ▽▽▽ ▽		
▽▽▽ ▽▽			＜＜ ▽▽▽ ▽▽			＜＜＜＜ ▽▽▽ ▽▽		
▽▽▽ ▽▽▽			＜＜ ▽▽▽ ▽▽▽			＜＜＜＜ ▽▽▽ ▽▽▽		
▽▽▽ ▽▽▽ ▽			＜＜ ▽▽▽ ▽▽▽ ▽			＜＜＜＜ ▽▽▽ ▽▽▽ ▽		
▽▽▽ ▽▽▽ ▽▽			＜＜ ▽▽▽ ▽▽▽ ▽▽			＜＜＜＜ ▽▽▽ ▽▽▽ ▽▽		
▽▽▽ ▽▽▽ ▽▽▽			＜＜ ▽▽▽ ▽▽▽ ▽▽▽			＜＜＜＜ ▽▽▽ ▽▽▽ ▽▽▽		
＜			＜＜＜			＜＜＜＜＜		
＜ ▽			＜＜＜ ▽			＜＜＜＜＜ ▽		
＜ ▽▽			＜＜＜ ▽▽			＜＜＜＜＜ ▽▽		
＜ ▽▽▽			＜＜＜ ▽▽▽			＜＜＜＜＜ ▽▽▽		
＜ ▽▽▽ ▽			＜＜＜ ▽▽▽ ▽			＜＜＜＜＜ ▽▽▽ ▽		
＜ ▽▽▽ ▽▽			＜＜＜ ▽▽▽ ▽▽			＜＜＜＜＜ ▽▽▽ ▽▽		
＜ ▽▽▽ ▽▽▽			＜＜＜ ▽▽▽ ▽▽▽			＜＜＜＜＜ ▽▽▽ ▽▽▽		
＜ ▽▽▽ ▽▽▽ ▽			＜＜＜ ▽▽▽ ▽▽▽ ▽			＜＜＜＜＜ ▽▽▽ ▽▽▽ ▽		
＜ ▽▽▽ ▽▽▽ ▽▽			＜＜＜ ▽▽▽ ▽▽▽ ▽▽			＜＜＜＜＜ ▽▽▽ ▽▽▽ ▽▽		
＜ ▽▽▽ ▽▽▽ ▽▽▽			＜＜＜ ▽▽▽ ▽▽▽ ▽▽▽			＜＜＜＜＜ ▽▽▽ ▽▽▽ ▽▽▽		
＜＜			＜＜＜＜					

CHAPTER 7 REVIEW

Section 7.1 Numeral Systems Based on Position

Definitions

Babylonian Numeral System: The Babylonian Numeration System for the numbers 1 through 60 is as follows. The Babylonian's system was additive and horizontal.

▽	1	∢ ▽	11	∢∢ ▽	21	∢∢∢ ▽	31	∢∢∢ ▽	41	∢∢∢ ▽	51
▽▽	2	∢ ▽▽	12	∢∢ ▽▽	22	∢∢∢ ▽▽	32	∢∢∢ ▽▽	42	∢∢∢ ▽▽	52
▽▽▽	3	∢ ▽▽▽	13	∢∢ ▽▽▽	23	∢∢∢ ▽▽▽	33	∢∢∢ ▽▽▽	43	∢∢∢ ▽▽▽	53
▽▽▽ ▽	4	∢ ▽▽▽ ▽	14	∢∢ ▽▽▽ ▽	24	∢∢∢ ▽▽▽ ▽	34	∢∢∢ ▽▽▽ ▽	44	∢∢∢ ▽▽▽ ▽	54
▽▽▽ ▽▽	5	∢ ▽▽▽ ▽▽	15	∢∢ ▽▽▽ ▽▽	25	∢∢∢ ▽▽▽ ▽▽	35	∢∢∢ ▽▽▽ ▽▽	45	∢∢∢ ▽▽▽ ▽▽	55
▽▽▽ ▽▽▽	6	∢ ▽▽▽ ▽▽▽	16	∢∢ ▽▽▽ ▽▽▽	26	∢∢∢ ▽▽▽ ▽▽▽	36	∢∢∢ ▽▽▽ ▽▽▽	46	∢∢∢ ▽▽▽ ▽▽▽	56
▽▽▽ ▽▽▽ ▽	7	∢ ▽▽▽ ▽▽▽ ▽	17	∢∢ ▽▽▽ ▽▽▽ ▽	27	∢∢∢ ▽▽▽ ▽▽▽ ▽	37	∢∢∢ ▽▽▽ ▽▽▽ ▽	47	∢∢∢ ▽▽▽ ▽▽▽ ▽	57
▽▽▽ ▽▽▽ ▽▽	8	∢ ▽▽▽ ▽▽▽ ▽▽	18	∢∢ ▽▽▽ ▽▽▽ ▽▽	28	∢∢∢ ▽▽▽ ▽▽▽ ▽▽	38	∢∢∢ ▽▽▽ ▽▽▽ ▽▽	48	∢∢∢ ▽▽▽ ▽▽▽ ▽▽	58
▽▽▽ ▽▽▽ ▽▽▽	9	∢ ▽▽▽ ▽▽▽ ▽▽▽	19	∢∢ ▽▽▽ ▽▽▽ ▽▽▽	29	∢∢∢ ▽▽▽ ▽▽▽ ▽▽▽	39	∢∢∢ ▽▽▽ ▽▽▽ ▽▽▽	49	∢∢∢ ▽▽▽ ▽▽▽ ▽▽▽	59
∢	10	∢∢	20	∢∢∢	30	∢∢∢∢	40	∢∢∢∢∢	50		

Mayan Numeral System: The Mayan Numeration System for the numbers 1 through 19 is as follows. The Mayan's system was additive and vertical.

0	1	2	3	4
〇	•	••	•••	••••
5	**6**	**7**	**8**	**9**
—	•	••	•••	••••
10	**11**	**12**	**13**	**14**
═	•	••	•••	••••
15	**16**	**17**	**18**	**19**
≡	•	••	•••	••••

Section 7.2 Early Numeral Systems

Definitions

Egyptian Numeral System: The Egyptian hieroglyphics that represent the digits and powers of 10 are given in Table 7.2.1. The Egyptian numeral system uses duplicate characters to represent multiples of numbers. The position of the characters does not play a role in the value of the characters.

Table 7.2.1: Hieroglyphics for Egyptian Numerals

Egyptian Hieroglyphic	Hieroglyphic Description	Hindu-Arabic Numeral
(vertical line)	Vertical line	1
(cattle hobble)	Cattle hobble	10
(coiled rope)	Coiled rope	100
(lotus plant)	Lotus plant	1000
(finger)	Finger	10,000
(frog)	Frog	100,000
(a god with his hands raised above his head)	A god with his hands raised above his head	1,000,000

Ionic Greek Numeral System: Table 7.2.2 shows the symbols used for numbers from 1 to 900 in the Ionic Greek Numeral System.

- Putting a comma before any of the first nine symbols multiplies the value by 1000.

- An Ionic Greek numeral written above M means that value is multiplied by 10,000.

- Characters written next to one another in the Ionic Greek system are added together.

Table 7.2.2: Ionic Greek Numerals

alpha	beta	gamma	delta	epsilon	vau	zeta	eta	theta
α	β	γ	δ	ε	ϛ	ζ	η	θ
1	2	3	4	5	6	7	8	9

iota	kappa	lambda	mu	nu	xi	omicron	pi	koph
ι	κ	λ	μ	ν	ξ	o	π	ϟ
10	20	30	40	50	60	70	80	90

rho	sigma	tau	upsilon	phi	chi	psi	omega	sampi
ρ	σ	τ	υ	ϕ	χ	ψ	ω	ϡ
100	200	300	400	500	600	700	800	900

Roman Numeral System: Table 7.2.4 shows the Roman numerals along with their values in Hindu-Arabic numerals.

Table 7.2.4: Roman Numerals Up to 1000

Value	Roman Numeral	Value	Roman Numeral	Value	Roman Numeral
One	I	Eleven	XI	Thirty	XXX
Two	II	Twelve	XII	Forty	XL
Three	III	Thirteen	XIII	Fifty	L
Four	IV	Fourteen	XIV	Sixty	LX
Five	V	Fifteen	XV	Seventy	LXX
Six	VI	Sixteen	XVI	Eighty	LXXX
Seven	VII	Seventeen	XVII	Ninety	XC
Eight	VIII	Eighteen	XVIII	One hundred	C
Nine	IX	Nineteen	XIX	Five hundred	D
Ten	X	Twenty	XX	One thousand	M

Chinese Numeral System: Table 7.2.5 shows the characters for numbers 1 through 10,000 in the Chinese numeral system.

Table 7.2.5: Chinese Numerals

Simple Character	Formal Character	Hindu-Arabic Numeral
一	零	1
二	壹	2
三	貳	3
四	叁	4
五	肆	5
六	伍	6
七	陸	7
八	柒	8
九	捌	9
十	玖	10
百	佰	100
千	仟	1000
万	萬	10,000

Properties

Addition and Subtraction Rules for Roman Numerals

1. A letter may be repeated up to 3 times to indicate multiple values. For example,

   III = 3
   CC = 200

2. If a smaller value is placed on the right of a character, that smaller value is added to the larger one. For example,

$$\overset{5\ \ 1}{VI} = 5 + 1 = 6$$

$$\overset{100\ 10\ 10}{C\,X\,X} = 100 + 10 + 10 = 120$$

3. If a smaller value is placed on the left of a character, that smaller value is subtracted from the larger one. For example,

$$\overset{1\ \ 5}{I\,V} = 5 - 1 = 4$$

$$\overset{100\ 1000}{C\ M} = 1000 - 100 = 900$$

4. A bar is placed above a symbol, or group of symbols, to indicate that the numeral is to be multiplied by 1000. For example,

$$\overline{D} = 500 \cdot 1000 = 500,000$$

$$\overline{VI} = 6 \cdot 1000 = 6000$$

Addition and Subtraction Rules for the Chinese Numeral System

1. A numeral to the right of a multiple of 10 (10, 100, 1000, …) indicates addition. For instance,
$$十六 = 10 + 6 = 16$$

2. A numeral to the left of a multiple of 10 (10, 100, 1000, …) indicates multiplication. For instance,
$$六十 = 6 \cdot 10 = 60$$

3. A position holder, 零 is used for when a zero occurs in the middle of a numeral. Only one is needed to show any number of consecutive zeros. For instance,
$$201 = 二百零一$$
$$2001 = 二千零一$$

Section 7.3 Working with Base Number Systems

Definition

Hexadecimal System: The digits and letters used for the hexadecimal system are shown in Table 7.3.1.

Table 7.3.1: Characters Used for Base 16 (the Hexadecimal System)

Base 10 Digit	0	1	2	3	4	5	6	7	8	9	10	11	12	13	14	15
Hexadecimal (Base 16) Digit	0	1	2	3	4	5	6	7	8	9	A	B	C	D	E	F

Section 7.4 The Metric System

Definition

Conversion Factor: A **conversion factor** is a fraction equal to 1 that contains different units in the numerator and denominator.

Conversion Tables

Table 7.4.1: Measurements Used in the US Customary System

Units of Length	Units of Mass
12 inches (in.) = 1 foot (ft)	16 ounces (oz) = 1 pound (lb)
36 inches = 1 yard (yd)	2000 pounds (oz) = 1 ton (T)
3 feet = 1 yard	
5280 feet = 1 mile (mi)	

Units of Capacity	Units of Time
8 fluid ounces (fl oz) = 1 cup (c)	60 seconds (sec) = 1 minute (min)
2 cups = 1 pint (pt) = 16 fl oz	60 minutes = 1 hour (hr)
2 pints = 1 quart (qt)	24 hours = 1 day
4 quarts = 1 gallon (gal)	7 days = 1 week

Table 7.4.2: Common Metric Prefixes for Any Base Unit

Prefix	Multiple	Example (Using Meters)
kilo (k)	Base unit × 1000	1 kilometer (km) = 1000 m
hecto (h)	Base unit × 100	1 hectometer (hm) = 100 m
deka (da)	Base unit × 10	1 dekameter (dam) = 10 m
	Base unit	1 meter (m)
deci (d)	Base unit × 0.1	1 decimeter (dm) = 0.1 m
centi (c)	Base unit × 0.01	1 centimeter (cm) = 0.01 m
milli (m)	Base unit × 0.001	1 millimeter (mm) = 0.001 m

Table 7.4.3: Common Units in the Metric System

Prefix	Length	Mass	Capacity
kilo- (1000)	kilometer (km) = 1000 meters	kilogram (kg)	kiloliter (kL)
hecto- (100)	hectometer (hm) = 100 meters	hectogram (hg)	hectoliter (hL)
deka- (10)	dekameter (dam) = 10 meters	dekagram (dag)	dekaliter (daL)
Base unit	meter (m)	gram (g)	liter (L)
deci- $\left(\frac{1}{10}\right)$	decimeter (dm) = 0.1 meters	decigram (dg)	deciliter (dL)
centi- $\left(\frac{1}{100}\right)$	centimeter (cm) = 0.01 meters	centigram (cg)	centiliter (cL)
milli- $\left(\frac{1}{1000}\right)$	millimeter (mm) = 0.001 meters	milligram (mg)	milliliter (mL)

Table 7.4.4: Measures of Land Area and Capacity

US Customary Measure of Land Area	Metric Measure of Land Area	US Customary Measure of Capacity	Metric Measure of Capacity
1 acre = 4840 yd^2	1 are (a) = 100 m^2	1 pint = 28.875 in^3	1 liter (L) = 1000 cm^3
1 acre = 0.0015625 mi^2	1 hectare (ha) = 100 a = 10000 m^2	1 pint = 0.0167101 ft^3	1 mL = 1 cm^3

Section 7.5 Converting between the US Customary System and the Metric System

Definition

Dimensional Analysis: Dimensional Analysis is a process by which the relationship between different physical quantities of the same dimension can be compared by converting between units of measurement using the rules of algebra.

US and Metric Equivalences

Table 7.5.1: Commonly Used Equivalences

	US to Metric	Metric to US
Length	1 in. = 2.54 cm	1 cm ≈ 0.394 in.
	1 ft ≈ 0.305 m	1 m ≈ 3.28 ft
	1 yd ≈ 0.914 m	1 m ≈ 1.09 yd
	1 mi ≈ 1.61 km	1 km ≈ 0.62 mi
Area (Length²)	1 in.² ≈ 6.45 cm²	1 cm² ≈ 0.155 in.²
	1 ft² ≈ 0.093 m²	1 m² ≈ 10.764 ft²
	1 yd² ≈ 0.836 m²	1 m² ≈ 1.196 yd²
	1 mi² ≈ 2.6 km²	
	1 acre ≈ 0.405 ha	1 ha ≈ 2.47 acres
Volume (Length³)	1 tsp ≈ 5 mL	1 L ≈ 61.02 in.³
	1 tbsp ≈ 15 mL	
	1 c ≈ 0.24 L	
	1 pint ≈ 0.47 L	
	1 qt ≈ 0.946 L	1 L ≈ 1.06 qt
	1 gal ≈ 3.785 L	1 L ≈ 0.264 gal
Mass	1 oz ≈ 28.35 g	1 g ≈ 0.035 oz
	1 lb ≈ 0.454 kg	1 kg ≈ 2.205 lb

Formulas

Temperature Formulas: The following formulas are used to convert temperature between degrees Celsius and degrees Fahrenheit, where F = Fahrenheit temperature and C = Celsius temperature.

$$F = \frac{9C}{5} + 32 \qquad C = \frac{5(F-32)}{9}$$

CHAPTER 8

Number Theory

📑 **SECTIONS**

Introduction

Number theory, once known as *higher arithmetic*, is the branch of mathematics that studies the relationships between different types of numbers. Today we think of arithmetic as the elementary calculations learned at a young age, but until the twentieth century, number theory was known as higher arithmetic and only studied for the "love of math." Carl Friedrich Gauss once said, "Mathematics is the queen of the science—and number theory is the queen of mathematics."

While we now recognize the usefulness of number theory and its application to everyday things, such as bar codes and encryption technology, it is a field full of interesting, unsolved problems. One such problem is Goldbach's conjecture. Goldbach's conjecture, originally proposed nearly 300 years ago, states that every even whole number greater than 2 is the sum of two prime numbers. While this conjecture has been shown to be true for all integers less than 4×10^{18}, it hasn't been proven yet. Once someone finally proves Goldbach's conjecture to be true, it will be true forever and mathematicians will be able to use it as a basis to create and explore new mathematical concepts.

Because the study of mathematics often leads to new and exciting breakthroughs, the Clay Mathematics Institute decided back in 2000 to offer a prize of 1 million dollars to anyone who solves one of the seven Millennium Problems. Two of the seven problems are famous unsolved number theory problems: the Birch and Swinnerton-Dyer conjecture and the Riemann Hypothesis. While the Goldbach Conjecture is not one of the seven, it is the oldest unsolved problem in all mathematics. To date only one of the seven Millennium Problems, the Poincaré conjecture, has been solved. The problem was solved by Grigori Perelman, who declined the prize money. Had you ever considered that you could become a millionaire by solving a single math problem?

8.1 PRIME NUMBERS

OBJECTIVES

1. Identify prime and composite numbers.

2. Find the prime factorization of numbers.

3. Find the greatest common divisor of a set of numbers.

OBJECTIVE 1 Prime and Composite Numbers

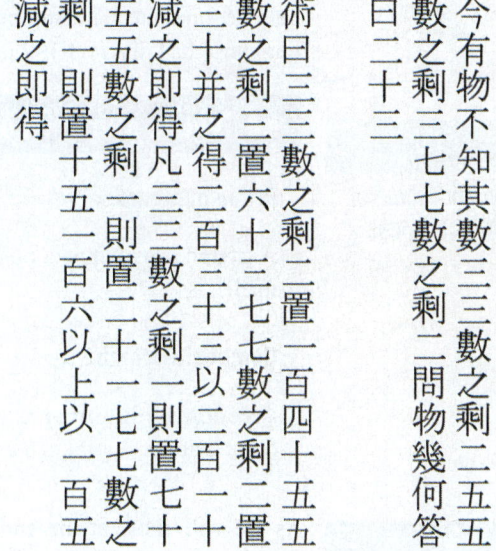

Figure 8.1.1: Sun Tzu Suan-Ching's Puzzle

The text in Figure 8.1.1 is from a classical Chinese text *Sun Tzu Suan-Ching* dating from the third or fourth century. Here is a translation.

We have things of which we do not know the number; if we count by threes the remainder is two; if we count by fives the remainder is three; if we count by sevens the remainder is two. How many things are there?

Sun Tzu goes on to give his method to solve this puzzle.

If you count by threes and have the remainder two, put 140. If you count by fives and have a remainder three, put 63. If you count by sevens and have the remainder two, put 30. Add these numbers and you get 233. From this subtract 210 and you have the result. For each unity as a remainder when counting by three, put 70. For each unity as a remainder when counting by fives, put 21. For each unity as a remainder when counting by sevens, put 15. If the sum is more than 106 subtract 105 from this and you get the result.

Perhaps this makes things even more puzzling! It is certainly not clear where all these numbers come from. Did Sun Tzu actually solve the puzzle? If he did, it is not clear how this method could be generalized to solve a similar problem that does not involve counting by threes, fives, and sevens.

We're all familiar with the concept of odd and even numbers, but understanding Sun Tzu's solution requires understanding how numbers can be characterized into more complex groups. This is where number theory comes in. Number theory looks at the properties of families of numbers and how they interconnect. One of the fundamental building blocks of this interconnection is the set of prime numbers. Just as all matter is built up from the elements of the periodic table, we will see that every number can be built up from the "mathematical periodic table" of prime numbers.

💬 **Think Back**

The set of integers includes all whole numbers, their negatives, and 0.

💬 **Think Back**

Remember that a *divisor*, or *factor*, of a number x is a number that divides x with a remainder of 0.

💬 **Fun Fact**

The only even prime number is 2.

💬 **Math Milestone**

Eratosthenes (276–194 B.C.) was a Greek librarian of Alexandria who is remembered as a scholar, poet, and inventor. He is known for his measurement of the circumference of the Earth as well as his estimates of the distances from the Earth to the sun and the moon. He is also credited with the discipline of geography as we know it, inventing the system of longitude and latitude.

We say that a positive integer is **prime** if it has precisely two divisors: 1 and itself. Since 0 is neither positive nor negative, it cannot be prime. The number 1 is also not considered prime because it does not have precisely two divisors. It has precisely one—itself. Therefore, the first prime number is 2. If a positive integer greater than 1 is not prime, it is **composite**. That is, a composite number has more than two divisors (or factors). Notice that 0 and 1 are also eliminated from being composite. The number 3 only has two divisors (1 and 3) so it is prime. The number 4 has divisors 1, 2, and 4, so it is the first composite number.

📖 **PRIME AND COMPOSITE NUMBERS**

Prime numbers

A **prime number** is a positive integer that has precisely two divisors: 1 and itself.

Composite Numbers

A **composite number is** a positive integer that has more than two divisors. Note that the numbers 0 and 1 are neither prime nor composite.

As we will show at the end of this section, there is an infinite number of both prime and composite numbers. The following is the beginning of the list of prime numbers.

$$2, 3, 5, 7, 11, 13, 17, 19, 23, \ldots$$

The following is the beginning of the list of composite numbers.

$$4, 6, 8, 9, 10, 12, 14, 15, 16, 18, 20, 21, 22, 24, \ldots$$

How do we find or determine prime numbers? A very early technique for finding primes is credited to Eratosthenes in the third century B.C. His idea was to organize the computations in order to find all the prime numbers smaller than a particular number N; a process commonly called the **sieve of Eratosthenes**.

≡ **SIEVE OF ERATOSTHENES**

Step 1: List all the positive integers between 2 and N.

Step 2: Highlight the smallest number (p) not crossed out or previously highlighted. This is a prime number.

Step 3: Cross out all multiples of p.

Step 4: Repeat steps 2 and 3 until all numbers are either highlighted or crossed out.

Let's look at how the sieve of Eratosthenes is used to find the prime numbers up to 101. Begin by writing out all the positive integers from 2 to 101. The number 2 is the smallest number not crossed or highlighted, so we highlight 2 (indicating that it is prime) and cross out all multiples of 2, which all have 2 as a divisor. Now the smallest number that isn't highlighted or crossed out is the number 3. Three has no smaller divisors since it is not a multiple of 2. Highlight 3 to indicate it is prime and cross out the multiples of 3. Notice that some multiples are already crossed out since they are also multiples of 2. Repeat the process until all numbers are either highlighted (prime numbers) or crossed out (composite numbers). Figure 8.1.2 shows the end result of this process, with the first twenty-six prime numbers highlighted.

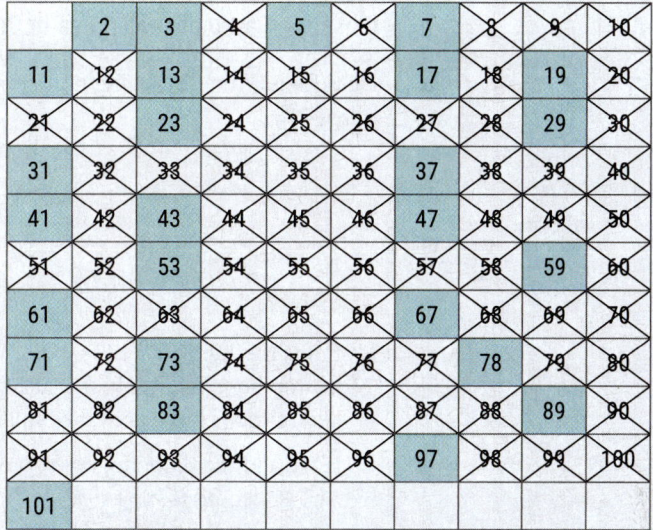

Figure 8.1.2: Sieve of Eratosthenes

One of the first questions that you might ask yourself when trying to decide if a number is prime or composite is "How do I know what the divisors of a number are?" Using the sieve of Eratosthenes is not a convenient nor efficient method for finding primes once the numbers become large. Fortunately, there is a clever way to check if any of the numbers 2, 3, 4, 5, 6, 7, 8, 9, or 10 is a divisor of a number. Table 8.1.1 lists these divisibility rules.

Table 8.1.1: Divisibility Rules

Divisor	Test	Example
2	The number is even.	The number 391,574 is divisible by 2 because it is even.
3	When the digits of the number are added together, the resulting number is divisible by 3.	87,408 is divisible by 3 because $8 + 7 + 4 + 0 + 8 = 27$ and 27 is divisible by 3.
4	The last 2 digits of the number form a number divisible by 4.	316 is divisible by 4 because 16 is divisible by 4.
5	The number ends in a 0 or a 5.	29,345 is divisible by 5 because it ends in a 5.
6	The number is divisible by both 2 and 3.	628,116 is divisible by 6 because it's divisible by 2, since it's even, and 3, since $6 + 2 + 8 + 1 + 1 + 6 = 24$ and 24 is divisible by 3.
7	Double the last digit, then subtract it from the remaining digits of the number. If the answer is divisible by 7, then so is the original number.	819 is divisible by 7 since $2 \cdot 9 = 18$ and $81 - 18 = 63$. 63 is divisible by 7, so 819 is also.
8	The last 3 digits of the number form a number divisible by 8.	2160 is divisible by 8 because 160 is divisible by 8.
9	When the digits of the number are added together, the resulting number is divisible by 9.	189 is divisible by 9 because $1 + 8 + 9 = 18$ and 18 is divisible by 9.
10	The number ends in a 0.	9,145,830 is divisible by 10 because it ends in a 0.

This list of divisibility rules will help us determine if a number has more than two divisors. While it isn't helpful in finding divisors bigger than 10, it's a good start.

Example 8.1.1

Classifying a Number as Prime or Composite

Determine if the following numbers are prime or composite using the divisibility rules.

a. 312

b. 101

c. 2,344,017

Solution

a. Since the number 312 is even, we know that it is divisible by 2. It is enough to stop here and know that the number is composite, but we'll continue on to determine if there are other small divisors of 312.

If we add the digits together, we get $3 + 1 + 2 = 6$. Because 6 is divisible by 3, 312 is also divisible by 3.

In addition, the last 2 digits form the number 12, which is divisible by 4, so we know that 312 is also divisible by 4.

Although we have plenty of examples to show us the number is composite, one last point to notice is that because 312 is divisible by 2 and by 3, it must also be divisible by 6.

b. Since 101 is odd, it is not divisible by any of the even divisors: 2, 4, 6, 8, or 10. $1 + 0 + 1 = 2$, which is not divisible by 3, so 101 is not divisible by 3 or 6. Since the sum of the digits, 2, is not divisible by 9, 101 is also not divisible by 9. That leaves 5 and 7 to check. 101 is not divisible by 5 because it does not end in 0 or 5. Lastly, $2 \cdot 1 = 2$ and $10 - 2 = 8$. 8 is not divisible by 7, so 101 is not either.

That eliminates all of the small divisors, but to establish that 101 is prime, we have to deal with all divisors smaller than 101. Notice that if both a and b are integers bigger than 10, then $a \cdot b \geq 121$. So if 101 had divisors a and b, one of them must be smaller than 10. Since we've already established that there are no divisors smaller than 10, 101 is prime.

While 101 was shown to be prime earlier in the text, testing for factors is a method for checking without making a sieve or memorizing the list of primes.

c. We know that the number 2,344,017 is not divisible by 2, 4, 6, 8, or 10 since it is not even. Adding the digits together gives us $2 + 3 + 4 + 4 + 0 + 1 + 7 = 21$, which is divisible by 3, so the number is divisible by 3. Again, although we now know the number is composite, we'll continue to show how to check for other small divisors.

2,344,017 is not divisible by 5 since it does not end in a 5 or 0.

Doubling the last digit we have $7 \cdot 2 = 14$. Then, $234,401 - 14 = 234,387$. This is still a large number, so we can apply the rule again to check if it's divisible by 7. Continue repeating this process until we are able to definitively say if a number is divisible by 7.

$$234,387$$
$$7 \cdot 2 = 14 \Rightarrow 23,438 - 14 = 23,424$$
$$4 \cdot 2 = 8 \Rightarrow 2342 - 8 = 2334$$
$$4 \cdot 2 = 8 \Rightarrow 233 - 8 = 225$$
$$5 \cdot 2 = 10 \Rightarrow 22 - 10 = 12$$

Since 12 is not divisible by 7, our original number is not divisible by 7. Although we used the divisibility rule for 7 here, it may have been quicker to use long division and see if the number has a remainder when divided by 7.

The only other number left in our list to check is 9. Adding the digits together, as we did earlier, we have 21. Since 21 is not divisible by 9, the number 2,344,017 is not divisible by 9.

However, we know that 2,344,017 is composite because it divisible by the number 3.

✔ Skill Check 8.1.1

Determine if 3,743,216 is prime or composite.

OBJECTIVE 2 Prime Factorizations

Now that we have a working idea of the definition of prime numbers, we can more carefully examine how prime numbers are the building blocks of all numbers. The following theorem proves that every number can be built from prime numbers in a unique way. This particular theorem was proven by Euclid around 300 B.C.

💬 Helpful Hint

A **mathematical theorem** is a statement or rule that can be, and has been, proven to be (logically) true.

🔑 FUNDAMENTAL THEOREM OF ARITHMETIC

The **fundamental theorem of arithmetic** states that every positive integer greater than 1 is either a prime number or can be written as a unique product of prime numbers. This unique product of prime numbers is called its *prime factorization*.

One method to find the prime factorization of a number is to use a *factor tree* like the one shown in Figure 8.1.3.

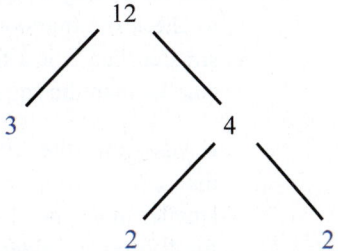

Figure 8.1.3: Factor Tree for 12

To find the prime factorization of a number N using this method, begin by choosing any pair of factors, other than 1 and N, that multiply together to give you N. In Figure 8.1.3, we started factoring 12 with the numbers 3 and 4. Then, find a pair of factors for each of these numbers. Continue until all factors are prime. Notice that in Figure 8.1.3, we continued to find factors for 4 but not for 3 since 3 is already prime. Once all factors are prime, we have found the prime factorization of the number. We call this method a factor tree because of the way we organize the pairs of factors. The original number N is placed at the top of the "tree" and then the factors "branch off" below.

Example 8.1.2

Using a Factor Tree to Find a Prime Factorization

Use a factor tree to determine the prime factorization of the number 84.

Solution

Start by choosing any pair of factors you like, other than 1 and 84. We'll use the pair 4 and 21, as shown here.

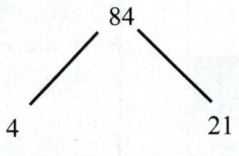

Figure 8.1.4

Helpful Hint

Although factor trees for the same number may not have identical branches, the prime factorization on the bottom row will always be the same.

Next, notice that both of the numbers 4 and 21 also have factors. Branching each off into its own factors, we have the following.

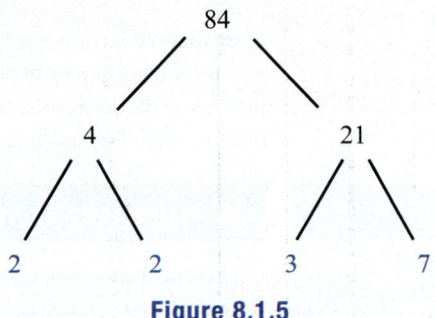

Figure 8.1.5

✔ Skill Check 8.1.2

Use a factor tree to find the prime factorization of 84 by starting with the pair of factors 2 and 42.

This time, as we look across the bottom row of factors, we can see that they are all prime. Thus, the prime factorization of 84 is $2 \cdot 2 \cdot 3 \cdot 7$.

One of the ways we can use prime factorization is to cut down on the work needed to check if a number N is prime. Instead of having to determine if every number smaller than N is a factor of N or not, we only need to check the prime numbers smaller than the square root of N.

If N is composite, then N can be written in the form $a \cdot b$. If both a and b are bigger than $\sqrt{N}$, then $a \cdot b$ will be bigger than N. Consequently, either a or b must be smaller than $\sqrt{N}$. For large numbers, this fact cuts down the work of checking for divisors considerably.

For example, is the number 9997 prime? To check, we don't need to test 9996 numbers, we only need to look at the primes smaller than $\sqrt{9997} \approx 99.98$. The sieve of Eratosthenes at the beginning of the section showed us that there are only 25 prime numbers smaller than 99.98. (Go back and check this for yourself.) This is a huge time saver!

Example 8.1.3

Classifying a Number as Prime or Composite

Determine whether 197 is prime or composite.

Solution

If 197 is composite, it must have a prime divisor that is less than $\sqrt{197} \approx 14.0357$. The prime numbers less than 14 are 2, 3, 5, 7, 11, and 13.

197 is not even, so it is not divisible by 2. $1 + 9 + 7 = 17$ and 17 is not divisible by 3, which eliminates 3. 197 doesn't end in 0 or 5, which eliminates 5, and $19 - (7 \cdot 2) = 5$ and 5 is not divisible by 7, which eliminates 7. All that remains is to check 11 and 13.

$$\frac{197}{11} = 17.\overline{90} \quad \text{and} \quad \frac{197}{13} \approx 15.1538$$

Therefore, 197 is prime.

OBJECTIVE 3 Greatest Common Divisor

Using prime factorizations, we can find the **greatest common divisor (GCD)** of any collection of numbers. The greatest common divisor of two numbers is the largest integer that divides both numbers without a remainder.

> 📖 **GREATEST COMMON DIVISOR (GCD)**
>
> The largest integer that divides two numbers without a remainder is called the **greatest common divisor (GCD)**. If the GCD = 1 for a pair of numbers, the numbers are said to be **relatively prime**.

For example, the GCD of 10 and 15 is 5 because 5 is the largest number that divides both 10 and 15. When the GCD of two numbers is less obvious, it can be determined using factor trees.

To find the GCD of two numbers, such as 24 and 16, using factor trees, begin by creating the factor tree of each number. The factor trees of 24 and 16 are shown in Figure 8.1.6.

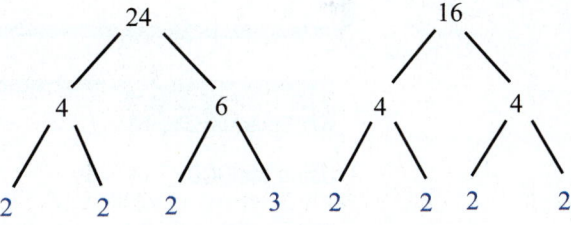

Figure 8.1.6

Once the trees are complete, the GCD of the two numbers is the product of all the prime factors that the two numbers have in common. In the case of 24 and 16, you can see that the prime number 2 appears three times in both of the prime factorizations. Therefore, the greatest common divisor for these numbers is $2 \cdot 2 \cdot 2 = 8$.

As we stated in the definition of the greatest common divisor, if there is no prime factor in common between two numbers, then the GCD is 1, and the numbers are said to be **relatively prime**, or *co-prime*. For instance, consider the numbers 10 and 21. Their factor trees are shown in Figure 8.1.7.

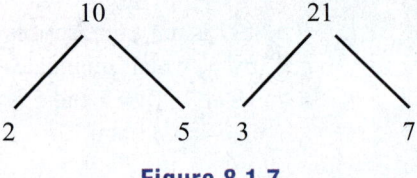

Figure 8.1.7

Since 10 and 21 have no common factors, they are relatively prime.

Example 8.1.4

Finding the GCD Using Factor Trees

Use factor trees to find the greatest common divisor of 40 and 60.

Solution

Begin by constructing the factor trees of 40 and 60 as shown.

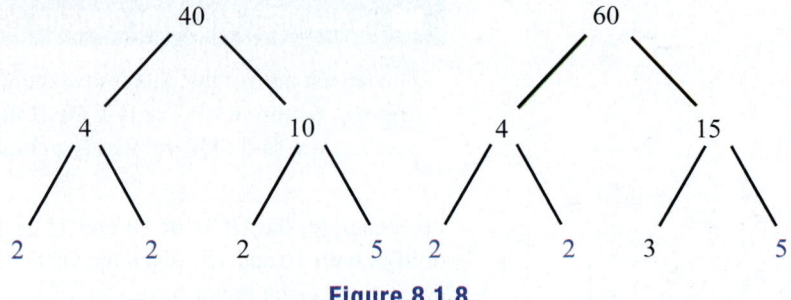

Figure 8.1.8

Notice that both 40 and 60 have the factors 2 and 5 in common. In fact, both numbers have two 2s in common as well as one 5. So, the GCD of 40 and 60 is $2 \cdot 2 \cdot 5 = 20$.

Example 8.1.5

Using the GCD

One way that the Manna Café Pantry serves the hungry of Clarksville, TN, is by distributing boxes of food each week to the local shelters. As part of the stipulation for receiving local governmental funds, all boxes must contain the same number of items from each of the following categories: pasta, canned vegetable, and canned meat. This week, the pantry has the following in supply: 360 pasta items, 540 canned vegetables, and 240 canned meats.

a. Using all of the food, what is the maximum number of food boxes that the Manna Café Pantry can distribute this week?

b. How many of each item will be in each box?

Solution

a. In order to distribute all of the food items evenly among the food boxes, the number of boxes must be a divisor of 360, 540, and 240. To find the *maximum* number of food boxes that can be made from all of the pantry supplies, we need to find the GCD of the three numbers. First, find the prime factorization of each number. We'll do this by constructing factor trees of each number.

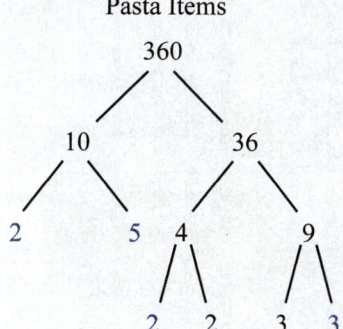

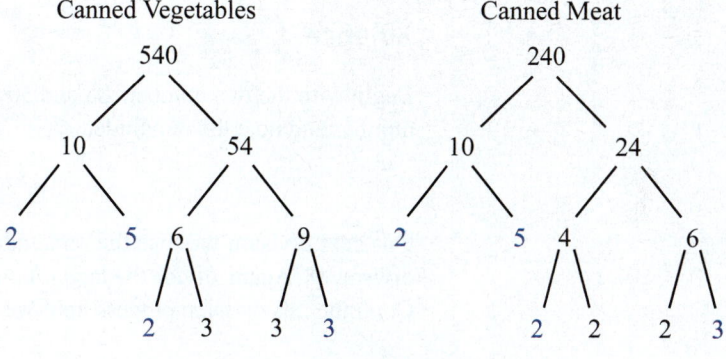

Figure 8.1.9

The GCD of 360, 540, and 240 is $2 \cdot 2 \cdot 3 \cdot 5 = 60$. Therefore, the pantry can give out 60 food boxes to the local shelters this week.

b. In order to find how many of each item will be in the boxes, we can use the prime factorizations that we found in part a. For each food category, if we remove the GCD from the prime factorization, the remaining factors will tell us how many of each item needs to go in a box. Each prime factorization with the GCD marked out is as follows.

Pasta items: $\cancel{2} \cdot \cancel{2} \cdot 2 \cdot \cancel{3} \cdot 3 \cdot \cancel{5}$

Canned vegetables: $\cancel{2} \cdot \cancel{2} \cdot \cancel{3} \cdot 3 \cdot 3 \cdot \cancel{5}$

Canned meat: $\cancel{2} \cdot \cancel{2} \cdot 2 \cdot 2 \cdot \cancel{3} \cdot \cancel{5}$

Therefore, we can see that there will be $2 \cdot 3 = 6$ pasta items, $3 \cdot 3 = 9$ canned vegetables, and $2 \cdot 2 = 4$ canned meats in each box.

There are other methods besides factor trees that can be used to find the GCD of groups of numbers. The main point is that the GCD can be found by knowing the prime factorization of each number. If we are trying to find the GCD of very large numbers, then finding the prime factors can get quite cumbersome. We once again turn to Euclid for a solution. Around 300 B.C., Euclid developed an algorithm that calculates the GCD of two numbers. This algorithm is referred to as **Euclid's Algorithm**, or the *Euclidian Algorithm*.

☰ STEPS FOR USING EUCLID'S ALGORITHM

1. Divide the larger of the given numbers by the smaller (the divisor) and note the remainder.

2. Divide the original divisor (the smaller of the given numbers) by the remainder found in Step 1.

3. Continue dividing the previous divisor by the previous remainder until the remainder is 0.

4. The last nonzero remainder is the GCD of the given numbers.

Example 8.1.6

Finding the GCD Using Euclid's Algorithm

Use Euclid's Algorithm to find the greatest common divisor of 88 and 300.

Solution

Begin with the two numbers 88 and 300. Divide the larger number by the smaller number and note the remainder.

$$88\overline{)300} \quad 3R36$$

Our next division will use the remainder that we just found, 36, and the original divisor, 88. Again, divide the larger number by the smaller and note the remainder. Continue this division process until we get a remainder of zero.

$$36\overline{)88} \quad 2R16$$

$$16\overline{)36} \quad 2R4$$

$$4\overline{)16} \quad 4R0$$

Since 4 is the last nonzero remainder, it is the GCD of 88 and 300.

Example 8.1.7

Using Euclid's Algorithm

Two college classes, one with 88 students and one with 17, need to be split into smaller groups of equal sizes. Use Euclid's Algorithm to determine what size the smaller groups could be, if it is possible.

Solution

Begin by dividing the larger number by the smaller number and note the remainder.

$$\overset{5R3}{17\overline{)88}}$$

Our next division will use the remainder we just found and the divisor. Repeat this process until the remainder is zero.

$$\overset{5R2}{3\overline{)17}}$$

$$\overset{1R1}{2\overline{)3}}$$

$$\overset{2R0}{1\overline{)2}}$$

The last nonzero remainder is 1, so the GCD of 88 and 17 is 1, which means that these numbers are relatively prime. Therefore, the two classes of 88 and 17 cannot be divided into smaller groups of equal sizes other than having 105 groups with 1 person each.

As we will see in Section 8.4, which covers public-key encryption, prime numbers are very useful. Euclid's prime factorization theorem also ensures that we will never run out of prime numbers—there are infinitely many. We'll finish this section with a generalization of Euclid's proof of this.

Euclid's Proof of Infinitely Many Primes (A Summary)

One of the consequences of Euclid's Algorithm is that there are infinitely many prime numbers. Let's think about why this is true. Suppose that $p_1, p_2, p_3, \ldots, p_k$ is a complete, finite list of all of the prime numbers that exist. Take a look at the number n, the number that you get by multiplying all of those primes together and adding 1.

$$n = p_1 \cdot p_2 \cdot p_3 \cdots p_k + 1$$

As we have seen in the previous equation, if n is composite, then it must have a prime divisor. Since $p_1, p_2, p_3, \ldots, p_k$ are all of the primes that exist, the divisor must be one of the numbers in the list. But, by our definition of n, there is a remainder of 1 when n is divided by any of the primes on the list. (Check this for yourself.) So the divisor cannot be on our finite list of primes. Therefore, n must be an even bigger prime that we left off the list. Because we can continue to play this game, there must be infinitely many primes.

Skill Check Answers

1. It's even, so it's composite. **2.** The prime factorization of 84 is $2 \cdot 2 \cdot 3 \cdot 7$.

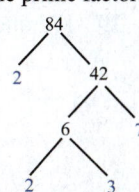

8.1 Exercises

✔ CONCEPT CHECK

1. If a number is divisible by 2 and 3, then it is also divisible by _____ .

2. If a number is even, then it is divisible by _____ .

3. True or False: 1 is a prime number.

4. True of False: A composite number is a positive integer with two or more divisors.

💡 PRACTICE

Determine whether each of the numbers is prime or composite.

5. 245

6. 939

7. 149

8. 4372

9. 68,045,800

10. 113

11. 51

12. 79

13. 91

14. 147

Determine the prime factorization of each number.

15. 16

16. 124

17. 162

18. 630

19. 625

20. 72

21. 105

22. 81

Use a factor tree to determine the greatest common divisor of each pair of numbers.

23. 35 and 14

24. 28 and 42

25. 90 and 225

26. 350 and 217

27. 36 and 72

28. 36 and 54

29. 175 and 245

30. 360 and 216

Use Euclid's Algorithm to find the greatest common divisor of each pair of numbers.

31. 357 and 217

32. 350 and 140

33. 1235 and 5687

34. 1556 and 236

35. 2216 and 1008

36. 288 and 672

37. 894 and 204

38. 225 and 121

Use Euclid's Algorithm to determine whether each pair of numbers is relatively prime.

39. 67 and 17

40. 1231 and 5673

41. 351 and 141

42. 289 and 169

43. 39 and 51

44. 133 and 153

45. 428 and 1284

46. 1084 and 1444

 **APPLICATIONS**

47. There are 16 cellists, 32 violinists, 24 flautists, and 16 violists at a summer classical music camp.

    a. What is the largest number of identical groups that can be made from all of the musicians?

    b. How many violinists will be in each group?

    c. How many campers will be left without a group?

48. Berrylin was asked to make flower arrangements for the tables at a sports banquet. She was given 36 tulips, 27 daisies, and 18 carnations.

    a. What is the largest number of identical arrangements that Berrylin can make from all of the flowers?

    b. How many of each flower will be put into each arrangement?

49. Phillip's Bakery gives away any remaining items at the end of the day to local hunger organizations. Today they had 39 loaves of bread and 86 cookies left. Bill is allowed to take some of the food home himself because he volunteers to box the goods each day. If he plans to take 3 loaves of bread and 10 cookies home, how many boxes will Bill fill given that each box must contain the same number of loaves of bread and cookies?

50. The library received 117 used books for their summer used book drive this year. There were 45 hardbacks and 72 paperbacks. The library delivers the books in bundles to local shelters.

    a. How many bundles can the library make if each bundle must have the same number of hardback and paperback books?

    b. How many paperback books will be in each bundle?

51. Two girl scout troops are planting gardens side-by-side on donated land. Based on the number of girls, Troop A is planning a garden that will cover 180 square feet, and Troop B's garden will be 204 square feet. The gardens must have a fence completely surrounding each one. The fencing is only sold in whole foot sections. In order to save money, the troops decide to maximize the shared amount of fence between the gardens.

    a. What is the greatest amount of fence the troops can share?

    b. What will the dimensions of each garden be if the shared fencing is maximized?

✏ **WRITING & THINKING**

52. Which possible divisors must be checked to see if 283 is prime?

53. Which possible divisors must be checked to see if 1291 is prime?

54. Suppose you are choosing the size of an elementary school class. It must be between 15 and 20 students. Knowing that teachers regularly break the class into smaller groups for projects and activities, answer the following questions.

    a. List the advantages and disadvantages for each of the possible class sizes between 15 and 20.

    b. Which size would you say is best?

8.1 PROJECT

PRIME FACTORIZATION AND THE NUMBER OF FACTORS

In Section 8.1, you learned about prime and composite numbers. In this project, you will explore a way to determine the number of factors a number has by using its prime factorization.

1. Find the prime factorization of 18. How many times does each prime factor appear in the prime factorization? Find all possible factors of 18. How many total factors are there?

2. Find the prime factorization of 100, and then find all factors of 100. How many times does each prime factor appear in the prime factorization? How many total factors are there?

It's possible that in parts 1 and 2 you determined the total number of factors by using a brute-force method to list each factor. Let's explore a systematic way to determine the total number of factors.

3. Does any factor in the list of factors of 18 have prime factors that are not listed in the prime factorization of 18? Similarly, does any factor in the list of factors of 100 have prime factors that are not listed in the prime factorization of 100? Explain why it is or isn't possible for the factors to have a prime factor that is not listed in the prime factorization.

Consider the number 72, which has a prime factorization of $2 \cdot 2 \cdot 2 \cdot 3 \cdot 3$. We can create the factors of 72 by choosing how many 2s and 3s we wish to have in the prime factorization of each individual factor.

4. Find the factors of 72 that have no 2s in their prime factorization.

5. Find the factors of 72 that have exactly one 2 in their prime factorization.

6. Find the factors of 72 that have exactly two 2s in their prime factorization.

7. Find the factors of 72 that have exactly three 2s in their prime factorization.

If you combine the factors found in parts 4 through 7 into a list, you will have all of the factors of 72.

8. How many factors appear in each set of factors in parts 4 through 7? How does this number compare to the number of times 3 appears in the prime factorization of 72?

9. How many sets of factors did we find in parts 4 through 7? How does this compare to the number of times 2 appears in the prime factorization of 72?

10. Use the answers to parts 8 and 9 to create an algebraic expression to describe how the total number of factors of 72 compares to the number of 2s and number of 3s in the prime factorization of 72.

11. Can you use the same logic to find an algebraic expression to describe how the total number of factors compares to the prime factorizations of 18 and 100? If so, write an expression for each.

12. Find a general rule that gives the number of factors of a given number based on its prime factorization. (**Hint:** Let p_1, p_2, and so on represent the prime factors in the prime factorization.)

13. Find the number of factors of 900 using the prime factorization. Compare using the formula to the brute-force method of listing every factor.

8.2 MODULAR ARITHMETIC

OBJECTIVES

1. Evaluate numbers using modular arithmetic.

2. Perform addition, subtraction, and multiplication with modular arithmetic.

3. Check bar codes using modular arithmetic.

OBJECTIVE 1 Introduction to Modular Arithmetic

In Section 8.1, we looked at the divisibility property of numbers. We introduced clever ways to test for divisibility by using the numbers 2 through 10. We also focused on numbers having precisely two divisors; that is, prime numbers.

Suppose we continue the idea of divisibility by creating a fictitious classification of numbers; let's call it "3visible." In our new classification, a number will be 3visible if it can be divided evenly by 3.

Here's a beginning list of 3visible numbers: 3, 6, 9, 12, 15, 18, … (notice they are the multiples of 3).

What if a number is not 3visible? If a number is not 3visible, then it must have a remainder when divided by 3. The number can be either 1 greater than 3visible or 2 greater than 3visible. But, if it's 3 greater than 3visible, then it is actually 3visible. Therefore, the remainder must either be a 1 or 2. Table 8.2.1 can help visualize this by listing numbers and their 3visible properties, starting with the number 3.

Table 8.2.1: 3visible Properties

Number	3visible	Remainder When Divided by 3
3	Yes	0
4	No	1
5	No	2
6	Yes	0
7	No	1
8	No	2

Let's consider the arithmetic properties of 3visible. If we add together two numbers that are 3visible, then the answer is also 3visible. If you add together two numbers that are 1 unit greater than 3visible numbers, then the answer is 2 greater than 3visible. For instance, add together the numbers 4 and 7, which both have a remainder of 1 when divided by 3.

$$4 + 7 = 11$$

$$11 \div 3 = 3 \quad \text{Remainder 2}$$

What do you think happens if you add a number that is 1 greater than 3visible to one that is 2 greater than 3visible? Try the numbers 4 and 5.

$$4 + 5 = 9$$

$$9 \div 3 = 3 \quad \text{Remainder 0}$$

Therefore, the sum of 4 and 5 is 3visible.

In fact, performing 3visible arithmetic is equivalent to adding the remainders together, as long as you remember to go back to 0 every time you get to 3. The classification we created works for 3visible, but what if we wanted to do the same for the number 4, or "4visible"? It would get a bit awkward to have an entire vocabulary of classifications that only we could understand. Thankfully, a system is already in place for this type of classification. It's called **modular arithmetic**.

Here's how it works. The operation modulo (or "mod" for short) is simply the remainder when dividing. For example, 4 mod 3 means "the remainder when 4 is divided by 3." Of course, the answer is 1 in this case, since we already know that when you divide 4 by 3, there is a remainder of 1. The following notation is used in modular arithmetic to denote this.

$$4 \bmod 3 = 1$$

📖 MODULAR ARITHMETIC

In **modular arithmetic**, a number n is congruent to the remainder r when it is divided by a fixed number m. We write $n \equiv r \pmod{m}$. Note that m is referred to as the **modulus**.

Sometimes we want to express that two numbers have the same remainder when divided by a modulus. To express this, we use an equivalence sign, ≡, and say that the two numbers are congruent. For instance, $5 \equiv 8 \pmod 3$ is read " 5 is congruent to 8 mod 3," which means 5 and 8 both have a remainder of 2 when divided by 3.

Now that we have the notation for modular arithmetic, we can understand 3visible more mathematically. Instead of 3visible we can now refer to this classification as modulo 3 (or mod 3, for short). Likewise, 4visible is simply mod 4. Table 8.2.2 is a modification of Table 8.2.1 with an added column for the congruence of modulo 3.

Table 8.2.2: 3visible Properties

Number	3visible	Remainder When Divided by 3	Congruence
3	Yes	0	$3 \equiv 0 \pmod 3$
4	No	1	$4 \equiv 1 \pmod 3$
5	No	2	$5 \equiv 2 \pmod 3$
6	Yes	0	$6 \equiv 0 \pmod 3$
7	No	1	$7 \equiv 1 \pmod 3$
8	No	2	$8 \equiv 2 \pmod 3$

Notice that in modulo 3, all numbers are congruent to either 0, 1, or 2. This cycling pattern is based on the possible remainders when divided by 3. In general, for arithmetic modulo m, the cycle will include the numbers $0, 1, 2, \ldots, m - 1$. For example, the cycle for modulo 6 will consist of the numbers 0, 1, 2, 3, 4, and 5.

Example 8.2.1

Completing a Modular Congruence Table

Complete the following table for modulo 5.

Table 8.2.3: Modulo 5

Number	Remainder When Divided by 5	Congruence
0	0	$0 \equiv 0 \pmod 5$
1	1	
2	2	
3	3	
4	4	
5	0	
6		
7		
8		
9		

Solution

The cycle of remainders for modulo 5 will consist of 0, 1, 2, 3, and 4. Because the cycle returns to 0 after reaching 4, we can complete the second column with the cycle. The last column of Table 8.2.4 is the notation that represents these remainders. Notice that until you reach the value of m, the remainder is the number itself.

Table 8.2.4: Modulo 5

Number	Remainder When Divided by 5	Congruence
0	0	$0 \equiv 0 \pmod 5$
1	1	$1 \equiv 1 \pmod 5$
2	2	$2 \equiv 2 \pmod 5$
3	3	$3 \equiv 3 \pmod 5$
4	4	$4 \equiv 4 \pmod 5$
5	0	$5 \equiv 0 \pmod 5$
6	1	$6 \equiv 1 \pmod 5$
7	2	$7 \equiv 2 \pmod 5$
8	3	$8 \equiv 3 \pmod 5$
9	4	$9 \equiv 4 \pmod 5$

It is not necessary to make a modular congruency table for all possible modular congruences of a particular number. Often we are asked to find the modular congruence of a single number. Remembering that we only need division, the process is straightforward.

Example 8.2.2

Finding Modular Congruence

Evaluate the following.

a. 15 mod 6

b. 72 mod 13

Solution

a. To find 15 mod 6, we need to divide 15 by 6 and find the remainder.

$$15 \div 6 = 2 \text{ remainder } 3$$

So $15 \equiv 3 \pmod 6$.

b. To find 72 mod 13, we need to divide 72 by 13 and find the remainder.

$$72 \div 13 = 5 \text{ remainder } 7$$

So $72 \equiv 7 \pmod{13}$.

OBJECTIVE 2 Operations with Modular Arithmetic

Most of us use modular congruence every day when we refer to the time of day. The military uses the 24-hour clock, but civilians normally use a 12-hour clock. At 1400 hours military time, we say it's 2:00 p.m. That's because $14 \equiv 2 \pmod{12}$. Similarly, 2300 hours is 11:00 p.m. because $23 \equiv 11 \pmod{12}$.

Modulo is more than a notation for writing congruent numbers. It extends standard arithmetic to modular arithmetic. It is possible to add, subtract, and multiply with modular congruence. Once again, you are familiar with adding and subtracting in modular arithmetic because of clock time. For instance, 2 hours after 11:00 p.m. is 1:00 a.m., not 13:00 p.m. The following example illustrates addition, subtraction, and multiplication in modular arithmetic.

Example 8.2.3

Evaluating Modular Operations

Evaluate each of the following.

a. $(12 + 7) \bmod 5$

b. $(21 - 18) \bmod 6$

c. $(35 \cdot 22) \bmod 10$

Solution

a. In order to calculate the sum of (12 + 7) mod 5, we can either calculate 19 mod 5, or we can calculate the individual pieces first and then add them together.

$$(12+7) \bmod 5 = (12 \bmod 5) + (7 \bmod 5)$$
$$\equiv 2 + 2 \ (\bmod \ 5)$$
$$\equiv 4 \ (\bmod \ 5)$$

Since the numbers in this example are small, there is little difference between calculating (12 mod 5) + (7 mod 5) and 19 mod 5, although both return the same result. However, as we will see later, this way of breaking things down is useful with larger numbers.

b. We can perform subtraction in the same manner as addition by calculating the individual pieces and then subtracting.

$$(21-18) \bmod 6 = (21 \bmod 6) - (18 \bmod 6)$$
$$\equiv 3 - 0 \ (\bmod \ 6)$$
$$\equiv 3 \ (\bmod \ 6)$$

c. Multiplication is consistent with both addition and subtraction, in that you can find the modular congruences and then multiply.

$$(35 \cdot 22) \bmod 10 = (35 \bmod 10) \cdot (22 \bmod 10)$$
$$\equiv 5 \cdot 2 \ (\bmod \ 10)$$
$$\equiv 10 \ (\bmod \ 10)$$
$$\equiv 0 \ (\bmod \ 10)$$

When performing any of these three operations with a particular modulus, we can either find the individual modulo first and then carry out the operation, or we can carry out the calculation on the numbers and then find the modular congruence. Both methods provide the same answer. Note, however, that these operations must use the same modulus all the way through the calculation.

Example 8.2.4

Evaluating Modular Addition

Calculate each of the following and compare the answers.

a. (10 mod 4) + (17 mod 4)

b. (10 + 17) mod 4

Solution

a. In order to find (10 mod 4) + (17 mod 4), we need to find each modulo first as we did in Example 8.2.3.

$$(10 \bmod 4) + (17 \bmod 4) \equiv 2 + 1 \ (\bmod \ 4)$$
$$\equiv 3 \ (\bmod \ 4)$$

b. For $(10 + 17) \bmod 4$, we can either add the numbers in the parentheses together first and then apply the modulus or we can find each individual modulus and then add them together as we did in part a. We will add the numbers first since they are small.

$$(10 + 17) \bmod 4 = 27 \,(\bmod\,4)$$
$$\equiv 3\,(\bmod\,4)$$

Comparing answers in parts a. and b., we can see that they provide equivalent solutions using modular arithmetic.

OBJECTIVE 3 Modular Arithmetic and Bar Codes

Finding modular arithmetic in action is as easy as picking up a book and finding its bar code. The bar code on a book contains a string of numbers called the International Standard Book Number (ISBN) that is unique to each book title and conveys such information as the publisher and country of origin. If the book was assigned an ISBN before 2007, the ISBN is 10 digits long. After 2007, the ISBN changed to 13 digits. Either way, the numbers in an ISBN are not assigned at random; modular arithmetic plays an integral part in the construction of either the 10-digit or 13-digit ISBN.

Figure 8.2.1: ISBN of a Textbook

Let's look at how the 10-digit ISBN is constructed. All 10-digit ISBNs are constructed from the numbers 0 through 9 as well as the letter X. The X stands for the number 10. The first nine digits are numbers that represent the language group, publisher, and title. The tenth digit is assigned such that when added to certain multiples of the other nine digits, the sum is 0 modulo 11. This tenth digit is referred to as the **check-sum digit**. Figure 8.2.2 shows the breakdown for a 10-digit ISBN.

Figure 8.2.2: ISBN Breakdown

In order to determine the check-sum digit for a particular 10-digit ISBN, a process of multiplication, addition, and modular arithmetic takes place. The following outlines how the process works once the initial nine digit are assigned.

Process for Finding the Check-Sum Digit for a 10-digit ISBN

1. Multiply the 1st digit by 10.

2. Multiply the 2nd digit by 9.

3. Multiply the 3rd digit by 8.

4. Multiply the 4th digit by 7.

5. Multiply the 5th digit by 6.

6. Multiply the 6th digit by 5.

7. Multiply the 7th digit by 4.

8. Multiply the 8th digit by 3.

9. Multiply the 9th digit by 2.

10. Add the multiples together.

11. The check-sum digit (or 10th digit) is then chosen so that the total of the products and the check-sum digit is 0 modulo 11.

Example 8.2.5 validates that a particular ISBN meets the constraints given by the process for finding the check-sum digit.

Example 8.2.5

Verifying the Validity of a 10-Digit ISBN

Verify the validity of the ISBN shown here by performing the process described for finding the check-sum digit.

ISBN 0-07-338125-X

9780073 381251

Figure 8.2.3

Solution

To verify the ISBN, we need to find the various multiples of the first 9 digits in the number, and add them together with the check-sum digit. Remember, the sum should be congruent to 0 mod 11.

First, we'll find the multiples of the first 9 digits.

$$(10 \cdot 0) + (9 \cdot 0) + (8 \cdot 7) + (7 \cdot 3) + (6 \cdot 3) + (5 \cdot 8) + (4 \cdot 1) + (3 \cdot 2) + (2 \cdot 5)$$
$$= 0 + 0 + 56 + 21 + 18 + 40 + 4 + 6 + 10$$
$$= 155$$

Next, we add 155 and 10, since X represents the number 10.

$$155 + 10 = 165$$

✔ Skill Check 8.2.2

Verify that the following is a valid 10-digit ISBN.

ISBN 0-306-40615-2

Finally, we need to find 165 (mod 11).

$$165 \equiv 0 \ (\text{mod } 11)$$

Because the sum is congruent to 0 (mod 11), the ISBN is indeed valid.

Example 8.2.6

Finding a Missing Digit of a 10-Digit ISBN

What should the third digit of the ISBN be for the following barcode? Notice that the check-sum number is 5.

ISBN 0-7?56-2153-5

9780735 621534

Figure 8.2.4

Solution

The ISBN we were given is 07?5621535.

We can set up the equation for the multiplication as follows, using x to represent the missing digit.

$$(10 \cdot 0) + (9 \cdot 7) + (8 \cdot x) + (7 \cdot 5) + (6 \cdot 6) + (5 \cdot 2) + (4 \cdot 1) + (3 \cdot 5) + (2 \cdot 3) + 5$$
$$\equiv 0 \ (\text{mod } 11)$$

Simplifying, we have

$$0 + 63 + 8x + 35 + 36 + 10 + 4 + 15 + 6 + 5 \equiv 0 \left(\text{mod } 11 \right)$$

$$8x + 174 \equiv 0 \left(\text{mod} 11 \right)$$

To make things easier on ourselves here, we can go ahead and evaluate 174 modulo 11 and substitute that value into our equation.

$$174 = 9 \ (\text{mod } 11)$$

So, now we have $8x + 9 = 0$ (mod 11).

This tells us that when the third digit of the ISBN, x, is multiplied by 8 and added to 9, the sum will be congruent to 0 (mod 11).

Now, we need to use a bit of trial and error to find the missing digit. We know that the digit must be a number between 0 and 10, inclusive. So we can begin trying numbers, starting with 0. Substituting $x = 0$ in our equation gives the following.

$$8x + 9 = 8(0) + 9$$
$$= 0 + 9$$
$$\equiv 0(\bmod 11) + 9(\bmod 11)$$
$$\equiv 9(\bmod 11)$$

Since this is not congruent to 0 (mod 11), the third digit cannot be a 0.

Next, suppose the third digit was a 1. Let's substitute $x = 1$ in our equation.

$$8x + 9 = 8(1) + 9$$
$$= 8 + 9$$
$$\equiv 8(\bmod 11) + 9(\bmod 11)$$
$$\equiv 17(\bmod 11)$$
$$\equiv 6(\bmod 11)$$

Since this is not congruent to 0 (mod 11), the third digit cannot be a 1.

Now, let $x = 2$.

$$8x + 9 = 8(2) + 9$$
$$= 16 + 9$$
$$\equiv 5(\bmod 11) + 9(\bmod 11)$$
$$\equiv 14(\bmod 11)$$
$$\equiv 3(\bmod 11)$$

Again, we do not have a value equivalent to 0 (mod 11), so we still have not found our number. Let $x = 3$.

$$8x + 9 = 8(3) + 9$$
$$= 24 + 9$$
$$\equiv 2(\bmod 11) + 9(\bmod 11)$$
$$\equiv 11(\bmod 11)$$
$$\equiv 0(\bmod 11)$$

Since 11 modulo 11 is congruent to 0, we have found our missing digit!

The third digit in the ISBN is 3.

One might wonder why an elaborate system of numbers, which includes a check-sum number and modular arithmetic, is needed for books. Although it might take us a few minutes to confirm the validity of an ISBN by hand, in our world of computers, this process only takes milliseconds to compute. A computer can quickly identify if a number has been entered incorrectly or even if a particular ISBN is legitimate. We can take one step further and apply this technique to credit card numbers since one of the 16 digits in a credit card number is a check-sum number. You can imagine how computers can immediately identify when you've accidentally typed in the wrong number.

Skill Check Answers

1. $(16 \bmod 2) \cdot (3 \bmod 2) \equiv 0 \cdot 1 (\bmod 2) \equiv 0 (\bmod 2)$

   $(16 \cdot 3) \bmod 2 \equiv 48 (\bmod 2) \equiv 0 (\bmod 2)$

2. First, find the multiples of the first 9 digits.

$$(10\cdot0)+(9\cdot3)+(8\cdot0)+(7\cdot6)+(6\cdot4)+(5\cdot0)+(4\cdot6)+(3\cdot1)+(2\cdot5)=130$$

Next, add 2 for the final digit of the ISBN. $130+2=132$

Finally, find $132\,(\text{mod}\,11)$.

$$132\equiv0\,(\text{mod}\,11)$$

Because the sum is congruent to 0 (mod 11), the ISBN is valid.

8.2   Exercises

✔ CONCEPT CHECK

1. The 10th digit of a 10-digit ISBN is referred to as the _____.

2. True or False: If a number is a multiple of n, that number will be congruent to 0 in modulus n.

3. True or False: The numbers in an ISBN are selected at random.

4. True or False: The check-sum digit in an ISBN must always be equal to 0.

💡 PRACTICE

Determine whether each statement is true or false.

5. $2^3\equiv3\,(\text{mod}\,5)$ 6. $\sqrt{81}\equiv3\,(\text{mod}\,2)$

7. $(3\cdot4)\equiv0\,(\text{mod}\,6)$

8. Any two even numbers x and y are equivalent to each other mod 2; that is, $x\equiv y\,(\text{mod}\,2)$.

9. Any two odd numbers a and b are equivalent to each other mod 2; that is, $a\equiv b\,(\text{mod}\,2)$.

10. $645,234\equiv111,111,111\,(\text{mod}\,3)$ (**Hint:** Two numbers are congruent mod 3 if both are divisible by 3. Use Section 8.1 methods to help you decide.)

11. $9,436,278,463,920\equiv764,283,237,885\,(\text{mod}\,5)$

Compute each value.

12. $12\equiv\underline{\quad}\,(\text{mod}\,5)$ 13. $13\equiv\underline{\quad}\,(\text{mod}\,4)$

14. $120\equiv\underline{\quad}\,(\text{mod}\,11)$ 15. $84\equiv\underline{\quad}\,(\text{mod}\,3)$

16. $5^2\equiv\underline{\quad}\,(\text{mod}\,2)$ 17. $4328\,\text{mod}\,10$

18. 60,002 mod 6

19. 311 mod 4

20. 44 mod 12

21. 113 mod 12

22. $3^4 \equiv$ __ (mod 3)

23. 812 mod 6

24. $72 \equiv$ ___ (mod 11)

25. 113 mod 8

Convert each 24-hour clock time to an equivalent 12-hour clock time.

26. 0800 hours

27. 1300 hours

28. 2100 hours

29. 0000 hours

Convert each 12-hour clock time to an equivalent 24-hour clock time.

30. 5:00 p.m.

31. 12:00 p.m.

32. 1:00 p.m.

33. 7:00 a.m.

Compute each value.

34. $(56+87) \bmod 13$

35. $\left[12(34+6)\right] \bmod 7$

36. $\left[128-(15+8)\cdot 23\right] \bmod 11$

37. $\left(2^5-17\right) \bmod 3$

38. $(11+39) \bmod 5$

39. $(91-16) \bmod 6$

40. $(27 \cdot 18) \bmod 7$

41. $[6(11+5)-21] \bmod 13$

42. $[3(15+17)+2(56-14)] \bmod 8$

43. $(9^2-3^4) \bmod 3$

🚀 **APPLICATIONS**

Determine whether each 10-digit ISBN is valid. If it is not, state what the correct check-sum number should be.

44. 0-392-31123-2

45. 1-103-24582-6

46. 0-332-15573-0

47. 1-02-345678-8

48. 0-022-44668-8

49. 2-013-31943-2

Determine the missing digit for each 13-digit ISBN.

For a 13-digit ISBN, multiply the first digit by 1, the second digit by 3, the third digit by 1, the fourth digit by 3, and so on, until you get the 12th digit, which you multiply by 3. These products are added and the 13th digit is chosen so that the total of all these numbers is 0 modulo 10.

50. *Emma* by Jane Austin: 978048640648**?**

51. *Harry Potter and the Sorcerer's Stone* by J.K. Rowling: 978059035342**?**

52. *The Hunger Games* by Suzanne Collins: 97**?**0439023528

53. *The Girl with the Dragon Tattoo* by Stieg Larsson: 9**?**80307949486

54. *The Way of Kings* by Brandon Sanderson: 978-0765376**?**71

55. *Furies of Calderon* by Jim Butcher: 978-**?**441012688

Decide whether each barcode is valid. If it is not, state what the correct check-sum number should be.

Each of the following bar codes consists of 12 digits. To find the check-sum digit, multiply the first digit by 3, the second digit by 1, the third digit by 3, the fourth digit by 1, and so on, until you get the 11th digit, which you multiply by 3. The final digit is chosen so that the sum of the first 11 products and the final digit is 0 modulo 10.

56.
100230 42041?

57.
374034 61005?

58.
102010 40302?

Calculate the correct first digit for each barcode.

59.
?00200 30X450

60.
?21045 729336

61.
?21323 435362

Credit card companies use the Luhn algorithm to help construct secure numbers. Use the following steps to determine whether each credit card number is valid. If it is not, find the correct check-sum number.

1. The check-sum digit is the last digit in the number, whether the card number is 13, 15, or 16 digits long. Working right to left, starting with the digit to the left of the check-sum number, double the value of every other digit.

2. If a number becomes a 2-digit number after doubling, treat each digit as an individual digit. Finally, sum the digits of all doubled numbers as well as the undoubled numbers including the check-sum number.

3. If the total is congruent to 0 (mod 10), then the number is valid according to the Luhn formula; otherwise it is not valid.

62. 3780 2850 1184 225   **63.** 300 9255 939 6891

64. 389 841 621 516 22   **65.** 4929 1175 0198 3180

8.2 PROJECT

AFFINE SHIFT CIPHERS

In Section 8.2, you learned about modular arithmetic. In this project, you will use modular arithmetic to encode and decode messages.

One method of encoding messages is to simply convert the letters of the alphabet to numbers by identifying A with 0, B with 1, and so on.

A	B	C	D	E	F	G	H	I	J	K	L	M	N	O	P	Q	R	S	T	U	V	W	X	Y	Z
0	1	2	3	4	5	6	7	8	9	10	11	12	13	14	15	16	17	18	19	20	21	22	23	24	25

A message encoded in this manner would be a string of numbers. We can make this encryption slightly more advanced by using a shift cipher, also called a Caesar cipher, which converts the message back to a string of letters. To use a shift cipher on the numerically encoded alphabet, add a fixed value (the "shift" value) to the number and then calculate the result modulo 26, then substitute the corresponding value into the encoded string.

For example, suppose we want to encode the letter T using a shift cipher with a shift of 9. First, we need to know that T corresponds to 19. Adding the shift value results in a value of 28, which is not a number modulo 26. We can calculate that 28 mod 26 = 2. The letter we'd use in the encoded message is C.

1. Encode the word NOTE using a shift cipher with a shift value of 11.

If the value of the shift is known, a shift cipher is easy for anyone to decode. In the example with a shift of 9, the encoded letter was C, which corresponds to a value of 2. To decode this letter, we subtract the shift value to get −7. Since this number is negative, we can add 26 to get a value between 0 and 25. In this case, we get 19, which corresponds to T.

2. Decode DLQP that was coded using a shift cipher with a shift of 11.

Since shift ciphers are relatively easy to decode, more advanced coding methods are used to keep information safe. One slightly more advanced method is an affine shift cipher. This method involves multiplying the letter value by a number relatively prime to 26, called the key, before adding the shift value. For example, suppose we want to encode the letter T using an affine shift cipher with a key of 7 and a shift of 9. We would compute $7(19)+9=142$. Next, we would calculate 142 mod 26 = 12. This means the encoded letter is M.

3. Encode the word CODE using an affine shift cipher with a key of 5 and a shift of 11.

To decode a message encoded with a shift cipher, the shift value is subtracted. To decode a message encoded with an affine shift cipher, the shift value is first subtracted and then division modulo 26 is performed. Division in modular arithmetic is a little tricky; it involves multiplying by the multiplicative inverse of the key modulo 26. For example, if the value of the key is k, we need to find a value d so that $kd \equiv 1 \pmod{26}$. This is the reason we want k to be relatively prime to 26: d is the multiplicative inverse of k modulo 26.

4. List out the numbers less than 26 that are relatively prime to 26.

5. Suppose you intercept a message and discover that it is encoded using an affine shift cipher with a key of 3 and a shift of 12. Determine the multiplicative inverse of the key modulo 26. (**Hint:** Use the list created in part 4 and determine the value of each product modulo 26.)

6. You intercept a message that you know has been encoded with an affine shift cipher that has a key of 3 and a shift of 12. Decode the following message.

OCO MOOKORMZSY LYIUKLYV

7. You are told that the following responses to such messages are common.

    • SENDING   • DECLINE   • UNCLEAR

Choose a response to the message and encode it using the same cypher.

8.3 FERMAT'S LITTLE THEOREM AND PRIME TESTING

OBJECTIVES

1. Verify Fermat's Little Theorem.

2. Verify numbers are prime using the contrapositive of Fermat's Little Theorem.

OBJECTIVE 1 Fermat's Little Theorem

Two 17th-century mathematicians, Pierre de Fermat and his friend and confidant Bernard Frénicle de Bessy, often corresponded by letter. In a letter dated October 18, 1640, Fermat tells Frénicle de Bessy about a marvelous discovery in number theory that he has come across. As he often did in his letters, Fermat provided no proof of his discovery. The proof of the comment in the letter took another 100 years to be rediscovered. Thankfully, it was published by yet another great mathematician, Leonhard Euler, and since then has been known as Fermat's Little Theorem. More than 400 years later, we continue to use this theorem to help us establish whether a number is prime. The theorem is stated as follows.

🔑 **FERMAT'S LITTLE THEOREM**

Let p be any prime number and x be any positive integer. Then, $x^p - x \equiv 0$ (mod p).

This theorem is a statement about a calculation in modular arithmetic. But rather than talking about addition or multiplication as we did in the previous section, this time exponents are involved. Fermat's Little Theorem says that if you multiply a number by itself p times and subtract off that same number, then the answer is a multiple of p; that is, the answer is congruent to 0 (mod p).

To try it out, we need a prime number p, such as 5, and another positive integer x, such as 2. Now let's carry out the calculation in the theorem with $p = 5$ and $x = 2$. We'll raise 2 to the 5th power and subtract 2, then show that the answer is a multiple of 5.

$$\begin{aligned}
x^p - x = 2^5 - 2 \\
= 2 \cdot 2 \cdot 2 \cdot 2 \cdot 2 - 2 \\
= 32 - 2 \\
= 30 \\
\equiv 0 \left(\text{mod } 5 \right)
\end{aligned}$$

Notice that 30 is congruent to 0 modulo 5 because 30 is a multiple of 5.

Let's try another example.

Example 8.3.1

Verifying Fermat's Little Theorem

Verify that $x^p - x \equiv 0$ (mod p) when $p = 7$ and $x = 4$.

Solution

$$4^7 - 4 = 4 \cdot 4 \cdot 4 \cdot 4 \cdot 4 \cdot 4 \cdot 4 - 4$$

We'll stop at this point and introduce a useful short cut. You might not be able to calculate 4^7 in your head, but the beauty of modular arithmetic is that you don't have to. Remember that we want our equation to work in modulo 7. So changing any number to its equivalent modulo 7 throughout the calculation is helpful in that it makes the numbers we are manipulating smaller. Notice that $4 \cdot 4 = 16 \equiv 2 \pmod{7}$. By replacing each pair of 4s that are multiplied together with a 2, we can make life easier.

$$4^7 - 4 = (4 \cdot 4) \cdot (4 \cdot 4) \cdot (4 \cdot 4) \cdot 4 - 4$$
$$\equiv 2 \cdot 2 \cdot 2 \cdot 4 - 4 \pmod{7}$$
$$\equiv 8 \cdot 4 - 4 \pmod{7}$$
$$\equiv 32 - 4 \pmod{7}$$
$$\equiv 28 \pmod{7}$$
$$\equiv 0 \pmod{7}$$

Therefore, we've shown that $4^7 - 4 \equiv 0 \pmod{7}$.

✔ **Skill Check 8.3.1**

Verify that $x^p - x \equiv 0 \pmod{p}$ when $p = 11$ and $x = 10$.

OBJECTIVE 2 Prime Testing

Recall that one method for establishing primes is to test for possible divisors. As we showed in previous examples, that method works very well for small numbers. But how would you like to test the divisors of the number 4,294,967,297 or the number 18,446,744,073,709,551,617? Although we only need to test the prime numbers as large as the square root of each number, for the second number, that's still more than two hundred million prime numbers to try out one-by-one.

Luckily, we can use Fermat's Little Theorem to provide a more useful method for testing large numbers. The theorem states that for any prime number p, $x^p - x \equiv 0 \pmod{p}$ is true. Recall from our discussion of logic in Chapter 3 that if a statement is true, then its contrapositive is also true. In other words, if a implies b is true, then not b implies not a is true. So, we can use logic to turn the theorem around and tell us something about prime numbers. Fermat's Little Theorem says that if you raise a number to a prime, then the result after subtraction of the same number is equivalent to 0 mod that prime. So the contrapositive says that if we perform the operation and *don't* get a value equivalent to 0 $\pmod{p}$, then p is not a prime number to begin with. The formal theorem is stated as follows.

🔑 **CONTRAPOSITIVE OF FERMAT'S LITTLE THEOREM**

Let x and n be positive integers. If $x^n - x \not\equiv 0 \pmod{n}$, then n is **not** a prime number.

Let's stop and emphasize that the contrapositive says that if we do *not* get 0, then the power we used is *not* prime. However, we must be careful to not misinterpret the theorem to imply that obtaining 0 means that the power must be a prime number. In other words, if we can find at least one value for x to use in the formula that results in a number other than 0, we've shown that the number n is not prime. So, this contrapositive allows us to test if a number is not prime simply by trying Fermat's Little Theorem with a single number. Let's try this out for some small manageable numbers before we move on to our bigger example of 4,294,967,297. We know that the number 8 is not prime, so we'll start with that.

Example 8.3.2

Prime Testing

Use the contrapositive of Fermat's Little Theorem to verify that the number $n = 8$ is not prime by using the number 2 for x.

Solution

Substituting in $n = 8$ and $x = 2$ into the contrapositive, we have the following. Note that 2^3 is congruent to 0 (mod 8).

$$x^n - x = 2^8 - 2$$
$$= (2 \cdot 2 \cdot 2) \cdot (2 \cdot 2 \cdot 2) \cdot (2 \cdot 2) - 2 \qquad \text{Factor } 2^8 \text{ and group.}$$
$$\equiv 0 \cdot 0 \cdot 4 - 2 \, (\text{mod } 8) \qquad \text{Substitute } 2^3 \equiv 0 \, (\text{mod } 8).$$
$$\equiv -2 \, (\text{mod } 8)$$
$$\equiv (-2 \bmod 8) + (8 \bmod 8) \qquad \text{Add 8 (mod 8) to remove the negative.}$$
$$\equiv -2 + 8 \, (\text{mod } 8)$$
$$\equiv 6 \, (\text{mod } 8)$$
$$\not\equiv 0 \, (\text{mod } 8)$$

Notice that this is the first time a negative has appeared at the front of an equivalence in our calculations. To simplify, we added 0 to the equation by adding 8 (mod 8). This allowed us to return to a positive equivalence that we are accustomed to.

Since raising 2 to the 8th power and subtracting 2 does not result in a multiple of 8, we have confirmed that 8 is not prime. It's worth noting that although we are told to start with $x = 2$, we could use any number to show that 8 is not a prime number.

> **Helpful Hint**
>
> To calculate negative equivalences, you can add any multiple of the modular base. In other words, −2 mod 8 is congruent to (−2 + 8) (mod 8), or 6 (mod 8).

To get the most out of this method, we need to learn how to calculate large exponents in modular arithmetic. Table 8.3.1 recalls a few rules of exponents.

> **✔ Skill Check 8.3.2**
>
> Simplify each of the following.
>
> **a.** $3^3 \cdot 3^4$
>
> **b.** $(4^2)^5$
>
> **c.** $\dfrac{2^{10}}{2^4}$

Table 8.3.1: Rules of Exponents

Product Rule	$x^a \cdot x^b = x^{a+b}$
Quotient Rule	$\dfrac{x^a}{x^b} = x^{a-b}$
Power Rule	$(x^a)^b = x^{ab}$

To use the contrapositive of Fermat's Little Theorem for a large number n, we will have to raise a number x to that large exponent. Using standard arithmetic, x to the power n will be a very difficult number to calculate. However, the beauty of using modular arithmetic is that we can always keep numbers relatively small; that is, less than n. The numbers might still be large, but they will be manageable. Let's try testing a slightly bigger number as an example of how we can use modular arithmetic to make calculations easier. We're going to use a squaring technique to break the exponent we want into manageable pieces.

Example 8.3.3

Prime Testing Using Modular Arithmetic

Verify that 39 is not prime using the contrapositive of Fermat's Little Theorem and $x = 2$.

Solution

We want to show that 39 is not prime using the number 2. To utilize the contrapositive of Fermat's Little Theorem we would have to establish that $2^{39} - 2$ is not congruent to 0 modulo 39. To do this directly we would have to calculate 2^{39}—a very large number. Instead, we can use a variant of the grouping idea that we have seen before to build up 2^{39} piece by piece. We will just be very strategic in how we group.

Notice that $2^{39} = 2^{32} \cdot 2^4 \cdot 2^2 \cdot 2 \cdot 1$ So if we can find these powers of 2, we can combine them to find 2^{39}. The following steps show the most efficient way to find these powers.

$$2^1 \equiv 2 \pmod{39}$$
$$2^2 \equiv 4 \pmod{39}$$
$$2^4 = 2^2 \cdot 2^2 \equiv 4 \cdot 4 \equiv 16 \pmod{39}$$
$$2^8 = 2^4 \cdot 2^4 \equiv 16 \cdot 16 = 256 \equiv 22 \pmod{39} \quad \text{Verify this yourself.}$$

> **Helpful Hint**
>
> Notice that each line uses the result of the previous line after the first congruence symbol. For instance, the third line uses 4 in place of 2^2. When an equal sign appears in the string of calculations, standard arithmetic was performed.

In standard arithmetic, $2^8 = 256$, and we would need to continue on with our calculations of 2^{39} using 256. However, in arithmetic modulo 39, $2^8 \equiv 22 \pmod{39}$. This smaller number of 22 will be much easier to work with. This pattern of reducing our calculations modulo 39 keeps the numbers manageable. Notice how the pattern continues in the next steps.

$$2^{16} = 2^8 \cdot 2^8 \equiv 22 \cdot 22 = 484 \equiv 16 \pmod{39} \quad \text{Again, verify this yourself.}$$
$$2^{32} = 2^{16} \cdot 2^{16} \equiv 16 \cdot 16 = 256 \equiv 22 \pmod{39}$$

From these powers of 2, we have all we need to calculate $2^{39} - 2$.

$$2^{39} - 2 = 2^{32} \cdot 2^4 \cdot 2^2 \cdot 2^1 - 2$$
$$\equiv 22 \cdot 16 \cdot 4 \cdot 2 - 2 \pmod{39}$$
$$\equiv 2816 - 2 \pmod{39}$$
$$\equiv 2814 \pmod{39}$$
$$\equiv 6 \pmod{39}$$

Verify for yourself that $2814 \equiv 6 \pmod{39}$. Once again, since this is not equivalent to 0 (mod 39), we have established that 39 is not prime. Notice that we did the calculation without ever multiplying two numbers that were bigger than 39.

So far, we have only confirmed that a few relatively small numbers are not prime, which may have been obvious to some. To demonstrate the real power of Fermat's Little of Theorem, we can show that the monstrous number 4,294,967,297 is not prime using the same technique we developed in Example 8.3.3. To illustrate this technique, we didn't choose 4,294,967,297 at random, but instead chose to use $2^{32+1} = 4,294,967,297$.

To use the Fermat's Little Theorem technique, we need to calculate

$$x^{4,294,967,297} - x \pmod{4,294,967,297}$$

for some value of x.

If we were able to establish that the answer isn't zero for some value of x, then we would know that 4,294,967,297 is not prime. Of course $x = 1$, doesn't work because $1^{4,294,967,297} - 1 = 0$. Unfortunately, there is no way we can get an answer by simply typing $x^{4,294,967,297}$ into a calculator or computer no matter what other value of x we choose. The answer is more than one billion digits long. (Think for a second what a number this large might look like.)

However, we can employ the successive squaring technique we have just developed by using exponent rules once again. This large number is 2^{32+1}, so we have the following.

$$x^{4,294,967,297} = x^{\left(2^{32}\right)+1} = x^{\left(2^{32}\right)} \cdot x^1$$

As we have already seen, we can calculate $x^{2^{32}} \pmod{4,294,967,297}$ by successive squaring. We will see the details in just a second.

Now, we need to choose the value for x. There's no reason not to choose the next smallest possibility after 1. Therefore, we'll let $x = 2$. Let's begin the process of squaring x.

$$2 \equiv 2 \pmod{4,294,967,297}$$

$$2^{\left(2^1\right)} = 2^2 = 2 \cdot 2 \equiv 4 \pmod{4,294,967,297}$$

$$2^{\left(2^2\right)} = 2^4 = 2^2 \cdot 2^2 \equiv 4 \cdot 4 \equiv 16 \pmod{4,294,967,297}$$

$$2^{\left(2^3\right)} = 2^8 = 2^4 \cdot 2^4 \equiv 16 \cdot 16 \equiv 256 \pmod{4,294,967,297}$$

$$2^{\left(2^4\right)} = 2^{16} = 2^8 \cdot 2^8 \equiv 256 \cdot 256 \equiv 65,536 \pmod{4,294,967,297}$$

$$2^{\left(2^5\right)} = 2^{32} = 2^{16} \cdot 2^{16} \equiv 65,536 \cdot 65,536 \equiv 4,294,967,296 \pmod{4,294,967,297}$$

$$2^{\left(2^6\right)} = 2^{64} = 2^{32} \cdot 2^{32} \equiv (4,294,967,296) \cdot (4,294,967,296)$$
$$= 18,446,744,073,709,551,616 \equiv 1 \pmod{4,294,967,297}$$

$$2^{\left(2^7\right)} = 2^{128} = 2^{64} \cdot 2^{64} \equiv 1 \cdot 1 \equiv 1 \pmod{4,294,967,297}$$

$$2^{\left(2^8\right)} = 2^{256} = 2^{128} \cdot 2^{128} \equiv 1 \cdot 1 \equiv 1 \pmod{4,294,967,297}$$

$$2^{\left(2^9\right)} = 2^{512} = 2^{256} \cdot 2^{256} \equiv 1 \cdot 1 \equiv 1 \pmod{4,294,967,297}$$

Now, we can see that the pattern developing means we will get a value equivalent to 1 (mod 4,294,967,297) every time we square from now on. We will get 1(mod 4,294,967,297) for 2^{1024}, 2^{2048}, …, and all the way up to $2^{4,294,969,296}$.

So now we are ready to calculate $2^{4,294,967,297} - 2 \pmod{4,294,967,297}$.

$$2^{4,294,967,297} - 2 = 2^{4,294,967,296+1} - 2$$
$$= 2^{4,294,967,296} \cdot 2^1 - 2$$
$$\equiv 1 \cdot 2^1 - 2 \pmod{4,294,967,297}$$
$$\equiv 2 - 2 \pmod{4,294,967,297}$$
$$\equiv 0 \pmod{4,294,967,297}$$

Now, let's think carefully about what we just found out. Remember, as we stated before, the theorem says that if we do *not* get 0 (mod n), then the number is *not* prime. It does **not** say that getting 0 (mod n) means the number is prime. So, we have not found out anything helpful in determining if our number is prime *yet*. We can try a different value for x to see if that helps. Let's try $x = 3$ instead. Don't get discouraged here. We know that these calculations are very large, but in bite-size steps they are manageable.

$$3 \equiv 3 \pmod{4,294,967,297}$$

$$3^{(2^1)} = 3^2 \equiv 3 \cdot 3 \equiv 9 \pmod{4,294,967,297}$$

$$3^{(2^2)} = 3^4 \equiv 9 \cdot 9 \equiv 81 \pmod{4,294,967,297}$$

$$3^{(2^3)} = 3^8 \equiv 81 \cdot 81 \equiv 6561 \pmod{4,294,967,297}$$

$$3^{(2^4)} = 3^{16} \equiv 6561 \cdot 6561 \equiv 43,046,721 \pmod{4,294,967,297}$$

$$3^{(2^5)} = 3^{32} \equiv 43,046,721 \cdot 43,046,721 \equiv 3,793,201,458 \pmod{4,294,967,297}$$

$$3^{(2^6)} = 3^{64} \equiv 3,793,201,458 \cdot 3,793,201,458 \equiv 1,461,798,105 \pmod{4,294,967,297}$$

$$3^{(2^7)} = 3^{128} \equiv 1,461,798,105 \cdot 1,461,798,105 \equiv 852,385,491 \pmod{4,294,967,297}$$

$$3^{(2^8)} = 3^{256} \equiv 852,385,491 \cdot 852,385,491 \equiv 547,249,794 \pmod{4,294,967,297}$$

$$3^{(2^9)} = 3^{512} \equiv 547,249,794 \cdot 547,249,794 \equiv 1,194,573,931 \pmod{4,294,967,297}$$

$$3^{(2^{10})} = 3^{1024} \equiv 1,194,573,931 \cdot 1,194,573,931 \equiv 2,171,923,848 \pmod{4,294,967,297}$$

$$\vdots$$

$$3^{(2^{31})} = 3^{2,147,483,648} \equiv 10,324,303 \pmod{4,294,967,297}$$

$$3^{(2^{32})} = 3^{4,294,967,296} \equiv 3,029,026,160 \pmod{4,294,967,297}$$

Notice that between 3^{1024} and $3^{2,147,483,648}$ we've skipped over several steps. In fact, we need to continue doubling 21 times between the steps listed. For sake of space, we've taken the liberty to do the work for you and leave you to verify these steps on your own.

Now to use the theorem, we calculate the following.

$$3^{4,294,967,297} - 3 = 3^{4,294,967,296+1} - 3$$
$$= 3^{4,294,967,296} \cdot 3^1 - 3$$
$$\equiv 3,029,026,160 \cdot 3 - 3 \,(\text{mod}\, 4,294,967,297)$$
$$\equiv 9,087,078,480 - 3 \,(\text{mod}\, 4,294,967,297)$$
$$\equiv 9,087,078,477 \,(\text{mod}\, 4,294,967,297)$$
$$\equiv 497,143,883 \,(\text{mod}\, 4,294,967,297)$$

Thankfully, this is not 0. So 4,294,967,297 is indeed not prime, but composite.

This is quite a lot of work to do by hand with a calculator, but certainly far less than the hundreds of millions of divisions we would otherwise have had to check by the divisor method. Even computers can't handle the divisor method for numbers this big, but they can use modular arithmetic to apply Fermat's Little Theorem. Now we have a healthy appreciation for what goes on behind the scenes for this computation.

Skill Check Answers

1. $10^{11} - 10 = (10 \cdot 10) \cdot (10 \cdot 10) \cdot (10 \cdot 10) \cdot (10 \cdot 10) \cdot (10 \cdot 10) \cdot 10 - 10$ **2. a.** 3^7 **b.** 4^{10} **c.** 2^6
$$\equiv 1 \cdot 1 \cdot 1 \cdot 1 \cdot 1 \cdot 10 - 10 \,(\text{mod}\, 11)$$
$$\equiv 1 \cdot 10 - 10 \,(\text{mod}\, 11)$$
$$\equiv 0 \,(\text{mod}\, 11)$$

8.3 Exercises

✔ CONCEPT CHECK

1. One can test for prime numbers using _____.

2. True or False: If a statement is true, then the contrapositive is also true.

3. True or False: If you calculate a 0 from the contrapositive of Fermat's Little Theorem, then the power must be a prime number.

💡 PRACTICE

Verify that Fermat's Little Theorem holds true for each prime number using the value of x given.

4. The prime number 11 and $x = 6$ 5. $x = 2$ with the prime 23

6. The third prime number and $x = 10$ 7. $p = 79$ and $x = 3$

8. The prime number 13 and $x = 3$

9. The prime number 17 and $x = 2$

10. $x = 3$ with the prime 19

11. $p = 31$ and $x = 4$

12. $p = 7$ and $x = 2$

Use the contrapositive of Fermat's Little Theorem to show that each given number is composite.

13. 63,571 using $x = 3$

14. 12,731 using $x = 3$

15. 65,476,751 using $x = 2$

17. 7802 using $x = 2$

18. 10,431 using $x = 3$

19. 103,322 using $x = 3$

20. 123,456,782 using $x = 2$

Use technology to confirm that each given number is composite using the contrapositive of Fermat's Little Theorem.

21. 2427

22. 24,271

23. 245,127

24. 10,050,207

25. 1963

26. 1649

27. 903,219

28. 10,203

29. 565,841

30. 106,057

31. 35,066

32. 14,772,361

8.3 PROJECT

FERMAT AND MERSENNE PRIMES

In Section 8.3, you learned about Fermat's Little Theorem and testing whether numbers are prime. In this project, you will explore two special types of numbers and test whether they are prime.

Very large prime numbers play a vital role in keeping information secure. There are two special types of numbers that have played a role in searching for large primes. They are called Fermat numbers and Mersenne numbers. A Fermat number is a number of the form $2^{2^n} + 1$, where $n = 0, 1, 2, \ldots$. For example, when $n = 2$, we have the Fermat number $2^{2^2} + 1 = 2^4 + 1 = 17$. Since this resulting number is prime, we say that 17 is a Fermat prime.

1. Find the Fermat numbers that correspond to $n = 0$, $n = 1$, and $n = 3$. Are these numbers Fermat primes? (Note that the Fermat number 65,537, which corresponds to $n = 4$, is a Fermat prime.)

2. Find the Fermat number that corresponds to $n = 5$.

3. Evaluate the expression $\left(2^9 + 2^7 + 1\right)\left(2^{23} - 2^{21} + 2^{19} - 2^{17} + 2^{14} - 2^9 - 2^7 + 1\right)$. Is the Fermat number that corresponds to $n = 5$ a prime number? Why or why not?

A Mersenne number is a number of the form $2^p - 1$, where p is a prime number. If the result is a prime, then the number is a Mersenne prime.

It is generally unknown which Fermat numbers and which Mersenne numbers are prime. It is believed that only the Fermat numbers corresponding to $n = 0$, 1, 2, 3, and 4 are prime and infinitely many Mersenne numbers are prime. In fact, since 1996, the Great Internet Mersenne Prime Search (GIMPS) has been testing larger and larger Mersenne numbers to determine which ones are prime. This collaborative effort uses the computing power of multiple volunteer computers connected to the internet and led to the discovery, in December 2018, of the largest known prime number, $2^{82,589,933} - 1$, which has 24,862,048 digits.

4. Find the Mersenne numbers that correspond to $p = 2$, $p = 3$, $p = 5$, and $p = 7$. Determine whether these are Mersenne primes.

5. Find the Mersenne number that corresponds to $p = 11$.

Recall that the contrapositive of Fermat's Little Theorem is stated as follows:

Let x and n be positive integers. If $x^n - x \not\equiv 0 \pmod{n}$, then n is not a prime number.

6. Use the contrapositive of Fermat's Little Theorem to show that the Mersenne number corresponding to $p = 11$ is not a Mersenne prime. (**Hint:** Rewrite the exponent as a sum of powers of 2.)

7. Find a prime factorization for the Mersenne number in the previous problem. How did you approach factoring it?

8.4 FERMAT'S LITTLE THEOREM AND PUBLIC-KEY ENCRYPTION

OBJECTIVE 1 Euler's Theorem

Secret codes, or encryptions, have been around since the beginning of history. One of the earliest and simplest was one that is said to have been used by Julius Caesar to communicate with his military. Now known as a Caesar cipher, or cipher shift, messages were encoded by substituting each letter in the alphabet by another letter some fixed number of letters down the alphabet. For example, in a right shift of 3, the letter A would be replaced by the letter D, the letter B would be replaced by E, and so on.

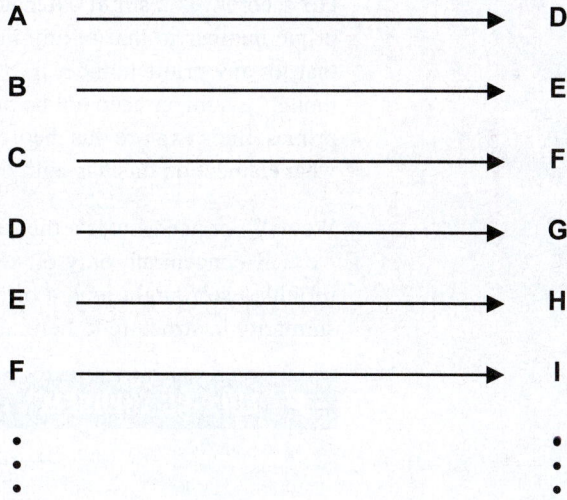

Figure 8.4.1: Cipher Shift

Helpful Hint

Secret codes, or encryptions, are a part of cryptography. Cryptography is the technique of transferring information through secure codes in such a way that only those intended to understand the message will be able to do so.

Once a message was sent to a general in the army, all the general needed was the size of the shift down the alphabet and he could know Caesar's wishes. Although effective, you can see that if someone knew the size (and direction) of the cipher shift, they could easily decode an intercepted message.

Over time, codes have become more and more advanced and harder to break. One of the most widely used coding techniques today is called **public-key encryption**. Its existence is crucial in our computer-driven world as it is the most widely used system for encrypting and protecting internet transactions. Its beauty lies in the fact that to "unlock" a message, two keys are required—one that is secret and another that is public—hence its name.

One of the best explanations for public-key encryption comes from an article in *The Chronicle of Higher Education* by Konrad M. Lawson. He tells a story of two professors, *Professor More* and *Professor Erasmus*, who want to exchange private documents via the public mail system. To do this, both professors buy a padlock. Then they each send the open padlocks (without keys) to the other. When it's time to send a private document, Professor More puts the document in a box and locks it with Professor Erasmus' lock. Upon receiving the package, Professor Erasmus can use his own key, which he kept, to unlock it. If he chooses to send something back, Professor Erasmus uses Professor More's lock to secure the document. Neither professor has risked someone intercepting the package or stealing the key, since the keys never left the possession of the owner. The padlocks are analogous to the "public key" that only allows someone to encode a message; that is, to "lock the box." The keys are analogous to the "private keys," which are the essential bits used to decode the message, or "unlock the box."[1]

The idea for a public-key cipher was first publicly described by three mathematicians and computer scientists: Ron Rivest, Adi Shamir, and Leonard Aldeman in 1977. This cipher, called RSA after the three men, is built into almost all secure computer systems in some way, and luckily for them, they patented their idea (US Patent 4,405,829). Despite their patent and independent work, they were not the first to come up with the idea. Clifford Cocks, an English mathematician discovered this idea in 1973. But because he was working for the UK secret service, GCHQ, at the time of his discovery, his work remained classified until 1998.

> **Fun Fact**
>
> An open problem in cryptography is determining whether factoring large numbers is the only way to break an RSA encryption. This is known as the RSA Problem. Perhaps it has already been solved, but is still classified information. Only time will tell.

The idea for the cipher revolves around a theorem attributed to Leonhard Euler. His theorem is a slight extension of Fermat's Little Theorem from the use of a prime number to that of any number. Remember, Fermat's Little Theorem says that for any prime number p, $x^p - x \equiv 0 \pmod{p}$ for any x. For Euler's Theorem, though, a number need not be prime; it just needs to be the product of at least two primes. Let's explore this theorem to understand what Euler is saying and then see what connection this has with encryptions.

We will not need Euler's theorem in its full generality for our purposes. Instead we will concentrate only on when n is a product of two primes. Although the formula itself might look a bit intimidating at first, see if you can recognize the similarity in structure to Fermat's Little Theorem.

🔑 **EULER'S THEOREM**

$$x^{a(p-1)(q-1)+1} - x \equiv 0 \pmod{n},$$

where p and q are prime numbers, $n = pq$, and x and a are any positive integers.

Did you notice that the basic format of the equation in Euler's Theorem looks similar to the one in Fermat's Little Theorem? Both are of the general form

$$x^{\text{number}} - x \equiv 0 \pmod{\text{modulus}}.$$

In Fermat's Little Theorem, the exponent and the modulus are the same prime number. In Euler's Theorem, that is not the case. Example 8.4.1, uses a set of prime numbers to demonstrate this.

Example 8.4.1

Verifying Euler's Theorem

Verify Euler's Theorem with the prime numbers $p = 3$ and $q = 5$. Let $x = 2$ and $a = 1$.

---

1   Konrad M. Lawson, "Secure Communication with Public-Key Encryption," *The Chronicle of Higher Education,* Published February 7, 2011. https://www.chronicle.com/blogs/profhacker/secure-communication-with-public-key-encryption

Solution

We are told to use the prime numbers 3 and 5. Therefore, $n = 3 \cdot 5 = 15$. We can substitute these values into the formula as follows.

$$2^{a(p-1)(q-1)+1} - 2 = 2^{1(3-1)(5-1)+1} - 2$$
$$= 2^9 - 2$$
$$= 512 - 2$$
$$= 510$$
$$\equiv 0 \,(\text{mod} \, 15)$$

Just as with Fermat's Little Theorem in the previous section, there is nothing special about the numbers we chose, other than the fact that 3 and 5 are both prime.

> ✔ **Skill Check 8.4.1**
>
> Verify Euler's Theorem with the prime numbers $p = 2$ and $q = 3$. Let $x = 2$ and $a = 1$.

OBJECTIVE 2 Public-Key Encryption

The way that we presented both Fermat's Little Theorem and Euler's Theorem illustrates how the two are similar. We've written them side by side here so that you can see for yourself.

Table 8.4.1: Fermat's Little Theorem vs. Euler's Theorem

Fermat's Little Theorem	Euler's Theorem
$x^p - x \equiv 0 \,(\text{mod} \, p)$	$x^{a(p-1)(q-1)+1} - x \equiv 0 \; (\text{mod} \, n)$
p is prime	p and q are prime, $n = pq$

As you can see, both theorems involve raising x to some power and then subtracting that same x. Both of these theorems depend on certain numbers being prime. Public-key encryption also relies on prime numbers.

Here's where Euler's Theorem comes into play. To help us better understand public encryption codes, we're going to rearrange Euler's Theorem a bit. Although we are not changing the meaning behind the theorem, changing the way it is written can help us see how to use it in encryption codes.

> 🔑 **EULER'S THEOREM**
>
> $$x^{a(p-1)(q-1)+1} \equiv x \; (\text{mod} \, n),$$
>
> where p and q are prime numbers, $n = pq$, and x and a are any positive integers.

Think of the new arrangement of the formula in the following way. Suppose you have any number and you raise that number to a special power involving two primes, p and q. What you get is congruent to itself, modulo the product of p and q. In other words, you can manipulate a number exponentially with primes and get right back where you started using modular arithmetic.

Using the formula in this format is beneficial in public encryption. Suppose we take your credit card number and encode it by raising it to a power involving two primes. Now we have your new encoded credit card number. If I wanted to, I could give your encoded number along with the directions of how I encoded it to anyone without resulting in a breach in your credit card security. Why? Because the directions tell you the power to use when encoding, but not the values of the two primes, I'm the only one who knows the exact two prime numbers that you need for decoding.

Now, you might think that it would be easy for someone to just "figure out" what the two prime numbers are. That, however, is not so easy if the prime numbers chosen are very, very large—say 200 digits each!

Remember, it is crucial to know the prime numbers p and q in order to decode the number. Usually for these purposes p, q, and n are very large numbers, often hundreds of digits long, and at present, no one knows how to carry out the search for prime numbers in any practical way. A good analogy is to suppose that I were to give you a 200-digit number n and its factorization into two 100-digit primes written down on a piece of paper. Then, you lost the primes by accidentally shredding the piece of paper. It would be quicker for you to search for the shreds of paper in the public dump and try to reconstruct the primes on the paper rather than to try to find the factors by trial and error. It's like finding a needle in a haystack.

Here's how the process works in encryption. The formula requires that we raise a number to a power and then perform a modular reduction to get back to the original number. For encryption purposes, the sender will do part of the raising to a power and the receiver will finish off the raising to a power. Both will perform modular arithmetic using the product of two prime numbers.

Let's use Samantha and Carlos to help us explain. After choosing two prime numbers, p and q, Samantha multiplies them together with another number a in the following manner so that she forms the exponent in the formula from Euler's Theorem.

$$a(p - 1)(q - 1) + 1$$

She then takes this number and factors it into two numbers e and d.

$$a(p - 1)(q - 1) + 1 = ed$$

She tells Carlos the numbers e and n and instructs him to encrypt a secret number by raising it to the power e and finding its congruence modulo n. (Notice that by telling Carlos e and n, where n is the product of p and q, Samantha has not actually given away what p and q are.)

So, Carlos does the following and sends it to Samantha.

$$(\text{secret number})^e \pmod{n}$$

Now, when Samantha gets the coded number from Carlos, she raises it to the power d.

$$(\text{secret number}^e)^d \pmod{n}$$

> **💬 Helpful Hint**
>
> To encode a number, we use $(\text{secret number})^e \pmod{n}$.
>
> To decode a number, we use $(\text{secret number}^e)^d \pmod{n}$.

Now Samantha has (secret number)^{ed}. Since Samantha chose e and d based on her prime number calculation, this is the same as (secret number)$^{a(p-1)(q-1)+1}$.

Because of Euler's Theorem, she can then perform modular arithmetic with modulus $n = pq$ and have the original secret number Carlos started with.

$$\left(\text{secret number}\right)^{ed} = \left(\text{secret number}\right)^{a(p-1)(q-1)+1}$$
$$\equiv \left(\text{secret number}\right)\left(\bmod\, n\right)$$

With Samantha and Carlos's help, we can see that the public part of the encryption code, called the public key, consists of the numbers e and n. The private part of the process, which is used to decode the number, is d along with p and q.

The following table gives a summary of the variables used in public-key encryption.

Table 8.4.2: Summary of the Variables Used in Public-Key Encryption

Variable	Definition
p and q	prime numbers
a	any positive integer
e and d	factors of $a(p-1)(q-1)+1$
n	the product pq
e and n	the public key
p, q, and d	the private key

The next examples take us through encoding and decoding a message using public-key encryption.

Example 8.4.2

Using Public-Key Encryption

Let $C = 2$ be a secret number. Encode C using the following public key. Let M be the new encoded number.

Public Key: $n = 115$ and $e = 17$

Solution

To encode C, we have to calculate C^e (mod n).

$$C^e\left(\bmod\, n\right) \equiv 2^{17}\left(\bmod\, 115\right)$$
$$\equiv 131,072\left(\bmod\, 115\right)$$
$$\equiv 87\left(\bmod\, 115\right)$$

So, our encoded number is now $M = 87$.

✔ Skill Check 8.4.2

Using the same public-key information in Example 8.4.2, encode the number $C = 3$.

Let's now try to decode the newly transmitted number from Example 8.4.2.

Example 8.4.3

Decoding with Private-Key Decryption

Decode the number $M = 87$, which is the secret number to the power e, using the private key $d = 57$ and $n = 115$.

Solution

To decode, we calculate the following.

$$M^d \,(\text{mod } n) = 87^{57}(\text{mod } 115)$$

As it stands, these numbers are too large for direct computation without a powerful computer. Verify for yourself each step of the process with modular arithmetic.

$$87^1 = 87\,(\text{mod } 115)$$
$$87^2 = 87^1 \cdot 87^1 = 87 \cdot 87 = 7569 \equiv 94\,(\text{mod } 115)$$
$$87^4 = 87^2 \cdot 87^2 \equiv 94 \cdot 94 = 8836 \equiv 96\,(\text{mod } 115)$$
$$87^8 = 87^4 \cdot 87^4 \equiv 96 \cdot 96 = 9216 \equiv 16\,(\text{mod } 115)$$
$$87^{16} = 87^8 \cdot 87^8 \equiv 16 \cdot 16 = 256 \equiv 26\,(\text{mod } 115)$$
$$87^{32} = 87^{16} \cdot 87^{16} \equiv 26 \cdot 26 = 676 \equiv 101\,(\text{mod } 115)$$

Now, we can once again use some rules of exponents to give us 87^{57}. Since we can use a combination of the numbers we have calculated, we have the following.

$$87^{57} = 87^{32} \cdot 87^{16} \cdot 87^8 \cdot 87^1$$
$$\equiv 101 \cdot 26 \cdot 16 \cdot 87\,(\text{mod } 115)$$
$$\equiv 3{,}655{,}392\,(\text{mod } 115)$$
$$\equiv 2\,(\text{mod } 115)$$

And, just as we knew, the original number is 2.

Although we have shown this long computation by hand, computers can calculate $87^{57}\,(\text{mod } 115)$ in one step.

Skill Check Answers

1. $2^{1(2-1)(3-1)+1} - 2 = 2^3 - 2$
   $= 8 - 2$
   $= 6$
   $\equiv 0\,(\text{mod } 6)$

2. $M = 108$

8.4 Exercises

✔ CONCEPT CHECK

1. In Euler's Theorem, the variables p and q are _____ numbers.

2. Public-key encryption relies on _____ numbers.

3. True or False: It was said that Julius Caesar communicated with his army using the Julius cipher.

💡 PRACTICE

Verify Euler's Theorem for each given number.

4. $n = 77, p = 7, q = 11, x = 2$, and $a = 1$

5. $n = 91, p = 13, q = 7, x = 3$, and $a = 2$

6. $n = 155, p = 31, q = 5, x = 4$, and $a = 6$

Encode each number using the given public key.

7. Encode the number $M = 14$ using the public key $n = 77$ and $e = 13$.

8. Encode the number $T = 49$ using the public key $n = 77$ and $e = 37$.

9. Encode the number $CC = 101$ using the public key $n = 119$ and $e = 11$.

10. Encode the number $K = 55$ using the public key $n = 119$ and $e = 35$.

11. Encode the number $M = 3$ using the public key $n = 105$ and $e = 10$.

12. Encode the number $A = 8$ using the public key $n = 91$ and $e = 4$.

13. Encode the number $BB = 21$ using the public key $n = 15$ and $e = 2$.

14. Encode the number $J = 100$ using the public key $n = 63$ and $e = 8$.

Decode each number using the given private key.

15. Decode the number $PR = 15$ using the private key $d = 53$ and $n = 77$.

16. Decode the number $TD = 38$ using the private key $d = 17$ and $n = 77$.

17. Decode the number $C = 64$ using the private key $d = 35$ and $n = 221$.

18. Decode the number $W = 219$ using the private key $d = 151$ and $n = 781$.

19. Decode the number $AC = 12$ using the private key $d = 7$ And $n = 77$.

20. Decode the number $B = 35$ using the private key $d = 3$ And $n = 99$.

21. Decode the number $RJ = 21$ using the private key $d = 31$ And $n = 65$.

22. Decode the number $M = 72$ using the private key $d = 11$ And $n = 57$.

Use technology to break each code.

23. Suppose $p = 23$ and $q = 71$. Calculate what d must be if $e = 51$ and $a = 5$.

24. Suppose $p = 17$ and $q = 61$. Calculate what d must be if $e = 71$ and $a = 23$.

25. Suppose $n = 187$ and $e = 19$. Calculate what d must be as well as the two primes p and q, if $a = 7$.

26. Suppose $n = 589$ and $e = 23$. Calculate what d must be as well as the two primes p and q, if $a = 2$.

Chapter 8 Exercises

Solve each problem.

1. The student government body consists of 16 females and 12 males. What is the greatest number of subgroups that can be made if each group must have the same number of males and females?

2. As part of a service project, Brooke, Madison, and Collin have collected school supplies (pens, pencils, glue, and notebooks) along with toothbrushes and toothpaste to send to school kids in Haiti. They collected the following.

   60 glue sticks   122 pencils   180 pens   90 notebooks
   70 toothbrushes   40 tubes of toothpaste

   a. If they want to make identical care packages using the supplies they collected, what is the maximum number of care packages they can make?

   b. How many of each item will be in each package?

3. What is the largest number that divides both 691 and 861 leaving a remainder of 11 each for each number.

4. The Food Mission is a local nonprofit organization that teaches high school students how to grow a sustainable garden. They donate all the food grown to local hunger organizations. The land they have must be divided into two pieces so that they can rotate through spring and summer crops. The first garden needs to cover 400 square feet, and the second garden needs to be 340 square feet. In order to stretch donated supplies, the shared boundary between the two gardens needs to be maximized.

   a. What is the maximum boundary the gardens can share?

   b. What will the dimensions of each garden be if the shared boundary is maximized?

5. Find the GCD of the numbers 1785 and 546. Are the numbers relatively prime?

Determine the missing digit for each 10-digit ISBN.

6. 0–337–25?16–9

7. 1–00–?82634–1

8. 0–07–7734?1–1

9. ?–271–53981–1

Decide whether each barcode is valid. If it is not, state what the correct check-sum number should be.

> Each of the following bar codes consists of 12 digits. To find the check-sum digit, the first 11 digits are multiplied by 3 and 1, alternatively. These multiples are summed with the last digit so that the result is 0 (mod 10).

10.
1 2 3 4 5 2 3 7 3 9 2 5

11.
3 2 8 9 5 4 3 1 0 0 4 3

12.
2 6 0 0 1 7 0 3 0 1 1 1

Find the missing digit for each credit card number.

13. 4485 4217 6305 8?81

14. 6011 5341 1874 008?

15. ?465 9337 2511 334

Encode each number using the given public key.

16. Encode the number $K = 21$ using the public key $n = 77$ and $e = 13$.

17. Encode the number $T = 28,462$ using the public key $n = 5,298,463$ and $e = 99$.

Decode each number using the given private key.

18. Decode the number 5,031,323 using the private key $d = 105,019$ and $n = 5,298,463$.

19. Decode the number 51,189,234 using the private key $d = 27,743$ and $n = 152,472,479$.

CHAPTER 8 PROJECT

INVESTIGATING NUMBER SEQUENCES

Number theory is sometimes studied simply for the love of math. There is a beauty in finding patterns in sequences of numbers and learning how these sequences overlap with different areas of study. In this project, you will investigate two sequences of numbers and learn how their patterns interact with each other and the golden ratio.

1. The Fibonacci numbers form a sequence of numbers that have a distinct pattern. The first six values of the Fibonacci numbers are as follows.

$$0, 1, 1, 2, 3, 5, \ldots$$

Write a description for the pattern that forms these Fibonacci numbers. (**Hint:** Write sentences to describe how to obtain the next four values in the sequence if the first two values are given.)

2. Using the pattern described in part 1, write the first 15 numbers of the Fibonacci numbers.

3. The notation for each value in the Fibonacci sequence is F_n where n is nth term of the sequence, starting at $n = 0$. This means that $F_0 = 0$ and $F_1 = 1$. One way to think of the Fibonacci sequence is that the pattern described in part 1 is "seeded" with the values $F_0 = 0$ and $F_1 = 1$. Describe the sequence Y_n that would be created with the same pattern if it were seeded with the values $Y_0 = 1$ and $Y_1 = 0$. (**Hint:** List out the first 10 digits of the sequence.)

4. The Lucas numbers L_n form a sequence of numbers that share the same pattern as the Fibonacci numbers, but the sequence is seeded with $L_0 = 2$ and $L_1 = 1$. List out the first 15 numbers of the Lucas numbers. (**Hint:** Use the pattern that was described in part 1.)

5. Compare the Fibonacci numbers with the Lucas numbers. Are any values the same? Do any values match up exactly between the two sequences (that is, does $F_n = L_n$ for any values of n)? Which sequence seems to grow in size the fastest?

6. The *golden ratio* is a mathematical property that has been studied since 300 B.C. Perform an internet search and write a short description of the golden ratio. Include the value of the golden ratio rounded to the nearest millionth. Be sure to describe the use of the golden ratio.

7. Find the ratio of the two largest values of the Fibonacci numbers that were listed in part 2. Find the ratio of the two largest Lucas numbers that were listed in part 4. Round the ratios to the nearest millionth. Compare these two ratios to the value of the golden ratio. Does either the Fibonacci ratio or the Lucas ratio match the golden ratio? How can we improve our investigation into whether each series satisfies the golden ratio?

8. Pick two new seeds for the pattern described in part 1 and list the first 15 values.

9. Compare the values of the new sequence of numbers you created in part 8 to both the Fibonacci numbers and the Lucas numbers. Are any values the same? Do any values match up exactly? Does the new sequence grow faster or slower than the Fibonacci numbers and Lucas numbers?

10. Find the ratio of the two largest values of the new number sequence you created in part 8. Round the ratio to the nearest millionth. Compare this value to the golden ratio. Does it match the golden ratio? Does it match the golden ratio any better or worse than the Fibonacci numbers and Lucas numbers?

11. Do you think that using different seed values for the pattern described in part 1 would result in a ratio that matches closer to the golden ratio? Explain your reasoning.

📋 CHAPTER 8 REVIEW

Section 8.1 Prime Numbers

Definitions

Prime numbers: A **prime number** is a positive integer that has precisely two divisors: 1 and itself.

Composite Numbers: A **composite number** is a positive integer that has more than two divisors. Note that the numbers 0 and 1 are neither prime nor composite.

Greatest Common Divisor (GCD): The largest integer that divides two numbers without a remainder is called the **greatest common divisor (GCD)**. If the GCD = 1 for a pair of numbers, the numbers are said to be relatively prime.

Procedures

Sieve of Eratosthenes

    Step 1: List all the positive integers between 2 and N.

    Step 2: Highlight the smallest number (p) not crossed out or previously highlighted. This is a prime number.

    Step 3: Cross out all multiples of p.

    Step 4: Repeat steps 2 and 3 until all numbers are either highlighted or crossed out.

Steps for Using Euclid's Algorithm

    1. Divide the larger of the given numbers by the smaller (the divisor) and note the remainder.
    2. Divide the original divisor (the smaller of the given numbers) by the remainder found in Step 1.
    3. Continue dividing the previous divisor by the previous remainder until the remainder is 0.
    4. The last nonzero remainder is the GCD of the given numbers.

Theorems

Fundamental Theorem of Arithmetic: The **fundamental theorem of arithmetic** states that every positive integer greater than 1 is either a prime number or can be written as a unique product of prime numbers. This unique product of prime numbers is called its prime factorization.

Section 8.2 Modular Arithmetic

Definition

Modular Arithmetic: In **modular arithmetic**, a number n is congruent to the remainder r when it is divided by a fixed number m. We write $n = r(\text{mod } m)$. Note that m is referred to as the **modulus**.

Section 8.3 Fermat's Little Theorem and Prime Testing

Theorems

Fermat's Little Theorem

Let p be any prime number and x be any positive integer. Then, $x^p - x \equiv 0 \pmod{p}$.

Contrapositive of Fermat's Little Theorem

Let x and n be positive integers. If $x^n - x \not\equiv 0 \pmod{n}$, then n is **not** a prime number.

Section 8.4 Fermat's Little Theorem and Public-Key Encryption

Theorems

Euler's Theorem

$$x^{a(p-1)(q-1)+1} - x \equiv 0 \pmod{n},$$

where p and q are prime numbers, $n = pq$, and x and a are any positive integers.

(Modified) Euler's Theorem

$$x^{a(p-1)(q-1)+1} \equiv x \pmod{n},$$

where p and q are prime numbers, $n = pq$, and x and a are any positive integers.

CHAPTER 9

Geometry

🔖 SECTIONS

Introduction

Why are Pringles packaged in a tube but Doritos are in a bag? Why is rice usually sold in a bag, but cereal is most often sold in a box? How do companies determine the size and shape of the containers they use for packaging? To the consumer, sometimes containers seem larger than necessary. We get a sense of feeling cheated that a bag of chips is only half full. But "slack fill," or extra space around the chips, is what keeps the chips from getting crushed during shipping.

In order to keep their products safe along with keeping the packaging cost effective, companies use geometric principles to optimize slack fill. A container that is too large can result in fewer items per box, which increases overall shipping costs and requires more shelf space in stores. A package that is too small can lead to potentially damaged products. These considerations and the math behind them have made the packaging industry a big business. In the US alone, the packaging business raked in nearly $180 billion in 2019, and the global packaging market is expected to reach $1.2 trillion by 2028.[1]

Another multibillion-dollar business is aircraft manufacturing, which requires a tremendous use of trigonometry, the basic concepts of which we'll cover in this chapter. One of the most significant changes in the aircraft industry is the improvements of technology for the navigation and targeting systems used by fighter pilots. Before aircrafts had computers, fighter pilots had to use trigonometry to create a "bombing triangle" and perform the necessary trigonometric calculations in their head while flying the plane.

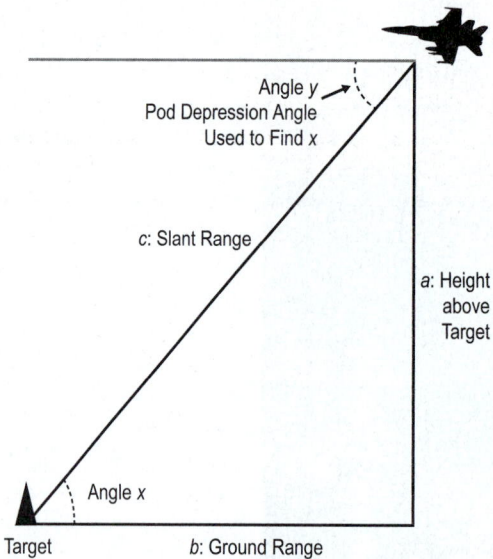

Thankfully, today these calculations no longer lay on the shoulders of the pilots. The technology used by the companies building these new aircrafts has been instrumental in increasing the standard for accuracy from 80% hit rate to 97%, giving fighter pilots extra mental space to concentrate on other challenges at hand.

---

1   "Four key trends that will shape the future of packaging to 2028," Smithers, Accessed October 27, 2021, https://www.smithers.com/re-sources/2019/feb/future-packaging-trends-2018-to-2028.

OBJECTIVES

1. Calculate perimeter and area.

2. Use trapezoids to estimate the area under a curve.

Fun Fact

The word geometry translates from Greek to mean *earth measure*.

Think Back

A polygon is a closed figure that is determined by three or more straight line segments.

OBJECTIVE 1 Area and Perimeter

The quest to fully understand the planet we live on is an ancient one—and it involves thinking in shapes and sizes. In other words, it involves exploring the **geometry** of things. You yourself do this on a daily basis. Our brains perform geometric spatial calculations for the most basic moments in life, such as setting your foot on the floor when you get out of bed in the morning, successfully walking down a flight of stairs, or parallel parking your car. Humanity also benefits from advanced geometric phenomena like space travel or robotic medical procedures. Virtually every realm of society is touched by geometry: art, architecture, engineering, robotics, astronomy, space, nature, sports, machines, cars, computer imaging, navigation, medical technology, and even the movie industry. By bringing the logic-driven, technical left side of the brain together with the creative, artistic right side, geometry helps to create balance in our world.

Precision with geometric shapes is vital when calibrating machines for CT scans or designing a skyscraper. However, our focus in this chapter will be more on the calculations and applications of geometric shapes rather than producing drawings and constructions. Concentrating on some of the practical uses that geometry offers in the routine of daily life, we'll apply the relationships between the shapes and sizes of surfaces and solids in the first two sections of this chapter.

Let's start with a brief recap of terms and formulas that we will use throughout the chapter. Recall that the distance around the outside of any two-dimensional shape is its **perimeter**. The perimeter of a circle has a special name; it's called the circle's *circumference*. The measure of the space inside the shape is its **area**. Table 9.1.1 provides a summary of some of the most common shapes alongside their respective formulas for perimeter and area.

Table 9.1.1: Summary of Perimeter and Area Formulas

Shape	Definition	Formula for Perimeter	Formula for Area
Rectangle	A **rectangle** is a quadrilateral with two pairs of parallel sides and four right angles. We denote the longer side as the length l and the shorter side as the width w.	$P = 2l + 2w$	$A = lw$
Square	A **square** is a quadrilateral with two pairs of parallel sides, four sides that are the same length, and four right angles. We denote the side length as s. Note that all squares are also rectangles.	$P = 4s$	$A = s^2$

Shape	Definition	Formula for Perimeter	Formula for Area
 Triangle	A **triangle** is a three-sided polygon that is identified by its base b and height h.	$P = \text{sum}$ of the side lengths	$A = \dfrac{1}{2}bh$
 Parallelogram	A **parallelogram** is a quadrilateral with two pairs of parallel sides. The base of the parallelogram is denoted by b and the height by h.	$P = \text{sum}$ of the side lengths	$A = bh$
 Trapezoid	A **trapezoid** is a quadrilateral with exactly one pair of parallel sides. The sides that are parallel are called bases and are denoted b_1 and b_2. The height is denoted by h.	$P = \text{sum}$ of the side lengths	$A = \dfrac{1}{2}(b_1 + b_2)h$
 Circle	A **circle** is the set of all points in a plane that are a certain distance from a fixed center point. The diameter, d, is the length of a line segment that passes through the center. The radius, r, is half of the diameter.	$C = 2\pi r = \pi d$	$A = \pi r^2$

These formulas not only enable us to find the perimeters and areas of standard shapes, but they also allow us to find the measurements of more complex shapes if we combine them. This simple idea has lots of practical applications, as we'll see throughout the examples in this section.

Example 9.1.1

Using Area of Rectangles

Agatho's High School is updating part of their cafeteria by painting an area of the floor in one of the school colors. The parent association was tasked with estimating the cost of the project. They must use commercial grade non-skid paint. Each gallon costs $135 and covers approximately 100 ft². Two coats are recommended, and this paint is only sold by the gallon. A group of parents in the association volunteered to provide the labor needed and measured the proposed area to be painted. The floor portion to be painted is shown in Figure 9.1.1. Estimate the cost of this project.

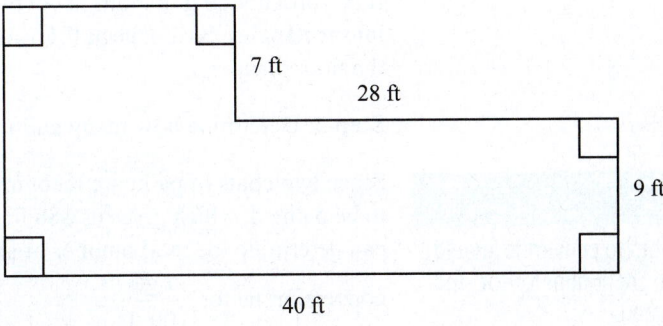

Figure 9.1.1

Solution

In order to estimate the cost of the project, the parents must do the following.

1. Calculate the area of floor that needs to be painted.

2. Determine how many gallons of paint are needed.

3. Calculate the cost of the paint.

Step 1: Calculate the area of floor that needs to be painted.

Although the area of the floor to be painted is not a standard shape that we have a formula for, we can divide the figure into shapes whose areas we can calculate more easily. In this case, the shape can be divided into rectangles using the known dimensions. The dashed line in Figure 9.1.2 divides the area into two rectangles using the measurements provided in the Figure 9.1.1. Now we can determine the missing lengths for each side of the new rectangles. The top side of Rectangle 1 has length 40 ft – 28 ft = 12 ft. The left side of Rectangle 1 has length 7 ft + 9 ft = 16 ft. While we will keep the labels of 7 ft and 40 ft on the figure, they will not be used when calculating the area.

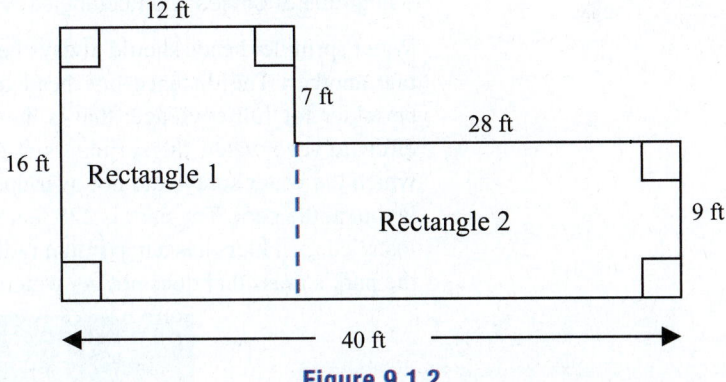

Figure 9.1.2

We know that the formula for the area of a rectangle is $A = lw$. Using this formula, the area for each rectangle is calculated as follows.

Area of Rectangle 1: $A_1 = lw = (16 \text{ ft})(12 \text{ ft}) = 192 \text{ ft}^2$

Area of Rectangle 2: $A_2 = lw = (28 \text{ ft})(9 \text{ ft}) = 252 \text{ ft}^2$

Adding the areas together, we have the total area of the floor to be painted.

Total Area of Painted Floor: $\text{Area}_1 + \text{Area}_2 = 192 \text{ ft}^2 + 252 \text{ ft}^2 = 444 \text{ ft}^2$

It is worth noticing that this is not the only way that the cafeteria could be divided into rectangles. Skill Check 9.1.1 asks that you determine another way to divide the floor plan.

Step 2: Determine how many gallons of paint are needed.

Since two coats of paint are recommended, we need to double the square footage to be painted, which gives us 888 ft². Since each gallon of paint covers 100 ft², we can determine the total number of gallons needed by multiplying this area by the conversion factor $\frac{1 \text{ gal}}{100 \text{ ft}^2}$.

$$\text{Gallons of Paint Needed} = 888 \text{ ft}^2 \cdot \frac{1 \text{ gal}}{100 \text{ ft}^2} = 8.88 \text{ gal}$$

Since this paint is only sold by the gallon, the school will need to purchase 9 gallons of paint in order to ensure full coverage.

Step 3: Calculate the cost of the paint.

We are told the paint costs \$135 per gallon. So 9 gallons of paint will be $9 \cdot \$135 = \$1215$. (Note that if the school doesn't qualify for a tax exemption, tax would need to be included. We'll assume this school qualifies for a tax exemption.)

Since the parents are providing the labor, the only other cost consideration will be painting supplies, such as rollers, tape, or cleaning materials. Therefore, a reasonable estimate for the project would be to round up to \$1300.

Example 9.1.2

Using Area of Circles and Rectangles

Water sprinkler heads should always be placed so that the sprays of water overlap one another. The distance one head throws should reach the head of the next sprinkler for full coverage; that is, head-to-head coverage. Unfortunately, in an effort to save water, the sprinklers at a local park were installed in a pattern in which the water sprays did not overlap. Figure 9.1.3 is a drawing of the sprinkler layout at the park. The park is 228 feet wide and 380 feet long, and each of the 15 installed sprinklers has a maximum radius of 38 feet. Determine the percentage of the park's grass that does not get watered with this sprinkler arrangement.

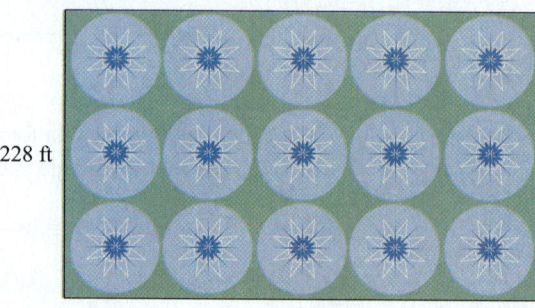

228 ft

380 ft

Figure 9.1.3

Solution

To determine how much of the grass is not getting water, we need to use a combination of the areas of rectangles and circles. We can find the area covered by the sprinklers and subtract that from the total grassy area of the park.

The park is a rectangle measuring 228 ft by 380 ft. Thus, the area of the grass can be found by multiplying the two dimensions.

$$A_{grass} = lw$$
$$= (228 \text{ ft})(380 \text{ ft})$$
$$= 86,640 \text{ ft}^2$$

Each sprinkler creates a circular pattern with a radius of 38 feet. Using the area formula for a circle ($A = \pi r^2$), we can calculate the area watered by each sprinkler.

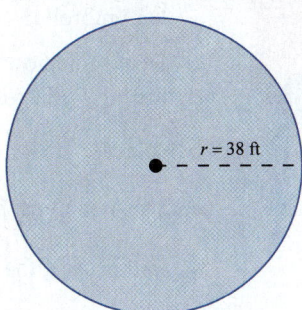

$r = 38$ ft

Figure 9.1.4

$$A_{sprinkler} = \pi r^2$$
$$= \pi (38 \text{ ft})^2$$
$$\approx 4536.459791 \text{ ft}^2$$

Helpful Hint

When calculations involve several steps, avoid rounding at intermediate calculations. If necessary, round intermediate calculations to at least six decimal places to avoid additional rounding error in subsequent calculations.

Since there are 15 sprinkler heads, multiplying the area by 15 gives us the total area of watered grass.

$$A_{watered\ grass} = 4536.459791 \text{ ft}^2 \cdot 15$$
$$= 68,046.896865 \text{ ft}^2$$

Subtracting the watered area from the total grassy area, we can find the area of grass that does not receive water.

$$A_{unwatered} = A_{grass} - A_{watered\ grass}$$
$$= 86,640 \text{ ft}^2 - 68,046.896865 \text{ ft}^2$$
$$= 18,593.103135 \text{ ft}^2$$

To determine the percentage of grass not watered, we divide the unwatered amount of grass by the total grass area.

Proportion of Unwatered Grass: $\dfrac{18,593.103135 \text{ ft}^2}{86,640 \text{ ft}^2} \approx 0.214602$

✔ **Skill Check 9.1.2**

Find the area and circumference of a circle with a diameter of 10 inches.

Therefore, the current layout of sprinklers means that approximately 21% of the park's grass is not being watered.

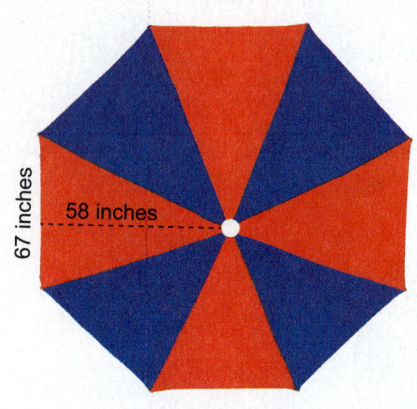

67 inches

58 inches

Figure 9.1.5

Example 9.1.3

Using Area of Triangles

Barbara has decided to personalize her patio umbrella with alternating colors. Her current umbrella has 8 triangular panels that measure 58 inches from the center to the outer edge, as shown in Figure 9.1.5. Each triangular panel has a base length of 67 inches.

a. Determine the minimum amount of material (in square yards) that Barbara will need in each color for her new design.

b. Barbara wants to waterproof the umbrella once it is finished. The can of water repellant states that a 12-ounce can will cover approximately 60 square yards, or 540 square feet. How many cans does Barbara need to buy to waterproof her umbrella?

c. Determine how many yards of trim Barbara would need if she decided to edge the umbrella with it.

Solution

a. The area of each triangular panel can be found by using the formula $A = \frac{1}{2}bh$.

In this case, the base of each triangle is 67 inches and the height is 58 inches.

$$\text{Area of Triangular Panel:} \quad A = \frac{1}{2}bh$$

$$= \frac{1}{2}(67 \text{ in.})(58 \text{ in.})$$

$$= 1943 \text{ in.}^2$$

Therefore, each panel has an area of 1943 square inches. Since Barbara needs 4 panels of each color, we multiply the area by 4.

$$\text{Area for Four Panels: } 1943 \text{ in.}^2 \cdot 4 = 7772 \text{ in.}^2$$

This means that Barbara needs a minimum 7772 square inches of material in each color. To calculate what this amount is in yards, we need to use a conversion factor for inches and yards, $\frac{1 \text{ yd}}{36 \text{ in.}}$. However, we need to square this conversion factor so that it contains the correct units. The conversion factor we need is $\left(\frac{1 \text{ yd}}{36 \text{ in.}}\right)^2 = \frac{1 \text{ yd}^2}{1296 \text{ in.}^2}$. Multiplying the area by the conversion factor gives us the following.

$$\text{Area for Four Panels in Square Yards: } 7772 \text{ in.}^2 \cdot \frac{1 \text{ yd}^2}{1296 \text{ in.}^2} \approx 6.0 \text{ yd}^2$$

So Barbara needs approximately 6 square yards of each color.

b. We are told that a 12-ounce can will cover approximately 60 square yards. Since Barbara will be using 6 square yards of each of the two colors, we know there will be 12 square yards of material total. She will only need one can to waterproof all 12 square yards of material.

c. The perimeter of the umbrella consists of the bases of each of the eight triangles. This means that the perimeter is 8 times the base length of 67 inches.

$$\text{Perimeter of the Umbrella: } 8(67 \text{ in.}) = 536 \text{ in.}$$

To determine the yardage, we need to multiply by the conversion factor for inches and yards, $\dfrac{1 \text{ yd}}{36 \text{ in.}}$.

$$\text{Perimeter in Yards: } 536 \text{ in.} \cdot \left(\frac{1 \text{ yd}}{36 \text{ in.}} \right) \approx 14.89 \text{ yd}$$

So Barbara needs approximately 15 yards of trim for the edge of the umbrella.

Example 9.1.4

Using Area of Parallelograms

The Internal Revenue Service allows qualifying taxpayers to claim a home office deduction on their taxes for any area of their home that they use for business purposes. Taxpayers are allowed to claim $5 per square foot for business use, with a maximum of 300 square feet. The Martins meet the requirements to deduct their home office space on their taxes. Calculate the amount they are allowed to deduct based on their home measurements.

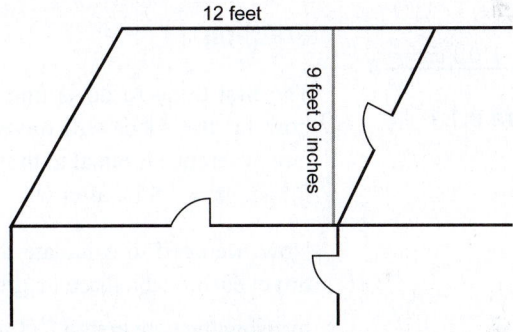

Figure 9.1.6

Solution

The home office area that the Martins are deducting is in the shape of a parallelogram. Recall that the area of a parallelogram is given by $A = bh$. We can see from the floor plan that the base measures 12 feet and the height is 9 feet 9 inches. Before we can multiply to find the area, we need to have both measurements in the same unit (either all feet or all inches). We will change the height measurement, which is given in feet and inches, to feet. The conversion factor to change the inches into feet is $\dfrac{1 \text{ ft}}{12 \text{ in.}}$. We convert 9 inches to feet by

multiplying; that is, we get $9 \text{ in.} \cdot \dfrac{1 \text{ ft}}{12 \text{ in.}} = 0.75 \text{ ft}$. Adding this to the 9 feet, we

have that the height measurement is 9.75 feet.

Now we can use the area formula for a parallelogram $A = bh$ to find the area of their home office.

$$A_{\text{home office}} = bh = (12 \text{ ft})(9.75 \text{ ft}) = 117 \text{ ft}^2$$

The Martins are allowed to deduct $5 per square foot of their home office, so we multiply the area by $5/ft^2.

$$\text{Deduction Amount: } 117 \text{ ft}^2 \cdot \frac{\$5}{1 \text{ ft}^2} = \$585$$

Thus, the Martins are allowed to deduct $585 on their taxes for their home office space.

Example 9.1.5

Using Circumference

Members of a robotics club are tasked with moving their robot from point A to point B. The robot understands the following commands.

• Rotate wheels forward x number of rotations.

• Rotate wheels backward x number of rotations.

• Turn 90° left. Turn 90° right.

If the wheels of the robot are 2.5 inches in diameter, determine the sequence of commands that the robot should be given.

Solution

The first thing to do is find the circumference of the wheels. This will tell us how far the robot will travel after one full rotation of the wheels. Recall that circumference is equal to the diameter d times π. This means that the robot travels 2.5 in. $\cdot \pi \approx 7.8$ in. after one rotation of the wheels.

Now, we need to calculate the number of rotations the wheels need to make to travel each of the three lengths. We can calculate the number of rotations needed by using the conversion factor $\dfrac{1 \text{ rotation}}{7.85 \text{ in.}}$. Notice that the second and third lengths will also need to be converted to inches for our calculations.

$$\text{Wheel Rotations to Travel 20 Inches: } (20 \text{ in.})\left(\frac{1 \text{ rotation}}{7.85 \text{ in.}}\right) \approx 2.55 \text{ rotations}$$

$$\text{Wheel Rotations to Travel 8 Feet: } (8 \text{ ft})\left(\frac{12 \text{ in.}}{1 \text{ ft}}\right)\left(\frac{1 \text{ rotation}}{7.85 \text{ in.}}\right) \approx 12.23 \text{ rotations}$$

Wheel Rotations to Travel 110.3 Centimeters:

$$(110.3 \text{ cm})\left(\frac{1 \text{ in.}}{2.54 \text{ cm}}\right)\left(\frac{1 \text{ rotation}}{7.85 \text{ in.}}\right) \approx 5.53 \text{ rotations}$$

Thus, the commands for the robot should be as follows.

1. Rotate wheels forward 2.55 rotations.

2. Turn 90° right.

3. Rotate wheels forward 12.23 rotations.

4. Turn 90° left.

5. Rotate wheels forward 5.53 rotations.

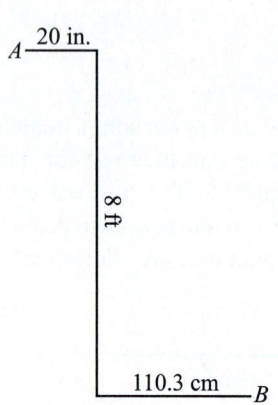

Figure 9.1.7

Figure 9.1.9

From 1966–1991, the title was held by the Astoria-Megler bridge in Oregon, measuring 376 meters.

Figure 9.1.10

Images Courtesy of Hiroshi Nakai and Ron Reiring, Creative Commons 2.0 Generic License (https://creativecommons.org/licenses/by/2.0/deed.en).

Example 9.1.6

Using Area of Trapezoids

A truss is a support framework that is used for structures like bridges and roofs. Typically, trusses are composed of triangles due to their stability. The Warren truss, popular in bridge construction, consists of equilateral triangles that form a trapezoid, as shown in Figure 9.1.8.

A covered bridge is to be built using two Warren trusses, one on each side, supporting the roof. Calculate the amount of cedar in square feet that will be needed to cover both sides of the bridge.

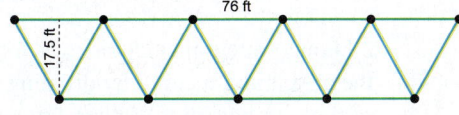

Figure 9.1.8

Solution

The face of the Warren truss forms a trapezoid, the area of which is found by the formula $A = \frac{1}{2}(b_1 + b_2)h$. From Figure 9.1.8, we can see that the bases measure 76 feet and 66 feet and the height is 17.5 feet.

$$A = \frac{1}{2}(b_1 + b_2)h$$
$$= \frac{1}{2}(76 \text{ ft} + 66 \text{ ft}) \cdot (17.5 \text{ ft})$$
$$= 1242.5 \text{ ft}^2$$

Since the bridge has two Warren trusses, the area to be covered in cedar is $2(1242.5 \text{ ft}^2) = 2485 \text{ ft}^2$.

OBJECTIVE 2 Area under a Curve

As we've seen, it's sometimes useful to divide a more complicated shape into simpler shapes. When a surface is covered using one or more geometric shapes so that there are no gaps or overlaps, the covering is called a **tessellation**. Tessellations have been used to tile walls and floors throughout time, even as far back as the 3rd millennium B.C., and were popularized thousands of years later by the Romans and Greeks. But we don't have to go to a museum to see tessellations, they're commonly used in our kitchen backsplashes and on our bathroom floors. We'll look more at tessellations in Section 9.3.

One way that tessellations are used is to help approximate the area under a particular curve. Calculating the precise area under a curve is one of the fundamental elements of integral calculus and has applications in the fields of physics, chemistry, economics, or statistics, among others. For instance, suppose the curve shown in Figure 9.1.12 is the speed of a Formula 1 car as it travels

Figure 9.1.11

around a segment of the Monte Carlo racecourse. By finding the area under the curve, we can determine the distance the car traveled during that time period.

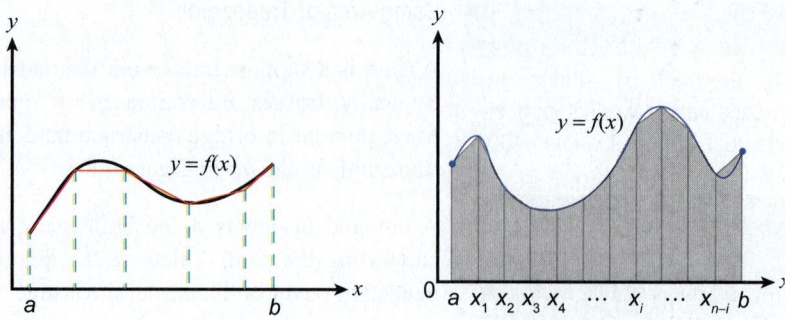

Figure 9.1.12

Although integral calculus is beyond the scope of this book, we can approximate the area under a curve by dividing it into a lot of trapezoids and adding the areas of those trapezoids together. Example 9.1.7 looks at that process, formally called the Trapezoidal Rule.

Example 9.1.7

Using Area of Trapezoids

The graph of the function $y = x^2$ is shown in Figure 9.1.13, where x and y are both measured in inches. Approximate the shaded area under the curve between $x = 1$ and $x = 5$ by finding the areas of 4 trapezoids.

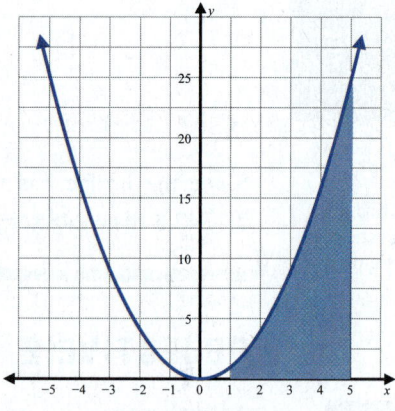

Figure 9.1.13

Solution

Begin by dividing the area we are interested in, the area between $x = 1$ and $x = 5$, into four trapezoids, T_1, T_2, T_3, and T_4.

Recall that the formula for the area of a trapezoid is $A = \dfrac{1}{2}(b_1 + b_2)h$. In Figure 9.1.14, for each individual trapezoid, h is the width of the trapezoid along the x-axis. The parallel side lengths (b_1 and b_2) are the vertical lengths of each trapezoid. Notice that all four of the trapezoids will have a width of $h = 1$.

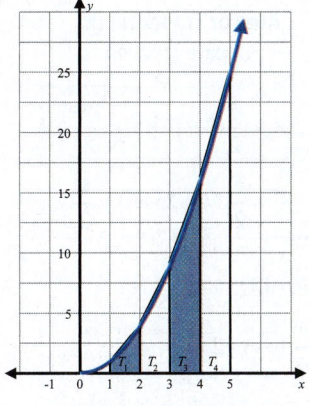

Figure 9.1.14

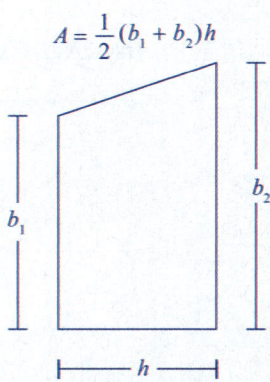

Figure 9.1.15

Since the curve is the graph of $f(x) = x^2$, we can calculate the lengths of the sides of each trapezoid by evaluating the function $f(x) = x^2$. Let's begin with the leftmost trapezoid, T_1, which is located between $x = 1$ and $x = 2$. The length of the left-hand side is the y-value of the function at $x = 1$, or $f(1)$, and the length of the right-hand side is the y-value of the function at $x = 2$, or $f(2)$ as shown in Figure 9.1.16.

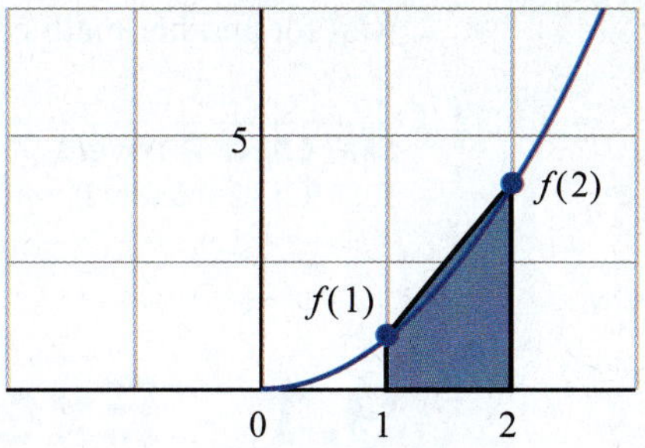

Figure 9.1.16

Evaluating each, we have the following for T_1.

Length of Left-Hand Side: $f(1) = (1)^2 = 1$

Length of Right-Hand Side: $f(2) = (2)^2 = 4$

Thus, the area of the first trapezoid, T_1, is $A_1 = \dfrac{1}{2}(1+4)(1) = \dfrac{5}{2}$.

The areas of the other three trapezoids are found in the same manner. Each of the trapezoids has width 1, and the side lengths are found by evaluating the function at the endpoints along the x-axis

The areas of the remaining trapezoids are shown in Table 9.1.2.

Table 9.1.2

Trapezoid	Length of Left-Hand Side (in inches)	Length of Right-Hand Side (in inches)	Area of Trapezoid (in square inches)
T_1	1	4	$\dfrac{5}{2}$
T_2	$f(2) = (2)^2 = 4$	$f(3) = (3)^2 = 9$	$A_2 = \dfrac{1}{2}(4+9)(1) = \dfrac{13}{2}$

Trapezoid	Length of Left-Hand Side (in inches)	Length of Right-Hand Side (in inches)	Area of Trapezoid (in square inches)
T_3	$f(3) = (3)^2 = 9$	$f(4) = (4)^2 = 16$	$A_3 = \dfrac{1}{2}(9+16)(1) = \dfrac{25}{2}$
T_4	$f(4) = (4)^2 = 16$	$f(5) = (5)^2 = 25$	$A_4 = \dfrac{1}{2}(16+25)(1) = \dfrac{41}{2}$

The Total Area of the Four Trapezoids:

$$\frac{5 \text{ in.}^2}{2} + \frac{13 \text{ in.}^2}{2} + \frac{25 \text{ in.}^2}{2} + \frac{41 \text{ in.}^2}{2} = \frac{84 \text{ in.}^2}{2} = 42 \text{ in.}^2$$

Thus, the approximate area under the curve is 42 square inches.

If we divided the area under the curve in the previous example into more and more and more trapezoids, we would get a better and better approximation for the area. This is exactly the foundation for integral calculus, but the details will have to wait for another math course on another day.

Skill Check Answers

1. 7 ft × 12 ft and 9 ft × 40 ft   2. Area = 78.5 in.²; Circumference = 31.4 in.

9.1   Exercises

✔ CONCEPT CHECK

1. The perimeter of a circle is called its _____.

2. A trapezoid is a quadrilateral that has exactly one pair of _____ sides.

3. A _____ is a pattern made of one or more geometric shapes so that there are no gaps or overlaps.

4. True or False: The formulas for the perimeter and area of a rectangle can be used to find the perimeter and area of a square.

🚀 APPLICATIONS

Round your answer to the nearest tenth, if necessary.

5. A rectangular pane of antique glass has dimensions of 25 inches by 48 inches. If glass sells for $0.50 per square inch, how much will each window cost?

6. Carpet Pro sells carpet for $1.50 per square foot. If you wish to carpet a room that measures 16 feet by 25 feet, how much will the carpet cost?

7. A 25-lb bag of grass seed will cover 1200 square feet. If your lawn is in the shape of a rectangle and measures 205 feet by 145 feet, how many bags of grass seed will you need to cover the lawn?

8. Devon wants to paint an accent wall in his condo. The wall measures 12 feet long and 8 feet high. If a quart of paint covers 100 square feet, will one quart be enough to paint the wall?

Find the area of the shaded region in each figure. Round your answer to the nearest hundredth, if necessary.

9.

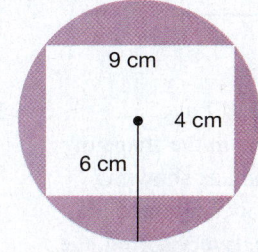

10.

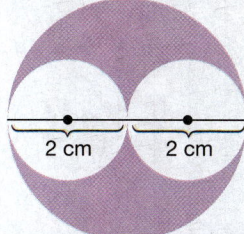

11.

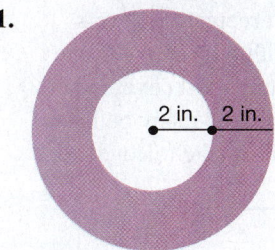

12.

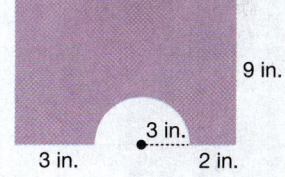

13. A swimming pool is in the shape of the following figure. The homeowner wishes to create a canvas cover for the pool. How many square feet of canvas would be required to make a cover for the pool?

14. A homeowner wants to create a canvas cover for the pool that is the shape of the following figure. How many square feet of canvas would be needed to cover the pool?

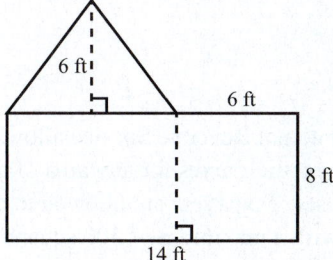

15. Gilbert is making a glass art project that requires two squares and two right triangles as shown. If the glass cost $0.24 per square inch, how much will the glass for the project cost?

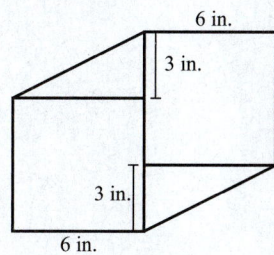

16. Shira needs to buy new siding for one end of her house, which is shown in the following figure. How many square meters of siding will she need?

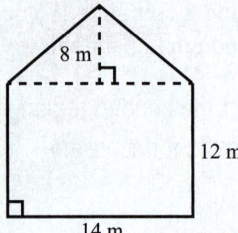

8 m

12 m

14 m

17. Micah is tiling his bathroom with parallelogram-shaped tiles as shown in the following figure. If the bathroom is 8 feet by 9 feet, how many boxes of tiles will he need if each box contains 26 tiles?

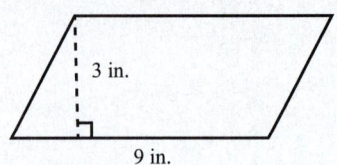

3 in.

9 in.

18. Each campsite at Lake Montgomery is in the shape of a parallelogram, as shown in the figure below, and there are 22 camp sites. Every spring, the department of recreation covers all 22 sites with pine straw. If one bail of pine straw will cover about 45 square feet, how many bails of pine straw will it take to cover all 22 sites?

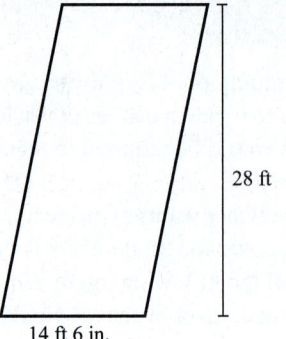

28 ft

14 ft 6 in.

19. Jackson is making a flower bed that will have a triangular portion in the middle and two parallelogram-shaped portions on either side, as shown in the figure. The parallelograms will be filled with shrubs, and each requires a minimum of 4 square feet. The triangle will be filled with flowers and each flower will need a minimum of 0.25 square feet. How many shrubs and how many flowers can Jackson put in the flower bed?

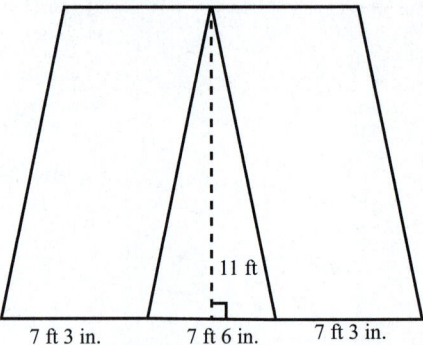

11 ft

7 ft 3 in.   7 ft 6 in.   7 ft 3 in.

20. The Internal Revenue Service allows qualifying taxpayers to claim a home office on their taxes for any area of their home that they use for business purposes. Taxpayers are allowed to claim $5 per square foot for business use, with a maximum of 300 square feet. The Hills meet the requirements to deduct their home office space on their taxes. Calculate the amount they are allowed to deduct based on their home measurements.

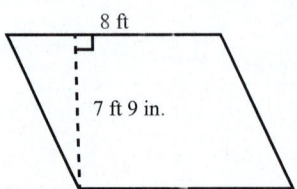

8 ft

7 ft 9 in.

21. Members of a robotics club are tasked with moving their robot from point A to point B. The robot understands the following commands.

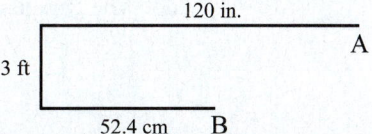

- Rotate wheels forward *x* number of rotations.

- Rotate wheels backward *x* number of rotations.

- Turn 90° left.

- Turn 90° right.

If the wheels of the robot have a 3-inch radius, determine the sequence of commands that the robot should be given.

22. Members of a robotics club are tasked with moving their robot from point A to point B. The robot understands the following commands.

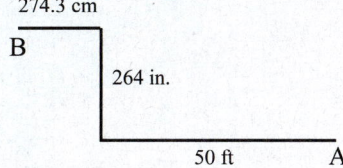

- Rotate wheels forward *x* number of rotations.

- Rotate wheels backward *x* number of rotations.

- Turn 90° left.

- Turn 90° right.

If the wheels of the robot have a diameter of 12 cm, determine the sequence of commands that the robot should be given.

23. Farah rode her bike from point A to point B. If it took her bike's wheels exactly 8 rotations, and if the distance between the two points is 54.5 feet, what is the radius of Farah's tires in inches?

24. In hockey, "the trapezoid" is the trapezoidal area behind the goal, as shown in the following figure, where the goalie is legally allowed to play the puck. What is the area of the trapezoid?

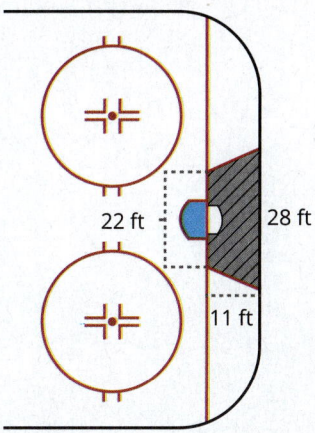

25. Bart built a workshop in the shape of a trapezoid as shown. Find the cost of the siding needed for the front of the shop if the siding cost $2.65 per square foot. The garage door should not be included.

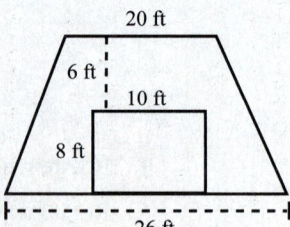

26. Luis is making trapezoidal tables for the preschool at his church. He plans to make 6 of them and will cover each table with laminate. If the laminate cost $23 per square foot, how much will the laminate for the 6 tables cost?

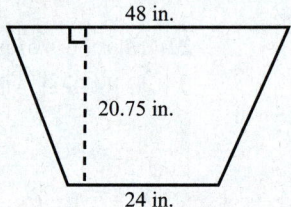

27. The trapezoidal window shown got broken during a storm. Find the cost if the replacement glass costs $12.97 per square foot.

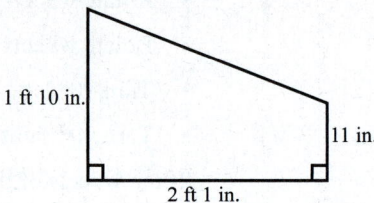

28. Find the perimeter and area of the floor tile represented by the parallelogram.

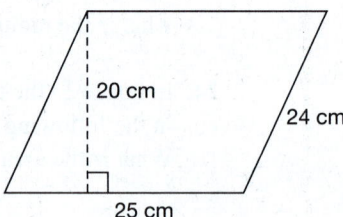

29. The side view of a wheel for a skate is shown. What is the circumference of the wheel? How many square inches of paint are required to paint the wheel?

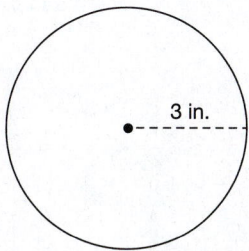

30. A piece of cake has the shape of the following figure. What is the area of the top of the cake slice?

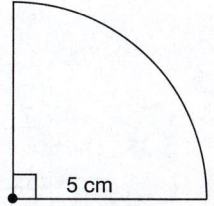

31. A window in the shape of the following figure is needed for a new home. If glass costs $0.85 per square foot, how much will the window cost?

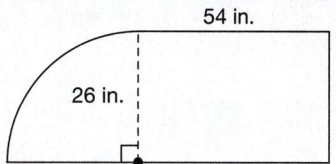

32. A rectangular piece of land has dimensions 1200 ft by 850 ft. Find the amount of fencing needed to construct a fence around the perimeter of this piece of land. Also find the area of the land to be enclosed.

33. A hot tub will have the footprint of a regular polygon, shown in the following figure. If your builder wishes to cover the area under the hot tub with plastic, how many square meters of plastic are needed?

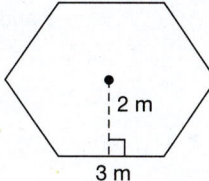

34. The graph of the function $y = \dfrac{1}{4}x^2$ is shown, where x and y are both measured in centimeters. Approximate the area under the curve between $x = 2$ and $x = 4$.

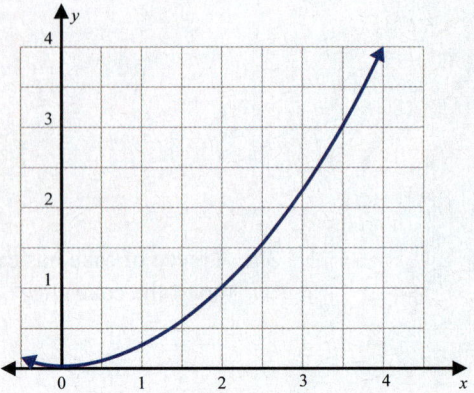

35. The graph of the function $y = 2\sqrt{x}$ is shown, where x and y are both measured in inches. Approximate the area under the curve between $x = 1$ and $x = 4$.

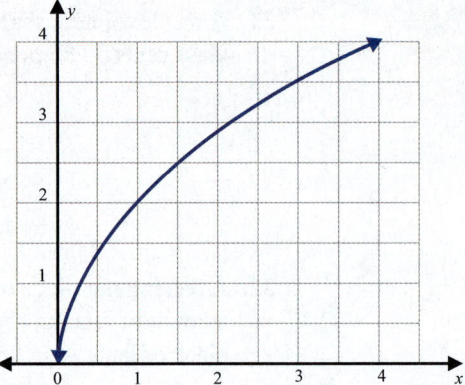

36. The graph of the function $y = \sqrt{-x}$ is shown, where x and y are both measured in centimeters. Approximate the area under the curve between $x = -1$ and $x = -4$.

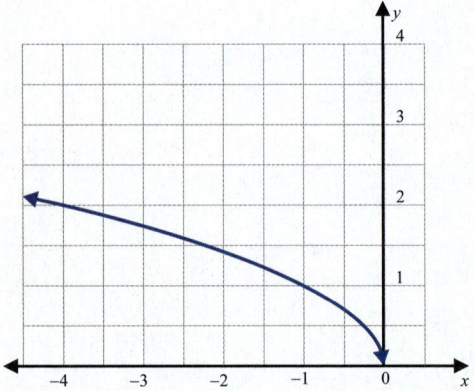

37. The graph of the function $y = x$ is shown, where x and y are both measured in meters. Find the area under the curve between $x = 1$ and $x = 3$.

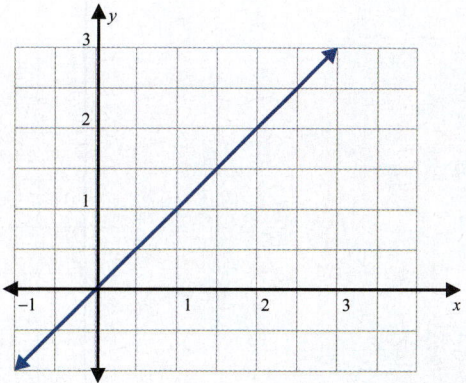

TETRIS-ROBOTICS

Have you ever played the game Tetris? Tetris is a tile-matching video game that was created by Alexey Pajitnov in 1984. The game has seven different pieces, called tetriminos.

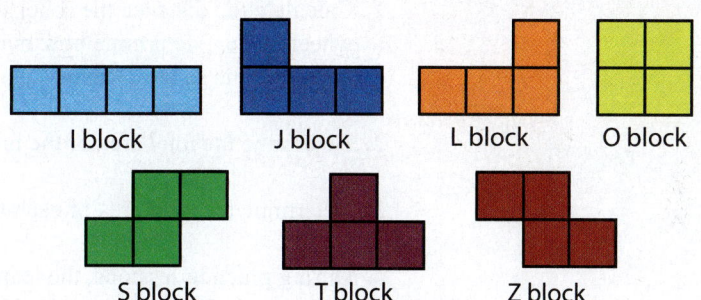

Suppose a science club is programming a robot to play a modified game of Tetris. In their game, each square that makes up a tetrimino has a side of 10 inches. For example, the O block, which is 2 squares wide by 2 squares tall, would be 20 inches wide and 20 inches tall while the dimensions of the I block would be 40 inches wide by 10 inches tall. The corresponding playfield is 10 squares wide by 16 squares tall. In this project, you will create a sequence of commands for a robot to navigate a specific path to place a tetrimino in the playfield.

The robot's wheels are 3 inches in diameter, and the robot understands the following commands:

- Rotate wheels forward n rotations, where n is a positive real number.
- Rotate wheels backward n rotations, where n is a positive real number.
- Turn 90° left.
- Turn 90° right.

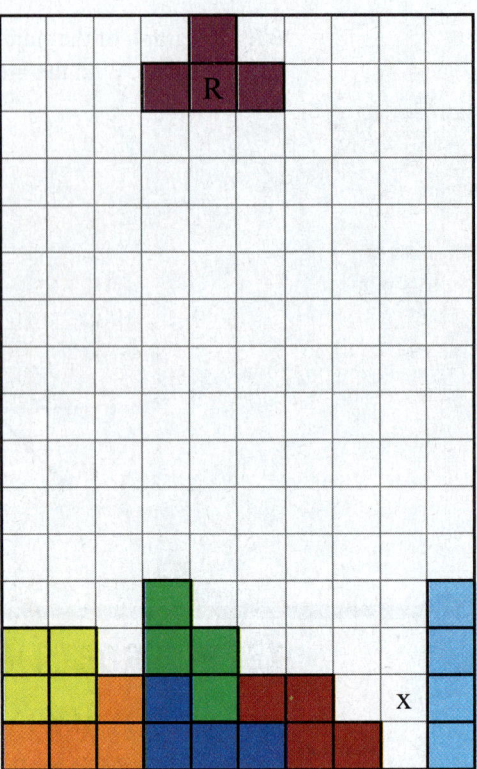

1. Calculate the distance the robot would travel after one full rotation of its wheels. Then, determine how many rotations are needed to travel the side length of one square. Round your answers to the nearest hundredth.

2. Determine the total area of the playfield.

3. Determine the total area of each tetrimino.

4. During practice sessions, the team had to figure out the basic moves needed to reposition the tetriminos. During one practice session, they worked through the following sequences. In each situation, the robot is centered under the square marked R.

   a. The robot starts as shown, facing toward the right of the playfield. Draw the Z block if the robot turns 90° left. Show R with the correct orientation.

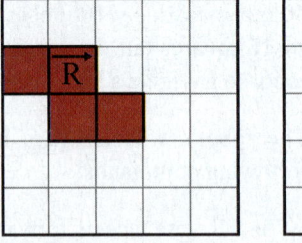 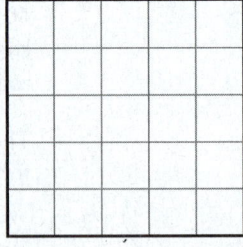

   b. The robot starts as shown, facing toward the top of the playfield. Draw the I block if the robot rotates its wheels backwards 1.06 rotations. Show R with the correct orientation.

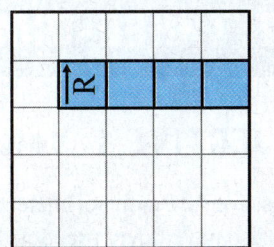
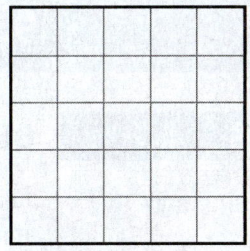

c. The robot starts as shown, facing toward the right of the playfield. Draw the J block if the robot rotates its wheels backwards 2.12 rotations and turns 90° right. Show R with the correct orientation.

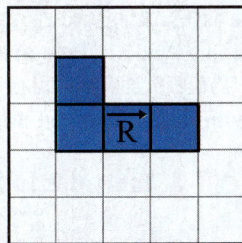
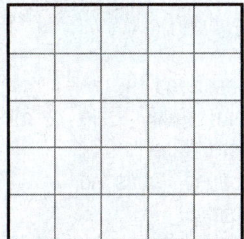

d. Give the directions needed if the robot starts out in the first position and ends in the second position.

First Position   Second Position

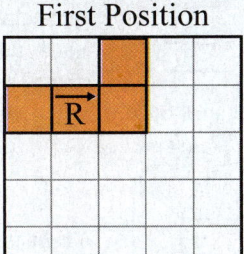
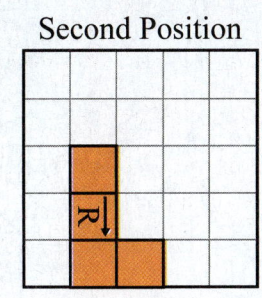

e. Give the directions needed if the robot starts out in the first position and ends in the second position.

First Position   Second Position

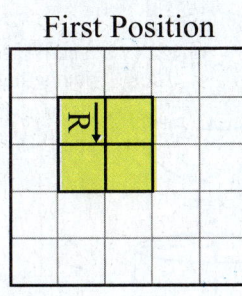
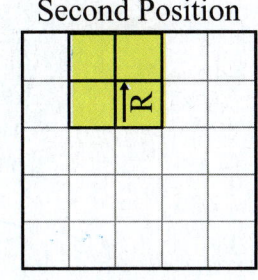

5. During a science fair, the first step of the team's demonstration is for the robot to start at the top left corner of the playfield and then travel around the edge of the entire field. How far will the robot travel?

6. The final demonstration of the robot is to move a T block from the top of the playing field to the bottom so that two full rows are completed. The robot is facing towards the right side of the playfield and centered under the square marked R. What commands should be given to the robot to properly place the block, which would result in the robot being centered in the square marked with an X?

OBJECTIVES

1. Calculate volume and surface area.

Fun Fact

A Klein bottle, first described by the German mathematician Felix Klein in 1882, is a unique structure in that it has no volume because it has no definable inside or outside.

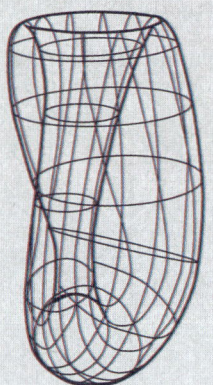

Figure 9.2.1

OBJECTIVE 1 Volume and Surface Area

Just as we saw with perimeter and area in the previous section, volume and surface area also have many practical applications. International shipping companies base their fees on the volume of goods to be shipped, scientists study the surface area of microorganisms for intestinal research, and conservationists look at human water consumption by volume to aid in environmental predictions.

Volume is the measure of the amount of space occupied by an object in three dimensions. And **surface area** is the measurement of the outside surface of the object. Table 9.2.1 provides a summary of some of the most common shapes along with their respective formulas.

Table 9.2.1: Volume and Surface Area Formulas

Shape	Definition	Volume Formula	Surface Area Formula
Rectangular Solid	A **rectangular solid** has 6 faces, all of which are rectangles. We denote the longer side of the base as length l and the shorter side as the width w. The height is h.	$V = lwh$	$SA = 2lw + 2lh + 2wh$
Cube	A **cube** is made of six congruent squares. The side length is denoted by s.	$V = s^3$	$SA = 6s^2$
Sphere	A **sphere** is a the set of all points that are a certain distance from a fixed center point. The radius is denoted by r.	$V = \dfrac{4}{3}\pi r^3$	$SA = 4\pi r^2$
Right Circular Cylinder	A **right circular cylinder** is a cylinder with circle bases of radius r. The bases are perpendicular to the height h of the cylinder.	$V = \pi r^2 h$	$SA = 2\pi r^2 + 2\pi rh$

Shape	Definition	Volume Formula	Surface Area Formula
Right Circular Cone	A **right circular cone** is a figure with a circular base of radius r which tapers to a point at a perpendicular height, h.	$V = \dfrac{1}{3}\pi r^2 h$	$SA = \pi r^2 + \pi r\sqrt{r^2 + h^2}$
Square Pyramid	A **square pyramid** is a figure with a square base with sides s and four triangular sides that taper to a point. The height is denoted by h.	$V = \dfrac{1}{3}s^2 h$	$SA = s^2 + 2s\sqrt{h^2 + \left(\dfrac{s}{2}\right)^2}$

Before we jump in to using the formulas from Table 9.2.1, let's start by distinguishing between scenarios that require surface area and those that require volume.

Example 9.2.1

Determining which Measurement to Calculate

In each of the following scenarios, consider the space being described and determine whether you should calculate the volume or surface area to resolve the issue.

a. The amount of packing space in a moving truck

b. The amount of fabric needed to re-cover a chair

c. The amount of wrapping paper needed for Christmas gifts

d. The amount of soil needed for a new planter

Solution

a. The space inside a moving truck requires that you know the **volume** the truck can hold.

b. The fabric will cover the outside of the chair, so **surface area** needs to be calculated.

c. Wrapping paper will cover the outside area of packages, and hence **surface area** is needed.

d. The soil will fill in the inside of a planter, so **volume** needs to be calculated.

As these examples show, being able to calculate volume and surface area has many practical uses. We can also combine the formulas for the volume and surface area of standard solids to enable us to find the volumes and surface areas of more complicated figures. We will explore this idea throughout the remainder of this section. Let's begin with the volume of a rectangular solid.

Example 9.2.2

Using the Volume of a Rectangular Solid

You want to build a concrete patio onto the back of your house. The patio will be 18 feet long, 12 feet wide, and 6 inches deep, and concrete installation (including the cost of the concrete) is $126 per cubic yard.

a. How many cubic yards of concrete will be needed to build the patio?

b. Determine the cost of installing the patio.

Solution

a. A concrete patio is a rectangular solid and the volume is determined by the formula $V = lwh$.

We know that the length l of the patio is 18 feet and the width w is 12 feet. However, the depth (which is the height) is given in inches. Because we should always work with dimensions that are in the same units, we need to first convert the depth to feet. A depth h of 6 inches is equivalent to 0.5 feet since there are 12 inches in a foot.

The volume of the patio is calculated as follows.

$$V = lwh$$
$$= (18 \text{ ft})(12 \text{ ft})(0.5 \text{ ft})$$
$$= 108 \text{ ft}^3$$

In order to determine the number of cubic yards of concrete that are required to build the patio, we need to use a version of the conversion factor for feet and yards, $\dfrac{1 \text{ yd}}{3 \text{ ft}}$. Since volume is in cubic units, we need to cube the conversion factor so that it contains the correct units. Therefore, the conversion factor we need is $\left(\dfrac{1 \text{ yd}}{3 \text{ ft}}\right)^3 = \dfrac{1 \text{ yd}^3}{27 \text{ ft}^3}$. Multiplying the volume by the conversion factor gives us the following.

$$\text{Volume in Cubic Feet: } (108 \text{ ft}^3)\left(\dfrac{1 \text{ yd}^3}{27 \text{ ft}^3}\right) = 4 \text{ yd}^3$$

Therefore, 4 cubic yards of concrete are required to build the patio.

b. We are told that the installation fee is $126 per cubic yard. So we can use the conversion factor $\dfrac{\$126}{1 \text{ yd}^3}$ to determine the installation cost for the patio.

$$\text{Installation Cost: } (4 \text{ yd}^3)\left(\dfrac{\$126}{1 \text{ yd}^3}\right) = \$504$$

Therefore, the concrete patio will cost $504 to install.

Example 9.2.3

Using the Surface Area of a Rectangular Solid

Aimee is icing a cake for her sister's birthday. It's a half-sheet cake and measures 12 in. by 18 in. by 2 in.

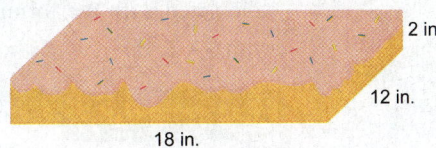

Figure 9.2.2

a. Find the surface area of the cake that will need icing.

b. If the cake is cut into pieces that are 2-inch squares, how many pieces of cake can be served?

Solution

a. The cake is a rectangular solid with length $l = 18$ in., width $w = 12$ in., and height $h = 2$ in. Using our formula, we calculate the surface area as follows.

$$\begin{aligned}
\text{SA} &= 2lw + 2lh + 2wh \\
&= 2(18 \text{ in.})(12 \text{ in.}) + 2(18 \text{ in.})(2 \text{ in.}) + 2(12 \text{ in.})(2 \text{ in.}) \\
&= 432 \text{ in.}^2 + 72 \text{ in.}^2 + 48 \text{ in.}^2 \\
&= 552 \text{ in.}^2
\end{aligned}$$

Therefore, the total surface area of the cake is a 552 square inches. However, remember the bottom of the cake will not be iced, so we need to subtract off the area of the bottom of the cake. The area for the bottom of the cake is a rectangle of length 18 inches and width 12 inches, which gives us $A_{\text{bottom}} = 18$ in. $\cdot$ 12 in. $= 216$ in.^{2} So the surface area of the cake that needs icing is calculated as follows.

$$552 \text{ in.}^2 - 216 \text{ in.}^2 = 336 \text{ in.}^2$$

b.

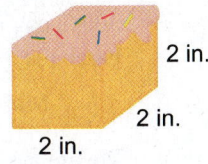

Figure 9.2.3

To find the number of 2-inch square servings in the cake, we only need to consider the top of the cake. We know that the top area of the cake is the same as the bottom, which we have already calculated as 216 in.^{2} So, we can divide the top area of the cake by the top area of a single slice, 2 in. $\cdot$ 2 in. $= 4$ in.^{2}, and obtain the following.

$$\frac{216 \text{ in.}^2}{4 \text{ in.}^2} = 54 \text{ slices}$$

✔ **Skill Check 9.2.1**

Find the volume and surface area for the given rectangular solid.

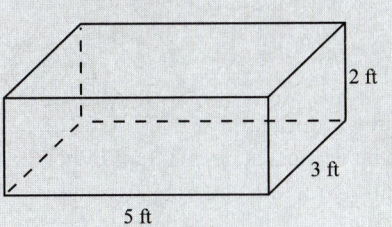

Figure 9.2.4

A sphere, such as a marble or a globe, is defined as the set of points in space that are all at an equal distance r from a given point. The given point is called the center of the sphere and we describe the distance from the center to any point on the surface of the sphere as the radius, r, of the sphere. In contrast to the other shapes we have looked at, the sphere has no edges or corners. To calculate its volume or surface area, we only need to know the radius. Spherical volume is calculated with the formula $V = \dfrac{4}{3}\pi r^3$ and a sphere's surface area is calculated with $SA = 4\pi r^2$.

Example 9.2.4

Using the Surface Area of a Sphere

Tyson's Athletic Gear Inc. manufactures 300 new baseballs a day. The cork center (or pill) is wrapped in rubber and then wool before having a cowhide cover applied. Each finished ball has a diameter of 2.9 inches and weighs between 5.0 and 5.25 ounces

a. Calculate the surface area of one baseball, rounded to two decimal places.

b. What is the minimum amount of cowhide needed (in square feet) to manufacture a week's worth of baseballs if the plant operates 5 days a week? Explain why this is a minimum amount.

Figure 9.2.5

> **Think Back**
>
> The diameter of a circle is twice the length of the radius.

Solution

a. The formula to calculate the surface area of a sphere is $SA = 4\pi r^2$. Since we were given the diameter, and we need the radius for the formula, we need to divide the diameter by two. Thus, the radius of each baseball is $2.9 \div 2 = 1.45$ in. Using the formula, we can calculate the surface area of the baseball.

$$SA = 4\pi r^2$$
$$= 4\pi \left(1.45 \text{ in.}\right)^2$$
$$\approx 26.42 \text{ in.}^2$$

So the surface area of one baseball is approximately 26.42 in.²

b. To find the minimum amount of cowhide needed to manufacture a week's worth of baseballs, we need to multiply the surface area we found in part a. by 300 (a day's worth of baseballs) and then by 5 since the plant is manufacturing baseballs 5 days a week.

Minimum Cowhide Needed: 26.42 in.²(300)(5) = 39,630 in.²

We are asked to provide the area in square feet, so we need to change square inches to square feet using the conversion factor $\left(\dfrac{1 \text{ ft}}{12 \text{ in.}}\right)^2 = \dfrac{1 \text{ ft}^2}{144 \text{ in.}^2}$.

$$39{,}630 \text{ in.}^2 \cdot \frac{1 \text{ ft}^2}{144 \text{ in.}^2} \approx 275.21 \text{ ft}^2$$

Therefore, a week's worth of baseballs requires a minimum of 275.21 ft² of cowhide. This is a minimum amount of cowhide because it does not allow for any wasted material that might occur when cutting out the coverings. We will look more at this type of challenge in Section 9.3.

Example 9.2.5

Using the Volume of a Sphere

Whitt is filling balloons with an air pump for his birthday party.

a. After 2 pumps of air, a balloon has a radius of 2 inches. Find the volume of air contained in the balloon.

b. Whitt wants to make the balloon bigger, so he puts in 2 more pumps of air and the volume of air doubles. What is the volume of air in the balloon now?

c. How big is the balloon's radius after the 4 pumps of air?

d. By what percentage does the radius of the balloon increase when Whitt goes from 2 pumps of air to 4 pumps?

Solution

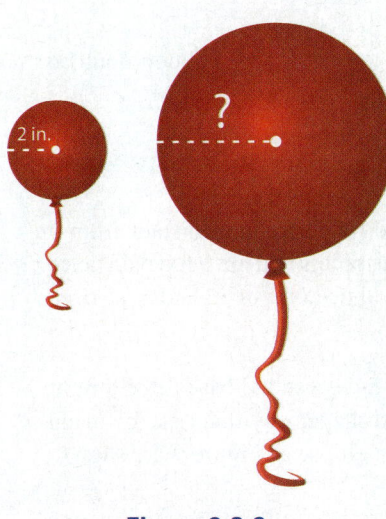

Figure 9.2.6

a. We know that the volume of a sphere can be determined by the formula $V = \frac{4}{3}\pi r^3$. After 2 pumps of air, the balloon has a radius of 2 inches. Substituting in the known values and simplifying gives the following.

$$V = \frac{4}{3}\pi r^3$$

$$V = \frac{4}{3}\pi (2 \text{ in.})^3$$

$$\approx 33.51 \text{ in.}^3$$

The balloon has a volume of approximately 33.51 in.³ after two pumps of air.

b. Since we are told that the volume of air doubles after two more pumps, the volume of air after 4 pumps is as follows.

$$2(33.51 \text{ in.}^3) = 67.02 \text{ in.}^3$$

c. In part b., we found the volume of the balloon after the 4 pumps of air. We will use that information to find the new radius of the balloon by substituting the known values into the formula for the volume and solving for the radius, r.

$$V = \frac{4}{3}\pi^3$$

$$67.02 \text{ in.}^3 = \frac{4}{3}\pi^3$$

$$\frac{3}{4\pi}\left(67.02 \text{ in.}^3\right) = r^3 \qquad \text{Isolate the variable.}$$

$$16.00 \text{ in.}^3 \approx r^3$$

$$2.53 \approx r \qquad \text{Find the cube root of both sides.}$$

Thus, the radius of the balloon after four pumps is approximately 2.52 inches.

d. Recall that percentage increase is found by dividing the absolute change in the measurements by the original amount, where absolute change is simply the difference between the two radii.

Percent increase:

$$\frac{\text{New Radius} - \text{Original Radius}}{\text{Original Radius}} \cdot 100\% = \frac{2.52 \text{ in.} - 2 \text{ in.}}{2 \text{ in.}} \cdot 100\% = 26\%$$

Thus, the radius increased by 26% when the volume of the balloon doubled (that is, when the volume increased by 100%).

💬 **Think Back**

Percentage Change $= \dfrac{\text{Absolute Change}}{\text{Original Amount}}$

For more on percent increase or decrease, see Section 4.2.

In contrast to a sphere, which consists of all the points equally distant from a single point, a right circular cylinder consists of the points that are a fixed distance from a line segment. If we revolve a rectangle about one of its sides, a right circular cylinder is formed.

However, a cylinder does not necessarily have to have a round base or be a right cylinder. As long as the ends are the same size and shape, a cylinder can be made from any shape. For instance, the base could be an ellipse as Figure 9.2.7 shows.

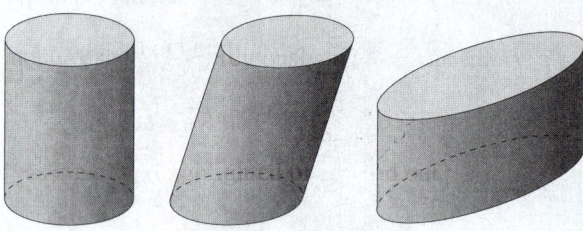

Figure 9.2.7

For our purposes, we will only consider right circular cylinders. The volume for a right circular cylinder is $V = \pi r^2 h$, where r is the radius of the circle and h is the height of the cylinder.

Example 9.2.6

Using the Volume of a Right Circular Cylinder

The sound a drum makes varies according to the diameter and height of the drum (assuming that the materials used to make the drum remain the same).

a. Find the volume of air, to the nearest tenth, in a snare drum that has a diameter of 14 inches and a height of 5 inches.

b. A tom-tom drum has a radius of 5 inches and a height of 8 inches. Find the volume of air, to the nearest tenth, in the tom-tom drum.

c. Vertical 5-inch brads (small nails) are placed around the edge of the snare drum. Calculate how many brads are on the drum if they are placed approximately 4 inches apart.

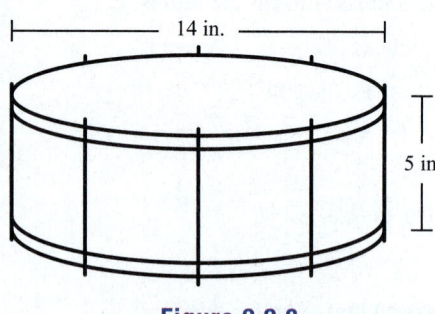

Figure 9.2.8

Solution

a. The drum forms a right cylinder, so we can use the formula for the volume of a right circular cylinder, $V = \pi r^2 h$. We are given the diameter, $d = 14$ in., so the radius is $r = \dfrac{14}{2}$ in. $= 7$ in. The height of the drum, h, is 5 inches. Substituting these values into the formula, we have the following.

💬 **Fun Fact**

As a general rule, the smaller the diameter of the drum is, the higher its pitch is. A drum with a 15" diameter will sound deep and fat, while a 10" diameter will produce a thinner, sharper sound.

$$V = \pi r^2 h$$

$$= \pi (7 \text{ in.})^2 (5 \text{ in.})$$

$$\approx 769.7 \text{ in.}^3$$

So the approximate volume of the snare drum is 769.7 in.^{3}

b. To find the volume of air in the tom-tom drum, we use the same formula, with $r = 5$ in. and $h = 8$ in.

$$V = \pi r^2 h$$

$$= \pi (5 \text{ in.})^2 (8 \text{ in.})$$

$$\approx 628.3 \text{ in.}^3$$

So the approximate volume of the tom-tom drum is 628.3 in.^{3}

c. The brads are placed along the circumference of the drum. To determine how many there are, we need to calculate the drum's circumference using the formula $C = \pi d$.

$$C = \pi (14 \text{ in.}) \approx 43.98 \text{ in.}$$

Since the brads are spaced 4 inches apart, we divide the circumference by four to determine the number of brads.

$$\text{Number of Brads: } \frac{43.98 \text{ in.}}{4 \text{ in.}} = 10.995$$

We can conclude that 11 brads are used and placed approximately 4 inches apart.

A cone is a three-dimensional shape that tapers smoothly from a flat circular base to the peak of the cone. If the axis forms a right angle with the base, it is a right circular cone; otherwise, it is an oblique cone. A right cone is generated by revolving a right triangle about one of its legs. For our purposes, we will only consider right circular cones. The volume is given by the formula $V = \frac{1}{3}\pi r^2 h$, where r is the radius of the circle base and h is the height of the cone (which is also the length of the axis).

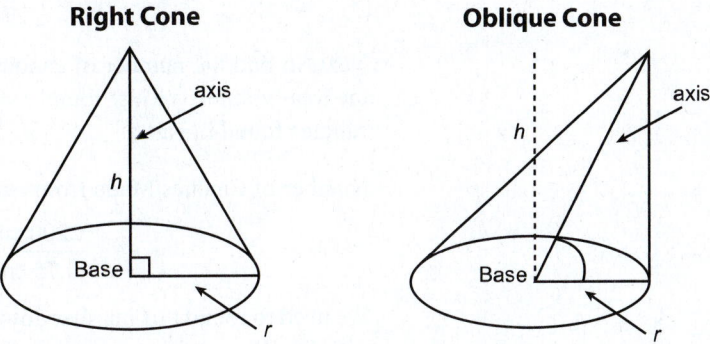

Figure 9.2.9

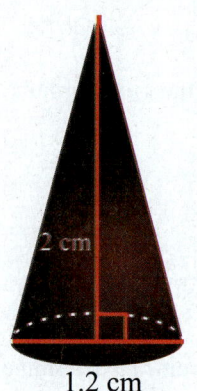

1.2 cm

Figure 9.2.10

Example 9.2.7

Finding the Volume of a Right Circular Cone

E.D.'s Candy Shoppe is creating a new solid chocolate in the shape of a right circular cone. The height of the chocolate cone is 2 cm and the diameter of the base is 1.2 cm.

a. Find the volume of the new chocolate candy.

b. A rectangular block of chocolate measuring 22 cm by 10 cm by 10 cm is melted down and used to make a batch of the new candies. Find the maximum number of candies that can be made from the block of chocolate.

Solution

a. The formula for the volume of a right circular cone is $V = \frac{1}{3}\pi r^2 h$. The radius of the base of the new chocolate is half of the diameter we were given, or $r = \frac{1.2 \text{ cm}}{2} = 0.6 \text{ cm}$. The height is 2 cm. Substituting these values into the formula, we have the following.

$$V_{\text{New Candy}} = \frac{1}{3}\pi r^2 h$$

$$= \frac{1}{3}\pi (0.6 \text{ cm})^2 (2 \text{ cm})$$

$$\approx 0.75 \text{ cm}^3$$

Thus, the volume of the new chocolate is approximately 0.75 cm³.

b. To find the maximum number of candies that can be made from a block of chocolate, we need to begin by finding the volume of the chocolate block. We know the volume of a rectangular solid is $V = lwh$ and that the block has a length of 22 cm, a width of 10 cm, and a height of 10 cm. Substituting these values into the formula, we have the following.

$$V_{\text{Chocolate Block}} = lwh$$

$$= (22 \text{ cm})(10 \text{ cm})(10 \text{ cm})$$

$$= 2200 \text{ cm}^3$$

Next, to find the number of candies that can be made from this block, divide the total volume we just found by the volume of chocolate of one of the new candies found in part a.

Number of Candies Made from One Block of Chocolate:

$$\frac{2200 \text{ cm}^3}{0.75 \text{ cm}^3} = 2933.33$$

We need to round this number down to the nearest whole number, since we are asked for the maximum number of chocolates that can be made from one block. So a maximum of 2933 candies can be made from the block of chocolate.

Example 9.2.8

Comparing Volumes

If the volume of a right circular cylinder is 600 cm³, find the volume of a right circular cone with the same height and base.

Solution

Let's begin by considering the volume formulas for both the right circular cylinder and the right circular cone.

$$V_{\text{cylinder}} = \pi r^2 h \quad V_{\text{cone}} = \frac{1}{3}\pi r^2 h$$

Notice that since the base and height are the same for each figure, r and h will be the same in each equation. From the equations, we can see that the cone will then contain one third of the volume of the cylinder. Thus, if the cylinder has a volume of 600 cm³, the cone's volume will be $\dfrac{600 \text{ cm}^3}{3} = 200 \text{ cm}^3$.

A square pyramid is defined as a three-dimensional figure that has a square base and isosceles triangles as faces with a volume of $V = \dfrac{1}{3}s^2 h$. It should be noted that a pyramid can have any polygon as its base. However, for the purposes of this discussion, we will only consider the case of the square pyramid.

Example 9.2.9

Using the Volume of a Square Pyramid

The Great Pyramid of Giza was originally about 480 ft tall with a square base whose sides measured about 755 ft long.

a. Find the volume of the Great Pyramid of Giza.

b. For a sandcastle competition, a scale model of the Great Pyramid of Giza is made. If the model has a ratio of 1 : 100, what is the volume of sand needed to build the model?

Solution

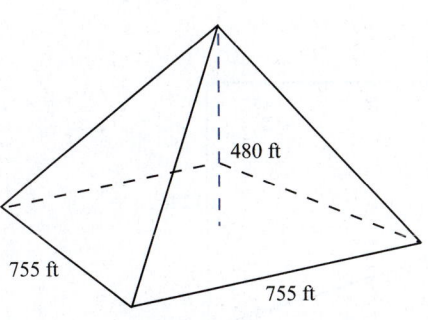

Figure 9.2.11

a. The formula for the volume of a square pyramid is $V = \dfrac{1}{3}s^2 h$. The Great Pyramid of Giza has a height of 480 feet and each side of the base is 755 feet long. Substituting these values into the formula, we have the following.

$$V = \frac{1}{3}s^2 h$$

$$= \frac{1}{3}(755 \text{ ft})^2 (480 \text{ ft})$$

$$= 91,204,000 \text{ ft}^3$$

Therefore, the volume of the Great Pyramid of Giza was about 91,204,000 ft³.

b. Using our knowledge of ratios, we know that if the scale model has a ratio of 1:100, then we need to divide the measurements by 100. We can do this in one of two ways. We can either divide each of the original length measurements by 100 and then multiply them together, or we can divide the volume by $100^3 = 1,000,000$, which is the method we will use. Either calculation will give the same result, and we'll leave the other method as an exercise. The volume of the model is calculated as follows.

$$V_{\text{scaled model}} = \frac{91,204,000 \text{ ft}^3}{1,000,000} = 91.204 \text{ ft}^3$$

Therefore, the model pyramid will require 91.204 ft³ of sand.

> ✔ **Skill Check 9.2.2**
>
> Show that dividing each of the original measurements in Example 9.2.9 by 100 before using the formula produces the same volume found in part b.

Example 9.2.10

Comparing Volumes

Suppose you want to build a swimming pool in your backyard. The city in which you live has a restriction on the volume of water that the pool can hold. Pools may contain at most 4800 cubic feet of water. Your plans are to a build a pool that is 16 feet wide, 24 feet long, and has a depth that varies from 2 feet to 12 feet on a constant slope. Does the pool meet the restrictions required by the city?

Solution

To begin, let's use some of the problem-solving techniques from Chapter 1. By drawing a picture of the pool with views from both the top and side, we can better understand what is being asked. Figure 9.2.12 shows the top view of the swimming pool while Figure 9.2.13 shows the side view of the pool sloping from a depth of 2 feet to a depth of 12 feet.

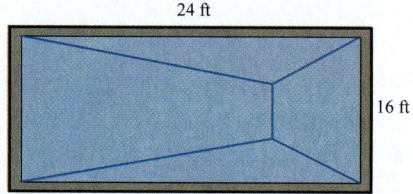

Figure 9.2.12: Top View of the Pool

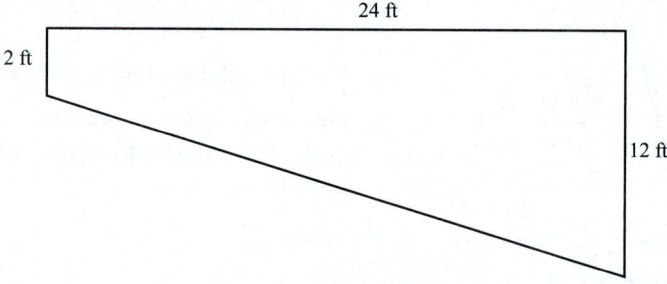

Figure 9.2.13: Side View of the Pool

Notice from the side view that the pool is actually in the shape of a trapezoid. Although Table 9.2.1 does not list a formula for the volume of a solid in the shape of a trapezoid, we can find the volume by finding the area of the trapezoid and multiplying it by the width of the pool (16 feet).

The formula for the area of a trapezoid is Area $= \frac{1}{2}(b_1 + b_2)h$, where b_1 and b_2 are the lengths of the bases and h is the height. In this case, the lengths of the bases are 2 feet and 12 feet and the height is 24 feet. Substituting these values in the area formula, we have the following.

$$A_{\text{pool side}} = \frac{1}{2}(b_1 + b_2)h = \frac{1}{2}(2 \text{ ft} + 12 \text{ ft})(24 \text{ ft}) = 168 \text{ ft}^2$$

The volume of the pool is then equal to the product of the area of the trapezoidal side view (168 ft²) and the width of the pool (16 ft).

$$V = A_{\text{pool side}} \cdot \text{Pool Width} = (168 \text{ ft}^2)(16 \text{ ft}) = 2688 \text{ ft}^3$$

Now we need to answer the initial question of whether a pool with these dimensions meets the restrictions set by the city. The answer is yes, the pool meets the restrictions because the volume of the pool (2688 ft³) is less than the maximum volume allowed by the city (4800 ft³).

Example 9.2.11

Finding the Dimensions of a Right Circular Cylinder

A farmer wishes to build a grain silo in the shape of a right circular cylinder capable of storing 400,000 cubic feet of grain. City ordinances say the silo can be no taller than 100 feet. Determine a set of dimensions for a silo that satisfies the city's restrictions and meets the needs of the farmer.

Solution

Note that we are asked to find *a* set of dimensions, not *the* set of dimensions. Since the radius and height of the silo can both vary, there are many differently sized silos that meet the criteria.

The farmer's desire to have a silo that holds a specific amount of grain means that we need to work with the volume formula for a right circular cylinder, $V = \pi r^2 h$, to help us with the dimensions. Notice that we have values for two of the three variables in the formula, the volume V and the height h. We are told that the volume needs to be 400,000 ft³ and the height can be no more than 100 ft. Although there are many possible heights less than the restriction, we are going to assume that the height will be the maximum, 100 feet.

Thus, we are left to find the radius of the base to determine the full dimensions of the silo. Using the formula for the volume of a cylinder, we can calculate the radius as follows.

$$V = \pi r^2 h$$

$$400,000 \text{ ft}^3 = \pi r^2 (100 \text{ ft}) \qquad \text{Substitute the known values.}$$

$$\frac{4000}{\pi} \text{ ft}^2 = r^2 \qquad \text{Isolate the variable.}$$

$$\sqrt{\frac{4000}{\pi} \text{ ft}^2} = \sqrt{r^2} \qquad \text{Take the square root of both sides.}$$

$$35.68 \text{ ft} \approx r$$

Therefore, at 100 feet tall, in order for the silo to hold 400,000 cubic feet of grain, the radius should be approximately 35.68 feet.

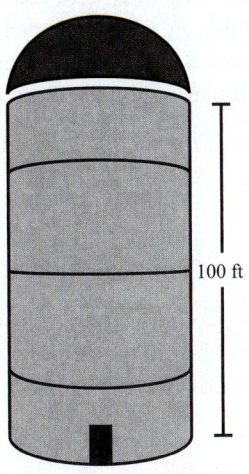

Figure 9.2.14

100 ft

> 💬 **Helpful Hint**
>
> Note that when we divide cubic feet by a foot, the result is in square feet.

> ✔ **Skill Check 9.2.3**
>
> Determine another set of dimensions for a silo with $V = 400,000$ ft³, as in Example 9.2.11, using a height of 85 feet.

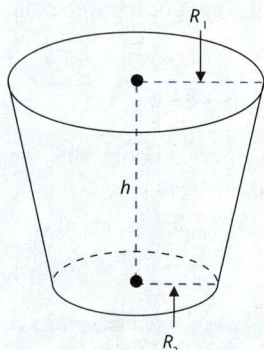

Figure 9.2.15

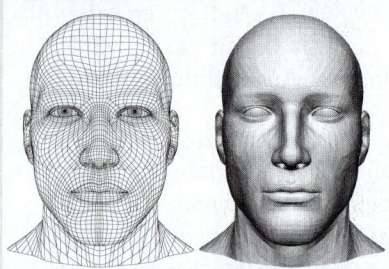

Figure 9.2.16

Figure 9.2.17

Example 9.2.12

Finding the Dimensions of a Conical Frustrum

The I Scream, U Scream Ice Cream shop is famous for their 0.5 L ice cream sundae. They had a special cup (instead of a cone) created just for them. The cup is in the shape of a conical frustum. (If we cut the top—or bottom depending on the orientation—off of a cone, we get a conical frustum, as shown in the figure, where the top radius is R_1, the bottom radius is R_2, and the height is h.)

The original cup has dimensions of $R_1 = 4.5$ cm, $R_2 = 3$ cm, and $h = 11$ cm. Recently, the owner decided that he wanted to break the world record for the size of a sundae, which currently is approximately 37,500 L of ice cream. He needs to enlarge his special cup in order to achieve the goal of breaking the world record. The cup for breaking the world record has a constraint that $R_1 = 2.5$ meters and $R_2 = 1$ meter. The owner wants the final amount of ice cream that will fit in the cup to be 38,000 L. If the volume of a frustum can be found using the formula $V = \frac{1}{3}\pi h\left(R_1^2 + R_1 R_2 + R_2^2\right)$, what should the dimensions of the enlarged cup be? (**Note:** 1000 L = 1 m³.)

Solution

Notice that with the constraints that were given ($R_1 = 2.5$ m, $R_2 = 1$ m, and $V = 38{,}000$ L), the only dimension that is unknown is the height (h) of the cup. We need to substitute these values into the formula for the volume and solve for h.

$$V = \frac{1}{3}\pi h\left(R_1^2 + R_1 R_2 + R_2^2\right)$$

$$38{,}000 \text{ L} = \frac{1}{3}\pi h\left((2.5 \text{ m})^2 + (2.5 \text{ m})(1 \text{ m}) + (1 \text{ m})^2\right)$$

$$38{,}000 \text{ L} = \frac{1}{3}\pi h\left(9.75 \text{ m}^2\right) \qquad \text{Simplify.}$$

$$\frac{3(38{,}000 \text{ L})}{(9.75 \text{ m}^2)\pi} = h \qquad \text{Isolate the variable.}$$

$$3721.78\,\frac{\text{L}}{\text{m}^2} \approx h$$

We now need to use a conversion factor for liters and meters: 1000 L = 1 m³. Placing liters in the denominator gives us the fraction $\dfrac{1 \text{ m}^3}{1000 \text{ L}}$.

$$h \approx \frac{3721.78 \text{ L}}{\text{m}^2} \cdot \frac{1 \text{ m}^3}{1000 \text{ L}} \approx 3.72 \text{ m}$$

Therefore, I Scream, U Scream Ice Cream will need to build a cup with a top radius of 2.5 meters, a bottom radius of 1 meter, and a height of about 3.72 meters. This will hold the desired 38,000 L of ice cream and become the new world record.

Skill Check Answers

1. $V = 30$ ft³; SA $= 62$ ft²   **2.** $\frac{1}{3}\left(\frac{755 \text{ ft}}{100}\right)^2\left(\frac{480 \text{ ft}}{100}\right) = \frac{1}{3}(7.55 \text{ ft})^2(4.8 \text{ ft}) = 91.204$ ft³
3. $r \approx 38.70$ ft, $h = 85$ ft

9.2   Exercises

✔ CONCEPT CHECK

1. Volume is measured in _____ dimensions while surface area is measured in _____ dimensions.

2. All points on the surface of a sphere are the same distance from a _____ point.

3. A square pyramid has four _____ sides.

4. True or False: A cube is a type of rectangular solid.

5. True or False: A cylinder can have ends that are different sizes and shapes.

💡 PRACTICE

In each of the following scenarios, consider the space being described and determine whether you should calculate the volume or surface area.

6. The amount of coffee in a can

7. How much shaved ice is needed for a snow cone

8. How many gallons of paint needed for the siding on your house

9. The amount of sand in a sandbox

10. How much cedar is needed to line a hope chest

11. The amount of leather needed to cover a basketball

🚀 APPLICATIONS

12. If the volume of a right circular cone is 360 ft³, find the volume of a right circular cylinder with the same radius and height.

13. If the volume of a right circular cylinder is 360 ft³, find the volume of a right circular cone with the same radius and height.

14. You want to make a fish tank in the shape of a rectangular solid that can hold 26,250 cubic centimeters of water. If the length is to be 50 cm and the height is to be 35 cm, then how wide is the tank? How many square centimeters of glass are needed to construct the tank?

15. The concrete foundation of a house requires 15 truckloads of concrete. The concrete was poured at a depth of 0.5 meters and covered an area of 540 square meters.

    a. How many cubic meters of concrete were used?

    b. If each truckload was the same size, how many cubic meters did each truck carry?

16. The surface of a house is covered with vinyl siding. The gable end (triangular portion) of the house and the roof will not need siding. Determine the amount of siding required.

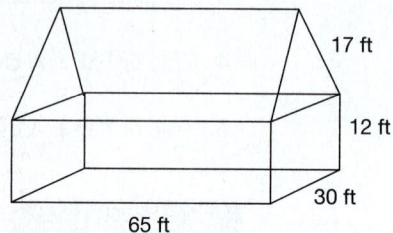

17. Montgomery has a box that is 24 inches by 36 inches by 12 inches and is filled with sand. He wants to use this sand to fill as many 4-inch cubes as he can. How many cubes can he fill?

18. A company is shipping a piece of equipment that requires a cubic box with a volume of 216 cubic feet. How many square feet of wood are required to construct the box?

19. Isaac made a rectangular shaped raised flower bed that is 12-feet long, 4-feet wide and 1.5-feet high. He wants to line the inside of the bed with landscape fabric. How many square feet of fabric will he need?

20. Emeril baked a sheet cookie that is 15 inches by 24 inches. If he wants to serve one piece to each of his 40 guests, what size squares should he cut?

21. A company needs cubic boxes that will hold 729 cubic inches. How many square inches of cardboard will be required to construct each box?

22. The Earth has a diameter of about 7900 miles.

    a. What is the approximate surface area of the Earth?

    b. What is the approximate volume of the Earth?

    c. If $\frac{2}{3}$ of the Earth's surface is covered with water, about how many square miles of the surface are covered by land?

23. The moon has a diameter of approximately 2160 miles.

    a. What is the approximate surface area of the moon?

    b. If one-third of the moon is covered in craters, about how many square miles of the moon's surface is crater free?

24. Lisa makes Christmas ornaments from Styrofoam balls with a 3-inch diameter.

    a. Calculate the surface area of one of the balls rounded the area to the nearest hundredth.

    b. What is the maximum number of balls Lisa can make from 1 square yard of fabric?

25. Patti is making 2 papier-mâché balls for a party. To make the balls, she will cover balloons that have a 12-inch diameter with the papier-mâché material.

    a. Calculate the surface area of one balloon rounded to the nearest hundredth.

    b. If each balloon must be covered five times to create a sturdy ball, how many pieces of 8.5 by 11 inch paper will be needed to cover the two balloons?

26. Helen is filling round balloons with helium for a wedding. Each balloon has a diameter of 12 inches. Her helium tank contains 14.9 cubic feet of helium.

    a. How many cubic feet of helium does one balloon hold when filled to a diameter of 12 inches?

    b. How many balloons can she fill with the amount of helium in the tank?

    c. If Helen wants to use more balloons in the decorations but doesn't want to purchase more helium, she can fill each balloon to an 11-inch diameter instead of 12. How many balloons can she fill now?

    d. By what percentage does the volume decrease when the diameter decreases from 12 inches to 11 inches?

27. A grain silo in the shape of a circular cylinder is 75 feet tall and has a diameter of 24 feet.

    a. What is the volume of the silo?

    b. If a bushel of grain is 1.25 cubic feet, then how many bushels of grain will the silo hold?

    c. If a truck can unload 200 cubic feet of grain per minute into the silo, how long will it take to fill the entire silo?

28. The town of Batesburg has a cylindrical water tower that has a height of 15 meters and a diameter of 8 meters.

    a. Find the volume of the water tower.

    b. If 1 cubic meter is equal to 264.172 gallons, how many gallons of water will the water tower hold?

    c. Suppose the water tower holds about a day's worth of water for the community when filled to full capacity. Approximately how many households are in Batesburg if the average household uses 400 gallons of water per day?

29. Maxine has two cylindrical vases. Vase A has a radius of 2.5 cm and a height of 25 cm, and vase B has a radius of 4 cm and a height of 10 cm.

    a. Find the volume of each vase.

    b. Which vase has a larger volume? What is the difference in volume between the two vases?

30. The manager of the Blizzard Snow Cone truck ordered a new style of paper cones. These new cones have a height of 12 cm and a radius of 8 cm. The dome of ice that goes on top of the cone will remain the same size, and the dome is made of 1072 cm³ of ice.

    a. Find the volume of the cone rounded to the nearest hundredth. (**Hint:** Do not include the volume of the dome.)

    b. The shaved ice is kept in a cooler that is 80 cm by 40 cm by 70 cm. How many snow cones can be made? (**Hint:** Do not forget to include the dome.)

31. A shaved ice stand bought a new ice shaving machine and received a new inventory of paper cones.

    a. Find the volume of a cone if the paper cup has a diameter of 9 cm and a height of 8 cm.

    b. The new ice shaving machine can shave about 50 cm³ per minute. About how many minutes will it take to shave enough ice to fill 12 snow cones if the domes of ice that go on top take an additional 190 cm³.

32. A reception will be held in the lobby of a building that has an inverted square pyramid fountain with a height of 150 inches and side-length of 120 inches. Lani is making table decorations for the reception by filling small replicas of the fountain with glass beads.

    a. Find the volume of the small vases if they are in a 1 : 20 ratio with the fountain.

    b. If each bag of beads contains 30 cubic inches of beads and there are 8 replicas, how many bags of beads will Lani need?

33. Bo built a sandcastle in the shape of a square pyramid. The side length of the pyramid was 8 feet and the height was 4 feet.

    a. How many cubic feet of sand did it take to build the castle?

    b. If Bo decides to build a small replica that has a ratio of 1 : 3, how much sand would be needed for the replica?

34. Selene's teacher made a square pyramid and gave the class an assignment to make replica pyramids out of clay with a ratio of 1 : 6. If the side length of the pyramid made by the teacher is 24 inches and the height is 30 inches, find the volume of the replica that Selene needs to make.

35. The Great Pyramid of Giza is the largest structure ever built by man. It is a square pyramid with a side length of approximately 230.36 meters and a height of approximately 146.59 meters. A gift shop sells replicas that has a ratio of 1: 200. Determine the volume of the replica.

36. Which swimming pool holds more water?

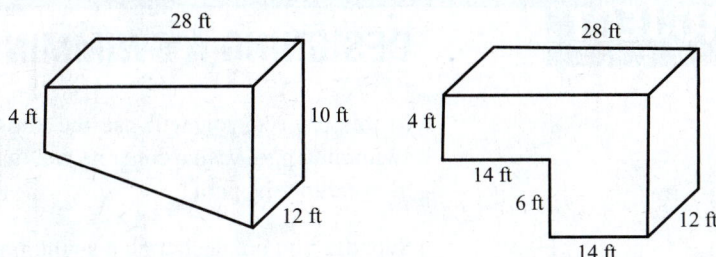

37. Michaela is making a cylindrical candle. She has 2 blocks of candle wax that will melt to give her a total of 400 cm3 of wax. She doesn't want the candle to be more than 20 cm tall. Find a set of dimensions that will meet the restrictions and will also use all of the candle wax.

38. Delilah is making a cylindrical candle. She has 3 blocks of candle wax that will melt to give her a total of 80 in.³ of wax. She doesn't want the radius of the candle to be smaller than 1 inch. Find a set of dimensions that will meet the restrictions and will also use all of the candle wax.

39. David is building a cylindrical rainwater harvesting storage tank in his backyard, and he wants it to hold 1000 gallons or about 134 cubic feet of water. His HOA will not allow any free-standing structures taller than 8 feet. Find a set of dimensions that will meet these restrictions.

40. Phillipe wishes to have a grain silo in the shape of a right circular cylinder that can store 200,000 cubic feet of grain. If he doesn't want the silo to be any taller than 50 feet, determine a set of dimensions that will meet his needs.

41. Refer to Example 9.2.12 of the section to solve the following problems.

  a. Find the surface area of the record-breaking cup, given that the surface area of a frustum can be calculated using the formula $SA = \pi\left(R_1 + R_2\right)\sqrt{\left(R_1 - R_2\right) + h^2} + \pi R_2^2 + \pi R_1^2$.

  b. Paper is not a viable option for the cup because it is not strong enough, so the owner has opted to use steel. If steel costs $7.25 per square meter, what is the cost of material to make the world record cup?

9.2 PROJECT

DESIGNING A SWIMMING POOL

In this project, you will use the knowledge gained in Section 9.2 to design a swimming pool with a concrete walkway and estimate some of the costs involved in installing the pool.

Suppose you are designing a swimming pool for an apartment complex. You are given a few guidelines and a specific location in which to build the pool, but you are able to create the pool however you wish. The overall design phase will consist of designing the pool, estimating costs, and reflecting on your choices.

Design

1. Determine the shape of the pool. The shape of the pool can be square, circular, or rectangular. If the pool is circular, the depth should be constant. If the pool is square or rectangular, the pool can have a depth that steadily increases from one side to the other to provide a deep end and a shallow end. Create a sketch of the pool.

2. Consider what a reasonable length, width, and depth would be for the pool. Add measurements to your sketch in feet.

3. The concrete walkway around the pool needs to be the same width all the way around and wide enough to allow the use of poolside lounge chairs and provide enough room to safely walk. Determine the width you would use for the walkway and add it to your sketch. The depth of the concrete should be 6 inches. (**Note:** Assume poolside lounge chairs have a length of 6 feet.)

Cost Estimation after the Pool is Installed

4. Once the pool is constructed, the interior surface will need to be painted with two coats of swimming pool paint. Calculate the surface area of the pool. The paint you will use costs $50 per gallon and covers 250 square feet per gallon. How many gallons will you need for the two coats of paint? How much will the paint cost?

5. Determine the volume of the pool in both cubic feet and gallons using the conversion factor 1 cubic foot ≈ 7.48 gallons. (Assume the pool is filled to the top.) To fill the swimming pool with water, it will cost $9 for every 1000 gallons of water. How much will the water cost to fill the pool one time, rounded to the nearest dollar?

6. Calculate the volume of concrete (in cubic feet) the walkway around the pool will require. It will cost $4 per cubic foot to have the concrete poured. How much will it cost to pour the concrete?

Analysis

7. Do any of the costs seem unreasonable? Explain your answer.

8. What can you change in your plan to decrease the overall cost after the pool is installed?

9. What are some cost factors that were not considered in this project for the design and installation of the pool?

OBJECTIVES

1. Calculate measures of angles.

2. Determine whether shapes can create a tessellation.

3. Use the Pythagorean Theorem.

4. Use basic trigonometric identities.

5. Use the Law of Sines and Cosines.

OBJECTIVE 1 Angles

The geometry of shapes isn't just limited to measurements of dimension in the form of perimeter, area, and volume. The study of angles is another pivotal aspect of geometry. The properties and uses of angles are embedded in many fields such as architecture (where exact drawings are crucial), land surveying (where precise angle measurement determines land boundaries), and camera design (where lenses create sharp images based on the angles of light reflections).

We begin our probe into the practical uses of angle measurement with the vocabulary used in the study of angles. When we talk about the measure of angles, there are three commonly used base units to work with: degrees, radians, and gradians. In a full circle, there are 360 degrees (360°), 2π radians, or 400 gradians. Although degrees are the most familiar units to the general population, engineers often use gradians, while radians are more common in computer programing or trigonometry. In this section, we will focus our attention on degree measurements.

There are several standard ways to label angles when we are referring to them. The symbol $\angle$, read "angle," is followed by one the following: a letter of the Greek alphabet (such as θ or φ); a single number or capital letter; or by three capital letters representing the points with the vertex as the middle letter. All of these labels are interchangeable and will be used throughout this section. When we refer to the measurement of the angle, we use the letter m in front of the angle symbol. For instance, $m\angle A = 30°$ is read "the measure of angle A is 30 degrees."

💬 Think Back

In geometry, the vertex is the common point where two or more lines, curves, or edges meet.

📖 LABELING ANGLES

There are three common ways of labeling angles.

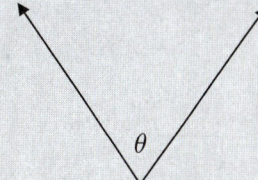

Figure 9.3.1: $\angle \theta$

Using a letter of the Greek alphabet such as θ or φ.

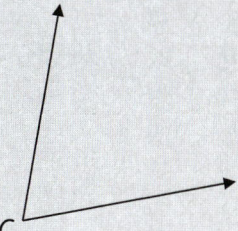

Figure 9.3.2: $\angle C$

Using a single number or capital letter.

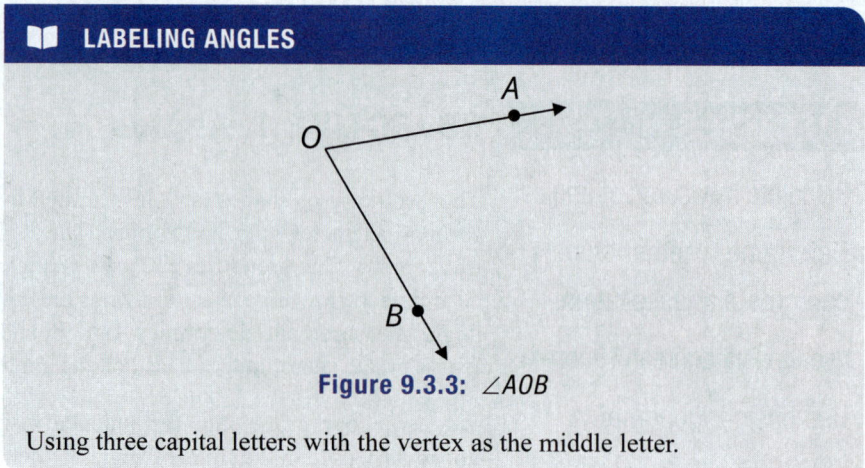

LABELING ANGLES

Figure 9.3.3: ∠AOB

Using three capital letters with the vertex as the middle letter.

Another way to refer to angles is by their properties. For instance, you might not know the exact measure of an angle needed in construction, but you can describe the constraints is has; for instance, that the angle must be *acute*, that an angle needs to be *supplementary* to another angle, or that the angle must be the ever-useful *right angle*. Some of the universal properties of angles and pairs of angles are listed in Table 9.3.1. We will use these attributes to work out various unknown angles in different situations.

Table 9.3.1: Angles Classified by Measure

Name	Measure	Illustrations with Notation
Actute	$0° < m\angle A < 90°$	∠A is an acute angle.
Right	$m\angle B = 90°$	∠B is a right angle.
Obtuse	$90° < m\angle C < 180°$	∠C is an obtuse angle.
Straight	$m\angle D = 180°$	∠D is a straight angle.
Complementary	$m\angle\alpha + m\angle\beta = 90°$	∠α and ∠β are complementary.

Name	Measure	Illustrations with Notation
Supplementary	$m\angle\alpha + m\angle\beta = 180°$	$\angle\alpha$ and $\angle\beta$ are supplementary.
Vertical	$m\angle A = m\angle B$	$\angle A$ and $\angle B$ are vertical angles. They are the opposite angles formed when two lines intersect.

Example 9.3.1

Calculating Measurements of Angles

In the figure shown, $\overleftrightarrow{AD}$ and $\overleftrightarrow{BF}$ are straight lines.

> **Think Back**
>
> A line may be denoted using any two of its points. For instance, the line passing through the points A and B is written as $\overleftrightarrow{AB}$.

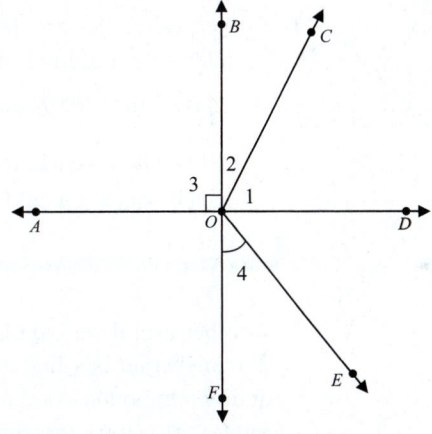

Figure 9.3.4

a. If $m\angle 1$ is $(26x - 7)°$ and $m\angle 2$ is $(2x + 13)°$, find the measures of $\angle 1$ and $\angle 2$.

b. The measure of $\angle BOE$ is 20° more than four times the measure of $\angle 4$. Find the measure of both angles.

c. Identify the two pairs of vertical angles.

Solution

a. Since $\overleftrightarrow{AD}$ is a straight line and $\angle 3$ is a right angle we know that $\angle BOD$ is also a right angle. This means that $\angle 1$ and $\angle 2$ are complementary and the sum of their angles is equal to 90°. Using this information, we can set up an equation to solve for the value of x and find the measures of each angle.

$$m\angle 1 + m\angle 2 = 90°$$
$$(26x - 7) + (2x + 13) = 90°$$
$$28x + 6 = 90°$$
$$x = 3°$$

Substituting this *x*-value into the original expressions for the angles, we have the following.

$$m\angle 1: (26x-7)° = (26(3)-7)° = 71°$$

$$m\angle 2: (2x+13)° = (2(3)+13)° = 19°$$

b. Begin by writing down ∠*BOE* in algebraic terms relating to ∠4. We will let $m\angle 4 = y$. We are told that if the $m\angle 4$ is multiplied by four and added to 20, we have the measure of ∠*BOE*. Thus, $m\angle BOE = 4y + 20$.

We also know that since ∠*BOE* and ∠4 are supplementary angles, the sum of their angles is equal to 180°. Using this information, we can set up an equation to determine the angle measures.

$$m\angle BOE + m\angle 4 = 180°$$
$$(4y+20) + y = 180°$$
$$5y + 20 = 180°$$
$$y = 32°$$

Thus, ∠4 measures 32° and ∠*BOE* measures 4(32°) + 20° or 148°.

c. Vertical angles are the opposite angles formed when two lines intersect. The intersection of lines $\overrightarrow{AD}$ and $\overrightarrow{BF}$ form two sets of vertical angles.

∠*AOF* and ∠*BOD* are vertical angles. ∠3 and ∠*DOF* are vertical angles.

Notice that we had to use three capital letters to identify ∠*AOF*, ∠*BOD*, and ∠*DOF* since it would be ambiguous to use only the letter *O*.

 Skill Check 9.3.1

In the figure, $\overrightarrow{MQ}$ is a straight line. Identify all pairs of complementary and supplementary angles.

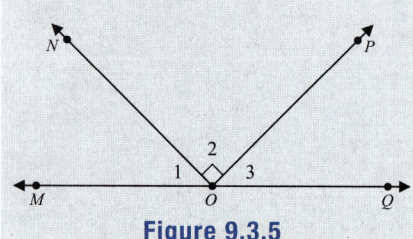

Figure 9.3.5

Another useful way to classify angles is by their position between parallel lines. A **transversal** is a line in a plane that intersects two or more lines in that plane at different points. A transversal and a pair of parallel lines create **corresponding angles**, **alternate interior angles**, and **alternate exterior angles**. The angles of each of these pairs are **congruent**; that is, they have the same measurement.

 Think Back

Two lines in the same plane are parallel if their slopes are equal, signifying that they never intersect one another.

 **TRANSVERSAL AND CONGRUENT ANGLES**

Transversal

A **transversal** is a line in a plane that intersects two or more lines in that same plane at different points.

Congruent Angles

Two angles of equal measure are said to be **congruent** to one another.

Table 9.3.2: Angles Formed by a Transversal

Name	Definition	Illustrations with Notation
Corresponding Angles	When two parallel lines are crossed by a transversal, the angles located in the same relative positions are called **corresponding angles** and are congruent.	$\angle A$ and $\angle B$ are one pair of corresponding angles.
Alternate Interior Angles	When two parallel lines are crossed by a transversal, the angles located between the parallel lines on opposite sides of the transversal are called **alternate interior angles** and are congruent.	$\angle C$ and $\angle D$ are one pair of alternate interior angles.
Alternate Exterior Angles	When two parallel lines are crossed by a transversal, the angles outside of the parallel lines on opposite sides of the transversal are called **alternate exterior angles** and are congruent.	$\angle E$ and $\angle F$ are one pair of alternate exterior angles.

> **Math Milestone**
>
> Greek astronomer and mathematician Eratosthenes calculated the earliest known circumference of the Earth. Although the method he used was lost, alternate interior angles are used to recreate our best guess at his accomplishment.

Example 9.3.2

Calculating Angles

In Figure 9.3.6, lines *m* and *n* are parallel. Find the measures of angles 1, 2, and 3.

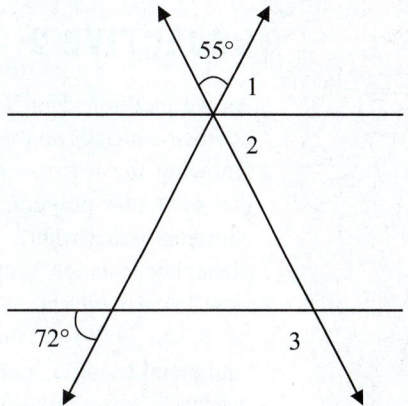

Figure 9.3.6

Solution

In order to find the measure of ∠1, we need to use some of the angles not yet labeled. Among the many different approaches, let's use the angle whose measurement is 72°. The corresponding angle on line *m* will be congruent and also measure 72° as shown in Figure 9.3.7.

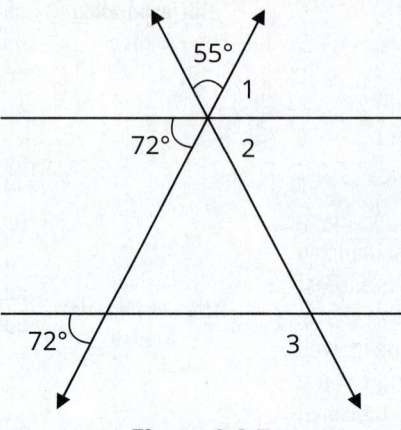

Figure 9.3.7

This angle forms a vertical angle with angle 1, and thus they are congruent. So $m\angle 1 = 72°$.

Notice that the angle with measurement 55°, ∠1, and ∠2 all form a straight line and thus the sum of their measurements is 180°. By subtracting, we can determine the measure of angle 2.

$$m\angle 2 = 180° - 55° - 72° = 53°$$

Lastly, notice that ∠3 is an alternate exterior angle to the angle formed by angle 55° and ∠1, which we will call angle β. From our previous calculations, we know that $m\angle\beta = 55° + 72° = 127°$. Thus, ∠3 also measures 127°.

In summary, $m\angle 1 = 72°$, $m\angle 2 = 53°$, and $m\angle 3 = 127°$.

OBJECTIVE 2 Tessellations

As we mentioned in Section 9.1, when we cover a surface using one or more geometric shapes so that there are no gaps or overlaps, it is called a **tessellation**. Knowing the interior angle measurements of polygons allows us to determine if a particular polygon will tessellate a surface. Geometric figures that have a common vertex where the sum of the vertex angles adds to 360° will tessellate a plane. For instance, four squares (each having a vertex angle measuring 90°) will tessellate a plane since 4 · 90° = 360°. The same is true for six equilateral triangles (6 · 120° = 360°) and three regular hexagons (3 · 120° = 360°). In fact, the square, equilateral triangle, and regular hexagon are the only three regular polygons that produce a tessellation when no other shape is used. Figure 9.3.8 shows the three regular polygon tessellations.

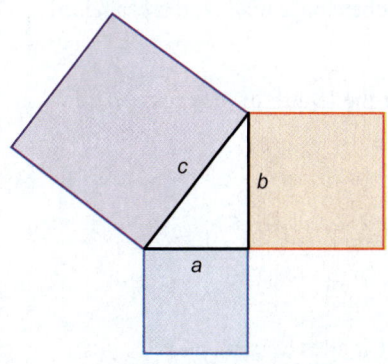

Figure 9.3.9

💬 Think Back

A regular polygon is a polygon where all sides have equal length and all angles are congruent.

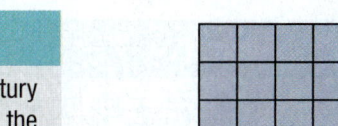 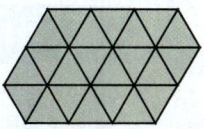 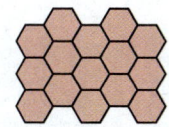

Figure 9.3.8

Example 9.3.3

Determining Whether Shapes Will Tessellate

A regular octagon has eight interior angles, each measuring 135°. Explain why a tessellation cannot be created using only regular octagons.

Solution

To tessellate a surface, the interior angles of the geometric figure at the common vertex must equal 360°. By showing that the interior angle of an octagon (135°) does not evenly divide 360°, we show that a tessellation cannot happen.

$$\frac{360°}{135°} \approx 2.667$$

From this we know that placing three octagons will create an overlap, while having only two octagons leaves a gap. Thus, regular octagons will not create a tessellation.

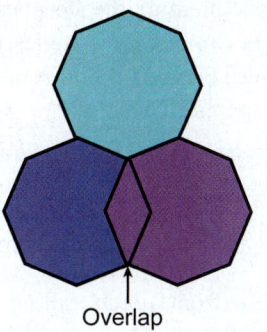 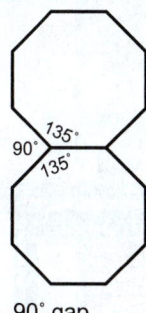

Overlap   90° gap

Figure 9.3.10

OBJECTIVE 3 Pythagorean Theorem

As the simplest of geometric shapes, triangles and their properties form the basic building blocks for much of plane geometry. There are connections between the angles and side lengths of triangles that despite being discovered long ago are still widely used today.

We will begin with a theorem that relates the sides of a right triangle—its two legs and the hypotenuse. Known as the **Pythagorean Theorem**, it states that in a right triangle, if squares are made on each of the three sides, as shown in Figure 9.3.10, the area of the biggest square has the exact same area as the other two squares added together. In algebraic terms, we write this as $a^2 + b^2 = c^2$.

Figure 9.3.11

PYTHAGOREAN THEOREM

In a right triangle, the square length of the hypotenuse is equal to the sum of the squares of the lengths of the other two sides.

$$a^2 + b^2 = c^2$$

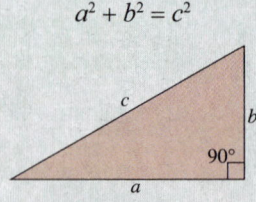

Figure 9.3.12

Here, a and b are the are the legs and c is the hypotenuse.

The Pythagorean Theorem is one of the most widely proven theorems, having over 350 different proofs! It's so widely known that novels, plays, and even cartoons have referenced it throughout history. Although the Pythagorean Theorem is usually stated in the way we presented it, its converse is also true. If a triangle has sides a, b, and c so that $a^2 + b^2 = c^2$, then the triangle is a right triangle. These two facts have useful implications in all types of arenas. In construction, workers can measure out the sides of a right triangle and know they are constructing the foundation of a building at a proper 90° angle. In surveying, right triangles provide ways to measure the steepness of mountains from a distance. And in navigation, sailors can use right angles to determine the shortest distance needed to travel between two points on the open sea. Let's look at a couple of practical uses of the Pythagorean Theorem.

Example 9.3.4

Using the Pythagorean Theorem

A new front porch railing is being installed and will look as pictured in Figure 9.3.13. The vertical beams, or balusters, will be 20 inches apart and 18 inches high. Determine the length of the boards that will make the crosses between the balusters.

Solution

Notice that the cross boards form right triangles with the baluster beams. Each triangle has legs that are 20 inches wide and 18 inches long, while the cross board is the hypotenuse.

We can use the Pythagorean Theorem to solve for the length of the cross board.

$$a^2 + b^2 = c^2$$
$$(20 \text{ in.})^2 + (18 \text{ in.})^2 = c^2$$
$$724 \text{ in.}^2 = c^2$$
$$c \approx 26.91 \text{ in.}$$

Each cross board needs to be approximately 26.91 inches long.

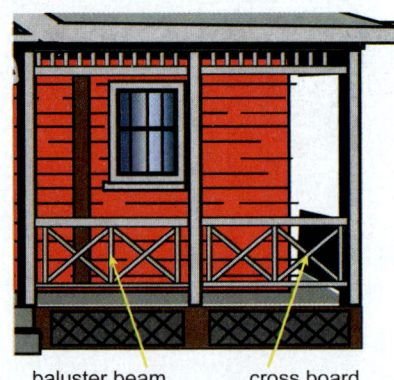

baluster beam   cross board

Figure 9.3.13

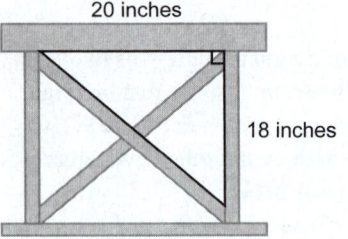

20 inches

18 inches

Figure 9.3.14

Example 9.3.5

Using the Pythagorean Theorem

A carpenter builds custom bookcases in a workshop and then installs them in customers' homes. Once built, a bookcase is brought into the home and laid front side down before being stood upright. The carpenter must ensure that once the bookcase is ready to install, it does not hit the ceiling when it is set upright in the room.

a. If the bookcase is 12 inches deep and the ceiling is 10 feet high, what is the maximum height the carpenter can build the bookcase and not hit the ceiling when it is set upright?

b. Determine a general formula that the carpenter can use when building bookcases so that he can always know the maximum height of any bookcase given its depth and the ceiling height.

Solution

a. Begin by sketching the situation.

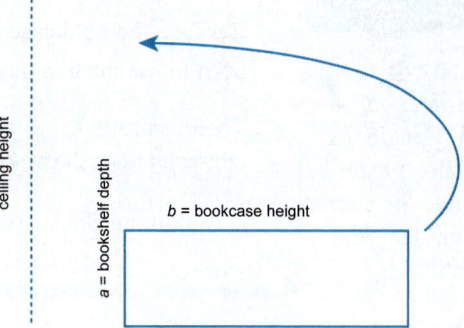

Figure 9.3.15

Notice that the longest part of the bookcase is not its height (b), but its diagonal across the side (c), as shown in Figure 9.3.16. The carpenter needs to make sure that the length of this diagonal is less than the height of the ceiling.

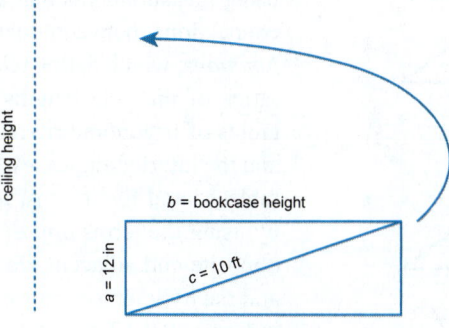

Figure 9.3.16

Using the Pythagorean Theorem, we can determine the maximum bookcase height. Before substituting the values into the formula though, we need to convert the measurements into the same units using the conversion factor $\dfrac{12 \text{ in.}}{1 \text{ ft}}$. Although it does not matter which measurement is changed, we will convert c to inches so that at the end of the calculation we can describe the height of the bookcase in feet and inches.

$$10 \text{ ft} \cdot \frac{12 \text{ in.}}{1 \text{ ft}} = 120 \text{ in.}$$

Now substitute into the formula and solve for b.

$$\left(12 \text{ in.}\right)^2 + b^2 = \left(120 \text{ in.}\right)^2$$
$$b^2 = 14{,}400 \text{ in.}^2 - 144 \text{ in.}^2$$
$$b = \sqrt{14{,}256 \text{ in.}^2} \approx 119.40 \text{ in.}$$

So the bookcase must be less than 119.4 inches tall, which is approximately 9 feet 5 inches.

b. We can rearrange the Pythagorean Theorem to solve for b (the bookcase height) in terms of the ceiling height and the depth of the bookcase.

$$a^2 + b^2 = c^2$$
$$b^2 = c^2 - a^2$$
$$b = \sqrt{c^2 - a^2}$$

Because the bookcase height must be less than the height of the ceiling, we need to use the less than inequality in the formula, $b < \sqrt{c^2 - a^2}$.

Therefore, we have the following general formula the carpenter can use to construct bookshelves that will not hit the ceiling when put upright.

$$\text{Maximum Bookcase Height} < \sqrt{\text{Ceiling Height}^2 - \text{Bookshelf Depth}^2}$$

> ✔ **Skill Check 9.3.2**
>
> Using the month and day of your birthday as the legs of a right triangle, find the length of the hypotenuse to the nearest hundredth. For instance, if your birthday is March 5th, the legs would be $a = 3$ and $b = 5$.

OBJECTIVE 4 Trigonometric Functions

The study of triangles is so profound it spawned the entire branch of mathematics called **trigonometry**. Trigonometry, the study of triangles, is both algebraic and geometric in nature. It allows us to fill in the missing information about a triangle using measurements that are known. The earliest results in this field observe the connections between mathematics and astronomy. Then came *trigonometric functions*, which define relationships between arcs of circles and their radii. Using ratios of the side lengths of triangles inside circles, mathematicians produced tables of trigonometric ratios. In any size right triangle, the lengths of the sides and the interior angles always satisfy the ratio functions **sine, cosine,** and **tangent** (abbreviated to sin, cos, and tan, respectively). We describe these relationships by using the terms *opposite side, adjacent side,* and *hypotenuse*. The descriptors opposite and adjacent are in reference to one of the acute angles in the triangle, and the hypotenuse, as we already know, is always the longest side. For instance, in Figure 9.3.17, you can see that in order to label the opposite side and adjacent side, we need to first establish the angle we are referencing.

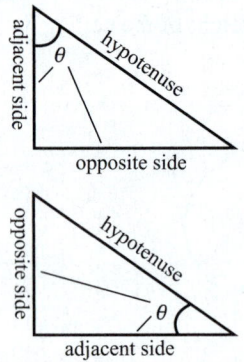

Figure 9.3.17

📖 **SINE, COSINE, AND TANGENT**

For an acute $\angle A$ in a right triangle,

$$\sin A = \frac{\text{Opposite}}{\text{Hypotenuse}} = \frac{a}{c},$$

$$\cos A = \frac{\text{Adjacent}}{\text{Hypotenuse}} = \frac{b}{c}, \text{ and}$$

$$\tan A = \frac{\text{Opposite}}{\text{Adjacent}} = \frac{a}{b}.$$

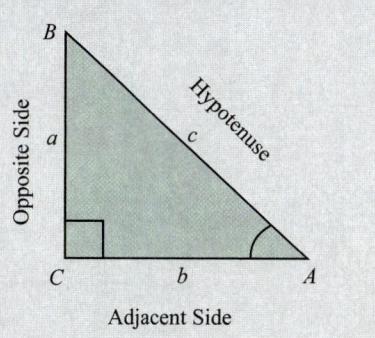

Figure 9.3.18

Example 9.3.6

Calculating Trigonometric Ratios

A right triangle with hypotenuse of length 26 cm and legs of lengths 10 cm and 24 cm is shown in Figure 9.3.19, with one of the acute angles labeled θ. Find the values of $\sin \theta$, $\cos \theta$, and $\tan \theta$. Leave your answer as a reduced fraction.

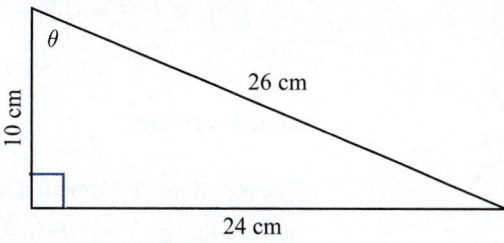

Figure 9.3.19

Solution

$$\sin \theta = \frac{\text{Opposite}}{\text{Hypotenuse}} = \frac{24}{26} = \frac{12}{13}$$

$$\cos \theta = \frac{\text{Adjacent}}{\text{Hypotenuse}} = \frac{10}{26} = \frac{5}{13}$$

$$\tan \theta = \frac{\text{Opposite}}{\text{Adjacent}} = \frac{24}{10} = \frac{12}{5}$$

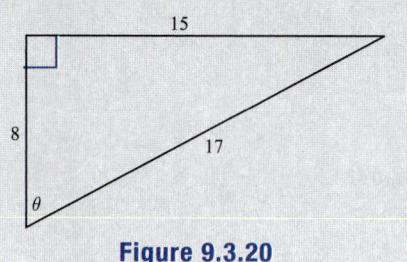
Example 9.3.7

Solving for Unknowns in a Right Triangle

A contractor is marking the property corners on a new construction site. She has marked the location of the first two corners, points A and B, which are 100 feet apart. The next corner will be 72 feet from point B at an angle of 55°. Using only a surveyor's rope and string, she needs to mark the 55° angle at point B. Determine how she can create this angle using a right triangle.

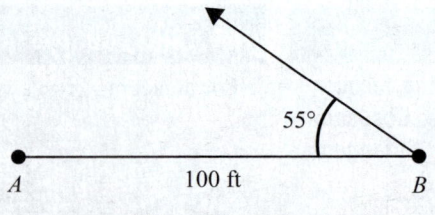

Figure 9.3.21

Solution

Begin by forming a right triangle with the right angle at point *A* as illustrated in Figure 9.3.22. Using the trigonometric ratios, we can find the lengths of the opposite side *x* and the hypotenuse *y*. The angle (55°) and the adjacent side (100 feet) are known.

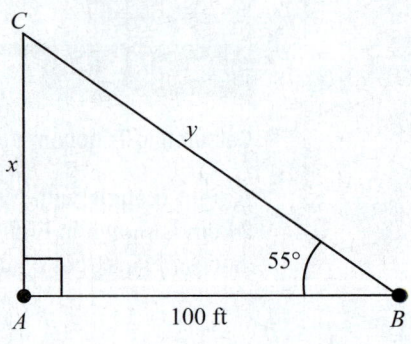

Figure 9.3.22

Length of Side *x*:

The trigonometric function involving the opposite side and the adjacent side is the tangent function, $\tan\theta = \dfrac{\text{Opposite}}{\text{Adjacent}}$.

So using the tangent function gives the following.

$$\tan 55^\circ = \frac{x}{100 \text{ ft}}$$

$$\left(\tan 55^\circ\right)100 \text{ ft} = x$$

$$x \approx 142.8 \text{ ft}$$

Length of Side *y*:

The trigonometric function involving the hypotenuse and the adjacent side is the cosine function, $\cos\theta = \dfrac{\text{Adjacent}}{\text{Hypotenuse}}$.

So using the cosine function gives the following.

$$\cos 55^\circ = \frac{100 \text{ ft}}{y}$$

$$y = \frac{100 \text{ ft}}{\cos 55^\circ}$$

$$y \approx 174.3 \text{ ft}$$

📈 **TECH TIP**

To find the values of trigonometric function of angles measured in degrees, the calculator must be set in degree mode.

Thus, to create a 55° angle, the contractor needs to make a right triangle with sides measuring 100 feet and 142.8 feet, and a hypotenuse measuring 174.3 feet. The contractor would then need to measure 72 feet along the hypotenuse to mark the next corner.

Many applications of right triangles refer to angles from the point of view of an observer's eye level. An **angle of elevation** is the angle formed by a horizontal line at your eye level and your line of vision as you look upward. Similarly, an **angle of depression** is formed when you look downward.

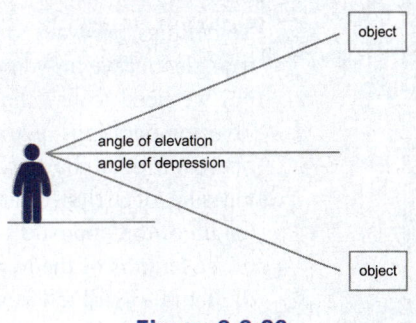

Figure 9.3.23

> ### 💬 Helpful Hint
>
> When two parallel lines are cut by a transversal, the angle of elevation and the angle of depression are equal to one another because they form alternate interior angles.
>
>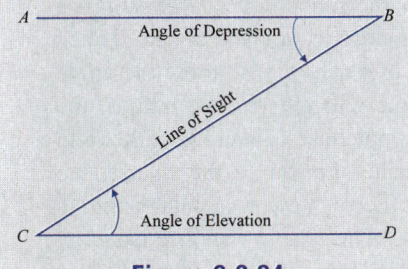
>
> **Figure 9.3.24**

Example 9.3.8

Using Angles of Depression

A ship's sonar locates a wrecked ship, which is on the ocean floor, at a 17° angle of depression. A diver is lowered 105 meters directly down to the ocean floor. Calculate how far the diver needs to travel along the ocean floor to get to the wreckage. Assume the ocean floor is parallel to the surface of the water. Round your answer to the nearest meter.

Solution

Begin by making a sketch of the situation.

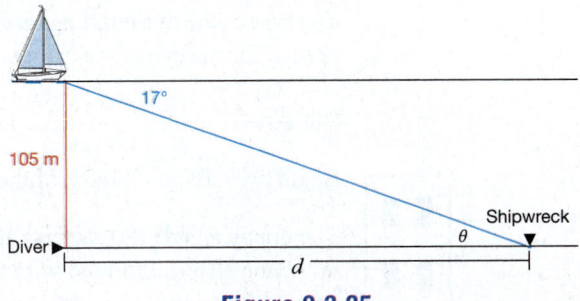

Figure 9.3.25

Notice that the angle of depression from the ship to the shipwreck (17°) is equal to the angle of elevation from the shipwreck to the ship, θ, because they are alternate interior angles. From the sketch, we see that the length of the opposite side is 105 meters and the unknown length d is the adjacent side. Thus, we should use the tangent function, $\tan \theta = \dfrac{\text{Opposite}}{\text{Adjacent}}$, to find d.

$$\tan 17° = \frac{105 \text{ m}}{d}$$

$$d = \frac{105 \text{ m}}{\tan 17°}$$

$$d \approx 343.44 \text{ m}$$

Therefore, the diver needs to travel approximately 343 meters along the ocean floor to reach the wreckage.

Sometimes, we want to find the measurement of one of the acute angles in a right triangle, but we only know the lengths of the sides. We can still use trigonometry, but we need to use the *inverse*, or opposite, of the functions. You already use inverse operations: addition and subtraction, or multiplication and division. Each operation does the opposite of its inverse. For instance, if you know the angle measure of a right triangle, the sine of that angle will tell you the ratio of the length of the opposite side to the length of the hypotenuse. If you know the ratio of the lengths of the hypotenuse to the side opposite of an angle, the inverse sine of that ratio will tell you the angle measurement. Fortunately, a calculator is able to perform this computation for us for any angle. The notation we use is $\sin^{-1}x$, $\cos^{-1}x$, and $\tan^{-1}x$.

> ### 💬 Helpful Hint
>
> Inverse trig functions can also be labeled using "arc."
>
> $$\sin^{-1} x = \arcsin x$$
> $$\cos^{-1} x = \arccos x$$
> $$\tan^{-1} x = \arctan x$$

📖 INVERSE TRIGONOMETRIC FUNCTIONS

The **inverse trigonometric functions** are used to find the angle from any trigonometric ratio. The common inverse trigonometric functions are $\theta = \sin^{-1}x$, $\theta = \cos^{-1}x$, and $\theta = \tan^{-1}x$.

Example 9.3.9

Finding the Angle of Depression

A surveillance camera is being installed on the top corner of a building that is 80 meters tall in order to monitor the exit door at the base of a neighboring building. The base of the two buildings are 128 meters apart. Calculate the angle of depression needed for the eye of the camera so that it points directly at the exit door.

Solution

Begin by making a sketch of the situation.

The opposite side (80 meters) and the adjacent side (128 meters) are both known in the top triangle formed with the angle of depression (θ). Thus, we will use the tangent function.

$$\tan \theta = \frac{80 \text{ m}}{128 \text{ m}}$$

To find the angle, we need to use the inverse tangent function.

$$\theta = \tan^{-1}\left(\frac{80 \text{ m}}{128 \text{ m}}\right)$$

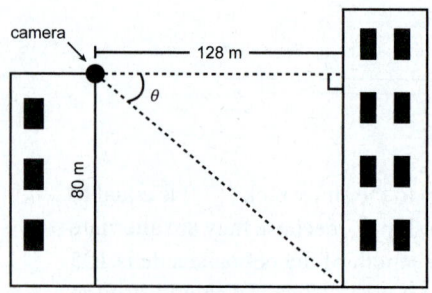

Figure 9.3.26

Using a calculator to solve, press 2nd tan. On your screen, tan⁻¹(will appear. Enter 80/128) and press enter. The calculator will return the value 32.00538321.

Find $m\angle\theta$.

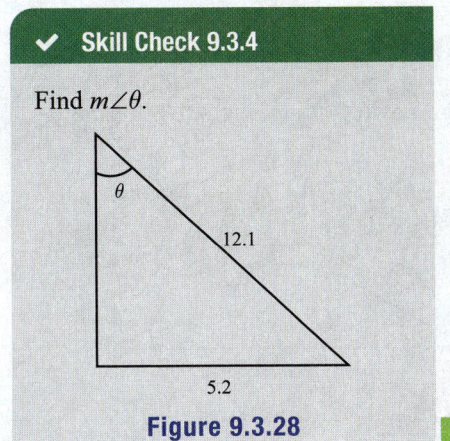

Figure 9.3.28

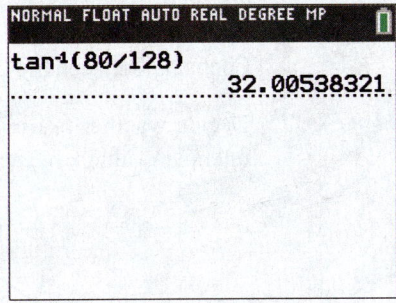

Figure 9.3.27

Thus, the angle of depression needed for the eye of the camera is approximately 32°.

OBJECTIVE 5 Law of Sines and Cosines

Of course, not every triangle is a right triangle, and often we need to calculate side lengths and angles in other triangles, too. We are able to extend the usefulness of sine and cosine functions so that they apply to any triangle with the Law of Sines and the Law of Cosines. The **Law of Sines** relates the sine of an angle and its corresponding opposite side to other angles and sides by the equation $\dfrac{a}{\sin A} = \dfrac{b}{\sin B} = \dfrac{c}{\sin C}$.

We can use the Law of Sines any time we know the measurements of two angles and one side. If we know the measurements of two sides, we can only use the Law of Sines if we also know the measurement of one of the angles that is not between the two known sides. In situations where we know the measurements of two sides and the angle between them, we need the Law of Cosines. The **Law of Cosines** states that given two sides, b and c, and the included angle A, the third side can be found by the equation $a^2 = b^2 + c^2 - 2bc \cos A$.

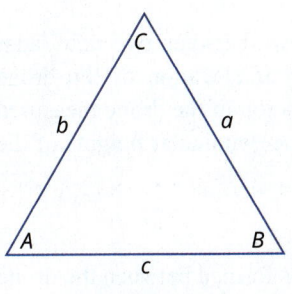

Figure 9.3.29

📖 LAW OF SINES AND LAW OF COSINES

For a triangle where a is the length of the side opposite angle A, b is the length of the side opposite angle B, and c is the length of the side opposite angle C, we have the following.

Law of Sines

$$\frac{a}{\sin A} = \frac{b}{\sin B} = \frac{c}{\sin C}$$

Law of Cosines

$$a^2 = b^2 + c^2 - 2bc \cos A$$

$$b^2 = a^2 + c^2 - 2ac \cos B$$

$$c^2 = a^2 + b^2 - 2ab \cos C$$

The usefulness of both laws is best seen by Example 9.3.10. Again, our goal is to find the missing measurements of the triangle.

Example 9.3.10

Choosing Between the Law of Sines and the Law of Cosines

Decide whether to use the Law of Sines or the Law of Cosines to solve for the unknown value x in each scenario.

a.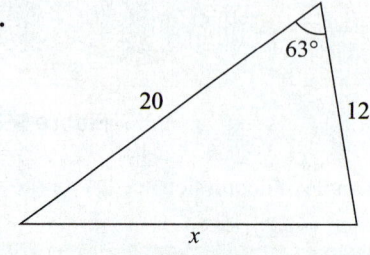

Figure 9.3.30

b.

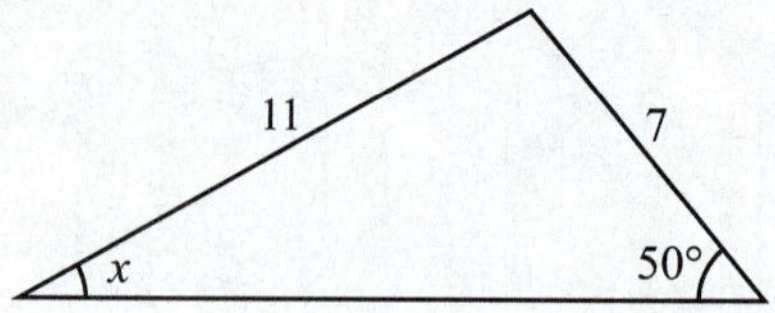

Figure 9.3.31

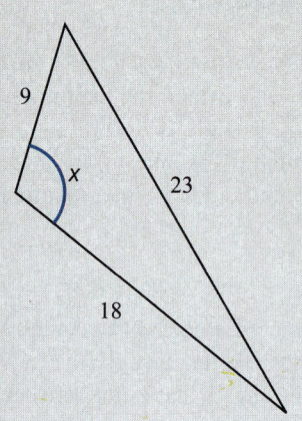

✔ Skill Check 9.3.5

Determine whether to use the Law of Sines or the Law of Cosines to find the value of x in the triangle.

Figure 9.3.32

Solution

a. We are given the lengths of two sides and their included angle, and we wish to know the length of the third side. This is enough information to use the Law of Cosines to solve for x.

b. In order to use the Law of Sines, you need to know at least three of the four parts in the proportional equation $\dfrac{a}{\sin A} = \dfrac{b}{\sin B}$. In other words, you need one *side* and its *opposite angle* in the triangle, in addition to one of the corresponding parts for the unknown pair.

We know the measure of an angle (50°) and its opposite side (11), and we want to calculate the measure of angle x, which is opposite of the side with length 7. So we can use the Law of Sines to solve for x.

Example 9.3.11

Using the Law of Sines

In order to test a new security system, a drone is flown between two new radar stations that are located 50 miles apart. The angle of elevation of the drone measured by the first station is 36°. The angle of elevation of the drone measured by the second station is 17°. Calculate the altitude (perpendicular height) of the drone when it was detected.

Solution

Begin by making a sketch of the situation. A triangle is formed between the drone and the two stations.

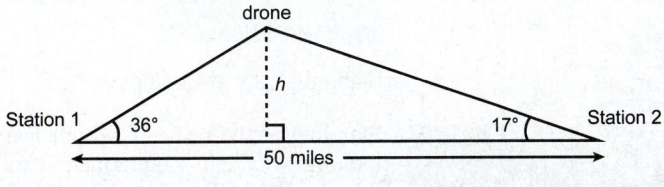

Figure 9.3.33

Because the triangle formed by the drone and the stations is not a right triangle, we cannot use the trigonometric functions to find the missing sides. However, we do know that the interior angles of a triangle sum to 180°. So the unknown interior angle is $\theta = 180° - 36° - 17° = 127°$.

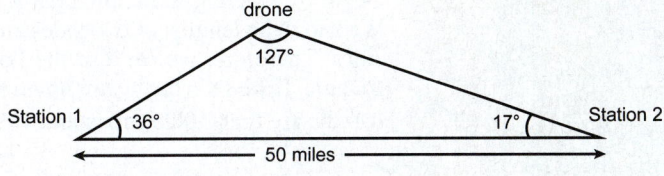

Figure 9.3.34

We now know an angle ($m\angle A = 127°$) and the length of the side opposite it ($a = 50$ miles). We also have the measure of the angle ($m\angle B = 17°$) opposite the unknown side b. This provides enough information to use the Law of Sines to find a missing side of the triangle.

$$\frac{\sin 127°}{50 \text{ mi}} = \frac{\sin 17°}{b}$$

$$b = \frac{\left(\sin 17°\right)\left(50 \text{ mi}\right)}{\sin 127°}$$

$$b \approx 18.304 \text{ mi}$$

The unknown altitude h forms two smaller right triangles. In the smaller right triangle formed with station 1, we know the length of the hypotenuse (18.304 miles) and the measure of an angle (36°). The missing altitude is the side opposite the angle.

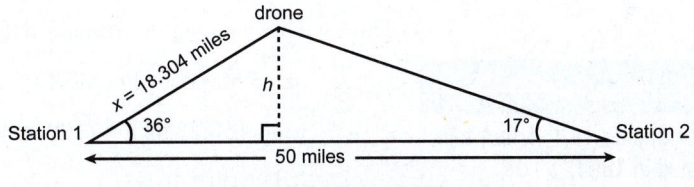

Figure 9.3.35

With this information, we can use the sine function to find the altitude h.

$$\sin 36° = \frac{h}{18.304 \text{ mi}}$$

$$18.304\left(\sin 36°\right) = h$$

$$10.759 \text{ mi} \approx h$$

Thus, the altitude of the drone was approximately 10.759 miles when it was detected.

💬 Helpful Hint

Sine, cosine, and tangent can only be used with right triangles.

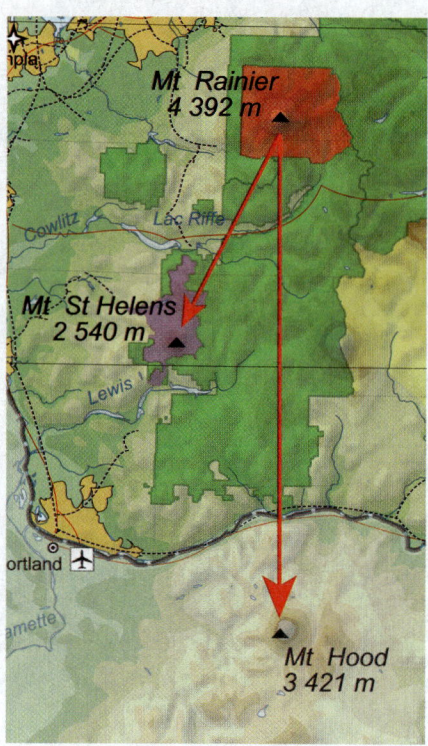

Figure 9.3.36

Example 9.3.12

Using the Law of Cosines

Standing at the peak of Mount Rainier (located in the state of Washington) and looking south, a mountaineer can see two other volcanos: Mount St. Helens (also in Washington) and Mount Hood (in Oregon). From her current location, the mountaineer knows that Mount St. Helens is 50.25 miles away while Mount Hood is 109 miles away. If she approximates the angle that she sees between the two volcanos as 20°, what is the approximate distance between Mount St. Helens and Mount Hood?

Solution

Begin by labeling the known parts of the triangle formed by the three volcanos. We know the lengths of two sides and the measure of the angle between those two sides. Therefore, we can use the Law of Cosines to find the missing side of the triangle. If we let a be the unknown side, then the measure of angle A is 20°. Sides b and c are then 50.25 miles and 109 miles, respectively.

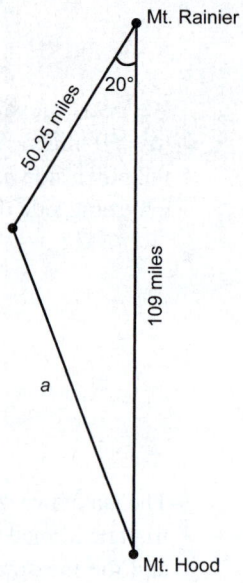

Figure 9.3.37

Thus, we have enough information to use the Law of Cosines.

$$a^2 = b^2 + c^2 - 2bc \cos A$$
$$a^2 = (50.25 \text{ mi})^2 + (109 \text{ mi})^2 - 2(50.25 \text{ mi})(109 \text{ mi})(\cos 20°)$$
$$a^2 \approx 4112.20 \text{ mi}^2$$
$$a \approx 64.13 \text{ mi}$$

Mount Hood and Mount St. Helens are approximately 64 miles apart.

Skill Check Answers

1. ∠1 and ∠3 are complementary; ∠*MOP* and ∠3 are supplementary; ∠*NOQ* and ∠1 are supplementary. **2.** Answers will vary. For March 5th, $c \approx 5.83$ **3.** $\sin \theta = \dfrac{15}{17}$, $\cos \theta = \dfrac{15}{17}$, $\tan \theta = \dfrac{15}{8}$ **4.** Approximately 25.5° **5.** Law of Cosines

9.3 Exercises

✔ CONCEPT CHECK

1. Two angles are supplementary angles if the sum of their measures equals _____ degrees.

2. Two angles are vertical angles if they are the _____ angles formed when two lines intersect.

3. The three basic trigonometric functions are _____, _____, and _____.

4. True or False: The Pythagorean Theorem can be used for any type of triangle.

5. True or False: The Law of Sines and Law of Cosines can be used interchangeably.

💡 PRACTICE

Determine whether each angle is acute, right, or obtuse.

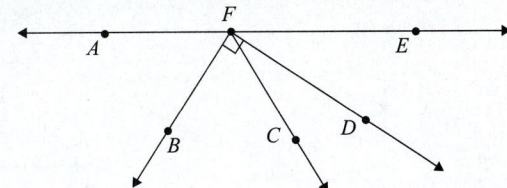

6. ∠BFD	7. ∠DFE	8. ∠AFD
9. ∠AFB	10. ∠EFC	11. ∠EFB

In the figure, $\overrightarrow{YU}$ bisects (divides into 2 equal parts) ∠WYV and $\overrightarrow{YT}$ bisects ∠XYV. Use the figure to answer each question.

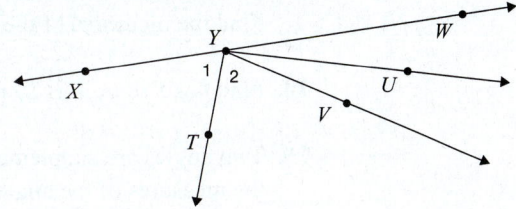

12. If $m\angle WYV = 36°$, find $m\angle WYU$.

13. If $m\angle 1 = 56°$, find $m\angle 2$.

14. If $m\angle VYW = 82°$ and $m\angle WYU = (4r + 25)°$, find r.

15. If ∠XYV has a measure of 162°, find $m\angle 1$.

Use the figure to answer each question.

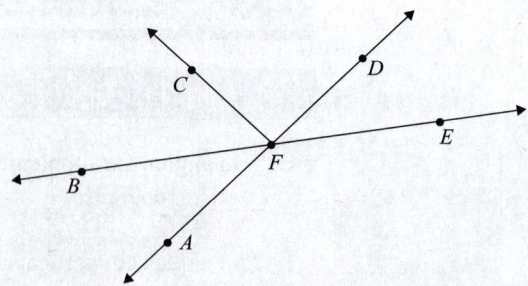

16. If $m\angle CFD = (15a + 45)°$, find a so that $\overrightarrow{FC} \perp \overrightarrow{FD}$.

17. If $m\angle AFB = (8x - 6)°$ and $m\angle BFC = (14x + 8)°$, find the value of x so that $\angle AFC$ is a right angle.

18. If $m\angle BFA = (3r + 12)°$ and $m\angle DFE = (-8r + 210)°$, find $m\angle AFE$. (**Hint:** Recall that lines that intersect at a common point share vertical angles that are congruent.)

In the following figure, lines *a* and *b* are parallel.

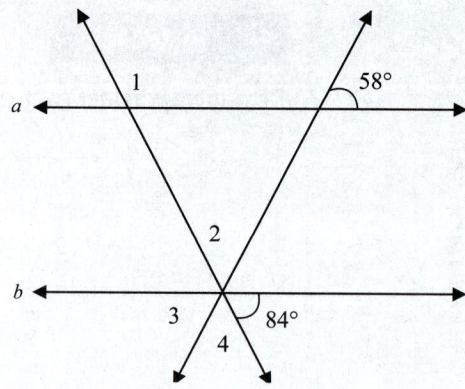

19. Find $m\angle 1$.

20. Find $m\angle 2$.

21. Find $m\angle 3$.

22. Find $m\angle 4$.

23. The measures of two complementary angles are $(16z - 1)°$ and $(12z + 7)°$. Find the measures of the angles.

24. Find $m\angle T$ if $m\angle T$ is 20 more than four times its supplement.

25. Two angles are supplementary. One angle is 12° more than the other. Find the measures of the angles.

26. A regular dodecagon has 12 interior angles, each measuring 150°. Can a tessellation be created using only regular dodecagons?

27. A regular nonagon has 9 interior angles, each measuring 140°. Can a tessellation be created using only regular nonagons?

28. Use △*ABC* to find sin *A*, cos *A*, tan *A*, sin *B*, cos *B*, and tan *B*. Round your answers to the nearest hundredth.

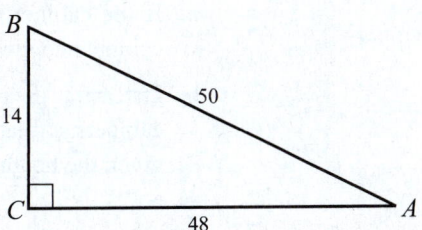

29. Use △ABC to find sin A, cos A, tan A, sin B, cos B, and tan B. Round your answers to the nearest hundredth.

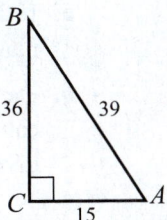

Decide whether to use the Law of Sines or the Law of Cosines to solve for the unknown value *x* in each scenario.

30.

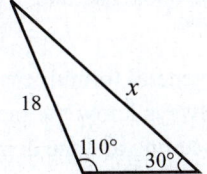

31.

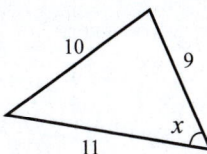

32.

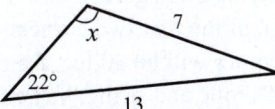

33.

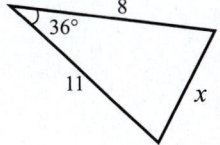

🚀 **APPLICATIONS**

34. The shortest path from A to B is across Lazy Lake, shown below. How far is this path?

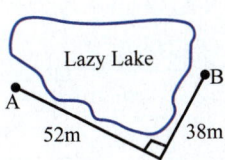

35. Find the value of *x* in the following diagram of a truss bridge.

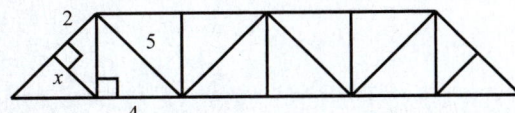

36. Jonathan is building a cabinet in his workshop and will be installing it in his den which has 7-foot-high ceilings.

   a. If the cabinet is 6-feet tall, what is the deepest that he can make the cabinet and clear the ceiling when it is stood upright?

   b. Determine a general formula that Jonathan can use when building cabinets so that he can always know the maximum depth of any cabinet given the height of the ceiling and the height of the cabinet.

37. Anastasia is designing a cabinet for a client whose living room has 9-foot ceilings. The client wants the cabinet to be as tall as possible. Anastasia will create the cabinet in her workshop and transport it to the client's home.

   a. If the depth of the cabinet is 18 inches, what is the tallest Anastasia can make the cabinet and clear the ceiling when it is stood upright?

   b. Determine a general formula that Anastasia can use when designing cabinets so that she can always know the maximum height of any cabinet given the height of the ceiling and the depth of the cabinet.

38. Bartholomew is designing a bookcase for the town library. The ceilings are 12-feet high.

   a. If he wants the bookcase to be as tall as possible, and if the depth of the bookcase must be 16 inches, what is the maximum height that he can make the bookcase and it still be able to clear the ceiling when stood upright?

   b. Determine a general formula that he can use when designing cabinets so that he can always know the maximum height of any cabinet given the height of the ceiling and the depth of the cabinet.

39. Calvin is laying out a garden based on an architectural plan. He has marked the location of the first two corners, points A and B, which are 60 feet apart. The next corner will be 30 feet from point B at an angle of 50°. Using only a surveyor's rope and string, he needs to mark the 50° angle at point B. Determine how he can create this angle using a right triangle.

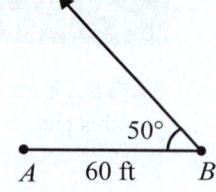

40. A drone is flying 2 meters directly above a flagpole that is located 6 meters from a building. If the angle of depression from the drone to the base of a building is 64°, find the height of the flagpole.

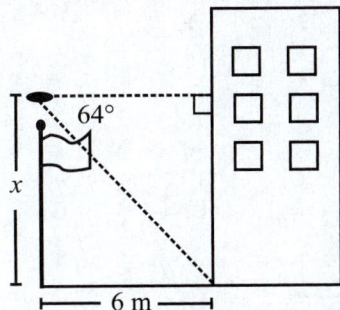

41. The sun hits a building of unknown height so that the building casts a shadow of 517 feet. If the angle of elevation from the tip of the shadow to the sun is 16°, how tall is the building?

42. A snow ski slope has a run of 1.5 miles with an angle of elevation of 24°. What is the length of the vertical drop of the ski slope?

43. A wheelchair ramp is to be constructed with a maximum incline of 6°. If the height of the ramp is to be 5 feet, how long should the ramp be after construction?

44. A section of highway has an incline of 10°. If the length of the incline is 5 miles, what is the horizontal distance of the incline?

45. A 16 ft ladder is leaning against a wall where it makes an angle of 68° with the ground and 22° with the building.

   a. How far is the ladder base from the building?

   b. How far does the ladder reach up the building?

46. Leo is flying his drone from his apartment balcony to watch the entrance for his friend's arrival. The drone is currently flying 2 meters directly above a 10-meter tall flagpole, which is located 6 meters from the entrance of the building. Find the angle of depression that the camera on the drone would need to use to focus on the base of the door of the building. Round your answer to the nearest tenth.

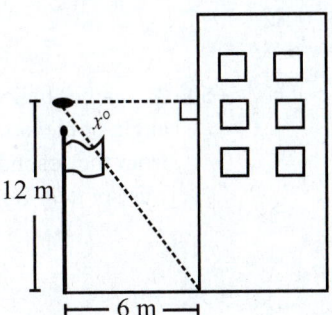

47. A plane is flying at an altitude of 2000 feet. It is on course maintain altitude and pass directly over a 150-foot tower. Determine the current angle of depression to the top of a 150-foot tower if the tower is 1000 feet away. Round your answer to the nearest tenth.

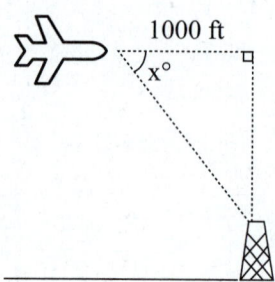

48. Selena spots a balloon that is 16 feet in the air. If she is 5-feet tall and the distance from the top of her head to the balloon is 15 feet, what is the angle of elevation from the top of her head to the balloon? Round your answer to the nearest tenth.

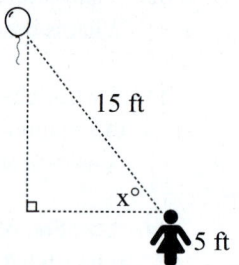

49. Two buildings are located 50 feet apart. One building is 100-feet tall and the other is 60-feet tall. A camera is placed the top edge of the shorter building to watch a bird nest located at the top edge of the taller building. What angle of elevation is needed to focus the camera from the top of the shorter building to the top of the taller building? Round your answer to the nearest tenth.

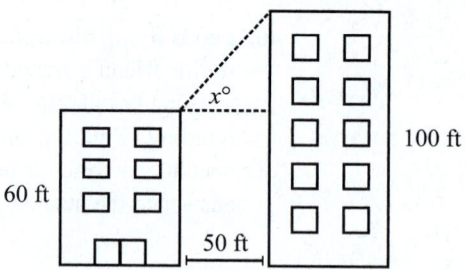

50. A weather balloon is spotted by two people who are 20 meters apart. The angle of elevation from the first person is 56° and the angle of elevation from the second person is 40°. Calculate the altitude of the balloon when it was spotted. Round your answer to the nearest tenth.

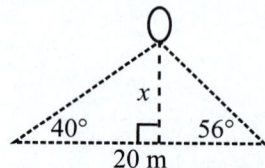

51. A person is standing 16 feet from the base of a building and looking at the top of a flagpole that is on top of the building. The flagpole is 10-feet tall. The angle of elevation from where the person is standing to the top of the flagpole is 75°. Determine the height of the building, rounded to the nearest tenth.

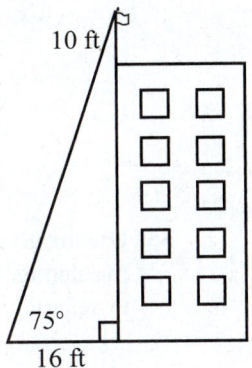

52. Patricia is currently located on a dock on an island. She wants to determine the route that would be fastest to travel from the island to dock B on the mainland. She has the options of rowing her boat to dock A and then riding a bike to dock B or of rowing her boat directly to dock B. She knows that the distance from the island dock to dock A is 1.2 miles and the distance from the island dock to dock B is 1.8 miles. The angle between docks A and B is 68 degrees. Patricia needs to determine the distance from dock A to dock B to help her determine which would be faster. How far is dock B from dock A?

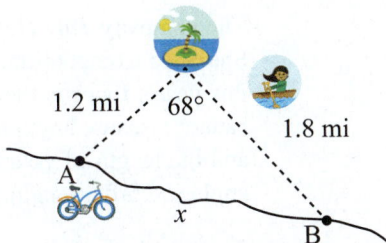

53. If you are standing at a certain spot on Lookout Mountain in Tennessee, you can see seven states. Of these seven, you can see Alabama, which is 25 miles away, and South Carolina, which is 80 miles away. The angle between these two states from your location is approximately 70°. Determine the distance between these two sites, rounded to the nearest mile.

54. Can you tessellate the plane with the following figure? Explain your answer.

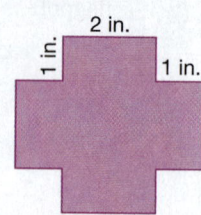

55. Use the given figure to create a tessellation by rotating, reflecting, or translating. Describe which transformations you used to create the tessellation.

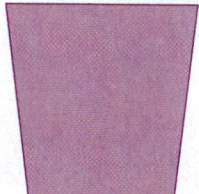

9.3 PROJECT

MAN CAVES AND SHE SHEDS

The company *Tiny House Livings* is gearing up for their annual Man Cave and She Shed Competition. Every year the competition focuses on different features; this year's focus is the roof. The main contest restrictions this year are that you cannot measure anything directly with a tape measure, you must use your height and the length of your hand to "measure," and then you must provide the pitch angle and rafter length.

Suppose that you have always dreamed of owning one but haven't had a chance to save up the money to buy one nor the tools and materials to build one. Thanks to the competition, you now have the chance to win one! Use your own hand length as the standard unit of measurement.

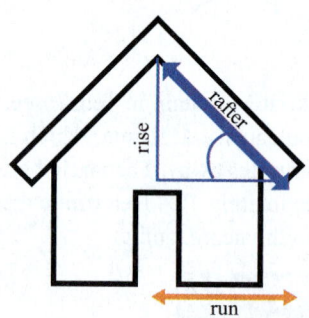

1. Refer to the sketch provided of a man cave or she shed. First, estimate your height in hands and then estimate that the end of the roof overhang is 4 hands above your head. How high is the end of the roof overhang from the ground?

2. You use the length of your hand to measure the run of one side of the shed roof, which is half the width of the shed. You discover the run is twenty "hands" long. What is the length of the run in inches?

3. You estimate the rise to be 8 "hands" high. What is the pitch angle of the roof? **Hint:** Pay attention to your units!

4. What is the length of the rafter?

5. What influences the accuracy of your calculations? What can you do to make the estimations more precise?

Chapter 9 Exercises

Find the area of the shaded region in each figure. Round your answer to the nearest hundredth, if necessary.

1.

11 in.

5 in.

8 in.

2.

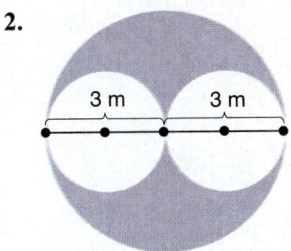

3 m   3 m

Solve each problem. Round your answer to the nearest hundredth, if necessary.

3. A window pane has dimensions of 30 inches by 22 inches. If glass sells for $0.80 per square inch, how much will each window cost?

4. A 50-pound bag of fertilizer will cover 5500 square feet. If your lawn measures 215 feet by 168 feet, how many pounds of fertilizer will you need to cover the lawn?

5. Civil War cannonballs were made out of cast iron. A collector found a set of three cannonballs for sale at an auction. The balls had diameters of 4.52 in., 4.346 in., and 3.908 in., respectively. Find the total volume of cast iron contained in the three Civil War cannonballs.

6. Gregory wanted a soccer-ball piñata for his birthday. Find the maximum amount of candy Gregory's mother could put in the piñata if the tag describes the ball as 12 inches in diameter.

7. A tube of lip balm has a height of 2.75 in. and a diameter of 0.75 in.

    a. If the cap on the tube is 0.20 inches high, find the maximum volume of lip balm that the tube can hold.

    b. What are the minimum dimensions of a box that can hold 30 tubes of lip balm, which are placed in six rows, with five tubes in each row?

8. A stick of butter measures 4.5 in. long, 1.25 in. wide, and 1.25 in. deep.

    a. How much butter is in each stick in cubic inches?

    b. What is the minimum surface area of a box containing 4 sticks of butter?

9. An 8-ounce tub of sour cream needs to contain 19.32 in.^{3} of sour cream. The tub must be in the shape of a right circular cylinder and leave a quarter inch between the top of the container and the sour cream. If the height of the container is 2 in., find the diameter of the tub of sour cream.

10. Oddie's Packaging Company always includes a serving size on its packages. The company has recently started making cheese balls and needs to estimate the number of servings in a ball with a diameter of three inches. Assuming the ball is a perfect sphere and that each serving is four fluid ounces, what should Oddie's estimate be? (**Hint:** There are 1.8 cubic inches in 1 fluid ounce.)

11. The measures of two complementary angles are $(18z + 5)°$ and $(2z + 5)°$. Find the measures of the angles.

12. Find $m\angle T$ if $m\angle T$ is 30 degrees larger than four times the measure of its supplement.

13. Two angles are supplementary. One angle measure is 24° larger than the other. Find the measures of the angles.

14. Use $\triangle ABC$ to find sin A, cos A, tan A, sin B, cos B, and tan B.

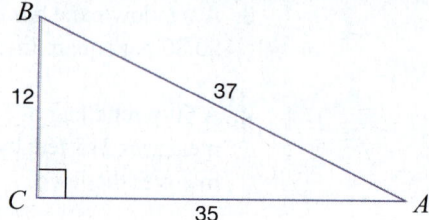

15. An 18-foot ladder is leaning against a wall where it makes an angle of 58° with the ground and 32° with the building.

    a. How far is the ladder base from the building?

    b. How far does the ladder reach up the building?

16. A plane is flying at an altitude of 15,000 feet and needs to fly an altitude of 25,000 feet. The angle of elevation that should be used is 4°. Over how many miles will it take the plane to reach the correct height? (**Hint:** There are 5280 feet in 1 mile.)

17. The sun hits a building of unknown height so that the building casts a shadow of 410 feet. If the angle of elevation of the sun is 16°, how tall is the building?

18. Suppose you can see two flags from the point where you are standing. One flag is 60 yards away from where you are standing and the other is 100 yards away from where you are standing. The angle between the flags is 40 degrees. Determine how far apart the flags are from each other, rounded to the nearest tenth.

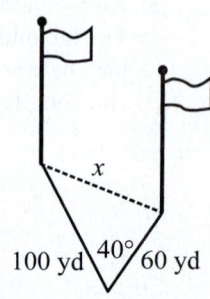

CHAPTER 9 PROJECT

It's All in the Packaging

As a member of the design team at Palisade Pharmaceuticals, you've been assigned the job of creating part of the packaging for the company's new syringe line. The packages will be sold to retailers in multiples of 10 syringes, and your job is to minimize the packaging for each bundle. Each plastic 1 mL syringe has a diameter of 6 mm and a length of 84 mm.

Part I—Minimum Surface Area

Begin by deciding on the best way to bundle the 10 syringes together; you want to have the smallest surface area in your design. We will try bundling the 10 syringes in both a rectangular format and a triangular format. For each possible configuration, calculate the minimum area that would be required to cover the top of that particular bundle of syringes. Be sure to show your calculations. Round your answers to the nearest hundredth.

A. Rectangle

The syringes are arranged so that the top of the syringes form the smallest possible rectangle. In order to find the surface area needed to package them, we need to find the area of this rectangle.

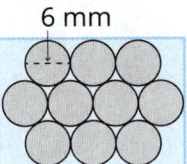

1. Let's start with the length L of the rectangle. This is the easier of the two measurements to calculate. On the following figure, mark the center of each syringe in the middle row. Then draw a line through the centers that represents the length we need to find.

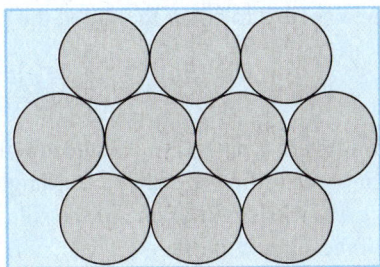

2. Using the line you just created, determine the general formula for the length L of the rectangle in terms of the diameter d of each syringe.

3. Now use your formula and the information given about the syringes to calculate the actual length of the rectangle needed to cover the syringes.

Next we'll calculate the width of the rectangle needed. As you are calculating this side, be careful to note that the circles are not sitting directly on top of one another. This allows for the smallest rectangle possible. Let's take a closer look.

Once again, mark the center of each syringe. By using the radius of each circle, connect the centers of the four syringes shown in such a way that you form two triangles sitting on top of each other. Label each radius *r*.

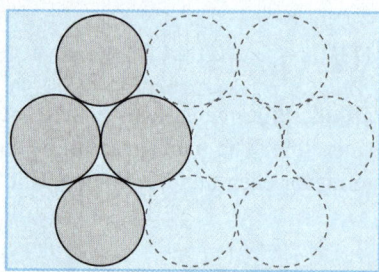

4. Use either trigonometry or the Pythagorean Theorem to find the height *h* of each triangle you drew based on the radius *r*. Show your work.

Now that you have the height of each of your triangles, you are ready to determine the width of the rectangle needed to cover the syringes. In the following figure, a line is drawn representing the width of the rectangle broken into lengths *a*, *b*, *c*, and *d*. You are now equipped to find each of these lengths in terms of *r*.

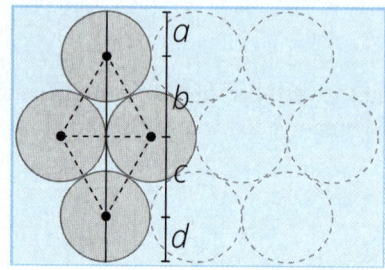

5. Create a formula for the width of the rectangle in terms of *r*.

6. Use the formula created in Step 5 and the information given to calculate the actual width of the rectangle covering the syringe bundle.

7. Finally, calculate the minimum area needed to cover the top of the rectangular bundle of syringes.

B. Equilateral Triangle

In the triangular configuration, the syringes are arranged so that the top of the syringes form the smallest possible triangle. In order to find the area the triangle covers, we need to know the height of the triangle and the length of one of the sides. Notice that this is an equilateral triangle, so any of the sides will do.

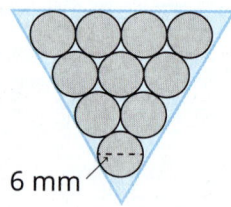

6 mm

Let's start by finding the length of the sides of the triangle. We've enlarged one of the sides for you in the following figure and indicated the centers of the syringes.

S

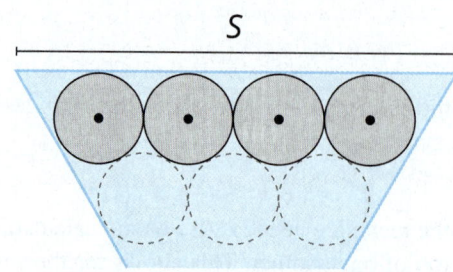

8. Notice in the next figure that we can determine part of the length of S by using the radius of each syringe. Give the length of x, shown in the figure, in terms of r.

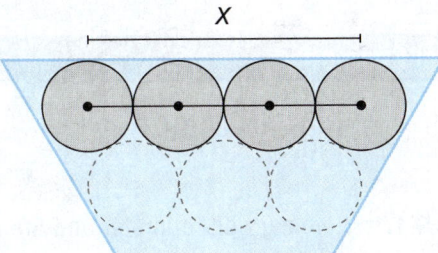

9. Now, let's find the remaining portion of side S. We'll zoom in closer to the right side. We can create a triangle in the corner using the radius of the circle as our base, as shown. (**Note:** This is the same triangle that we found the height of in the first layout design. So the length of y is the same as the earlier height h.) Determine the length of y in terms of r.

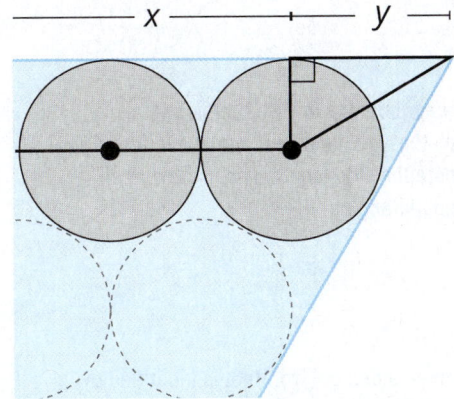

10. Now, determine the total length of a side of the triangle in terms of r.

11. Since this is an equilateral triangle, you can now use the Pythagorean Theorem to find the height of the triangle. Round your answer to the nearest hundredth. Show your work.

12. Using the lengths of the sides and height of the triangle, calculate the minimum area needed to cover the top of the triangular bundle.

13. Which configuration in Part 1, the rectangle or triangle, requires the smaller area to cover the top of a bundle of 10 syringes?

Part II—Packaging

14. For each bundle configuration, calculate the minimum total surface area of a three-dimensional package needed for the 10 syringes. Show your calculations for **a.** the rectangle and **b.** the triangle.

15. Which of the configurations would you recommend to the design team at Palisade Pharmaceuticals? Explain your reasoning.

Part III—For Consideration

16. Suppose you changed the diameter of the syringes. Would the most efficient packaging stay the same? Why or why not?

17. Consider other configurations for the bundles. Can you find a way to package them so that it uses less surface area? Show your work.

☰ CHAPTER 9 REVIEW

Section 9.1 Two-Dimensional Geometry

Formulas

Perimeter and Area

Table 9.1.1: Summary of Perimeter and Area Formulas

Shape	Definition	Formula for Perimeter	Formula for Area
Rectangle (with length l and width w)	A **rectangle** is a quadrilateral with two pairs of parallel sides and four right angles. We denote the longer side as the length l and the shorter side as the width w.	$P = 2l + 2w$	$A = lw$
Square (with side s)	A **square** is a quadrilateral with two pairs of parallel sides, four sides that are the same length, and four right angles. We denote the side length as s. Note that all squares are also rectangles.	$P = 4s$	$A = s^2$
Triangle (with height h and base b)	A **triangle** is a three-sided polygon that is identified by its base b and height h.	$P =$ sum of the side lengths	$A = \dfrac{1}{2}bh$
Parallelogram (with height h and base b)	A **parallelogram** is a quadrilateral with two pairs of parallel sides. The base of the parallelogram is denoted by b and the height by h.	$P =$ sum of the side lengths	$A = bh$
Trapezoid (with bases b_1, b_2 and height h)	A **trapezoid** is a quadrilateral with exactly one pair of parallel sides. The sides that are parallel are called bases and are denoted b_1 and b_2. The height is denoted by h.	$P =$ sum of the side lengths	$A = \dfrac{1}{2}(b_1 + b_2)h$
Circle (with diameter d and radius r)	A **circle** is the set of all points in a plane that are a certain distance from a fixed center point. The diameter, d, is the length of a line segment that passes through the center. The radius, r, is half of the diameter.	$C = 2\pi r = \pi d$	$A = \pi r^2$

Section 9.2 Three-Dimensional Geometry

Formulas

Volume and Surface Area

Table 9.2.1: Volume and Surface Area Formulas

Shape	Definition	Volume Formula	Surface Area Formula
 Rectangular Solid	A **rectangular solid** has 6 faces, all of which are rectangles. We denote the longer side of the base as length l and the shorter side as the width w. The height is h.	$V = lwh$	$SA = 2lw + 2lh + 2wh$
 Cube	A **cube** is made of six congruent squares. The side length is denoted by s.	$V = s^3$	$SA = 6s^2$
 Sphere	A **sphere** is a the set of all points that are a certain distance from a fixed center point. The radius is denoted by r.	$V = \frac{4}{3}\pi r^3$	$SA = 4\pi r^2$
 Right Circular Cylinder	A **right circular cylinder** is a cylinder with circle bases of radius r. The bases are perpendicular to the height h of the cylinder.	$V = \pi r^2 h$	$SA = 2\pi r^2 + 2\pi rh$
 Right Circular Cone	A **right circular cone** is a figure with a circular base of radius r which tapers to a point at a perpendicular height, h.	$V = \frac{1}{3}\pi r^2 h$	$SA = \pi r^2 + \pi r\sqrt{r^2 + h^2}$
 Square Pyramid	A **square pyramid** is a figure with a square base with sides s and four triangular sides that taper to a point. The height is denoted by h.	$V = \frac{1}{3}s^2 h$	$SA = s^2 + 2s\sqrt{h^2 + \left(\frac{s}{2}\right)^2}$

Section 9.3 Angles and Trigonometry

Definitions

Transversal Angles: A **transversal** is a line in a plane that intersects two or more lines in that same plane at different points.

Congruent Angles: Two angles of equal measure are said to be **congruent** to one another.

Sine, Cosine, and Tangent

For an acute $\angle A$ in a right triangle,

$$\sin A = \frac{\text{Opposite}}{\text{Hypotenuse}} = \frac{a}{c},$$

$$\cos A = \frac{\text{Adjacent}}{\text{Hypotenuse}} = \frac{b}{c}, \text{ and}$$

$$\tan A = \frac{\text{Opposite}}{\text{Adjacent}} = \frac{a}{b}.$$

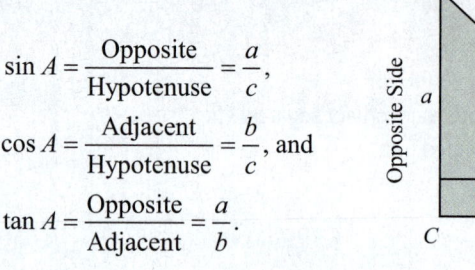

Figure 9.3.18

Inverse Trigonometric Functions

The inverse trigonometric functions are used to find the angle from any trigonometric ratio. The common inverse trigonometric functions are $\theta = \sin^{-1}x$, $\theta = \cos^{-1}x$, and $\theta = \tan^{-1}x$.

Law of Sines and Law of Cosines

For a triangle where a is the length of the side opposite angle A, b is the length of the side opposite angle B, and c is the length of the side opposite angle C, we have the following.

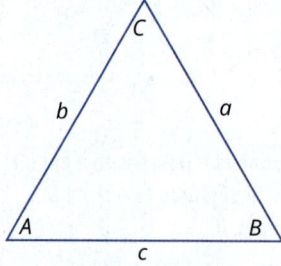

Figure 9.3.29

Law of Sines

$$\frac{a}{\sin A} = \frac{b}{\sin B} = \frac{c}{\sin C}$$

Law of Cosines

$$a^2 = b^2 + c^2 - 2bc \cos A$$

$$b^2 = a^2 + c^2 - 2ac \cos B$$

$$c^2 = a^2 + b^2 - 2ab \cos C$$

Properties

Labeling Angles

There are three common ways of labeling angles.

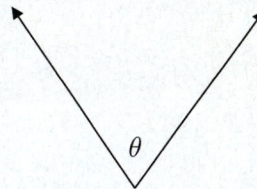

Figure 9.3.1: $\angle\theta$

Using a letter of the Greek alphabet such as θ or φ.

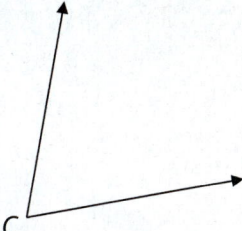

Figure 9.3.2: $\angle C$

Using a single number or capital letter.

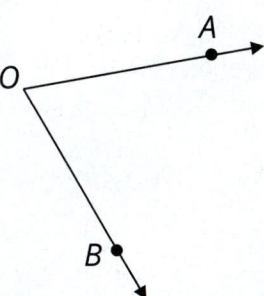

Figure 9.3.3: $\angle AOB$

Using three capital letters with the vertex as the middle letter.

Theorem

Pythagorean Theorem

In a right triangle, the square length of the hypotenuse is equal to the sum of the squares of the lengths of the other two sides.

$$a^2 + b^2 = c^2$$

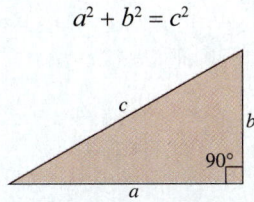

Figure 9.3.12

Here, a and b are the are the legs and c is the hypotenuse.

CHAPTER 10

Probability

▌ SECTIONS

Introduction

Monty Hall was the creator of a game show called Let's Make a Deal and hosted it for 30 years. The heart of the show was based on a probability problem that came to be known as the Monty Hall Problem—having made a choice among three options, are you more likely to win the big prize if you stick with it or change your decision? Let's walk through the scenario and see what you decide.

Suppose you are presented with three doors to choose from. You are told that there is a car behind one of the doors and a teddy bear behind the other two doors. You choose one of the three doors. After you make your selection, one of the two doors that you didn't chose is opened to reveal a teddy bear. Do you now stay with your first choice or switch to the other unopened door?

It may surprise you, but if you want to increase your chance of winning, you should switch doors. If you switch, the only way you will lose is if you were correct when you made your first selection, and the odds are against that. Initially you only had a one-in-three chance of being correct. That is, when making your first choice you had one chance of being right and two chances of being wrong. So it is more likely that your first guess was wrong. Let's assume that you chose the first door and look at the possible scenarios.

	1st Door	2nd Door	3rd Door	
1.	🚗	🧸	🧸	Switch → Lose
2.	🧸	🧸	🚗	Switch → Win
3.	🧸	🚗	🧸	Switch → Win

Until you look at the possible scenarios, it might seem like your chance of winning is the same regardless of whether you switch after the reveal or stay with your first choice. But as you can see, that isn't the case. Notice that if you switch, the first scenario is the only scenario in which you lose the car. This means that if you make the switch, you win the car in both the second and third scenarios. The probability of winning if you stay with the door you first chose is one out of three, and the probability of winning if you switch is two out of three. Do you see why you should switch? In this chapter, you will learn about probablity to help you better understand scenarios like this one.

10.1 INTRODUCTION TO PROBABILITY

OBJECTIVES

1. Determine sample spaces.

2. Calculate experimental probability.

3. Calculate classical probability.

💬 **Helpful Hint**

The word inclusive indicates that the end points are included in a range of numbers. For example, "natural numbers 1 to 3, inclusive," denotes 1, 2, and 3.

OBJECTIVE 1 Probability and Sample Spaces

Consider how likely it would be for you to…

- Be struck by lightning
- Win the lottery
- Play football in the NFL
- Become President of the United States
- Be in a car accident
- Die of a spider bite
- Get married
- Get divorced
- Throw two sixes on a set of dice

We have a natural understanding of likelihood. Certainly, if you were asked to categorize these occurrences listed above as likely to happen, unlikely to happen, or very unlikely to happen, you would be able to. If instead the task was to put them in order of increasing likelihood, that would be much more of a challenge. Getting married is probably more likely than being struck by lightning, but is winning the lottery more or less likely than being President of the United States? And how likely are you to die from a spider bite?

The study of *probability* is the mathematical approach to analyzing how likely things are to happen. In a nutshell, the probability of something occurring is a number between 0 and 1, inclusive, that represents how likely the event is to happen. If an event *cannot* occur, then its probability is 0; if an event is *certain* to occur, its probability is 1. All other probabilities fall somewhere between those two possibilities.

The closer a probability is to 0, the less likely the event is to happen; the closer the probability is to 1, the more likely the event is to happen. A probability of 0.5, or $\frac{1}{2}$, means an event is just as likely to happen as to not happen. Figure 10.1.1 illustrates this for us.

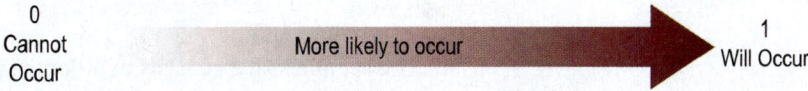

0
Cannot Occur
More likely to occur
1
Will Occur

Figure 10.1.1: Numerical Property of Probability

In this chapter, we'll look at terminology used in the study of probability, some different techniques to help us calculate probabilities, and how to use probabilities to make informed decisions.

To begin, we need to define a few terms. First, a *probability experiment*, or a *trial*, is any process with a result determined by chance, such as flipping a coin, drawing a number from one to ten out of a hat, or having a computer randomly generate a three-digit number. Each of these examples illustrates the possible individual results, called *outcomes*, in a trial. For instance, when you flip a coin, the possible outcomes are either a head or a tail. When you draw a number from one to ten out of a hat, four is a possible outcome, as is nine. In fact, all the numbers from

💬 **Helpful Hint**

The outcomes in a sample space can be listed between curly brackets and separated by commas. For instance, the sample space for a coin flip is {head, tail}.

one to ten are outcomes. The set of all possible outcomes from a given probability experiment is called the *sample space* of that experiment. We've already seen that the sample space for a coin flip is {head, tail}. The sample space for the computer example is all possible three-digit numbers.

📖 PROBABILITY TERMINOLOGY

Probability

The **probability** of something occurring is a number between 0 and 1, inclusive, that represents how likely the event is to occur. If an event cannot occur its probability is 0. If an event is certain to happen its probability is 1.

Probability Experiment

A **probability experiment**, or a *trial*, is any process with a result determined by chance.

Outcome

Each individual result that is possible for a probability experiment is an **outcome**.

Sample Space

The **sample space** is the set of all possible outcomes for a given probability experiment.

Example 10.1.1

Determining Sample Spaces

Identify the sample space for each of the following experiments.

a. Rolling a single fair six-sided die

b. Seat assignments on an airplane where the row options are aisle (A) or window (W), and the section options are business class (B) or economy class (E)

Solution

💬 **Helpful Hint**

The sample space is the set of outcomes for an experiment, not the number of outcomes.

a. The sample space consists of all of the possible outcomes of rolling a fair die. It can land on any of the six sides of the die. Therefore, the sample space is the following.

$$\text{Sample Space} = \left\{ \boxed{\cdot}, \boxed{\because}, \boxed{\therefore}, \boxed{::}, \boxed{\because\cdot}, \boxed{:::} \right\}$$

b. The sample space consists of all possible seat assignments. Each outcome is comprised of a seat position in the row and a section in the plane. For example, an aisle seat (A) in the business class (B) could be denoted AB. Using this notation, the sample space can be written as follows.

$$\text{Sample Space} = \{AB, AE, WB, WE\}$$

✔ **Skill Check 10.1.1**

Identify the sample space for tossing two coins together, where H is heads and T is tails.

Sometimes it is easy to list all the outcomes in a sample space as we did in Example 10.1.1. However, sometimes the sample space is large, and other techniques should be used to ensure that no outcomes are omitted. When an experiment consists of several stages, using a *tree diagram* is a great way to stay organized. Just as its name suggests, a tree diagram is a visual tool that indicates the possible outcomes at each stage in a multistage experiment. A tree diagram begins with one stage and then branches off into the possible outcomes of that stage. Branches extend off of each outcome to indicate the next stage of the experiment. To identify a specific outcome in the multistage experiment, list the sequence of outcomes that describes a path from the top of the tree to the bottom of the tree. Note that the number of branch paths on the tree is equal to the number of outcomes in the sample space.

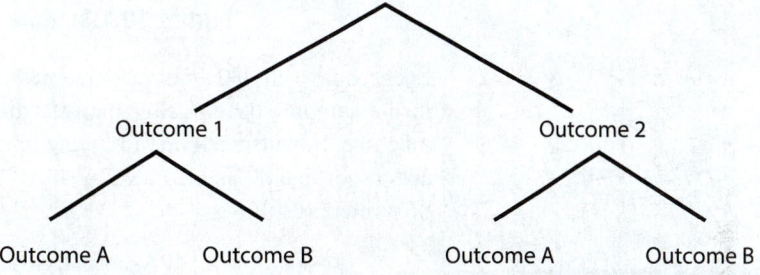

Sample Space = {1A, 1B, 2A, 2B}

Figure 10.1.2: Example of a Tree Diagram

📖 **TREE DIAGRAM**

A **tree diagram** uses branches to indicate all possible outcomes at each stage for an experiment. Each path of branches in a tree diagram indicates a single possible outcome for the experiment.

Example 10.1.2

Using a Tree Diagram to Determine a Sample Space

You are considering buying a new smartwatch. Suppose you have the following options to choose from.

- Face Size: 40 mm, 44 mm
- Case: Aluminum, Stainless Steel, Titanium, Ceramic
- Band: Rubber Sports Band, Fabric Sports Band, Leather, Stainless Steel

Use a tree diagram to determine the sample space for possible watch choices.

Solution

The tree in Figure 10.1.3 begins with the two possibilities for the face size—40 mm or 44 mm. It then branches out for each case option in the second row and then again in the third row with the type of band.

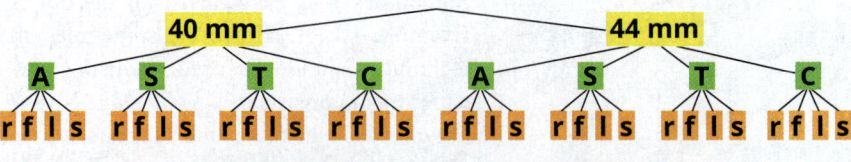

Key:

Cases	Bands
A = Aluminum	**r** = Rubber Sports Band
S = Stainless Steel	**f** = Fabric Sports Band
T = Titanium	**l** = Leather
C = Ceramic	**s** = Stainless Steel

Figure 10.1.3: Smartwatch Tree Diagram

Each path to the 32 bottom outcomes on the tree represents a unique outcome in the sample space. Using the tree diagram as a guide, we can identify each outcome. For instance, the first outcome—a 40 mm face with an aluminum case and a sport band—is indicated by 40Ar. Using this notation, the sample space can be written as follows.

$$\text{Sample space} = \left\{ \begin{array}{l} \text{40Ar, 40Af, 40Al, 40As, 40Sr, 40Sf, 40Sl, 40Ss,} \\ \text{40Tr, 40Tf, 40Tl, 40Ts, 40Cr, 40Cf, 40Cl, 40Cs,} \\ \text{44Ar, 44Af, 44Al, 44As, 44Sr, 44Sf, 44Sl, 44Ss,} \\ \text{44Tr, 44Tf, 44Tl, 44Ts, 44Cr, 44Cf, 44Cl, 44Cs} \end{array} \right\}$$

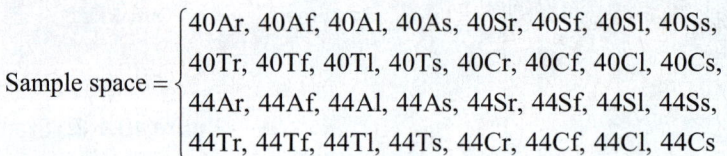

✔ Skill Check 10.1.2

Draw a tree diagram for an experiment consisting of 2 stages where each stage has 3 possible outcomes.

Often in probability we will be interested in grouping together outcomes in the sample space. A group, or subset, of outcomes in the sample space is called an *event*. An event may include one, some, or all the members of the sample space. For example, consider Example 10.1.1 where we rolled a fair single six-sided die. The sample space is the numbers 1 through 6. The event "rolling an even number" is the subset of outcomes {2, 4, 6}. On the other hand, the event "rolling a number less than 10" is the set {1, 2, 3, 4, 5, 6}.

📖 EVENT

An **event** E is a group, or subset, of outcomes in the sample space.

Informally, when we talk of the probability or likelihood of an event, we are often referring to either a precise theoretical probability, called *classical probability*, or more of an investigational approach, called *empirical probability* or *experimental probability*. We'll start with the less precise method of calculating probability using experimental probability.

OBJECTIVE 2 Experimental Probability

As its name suggests, *experimental probability*, or *empirical probability*, is found by performing an experiment. Specifically, the experimental probability that an event occurs is calculated by dividing the number of times an event occurs by the total number of trials performed.

EXPERIMENTAL PROBABILITY

In **experimental probability**, if E is an event, then $P(E)$, which is read "the probability that E occurs," is given by

$$P(E) = \frac{f}{n}$$

where f is the number of times the event occurs (the *frequency* of the event) and n is the total number of times the experiment is performed. The value of $P(E)$ will always be a real number between 0 and 1, inclusive.

For example, suppose we wanted to know the probability that a customer at a concession stand orders a lemonade. As an experiment, we can record the number of drink orders made over the next two hours, noting how many of those included a lemonade. Suppose that we found there were 34 drink orders made and 7 of those included lemonade. We can then calculate the experimental probability of ordering a lemonade as follows.

$$P(\text{Lemonade}) = \frac{\text{Number of Drink Orders That Include Lemonade}}{\text{Total Number of Drink Orders}}$$

$$= \frac{7}{34}$$

$$\approx 0.2059$$

This is a reasonable estimate based on the two hours of ordering at the concession stand. Suppose we recorded the same information over a longer period of time, say four hours. How do you think our estimate would change? What if we collected the information for several days? Or months? The **Law of Large Numbers** says that the greater the number of trials, the closer the experimental probability will be to the *classical probability*. In other words, the more you perform an experiment, the more accurate the estimated probability.

Example 10.1.3

Calculating Experimental Probability

For her elementary school science fair project, Libby is conducting research on the accuracy of the weather prediction from her local news channel. She recorded the forecast and the actual weather for two weeks. The following table shows her results.

Table 10.1.1: Accuracy of Weather Predictions

Forecast	Actual Weather
Rain	Rain
Snow	Rain
Snow	Snow
Cloudy	Clear
Cloudy	Cloudy
Rain	Rain
Clear	Drizzling rain
Clear	Clear
Cloudy	Clear

Forecast	Actual Weather
Snow	Cloudy
Clear	Clear
Clear	Clear
Rain	Rain
Rain	Cloudy

a. Using experimental probability, what is the probability that the news channel accurately predicts the next day's weather?

b. Discuss what might happen to the calculated probability if Libby records the weather prediction and outcome for another 14 days.

Solution

a. For Libby to calculate the probability, she needs to count the number of days the weatherman correctly predicted the weather and divide it by 14 (the total number of days she did the experiment).

$$P(\text{Correct Prediction}) = \frac{\text{Number of Times the Forecast Was Correct}}{\text{Total Number of Times the Weather Was Recorded}}$$

$$= \frac{8}{14}$$

$$\approx 0.5714$$

Because Libby can now estimate that the news channel correctly predicts the weather approximately 57% of the time, she knows that this is also the probability that the prediction for the following day's weather will be correct.

b. With more data points to include in the probability calculation, Libby's experimental probability will get closer to the true probability of the news channel correctly predicting the weather. If the prediction was correct more often in the next 14 days, the calculated probability of a correct prediction would be higher; if the prediction was correct less often, the probability would be lower. Either way, the experimental probability would get closer to the true probability.

> **Think Back**
>
> Remember, when changing from a decimal to a percent, you move the decimal two places to the right and add the % sign.

OBJECTIVE 3 Classical Probability

What then is the *true probability* that we are aiming to find when we calculate experimental probability? This theoretical probability is called *classical probability*. This is the most precise method for calculating probability because it takes *all* possible outcomes for an experiment into account. For instance, the probability that you roll the number three on a fair six-sided die is $\frac{1}{6}$ because there is only one way to roll a three out of the six numbers on the die.

Classical probability says that if all outcomes are equally likely, then the probability that an event occurs is equal to the number of outcomes included in the event divided by the total number of outcomes in the sample space. Be careful; the term *equally likely* here is important. There is an obvious flaw in the statement, "There's a 50% chance I'll win the lottery tomorrow." Even though the sample space contains the two outcomes {win the lottery, do not win the lottery}, the probability of winning the lottery is not $\frac{1}{2}$ because the outcomes are definitely not equally likely here!

CLASSICAL PROBABILITY

In **classical probability**, if all outcomes are equally likely to occur, then $P(E)$, which is read "the probability that E occurs," is given by

$$P(E) = \frac{n(E)}{n(S)}$$

where $n(E)$ is the number of outcomes in the event and $n(S)$ is the number of outcomes in the sample space. The value of $P(E)$ will always be a real number between 0 and 1, inclusive.

Example 10.1.4

Calculating Classical Probability

Suppose you were asked to draw a card from a standard deck of 52 cards. A standard deck of cards contains the following cards.

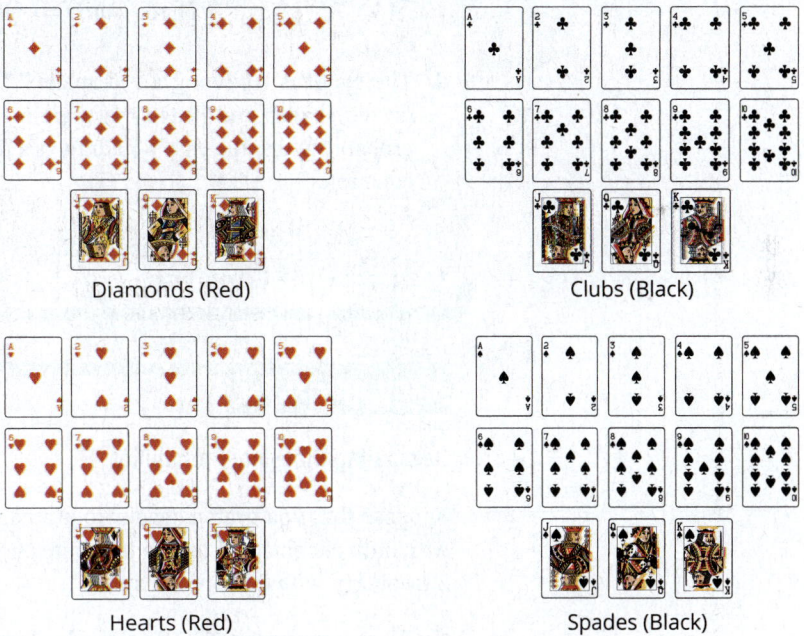

Diamonds (Red)   Clubs (Black)

Hearts (Red)   Spades (Black)

Figure 10.1.4: Cards in a Standard Deck

a. What is the probability that the card you draw is red?

b. What is the probability that the card you draw is a diamond?

c. What is the probability that the card you draw is a face card (king, queen, or jack)?

d. What is the probability of drawing a red spade?

Solution

In all of these questions, we are drawing from a standard deck of cards, which means that each card has the same probability of being chosen and that the sample space contains 52 cards; that is, $n(S) = 52$.

a. The event we are interested in is "drawing a red card." Since half of the cards are red, there are 26 red cards; that is, the number of outcomes in the event is 26. So we calculate the probability that the card you draw is red as follows.

$$P(\text{Red Card}) = \frac{n(E)}{n(S)} = \frac{26}{52} = 0.5$$

b. The event "drawing a diamond" has 13 outcomes, since there are 13 cards in the diamond suit. So the probability that the card you draw is a diamond is found using the following equation.

$$P(\text{Diamond}) = \frac{n(E)}{n(S)} = \frac{13}{52} = 0.25$$

c. The event "drawing a face card" means that we must consider all four suits. Each suit has three face cards—a king, a queen, and a jack. So there are $4 \cdot 3 = 12$ cards in the event. The probability that the card you draw is a face card is calculated as follows.

$$P(\text{Face Card}) = \frac{n(E)}{n(S)} = \frac{12}{52} \approx 0.2308$$

d. The event is "drawing a red spade." Because all spades are black, there are no red spades and hence there are no outcomes in the event. Therefore, the probability that the card you draw is a red spade is 0, as shown in the following equation.

$$P(\text{Red Spade}) = \frac{n(E)}{n(S)} = \frac{0}{52} = 0$$

Example 10.1.5

Calculating Classical Probability

Suppose that you grab a snack from a bag of chocolates that contains 4 caramel with milk chocolate, 4 mint with white chocolate, 6 mint with dark chocolate, and 2 raspberry with dark chocolate.

a. What is the probability that you randomly grab a raspberry with dark chocolate for your snack?

b. What is the probability that you get a chocolate that does not contain dark chocolate?

c. Determine the probability that the snack you grab contains chocolate.

Solution

a. We first need to know the number of outcomes in the sample space. The sample space consists of the number of all chocolates in the bag, which is $4 + 4 + 6 + 2 = 16$ different chocolates.

There are 2 outcomes in the event "choosing a raspberry with dark chocolate." The probability is then calculated as follows.

$$P\left(\text{Raspberry with Dark Chocolate}\right) = \frac{n(E)}{n(S)} = \frac{2}{16} = 0.125$$

b. The event "does not contain dark chocolate" includes the flavors caramel with milk chocolate and mint with white chocolate. Thus, there are $4 + 4 = 8$ pieces that do not contain dark chocolate. The probability is then calculated as follows.

$$P\left(\text{Not Dark Chocolate}\right) = \frac{n(E)}{n(S)} = \frac{8}{16} = 0.5$$

c. Since there are only chocolates in the bag, choosing a chocolate is a certainty. Therefore, the probability is 1.

$$P\left(\text{Chocolate}\right) = \frac{n(E)}{n(S)} = \frac{16}{16} = 1$$

> ✔ **Skill Check 10.1.3**
>
> Determine the probability of rolling a four on a fair six-sided die.

Example 10.1.6

Calculating Classical Probability

Consider a tennis game between two equally matched players in different age groups. The outcome of each point is that either the older player wins (O) or the younger player wins (Y), and since the players are equally matched, we'll assume these two outcomes are equally likely. What is the probability that the older player will win the first three points?

Solution

Since we are told both outcomes O and Y are equally likely, we can use classical probability. To determine how many outcomes are in the sample space, we can use either a pattern or a tree diagram. Let's use a tree diagram.

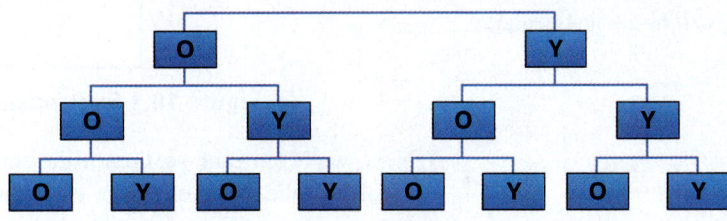

Figure 10.1.5

This gives 8 possible outcomes for the first three points: {OOO, OOY, OYO, OYY, YOO, YOY, YYO, YYY}. Notice that only 1 of the outcomes consists of the older player winning all three times (OOO). Thus, the probability of the older player winning all three of the points is calculated as follows.

$$P\left(\text{Older Player Winning All Three Points}\right) = \frac{n(E)}{n(S)} = \frac{1}{8} = 0.125$$

Notice that the probability of the younger player winning all three points is also $\frac{1}{8}$.

Helpful Hint

A gene consists of two alleles, one from each parent.

Example 10.1.7

Calculating Classical Probability

In biology we learn that many traits are genetic. One example is the shape of our hairline, which is either straight or has a V-shape, called a widow's peak. Each person has two alleles that make up their hairline shape gene—either a widow's peak allele or a straight hairline allele, one inherited from each parent. Suppose a child has a mother with a widow's peak allele and a straight hairline allele, and a father who has two straight hairline alleles. What is the probability that the child will have a straight hairline if the widow's peak allele is always dominant over the straight hairline allele? (Note that since the widow's peak is dominant, any gene containing the widow's peak allele will result in having a widow's peak.)

Solution

So far, we have listed the outcomes of an experiment in an orderly fashion by using a tree diagram. In biology, a Punnett square is commonly used to help list the outcomes of a genetic experiment. The mother's two alleles of a specific gene are listed along one side of a square, and the father's two alleles of the gene are listed along an adjacent side. The square is then filled in by writing the four possible combinations of alleles inherited—one from each parent.

For this experiment, we will label a widow's peak allele with an uppercase W and a straight hairline allele with a lowercase w. The mother's gene will be labeled Ww since she has both a widow's peak and a straight hairline allele. The father, with two straight hairline alleles, will be labeled ww. These go on the outside of the Punnett square as shown.

Math Milestone

Reginald Punnett (1875–1967) was a British geneticist who developed the diagram discussed in Example 10.1.7, which is named after him, called the Punnett square. He is also known for cofounding the Journal of Genetics and writing one of the first textbooks on genetics.

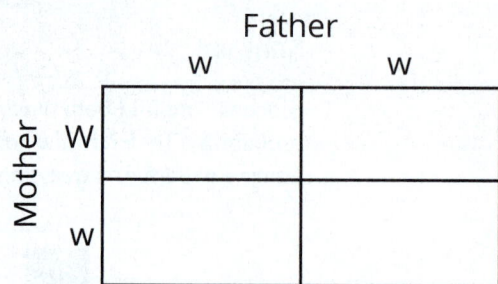

Figure 10.1.6: Punnett Square for Hairline Genes

Next, we fill in each section of the square by writing the possible combinations of alleles inherited—one from each parent. For instance, the top left section will have a W from the mother and a w from the father.

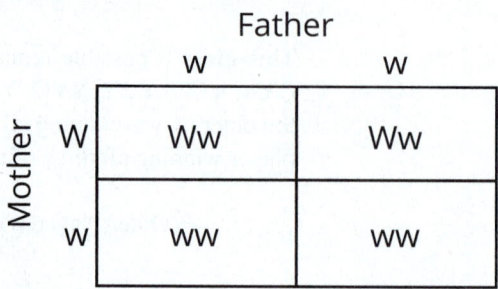

Figure 10.1.7: Punnett Square for Hairline Genes

So the four possibilities for the child's genes are {Ww, Ww, ww, ww}. We are told that the widow's peak is dominant, so any gene containing the widow's peak allele, W, will result in having a widow's peak. That leaves two with only the straight hairline allele, {ww, ww}. Thus, if E is the event of the child having a straight hairline, then the probability that E occurs is calculated as follows.

$$P(\text{Straight Hairline}) = \frac{n(E)}{n(S)} = \frac{2}{4} = 0.5$$

This means that the child has a 50% chance of having a straight hairline.

The next example looks at distinguishing between the two types of probability that we've discussed: experimental and classical. Remember, experimental probability is based on the results of an experiment whereas classical probability is an exact calculation that can be made when all possible outcomes of an event are known and equally likely.

Example 10.1.8

Classical vs. Experimental Probability

Determine if the scenarios given are examples of classical or experimental probability.

a. Katie is curious about her chances of winning an e-reader from the student government association. She polled her friends to find out how many of them filled out the survey to be entered in the contest.

b. Tristan is interested in his chances of winning at the blackjack table. He determines the probability of what his next card will be by knowing the cards that have already been played.

c. Based on the recent United States Census, the local government estimates the amount of growth the community will experience in the coming years.

Solution

a. Because Katie is conducting an informal survey and not all students are included, the probability is experimental.

b. This is an example of classical probability since *all* cards have an equal chance of being dealt at the beginning, and Tristan adjusts his chances by accounting for those cards that have already been drawn.

c. Since the United States Census is actually an incomplete count, any probability calculated from it would be experimental.

Skill Check Answers

1. {HH, HT, TT, TH}   **2.**   **3.** $\frac{1}{6}$

10.1   Exercises

✔ CONCEPT CHECK

1. A probability experiment, or a _____, is any process with a result determined by chance.

2. The _____ is the set of all possible outcomes for a given probability experiment.

3. Each path of branches in a _____ indicates a single possible outcome.

4. The Law of Large Numbers says that the greater the number of trials, the closer the _____ probability will be to the _____ probability.

5. True or False: The probability of something occurring can never be 1.

♀ PRACTICE

Create a tree diagram to list the outcomes in each sample space.

6. Find the sample space for the gender of each child in regard to birth order for a family with three children.

7. In picking out a new car, there are several choices to make. The color can be red, white, or silver. The seats can be cloth or leather. Finally, it can have a sunroof, a moonroof, or neither. Find the sample space for the possible new car combinations using a tree diagram.

8. Use a tree diagram to find the sample space for tossing a coin three times.

9. Four students are randomly selected from an algebra class and asked whether they suffer from math anxiety. Find the sample space for the possible outcomes of the survey using a tree diagram.

Identify the sample space for each of the following experiments.

10. A coin is flipped and then a single die is rolled.

11. Choosing a number from all positive two-digit integers where the digits are repeated (for example, 11).

12. Five marbles are in a bag, one of each of the following colors: blue (B), clear (C), green (G), yellow (Y), and red (R). Two marbles are drawn consecutively. Assume that the first marble is not put back in the bag before the second marble is drawn. Order of the selection matters. In other words, BG and GB are two different selections.

13. When choosing an outfit, you have a choice of 3 shirts: white (W), black (B), or patterned (P); a choice of 2 types of jeans: faded (F) or dark wash (D); and a choice of 4 pairs of shoes: sandals (S), running shoes (R), climbing boots (C), or mules (M). List the sample space in regard to the outfits (combination of shirt, jeans, and shoes) you could pick from.

14. When ordering pizza for a party, you have the choice between three restaurants: Dominos (D), Pizza Hut (P), and Papa John's (J); a choice of delivery (d) or store pickup (p); and side options of breadsticks (b), wings (w), or salad (s).

15. Tickets for an outdoor orchestra concert can be purchased for pavilion seating (P) or lawn seating (L). When purchasing tickets, you can choose to pay extra for a parking pass for the close lot (C) or opt for the free general parking pass (G). You can also add on meal and drink vouchers (M) or choose not to purchase any food vouchers (N).

16. You are purchasing a jigsaw puzzle from a local bookstore. The puzzles come in four sizes: 200 piece (2), 500 piece (5), 1000 piece (10), and 1200 piece (12). Each puzzle size has five images to choose from: space (S), landscapes (N), cars (C), animals (A), and bridges (B).

17. A small company is choosing packaging materials for a new product. For the box, the options are a recycled cardboard box (C), a recycled plastic box (P), or a box made from new materials (N). For the instruction manual, the options are a card with a link to online instructions (O) or a fully printed manual (M). Finally, the box can be sealed with temper resistant stickers (T) or staples (S).

Determine whether each probability is experimental or classical.

18. In order to find the percentage of bass that Troy had in his pond this spring, he decided to spend three days catching a total of 15 fish each day and counting how many of those were bass.

19. Jason wants to know how likely he is to win a raffle if he bought 3 of the 1000 tickets that were sold.

20. Virginia wants to know how likely it is for her to win a backgammon game if she only needs to roll a double six to win.

21. Emre wants to see how many students own a smartphone. He surveys 100 college freshmen at a Winter Welcome event and asks what kind of phone they carry.

22. Richard wants to determine the probability that a red Skittle will appear in a bag of regular Skittles, so he purchases 20 bags of Skittles and counts how many red Skittles are in the bags.

23. Rita volunteers at a local shelter and wonders what the probability is that someone chooses to adopt a cat from the local shelter today, so she counts how many total animals they have for adoption and the total number of cats available for adoption.

🚀 APPLICATIONS

Calculate each experimental probability. Round your answer to the nearest millionth, if necessary.

24. A news organization asked a selection of voters exiting a polling place their age bracket in order to paint a picture of turn out on election day. Here's the record of the results collected so far.

Voting Age			
17–29	**30–44**	**45–64**	**65 and Older**
9	8	32	15

a. What is the probability that the next voter to exit will be between 30 and 44?

b. What is the probability that the next voter to exit will be in either of the youngest two age groups?

c. What is the probability that the next voter to exit will be under 65?

25. The blood types of 200 people are collected at a doctor's office. The table shows the breakdown of patients per blood type. If a person from this group is selected at random, what is the probability that this person has type O blood?

Blood Type Survey Results	
Blood Type	**Number of Patients**
A	50
B	65
O	70
AB	15

26. As students were exiting the student center on campus, Vicki took note whether they were listening to headphones. The table shows results that she collected.

Based on Vicki's data, what is the probability that a randomly selected student will have headphones in their ears when exiting the student center?

Data for Students Exiting the Student Center	
	Number of Students
Headphones	33
No Headphones	51

27. A sample of 500 active-duty military showed that 82 of them suffered from Post Traumatic Stress Disorder (PTSD). However, 95 of those studied reported that they would be too embarrassed to seek mental health services.

a. Based on this sample, if a soldier is chosen at random, what is the probability that he/she suffers from PTSD?

b. Based on this sample, if a soldier is chosen at random, what is the probability that he/she would be willing to seek mental health services?

28. The owner of a parking lot takes an inventory of the 150 cars parked in the lot one evening. The table shows the breakdown of cars by brand. If a car from this group is randomly selected, what is the probability that it is a Nissan?

Cars Parked in Parking Lot

Brand	Number of Cars
Chevrolet	20
Ford	18
Honda	42
Hyundai	26
Nissan	44

29. The favorite donut flavors of 225 people are collected at a donut shop. The table shows the breakdown of the number of customers that prefer each flavor. If a flavor is selected at random, what is the probability that it is maple dipped?

Donut Flavor Survey Results

Flavor	Number of Customers
Chocolate Sprinkle	57
Cream-filled	31
Cruller	36
Glazed	51
Jelly-Filled	32
Maple Dipped	18

30. A librarian walked through the quiet area of the university library and recorded whether students were studying or looking at their phones. The table shows the data that was collected. Based on this data, what is the probability that a randomly selected student was looking at their phone instead of studying?

Data for Students in the Library

	Number of Students
Studying	47
Looking at Phone	22

31. A restaurant at a mall gives out free samples in hopes of gaining new customers. The employee giving out samples for an hour recorded whether the people who took a sample ordered food from the restaurant or not. The table shows the data that was collected. Based on this data, what is the probability that a person who takes a sample will order food from the restaurant?

Data for People Who Took a Sample

	Number of People
Ordered Food	38
Did Not Order Food	77

For each situation, draw a tree diagram to determine the sample space and then calculate the given probability.

32. When ordering the three-course lasagna special at a restaurant, customers have the choice between soup (S) or salad (A) for an appetizer, meat lasagna (M) or vegetarian lasagna (V) for the main course, and cake (C) or pie (P) for dessert. What is the probability of someone ordering salad, vegetarian lasagna, and pie?

33. When picking dinnerware to add to their wedding registry, a couple decides they want white dinner plates. The couple can choose between square plates (S) or round plates (R); gold-banded plates (G), silver-banded plates (B), or plain plates (P), and glazed earthenware (E) or ceramic (C). What is the probability the couple will pick any type of glazed earthenware plates?

Calculate each classical probability. Assume individual outcomes are equally likely. Round your answer to the nearest millionth, if necessary.

34. What is the probability that, out of 235 attendees (including yourself) at a conference, you are selected to win the door prize at the opening session?

35. On an American roulette wheel, there are 18 red pockets, 18 black pockets, and 2 green pockets. What is the probability of landing on a red pocket?

36. What is the probability that a card drawn randomly from a standard deck of cards will be an ace?

37. A book contains 321 pages numbered 1, 2, 3, . . . , 321. If a student randomly opens the book, what is the probability that the page's number has all of its digits the same (ignoring single-digit page numbers)?

38. Consider a cat with a litter of four kittens. Find the probability that all four kittens are male.

39. Mason has 213 songs in a playlist. He's categorized them in the following manner: 20 from soundtracks, 8 spiritual, 31 jazz, 16 Latin American, 27 R&B, 47 rock, and 64 pop. If Mason puts his playlist on shuffle, what is the probability that the first song played is a pop song?

40. Landon decided to play a joke on his friends at work. When he stocked the vending machine with sodas, he randomly put the drinks into the different slots. If he put in 15 Diet Coke, 15 Coke, 12 Sprite, 17 Dr. Pepper, and 10 Fanta cans, what is the probability that the next person will get a Fanta drink when they put their money into the machine?

41. A shelf of fiction books contains 15 fantasy books, 12 science-fiction books, and 9 historical fiction books. What is the probability that a book selected at random is science-fiction?

42. A bag of apples picked from an orchard contains 8 Honeycrisp, 9 Gala, 7 Red Delicious, and 10 Granny Smith. If an apple is selected at random, what is the probability that it will be a gala apple?

43. To divide a class into groups for a project, numbers are placed into a box. Each student selects a number which assigns them to a group. In the box, there are six 1s, six 2s, six 3s, and six 4s. The first student selects a number at random. What is the probability that the student draws a 3?

44. An American roulette wheel has slots marked with a number 1 through 36 along with 0 and 00. The 0 and 00 slots are colored green. Half of the remaining slots are black and the other half are red. If a roulette ball randomly stops on a slot, what is the probability that the slot is red?

45. Lunch is served at a seminar and consists of a variety of sub sandwiches boxed with potato chips and a chocolate chip cookie. There are 14 turkey sandwiches, 16 ham sandwiches, and 8 roast beef sandwiches.

    a. What is the probability that a randomly selected lunch box contains a ham sandwich?

    b. What is the probability that a randomly selected lunch box contains a chocolate chip cookie?

    c. What is the probability that a randomly selected lunch box does not contain a roast beef sandwich?

46. A store sets up a display featuring the favorite movies of the employees. The display contains 12 action movies, 8 romances, 5 science-fiction movies, and 7 dramas.

    a. What is the probability that a randomly selected movie is a science-fiction?

    b. What is the probability that a randomly selected movie is a documentary?

    c. What is the probability that a randomly selected movie is not an action movie?

✏ WRITING & THINKING

47. Compare and contrast classical probability and experimental probability.

10.1 PROJECT

HOW HARD IS BASEBALL COMPARED TO OTHER SPORTS?

Many say that baseball, which is commonly called America's national pastime, is a game with a low success rate. Consider Mookie Betts, one of the greatest players the sport has ever seen. He had a batting average (BA) of .346 during the 2018 season. His batting average was the highest of all players for that year.

A player's batting average is computed by dividing the number of times a player has a hit by the player's total number of at-bats.

$$BA = \frac{\text{Number of Hits}}{\text{Number of At-Bats}}$$

Notice that this formula looks a lot like a probability where the sample space is the set of all at-bats and the event is getting a hit.

When fans talk about batting averages, they commonly think of it in terms of probability. That is, if a player is batting .250, then there is a 25% chance that they will get a hit at their next at-bat.

1. Do you agree with the interpretation of the batting average as a probability? Is it possible to compute the theoretical probability that a player will get a hit when at bat? Explain your answer.

2. Let's return our attention to Mookie Betts. Betts had 520 at-bats during the 2018 season; how many hits did he have that season? Round your answer to the nearest whole number.

3. Suppose that Mookie Betts had another 20 at-bats during the 2018 season. Assume he had a BA of .500 for those extra 20 at-bats. What would his new 2018 batting average be with these extra 20 at-bats?

During the 2018 season, the entire Major League Baseball batting average was .248. That is, on average, a player would get a hit 24.8% percent of the time at bat.

4. Perform an internet search to find the definition of Field Goal Percentage (FG%) in basketball. Find the National Basketball Association average FG% in the 2018–19 season. Explain this number in terms of players and field goals.

5. Using this information, decide which sport has a higher success rate. Explain your answer.

10.2 COUNTING OUTCOMES

OBJECTIVES

1. Use the Fundamental Counting Principle.

2. Calculate factorials.

3. Calculate combinations and permutations.

As you might imagine, carrying out trials and surveys to accurately estimate probabilities using the experimental approach we discussed previously can be very time consuming and potentially costly. Consequently, although using experimental probability may sometimes be unavoidable, it is certainly preferable to use classical probability whenever it can be obtained. In order for us to calculate classical probability we need to develop some techniques to count the number of outcomes in a sample space or an event. We'll spend this section looking at ways to count outcomes, and then in a later section we'll begin calculating more involved classical probabilities.

OBJECTIVE 1 The Fundamental Counting Principle

The first technique we'll look at is the Fundamental Counting Principle. When there are several outcomes at each stage of an experiment, the Fundamental Counting Principle states that you can multiply together the number of possible outcomes at each stage of the experiment in order to obtain the total number of outcomes for the experiment.

📖 FUNDAMENTAL COUNTING PRINCIPLE

The **Fundamental Counting Principle** states that for an experiment with a sequence of n stages where the first stage has k_1 outcomes, the second stage has k_2 outcomes, the third stage has k_3 outcomes, and so forth, the total number of possible outcomes for the experiment is calculated as follows.

$$(k_1) \cdot (k_2) \cdot (k_3) \cdot \ldots \cdot (k_n)$$

For instance, recall that in Example 10.1.2 we used a tree diagram to methodically list out all of the elements in the sample space of smartwatches having a choice of 2 face sizes, 4 case styles, and 4 band styles. We showed that the sample space consisted of 32 different smartwatch designs. If we think of each design choice as a different stage, we can multiply the numbers of choices together to come to the same conclusion; $2 \cdot 4 \cdot 4 = 32$. Note that by using the Fundamental Counting Principle like this, we only calculate the *size* of the sample space; we do not determine the actual elements in the sample space as we did with the tree diagram.

When counting outcomes in multistage experiments, we need to determine if repetition of outcomes is permitted, also known as an experiment *with replacement*, or if each outcome can only be used once, known as an experiment *without replacement*. For example, suppose you are choosing several cards from a deck, one at a time. If the card is placed back in the deck after each choice, then that experiment is *with replacement*. If the card is not returned to the deck after each choice, then the experiment is *without replacement*.

REPLACEMENT

When counting possible outcomes **with replacement**, objects are placed back into consideration for the following choice.

When counting possible outcomes **without replacement**, objects are not placed back into consideration for the following choice.

Example 10.2.1

Using the Fundamental Counting Principle

In order to log in to your new email account, you must create a password. The requirements are that the password needs to be 8 characters long, consisting of 5 lowercase letters and followed by 3 numbers.

a. If you are allowed to use a character more than once, that is, *with replacement*, how many different possibilities are there for passwords?

b. How many passwords are possible if the characters may not be duplicated, that is *without replacement* or *without repetition*.

Solution

a. If we think about each character in the password as a slot to fill, then we have 8 slots that need to be filled. The first 5 can be filled with letters and the last 3 with digits, as shown in Figure 10.2.1.

Slot 1	Slot 2	Slot 3	Slot 4	Slot 5	Slot 6	Slot 7	Slot 8
a b c ... x y z	a b c ... x y z	a b c ... x y z	a b c ... x y z	a b c ... x y z	0 1 2 ... 7 8 9	0 1 2 ... 7 8 9	0 1 2 ... 7 8 9
26 Options	26 Options	26 Options	26 Options	26 Options	10 Options	10 Options	10 Options

Figure 10.2.1

The first 5 slots contain 26 possibilities each, one for each letter of the alphabet. Notice that since we are allowed to have repeating characters in the password, we use the number 26 for each of the five slots and not just the first one. Similarly, the last 3 slots have 10 possible possibilities each, one for each digit 0 through 9. Using the Fundamental Counting Principle, we multiply the numbers of possibilities together to get the total number of possible passwords for the new account.

Possible Passwords: $26 \cdot 26 \cdot 26 \cdot 26 \cdot 26 \cdot 10 \cdot 10 \cdot 10 = 11{,}881{,}376{,}000$

Therefore, when characters are allowed to be used more than once, there are 11,881,376,000 possible passwords.

b. This time the parameters change slightly. The first slot still has a possibility of 26 letters, but the second slot now only has 25 choices since we used one letter for the first slot and are not allowed to reuse it. Similarly, the third slot has 24 choices, and so forth. A similar thought process is used to fill in the numbers of choices for the digits in the last 3 slots.

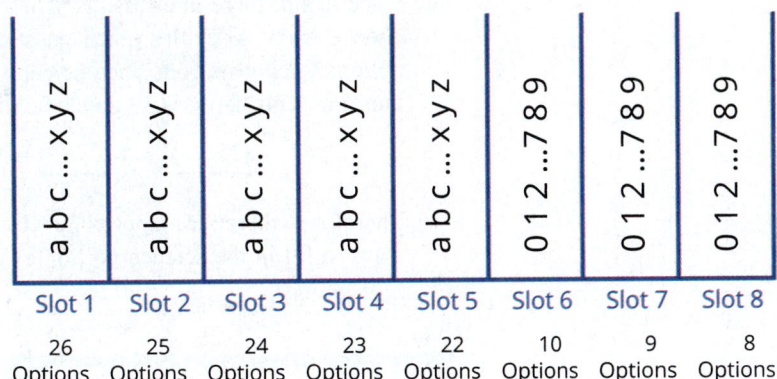

Slot 1	Slot 2	Slot 3	Slot 4	Slot 5	Slot 6	Slot 7	Slot 8
26 Options	25 Options	24 Options	23 Options	22 Options	10 Options	9 Options	8 Options

Figure 10.2.2

Using the Fundamental Counting Principle, we multiply the numbers of options together once again.

Possible Passwords: $26 \cdot 25 \cdot 24 \cdot 23 \cdot 22 \cdot 10 \cdot 9 \cdot 8 = 5{,}683{,}392{,}000$

Therefore, when we are not allowed duplicate characters there are about half as many possible passwords, even though 5,683,392,000 is still a lot of choices.

✔ **Skill Check 10.2.1**

How many 3-digit pin numbers are available using the numbers 1 through 9 if repetition is allowed?

Example 10.2.2

Using the Fundamental Counting Principle

The science section of the ACT test has 40 multiple choice questions, each with answer options A through D.

a. How many different ways are there to fill in the answer sheet for the science portion of the ACT if all the questions are answered?

b. Suppose a test taker decides to fill in the answers to the science portion randomly. The only restriction he puts on himself is that he can't use the same answer choice back-to-back. For example, if he chooses A for question 1, he can't use A for question 2, but he can use it again for question 3. How many possible ways are there for him to fill in the answer sheet for the science section?

Solution

a. The test has 40 questions, each with 4 possible answer options. You can think of this as counting "with replacement" since each of the 4 choices can be used for every question. So there are 40 "slots" to fill, each with 4 possible answers. Using the Fundamental Counting Principle, we multiply 4 by itself 40 times. (Notice that we do not multiply 4 times 40.)

$$\underbrace{4 \cdot 4 \cdot 4 \cdot 4 \cdot 4 \cdot 4 \cdot 4 \cdots \cdots 4}_{40} = 4^{40}$$

This number is so large that most of our calculators struggle to display it. Instead, many calculators give the answer in scientific form: 1.208926×10^{24}. This means there are approximately 1,208,926,000,000,000,000,000,000 ways to fill in just the science portion of the ACT.

b. Once again, there are 40 slots. The first slot will have 4 answer options to choose from. After that, each question will essentially have only 3 answer options to choose from since he can never use the answer from the previous question. This happens 39 times. So the multiplication will be as follows.

$$\underbrace{4 \cdot 3 \cdot 3 \cdot 3 \cdot 3 \cdot 3 \cdot 3 \cdot \cdots \cdot 3}_{39} = 4 \cdot 3^{39} \approx 1.621022 \times 10^{19}$$

Therefore, there are approximately 16,210,220,000,000,000,000 possible ways to fill in the science portion of the ACT if no answer choices are used back-to-back.

OBJECTIVE 2 Factorials

Notice that sometimes we need to multiply successive decreasing numbers together as we did in Example 10.2.1. This is a common occurrence when solving this type of problem. Mathematically, we can represent this by using *factorials*. A **factorial** is the product of all positive integers less than or equal to a given positive integer, n. Symbolically, a factorial is written as $n!$, which is read "n factorial."

$f(x)$ N FACTORIAL

In general, $n!$ (read "n factorial") is the product of all the positive integers less than or equal to n, where n is a positive integer.

$$n! = n(n-1)(n-2)(n-3) \cdots (2)(1)$$

The expression $0!$ is defined to be 1.

Thus, $4!$ is the product of all positive integers less than or equal to 4, which equals $4 \cdot 3 \cdot 2 \cdot 1 = 24$. One important factorial to memorize is $0!$, which is equal to 1 and appears in the denominator of many calculations.

Since factorials equal the products of strings of positive numbers, their values get very big, very quickly. While it was easy to calculate $4!$ without the assistance of a calculator, expressions such as $100!$ are much too large even for ordinary calculators to display. (Check and see how large a factorial your calculator can display.) Hence when calculating factorials by hand, it's convenient to use some methods for reducing factorials at the start of a calculation if at all possible. When using technology, intermediate calculations are done for us, as you will see in the following examples.

Example 10.2.3

Calculating Factorials

Calculate the values of the following factorial expressions.

a. $8!$

b. $\dfrac{3!}{0!}$

c. $\dfrac{89!}{87!}$

d. $\dfrac{7!}{(5-1)!}$

e. $\dfrac{5!}{3!(4-2)!}$

Solution

a. The expression $8!$ indicates that we need to multiply together each positive integer less than or equal to 8.

By Hand

$$8! = 8 \cdot 7 \cdot 6 \cdot 5 \cdot 4 \cdot 3 \cdot 2 \cdot 1 = 40{,}320$$

TI-83/84 Plus

To calculate a factorial of the form $n!$, enter the value of n in the calculator and press math . Highlight **PROB** to go to the probability menu and select **!**. The calculator returns 40320.

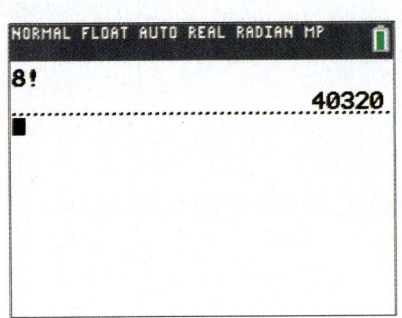

Figure 10.2.3

Microsoft Excel

We can use the built-in factorial function =**FACT(number)** to evaluate factorials. In an empty cell, type "=**FACT(8)**" and press Enter. The value is given as 40320.

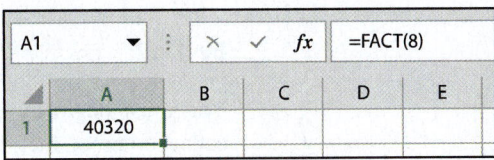

Figure 10.2.4

b. $\dfrac{3!}{0!}$

Remember that $0! = 1$, so we are not dividing by 0.

By Hand

Calculate each factorial and then divide.

$$\frac{3!}{0!} = \frac{3 \cdot 2 \cdot 1}{1} = 6$$

TI-83/84 Plus

We can do this calculation in one step with the calculator by typing in $3!/0!$. The calculator returns 6.

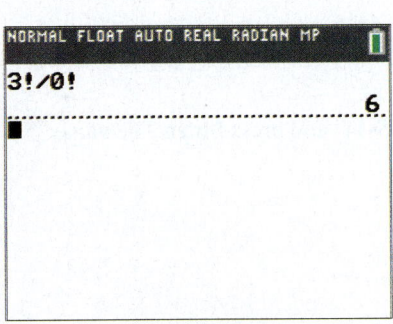

Figure 10.2.5

Microsoft Excel

In an empty cell, type "=**FACT(3)/FACT(0)**" and press Enter. The value returned is 6.

c. $\dfrac{89!}{87!}$

By Hand

It would be very cumbersome to multiply out 89! and 87! and then divide. Instead, we will cancel out some of the common factors that appear in both the numerator and denominator first. Begin by expanding the numerator and then canceling.

$$\frac{89!}{87!} = \frac{89 \cdot 88 \cdot \cancel{87} \cdot \cancel{86} \cdots \cancel{2} \cdot \cancel{1}}{\cancel{87} \cdot \cancel{86} \cdots \cancel{2} \cdot \cancel{1}} = 89 \cdot 88 = 7832$$

TI-83/84 Plus

Notice that if you type 89!/87! into a TI-83/84 Plus, you will get an overflow error message. Even though the expression $\dfrac{89!}{87!}$ simplifies to $89 \cdot 88 = 7832$, which can easily be done in a calculator, the initial value of 89! is too large for the TI-83/84 Plus to compute. Ironically, most modern smartphone calculators are able to do this calculation without simplifying first!

Microsoft Excel

In an empty cell, type "=**FACT(89)/FACT(87)**" and press Enter. The value is given as 7832.

d. $\dfrac{7!}{(5-1)!}$

By Hand

You must begin by evaluating the expression in the parentheses. Next, cancel common factors in the numerator and denominator and then simplify.

$$\frac{7!}{(5-1)!} = \frac{7!}{4!} = \frac{7 \cdot 6 \cdot 5 \cdot \cancel{4} \cdot \cancel{3} \cdot \cancel{2} \cdot \cancel{1}}{\cancel{4} \cdot \cancel{3} \cdot \cancel{2} \cdot \cancel{1}} = 7 \cdot 6 \cdot 5 = 210$$

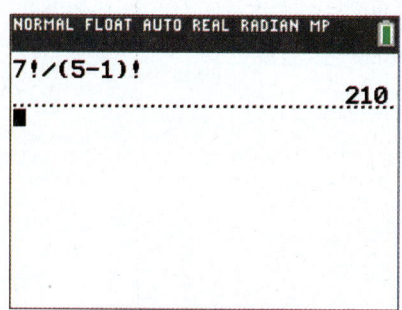

Figure 10.2.6

TI-83/84 Plus

Notice that you can type in the entire expression without simplifying first. Be sure to place the parentheses in the denominator. The calculator returns 210.

Microsoft Excel

In an empty cell, type "=**FACT(7)/FACT(5-1)**" and press Enter. The value returned is 210.

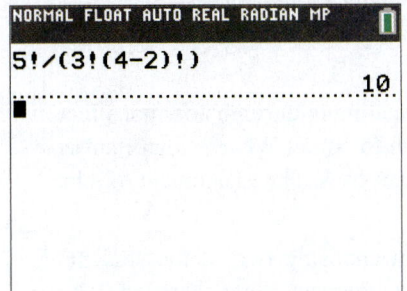

Figure 10.2.7

e. $\dfrac{5!}{3!(4-2)!}$

By Hand

Make sure that you begin by subtracting to simplify the expression in parentheses.

$$\frac{5!}{3!(4-2)!} = \frac{5!}{3!2!} = \frac{5 \cdot 4 \cdot \cancel{3} \cdot \cancel{2} \cdot \cancel{1}}{(\cancel{3} \cdot \cancel{2} \cdot \cancel{1})(2 \cdot 1)} = \frac{5 \cdot 4}{2 \cdot 1} = 10$$

TI-83/84 Plus

Once again, we can enter the entire original expression in one line. However, we must place parentheses around the entire expression in the denominator, which produces the correct solution of **10**.

Microsoft Excel

In an empty cell, type "**=FACT(5)/(FACT(3)*FACT(4-2))**". Be sure to put parentheses around the entire expression in the denominator, which results in the correct solution of 10.

OBJECTIVE 3 Combinations and Permutations

We often want to be able to count the number of ways that we can choose members from a group of objects. For instance, how many different sandwiches can be made with the ingredient choices at a sandwich shop? Or how many ways can the top three spots be filled at the end of a race with 140 runners? These types of scenarios can be represented by using either a combination or a permutation. The important distinction is whether the order of choice is important.

For instance, we'll contend that the order in which you choose the ingredients to make a sandwich does not affect the sandwich itself. A bacon, lettuce, tomato sandwich is the same as a lettuce, tomato, bacon one. This is a *combination* scenario since order is not critical. However, if you are determining the top three spots for winners of a race, there is a difference between Chloë finishing first with Blake and Crawford second and third, respectively, or Blake getting the gold, Crawford the silver, and Chloë the bronze. In this case, order is key, so we use a *permutation*.

📖 COMBINATIONS AND PERMUTATIONS

Combinations

A **combination** is a selection of objects from a group without regard to their arrangement; that is, their order *is not* important.

Permutations

A **permutation** is a selection of objects from a group where the arrangement is specific; that is, their order *is* important.

Example 10.2.4

Comparing Combinations and Permutations

Decide whether you would use a permutation or combination to count the number of outcomes for each of the following scenarios.

a. In how many ways can a club elect a president, vice president, and treasurer if there are 30 students participating?

b. If each department needs two student representatives from each major on a campus committee, how many ways can the biology department choose the representatives from the 45 students who are majoring in biology?

Solution

a. First notice that the number of students participating in the club does not affect whether we use a permutation or combination to count. What is important is the arrangement of the officers for counting purposes. The assignment of jobs makes a difference, so this is a permutation.

b. In this second scenario, we are interested in choosing two of the students from those majoring in biology, where the arrangement is not specified to be important. Therefore, we use a combination to count the number of outcomes.

✔ **Skill Check 10.2.2**

Would you use a combination or a permutation to count the number of ways that five people can line up to buy tickets to a concert?

Once you've determined whether the order in which objects are chosen is important, you can apply the appropriate formula to calculate the total number of possibilities. The formulas for combinations and permutations are given below.

💬 **Helpful Hint**

The following are all alternate notations for combinations and permutations.

Combinations

$$_nC_r = C(n,r)$$
$$= C_{n,r}$$
$$= {}^nC_r$$
$$= \binom{n}{r}$$

Permutations

$$_nP_r = P(n,r)$$
$$= P_{n,r}$$
$$= {}^nP_r$$

ƒ(×)   FORMULAS FOR COMBINATIONS AND PERMUTATIONS

The following formula is used to calculate the number of combinations.

$$_nC_r = \frac{n!}{r!(n-r)!}$$

The following formula is used to calculate the number of permutations.

$$_nP_r = \frac{n!}{(n-r)!}$$

In both of these formulas, r objects are selected from a group of n distinct objects, so r and n are both positive integers with $r \le n$.

The number of combinations is also commonly written as $\binom{n}{r}$. However, we will use the $_nC_r$ format throughout our text. Additional symbolic notations for combinations and permutations are given in the margin. A permutation is read "n things permuted r at a time," whereas a combination can be read in the shorter form "n choose r."

Example 10.2.5

Calculating Numbers of Combinations and Permutations

Consider a group of 15 candidates auditioning for a talent search firm.

a. How many ways can you arrange the sequence in which the candidates audition in the first round?

b. How many ways can you choose 5 of the candidates to move on to the next round?

Solution

a. First, notice that the arrangement of the 15 candidates is important for counting purposes because being the first to audition is certainly not the same as being the fifteenth. Therefore, this is a permutation. There are 15 candidates to choose from, so $n = 15$, and we want to arrange all of them, so r is also fifteen. The number of permutations of 15 things permuted 15 at a time is calculated as follows.

By Hand

$$
\begin{aligned}
{}_{15}P_{15} &= \frac{15!}{(15-15)!} \\
&= \frac{15!}{0!} \\
&= \frac{15 \cdot 14 \cdot 13 \cdots 2 \cdot 1}{1} \\
&= 1,307,674,368,000
\end{aligned}
$$

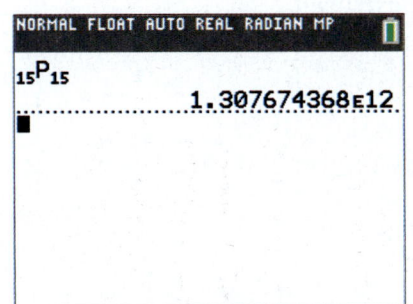

Figure 10.2.8

TI-83/84 Plus

To calculate a permutation of the form ${}_nP_r$, first enter the n value, 15. Then press [math], highlight **PROB** to go to the probability menu, and select nPr. Back on the home screen, enter in the r value, 15. The value is given as 1.3Ø7674368E12. This value is given in scientific notation and is equivalent to 1,307,674,368,000.

Microsoft Excel

We can use the built-in permutation function, **PERMUT(number, number_chosen)**, to evaluate permutations. In an empty cell, type "=PERMUT(15,15)" and press Enter. The value is given as 1,307,674,368,000. (Note that the value is initially given as 1.30767E+12. To see what this value is in expanded form instead of in scientific notation, right-click on the cell and select "Format cells". On the "Number" tab, select "Number" for the category, and click on "Okay".)

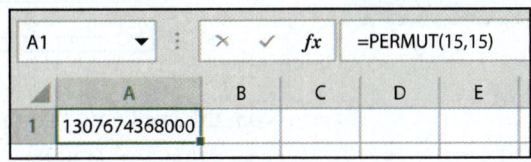

Figure 10.2.9

So there are 1,307,674,368,000 possible choices for the candidate lineup. Notice that the result of ${}_{15}P_{15}$ is the same as 15!

Helpful Hint

In the special permutation where $n = r$, the result will always equal n!

That is, ${}_nP_n = n!$

b. In the second part, we are interested in choosing 5 of the 15 candidates to move on to the next round. The order of the chosen candidates is not important since each one receives the same prize of moving on to the next round, so we are counting the number of combinations of 5 people from a group of 15; that is, $n = 15$ and $r = 5$.

By Hand

$$_{15}C_5 = \frac{15!}{5!(15-5)!}$$

$$= \frac{15!}{5!\,10!}$$

$$= \frac{15 \cdot 14 \cdot 13 \cdot 12 \cdot 11 \cdot 10 \cdot 9 \cdots 2 \cdot 1}{(5 \cdot 4 \cdot 3 \cdot 2 \cdot 1)(10 \cdot 9 \cdots 2 \cdot 1)}$$

$$= \frac{14 \cdot 13 \cdot 12 \cdot 11}{4 \cdot 2 \cdot 1}$$

$$= 3003$$

TI-83/84 Plus

Calculating a combination of the form $_nC_r$ is similar to calculating a permutation. First, enter the n value, 15. Then press [math], highlight **PROB**, and select **nCr**. Back on the home screen, enter in the r value, 5. The calculator returns **3003**.

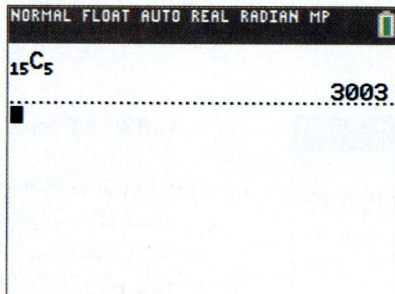

Figure 10.2.10

Microsoft Excel

Just as there is a formula for permutations, there is a built-in formula that is used to evaluate combinations, **COMBIN(number, number_chosen)**. In an empty cell, type "=COMBIN(15,5)" and press Enter. The value is given as 3003.

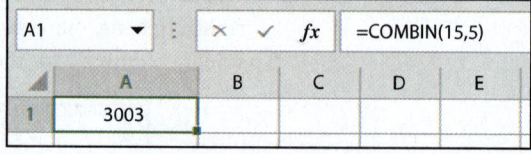

Figure 10.2.11

From this we know that there are 3003 distinct ways that 5 candidates can be chosen to move on to the next round.

✔ **Skill Check 10.2.3**

Use the appropriate formula to calculate each of the following.

a. 10 things permuted 3 at a time

b. 8 choose 2

In the formulas for combinations and permutations, it's important to note that a *distinct* group of objects is required. In other words, none of the objects in each group were the same—no two runners, no two students, and no two candidates. However, suppose the objects we have to choose from contain some identical members (that is, objects are repeated within the group) and we'd like to count how many distinct ways we can arrange, or permute, the objects. For instance, how many ways are there to arrange the letters in the word Mississippi? In order to count these arrangements, we must use a slightly different permutation formula that takes into consideration the repeated objects.

---

f(x) PERMUTATIONS WITH REPEATED OBJECTS

The number of distinguishable permutations of n objects, of which k_1 are all alike, k_2 are all alike, and so forth, is given by the following formula.

$$\frac{n!}{(k_1!)(k_2!)(k_3!)\cdots(k_p!)}$$

Here, $k_1 + k_2 + \cdots + k_p = n$.

To account for repetitions of objects when counting distinct permutations, we divide by the factorial representing the number of times each object is duplicated. Example 10.2.6 demonstrates this.

Example 10.2.6

Calculating Numbers of Permutations with Repeated Objects

Consider a football team with an 11-game schedule. Each game ends with a win (W), a loss (L), or a tie (T).

a. How many ways can the 11-game schedule be arranged if the last game is always with the local rival team?

b. Once the schedule is set, how many ways can the schedule end with 6 wins, 4 losses, and 1 tie?

Solution

a. First, notice that the arrangement of the 11 games is important for counting purposes, so this is a permutation. Since the last game is always with the rival team, we really only need to find how many ways the first 10 games can be arranged, so n is 10, and we want to arrange all of them, so r is also 10. Therefore, the number of permutations of 10 things permuted 10 at a time is calculated as follows.

By Hand

$$_{10}P_{10} = \frac{10!}{(10-10)!}$$

$$= \frac{10!}{0!}$$

$$= \frac{10 \cdot 9 \cdot 8 \cdot \ldots \cdot 2 \cdot 1}{1}$$

$$= 3,628,800$$

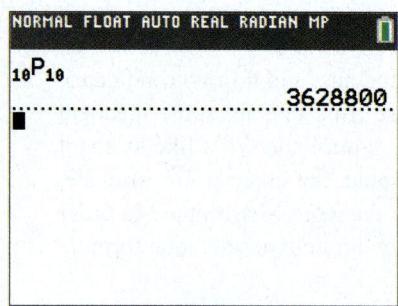

Figure 10.2.12

TI-83/84 Plus

The computation will have the form $_nP_r$, with $n = 10$ and $r = 10$. The value given is 3628800.

Therefore, there are 3,628,800 ways to arrange the 11-game schedule such that the last game is always against the rival team.

b. In the second part, we can think of this as arranging wins (W), losses (L), and ties (T). Although the team might have a different opinion, we can make no distinction between the 6 wins or the 4 losses, so we will consider these to be repeated objects. Since we want to arrange 6 wins, 4 losses, and 1 tie, $n = 11$, $k_1 = 6$, $k_2 = 4$, and $k_3 = 1$. Substituting these values into the formula for permutations with repeated objects, $\dfrac{n!}{(k_1!)(k_2!)(k_3!)\cdots(k_p!)}$, gives the answer.

By Hand

$$\frac{n!}{(k_1!)(k_2!)(k_3!)\cdots(k_p!)} = \frac{11!}{6!4!1!}$$

$$= \frac{11\cdot10\cdot9\cdot8\cdot7\cdot\overline{6\cdot5\cdot4\cdot3\cdot2\cdot1}}{(6\cdot5\cdot4\cdot3\cdot2\cdot1)(4\cdot3\cdot2\cdot1)(1)}$$

$$= \frac{11\cdot10\cdot9\cdot7}{3\cdot1\cdot1}$$

$$= 2310$$

TI-83/84 Plus

Notice that we do not need to simplify first when using the calculator. We simply enter $11!/(6!\ 4!\ 1!)$. Be sure to include parentheses around the denominator. The result is 2310.

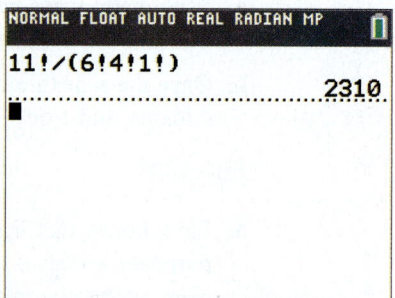

Figure 10.2.13

Thus, there are 2310 ways that the wins, losses, and ties can be arranged.

✔ **Skill Check 10.2.4**

How many different ways can you arrange the letters in the word Mississippi?

Skill Check Answers

1. 729 **2.** Permutation **3. a.** 720 **b.** 28 **4.** 34,650

10.2 Exercises

1. When there are several outcomes at each stage of an experiment, as we saw with the start watch example in the first section, you can _____ together the number of possible outcomes at each stage of the experiment in order to obtain the total number of outcomes for the experiment.

2. An experiment where repetition of outcomes is permitted is considered to be an experiment _____.

3. The factorial of a number is the _____ of all the positive integers less than or equal to that number.

4. When a selection of objects from a group is being made and order is not important, this is a _____. If order is important, it is a _____.

5. True or False: 0! = 1.

💡 **PRACTICE**

Evaluate each factorial expression.

6. $5!$

7. $9!$

8. $\dfrac{8!}{6!}$

9. $1!$

10. $0!$

11. $\dfrac{10!}{2!\,4!}$

12. $\dfrac{6!}{2!\,3!}$

13. $\dfrac{7!}{4(9-6)!}$

14. $\dfrac{12!}{10!(12-10)!}$

15. $\dfrac{17!}{12!(15-10)!}$

Evaluate each permutation or combination.

16. $_8P_2$

17. $_7P_4$

18. $_5P_1$

19. $_4P_4$

20. $_3C_2$

21. $_{30}C_1$

22. $\dfrac{_3C_2}{_3P_2}$

23. $\dfrac{_5P_3}{_5C_3}$

24. $\dfrac{_5C_3}{_5C_2}$

25. $\dfrac{_5P_3}{_5P_2}$

26. $\dfrac{_6P_4}{_6C_2}$

27. $\dfrac{_7C_3}{_7P_3}$

28. $_7C_4 + {_7C_3} + {_7C_2} + {_7C_1}$

29. $_7P_4 + {_7P_3} + {_7P_2} + {_7P_1}$

30. How many three-digit area codes can be made from the digits 0 through 9? Assume that the digits may repeat and that area codes beginning with 0 are allowed.

31. How many three-digit area codes can be made from the digits 0 through 9 if the first digit is not allowed to be a 0 and the digits are allowed to repeat?

32. How many six-character password codes are possible if you are allowed both lowercase letters and the digits 0 through 9, and repeating characters are allowed?

33. When ordering a new e-reader, you have several choices to make. You can choose from five price ranges, decide between Wi-Fi and 3G, choose to have a one-year or three-year warranty, and pick a black, white, or silver casing. How many possible e-readers are there for you to choose from?

34. The Youngs are planning their next family night. They always have dinner out somewhere and then do something fun together. There are two adults and four boys in the family. Each family member is allowed two meal suggestions, and each boy is allowed three activity suggestions. Assuming no family members choose the same thing, how many different family night possibilities are there?

35. If there are seven children lining up for recess, how many different ways can they line up?

36. The college's soccer team will play 11 games next fall. Each game can result in one of three outcomes: a win, a loss, or a tie. Find the total possible number of outcomes for the season record.

37. Harper is deciding on her schedule for next semester. She must take each of the following classes: English 102, College Algebra, History 102, and Biology 101. If there are 16 sections of English 102, 14 sections of College Algebra, 7 sections of History 102, and 12 sections of Biology 101, how many different possible schedules are there for Harper to choose from? Assume there are no time conflicts between the different classes.

38. If there were no restrictions on the use of the number zero, in theory, how many seven-digit telephone numbers are possible?

39. How many four-digit even whole numbers exist?

40. In the new ice creamery, several choices need to be made before tasting "a little bit of heaven," as the advertisement suggests. First, there are three cup sizes to choose from, then 39 different ice cream flavors to decide from, and finally several mix-ins to choose from: 10 candy bars, 8 fruits, 8 nuts, and 10 cookies or cakes. If you want a medium cup with one flavor of ice cream and two mix-ins, one candy bar, and one nut, how many possible choices are there for you to have your "taste of heaven"?

41. As part of a fundraising campaign, a city's parks and recreation department is selling bicycle license plates featuring the blueprint of a future park. Each license plate has unique and consists of 2 capital letters followed by 2 numbers (from the digits 0 through 9).

    a. Determine the number of possible unique license plates if letters and numbers are allowed to be duplicated.

    b. In an effort to decrease the number of license plates available, a decision was made to not allow any duplicates of letters or numbers on the license plates. Determine how many unique license plates are possible under this new rule.

42. A 5th grade class is choosing books to read for upcoming book reports. Each student has to complete two book reports. They must select two books from a selection of 20 books, then they must pick two book report formats from the following options: written report, poster, or shoebox diorama.

    a. If the teacher does not specify that the books and report types must be unique, how many possible combinations are there for a student to choose from?

    b. If the teacher specifies that the books and the report types must be unique, how many possible combinations are there for a student to choose from?

43. A bakery has 10 different cupcake flavors and sells boxes that hold 4 cupcakes. How many different ways are there to fill a box with 4 cupcakes?

44. Suppose a bakery has 10 different cupcake flavors and sells variety boxes of 4 cupcakes. The variety boxes have the restriction that the same flavor cannot be added to the box back-to-back. (For example, if a chocolate cupcake with vanilla frosting is the first cupcake placed into the box, then the next cupcake placed into the box cannot be a chocolate cupcake with vanilla frosting.) How many different variety boxes of cupcakes can be created?

45. To determine the theme for a short story, an author places 15 different plot ideas into a box and draws out 3, placing each idea in order on a table. The author then writes a story that progresses through those 3 ideas. Determine how many possible short story topics can be created using this method.

46. A personal trainer creates a challenge for his clients where each client must choose one of five exercises to perform each day for three weeks (21 days). How many different ways can a client complete the month-long challenge?

Determine whether to use a permutation or combination to answer each question, and then determine the total number of outcomes.

47. There are 10 board members on the Community Arts Council. In how many ways can a president and treasurer be chosen? Assume that no member can hold both positions at the same time.

48. In how many ways can a committee of five people be chosen from a pool of 120 employees?

49. If there are 84 runners in a race, in how many ways can 1st, 2nd, and 3rd place ribbons be given out?

50. Elliot has to submit three photographs for the school art show. This semester he has taken 29 photographs that he thinks are show-worthy. In how many ways can he choose the photographs to submit?

51. Avery was born on 10/15/1995. How many eight-digit codes could she make using the digits in her birthday?

52. There are 18 tenured faculty in the biology department on campus. The department needs one tenured faculty member to facilitate undergraduate research, one member to supervise graduate advising, and one to coordinate grant proposals. In how many ways can these tasks be assigned, if a member may be appointed to only one duty?

53. A service organization on campus needs a group of six students from their organization's membership of 123 students to serve as program attendants at graduation. In how many ways can the attendants be chosen?

54. In how many ways can the letters in the word STATISTICS be arranged?

55. In how many ways can the letters in the word TENNESSEE be arranged?

56. A bookstore has a buy three get one free sale on used books. If the current selection of used books contains 18 unique books, how many different groups of four books can be chosen to qualify for the sale?

57. If a talent show has 12 contestants, in how many ways can the prizes for first, second, and third place be awarded?

58. A company is considering using a new productivity management system and wants to test out the latest system on a group of four employees. If the company has 50 employees, how many ways can the test group be selected?

59. A designer needs to select four colors to use to create a poster for an upcoming event. After some brainstorming, she comes up with eight different colors to use, all of which go together. How many different ways can she choose four colors from the eight possible colors?

60. An ice cream store can create 25 different flavors onsite but their display freezer limits the store to only selling 10 flavors per day. How many different flavor combinations can be created per day?

61. A movie theater's newest promotion is an early-bird classics week where a different classic movie is played each morning for a week (seven days). If the theater has the rights to show twenty different classic movies, how many different ways can seven movies be shown during the week?

62. The Renfield family is staying at a resort for their annual vacation. As part of the resort package, the family can choose three of twenty different excursions to go on from options such as zipline tours, guided hikes, and sundown bat watching. How many ways can the family choose which excursions to go on?

63. An instructor is creating a 10-question quiz where 6 questions are true and 4 are false. How many different ways can the questions be arranged in the quiz?

64. Naval signs are made by arranging colored flags in a vertical line. Using 6 flags, how many distinct symbols can be made if you have 2 red flags, 3 green flags, and 1 blue flag?

65. Allison purchases a variety pack of mochi ice cream that contains the following flavors: 2 salted caramel, 2 mango, 4 vanilla, and 4 chocolate. If she plans to eat one mochi per day, how many flavor arrangements are possible?

✏ WRITING & THINKING

66. Without calculating the permutations, decide which of the following words would produce the greatest number of four-letter arrangements.

   a. PASS

   b. TEST

   c. FAIR

   d. FREE

67. Determine how many ways a group of three students from your class could be selected.

10.2 PROJECT

PASSWORDS AND SECURITY: DO HACKERS TEST ALL POSSIBILITIES?

Mark Zuckerberg, the founder and CEO of Facebook, is known to have kept "dadada" as his Twitter and Instagram passwords. Simple passwords of this type are a treat for hackers trying to gain access to an account. These passwords can be guessed in a short amount of time by what is called a dictionary attack. In this case, an intruder tests common words and their combinations until the correct password is found.

What about more secure passwords? Is it possible to simply test all possibilities for a password and gain access to someone's account? In this activity, you will estimate the time it would take to correctly guess a social media account password.

Assume that a social media platform requires your password to have exactly 8 characters selected from any of the 26 letters of the alphabet, the numbers 0 through 9, and any of the characters #, $, %, or &. In order to simplify our computations, let's assume repetitions are allowed.

1. Determine the number of distinct passwords that can be created using the requirements described above. The number is very large, so you should use a calculator or computer for your computation.

Now, let's assume that it takes 0.02 seconds to check a single possible password.

2. How many seconds would it take to check every single possible password? This approach is called a "brute force" attack.

3. How many seconds are in a year?

4. How many years would it take to check every single possible password? Round your answer to the nearest whole year.

5. Explain why it is really a worst-case scenario for the hacker for the brute force attack to take this long.

6. Do you believe hackers use the brute force attack often? Explain your reasoning.

10.3 PROBABILITY OF SINGLE EVENTS

OBJECTIVES

1. Calculate classical probability using combinations and permutations.

2. Calculate classical probability using complements.

3. Calculate odds.

OBJECTIVE 1 Probability with Combinations and Permutations

Recall that classical probability is more precise than experimental probability, and hence preferable to use whenever it can be obtained. Now that we've looked at several methods of counting the outcomes in a sample space, we can use those methods to aid in calculating classical probabilities.

Example 10.3.1

Calculating Classical Probability Using Combinations

Suppose that as one of the 20 graduate students in the physics department, you have a chance of being selected for one of the three student spots for a conference trip to Cancun. If the names of all the graduate students are put in a hat and three are drawn, what is the probability that you and your two friends, Leonard and Sheldon, end up being chosen?

Solution

The first thing we need to do is count the number of ways that the three student spots on the trip can be filled. Because the order in which the students are chosen is not specified to be important, we can count the outcomes using the combination formula. We are choosing 3 students from the 20 eligible graduate students. Thus, $n = 20$ and $r = 3$. The number of combinations of 3 things from a group of 20 is calculated as follows.

$$_{20}C_3 = \frac{20!}{3!(20-3)!} = 1140$$

This means that there are 1140 possible ways to choose 3 students from 20 for the trip. Note that there is only one way in which to choose you and your two friends as the students who go on the trip. The probability of this event happening is calculated as follows.

$$P(\text{You, Leonard, and Sheldon in Cancun}) = \frac{1}{1140} \approx 0.000877$$

In other words, it is very unlikely that the three of you would randomly be chosen to go on the conference trip to Cancun together.

Example 10.3.2

Calculating Classical Probability Using Permutations

Let's change the previous example slightly. Suppose that as one of the 20 graduate students in the physics department, you have a chance of being selected for one of the three student spots for a conference in Cancun. The names of all the graduate students are put in a hat and three are randomly drawn. However, if your name is drawn first, you get all expenses paid. If you are chosen second, everything is paid for except meals. And if you're chosen third, you must pay for your own meals and hotel. (This means that the department still picks up the tab for the flight and conference fees of the three lucky students, so it's not a bad deal!) What is the probability that you and your two friends, Leonard and Sheldon, all end up being chosen, and that your name is drawn first?

Solution

This time, the order in which the three students are chosen does make a difference when we are counting, so we'll use a permutation. Note that n is still 20 and r is still 3.

$$_{20}P_3 = \frac{20!}{(20-3)!} = 6840$$

Thus, there are 6840 possible ways to choose the three lucky students for the trip. However, let's consider how many outcomes are in the event that you, Sheldon, and Leonard are chosen, and that your name is drawn first; we'll call this event E.

There are only two ways that your name can be drawn first followed by Leonard and Sheldon. The possibilities are {you, Leonard, Sheldon} or {you, Sheldon, Leonard}.

Therefore, the probability that the event described occurs is calculated as follows.

$$P(E) = \frac{2}{6840} \approx 0.000292$$

It seems that there is an even smaller chance of this happening, so it's better not to be greedy and wish for the top spot!

✔ Skill Check 10.3.1

Using the scenario from Examples 10.3.1 and 10.3.2, find the probability that you, Sheldon, and Leonard are all chosen for the department trip and you are the third to be chosen.

Example 10.3.3

Calculating Classical Probability

A university is putting together a hiring committee of 6 members for the next university provost. The eligible pool of members consists of 8 administrative personnel, 9 faculty, and 7 students.

a. In how many ways can the committee be formed?

b. In how many ways can the committee be formed if there must be 2 members chosen from each of the 3 subgroups?

c. Suppose the committee is chosen at random and with no restrictions. What is the probability that each subgroup has 2 members on the committee?

Solution

a. Since there are no stipulations as to how the committee is to be formed, nor is the order specified to be important, we simply need to calculate the number of combinations by choosing 6 committee members from the $8 + 9 + 7 = 24$ people in the pool.

$$_{24}C_6 = \frac{24!}{6!(24-6)!}$$

$$= 134,596$$

Thus, there are 134,596 ways in which a committee of 6 can be formed.

b. This time we have the stipulation that there must be 2 members from each subgroup on the committee of 6 members. This will require a two-stage solution. First, we need to use the combination formula to count the number of ways we can choose 2 people from each subgroup. Then, we can use the Fundamental Counting Principle and multiply those numbers of combinations together to find out the total number of ways to form the committee.

Begin by finding the number of combinations specific to each subgroup.

Number of Administrative Personnel Combinations:

$$_8C_2 = \frac{8!}{2!(8-2)!} = \frac{8!}{2!6!} = 28$$

Number of Faculty Combinations:

$$_9C_2 = \frac{9!}{2!(9-2)!} = \frac{9!}{2!7!} = 36$$

Number of Student Combinations:

$$_7C_2 = \frac{7!}{2!(7-2)!} = \frac{7!}{2!5!} = 21$$

Now we multiply these numbers of combinations together.

$$28 \cdot 36 \cdot 21 = 21{,}168$$

Thus, there are 21,168 ways to choose the committee if there must be 2 members chosen from each of the 3 subgroups.

c. The final part of this question asks us to find the probability that a committee chosen at random would meet the requirements given in part b. That is, each subgroup would have 2 members on the committee. We can combine our answers from parts a. and b. to find the answer to this probability question.

$$P\left(\begin{array}{l}\text{Committee Has 2 Members} \\ \text{from Each Subgroup}\end{array}\right) = \frac{\begin{array}{c}\text{Number of Possible Committees} \\ \text{with 2 Members from Each Subgroup}\end{array}}{\text{Number of Possible Committees of 6 Members}}$$

$$= \frac{21{,}168}{134{,}596}$$

$$\approx 0.1573$$

In other words, if the members of the committee are randomly selected, there is a 15.73% chance that the committee will be made up of 2 administrative personnel, 2 faculty, and 2 students.

Example 10.3.4

Calculating Classical Probability

An employee is setting up a new password for his in-house messaging account. The password must contain at least 4 but no more than 6 characters using only lowercase letters and the digits 0 through 9, and each character can only be used once. Suppose he lets the computer randomly set his password using all of the restrictions listed. Use the following steps to calculate the probability that the password generated by the computer would contain all the letters in his dog's name—Bailey.

Step 1: Begin by determining how many different passwords are possible given the restrictions.

Step 2: Determine the number of possible passwords that contain all the letters in the dog's name.

Step 3: Calculate the probability using the values from Steps 1 and 2.

Solution

Step 1:

First notice that for a password, the order of characters is important, so we are counting permutations. The key words *at least* and *no more than* tell us that the password can be 4, 5, or 6 characters in length. So we will need to count the number of permutations for each length of password and then add them together.

There are 26 letters and 10 digits, which means that there are 36 possible characters to choose from no matter the length of the password, so $n = 36$ in all cases.

Begin by calculating each of the numbers of permutations for the different lengths of password.

Number of Permutations of 4 Characters:

$$_{36}P_4 = \frac{36!}{(36-4)!} = \frac{36!}{32!} = 1,413,720$$

Number of Permutations of 5 Characters:

$$_{36}P_5 = \frac{36!}{(36-5)!} = \frac{36!}{31!} = 45,239,040$$

Number of Permutations of 6 Characters:

$$_{36}P_6 = \frac{36!}{(36-6)!} = \frac{36!}{30!} = 1,402,410,240$$

Now we add all of the numbers of permutations together.

$$1,413,720 + 45,239,040 + 1,402,410,240 = 1,449,063,000$$

Therefore, there are a total of 1,449,063,000 different possibilities for the password if each character can only be used once. Imagine how many possibilities there would be if the password could use up to 9 or 10 characters!

Step 2:

There are 6 letters in the dog's name, Bailey. Finding the number of passwords that can be made using all of these letters requires we need to count the number of permutations of 6 things from a set of 6.

$$_6P_6 = \frac{6!}{(6-6)!} = 6! = 720$$

So there are 720 possible passwords using the 6 letters in Bailey's name.

💬 **Think Back**

Remember that in permutations, if $n = r$, then you can simply calculate $n!$.

Step 3:

To find the probability that a randomly generated password would contain the 6 letters from Bailey's name, we need the total number of possible passwords made from the letters in Bailey's name, which we calculated in Step 2, divided by the total number of possible passwords, which we calculated in Step 1.

Thus, the probability is calculated as follows.

$$P\left(\begin{array}{l}\text{Password Containing the Letters}\\ \text{b, a, i, l, e, y}\end{array}\right) = \frac{\begin{array}{c}\text{Number of Possible Passwords}\\ \text{Using b, a, i, l, e, y}\end{array}}{\text{Number of Possible Passwords}}$$

$$= \frac{720}{1,449,063,000}$$

$$\approx 0.0000004969$$

Therefore, the probability that the computer randomly sets the password to contain the letters in the dog's name is approximately 0.00005%, an extremely unlikely possibility.

OBJECTIVE 2 Probability with Complements

Sometimes, when we calculate probabilities, it is easier to count the outcomes that are not in an event rather than the ones that are. This might sound strange, but consider the following situation. You are the keynote speaker with a room full of 500 participants. From the podium you ask, "Who would like a free gift? Raise your hand." Instead of counting all those hands, it might be easier to ask, "Who doesn't want the free gift?" There will likely be many fewer hands to count this way. Since you know there are 500 people in attendance, you can just subtract those that don't want the free gift from 500 to find out how many people do want the gift. This is the idea behind *complements*.

📖 COMPLEMENT

The **complement** of event E, denoted by E^c, consists of all outcomes in the sample space that are not in event E.

By definition, if you combine the outcomes from E and the outcomes from E^c, you will have the entire sample space. That is, $E + E^c = S$, as illustrated in Figure 10.3.1.

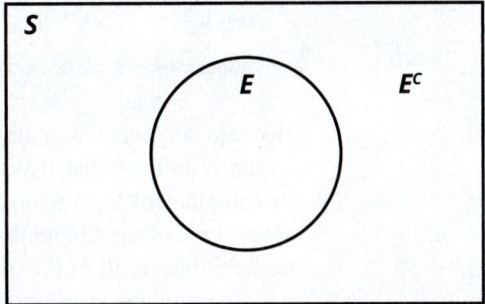

Figure 10.3.1: Venn Diagram of a Sample Space

In Figure 10.3.1, the entire rectangle represents the outcomes that make up the entire sample space. The region inside the circle represents the outcomes that are part of event E. The region outside the circle represents the outcomes that are in E^c, the complement of E.

As an example, consider the experiment of rolling a fair six-sided die. If E is the event of rolling an even number, $E = \{2, 4, 6\}$, then its complement is rolling an odd number, $E^c = \{1, 3, 5\}$. Notice that $E + E^c = S$; that is,

$$\{2, 4, 6\} + \{1, 3, 5\} = \{1, 2, 3, 4, 5, 6\}.$$

An illustration of this die example is shown in Figure 10.3.2.

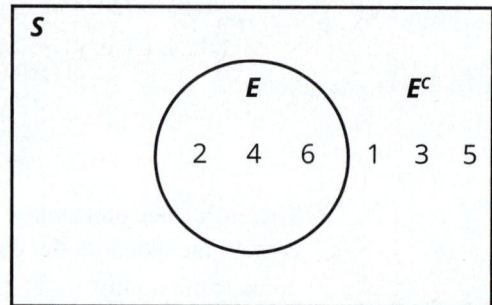

Figure 10.3.2: Venn Diagram of the Complement of an Event

Example 10.3.5

Finding the Complement of an Event

Describe the complement for each of the following events.

a. Choosing a vowel (a, e, i, o, u) from the English alphabet.

b. Choosing a number that doesn't end in 1 from all positive two-digit whole numbers.

c. From a class of 52 students, choosing a student who is over 21 years old.

Solution

a. The English alphabet contains 26 letters, 5 of which are vowels (a, e, i, o, u). Since the 5 vowels are contained in the event, the set of remaining 21 letters (consonants) are the complement.

b. The set of all positive two-digit whole numbers includes the numbers 10 through 99. The complement of the event contains all two-digit numbers that do end in 1 (that is, 11, 21, 31, 41, 51, 61, 71, 81, and 91).

c. The complement consists of the students in the class who are 21 years old or younger.

✔ **Skill Check 10.3.2**

Let the event E be an odd sum when a pair of dice is rolled. For each of the following, determine if the outcome could be in the complement of E.

a. A sum greater than 8

b. A sum that is an even number

c. A sum less than 5

d. A sum that is a multiple of 3

e. All of the above

Because an event and its complement comprise the entire sample space, it naturally follows that if you add the probability of event E to the probability of its complement E^c, you have accounted for all possible probabilities in the sample space. One of the fundamental properties of probability is that the sum of all the probabilities of all of the outcomes in the sample space must equal 1. Therefore, we can write this property mathematically in the following ways.

💬 **Helpful Hint**

$E + E^c = S$

That is, the outcomes in the event E plus the outcomes in the complement E^c equal the entire possible set of outcomes, which is the sample space S.

f(x) COMPLEMENT RULES OF PROBABILITY

1. $P(E) + P(E^c) = 1$
2. $P(E) = 1 - P(E^c)$
3. $P(E^c) = 1 - P(E)$

Recall that it is sometimes it is easier to find the probability of an event by calculating the probability of its complement rather than the probability of the event itself. This is demonstrated in the following examples.

Example 10.3.6

Calculating Probability Using Complements

💬 **Helpful Hint**

Some of the more common complement phrases and their opposites are as follows.

	Not all are
All are	Some are not
	At least one is not
Some are	None are
Some are not	All are
None are	At lease one is

a. Suppose you order a new phone charger online. If there is a 0.5% probability of receiving a defective charger, what is the probability that you will receive a charger that is not defective?

b. If there is a 3% chance that none of the books you would like to check out of the library are available, what is the probability that at least one of the books you want is available?

Solution

a. The complement to the outcome of receiving a charger that is not defective is receiving a defective one. Thus, the probability of receiving a charger that is not defective can be calculated as follows.

$$P(\text{Charger That Is Not Defective}) = 1 - P(\text{Defective Charger})$$
$$= 1 - 0.005$$
$$= 0.995$$
$$= 99.5\%$$

So the good news is that there is a 99.5% chance of getting a charger that works.

b. The complement to having *none* of something is having *at least one* of that thing. Thus, the probability can be calculated as follows.

$$P(\text{At Least One Book Available}) = 1 - P(\text{No Books Available})$$
$$= 1 - 0.03$$
$$= 0.97$$
$$= 97\%$$

Thus, there is a 97% chance that at least one of the books you want is available.

Example 10.3.7

Calculating Probability Using Complements

Using the data in Table 10.3.1, find the following probabilities involving nuts imported into the United States from 2006 to 2011.

Table 10.3.1: US Import Sources by Weight (in Pounds) for Fresh or Dried Walnuts and Pistachios, 2006–2011

Walnuts	India	4,373,000
	Mexico	1,268,000
	Spain	5,239,000
	China	1,533,000
	Austria	1,938,000
	Other countries	4,157,000
Pistachios	Iran	2,012,000
	Turkey	2,030,000
	Hong Kong	262,000
	Switzerland	64,000
	Italy	115,000
	Other countries	323,000

Source: USDA. "Fruit and Tree Nut Data." http://www.ers.usda.gov/data-products/fruit-and-tree-nut-data/data-by-commodity.aspx

a. Calculate the probability that imported walnuts purchased in the United States during this period were from Austria.

b. Calculate the probability that imported walnuts purchased in the United States during this period were from somewhere other than Austria.

c. Calculate the probability that imported pistachios purchased in the United States during this period were from somewhere other than Italy or Switzerland.

Solution

a. The probability that imported walnuts purchased in the United States during this period were from Austria is found by dividing the weight of walnuts imported from Austria by the total weight of walnuts imported during this time period. The weight of walnuts imported from Austria during this time is given in the table as 1,938,000 pounds. To find the total weight of walnuts imported, we need to add together all of the weights of walnuts imported.

$$\text{Total Walnut Weight} = 4{,}373{,}000 \text{ lb} + 1{,}268{,}000 \text{ lb} + 5{,}239{,}000 \text{ lb}$$
$$+ 1{,}533{,}000 \text{ lb} + 1{,}938{,}000 \text{ lb} + 4{,}157{,}000 \text{ lb}$$
$$= 18{,}508{,}000 \text{ lb}$$

The probability that walnuts purchased during this time period were from Austria is found as follows.

$$P(\text{Walnuts from Austria}) = \frac{1{,}938{,}000 \text{ lb}}{18{,}508{,}000 \text{ lb}} \approx 0.1047$$

Therefore, the probability that imported walnuts purchased during this time period were from Austria is approximately 10.47%.

b. We could find the probability that imported walnuts purchased in the United States came from somewhere other than Austria by combining all the remaining places together. However, given that we just calculated the probability that the walnuts were from Austria, it is easier to use the complement.

$P(\text{Not from Austria}) = 1 - P(\text{Walnuts from Austria}) \approx 1 - 0.1047 = 0.8953$

Therefore, the probability that imported walnuts purchased during this time were from somewhere other than Austria is approximately 89.53%.

   c. Again, it will be easier to use the complement here. First, we'll need to know the total weight of imported pistachios and the weight of the pistachios imported from Italy and Switzerland.

$$\text{Total Pistachio Weight} = 2,012,000\,\text{lb} + 2,030,000\,\text{lb} + 262,000\,\text{lb}$$
$$+ 64,000\,\text{lb} + 115,000\,\text{lb} + 323,000\,\text{lb}$$
$$= 4,806,000\,\text{lb}$$

Now, calculate the probability that the imported pistachios purchased in the United States came from either Italy or Switzerland.

$$P\left(\text{From Italy or Switzerland}\right) = \frac{\text{Italy} + \text{Switzerland}}{\text{Total Pistachio Imports}} = \frac{115,000 + 64,000}{4,806,000}$$
$$\approx 0.0372$$

To calculate the probability that the pistachios came from somewhere other than Switzerland or Italy, we'll find the probability of the complement.

$$P\left(\text{Not from Italy or Switzerland}\right) = 1 - P\left(\text{From Italy or Switzerland}\right)$$
$$\approx 1 - 0.0372$$
$$= 0.9628$$

Therefore, the probability that the pistachios came from somewhere other than Switzerland or Italy is approximately 96.28%.

OBJECTIVE 3 Odds

The concept of *odds* as you might see at a casino, racetrack, or any gambling establishment is often misinterpreted. Although it is a common mistake, the words "odds" and "probability" are not interchangeable. First of all, there are two ways that odds are given: *odds against* and *odds in favor*. Both are ways of conveying the same information, but odds against are the most commonly used. Either way, odds are a way for a bookie to indicate how likely they think something is to happen and how that affects how much they are willing to pay out. Let's look what odds tell us about probability.

Consider the betting phrase "odds of 3 : 1 against." Recall that this is a ratio and we read 3 : 1 as "three to one." So if it is a ratio, what is it a ratio of? When you are given odds against (when it comes to betting) you are being told the ratio of the probability of losing money to the probability of winning money. With that in mind, we can now use the ratio to calculate each of these probabilities.

One of the foundational features of ratios is that the sum of all of the components of a ratio determines the total number of equal parts in the whole. So the ratio 3 : 1 means that there are 4 (that is, 3 + 1) equal parts to consider: 3 against and 1 for. If we consider the probability implications, this means that by offering odds of 3 : 1, the bookie is saying that they think the probability of your losing money (you paying the bookie) is $\frac{3}{4}$ and the probability of your winning money (the bookie paying you) is $\frac{1}{4}$.

Practically speaking, odds against also indicate how much money you can win or lose. Having the odds 3 : 1 against in the context of horse races means that for every $1.00 you place on the bet, you win $3.00 if your chosen horse wins; but you lose $1.00 if your chosen horse does not win. Since it is the most convenient way to understand the payout if the bet is successful for the player, the "odds against" are most commonly given instead of "odds in favor."

📖 **ODDS**

The **odds against** an event E are given as follows.

$$\frac{P(\text{Losing})}{P(\text{Winning})} \text{ or } a : b$$

The **odds in favor** of an event E are given as follows.

$$\frac{P(\text{Winning})}{P(\text{Losing})} \text{ or } b : a$$

Here, a is the number of unfavorable outcomes and b is the number of favorable outcomes.

Example 10.3.8

Determining Probability From Odds

The Women's National Basketball Association (WNBA) was comprised of 12 teams in 2020. During the playoffs in late September, a week before the finals were set to begin, an online betting site gave the following odds to use when betting on the winner of the 2020 WNBA Finals. Use Table 10.3.2 to answer the following questions. Note that the odds in Table 10.3.2 are the odds against each team winning the finals.

Table 10.3.2: Odds to Win 2020 WNBA Finals (9/30/2020)

Team	Odds
Seattle Storm	1 : 2
Las Vegas Aces	9 : 2
Connecticut Sun	11 : 2
Minnesota Lynx	18 : 1

Source: "WNBA Futures Odds: 2020 WNBA Finals Odds," VegasInsider.com, accessed September 30, 2020, https://www.vegasinsider.com/wnba/odds/futures/.

a. Determine the probability of losing money on a bet for each of the teams listed.

b. Based on these odds, which team was most likely to win the 2020 WNBA finals at this point? Explain what the odds mean for this team in terms of a bet.

c. How much money could be won with a $50 bet placed on the Connecticut Suns?

Solution

a. The probability of losing money is the left side of the ratio divided by the sum of the parts; for example, the ratio 1 : 2 has 3 parts and the probability of losing is $\frac{1}{3} \approx 0.3333$.

Table 10.3.3: 2020 WNBA Finals Odds and Probability

Team	Odds	Probability of Losing Money
Seattle Storm	1 : 2	$\dfrac{1}{3} \approx 0.3333$
Las Vegas Aces	9 : 2	$\dfrac{9}{11} \approx 0.8182$
Connecticut Sun	11 : 2	$\dfrac{11}{13} \approx 0.8462$
Minnesota Lynx	18 : 1	$\dfrac{18}{19} \approx 0.9474$

Notice that these probabilities add to more than 1 and we've only included 4 of the 12 teams. Remember that this is not the probability that a team will win the WNBA finals, but an assessment by the bookie of the probability that you will lose a money on a bet for that team.

b. The team that has the smallest probability of losing money for a bet is the team most likely to win. Therefore, the Seattle Storm was the team most likely to win the 2020 WNBA finals. Their odds of 1 : 2 mean that for each $2 bet, you would win $1.

c. Since the odds for the Connecticut Sun are 11 : 2, for every $2 bet, a bet would win $11. Therefore, we need to divide the bet by $2 and then multiply by $11 to determine the amount that would be won.

Winnings for a $50 Bet on the Connecticut Sun: $\dfrac{\$50}{\$2} \cdot \$11 = \$275$

> ✔ **Skill Check 10.3.3**
>
> If the odds on a bet are 4 : 1 against, what is the probability of winning?

Skill Check Answers

1. $P(E) \approx 0.000292$ **2.** e **3.** $\dfrac{1}{5}$ or 20%

10.3 Exercises

✔ CONCEPT CHECK

1. If you combine the outcomes from a set and its _____, you will have the entire sample space.

2. The sum of all the probabilities of all the outcomes in a sample space must equal _____.

3. There are two ways that odds are given: odds _____ and odds _____.

4. When you are given the odds _____ (when it comes to betting) you are being told the ratio of the probability of losing money to the probability of winning money.

5. True or False: If the odds of something happening are 1:2, that means that the probability of it happening are 50%.

6. Find the probability that three people randomly line up to buy tickets in order of their height (tallest, middle, shortest). Assume that no two people in the line are of the exact same height.

7. One option to play the lottery is called "3-way any order." In order to play this method, you select three digits, from 0 to 9, such that precisely two of the digits are the same (for example, 1, 1, 2). You're a winner if your three digits show up in any order in the lottery's three randomly chosen digits. Digits may be repeated when the lottery chooses the winning number. Find the probability of winning with the "3-way any order" method.

8. Another option of playing the lottery is to choose three numbers (allowing repetition) in the exact order they will appear. Find the probability of winning the lottery with one ticket.

9. Ian is playing Scrabble. What is the probability that the next three letters he draws from the bag spell out his name in the order that he draws them? Assume there is one of each letter in the alphabet left in the bag.

10. There are two sets of balls numbered 1 through 5 placed in a bowl. If two balls are randomly chosen without replacement, find the probability that the balls have the same number.

11. William and Gavin are going to play video games after work. Together they have 48 games. If they decide to randomly choose two games to play, what is the probability that the two games they choose consist of William's favorite game and Gavin's favorite game? Assume they have different favorites.

12. A local pizza parlor has the following list of toppings available for selection. The parlor is running a special to encourage patrons to try new combinations of toppings. They list all possible three-topping pizzas (three distinct toppings) on individual cards and give away a free pizza every hour to a lucky winner.

Pizza Toppings

Green Peppers	Onions	Pepperoni	Sausage
Baby Portabello Mushrooms	Black Olives	Ham	Spicy Italian Sausage
Roma Tomatoes	Pineapple	Beef	Grilled Chicken
Jalapeño Peppers	Banana Peppers	Bacon	Extra Cheese

a. How many three-topping pizza cards are there?

b. Find the probability that the first winner randomly selects the card with the pizza containing green peppers, ham, and bacon on it.

13. A combination padlock is a lock in which a sequence of numbers is used as the "key" to open the lock. Suppose a combination padlock has 10 digits to choose from for each of the four sections of the lock.

a. Does the "key" for a combination padlock involve permutations or combinations?

b. How many possible "keys" are there for the combination padlock?

c. What is the probability that you randomly buy one of these locks whose "key" is made of 4 of the same digit?

14. Four students, three girls and a boy, have arranged to meet on the first day of class and sit in the front row. Suppose they agree to sit in the first four seats in the order that they arrive.

    a. How many possible seating arrangements are there for the four friends?

    b. What is the probability that all three girls end up sitting next to one another?

15. A bakery has 20 unique donut flavors but only sells 5 flavors per day. If the flavors are chosen at random, determine the probability that the flavors selected for today are s'mores, crème brûlée, very vanilla, strawberry shortcake, and chocolate dream.

16. An introduction to rock climbing class is divided up into groups of 3 for rappelling practice. Out of the 12 people in the class, 3 are complete beginners. What is the probability that all three beginners are placed in the same group?

17. An instructor is creating the syllabus for the upcoming semester and must choose 5 chapters to cover out of a 10-chapter math textbook. What is the probability that the instructor chooses to cover topics in the following order: set theory, logic, finance, probability, and statistics?

18. Three friends, Allen, Paul, and Trey, are sitting in a room with nine other people waiting for their turn to audition for an upcoming reality show. What is the probability that the next three people called to audition are Allen, Paul, and then Trey?

19. In the game of Texas hold'em poker, a hand is made up of two cards from a deck of 52. What is the probability of being dealt a 2 of clubs and a 7 of hearts?

20. The drama department at a school is selecting plays to perform for the year. They can choose one play for the fall and another play for the spring. If the department holds the rights to perform 22 different plays, what is the probability that they select Romeo and Juliet and Clue?

21. As a sandwich aficionado, you believe that the order in which the ingredients are placed influences the overall flavor of the sandwich. If you order a turkey club sandwich, what is the probability that the ingredients are, from bottom to top, turkey, cheese, bacon, lettuce, tomato, and mayo?

22. The producers for an awards show plan to have a commercial break after every two awards presented. During the first hour, they decide to present the awards for best leading actor, best supporting actor, best leading actress, best supporting actress, best overall cast, and best script. What is the probability that the awards for best overall cast followed by best supporting actress are presented before the first commercial break?

23. A committee of four is being formed randomly from the employees at a school: 5 administrators, 37 teachers, and 4 staff.

    a. How many ways can the committee be formed?

    b. What is the probability that all four members are staff?

    c. What is the probability that no member is an administrator?

24. A hand of poker is made up of five cards from a standard deck of cards.

    a. How many possible hands of poker are there in a standard deck of 52 cards?

    b. A royal flush consists of the cards Ace, King, Queen, Jack, and ten, all in the same suit. What is the probability of being dealt a royal flush?

25. Matthew needs to set the pass code on his smartphone. It must be a four-digit number and repeated digits are allowed.

    a. How many possible pass codes are there for Matthew to choose from?

    b. How many possible pass codes are there for Matthew if he decides to choose four distinct numbers?

    c. A spy sneaks a look at Matthew's phone and sees his fingerprints on the screen over four numbers. What is the probability that the spy is able to unlock the phone on his first try?

    d. The spy knows the fingerprint trick and so on his phone he uses a repeated digit in his code. If you could see the three fingerprints on the spy's phone, what is the probability that you could unlock the phone on your first attempt?

    e. Based on parts **c.** and **d.**, is it better to repeat a digit or have four distinct digits in the code on your phone for security purposes?

26. A hand of blackjack consists of two cards. The dealer deals you a hand from a fresh deck.

    a. What is the probability that the two cards have the same face value, for instance, both cards are Kings or both cards are 5s?

    b. If aces count 1 or 11, picture cards count 10, and card numbers 2 through 10 are equal to their face value, what is the probability that the two cards sum to 21?

27. For a pick-up game of basketball, jerseys are in a box and people start grabbing them. The box contains three extra-large, seven large, and four medium jerseys. If you are first to the box and grab two jerseys, what is the probability that you randomly grab two extra-large jerseys?

28. A junk drawer at home contains a half-dozen pens, two of which work. What is the probability that you randomly grab two pens from the drawer and don't end up with a pen that works?

29. A city activity committee is planning to show one movie per week at a local park for five weeks during the summer. The committee decides that this year's theme is comedy movies and can choose from five PG rated comedies and six G rated comedies.

    a. In how many ways can the movies be selected?

    b. In how many ways can the movies be selected if three must be G rated and the other two must be PG rated?

    c. Suppose that the movies are selected at random with no restrictions. What is the probability that three G rated movies are selected and two PG movies are selected?

30. A book club reads one book per month. The book selection committee of the club is assigning the books they will read over the next six months. The overall book selection includes six fiction books and seven nonfiction books.

    a. In how many ways can the books be selected?

    b. In how many ways can the books be selected if three books must be fiction and three books must be nonfiction.

    c. Suppose the books are selected at random with no restrictions. What is the probability that the selection contains three fiction and three nonfiction books?

31. A bag contains each letter of the alphabet. Find the probability that a randomly selected letter from the bag will not be one of the five vowels.

32. Find the probability of randomly choosing a letter other than the letter O from a bag that contains the eighteen letters of the Italian city GUIDONIA MONTECELIO.

33. Using the table containing the breakdown of all employees on nonfarm payrolls in the United States during March 2014, find the probability that a randomly selected US worker was not in either retail trade or wholesale trade.

Employees on Nonfarm Payrolls (in Thousands), March 2014

	Area of Employment	Number of Employees (in Thousands)
Private Sector	Goods-Producing	18,558.2
	Wholesale Trade	5803.7
	Retail Trade	15,004.0
	Transportation and Warehousing	4524.8
	Utilities	550.3
	Information	2653.0
	Financial Activities	7870.0
	Professional and Business Services	18,832.0
	Education and Health Services	21,481.0
	Leisure and Hospitality	14,143.0
	Other Private Service-Providing Services	5464.0
Public Sector	Federal Government	2705.0
	State Government	5217.0
	Local Government	14,341.0
	Total Nonfarm Employees	137,147.0

Source: Bureau of Labor Statistics. "Table B-1. Employees on nonfarm payrolls by industry sector and selected industry detail." Accessed June 2014. http://www.bls.gov/news.release/empsit.t17.htm

34. In June 2011, the week of the final mission of the US space shuttle program, a Pew Research poll asked 1502 US adults whether the United States must continue to be a world leader in space exploration. The following table gives a breakdown of their opinions.

The United States Continuing to be a World Leader in Space Exploration is. . .

Essential	Not Essential	Don't Know
871	571	60

Source: Pew Research Center. "Majority Sees U.S. Leadership in Space Essential." July 5, 2011. http://www.people-press.org/2011/07/05/majority-sees-u-s-leadership-in-space-as-essential/

   a. Find the probability that someone responded "essential."

   b. Find the probability that someone did not respond "essential."

35. Find the probability of rolling two dice and not getting the same number on both dice.

36. Suppose a family has five pets. Find the probability that at least one of the pets is male.

37. The weather report says there is a 25% chance of rain this evening. What is the probability that it does not rain this evening?

38. There is a 12% chance that all of the strawberries in a container are moldy. What is the probability that some of the strawberries are not moldy?

Use the data in the table to determine the following probabilities.

US Import Sources by Weight (in Pounds) for Frozen Blackberries in 2019

Chile	17,303,000
Mexico	4,542,000
Serbia	3,705,000
China	169,000
Other Countries	514,000

Source: USDA. "Fruit and Tree Nut Data." http://www.ers.usda.gov/data-products/fruit-and-tree-nut-data/data-by-commodity.aspx

39. What is the probability that frozen blackberries purchased in the United States during this period were from Serbia or China?

40. What is the probability that frozen blackberries purchased in the United States during this period were not from Chile?

41. If the odds on a bet are $6:1$ against, what is the probability of winning?

42. Suppose the probability of a football team winning a playoff game is 0.25. What are the odds of winning?

43. The odds of a teenage male having an accident are $2:3$. What is the probability of a teenage male having an accident?

44. An insurance company claims the probability of surviving a certain type of cancer is 95%. What are the odds of surviving?

45. The UVest investment company publishes that the odds of increasing your wealth with their company is $5:2$. What is the probability of UVest increasing your investment?

46. Odds against being struck by lightning in one year are 1,000,000 to 1.

    a. If you live to be 80, what are the odds against being struck by lightning over your lifetime? Assume each year has the same probability.

    b. The National Weather Service gives the odds against being struck by lightning over an 80-year lifetime as 10,000 to 1. Why do you think this is different from the answer you got in part **a.**?

47. Overall odds in favor of winning in a state lottery game are $4.63:1$.

    a. Find the probability of winning in the lottery game.

    b. The prize for this lottery game is $100. If the cost to play the game is $2.00, what is the expected value for playing this game?

48. Suppose the odds on a bet are $10:1$ against. Your friend tells you he thinks the odds are too generous. Odds are considered less generous if the probability of losing is greater. Write down some less generous odds.

49. Some of the odds to win the 2021 Palmetto Championship are listed in the table.

Odds to Win 2021 Palmetto Championship	
Player	**Odds Against**
Brooks Koepka	$9:1$
Dustin Johnson	$8:1$
Matthew Fitzpatrick	$16:1$
Sungjae Im	$20:1$
Tyrrell Hatton	$14:1$

Source: "Golf Futures Betting Odds," Vegas Insider, accessed June 11, 2021, https://www.vegasinsider.com/golf/odds/futures/

    a. Determine the probability of losing money on a bet for each of the players listed.

    b. Based on these odds, which player was most likely to win the 2021 Palmetto Championship?

50. Some of the odds to win the 2021 Kentucky Derby are listed in the table.

Odds to Win 2021 Kentucky Derby	
Horse	**Odds Against**
Rock Your World	9 : 2
Essential Quality	3 : 1
Medina Spirit	12 : 1
Known Agenda	10 : 1
Hot Rod Charlie	6 : 1

Source: "2021 Kentucky Derby Betting Odds & Results," Vegas Insider, accessed June 11, 2021, https://www.vegasinsider.com/horse-racing/odds/kentucky-derby/

a. Determine the probability of losing money on a bet for each of the horses listed.

b. Based on these odds, which horse was most likely to win the 2021 Kentucky Derby?

✎ WRITING & THINKING

51. Describe the complement of the set of odd numbers greater than 0 within the set of positive integers.

52. Let the event E be the sum of a pair of dice that is divisible by 3. List the events in E^c.

53. The following is a table of the ages of boys on a soccer team.

Let $A = \{$soccer players older than 9$\}$. How many players are in the complement of A?

Ages of Boys on Soccer Team	
Age	**Number of Boys**
8	3
9	6
10	7
11	2

54. Describe the complement of the set of face cards in a standard deck of cards.

55. In a company, all employees who have worked there for more than five years receive a gift. Describe the complement of this group of employees.

56. In a bookstore, all books released in the past year that are currently on the best sellers list are placed on a table near the entrance. Describe the complement of this set of books.

10.3 PROJECT

THE BIRTHDAY PARADOX

It is often the case that we can gauge how likely an event is by simply thinking of our past experiences and comparing it to other events in our daily lives. For instance, most people would agree that the probability of buying one lottery ticket and winning the jackpot is much lower than the probability of rolling one die and getting a six, which is 1 in 6. We could use data or computations to confirm our intuition about these situations. In fact, if a lottery was set up where you choose 6 numbers (in any order) from a possible pool of 49 numbers, your chances of winning the jackpot are 1 in 13,983,816.

Sometimes, however, our intuition betrays us. In this activity, we will investigate a classic probability problem called the birthday paradox.

Consider a room that has 25 people in it.

1. Do you think that the probability of at least two people sharing a birthday (same month and same day) is above or below 50%? Would you say the probability is below 10% or above 90%? Explain your reasoning.

2. If we assume that there are 365 days in a year, what is the smallest number of people in a room that would guarantee that at least two people in the room share a birthday?

If we have two people in a room, they either share a birthday or they don't. These are complementary events, and we can write the following equation.

$$P(\text{Same Birthday}) = 1 - P(\text{Different Birthday})$$

Since there are 365 days in a year, there are $365(364) = 132,860$ different ways that the two people can have different birthdays. There are also $365(365) = 133,225$ possible pairs of birthdays. From this information, we get the following probability.

$$P(\text{Same Birthday}) = 1 - \frac{\text{Different Birthdays}}{\text{All Possible Birthdays}} = 1 - \frac{365(364)}{365(365)} \approx 0.0027$$

Hence, with two people in a room, the probability that they share a birthday is 0.27%.

In a room with three people, we can use the same argument. Either no birthday is shared or at least two people share a birthday. We would have the following probability.

$$P(\text{At Least Two People with the Same Birthday}) = 1 - \frac{365(364)(363)}{365(365)(365)}$$

3. Complete the computation above. Did the probability increase or decrease by adding just one extra person? By how much?

4. Now, follow the same reasoning and write an expression for the probability of at least two people sharing a birthday in a room with 25 people.

5. Use a computer to determine the value of the expression you found in part 4. (**Hint:** The website wolframalpha.com has a good computation engine.)

6. How does the answer found in part 5 compare to your answer from part 1?

10.4 ADDITION AND MULTIPLICATION RULES OF PROBABILITY

OBJECTIVES

1. Calculate probability for dependent events.

2. Identify mutually exclusive events.

3. Calculate probability for independent events.

4. Use Bayes' Theorem.

So far in the chapter, we've calculated the probabilities of single events using counting methods and the idea of complements to help us. But often situations call for analyzing probabilities that involve combinations of events. Let's turn our attention to how we might deal with those situations. There are basically two categories to consider:

1. Event A happening *or* event B happening
2. Event A happening *and* event B happening

There are some subtleties of distinction that we will need to take note of as we go along.

OBJECTIVE 1 Event *A* Happening or Event *B* Happening

Let's start with the *or* events. Consider a cashier giving change to a customer. They might say, "would you like 1 ten or 2 fives?" when handing back change. In this case, it's understood that the customer will get either 1 ten or 2 fives, but not both, because that would result in an overpayment. In contrast, when we use *or* in probability to say "event A or event B will occur," we mean that event A will occur, event B occur, or both events will occur; that is, at least one event will occur.

Let's imagine calculating the probability of selecting a king or a spade from a standard deck of cards. The probability of selecting a king is $\frac{4}{52}$ and the probability of selecting a spade is $\frac{13}{52}$.

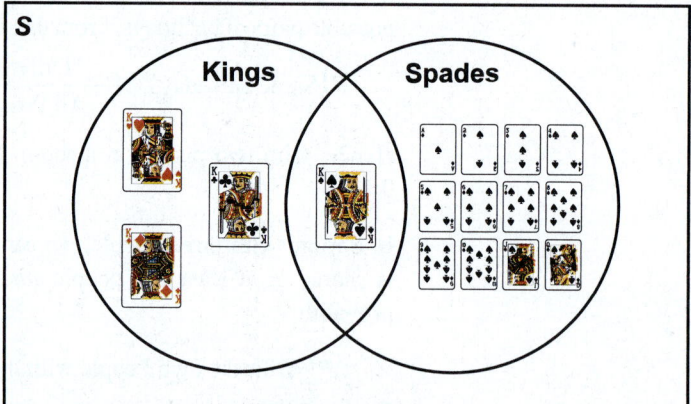

Figure 10.4.1

It's tempting to want to just add the two probabilities together. However, think about the card that is both a king and a spade. The king of spades is an outcome in both of the events, so we would have actually counted it twice by adding the fractions together. We can adjust for the overcounting by using subtraction. The probability of choosing a king or a spade from a standard deck of cards is calculated as follows.

$$P(\text{king } or \text{ spade}) = P(\text{king}) + P(\text{spade}) - P(\text{king } and \text{ spade})$$

$$= \frac{4}{52} + \frac{13}{52} - \frac{1}{52}$$

$$= \frac{16}{52}$$

$$\approx 0.3077$$

If A and B are events that have some outcomes in common, then the probability that A or B will happen is calculated by adding the individual probability of each and then subtracting the probability that both events occur simultaneously. This is referred to as the *Addition Rule of Probability*.

f(x) ADDITION RULE OF PROBABILITY

The **Addition Rule of Probability** states that the probability of event A happening *or* event B happening is given by the following formula.

$$P(A \text{ or } B) = P(A) + P(B) - P(A \text{ and } B)$$

Example 10.4.1

Using the Addition Rule of Probability

Suppose that a student is chosen at random to receive a gift card for filling out a survey. The following table shows a breakdown of who filled out the survey.

Table 10.4.1: Breakdown of Survey Takers

Class	Student Government Member	Not a Student Government Member	Total
Freshman	3	15	18
Sophomore	1	11	12
Junior	2	7	9
Senior	4	3	7
Total	**10**	**36**	**46**

What is the probability that the winner is either a freshman or a member of student government?

Solution

The key word *or* tells us we will be using the Addition Rule of Probability to find the solution.

$$P(\text{Freshman } or \text{ Student Government}) = P(\text{Freshman}) + P(\text{Student Government})$$
$$- P(\text{Freshman } and \text{ Student Government})$$

We begin by finding the probability of choosing each of the individual criteria. We can see from the table that there were a total of 46 students who filled out the survey. The number of freshmen who filled it out was 18. So the probability of choosing a freshman is calculated as follows.

$$P(\text{Freshman}) = \frac{18}{46}$$

In previous sections, we converted our probability answers from fractions to decimals. However, since we are going to use the Addition Rule of Probability, which requires us to add probabilities together, it's better to leave them as nonreduced fractions so that we avoid any error in rounding.

There were 10 members of the student government who filled out the survey, so the probability of choosing a member of the student governing board is as follows.

$$P(\text{Student Government}) = \frac{10}{46}$$

There are 3 students who are both freshmen and members of the student government, so the probability of choosing a student who is in both groups is calculated as follows.

$$P(\text{Freshman } and \text{ Student Government}) = \frac{3}{46}$$

We can now use the Addition Rule of Probability to calculate the probability of the winner being either a freshman or a member of student government.

$$P(\text{Freshman } or \text{ Student Government}) = \frac{18}{46} + \frac{10}{46} - \frac{3}{46}$$
$$= \frac{25}{46}$$
$$\approx 0.5435$$

Thus, the probability that the winner of the gift card is either a freshman or a member of student government is about 54.35%.

✔ **Skill Check 10.4.1**

Using Table 10.4.1 in the previous example, calculate the probability that the winner was either a senior or not a member of the student government.

Example 10.4.2

Using the Addition Rule of Probability

The tree diagram in Figure 10.4.2 illustrates the makeup of students in Art History 202. The number of students fitting each descriptor is given in the parentheses. Suppose a student is chosen at random from the class to attend the annual regional art conference. Use the diagram to answer the following questions.

a. What is the probability that the student chosen is a full-time senior?

b. What is the probability that the student chosen is either a full-time student or a senior?

c. What is the probability that the student chosen is neither a full-time student nor a senior?

Solution

a. Notice that the first question is a simple probability question involving only one event—choosing a full-time senior. Since there are 39 students to choose from and 2 of them are full-time seniors, we calculate the probability as follows.

$$P(\text{Full-Time Senior}) = \frac{2}{39} \approx 0.0513$$

Thus, the probability of randomly choosing a full-time senior is approximately 5%.

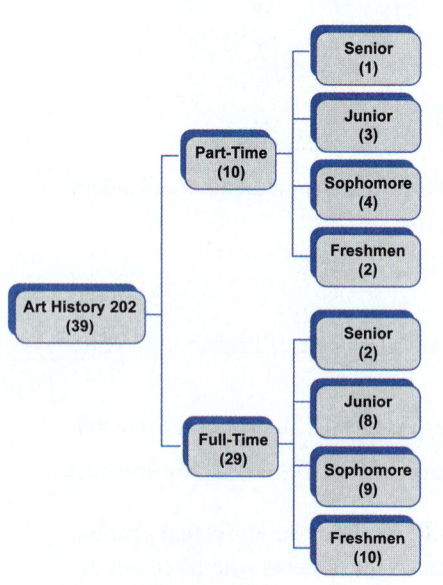

Figure 10.4.2

b. This time, the key word *or* implies two events—choosing a full-time student *or* choosing a senior. The Addition Rule of Probability tells us we find the probability as follows.

$$P(\text{Full-Time or Senior}) = P(\text{Full-Time}) + P(\text{Senior}) - P(\text{Full-Time Senior})$$

We calculated the last part of this formula in part a., so we only need to calculate the other two probabilities. The chart shows that there are 29 full-time students.

$$P(\text{Full-Time}) = \frac{29}{39}$$

We need to include all of the senior students, both full-time and part-time, when calculating the probability of choosing a senior. There are $2 + 1 = 3$ seniors in the class.

$$P(\text{Senior}) = \frac{3}{39}$$

Substituting these into the formula, we can calculate the probability.

$$P(\text{Full-Time } or \text{ Senior}) = \frac{29}{39} + \frac{3}{39} - \frac{2}{39} = \frac{30}{39} \approx 0.7692$$

Thus, the probability of choosing either a full-time student or a senior is approximately 77%.

c. Finally, we are asked to find the probability that the student is not a full-time student nor a senior. We can either determine the number of people who fall into this category or we can use the complement rule and simply subtract the answer we found part b. from 1.

$$P(\text{Not Full-Time } Nor \text{ Senior}) = 1 - P(\text{Full-Time } or \text{ Senior})$$
$$\approx 1 - 0.7692 = 0.2308$$

Thus, the probability of choosing a student who is neither a full-time student nor a senior is approximately 23.08%.

OBJECTIVE 2 Mutually Exclusive Events

What about the case when the two events do not have any outcomes in common? Think about the following choices:

- Rolling a 1 or a 6 on a single roll of a fair die

- Living in the city or the country

- Buying a red car or a black truck as your first vehicle

- Going to Hawaii or Sweden for your one week of vacation

- Building a 1500-square-foot home or a 2100-square-foot home on your new single-home property

All of these examples have no outcomes in common. We refer to events such as these as *mutually exclusive events*.

📖 **MUTUALLY EXCLUSIVE EVENTS**

Mutually exclusive events are events that have no outcomes in common.

Example 10.4.3

Identifying Mutually Exclusive Events

Determine if the following pairs of events are mutually exclusive.

a. Let the first event consist of randomly selecting an adult from the mall who has shopped online at least once in the past 6 months. Let the second event consist of randomly selecting an adult from the mall who has never shopped online.

b. Let event *A* consist of selecting an even number and event *B* consist of selecting a prime number.

Solution

a. Notice that it is impossible for a person to be in both of the events "shopped online at least once in the past 6 months" and "never shopped online." Therefore, these are mutually exclusive events.

b. There exists one even number that is also prime, namely 2. Therefore, events *A* and *B* are not mutually exclusive events.

If we think about the formula for the Addition Rule of Probability, the formula adjusts for overcounting duplicates by subtracting outcomes that are in both events. Because there is no need to worry about overcounting when events have no outcomes in common, we will single out this type of probability by calling the formula the *Addition Rule of Probability for Mutually Exclusive Events*. However, it is important to realize that this is simply a special case of the Addition Rule.

𝑓(x)   ADDITION RULE OF PROBABILITY FOR MUTUALLY EXCLUSIVE EVENTS

The **Addition Rule of Probability for Mutually Exclusive Events** states that the probability of event *A* happening or event *B* happening when *A* and *B* have no outcomes in common is given by the following formula.

$$P(A \text{ or } B) = P(A) + P(B)$$

Example 10.4.4

Applying the Addition Rule of Mutually Exclusive Events

Suppose that you have decided you'd like to get a pet. Your renter's agreement allows you only one pet, and you decide to go to the local animal shelter to adopt one of their animals. Because you can't decide between a dog and a cat, you've left the choice up to chance. You're going to run your finger down the list of available animals without looking and let the lucky pet be the one you stop on. The graph in Figure 10.4.3 shows the distribution of animals currently at the shelter. Unfortunately, you didn't consider that other kinds of animals might be on the shelter's list. What is the probability that you randomly choose either a cat or a dog to take home with you?

Available Animals for Adoption

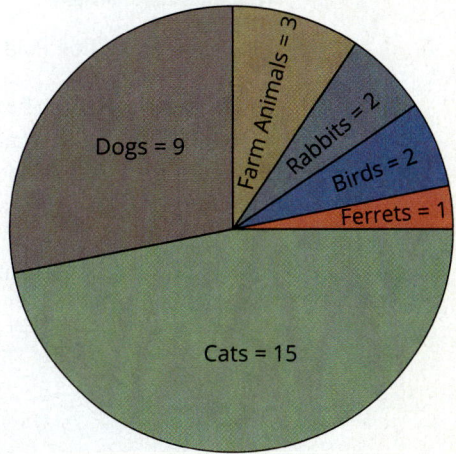

Figure 10.4.3: Available Animals for Adoption

Solution

The key word *or* implies that we need to use the Addition Rule of Probability. However, because an animal cannot be both a cat and a dog at the same time, these are mutually exclusive events, and thus have no outcomes in common. So we can use the Addition Rule of Mutually Exclusive Events, which gives us the following equation.

$$P(\text{Cat or Dog}) = P(\text{Cat}) + P(\text{Dog})$$

By adding the numbers all of the different animals together, we know that there are currently 32 animals available at the shelter.

The probability of choosing a cat is found by dividing the number of cats by the total number of animals.

$$P(\text{Cat}) = \frac{15}{32}$$

The probability of choosing a dog is found by dividing the number of dogs by the total number of animals.

$$P(\text{Dog}) = \frac{9}{32}$$

Substituting these probabilities into the formula, we have the following.

$$P(\text{Cat } or \text{ Dog}) = \frac{15}{32} + \frac{9}{32}$$
$$= \frac{24}{32}$$
$$= 0.75$$

Even though you forgot about the fact that other animals might be at the shelter, you still have a 75% chance of randomly choosing a cat or a dog. However, that means there is also a 25% chance you will select one of the other animals, some of which your landlord might not approve of!

Example 10.4.5

Applying the Addition Rule of Mutually Exclusive Events

Choosing a college can be an exciting and nerve-racking experience in the life of a high school student. Emma has finally narrowed down her choices to the top four. She's also given each school a probability based on certain characteristics. What is the probability that Emma ends up at University B or University D?

Table 10.4.2: University Probabilities

University	Characteristic	Probability
A	Closest to home	$P(A) = 0.25$
B	Best sports	$P(B) = 0.10$
C	Her best friend's choice	$P(C) = 0.30$
D	Best academic program of her choice	$P(D) = 0.35$

Solution

Since Emma will choose one or the other, but not both at the same time, these events are mutually exclusive. So we just need to add the probability of her choosing University B to the probability of choosing University D.

$$P(\text{University B or University D}) = 0.10 + 0.35 = 0.45$$

Thus, there is a 45% probability that Emma will attend University B or University D.

OBJECTIVE 3 Event *A* Happening and Event *B* Happening

Now we'll turn our attention to the possibility of two events both happening. In other words, one event occurs *and* a second event occurs. For example, winning the lottery and getting a job promotion on the same day, or rolling a five and a three on a pair of dice, or randomly choosing a teenager and a grandparent from a group of people of varying ages. The key to calculating probabilities involving two events both happening is determining if the occurrence of one event influences the probability of the other event happening. Sometimes it does and sometimes it doesn't.

Does rolling a six on a fair six-sided die affect the probability of drawing an ace from a deck of cards? No, but how about being dealt the queen of hearts and then a red face card from a deck of cards? Here, the first event does influence the probability of the second event. If you were dealt the queen of hearts first, the probability of getting a red face card would decrease for the second card because there would be one fewer red face card in the deck to choose from.

Let's stop here and make a distinction between the two situations. We say two events are *independent* when the occurrence of one event *does not* influence the probability of the other event happening. If the result of one event *does* influence the probability of the second, we say that the two events are *dependent*.

📖 INDEPENDENT EVENTS AND DEPENDENT EVENTS

Independent Events

Independent events are events where the result of one event does not influence the probability of the other event happening.

Dependent Events

Dependent events are events where the result of one event influences the probability of the other event happening.

Example 10.4.6

Identifying Independent and Dependent Events

Determine if the following pairs of events are independent

a. Event *A*: Eating a red candy from a new bag of Skittles

   Event *B*: Pulling a second Skittle from the same bag that is also red

b. Event *A*: A woman giving birth to a daughter

   Event *B*: The same woman's second child is also a girl

c. Event *A*: Tina is the first woman to finish the 2030 Boston Marathon.

   Event *B*: Tina is the first woman to finish the 2031 New York Marathon.

Solution

a. These events are dependent. The chances of drawing a second red Skittle from the bag decreases after the first one is drawn, because it obviously was not replaced in the bag. It was eaten.

b. Although these events might appear dependent on one another, the probability that a child is a girl is the same for any given pregnancy. It is not affected by the gender of a child from any previous pregnancy. Therefore, these events are independent.

c. At first glance these events might seem independent of one another. You might think that winning one race has no effect on winning a second race. However, the best starting positions for runners in large marathons are given to winners of previous races. This means that the first race does have an effect on the second race. At the same time, running two marathons in consecutive years will have an effect on your body when training for and running the second one (whether that effect is positive or negative). Therefore, these events are dependent.

This example highlights the fact that the dependence of events isn't always obvious. Sometimes our own personal knowledge or experiences are not enough to rely on, and we need to look to experts in other fields to help us determine whether events are dependent or not.

✔ Skill Check 10.4.2

Determine if the following events are independent or dependent.

a. Finding the batteries in your calculator are dead and then finding that the battery in your car is dead.

b. Parking your car illegally and then getting a parking ticket.

Let's focus on finding the probabilities of independent events first. If two events are independent, we can find the probability of both events occurring by multiplying the individual probabilities together. This is referred to as the **Multiplication Rule of Probability for Independent Events**. In fact, satisfying the multiplication rule defines independent events.

---

f(x) MULTIPLICATION RULE OF PROBABILITY FOR INDEPENDENT EVENTS

The **Multiplication Rule of Probability for Independent Events** states that for two independent events, A and B, the probability of event A happening *and* event B happening is given by the following formula.

$$P(A \text{ and } B) = P(A) \cdot P(B)$$

Note that it doesn't matter how many events are in question; if each one is independent of the others, then we can multiply the individual probabilities.

Example 10.4.7

Using the Multiplication Rule of Probability for Independent Events

Given a fair six-sided die and a standard deck of 52 cards, find the probability of rolling a 6 and drawing an ace.

Solution

Because the number rolled on the die does not affect the card drawn from the deck, and vice versa, the events here are independent. Using the Multiplication Rule of Probability for Independent Events, we have the following.

$$P(6 \text{ and Ace}) = P(6) \cdot P(\text{Ace}) = \frac{1}{6} \cdot \frac{4}{52} = \frac{4}{312} \approx 0.0128$$

Thus, the probability of rolling a 6 and drawing an ace is approximately 1.28%.

Example 10.4.8

Using the Multiplication Rule of Probability for Independent Events

Suppose we know the following breakdown for internal medicine hospitalists who work at Madison Regional Hospital and the ages of their patients on a given day.

Table: 10.4.3: Hospitalists at Madison Regional Hospital

		Number
Years of Experience	Less than two years	6
	Greater than two years	7
Patient Age	Less than 35	16
	35 to 55	35
	Greater than 55	21

What is the probability that the first patient treated is over 55 years old and treated by a hospitalist with less than two years of experience at Madison Regional Hospital?

Solution

Because the age of the patient and the experience of the hospitalist have no effect on one another, these events are independent. So we use the Multiplication Rule of Probability for Independent Events and multiply the individual probabilities together. Let's find the individual probabilities first.

$$P\left(\text{Patient Age} > 55\right) = \frac{21}{72}$$

$$P\left(\text{Less Than Two Years Experience}\right) = \frac{6}{13}$$

Using the Multiplication Rule of Probability for Independent Events, we have the following equation.

$$P\left(\begin{array}{l}\text{Patient Age} > 55 \text{ and Hospitalist with} \\ \text{More Than Two Years Experience}\end{array}\right) = \frac{21}{72} \cdot \frac{6}{13}$$

$$= \frac{126}{936}$$

$$\approx 0.1346$$

Therefore, the probability of the first patient being over the age of 55 and treated by a hospitalist with less than two years of experience is approximately 13.5%.

As we defined earlier, when two events are not independent, it means that the occurrence of one influences the probability that the other occurs. For example, consider choosing chocolates from a variety box. Each time you remove a chocolate, the chances of drawing out a certain flavor next will change. These events are said to be *dependent*. To calculate probability for dependent events, we multiply the probabilities at each stage, taking care to adjust each probability based on the previous stage. In general, this type of probability is called **conditional probability**, denoted as $P(B|A)$ and read "the probability of B given A."

📖 CONDITIONAL PROBABILITY

Conditional probability, denoted $P(B|A)$ and read "the probability of B given A," is the probability of event B occurring given that event A occurs first.

Conditional probability is necessary for finding the probability of two dependent events. We multiply the probability of one event by the conditional probability of the other event, as shown in the following formula, which is the **Multiplication Rule of Probability for Dependent Events**.

f(x) MULTIPLICATION RULE OF PROBABILITY FOR DEPENDENT EVENTS

The **Multiplication Rule of Probability for Dependent Events** states that if A and B are dependent events, the probability of A and B occurring is given by the following formula.

$$P(A \text{ and } B) = P(A) \cdot P(B|A)$$

Example 10.4.9

Using the Multiplication Rule of Probability for Dependent Events

Find the probability that, from a standard deck of cards, you're dealt two particular cards in a row: the queen of hearts and then a face card. Assume the cards are chosen without replacement.

> **💬 Helpful Hint**
>
> A standard deck of cards is 52 cards with four suits (hearts, diamonds, spades, and clubs) and 13 cards in each suit. The cards in each suit are ace, king, queen, jack, 10, 9, 8, 7, 6, 5, 4, 3, and 2. The king, queen, and jack are called face cards.

Solution

Recall that without replacement means the first card is not placed back into consideration before drawing the next card. Therefore, being dealt the queen of hearts for the first card reduces the number of face cards as well as the total number of cards left in the deck for the second card. That means that these events are dependent and we need to use the Multiplication Rule for Dependent Events. We begin by calculating the probability of first being dealt the queen of hearts. Note that all cards are available.

$$P\left(\text{Queen of Hearts}\right) = \frac{1}{52}$$

When the second card is dealt, there are no longer 12 face cards in the deck. Only 11 face cards remain in a deck of 51 cards (remember that there is one fewer card). So the probability of drawing a face card given that the queen of hearts has already been drawn is calculated as follows.

$$P\left(\text{Face Card|Queen of Hearts}\right) = \frac{11}{51}$$

Finally, substitute these values into the Multiplication Rule for Dependent Events.

$$P\left(\begin{array}{c}\text{Queen of Hearts}\\\text{and Face Card}\end{array}\right) = P\left(\text{Queen of Hearts}\right) \cdot P\left(\text{Face Card|Queen of Hearts}\right)$$

$$= \frac{1}{52} \cdot \frac{11}{51}$$

$$= \frac{11}{2652}$$

$$\approx 0.0041$$

Thus, the probability of being dealt the queen of hearts and then a face card from a standard deck of cards is approximately 0.41%, which is not very likely.

Example 10.4.10

Calculating Conditional Probability

Suppose 290 students were randomly selected and asked about their satisfaction with their interactions with the financial aid office on campus. Their responses are given in the following table.

Table 10.4.4: Financial Aid Office Satisfaction

Class	Satisfied	Dissatisfied	Did Not Use
Freshman	55	21	13
Sophomore	15	33	24
Junior	48	6	8
Senior	22	18	3
Graduate	4	1	19

If one response was selected at random, find the probability that the following occurred.

a. The response indicated that the student was satisfied with their experience.

b. The response indicated that the student was satisfied given that they were a senior.

c. The response indicated that the student was dissatisfied given that they were a freshman or sophomore.

d. The response indicated that the student was a graduate student given that they did not use the financial aid office.

Solution

a. The probability of selecting a response from a student who responded they were satisfied with their experience is found by dividing the number of students who said they were satisfied by the total number of students surveyed. There are no restrictions given, so this calculation does not require conditional probability. To find the total number of students who responded that they were satisfied, we need to add the values in the first column together.

Students Who Responded They Were Satisfied = 55 + 15 + 48 + 22 + 4 = 144

To find the probability that a randomly selected response is one of these, we divide the number of satisfied responses by the total number of responses.

$$P(\text{Satisfied}) = \frac{144}{290}$$
$$\approx 0.4966$$

Therefore, the probability of randomly selecting a student who responded that they were satisfied with their financial aid experience is approximately 49.66%.

b. In this second scenario, we do have conditions to take into account in order to calculate the probability, namely that the respondents are seniors. To find the probability of selecting a response from a student who responded they were satisfied given that they were a senior, we need to limit our satisfied responses to those of senior students only. First, add all of the numbers in the senior row of the table to find out how many of the 290 total students were seniors.

Senior Students = 22 + 18 + 3 = 43

Now, divide the number of seniors who responded that they were satisfied by the total number of seniors.

$$P(\text{Satisfied|Senior}) = \frac{22}{43}$$
$$\approx 0.5116$$

Thus, the probability of randomly selecting a response from a student who responded they were satisfied given that they were a senior is approximately 51.16%.

c. Once again, the restrictions imposed mean that we are calculating a conditional probability. In this case, we need to limit our responses to only freshmen and sophomores. The number of respondents in these two classes is found by adding all the numbers from the appropriate rows.

Freshman and Sophomore Students = 55 + 21 + 13 + 15 + 33 + 24 = 161

This time we're interested in the number of dissatisfied responses. Looking in the dissatisfied column, we see that there were 21 + 33 = 54 dissatisfied freshmen and sophomores. We can then divide to find the probability that we randomly chose one of these responses.

$$P\left(\text{Dissatisfied}|\text{Freshman or Sophomores}\right) = \frac{54}{161}$$
$$\approx 0.3354$$

Therefore, the probability that a randomly chosen response was from a student who responded that they were dissatisfied given that they were either a freshman or sophomore is approximately 33.54%.

d. This final conditional probability requires us to consider only the column of responses from students who did not use the financial aid office.

Students Who Did Not Use the Financial Aid Office = 13 + 24 + 8 + 3 + 19 = 67

Of these 67 responses, 19 of them were from graduate students. Thus, the probability that we randomly chose the response of a graduate student given that they did not use the financial aid office is calculated as follows.

$$P\left(\text{Graduate Student}|\text{Did Not Use Financial Aid Office}\right) = \frac{19}{67}$$
$$\approx 0.2836$$

There is approximately a 28.36% chance of randomly choosing a response from a graduate student given that they did not use the financial aid office.

OBJECTIVE 4   Bayes' Theorem

 Math Milestone

Likely born in 1701 or 1702, Reverend Thomas Bayes was an English statistician, philosopher, and Presbyterian minister. He died before ever publishing what would become his most famous work.

Very often we have the ability to calculate the conditional probability $P(A|B)$, but what we really want to know is the reverse of that, $P(B|A)$. Despite the fact that these two formulas look quite similar, they can represent remarkably different situations. Imagine, for instance, that you take a medical test for a disease that comes back positive. If you know that the test is 90% accurate, then what you know is $P(\text{Positive Test}|\text{You Have the Disease}) = 0.90$. However, what you probably want to know how likely it is that you have the disease now that you have tested positive; that is, $P(\text{You Have the Disease}|\text{Positive Test})$. A mathematician named Thomas Bayes is credited with a theorem simplifying the relationship between the two conditional probabilities. His theorem, known as **Bayes' Theorem**, shows how one relates to the other.

> 🔑 **BAYES' THEOREM**
>
> The probability of event A occurring given that event B occurred is calculated as follows.
>
> $$P(A|B) = \frac{P(B|A) \cdot P(A)}{P(B)}$$
>
> Here, $P(B) \neq 0$.

Example 10.4.11 examines the usefulness of this theorem.

Example 10.4.11

Calculating Probability Using Bayes' Theorem

According to statistics from SEER, men have a 1 in 8 (or about 12.5%) chance of developing prostate cancer. A PSA test is used to detect prostate cancer and can give either a positive or negative result. Studies have shown that of those men who have developed prostate cancer, 15% receive a negative PSA test result. Studies also show that 58.5% of all men receive a positive PSA test. Based on these statistics, calculate the probability that a man who has a positive PSA test result actually has prostate cancer. Use parts a. through d. as an aid to determining this value in part e.

a. What is the probability that a man develops prostate cancer?

b. What is the probability that a man with prostate cancer has a negative PSA test?

c. What is the probability that a man with prostate cancer has a positive PSA test?

d. What is the probability that a man has prostate cancer and a positive PSA test?

e. What is the probability that a man with a positive PSA test has prostate cancer?

Solution

a. We are asked to calculate the probability that a man develops prostate cancer. We are actually told this probability in the information given.

$$P(\text{Prostate Cancer}) = \frac{1}{8} = 0.125$$

b. This time we are asked about the probability that a man with prostate cancer has a negative PSA test; that is, the probability of a negative test given that a man has prostate cancer, written $P(\text{Negative}|\text{Cancer})$. Once again, we do not need to calculate this because this information is given.

$$P(\text{Negative}|\text{Prostate Cancer}) = 15\% = 0.15$$

c. Here, we want to know the probability of a positive test, given that a man has cancer; that is, $P(\text{Positive}|\text{Cancer})$. Since the information we were given tells us that men with cancer have a negative PSA test 15% of the time, we can use the complement to find the probability of men with cancer having a positive test.

$$P(\text{Positive}|\text{Prostate Cancer}) = 1 - 0.15 = 0.85$$

d. Notice that the wording is different this time. We are asked to find the probability that a man has prostate cancer *and* a positive PSA test result; that is, *P*(Prostate Cancer and Positive). We know from the Multiplication Rule for Dependent Events that we can calculate this as follows, using the results we found in parts a. and c.

$$P\left(\begin{array}{c}\text{Prostate Cancer}\\\text{and Positive}\end{array}\right) = P(\text{Prostate Cancer}) \cdot P(\text{Positive}|\text{Prostate Cancer})$$

$$= 0.125 \cdot 0.85$$
$$= 0.10625$$

Therefore, the probability that a man has both prostate cancer and a positive test is 10.625%.

e. Be careful how you read this question. It's slightly different than the one in part d. This time we want to know the probability of a man actually having prostate cancer, given that he has a positive test; that is, *P*(Prostate Cancer|Positive). We can use Bayes' Theorem along with our previous calculations to find this probability. Bayes' Theorem is written again for you here.

$$P(A|B) = \frac{P(B|A) \cdot P(A)}{P(B)}$$

Thus, we are looking to calculate the following probability.

$$P(\text{Prostate Cancer}|\text{Positive}) = \frac{P(\text{Positive}|\text{Prostate Cancer}) \cdot P(\text{Prostate Cancer})}{P(\text{Positive})}$$

We know the numerator of the fraction from parts a. and c.:

P(Prostate Cancer) = 0.125 and *P*(Positive|Prostate Cancer) = 0.85. From the problem statement, we also know that 58.5% of all men receive a positive PSA test; that is, *P*(Positive) = 0.585. Substituting these values into Bayes' Theorem, we have the following equation.

$$P(\text{Prostate Cancer}|\text{Positive}) = \frac{P(\text{Positive}|\text{Prostate Cancer}) \cdot P(\text{Prostate Cancer})}{P(\text{Positive})}$$

$$= \frac{0.85 \cdot 0.125}{0.585}$$
$$\approx 0.1816$$

This means if a man has a positive PSA test, the probability that he actually has prostate cancer is approximately 0.1816, or 18.16%.

Take a moment and think about the difference between parts d. and e. In part d., we found the probability that a random man has prostate cancer and a positive PSA test. This is not as useful to an individual male who receives a positive PSA test. The conditional probability we found in part e.—the probability that he actually has prostate cancer given that his PSA test came back positive—is more helpful when considering test results.

Skill Check Answers
1. 86.96% **2. a.** Independent **b.** Dependent

10.4 Exercises

✔ CONCEPT CHECK

1. If A and B are events that have some outcomes in common, then the probability that A or B will happen is calculated by adding the individual probability of each and then _____ the probability that both events occur simultaneously.

2. If two events have ___ outcomes in common, they are called mutually exclusive events.

3. The probability of events A and B happening is equal to the probability of A happening times the probability of B happening IF the two events are _____.

4. _____ probability is the probability of one event happening given that another event has already happened.

5. True or False: If the result of one event influences the probability of a second event, the two events are independent.

💡 PRACTICE

Determine whether each situation contains mutually exclusive events.

6. Let event A consist of numbers that are prime. Let event B consist of numbers that are multiples of 5.

7. Let event A consist of the multiples of 3. Let event B consist of the factors of 10.

8. Let event A consist of randomly selecting a member of a concert audience who can play the guitar. Let event B consist of randomly selecting a member of a concert audience who cannot play the guitar.

9. Let event A consist of rolling a die and getting a 3. Let event consist of rolling a die and getting a 6.

Determine whether each situation contains independent events.

10. The color of car driven by three randomly chosen classmates.

11. A password must be six characters long with no repeated characters. Are the choices of consecutive characters independent?

12. There are 15 board members, of which seven are men and eight are women. Two randomly chosen members will serve on the United Way campaign committee. If you wish to find the probability that both members chosen are the same sex, do you treat these selections as independent events?

13. Are receiving a bill in Monday's mail and receiving a letter from your grandparents in Monday's mail independent events?

14. Naomi and Amelia both put two business cards into the basket at a coffee shop. The shop owner selects three cards from the basket. Are the two events that Naomi's card is chosen and Amelia's card is chosen independent?

🚀 APPLICATIONS

Calculate the probability of each set of events that are not mutually exclusive. Round your answer to the nearest millionth, if necessary.

15. A pair of dice is rolled. What is the probability that the sum of the numbers is an even number or a multiple of 3?

16. A bag of eleven marbles contains five marbles with red on them, three with green on them, seven with black on them, and four with black and red on them. What is the probability that a randomly chosen marble has either black or red on it?

17. What is the probability that a card selected from a deck will be either an ace or a spade?

18. The following is a table showing the results of a poll taken on campus.

Will You Vote in the Upcoming Election?

	Male	Female
Yes	16	24
No	19	11
Not decided	21	22

a. What is the probability that a randomly selected student from this poll would be a male who has not decided whether he will vote in the upcoming election?

b. What is the probability that a randomly selected student from this poll is female or will not vote in the upcoming election?

c. What is the probability that a randomly selected student from this poll has decided to vote in the upcoming election?

19. Out of a class of 30 students, there are 16 students who study Latin, 21 who study German, and 7 who study both. What is the probability that a randomly selected student from the class will study only Latin?

20. Of the 11 instructors in the English department, four are new to the department and three are female. However, there is only one who fits all of the descriptions. Find the probability that if you randomly choose a course taught by these instructors, you get either a new instructor or a female instructor.

21. The following is a table representing the students who are on the Student Government Board.

Students on the Student Government Board

	On-Campus Housing	Off-Campus Housing
Freshman	3	1
Sophomore	3	2
Junior	2	3
Senior	0	3
Graduate Student	0	2

Find the probability that a randomly chosen member of the Student Government Board is either a sophomore or lives in on-campus housing.

22. A box of a dozen chocolates contains 8 chocolates that contain nougat, 6 that contain caramel, and 2 that contain both caramel and nougat. Determine the probability that a randomly selected chocolate will contain only nougat.

23. A pair of dice is rolled. What is the probability that the sum of the numbers is divisible by 5 or a prime number?

24. A bag of meeple game pieces contains 10 meeples with yellow on them, 20 meeples with blue on them, 12 meeples with green on them, and 6 meeples that are striped yellow and blue. What is the probability that a randomly chosen meeple has either yellow or blue on it?

25. What is the probability that a card randomly selected from a standard deck will be either a face card or a diamond?

Calculate the probability of each set of mutually exclusive events. Round your answer to the nearest millionth when necessary.

26. Suppose that the probability of obtaining zero defective items in a sample of 50 items off the assembly line is 0.34 while the probability of obtaining 1 defective item in the sample is 0.46. What is the probability of the following?

    a. Obtaining no more than one defective item in a sample.
    b. Obtaining more than one defective item in a sample.

27. A pair of dice is rolled. What is the probability that the sum of the numbers is either 7 or 11?

28. A single letter from the word MISSISSIPPI is chosen. What is the probability of choosing an S or an I?

29. What is the probability that a card selected from a deck will be either an ace or a queen?

30. A reporter for an international newspaper is given an assignment that is randomly chosen from the following destinations worldwide: 13 continental United States assignments, 7 South American assignments, 21 European Union assignments, and 5 Asian assignments. Find the probability that he gets an assignment in Asia or South America.

31. The following table shows the breakdown of opinions for both faculty and students in a recent survey about the new restructuring of the campus to be a walking campus.

Survey Results on Restructuring Campus to a Walking Campus

	Favor	Oppose	Neutral	Total
Faculty	12	4	3	19
Student	33	57	28	118
Total	45	61	31	137

   a. Find the probability that a randomly selected person is either a faculty member in favor of the change or a student who has an opinion either for or against.

   b. Find the probability that a randomly selected person is either neutral or in favor of the restructuring.

32. The probability of the stoplight being green at the intersection of Meeting Street and Main Street is 0.55, while the probability of it being yellow is 0.15. Find the probability that the light is red when you get to the intersection of Meeting Street and Main Street. Assume that the light will be working and will be a solid color: red, yellow, or green.

33. In a box of pens and pencils, the probability of randomly choosing a sharpened pencil is 0.54 and the probability of randomly choosing a pen from the box is 0.39. Find the probability of randomly selecting either an unsharpened pencil or a pen from the box.

34. A couple is deciding where to spend their honeymoon. They've narrowed down their options to 4 locations in the United States, 3 locations in Mexico, and 3 locations in the United Kingdom. What is the probability of the couple choosing a honeymoon location in Mexico?

35. A community extracurricular soccer league uses volunteers for the coaching positions. This year, only 7 of the 16 volunteers have previous coaching experience. What is the probability that a team will have a coach with no previous coaching experience?

Calculate the probability of each set of independent events. Round your answer to the nearest millionth, if necessary.

36. Suppose the probability that my pet will be alive in five years is 0.65 and the probability that my cousin's pet will be alive in five years is 0.48. Find the probability that both of these pets will be alive in five years assuming that they are independent events.

37. Two dice are thrown. Find the probability of getting an even number on the first die and an odd number on the second die.

38. The following table shows the student demographics for a sociology class.

Sociology 101 Student Demographics

	Male	Female
Freshman	3	11
Sophomore	4	9
Junior	0	3
Senior	1	0

    a. Find the probability that a randomly selected student from the class is a male.

    b. Find the probability that if two students are randomly selected, without replacement, the first is a female junior and the second is a male sophomore.

39. Find the probability of choosing a heart and then an ace from a standard deck of cards with replacement.

40. On any given day at the beach, there is a 49% chance of precipitation. What is the probability that you will get precipitation for three days in a row on your beach vacation? Assume that the weather on a particular day at the beach is independent of the weather the day before.

Calculate the probability of each set of dependent events. Round your answer to the nearest millionth, if necessary.

41. A bag of marbles contains 7 marbles with green on them, 10 marbles with pink on them, 6 marbles with blue on them, and 3 marbles with both pink and blue on them. Two marbles are selected from the bag in a row without replacement. Determine the probability that a marble with blue and pink on it was selected first and a marble with blue on it was selected second.

42. A group of candidates are sitting in a lobby waiting to be interviewed. Six of the candidates have previous experience related to the job, eight of the candidates have a bachelor's degree, and three of the candidates have both previous experiences and a bachelor's degree. Determine the probability that the next candidate called has a bachelor's degree and previous experience followed by a candidate that only has a bachelor's degree.

43. A group of friends are deciding which two movies to watch for movie night. They narrowed their options down to 6 comedies, 5 romances, 4 action films, and 3 romantic comedies. What is the probability that the first movie they select is a comedy and the second movie they select is a romantic comedy? (Assume they will not watch the same movie twice in a row.)

44. Suppose you are trying out different types of mouthwash and the store you visit has 10 mouthwashes that contain alcohol, 7 mouthwashes that contain fluoride, 3 mouthwashes that contain neither alcohol nor fluoride, and 5 mouthwashes that contain both alcohol and fluoride. If you choose two different mouthwashes at random, what is the probability that the first mouthwash you select contains both alcohol and fluoride and the second mouthwash you select contains fluoride?

Calculate each conditional probability. Round your answer to the nearest millionth, if necessary.

45. A swim team consists of four boys and three girls. A relay team of four swimmers is chosen at random from the team members. What is the probability that there are two boys on the relay team given that there are two girls on the relay team?

46. Emma is playing Monopoly, a game played with two dice. What is the probability that the sum of the two dice she rolls is less than 4 given that she rolls an odd number?

47. Hunter bets his friend that he can draw two aces in a row from a standard deck of cards. What is the probability that Hunter draws a second ace given that his first card was an ace?

48. The probability that a student passes Intermediate Algebra is 0.55. The probability that a student passes College Algebra given that they pass Intermediate Algebra is 0.70. What is the probability that a student passes both College Algebra and Intermediate Algebra?

49. On each point in racquetball, a player is allowed two serves. Suppose while playing racquetball, Tim gets his first serve in about 75% of the time. He gets his first serve in and wins the point about 50% of the time. What is the probability that he wins the point, given that he gets his first serve in?

50. Suppose 170 people were randomly selected and asked whether their job earnings during the past month were less than $4000 or greater than or equal to $4000. Their responses are given in the following table.

| | **Earnings** | | |
Age	<$4,000	≥$4,000	Unemployed
18–27	14	7	9
28–37	16	17	3
38–47	4	31	0
48–57	6	28	2
>58	12	10	11

If one response was selected at random, find the probability that it indicated that the person earned more than $4,000.

51. Suppose 170 people were randomly selected and asked whether their job earnings during the past month were less than $4000 or greater than or equal to $4000. Their responses are given in the following table.

Age	Earnings		
	<$4,000	≥$4,000	Unemployed
18–27	14	7	9
28–37	16	17	3
38–47	4	31	0
48–57	6	28	2
>58	12	10	11

If one response was selected at random, find the probability that it indicated that the person earned at least $4,000 given that they were of age 38 to 47.

52. Suppose 170 people were randomly selected and asked whether their job earnings during the past month were less than $4000 or greater than or equal to $4000. Their responses are given in the following table.

Age	Earnings		
	<$4,000	≥$4,000	Unemployed
18–27	14	7	9
28–37	16	17	3
38–47	4	31	0
48–57	6	28	2
>58	12	10	11

If one response was selected at random, find the probability that it indicated that the person was of age 18 to 27 given that they were unemployed.

Use Bayes' Theorem to calculate each probability. Round your answer to the nearest millionth, if necessary.

53. One of the world's most common parasites is Toxoplasma gondii which causes a disease called Toxoplasmosis. It is possible that 50% of the world's population is infected by toxoplasmosis. Cats spread Toxoplasma, and about 26% of the world population own cats. If 42% of people infected by toxoplasmosis own a cat, what is the probability that a cat owner is infected by toxoplasmosis?

54. The company expects the probability of economic recovery to be 4% and the probability of the company's revenue growth due to economic recovery to be 78%. Regardless of whether the economy grows, the company's revenue will grow with a 32% probability. What is the probability that the economy has risen if the company's revenue has grown?

55. Suppose 34% of students at a certain school know French and 23% know German. The share of French-speaking students who also know German is 12%. What is the probability that a student speaks French given that he or she speaks German?

56. Major depressive disorder affects approximately 7.1% of the US population aged 18 and older. Moreover, depression is the cause approximately two-thirds of the reported suicides in the US every year. Given that 0.015% of Americans die by suicide each year, what is the probability of a person committing suicide if they have depression?

10.4 PROJECT

THE PROBABILITY OF SPAM FILTERING

According to the website statista.com, 28.5% of all email traffic in 2019 was made up of spam—those pesky, useless, and potentially dangerous messages that just clog our email inboxes. Most email servers these days can filter spam automatically. Spam messages often have certain suspicious phrases in the subject lines. For example, "You Have Been Selected" is one such phrase.

An incoming email is checked for key elements, such as this phrase, then the server decides whether to put the email in your mailbox or send it to the spam folder.

In this activity, you will estimate the probability that an email with a specific subject line is classified as spam.

Let $P(S)$ be the probability that an email you have received is spam and $P(S^c)$ be the probability that the email is not spam.

1. According to statista.com, what were the values of $P(S)$ and $P(S^c)$ in 2019?

Let's assume that 10% of all spam messages contain the word *selected* in the subject line. In order to simplify our notation, we will name the events as follows.

   S = email is spam

   S^c = email is not spam

   W = subject line contains the word *selected*

   W^c = subject line does not contain the word *selected*

2. Express the statement "10% of all spam messages contain the word *selected* in the subject line" as a conditional probability.

3. We also will assume that 0.5% of all nonspam messages also contain *selected* in the subject line. Express the previous statement as a conditional probability.

Since every message can be classified as either spam or not spam, the probability that any message has *selected* in the subject line is the following.

$$P(W) = P(W \mid S)P(S) + P(W \mid S^c)P(S^c)$$

4. Compute the value of $P(W)$.

5. Finally, determine the probability that an email is spam, knowing it has the word *selected* in the subject line. (**Hint:** Use Bayes' Theorem.)

OBJECTIVES

1. Determine whether an experiment is a binomial experiment.

2. Calculate binomial probability.

3. Calculate cumulative binomial probability.

OBJECTIVE 1 Binomial Experiments

Let's turn our attention to situations where we want to calculate the probability of events that involve only two outcomes. Although this might seem restrictive, many situations can be thought of this way. For instance, all of the following scenarios have essentially two outcomes: a patient being treated for asthma either finds relief from the treatment or does not; when you answer a true or false question, either the answer is correct or not; a baseball batter will either get on base or not when he's at bat; or you flip a coin and either win the toss or not.

The scenarios we've listed, in which the outcome has two possible values, all qualify as *binomial experiments*. A binomial experiment is one in which there are a fixed number of identical trials and each trial is independent of the others. The two outcomes in a binomial experiment are often labeled *success* and *failure*, where success indicates the outcome of interest that we want to count. Note that although it might sound strange, labeling an outcome a success in an experiment is simply a method of record keeping, not a judgement on the outcome. For instance, in some medical areas the outcome to be counted is the occurrence of a disease, which is obviously not really a success. This terminology is typically used when discussing binomial experiments. As a result, it's important that we clearly specify which outcome is the success and which is the failure. In fact, all of the following guidelines must be met for an experiment to qualify as binomial.

📖 BINOMIAL EXPERIMENT

A **binomial experiment** is a probability experiment that satisfies the following assumptions:

1. The experiment consists of a fixed number n of identical trials.

2. Each trial is independent of the others.

3. For each trial, there are only two possible outcomes. For counting purposes, one outcome is labeled a *success* and the other a *failure*.

4. For every trial, the probability p of getting a success remains the same.

Example 10.5.1

Determining If an Experiment Is Binomial

Determine whether the given procedure meets the criteria of a binomial experiment. If not, identify at least one requirement that is not satisfied.

a. Surveying the student ratings of the food in the campus dining hall on a scale of 1 to 10.

b. Counting the number of times a professional basketball player makes a free-throw shot during a game.

c. Randomly choosing 3 seniors from a mixed group of 40 students for a panel of judges.

Solution

a. No; this experiment does not meet the criteria of a binomial experiment. There are ten outcomes for each trial instead of just two (the ratings 1, 2, 3, 4, 5,6, 7, 8, 9, and 10). Also, there is not a fixed number of trials.

b. Yes; this experiment does meet the criteria of a binomial experiment. There are two outcomes: makes a shot or does not make a shot. Each shot is an independent trial. There are a fixed number of trials (the number of free-throw shots the player takes in the game) and the probability of success is the same for every free-throw shot.

c. No; this experiment does not meet the criteria of a binomial experiment. Although there are only two possible outcomes (choosing a senior or not) and there are a fixed number of trials (3), the probability of choosing a senior does not stay the same for each trial. After each student is chosen for the panel, the probability of choosing a senior changes for the next student selection.

OBJECTIVE 2   Binomial Probability

If an experiment meets all the binomial requirements, we can compute the probability of observing a specified number of successes in the experiment by using the following formula.

f(x)   BINOMIAL PROBABILITY FORMULA

The probability of obtaining x successes in n independent trials of a binomial probability experiment is given by

$$P(x) = {_nC_x} \cdot p^x (1-p)^{(n-x)},$$

where x is the number of successes,

n is the number of trials, and

p is the probability of getting a success on any trial.

Coin tosses easily lend themselves to binomial experiments since it's obvious that there are only two outcomes (heads or tails), each toss is independent of another toss, and the probability of getting a head (or tail) is always the same (0.5). In Example 10.5.2, we consider a coin toss to illustrate how the binomial probability formula is used. In the remainder of the section, we explore other scenarios involving binomial experiments.

Example 10.5.2

Calculating Binomial Probability

What is the probability of getting exactly six heads in 10 coin tosses?

Solution

As we said earlier, coin tosses meet the criteria of a binomial experiment. And since we have a fixed number of tosses, we can use the binomial probability formula to calculate this probability. We will define a success as getting a head. Therefore, we want the probability of *exactly six* successes, so $x = 6$. The probability of flipping a head in one coin toss is 0.5, which means that $p = 0.5$. There are 10 coin tosses, so $n = 10$.

By Hand

Substituting these values into the binomial probability formula gives us the following.

$$P(x) = {}_nC_x \cdot p^x (1-p)^{(n-x)}$$
$$P(6) = {}_{10}C_6 \cdot (0.5)^6 (1-0.5)^{(10-6)}$$
$$= \frac{10!}{6!4!}(0.5)^6 (0.5)^4$$
$$\approx 0.2051$$

TI-83/84 Plus

The TI-83/84 Plus has a built-in binomial probability formula that can be accessed by pressing [2nd] [vars] to open the distr menu, scrolling down, and selecting binompdf(. You will be prompted to enter the following values.

trials is the number of trials in the experiment.

p is the probability of getting a success on any trial.

x value is the number of successes.

We will enter 10 for trials since there are 10 coin tosses. The probability of flipping a head in one coin toss is 0.5, so we'll enter 0.5 for p. Since we are finding the probability of exactly six successes, we'll enter 6 for x value. Move the cursor down to highlight Paste and press [enter] twice. The calculator will give the result 0.205078125.

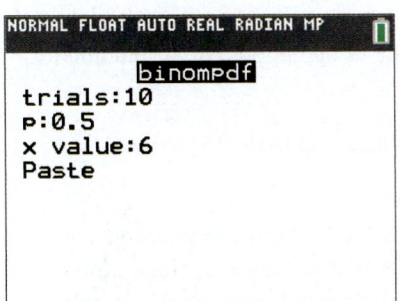

Figure 10.5.1

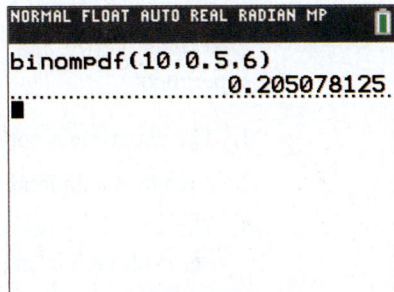

Figure 10.5.2

Microsoft Excel

We can use the built-in binomial distribution function, **BINOM. DIST(number_s, trials, probability_s, cumulative)**.

number_s is the number of successes.

trials is the number of trials.

probability_s is the probability of a success on any trial.

cumulative should be TRUE if you are finding the probability of at most *x* successes (which is called the cumulative probability and will be discussed later in the section) and it should be FALSE if you are finding the probability of exactly *x* successes. Default is FALSE.

Since we want to know the probability of getting *exactly* six heads, we want to use the probability mass function, which means we should either type FALSE for the last entry or leave it blank. In an empty cell, type "=BINOM.DIST (6, 10, 0.5, FALSE)" and press Enter. The value given is 0.205078125.

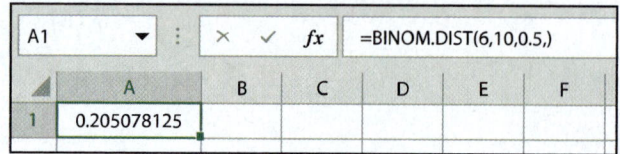

Figure 10.5.3

Therefore, the probability of getting exactly six heads in 10 coin tosses is approximately 20.51%.

Example 10.5.3

Calculating Binomial Probability

Suppose that 90% of all adults with allergies report that they get symptomatic relief with a specified medication. If the medication is given to a sample of 15 adults with allergies, what is the probability that exactly two *will not* experience symptomatic relief?

Solution

First let's verify that this process satisfies the assumptions of a binomial experiment.

1. The experiment consists of a fixed number of identical trials, 15, so $n = 15$.

2. Each trial is independent of the others.

3. There are only two possible outcomes for each trial: relief from symptoms or not. Although it might seem harsh, we will consider *not* experiencing relief from allergies as a success as that is our focus. (Remember, *success* is simply a label to help us count). Since we are looking for the probability of exactly two adults not getting relief, $x = 2$.

4. For every trial, the probability of getting a success remains the same. Since 90% get symptomatic relief, 10% do not report relief. Therefore, $p = 0.1$.

By Hand

Using the binomial probability formula for our solution gives us the following equation.

$$P(x) = {}_nC_x \cdot p^x (1-p)^{(n-x)}$$
$$P(2) = {}_{15}C_2 \cdot (0.1)^2 (1-0.1)^{(15-2)}$$
$$= \frac{15!}{2!13!}(0.1)^2 (0.9)^{13} \approx 0.2669$$

TI-83/84 Plus

As in the previous example, we can use the `binompdf(` formula from the `distr` menu. This time, `trials = 15`, `p = 0.1`, and `x value = 2`. The calculator will give the result `0.266895912`.

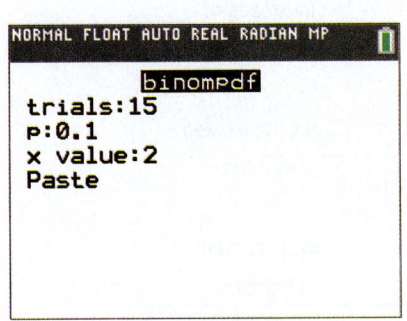

Figure 10.5.4

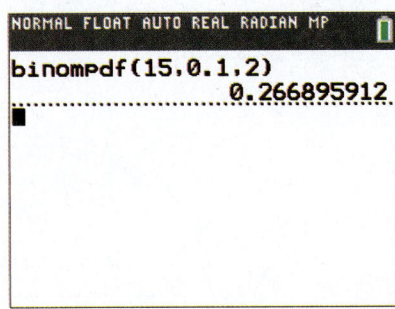

Figure 10.5.5

Microsoft Excel

Again, we can use the binomial distribution function, **BINOM. DIST(number_s, trials, probability_s, cumulative)**.

Since we want to know the probability that *exactly* two people do not experience symptomatic relief, we want to use the probability mass function, which means we should either type FALSE for the last entry or leave it blank. In an empty cell, type "BINOM.DIST(2, 15, 0.1, FALSE)" and press Enter. The value given is 0.266895912.

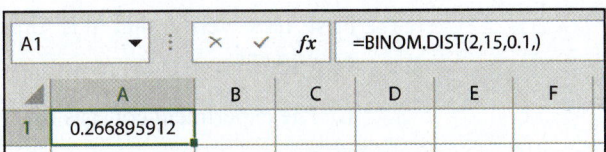

Figure 10.5.6

Therefore, the probability that exactly two out of the 15 adults will not experience symptomatic relief is 26.69%.

OBJECTIVE 3 Cumulative Binomial Probability

So far, we have calculated the probability of getting *exactly* a certain number of successes in a binomial experiment. Suppose we are interested in computing the probability of a range of outcomes instead. For instance, we can find the probability that *at least* two adults in the previous example experience symptomatic relief, or we could find the probability that *no more than* eight adults report relief. This is known as **cumulative probability**.

The key to identifying the proper range of successes to be counted in cumulative probability is the phrasing cues in a scenario. Some key phrases are provided in Table 10.5.1.

Table 10.5.1: Key Phrases

Symbol	Phrase	
<	Less than	Up to
	Fewer than	Under
>	Greater than	In excess of
	More than	Over
	Exceeds	
≤	Less than or equal to	Does not exceed
	No more than	At most
	Maximum	
≥	Greater than or equal to	No less than
	At least	Minimum

The following examples look at how to calculate cumulative binomial probability.

Example 10.5.4

Calculating Cumulative Binomial Probability

A quality control expert at a large factory estimates that 8% of all the batteries produced at the factory are defective. If he takes a random sample of 20 batteries, what is the probability that no more than two are defective?

Solution

First let's verify that this process satisfies the assumptions of a binomial experiment.

1. The experiment consists of a fixed number of identical trials, 20, so $n = 20$.

2. Each trial is independent of the others.

3. There are only two possible outcomes for each trial, either the battery is defective or it is not. Though it may sound strange, we will consider a defective battery as a success since that is what we are interested in counting. We want to know the probability that *no more than two* are defective; that is, two are defective, one is defective, or none are defective. Therefore, we need to find the probabilities for $x = 2$, $x = 1$, and $x = 0$.

4. For every trial, the probability of getting a success remains the same. Since 8% of all batteries produced are defective, the probability of getting an individual success is $p = 0.08$.

By Hand

We can find the probability of no more than two by adding together the three individual probabilities. Using the binomial probability formula for our solution gives us the following expression.

$$P(\text{No More Than 2 Defects}) = P(2) + P(1) + P(0)$$

$$= {}_{20}C_2 \cdot (0.08)^2 (0.92)^{18} + {}_{20}C_1 \cdot (0.08)^1 (0.92)^{19}$$

$$+ {}_{20}C_0 \cdot (0.08)^0 (0.92)^{20}$$

$$\approx 0.271091 + 0.328162 + 0.188693$$

$$= 0.7879$$

TI-83/84 Plus

Since we are not trying to find the probability of exactly *x* successes, we will not use the `binomialpdf(` formula. In addition to the binomial probability formula, the TI-83/84 Plus has a built-in cumulative probability formula. This formula can be used to find the probability of at most *x* successes. It can be accessed by pressing [2nd] [vars] to open the `distr` menu, scrolling down and selecting `binomcdf(`. Enter 20 for `trials`, 0.08 for `p`, and 2 for `x value`. Using an *x*-value of 2 with cumulative probability means finding the probability that there are no more than 2 successes. Once `Paste` is highlighted, press [enter] twice. The calculator will give the result 0.7879462459.

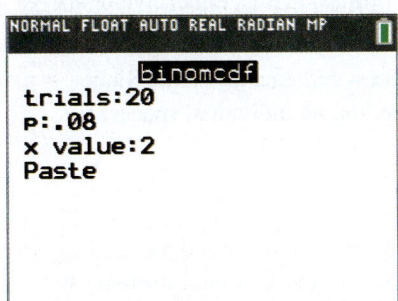

Figure 10.5.7

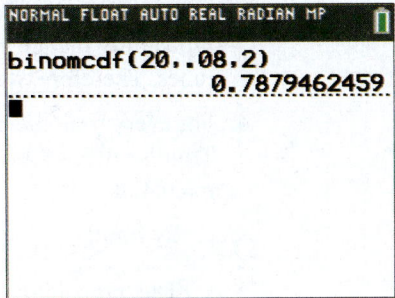

Figure 10.5.8

Microsoft Excel

As in the previous two examples, we can use the binomial distribution function, **BINOM.DIST(number_s, trials, probability_s, cumulative)**.

This time, however, since we are looking for the probability of no more than 2 successes, we want to use the cumulative mass function. This means we should use TRUE for the last entry. In an empty cell, type "=BINOM.DIST(2, 20, 0.08, TRUE)" and press Enter. The value given is 0.787946246.

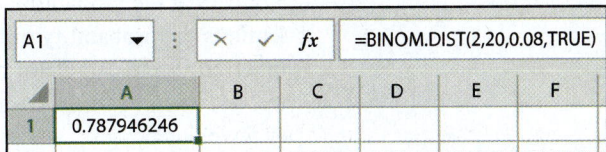

Figure 10.5.9

Therefore, the probability that no more than two out of the 20 batteries are defective is approximately 78.8%.

Example 10.5.5

Calculating Cumulative Binomial Probability

A baseball player's on-base percentage (OBP) is a measure of the likelihood that a runner will reach base safely each time at bat, taking into account things such as hitting the ball, being walked by the pitcher, or even being hit by a pitch. In 2019, Mike Trout had the leading OBP for Major League Baseball at 43.8%. This means that he reached first base 43.8% of the times he came up to bat. Assuming that each of his six at bats in a game is independent of any other at bat, what is the probability that he gets on base more than four times?

Solution

First let's verify that this process satisfies the assumptions of a binomial experiment.

1. The experiment consists of a fixed number of identical trials, six, so $n = 6$.

2. We are told that each trial is independent of the others.

3. There are only two possible outcomes for each trial: either Mike Trout reaches first base or he does not. We will consider reaching base a success since that is what we are interested in counting. We want to know the probability that he gets on base more than four times; that is, either five or six times. Therefore, we need to find the probabilities for $x = 5$ and $x = 6$.

4. For every trial, the probability of getting a success remains the same. Since Trout's OBP is 43.8%, the probability of his getting an individual success is $p = 0.438$.

By Hand

The probability of *more than four* successes can be found by adding together the two individual probabilities. Using the binomial probability formula for our solution gives us the following equation.

$$P(\text{More Than 4 on Base}) = P(5) + P(6)$$
$$= {}_6C_5 \cdot (0.438)^5 (0.562)^1 + {}_6C_6 \cdot (0.438)^6 (0.562)^0$$
$$\approx 0.054357 + 0.007061$$
$$= 0.0614$$

TI-83/84 Plus

Notice that the probability of *more than four* successes is the same thing as 1 minus the probability of four or fewer successes. Since the `binomcdf(` function in the calculator returns the probability of x or fewer successes, we will need to find the value of 1 minus the cumulative probability of 4 or fewer successes.

First, enter `1-` into the calculator. Then access the `distr` menu, scroll down, and select `binomcdf(`. Enter 6 for `trials`, `0.438`, for p, and 4 for the `x value`. Highlight `Paste` and press enter twice. The calculator will give the result `0.0614179799`.

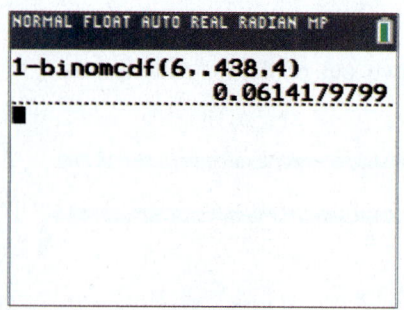

Figure 10.5.10

Microsoft Excel

We can again use the binomial distribution function, **BINOM. DIST(number_s, trials, probability_s, cumulative)**, to find the probability of 4 or fewer hits. This number subtracted from 1 is the probability of more than 4 hits.

Note that we'll be using TRUE for the cumulative argument since we are not finding the probability of exactly 4 hits.

In an empty cell, type "=1-BINOM.DIST(4, 6, 0.438, TRUE)" and press Enter. The value given is 0.06141798.

A1	▼	⋮	×	✓	fx	=1-BINOM.DIST(4,6,0.438,TRUE)		
◢		A		B	C	D	E	F	G
1		0.06141798						

Figure 10.5.11

Therefore, the probability that Mike Trout reaches base more than four times out of six at bats is approximately 6.14%.

Example 10.5.6

Calculating Cumulative Binomial Probability

Suppose 44.1% of all voters in the last election supported the current governor. A telephone survey contacts 200 voters from the last election and asks if they voted for the current governor. What is the probability that at least half of the voters contacted supported the current governor in the last election?

Solution

First let's verify that this process satisfies the assumptions of a binomial experiment.

1. The experiment consists of a fixed number of identical trials, 200, so $n = 200$.

2. Each trial is independent of the others.

3. There are only two possible outcomes for each trial, either they voted for the governor or they did not. We will consider voting for the governor a success since that is what we are interested in counting. We want to know the probability that *at least half* of the trials were successes; that is, 100, 101, 102, …, or 200 successes. Therefore, we need to find the probabilities for $x = 100$, $x = 101$, $x = 102$, and so forth all the way to $x = 200$.

4. For every trial, the probability of getting a success remains the same. Since 44.1% of all voters in the last election supported the current governor, the probability of getting an individual success is $p = 0.441$.

As you can see, we would need to find 101 individual probabilities if we were to try and calculate this by hand. As this is unreasonable, we will show the calculator method as well as how to use Microsoft Excel to calculate this probability.

TI-83/84 Plus

The probability that *at least half* of the survey respondents voted for the current governor is the same thing as 1 minus the probability that *less than half* of the respondents voted for the current governor. So we will use the calculator to find 1 minus the cumulative probability of 99 or fewer successes.

First, enter 1− into the calculator. Then access the distr menu, scroll down, and select binomcdf(. Enter 200 for trials, 0.441, for p, and 99 for the x value. Highlight Paste and press [enter]. Press [enter] again. The calculator will give the result 0.0541418839.

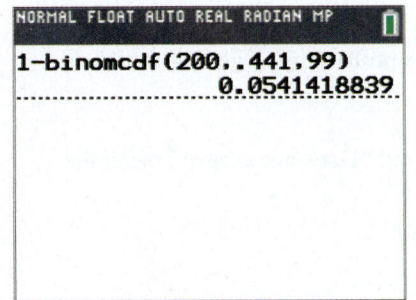

Figure 10.5.12

Microsoft Excel

We can use the binomial distribution function, **BINOM.DIST(number_s, trials, probability_s, cumulative)** to find the probability that less than half of the survey respondents voted for the current governor. This number subtracted from 1 is the probability that at least half of the respondents voted for the current governor.

In an empty cell, type "=1-BINOM.DIST(99, 200, 0.441, TRUE)" and press Enter. The value given is 0.054141885.

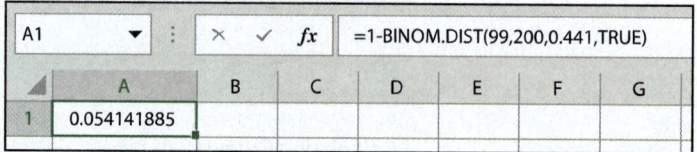

Figure 10.5.13

Therefore, the probability that more than half of the survey respondents voted for the current governor in the last election is approximately 5.4%.

Skill Check Answers

1. **a.** Yes **b.** No; there are ore than two outcomes. **c.** No; there is not a fixed number of trials, and the probability of getting an ace changes with each draw unless the card is replaced each time.

2. $P(\text{At Least 4 Sixes}) = P(4 \text{ Sixes}) + P(5 \text{ Sixes}) + P(6 \text{ Sixes}) + P(7 \text{ Sixes})$
$$+ P(8 \text{ Sixes}) + P(9 \text{ Sixes}) + P(10 \text{ Sixes})$$

10.5   Exercises

✔ CONCEPT CHECK

1. In a binomial experiment, the two possible outcomes are labeled as "_____" and "_____."

2. The probability of at least 3 successes in a binomial experiment is an example of _____ probability.

3. True or False: In a binomial experiment, the probability of getting a success must be the same for every trial.

4. True or False: The phrases "Greater than" and "Does not exceed" mean the same thing.

🔅 PRACTICE

5. Counting the number of broken lightbulbs in a case of lightbulbs.

6. Surveying whether restaurant customers prefer mild, medium, or hot salsa on their tacos.

7. Surveying whether customers at a local ice cream shop prefer their ice cream served in a cup or a cone.

8. Counting the number of male and female lions out of the six lions at a local zoo.

9. Randomly selecting five theater majors to participate in an upcoming play from a group of 40 students comprised of theater majors and non-theater majors.

10. Asking a class of 32 students whether they would prefer the final exam to be free answer or multiple choice.

11. Asking the employees of a small business whether they'd prefer a catered lunch of tacos, pizza, or neither.

12. Asking each audience member in a theater whether they enjoyed the movie they just watched or not.

13. Determining the number of pickles in each jar of a case of 16 jars of pickles.

14. Asking the 140 guests at a wedding whether they would prefer chicken or steak for their dinner entree.

15. Counting the number of true questions on a 20-question true-or-false test.

16. Tracking whether patients entering the ER during an afternoon are having chest pain or not.

🚀 APPLICATIONS

Calculate the binomial probability. Round your answer to the nearest ten-thousandth, if necessary.

17. What is the probability of getting exactly 4 heads in 8 coin tosses?

18. When rolling a standard 6-sided die 9 times, what is the probability of the die landing on an even number exactly 2 times.

19. A spinner for a board game has 10 wedges, 4 of which are blue while the other wedges are red. What is the probability of the spinner landing of red exactly 4 times if it is spun 5 times.

20. A spinner at a carnival has 8 wedges. Landing on one specific wedge wins a prize while landing on any of the other wedges does not win a prize. If the spinner is spun 5 times, what in the probability that exactly zero prizes are won.

21. Data collected from parents indicates that 12% of children attending a summer camp are allergic to bees. If 10 children are stung by bees during the summer camp, what is the probability that exactly 8 of the children are not allergic to bees.

22. Data pulled from the 72 resumes for a data entry position indicates that 75% of applicants have previous experience for the job. If 6 resumes are pulled at random for the first round of interviews, what is the probability that exactly 4 of the candidates have previous experience?

23. According to a survey of 75 dentists, 84% recommend that their patients with gum inflammation use a pH-balanced mouthwash. If 5 of these dentists are asked at random, what is the probability that exactly 2 of them recommend a pH-balanced mouthwash?

24. Tea Leaves magazine surveyed its readers and discovered that 62% of them prefer milk in their tea instead of sugar. If 20 readers were surveyed at random, what is the probability that exactly 12 of them would prefer milk in their tea?

25. The manager of a bookstore discovers that 16% of customers enjoy sitting in the bookstore café to read. What is the probability that exactly 9 out of 10 randomly selected customers do not enjoy sitting in the bookstore café to read?

26. Human resources learns that 54% of all workplace accidents happen during third shift. If 9 accidents are reported in one week, what is the probability that 4 of them did not happen during third shift?

27. The local hospital's data indicate that 52.5% of newborn babies are boys. What is the probability that of the next 7 newborn babies, less than 3 will be boys?

28. The probability of certain flower seeds sprouting is estimated to be 86%. If 20 seeds are sowed, what is the probability that no more than 3 will not sprout?

29. Suppose that 60% of tickets for a certain lottery win a prize and the remaining 40% do not win anything. If Carol buys twenty lottery tickets, what is the probability that the number of winning tickets will not exceed ten?

30. It is believed that about 10% of people in the world have blue eyes. What is the probability that out of 30 randomly chosen people, at most 5 will have blue eyes?

31. About 7% of people in the world use Spanish as their native language. If 50 random people are selected, what is the probability that fewer than seven of them are Spanish native speakers?

32. Suppose that 1% of juice cans come underfilled from the filling machine. What is the probability that in a batch of 100 juice cans, no more than two will be underfilled?

33. Heating is one of the ways to improve the quality of a natural gem. The gem is put into a high-temperature flame for a specific time and then allowed to cool slowly. This process can enhance the color of the gem and remove the inclusions. If we assume that about 3% of the gems get significantly better color and clarity after the heat treatment, what is the probability that of 10 gemstones, at least 2 will become significantly better?

34. If an archer has an 80% chance to hit the target, what is the probability that he will hit it at least 8 times in 10 tries?

35. A machine includes a set of 8 identical nodes, and it can continue working while at least 6 nodes are functioning correctly. If the probability of each node failing at any particular moment is 0.6%, what is the probability that the machine will stop working?

36. As a part of an advertising campaign, each chocolate bar of a particular company comes with a sticker hidden inside. Jacob wants to collect all 15 different stickers, and now he is missing only one. If the chance of getting each sticker is the same and Jacob buys five chocolate bars, what is the probability that he will find more than one copy of the missing sticker?

37. Suppose eight cards are drawn with replacement from a standard deck of cards. What is the probability that the number of red cards drawn is no less than six?

38. Nine fair six-sided dice are thrown. What is the probability of getting a 1 on more than two dice?

39. The probability of winning when placing a bet on a single color (either red or black) in a game of roulette is 18/38. What is the probability of winning more than half of such bets in 20 roulette games?

40. A recent survey showed that about 9% of adult people in the US do not have a car in the household. If 40 adult US people are chosen at random, what is the probability that a minimum of six of them do not have a car?

41. Suppose that 12% of people in a certain population suffer from migraines. If 40 people are selected at random from this population, what is the probability that at least 5 of them suffer from migraines?

42. In a recent survey conducted among young adults, about a quarter of respondents stated that they are indifferent toward the political life of their country. If a group of 30 young adults is selected at random, what is the probability that the number of people indifferent to politics will be greater than or equal to 7?

43. About 45% of the US population have blood type O. What is the probability that at least 10 out of 20 randomly selected US people have blood type O?

44. According to the 2019 National Survey on Drug Use and Health, about 55% of US adults aged 26 or older drank alcohol in the latest month. If 25 US adults are chosen at random, what is the probability that at least 15 of them have consumed alcohol in the latest month?

45. There are three major colors of Labrador Retrievers: black, brown, and yellow. The individual color depends on the combination of several specific genes. When two black Labradors with a particular genetic code are mated together, there is about a 25% chance for each of their puppies to have a yellow coat. If there are 7 puppies in a litter of two such black Labradors, what is the probability that more than 3 of them are yellow?

46. In a certain survey, 26% of adult respondents indicated that they did not have any physical activity during the week preceding the survey. If 15 adult people are randomly selected from the population being surveyed, what is the probability that at least 10 of them have exercised in the past week?

10.5 PROJECT

AMERICA'S THIRD FAVORITE PET: CAN WE PREDICT THEIR FUR COLOR?

According to the Humane Society of the United States, rabbits are the third most popular type of mammal owned as pets, with cats and dogs ranking higher. In this activity, you will calculate the probability of seeing a particular fur color when breeding rabbits.

Suppose that a particular breed of rabbit can be one of two colors: gray or white. The color of this breed of rabbit is determined by whether it has at least one dominant allele G in its genotype or whether it has only the recessive allele g. Out of the three possible genotypes (GG, Gg, and gg), rabbits with genotypes GG and Gg exhibit gray fur while rabbits with genotype gg exhibit white fur. Assume that each parent passes on one of its two color alleles to each offspring with equal probability.

If we breed two gray rabbits of genotype Gg, the possibilities for the fur color of the offspring are given in the following table.

1. Use the table to determine the probability of a rabbit being gray (GG or Gg alleles) and determine the probability of the rabbit being white (gg alleles) as decimal values.

2. Construct a table for the breeding of a white rabbit (gg) and a gray rabbit (Gg).

3. Use your table from part 3 to determine the offspring coloring probability for the breeding of a white rabbit (gg) and a gray rabbit (Gg).

   a. What is the probability that an offspring will be white (gg)?

   b. What is the probability that an offspring will be gray with genotype Gg?

   c. What is the probability that an offspring will be gray with genotype GG?

4. Use the binomial distribution to determine the probability that a litter of 10 offspring from a white rabbit (gg) and a gray rabbit (Gg) will contain exactly 6 white rabbits.

5. Use the binomial distribution to determine the probability that a litter of 10 offspring from a white rabbit (gg) and a gray rabbit (Gg) will contain at least 1 gray rabbit.

6. Without doing any computations, explain what the color of the parents must be to guarantee 100% of white rabbits as offspring.

7. Explain what the color of the parents must be to guarantee 100% of gray rabbits as offspring.

10.6 EXPECTED VALUE

OBJECTIVES

1. Calculate expected value for a random variable.

OBJECTIVE 1 Expected Value for a Random Variable

As we've seen through all the sections thus far in this chapter, probability is a way for us to describe how likely something is to happen. For example, we learned how to find the probability that you win a trip to Cancun with friends and even the probability of having prostate cancer given you have a positive test. In this section, we turn our attention to scenarios that involve things like predicting the average winnings for a state lottery ticket or the average expense a business should expect to incur for a customer loyalty program. Let's begin by defining some concepts needed to solve these types of problems.

We refer to the random numerical outcomes of an event as values of a *random variable*. For instance, suppose we roll a six-sided die. We know that one of the numbers one through six will appear. The roll of a die is an example of a random variable; that is, a variable whose value is determined by chance. (This is unlike the term *variable* in an algebraic equation, whose value we can systematically solve for.) Or suppose we look at the amount of screen time you log on your phone each day. Again, it is a random variable because we cannot know beforehand what the amount of time will be. Formally, a random variable is usually denoted with a capital letter, such as X, Y, or Z.

📖 RANDOM VARIABLE

A **random variable** is a variable whose numeric value is determined by the outcome of a probability experiment.

Let's go back to our die-rolling example. Suppose you roll a fair six-sided die, and you win $1.00 times the number you roll. If you roll a six, you win $6.00. If you roll a two, you win $2.00. You might begin to ask yourself, "how much should I expect to win when I roll the die?" We can calculate your expected winnings by knowing the probability of each outcome (which happens to be $\frac{1}{6}$ for all of the numbers on a fair die), and then multiplying each probability by its outcome.

$$\text{Expected Winnings from a Die Roll} = \$1 \cdot \frac{1}{6} + \$2 \cdot \frac{1}{6} + \$3 \cdot \frac{1}{6} + \$4 \cdot \frac{1}{6} + \$5 \cdot \frac{1}{6} + \$6 \cdot \frac{1}{6}$$

$$= \$3.50$$

This quantity is referred to as the *expected value E* of a random variable. Expected value means just what you intuitively think it means—the value we should expect the random variable to have on average. You might be thinking to yourself, "but you can't win $3.50 on a single die roll because there isn't a 3.5 on a die!" and you'd be right. Expected value is a *long-term average*. So if you rolled the die 100 times, you should expect to win an average of $3.50 per roll, or $350 at the end of the 100 rolls.

EXPECTED VALUE

The **expected value** E of a random variable W is calculated by multiplying each possible outcome by the probability of it occurring and then adding these products together.

$$E(W) = w_1 \cdot P(w_1) + w_2 \cdot P(w_2) + \cdots + w_n \cdot P(w_n)$$

Here, $w_1, w_2, w_3, \ldots, w_n$ are possible outcomes of random variable W with respective probabilities of occurring $P(w_1), P(w_2), P(w_3), \ldots, P(w_n)$.

We'll use a simple example of a carnival game to walk through using the expected value formula first, and then we'll move on to using expected values to analyze financial scenarios.

Example 10.6.1

Calculating Expected Value

Figure 10.6.1 shows a wheel used in a carnival game at a children's party where a player spins and wins the amount of candy the wheel lands on. Table 10.6.1 lists the possible outcomes along with their respective probabilities for the wheel.

Helpful Hint

Expected value is a *long-term average* of random numerical outcomes.

a. Use the wheel to determine the expected number of pieces of candy each player wins.

b. The host of the party does not want to run out of candy. If 15 children each play the game 5 times, how many pieces of candy should the host expect to give out?

c. What are the maximum and minimum amounts of candy that the host should expect to hand out at this game?

Table 10.6.1: Outcomes and Probabilities for Carnival Game

Outcome, x_i	Probability, $P(x_i)$	$x \cdot P(x_i)$
1	0.5	
2	0.25	
5	0.125	
10	0.125	

Figure 10.6.1: Carnival Spinner

Solution

a. In order to calculate the expected value, we need to multiply each outcome by its probability, and then add the resulting products.

By Hand

The third column shows each outcome multiplied by its probability.

Table 10.6.2: Outcomes and Probabilities for Carnival Game

Outcome, x_i	Probability, $P(x_i)$	$x \cdot P(x_i)$
1	0.5	0.5
2	0.25	0.5
5	0.125	0.625
10	0.125	1.25

The expected value is then the sum of the values in the third column.

$$E(x) = 0.5 + 0.5 + 0.625 + 1.25 = 2.875$$

TI-83/84 Plus

We can store information as lists in the TI-83/84 Plus and then use built-in functions to analyze the data. Specifically, we can find the expected value.

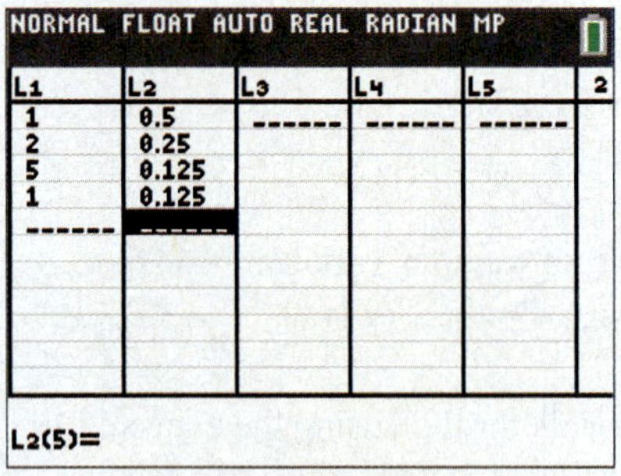

Press stat and then EDIT to access lists. Enter the outcomes, x, in L1 and the probabilities, $P(x)$, in L2. Now that you have two lists, press stat again and highlight CALC. Select 1-Var Stats. L1 should automatically appear for List, but you'll need to input L2 for FreqList. To do this, press 2nd 2. Move the cursor down to highlight Calculate and press enter. The first number listed, $\overline{x}$, is the expected value and is given as 2.875.

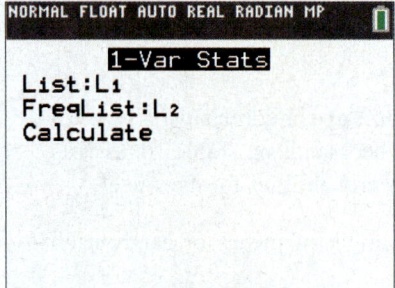

Figure 10.6.2

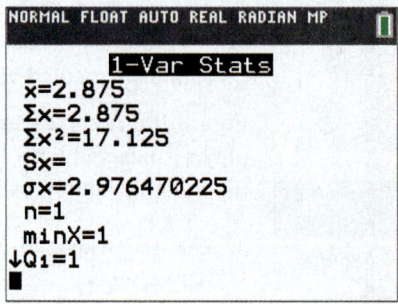

Figure 10.6.4

Figure 10.6.3

Microsoft Excel

We can also use Excel to calculate the expected value. For clarity, we'll begin by labeling the columns "Outcome, x", "Probability P(x)", and "x*P(x)" in A1 through C1, respectively. Next enter the necessary data into the first two empty columns; that is, enter the values of x in column A and their probabilities in column B. In cell C2, type the formula for the product, "=A2*B2", and press Enter. The value will appear as 0.5.

Drag the bottom right-hand corner of cell C2 down through cell C5. This will copy the formula into cells C3, C4, and C5 while updating the referenced cells. The values 0.5, 0.625, and 0.125, will appear in cells C3, C4, and C5, respectively.

The expected value is the sum of these products, which are now all located in column C. We can use the **SUM(number1, [number2],...)** formula to find this sum. In cell C6, type "=sum(" and then highlight the four cells we need to add: C2, C3, C4, and C5. Then type the closing parenthesis and press Enter. Alternatively, you could type "=sum(C1, C2, C3, C4)" or "=sum(C1:C4)". The resulting value is 2.875.

C5 | fx | =A5*B5

	A	B	C
1	Outcome, x	Probability, P(x)	x*P(x)
2	1	0.5	0.5
3	2	0.25	0.5
4	5	0.125	0.625
5	10	0.125	1.25

Figure 10.6.5

C6 | fx | =SUM(C2:C5)

	A	B	C	D
1	Outcome, x	Probability, P(x)	x*P(x)	
2	1	0.5	0.5	
3	2	0.25	0.5	
4	5	0.125	0.625	
5	10	0.125	1.25	
6			2.875	

Figure 10.6.6

Thus, the expected value of spinning the wheel at the party is 2.875. In other words, over time, the average spin will win 2.875 pieces of candy, even though a single spin will never win that amount.

b. If each of the 15 children spin the wheel 5 times, there will be $15 \cdot 5 = 75$ spins on the wheel. To find the amount of candy the host should expect to give out, we multiply the total number of spins by the expected value of 2.875.

$$\text{Total Candy Expected to Be Given Out} = 2.875 \cdot 75 = 215.625$$

Therefore, the host should expect to give out about 216 pieces of candy.

c. The host would give out the maximum amount of candy if each of the 75 spins landed on 10 pieces. Thus, $75 \cdot 10 = 750$ pieces of candy would be given out. The minimum would happen if all spins landed on 1 piece, in which case only 75 pieces of candy would be given out. This large difference in candy amounts shows why the expected value is a better gage of the amount of candy the host will need. This way the host is less likely to buy too much or too little candy. Notice that the expected value is not the average of the minimum and maximum amounts. We leave this for you to verify for yourself.

✔ Skill Check 10.6.1

Given the following table of outcomes for an event and their respective probabilities, complete the table and then determine the expected value of the event.

Table 10.6.3

Outcome, x_i	$P(x_i)$	$x_i \cdot P(x_i)$
3	0.45	______
6	0.3	______
9	0.25	______
		______

Example 10.6.2

Calculating Expected Value

The campus dining services at State University is preparing for the upcoming fall semester. Based on past years, they have observed the following data on the probability of selling different meal plans during the fall semester. Each plan consists of a set number of meals along with Plus Dollars, which can be used anywhere on campus but are restricted to food purchases. The university predicts an enrollment of 8421 students in the coming fall semester. Use Table 10.6.4 to answer questions a. through c.

Table 10.6.4: Meal Plans

Plan	Description	Price	Probability
1	19 meals per week and \$100 Plus Dollars for the fall semester	\$1245	$\frac{1}{10}$
2	14 meals per week and \$250 Plus Dollars for the fall semester	\$1245	$\frac{1}{15}$

Plan	Description	Price	Probability
3	10 meals per week and $350 Plus Dollars for the fall semester	$1245	$\frac{1}{30}$
4	$815 Plus Dollars to be used anytime during the fall semester (no meals)	$815	$\frac{1}{10}$
5	90 meals to be used anytime during the fall semester and $300 Plus Dollars	$870	$\frac{1}{12}$

Note: Students have the option of choosing to not buy a meal plan.

On average, how much does one student spend on a meal plan for the fall semester?

d. Suppose that for each meal plan sold, dining services makes $510 from Plan 1, $550 from Plan 2, $600 from Plan 3, $220 from Plan 4, and $270 from Plan 5. What is the expected profit per student for dining services in the upcoming semester?

e. What overall profit can dining services expect from student meal plans in the upcoming semester?

Solution

a. The question is asking us to find the average amount that students spend on meal plans during the fall semester; that is, the expected value of the price for a meal plan. To find the expected value, we need to multiply each meal plan price by the probability of a student purchasing that meal plan and then add these products together.

We also need to take into consideration the note at the bottom of Table 10.6.4 that says that students may choose not to buy a meal plan. This option must also be included in the calculation even though students who choose this option spend $0 on a meal plan. The probability that a student will choose no meal plan at all can be found by knowing that the probabilities in the sample space must add up to 1. Using subtraction, we can find the unknown probability as follows.

$$P\left(\text{No Meal Plan}\right) = 1 - \left(\frac{1}{10} + \frac{1}{15} + \frac{1}{30} + \frac{1}{10} + \frac{1}{12}\right) = \frac{37}{60}$$

Now that we know the probabilities for all possible amounts a student can spend on meal plans, we can calculate the amount we should expect one student to spend on average on a meal plan.

By Hand

Substituting these values into the expected value formula, we have the following equation.

$$E\left(\text{Meal Plan Cost}\right) = \frac{1}{10}(\$1245) + \frac{1}{15}(\$1245) + \frac{1}{30}(\$1245) + \frac{1}{10}(\$815)$$
$$+ \frac{1}{12}(\$870) + \frac{37}{60}(\$0)$$
$$= \$124.50 + \$83.00 + \$41.50 + \$81.50 + \$72.50 + \$0.00$$
$$= \$403.00$$

Helpful Hint

Recall that the sum of all the probabilities of all the outcomes in a sample space must equal 1.

TI-83/84 Plus

We can again use the calculator's built-in functions to find the expected value. Press [stat] and EDIT to access lists. Enter the price of each meal plan in L1 and the probabilities in L2. (Notice that the fractions entered in L2 are converted to decimal form.) Press [stat] again and highlight CALC. Select 1-Var Stats. Use L1 for List and L2 for FreqList. Move the cursor down to highlight Calculate and press [enter]. The first number listed, $\bar{x}$, is the expected value and is given as 403.

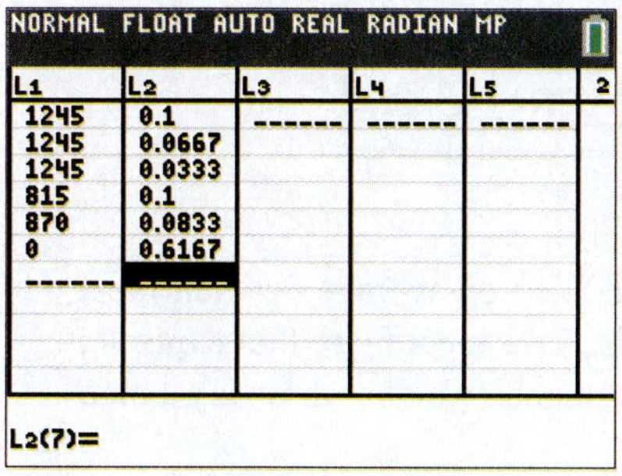

Figure 10.6.7

Figure 10.6.8

Microsoft Excel

We can also use Excel to calculate the expected value. For clarity, we'll begin by labeling columns "Price, x", "Probability P(x)", and "x*P(x)" in A1 through C1, respectively. Enter the price of each plan into column A and the probability that each plan will be chosen into column B. (Notice that the fractions entered in column B are converted to decimal form.) In cell C2, type the formula for the product of the first row, "=A2*B2" and press Enter. The value in C2 will appear as 124.5. To copy this formula down column C, drag the lower right-hand corner of the cell down the column. The row references will update in the formula and the values 83, 41.5, 81.5, 72.5, and 0 will appear in cells C3 through C7, respectively.

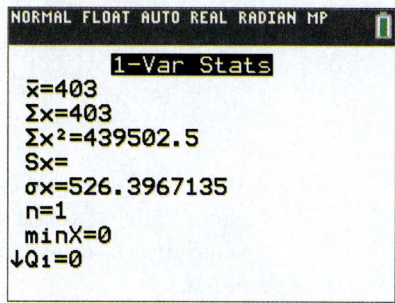

C7	▼	⋮	× ✓ fx	=A7*B7	
	A		B	C	D
1	Price, x		Probability, P(x)	x*P(x)	
2	1245		0.1	124.5	
3	1245		0.066666667	83	
4	1245		0.033333333	41.5	
5	815		0.1	81.5	
6	870		0.083333333	72.5	
7	0		0.61666666	0	

Figure 10.6.9

Next, use the **SUM(number1, [number2],...)** formula in cell C8 to determine the sum of the values in column C. The resulting value is 403.

Figure 10.6.10

Therefore, we would expect one student to spend $403 on average for their meal plan. While no single student actually pays $403 for a meal plan, this expected value gives us an average per student across all students, whether they buy a plan or not.

b. The profits per meal plan are summarized in Table 10.6.5.

Table 10.6.5: Meal Plan Profits

Plan	Probability per Meal Plan	Profit per Meal Plan
1	$\dfrac{1}{10}$	$510
2	$\dfrac{1}{15}$	$550
3	$\dfrac{1}{30}$	$600
4	$\dfrac{1}{10}$	$220
5	$\dfrac{1}{12}$	$270
No Plan	$\dfrac{37}{60}$	$0

In order to calculate the expected profit per student, multiply the estimated profit for each meal plan by the probability of that meal plan being chosen.

By Hand

Substituting these values into the expected value formula, we have the following equation.

$$E\left(\text{Profit per Student}\right) = \frac{1}{10}\left(\$510\right) + \frac{1}{15}\left(\$550\right) + \frac{1}{30}\left(\$600\right) + \frac{1}{10}\left(\$220\right)$$

$$+ \frac{1}{12}\left(\$270\right) + \frac{37}{60}\left(\$0\right)$$

$$= \$51.00 + \$36.67 + \$20.00 + \$22.00 + \$22.50 + \$0$$

$$= \$152.17$$

TI 83/84 Plus

Press … and EDIT to access the list function. Enter the profit per meal plan in L1 and the probabilities in L2. Press y z to return to the home screen. Press … again and highlight CALC. Select 1-Var Stats. Use L1 for List and L2 for FreqList. Move the cursor down to highlight Calculate and press Í. The first number listed, $\bar{x}$ =152.1666667 is the expected value.

NORMAL FLOAT AUTO REAL RADIAN MP

L1	L2	L3	L4	L5	2
510	0.1	------	------	------	
550	0.0667				
600	0.0333				
220	0.1				
270	0.0833				
0	0.6167				
------	------				

L2(7)=

Figure 10.6.11

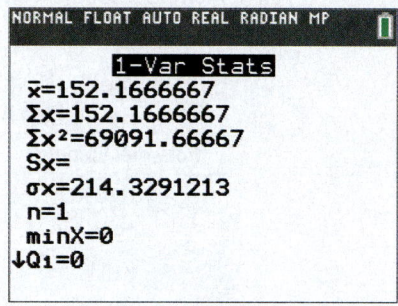

NORMAL FLOAT AUTO REAL RADIAN MP

1-Var Stats
$\bar{x}$=152.1666667
Σx=152.1666667
Σx^2=69091.66667
Sx=
σx=214.3291213
n=1
minX=0
↓Q₁=0

Figure 10.6.12

Microsoft Excel

Begin by labeling columns "Profit, x", "Probability, P(x)", and "x*P(x)" in cells A1 through C1. Enter the profit per meal plan into column A and the probabilities into column B. In cell C2, type the formula for the product of the first row, "=A2*B2", and press Enter. The value in C2 will appear as 51. To copy this formula down column C, drag the lower right-hand corner of the cell down the column. The row references will update in the formula and the values 36.66667, 20, 22, 22.5, and 0 will appear in cells C3 through C7, respectively. (Notice that the value in C3 may vary in length depending on the width of the cell.)

Next, use the =SUM(number1, [number2],…) formula in cell C8 to determine the sum of the values in column C. The resulting value is 152.1667.

C8	▼	⋮	× ✓	fx	=SUM(C2:C7)

	A	B	C	D
1	Profit, x	Probability, P(x)	x*P(x)	
2	510	1/10	51	
3	550	1/15	36.66667	
4	600	1/30	20	
5	220	1/10	22	
6	270	1/12	22.5	
7	0	37/60	0	
8			152.1667	

Figure 10.6.13

Therefore, dining services can expect an average profit of approximately $152.17 per student next semester.

c. Since the university predicts an enrollment of 8421 students next semester, dining services can multiply the expected profit for each student by the number of students to find their overall expected profit as follows.

Overall Expected Profit for Dining Services = $152.17 · 8421 = $1,281,423.57

Example 10.6.3

Calculating Expected Value

Suppose you are trying to decide between two different investment opportunities. The two plans are summarized in Table 10.6.6. The left column for each plan gives the potential earnings, and the right columns give their respective probabilities. Which plan should you choose?

Table 10.6.6: Investment Plans

Plan A		Plan B	
Potential Earnings	**Probability**	**Potential Earnings**	**Probability**
$1200	0.1	$1500	0.3
$950	0.2	$800	0.1
$130	0.4	− $100	0.2
− $575	0.1	− $250	0.2
− $1400	0.2	− $690	0.2

Solution

It is difficult to determine which plan will yield the higher return simply by looking at the table. However, we can calculate the expected value for each plan in order to compare them. We will calculate the expected value using Microsoft Excel.

For investment Plan A, begin by labeling the columns "Earnings, x", "Probability, P(x)", and "x*P(x)". Enter the potential earnings of Plan B into column A and the probabilities into column B. In cell C2, type the formula for the product of the earnings and the probability for the first row of the table, "=A2*B2" and press Enter. The value in C2 will appear as 120. To copy this formula down column C, drag the lower right-hand corner of cell C2 down the column. The row references will update in the formula and the values 190, 52, 57.5, and –280 will appear in cells C3 through C6, respectively.

Finally, use the **SUM(number1, [number2], ...)** formula in cell C7 to determine the sum of the values in column C. The resulting value is $24.5.

C7	▼ ⋮	× ✓	fx	=SUM(C2:C6)	
◢	**A**	**B**	**C**	**D**	
1	Earnings, x	Probability, P(x)	x*P(x)		
2	1200	0.1	120		
3	950	0.2	190		
4	130	0.4	52		
5	-575	0.1	-57.5		
6	-1400	0.2	-280		
7			24.5		

Figure 10.6.14: Plan A

Thus, the expected value for Plan A is $24.5.

Repeat this process in Excel for Plan B.

	A	B	C	D
	C7	▼ ⋮	× ✓ *fx*	=SUM(C2:C6)
1	Earnings, x	Probability, P(x)	x*P(x)	
2	1500	0.3	450	
3	800	0.1	80	
4	-100	0.2	-20	
5	-250	0.2	-50	
6	-690	0.2	-138	
7			322	

Figure 10.6.15: Plan B

Thus, the expected value for Plan B is $322.00.

From these calculations, we see that the expected value of Plan A is $24.50 and the expected value of Plan B is $322.00. Therefore, Plan B appears to be the wiser investment option over time.

Casinos rely heavily on expected values in order to turn a profit. Their entire business depends on creating games where the experience is appealing enough for you to play the game, but the probabilities end up in the casino's favor in the long run. Hence, the phrase "the house always wins." Let's look at a simple example.

Example 10.6.4

Calculating Expected Winnings

In American roulette, the wheel contains the numbers 1 through 36, alternating between black and red. There are two green spaces numbered 0 and 00.

a. Calculate the probability of the roulette ball landing on a red pocket (which will be the same as the probability of landing on a black pocket, since the number of red and black pockets is the same).

b. Calculate the probability of the ball not landing on a red pocket.

c. A player places a bet of $1.00 on red to play the game. If the ball lands on red, the player gets $1.00 for winning and receives the original dollar back. If the ball does not land on red, then the player simply loses the $1.00 placed on the bet. Calculate the expected winnings on one game.

d. If the player places a $1.00 bet on red 10 times, what is the player's expected winnings?

Solution

a. First note that there are 38 possible pockets on the wheel—one for each of the numbers 1 through 36 and the two green pockets containing 0s. Of those, 18 are red pockets. So the probability of the ball landing on a red pocket is the number of red pockets divided by the total number of pockets.

$$P(\text{Red}) = \frac{18}{38}$$

Figure 10.6.16:
American Roulette Wheel

b. The event that the ball does not land on a red pocket is the complement of the ball landing on a red pocket, so we can calculate the probability as follows.

$$P(\text{Not Red}) = 1 - P(\text{Red})$$

$$= 1 - \frac{18}{38}$$

$$= \frac{20}{38}$$

c. If a player bets $1.00 on red, the chance of winning a dollar ($1.00) is $\frac{18}{38}$ and the chance of losing a dollar (effectively winning $- \$1.00$) is $\frac{20}{38}$. Multiply each dollar amount by its respective probability and then add the products to find the player's expected winnings.

$$E(\text{Winnings}) = \frac{18}{38}(\$1.00) + \frac{20}{38}(-\$1.00)$$

$$\approx -\$0.05$$

This means that the player can expect to lose about 5¢ on average for every dollar bet on red while at the roulette table. Of course, a player never actually loses 5¢ on a single bet, but this is the expected average outcome for multiple trials.

d. If the player made the same bet (that is, $1.00 on red) 10 times, then we multiply the expected winnings by 10 to determine his anticipated result.

$$\text{Overall Winnings} = 10(-\$0.05) = -\$0.50$$

So after 10 bets of $1.00 on red, the player should expect to lose 50¢.

Let's take a moment and consider the casino's point of view. Gaining 50¢ from one player doesn't sound very profitable, but if the casino has millions of roulette players every year, they can expect to make millions of dollars over time. Think about how this advantage changes if you place the same bet on a European roulette wheel, which mimics the American wheel but only has one green pocket. Determining the difference in advantage is left as an exercise.

Skill Check Answers

1. $E(X) = 1.35 + 1.8 + 2.25 = 5.4$

10.6 Exercises

✔ CONCEPT CHECK

1. A random variable's numeric value is determined by the outcome of a _____ _____.

2. Expected value is a _____ average.

3. True or False: In the scenario where you win $1 if a fair six-sided die lands on a 1, $2 if it lands on a 2, and so on, the expected value must be an integer 1 through 6.

4. True or False: The expected value of a game is the average of the maximum and minimum outcomes.

💡 PRACTICE

Calculate the expected value of each scenario. Round your answer to the nearest hundredth, if necessary.

5.

x_i	$P(x_i)$
1	0.21
2	0.58
3	0.06
4	0.15
5	0.0

6.

x_i	$P(x_i)$
−$1.50	0.3
$0.00	0.5
$2.75	0.1
$5.00	0.1

7.

x_i	$P(x_i)$
25	$\frac{1}{3}$
15	$\frac{2}{5}$
10	$\frac{1}{15}$
5	$\frac{1}{5}$

8.

x_i	$P(x_i)$
1	0.12
2	0.09
3	0.33
4	0.14
5	0.04
6	0.28

9.

x_i	$P(x_i)$
$2.25	0.1
−$1.43	0.2
$4.67	0.3
$3.11	0.3
$0.56	0.1

10.

x_i	$P(x_i)$
10	$\dfrac{4}{21}$
5	$\dfrac{1}{7}$
35	$\dfrac{1}{6}$
15	$\dfrac{3}{10}$
20	$\dfrac{1}{5}$

11. The number of even numbers showing when a pair of standard six-sided dice are rolled.

x_i	$P(x_i)$
0	$\dfrac{9}{36} = \dfrac{1}{4} = 0.25$
1	$\dfrac{18}{36} = \dfrac{1}{2} = 0.5$
2	$\dfrac{9}{36} = \dfrac{1}{4} = 0.25$

12. The number of odd numbers showing when three standard six-sided dice are rolled.

x_i	$P(x_i)$
0	0.125
1	0.375
2	0.375
3	0.125

 APPLICATIONS

13. Alex wanted to learn how to play guitar, so his father bought him an inexpensive model for training. The number of strings x_i that will require a replacement within a year is distributed as follows.

x_i	$P(x_i)$
0	0.262
1	0.393
2	0.246
3	0.082
4	0.015
5	0.002
6	0

Find the average number of strings to be replaced in a year.

14. A shooter fires at 5 targets. The number of hits x_i is distributed as follows.

x_i	$P(x_i)$
0	0.01
1	0.08
2	0.23
3	0.34
4	0.26
5	0.08

Find the average number of targets the shooter will hit.

15. An insurance policy for a car costs $1500 per year. According to the past data, the probability of having a minor accident during a year is 0.2, and the average amount of insurance claim is $5000. The probability of having a serious accident is 0.01, and the average claim is $30,000.

   a. Find the average amount of the insurance claim.

   b. Find the average income of an insurance company from the sale of one policy.

16. A horse in a circus is trained to jump over obstacles. It will jump over one obstacle with a probability of 60%, over two obstacles with a probability of 30%, and over three obstacles with a probability of 10%. Find the expected number of obstacles that the horse will jump over.

17. A student bought a second-hand laptop. During the first year of its usage, he will have to spend $100 on repairs with the probability of 0.45 and $200 with the probability of 0.35. Find the average amount the student will have to spend on repairs in the first year.

18. Suppose that there is a 60% chance that Michael will go to the gym twice a week, a 35% chance that he will go once, and the remaining 5% that he will not go to the gym.

   a. What is the expected value for the number of times Michael goes to the gym in a week?

   b. Assuming there are 4 weeks in a month, what is the average number of times Michael goes to the gym in a month?

19. A coffee shop analyzed the sales for the latest month and got the following data. 55% of customers spent about $5, 25% spent about $9, and the remaining 20% spent about $11.

   a. What is the expected amount that a customer will spend?

   b. If the average number of customers per day is 80, what is the average daily revenue of the coffee shop?

20. A device consists of four independently operating nodes. The probability that exactly one node will fail in an experiment is 0.3. The probability of two nodes failing at once is 0.05. The probability of three nodes failing simultaneously is 0.03. The probability of all four nodes failing at the same time is negligible.

   a. Find the expected number of failing nodes in the experiment.

   b. If a similar test is performed simultaneously on three devices, what is the expected number of failing nodes?

21. A large entrepreneur plans to invest in a certain project. According to forecasts, the profit from the project will be $20,000 with a probability of 0.5, $30,000 with a probability of 0.2, or $25,000 with a probability of 0.3. What is the expected profit on this investment?

22. Suppose Piper eats out twice a week 15% of the time, she eats out once a week 35% of the time, and she doesn't eat out anytime during the week 50% of the time. What is the expected value for the number of times Piper eats out during a week?

23. Suppose that you and a friend are playing cards and decide to make a bet. If you draw two aces in succession from a standard deck of cards without replacing the first card, you win $50.00. Otherwise, you pay your friend $10.00.

   a. What is the expected value of your bet?

   b. If the same bet was made 25 times, how much would you expect to win or lose?

24. A European roulette wheel has only one green slot instead of two. Using Example 2 from this section as a guide, calculate the expected winnings on a European roulette wheel if a player bets $1.00 on red to play the game.

25. Jim likes to day-trade on the Internet. On a good day, he averages a $1100 gain. On a bad day, he averages a $900 loss. Suppose that he has good days 25% of the time, bad days 35% of the time, and the rest of the time he breaks even.

   a. What is the expected value for one day of Jim's day-trading hobby?

   b. If Jim day-trades every weekday for three weeks, how much money should he expect to win or lose?

26. A university in town is raffling off $20,000 for student scholarships. You can buy one ticket for $10, three tickets for $25, or five tickets for $40. Assume that the university sells 10,000 tickets.

   a. Find the expected value for each of the three ticket options: purchasing just one ticket, purchasing three tickets, or purchasing five tickets.

   b. Should you buy one, three, or five tickets in order to maximize the money you expect to have at the end of the raffle?

27. You need to borrow money from your sister. She's feeling quirky on the day you ask and says she wants you to flip a coin. Heads, you get $15, tails you get $5. Thinking this is weird, you ask your mother for money instead. She says she'll let you roll a die and she'll give you $2 times the number that appears on the die. Before agreeing to either of these unique offers from the "mathy" folk in your family, you decide to see which is the better offer by calculating the expected value for each method (realizing that you too fit the bill of a "mathy" member of your family). Which offer should you take? Explain your reasoning.

28. Assume that stock in Degree Compass, a predictive analytics company in higher education, returns the percentages shown in the table.

Degree Compass Stock Returns

Annual Return Rate	Probability
15%	0.17
30%	0.51
45%	0.32

Calculate the expected value of the return rate for stock in Degree Compass.

29. During the NCAA basketball tournament season, affectionately called *March Madness*, part of one team's strategy is to always foul their opponent's tall forward. Because he is so tall, he makes 57% of shots he takes close to the basket. However, when he is fouled, his free-throw shooting percentage is only 51.5%. The shots he makes close to the basket are worth two points and each of the two free throw shots after being fouled are worth one point.

   a. Calculate the expected value of the number of points the forward makes when he takes a shot close to the basket.

   b. Calculate the expected value of the number of points the forward makes when he shoots two foul shots.

   c. Based on these expected values, is fouling the tall forward a good strategy? Explain your answer.

30. On your next multiple-choice test, each question has four incorrect answers and one correct answer to choose from. Your professor tells you that each correct answer you make, you receive 1 point, but you lose $\frac{1}{4}$ point for each incorrect answer.

a. What is your expected gain or loss on a question if you have no idea of the correct answer and end up simply guessing?

b. What is your expected gain or loss if you guess on all 25 questions?

10.6 PROJECT

EXPECTED VALUE AND INSURANCE PREMIUMS

When you purchase a homeowners insurance policy, you pay a certain amount of money (the premium) to the insurance company. If nothing happens to your home, the insurance company keeps the money. If you make a claim, the company will pay to fix your home, usually spending a lot more than what your premium was. Due to this set up, the insurance company makes money on some policies and loses money on others.

In this activity, we will investigate the way an insurance company computes insurance premiums.

Suppose that in a certain neighborhood, an insurance company has used historical data to determine that the probability of a house fire occurring at a home over the course of one year is 0.02%.

1. What is the probability that there will not be a fire at a house in the neighborhood over the course of one year?

Assume that the insurance company charges a $300 annual premium for fire insurance. If there is a fire, the insurance company will pay out $200,000 to the homeowner.

2. Fill in the table below. What happens when there is no fire? Explain why this is.

Probability and Payout for Insurance Company

Event	Probability	Payout – Premium
Fire		
No Fire		

3. Determine the expected value that the insurance company will pay out.

4. What does the expected value say about the average loss or gain for the insurance company on each policy they sell?

5. How much would the insurance company expect to earn on average if it sold 10,000 policies in a year?

Chapter 10 Exercises

1. Ali is interested in the chance of someone buying vanilla ice cream, so he counts how many people order vanilla ice cream during two hours at a local ice cream shop. Determine whether the probability is experimental or classical.

2. Identify the sample space for Mrs. Okeela spinning the wheel once.

3. Describe how you would create the sample space for spinning the wheel twice.

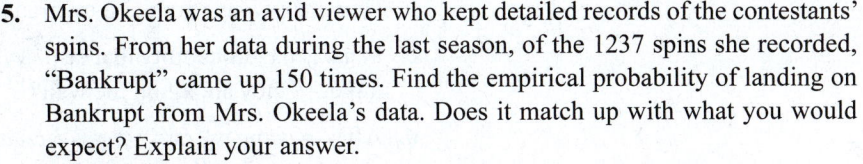

4. Find the classical probability of landing on

   a. $300

   b. Bankrupt

   c. $500

   d. A tropical vacation

5. Mrs. Okeela was an avid viewer who kept detailed records of the contestants' spins. From her data during the last season, of the 1237 spins she recorded, "Bankrupt" came up 150 times. Find the empirical probability of landing on Bankrupt from Mrs. Okeela's data. Does it match up with what you would expect? Explain your answer.

6. Dr. Gladden was interested in studying whether gender affected students' preferred social media outlet. A random sample of students was selected from a sociology class and asked to choose their top social media site out of Facebook, Twitter, or Pinterest. Their social media choice, gender, and education level (freshman, sophomore, junior, or senior) were recorded.

   a. Find the sample space for the possible outcomes of the survey using a tree diagram.

   b. Dr. Gladden decides to add Instagram and Tumblr to the choices for preferred social media sites. She also decides to include high school students as an additional education level. Use the Fundamental Counting Principle to establish how many possible outcomes are now in the sample space.

   c. The students in the survey are asked to rank their top two social media sites from the five possibilities. How many ways are there for students to rank their top two favorites?

7. Mrs. Darcy is asked to create a 5-digit pin code for an account and decides to use the digits in her anniversary, 1/28/13. How many different codes can she create?

8. Billy got a new tablet computer for Christmas, but his envious sister Meg set a passcode on it without his knowing. The passcode is a sequence of five digits.

   a. What is the probability that Billy can guess the correct code the first time by entering five random digits?

   b. Frustrated, Billy looks at the screen much more carefully and sees Meg's sticky fingerprints on five of the number-pad keys. What is the probability that Billy can correctly guess the code the first time now?

   c. Under duress, Meg admits that she only actually used four of the five smudged digits, but that she did keep them in increasing order. Using this information, what is the probability that Billy can correctly guess the code on his first attempt?

9. Emre was delighted with his BB gun for Christmas and couldn't wait to get outside to try out his marksmanship. He grabbed six bottles and cans to shoot at as he ran from the house. Unfortunately, no one noticed that he grabbed one of his Nana's antique glass tumblers as one of the six targets to shoot at.

   a. Initially he decides to only set up four of the targets. What is the probability that the antique tumbler is not one of them?

   b. How many ways are there to arrange all six targets on the wall?

   c. What is the probability that he places the antique tumbler as the left-most target, if they are all on the wall?

   d. After placing all six targets and shooting from left to right, he had already downed four of the targets by the time an adult discovered the tumbler was missing from the cabinet. What is the probability that the antique survived?

10. A single letter from the word MISSISSIPPI is chosen. What is the probability of choosing an S or a P?

11. An experiment consists of tossing a coin and then rolling a die. List the sample space for this experiment.

An urn contains 100 marbles. Fifty are purple, ten are green, fifteen are red, and twenty-five are orange. Two marbles are drawn at random one after another with replacement. Calculate each probability.

12. $P(\text{purple and orange})$ 13. $P(\text{green and green})$

14. $P(\text{same color})$

An urn contains 100 marbles. Fifty are purple, ten are green, fifteen are red, and twenty-five are orange. Two marbles are drawn at random one after another without replacement. Calculate each probability.

15. $P(\text{purple and orange})$ 16. $P(\text{green and green})$

17. $P(\text{same color})$

Two dice are rolled. Calculate each probability.

18. $P(\text{sum of 8})$

19. $P(\text{sum} > 7)$

20. $P(\text{sum is odd})$

21. $P(\text{sum is 6 or 8 or 10})$

One card is drawn from an ordinary deck of cards. Calculate each probability. Round your answer to the nearest millionth, if necessary.

22. $P(\text{heart or a face card})$

23. $P(\text{ace and a spade})$

24. $P(\text{face card and an ace})$

25. Apple's iPhone takes the leading place in the smartphone market. About 13% of people worldwide use iPhones, and about 45% of smartphone users in the United States use an iPhone. Given that 4% of world population is American, what is the probability that a person uses iPhone given that he or she is not American?

26. What is the probability of getting exactly 3 heads in 9 coin tosses?

27. Suppose that 65% of visitors to a chocolate factory prefer milk chocolate. If 10 visitors are randomly selected from a tour group, what is the likelihood that exactly 6 of them prefer milk chocolate?

28. Suppose that 20% of people trying a certain computer program buy it after the trial period expires. What is the probability that of the next 50 people who will try the program, no less than 15 will buy it?

29. A survey showed that 30% of people in a certain country are satisfied with their salary. If 10 citizens of this country are chosen at random, what is the probability that the number of satisfied people among them exceeds 3?

30. A study of attendance at a football game, based on weather, shows the following pattern. Find the expected value of the attendance at a football game.

Football Attendance Based on Weather

Weather	Attendance	Probability
Extremely Cold	10,000	0.05
Cold	15,000	0.30
Moderate	30,000	0.40
Warm	35,000	0.25

31. The odds in favor of Fast Enough winning the horse race are 6 to 5. If you place a bet of $2500 on Fast Enough and he wins the race, you win $6500.

   a. What is the expected value of your bet?

   b. Suppose you win. What are your winnings to take home, that is, how much ahead are you after the win?

32. On a roulette wheel, the slots are numbered 1 through 36, 0, and 00 for a total of 38 slots. Assume you bet on a single slot. The payout for landing on a single slot is 35 to 1. If a chip is worth $5, what is the expected value for a player that plays the number 12 repeatedly?

33. In a raffle, a ticket costs $50. If there are 5000 tickets sold in the raffle and the payout is $10,000, what is the expected value of purchasing one ticket?

34. In a lottery, 5% of tickets win $100, 1% of tickets win $500, 0.1% of tickets win $1000, and the rest do not win anything. Find the average winnings per ticket for this lottery.

35. A weather forecast gives the following probabilities for the number of rainy days x_i on the next week.

x_i	$P(x_i)$
0	0.002
1	0.017
2	0.077
3	0.194
4	0.290
5	0.261
6	0.131
7	0.028

Find the average number of rainy days the forecast predicts.

CHAPTER 10 PROJECT

Benford's Law: What Do Electricity Bills, House Prices, Population Numbers, Death Rates, and the Lengths of Rivers Have in Common?

Benford's law, also known as the Newcomb–Benford law, states that the first digit in real-life data is more likely to be a small number (such as 1) than it is to be a large number (such as 9). Simon Newcomb first observed the phenomenon in 1881 (with Frank Benford independently rediscovering it in 1938) after testing data from 20 different fields of study, including the surface areas of rivers, physical constants, and molecular weights. In this project, you will explore some applications of Benford's law.

Suppose we wrote each 4-digit number on separate pieces of paper and put them in a hat. (Notice that the first digit must be 1, 2, 3, 4, 5, 6, 7, 8, or 9 while the remaining digits could also include 0.)

1. How many numbers in the hat start with the digit 1?

2. What is the probability that a randomly selected number from the hat has 1 as its first digit?

3. Suppose the hat contained every 12-digit number. What would be the probability of randomly selecting a number with the first digit equal to 1?

Now, let's see if switching the first digit makes any difference in the probability.

4. How many 4-digit numbers in the hat start with the digit 9?

5. What is the probability that a randomly selected number from the hat has 9 as its first digit?

6. Suppose the hat contained every 12-digit number. What would be the probability of randomly selecting a number with the first digit equal to 9?

The investigation in parts 1 through 6 illustrate what we call a *uniform* distribution. It seems like a reasonable thing to accept; there shouldn't be any reason why 1 would be more or less likely to occur as the first digit of a randomly generated number than 9.

7. Suppose you are given a list of 1000 numbers. If we assume there is a uniform distribution of numbers within the list, how many of those numbers would begin with 1?

Benford's law, strangely enough, states that the distribution of first digits is indeed not uniform but it follows the probabilities in the following tables.

Digit	Probability of Being the First Digit
1	30.1%
2	17.6%
3	12.5%
4	9.7%
5	7.9%
6	6.7%
7	5.8%
8	5.1%
9	4.6%

8. Suppose that you were presented with a set of 1000 numbers. Use Benford's law to calculate how many numbers would you expect to start with 1.

9. Use Benford's law to calculate how many numbers would you expect to start with 9.

10. Compare the values found in parts 8 and 9. Explain what the values mean.

Let's test Benford's law using a few examples.

11. Write a list of the first 30 powers of 2. That is, write the numbers that represent 21, 22, 23, ..., 230. (Hint: The use of Microsoft Excel or a calculator will help with the calculations!)

12. How many numbers in the first 30 powers of 2 begin with the number 1?

13. What is the percentage of numbers that start with the digit 1 in the first 30 powers of 2?

14. What is the probability that a randomly selected power of 2 starts with the digit 1 if it is chosen from the list of the first 30 powers of 2? Does this result align with Benford's Law?

15. Next, perform an internet search to find a list of counties in South Carolina with a population under 250,000. What is the percentage of counties whose population starts with the digit 1? How does this compare to Benford's law?

16. Perform an internet search to learn about different ways Benford's law can be used to detect fraud. Describe one instance where Bendford's law detected fraud.

CHAPTER 10 REVIEW

Definitions

Probability: The **probability** of something occurring is a number between 0 and 1, inclusive, that represents how likely the event is to occur. If an event cannot occur its probability is 0. If an event is certain to happen its probability is 1.

Probability Experiment: A **probability experiment**, or a *trial*, is any process with a result determined by chance.

Outcome: Each individual result that is possible for a probability experiment is an **outcome**.

Sample Space: The **sample space** is the set of all possible outcomes for a given probability experiment.

Tree Diagram: A **tree diagram** uses branches to indicate all possible outcomes at each stage for an experiment. Each path of branches in a tree diagram indicates a single possible outcome for the experiment.

Event: An **event** E is a group, or subset, of outcomes in the sample space.

Formulas

Experimental Probability: In **experimental probability**, if E is an event, then $P(E)$, which is read "the probability that E occurs," is given by

$$P(E) = \frac{f}{n}$$

where f is the number of times the event occurs (the *frequency* of the event) and n is the total number of times the experiment is performed. The value of $P(E)$ will always be a real number between 0 and 1, inclusive.

Classical Probability: In **classical probability**, if all outcomes are equally likely to occur, then $P(E)$, which is read "the probability that E occurs," is given by

$$P(E) = \frac{n(E)}{n(S)}$$

where $n(E)$ is the number of outcomes in the event and $n(S)$ is the number of outcomes in the sample space. The value of $P(E)$ will always be a real number between 0 and 1, inclusive.

Definitions

Fundamental Counting Principle: The **Fundamental Counting Principle** states that for an experiment with a sequence of n stages where the first stage has k_1 outcomes, the second stage has k_2 outcomes, the third stage has k_3 outcomes, and so forth, the total number of possible outcomes for the experiment is calculated as follows.

$$(k_1) \cdot (k_2) \cdot (k_3) \cdot \ldots \cdot (k_n)$$

Replacement: When counting possible outcomes **with replacement**, objects are placed back into consideration for the following choice.

When counting possible outcomes **without replacement**, objects are not placed back into consideration for the following choice.

***n* Factorial:** In general, $n!$ (read "n factorial") is the product of all the positive integers less than or equal to n, where n is a positive integer.

$$n! = n(n-1)(n-2)(n-3)\cdots(2)(1)$$

The expression $0!$ is defined to be 1.

Combinations: A **combination** is a selection of objects from a group without regard to their arrangement; that is, their order is not important.

Permutations: A **permutation** is a selection of objects from a group where the arrangement is specific; that is, their order is important.

Formulas

Combinations: The following formula is used to calculate the number of combinations.

$$_nC_r = \frac{n!}{r!(n-r)!}$$

Permutations: The following formula is used to calculate the number of permutations.

$$_nP_r = \frac{n!}{(n-r)!}$$

In both of these formulas, r objects are selected from a group of n distinct objects, so r and n are both positive integers with $r \le n$.

Permutations with Repeated Objects: The number of distinguishable permutations of n objects, of which k_1 are all alike, k_2 are all alike, and so forth, is given by the following formula.

$$\frac{n!}{(k_1!)(k_2!)(k_3!)\cdots(k_p!)}$$

Here, $k_1 + k_2 + \cdots + k_p = n$.

Section 10.3 Probability of Single Events

Definitions

Complement: The **complement** of event E, denoted by E^c, consists of all outcomes in the sample space that are not in event E.

Odds: The **odds against** an event E are given as follows.

$$\frac{P(\text{Losing})}{P(\text{Winning})} \quad \text{or } a : b$$

The **odds in favor** of an event E are given as follows.

$$\frac{P(\text{Winning})}{P(\text{Losing})} \quad \text{or } b : a$$

Here, a is the number of unfavorable outcomes and b is the number of favorable outcomes.

Properties

Complement Rules of Probability

1. $P(E) + P(E^c) = 1$
2. $P(E) = 1 - P(E^c)$
3. $P(E^c) = 1 - P(E)$

Section 10.4 Addition and Multiplication Rules of Probability

Definitions

Mutually Exclusive Events: Mutually exclusive events are events that have no outcomes in common.

Independent Events: Independent events are events where the result of one event does not influence the probability of the other event happening.

Dependent Events: Dependent events are events where the result of one event influences the probability of the other event happening.

Conditional Probability: Conditional probability, denoted $P(B|A)$ and read "the probability of B given A," is the probability of event B occurring given that event A occurs first.

Formulas

Addition Rule of Probability: The **Addition Rule of Probability** states that the probability of event A happening *or* event B happening is given by the following formula.

$$P(A \text{ or } B) = P(A) + P(B) - P(A \text{ and } B)$$

Addition Rule of Probability for Mutually Exclusive Events: The **Addition Rule of Probability for Mutually Exclusive Events** states that the probability of event A happening or event B happening when A and B have no outcomes in common is given by the following formula.

$$P(A \text{ or } B) = P(A) + P(B)$$

Multiplication Rule of Probability for Independent Events: The **Multiplication Rule of Probability for Independent Events** states that for two independent events, A and B, the probability of event A happening *and* event B happening is given by the following formula.

$$P(A \text{ and } B) = P(A) \cdot P(B)$$

Multiplication Rule of Probability for Dependent Events: The **Multiplication Rule of Probability for Dependent Events** states that if A and B are dependent events, the probability of A and B occurring is given by the following formula.

$$P(A \text{ and } B) = P(A) \cdot P(B|A)$$

Theorem

Bayes' Theorem: The probability of event A occurring given that event B occurred is calculated as follows.

$$P(A|B) = \frac{P(B|A) \cdot P(A)}{P(B)}$$

Here, $P(B) \neq 0$.

Section 10.5 Binomial Probability

Definition

Binomial Experiment: A **binomial experiment** is a probability experiment that satisfies the following assumptions:

1. The experiment consists of a fixed number n of identical trials.

2. Each trial is independent of the others.

3. For each trial, there are only two possible outcomes. For counting purposes, one outcome is labeled a *success* and the other a *failure*.

4. For every trial, the probability p of getting a success remains the same.

Formula

Binomial Probability Formula: The probability of obtaining x successes in n independent trials of a binomial probability experiment is given by

$$P(x) = {}_nC_x \cdot p^x (1-p)^{(n-x)},$$

where x is the number of successes,

n is the number of trials, and

p is the probability of getting a success on any trial.

Section 10.6 Expected Value

Definition

Random Variable: A **random variable** is a variable whose numeric value is determined by the outcome of a probability experiment.

Formula

Expected Value: The **expected value** E of a random variable W is calculated by multiplying each possible outcome by the probability of it occurring and then adding these products together.

$$E(W) = w_1 \cdot P(w_1) + w_2 \cdot P(w_2) + \cdots + w_n \cdot P(w_n)$$

Here, w_1, w_2, w_3, ... , w_n are possible outcomes of random variable W with respective probabilities of occurring $P(w_1)$, $P(w_2)$, $P(w_3)$, ... , $P(w_n)$.

CHAPTER 11

Statistics

📑 **SECTIONS**

Introduction

Researchers at the University of Oregon[1] concluded that "by the time children are 3-years old, they already have an adult-like preference for visual fractal patterns commonly seen in nature." You might be thinking, "How can they know this?" The researchers didn't simply survey a lot of babies and adults to ask them what shapes they liked best. The researchers conducted a well-thought-out scientific study that gathered data, analyzed the data, and drew statistical conclusions. The study, published in the nature journal *Humanities and Social Sciences Communication* in November 2020, explains how the researchers went about conducting the research. They exposed participants, both young and old, to images of fractal patterns and recorded their preferences. In the publication, the researchers described the statistical methods used and how confident they are in their research.

Statistical research like that done in Oregon isn't limited to any one field. A quick glance at the headlines on the website ScienceDaily reveals the vast array of topics that humans are interested in studying. For instance, the top four articles listed on their site from September 29, 2021, cover medical technology, archaeology, the environment, and pain levels.

Contact-Tracing Apps Could Improve Vaccination Strategies

A Statistical Fix for Archaeology's Dating Problem

Threat of Catastrophic Supervolcano Eruptions

Standard Vital Signs Could Help Estimate People's Pain Levels

All of these studies seek to provide insight into our world by using established statistical methods, similar to the researchers at the University of Oregon. To look at how this is done, in this chapter we'll discuss different ways to gather data, describe the data we collect, effectively display the data so that others can comprehend it, and determine what information we can (and cannot) gain from the data.

---

1   University of Oregon, "Like adults, children by age 3 prefer seeing fractal patterns: Study concludes that preference for common natural patterns may develop early in life," ScienceDaily, December 11, 2020, https://www.sciencedaily.com/releases/2020/12/201211135507.htm

OBJECTIVE 1 Parts of a Statistical Study

Whether reviewing data from a medical trial for a new drug, federal election polls, or even baseball batting averages, we encounter statistics everywhere. As informed citizens and discerning consumers, it is crucial that we understand exactly what these statistics mean. The goal of statistics, a branch of mathematics, is to provide information so that informed decisions can be made. The word **statistics** itself may refer to either the science of gathering, describing, and analyzing data *or* to the actual numerical descriptions of sample data. To illustrate the difference in meanings, note how the word is used in the following two sentences.

- Section 11.1 is the first lesson in *statistics* (the science).

- Throughout this chapter, you will learn how to properly collect and analyze *statistics* (actual data).

📖 STATISTICS

Statistics is the science of gathering, describing, and analyzing data.

or

Statistics is the actual numerical description of sample data.

Most statistical studies are brought about by the desire to know the answer to a particular question. For instance, you might want to know which car gets the best gas mileage, which university is the best value for the money, or how often you might expect to be called in for jury duty. The search for the answers to many questions, like the ones just discussed, is helped along by a statistical study that involves collecting data, analyzing it, and then drawing conclusions.

When you read a statistical study, it is up to you to decide on the validity of the researcher's answer to the question being asked. Therefore, being aware of how to design a good study helps in critiquing the results of studies you encounter in professional journals or the news. The basic process of a statistical study is to state a question, collect data regarding the question, and then organize and analyze the data in order to answer the question.

☱ CONDUCTING A STATISTICAL STUDY

1. Design the study.

   a. State the question to be studied.

   b. Determine the population and variables.

   c. Determine the type of study: observational or experimental.

2. Collect the data.

3. Organize the data.

4. Analyze the data to answer the question.

This section focuses on the methods used in Step 1—the design of a statistical study. Each statistical study begins with a thoughtful design. What knowledge do you wish to gain? Where and how will you collect data to gain this knowledge? With these things in mind, we define the **population**, or the particular group we are interested in. A study's population consists of all persons or things about which we are trying to make an inference or decision.

Let's consider the question of which car gets the best gas mileage. The first assumption might be that the desired population for the study is "all cars." However, stop for a moment and think about the implications of trying to study all makes and models of all cars, in all parts of the world! Even if we were able to do it, this might not be very useful. What if the car with the best gas mileage was only available in Romania 20 years ago? As statisticians, it is imperative that we carefully define the population we are studying. By being more precise about how we define the population of interest, we narrow our potential population to only what we are truly interested in studying. In most cases, this also creates a more manageable population to work with. When defining the statistical group of interest, it is impossible to overclarify the population you wish to study. Therefore, we could narrow our attention to all passenger cars manufactured in the United States in 2021.

When studying a population, the values that change among members of the population are called **variables**. The information gathered about a specific variable is collectively called **data** and it may be in a variety of forms including counts, measurements, or observations.

> **Helpful Hint**
>
> The singular form of data is datum.

POPULATION, VARIABLES, AND DATA

Population

A **population** is the particular group of interest in a statistical study.

Variable

A **variable** is a value or characteristic that changes among members of the population.

Data

Data are the counts, measurements, or observations gathered about a specific variable in a population in order to study it.

The question being asked in the study determines the population and the variables. For instance, if the question is "Does taking the drug levodopa reduce symptoms such as slow movements and stiff, rigid body parts for those with Parkinson's disease?" then the population of interest would be people with Parkinson's disease. The variables would be the amount of levodopa taken and the occurrence of the symptoms. Notice how the question tells us both the population and the variables that are being studied.

Example 11.1.1

Identifying Population and Variables

Scientists want to determine if the existence of certain types of cells can predict longer overall survival in patients with a particular type of lung cancer. The goal of the study is to see if high levels of TGF-β (proteins produced by white blood cells) and Treg cells (suppressive immune cells) are associated with favorable clinical outcomes for patients with non-small cell lung cancer. Identify the population of interest and the variables in this study.

Solution

Since the study seeks to look at the outcomes for patients with non-small cell lung cancer, the population of the study is limited to patients with this specific type of lung cancer. The variables of interest are the amounts of TGF-β and Treg cells in patients along with the survival lengths for the patients.

✔ Skill Check 11.1.1

Identify the population of interest and the variables for the following scenario.

Real estate agencies in Kansas City, Missouri, are interested in how the housing market has changed over the past three years. The goal of the study is to determine any change in the average price per square foot for single-family homes as well as the average amount of time these homes stay on the market from listing to closing.

Collecting data, or information, from each member of the population always results in the best analysis we can have of a particular group. When we are able to achieve this, the data we've gathered is called a **census**. A census gives us the ability to speak with 100 % confidence about the data we gathered from the group as a whole.

We can identify population parameters from data gathered from a census. A **parameter** is the numerical description of a particular population characteristic. For instance, suppose the average gas mileage for all passenger cars manufactured in the United States in 2021 was 23.6. Then, 23.6 mpg is a population parameter. All populations have fixed parameters that numerically describe the structure of the population. The goal of a statistical study is to identify, as best we can, the population parameter we are interested in.

💬 Fun Fact

The United States Census happens once every 10 years, as required by the constitution; the most recent being in 2020. Although statistical data from the census is released as soon as they are available, individual specific records are sealed for 72 years—a number slightly higher than the average female life expectancy of 71.6 years.

Although conducting a census is the ideal way to identify population parameters, this is often an unrealistic goal. Sometimes the population is too big or too hard to collect data from, such as if we were to consider the gas mileage from all cars around the globe. That's a lot of cars! Even with a small population, data from every member may not be available to you; for example, obtaining the gas mileage from cars of *all* students registered in your class right now. When a census cannot be obtained, we gather information from a subset of the population, or a **sample**. The numeric descriptions of particular sample characteristics are called **sample statistics**. For instance, suppose you collected data on the gas mileage of the cars driven by the students who are in attendance in your class today and found that the average mpg was 19.1. Then, 19.1 mpg is a sample statistic.

📖 TERMS RELATED TO POPULATIONS AND SUBSETS

Census

A **census** is a study involving data obtained from every member of the population.

Parameter

A **parameter** is the numerical description of a particular population characteristic.

Sample

A **sample** is a subset of a population from which data are collected in a statistical study.

Sample Statistic

A **sample statistic** is the numerical description of a particular sample characteristic.

It's important to keep in mind the relationship between a population and a sample. Figure 11.1.1 and Table 11.1.1 help visualize this relationship and illustrate the differences between populations and samples. The large oval represents the entire population, and the smaller oval represents the sample chosen from the population.

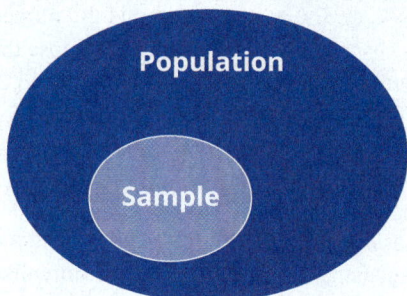

Figure 11.1.1

Table 11.1.1: Population vs. Sample

Population	Sample
Whole group	Part of a group
Group I want to know about	Group I do know about
Numerical descriptions of characteristics are called parameters	Numerical descriptions of characteristics are called statistics
Parameters are generally unknown	Statistics are always known
Parameters are fixed	Statistics vary with the sample chosen

💬 Helpful Hint

In set notation, the population is referred to as the universal set and the sample is a subset.

Example 11.1.2

Identifying Populations, Samples, Parameters, and Statistics

Two shortened survey reports are given. In each report, identify the following: the population, the sample, the results, and whether the results represent a sample statistic or a population parameter.

a. A headline about the rising obesity among young people led a school board to survey local high school students. Out of 231 students surveyed, 58% reported eating a "high fat" snack at least 4 times a week.

b. A nonprofit organization interviewed 618 adult shoppers at malls across Louisiana about their views on obesity in young people. The resulting report stated that an estimated 48% of Louisiana adults are in favor of government regulation of "high fat" fast food options.

Solution

a. Population: local high school students

Sample: the 231 students who were surveyed

Results: 58% of students surveyed eat a "high fat" snack at least 4 times a week.

The result refers to only those students who were surveyed, thus the result is a sample statistic.

b. Population: Louisiana adults

Sample: the 618 adult Louisiana mall shoppers who were surveyed

Results: 48% of Louisiana adults are in favor of government regulation of "high fat" fast food options.

The results refer to all Louisiana adults, thus this is a population parameter. This population parameter is an estimate based on the sample statistics, which were not reported.

✔ **Skill Check 11.1.2**

Determine if the result given represents a sample statistic or a population parameter.

In 2020, a survey by the US Census Bureau sampled about 60,000 households and found 52% of those households had a young adult living with a parent.

When designing a statistical study, we also need to determine the type of study that will be conducted. In other words, we need to determine how the data will be collected. One way to gather data is through observing data that already exists. Studies that use this method to gather data are called **observational studies**. Alternatively, we can generate data by performing an **experiment** or *simulation*. In situations where we wish to determine if one thing causes another to happen, it is best to use an experiment. Although observational studies can show that two variables are related in some meaningful way, they cannot be used to determine causality in the way that an experiment can.

The question that we are trying to answer is often the best indicator of how to collect the data for the study. For the research question "Which passenger car manufactured in the United States in 2021 gets the best gas mileage?" it would make sense to observe the car data that already exists. Thus, we would perform an observational study. For our question earlier about levodopa reducing symptoms in patients with Parkinson's disease, it would be better to compare two groups—one that takes the levodopa and one that does not. This scenario would be an experiment, which is necessary to draw the conclusion that one variable causes the change in the other variable.

> **📖 OBSERVATIONAL STUDY AND EXPERIMENT**
>
> **Observational Study**
>
> An **observational study** observes existing data.
>
> **Experiment**
>
> An **experiment** generates data to help identify cause-and-effect relationships.

Example 11.1.3

Identifying Observational Studies and Experiments

Given each scenario, which type of study would you conduct: an observational study or an experiment?

a. Researchers wish to determine if a vaccine is effective in preventing the spread of COVID-19.

b. You wish to determine the average salary of college graduates across your state during their first year after graduation.

Solution

a. An experiment would need to be used in order to establish a cause-and-effect relationship between the vaccine and COVID-19 spread prevention.

b. An observational study would be used since you only need to consider existing records of college graduates in your state and determine the average salary.

OBJECTIVE 2 Observational Studies

When we say observational studies observe existing data, we mean that we are not creating any new data. Instead, we are examining information that already exists, such as the results of an election or scores on standardized tests. No manipulation of the sample is performed by the researchers. One limitation that we've already pointed out is that no cause-and-effect relationship can be determined between the variables in this type of study. For example, if we collect data on GRE scores and the undergraduate GPAs of students, we might find a relationship between the variables, but it would be inappropriate to conclude that a cause-and-effect relationship exists.

So how do we go about collecting data from a population? Since we've already noted that obtaining the values of population parameters by conducting a census is often unrealistic, we will focus our discussion on samples and how their characteristics help us know more about the population. It's important to stop here and note that the way we choose a sample from the population is important. Care must be taken in choosing your sample so that the population is well represented and the results of the study are meaningful. For instance, consider our earlier example of a study wanting to know the average gas mileage for cars in the United

States. If the data you gathered were from half the students in your class, would this be an accurate picture of the population of cars driven in the United States? What if we only chose cars of people who lived within the city limits of a large city? Would that be any different from cars driven in only rural areas of the country?

When choosing a sample from the population, it's important that the sample is **representative** of the entire population. Without a representative sample, it may not be possible to accurately generalize the results to the population as a whole. If the sample is poorly chosen, a form of **bias** (a favoring of a certain outcome) might occur. Consider the sample of cars driven by residents of the city of San Diego. Because of its dense population, hilly terrain, and popularity with a young affluent crowd, cars driven there tend to be either compact cars, which are more fuel efficient, or luxury vehicles. This sample would certainly favor a narrow sector of cars for our study on average miles per gallon, which means it has a bias.

📖 REPRESENTATIVE SAMPLE AND BIAS

Representative Sample

A **representative sample** is one that has the same relevant characteristics as the population and does not favor one group of the population over another.

Bias

A **bias** is a favoring of a certain outcome in a study.

There are several standard statistical methods for choosing samples that allow for the best possible representative samples. As you might imagine, some methods are better suited to particular situations than others. We will highlight certain advantages and disadvantages of the different sampling methods as we describe each of them.

Random Sampling: A **random sample** is one in which every member of the population has an equal chance of being selected. Drawing for door prizes by placing all of the ticket stubs in a basket and mixing them up is a very common example of random sampling. A random sampling can also be performed by assigning identification numbers to each member of the population and using a random number generator, such as a Microsoft Excel function or an app for smartphones, to choose ID numbers. This basic sampling technique forms a basis for some of the other methods we will discuss.

Figure 11.1.2: Random Sample

A note of caution though. You should always be careful when letting a human choose a "random" sample. In reality, it is against our human nature to choose members of the population entirely at random. Take a moment to choose 3 random digits between 0 and 9. Try to make sure they are truly random. Now, consider these questions.

• Are the numbers spaced out or grouped closely together—probably spaced out, right?

• Are they all even, odd, or some of both?

• Did you have a reason for choosing any of the numbers—maybe your favorite?

• Did you alter any of your original responses. If so, why?

• Did you repeat any digits? If not, why? It wasn't against the rules.

• Are your digits consecutive, such as 1, 2, 3?

If any factors influenced your choice of digits, then your sample is not random. Because it is hard for humans to completely eliminate influences, it's best to take the human element out of it and allow technology to choose a random sample for us.

Stratified Sampling: A **stratified sample** is one in which members of the population are divided into two or more subgroups, called **strata**, that share similar characteristics like age, gender, or ethnicity. A random sample from each stratum is then drawn. Stratified sampling is used when we want to be confident that certain subgroups of a population are represented in the sample. For instance, if we divide the population of college students into strata based on the number of courses completed, then our strata would be freshmen, sophomores, juniors, and seniors. We would then select our sample by choosing a random sample of freshmen, a random sample of sophomores, and so on. This type of sampling is also called **quota sampling**.

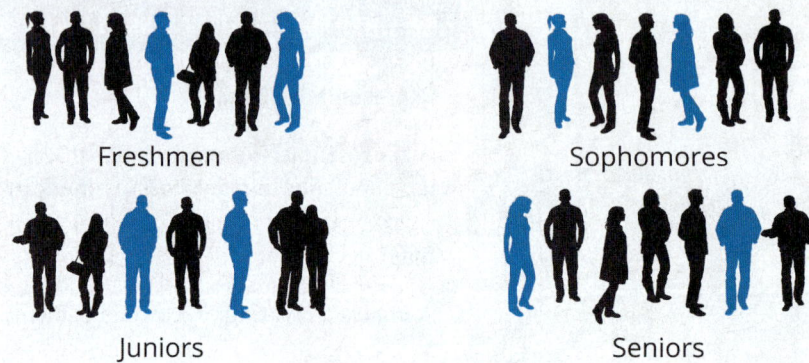

Freshmen   Sophomores

Juniors   Seniors

Figure 11.1.3: Stratified Sample

Stratified sampling is one of the best ways to enforce representativeness of particular characteristics on a sample. Since a random sample of all students cannot guarantee that any sophomores would be chosen, we would use a stratified sample if it were important that sophomores be included in our sample. Furthermore, by using stratified sampling you can preserve certain characteristics of the population. For instance, if freshmen make up 40% of the population, then we could choose 40% of the sample from the freshmen stratum to maintain this characteristic.

Cluster Sampling: Cluster sampling is similar to stratified sampling in that groups of the population are considered when choosing a sample. A **cluster sample** is one chosen by dividing the population into groups, called **clusters**, that are each similar to the entire population. The researcher then randomly selects some of the clusters for data collection. Unlike in stratified sampling, the sample consists of the data collected from *every* member of only the clusters selected. Populations often naturally lend themselves to subgroups from which this type of sampling might be done. For instance, if you wanted to know the average grade on the first test of all students in English 101, you could divide the students up by the section they are in. For cluster sampling, you would then randomly choose several sections and collect the grades from all students in each of the chosen section. Note that you must use caution when choosing clusters to ensure that each of the clusters is representative of the entire population.

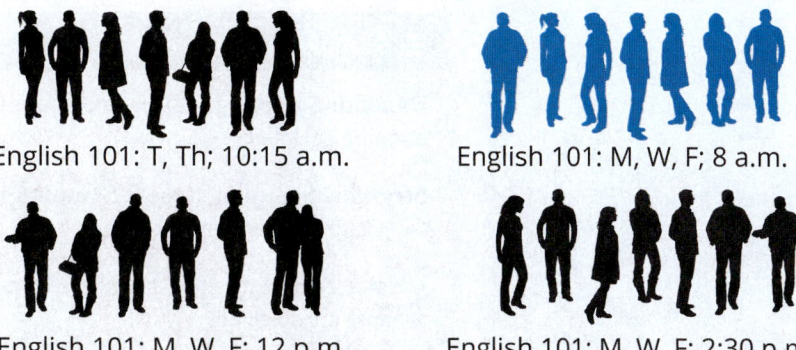

English 101: T, Th; 10:15 a.m.

English 101: M, W, F; 8 a.m.

English 101: M, W, F; 12 p.m.

English 101: M, W, F; 2:30 p.m.

Figure 11.1.4: Cluster Sample

Systematic Sampling: A **systematic sample** is one chosen by selecting every nth member of the population. Assembly lines in a factory are often tested for quality control using systematic sampling. If every 10th product is checked for defects, then the manufacturer can feel confident that the plant is producing quality products. This sampling technique provides a nice way to systematically choose the sample (as the name suggests), but there can be disadvantages. If the population has a pattern to it, systematic sampling can easily produce a sample that does not represent the population. For instance, as a manager of the production line, you would want to ensure that there was no obvious bias in choosing every 10th product as opposed to every 9th product. If there was a unique characteristic about every 10th product, such as they came only from Machine A, you would not want to choose 10 as your counting interval.

Every 10th bottle

Figure 11.1.5: Systematic Sample

Convenience Sampling: A **convenience sample** is one in which the sample is "convenient" to select. It is so named because it is convenient for the researcher. Although convenience sampling is the least reliable for obtaining a representative sample, in some instances it is all the data we can obtain, and the "some is better than none" factor comes into play. For instance, suppose you need 20 students to fill out a survey for your sociology project. Because of time constraints, you ask students in your English 101 class to participate. This sample of convenience is better than no sample at all. On the whole, however, caution should be used when results are obtained merely from convenience sampling, as this method often leads to members of the sample all having a similar characteristic and hence is prone to creating nonrepresentative samples.

☰ SAMPLING METHODS

Random Sampling—Every member of the population has an equal chance of being selected.

Stratified Sampling (Quota Sampling)—A few members from each stratum (or group) are randomly chosen.

Cluster Sampling—All members from a few randomly chosen clusters (or groups) are selected.

Systematic Sampling—Every nth member of the population is chosen.

Convenience Sampling—The sample is chosen because it is convenient for the researcher.

Example 11.1.4

Identifying Sampling Techniques

Identify the sampling technique used to obtain a sample in each of the following situations.

a. To conduct a survey on collegiate social life, you knock on every 5th dorm room door on campus.

b. Student ID numbers are randomly selected from a computer printout for free tickets to the championship game.

c. Fourth grade reading levels across the county were analyzed by the school board by randomly selecting 25 fourth graders from each school in the county district.

d. In order to determine which ice cream flavors would sell best, a grocery store polls shoppers that are in the frozen foods section.

e. To determine the average number of cars per household, each household in 4 of the 20 local counties were sent a survey regarding car ownership.

Solution

a. Because the sample is obtained by choosing every nth dorm room, this is systematic sampling. This is a representative sample, as long as students were randomly assigned to dorm rooms and there are no hidden potential biases, like only males live in every nth room.

b. Since every member has an equal chance of being selected, this is random sampling.

c. The students were divided into strata based on their schools and then a random sample from each school was chosen. This is stratified sampling.

d. Because of the ease of choosing shoppers right in their own store, this is convenience sampling. In this case, convenience sampling is a viable method for gaining a representative sample since the store would be interested in knowing the thoughts of their customers.

e. Cluster sampling was used here because the counties are the natural clusters and all of the households in some of the counties received the surveys.

OBJECTIVE 3 Experiments

As we stated earlier, an experiment seeks to create data to help identify cause-and-effect relationships, as opposed to observational studies that observe data. In an experiment, researchers apply a **treatment** to a group of people or things, called **subjects**, and measure the response. (If the group is made up of people, they can also be referred to as **participants**.) A treatment is simply a certain condition that is applied to a group of subjects for experimental purposes, such as asking one group of people to adhere to a specific diet or adding fertilizer to a crop of vegetables.

The variable that measures any response to the treatment in an experiment is called the **response variable**. The variable that causes the change in the response variable is referred to as the **explanatory variable**. For instance, if an experiment is designed to determine if a dietary regimen will lower cholesterol, then the dietary regimen is the explanatory variable and the cholesterol level is the response variable. If an experiment is designed to determine the growing effects when fertilizer is added to a crop, the fertilizer is the explanatory variable and the growth rate is the response variable.

PARTS OF AN EXPERIMENT

Treatment

A **treatment** is a certain condition that is applied to a group of subjects in an experiment.

Subjects

Subjects are the people or things being studied in an experiment.

Participants

Subjects in an experiment that consist solely of people are **participants**.

Response Variable

The **response variable** is the variable in an experiment that measures any response to the treatment.

Explanatory Variable

The **explanatory variable** is the variable in an experiment that causes the change in the response variable.

While there are many ways to set up an experiment, the goal of any experiment is to compare the results of the treatment to a control in order to establish a cause and effect. The **treatment group** is the group of subjects the treatment is applied to, while the **control group** does not receive the treatment. In order to generalize the results, it is important that the two groups are as similar as possible so that any difference between the groups can be attributed to the treatment. In some instances, the control and treatment groups may actually be the same group of participants measured before and after the treatment is applied, such as in a pre-test/post-test scenario.

Just as with observational studies, the design of any experiment determines how reliably the results can be used to make inferences about the population. The three main principles in experimental design are to **randomize**, **control**, and **replicate**.

> ### ⬚ PRINCIPLES OF EXPERIMENTAL DESIGN
>
> 1. Randomize the control and treatment groups.
> 2. Control for outside effects on the response variable.
> 3. Replicate the experiment a significant number of times to see meaningful patterns.

Randomize

Researchers cannot randomly select participants for an experiment as they would for an observational study. Generally, participants volunteer to be part of an experiment. Therefore, techniques must be established to create two similar groups from the available participants. One method that researchers use to create similar groups is to randomly split the volunteers into two groups. Experience has shown that groups created in this way usually have similar characteristics. Alternatively, researchers create the groups by purposely assigning volunteers based on important characteristics. For instance, they might assign an equal number of participants based on age, gender, disabilities, smoking habits, and so forth, to each group.

Control

Factors other than the treatment that cause an effect on the groups are called **confounding variables**. Researchers need to "control" for as many of these as possible so that experiments provide the best results possible. Confounding variables that are not taken into account might suggest that there is a cause-and-effect relationship when in fact there is not one. When researchers directly assign participants to the various groups, they are controlling for different confounding variables.

Researchers must also control for a phenomenon called the **placebo effect**. Because people respond to suggestion, giving someone a drug for the common cold and telling them that it will cure them will often produce effects caused by the suggestion alone and not the drug itself. This response is known as the placebo effect. To counteract this, subjects in the control groups are sometimes given a **placebo**, which appears identical to the treatment, but contains no intrinsic beneficial elements. For example, when testing a new drug, both groups would be given an identical pill to take, but only the treatment group's pill would have active ingredients in it. The control group would get the placebo pill with no active ingredients.

In experiments that use a placebo, the subjects do not know if they are in the control group or the treatment group. This is called a **single-blind** experiment. In a single-blind experiment, the people interacting with the subjects know which group each subject has been placed in. However, in a double-blind experiment, neither the subjects nor the people interacting with the subject know who is getting the placebo. This avoids the researchers subconsciously influencing the results of the study through their interactions with the participants. This **double-blind** design is preferable when the researchers have to make subjective decisions regarding the outcomes of the experiment, such as in clinical trials for depression or schizophrenia.

Replicate

Lastly, when setting up an experiment, the design should be described in such a way that future researchers can validate the findings of the study by replicating the original study. Additionally, it is important to make sure that the original experiment is replicated a sufficient number of times if the sample size is small. Performing a single experiment on just two subjects is unlikely to yield meaningful results that can be generalized to any population. However, multiple iterations that produce the same results can be meaningful.

📖 TERMS RELATED TO EXPERIMENTS

Control Group

A **control group** is a group of subjects to which no treatment is applied in an experiment.

Treatment Group

A **treatment group** is a group of subjects that researchers apply a treatment to in an experiment.

Confounding Variables

Factors other than the treatment that cause an effect on the subjects of an experiment are called **confounding variables**.

Placebo Effect

The **placebo effect** is a response to the power of suggestion, rather than the treatment itself, by participants of an experiment.

Placebo

A **placebo** is a substance that appears identical to the actual treatment but contains no intrinsic beneficial elements.

Single-Blind

In a **single-blind experiment**, subjects do not know if they are in the control group or the treatment group, but the people interacting with the subjects in the experiment know which group each subject has been placed in.

Double-Blind

In a **double-blind experiment**, neither the subjects not the people interacting with the subjects know group each subject belongs to.

💬 Helpful Hint

Researchers must get approval for an experiment from an Institutional Review Board (IRB) before beginning. The IRB reviews the design of the study to make sure it is appropriate and no unnecessary harm will come to the subjects involved.

Example 11.1.5

Identifying Parts of an Experiment

Consider a statistical study in which researchers want to determine if taking doses of the chemical salicin, found in willow bark, will reduce inflammation in patients. Suppose that the study is focusing only on patients with osteoarthritis, a common chronic joint condition. After study approval, volunteers are randomly assigned to two groups, each having 30 participants. Participants in Group A are administered doses of salicin, and their inflammation is tracked. Participants in Group B are administered doses of saline (which has no active ingredients), and their inflammation is also tracked. The patients are not told which of the two groups they are in; however, the nurses administering the salicin are aware of the group assignments. After a predetermined length of time, the inflammation reported by the separate groups are compared to determine if salicin reduces inflammation for those with osteoarthritis.

a. Identify the explanatory and response variables.

b. Identify the treatment in the experiment.

c. Determine which group is the treatment group and which group is the control group.

d. What is the purpose of administering saline to Group B?

e. State whether the experiment is a single-blind or double-blind study. Do you think this is the best choice for this study?

Solution

a. The explanatory variable is what "explains" the changes in the response variable. Since the researchers are trying to determine if salicin can reduce inflammation, the explanatory variable is salicin and the response variable is the amount of inflammation recorded for each patient.

b. The treatment is what is being applied to the group. Thus, the treatment is salicin.

c. Group A is the treatment group since they are receiving salicin. Group B is the control group since they did not receive the treatment.

d. The saline that is administered to Group B is a placebo and is administered to compensate for the placebo effect so that all patients are responding to the same suggestion that they might be receiving the treatment.

e. Since only the patients are unaware of the group assignments, this is a single-blind study. Since inflammation is not a subjective measure, this is an appropriate choice for the study.

> ✔ **Skill Check 11.1.4**
>
> A dentist wants to look at the effects of a new dental material he has patented to be used for fillings. Should the experiment be run as a single-blind or double-blind experiment? Explain your choice.

OBJECTIVE 4 Critiquing a Published Study

As you evaluate statistical studies, employing critical thinking skills based on your knowledge of the principles of a good study can help you better identify valid conclusions or spot questions that might need further research. Let's take a look at a few filters you should consider when critiquing a study.

Consider the *Source*

The source of information is a clue about its reliability. As you read a statistical study—or a news article for that matter—it's important to ask questions about the source of the study such as the following.

• Who paid for the study?

• Where was the data collected?

- When was the data collected?
- Do the organizations funding or publishing the study have a vested interest in the results?

It's not that there is a wrong answer to these questions, but the answers help you put the conclusions that are drawn into a proper perspective and note any inherent biases. For instance, suppose a chicken restaurant reports that 8 out of 10 fast food customers prefer chicken sandwiches over hamburgers. In this case, the restaurant clearly has some benefit to reporting this finding, so we'd need to look deeper into the data. The truth is that we do need organizations and groups promoting different views, and it's perfectly reasonable for them to use statistics to support their viewpoints.

The same can be said for data collected from individual opinions. If a population is not qualified to answer a question, you should be wary of placing value in their collective opinion. However, if the population being surveyed are experts in their fields, we can place a high value on their opinions.

Consider the *Variables*

Some variables are easy to measure, like the average weight of newborn babies. On the other hand, some variables require more carefully defined explanations. For instance, suppose you surveyed locals on the "best restaurant in town." You would need to define what criteria should be considered for "best"—best service, best location, fastest service, best taste, largest wine menu, etc. You get the idea. You should not assume that your definition of best or worst is automatically the one being measured. There is not necessarily only one correct way to measure criteria such as these, but the point is that you as a reader should understand which definition is being used. Along with well-defined variables, we also need to consider any confounding variables and how they were accounted or controlled for.

Consider the *Setup*

A good critique also considers the study's general setup. Is the question being answered well defined? Is the sample size large enough? Is the sample representative of the population or is there inherent bias in the method of sampling? In other words, does the method used to select the sample favor any particular outcome. It's also good to consider questions that assess the potential for errors in processing or participation. If it is an experiment, was it replicated enough?

Consider the *Conclusions*

In the end, the conclusion of the study is what we should be most prudent about. If a study is well constructed and unbiased, does the data set support the conclusions being made? Most often the answer is yes, but it's wise to consider what might not be reported, as well as what is being reported. For instance, consider the headline "Most Adults Don't Like Scented Toilet Paper." The results in the article state that of those who responded, 60% do not like it and 40% don't mind. What isn't stated is the 90% of the people surveyed couldn't be bothered to answer the question! (Not hard to imagine, is it?) The point is to consider if the results present the whole picture or if there could be other conclusions.

Example 11.1.6

Critiquing a Study

Consider the headline from a study published by the Pew Research Center in October 2020.

> 64% of Americans say social media have a mostly negative effect on the way things are going in the US today

a. List some of the questions that the article could address that would help you determine the validity and applicability of that claim as you read the article.

b. Describe as many potential sources of bias as you can.

Solution

a. There is not a single set of correct questions to ask; however, some possible questions might include the following.

- How big was the sample?
- What age range was sampled?
- How did they measure "negative effect"?
- Was "social media" defined?

b. Potential biases might include the following.

- Sampling method
- Source of the study and their interests in the results
- Ill-defined terms
- Nonresponses

For more information on the survey, see https://www.pewresearch.org/fact-tank/2020/10/15/64-of-americans-say-social-media-have-a-mostly-negative-effect-on-the-way-things-are-going-in-the-u-s-today/.

Skill Check Answers

1. The population of interest is single-family homes in Kansas City, Missouri. The variables of interest are the price per square foot and listing times.   **2.** 52% represents a sample statistic   **3.** Explanatory variable is the drug Chantix and the urge to smoke is the response variable.   **4.** Because there is potential researcher bias from the dentist since it is his patent, a double-blind experiment is best.

11.1 Exercises

✔ CONCEPT CHECK

1. Data are the _____, _____, or _____ gathered about a specific variable in a population.

2. During a census, data is obtained from _____ of the population being studied.

3. When a study has _____, it favors a certain outcome.

4. Stratified sampling is also referred to as _____ sampling.

🚀 APPLICATIONS

5. A study was taken to determine whether insulin level influences the course of polycystic ovary syndrome (PCOS). A group of women with PCOS was selected, and half of them were treated with metformin (a drug commonly used to treat diabetes) while the other half was in a control group. Researchers compared the symptoms of PCOS in the two groups. Identify the population of interest and the variables in this study.

6. Lactobacillus is very important for human health, specifically gut health. A group of researchers decided to study how the amount of sugar in a diet could affect the activity of lactobacillus in the gut. They selected a group of people who did not suffer from any gut diseases and divided them into three subgroups. The first subgroup got a diet with high sugar content, the second got a diet with low sugar content, and the third did not change their diet. Identify the population of interest and the variables in this study.

7. An experiment was designed to study the influence of noise level on people's stress level. Several volunteers of differing ages and social groups were subjected to noises of various types and intensities. During the study, researchers measured participants' level of cortisol, which is a hormone that is produced in response to stress. Identify the population of interest and the variables in this study.

8. Social research was conducted to determine if there is a correlation between annual income and the number of children in families. Researchers surveyed two-parent households where both parents were from 25- to 40-years old. Identify the population of interest and the variables in this study.

Decide if each numerical value from a statistical study is a population parameter or sample statistic.

9. After the presentation by the United Way representative, 67 of the night's 300+ attendees were interviewed about their reaction. Of the 67 people interviewed, **79%** said they were motivated to donate more of their time and/or money to helping nonprofit organizations.

10. The Nashville Country Music Marathon reported an average finish time of 4:46:28 for the 4082 finishers.

11. A medical study showed that the heart-rate profile during exercise and recovery is a predictor of sudden death. Of the 5713 men studied, the mean maximum heart rate (expressed as the percentage of the predicted maximum heart rate) during exercise was found to be 96% in subjects who died suddenly from cardiac causes during the 23 years when the follow-up was conducted.[2]

12. An estimated 65% of drivers in rural or town areas view traffic conditions in a good light versus 39% of drivers in the city or suburbs based on a recent news report that aired nationally.

In each scenario, identify the population, the sample, and any population parameters or sample statistics that are given.

13. A medical study followed 675 cancer patients who received either the new drug *Lafent* or a standard chemotherapy drug. 76% of those who received *Lafent* were still living after eight months, compared to only 63% of those who received the standard chemotherapy drug.

14. US realtors across the country are encouraged by the latest reports on the real estate market. Taking into account seasonal factors, sales rose by nearly 9% in the West, 3.5% in the South, 3.4% in the Northeast, and 1% in the Midwest.

15. The online version of *Health* magazine, Health.com, reported the following: "Data from the Women's Health Study . . . examined the medical records of more than 36,000 women who had no history of depression at the start of the study. Roughly 18% of the women were experiencing some form of migraine or had suffered from the headaches in the past. Over the next 14 years, 11% of the study participants received a depression diagnosis." One of the conclusions of the study was that middle-aged women who experience migraines are 40% more likely to become depressed.[3]

16. In a recent study, US graduates of for-profit higher education institutions were between 4.8 and 6.7 percentage points more likely to be unemployed than those at nonprofit institutions and community colleges.[4]

2   Xavier Jouven et al., "Heart-Rate Profile during Exercise as a Predictor of Sudden Death."
3   Matt McMillen, *Health*, http://www.health.com
4   Dan Berrett, *The Chronicle of Higher Education*, http://www.chronicle.com

17. Java Express is a company that sells high-end coffee appliances and gourmet coffee. Due to the current economic conditions, the company is looking into ways to change its strategy in order to appeal to more frugal consumers.

    Before investing in costly setup expenses, Java Express needs to know the percentage of coffee consumers who would buy from a new line of less expensive products. To estimate this percentage, the company chooses 5 of its largest markets and surveys every registered customer from each of these chosen markets. A total of 6195 customers complete the survey.

    The results show that 45% of those surveyed are very interested in the new line of products, 32% are somewhat interested, and the remaining 23% are not interested in the new products at all. Based on these results, the researchers estimate that approximately 77% of coffee consumers will be receptive to the new line of products. Java Express decides that it is worthwhile to introduce a new line of less expensive products to the market.

18. A study was conducted to investigate whether or not exercise is a factor in the overall outlook and positive spirit of Americans.

    For the study, 400 Americans were selected at random from each of the following two categories: those who exercise on a regular basis and those who do not. For the purpose of the study, regular exercise was defined to be light to moderate activity at least three times per week. A total of 800 Americans completed the brief telephone survey. Of those who exercise regularly, 74% said that their overall outlook on life is positive. Of those who do not exercise regularly, 68% said that their overall outlook on life is positive.

    Though not a definitive study, the researchers concluded that exercise does appear to influence the attitudes of Americans. In addition, the researchers estimated that approximately 71% of all Americans have a positive outlook. Further research would likely be conducted on the subject at a later time.

Determine which type of study should be conducted: an observational study or an experiment.

19. A football coach wants to know the average weight of his offensive linemen.

20. A doctor wants to study the effect of ginseng on patients' memories.

21. A city planner wants to know the average number of vehicles parked in downtown parking lots on any given business day.

22. A cell phone company wants to know the average total length of time teenagers spend on the phone each day.

23. A dentist wants to look at the effects of a new dental material used for fillings.

24. A financial institution wants to know if customer behavior will change significantly when interacting with a new mobile app feature.

Identify the sampling technique used in each scenario.

25. To ensure the quality of its product, a company tests every 15th item off the assembly line.

26. A student committee in the biology department was formed by randomly choosing three biology majors from every level of student, that is, 1st year, 2nd year, etc.

27. In order to choose the winners of free tickets to the campus concert, student housing printed out all 60 pages of student ID's and then chose three of the pages randomly.

28. A candidate for the local school board surveyed parents picking up their children from the public school on three school days in the last month.

29. Surveying movie goers as they exited the late movie, researchers determined that only young adults under the age of 22 see movies in the theater anymore.

30. In order to sample the production yield of milk from the farm's herd of cows, random samples are taken from each of the five different types of cows.

31. Using Excel to generate an arbitrary list of customers, the marketing department sent promotional materials to the top 250 names on the list.

Determine an appropriate sampling method for each scenario. Give your reasons for choosing the method. Describe any potential biases that the researcher might need to consider.

32. A state senator wishes to survey his constituents on issues in the upcoming legislature session.

33. A pharmaceutical company wishing to test the outcomes of a new drug for a particular skin disease finds 1200 patients willing to participate in the study. They can only choose 100 of them for the actual study.

34. Medical scientists wish to evaluate the effects of vitamin E and low-dose aspirin in primary prevention of cardiovascular disease and cancer in apparently healthy women.

35. Researchers wish to study the effects of watching too much television at a young age for children diagnosed with autism.

36. A pharmaceutical company claims their new antacid has fewer side effects than the leading drug and is more effective in relieving the pain of occasional heartburn in patients with reflux disease. To prove this claim, the company conducted a clinical trial on 50 patients with reflux disease. Half of them got the new antacid, and the other half got the leading drug. After a predetermined period of time, the patients were surveyed on their experience. The researchers, unlike the patients, knew what drug each patient was taking.

    a. Identify the explanatory and response variables.

    b. Identify the treatment in the experiment.

    c. Determine which group is the treatment group and which group is the control group.

    d. State whether the experiment is a single-blind or double-blind study. Do you think this is the best choice for this study?

37. A study was conducted to prove that a new virus-suppressing drug was more effective than the conventional antiviral agent. Patients susceptible to viral infection were randomized into two groups. Participants in Group A were given the new virus-suppressing drug and Participants in Group B were given the conventional antiviral agent. The patients were not informed which drug they were taking, but the researchers were. A viral load was calculated by estimating the number of viral particles per milliliter of blood plasma. The viral load was measured before and three weeks after the course of treatment.

    a. Identify the explanatory and response variables.

    b. Identify the treatment in the experiment.

    c. Determine which group is the treatment group and which group is the control group.

    d. State whether the experiment is a single-blind or double-blind study. Do you think this is the best choice for this study?

38. A clinical trial was aimed at studying a drug for chronic progressive degenerative disease. Researchers expected the drug to slow down or even reverse the disease's degenerative processes. A group of people with this disease was selected, and some of them were injected with the drug while some were given a placebo in the form of saline. Neither the subjects nor the researchers knew who was receiving the drug and who was receiving the placebo. Before, during, and after the trial, patients underwent various tests to identify and evaluate degenerative processes.

    a. Identify the explanatory and response variables.

    b. Identify the treatment in the experiment.

    c. Determine which group is the treatment group and which group is the control group.

    d. State whether the experiment is a single-blind or double-blind study. Do you think this is the best choice for this study?

39. To test the hypothesis that vitamin C increases the absorption of iron, researchers recruited a group of people with iron deficiency. Some of the people were prescribed iron supplements, and the others were given iron with vitamin C. Both the researchers and the subjects knew what they were taking. After a month of taking the drugs, the patients' iron levels were remeasured.

   a. Identify the explanatory and response variables.

   b. Identify the treatment in the experiment.

   c. Determine which group is the treatment group and which group is the control group.

   d. State whether the experiment is a single-blind or double-blind study. Do you think this is the best choice for this study?

40. *Self-selected* samples are a type of convenience sampling in which the participants volunteer to participate in the study rather than being chosen by the researcher. One issue with this type of sampling is that people who "self-select" to be in a survey often either have very strong opinions about an issue or desire monetary rewards for their participation. For instance, consider online polls regarding political views that are conducted by popular news outlets. Describe some of the potential responses from those who take the time to log on and respond. Is it reasonable to generalize the results from a study like this to include all of the American public? Why?

41. Often words like "best" and "worst" are used in reporting data. However, descriptive words like these are hard to measure in a concrete way. Consider how you could measure small towns to compile a list of "America's 50 Best Small Towns to Live In." Name at least five distinct measurements.

42. Researchers try to eliminate as many biases in studies as they can in order to get a true picture of the population they are studying. Describe as many potential sources of bias as you can if you were asked to study the effects of alcohol on college campuses.

✎ **WRITING & THINKING**

43. Explain how cluster sampling is different from stratified sampling.

11.1 PROJECT

MODIFYING THE FOCUS OF A STUDY

When presented with the results of a study, it's important to consider the study as a whole before deciding what to do with the information presented. Determining the population and variables of a study will help you understand what was being analyzed. Understanding the results that are presented will help you know whether the type of information used was selected for a reason and whether it might skew the perceived outcome. In this project, you will explore adjusting the study to create more useful results.

1. A study performed by the Refuel Agency[5] determined that college students spent $39.6 billion on food in 2020. Identify the population and the variable in this study.

2. Consider the population and variable from the study found in part 1. Can this information be used to say anything specific about the population at the college you attend? Explain why or why not.

3. Let's create a more focused population. List three characteristics or restrictions that can be used to define a smaller population that is a subset of the original population. Then, use one or two of these characteristics to create a more specific population description.

4. Now, let's create a more manageable variable. List three characteristics or restrictions that can be used to more clearly define a narrower variable that is a subset of the original variable. Then, use one or two of these characteristics to create a more specific variable description.

5. Write a new research question using the population from part 3 and the variable from part 4.

Now that we have a new research question, it's time to decide which sampling method to use. We will assume that the population chosen is too large to get data from every single student in it. (If the population is small enough to get data from every person in it, expand it before continuing to the next steps.)

6. Create a short survey that could be used to collect the data that you need to answer your research question.

7. Select one of the sampling methods and explain why you chose it.

8. Explain why the remaining sampling methods were not chosen.

9. Identify two possible types of bias that your choice of sampling may have.

5 "Is Your Brand Effectively Marketing to College Undergrads?" Refuel Agency, Last Accessed November 9, 2021, https://www.refuelagency.com/blog/market-to-college-students/

11.2 DISPLAYING DATA

OBJECTIVES

1. Construct and interpret frequency distributions.

2. Construct and interpret grouped frequency distributions.

3. Interpret pie charts, bar graphs, histograms, and line graphs.

4. Analyze graphs for misleading information.

OBJECTIVE 1 Frequency Distributions

Once we have designed a study and collected the data, our attention turns to the third step of the process: organizing the data. If data were simply presented in its raw form, it would neither elicit attention nor allow those who look at it to fully investigate the information. As a result, it's important to arrange and display the data in a way that is well organized and easily understandable. For instance, consider the 50 final grades from a statistics class shown below.

Final Grades in Statistics

C, C, D, A, A, C, D, C, C, C, D, C, B, A, A, C, C, C, C, C, C, D, C, C, C, C, C,
B, C, C, C, B, C, C, C, C, C, C, C, C, C, D, F, C, C, C, C, C, F, F, C, C, D, A

Although we have all of the grades for the class, it's not easy to do anything with a big list of data like this. Organizing the grades into a table is an easy and natural way to begin. A **frequency distribution** (or *frequency table*) is a table that displays each observed data value along with how often it appears. For instance, the frequency distribution in Table 11.2.1 shows the grade distribution for students in the statistics class. Because of the organized manner of the table, we can easily see that there were 50 students who received grades in the class, and of those, 33 students earned a C. We can also deduce that there was the same number of Bs as there were Fs and that similar numbers of people earned As and Ds. By using the frequency distribution, we can glean information considerably faster than sorting through the list of all 50 grades to determine these facts.

Table 11.2.1: Frequency Distribution for Final Grades in Statistics

Final Grade	Frequency
A	5
B	3
C	33
D	6
F	3
Total	**50**

A frequency distribution often includes columns containing other pertinent information about the data. One such statistic is the **relative frequency** of a data value. The relative frequency is the percentage of the data set that falls into a particular **class**, or category of data. Relative frequency is calculated by dividing the frequency of the class by the total number of members in the data set, n. (Relative frequency can be displayed as a percentage or a fraction.) Table 11.2.2 includes the relative frequency for the statistics grades from Table 11.2.1. Now we can see that 66% of the class received a C.

Table 11.2.2: Frequency Distribution for Final Grades in Statistics

Final Grade	Frequency	Relative Frequency
A	5	$\dfrac{5}{50} = 10\%$
B	3	$\dfrac{3}{50} = 6\%$
C	33	$\dfrac{33}{50} = 66\%$
D	6	$\dfrac{6}{50} = 12\%$
F	3	$\dfrac{3}{50} = 6\%$
Total	**50**	

Notice that the sum of the percentages in the relative frequency column should be 100% (or very close to it, allowing for rounding errors). Check for yourself that the sum of the percentages in Table 11.2.2 does in fact equal 100%.

FREQUENCY DISTRIBUTION, CLASS, AND RELATIVE FREQUENCY

Frequency Distribution

A **frequency distribution** is a display of the values that occur in a data set and how often each value, or range of values, occurs.

Class

In a frequency distribution, a **class** is a category of data.

Relative Frequency

The **relative frequency** is the percentage of the data set that falls into a particular class.

$$\text{Relative Frequency} = \frac{f}{n}$$

Here, f is the class frequency and n is the total number of members in the data set.

The following example looks at interpreting a frequency distribution.

Example 11.2.1

Interpreting a Frequency Distribution

In the state of Washington, researchers hoped to better understand the effects of roadside objects on vehicle accident frequency and severity. Table 11.2.3 shows data collected on the number of accidents involving roadside objects along State Route 3 in the state of Washington. Answer the following questions based on the table.

Table 11.2.3: Vehicle Accidents with Roadside Objects Along State Route 3, Washington

Roadside Object	Number of Crashes (% of total)
Guardrail	57 (15.36)
Earth Bank	55 (14.82)
Ditch	42 (11.32)
Tree	42 (11.32)
Concrete Barrier	38 (10.24)
Over Embankment	31 (8.36)
Utility Pole	20 (5.39)
Wood Sign Support	19 (5.12)
Bridge Rail	17 (4.58)
Culvert	7 (1.89)
Boulder	6 (1.62)
Luminaire	6 (1.62)
Mailbox	5 (1.35)
Fence	5 (1.35)
Building	5 (1.35)
Other Object	16 (4.31)

Source: Jinsun Lee and Fred Manning. "Analysis of Roadside Accident Frequency and Severity and Roadside Safety Management." Washington State Transportation Commission. December 1999. http://www.wsdot.wa.gov/research/reports/fullreports/475.1.pdf

a. What do the numbers in the parentheses represent?

b. Which roadside object was involved in the most accidents?

c. How many total accidents did the survey cover?

Solution

a. The numbers in parentheses represent the relative frequency of each class; that is, the number of accidents in a particular category as a percentage of the total number of accidents. For example, the table shows that there were 57 accidents involving a guardrail and that those accidents account for 15.36% of the total accidents along State Route 3.

b. Because the guardrail category has the highest frequency listed, we know that guardrails were involved in the most accidents.

c. By adding up the frequencies of all the different categories, we can see that there were 371 accidents involving roadside objects.

> ✔ **Skill Check 11.2.1**
>
> Which roadside object in Example 11.2.1 was involved in the smallest number of crashes?

OBJECTIVE 2 Grouped Frequency Distributions

You can imagine that if you had 500 unique possible data values, a frequency distribution would be tedious and not much more help than the original list. Consequently, sometimes it is more helpful to group the data into ranges of values, thus allowing for fewer classes and a more manageable frequency distribution. A frequency distribution of this type, where the classes are ranges of possible values, is called a **grouped frequency distribution**.

For example, consider the grouped frequency distribution in Table 11.2.4. It displays the ages (in years) of children who attended a program at an aquarium last week.

Table 11.2.4: Ages of Children for Aquarium Program

Age	Number of Children
0–5	10
6–11	14
12–17	7

Notice the first class contains the ages 0 through 5. There are six distinct ages possible in the first class: 0, 1, 2, 3, 4, and 5. The second class contains the next six distinct ages: 6, 7, 8, 9, 10, and 11. Then the final class contains the remaining possible ages: 12, 13, 14, 15, 16, and 17. Note that by adding the frequencies in the column, we know that there were $10 + 14 + 7 = 31$ children at the program.

The numbers 0, 6, and 12 (found on the left-hand side of the class ranges) are referred to as the **lower class limits** and the numbers 5, 11, and 17 are the **upper class limits**. The **class width** is equal to the difference between either consecutive lower class limits or consecutive upper class limits. For instance, the difference between the first two consecutive lower class limits is $6 - 0 = 6$, and the difference between consecutive upper class limits is also 6. (Check this for yourself.) The class widths must be the same for all classes in a grouped frequency distribution, and the classes should never overlap. In other words, a particular piece of data must have only one possible category it could fall into.

📖 GROUPED FREQUENCY DISTRIBUTIONS

Grouped Frequency Distribution

A frequency distribution in which each category represents a range of values is call a **grouped frequency distribution**.

Class Width

The **class width** is the difference between the lower limits or upper limits of two consecutive classes of a frequency distribution.

Lower Class Limit

The **lower class limit** is the smallest number that can belong to a particular class.

Upper Class Limit

The **upper class limit** is the largest number that can belong to a particular class.

The arbitrary choice of the first class width in the example about the children's ages dictated the layout for the classes that followed. However, the width and the number of classes for any grouped frequency distribution can be adjusted as long as we adhere to the guidelines that the classes maintain equal widths and do not overlap with one another. In general, the procedure for creating a grouped frequency distribution (which from this point on we will refer to simply as a *frequency distribution*) is outlined as follows.

CONSTRUCTING A FREQUENCY DISTRIBUTION

1. **Decide on the number of classes for the distribution**. There are typically between 5 and 20 classes. Having too many class divisions does not add any value over simply listing every data point and can be overwhelming. Having too few class divisions results in the data being lumped together and lacks detail.

2. **Choose an appropriate class width**. In some cases, the data set lends itself to natural divisions, such as decades or years. In other cases, you need to choose an appropriate class width so that the classes formed present a clear representation of the data and include all members of the data set. One method of finding the class width is to subtract the lowest number in the data set from the highest number and divide the difference by the number of classes.

3. **Find the class limits**. Begin by determining the lower class limit of the first class. You should choose the first lower limit so that reasonable classes will be produced. Using the minimum data value is a good place to start.

4. **Determine the frequency of each class**. Make a tally mark for each data value in the appropriate class. Count the marks to find the total frequency for each class.

> 💬 **Helpful Hint**
>
> Although choosing the smallest data value as the first lower class limit is easy, it's not necessary to start there. Classes can begin with any number at or below the smallest data value.

The central idea to keep in mind when creating a frequency distribution is to consider the data given and think about how you would divide up the classes so that they make the data easier to digest and analyze. There is really no one correct answer as long as you follow the guidelines we mentioned. Of course, there will always be cases where the data set divides into classes more naturally than others.

Example 11.2.2

Constructing a Grouped Frequency Distribution

The number of text messages used per month for 52 individuals was recorded and is listed. Construct a grouped frequency distribution for the data using six classes.

357	239	379	381	298	377	130	312	333	287	389	357	328
345	301	278	222	388	391	295	342	327	276	289	367	399
265	298	326	301	333	344	313	389	372	369	328	337	317
352	189	355	318	272	361	340	322	371	358	266	345	207

Solution

The first thing we need to do is identify the smallest and largest pieces of data to ensure that our distribution will include all of the data values. The smallest number of text messages used was 130 and the largest was 399. Because we were told how many classes to include, we will begin with Step 2 in the procedure: choosing an appropriate class width. Subtract the lowest data value from the highest and divide by the number of classes, as shown.

$$\frac{399 - 130}{6} \approx 44.83$$

This would give us a class width of approximately 45. We will stop here and consider some options. Choosing a class width of 45 is reasonable from a theoretical point of view. However, rounding the number up to 50 makes the arithmetic more convenient and the frequency distribution more easily understood. Therefore, we will choose a class width of 50.

If our classes each have a width of 50, where should the first class start? Let's try with the smallest data value. By adding the width to 130, we can find each of the other five lower class limits.

Table 11.2.5: Number of Text Messages Used per Month

Number of Text Messages Used per Month	Frequency
130–	
180–	
230–	
280–	
330–	
380–	

Next, in order to find upper class limits, we can use the lower limits. For the first upper limit, simply count backward one unit from the lower limit of the second class, making sure the classes don't overlap with one another. For example, the first upper class limit will be $180 - 1 = 179$. We could continue this same process of subtraction to obtain the remaining upper limits. Alternatively, we can add the class width of 50 to the upper limit we just found. Either method produces identical upper limits. We then have the following table.

Table 11.2.6: Number of Text Messages Used per Month

Number of Text Messages Used per Month	Frequency
130–179	
180–229	
230–279	
280–329	
330–379	
380–429	

Finally, we tabulate the number of data values that occur in each class. This produces the following frequency table.

Table 11.2.7: Number of Text Messages Used per Month

Number of Text Messages Used per Month	Frequency
130–179	1
180–229	3
230–279	6
280–329	16
330–379	20
380–429	6

✔ **Skill Check 11.2.2**

Identify the lower and upper class limits of the 4th class in Example 11.2.2

When you complete your grouped frequency distribution, the sum of the frequencies should add up to the total number of data values. Check for yourself that it does in Example 11.2.2.

As consumers of information, we're likely to be asked to interpret information in the form of a table more often than we'd be asked to construct one. Interpreting data correctly is a crucial step in making informed decisions. Example 11.2.3 looks at reading and interpreting a frequency distribution.

Example 11.2.3

Interpreting a Grouped Frequency Distribution

In the Fall of 2017, the US Energy Information Administration produced a report entitled International Energy Outlook 2017. Among other things, it predicted the world's energy use through 2050. The predictions are given in quadrillion BTUs (quad BTU) by region. A quad BTU is approximately equal to the amount of energy in 45 million tons of coal, or 1 trillion cubic feet of natural gas, or 170 million barrels of crude oil. The predictions for both 2020 and 2030 are given in Table 11.2.8 and a frequency distribution of the results are shown in Table 11.2.9. Use the tables to answer the following questions.

Table 11.2.8: Raw Data of World Total Primary Energy Consumption by Region (in Quad BTU)

Region	2020	2030
Australia and New Zealand	7.2	8.4
Mexico and Chile	9.1	10.2
South Korea	12.8	14.7
Brazil	14.8	17.5
Canada	15.1	15.5
Other Non-OECD Americas	16.2	17.9
Other Non-OECD Europe and Eurasia	17.7	17.7
Japan	19.2	19.4
Africa	24.2	27.9
Russia	29.8	30
India	33.3	44.3
Middle East	36.4	41.8
Other Non-OECD Asia	42.2	54.8
OECD Europe	80.5	83
United States	100.2	99.1
China	146.4	161.2
Total World	**604.9**	**663.2**

Table 11.2.9: Frequency Distribution of Quad BTU Predicted Use by Region

Quad BTU	2020	2030
0.1–10	2	1
10.1–20	6	7
20.1–30	2	1
30.1–40	2	1
40.1–50	1	2
50.1–60	0	1
60.1–70	0	0
70.1–80	0	0
80.1–90	1	1
90.1–100	0	1
100.1–110	1	0
110.1–120	0	0
120.1–130	0	0
130.1–140	0	0
140.1–150	1	0
150.1–160	0	0
160.1–170	0	1

a. How many classes are there in the frequency distribution?

b. What is the class width of the frequency distribution?

c. How many regions are represented in the report? In which table is this information found?

d. Which table would be best to use if you wanted to know which particular region used the most or least BTUs? Explain your choice.

e. Is the statement "Most regions are predicted to use between 10.1 and 20.0 quad BTUs in 2020" a true statement? Explain your answer and indicate which table you used to find your answer.

f. Is it appropriate to remove the six classes that have no data points in either 2020 or 2030 so that the frequency distribution is easier to read? Explain your answer.

Solution

a. The number of classes in the frequency distribution is the number of categories in the distribution table. There are 17 classes in this distribution.

b. The class width is the difference between consecutive class limits—either lower or upper. We can use the first two upper class limits and subtract to find the class width.

$$\text{Class Width} = 20 - 10 = 10$$

c. To determine how many regions are represented, we can use either table. In the first table, the regions are listed individually, so we add the number of listings. Thus, there are 16 regions.

In the frequency distribution, we add the frequencies for one of the years, using either the 2020 column or the 2030 column. Again, there are 16 regions represented in the distribution.

Notice that the sum of each column (2020 and 2030) is 16, which matches the number of regions in Table 11.2.8.

d. By design, a frequency distribution loses some of the uniqueness of each data point because we group them together into categories. For instance, we can know that two regions fell in the lowest usage category in the distribution, but we are not able to tell which regions they are. If we want to know which particular region used the most or least BTUs, we need to use the raw data that lists the regions individually, which is found in Table 11.2.8.

e. Although the class of 10.1–20 quad BTUs is the class with the most data points in the frequency distribution, that does not mean that "most regions" fall into this class. In order to have the most regions, the class would need to contain more than half of all the regions. We know that there are 16 regions represented, which means 6 is not more than half of the regions. Therefore, the statement is not true. However, it is true that the class 10.1–20 contains the most regions in the distribution.

f. It would not be appropriate to remove the classes that have no data points. Removing them would create a distortion of the gap between the data.

OBJECTIVE 3 Graphical Displays of Data

Although a frequency distribution is a reasonable place to start for displaying data, sometimes even a table can contain an overwhelming amount of information. That's when graphical displays such as pie charts, bar graphs, histograms, and line graphs help to convey the data more clearly. A **graph** is a picture of the data that allows us to view patterns at a glance, and graphs are more visually appealing than a table or list.

Because technology helps us make these types of graphs with relative ease, we'll focus our attention on interpreting the graphs. Every well-made graph should have a descriptive title as well as labels for any axes. A footnote usually lists the source of the data, and the date should be included whenever possible since data can change dramatically over time. When appropriate, a **legend** should be provided that labels the different data categories being compared. As a general rule, a graph should be labeled well enough that it is able to stand alone, without the original data.

> **📖 GRAPH AND LEGEND**
>
> **Graph**
>
> A **graph** is a picture of the data that allows us to view patterns at a glance.
>
> **Legend**
>
> A **legend** is a description of how each data category is identified in a graph.

Pie Charts

When we want to show a comparison between parts of the data set and the whole data set, we use a **pie chart**, also called a *circle graph*. A pie chart is a circular graph that represents the size of each category in relation to the whole. To make a pie chart, we can use the relative frequency distribution to divide the "pie" into different-sized wedges. The size of each wedge in the pie chart is proportional to the relative frequency of the category. Thus, a category that contains 20% of the data values would be represented by 20% or $\frac{1}{5}$ of the pie.

> **Helpful Hint**
>
> When constructing a pie chart by hand, the measure of the central angle of a particular wedge is calculated by multiplying the relative frequency of the class by 360° and rounding to the nearest whole degree.

> **📖 PIE CHART**
>
> A **pie chart** is a circular graph that depicts parts of a whole and shows the size of each category of data in relation to the whole.

Consider the following pie chart of religious affiliations of adults in the United States, which was made from data gathered by the Pew Research Center.[6] Notice that although there is quite a bit of data being displayed, the information is labeled clearly and easy to read.

Religious Afiiliations of Adults in the US

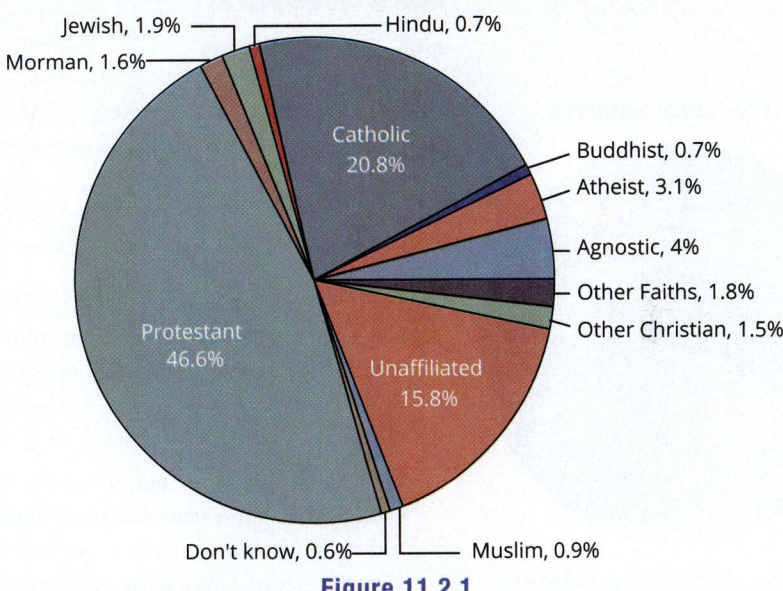

Jewish, 1.9% — Hindu, 0.7%
Morman, 1.6%
Catholic 20.8%
Buddhist, 0.7%
Atheist, 3.1%
Agnostic, 4%
Other Faiths, 1.8%
Other Christian, 1.5%
Protestant 46.6%
Unaffiliated 15.8%
Don't know, 0.6% — Muslim, 0.9%

Figure 11.2.1

Using the graph, we can easily answer questions along the lines of determining which religion has the largest US adult following. Visually, we can see that it is Protestant since this religion is represented by the largest pie slice. The next two largest slices are the categories Catholic and unaffiliated. By looking at their percentages, we know that a smaller amount of adults identify as unaffiliated (15.8%) compared to the amount of adults that identify with the Catholic faith (20.8%).

As we critically look at the graph, one question you might consider is how large of a study was done? Unfortunately, the pie chart does not tell us this information. However, the Pew Research Center gives an exact description of their survey methodology on their website.

---

6   "Religious Landscape Study," Pew Research Center, Accessed June 30, 2021, https://www. pewforum.org/religious-landscape-study/

"The centerpiece of Pew Research Center's 2014 Religious Landscape Study is a nationally representative telephone survey conducted June 4–Sept. 30, 2014, among a sample of 35,071 U.S. adults. Approximately 60% of the interviews were conducted with respondents reached on cellphones ($n = 21,160$) and 40 % were completed on landlines ($n = 13,911$). A minimum of 300 interviews were conducted in every state and the District of Columbia. Interviewing was conducted in English and Spanish. The survey is estimated to cover 97% of the non-institutionalized U.S. adult population; 3% of U.S. adults are not reachable by telephone or do not speak English or Spanish well enough to participate in the survey. No adjustments have been made to the survey's estimates of the religious composition of the U.S. population to attempt to account for the small amount of non-coverage."

We could have conveyed some of this information from the methodology on the pie chart by stating how many adults were surveyed. Adding that information, however, might make the graph look too cluttered and hard to read. There is a fine line between keeping the information displayed as clear as possible and providing too little information. As a general rule, more information is always better than less. But together, the pie chart and survey methodology provide a strong picture of the study that was done.

Example 11.2.4

Interpreting a Pie Chart

Entry-level salaries for a variety of professions are shown in the pie chart in Figure 11.2.2. Use the chart to answer the following questions.

a. Which salary level occurred the most?

b. How many of the entry-level salaries were $30,000?

c. What percentage of salaries were above $25,000?

d. Is the statement "Most jobs in the study had entry-level salaries that were more than $25,000" true? Explain your answer.

Solution

a. At a glance, we know that it's either the wedge for $30,000 or the wedge for $38,000 that represents the most common salary. In order to determine the underlying frequencies of the original data, we need to look at the percentages of each category. We can see that the $38,000 salary occurred the most.

b. The key at the bottom of the graph tells us that there were 541 salaries in the study. To calculate how many data points were at the $30,000 level, multiply 541 by the appropriate percentage for $30,000 (which is 21%).

$$541(21\%) = 541(0.21) = 113.61$$

Consider this answer. We know that there couldn't have been a decimal number of salaries in this category, so we can conclude that the percentages were rounded off when constructing the pie chart. We'll do the same with our answer and estimate that there were 114 salaries that fell into the $30,000 category.

c. To determine what percentage of salaries were above $25,000, we can simply add together the percentages for all categories that are above $25,000 (that is, add the percentages for the $30,000 and $38,000 categories).

$$21\% + 23\% = 44\%$$

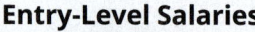

Entry-Level Salaries

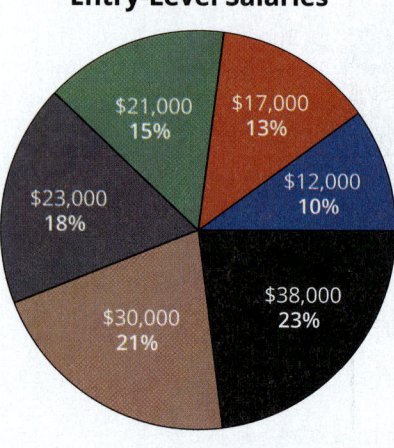

$21,000 15%
$17,000 13%
$23,000 18%
$12,000 10%
$30,000 21%
$38,000 23%

Total Number of Professions Surveyed = 541

Figure 11.2.2

Therefore, 44 % of salaries were above the $25,000 level in the study.

d. Because the percentage we calculated in part c. is not above 50%, it is incorrect to say that "most salaries" were more than $25,000. However, it would be correct to say that "almost half" of the salaries were more than $25,000.

Bar Graphs

One of the most common ways to display data that was collected in different categories is to use a bar graph. Like a pie chart, a bar graph represents the amount of data in each category. However, it uses either vertical or horizontal bars instead of parts of a circle, where the height (or length) of the bar represents the amount of data in each category. For example, Figure 11.2.3 uses a bar chart to display the results of a survey about the most popular pastries in the world. The bar chart makes determining the most popular pastry (croissant) quick and easy.

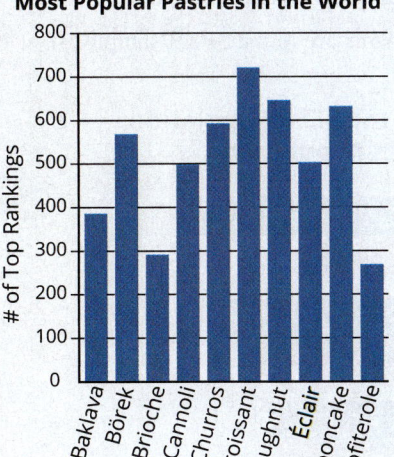

Most Popular Pastries in the World

Figure 11.2.3

> 📖 **BAR GRAPH**
>
> A **bar graph** is a graph that uses bars to represents the amount of data in each category.

One advantage of bar graphs over pie charts is the ability to see small differences between individual categories. Thus, we might choose to use a bar graph instead of a pie chart if we wish to compare different categories that have similar amounts of data in them. It's also a useful tool if we are more interested in the frequencies in individual categories rather than how the categories compare to the whole.

In a bar graph, one axis displays the categories of data and the other axis displays the frequencies. Traditionally, for vertical bar graphs, the horizontal axis contains the categories and the vertical axis represents the frequencies. Because each bar represents a category, the width of the bar along the horizontal axis does not relay anything about the data itself. However, the height of the bar should equal the frequency (or relative frequency) of the category. Similarly, bars touching or overlapping in some way might suggest that the categories are not completely separate from one another.

Bar graphs are also used to show a comparison of categories between different groups. We can either use a **side-by-side bar graph** or a **stacked bar graph** to show these comparisons. A side-by-side bar graph is just as the title suggests. The bars are placed next to one another to show the similarities and differences between populations. A stacked bar graph places the bars from each population on top of one another. This allows the reader to see a combined total for each category, yet at the same time, see the individual frequencies for each population.

> 💬 **Helpful Hint**
>
> Although there is no difference between using vertical bars versus horizontal bars in a graph, Microsoft Excel refers to a vertical bar graph as a "column graph" since the bars in the graph look like vertical columns. It uses horizontal bars to create what it calls a "bar graph."

Example 11.2.5

Interpreting a Side-by-Side Bar Graph

The Surveillance, Epidemiology, and End Results (SEER) Program of the National Cancer Institute works to provide information on cancer statistics in an effort to reduce the burden of cancer among the US population[7]. In order for the results of their studies to have the best impact, SEER researchers need representative samples of the US population. The following side-by-side bar graph shows the demographic breakdown of participants in a study by the institute versus the same demographic breakdown in the overall US population. Use the graph to answer the following questions.

a. Approximately what percentage of the United States population is foreign born? Compare this to the percentage of the SEER study participants with the same characteristic.

b. Which subpopulation is slightly underrepresented in the SEER sample?

c. Is it possible to determine how many participants are in the SEER sample? If so, estimate that number.

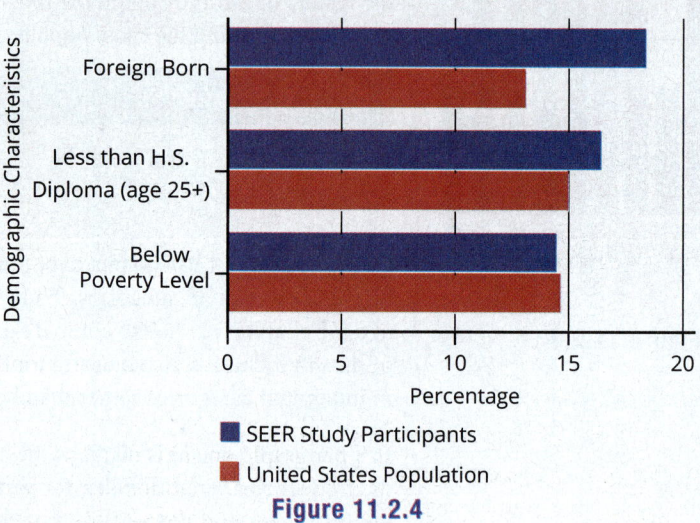

Figure 11.2.4

Solution

a. Based on the bar graph, approximately 12.5% of the United States population is foreign born compared to approximately 17.5% of the SEER study participants.

b. The bars are obviously longer for the SEER study participants than for the United States population in both the Foreign Born category and the High School Diploma category, meaning that these categories have more representation in the study group than in the general US population. However, if we look carefully at the Poverty Level category, although the percentages are very close, the SEER participants slightly underrepresented this demographic.

c. Unfortunately, it is not possible to determine how many people are in the SEER study using the graph. However, by looking at the SEER website, which is located in the footnote of the graph, we know that there were over 86,000,000 participants in study. When producing any type of graph, it's always a good idea to make sure that pertinent information is included for the reader; for example, the size of the sample.

---

7   "Population Characteristics," Surveillance, Epidemiology, and End Results Program, accessed June 10, 2021, https://seer.cancer.gov/registries/characteristics.html

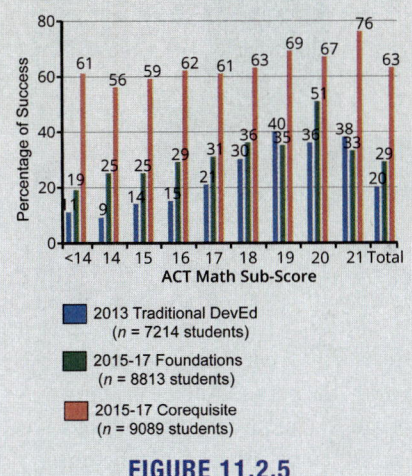

✔ Skill Check 11.2.3

For students with an ACT score of 17, which method of teaching provided the most success in a gateway math course?[8]

FIGURE 11.2.5

A stacked bar graph not only compares different categories, but also compares how the whole of each category is broken down. For instance, a college administrator might want to know how many spaces are available for the most popular classes on campus in the coming spring semester. The graph in Figure 11.2.6 shows not only how many total seats will be available for each class, but also how many are already filled. Notice the other categories that each class is broken down into.

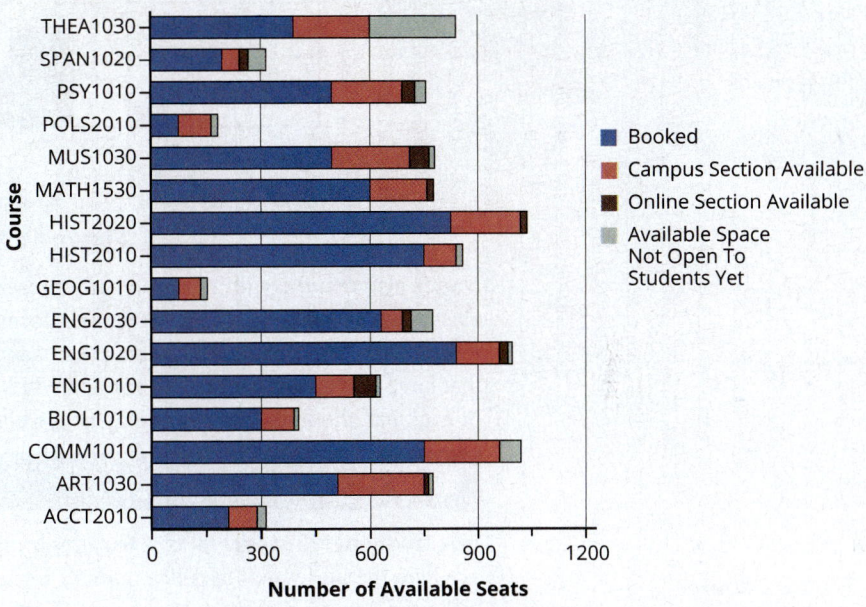

Figure 11.2.6

Let's take a closer look at the course Eng1010. This is a freshman-level English course required by all liberal arts majors at the college. We can see that there are approximately 625 seats that could be made available for this class for the spring semester. Of those 625 seats, the first portion of the bar indicates that a little more than 450 are already booked. The second portion tells us the available seats in physical classes that are offered on the campus; there are about 100 of these left. The third portion the number of spaces available in online classes, which is approximately 50. Finally, we can see by the last portion that there are very few spaces students do not already have access to.

A graph that contains this much information for each course would be good for an administrator who is trying to get an overall picture of the enrollment at his institution. However, if he required the exact numbers of spaces in the courses, this may not be the best way to convey the information.

Histograms

A special type of bar graph that displays the frequency distribution of numerical classes is called a frequency histogram, or simply a **histogram**. Because the bars represent classes and class limits are consecutive numbers, the bars touch in a histogram unlike in a bar graph, as shown in Figure 11.2.7.

---

8   National Academies of Sciences, Engineering, and Medicine. (2019). *Increasing Student Success in Developmental Mathematics: Proceedings of a Workshop.* Washington, DC: The National Academies Press. https://doi.org/10.17226/25547.

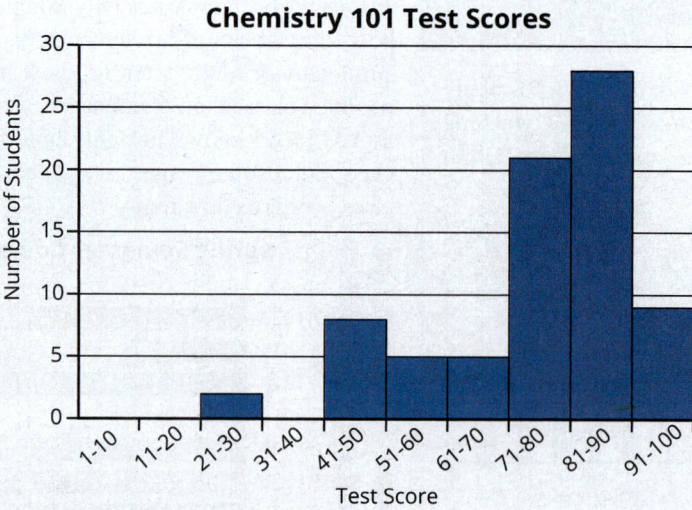

Figure 11.2.7

Notice also that the width of each bar represents the width of each class. Since the classes in a frequency distribution are uniform in width, the bars should be uniform in width as well. Thus, histograms are not appropriate for frequency distributions that have classes with an undetermined width, such as "20 or greater." The height of each bar should be the frequency of the class it represents.

> 📖 **FREQUENCY HISTOGRAM**
>
> A **frequency histogram** is a bar graph displaying a frequency distribution where the horizontal axis is a number line. The height of each bar represents the frequency of each class.

Example 11.2.6

Interpreting a Histogram

The following histogram displays data collected from delivery truck drivers. They were asked to record their wait times when an address had a closed entrance gate. Use the histogram to answer the questions.

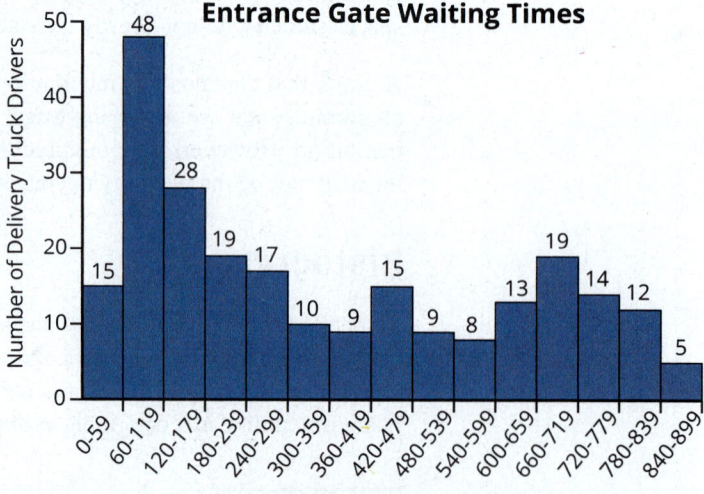

Figure 11.2.8

a. Determine how many wait times were recorded in the survey.

b. How many wait times were between 60 seconds and 179 seconds?

c. Approximately what percentage of wait times were at least 11 minutes?

d. What was the most common wait time for drivers?

e. The middle bar represents 420–479 seconds. Is it accurate to say that about half of the wait times were more than 420 seconds? Why or why not?

Solution

a. To determine the number of wait times recorded, add together each of the frequencies of the different classes.

$$15 + 48 + 28 + 19 + 17 + 10 + 9 + 15 + 9 + 8 + 13 + 19 + 14 + 12 + 5 = 241$$

Therefore, there were 241 wait times recorded in the survey.

b. To determine how many wait times were between 60 and 179 seconds, we need to add together the frequencies of each of the bars between those times. We know that 48 drivers waited between 60 and 119 seconds, and 28 drivers waited between 120 and 179 seconds. Therefore, $48 + 28 = 76$ wait times were between 60 seconds and 179 seconds.

c. To determine the percentage of wait times that were at least 11 minutes, we divide the number of drivers who waited 11 minutes or more by the total number of drivers in the survey. Since there are 60 seconds in one minute, 11 minutes is 660 seconds. So the last four bars represent drivers waiting at least 11 minutes. Thus, we have the following.

$$\frac{19 + 14 + 12 + 5}{241} \approx 0.2075$$

Therefore, approximately 21% of wait times at entrance gates were at least 11 minutes.

d. The class with the highest bar represents the most common range of wait times. Therefore, most wait times were between 60 and 119 seconds at the entrance gates.

e. Although 420 seconds represents the middle waiting time for the classes, it is inaccurate to say that half of the recorded times were less than this time and half were longer times. Instead, we need to think about the numbers of wait times. Since there are 241 wait times, we know that half of the wait times is approximately 121. We can add together the class frequencies until we reach approximately 120 and then note the waiting time.

The first two classes added together are $15 + 48 = 63$.

Adding the third class gives us $63 + 28 = 91$.

Continuing with the fourth class, we get $91 + 19 = 110$.

Adding the fifth class as well, we get $110 + 17 = 127$.

Now that we have passed 120 data points, we can stop. The fifth class has a lower limit of 240 seconds, so it is more accurate to say that about half of the wait times were more than 240 seconds at entrance gates.

Line Graphs

All of the previous graphs we've looked at depict data at one point in time. When your data consists of measurements over time, it is best to display these data using a **line graph**. In a line graph, the horizontal axis represents the continuous variable. The vertical axis represents the variable being measured. Each point on a line graph represents a data value and the time period it was observed. By joining the points together with line segments, changes over time are more easily observed. It is these line segments that give the graph its name.

> 📖 **LINE GRAPH**
>
> A **line graph** is a graph that uses points to represent data values at a particular time in history and then joins the points with line segments. Line graphs display changes in a variable over time.

Example 11.2.7

Interpreting a Line Graph

Figure 11.2.9 shows the approximate size of the internet based on the number of pages from August 2020 to November 2020. Use the graph to answer the following questions about the internet at that time.

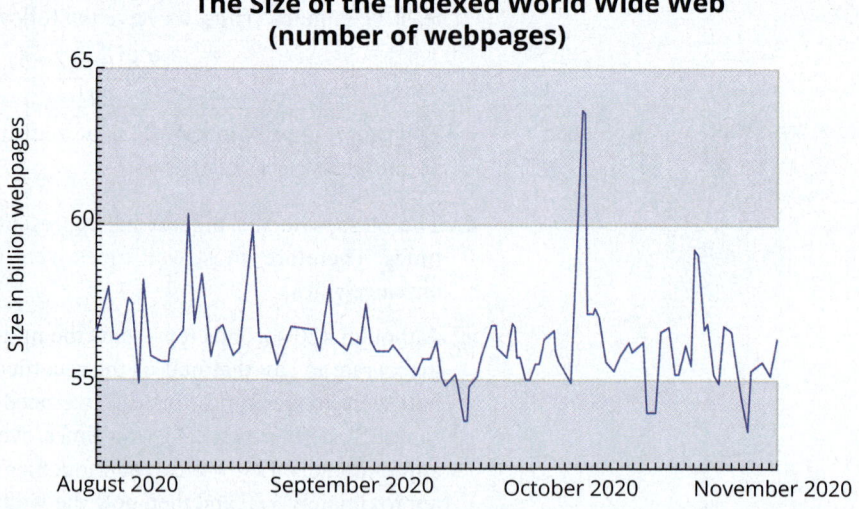

Figure 11.2.9

a. Approximately when did the number of pages peak during this time period?

b. Suppose you are writing an article about the size of the internet and referencing this graph. Would the headline "The Internet is Shrinking" be a fair representation of the information contained in this graph?

Solution

a. Determining this date is not as easy as you might think. Although it's easy enough to see the peak, the way the horizontal tick marks are displayed on the graph makes it difficult to easily identify the date of the peak. A good guess would be somewhere around October 10th, 2020. Making every 7th tick mark (a week) bold or identifying the first day of the new month on the tick marks would help tremendously with reading and interpreting the graph. All of these small details are helpful to remember when you're the one constructing the graph.

b. Based on the graph shown, it is true that the actual number of pages has decreased during this time period. However, we need to consider other factors along with a larger window of time. Consider that there are trends that sway the number of webpages that last longer than a 3-month period—holidays, political campaigns, and world events are all examples. Having data that covered a longer period of time would give more insight into the large-scale trends and thus the strength of the headline.

OBJECTIVE 4 Analyzing Graphs

As we said at the beginning of the section, it's important to display data so that it is organized clearly and effectively conveys the intended message. Ideally, graphs should be able to stand alone without the need for additional information in order to be understood. A graph should be properly labeled having a title that informs the reader of what type of information is being given. Each axis should be appropriately labeled, and the source and date of the information should be noted.

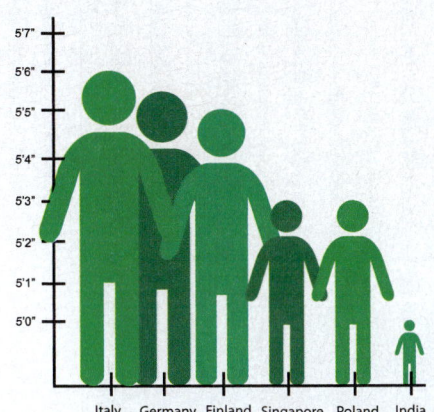

Figure 11.2.10: Average Female Height per Country

Sometimes graphs either intentionally or unintentionally convey the wrong message about data or are not quite clear enough to get their message across. It's important to be aware that there are visually misleading or ambiguous graphs out there. For example, making the bars different widths on a bar chart might imply that one category is somehow larger than another. So that we don't cause misunderstandings about the data, the bars should be of uniform width and not touching one another. Consider, the graph in Figure 11.2.10. It uses silhouettes to show average female heights in different countries. Without looking at the scale, what do you think the difference in average height is between the two extremes depicted, Italy and India? Although the "bars" make it seem like an enormous difference, in actuality, there are only 6 inches separating their average heights. Yet, the distance from 0 to 5′0″ ft is nearly the same as the distance from 5′0″ to 5′1″! Notice that the y-axis is also not uniform, making the graph even more misleading.

Spirit Week Participation

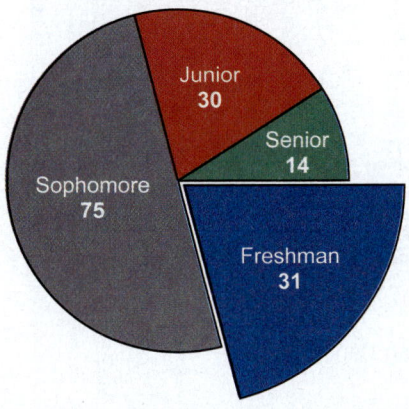

Figure 11.2.11: Misleading Pie Chart

Similarly, a distorted piece of a pie graph might inaccurately lead the reader to assume that one section of data is larger than another. An example of this is the graph in Figure 11.2.11. In the graph, the freshmen wanted to emphasize the fact that they came in second place for spirit week—although they just grabbed second place by a tiny margin. If the exact numbers were not on the graph, the emphasis on the freshmen piece of pie might visually imply that the freshman wedge is considerably larger than the junior wedge.

Making sure a graph is scaled properly is also important. If you stretch or shrink the scale on the y-axis, the shape of the graph may change dramatically. A line that rises gently on one scale might look very steep with a different scale. When analyzing a graph, check to make sure that the scale represents the data well. If there are large differences between data, then the graph should accurately reflect those differences. On the other hand, if the differences between the data values are small, then the graph should reflect this as well.

Example 11.2.8

Enhancing Graphs

Consider the following two graphs carefully.

a. Describe how the following graph could be improved for the reader.[9]

Figure 11.2.12: Census Data

b. These graphs appeared in the online text of a report by the Pew Research Center that contained the relevant information about how and when the data was collected.[10] Describe how the graphs could be improved for the reader.

Women leaders in national government

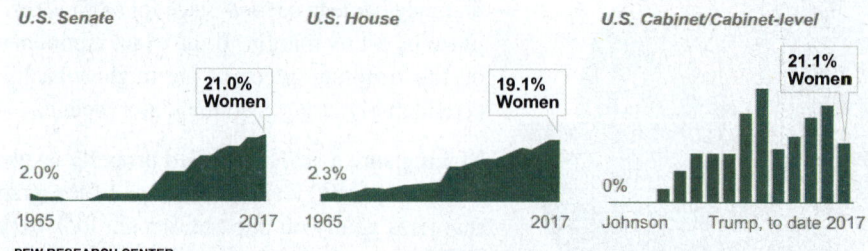

Figure 11.2.13: Pew Research Center Data

---

9   "Government Revenue from Airports," United States Census Bureau, last modified October 27, 2020, https://www.census.gov/library/visualizations/2020/comm/revenue-airports.html

10   Anna Brown, "Despite gains, women remain underrepresented among U.S. political and business leaders," Pew Research Center, last modified March 20, 2017, https://www.pewresearch.org/fact-tank/2017/03/20/despite-gains-women-remain-underrepresented-among-u-s-political-and-business-leaders/

Solution

a. Although the graph was on the United States Census website with relative source information noted, the graphic could be improved. The bold subtitle stating that the government receives a total of $24.9 billion dollars in revenue can be misleading for the data bars that are shown. The bars don't actually represent the entirety of that money. They only show the top 13 states, which represent a total of 76.4% of the revenue. A horizontal axis with markings might also help readers.

b. We are told that the graphs are included in the text, along with the data they represent. However, since the graphs are displayed side-by-side, having consistency with the horizontal labels as well as the type of graph would be an improvement. Notice that dates are given in the first two graphs while presidents' names are given in the third graph. It's also worth noting that 21.0% in the first graph is visually higher than 21.1% in the last graph.

Skill Check Answers
1. Mailboxes, fencing, and buildings were all tied for the smallest number of crashes at 5 each. 2. Lower class limit of the 4th class: 280; Upper class limit of the 4th class: 329
3. Corequisite

11.2   Exercises

✔ CONCEPT CHECK

1. A _____ is a table that displays how often each value in the data set occurs.

2. A _____ is a circular graph that illustrates the size of each category in relation to the whole data set.

3. In a _____, the height of each bar represents the frequency of each class.

4. True or False: The class width is equal to the lower class limit minus the upper class limit.

5. True or False: The width of the bars in a bar graph should be equal in size.

💡 PRACTICE

6. The grades on the first statistics test for Ms. Seago's class are listed in the following table. Construct a frequency distribution for the grades.

Grades on Statistics Test 1

A	C	F	C	C	D	F
B	D	F	B	A	A	F
B	C	C	A	B	F	D

7. Sisscon is a phone answering service. The following data are the numbers of calls per day reported by the company for the last month.

10	72	64	32	78	62	11
37	45	32	52	38	70	66
13	21	14	13	39	73	62
41	63	44	23	27	22	21
55	24	53	43	20	16	22

a. Create a grouped frequency distribution for the data using 8 classes and then use it to answer the following questions. Let the first lower class limit be 0 and the class width equal 10.

b. Calculate the relative frequencies for each class. Give your answer as a percentage rounded to the nearest tenth.

c. For what percentage of the days is the number of calls between 40 and 49?

d. For what percentage of the days is the number of calls in the single digits?

e. What is the most common range for the number of calls per day?

Create a frequency distribution with the indicated number of classes for each set of data. Include the frequency and relative frequency of each class.

8. The following data represent the numbers of curl-ups completed in 60 seconds for a group of 16 eight-year-old children. Use six classes that have a class width of 5. Begin with a lower class limit of 15.

31	34	41	36	27	29	18	33
31	28	34	22	26	28	36	42

9. The following data represent the caloric intakes in one day for a group of 15 people between the ages of 20 and 39. Use five classes that have a class width of 400. Begin with a lower class limit of 1800.

2700	2200	2500	2800	2600
3000	2600	2200	3100	2800
1800	3500	2500	3000	2900

10. A local book shop analyzed their sales to determine the most popular genres. The following table shows data collected on the number of books of each genre sold in the latest quarter. Answer the following questions based on the table.

Grades on Statistics Test 1 Book Genre	Number of Items Sold (% of total)
Romance	327 (25.85)
Detective and Mystery	189 (14.94)
Fantasy	153 (12.09)
Action and Adventure	128 (10.12)
Science Fiction	96 (7.59)
Suspense and Thrillers	91 (7.19)
Classics	77 (6.09)
Comic Book or Graphic Novel	75 (5.93)
Short Stories	62 (4.90)
Poetry	24 (1.90)
Other	43 (3.40)

   a. What do the numbers in the parentheses represent?

   b. Which genre was sold the most?

   c. How many books were sold by the shop in the last quarter?

11. Sofie keeps track of her expenses to make budget planning more transparent and understandable. The following table shows her expenses in each category for the last month.

Category	Amount Spent (% of total)
Housing and Utilities	$752.08 (25.53)
Tax and Insurance	$544.76 (18.49)
Food	$406.33 (13.79)
Household Expenses	$327.12 (11.10)
Wellness and Beauty	$217.64 (7.39)
Transport	$184.09 (6.25)
Entertainment	$174.85 (5.93)
Clothes and Accessories	$108.25 (3.67)
Other	$231.18 (7.85)

   a. What do the numbers in the parentheses represent?

   b. Which category required the most spending?

   c. What are Sophie's total expenses for the last month?

12. The following table gives the grouped frequency distribution of weights of 194 babies in kilograms. Answer the questions that follow based on the distribution.

Weights of Babies

Birth Weight (kg)	0.00–0.99	1.00–1.99	2.00–2.99	3.00–3.99	4.00–4.99	5.00–5.99
Frequency	2	17	39	89	46	1

a. How many classes are in the grouped frequency distribution?

b. What is the class width?

c. What is the value of the lower class limit of the 3rd class?

d. What is the value of the upper class limit of the 5th class?

e. What is the relative frequency of the 4th class? Give your answer as a percentage rounded to the nearest tenth.

13. The following table represents a grouped frequency distribution of the number of hours spent on the computer per week for 55 students.

Hours	Number of Students
0.0–4.4	9
4.5–8.9	14
9.0–13.4	21
13.5–17.9	11

a. Calculate the relative frequencies (as percentages rounded to the nearest tenth) for each class.

b. What percentage of the students used the computer between 9 and 13.4 hours per week?

c. What percentage of the students used the computer less than 9 hours per week?

14. The following pie chart shows destinations of recent University of Kent mathematics graduates, including business, financial math, and statistics.[11]

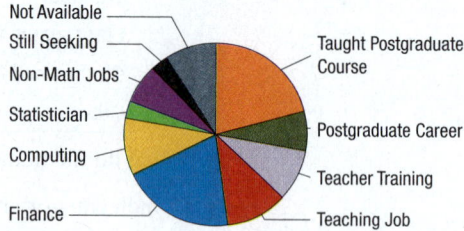

Destinations of Recent University of Kent Mathematics Graduates

a. Does it appear that any destination category accounts for more than 25% of the graduates?

b. Which pairs of destinations appear to have similar percentages of graduates?

c. How many mathematics graduates from the University of Kent were surveyed?

d. Is the graph misleading in any way?

11   University of Kent, http://www.kent.ac.uk

15. The following pie chart shows Apple's quarterly revenue[12] by category for the fiscal second quarter of 2021. Use the chart to answer the following questions.

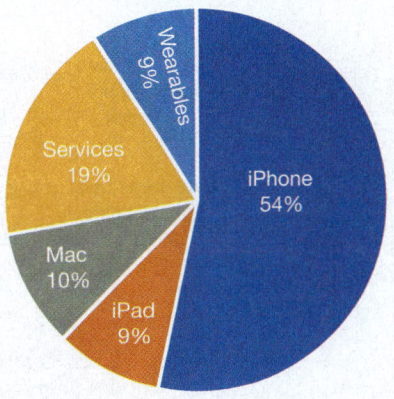

a. What product generated the most revenue?

b. What percentage of the sales were services?

c. What percentage of sales came from Mac, iPad, and Wearables?

d. Is the statement "Most revenue was generated from Mac, iPad, and Wearables" true? Explain.

16. Consider the bar graph of the predicted fastest growing occupations between 2010 and 2020.[13]

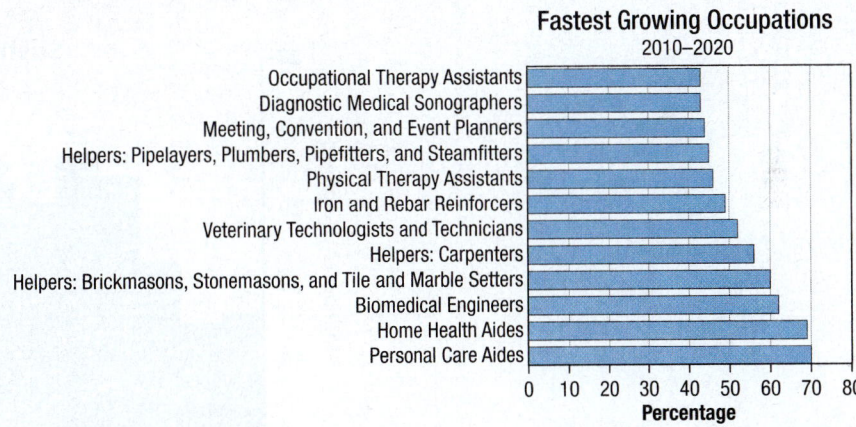

a. At a quick glance, which occupation is predicted to grow the most between 2010 and 2020? What is the predicted amount of growth?

b. How many new jobs will be available in the fastest-growing occupation in 2020?

12   "Investor Updates," News and Results, Apple, accessed January 4, 2022, https://investor.apple. com/investor-relations/default.aspx.

13   BLS Occupational Outlook Handbook, http://www.bls.gov/ooh

17. The stacked bar graph shows the average number of hours that married people in Japan spend each day doing various activities.[14]

Average Time Spent on Activities Each Day

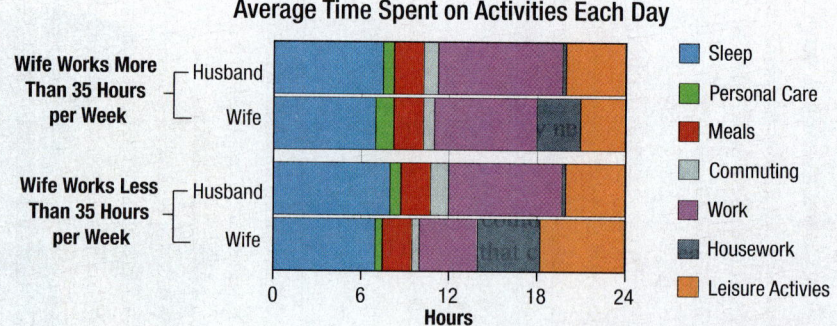

a. For wives who spend less than 35 hours per week working, how many hours on average are spent each day for leisure activities?

b. For husbands whose wives work more than 35 hours per week, approximately how many hours on average are spent on sleep?

c. Compare the number of hours spent sleeping for wives in each category.

d. What type of graph could be used to represent these data in a clearer fashion?

18. The following histogram displays data of wait times for delivery from a local sushi restaurant.

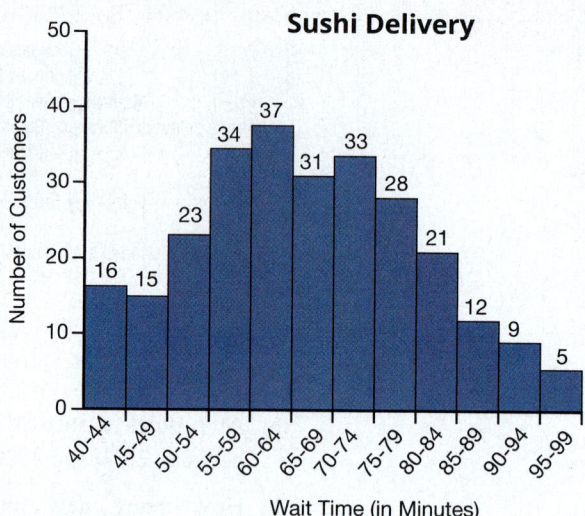

a. Determine how many wait times were recorded in the survey.

b. How many customers had to wait between 60 minutes and 74 minutes?

c. Approximately what percentage of customers had to wait at least one hour?

d. What was the most common wait time for customers?

e. The middle bars represent 65–69 minutes and 70–74 minutes. Is it accurate to say that about half of the wait times were more than 70 minutes? Why or why not?

14 Statistics Bureau (Japan), http://www.stat.go.jp

19. The following histogram displays data on the number of words in essays written by students at a high school.

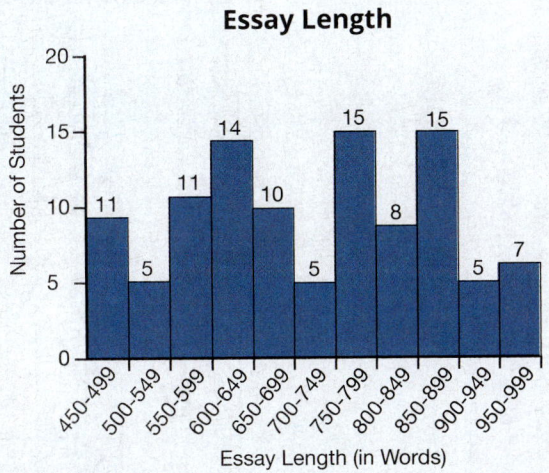

Essay Length

a. Determine how many essays were surveyed.

b. How many essays had the number of words between 700 and 899?

c. Approximately what percentage of essays had at least 800 words?

d. What was the most common numbers of words in essays?

e. The middle bar represents 700–749 words. Is it accurate to say that about half of the essays had 700 or more words? Why or why not?

20. The following graph shows the US unemployment rate from February 2011 to February 2013.[15]

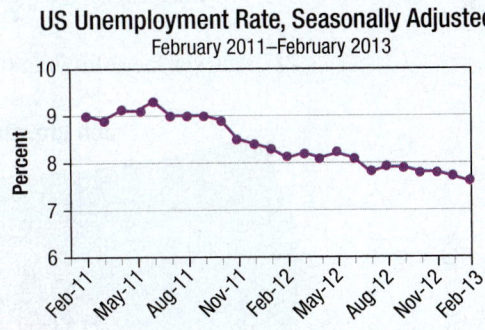

US Unemployment Rate, Seasonally Adjusted
February 2011–February 2013

a. Describe the trend of the unemployment percentage from February 2011 to February 2013.

b. Approximate the month and rate of the highest unemployment during this time period.

c. Approximate the month and rate of the lowest unemployment during this time period.

d. Is the graph misleading in any way?

---

15   Bureau of Labor Statistics, http://www.bls.gov

21. An online learning platform presented their analytics on average daily usage by their users. The data for June 2021 is shown in the following graph. Use the graph to answer the following questions.

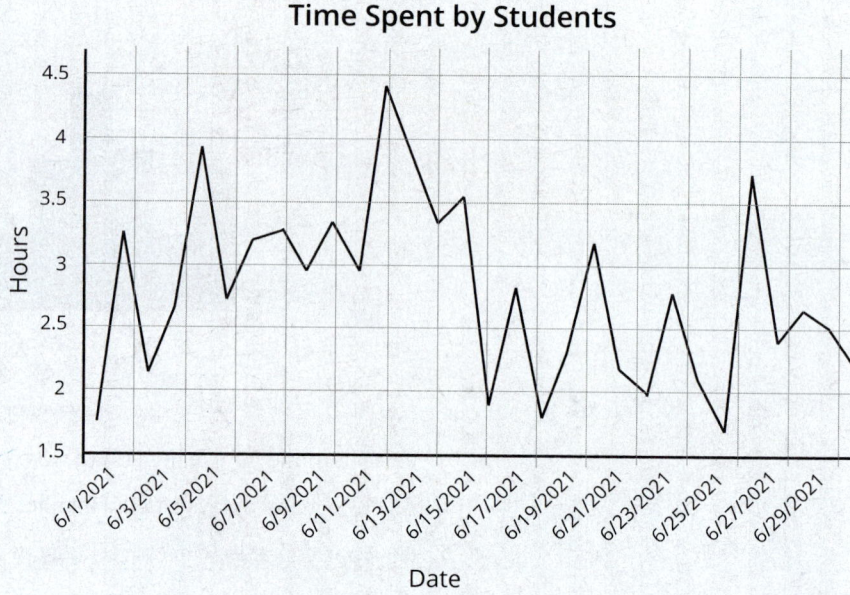

Time Spent by Students

a. Approximately when did the amount of time spent by students on the learning platform peak during this period?

b. Suppose you are writing an article about the usage of this platform and referencing this graph. Would the headline "Learning Engagement is Falling" be a fair representation of the information contained in this graph?

22. Answer the following questions about the Job Growth graph.[16]

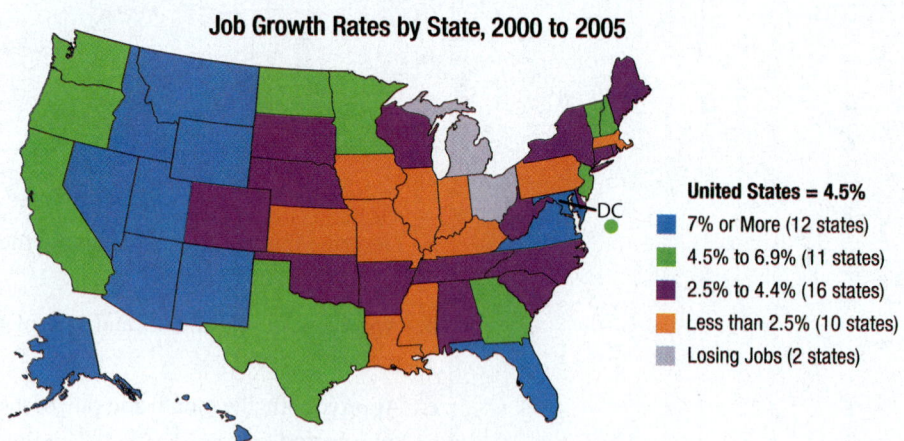

Job Growth Rates by State, 2000 to 2005

United States = 4.5%
- 7% or More (12 states)
- 4.5% to 6.9% (11 states)
- 2.5% to 4.4% (16 states)
- Less than 2.5% (10 states)
- Losing Jobs (2 states)

a. How many states lost jobs between 2000 and 2005?

b. In what part of the country are jobs growing the most in this time period?

23. What errors occur in the following histogram?

Hurricanes along the US Atlantic Coast

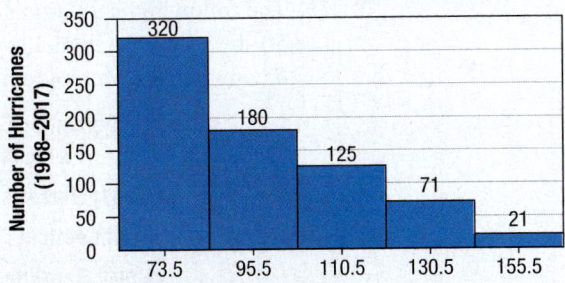

Source: Wikipedia. Atlantic hurricane season. https://en.wikipedia. org/wiki/Atlantic_hurricane_season#Number_of_storms_of_each_ strength_since_the_satellite_era (18 July 2019).

24. Consider the following excerpt from an online publication. Is the graph correctly labeled? If not, identify the corrections needed.

Stem-and-Leaf Plot

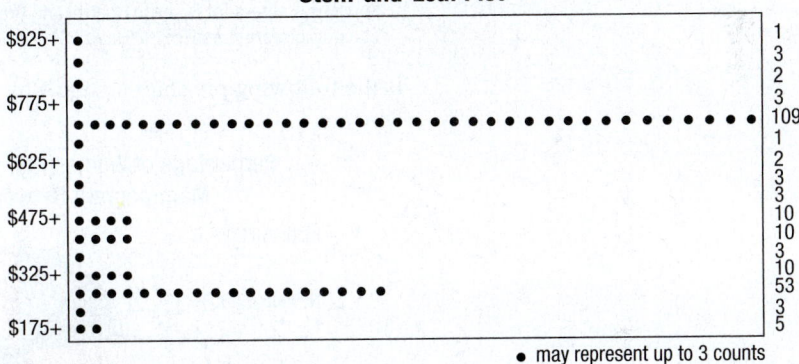

Source: NHHealthCost.org. "Health Costs for Consumers - Methodology." 16 Feb. 2007. http:// www.nhhealthcost.org/method.aspx (24 Jan. 2012).

25. The following is a portion of a table about state health facts. It lists 8 of the 50 states along with the percentage of women age 50 and older who report having had a mammogram between 2008 and 2010.

Percentage of Women Age 50 and Older Who Had a Mammogram Between 2008 and 2010

Massachusetts	87.50%
Connecticut	83.80%
North Carolina	81.20%
Virginia	79.10%
Alabama	77.60%
Mississippi	70.90%
Nevada	69.90%
Idaho	68.30%

Source: Centers for Disease Control and Prevention (CDC). Behavioral Risk Factor Surveillance System Survey Data. Atlanta, Georgia: U.S. Department of Health and Human Services, Centers for Disease Control and Prevention, 2010, available at http://apps.nccd.cdc.gov/brfss/list.asp?cat=WH&yr=2010&qkey=4427&state=All

Is the following pie chart a good way to display this data? Explain why or why not.

Percentage of Women Age 50 and Older Who had a Mammogram Between 2008 and 2010

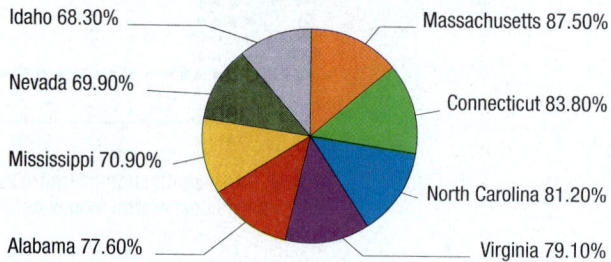

11.2 PROJECT

FOLLOWING THE BOOMERS

The US population is comprised of seven living *generations*: the Greatest Generation (born 1901–1927), the Silent Generation (1928–1945), Baby Boomers (1946–1964), Generation X (1965–1980), Millennials (1981–1995), Generation Z (1996–2010), and Generation Alpha (2011–2025). In this project, you will construct a series of stacked, side-by-side bar charts that demographers call an *age pyramid* and focus on how one of these generations affects the pyramid. This pyramid distribution for human population has been observed throughout the history of mankind but, since the 1950s, an interesting trend has developed. Our goal is to identify that trend and offer a reasonable explanation for its occurrence. In the table[17] below, we have the distribution of the percentage of the US population

---

17   Paul Taylor, "The Next America." Pew Research Center, Washington, D.C. (April 10, 2014) https://www.pewresearch.org/next-america/#Two-Dramas-in-Slow-Motion. Pew Research Center bears no responsibility for the analyses or interpretations of the data presented here. The opinions expressed herein, including any implications for policy, are those of the author and not of Pew Research Center.

by age groups for the years 1950, 1960, 1970, 1980, 1990, 2000, 2010, and 2020. The age groups are further broken down into sex as recorded at birth. The shaded cells roughly represent the portion of the population that was born between 1946 and 1964 (the Baby Boomers), with some overlap into the Silent Generation and Generation X.

Percentage of US Population

Age Group	1950 Male	1950 Female	1960 Male	1960 Female	1970 Male	1970 Female	1980 Male	1980 Female
0–9	9.96	9.60	11.01	10.66	9.22	8.86	7.42	7.09
10–19	7.28	7.10	8.53	8.29	9.97	9.63	8.82	8.48
20–29	7.80	8.00	6.08	6.13	7.55	7.54	9.12	9.05
30–39	7.43	7.68	6.67	6.90	5.46	5.63	6.94	7.08
40–49	6.36	6.40	6.15	6.35	5.71	6.06	4.89	5.12
50–59	5.12	5.12	4.93	5.10	4.95	5.37	4.88	5.36
60–69	3.62	3.71	3.52	3.92	3.51	4.15	3.78	4.54
70–79	1.74	1.94	1.98	2.38	1.90	2.65	2.08	3.04
80+	0.50	0.65	0.57	0.83	0.67	1.16	0.75	1.55
Total %	49.8	50.20	49.44	50.56	48.94	51.05	48.68	51.31

Percentage of US Population

Age Group	1990 Male	1990 Female	2000 Male	2000 Female	2010 Male	2010 Female	2020 Male	2020 Female
0–9	7.56	7.21	7.19	6.86	6.70	6.41	6.60	6.31
10–19	7.19	6.82	7.45	7.06	7.06	6.73	6.35	6.06
20–29	8.27	8.01	6.91	6.69	7.02	6.82	6.60	6.57
30–39	8.38	8.44	7.66	7.64	6.48	6.50	6.35	6.63
40–49	6.23	6.42	7.49	7.66	6.97	7.10	6.92	6.09
50–59	4.21	4.53	5.40	5.70	6.64	6.98	6.78	6.44
60–69	3.78	4.49	3.39	3.84	4.54	4.99	6.01	6.13
70–79	2.33	3.33	2.47	3.30	2.41	2.97	6.20	4.02
80+	0.89	1.91	1.10	2.18	1.32	2.31	5.58	2.10
Total %	48.84	51.16	49.06	50.93	49.10	50.81	3.39	50.65

1. For each year that is listed at the top of the table, construct a stacked, side-by-side bar chart by performing the following steps. (If time allows, construct your stacked, side-by-side bar charts on note cards. Be sure the horizontal axis stays the same for each graph. Stack the cards in order and flip through them to create an animated bar chart.)

    a. Center the horizontal axis at 0% and mark increments of 5% to the left and to the right.

    b. Beginning at the bottom of the chart with the 0–9 age group, extend a bar to the right of center indicating the appropriate percentage of the population that were female. Extend a bar to the left of center indicating the appropriate percentage of the population that were male.

c. Continue this process for each age group, stacking the bars on top of one another and shading the Baby Boomer bars a different color than the other bars. It should look similar to the following stacked, side-by-side bar chart. (The following chart is a visual aid only. Notice that the percentages and age groups are different than those given in the table.)

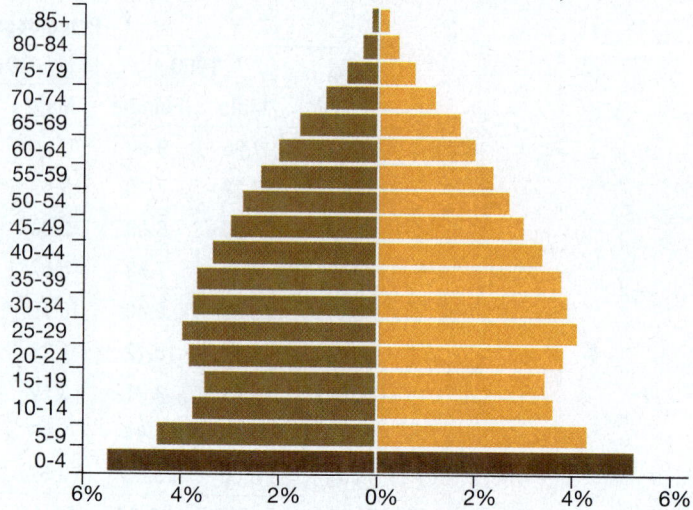

2. Compare the eight stacked, side-by-side bar charts. How does the shape of the age distribution change over time? What factors might be contributing to this changing shape?

3. What would you predict the stacked, side-by-side bar charts to look like in 2030, 2040, and 2050?

4. What additional information is gained by separating the population into sex?

OBJECTIVES

1. Calculate the measures of center of a data set.

2. Calculate the measures of dispersion of a data set.

3. Calculate measures of relative position.

As you can recall from the previous section, displaying data in a clear and informative way is certainly an important and necessary step in research. Just as important is describing the data set numerically so that it can be compared and analyzed. Imagine if the only way to compare data sets containing thousands of data points was to list each individual point. We would all throw our hands in the air and scream in frustration. The data would be cumbersome and useless. However, because of statistical tools like measures of central tendency and measures of dispersion (all of which we'll cover in this section), we can begin to get a better picture of what the data set is able to tell us by using summary statistics.

OBJECTIVE 1 Measures of Center

One of the first things that we are interested in knowing about a set of data is what the "average" data point looks like, or what best represents a typical value in the data set. We'll begin our discussion with the most common ways to describe the "center" or "average" of the data: the mean, the median, and the mode.

Mean

Often, we hear people refer to the "average" of a set of data, as in the average age of first-time parents or the average score on a test. Using the term average can refer to any of three measures of central tendency: the mean, the median, or the mode. Most commonly, the use of the word average is in reference to the **arithmetic mean**, often simply called the *mean*. The mean is the sum of all of the data values divided by the number of data values.

> **Helpful Hint**
>
> When calculating the mean, round to one more decimal place than the largest number of decimal places given in the data. Occasional exceptions to this rule can be made when the type of data lends itself to a more natural rounding method such as rounding currency values to two decimal places.

> $f(x)$ **ARITHMETIC MEAN**
>
> The **arithmetic mean**, or *mean*, is the sum of all of the data points divided by the number of data points. Formally, the formula for **population mean** μ is as follows.
>
> $$\mu = \frac{x_1 + x_2 + \cdots + x_N}{N}$$
>
> The formula for the **sample mean** $\bar{x}$ is as follows.
>
> $$\bar{x} = \frac{x_1 + x_2 + \cdots + x_n}{n}$$
>
> Here, x_i is the ith data value, N is the number of data values in the population, and n is the number of data values in the sample.

It's important to note that it is meaningless to find the mean of data that have no measurable values, such as the ratings of customer service: "very satisfactory," "satisfactory," or "needs improvement." Because there is not a defined measurable difference between each of the ratings, assigning a number value to each and taking the mean carries little meaning. However, you might often see such calculations being used. Later in this section, we'll show you a more appropriate descriptor to calculate this "average."

Example 11.3.1

Calculating the Mean

A sample of the number of sick days taken by employees at Witt's Insurance Agency during the last year is listed below. Calculate the mean of the sample data.

$$14, 5, 7, 11, 9, 7, 12, 6$$

Solution

Because we are given a sample of employee sick days, we are calculating a *sample* mean.

By Hand

Add all the values together and divide by 8, which is the number of data values in the sample.

$$\bar{x} = \frac{14+5+7+11+9+7+12+6}{8} = 8.875$$

TI-83/84 Plus

To calculate the mean using a TI-83/84 Plus calculator, we first need to input the data into the list function. Press $\boxed{\text{stat}}$, select `Edit…`, and enter the values into `L1`. Press $\boxed{\text{stat}}$ again, but this time select `CALC` and choose `1-Var Stats`. Be sure that `L1` is in the entry for `List:` and that the entry for `FreqList:` is blank. Highlight `Calculate` and press $\boxed{\text{enter}}$. The first value that the calculator returns is $\bar{x} = 8.875$.

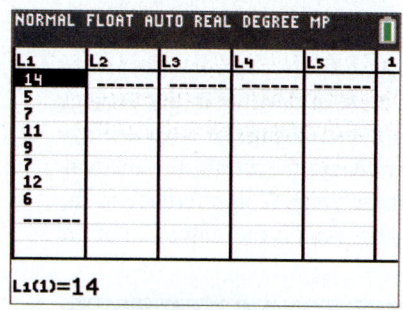

Figure 11.3.1

```
NORMAL FLOAT AUTO REAL DEGREE MP

         1-Var Stats
  x̄=8.875
  Σx=71
  Σx²=701
  Sx=3.181980515
  σx=2.976470225
  n=8
  minX=5
 ↓Q₁=6.5
```

Figure 11.3.2

Microsoft Excel

Excel has a built-in **AVERAGE** function that we can use to find the mean of a set of numbers. Enter the values into an empty column and in another cell, type "=average(". Then highlight the cells containing the values and press Enter. The value will appear as 8.875.

A9	▼	⋮	✕ ✓ fx	=AVERAGE(A1:A8)

	A	B	C	D	E	F
1	14					
2	5					
3	7					
4	11					
5	9					
6	7					
7	12					
8	6					
9	8.875					

Figure 11.3.3

> **📈 TECH TIP**
>
> If there are already values stored in this list, you can clear them out by pressing the up arrow to highlight the column name, **L1**, and pressing $\boxed{\text{clear}}$.

At the end of the calculation, round to one decimal place since the data values are whole numbers. Therefore, the mean for the number of sick days employees took at Witt's Insurance Agency last year is approximately 8.9.

Notice that the mean number of sick days in Example 11.3.1 isn't actually a member of the sample set. No employee took 8.9 sick days last year. However, it is a description of all the data points. It is an especially useful summary statistic when data sets are large and you may not be able to examine all the pieces of data individually.

A good way to visualize the mean is with a picture of a seesaw or teeter-totter. If all the data values were placed on the seesaw with even weights and the pivot point was at the mean, the seesaw would balance evenly. We've illustrated this in Figure 11.3.4 with the data from the previous example.

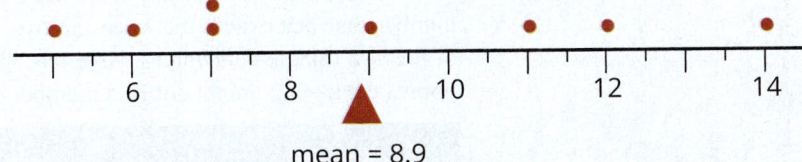

mean = 8.9

Figure 11.3.4: The Mean as a Seesaw Pivot Point

Example 11.3.2

Using the Mean to Find a Data Point

A plane that ferries visitors to a small resort island has strict guidelines on the weight allowed for passenger luggage. Consequently, the six passengers are limited to a maximum average luggage weight of 35 pounds (lb). The following are the weights of four out of six pieces of luggage: 39 lb, 32 lb, 35 lb, 37 lb. If the two pieces of luggage that haven't been weighed evenly split the remaining weight allowance, determine the weight for each remaining bag.

Solution

Since we are given the value of the population mean (35 pounds), we can use algebra techniques paired with the mean formula to find the remaining weight to be split. Substituting all the values we are given into the formula for the mean, we have the following.

$$\mu = \frac{x_1 + x_2 + x_3 + x_4 + x_5 + x_6}{N}$$

$$35 = \frac{39 + 32 + 35 + 27 + x_5 + x_6}{6}$$

$$6 \cdot 35 = \frac{133 + x_5 + x_6}{6} \cdot 6$$

$$210 = 133 + x_5 + x_6$$

$$77 = x_5 + x_6$$

Therefore, the two remaining pieces of luggage must evenly split the 77 pounds. In other words, each bag will weigh $\frac{77}{2} = 38.5$ pounds.

Median

Another measure of the center of a data set is the **median**. The median is the middle value in an ordered list of data. For this calculation to make sense, the data must be naturally orderable, such as counts or measurements. The median would not make sense for data that are simple labels such as "vanilla," "chocolate," or "strawberry."

Once the data is either in ascending or descending order, the median is the middle value of the list. If there is not a single middle value, then the median is the number that lies exactly between the two middle data values; that is, the mean of the two middle data values. Note that when there are an even number of data points, the median might not be a member of the data set.

≡ MEDIAN

The **median** of a data set is the middle value in an ordered list of the data. To find the median, follow the following steps.

1. List the data in ascending or descending order.
2. If the data set contains an *odd* number of values, the median is the middle value in the ordered list.
3. If the data set contains an *even* number of values, the median is the arithmetic mean of the two middle values in the ordered list. Notice that this implies that the median might not be a value in the data set.

In circumstances where the data points are tightly grouped values except for one or a few values, the median is a better choice for describing the "average" member of the data set. This is because extreme values do not affect the median in the way that they affect the mean. For example, at one time the highest mean salary earned by University of North Carolina graduates was not earned by the graduates in accounting, law, or medicine as may be expected, but by geography majors. This is because Michael Jordan was a geography major at the University of North Carolina. His salary alone was enough to outweigh the influence of all of the other salary points.

Example 11.3.3

Finding the Median

A VO_2 max score is the maximum amount of oxygen that one's body can transport and use during exercise. It is measured in liters of oxygen per minute (L/min). Given the following VO_2 max scores for 12 athletes, find the median score.

28.3, 27.7, 23.0, 25.5, 27.1, 26.94, 27.0, 27.52, 26.8, 27.2, 26.97, 27.53

Solution

By Hand

First, put the data in ascending numerical order.

23.0, 25.5, 26.8, 26.94, 26.97, 27.0, 27.1, 27.2, 27.52, 27.53, 27.7, 28.3

Since there are an even number of data points (12), the median will be the value between the middle two data points, 27.0 and 27.1. To find this, add the two together and divide by two.

$$\text{Median} = \frac{27.0 + 27.1}{2} = 27.05$$

TI-83/84 Plus

To calculate the median using a TI-83/84 calculator, we need to first enter the data by pressing [stat] and select Edit… to input the data as L1 in the list function. Press [2nd][mode] to return to the home screen. From here, there are two ways to find the median.

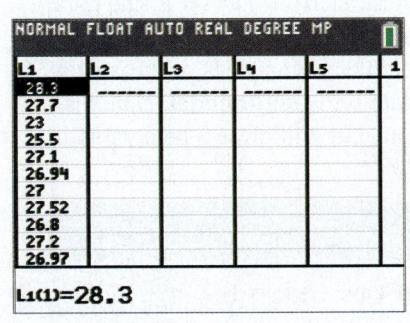

Figure 11.3.5

One way is to press [stat] again, this time selecting CALC and choosing 1-Var Stats, as we did when finding the mean. Highlight Calculate and press [enter]. Using the down arrow key, you will find that Med = 27.05.

Alternatively, once you've input the data as and returned to the home screen, you can press [2nd][stat] and arrow over to the MATH menu. Highlight median(and press [enter]. Press [2nd][stat] again and select L1. Back on the main screen, press [)] and [enter]. The calculator returns a value of 27.05.

Microsoft Excel

Excel has a built-in **MEDIAN** function that we can use to find the median of a set of numbers. Enter the values into an empty column and in an empty cell, type "=median(". Then highlight the cells containing the values and press Enter. The value will appear as 27.05.

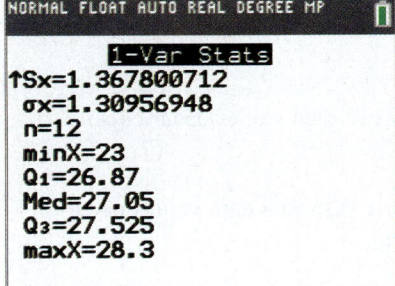

Figure 11.3.6

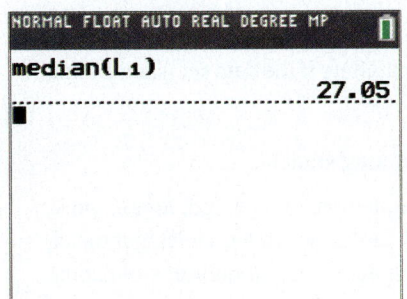

Figure 11.3.7

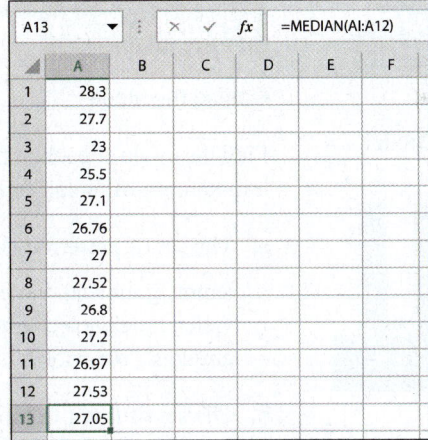

Figure 11.3.8

Thus, the median VO$_2$ max score is 27.05. Notice that the value of this "average" is not a member of the data set. However, it is a typical value in the sense that it is located in the middle of the data set when it is arranged numerically.

Mode

The last measure of center we will discuss is the mode. Sometimes a data set will not lend itself to numerical calculations such as the mean. For instance, if you surveyed students on the color of cell phone case they prefer, you would have a list of colors, which could not be added, subtracted, or put in ascending or descending order.

The **mode** is the value in the data set that occurs most frequently. If all the data values occur only once, or they each occur an equal number of times, we say that there is **no mode**. If only one value occurs the most, then the data set is said to be **unimodal**. If exactly two values occur equally often and more than any other data value, the data set is said to be **bimodal**. If more than two values occur equally often and more than any other data value, the data set is **multimodal**. Note that, unlike the mean and the median, if there is a mode, it will always be a value in the data set.

> 💬 **Helpful Hint**
>
> To find the mode of a data set, it is often helpful to first put the data in an ordered list.

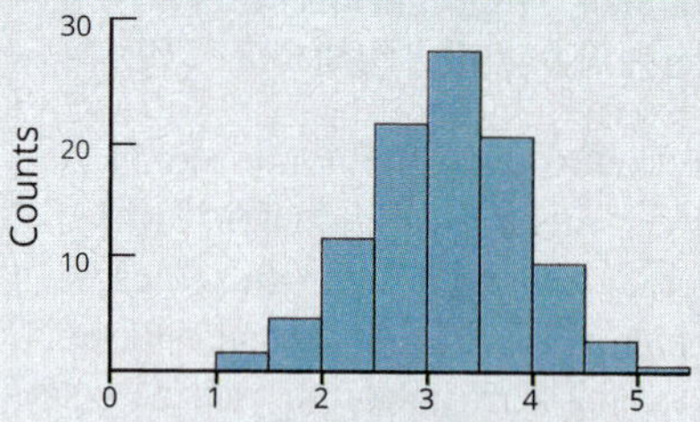

⚙ **MODE**

The **mode** of a data set is the value that occurs most frequently.

No mode describes a data set in which all of the data values occur only once or each value occurs an equal number of times.

Unimodal describes a data set in which only one data value occurs most often.

Bimodal describes a data set in which exactly two data values occur equally often and more than any other data value.

Multimodal describes a data set in which more than two data values occur equally often and more than any other data value.

> 💬 **Helpful Hint**
>
>
> **Figure 11.3.9:** Unimodal
>
>
> **Figure 11.3.10:** Bimodal
>
>
> **Figure 11.3.11:** Multimodal

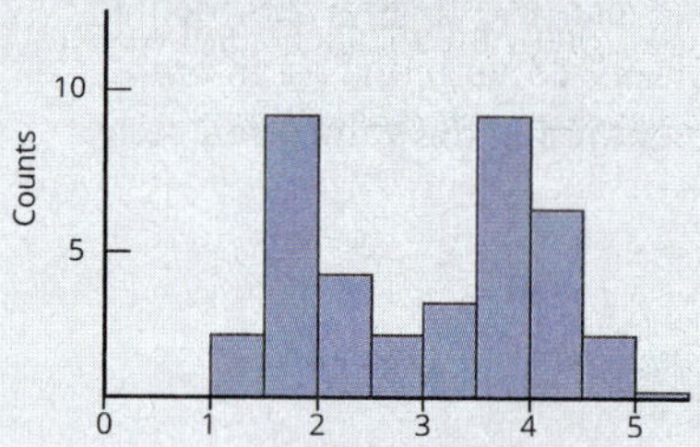

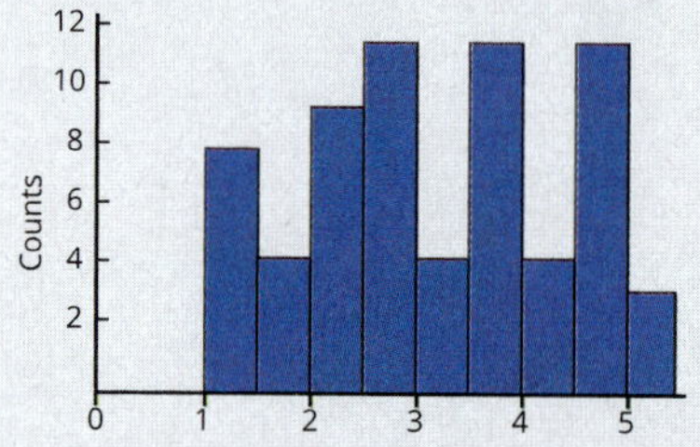

Example 11.3.4

Finding the Mode

Find the mode of each of the following sets of data. State if the data set is unimodal, bimodal, multimodal, or has no mode.

a. The set of preferred cell phone case colors among students

lemon, gunmetal, violet, turquoise, lime, violet, lemon, orange, red, lemon, pink, violet, lime, violet, lemon, pink, gunmetal, red, turquoise, violet, violet, gunmetal, turquoise, red, violet, turquoise, orange, pink, violet, violet, turquoise, violet, pink

b. The set of favorite football jersey numbers

32, 18, 99, 12, 7, 10, 28, 56, 13, 16, 19, 51, 23, 78

c. The set of ages of children at the community playground one afternoon

12, 4, 2, 7, 8, 4, 10, 6, 5, 7, 7, 4, 3

d. The set of the number of ATM withdrawals per hour at the downtown branch of University Bank

10, 13, 9, 13, 9, 14, 10, 14

Solution

a. The color violet occurs more than any other color, so the mode is violet. This data set is unimodal.

b. Each value occurs only once, so there is no mode.

c. The values 4 and 7 both occur three times, which is more than any other value. Thus, the set is bimodal with the modes 4 and 7.

d. Be careful here. Since each value occurs the same number of times, there is no mode in this data set.

For a given data set, which measure of center best describes the "average"? Well, it depends on the type of data and what we might mean by "average." For instance, the average color of car would refer to the most typical color sold, which is the mode of that data set. However, the mean should be used with data consisting of counts and measurements when the data set doesn't include any **outliers**. An outlier is a data value that is extreme compared with the rest of the data values in the set. It will influence the value of the mean of a data set but will not affect the median or mode. Because it is an extreme value, an outlier drags the value of the mean toward itself. Outliers cause the graph of the distribution to be *skewed* and not symmetrical. For example, suppose that we graphed a group of test scores for a test where most students in the class scored really low except for one student who earned a perfect score. Everyone in the class would say that this student "blew the curve," wouldn't they? Indeed, the graph created for this scenario would be skewed to the right; that is, it would be pulled toward that high grade.

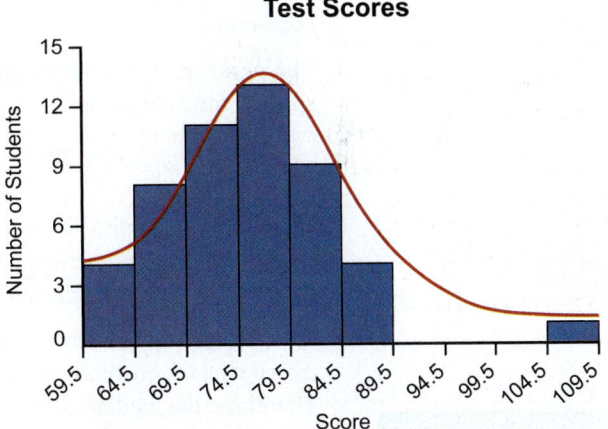

Figure 11.3.12: Skewed to the Right

We say that a graph is **skewed to the right** when the majority of the data points fall on the left side of the distribution. The "tail" of the distribution, so to speak, is on the right. Similarly, a graph is **skewed to the left** when the majority of the data fall on the right side of the distribution. A skewed graph is a good indicator that an outlier exists, although a distribution can be skewed without an outlier as well.

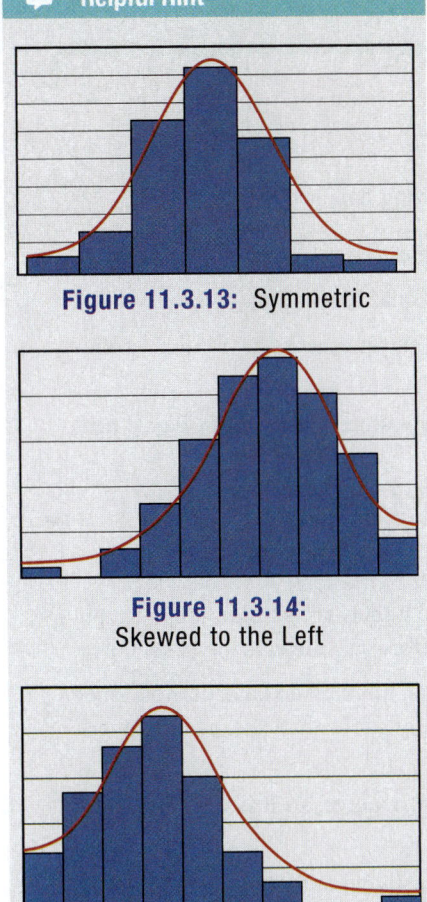

Helpful Hint

Figure 11.3.13: Symmetric

Figure 11.3.14: Skewed to the Left

Figure 11.3.15: Skewed to the Right

OUTLIERS AND SKEWED DATA SETS

Outlier

An **outlier** is a data value that is extreme compared with the rest of the data values in the set.

Skewed to the Right

A graph's shape is **skewed to the right** when a majority of the data fall on the left side of the distribution.

Skewed to the Left

A graph's shape is **skewed to the left** when a majority of the data fall on the right side of the distribution.

Example 11.3.5

Choosing the Most Appropriate Measure of Center

Choose the best measure of center for the following data sets.

a. Home prices for a subdivision of similar homes

b. Salaries for the Philadelphia Phillies professional baseball players

c. Lists of home countries by attendee at an international foreign policy conference

d. The number of people shopping in a grocery store on a certain day during specific time periods:

$$\{16, 44, 15, 48, 14, 57, 11, 54, 26, 15\}$$

Solution

a. Home prices are numerical counts. Since the homes are similar, there are not likely to be outliers in the subdivision. Therefore, the mean is the best choice.

b. The players' salaries are also numerical counts, but they do have outliers since the superstars on the team make substantially more than the other players. Therefore, the median is the best choice.

c. Countries are labels, and not counts or measurements, so it is best to use the mode as a measure of center.

d. Although there is a mode of 15, there is not really an outlier in the data, so the mean would be the best measure of center for this data.

✔ **Skill Check 11.3.1**

Find the mean, median, and mode of the following data set.

$$\{8, 12, 10, 11, 13, 12, 15, 9, 11, 16\}$$

OBJECTIVE 2 Measures of Dispersion

The numerical descriptors mean, median, and mode tell us where a data set is centered on a number line. Measures of dispersion, like the range and standard deviation, describe the "spread" of the data. In other words, they tell us whether the data values are all very similar or if they cover a wide section of the number line. For instance, we all know what it's like to be put on hold for "the next available customer service representative." If you are told the average wait time, knowing the variation in wait times might affect whether you stay on the line. Is the average the mean or the median or the mode? How would that affect how long you "hold on the line"? Measures of dispersion help us better analyze scenarios such as this.

Range

One of the simplest measures of dispersion is the **range** of the data. The range is the difference between the largest and smallest values in the data set. It tells you how much the data is spread out.

📖 RANGE

The **range** is the difference between the largest and smallest values in a data set, which tells you the distance covered on the number line between the two extremes.

$$\text{Range} = \text{Maximum Data Value} - \text{Minimum Data Value}$$

Example 11.3.6

Finding the Range

Find the range of the following sets of data.

a. The number of students enrolled as computer science majors over the past 12 semesters:

$$5, 21, 54, 33, 12, 14, 36, 40, 27, 29, 37, 22$$

b. The number of shoppers at a gas station downtown Monday through Sunday of one week:

$$1007, 1010, 1006, 1005, 1054, 1021, 1005$$

Solution

a. The maximum value for the data set is 54 and the minimum value is 5, so the range is calculated as follows.

$$54 - 5 = 49$$

b. The maximum value for the data set is 1054 and the minimum value is 1005, so the range is calculated as follows.

$$1054 - 1005 = 49$$

Although calculating the range is easy, it is not as descriptive as other measures of dispersion. Consider the two data sets in the previous example. Notice that both data sets have the same range. However, almost all of the values in the second data set are similar while the values in the first data set are more spread out. Additionally, the range can be affected by outliers, just as we saw with the mean. For instance, the data set {1, 10, 10, 10} has a range of 9, but this is a poor representation of the dispersion of the data. To help us shed light on situations such as these, we must use another measure of dispersion.

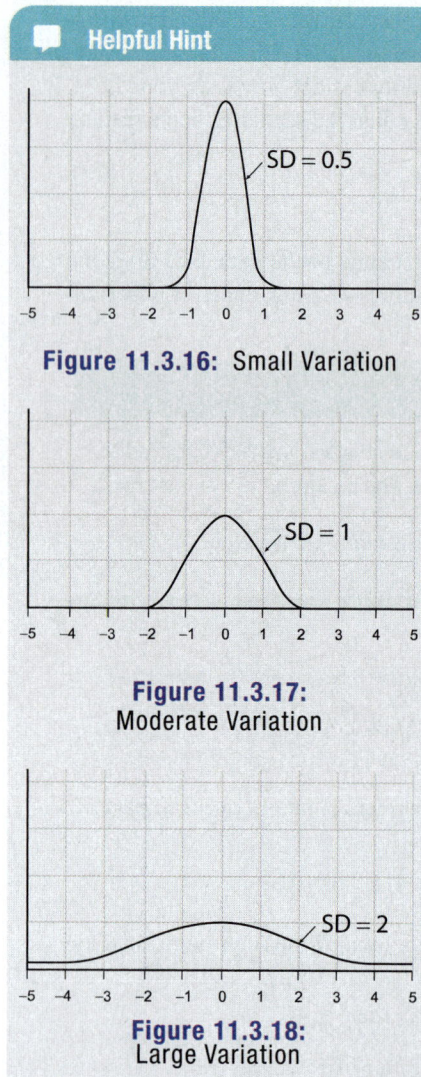

Figure 11.3.16: Small Variation

Figure 11.3.17:
Moderate Variation

Figure 11.3.18:
Large Variation

Standard Deviation

The **standard deviation** is a measure of how much we might expect a member of the data set to differ from the mean. The greater the standard deviation, the more the data values are spread out. Similarly, a smaller standard deviation indicates that the data values lie closer together. The standard deviation is always a number greater than or equal to zero. The standard deviation is equal to zero only when all of the data points are the same value.

The standard deviation allows us to interpret the variation of the data with a sense of scale. Two standard deviations that are exactly the same number don't necessarily represent the same variation for different populations. For instance, a standard deviation of 3.50 years when referring to an adult's age is not a large variation. However, if the standard deviation for gas prices at local gas stations had a value of $3.50, we would say that the data had a huge variation!

Standard deviations can also be used as a tool to compare two data sets. For example, suppose you received a bonus at work and are considering investing a portion of that money into the stock market. Company A touts that their stock prices over the past year have been consistent and have had a standard deviation of only $1.02. Company B, however, brags that their stock has been competitive over the past year with a standard deviation of $9.85. Where should you put your money? If you want a stable long-term investment, Company A appears to be the better choice since its standard deviation is smaller, which indicates that the stock price has not varied much. However, if you are looking to make a quick profit and are willing to take the risk, then Company B would seem to better suit your needs. Its price has varied much more over the past year, indicating an opportunity for potential gain, but also a potential loss.

Recall that we presented two formulas for the mean: the population mean and the sample mean. We also have two formulas for calculating standard deviation: the population standard deviation and the sample standard deviation. As with the mean, when the data set consists of the entire population, we use the population standard deviation; otherwise, we use the sample standard deviation.

𝑓(𝑥) STANDARD DEVIATION

The **standard deviation** is a measure of how much we might expect a member of the data set to differ from the mean.

The formula for finding the **population standard deviation** σ is as follows.

$$\sigma = \sqrt{\frac{\sum(x_i - \mu)^2}{N}}$$

Here, x_i is the ith data value, μ is the population mean, and N is the size of the population. The formula for finding the **sample standard deviation** s is as follows.

$$s = \sqrt{\frac{\sum(x_i - \overline{x})^2}{n-1}}$$

Here, x_i is the ith data value, $\overline{x}$ is the sample mean, and n is the sample size.

Example 11.3.7

Calculating Standard Deviation

The SAT Critical Reading scores for 12 of the members of the senior class at Richmond Prep High School are given in Table 11.3.1.

a. Calculate the appropriate standard deviation for the scores.

b. Describe what the standard deviation indicates about the SAT scores.

c. If you were part of the administration team at Richmond Prep High School, explain if you would rather see a smaller standard deviation on these scores or a larger one.

Table 11.3.1: SAT Critical Reading Scores

640	750	560	630
600	590	610	710
600	640	550	610

Solution

a. First note that the data given is a sample, because it consists of 12 of the members of the senior class, not the entire class. So we will use the formula for sample standard deviation given here.

Sample Standard Deviation

$$s = \sqrt{\frac{\sum(x_i - \bar{x})^2}{n-1}}$$

By Hand

When calculating the standard deviation by hand, the variables we need are the sample size n, the sample mean $\bar{x}$, and each of the data points x_i.

We are told $n = 12$, and we calculate that the sample mean is $\bar{x} \approx 624.17$.

Next, we subtract the mean from each data point, $x_i - \bar{x}$. This is the *deviation* of each score from the mean. Then, we square each of the deviations to find the *squared deviations* of the data points, $(x_i - \bar{x})^2$. Finally, we find the sum of the squared deviations, $\sum(x_i - \bar{x})^2$, which is the numerator of the fraction.

When calculating standard deviation by hand, it's helpful to use a table like Table 11.3.2 to keep everything organized.

Table 11.3.2: Deviations and Squared Deviations of the Data

Data, x_i	$(x_i - \bar{x})$	$(x_i - \bar{x})^2$
640	15.83	250.5889
600	−24.17	584.1889
600	−24.17	584.1889
560	−64.17	4117.7889
610	−14.17	200.7889

Data, x_i	$(x_i - \bar{x})$	$(x_i - \bar{x})^2$
550	−74.17	5501.1889
750	125.83	15,833.1889
590	−34.17	1167.5889
640	15.83	250.5889
630	5.83	33.9889
710	85.83	7366.7889
610	−14.17	200.7889

Finally, substitute the appropriate values into the formula for population standard deviation.

$$s = \sqrt{\frac{\sum(x_i - \bar{x})^2}{n-1}}$$

$$\approx \sqrt{\frac{36091.6668}{12-1}}$$

$$\approx 57.2805$$

> **Helpful Hint**
>
> In an effort to minimize any error introduced from rounding, it is helpful to leave intermediate calculations with more decimal places.

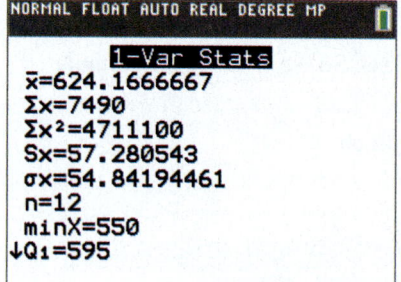

Figure 11.3.19

TI-83/84 Plus

In order to calculate the sample standard deviation using a TI-83/84 Plus calculator, begin by pressing [stat], selecting Edit..., and entering the data values in L1. Press [stat] again, scroll over to the CALC menu and select 1-Var Stats. Highlight Calculate and press [enter]. Notice that this is the same screen we've seen before when finding other statistics such as the mean. The sample standard deviation is given as Sx = 57.280543.

Microsoft Excel

Excel has built-in function for standard deviation of a set of values. For sample standard deviation the function is **STDEV.S.** Enter the values into empty cells. In another cell, type "=stdev.s(". Then highlight the cells containing the data points and press Enter. The sample standard deviation will appear as 57.2805.

> **Helpful Hint**
>
> On the TI-83/84 Plus, the sample standard deviation is **Sx** and the population standard deviation is σ**x**.

C8		▼	⋮	×	✓	fx	=STDEV.S(A1:B6)	
	A	B	C	D	E	F	
1	640	750					
2	600	590					
3	600	640					
4	560	630					
5	610	710					
6	550	610					
7							
8			57.2805				

Figure 11.3.20

Therefore, the sample standard deviation for the SAT Critical Reading scores is approximately 57.3 points.

b. The standard deviation indicates that the SAT scores differ from the mean score on average by 57.3 points. Notice that the mean is approximately 624.17. Although there is not an actual score of 624.17, you can see that many of the student's scores fall within about 60 points (or 1 standard deviation) of that mean, either larger or smaller.

c. As an administrator, a small standard deviation is a sign that students are preforming similarly. If the SAT scores are on the high end, this is a good fact. However, if the SAT scores are low, then a small standard deviation indicates that the students are all scoring poorly and would raise concern. If the standard deviation was relatively large, this is an indicator that SAT scores are more varied. As an administrator you would want to look at the data individually to determine any problem areas in your classes.

✔ **Skill Check 11.3.2**

Find the population standard deviation of the following data. Round to the nearest tenth.

8, 12, 10, 11, 13, 12, 15, 9, 11, 16

OBJECTIVE 3 Measures of Relative Position

Percentiles and Quartiles

Another way that we can describe the data set is to evaluate the position of a data value in relation to the other data points in the set. Often it is useful to know exactly where a particular value is located within a set, or how many data points are either above it or below it. Descriptors like **percentiles** and **quartiles** are measures of relative position. They are each based on dividing the data set into sections and then referencing the section in which the data value falls. Percentiles divide the data into 100 equal parts while quartiles divide the data into four parts.

💬 **Helpful Hint**

To divide something into 4 parts, you only need 3 dividers, as the figure illustrates.

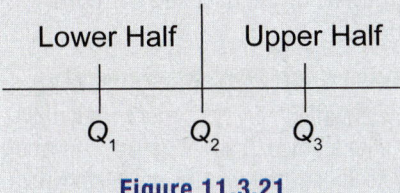

Figure 11.3.21

📖 **PERCENTILES AND QUARTILES**

Percentiles

Percentiles divide an ordered list into 100 equal parts and tell you approximately what percentage of the data lies at or below a given value.

Quartiles

Quartiles are values that divide an ordered list of data into four equal parts; equivalent to the 25th, 50th, and 75th percentiles.

Q_1 = First Quartile = 25th percentile; that is, 25% of the data is less than or equal to this value.

Q_2 = Second Quartile = 50th percentile; that is, 50% of the data is less than or equal to this value.

Q_3 = Third Quartile = 75th percentile; that is, 75% of the data is less than or equal to this value.

By definition, Q_2 is the same as the median.

To have a data point at the 81st percentile means that 81% of the population is at or below that data value. It's important to note that percentiles don't convey anything about the data value itself, only its relative position to the data set as a whole. One of the most prominent places you might have heard of percentiles would be in connection with any standardized tests you've taken, such as the ACT or SAT college entry exams. Birth measurements are also commonly reported in terms of percentiles.

Example 11.3.8

Interpreting Percentiles

Sierra received her scores from taking a mathematics placement test for her chosen university. Choose the best explanation for what it means for her to be in the 61st percentile.

a. She correctly answered 61% of the answers on the test.

b. 61% of people taking the test scored the same as Sierra.

c. Sierra's score was at least as good as 61% of the people taking the test.

d. Sierra missed 39% of the test questions.

Solution

The correct interpretation of her score is c.: "Sierra's score was at least as good as 61% of the people taking the test." Both a. and d. are incorrect because they refer to how many questions she answered correctly on the test and not how she did in comparison to others taking the test. Option b. is not quite correct because percentiles tell you the percentage that scored at or below you. They are not all necessarily the same score as Sierra's.

Notice that the quartiles are equivalent to percentiles. To find a rough approximation by hand for the quartiles, we can order the data set and determine the median. Then use the median to divide the data set into an upper half and a lower half. For an odd number of data values, do not include the median in either half. For an even number of data values, the data set will divide into a lower and upper half evenly. The first quartile is the median of the lower half of the data and the third quartile is the median of the upper half of the data.

Example 11.3.9

Interpreting Quartiles

On Karl's recent standardized test results, the picture graph of his score showed he was above the third quartile in language arts. His classmate, Asher, said his score was at the 70th percentile, while Rylie said hers was at the 79th percentile. Which of the three had the best language arts test score?

Solution

We know the percentile ranks of both Asher and Rylie are the 70th and 79th respectively. What we know about Karl's score is that it was above the third quartile. Since the third quartile is the same as the 75th percentile, we know that his score was somewhere at or above the 75th percentile. We can conclude that he did better than Asher, whose score was at the 70th percentile, but we can make no definite comparison with Rylie, whose score was at the 79th percentile, because we do not know for sure which one had the best language arts score.

✔ **Skill Check 11.3.3**

Calculate the three quartiles for the following set of data.

60, 62, 63, 65, 65, 67, 70, 71, 71, 75, 78, 79, 80, 81

Five-Number Summary and Box Plots

Quartiles are used in a numerical summary of a data set aptly called the **five-number summary** because it contains five numbers: the minimum value, the first quartile, the median (or second quartile), the third quartile, and the maximum value. The five-number summary is made up of these five numbers listed in order from smallest to largest.

Most commonly, we display the five-number summary graphically using a **box plot**, or *box-and-whiskers plot*. As Figure 11.3.22 illustrates, the endpoints represent the minimum and maximum data values, while the lines sectioning off the box in the middle represent each of the quartiles. This *box* represents the **interquartile range**, or **IQR**, which is the range of the middle 50% of the data. The IQR is equal to the difference between the third quartile and the first quartile. The *whiskers* are the lines that extend to reach the minimum and maximum values.

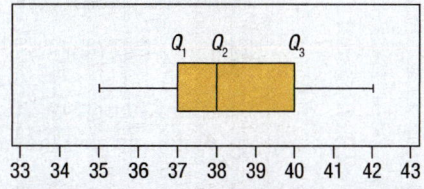

Figure 11.3.22: Box Plot Example

📖 FIVE-NUMBER SUMMARY AND BOX PLOTS

Five-Number Summary

The **five-number summary** is a numerical description of a data set that lists the minimum value; the first quartile, Q_1; the median or second quartile, Q_2; the third quartile, Q_3; and the maximum value in order from smallest to largest.

Box Plot

A **box plot**, or *box-and-whiskers plot*, is a graphical representation of a five-number summary.

Interquartile Range (IQR)

The **interquartile range** is the range of the middle 50% of the data, given by

$$IQR = Q_3 - Q_1,$$

where Q_3 is the third quartile and Q_1 is the first quartile.

To create a box plot by hand, follow the steps outlined below.

☰ CREATING A BOX PLOT

1. Draw a horizontal (or vertical) number line that contains the five-number summary.

2. Draw a small line segment above (or next to) the number line to represent each of the numbers in the five-number summary.

3. Connect the line segment that represents the first quartile to the line segment representing the third quartile, forming a box with the median's line segment in between.

4. Connect the "box" to the line segments representing the minimum and maximum values to form the "whiskers."

Example 11.3.10

Creating a Box Plot

A random sample of 20 adults was taken at an athletic training facility. Create a box plot to display the training heart rates of this sample.

Table 11.3.3: Training Heart Rates

130	124	125	133	129
126	126	130	124	121
136	116	113	128	127
119	126	127	132	131

Solution

The first thing we need to do is calculate the five-number summary.

By Hand

Begin by ordering the data set, which is essential to finding the values for the five-number summary by hand.

113	116	119	121	124	124	125	126	126	126
127	127	128	129	130	130	131	132	133	136

The minimum value and maximum value are then easy to identify: minimum = 113; maximum = 136.

The next step is to identify the quartiles. Begin with the median. Since $n = 20$, the median is the arithmetic mean of the values in the 10th and 11th positions.

$$Q_2 = \frac{126 + 127}{2} = 126.5$$

To find Q_1 and Q_3, we find the median of the lower half of data as well as the median of the upper half. Both halves contain 10 data values, so we need the arithmetic mean of the values in the 5th and 6th positions for Q_1 and the arithmetic mean of the values in the 15th and 16th positions for Q_3.

$$Q_1 = \frac{124 + 124}{2} = 124$$

$$Q_3 = \frac{130 + 130}{2} = 130$$

Stop and note that although we did the calculations for Q_1 and Q_3, the arithmetic mean of two identical numbers is always equal to that number.

TI-83/84 Plus

The first step to calculating the five-number summary on a TI-83/84 Plus is to store our data by pressing [stat], selecting Edit..., and entering the data values in L1. Once the values have been entered, press [stat] again, arrow over to the CALC menu, and select 1-Var Stats. Highlight Calculate and press [enter]. As we've seen before, 1-Var Stats provides lots of information about the data set, including all of the values for the 5-number summary. Using the down arrow key, you will find that minX = 113, Q_1 = 124, Med = 126.5, Q_3 = 130, and maxX = 136.

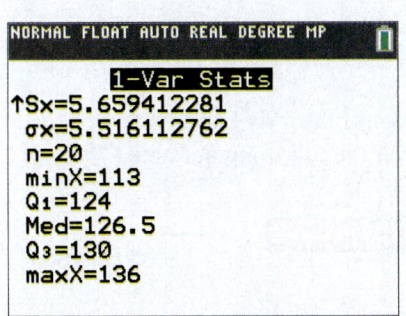

Figure 11.3.23

Thus, the five-number summary is 113, 124, 126.5, 130, and 136.

To create the box plot from the five-number summary, we'll follow the procedure steps given for creating a box plot.

Step 1:

Label the horizontal axis at even intervals, choosing a range that is large enough to include both the minimum value and the maximum value of the five-number summary.

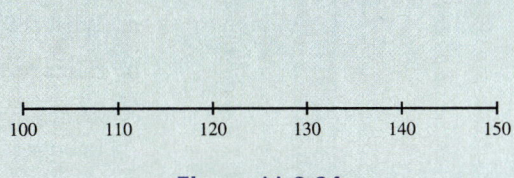

Figure 11.3.24

Step 2:

Place a small line segment above each of the numbers in the five-number summary.

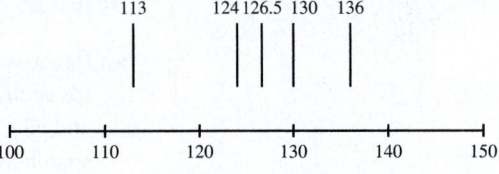

Figure 11.3.25

Step 3:

Connect the line segment that represents Q_1 to the line segment that represents Q_3, forming a box with the median's line segment in between.

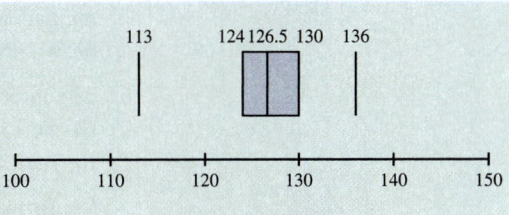

Figure 11.3.26

Step 4:

Connect the "box" to the line segments representing the minimum and maximum to form the "whiskers."

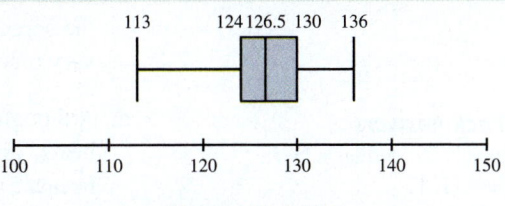

Figure 11.3.27

Because they are very easy to read and are effective at summarizing a data set, box plots are often used for the sake of comparison. Multiple sets of data can be displayed on a single graph to aid in contrasting and comparing the data sets. Example 11.3.11 looks at interpreting box plots.

Example 11.3.11

Interpreting Box Plots

The box plots in Figure 11.3.28 show the life spans of grizzly bears compared to the life spans of brown bears. Use them to answer the following questions.

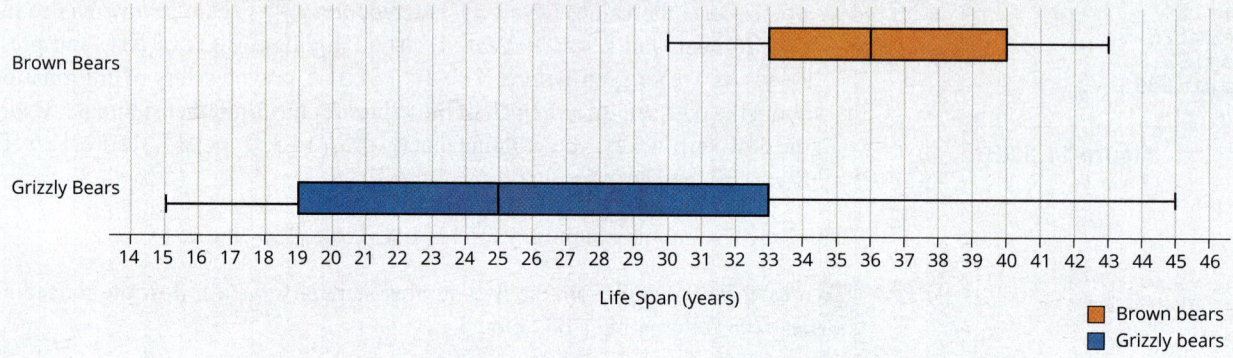

Figure 11.3.28

a. What is the range of life spans for each type of bear?

b. Based on the box plots, how long does the average brown bear live? Grizzly bear?

c. Which type of bear has the largest interquartile range for their life span? Interpret your answer.

d. Based on the box plots, which bears live longer? Explain your answer.

Solution

a. Because the ends of the whiskers identify the minimum and maximum life span for each type of bear, we can see that brown bears have a life span between 30 and 43 years, which is a range of $43 - 30 = 13$ years. Grizzly bears have a life span between 15 and 45 years, which is a range of $45 - 15 = 30$ years.

b. We can use the median as the "average" life span for each type of bear. The median is represented by the line segment in the middle of each box. Thus, brown bears live 36 years on average and the average grizzly bear lives 25 years.

c. The interquartile range (IQR) is the difference between Q_1 and Q_3 represented by the line segments at the end of the boxes. Thus, we can calculate the IQR for each type of bear as follows.

Brown Bear IQR $= 40 - 33 = 7$ Grizzly Bear IQR $= 33 - 19 = 14$

Grizzly bears have the larger interquartile range. Because the IQR represents the spread of the middle 50%, we can deduce that life spans of grizzly bears vary much more than those of brown bears.

d. Although the maximum life span for grizzly bears is longer than that for brown bears, one can argue that as a whole, brown bears live longer. We see this because the top 75% of life spans for brown bears lie in the same range as the top 25% of life spans for grizzly bears.

Skill Check Answers

1. Mean $= 11.7$; Median $= 11.5$; Mode $= 11, 12$

2. 2.5 **3.** $Q_1 = 65, Q_2 = 70.5, Q_3 = 78$

11.3 Exercises

✔ CONCEPT CHECK

1. The arithmetic mean is the _____ of all the data points _____ by the number of data points.

2. If a data set contains an _____ amount of values, the median is the middle value in the ordered list.

3. If a graph is skewed to the right, the majority of the data points fall on the _____ side of the distribution.

4. True or False: If all values in a data set are equal, the data set has no mode.

5. True or False: The values in a data set with a standard deviation of 3.5 is more spread out than the values in a data set with a standard deviation of 1.2.

💡 PRACTICE

Find the mean, median, mode, range, and standard deviation for each data set. When applicable state whether the data set is unimodal, bimodal, or multimodal. Round answers to one more decimal place than the largest number of decimal places given in the data. All data sets are samples unless stated otherwise.

6. 19, 32, 15, 21, 25, 22, 22, 28, 27, 27, 26

7. $11.40, $32.00, $22.50, $12.01, $10.08, $18.30, $18.40, $32.00

8. 45, 21, 26, 26, 45, 37, 22, 33, 26, 21, 42, 37, 41, 43, 46, 35, 31, 29, 46

9. 310, 310, 310, 310, 310, 310

10. 9, 3, −5, −3, −7, 3, 0, 6, −9, −7, −3, −8

11. The following data represent sample ACT scores from students at a local high school.

ACT Scores	
13	26
10	20
24	30
25	31
6	24
35	35
26	15

12. The following are lengths of each movie in the complete Harry Potter film series. Note that because these include all of the films in the series, this is a population.

Time Lengths for Harry Potter Film Series

Movie Title	Time (in Minutes)
Harry Potter and the Philosopher's Stone (2001)	152 minutes
Harry Potter and the Chamber of Secrets (2002)	161 minutes
Harry Potter and the Prisoner of Azkaban (2004)	141 minutes
Harry Potter and the Goblet of Fire (2005)	157 minutes
Harry Potter and the Order of the Phoenix (2007)	138 minutes
Harry Potter and the Half-Blood Prince (2009)	153 minutes
Harry Potter and the Deathly Hallows: Part 1 (2010)	146 minutes
Harry Potter and the Deathly Hallows: Part 2 (2011)	130 minutes

For each data set, determine the most appropriate measure of center.

13. Styles of houses in a suburb: ranch, colonial, bungalow, etc.

14. Grades on the final in Biology 210 at State University.

15. The ratings on a customer satisfaction survey: strongly disagree, disagree, neither agree nor disagree, agree, and strongly agree.

16. Salaries for janitorial staff at the state governmental buildings that include the Director of Sanitation's salary.

 **APPLICATIONS**

Use the formula for the mean to find the missing piece of data.

17. John knows that his first 4 tests grades were 84, 79, 82, and 88. Find John's grade on the fifth test if his average was 83.8.

18. A family has four children. Edward has a height of 58 inches, David 52 inches, and Chloe 44 inches. How tall is Benjamin if the average height of the children is 52.5 inches?

19. Ella went to a butcher shop to buy some meat. She chose five pieces of meat, three of which weighed 960 g, 840 g, and 1120 g. If Ella needs at least 5 kilograms of meat, how much should the other two pieces weigh, on average?

20. Jacob and his five friends choose what drinks to order at a coffee shop. The first three people want a cappuccino for $4.50, a latte for $4.25, and hot tea for $3.50. How much, on average, can the next three people spend on each drink if the total check should not exceed $25?

21. Marcel scored in the 91st percentile on the MCAT (Medical College Admissions Test). The medical school he is applying to only accepts students who score in the top 10% on the MCAT. Did Marcel score well enough to be considered for his school of choice?

22. Suppose that 110 male students are surveyed and that 52% have a height less than 1.776 m.

    a. True or False: Of those surveyed, the mean height must be under 1.776 m.

    b. True or False: Of those surveyed, the median height must be under 1.776 m.

23. If we know that a salary of $65,300 was in the 67th percentile in a company survey, can we determine how many employees were in the sample? Why or why not?

24. People who are 6 feet and 1 inch tall are in the 90th percentile among American adults. Which of the following is the most likely percentile for people who are 6 feet and 7 inches tall: 80th, 85th, 90th, 95th, or 99th?

25. The average annual temperature in Georgia is at the 90th percentile among all the US states, while the temperature in North Carolina is at the third quartile. Which state has the higher average annual temperature?

26. Ethan's salary is at the third quartile at the company he works. His wife's salary is at the second quartile at the company she works. Are we able to determine which of them has a higher salary? Explain.

27. A doctor told Amelia that her child's weight was at the 30th percentile in the average growth chart. Does the child weigh more or less than the median weight for his age?

28. Last year, the Smiths were at the 72nd percentile of the nation's income distribution. This year, they are at the third quartile. Did their position go up or down?

29. Consider the prices of 20 used cars sold on an online platform.

$24,400	$25,270	$27,570	$33,430	$28,500
$33,340	$33,760	$37,240	$23,270	$24,790
$21,590	$21,870	$33,650	$35,890	$30,560
$34,910	$38,680	$36,310	$36,640	$36,030

    a. Find the five-number summary of the used car prices.

    b. What percentage of used car prices are at or below $23,270?

    c. What is the range of used car prices on the platform?

30. Consider the given box plot.

Life Expectancy at Birth

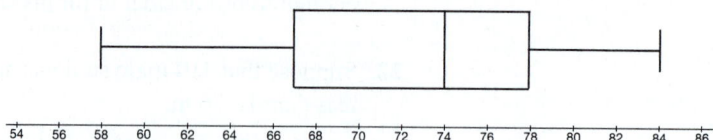

a. Based on the boxplot, estimate each of the values in the five-number summary.

b. The midrange is the average of the minimum and maximum data values. Estimate the mid-range from the box plot.

31. Calculate the values needed to construct a box plot for the following data. Sketch a graph of the box plot.

26	26	26	34	40	46	54
58	64	74	74	76	82	82
86	96	100	110	110	112	112
116	120	130	158	162	166	168

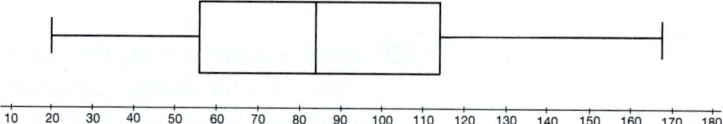

32. Given the five-number summary for three data sets, sketch a box plot for each, side-by-side on the same graph. Then answer the following questions based on your box plots.

Weight of Adult Male Dogs by Breed (in pounds)

	English Mastiffs	Golden Retrievers	Rottweilers
Min	190	55	95
Q_1	205	65	105
Q_2	220	68	112
Q_3	225	70	120
Max	240	85	135

a. Which breed has the largest range of weight?

b. Which breed has the least variation in weight?

c. Which breed has the largest median weight?

33. The following are house prices in one neighborhood.

$181,865	$119,442	$152,750	$100,960	$159,635
$150,963	$133,702	$149,788	$145,495	$182,500
$112,021	$120,900	$145,850	$164,590	$144,413

   a. Find the five-number summary of the house prices.

   b. What percentage of house prices is at or below $159,635?

   c. What is the range of house prices for this neighborhood?

34. The following graph contains a box plot.

Birth Weeks

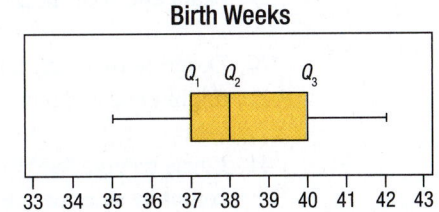

   33 34 35 36 37 38 39 40 41 42 43

   a. Based on the box plot, estimate each of the values in the five-number summary.

   b. The midrange is the average of the minimum and maximum data values. Estimate the midrange from the box plot.

35. Calculate the values needed to construct a box plot for the following data. Sketch a graph of the box plot.

310	320	450	460	470	500	520	540
580	600	650	700	710	840	870	900
1000	1200	1250	1300	1400	1720	2500	3700

36. Given the five-number summary for three data sets, sketch a box plot for each, side-by-side on the same graph. Then answer the following questions based on your box plots.

Committees and the Ages of Members

	Membership	Finance	Publicity
Min	23	26	25
Q_1	27	32	26
Q_2	29	38	27
Q_3	33	44	29
Max	35	46	33

   a. Which committee has the largest range of ages?

   b. Which committee has the least variation in the ages?

   c. Which committee has the smallest median?

✏ WRITING & THINKING

37. Accounting 101 has five class sections. All five classes took the same final. The mean scores on the final for each class were 72, 78, 76, 74, and 79. Can the mean final score for all students in Accounting 101 be found by averaging the mean scores in each class? Explain your answer.

38. Does the standard deviation of a data set equaling zero imply that all entries in the data set equal zero?

39. Explain the difference between Amelia making an 82 on her pre-calculus exam and scoring in the 82nd percentile in mathematics on the ACT test.

40. Describe two data sets, one that might have a large variation and one that might have a small variation.

41. Lucas received an e-mail containing the five-number summary for the company sales data that he asked for. Unfortunately, the e-mail cut off the summary labels and scrambled their order. Can you still determine which number is the first quartile, Q_1? Explain your answer.

five-number summary: 11, 17.5, 9, 13.5, 19

11.3 PROJECT

AVERAGES AND OUTLIERS BETWEEN FRIENDS

Suppose two competitive friends, Student A and Student B, are both finishing up a course on technical writing at different colleges. At the end of the semester, they compare their grades to see who did better in this course. Each friend had to complete a different number of graded assignments and exams, where each grade was given equal weight. The grades earned by the two students on all graded assignments and exams are as follows.

Student A: 78, 89, 95, 64, 98, 0, 87, 84, 76, 93, 89, 77, 61

Student B: 87, 79, 88, 91, 89, 77, 86, 93, 105, 89

In this project, you will use the skills learned in Section 11.3 to determine which friend earned the higher grade in the technical writing course.

1. For each set of grades, determine the mean, median, mode, range, and standard deviation. Round the answers to the nearest tenth, if necessary.

Comparison of Students' Grades

	Student A	Student B
Mean		
Median		
Mode		
Range		
Standard Deviation		

2. Compare the grades of the two students. Which value(s) did you use in your comparison? Explain your reasoning.

3. Determine which value(s) in each student's grades are outliers and remove the outlier(s) from the data sets. Explain why each data point you removed is an outlier.

4. For each modified data set, determine the mean, median, mode, range, and standard deviation. Round the answers to the nearest tenth, if necessary.

Comparison of Students' Grades with Outliers Removed

	Student A	Student B
Mean		
Median		
Mode		
Range		
Standard Deviation		

5. Do these new values change your mind of which student performed better in the technical writing course? Explain why or why not.

6. Do you think outliers should be removed when comparing grades between students? Explain your reasoning.

OBJECTIVES

1. Identify the properties of a normal distribution.

2. Use the Empirical Rule.

3. Find areas under the standard normal distribution.

4. Calculate probabilities for a normal distribution.

Consider outcomes that happen by chance—rolling a die, winning the lottery, or even the number of customers who enter a store on a given day. These numerical outcomes are called *random variables* because we cannot know the value that the variable takes before measuring due to its random nature. Even though we cannot know what value the variable will take, that doesn't mean that we cannot say anything about that random value itself. In a previous chapter, we investigated *algebraic* variables and saw that by writing down algebraic equations or inequalities to describe a particular scenario, we were able to conclude information about otherwise unknown values. Similarly with random variables, we can draw conclusions about a random outcome before it happens by recognizing the statistical properties of the scenario; but this time, our predictions will happen with high probability rather than with certainty, as in algebra.

In this section, we will consider statistical properties for random variables and what kinds of conclusions statistical techniques can provide.

OBJECTIVE 1 Normal Distributions

The fundamental property of a random variable is that although we don't know which of its possible values it will take, we do know that there is a predetermined probability that the variable will be each possible value. For instance, consider the random variable that consists of the roll of a single fair die. We know that each value 1, 2, 3, 4, 5, and 6 has the same chance of being rolled: $\frac{1}{6}$. For two dice, the sums of the values on the faces of the die are 2, 3, 4, …, 10, 11, and 12. As I'm sure you've experienced if you have ever played a board game, the sums from rolling two dice do not all have the same likelihood. In fact, rolling a sum of 7 is most likely and is three times more likely than rolling a sum of 3 or 11.

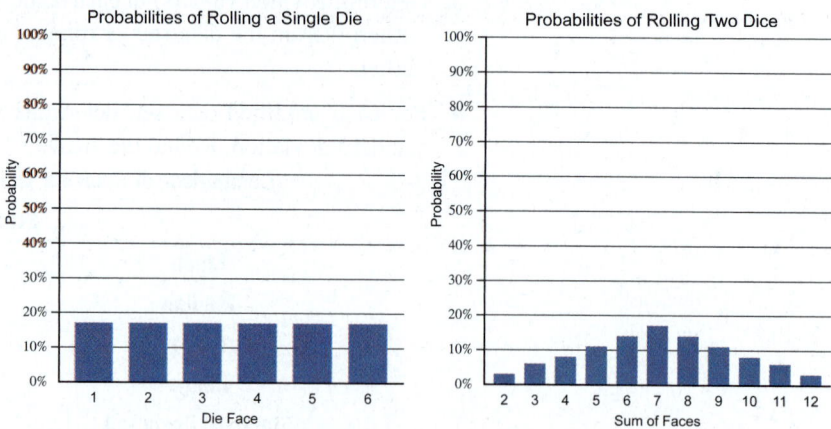

Figure 11.4.1: Probabilities of Rolling One Die and Two Dice

To describe the pattern of these predetermined probabilities, we say that a random variable has a distribution. Recall that a distribution is a way to describe the structure of a particular data set or population. Properties of the distribution can tell us a lot about the outcome of a random variable and allow us to draw conclusions about the outcome of a probabilistic scenario before it happens.

Distributions are classified into two major families based on the type of random variable involved: discrete and continuous.

Distributions categorized as discrete involve **discrete variables**. These types of variables can take on only particular values and cannot take on the values in between. Thus, discrete variables are often counts or categories. For example, the number of pets you have would be discrete because you can have either 2 pets or 3 pets, but not 2.75 pets. Likewise, your blood type is one of 8 distinct blood groups, A, B, AB, or O, each of which is either RhD+ or RhD−. You can't have a mixture of two or three blood types.

From our study of probability in Section 10.5, we know there are many scenarios that are collections of events with only two outcomes: yes or no; true or false; live or die. These binomial events are precisely what statisticians need to analyze when they study the number of foul shots a player will make at the line or whether a new vaccine is effective. Scenarios of this type of discrete variable are modeled by the *binomial distribution.*

However, suppose we want to model scenarios that are analogous to the number of pages in a random stack of papers, the number of people in line at a checkout, the number of letters a particular address receives each day, or the number of phone calls a call center receives in an hour. These discrete random variables measure the number of events in a fixed interval of time or space and are all modeled by the *Poisson distribution.*

The binomial and Poisson distributions are both examples of discrete distributions since they each only take values that are nonnegative integers. We will leave these distributions, and others like them, to be examined in more detail in most introductory statistics courses. We will focus our study on continuous distributions; namely, the normal distribution.

Distributions classified as continuous involve variables that can take on any value in a given range of numbers. **Continuous variables** are often measurements of some quantity. For example, the heights of trees, the lengths of newborn babies, or the ages of humpback whales can all be modeled by continuous distributions.

Helpful Hint

Common continuous distributions include

- Normal distribution
- Chi-square distribution
- *F*-distribution

📖 DISCRETE AND CONTINUOUS VARIABLES

Discrete Variable

A **discrete variable** is a variable whose numeric value can only take on particular values and is usually a count.

Continuous Variable

A **continuous variable** is a variable whose numeric value can take on any value in a given interval and is usually a measurement.

When the data are discrete, a distribution is often shown in a frequency table, as we saw earlier in the chapter. However, listing all the possibilities for a continuous distribution isn't feasible. For instance, think about listing all the possible lengths of newborn babies. Although it is common to list feet and inches to the nearest whole inch, there are actually infinitely many lengths that cannot be listed in a table. So as an alternative, we define continuous distributions by specific parameters that vary by the type of distribution.

We're going to focus in on the most prevalent continuous distribution, as its name suggests—the normal distribution. A **normal distribution** is a continuous probability distribution for a continuous random variable defined completely by its mean and standard deviation, and it has a very particular shape. In fact, there are four properties that define a normal distribution. These are listed in the definition box along with a general graph of a normal distribution.

✿ NORMAL DISTRIBUTION

A **normal distribution** is a probability distribution for a continuous random variable defined completely by its mean and standard deviation such that the following properties are true:

1. It is bell-shaped and symmetrical about its mean.
2. It is completely defined by its mean μ and standard deviation σ.
3. The total area under the curve is equal to 1.
4. The x-axis is the horizontal asymptote for a normal distribution curve.

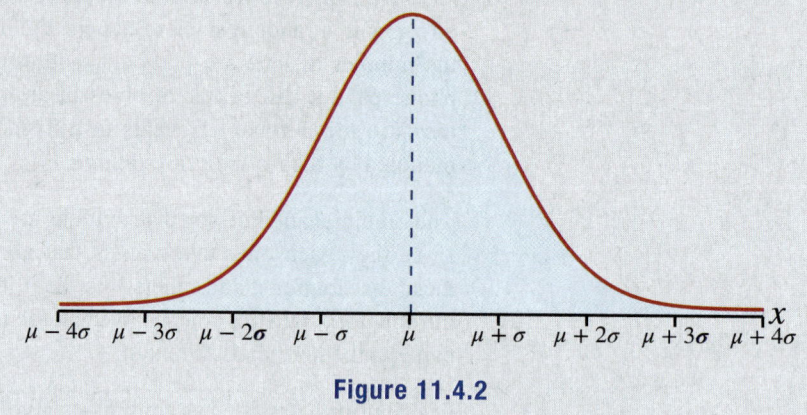

Figure 11.4.2

As an example, consider the data set that consists of the heights of 1000 randomly selected men in the United States. Heights of adult men follow a normal distribution—most adult men are fairly close to the same height, give or take a few inches. Very tall and very short men are rare. Figure 11.4.3 is a graph of the heights of a random sample of 1000 men, where the mean is close to 69.2 inches. Notice that the distribution has the shape of a bell with its center at the mean, and that the x-axis acts as a boundary, or *horizontal asymptote*, for the curve.

> **Think Back**
>
> A horizontal asymptote is a horizontal line that a graph approaches but never touches or crosses.

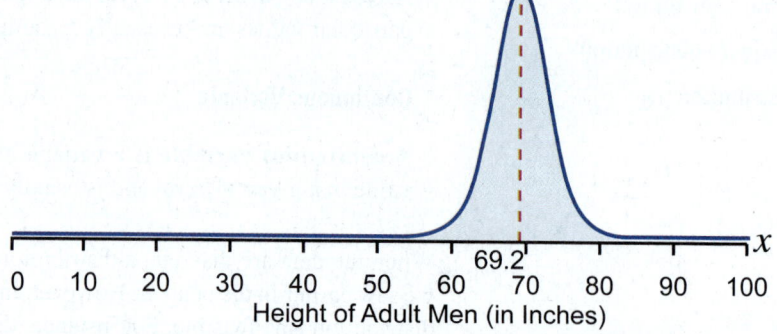

Figure 11.4.3: Example of Normally Distributed Data

Along with heights, other measurements such as human body temperatures, weights, and blood pressures are common examples of data that are normally distributed over a large randomly selected sample.

Example 11.4.1

Identifying Properties of a Normal Curve

Below are three normal distributions drawn on the same graph. Use them to answer the following questions.

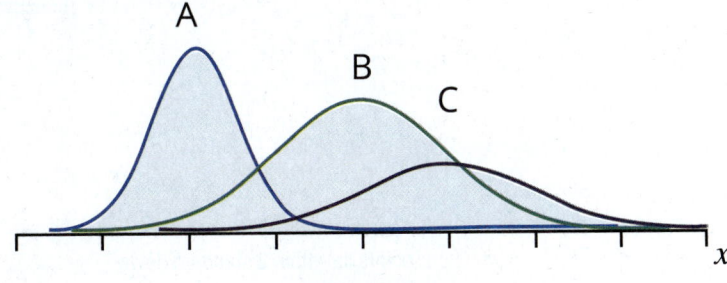

Figure 11.4.4

a. Determine which distribution has the largest mean and which has the smallest.

b. Determine which distribution has the largest standard deviation and which has the smallest.

c. What does the position of curve B in relation to curve A tell you about the data of the distribution?

Solution

a. We know that normal distributions curves are centered at their means. Therefore, we can determine the mean of each distribution by identifying where its peak is located along the *x*-axis. Even though the *x*-axis is not labeled, we can tell from the location of the peaks which mean is larger or smaller in relation to the other curves. Curve C has the largest mean and curve A has the smallest mean.

b. Normal distributions with a larger standard deviation will have a curve that appears more spread out from the mean. On the other hand, smaller deviations result in graphs that appear narrower and more concentrated around the mean. Therefore, curve C appears to have the largest standard deviation, whereas curve A appears to have the smallest standard deviation.

c. Because most of curve B is to the right of curve A along the *x*-axis, curve B appears to have larger data values than curve A does.

OBJECTIVE 2 The Empirical Rule

As we discussed at the beginning of this section, we can't be certain what values a random variable will take, but we can estimate how likely it is to fall within a range of values if we know its distribution. When the distribution of a set of data is approximately normal, we can use a rule of thumb called the **Empirical Rule** to estimate the percentage of data values that fall within a few standard deviations of the mean. The Empirical Rule states that approximately 68% of data will lie within one standard deviation of the mean, 95% of the data will be within two standard deviations of the mean, and 99.7% of the data will lie within three standard deviations of the mean. Notice this indicates that only a very small percentage of normally distributed data, only 0.3%, lie more than three standard deviations away from the mean.

Approximately 68% of all data points lie within 1 standard deviation above and below the mean.

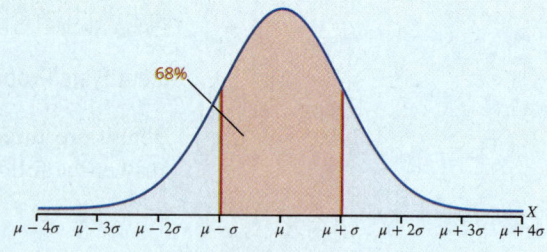

Figure 11.4.5

Approximately 95% of all data points lie within 2 standard deviations above and below the mean.

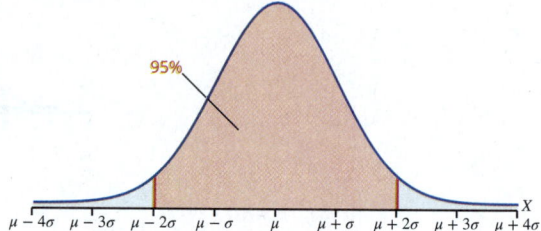

Figure 11.4.6

Approximately 99.7% of all data points lie within 3 standard deviations above and below the mean.

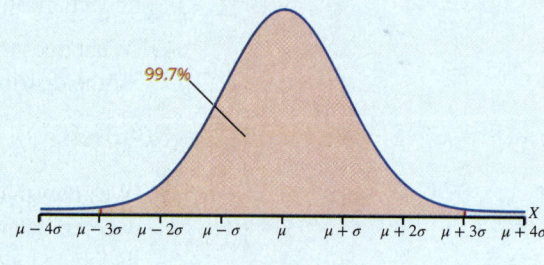

Figure 11.4.7

> 💬 **Helpful Hint**
>
> The Empirical Rule is also referred to as the Three Sigma Rule or the 68–95–99.7 Rule.

Example 11.4.2

Using the Empirical Rule

Emergency room waiting times for a local hospital are approximately normal. The data have a reported mean of 166.9 minutes and a standard deviation of 55.1 minutes.

a. Identify the range of waiting times that 68% of patients are likely to experience.

b. Estimate the percentage of patients that will wait less than 3 hours and 42 minutes.

Solution

a. By the Empirical Rule, we know that 68% of the data values lie within one standard deviation of the mean. The standard deviation of this distribution is 55.1; thus, by adding and subtracting 55.1 from the mean, we will get the range that contain the middle 68% of emergency room waiting times at the hospital.

$$\text{Upper Endpoint: } 166.9 + 55.1 = 222$$

$$\text{Lower Endpoint: } 166.9 - 55.1 = 111.8$$

Thus, we can estimate that 68% of emergency room waiting times for the local hospital are between 111.8 minutes and 222 minutes. We've sketched this on a normal curve in Figure 11.4.8.

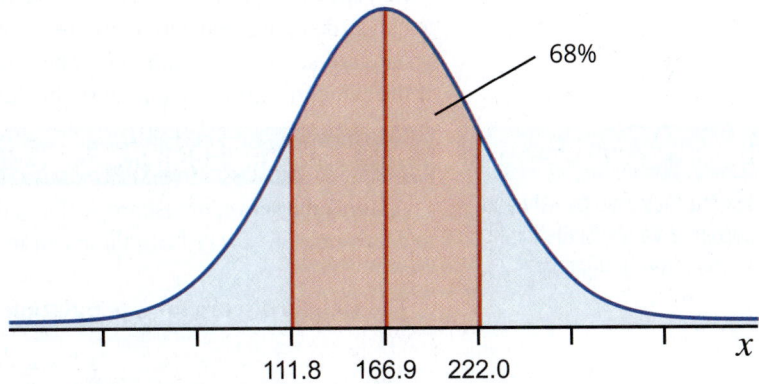

Figure 11.4.8: Emergency Room Wait Times (in Minutes)

b. To answer this question, we'll need to first convert 3 hours and 42 minutes into minutes only, since our mean and standard deviation are both in minutes. We know 3 hours is 180 minutes; we can then add the 42 minutes to get a total time of 222 minutes.

In part a., we determined that 222 minutes falls one standard deviation above the mean and that 68% of data fall between the endpoints of 111.8 minutes and 222 minutes. Because of the symmetry of the distribution, half of the 68% (that is, 34%) is above the mean and half is below. We also know that 50% of the data lies below the mean. Putting the upper 34% together with the 50% of the data that is below the mean, we have that approximately $50\% + 34\% = 84\%$ of patients wait less than 3 hours and 42 minutes. Figure 11.4.9 illustrates this.

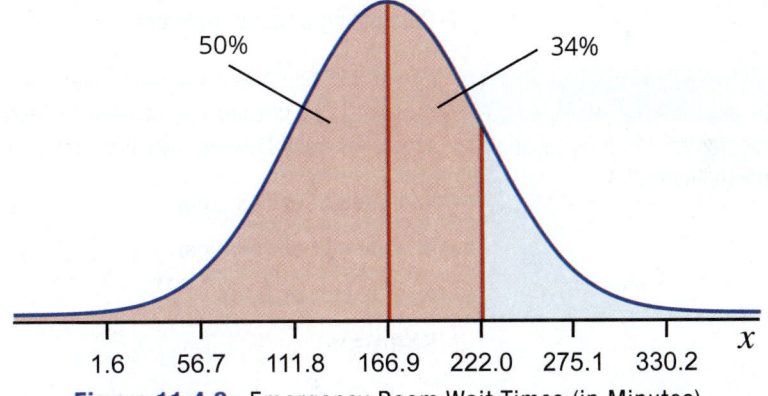

Figure 11.4.9: Emergency Room Wait Times (in Minutes)

✔ **Skill Check 11.4.1**

Estimate the range of emergency room waiting times that 95% of patients are likely to experience based on Example 11.4.2.

OBJECTIVE 3 Area Under a Normal Curve

One of the most important applications of a normal distribution, or any continuous probability distribution, is that the area under any part of the curve is equal to the *probability* of the random variable falling within that region. Thinking back to the last example, because we know that approximately 84% of the data is below 222 minutes, we also know that the probability that a randomly chosen patient will wait less than 222 minutes is approximately 84%.

Although the Empirical Rule is a good rule of thumb, there are techniques for calculating areas under the normal curve more precisely. One method is to use a **standard score**, or **z-score**. A standard score tells us how far a value lies from the mean, specifically, how many standard deviations it is away from the mean. So if a standard score is a positive number, the data value is greater than the mean μ; if a standard score is a negative number, the data value is less than the mean μ; and if the standard score is equal to 0, the data value is equal to the mean μ.

> **Helpful Hint**
>
> A standard score may also be referred to as a z-score or z-value. Statisticians use these terms interchangeably.

$f(x)$ STANDARD SCORE (z-SCORE)

A **standard score**, or z-score, is the number of standard deviations a particular data value lies away from the mean in a normal distribution.

The **standard score for a population** value is given by

$$z = \frac{x - \mu}{\sigma},$$

where x is the value of interest from the population, μ is the population mean, and σ is the population standard deviation.

The **standard score for a sample** value is given by

$$z = \frac{x - \overline{x}}{s},$$

where x is the value of interest from the sample, $\overline{x}$ is the sample mean, and s is the sample standard deviation.

Example 11.4.3

Calculating a Standard Score

> **Helpful Hint**
>
> Standard scores are normally rounded to the nearest hundredth.

Given that heights of Canadian women are normally distributed with a population mean of 159.5 cm and a population standard deviation of 7.1 cm, calculate the standard scores for the following measurements from randomly chosen Canadian women.

a. A height of 148.2 cm

b. A height of 160.3 cm

c. A height of 1.7 m

Solution

a. Notice that the height given is smaller than the mean, so we should expect the standard score to be negative; data values below the mean will always have a negative standard score. Using a height of 148.2 cm in the formula, we have the following.

$$z = \frac{x - \mu}{\sigma} = \frac{148.2 - 159.5}{7.1} \approx -1.59$$

What this tells us is that a height of 148.2 cm is 1.59 standard deviations below the mean.

b. A data value greater than the mean will always have a positive standard score; thus, we should expect a positive number. The standard score for a height of 160.3 cm is calculated as follows.

$$z = \frac{x - \mu}{\sigma} = \frac{160.3 - 159.5}{7.1} \approx 0.11$$

This tells us that a height of 160.3 cm, which we know is only slightly larger than the mean, is 0.11 standard deviations above the mean.

c. Before we can use the formula to compute the standard score, we must first convert the data into the same unit of measurement. That is, the measurements should be in either meters or centimeters. Since the previous measurements were in centimeters, we'll change 1.7 meters to centimeters. Note that you would get exactly the same standard score if you changed the mean and the standard deviation to meters instead. To change meters to centimeters, simply multiply by 100. So we have 1.7 m = 170 cm. Substituting into the formula we have the following.

$$z = \frac{x - \mu}{\sigma} = \frac{170 - 159.5}{7.1} \approx 1.48$$

So a height of 1.7 m is 1.48 standard deviations above the mean.

✔ **Skill Check 11.4.2**

Given that the heights of women in Malaysia are normally distributed with a mean of 154.7 cm and a standard deviation of 6.46 cm, calculate the standard score of a Malaysian woman who is 159.1 cm tall.

Standard scores also give us a way to compare two data values from different populations with different means and standard deviations. For example, you could determine if you scored better on the ACT exam or the SAT exam by comparing your standard score from each. Example 11.4.4 looks at this type of comparison.

Example 11.4.4

Comparing Standard Scores

On Karl's recent standardized test results, he scored an 18 in language arts. The mean was 15 and the standard deviation was 1.4. His classmate, Asher, took the test on a different date and also scored an 18 in language arts, but on that day the mean for the test was 17 and the standard deviation was 1.9. Another friend, Rylie, took the test in a different state, where the mean was 15.46 and the standard deviation was 0.8, and scored a 16 in language arts. Which of the three students actually performed the best on the language arts portion of the test compared to the others?

Solution

In order to compare the language arts scores, we need to calculate the standard score for each student using the appropriate mean and standard deviation for each.

$$\text{Karl: } z = \frac{18 - 15}{1.4} \approx 2.14$$

$$\text{Asher: } z = \frac{18 - 17}{1.9} \approx 0.53$$

$$\text{Rylie: } z = \frac{16 - 15.46}{0.8} \approx 0.68$$

💬 **Think Back**

In Example 11.3.9, we compared standardized test scores by looking at their percentiles. In this example, we make similar comparisons of standardized tests using standard scores.

Comparing each of the respective standard scores, we can see that Karl did the best on his test, even though his raw score was the same as Asher's. Karl's score of 18 was 2.14 standard deviations above the mean. Asher and Rylie's scores were fewer standard deviations above their respective means. The larger the standard score, the further above the mean a data point lies.

OBJECTIVE 4 Percentages and Probabilities with a Normal Distribution

As we stated earlier, the area under the normal curve is what we are most interested in, as it is equal to the probability of a random variable falling within that region. Calculating the area under the curve of any normal distribution requires integral calculus using techniques beyond the scope of this course. Fortunately, we only need the results of those calculations to determine probabilities using the normal curve. Around the 19th century, tables were created so that the calculus needed to calculate the area under a standard normal curve did not have to be repeated. All that was needed was to read the tables in order to find the desired area under a normal distribution curve. Because of the ease of reading the tables, this became the most widely used method for finding area under a normal curve and remained so until the mass distribution of technology. We will present both the table method and a technology method to find the area under the normal curve.

There are several different standard normal distribution tables that statisticians use, each one giving us a way to find the same area under the normal curve. We will use the **cumulative normal distribution tables**, although you can use any of the different tables to produce the same results. Table A and Table B in Appendix A give the area under the standard normal curve to the left of the given standard score z. A small excerpt of Table B from Appendix A is shown in Table 11.4.1.

Table 11.4.1: Excerpt from Table B: Standard Normal Distribution

z	0.00	0.01	0.02	0.03	0.04
0.0	0.5000	0.5040	0.5080	0.5120	0.5160
0.1	0.5398	0.5438	0.5478	0.5517	0.5557
0.2	0.5793	0.5832	0.5871	0.5910	0.5948
0.3	0.6179	0.6217	0.6255	0.6293	0.6331
0.4	0.6554	0.6591	0.6628	0.6664	0.6700
0.5	0.6915	0.6950	0.6985	0.7019	0.7054
0.6	0.7257	0.7291	0.7324	0.7357	0.7389
0.7	0.7580	0.7611	0.7642	0.7673	0.7704
0.8	0.7881	0.7910	0.7939	0.7967	0.7995

Notice that the z-values given in the table excerpt are rounded to two decimal places. The first decimal place of each z-value is listed in the left column, with the second decimal place listed along the top row. Where the appropriate row and column intersect, we find the amount of area under the standard normal curve to the *left* of that particular z-value.

Before we begin using the tables, we want to pause and emphasize how important it is to think about what you should expect to get when you look up a standard score in the tables. One of the best ways to help you do this is to always sketch a graph of the normal curve and indicate the area that you are interested in finding.

Recall that the area under the entire curve is equal to 1 and, due to its symmetry, half of that (0.5) falls to the left of the mean and half falls to the right. Positive standard scores are to the right of the mean, so we should expect that the area under the curve to the left of a positive standard score will be greater than half of the total area. A negative standard score is to the left of the mean and indicates that the area under the curve to the left of the value will be less than half (0.5). Both of these scenarios are shown in Figure 11.4.10.

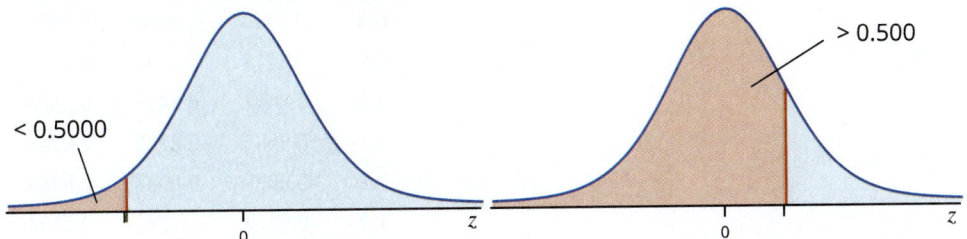

Figure 11.4.10: Area under a Standard Normal Curve in Relation to the Mean

Let's now look at an example of using the tables to find the area to the left of a particular value of z. At the same time, we will see how to find the area using technology.

Example 11.4.5

Finding the Area to the Left of a z-Value

Find the area under the normal curve to the left of each of the standard scores.

a. $z = 1.04$

b. $z = -2.11$

Solution

a. Begin by considering what value we expect to get when we look up the standard score. Since the standard score is *positive*, we should expect that the area under the curve to the left of this value will be greater than half of the total area (0.5000) as shown in the following figure.

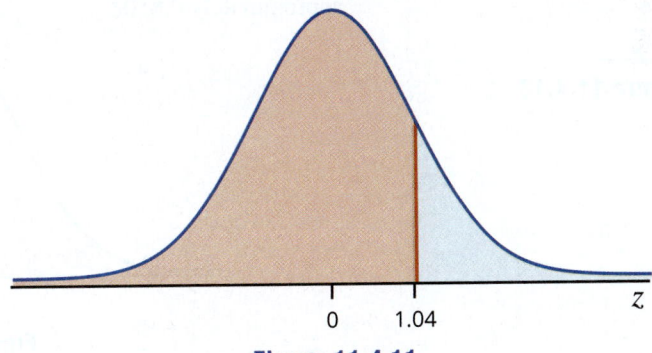

Figure 11.4.11

Tables

To use the table, we must break the given z-value (1.04) into two parts: one containing the first decimal place (1.0) and the other containing the second decimal place (0.04). Because the z-value is positive, we use Table B from Appendix A. Look across the row labeled 1.0 and down the column labeled 0.04. The row and column intersect at 0.8508. Notice that this value aligns with what we were expecting.

Table 11.4.2: Excerpt from Table B: Standard Normal Distribution

z	0.02	0.03	0.04	0.05	0.06
0.8	0.7939	0.7967	0.7995	0.8023	0.8051
0.9	0.8212	0.8238	0.8264	0.8289	0.8315
1.0	0.8461	0.8485	0.8508	0.8531	0.8554
1.1	0.8686	0.8708	0.8729	0.8749	0.8770
1.2	0.8888	0.8907	0.8925	0.8944	0.8962
1.3	0.9066	0.9082	0.9099	0.9115	0.9131

TI-83/84 Plus

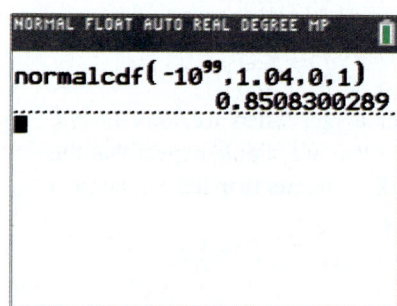

Figure 11.4.12

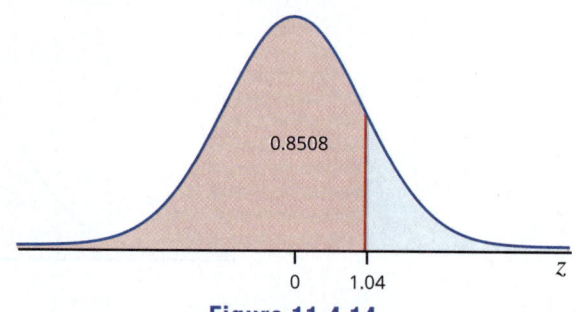

The TI-83/84 Plus has a built-in `normalcdf` function that can be used to find the area under the normal curve to the left of a given standard score. Press 2nd vars to access the DISTR menu and select `normalcdf(`. You will be prompted to enter values for the following items.

`lower` is the lower bound. Since we are asked to find the area to the left of z, the lower bound is $-\infty$. We cannot enter $-\infty$ into the calculator, so we will enter a very small value for this endpoint to act as an approximation of $-\infty$, such as `-10^99`. This number might appear as `-1E99` in the calculator.

`upper` is the upper bound, which is our z-value, `1.04`.

The calculator's default mode is the standard normal curve where $\mu = 0$ and $\sigma = 1$, so we do not need to enter μ or σ.

Highlight `Paste` and press enter. Back on the home screen, press enter again, and the calculator returns `0.8508300289`.

Thus, the area under the standard normal curve to the left of $z = 1.04$ is approximately 0.8508.

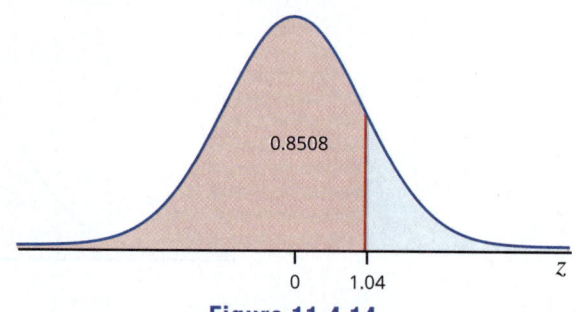

Figure 11.4.14

b. Begin by noting that since we are asked to find an area to the left of a *negative* z-value, we should expect an area under the curve that is less than 0.5000.

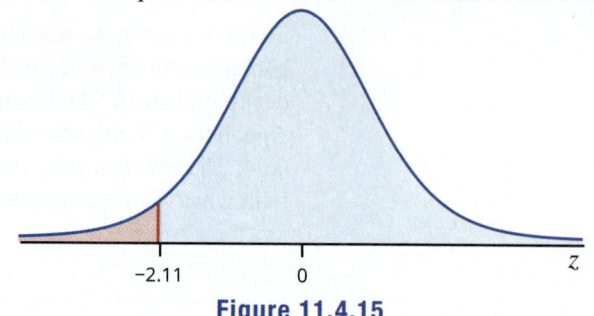

Figure 11.4.15

Tables

Notice that the first part of the z-value is –2.1 and the second part is 0.01. This time, use Table A from Appendix A since the z-value is negative. On the chart, look across the row labeled –2.1 and down the column labeled 0.01 (be careful to get the correct column here). The row and column intersect at 0.0174.

Table 11.4.3: Excerpt from Table A: Standard Normal Distribution

z	0.04	0.03	0.02	0.01	0.00
–2.4	0.0073	0.0075	0.0078	0.0080	0.0082
–2.3	0.0096	0.0099	0.0102	0.0104	0.0107
–2.2	0.0125	0.0129	0.0132	0.0136	0.0139
–2.1	0.0162	0.0166	0.0170	0.0174	0.0179
–2.0	0.0207	0.0212	0.0217	0.0222	0.0228
–1.9	0.0262	0.0268	0.0274	0.0281	0.0287

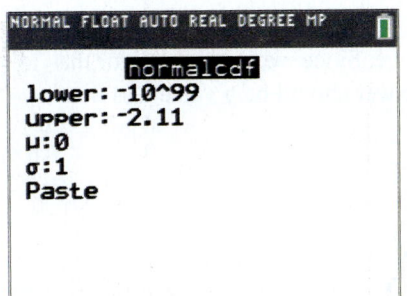

Figure 11.4.16

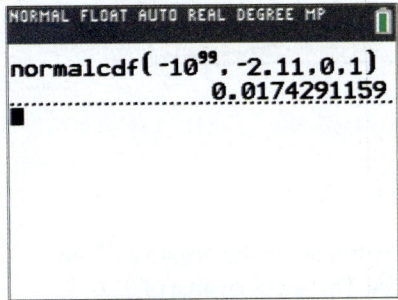

Figure 11.4.17

✔ **Skill Check 11.4.3**

Find the area under the standard normal curve to the left of $z = 1.76$.

TI-83/84 Plus

The built-in `normalcdf` function on a TI-83/84 Plus can again be used to find the area under the normal curve to the left of a given standard score. Press 2nd vars to access the DISTR menu and select `normalcdf(`. Since the lower bound is –∞, which we can't enter into the calculator, use `-10^99` for `lower`. The upper bound is the given z-score, –2.11, so use `-2.11` for `upper`.

Highlight **Paste** and press enter. Back on the home screen, press enter again, and the calculator returns Ø.Ø174291159.

Thus, the area under the standard normal curve to the left of $z = -2.11$ is approximately 0.0174.

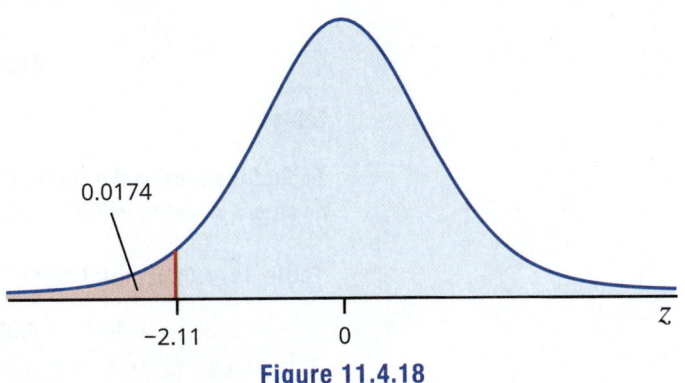

Figure 11.4.18

The cumulative normal curve tables we are using only give the area to the *left* of a given z-value, but we can use the tables, along with the properties of the standard normal distribution, to find other areas as well. Consider the area to the right of a given z-value. In terms of area under the curve, the curve's symmetry means that the area to the right of z is equal to the area to the left of –z. As a result, to find the area to the right of any z-value using the tables, we can simply look up the area to the left of its negative, –z. In other words, we can change the sign of the z-value and then look it up.

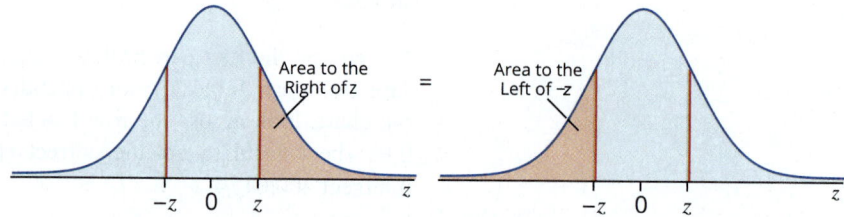

Figure 11.4.19

Example 11.4.6

Finding the Area to the Right of a *z*-Value

Find the area under the normal curve to the right of each of the *z*-scores.

a. $z = 0.68$

b. $z = -1.32$

Solution

a. Once again, consider how large the area will be. Since we are looking for the area to the right of a positive *z*-value, our answer should be a value less than 0.5000, as illustrated in Figure 11.4.20.

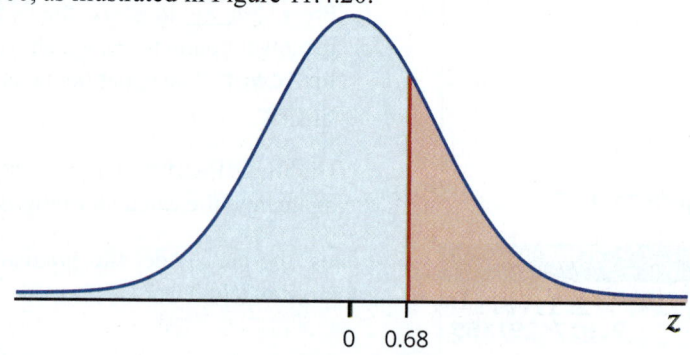

Figure 11.4.20

Tables

To find the area to the right of $z = 0.68$, we will look up the negative of the *z*-value, $z = -0.68$, in Table A from Appendix A. This gives an area of 0.2483.

Table 11.4.4: Excerpt from Table A: Standard Normal Distribution

z	0.09	0.08	0.07	0.06	0.05
−0.9	0.1611	0.1635	0.1660	0.1685	0.1711
−0.8	0.1867	0.1894	0.1922	0.1949	0.1977
−0.7	0.2148	0.2177	0.2206	0.2236	0.2266
−0.6	0.2451	0.2483	0.2514	0.2546	0.2578
−0.5	0.2776	0.2810	0.2843	0.2877	0.2912
−0.4	0.3121	0.3156	0.3192	0.3228	0.3264

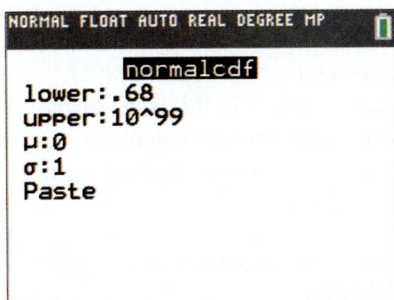

Figure 11.4.21

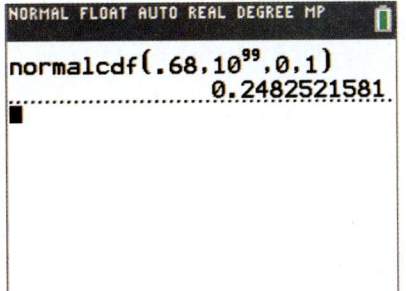

Figure 11.4.22

TI-83/84 Plus

When using technology to find the area to the right of a *z*-value, we don't have to change the sign of the *z*-value. The calculator just uses the upper and lower bounds as before. Press 2nd vars to access the DISTR menu and select normalcdf(. Because we want the area to the right of *z*, the *z*-value, 0.68, is the lower bound. The upper bound is ∞, which we can't enter into the calculator, so use we need to use a sufficiently large number for the upper bound, such as 10^99. (This value might appear as 1E99 in the calculator.)

Highlight Paste and press enter. Back on the home screen, press enter again, and the calculator returns 0.2482521581.

Thus, the area under the standard normal curve to the right of $z = 0.68$ is approximately 0.2483.

b. Begin by considering how much area is to the right of a negative *z*-value. The figure below illustrates that the area we are looking for will be larger than 0.5000.

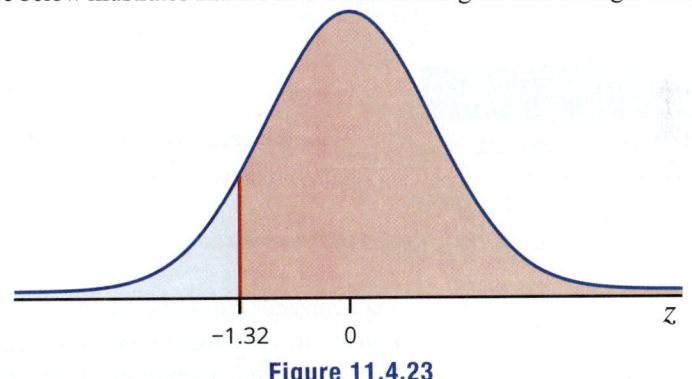

Figure 11.4.23

Tables

Look up the negative of the *z*-value, which is $z = 1.32$, in Table B from Appendix A. This gives an area of 0.9066.

Table 11.4.5: Excerpt from Table B: Standard Normal Distribution

z	0.01	0.02	0.03	0.04	0.05
0.9	0.8186	0.8212	0.8238	0.8264	0.8289
1.0	0.8438	0.8461	0.8485	0.8508	0.8531
1.1	0.8665	0.8686	0.8708	0.8729	0.8749
1.2	0.8869	0.8888	0.8907	0.8925	0.8944
1.3	0.9049	0.9066	0.9082	0.9099	0.9115
1.4	0.9207	0.9222	0.9236	0.9251	0.9265

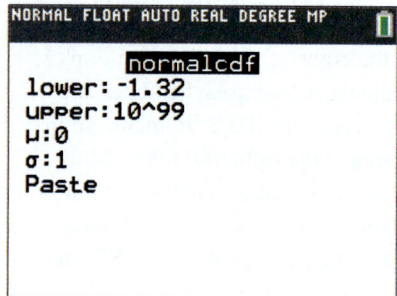

Figure 11.4.24

TI-83/84 Plus

Press 2nd vars to access the DISTR menu and select normalcdf(. Because we want the area to the right of z, the z-value, -1.32, is the lower bound. The upper bound is ∞, which we can't enter into the calculator, so use we need to use a sufficiently large number for the upper bound, such as 10^99.

Highlight Paste and press enter. Back on the home screen, press enter again, and the calculator returns 0.9065824276.

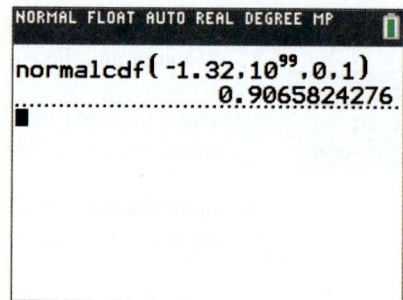

Figure 11.4.25

Thus, the area under the standard normal curve to the right of $z = -1.32$ is approximately 0.9066.

✔ **Skill Check 11.4.4**

Find the area under the standard normal curve to the right of $z = -0.62$.

Let's consider the area under the standard normal curve between two z-values. To find the area between two values of z using tables, we look up the area to the left of each z-value and then subtract the smaller area from the larger area, as illustrated in Figure 11.4.26.

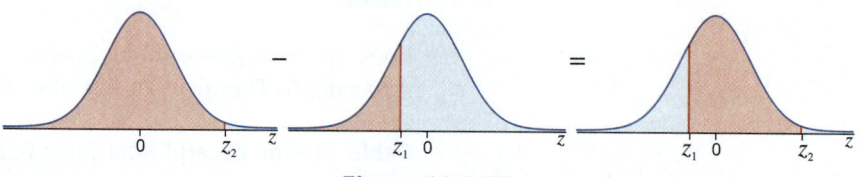

Figure 11.4.26

Example 11.4.7

Finding the Area between Two z-Values

Find the area under the normal curve that is between $z_1 = -0.56$ and $z_2 = 2.22$.

Solution

Tables

Begin by looking up the area to the left of each z-value using the tables in Appendix A.

Area to the left of $z_1 = -0.56$ is 0.2877.

Area to the left of $z_2 = 2.22$ is 0.9868.

Table 11.4.6: Excerpt from Table A: Standard Normal Distribution

z	0.09	0.08	0.07	0.06	0.05
−0.9	0.1611	0.1635	0.1660	0.1685	0.1711
−0.8	0.1867	0.1894	0.1922	0.1949	0.1977
−0.7	0.2148	0.2177	0.2206	0.2236	0.2266
−0.6	0.2451	0.2483	0.2514	0.2546	0.2578
−0.5	0.2776	0.2810	0.2843	0.2877	0.2912
−0.4	0.3121	0.3156	0.3192	0.3228	0.3264

Table 11.4.7: Excerpt from Table B: Standard Normal Distribution

z	0.01	0.02	0.03	0.04	0.05
2.0	0.9778	0.9783	0.9788	0.9793	0.9798
2.1	0.9826	0.9830	0.9834	0.9838	0.9842
2.2	0.9864	0.9868	0.9871	0.9875	0.9878
2.3	0.9896	0.9898	0.9901	0.9904	0.9906
2.4	0.9920	0.9922	0.9925	0.9927	0.9929
2.5	0.9940	0.9941	0.9943	0.9945	0.9946

Then, subtract the smaller from the larger: $0.9868 - 0.2877 = 0.6991$.

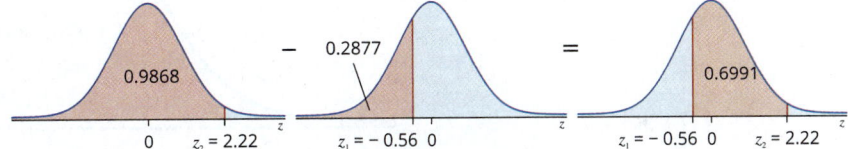

Figure 11.4.27

TI-83/84 Plus

Instead of having to look up two values and subtract them as we would need to do when using tables, the calculator finds the area between two z-values in one step. Press 2nd vars to access the DISTR menu and select normalcdf(. Because we want the area between two z-values, the smaller z-value will be the lower bound and the larger z-value will be the upper bound. So use -0.56 for lower and 2.22 for upper.

Highlight Paste and press enter. Back on the home screen, press enter again, and the calculator returns 0.6990509792.

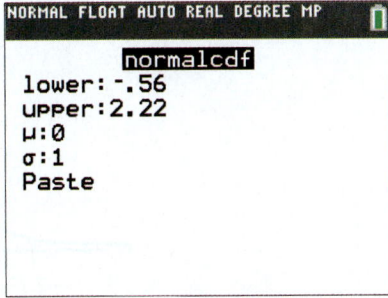

Figure 11.4.28

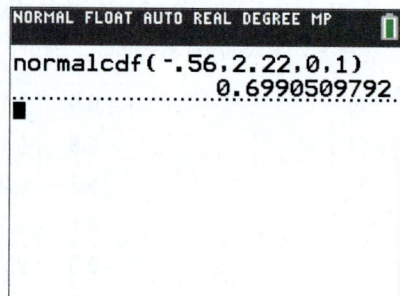

Figure 11.4.29

Thus, the area between the two z-values is approximately 0.6991.

The area that is not between two z-values, but instead is outside the z-values, is referred to as the "area in the tails" of the curve. To calculate the area in the tails of a curve, we simply find each individual area using the methods we've already discussed and then add them together. Figure 11.4.30 illustrates this. Notice that the area in the tails is also equal to one minus the area between the two z-values.

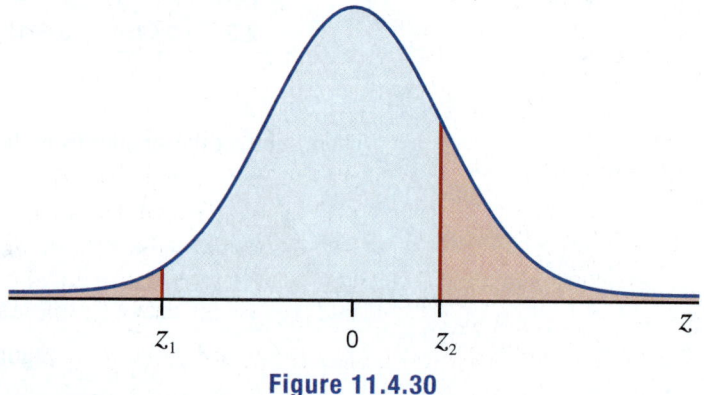

Figure 11.4.30

Example 11.4.8

Finding the Area in the Tails for Two z-Values

Find the total of the areas under the normal curve to the left of $z_1 = -2.73$ and to the right of $z_2 = 2.38$.

Solution

The area we are interested in is shown in the following figure.

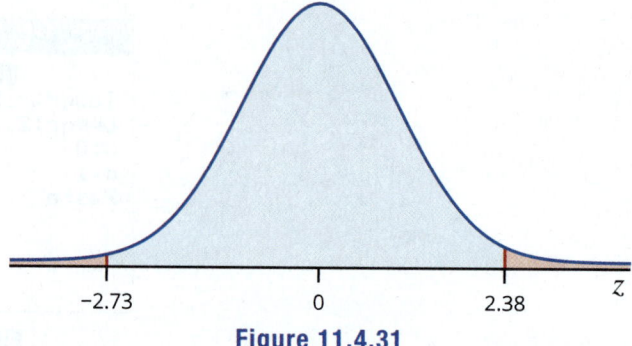

Figure 11.4.31

To find the total area we are interested in, we need to find each of the areas separately, and then add them together.

Tables

Find the area to the left of $z_1 = -2.73$ by looking it up in Table A. This gives us 0.0032.

Find the area to the right of $z_2 = 2.38$ by looking up its negative value, $z_2 = -2.38$. This is equal to 0.0087.

Table 11.4.8: Excerpt from Table A: Standard Normal Distribution

z	0.08	0.07	0.06	0.05	0.04	0.03
− 2.7	0.0027	0.0028	0.0029	0.0030	0.0031	0.0032
− 2.6	0.0037	0.0038	0.0039	0.0040	0.0041	0.0043
− 2.5	0.0049	0.0051	0.0052	0.0054	0.0055	0.0057
− 2.4	0.0066	0.0068	0.0069	0.0071	0.0073	0.0075
− 2.3	0.0087	0.0089	0.0091	0.0094	0.0096	0.0099
− 2.2	0.0113	0.0116	0.0119	0.0122	0.0125	0.0129

Adding these two values gives us $0.0032 + 0.0087 = 0.0119$.

TI-83/84 Plus

We will again use the `normalcdf` function of the TI-84 Plus. We could use this function once to find the area to the left of z_1 and again to find the area to the right of z_2, and then add those values together. An alternative is to find the area between the two z-values in one step, subtracting this area from one to find the area that is not between the two z-values.

Begin by typing 1 − on the home screen. Press 2nd vars to access the DISTR menu and select `normalcdf(`. Here we are finding the area between two z-values, so the smaller z-value will be the lower bound and the larger z-value will be the upper bound. Use `-2.73` for `lower` and `2.38` for `upper`. Highlight Paste and press enter. Back on the home screen, press enter again, and the calculator returns Ø.Ø118230769.

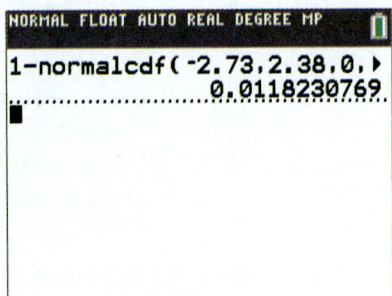

Figure 11.4.32

Thus, the total area in the tails is approximately 0.0119.

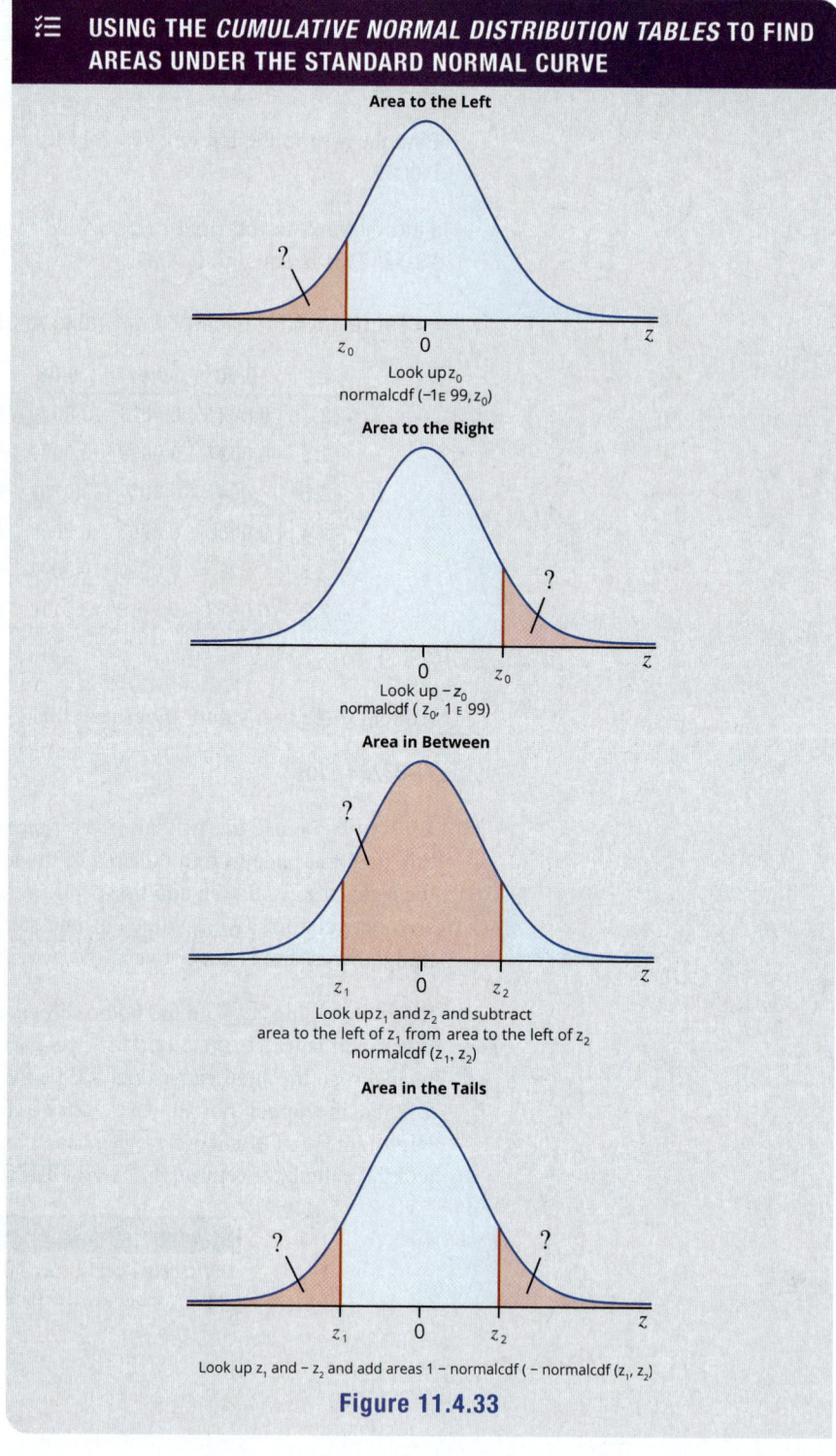

Figure 11.4.33

Once we know how to determine the area under a standard normal curve, we know how to find the probability that a value of a normally distributed random variable will occur within a specific range of values. For example, finding the area to the left of $z = 1.37$ is the same as calculating the probability that the z-score is less than or equal to 1.37. So now we are ready to find the probability of an event happening or answer questions about certain proportions of a population.

Example 11.4.9

Finding Percentages Using the Normal Distribution

Adult body temperatures are normally distributed with a mean of 98.6 °F and a standard deviation of 0.73 °F. Sometimes our bodies increase in temperature and we say we have a fever. Extreme fluctuations might signal that we need to seek medical help. With this in mind, determine what percentage of healthy adults have a body temperature that is normally above 100 °F.

Solution

We are in a position to use the normal curve to answer this question since we are told that body temperature for adults is normally distributed. Begin by sketching a normal curve to represent the distribution of body temperature, indicating the mean of 98.6 °F and the x-value we are interested in, 100 °F. Because the temperature we are interested in is a lower limit for the normal temperature, we need to find the area that is more than our x-value. Therefore, we shade the area to the right of 100 °F.

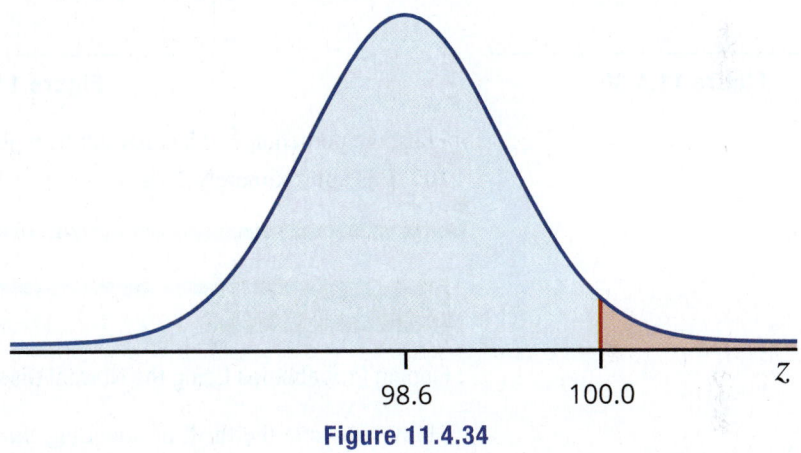

Figure 11.4.34

Notice that we can expect the value we find to be smaller than 0.5000 based on our normal curve drawing.

Next, find a standard score for the data point we are interested in, 100 °F, so that we can calculate the area under the standard normal curve. Substituting into the formula, we have the following.

$$z = \frac{(100 - 98.6)}{0.73} \approx 1.92$$

This z-value indicates that a normal temperature of 100.0 is 1.92 standard deviations above the mean normal temperature. Using either tables or technology, we now need to find the value to the right of $z = 1.92$.

Tables

Because we want the area to the right, we will look up the negative of the z-value in Table A, which is $z = -1.92$. This gives us a value of 0.0274.

TI-83/84 Plus

Press [2nd][vars] to access the DISTR menu and select normalcdf(. We are looking for the area to the right of $z = 1.92$, so that will be our lower bound. The upper bound is ∞, so we will use 10^{99}. Highlight Paste and press [enter]. Back on the home screen, press [enter] again, and the calculator returns 0.0274288813.

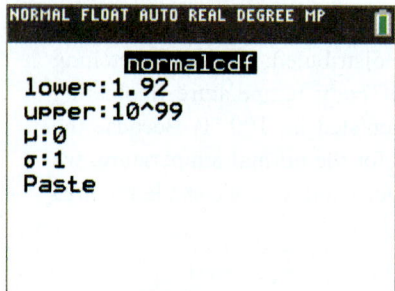

Figure 11.4.35

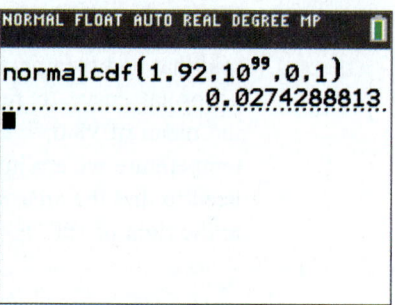

Figure 11.4.36

Thus, the percentage of healthy adults who have a body temperature of more than 100 °F is approximately 2.74%.

Example 11.4.10

Finding Probabilities Using the Normal Distribution

Many kids love the thrill of attending the state fair and riding rides of all types. However, not everyone gets to ride every attraction because there are height restrictions on some of the rides. Suppose for one ride guests must be at least 40 inches tall to ride. If heights for 6-year-olds are normally distributed with a mean of 3 feet 9 inches and a standard deviation of 2 inches, determine the probability that a random 6-year-old is shorter than the height restriction and is not allowed on the ride.

Solution

Note that we are told heights are normally distributed, and we can, therefore, use the normal distribution to answer the question. Since the height restriction is given in inches, we will first convert the mean into inches.

$$\mu = 3 \text{ ft } 9 \text{ in.} = 45 \text{ in.}$$

Draw a normal curve to represent the distribution of the heights of 6-year-olds with a mean of 45 inches and denote 40 inches on the x-axis to indicate the height restriction. Because this is an upper limit for heights, we are interested in finding the probability that a random 6-year-old's height is less than 40 inches. Therefore, we shade the area to the left of 40 on the curve.

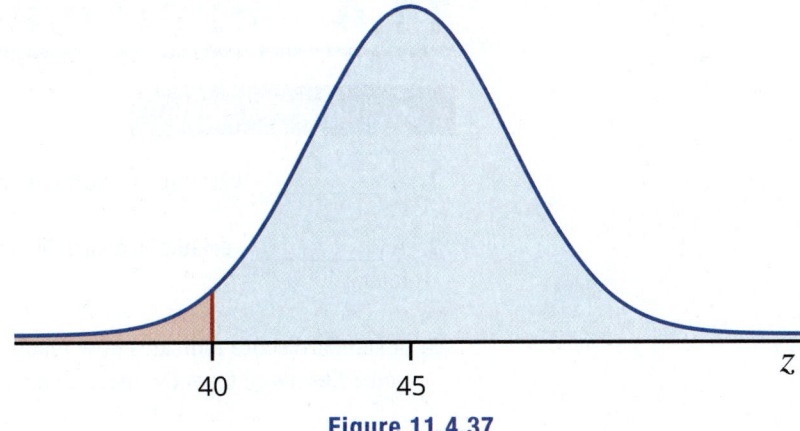

Figure 11.4.37

Notice that the area shaded is less than half of the area under the curve, so we can expect the value we find to be less than 0.5000.

Begin by finding a z-score for the data point we are interested in: 40 inches. Substituting into the formula, we have the following.

$$z = \frac{40 - 45}{2} = -2.5$$

Using either tables or technology, find the value to the left of $z = -2.5$.

Tables

Looking up $z = -2.5$ in Table A we see that the shaded region has an area of approximately 0.0062.

TI-83/84 Plus

Press 2nd vars to access the **DISTR** menu and select `normalcdf(`. We are looking for the area to the left of $z = -2.5$, so the lower bound is $-\infty$. Since we can't enter that into the calculator, we will use -10^{99} to approximate the lower bound. For the upper bound, use $z = -2.5$. Highlight **Paste** and press enter. Back on the home screen, press enter again, and the calculator returns `0.0062096799`.

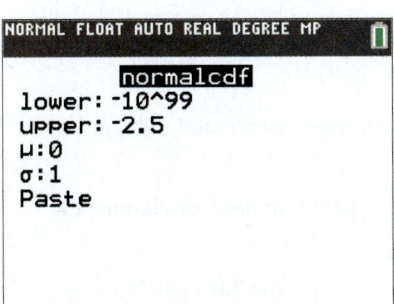

Figure 11.4.38

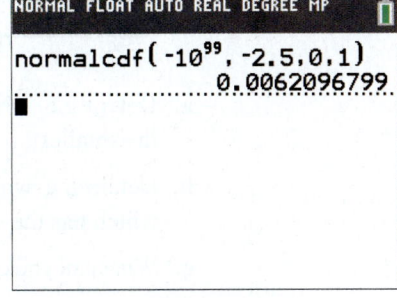

Figure 11.4.39

Thus, the probability of a 6-year-old having a height of less than 40 inches, and therefore not being allowed on the ride, is approximately 0.0062. We can also interpret this to mean that 0.62% of 6-year-olds have a height of less than 40 inches. Therefore, only a small percentage of 6-year-olds (less than 1%) are excluded from riding this ride.

Skill Check Answers
1. Between 56.7 minutes and 277.1 minutes
2. $z \approx 0.68$ 3. 0.9608   4. 0.7324

11.4   Exercises

✔ CONCEPT CHECK

1. A _____ variable is a variable that can only take on certain values.

2. A _____ variable is a variable that can take on any value in a given interval.

3. A standard score indicates how many _____ a data value lies away from the mean in a normal distribution.

4. True or False: According to the Empirical Rule, 99.7% of all data points in a data set lie within 2 standard deviations of the mean.

5. True or False: To find the area under a normal curve that lies to the right of a given z-value, look up the negative of the z-value in the Standard Normal Distribution table.

💡 PRACTICE

6. Below are three normal distributions representing the height of adults in China, France, and Norway. Use the graph to answer the following questions.

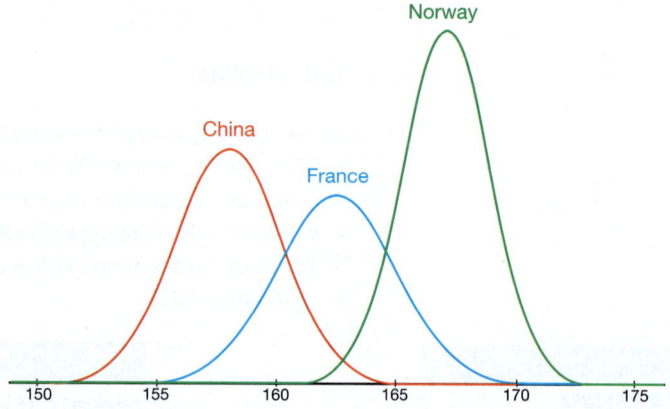

a. Determine which distribution has the largest mean and which has the smallest.

b. Determine which distribution has the largest standard deviation and which has the smallest.

c. What can you conclude about the heights of people in China and Norway?

7. Below are three normal distributions representing the prices of BMW X1, X2, and X3 models. Use the graph to answer the following questions.

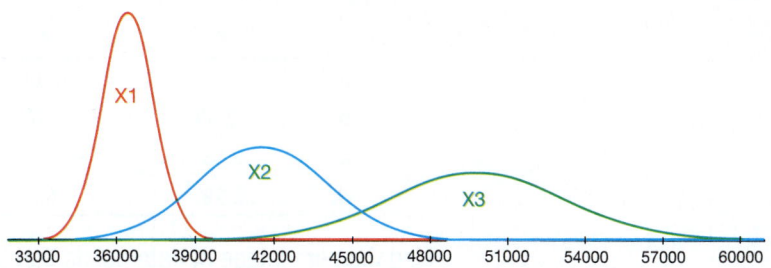

a. Determine which distribution has the largest mean and which has the smallest.

b. Determine which distribution has the largest standard deviation and which has the smallest.

c. What can you conclude about the prices of BMW X1 and X2 models?

Calculate the standard score for each given value. Round your answer to the nearest hundredth.

8. $\mu = 57, \sigma = 11$

   a. $x_1 = 63$

   b. $x_2 = 38$

   c. $x_3 = 58$

9. $\bar{x} = 1123, s = 241$

   a. $x_1 = 1284$

   b. $x_2 = 900$

   c. $x_3 = 1364$

   d. $x_4 = 1123$

10. $\bar{x} = 3.19, s = 0.06$

   a. $x_1 = 3.13$

   b. $x_2 = 3.22$

   c. $x_3 = 3.00$

11. $\mu = 178.15, \sigma = 49.3$

   a. $x_1 = 73.9$

   b. $x_2 = 267.3$

   c. $x_3 = 199.5$

Use the z-score formula to complete each table.

12. Find the missing value in each row of the table. Round answers to the same number of decimal places given in the table.

	z	x	μ	σ
a.		82.1	74.0	6.3
b.	1.05	162.3		8.9
c.	3.04		34.5	5.02
d.	−2.73	379	634	

13. Find the missing value in each row of the table. Round answers to the same number of decimal places given in the table.

	z	x	μ	σ
a.		4.33	6.10	2.04
b.	−2.39	−57		139.8
c.	0.58		118	21.2
d.	2.78	68	43	

Find the percentage of data points that lie below each z-score.

14. $z = -0.19$

15. $z = 1.46$

16. $z = 3.07$

17. $z = -2.22$

18. $z = 0$

Find the percentage of data points that lie above each z-score.

19. $z = 1.03$

20. $z = -1.87$

21. $z = -3.10$

22. $z = 2.84$

23. $z = 0$

Find the percentage of data points that lie between each pair of z-scores.

24. $z_1 = -1.00$
$z_2 = 1.00$

25. $z_1 = -2.40$
$z_2 = 1.73$

26. $z_1 = 2.00$
$z_2 = 3.00$

27. $z_1 = -3.01$
$z_2 = -0.56$

28. $z_1 = 0$
$z_2 = 2.61$

Find the percentage of data points that lie below z_1 and above z_2.

29. $z_1 = -1.10$
$z_2 = 1.10$

30. $z_1 = -2.84$
$z_2 = 2.84$

31. $z_1 = -1.75$
$z_2 = 0.53$

32. $z_1 = 1.09$
$z_2 = 2.88$

33. $z_1 = -0.01$
$z_2 = 0.02$

🚀 APPLICATIONS

34. Scores on a test have a mean of 73 and a standard deviation of 11. Steve has a score of 68. Convert Steve's score to a z-score.

35. The annual rainfall in a town has a mean of 47.22 inches and a standard deviation of 10 inches. Last year there was 51 inches of rain. How many standard deviations from the mean is that?

36. Mason's weekly poker winnings have a mean of $144 and a standard deviation of $51. Last week he won $165. How many standard deviations from the mean is that?

37. The mean score for a set of data is marked by the dotted line on the following graph. Which value is a likely *z*-score for the indicated value? Choose from **a.** –2.1, **b.** 0, or **c.** 2.7.

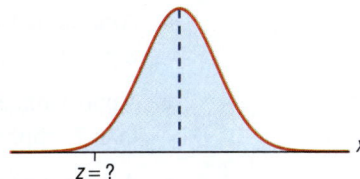

38. The average IQ score for adults is 100 with a standard deviation of 15. Assume that the distribution of IQ scores is approximately normal.

    a. Find the percentage of adults who have an IQ score less than 90.

    b. Find the percentage of adults who have an IQ score which exceeds the mean by at least 15 points.

    c. Find the percentage of adults who have an IQ score between 100 and 120.

    d. Find the percentage of adults who have an IQ score less than 55 or more than 145.

39. Assume the weights of offensive linemen in the NFL follow a normal distribution with a mean of 300 pounds and a standard deviation of 12.3 pounds.

    a. Find the percentage of linemen in the NFL who weigh more than 320 pounds.

    b. Find the percentage of linemen in the NFL who weigh between 275 and 325 pounds.

    c. Find the percentage of NFL linemen who weigh at least 260 pounds.

    d. Find the percentage of NFL linemen who weigh at most 315 pounds.

40. Although there is some controversy around the precise average body temperature of adults, new data suggest that the mean is 98.2° with a standard deviation of 0.6° and has a bell-shaped distribution.

    a. According to this distribution, approximately what percentage of body temperatures are between 97° and 99.4°?

    b. Approximately what percentage of temperatures are greater than 98.8°?

    c. Approximately what percentage of temperatures are no more than 98.2°?

41. In 2011, high school seniors had the following mean and standard deviation on the mathematics portion of the SAT exam: $\mu = 514$ and $\sigma = 117$.

    a. Approximately what percentage of scores were greater than 397 but less than 631?

    b. What two scores have approximately 95% of the data between them?

    d. Describe where the best and worst 0.3% of scores lie.

42. The speeds of vehicles on a certain part of a highway are normally distributed with a mean of 60 mph and a standard deviation of 6 mph. Use the Empirical Rule to answer the following questions.

    a. Approximately what percentage of vehicles travel at a speed between 54 mph and 66 mph?

    b. Approximately what percentage of vehicles travel at a speed greater than 72 mph?

    c. Approximately what percentage of vehicles travel at a speed not greater than 78 mph?

43. A machine is set to pour a mean of 497 mL of spring water in each bottle with a standard deviation of 2 mL. Assume that the distribution of the amount of water is bell-shaped. Use the Empirical Rule to answer the following questions.

    a. Approximately what percentage of bottles filled by the machine contain more than 491 mL and less than 503 mL of water?

    b. Between what two amounts of water will 68% of all bottles contain?

    c. How much water will the top 2.5% of bottles contain? How much water will the bottom 2.5% of bottles contain?

44. Ava scored a 92 on a test with a mean of 71 and a standard deviation of 15. Charlotte had a score of 688 on a test with a mean of 493 and a standard deviation of 150. Which score was better with respect to their test?

45. Avery started training to run a 5K. Her first race was a 5K for charity. She finished in 37.3 minutes. The average race time for the charity run was 36.42 with a standard deviation of 1.73 minutes. In her second race, Avery finished in 36.5 minutes. The race had a mean time of 33.02 minutes with a standard deviation of 2.45 minutes. In which race did Avery place higher in the list of finishers?

46. Michael works at a company where the mean annual salary is $37,000 with a standard deviation of $6800. Daniel works at a company where the mean annual salary is $39,000 with a standard deviation of $8200. If they both have a salary of $35,000, whose salary is better compared to the other people in their respective companies? Assume that the salaries are normally distributed.

47. The average height of students in Class A is 59.6 inches with a standard deviation of 2.7 inches. The average height of students in Class B is 57.2 inches with a standard deviation of 2.4 inches. Amanda is in Class A and is 62.2 inches tall. Sofia is in Class B and is 60.1 inches tall. Which girl is taller compared to the rest of their respective classes?

48. The Intelligence Quotient (IQ) score is one way to measure human intelligence. It is determined using a set of tests and then standardized to have a normal distribution with the mean of 100 and the standard deviation of 15. What percentage of people have an IQ score higher than 125?

49. The average weight of newborn boys in the US is 3530 grams with a standard deviation of 577 grams. Assuming that the distribution of weights is normal, what percentage of newborn boys weigh less than 3 kilograms?

50. Suppose you want to determine the voltage in a circuit, and your voltmeter allows you to find the voltage with a standard deviation of 0.65 volts. If the exact voltage is 95 volts, what is the probability that you obtain a value greater than or equal to 96 volts?

51. The systolic blood pressure of healthy adults is normally distributed with a mean of 116 mm Hg and a standard deviation of 23 mm Hg. However, if the blood pressure is higher than 130 mm Hg at rest, it can be a sign of hypertension. With this in mind, what percentage of healthy adults have blood pressure higher than 130 mm Hg?

52. Assume that the retirement age of people in a certain country is normally distributed with a mean of 62.4 years and a standard deviation of 2.7 years. What percentage of people in the country retire before the age of 60?

53. Assume that the daily milk yield of a cow has a normal distribution with a mean of 25 liters and a standard deviation of 7.5 liters. What is the probability that the cow will produce less than 20 liters of milk in a day?

✎ WRITING & THINKING

54. What is the minimum z-score that a piece of data would need to have in order to be in the top 10% of a normally distributed set of data?

55. What is an "average" z-score? Explain your answer.

56. What z-score represents the 1st quartile? 2nd quartile? 3rd quartile?

11.4 PROJECT

IS THAT NORMAL?

In Section 11.4, you learned how to use a normal distribution to calculate z-scores and answer questions about probability. We found that variables such as height, body temperature, weight, and blood pressure are common examples of data that are normally distributed. Because of the relative ease of calculating probabilities using normal distributions, it's tempting to assume, when collecting and analyzing data, that the variable or characteristic in question has a normal distribution. However, if the data set doesn't have a normal distribution, any conclusions we draw using standard scores is suspect at best. Fortunately, there is a method available to check the data beforehand and determine whether the population from which the data are taken has a normal distribution. It is called a *normality test*.

Normality Test

Step 1: Sort the data values so they are arranged from smallest to largest, and assign a rank to each value with 1 corresponding to the smallest value and n corresponding to the largest value, where n is equal to the sample size.

Step 2: Once sorted, compute the percentile occupied by each value using the formula

$$\frac{k - 0.5}{n},$$

where k is the rank.

Step 3: Using the percentile computed in the previous step, find the corresponding z-score for each value.

Step 4: Graph the points of the observed values from the data set on the horizontal axis and the corresponding expected z-scores from Step 3 on the vertical axis. This is called a *normal probability plot*.

Step 5: If the plot has a linear pattern, we can conclude the data values were drawn from a normal distribution. If the plot shows no pattern or some other trend rather than linear, we can conclude the data are not drawn from a normal distribution.

Note: The data should have no more than one *outlier*, an extremely large or extremely small value not typical of the other values in the data set. If there are two or more outliers, it is unlikely the data are drawn from a normal distribution.

The Normality Test is straightforward except for Step 3. To find z-scores from percentiles, you are essentially "working backwards" and will need appropriate technology to obtain them. For example, these are the steps when finding the corresponding z-score for the 25th percentile (0.25) using a TI-83/84 Plus calculator.

Step 1: Choose DISTR by pressing [2nd] then [vars].

Step 2: Select invNorm(.

Step 3: Press [enter].

Step 4: Input the area to the left of z (that is, the 25th percentile in decimal form), 0.25.

Step 5: Press [enter] five times.

Step 6: The z-score will be displayed. To two decimal places, it is −0.67. (Note that the negative sign indicates the z-score is in the lower half of the standard normal distribution and lies to the left of the standard mean of zero.)

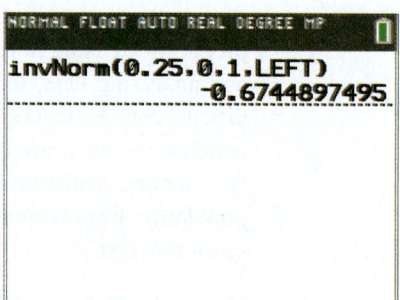

If technology is not readily available, the z-scores corresponding to the percentiles in a data set of sample size $n = 10$ are provided in the following table.

Rank	Percentile	z-Score
1	0.05	−1.645
2	0.15	−1.036
3	0.25	−0.674
4	0.35	−0.385
5	0.45	−0.126
6	0.55	0.126
7	0.65	0.385
8	0.75	0.674
9	0.85	1.036
10	0.95	1.645

Let's now test a data set to see if we can conclude that the population from which it is taken is normally distributed.

1. Randomly select 10 classmates and record their heart rate in beats per minute (bpm). You can count the rate for 15 seconds and multiple by four to save time, or you can count for the entire 60 seconds. The counting method is your choice.

2. Carry out Steps 1 through 4 in the Normality Test. If some heart rates appear more than once, each heart rate will still get its own distinct ranking. For example, if the three lowest rates are all 65 bpm, then followed by 66 bpm, the rankings would be 1: 65, 2: 65, 3: 65, 4: 66, and so on. Are there any outliers in your data set? If so, identify them.

3. Does your graph follow closely enough to a linear pattern for you to conclude the population from which it is drawn is normally distributed? If not, what might be some reasons why? What is the population? What is your conclusion about the distribution of heart rates in the population?

4. If time permits, repeat the test with another 10 randomly selected classmates and compare results.

5. If time permits and you have access to technology to calculate the z-score for any given percentile, choose a population of your own that you wish to test for normality and randomly select a sample of arbitrary sample size to conduct your test. Brainstorm with your project team members and instructor to come up with ideas.

11.5 CONFIDENCE INTERVALS

OBJECTIVES

1. Find a point estimate for a population parameter.

2. Interpret and construct confidence intervals.

3. Interpret margin of error.

> ### Helpful Hint
>
> A sample proportion is the fraction of a sample that has a certain characteristic.

> ### Fun Fact
>
> Article 1, Section 2, of the US Constitution mandates that the country conduct a count of its population once every 10 years. The 2020 Census marked the 24th time that the country has counted its population; the first was in 1790.

OBJECTIVE 1 Point Estimates

So far in this chapter we have looked at collecting, displaying, and describing data—all parts of the discipline of statistics. The most powerful part of statistics is how we use these data techniques to provide insights and draw conclusions about a whole population based on the sample we have collected. We refer to this process as **inferential statistics**.

> ### 📖 INFERENTIAL STATISTICS
>
> The branch of **inferential statistics**, as a science, involves using descriptive statistics to estimate population parameters.

So how does one go about estimating the unknown? Suppose we want to estimate the proportion of adults in the United States who speak two or more languages. It's unreasonable to expect that we could survey everyone. Instead, we can use a sample to arrive at a **point estimate**. A point estimate is a single value used to estimate a population parameter, such as the population mean, population proportion, or population percentage.

> ### 📖 POINT ESTIMATE
>
> A **point estimate** is a single-number estimate of a population parameter.

Understandably, we want to use the best point estimate for a population parameter that we can. When choosing a point estimate, the "best" point estimate is one that is *unbiased*—that is, a sample statistic that does not consistently underestimate or overestimate the population parameter. The sample mean is an unbiased estimator for the population mean and hence is the best point estimate to use. Similarly, a sample proportion is the best point estimate for a population proportion. We will look at methods for using a sample proportion to make an inference about a population proportion, but the same calculations can also be performed with probabilities or percentages. When working with percentages, we will use the equivalent proportion value.

Example 11.5.1

Finding a Point Estimate for a Population Mean

Find the best point estimate for the population mean of test scores on a standardized psychology final exam. The following is a simple random sample taken from the population of test scores.

68	83	68	90	91	72
86	92	85	93	45	69
76	81	76	71	97	70
99	100	96	88	84	89
100	91	55	75	74	85

Solution

The best point estimate for the population mean is the sample mean because it is an unbiased estimator. The sample mean for the given sample of test scores can be found as follows.

$$\bar{x} = \frac{\sum x_i}{n} = \frac{2449}{30} \approx 81.6$$

Thus, the best point estimate for the population mean of test scores on this standardized exam is 81.6.

> **✔ Skill Check 11.5.1**
>
> Out of 432 randomly selected students, 301 live in campus housing. Use the best point estimate to estimate the proportion of all students who live in campus housing.

OBJECTIVE 2 Confidence Intervals

As you can imagine, using only one value for estimating the population mean, or any other population parameter, leaves room for error. If we took a different sample, it would probably produce a different point estimate. If the methods used to sample the population were done appropriately, both point estimates would be good approximations. In theory, we could continue taking samples of the population and finding different point estimates indefinitely. Instead, statisticians look to identify a range, or interval, of values that provides a better guesstimate. This range of possible values is called an **interval estimate.**

We can use statistical methods to form an interval estimate with a specified level of confidence. This **level of confidence**, denoted by the letter c, is defined as the percentage of all possible samples of a given size that will produce interval estimates that contain the true population parameter. An interval estimate associated with a certain level of confidence is then called a **confidence interval.**

> ### 📖 CONFIDENCE INTERVALS
>
> **Interval Estimate**
>
> An **interval estimate** is a range of possible values for a population parameter.
>
> **Level of Confidence**
>
> The **level of confidence**, denoted by c, is the percentage of all possible samples of a given size that will produce interval estimates that contain the actual parameter.
>
> **Confidence Interval**
>
> A **confidence interval** is an interval estimate associated with a certain level of confidence.

Let's take a moment and note what a confidence interval conveys. Suppose a confidence interval for a population parameter has a 90% level of confidence. This means that 90% (9 out of 10) of all possible interval estimates will contain the true population parameter. Remember, the parameter we are trying to estimate is a fixed value and it is either in the confidence interval or it is not. Thus, the probability we are stating is about our ability to estimate the parameter, not the probability of the parameter changing. Figure 11.5.1 illustrates this concept around estimating μ. It demonstrates that we expect about 9 out of every 10 confidence intervals produced to contain the true population mean.

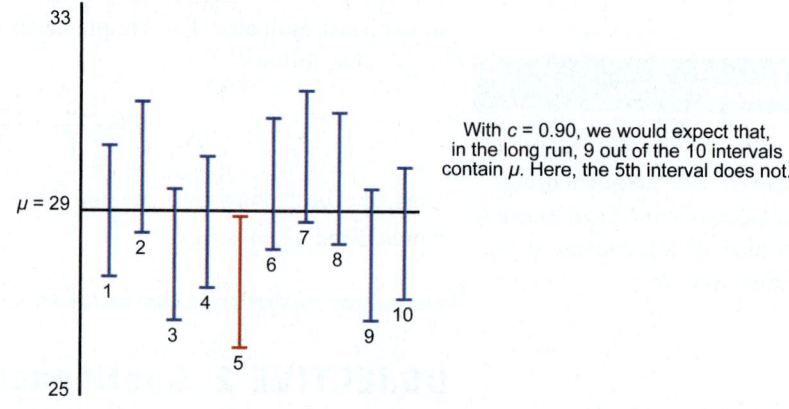

Figure 11.5.1: 90% Confidence Intervals for μ from 10 Random Samples

So what methods do we use to create a confidence interval? We begin with the best point estimate we can find and construct the interval around it. The next question is, *how big should the interval be?* Obviously, we could make the interval so big that it includes all the data points, and hence captures the parameter. But a large interval does not help us pinpoint the parameter's value. For instance, I can be 99.99% confident that the current average age of the reader of this text is between 1 year old and 100 years old. That's a very high level of confidence, but it really doesn't tell us much about the average age of a reader of this text. Using statistical techniques, our goal is to simultaneously shrink the interval but keep our confidence level relatively high. Generally, at the same level of confidence, a sample with a larger number of data values will produce a narrower confidence interval than another with fewer data values.

Example 11.5.2

Interpreting Confidence Intervals

During the 2020 global pandemic of COVID-19, the reproduction rate of the virus (R_t) was a measure of how fast the virus was spreading in the United States[18]. That is, R_t measured the average number of people who became infected by an infectious person. When R_t was below 1.0, the virus was not spreading. Below is an excerpt from November 2020, which shows 80% confidence intervals for each state's R_t value. Each state's initials inside the oval represent the sample mean. Use the excerpt to answer the following questions.

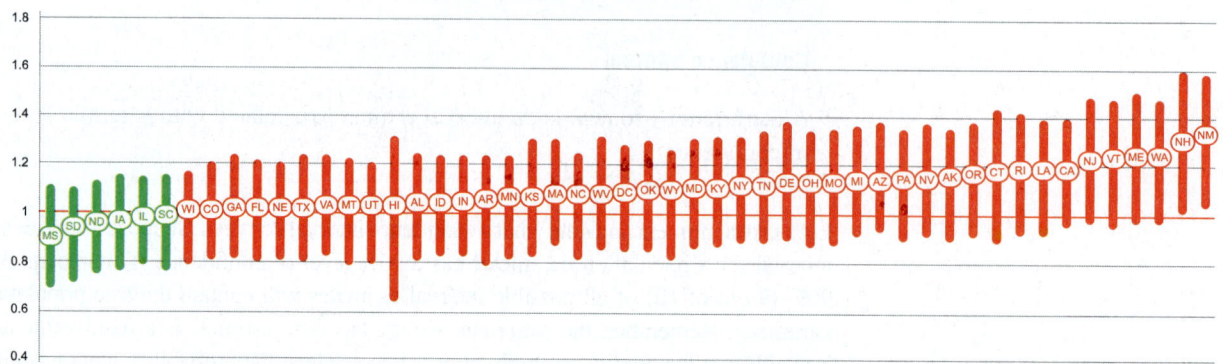

Figure 11.5.2

18   Rt: Effective Reproduction Number, last accessed November 23, 2020, https://rt.live/.

a. Approximate the 80% confidence interval for New Mexico's (NM) R_t value in November 2020. Based on this, what can you say about the spread of the virus in New Mexico at this point?

b. Can we say with certainty that Mississippi's (MS) R_t value is below one? Explain your answer.

c. Is there a state that we can be certain has an R_t value above one?

d. Compare the R_t values for Georgia (GA) and New Hampshire (NH).

Solution

a. Based on the graph, the 80% confidence interval for New Mexico's R_t value is approximately (1.05, 1.55). Therefore, we can be 80% confident that the R_t is above one, and thus the virus was indeed spreading in New Mexico during November 2020.

b. Although the sample mean for Mississippi falls below one, the confidence interval still contains one. Therefore, we cannot conclude with 80% confidence that the actual R_t value is less than one.

c. Because these are 80% confidence intervals, it is not possible to be 100% certain for any of the states. However, the two states whose intervals are completely above one, New Hampshire and New Mexico, are the most likely to have R_t values above one.

d. Although the sample mean for Georgia is smaller than that of New Hampshire, their confidence intervals overlap. Therefore, we cannot say that there is a statistically significant difference between the two states' R_t values.

OBJECTIVE 3 Margin of Error

The **margin of error**, or *maximum error of estimate*, E, is the largest possible distance away from the point estimate that the confidence interval will cover. In other words, it is the amount of error allowed around the point estimate, both above it and below it. To construct confidence intervals for the most common population parameters, we simply subtract the margin of error from the point estimate to obtain the lower endpoint and then add the margin of error to the point estimate to obtain the upper endpoint, as illustrated in Figure 11.5.3.

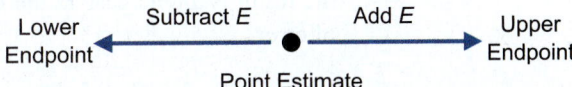

Figure 11.5.3: Constructing a Confidence Interval

📖 MARGIN OF ERROR

The **margin of error**, or *maximum error of estimate*, E, is the largest possible distance from the point estimate that a confidence interval will cover.

Example 11.5.3

Constructing a Confidence Interval with a Given Margin of Error

Suppose we wish to estimate the mean systolic blood pressure in adults. A sample of 3534 participants had a sample mean of 127.3 mm Hg (millimeters of mercury) with a standard deviation of 19.0 mm Hg. If the margin of error for the data using a 95% level of confidence is $E = 0.63$, construct a 95% confidence interval for the mean systolic blood pressure. Interpret your results.

Solution

To construct a confidence interval, we need the best point estimate for the population mean, which is the sample mean. So we will use $\bar{x} = 127.3$ mm Hg as the point estimate for this population parameter. Since we are given the value of the margin of error, we need to find the lower endpoint of the interval by subtracting E from the point estimate and the upper endpoint by adding E to the point estimate.

$$\text{Lower Endpoint: } \bar{x} - E = 127.3 - 0.63 = 126.67 \text{ mm Hg}$$

$$\text{Upper Endpoint: } \bar{x} + E = 127.3 + 0.63 = 127.93 \text{ mm Hg}$$

Therefore, the confidence interval ranges from 126.67 mm Hg to 127.93 mm Hg. The confidence interval can be written mathematically using either inequality symbols or interval notation.

$$126.67 < \bar{x} < 127.93$$

or

$$(126.67, 127.93)$$

The interpretation of our confidence interval is that we are 95% confident that the true population mean for systolic blood pressure in adults is between 126.67 mm Hg and 127.93 mm Hg.

> **✔ Skill Check 11.5.2**
>
> A sample of 3076 Americans 18 and older found that 29% of participants were unmarried. Construct a 95% confidence interval for the percentage of American adults who are unmarried if the margin of error is $E = 1.8\%$.

Note that the margin of error in the previous example is very small because of the large sample size that was taken. When our estimates are based on more data, the result is an increase in the level of confidence without increasing the margin of error.

Formulas for E will vary depending on the parameter being estimated and the size of the sample. However, using the Empirical Rule, we can approximate the margin of error percentage for a survey at a 95% level of confidence. If a sample of size n is taken, the margin of error for a 95% level of confidence is approximately equal to $\frac{1}{\sqrt{n}} \cdot 100\%$.

$f(x)$ RULE OF THUMB FOR MARGIN OF ERROR IN A SURVEY

With a 95% level of confidence, the margin of error, E, is approximately

$$\frac{1}{\sqrt{n}} \cdot 100\%$$

for a random sample of size n.

Example 11.5.4

Constructing a Confidence Interval for a Population Proportion

In a random sample of 1506 American adults in 2020, $\dfrac{51}{100}$ of the adults indicated that they received a text message about the upcoming presidential election.

a. Use the rule of thumb to calculate the margin of error for a 95% confidence interval. Round to the nearest tenth.

b. Write a statement about the percentage of American adults who received a text message about the election in 2020.

Solution

a. The sample size given is $n = 1506$. Thus, the approximate margin of error is calculated as follows.

$$E = \frac{1}{\sqrt{n}} \cdot 100\%$$

$$= \frac{1}{\sqrt{1506}} \cdot 100\% \approx 2.6\%$$

b. The best point estimate for a population proportion is a sample proportion, $\hat{p}$. Thus, we can use the sample proportion given, which is $\hat{p} = \dfrac{51}{100}$, to find the 95% confidence interval. The lower endpoint is found by subtracting the margin of error from the point estimate and the upper endpoint is found by adding the margin of error to the point estimate. We will use the equivalent percentage of 51% as the point estimate.

Lower Endpoint: $\hat{p} - E = 51\% - 2.6\% = 48.4\%$

Upper Endpoint: $\hat{p} + E = 51\% + 2.6\% = 53.6\%$

Therefore, we can be 95% confident that between 48.4% and 53.6% of American adults received a text message about the 2020 election cycle.

> **✔ Skill Check 11.5.3**
>
> A total of 937 university students were polled on their eating habits. Find the margin of error for this survey using the rule of thumb.

Example 11.5.5

Interpreting Margin of Error

The Pew Research Center surveyed American households in 2019 and reported the following.

> "37% of Americans now go online mostly using a smartphone, and these devices are increasingly cited as a reason for not having a high-speed internet connection at home."

Their website provides a breakdown of the sample size along with the margin of error for a 95% confidence interval for the overall survey as well as for different subgroups of data.

Table 11.5.1: Sample Breakdown

Group	Unweighted Sample Size	Margin of Error
Total Sample	1502	2.8 percentage points
18–29	236	7.2 percentage points
30–49	395	5.6 percentage points
50–64	424	5.4 percentage points
65+	391	5.6 percentage points
Less than $30,000	348	5.9 percentage points
$30,000–$74,999	400	5.5 percentage points
$75,000 or more	526	4.8 percentage points
High school or less	461	5.1 percentage points
Some college	375	5.7 percentage points
College+	656	4.3 percentage points
Urban	505	4.9 percentage points
Suburban	636	4.4 percentage points
Rural	258	6.9 percentage points
Smartphone users	1219	3.2 percentage points
Non-broadband users	400	5.5 percentage points
Smartphone only internet users	248	7.0 percentage points

Source: Monica Anderson, "Methodology," Pew Research Center, June 13, 2019, https://www.pewresearch.org/internet/2019/06/13/mobile-technology-and-home-broadband-methodology/

a. Use the data in Table 11.5.1 to create a graph of the sample sizes of each category versus the margin of error percentage points. For each point (x, y), let x represent the sample size and y represent the margin of error.

b. Based on the graph from part a., what can you say about the margin of error compared with the sample size?

c. Is it possible to create confidence intervals for the population percentages in each category using the data from Table 11.5.1?

Solution

a. Using the data from Table 11.5.1, where x is the sample size and y is the margin of error, we can plot each point on the graph.

Margin of Error vs. Sample Size, 95% Confidence

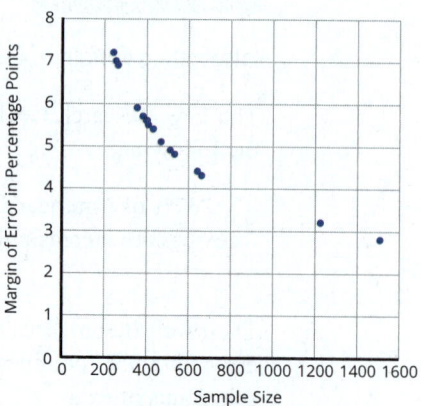

Figure 11.5.4

b. The graph shows us that as the sample size increases, the margin of error gets smaller. This means that with the same level of confidence we can create a tighter confidence interval, which gives a better approximation of the population parameter.

c. A confidence interval for a population parameter is constructed using a sample statistic and a margin of error. Since we are only given the margin of error and not the sample statistics, we do not have enough information to construct the confidence intervals.

Example 11.5.6

Interpreting Margin of Error

A 2019 survey of American households found that "most Americans were aware of facial recognition technology." Use the information in Table 11.5.2 to answer the following questions.

a. Construct a 95% confidence interval for the proportion of American adults who would say they have "a little" awareness of facial recognition technology.

b. Is it correct to say that 18- to 29-year-olds are more likely to say they know "a lot" about facial recognition technology than those 65 years old and older? Explain why or why not.

Table 11.5.2: Awareness of Facial Recognition Technology

Group	A Lot	A Little	Not at All	Unweighted Sample Size	Margin of Error
Total Sample	25%	61%	13%	4272	1.9 percentage points
18–29	29%	55%	15%	671	4.8 percentage points
30–49	27%	61%	12%	1314	3.3 percentage points
50–64	23%	63%	14%	1308	3.4 percentage points
65+	20%	67%	13%	977	3.8 percentage points

Source: Aaron Smith, "More Than Half of U.S. Adults Trust Law Enforcement to Use Facial Recognition Responsibly," Pew Research Center, September 5, 2019, https://www.pewresearch.org/internet/2019/09/05/more-than-half-of-u-s-adults-trust-law-enforcement-to-use-facial-recognition-responsibly/

Solution

a. To construct a confidence interval for the percentage of adults who say they have "a little" awareness of facial recognition technology, add the margin of error to the sample percentage found for the upper endpoint and subtract the margin of error from the sample percentage for the lower endpoint. From Table 11.5.2, we see that the sample percentage of adults who responded with "a little" is 61%, with a margin of error of 1.9%. Notice that we use the overall margin of error since we are interested in all American adults, and not a subpopulation.

$$\text{Lower Endpoint: } 61\% - 1.9\% = 59.1\%$$

$$\text{Upper Endpoint: } 61\% + 1.9\% = 62.9\%$$

Thus, the 95% confidence interval for the percentage of American adults who say they have "a little" awareness of facial recognition technology is (59.1%, 62.9%).

b. To establish if there is a statistical difference between the two subgroups, we can look at the confidence intervals for each of them and determine if the intervals overlap. If they do overlap, we cannot say that there is a statistical difference between the two. If they do not overlap, we can be 95% confident that one population parameter is different than the other. Begin by constructing the confidence intervals in the same manner as in part a., using the margin of errors given for each subpopulation.

Table 11.5.3

Age Group	Sample Proportion Who Responded "A Lot"	Margin of Error	95 % Confidence Interval for the Population Percentage
18- to 29-year-olds	29%	4.8%	(24.2%, 33.8%)
65+	20%	3.8%	(16.2%, 23.8%)

Notice that the confidence intervals for the two age groups do not overlap. Therefore, we can say that we are 95 % confident that 18- to 29-year-olds are more likely to say they know "a lot" about facial recognition technology than those 65 years old and older.

Skill Check Answers

1. $\dfrac{301}{432} \approx 0.697$ **2.** (27.2%, 30.8%)   **3.** 3.3%

11.5   Exercises

✔ CONCEPT CHECK

1. A _____ is a single-number estimate of a population parameter.

2. The margin of error is the largest possible distance from the _____ that a confidence interval will cover.

3. The range of all possible values for a population parameter is called the _____.

4. True or False: The level of confidence describes the likelihood that the data value in question will fall within the corresponding confidence interval.

🚀 **APPLICATIONS**

Find each specified point estimate or confidence interval.

5. A survey of 42 randomly selected teachers finds that they spend a mean of $18 per week on lunch. What is the best point estimate for the mean amount of money spent per week on lunch for all teachers?

6. The mean number of pets per student for a random sample of 47 students at Brown Elementary is 2.5 pets. What is the best point estimate for the mean number of pets per student for all students at Brown Elementary?

7. A survey of teachers reports that a point estimate for the mean amount of money spent each week on lunch is $18.00. If the margin of error for a 95% confidence interval for the mean amount of money spent each week on lunch by all teachers is $1.70, construct a 95% confidence interval for the mean amount of money spent each week on lunch for all teachers.

8. The mean batting average for a random sample of 35 professional baseball players is .283. If the margin of error for the population mean with a 99% level of confidence is .051, construct a 99% confidence interval for the mean batting average for professional baseball players.

9. The following graph[19] shows the 95% confidence intervals for mortality rates across several countries from 2010 to 2016. The middle point of each segment indicates the mean. Use the graph to answer the following questions.

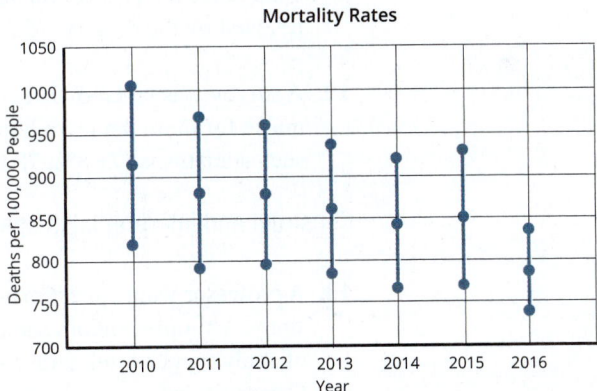

a. Approximate the 95% confidence interval for 2010. Based on this, what can you say about the mortality at this point?

b. Can we say with certainty that the 2012 mortality rate is below 900 deaths per 100,000 people? Explain your answer.

c. Is there a year that we can be certain has a mortality rate below 850 deaths per 100,000 people?

d. Compare the mortality rates for 2015 and 2016.

19   "Health Status," Organisation for Economic Co-Operation and Development, accessed January 4, 2022, https://stats.oecd.org/Index.aspx?DatasetCode=HEALTH_STAT.

10. The following graph[20] shows the 95% confidence intervals for the number of cancer cases in several countries in 2000, 2002, 2008, and 2012. The middle point of each segment indicates the mean. Use the graph to answer the following questions.

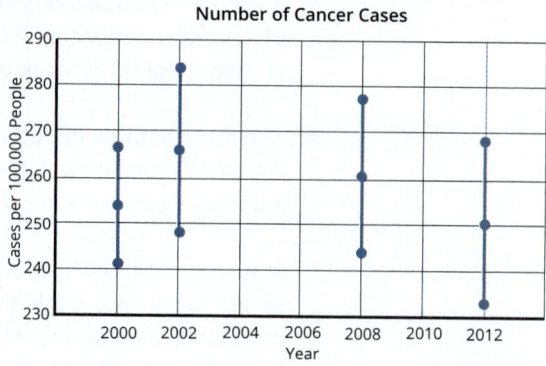

a. Approximate the 95% confidence interval for 2000. Based on this, what can you say about the number of cancer cases in 2000?

b. Can we say with certainty that there were fewer than 280 cancer cases per 100,000 people in 2002? Explain your answer.

c. Is there a year that we can be certain had fewer than 270 cancer cases per 100,000 people?

d. Compare the number of cancer cases in 2000 and 2002.

11. While testing a new anesthetic drug, the researchers found that the 90% confidence interval for the percentage of patients whose pain is completely relieved by the drug is (76.2%, 81.4%). Interpret their results.

12. A survey was taken of 1028 high school students asking if they listen to music for more than two hours per day. The 95% confidence interval for such students is (71.8%, 78.0%). Interpret the result of the survey.

Construct and interpret each specified confidence interval.

13. A professor wants to estimate how many hours per week her students study. A simple random sample of 78 students had a mean of 15.0 hours of studying per week with a standard deviation of 2.3 hours per week. Construct and interpret a 90% confidence interval for the mean number of hours a student studies per week if the margin of error for the data using a 90% level of confidence is $E = 0.4$.

14. A faculty advocacy group is concerned about the amount of time teachers spend each week doing schoolwork at home. A simple random sample of 56 teachers had a mean of 8.0 hours per week working at home after school with a standard deviation of 1.5 hours per week. Construct and interpret a 95% confidence interval for the mean number of hours per week a teacher spends working at home if the margin of error for the data using a 95% level of confidence is $E = 0.4$.

20   "Health Status," Organisation for Economic Co-Operation and Development, accessed January 4, 2022, https://stats.oecd.org/Index.aspx?DatasetCode=HEALTH_STAT.

15. A writer for a computer magazine is working on an article about computer usage in American households. A simple random sample of 120 American households has a mean computer usage time of 19.2 hours per week with a standard deviation of 3.3 hours per week. Construct and interpret a 95% confidence interval for the mean computer usage time per week for all American households if the margin of error for the data using a 95% level of confidence is $E = 0.6$.

16. A survey of 85 randomly selected homeowners finds that they spend a mean of $67 per month on home maintenance with a standard deviation of $14 per month. Construct and interpret a 99% confidence interval for the mean amount of money spent per month on home maintenance by all homeowners if the margin of error for the data using a 99% level of confidence is $E = 4$.

17. A survey of 97 randomly selected homeowners found that the mean amount spent on lawn service was $720 per year with a standard deviation of $123 per year. Construct and interpret a 98% confidence interval for the mean amount of money spent on lawn service per household each year if the margin of error for the data using a 98% level of confidence is $E = 29$.

18. A survey of a simple random sample of 140 dieters revealed that the numbers of times they "cheated" on their diets had a mean of 7.0 times per week with a standard deviation of 1.5 times per week. Construct and interpret a 99% confidence interval for the mean number of times dieters "cheat" on their diets each week if the margin of error for the data using a 99% level of confidence is $E = 0.3$.

19. A physical therapist is investigating the mean recovery time after ACL surgery for patients involved in a new therapy regimen. For the purpose of the study, a successful recovery was defined to be the ability to walk without crutches. For 38 randomly selected patients, the mean recovery time after ACL surgery was found to be 22.6 days with a standard deviation of 3.7 days. Find and interpret a 99% confidence interval for the mean recovery time for all ACL surgery patients undergoing the same new therapy if the margin of error for the data using a 99% level of confidence is $E = 1.5$.

20. The manufacturers of Caudill automotive oil wish to estimate the mean number of miles that motorists drive between oil changes. A random sample of 54 motorists has a mean of 5900 miles driven between oil changes with a standard deviation of 1350 miles. Construct and interpret a 95% confidence interval for the mean number of miles driven between oil changes for all motorists if the margin of error for the data using a 95% level of confidence is $E = 360$.

21. Thirteen out of 147 randomly selected faculty members who were surveyed at a community college know sign language. Use the rule of thumb to construct and interpret a 95% confidence interval for the percentage of all faculty members at the community college who know sign language. If necessary, round your upper and lower endpoints to the nearest hundredth.

22. A random sample of 200 computer chips is obtained from one factory and 4% are found to be defective. Construct and interpret a 95% confidence interval for the proportion of all computer chips from that factory that are defective if the margin of error for the data using a 95% level of confidence is $E = 2.7\%$.

23. Out of 140 randomly selected kindergartners who were surveyed, 32 said that pancakes are their favorite breakfast food. Use the rule of thumb to construct and interpret a 95% confidence interval for the percentage of all kindergartners who say pancakes are their favorite breakfast food. If necessary, round your upper and lower endpoints to the nearest hundredth.

24. Out of 543 randomly selected patients of a local hospital who were surveyed, 492 reported that they were satisfied with the care they received. Use the rule of thumb to construct and interpret a 95% confidence interval for the percentage of all patients satisfied with their care at that hospital. If necessary, round your upper and lower endpoints to the nearest hundredth.

25. A survey of 145 randomly selected students at one college showed that only 87 checked their campus e-mail account on a regular basis. Use the rule of thumb to construct and interpret a 95% confidence interval for the percentage of students at that college who do not check their e-mail account on a regular basis. If necessary, round your upper and lower endpoints to the nearest hundredth.

26. A random sample of 200 computer chips is obtained from one factory and 4% are found to be defective. Construct and interpret a 99% confidence interval for the percentage of all computer chips from that factory that are *not* defective if the margin of error for the data using a 99% level of confidence is $E = 3.6\%$.

27. The Pew Research Center surveyed pairs of US teens and one of their parents in 2019. Their website provides a breakdown of the sample size along with the margin of error for a 95% confidence interval for the overall survey as well as for different subgroups of data.

Teen Group	Unweighted Sample Size	Margin of Error
Total teen Sample	1811	3.1 percentage points
All religiously affiliated teens	1246	3.7 percentage points
Christian	1179	3.9 percentage points
Protestant	680	5.2 percentage points
Evangelical	407	6.6 percentage points
Mainline	208	9.0 percentage points
Catholic	428	6.4 percentage points
Unaffiliated	555	5.7 percentage points

Source: "Methodology," Pew Research Center, September 8, 2020, https://www.pewforum.org/2020/09/10/methodology-34/.

a. Use the data in the table to create a graph of the sample sizes of each category versus the margin of error percentage points. For each point (x, y), let x represent the sample size and y represent the margin of error.

b. Based on the graph from part a., what can you say about the margin of error compared with the sample size?

c. Is it possible to create confidence intervals for the population percentages in each category using the data from the table?

28. The Pew Research Center surveyed US adult Twitter users in 2021 and reported the following.

   "Seven in ten US adult Twitter news consumers have followed breaking news there."

   Their website provides a breakdown of the sample size along with the margin of error for a 95% confidence interval for the overall survey as well as for different subgroups of data.

Group	Unweighted Sample Size	Margin of Error
Total Sample	2548	3.4 percentage points
Gets news on Twitter	1783	4.0 percentage points
Does not get news on Twitter	751	6.1 percentage points
Among those who get news on Twitter		
Rep / Lean Rep	452	7.7 percentage points
Dem / Lean Dem	1311	4.8 percentage points

   Source: "Methodology," Pew Research Center, November 15, 2021, https://www.pewresearch.org/journalism/2021/11/15/new-on-twitter-methodology/.

   a. Use the data in the table to create a graph of the sample sizes of each category versus the margin of error percentage points. For each point (x, y), let x represent the sample size and y represent the margin of error.

   b. Based on the graph from part a., what can you say about the margin of error compared with the sample size?

   c. Is it possible to create confidence intervals for the population percentages in each category using the data from the table?

29. The Pew Research Center reported the following based on a survey conducted among 4860 US adults.

   "47% of Americans say dating is now harder than it was 10 years ago, while 19% say it's easier and 33% say it's about the same."

   Their website provides the margin of error of 2.1 percentage points for a 95% confidence interval for the overall survey. Interpret the provided margin of error.

30. A 2021 survey of US adults found that "Many Americans continue to experience mental health difficulties as pandemic enters second year. […] Psychological distress is especially common among adults with lower family incomes." Use the information in the table to answer the following questions.

Group	High Distress	Medium Distress	Low Distress	Unweighted Sample Size	Margin of Error (for 95% Confidence)
Upper Income	13%	22%	65%	2767	2.7%
Middle Income	18%	24%	56%	4810	2.2%
Lower Income	31%	25%	42%	2119	3.4%

   Source: Pew Research Center, February 2021, https://www.pewresearch.org/facttank/2021/03/16/many-americans-continue-to-experience-mental-health-difficulties-as-pandemic-enters-second-year/.

a. Construct a 95% confidence interval for the proportion of American adults with middle income who would say they have high distress.

b. Is it correct to say that people with lower income are more likely to say they are highly distressed than those with upper income? Explain why or why not.

31. A 2021 survey of US adults found that "About seven-in-ten US adults say major tech companies have too much economic power and influence." Use the information in the table to answer the following questions.

Group	Too much	About the Right Amount	Not Enough	Unweighted Sample Size	Margin of Error (for 95% Confidence)
US Adults	68%	25%	4%	4623	2.2%
Rep/Lean Rep	74%	22%	2%	1899	3.3%
Dem/Lean Dem	64%	29%	5%	2603	3.0%

Source: Pew Research Center, July 20, 2021, https://www.pewresearch.org/facttank/2021/07/20/56-of-americans-support-more-regulation-of-major-technology-companies/.

a. Construct a 95% confidence interval for the proportion of American adults who would say major technology companies have too much power and influence in today's economy.

b. Is it correct to say that Republicans are more likely to say technology companies have too much power than Democrats? Explain why or why not.

32. Suppose a poll was taken of 3196 high-school seniors to investigate how many of them plan to go to college. The researchers learned that 68.7% plan to go to college.

a. Use the rule of thumb to calculate the margin of error for a 95% confidence interval.

b. Interpret the obtained margin of error and the interval.

33. According to the 2019–2020 National Pet Owners Survey, 67% of US households own a pet. Suppose another survey was taken of 1352 US households. The researchers determined that 59.3% of them own a pet.

a. Use the rule of thumb to calculate the margin of error for a 95% confidence interval.

b. Interpret the obtained margin of error and the interval.

11.5 PROJECT

IN THE BLINK OF AN EYE

In Section 11.5, you learned that confidence intervals can be constructed using sample data in order to estimate a population parameter with a certain level of confidence. In this project, you will work with your classmates to collect sample data and construct confidence intervals for the number of times a person blinks in one minute in various circumstances.

Blinking our eyes is something we never think about until, well, we think about it. So let's think about it. Various sources maintain that an individual will, on average, blink somewhere between 15 and 20 times per minute unless they are either engaged in a conversation or focusing intently on a task. When carrying on a conversation, blinks per minute (bpm) tend to increase to between 19 and 26. The rate of blinking decreases to approximately 4.5 when focusing intently.

The goal of this project is to randomly select ten classmates, record how many times they blink in one minute for each of the three scenarios mentioned above, construct 95% confidence intervals from the data collected, and see if the reported average blinks per minute for each scenario falls within the interval you construct.

To do so, complete the following steps.

1. Form a group that consists of at least ten classmates.

2. For each classmate selected to have their blinks counted, have one group member observe, count, and record the number of blinks, while another group member watches a timer to start and stop the one-minute count. (Multiple experiments should be taking place within the group until the data collection is complete.)

3. Repeat this process until you have ten data points for each of the following three scenarios.

    a. Have the classmate relax and casually glance around the room, look at their fingernails, flip through a book, etc.

    b. Have the classmate carry on a conversation with another classmate.

    c. Have the classmate focus on something, such as playing a game on a cell phone. Tetris is an excellent choice, but any game will work.

4. Using the numbers of blinks per minute that your group observed, calculate the sample mean for each of the three scenarios.

5. Calculate the standard deviation for each sample either by using technology or by applying the following formula, where $\bar{x}$ is the sample mean and x_i is the ith value in the data set as i ranges from 1 to 10. Round each value to the nearest ten thousandth.

$$s = \text{sample standard deviation} = \sqrt{\frac{\sum_{i=1}^{10}(x_i - \bar{x})^2}{9}}$$

6. Using the following formulas, construct a confidence interval for each scenario (relaxed, during conversation, and playing a game). You may use a TI-83/84 Plus calculator or the following formulas. (**Note:** If you use a calculator, be sure to find the TInterval, not the ZInterval.) Round each endpoint to the nearest hundredth.

$$\text{Lower Endpoint: } \bar{x} - \left(2.262 \cdot \frac{s}{\sqrt{10}} \right)$$

$$\text{Upper Endpoint: } \bar{x} + \left(2.262 \cdot \frac{s}{\sqrt{10}} \right)$$

7. Do the reported averages (15–20 bpm when relaxed, 19–26 during conversation, and 4.5 while focused) fall within the confidence intervals you found? If not, list some factors that you think may have contributed to the reported average being outside of your confidence interval.

8. Do you think the experiments used to produce the data for this project are valid ways to measure the numbers of blinks per minute for people at rest, in conversation, and while focusing? Why or why not? In what ways do you think the method of data collection used in these experiments might result in inaccurate estimates of the population parameters?

9. Draw a general conclusion from your results.

Chapter 11 Exercises

In each scenario, identify the population, the sample, and any population parameters or sample statistics that are given.

1. The report titled *The American Freshman* looks at national norms of college freshman by analyzing the responses of 203,967 first-time, full-time freshmen entering 270 baccalaureate institutions across the United States. It found that over the past two years, the percentage of students who describe themselves as "liberal" (27.6%) or "conservative" (20.7%) has not changed significantly. The percentage of those identifying as "middle of the road" (47.4%) has risen slightly.[21]

2. *Social Metadata for Libraries, Archives, and Museums*: Executive Summary reported on work done in 2009–2010 by a 21-member RLG Partner Social Media Working Group from five countries. Their work reviewed 76 online sites that were relevant to libraries, archives, and museums. The sites supported social media features such as tagging, comments, reviews, etc. Among their findings, they reported that more than 70% of library sites have been offering social media features for 2 years or less and that 83% of respondents add new content at least monthly. Most of the respondents manage their own sites rather than use hosted services.[22]

Answer each question thoughtfully.

3. Suppose you want to know the proportion of students who wear glasses on your campus. Is simply surveying your current class a good sample to choose? Why or why not?

4. Suppose that in a race, male participants are given even entry numbers and females are given odd entry numbers. Would choosing every 10th runner to answer a survey give a representative sample of all racers? What about every 5th runner? Why?

5. You are assigned the task of determining the average age of people who shop at the local mall.

   a. Describe at least three different methods of sampling that could be used for the task. Which method would you say is the best one to use? Why?

   b. Describe any potential for bias in your method.

21   John H. Pryor et al., *The American Freshman*
22   Karen Smith-Yoshimura, *Social Metadata for Libraries, Archives, and Museums*

6. Determine which type of graph would most clearly depict the data described.

   a. The enrollment size at Austin Peay State University (APSU) over the past decade

   b. The size of each class—freshman, sophomore, junior, senior—for the current year at APSU

   c. The size of each class—freshman, sophomore, junior, senior—for the current year at APSU, specifically comparing the genders of students in each category

   d. The enrollment sizes of all state universities across the country

7. A clinical trial was designed to check the side effects of the drug for insomnia. Researchers gathered a group of people who suffered from insomnia and had no other pronounced diseases. Some of the people were given the drug, and the others were given a placebo. The patients didn't know whether they were taking the drug or placebo, but the researchers did know. After two weeks, the subjects were surveyed about side effects they noticed.

   a. Identify the explanatory and response variables.

   b. Identify the treatment in the experiment.

   c. Determine which group is the treatment group and which group is the control group.

   d. State whether the experiment is a single-blind or double-blind study. Do you think this is the best choice for this study?

Solve each problem.

8. Use the stacked bar graph on US defense spending trends from 2000–2011 to answer the following questions.[23]

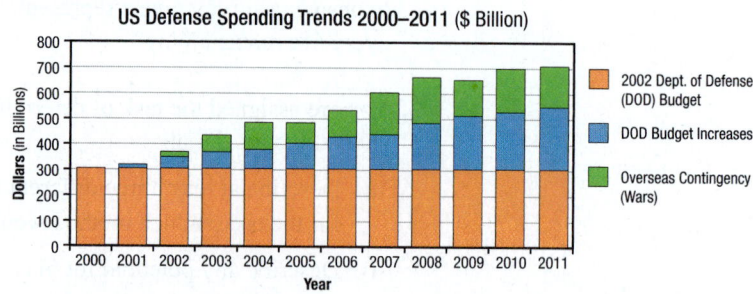

   a. How much is the estimated total spending for 2011?

   b. How much has spending increased from 2000 to 2011 in the Overseas Contingency category?

   c. By how much has the Department of Defense (DOD) budget increased from 2000–2011?

   d. Would you say that the stacked bar graph is a good way to represent the data here? Why or why not?

---

23 Emily Skarbek, "How Would You Cut Defense Spending?" http://www.mygovcost.org

9. The US Department of Agriculture estimates that between 2010 and 2015, the US economy will generate 54,000 annual openings requiring baccalaureate or higher degrees in food, renewable energy, and environmental specialties. Use the graph about employment opportunities to answer the following questions.[24]

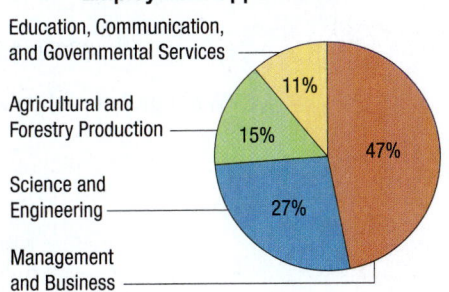

Employment Opportunities

a. What percentage of the jobs are expected to be in the business and science occupations?

b. Over the 5 years, how many total jobs openings are expected in education, communication, and governmental services?

c. How many annual openings are there expected to be in agricultural and forestry production?

d. Would you say that the pie chart is a good way to represent the data given? Why or why not?

10. Consider the following graph showing the results of a personal entertainment survey.

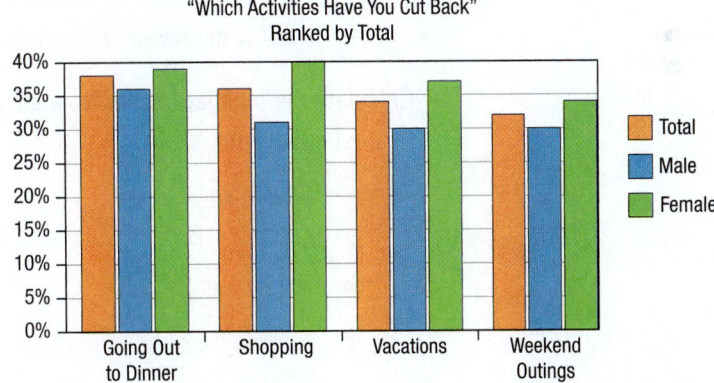

Activities That Have Decreased
"Which Activities Have You Cut Back"
Ranked by Total

a. Name at least two things that are missing from the graph for it to be an effective means of informing people about the data.

b. List ways in which a reader might be unclear or misled about what the graph shows.

For each data set, calculate the following numerical descriptors: mean, median, mode, range, sample standard deviation, and five number summary.

11. Area prices for unleaded gasoline for the week of 1/25/2013 are found to be the following.

$3.07 $3.49 $3.05 $3.35 $3.29 $3.06 $3.18 $3.29

$3.21 $3.29 $3.29 $3.31 $3.05 $3.21 $3.19

12. Whitt gathered the following data from students living in his dorm.

Daily Soda Consumption
Number of Cans per Day

2	0	1
0	1	0
6	1	1
0	0	2
0	3	0
4	0	1

13. Consider the weight in grams of 15 soda cans chosen at random from a batch by a quality control inspector.

343	335	338	344	333
345	324	329	322	323
342	341	326	334	335

a. Find the five-number summary of the weights.

b. What percentage of weights is at or below 333 grams?

c. What is the range of weights in the chosen batch?

Calculate the standard score for each given value. Round your answer to the nearest hundredth.

14. $\mu = 20, \sigma = 2$

a. $x_1 = 22$

b. $x_2 = 19$

c. $x_3 = 25$

15. $\bar{x} = 16.9, s = 0.04$

a. $x_1 = 17.15$

b. $x_2 = 16.84$

c. $x_3 = 17.4$

Solve each problem.

16. Scores on a test have a mean of 86 and a standard deviation of 13. Jerrica has a score of 94. Convert Jerrica's score to a z-score.

17. Hasef scored a 55 on a test with a mean of 47 and a standard deviation of 9. Kimberly had a score of 168 on a test with a mean of 145 and a standard deviation of 26. Which score was better with respect to the mean score on the given test?

Use the z-score formula to complete the table.

18. Find the missing value in each row of the table. Round your answers to the nearest hundredth when necessary.

	z	x	μ	σ
a.		26.2	35.0	4.3
b.	−1.24	152.2		12.2
c.	2.60		55.6	6.04
d.	−2.73	1250	1735	

Find the percentage of data points that lie below each z-score.

19. $z = -0.29$ **20.** $z = 1.16$

21. $z = 3.21$ **22.** $z = -2.75$

Find the percentage of data points that lie above each z-score.

23. $z = 1.17$ **24.** $z = -1.34$

25. $z = -3.25$ **26.** $z = 2.14$

Find the percentage of data points that lie between each pair of z-scores.

27. $z_1 = -2.00, \ z_2 = 2.00$ **28.** $z_1 = -1.40, \ z_2 = 2.56$

29. $z_1 = -2.00, \ z_2 = -1.00$ **30.** $z_1 = 0.62, \ z_2 = 3.1$

Solve each problem.

31. Suppose the length of Ethan's newborn son had a standard score of 0.48. True or false: Ethan's son is longer than the average newborn.

32. Salespeople for a car sales company have an annual sales average of $225,000, with a standard deviation of $18,000. What percentage of the salespeople will make the following sales? Assume the sales follow a normal distribution.

    a. More than $250,000   **b.** Less than $190,000

    c. Between $200,000 and $240,000

33. A potato chip manufacturer processes bags of chips with an average weight of 6.5 ounces and a standard deviation of 0.8 ounces. What percentage of the bags of chips will have the following weights? Assume the weights follow a normal distribution.

    a. More than 6.7 ounces   **b.** Less than 6.4 ounces

    c. Between 6.4 and 6.7 ounces

34. For extra credit, a high school statistics student found that the 95% confidence interval for the average weight of high school football players in the US was (180 lb, 210 lb). Interpret this confidence interval.

CHAPTER 11 PROJECT

Setting the Curve

Dr. Romero, a professor in the School of Computing at Klaggen University, is using a standardized final exam that is "nationally normed" for his Computer Science II class. Nationally normed implies that the normal distribution is an appropriate approximation for the probability distribution of students' scores on the exam. The probability distribution of students' scores on this standardized exam can be estimated using the normal distribution shown below.

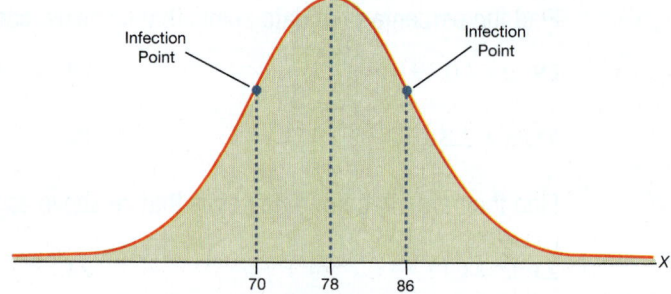

1. State the mean of the distribution of the computer science exam scores based on the figure above.

2. State the standard deviation of the computer science exam scores based on the figure above.

At some point in our academic pursuit, we all think that we would like our professors to curve our grades. Especially if we're the ones setting the curve. After considering his students' request to curve the grades on the final exam, Dr. Romero has come up with two options to use if he decides to curve the grades.

Follow the steps listed to determine the grading scale for each option.

Curved Grading Option #1

- Students whose raw scores are at or above the 90th percentile will receive an A.
- Students whose raw scores are in the 80th–89th percentile will receive a B.
- Students whose raw scores are in the 70th–79th percentile will receive a C.
- Students whose raw scores are in the 60th–69th percentile will receive a D.
- Students whose raw scores are below the 60th percentile will receive an F.

3. In order to know the cutoff exam scores required for Curved Grading Option #1, Dr. Romero needs to know what z-scores correspond to the upper limit percentiles. Find each z-score that corresponds to the following percentiles. Round your z-scores to the nearest thousandth. We've found the z-score of the 90th percentile for you.

90th percentile ___1.282___

80th percentile _____

70th percentile _____

60th percentile _____

4. Using the *z*-scores found in Step 3, find the exam scores that correspond to Curved Grading Option #1. Assume that the exam scores range from 0 to 100. (Round to the nearest whole number.)

Curved Grading Option #1

A: _____ – _100_

B: _____ – _____

C: _____ – _____

D: _____ – _____

F: _0_ – _____

Curved Grading Option #2

The second option for curving the grades is as follows:

- Students whose raw scores are at least two standard deviations above the mean of the standardized test will receive an A.

- Students whose raw scores are from one up to two standard deviations above the mean of the standardized test will receive a B.

- Students whose raw scores are from one standard deviation below the mean up to one standard deviation above the mean of the standardized test will receive a C.

- Students whose raw scores are from two standard deviations below the mean up to one standard deviation below the mean of the standardized test will receive a D.

- Students whose raw scores are more than two standard deviations below the mean of the standardized test will receive an F.

5. Using the information above, find the exam scores that correspond to Curved Grading Option #2. Assume that the exam scores range from 0 to 100. (Round to the nearest whole number.)

Curved Grading Option #2

A: _____ – _100_

B: _____ – _____

C: _____ – _____

D: _____ – _____

F: _0_ – _____

6. Using the grading scales you just created in Steps 4 and 5, complete the following table of the partial list of grades and find the new curved letter grades that the students would receive in each of the curving options given their raw scores on the exam.

Computer Science II Final Exam Scores

Name	Raw Score/Uncurved Grade	Option #1 Curved Grade	Option #2 Curved Grade
J. Alexander	79/C		
W. Thouy	69/D		
C. Bradford	88/B		
S. Nance	66/D		
A. Moore	75/C		
K. Pinkston	86/B		
C. Navas	91/A		
R. Alexandru	77/C		
S. Garcia	82/B		

After reviewing the grades for each student using the two optional curving methods, answer the following questions.

7. Who do you think benefits the *most* from Curved Grading Option #1? Explain your reasoning.

8. Who is likely to disapprove of Curved Grading Option #1? Why?

9. Who do you think benefits the *most* from Curved Grading Option #2? Why?

10. Who is likely to disapprove of Curved Grading Option #2? Why?

11. Which grading scale do you feel is *most* fair? Explain why?

CHAPTER 11 REVIEW

Section 11.1 Statistical Studies

Definitions

Statistics: Statistics is the science of gathering, describing, and analyzing data.

or

Statistics is the actual numerical description of sample data.

Population: A **population** is the particular group of interest in a statistical study.

Variable: A **variable** is a value or characteristic that changes among members of the population.

Data: Data are the counts, measurements, or observations gathered about a specific variable in a population in order to study it.

Representative Sample: A **representative sample** is one that has the same relevant characteristics as the population and does not favor one group of the population over another.

Bias: A **bias** is a favoring of a certain outcome in a study.

Terms Related to Populations and Subsets

Census: A **census** is a study involving data obtained from every member of the population.

Parameter: A **parameter** is the numerical description of a particular population characteristic.

Sample: A **sample** is a subset of a population from which data are collected in a statistical study.

Sample Statistic: A **sample statistic** is the numerical description of a particular sample characteristic.

Observational Study: An **observational study** observes existing data.

Experiment: An **experiment** generates data to help identify cause-and-effect relationships.

Parts of an Experiment

Treatment: A **treatment** is a certain condition that is applied to a group of subjects in an experiment.

Subjects: Subjects are the people or things being studied in an experiment.

Participants: Subjects in an experiment that consist solely of people are **participants**.

Response Variable: The **response variable** is the variable in an experiment that measures any response to the treatment.

Explanatory Variable: The **explanatory variable** is the variable in an experiment that causes the change in the response variable.

Terms Related to Experiments

Control Group: A **control group** is a group of subjects to which no treatment is applied in an experiment.

Treatment Group: A **treatment group** is a group of subjects that researchers apply a treatment to in an experiment.

Confounding Variables: Factors other than the treatment that cause an effect on the subjects of an experiment are called **confounding variables**.

Placebo Effect: The **placebo effect** is a response to the power of suggestion, rather than the treatment itself, by participants of an experiment.

Placebo: A **placebo** is a substance that appears identical to the actual treatment but contains no intrinsic beneficial elements.

Single-Blind: In a **single-blind experiment**, subjects do not know if they are in the control group or the treatment group, but the people interacting with the subjects in the experiment know which group each subject has been placed in.

Double-Blind: In a **double-blind experiment**, neither the subjects not the people interacting with the subjects know group each subject belongs to.

Procedures

Conducting a Statistical Study

1. Design the study.
   a. State the question to be studied.
   b. Determine the population and variables.
   c. Determine the type of study: observational or experimental.
2. Collect the data.
3. Organize the data.
4. Analyze the data to answer the question.

Sampling Methods

Random Sampling – Every member of the population has an equal chance of being selected.

Stratified Sampling (Quota Sampling) – A few members from each stratum (or group) are randomly chosen.

Cluster Sampling – All members from a few randomly chosen clusters (or groups) are selected.

Systematic Sampling – Every nth member of the population is chosen.

Convenience Sampling – The sample is chosen because it is convenient for the researcher.

Principles of Experimental Design

1. Randomize the control and treatment groups.
2. Control for outside effects on the response variable.
3. Replicate the experiment a significant number of times to see meaningful patterns.

Section 11.2 Displaying Data

Definitions

Frequency Distribution: A **frequency distribution** is a display of the values that occur in a data set and how often each value, or range of values, occurs.

Class: In a frequency distribution, a **class** is a category of data.

Relative Frequency: The **relative frequency** is the percentage of the data set that falls into a particular class.

$$\text{Relative Frequency} = \frac{f}{n},$$

Here, f is the class frequency and n is the total number of members in the data set.

Grouped Frequency Distribution: A frequency distribution in which each category represents a range of values is call a **grouped frequency distribution**.

Class Width: The **class width** is the difference between the lower limits or upper limits of two consecutive classes of a frequency distribution.

Lower Class Limit: The **lower class limit** is the smallest number that can belong to a particular class.

Upper Class Limit: The **upper class limit** is the largest number that can belong to a particular class.

Graph: A **graph** is a picture of the data that allows us to view patterns at a glance.

Legend: A **legend** is a description of how each data category is identified in a graph.

Pie Chart: A **pie chart** is a circular graph that depicts parts of a whole and shows the size of each category of data in relation to the whole.

Bar Graph: A **bar graph** is a graph that uses bars to represents the amount of data in each category.

Frequency Histogram: A **frequency histogram** is a bar graph displaying a frequency distribution where the horizontal axis is a number line. The height of each bar represents the frequency of each class.

Line Graph: A **line graph** is a graph that uses points to represent data values at a particular time in history and then joins the points with line segments. Line graphs display changes in a variable over time.

Procedure

Constructing a Frequency Distribution

1. **Decide on the number of classes for the distribution**. There are typically between 5 and 20 classes. Having too many class divisions does not add any value over simply listing every data point and can be overwhelming. Having too few class divisions results in the data being lumped together and lacks detail.

2. **Choose an appropriate class width**. In some cases, the data set lends itself to natural divisions, such as decades or years. In other cases, you need to choose an appropriate class width so that the classes formed present a clear representation of the data and include all members of the data set. One method of finding the class width is to subtract the lowest number in the data set from the highest number and divide the difference by the number of classes.

3. **Find the class limits**. Begin by determining the lower class limit of the first class. You should choose the first lower limit so that reasonable classes will be produced. Using the minimum data value is a good place to start.

4. **Determine the frequency of each class**. Make a tally mark for each data value in the appropriate class. Count the marks to find the total frequency for each class.

Section 11.3 Describing and Analyzing Data

Definitions

Outlier: An **outlier** is a data value that is extreme compared with the rest of the data values in the set.

Skewed to the Right: A graph's shape is **skewed to the right** when a majority of the data fall on the left side of the distribution.

Skewed to the Left: A graph's shape is **skewed to the left** when a majority of the data fall on the right side of the distribution.

Range: The **range** is the difference between the largest and smallest values in a data set, which tells you the distance covered on the number line between the two extremes.

$$\text{Range} = \text{Maximum Data Value} - \text{Minimum Data Value}$$

Percentiles: Percentiles divide an ordered list into 100 equal parts and tell you approximately what percentage of the data lies at or below a given value.

Quartiles: Quartiles are values that divide an ordered list of data into four equal parts; equivalent to the 25th, 50th, and 75th percentiles.

Q_1 = First Quartile = 25th percentile; that is, 25% of the data is less than or equal to this value.

Q_2 = Second Quartile = 50th percentile; that is, 50% of the data is less than or equal to this value.

Q_3 = Third Quartile = 75th percentile; that is, 75% of the data is less than or equal to this value.

By definition, Q_2 is the same as the median.

Five-Number Summary: The **five-number summary** is a numerical description of a data set that lists the minimum value; the first quartile, Q_1; the median or second quartile, Q_2; the third quartile, Q_3; and the maximum value in order from smallest to largest.

Box Plot: A **box plot**, or *box-and-whiskers plot*, is a graphical representation of a five-number summary.

Interquartile Range (IQR): The **interquartile range** is the range of the middle 50% of the data, given by

$$\text{IQR} = Q_3 - Q_1,$$

where Q_3 is the third quartile and Q_1 is the first quartile.

Formulas

Arithmetic Mean: The **arithmetic mean**, or *mean*, is the sum of all of the data points divided by the number of data points. Formally, the formula for **population mean** μ is as follows.

$$\mu = \frac{x_1 + x_2 + \cdots + x_N}{N}$$

The formula for the **sample mean** $\overline{x}$ is as follows.

$$\overline{x} = \frac{x_1 + x_2 + \cdots + x_n}{n}$$

Here, x_i is the ith data value, N is the number of data values in the population, and n is the number of data values in the sample.

Standard Deviation: The **standard deviation** is a measure of how much we might expect a member of the data set to differ from the mean.

The formula for finding the **population standard deviation** σ is as follows.

$$\sigma = \sqrt{\frac{\sum (x_i - \mu)^2}{N}}$$

Here, x_i is the ith data value, μ is the population mean, and N is the size of the population. The formula for finding the **sample standard deviation** s is as follows.

$$s = \sqrt{\frac{\sum\left(x_i - \overline{x}\right)^2}{n-1}}$$

Here, x_i is the ith data value, $\overline{x}$ is the sample mean, and n is the sample size.

Properties

Mode: The **mode** of a data set is the value that occurs most frequently.

No mode describes a data set in which all of the data values occur only once or each value occurs an equal number of times.

Unimodal describes a data set in which only one data value occurs most often.

Bimodal describes a data set in which exactly two data values occur equally often and more than any other data value.

Multimodal describes a data set in which more than two data values occur equally often and more than any other data value.

Procedures

Median: The **median** of a data set is the middle value in an ordered list of the data. To find the median, follow the following steps.

1. List the data in ascending or descending order.

2. If the data set contains an odd number of values, the median is the middle value in the ordered list.

3. If the data set contains an even number of values, the median is the arithmetic mean of the two middle values in the ordered list. Notice that this implies that the median might not be a value in the data set.

Creating a Box Plot

1. Draw a horizontal (or vertical) number line that contains the five-number summary.

2. Draw a small line segment above (or next to) the number line to represent each of the numbers in the five-number summary.

3. Connect the line segment that represents the first quartile to the line segment representing the third quartile, forming a box with the median's line segment in between.

4. Connect the "box" to the line segments representing the minimum and maximum values to form the "whiskers."

Section 11.4 The Normal Distribution

Definitions

Discrete Variable: A **discrete variable** is a variable whose numeric value can only take on particular values and is usually a count.

Continuous Variable: A **continuous variable** is a variable whose numeric value can take on any value in a given interval and is usually a measurement.

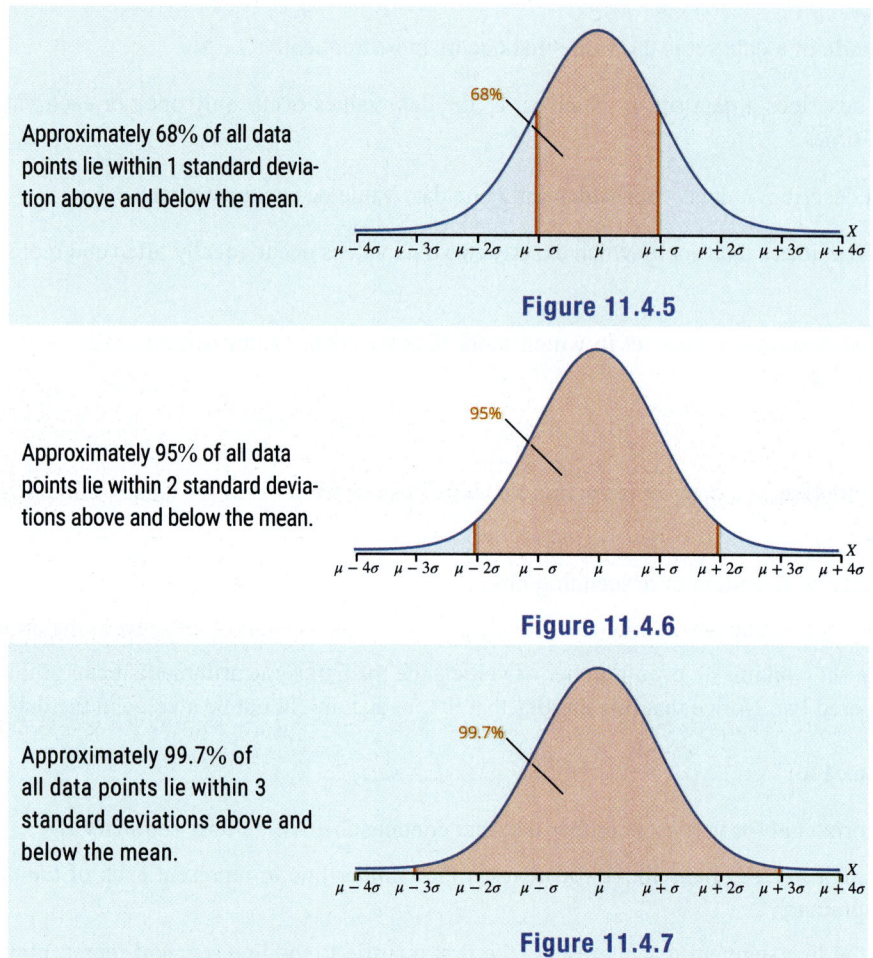

Approximately 68% of all data points lie within 1 standard deviation above and below the mean.

Figure 11.4.5

Approximately 95% of all data points lie within 2 standard deviations above and below the mean.

Figure 11.4.6

Approximately 99.7% of all data points lie within 3 standard deviations above and below the mean.

Figure 11.4.7

Formulas

Standard Score (z-Score): A **standard score**, or z-score, is the number of standard deviations a particular data value lies away from the mean in a normal distribution.

The **standard score for a population** value is given by

$$z = \frac{x - \mu}{\sigma},$$

where x is the value of interest from the population, μ is the population mean, and σ is the population standard deviation.

The **standard score for a sample** value is given by

$$z = \frac{x - \bar{x}}{s},$$

where x is the value of interest from the sample, $\bar{x}$ is the sample mean, and s is the sample standard deviation.

Properties

Normal Distribution: A **normal distribution** is a probability distribution for a continuous random variable defined completely by its mean and standard deviation such that the following properties are true:

1. It is bell-shaped and symmetrical about its mean.

2. It is completely defined by its mean μ and standard deviation σ.

3. The total area under the curve is equal to 1.

4. The x-axis is the horizontal asymptote for a normal distribution curve.

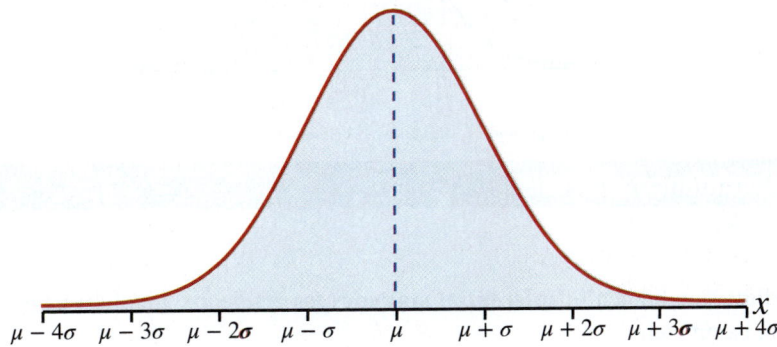

Figure 11.4.2

Procedure

Using the *Cumulative Normal Distribution Tables* to Find Areas under the Standard Normal Curve

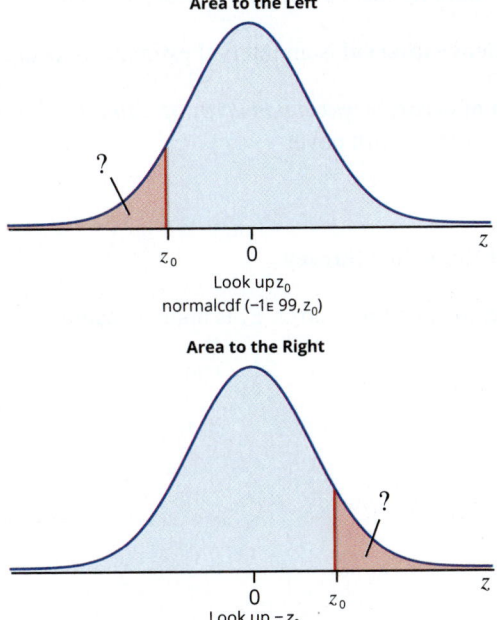

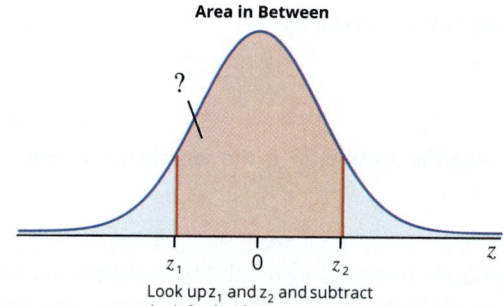

Area in Between

z_1 0 z_2

Look up z_1 and z_2 and subtract
area to the left of z_1 from area to the left of z_2
normalcdf (z_1, z_2)

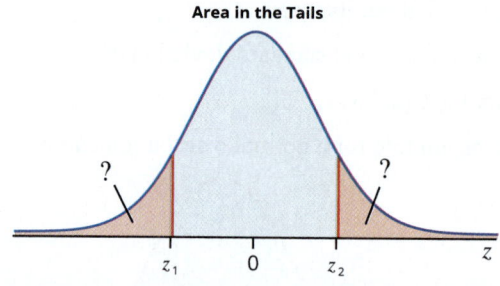

Area in the Tails

z_1 0 z_2

Look up z_1 and $-z_2$ and add areas $1 -$ normalcdf $(-$ normalcdf (z_1, z_2)

Section 11.5 Confidence Intervals

Definitions

Inferential Statistics: The branch of **inferential statistics**, as a science, involves using descriptive statistics to estimate population parameters.

Point Estimate: A **point estimate** is a single-number estimate of a population parameter.

Interval Estimate: An **interval estimate** is a range of possible values for a population parameter.

Level of Confidence: The **level of confidence**, denoted by c, is the percentage of all possible samples of a given size that will produce interval estimates that contain the actual parameter.

Confidence Interval: A **confidence interval** is an interval estimate associated with a certain level of confidence.

Margin of Error: The **margin of error**, or *maximum error of estimate*, E, is the largest possible distance from the point estimate that a confidence interval will cover.

Formulas

Rule of Thumb for Margin of Error in a Survey

With a 95% level of confidence, the margin of error, E, is approximately

$$\frac{1}{\sqrt{n}} \cdot 100\%$$

for a random sample of size n.

CHAPTER 12

Data Science

SECTIONS

Introduction

Every action you take in life has the potential to produce data. The products you purchase and how often you purchase them produces data for both the stores that stock the products and the businesses that create the products. Companies that create dating apps to match singles use data science to craft an algorithm that is based on swiping history to prioritize matches. Sports teams, like the Oakland Athletics, have effectively used data science to predict a player's potential in helping to lead them to the playoffs. The NBA and college teams are now using sensors on basketball hoop rims to gather data that displays shot details in real time to predict which shot a player is more likely to make.

How is all of this data analyzed? That job falls to data scientists. Data scientists are employed in almost all areas of business, public and private. They are employed by the government, Home Depot, Instagram, IBM, Uber, and Capital One, just to name a few. The list is long and inclusive, growing daily, and showing no signs of slowing down in the future.

This chapter will help introduce you to some of the processes and uses of data science. These processes include steps such as posing a question related to a field of study, gathering and cleaning data, and extracting information from the data to answer the question. Then we'll look at organizing and presenting the data to the people interested in the results, such as CEOs of companies, inventory specialists, or politicians.

12.1 THE SCIENCE OF DATA

OBJECTIVE 1 The Scientific Method for Data Science

To say we're living in the era of big data is a bit of an understatement. The infographic in Figure 12.1.1 shows the approximate amount of data that is generated every minute of every day by just a handful of companies at the end of 2021. The amount is mind blowing.

Figure 12.1.1

In 2020, 90% of the world's existing data was created in just the previous two years. Stop and consider that fact for a moment. That is an incredible rate of growth. Notice that most of the data was created through online sources such as Instagram posts, LinkedIn job applications, or streaming Netflix. Since almost 60% of the world's population already has access to the internet, this trend is likely to continue. It's estimated that in 2021 there were over 12.3 billion active connections in the IoT (which stands for Internet of Things). The IoT is a system of smart devices (such as smart speakers, smartwatches, and smart refrigerators) that communicate with other smart devices, all contributing to the data stockpile every second of the day. All of this "big data" is waiting, as the Oxford English Dictionary defines it, to be "analyzed computationally to reveal patterns, trends, and associations, especially relating to human behavior and interactions."

So what do we *do* with all of this information? We turn to the emerging field of data science. Still a relative newcomer in the world of sciences, the interdisciplinary field of data science applies the scientific method to big data, turning it from a large collection of data points into valuable and useful information.

Similar to any other scientific research, research in data science starts with asking a question or defining a problem. Data is then collected (or organized) and analyzed to produce an answer or solution to the defined question or problem. These steps are then repeated as needed to improve the results or to answer other questions. Of course, as you move through the process, new information gives rise to new questions, which in turn could necessitate backing up a step or two to refine

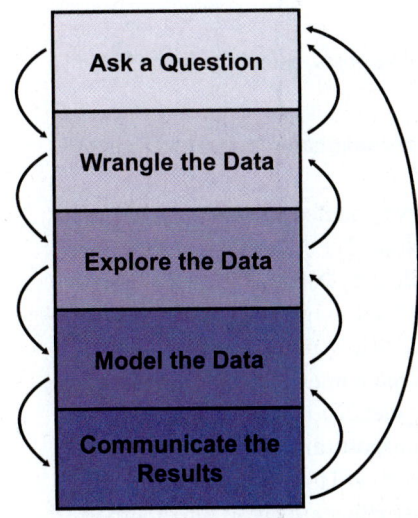

Figure 12.1.2: The Scientific Method for Data Science

the method of analysis or try a different approach. The arrows in Figure 12.1.2 indicate the inherent fluidness of using the scientific method on big data. As we move through the sections in this chapter, we will focus on different techniques and kills used in each of these components.

Example 12.1.1

Using the Scientific Method for Data Science

As part of a research project, you need to analyze the effects of social media on the mental health of teenagers. Answer the following questions to outline how you would work through the scientific method for data science.

a. What question can you pose to start your investigation?

b. Where might you find the data related to your investigation?

c. What are some ways you might explore the data to notice patterns or trends?

d. What are some ways you can model the data once it is cleaned up?

e. What might be an engaging way to communicate the results given the topic of the project?

Solution

a. The question we pose should result in a clear answer but should not be biased in any way. The question will also depend on which aspect of mental health the study will investigate. Some potential questions are as follows.

   1. Does the use of social media affect the rates of depression in teenagers?

   2. Does the use of social media affect the size of the social groups that teenagers interact with offline?

b. Data related to the study can be found by surveying teenagers on their social media usage, finding usage data posted by social media companies, and finding previously researched mental health data on teenagers.

c. You could create a scatter plot or order the data in a table to look for patterns and trends.

d. The data could be modeled by displaying the results graphically or in a summary table.

e. The results could be communicated through blog posts, social media posts, or online presentations.

Back in 2012, the occupation of *data scientist* was named the "Sexiest Job of the 21st Century" by the Harvard Business Review. But a decade later, the job description of a data scientist is still an evolving picture. Data scientists are tasked with everything from the very broad mission of "Tell us what the data says" to very precise predictions about the future, such as "How much inventory do we need to keep in stock?" They are asked to make recommendations to customers (such as Netflix or Amazon recommendations), identify behavioral patterns in society (which can be used to construct public transport timetables), or automate decision-making (such as processing debit and credit card transactions approvals). They optimize processes (such as finding the most efficient way to direct airport travelers through busy airports), increase revenue streams, or even detect fraud. The field of data science is both expansive and focused.

Despite the broad job description, what is clear is that a data scientist should be strong in at least four areas: Domain (or Business) Knowledge, Computer Science, Statistics and Probability, and Communication.

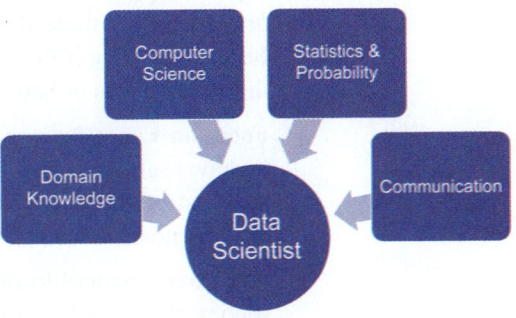

Figure 12.1.3

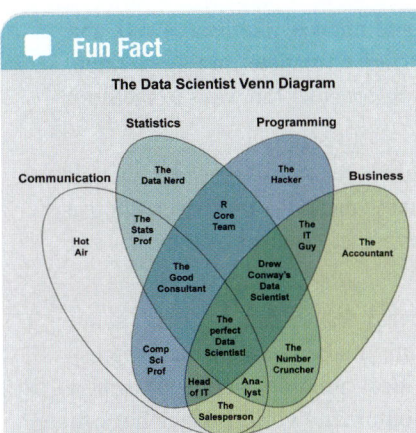

Figure 12.1.4

Stephan Kolassa designed the Venn diagram in Figure 12.1.4 to try and visually capture the data scientist. One online commenter is quoted as saying, "The perfect data scientist from Kolassa's Venn diagram is a mythical sexy unicorn ninja rock star who can transform a business just by thinking about its problems."

In this section, we will look at **domain knowledge**—the understanding of a particular field of interest such as insurance, retail sales, or nonprofit organizations. This understanding of the field of interest gives a deeper insight into issues and solutions that may be presented by the data. Being well versed in domain knowledge aids in refining the initial question and deciphering the results. Drawing upon strong **computer science** skills is essential in the second step of the scientific method. In Section 12.2, we will use Microsoft Excel to explore different types of data wrangling. In Section 12.3, we will focus on the mathematics of data science—using **statistics** and **probability** to explore and model the data. And finally, in Section 12.4, we will examine the strong **communication skills** data scientists need to effectively convey the results and set up ways to monitor their consequences. You can see how a strong background in each of these would definitely earn you the badge of "sexy data scientist."

OBJECTIVE 2 Domain Knowledge

> *In mathematics, the art of asking questions is more*
> *valuable than solving problems.*
>
> —Georg Cantor

Domain knowledge, or *business insight*, implies an understanding of how a particular business works in general and how a company's data can be used to achieve aspirational business goals. Data scientists need to be curious and results-oriented about the business or industry they find themselves in. They examine which questions need to be answered and determine where to find the applicable data. This requires the shrewdness to know where to look and what to look for in a particular setting. For instance, knowing the ins and outs of a production line can add insight on where a potential bottleneck might occur. Similarly, knowing about the recruiting process can help find ways to improve a company's hiring process. Often the data scientist is the person doing cross-departmental work, the proverbial bridge between departments such as IT, marketing, finance, operations, and management. Thus, a data scientist's domain knowledge might require insight into all of these fields.

Regardless of the domain you are working in, there are some important things to keep in mind when forming the question(s) to be answered.

- **Use specific well-defined terms and concepts.** Avoid vague language and sweeping terms like *all* or *most of*. For instance, rather than asking "Are adults lazy?" ask "How many hours of physical activity do adults between the ages of 30 and 40 get per day?"

- **Avoid subjective judgements.** Avoid words and phrases like *best, worst,* and *better than,* as these don't have universal criteria for determining a value. Instead of asking if Plan A is better than Plan B, consider the ways to measure the effectiveness of both plans.

- **Focus on a specific area of concern.** Make sure your study is narrow enough to draw useful conclusions. If there are too many questions to answer, choose the most central question and start there. Having too many focal points can result in no clear answer to any of them.

- **Ask questions within the data you have access to.** It's fine to dream about solving the world's problems, but you can only make constructive progress with the data you are given. For instance, a local nonprofit company cannot hope to answer the question of how best to eliminate world hunger but can instead look for ways to reduce hunger within their city or district.

Let's look at some examples on how to turn weak questions into stronger ones.

Question	Explanation
Weak Who has the best doctors? **Strong** Which hospital in the state has the lowest death rate along with a high patient satisfaction rating among people who were admitted to the critical care unit?	The first question is too broad and subjective. There are no defined criteria for "best." The second question uses clearly defined terms and narrows its focus to a specific group.
Weak What effect does social media have on sales? **Strong** How do sales before advertising on Instagram compare with sales in the three months following the initial advertising campaign?	The first question is not specific enough (which type of social media and what kinds of effects?) The second question defines the boundaries and particular terms more clearly.
Weak What do college graduates do after graduation? **Strong** Six months after graduating from a university nursing program, what proportion of graduates are practicing nurses?	The first question is about a very large population of which all the data may not be accessible. It is also unclear what the question is trying to measure; it could mean physically, socially, academically, or career-wise. The second question identifies a specific program and measurable quantity pertaining to the graduates.

Example 12.1.2

Strengthening Research Questions

Identify the weakness in each research question. Give three stronger, alternative questions.

a. Which toothpaste is the worst for your health?

b. Should cell phones be banned in schools?

Solution

a. Which toothpaste is the worst for your health?

The term "worst" is too subjective and not clearly defined. The same is true of the phrase "your health," which is too broad to measure. Stronger questions will define a measurement about a particular population.

Alternative Question 1:

For adults between the ages of 25–40, which toothpaste resulted in users experiencing the most oral bacteria after a period of 3 months?

This question provides a measurable quantity, *amount of oral bacteria*, and makes clear what is classified as "worse;" that is, more oral bacteria is worse. It also identifies a specific population—adults between the ages of 25 and 40.

Alternative Question 2:

What effect does a particular toothpaste have on the number of cavities in teenagers?

Again, the question is focused on a specific population (teenagers). It also identifies a measurable quantity (the number of cavities).

Alternative Question 3:

Is there a difference in the overall dental bill for users of different toothpastes over a period of 2 years?

Although the population is not narrowed down in this question, the emphasis here uses monetary impacts as a way to identify consequences.

b. Should cell phones be banned in schools?

The question is only subjective and does not provide any specific parameters for the researcher to measure. A stronger question will outline particular areas of concern around cell phones and schools for the researcher to investigate.

Alternative Question 1:

How do students use cell phones during the school day?

This question provides a quantifiable measurement for researchers to examine. The responses to the question will help frame what "banning cell phones" might entail and clarifies the possible impact a cell phone ban might have on students and their parents.

Alternative Question 2:

How does restricting cell phone use in school affect a student's academic performance?

This question ties the use of cell phones to academic performance. A researcher might be able to connect minutes of cellphone activity during the school day with academic grade performance.

Alternative Question 3:

What proportion of schools in the district currently allow students to carry a cell phone to school?

Phrasing the research question in this way provides a different angle to the potential issue. It provides a measurable quantity, allows the researcher to assess the prevalence of current cellphone use during the school day, and will help determine whether further investigation is needed around banning cell phones.

Strong research questions are the basis of fruitful outcomes for a data scientist. And domain knowledge is pivotal to the process. Each of the questions in the previous example were strengthened by this insight. In the case of the toothpaste question, the first question revision understands that the use of toothpaste is to decrease bacteria, the second recognizes the user base of toothpaste, and the third recognizes that the desired outcome of using toothpaste means better dental checkups. However, sometimes as a data scientist you won't have the luxury of strengthening the research question by simply substituting one weak question for a stronger one of your own choosing, as we did in the previous example. For instance, suppose you are lacking domain knowledge for a particular project, and the question posed to you is "How can we improve sales?" Because you recognize this as a weak research question, you might strengthen it by asking, "Does increasing store hours have a positive monetary effect on sales?" This is indeed a stronger research question and one which can be evaluated. Unfortunately, if the company is already open 24 hours a day, your stronger question is no better than the weaker one since the store hours cannot be increased. To find the right kind of stronger question, you need to work with colleagues who have domain knowledge (such as knowledge of the store hours and operations) to help them formulate the question that they really need to answer.

When the research question is determined by others, it is your job as the scientist to ascertain the goal of the project. Instead of guessing, you can ask leading questions and determine the key indicators that need to be assessed in the data. With this type of effective guidance, you can assist those who are initiating the research process in posing the strongest questions for their purposes. Helping clients articulate their concerns and see the opportunities the data can provide is a valuable skill.

So how do you help others formulate appropriate questions? The key is to encourage curiosity and continually look for ways to measure outcomes. For example, if the research question is "How can we spot whether a valuable employee is at risk of leaving the organization?" your job is to have the person who posed the question discern which measurable indicators could be used as clues. They need to articulate what it means to be a valuable employee and how that aspect can be measured. They also might need to identify early signs that someone may be considering leaving based on past experiences and determine how that data can be collected.

When searching for ways to explore the data and find solutions, there are no "wrong" questions to ask. In fact, asking without boundaries often inspires creative thought that can lead to new approaches. In this way, the freedom to think outside the box can be an advantage—not a weakness—of limited domain knowledge.

Example 12.1.3

Strengthening Research Questions

The sales and marketing team has proposed the following research question.

> How can we optimize the retail display area per location?

Identify questions that a data scientist who is unfamiliar with the specific domain knowledge of the sales and marketing department can ask the team to help strengthen their research process.

Solution

Without specific domain knowledge, a data scientist will need to ask questions that encourage the team to characterize what constitutes an "optimized retail display" and the key indicators involved in measuring it. You might begin with questions that determine how the sales and marketing team gauges retail display currently. Remember, you are looking for data that can be gathered and studied. There are no wrong questions in this exploratory stage. Questions may include the following.

- How much space is currently used for retail display?
- How much merchandise is currently displayed in the space?
- What proportion of each display is unused currently?
- How often are displays changed or updated?
- Where is the retail display in the store?
- What are current sales at each retail location?

As you help structure the appropriate question to investigate, it's important to remember that it's not your job as the data scientist to avoid the truth of the data. The truth isn't always pretty or what people expect. As a data science, preparing your audience at the *beginning* of the project will help with their understanding of the results. You can do this by stating up front that you don't know what the outcome will be and that the insights may not be what the audience expects.

12.1   Exercises

1. _____ is the understanding of the particular field of interest for the question being investigated.

2. When forming the questions to be answered, it is important to avoid subjective _____.

3. True or False: When working through the steps of the scientific method for data science, you cannot backtrack to a previous step.

4. True or False: In data science, a broad and less-defined question will yield more useful results than a narrow and well-defined question.

🚀 APPLICATIONS

5. As part of a research project, you need to analyze the effects of caffeine on the productivity of nightshift workers. Answer the following questions to outline how you would work through the scientific method for data science.

   a. What question can you pose to start your investigation?

   b. Where might you find the data related to your investigation?

   c. What are some ways you might explore the data to notice patterns or trends?

   d. What are some ways you can model the data once it is cleaned up?

   e. What might be an engaging way to communicate the results given the topic of the project?

6. As part of a research project, you need to help a local restaurant determine what type of weekly special will increase sales. Answer the following questions to outline how you would work through the scientific method for data science.

   a. What question can you pose to start your investigation?

   b. Where might you find the data related to your investigation?

   c. What are some ways you might explore the data to notice patterns or trends?

   d. What are some ways you can model the data once it is cleaned up?

   e. What might be an engaging way to communicate the results given the topic of the project?

7. As part of a research project, you need to determine the minimum inventory a store should keep in stock to be prepared for Thanksgiving holiday shopping. Answer the following questions to outline how you would work through the scientific method for data science.

   a. What question can you pose to start your investigation?

   b. Where might you find the data related to your investigation?

   c. What are some ways you might explore the data to notice patterns or trends?

d. What are some ways you can model the data once it is cleaned up?

e. What might be an engaging way to communicate the results given the topic of the project?

Identify a weakness in each research question. Then, give three stronger alternative questions.

8. When is the best time to post on social media?

9. How should the next iteration of the flagship smartphone be updated?

10. How does the consumption of aspartame affect overall health in comparison to the consumption of sugar?

11. Is it better to brush your teeth before flossing or better to floss before brushing your teeth?

12. Will an afterschool tutoring program improve student performance?

13. Does meditating improve focus?

For each situation, identify three questions that a data scientist who is unfamiliar with the specific domain knowledge indicated can ask their team to help strengthen their research process.

14. A sales and marketing department want to know how they can improve packaging to make a low-selling product look more enticing to customers.

15. The owner of a bakery wants to know the best way to increase sales of beverages that are sold along with the baked goods.

16. A nonfiction book publisher wants to know if they should extend their product lines into historical nonfiction books.

17. The human resources department of a company is interested in learning what steps can be taken to increase employee retention.

18. A small business wants to know how to decrease the time spent processing orders for shipment.

19. The owner of a museum wants to know the best way to redesign the ticket lobby so there is less congestion between the ticket lines and the entrance and exit of the museum.

OBJECTIVES

1. Apply methods of data acquisition.

2. Apply methods of data cleaning.

3. Apply methods of data parsing.

OBJECTIVE 1 Data Acquisition

Seventy percent of the work a data scientist does involves acquiring large amounts of "raw" data, then cleaning it, organizing it, and transforming it to make the data ready for analysis. This work is often picturesquely called *wrangling with data*. Data wrangling, also referred to as *data munging*, makes the data more user-friendly and thus more valuable. The process of data wrangling often requires some application of computer programing knowledge, including the skills needed to obtain the data—everything from downloading files to scraping websites for data. It also involves using the programming language R or statistical computing software programs such as Microsoft Excel and gretl. On the more advanced end of data wrangling, fluency in computer programming languages such as Python, Julia, or SQL is needed. Regardless of the skills used to transform raw data into useable data, this component of data science ensures that quality data exists for analysis in the statistical phase.

All of this may seem intimidating to a beginner in the field of data science. However, knowledge of computer languages and advanced data collecting techniques isn't required in this course. For our purposes, we'll cover simple methods of acquiring large amounts of data and explore how to clean this data with Excel.

> **📖 DATA WRANGLING**
>
> **Data wrangling**, or *data munging*, is the process of acquiring raw data, cleaning it, organizing it, and transforming it so that the data is user-friendly.

Acquiring data can happen in a number of ways.

1. Data can be downloaded directly from a source in a variety of formats, such as Excel files, text files, or zip files.

2. Data can be acquired with the assistance of an Application Programming Interface, or API. These are the middleman software packages that allow two applications to "talk" with one another. For instance, an API is employed every time you click the share icon or link in an article to share it to your preferred social media app. APIs also allow companies like Yelp to locate nearby restaurants or businesses. Any time a machine or system wants to interact with a different machine or system, an API is used.

3. Data can be acquired by *scraping* it from a web page. This refers to the process of writing code to pull information directly from a particular website, or even multiple websites, when it is not in a format that can be easily downloaded.

4. Data can be acquired in formats that require some work before we can manipulate it. Files like PDFs (Portable Document Format) or JPEGs (Joint Photographic Experts Group) are more like pictures of information and not the raw information itself. Pulling data from these types of files requires the use of applications that can export the data for you.

Our focus of data acquisition will be on one of the most widely used methods—downloading files in a CSV format. CSV stands for Comma-Separated Values. CSV files are commonly used to export complex data from one application and import it into another application. And just as the name suggests, the data is separated with commas (or sometimes other characters), and each row represents a data record. These types of files are similar to spreadsheets in that they store data in rows, but unlike a spreadsheet, the data look similar to a text file and cannot be manipulated with formulas. For example, suppose you have a list of students enrolled in this course and this list includes their student ID, classification, and email. Exported as a CSV file, your list would look similar to Figure 12.2.1.

```
File   Edit   Format   View   Help
Name,Student ID,Classification,Email
Victor Regan,33823,Sophomore,vregan@example.edu
Leon Travis,23453,Sophomore,ltravis@example.edu
Megan Smith,34224,Freshman,msmith@example.edu
```

Figure 12.2.1: Enrolled Student CSV File

While CSV files all have elements in common, some may be more complicated with thousands of lines, more entries on each line, or long strings of text. Some CSV files may not even have the headers at the top to identify the data, and some may use quotation marks to surround each piece of data. However, the basic format is the same in its simplicity. Figure 12.2.1 shows a CSV file in a plain text editor program such as Notepad, TextEdit, Evernote, or Atom. These types of editors show exactly what is contained in the CSV file, line by line. However, manipulating the data in this form is not feasible. We need to import the data into a different program in order to manipulate it.

Spreadsheet programs such as Excel are typically able to open the contents of a CSV file as if it were a spreadsheet, sorting the data into columns using the commas as a guide. All you have to do is open the file using Excel or right-click on the file and select Open With > Excel. Figure 12.2.2 shows a CSV file when opened in a plain text editor versus Excel.

```
File   Edit   Format   View   Help
Number,last_name,first_name,joined,state,email
1,Davis,Allie,20711022,OK,govmibrew@amo.et
2,Guzman,Etta,20570327,MT,ke@sauje.gy
3,Reyes,Trevor,20800610,IN,hebwah@jahtu.bz
4,McKenzie,Cecilia,20540620,VA,oksov@hatow.pm
5,Peters,Adam,21150818,TN,ibrev@efucelen.sl
6,Kelly,Lydia,20340522,DC,cibupfi@takownu.ss
7,Fields,Ellen,20360710,ME,bude@puekpi.ly
8,Harris,Nelle,20950629,WA,ofogub@vezerge.bs
9,Figueroa,Marguerite,20631126,NV,jehehduv@sepub.kh
10,Roberson,Ray,21200918,ME,uzudilu@hoos.ga
11,Gill,Benjamin,20400806,VT,lihobo@co.bg
12,Davis,Bryan,20460810,IL,eszavboz@eri.vi
13,Chambers,Ryan,20221101,IL,unucif@opca.is
14,Webb,Adrian,20920708,NY,vo@diwebawem.jm
15,Sanchez,Owen,20660913,OR,icezaku@leh.am
```

	A	B	C	D	E	F
1	Number	last_name	first_name	joined	state	email
2	1	Davis	Allie	20711022	OK	govmibrew@amo.et
3	2	Guzman	Etta	20570327	MT	ke@sauje.gy
4	3	Reyes	Trevor	20800610	IN	hebwah@jahtu.bz
5	4	McKenzie	Cecilia	20540620	VA	oksov@hatow.pm
6	5	Peters	Adam	21150818	TN	ibrev@efucelen.sl
7	6	Kelly	Lydia	20340522	DC	cibupfi@takownu.ss
8	7	Fields	Ellen	20360710	ME	bude@puekpi.ly
9	8	Harris	Nelle	20950629	WA	ofogub@vezerge.bs
10	9	Figueroa	Marguerite	20631126	NV	jehehduv@sepub.kh
11	10	Roberson	Ray	21200918	ME	uzudilu@hoos.ga
12	11	Gill	Benjamin	20400806	VT	lihobo@co.bg
13	12	Davis	Bryan	20460810	IL	eszavboz@eri.vi
14	13	Chambers	Ryan	20221101	IL	unucif@opca.is

Figure 12.2.2: CSV File Opened in a Plain Text Editor and a Spreadsheet

However, sometimes you need to copy and paste the raw data into a spreadsheet. To split up the data, the **Text to Columns** tool in Excel can be used to separate the data into columns based on certain criteria. Example 12.2.1 illustrates how to use this command.

Example 12.2.1

Importing a CSV file into Microsoft

Figure 12.2.3 shows a sample of raw data containing information from a police department regarding traffic accidents. When you click on the CSV link on the county's website, the data is displayed in the following format. Import the data into Microsoft Excel, separating each detail into its own column. (The full data set is available for download from math.hawkeslearning.com.)

```
Master Record Number,Year,Month,Day,Weekend?,Hour,Collision Type,Injury Type,Primary Factor,Reported_Location
902273125,2022,1,1,Weekday,0,2-Car,No injury,OTHER (DRIVER),1ST & MAIN
902305092,2022,1,2,Weekend,1500,2-Car,No injury,FOLLOWING TOO CLOSELY,2ND & COLLEGE
902328674,2022,1,4,Weekday,2300,2-Car,Non-incapacitating,DISREGARD SIGNAL/REG SIGN,WOOD & BLOOMFIELD
902342229,2022,1,5,Weekend,900,2-Car,Non-incapacitating,FAILURE TO YIELD RIGHT OF WAY,JACOBS & PEACHTREE
902331335,2022,1,4,Weekend,1100,2-Car,Non-incapacitating,FAILURE TO YIELD RIGHT OF WAY,W 3RD
902300774,2022,1,4,Weekend,1800,1-Car,No injury,DRIVER DISTRACTED,9TH AND WALNUT
902333574,2022,1,2,Weekend,1200,2-Car,No injury,ENGINE FAILURE OR DEFECTIVE,S CURIN PIKE
902349710,2022,1,6,Weekend,1400,1-Car,Incapacitating,FOLLOWING TOO CLOSELY,N LONDON RD
902319017,2022,1,1,Weekend,1600,2-Car,No injury,RAN OFF ROAD RIGHT,LIBERTY & W 3RD
902306318,2022,1,6,Weekday,1500,2-Car,No injury,UNSAFE BACKING,PATTERSON & W 3RD
902360989,2022,1,1,Weekend,1600,2-Car,No injury,FAILURE TO YIELD RIGHT OF WAY,S PEIDMONT
902321293,2022,1,3,Weekend,1700,1-Car,No injury,UNSAFE BACKING,E 6TH & MAIN
902273032,2022,1,2,Weekend,1800,1-Car,No injury,FAILURE TO YIELD RIGHT OF WAY,HEATHER & COLIN SPRINGS
902353266,2022,1,9,Weekend,1300,3+ Cars, ,Incapacitating,RAN OFF ROAD RIGHT,W PATTERSON & LILY
902310869,2022,1,9,Weekday,1500,2-Car,No injury,DISREGARD SIGNAL/REG SIGN,S WALNUT & 42ND
902319881,2022,1,4,Weekday,1800,1-Car,No injury,FOLLOWING TOO CLOSELY,17TH & MONROE
902348486,2022,1,9,Weekday,1800,2-Car,No injury,ROADWAY SURFACE CONDITION,N GRANT
902275376,2022,1,3,Weekday,1600,2-Car,No injury,FOLLOWING TOO CLOSELY,SW PEIDMONT
902288052,2022,1,2,Weekday,1600,1-Car,No injury,RAN RED LIGHT,N GRANT & HOUSTON
902285073,2022,1,2,Weekend,1800,1-Car,No injury,RAN OFF ROAD RIGHT,DILLMAN & SR37S
902282788,2022,1,8,Weekend,1200,2-Car,No injury,RAN OFF ROAD RIGHT,S CURIN PIKE
902273726,2022,1,9,Weekend,700,2-Car,No injury,FOLLOWING TOO CLOSELY,446 & S MAIN
902353669,2022,1,6,Weekday,700,2-Car,No injury,FAILURE TO YIELD RIGHT OF WAY,CHAPEL HILL & PINETREE
902264082,2022,1,7,Weekday,800,1-Car,No injury,SPEED TOO FAST FOR WEATHER CONDITIONS,DALE & PEACHTREE DR
```

Figure 12.2.3: Raw Data for County Police Department

Solution

We know that the data appears in comma-separated form since we downloaded a CSV file. To put it in a usable format, we need to copy the data and paste it in cell A1 of a new Excel spreadsheet.

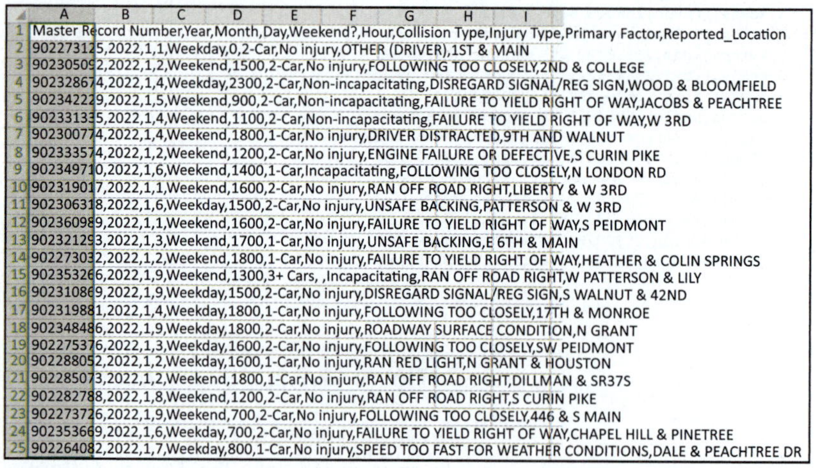

	A
1	Master Record Number,Year,Month,Day,Weekend?,Hour,Collision Type,Injury Type,Primary Factor,Reported_Location
2	902273125,2022,1,1,Weekday,0,2-Car,No injury,OTHER (DRIVER),1ST & MAIN
3	902305092,2022,1,2,Weekend,1500,2-Car,No injury,FOLLOWING TOO CLOSELY,2ND & COLLEGE
4	902328674,2022,1,4,Weekday,2300,2-Car,Non-incapacitating,DISREGARD SIGNAL/REG SIGN,WOOD & BLOOMFIELD
5	902342229,2022,1,5,Weekend,900,2-Car,Non-incapacitating,FAILURE TO YIELD RIGHT OF WAY,JACOBS & PEACHTREE
6	902331335,2022,1,4,Weekend,1100,2-Car,Non-incapacitating,FAILURE TO YIELD RIGHT OF WAY,W 3RD
7	902300774,2022,1,4,Weekend,1800,1-Car,No injury,DRIVER DISTRACTED,9TH AND WALNUT
8	902333574,2022,1,2,Weekend,1200,2-Car,No injury,ENGINE FAILURE OR DEFECTIVE,S CURIN PIKE
9	902349710,2022,1,6,Weekend,1400,1-Car,Incapacitating,FOLLOWING TOO CLOSELY,N LONDON RD
10	902319017,2022,1,1,Weekend,1600,2-Car,No injury,RAN OFF ROAD RIGHT,LIBERTY & W 3RD
11	902306318,2022,1,6,Weekday,1500,2-Car,No injury,UNSAFE BACKING,PATTERSON & W 3RD
12	902360989,2022,1,1,Weekend,1600,2-Car,No injury,FAILURE TO YIELD RIGHT OF WAY,S PEIDMONT
13	902321293,2022,1,3,Weekend,1700,1-Car,No injury,UNSAFE BACKING,E 6TH & MAIN
14	902273032,2022,1,2,Weekend,1800,1-Car,No injury,FAILURE TO YIELD RIGHT OF WAY,HEATHER & COLIN SPRINGS
15	902353266,2022,1,9,Weekend,1300,3+ Cars, ,Incapacitating,RAN OFF ROAD RIGHT,W PATTERSON & LILY
16	902310869,2022,1,9,Weekday,1500,2-Car,No injury,DISREGARD SIGNAL/REG SIGN,S WALNUT & 42ND
17	902319881,2022,1,4,Weekday,1800,1-Car,No injury,FOLLOWING TOO CLOSELY,17TH & MONROE
18	902348486,2022,1,9,Weekday,1800,2-Car,No injury,ROADWAY SURFACE CONDITION,N GRANT
19	902275376,2022,1,3,Weekday,1600,2-Car,No injury,FOLLOWING TOO CLOSELY,SW PEIDMONT
20	902288052,2022,1,2,Weekday,1600,1-Car,No injury,RAN RED LIGHT,N GRANT & HOUSTON
21	902285073,2022,1,2,Weekend,1800,1-Car,No injury,RAN OFF ROAD RIGHT,DILLMAN & SR37S
22	902282788,2022,1,8,Weekend,1200,2-Car,No injury,RAN OFF ROAD RIGHT,S CURIN PIKE
23	902273726,2022,1,9,Weekend,700,2-Car,No injury,FOLLOWING TOO CLOSELY,446 & S MAIN
24	902353669,2022,1,6,Weekday,700,2-Car,No injury,FAILURE TO YIELD RIGHT OF WAY,CHAPEL HILL & PINETREE
25	902264082,2022,1,7,Weekday,800,1-Car,No injury,SPEED TOO FAST FOR WEATHER CONDITIONS,DALE & PEACHTREE DR

Figure 12.2.4: CSV Police Data Imported into Excel

Notice that although the lines of characters appear to run across multiple columns, the data is actually all in the first cell of each row. We need to distribute the data into columns.

First, select the cells that you wish to distribute. Then, under the Data menu, click Text to Columns. In the pop-up window that appears, select Delimited and then click Next.

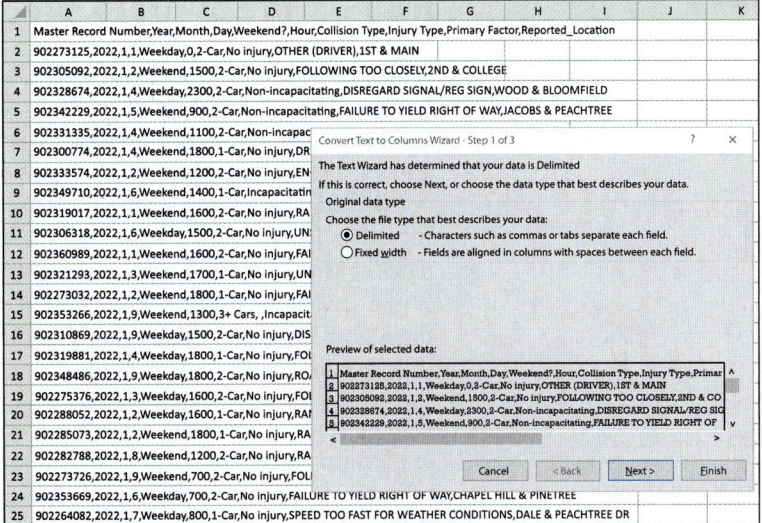

Figure 12.2.5

In the window that appears, check the box next to Comma as the delimiter and make sure all other boxes are unchecked. (Note that if your data file used a different separator character, a different character would be selected as the delimiter.) Click Next again. If you need the data to appear in specific columns, adjust the Destination field; we will keep this as it is. Click Finish. The spreadsheet then appears as shown in Figure 12.2.6.

Master Record Number	Year	Month	Day	Weekend?	Hour	Collision Type	Injury Type	Primary Factor	Reported_Location
902273125	2022	1	1	Weekday	0	2-Car	No injury	OTHER (DRIVER)	1ST & MAIN
902305092	2022	1	2	Weekend	1500	2-Car	No injury	FOLLOWING TOO CLOSELY	2ND & COLLEGE
902328674	2022	1	4	Weekday	2300	2-Car	Non-incapacitating	DISREGARD SIGNAL/REG SIGN	WOOD & BLOOMFIELD
902342229	2022	1	5	Weekend	900	2-Car	Non-incapacitating	FAILURE TO YIELD RIGHT OF WAY	JACOBS & PEACHTREE
902331335	2022	1	4	Weekend	1100	2-Car	Non-incapacitating	FAILURE TO YIELD RIGHT OF WAY	W 3RD
902300774	2022	1	4	Weekend	1800	1-Car	No injury	DRIVER DISTRACTED	9TH AND WALNUT
902333574	2022	1	2	Weekend	1200	2-Car	No injury	ENGINE FAILURE OR DEFECTIVE	S CURIN PIKE
902349710	2022	1	6	Weekend	1400	1-Car	Incapacitating	FOLLOWING TOO CLOSELY	N LONDON RD
902319017	2022	1	1	Weekend	1600	2-Car	No injury	RAN OFF ROAD RIGHT	LIBERTY & W 3RD
902306318	2022	1	6	Weekday	1500	2-Car	No injury	UNSAFE BACKING	PATTERSON & W 3RD
902360989	2022	1	1	Weekend	1600	2-Car	No injury	FAILURE TO YIELD RIGHT OF WAY	S PEIDMONT
902321293	2022	1	3	Weekend	1700	1-Car	No injury	UNSAFE BACKING	E 6TH & MAIN
902273032	2022	1	2	Weekend	1800	1-Car	No injury	FAILURE TO YIELD RIGHT OF WAY	HEATHER & COLIN SPRINGS
902353266	2022	1	9	Weekend	1300	3+Car	Incapacitating	RAN OFF ROAD RIGHT	W PATTERSON & LILY
902310869	2022	1	9	Weekday	1500	2-Car	No injury	DISREGARD SIGNAL/REG SIGN	S WALNUT & 42ND
902319881	2022	1	4	Weekday	1800	1-Car	No injury	FOLLOWING TOO CLOSELY	17TH & MONROE
902348486	2022	1	9	Weekday	1800	2-Car	No injury	ROADWAY SURFACE CONDITION	N GRANT
902275376	2022	1	3	Weekday	1600	2-Car	No injury	FOLLOWING TOO CLOSELY	SW PEIDMONT
902288052	2022	1	2	Weekday	1600	1-Car	No injury		N GRANT & HOUSTON
902285073	2022	1	2	Weekend	1800	1-Car	No injury	RAN OFF ROAD RIGHT	DILLMAN & SR37S
902282788	2022	1	8	Weekend	1200	2-Car	No injury	RAN OFF ROAD RIGHT	S CURIN PIKE
902273726	2022	1	9	Weekend	700	2-Car	No injury	FOLLOWING TOO CLOSELY	446 & S MAIN
902353669	2022	1	6	Weekday	700	2-Car	No injury	FAILURE TO YIELD RIGHT OF WAY	CHAPEL HILL & PINETREE
902264082	2022	1	7	Weekday	800	1-Car	No injury	SPEED TOO FAST FOR WEATHER C	DALE & PEACHTREE DR

Figure 12.2.6

One important feature to note is that the Text to Columns option in Excel deletes the delimiter character once it separates the data. Notice that the commas are no longer part of the traffic data in the spreadsheet.

✔ **Skill Check 12.2.1**

Determine how many columns Excel will use to distribute the following when the Text to Columns feature is used.

```
LF 22-003,FULERTON COUNTY FIRE PROTECTION NO. 8 PROPOSITION NO. 1,46,1,7,Total,NP,Registered Voters,464
LF 22-003,FULERTON COUNTY FIRE PROTECTION NO. 8 PROPOSITION NO. 1,46,1,7,Total,NP,Times Counted,242
LF 22-003,FULERTON COUNTY FIRE PROTECTION NO. 8 PROPOSITION NO. 1,46,1,7,Total,NP,Times Under Voted,0
LF 22-003,FULERTON COUNTY FIRE PROTECTION NO. 8 PROPOSITION NO. 1,46,1,7,Total,<NP>,Yes,195
LF 22-003,FULERTON COUNTY FIRE PROTECTION NO. 8 PROPOSITION NO. 1,46,1,7,Total,<NP>,No,47
LF 22-003,FULERTON COUNTY FIRE PROTECTION NO. 8 PROPOSITION NO. 1,46,1,7,Total,NP,Times Over Voted,0
```

Figure 12.2.7

OBJECTIVE 2 Data Cleaning

Once data has been acquired, it's time to clean it up. Rarely are we presented with flawless data, especially when working with constantly evolving information. The fact is that data is messy. And by messy, we mean that it doesn't meet the required structure to be able to perform analysis on it. For example, a dataset would be considered messy if it has missing entries or if the entries need to be standardized; for example, if "United States" is spelled out when it should be "US" or if the dates consist of 6 digits when they should be 8 digits long. It's not that one form is correct and the other is wrong, it's that we need data to be consistent. Because there is no one standard for data structure, we rely on project-level specifications to determine which portion of the data needs to be cleaned.

Data cleaning is important because a computer program will follow our instructions to perform calculations, but it won't try to interpret information and make sense of it. Since computers can't yet study the world on their own, we must describe to them exactly which steps to take. For example, suppose we instruct the computer to count how many times the letter S appears in the following sentence.

Suzy sells seashells down by the seashore.

Should the computer respond eight? Without clear guidelines, the computer might return a count of 1. Technically, there is only one *capital* S in the sentence and that is what we asked for. (Check for yourself.) It's situations like this that we have to be aware of and prevent by confirming project specifications and cleaning up the dataset.

When using spreadsheets to manipulate data, there is some common terminology that we need. Spreadsheet programs refer to a group of characters used as data as a **text string**, or a *string* or just *text*. Text strings can contain any combination of words, letters, numbers, special characters, or spaces. Unlike the human brain, computers read a space as a character. That means it will treat " hi" differently than "hi". Can you see the difference? Having extra spaces appear at the beginning, ending, or in the middle of text strings is considered *messy data*. As we will see in Example 12.2.2, Excel has a **TRIM** function that removes all spaces from a text string except single spaces between words.

📖 TEXT STRING

A **text string**, or *string*, is a group of characters used as data.

Example 12.2.2

Eliminating Spaces in Excel

The text from *Alice's Adventures in Wonderland* by Lewis Carroll has been loaded into an Excel file in column A. Clean up the table of contents by removing any extra spaces in lines 46 through 57. (The full data set is available for download from math.hawkeslearning.com in Microsoft Excel format.)

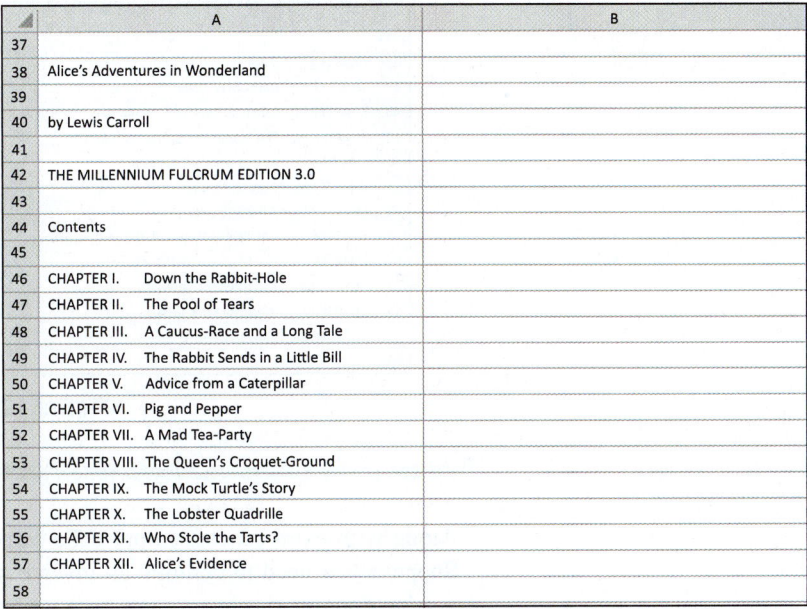

Figure 12.2.8

Solution

We can remove all extra spaces in a cell by using the **TRIM([text])** function. The text argument can be either the actual text to be trimmed or a cell reference to the text you want trimmed.

We will place the trimmed table of contents in column B. In cell B46, type "=TRIM(A46)" and then press Enter. The contents in cell B46 will be the chapter title with only single spaces between words.

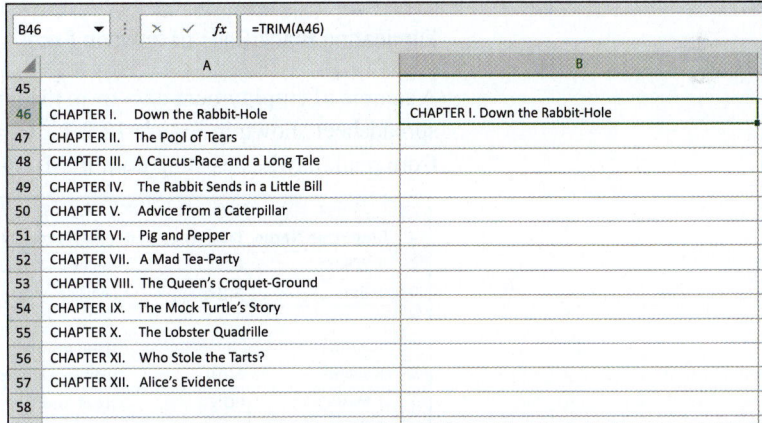

Figure 12.2.9

Next, drag the corner of cell B46 down to copy the formula to trim the extra spaces from the remaining titles. The resulting cleaned up table of contents is shown in Figure 12.2.10.

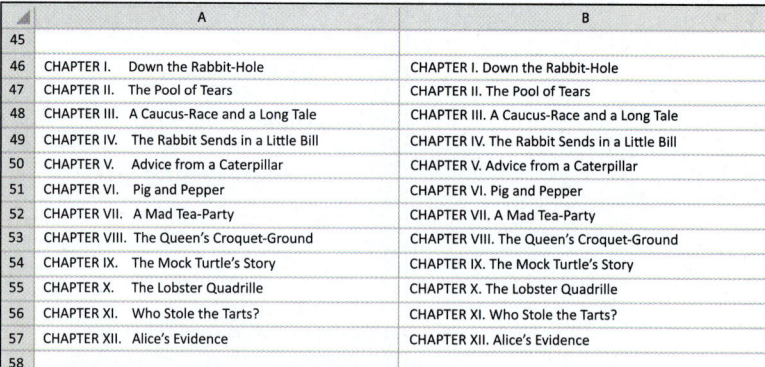

	A	B
45		
46	CHAPTER I. Down the Rabbit-Hole	CHAPTER I. Down the Rabbit-Hole
47	CHAPTER II. The Pool of Tears	CHAPTER II. The Pool of Tears
48	CHAPTER III. A Caucus-Race and a Long Tale	CHAPTER III. A Caucus-Race and a Long Tale
49	CHAPTER IV. The Rabbit Sends in a Little Bill	CHAPTER IV. The Rabbit Sends in a Little Bill
50	CHAPTER V. Advice from a Caterpillar	CHAPTER V. Advice from a Caterpillar
51	CHAPTER VI. Pig and Pepper	CHAPTER VI. Pig and Pepper
52	CHAPTER VII. A Mad Tea-Party	CHAPTER VII. A Mad Tea-Party
53	CHAPTER VIII. The Queen's Croquet-Ground	CHAPTER VIII. The Queen's Croquet-Ground
54	CHAPTER IX. The Mock Turtle's Story	CHAPTER IX. The Mock Turtle's Story
55	CHAPTER X. The Lobster Quadrille	CHAPTER X. The Lobster Quadrille
56	CHAPTER XI. Who Stole the Tarts?	CHAPTER XI. Who Stole the Tarts?
57	CHAPTER XII. Alice's Evidence	CHAPTER XII. Alice's Evidence
58		

Figure 12.2.10

Along with extra spaces, empty cells can also cause problems in data analysis. Suppose you need to compile an inventory based on previous data, but the data you have acquired is incomplete because several cells have been left blank. Filling the empty cells with zeros (assuming this was an oversight) or inserting the phrase "not available" are among the options to help organize and investigate the data set. However, filling in each empty cell individually is potentially very tedious depending on the size of the data set. The **Find and Replace** tool in Excel can simultaneously change any number of cells with one command. In particular, we can use it to find all empty cells in the data and replace their content with "0" or "not available". Example 12.2.3 illustrates the removal of both empty rows and empty cells.

Example 12.2.3

Eliminating Blank Cells or Rows in Excel

A sample of employment data from the human resource department is found in the spreadsheet shown in Figure 12.2.11. (The full data set is available for download from math.hawkeslearning.com in Microsoft Excel format.)

	A	B	C	D	E	
1	**Employee Name**	**Employee Code**	**Employee Designation**	**Employee Team**	**Employee Salary**	
2	J. Williams	GOTA	Software Engineer	Engineering	$ 65,000.00	
3	E. Rice	GOTB	Analyst	Production	$ 62,000.00	
4						
5	L. Smith		Lead Associate	Production		
6	R. Brown	GOTD	Software Engineer	Engineering	$ 64,000.00	
7	J. Messi	GOTE	Lead Associate	Production	$ 66,000.00	
8	P. Neymar	GOTF	Analyst	Production	$ 63,000.00	
9	A Davis	GOTG	Lead Associate	Production	$ 61,000.00	
10	N Martinez		Software Engineer		$ 60,000.00	
11	A. Lopez	GOTI	Analyst	Production	$ 58,000.00	
12	K. Sansa	GOTJ	Software Engineer	Engineering	$ 70,000.00	
13	T. Taylor	GOTK		Production	$ 55,000.00	
14	Moore	GOTL	Analyst	Production	$ 57,000.00	
15						
16	L. Lee	GOTM	Software Engineer	Engineering	$ 59,000.00	
17	M. Moore	GOTN	Lead Associate	Production	$ 68,000.00	
18	M. Perez	GOTO	Business Developer	Engineering	$ 72,000.00	
19						

Figure 12.2.11

Clean up the data set using the following steps.

a. Delete empty rows.

b. Use the Find and Replace tool to enter "Not Available" in each cell that is missing a data entry.

Solution

a. Before deleting any rows, we will begin by numbering the data entries so that we can return to this initial ordering of the data at any point in the wrangling phase. First add a new column to the left of the Employee Name by right-clicking on the column heading A and select "insert" from the menu. In the empty first column, enter the numbers 1 through 17, starting in the second row, next to the first employee's name.

> **TECH TIP**
>
> When inserting new columns in an excel spreadsheet, the new columns will always be inserted to the left. In this case, the new column to the left of column A becomes the new column A, while the existing column A becomes column B.

Next, we will format our data as a table so we can more easily sort it. Select all of the data, starting with the top of new column A and going to the last row of the last column. Click on the Insert tab and select Table. Since your data already includes headers in the first row, make sure that the box for "My table has headers" is checked. Click OK.

	A	B	C	D	E	F
1	Column1	Employee Name	Employee Code	Employee Designation	Employee Team	Employee Salary
2	1	J. Williams	GOTA	Software Engineer	Engineering	$ 65,000.00
3	2	E. Rice	GOTB	Analyst	Production	$ 62,000.00
4	3					
5	4	L. Smith		Lead Associate	Production	
6	5	R. Brown	GOTD	Software Engineer	Engineering	$ 64,000.00
7	6	J. Messi	GOTE	Lead Associate	Production	$ 66,000.00
8	7	P. Neymar	GOTF	Analyst	Production	$ 63,000.00
9	8	A Davis	GOTG	Lead Associate	Production	$ 61,000.00
10	9	N Martinez		Software Engineer		$ 60,000.00
11	10	A. Lopez	GOTI	Analyst	Production	$ 58,000.00
12	11	K. Sansa	GOTJ	Software Engineer	Engineering	$ 70,000.00
13	12	T. Taylor	GOTK		Production	$ 55,000.00
14	13	Moore	GOTL	Analyst	Production	$ 57,000.00
15	14					
16	15	L. Lee	GOTM	Software Engineer	Engineering	$ 59,000.00
17	16	M. Moore	GOTN	Lead Associate	Production	$ 68,000.00
18	17	M. Perez	GOTO	Business Developer	Engineering	$ 72,000.00
19						

Figure 12.2.12: Data in a Table with a Numbered Column

Next, click the drop-down arrow next to "Employee Name" and select "Sort A to Z." Once sorted, the empty cells in this column will appear at the bottom of the data set.

	A	B	C	D	E	F
1	Column1	Employee Name	Employee Code	Employee Designation	Employee Team	Employee Salary
2	8	A Davis	GOTG	Lead Associate	Production	$ 61,000.00
3	10	A. Lopez	GOTI	Analyst	Production	$ 58,000.00
4	2	E. Rice	GOTB	Analyst	Production	$ 62,000.00
5	6	J. Messi	GOTE	Lead Associate	Production	$ 66,000.00
6	1	J. Williams	GOTA	Software Engineer	Engineering	$ 65,000.00
7	11	K. Sansa	GOTJ	Software Engineer	Engineering	$ 70,000.00
8	15	L. Lee	GOTM	Software Engineer	Engineering	$ 59,000.00
9	4	L. Smith		Lead Associate	Production	
10	16	M. Moore	GOTN	Lead Associate	Production	$ 68,000.00
11	17	M. Perez	GOTO	Business Developer	Engineering	$ 72,000.00
12	13	Moore	GOTL	Analyst	Production	$ 57,000.00
13	9	N Martinez		Software Engineer		$ 60,000.00
14	7	P. Neymar	GOTF	Analyst	Production	$ 63,000.00
15	5	R. Brown	GOTD	Software Engineer	Engineering	$ 64,000.00
16	12	T. Taylor	GOTK		Production	$ 55,000.00
17	3					
18	14					
19						

Figure 12.2.13: Sorted Data Set

You will notice that sorting this way has moved the empty rows to the bottom of the data set. Select the empty the rows, right-click anywhere in the selected region, and click Delete. Because we sorted the second column alphabetically, the rows of data are no longer in their original order. If the data needs to be sorted in the original order, you can select the drop-down arrow in our numbering Column1 and select "Sort smallest to largest."

	Column1	Employee Name	Employee Code	Employee Designation	Employee Team	Employee Salary
1						
2	1	J. Williams	GOTA	Software Engineer	Engineering	$ 65,000.00
3	2	E. Rice	GOTB	Analyst	Production	$ 62,000.00
4	4	L. Smith		Lead Associate	Production	
5	5	R. Brown	GOTD	Software Engineer	Engineering	$ 64,000.00
6	6	J. Messi	GOTE	Lead Associate	Production	$ 66,000.00
7	7	P. Neymar	GOTF	Analyst	Production	$ 63,000.00
8	8	A Davis	GOTG	Lead Associate	Production	$ 61,000.00
9	9	N Martinez		Software Engineer		$ 60,000.00
10	10	A. Lopez	GOTI	Analyst	Production	$ 58,000.00
11	11	K. Sansa	GOTJ	Software Engineer	Engineering	$ 70,000.00
12	12	T. Taylor	GOTK		Production	$ 55,000.00
13	13	Moore	GOTL	Analyst	Production	$ 57,000.00
14	15	L. Lee	GOTM	Software Engineer	Engineering	$ 59,000.00
15	16	M. Moore	GOTN	Lead Associate	Production	$ 68,000.00
16	17	M. Perez	GOTO	Business Developer	Engineering	$ 72,000.00
17						

Figure 12.2.14

b. To fill in the remaining empty cells, we will use the Find and Replace tool. First, select the entire table to tell Excel where to look. On the Home tab, in the Editing section, click the Find & Select button. From the drop-down menu, choose Replace…. In the window that appears, leave the "Find what:" field empty and in the "Replace with:" field, type "Not Available", and click Replace All. A new window will appear telling you how many replacements were made. Click OK in that window and click Close in the Find and Replace window.

	Column1	Employee Name	Employee Code	Employee Designation	Employee Team	Employee Salary
1						
2	1	J. Williams	GOTA	Software Engineer	Engineering	$ 65,000.00
3	2	E. Rice	GOTB	Analyst	Production	$ 62,000.00
4	4	L. Smith	Not Available	Lead Associate	Production	Not Available
5	5	R. Brown	GOTD	Software Engineer	Engineering	$ 64,000.00
6	6	J. Messi	GOTE	Lead Associate	Production	$ 66,000.00
7	7	P. Neymar	GOTF	Analyst	Production	$ 63,000.00
8	8	A Davis	GOTG	Lead Associate	Production	$ 61,000.00
9	9	N Martinez	Not Available	Software Engineer	Not Available	$ 60,000.00
10	10	A. Lopez	GOTI	Analyst	Production	$ 58,000.00
11	11	K. Sansa	GOTJ	Software Engineer	Engineering	$ 70,000.00
12	12	T. Taylor	GOTK	Not Available	Production	$ 55,000.00
13	13	Moore	GOTL	Analyst	Production	$ 57,000.00
14	15	L. Lee	GOTM	Software Engineer	Engineering	$ 59,000.00
15	16	M. Moore	GOTN	Lead Associate	Production	$ 68,000.00
16	17	M. Perez	GOTO	Business Developer	Engineering	$ 72,000.00
17						

Figure 12.2.15

A frequent issue with data, whether it is entered manually or by multiple sources, is in its formatting. For instance, when writing the date, some people prefer to use the full year (2022) and others prefer an abbreviated version (22). The same is true for addresses (Drive vs. Dr.) or states (OH vs. Ohio) or days of the week (Tuesday vs. Tues). Again, one form is not necessarily better than the other, but consistency is key when analyzing the data.

In Example 12.2.3, we used the Find and Replace tool in Excel to replace empty cells with a specific common phrase. We can also use this tool to find particular number or text strings and change them to something else. You can search in particular rows, columns, within spreadsheets, or entire workbooks for particular strings you want to change.

Example 12.2.4

Formatting Data

Lauren started a book club and asked members to send their contact information so she could start a group chat. Figure 12.2.16 is a list of the phone numbers in their original format. Use Excel's Find and Replace tool to standardize each phone number so that they are all in the form XXX-XXX-XXXX. (The full data set is available for download from math.hawkeslearning.com in Microsoft Excel format.)

	A	B
1	**Member**	**Phone Number**
2	Lauren	312.555.0179
3	Dominique	773-555-0101
4	Katie	(312) 555-0199
5	Kiyah	708.555.0150
6	Lindsay	639-555-0125
7	Charnelle	847-555-0110
8	Joy	(773) 555-0164
9	Dannetta	(708) 555-0186
10	Cecelia	312 555 0137
11		

Figure 12.2.16

Solution

Notice that some of the phone numbers include parentheses, some have hyphens, and some use periods. First, we will replace any periods with a hyphen. Select all of the phone numbers. On the Home tab, in the Editing section, click the Find & Select button. Choose Replace… from the drop-down menu. In the window that appears, type a period in the "Find what:" field and type a hyphen in the "Replace with:" field. Then click Replace All.

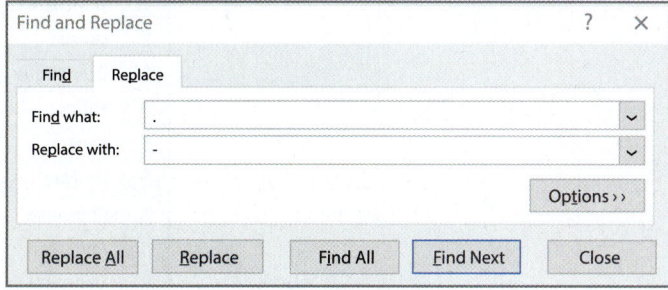

Figure 12.2.17

A pop up will appear alerting you that four replacements were made.

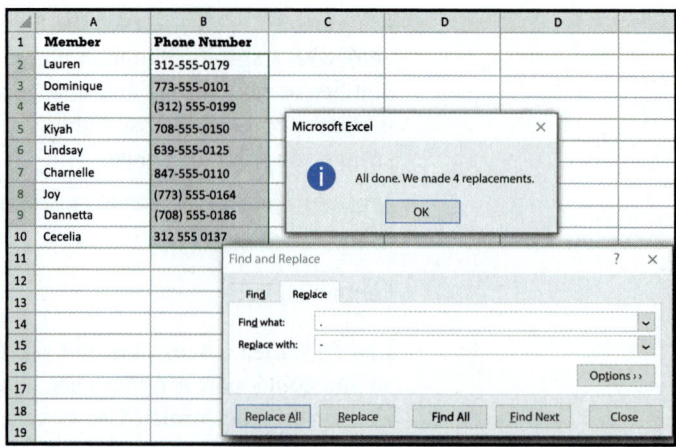

Figure 12.2.18

Next, we will fix the phone numbers that have parentheses around the area code. Since the area codes vary, we will have to use Replace twice: once for the opening parenthesis and once for the closing parenthesis.

Once again, open the Find and Replace tool. In the "Find what:" field, type an opening parenthesis. This time, the "Replace with:" field should be empty. Click Replace All. Repeat the same steps, this time with a closing parenthesis in the "Find what:" field.

Now we need to find any blank spaces between numbers and replace those spaces with hyphens. Using the Find and Replace tool once more, type a single space in the "Find what:" field and a hyphen in the "Replace with:" field. Click Replace All.

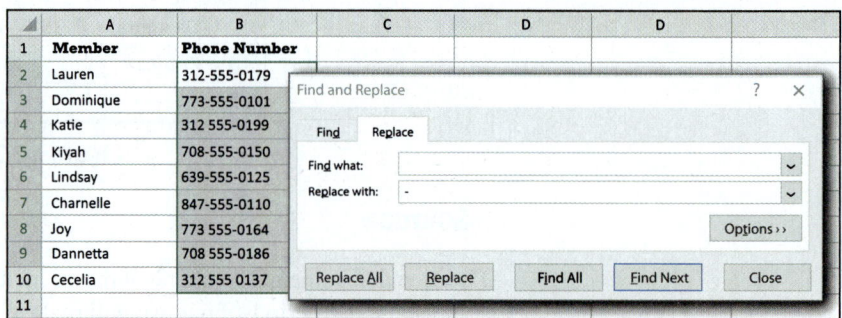

Figure 12.2.19

Now all of the phone numbers have the same format of XXX-XXX-XXXX.

	A	B
1	**Member**	**Phone Number**
2	Lauren	312-555-0179
3	Dominique	773-555-0101
4	Katie	312-555-0199
5	Kiyah	708-555-0150
6	Lindsay	639-555-0125
7	Charnelle	847-555-0110
8	Joy	773-555-0164
9	Dannetta	708-555-0186
10	Cecelia	312-555-0137
11		

Figure 12.2.20

✔ Skill Check 12.2.2

Which of the following data will Excel find if you run the command "Find ?et" and select the "Match case" option?

- bets
- meet
- pet
- etc.

When looking for a specific string using the Find and Replace tool, you can find exact matches by selecting the Match Case option, or you can use **wildcard characters** if you want to look for unknowns or multiple items. Excel considers the question mark (?) and the asterisk (*) to be wildcards, each with a different function. Table 12.2.1 shows the function of each wildcard when you select the Match case option or Match entire cell contents option.

Table 12.2.1: Excel Wildcards

Wildcard	Function	Example
Question mark (?)	Finds any single character.	p?n finds "pen" and "pan".
Asterisk (*)	Finds any number of characters in a row.	p*n finds both "pin" and "pontoon".

The Find and Replace tool can also be used to proactively avoid complications when cleaning data. For instance, notice how the chapter titles in the Alice and Wonderland table of contents, shown in Figure 12.2.21, use a period after the chapter numbers.

38	Alice's Adventures in Wonderland
39	
40	by Lewis Carroll
41	
42	THE MILLENNIUM FULCRUM EDITION 3.0
43	
44	Contents
45	
46	CHAPTER I. Down the Rabbit-Hole
47	CHAPTER II. The Pool of Tears
48	CHAPTER III. A Caucus-Race and a Long Tale
49	CHAPTER IV. The Rabbit Sends in a Little Bill
50	CHAPTER V. Advice from a Caterpillar
51	CHAPTER VI. Pig and Pepper
52	CHAPTER VII. A Mad Tea-Party
53	CHAPTER VIII. The Queen's Croquet-Ground
54	CHAPTER IX. The Mock Turtle's Story
55	CHAPTER X. The Lobster Quadrille
56	CHAPTER XI. Who Stole the Tarts?
57	CHAPTER XII. Alice's Evidence
58	

Figure 12.2.21: Table of Contents from Alice's Adventures in Wonderland

Suppose we wanted to use the Text to Columns tool to separate the chapter number from the chapter title. We could do this by using periods as the delimiter characters. Recall, however, that the Text to Columns tool deletes the delimiter character in each text string, so we will lose the period in the chapter title and create potential problems where the table of contents labels do not exactly match the chapter titles in the rest of the file. To avoid this potential issue, we need to keep the periods after the Roman numerals when applying the Text to Columns tool. By replacing the periods in the table of contents with ".$", we can then use $ as the delimiter. This way, the Text to Columns tool would only delete the $, leaving the original period intact. We'll leave this for you to investigate on your own.

Another common data cleaning technique involves duplicated data within a file. Duplicates are not inherently bad. Useful duplicates can be in the form of multiple sales made by the same salesperson, or an identical model of car released in different years. However, in some cases, a duplicate of a particular data point is undesirable. For instance, suppose there are 142 courses of a particular class taught by 39 instructors, and we want to know instructor-specific attributes, such as the average age or average salary. We can infer from the number of courses versus the number of instructors that some of the instructors taught more than one course. Therefore, the duplicate instructor names need to be removed from the list before we can analyze the data. Excel provides several ways to identify and to delete duplicates in a spreadsheet, as we will see in the next example.

Example 12.2.5

Identifying Duplicates in Excel

The following is a sample of data from a website submission. Users were asked to submit information to enter for a chance to win a drawing. Users were only allowed to be entered in the drawing once. However, some users hit the submit button twice thinking the form had not actually submitted. (The full data set is available for download from math.hawkeslearning.com in Microsoft Excel format.)

a. Highlight any duplicates in the data to determine if duplicates should be removed.

b. Delete any duplicate entries in the data.

	A	B	C	D
1	**Last Name**	**Street Name**	**Zip Code**	**State**
2	Fletcher	3028 W 6th St	30062	GA
3	Ward	84 Rose Circle SW	95492	CA
4	Grisham	4726 Sarah Ln	38655	MS
5	Allgood	427 Camden Drive NE	32765	FL
6	Meares	654 Bloomfield Lane	30215	GA
7	Headlee	190 Elizabeth Street NE	46077	IN
8	Schutz	20 Colonial Circle	30075	GA
9	Dobson	4469 E. Paradise Village Pkwy	37214	TN
10	Edlin	81 Paseo Manuel Girona	10468	NY
11	Jureidini	57 Midway Circle	37217	TN
12	Hutcheson	38 Bunker Hill	78705	TX
13	Hall	4436 Bear Creek Pike	80238	CO
14	Hilson	36 Albee Square Apt 25B	72034	AR
15	Youman	9315 Hwy 6 East	22801	VA
16	Ward	84 Rose Circle SW	95492	CA
17	Meares	654 Bloomfield Lane	30215	GA

Figure 12.2.22

Solution

a. To highlight any duplicates in the data, we will use Conditional Formatting. First, select the data. On the Home tab, click Conditional Formatting. In the drop-down menu, under Highlight Cell Rules, choose Duplicate Values. A pop-up window will appear. Choose Duplicate for the first drop-down menu. For the second drop-down menu, you can choose your preferred option, but "Light Red Fill with Dark Red Text" is the standard choice. Click OK.

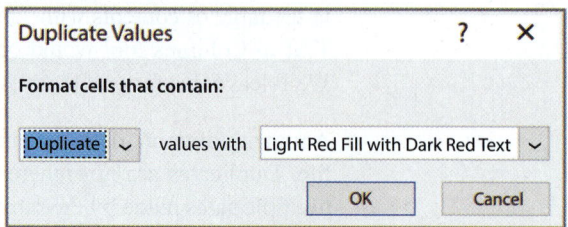

Figure 12.2.23

Notice in Figure 12.2.24 that the Conditional Formatting has flagged both entire rows that are identical as well as individual cells that are repeated within a column. This feature allows us to quickly check the data and confirm that we do want to remove duplicate rows in our data.

	A	B	C	D
1	**Last Name**	**Street Name**	**Zip Code**	**State**
2	Fletcher	3028 W 6th St	30062	GA
3	Ward	84 Rose Circle SW	95492	CA
4	Grisham	4726 Sarah Ln	38655	MS
5	Allgood	427 Camden Drive NE	32765	FL
6	Meares	654 Bloomfield Lane	30215	GA
7	Headlee	190 Elizabeth Street NE	46077	IN
8	Schutz	20 Colonial Circle	30075	GA
9	Dobson	4469 E. Paradise Village Pkwy	37214	TN
10	Edlin	81 Paseo Manuel Girona	10468	NY
11	Jureidini	57 Midway Circle	37217	TN
12	Hutcheson	38 Bunker Hill	78705	TX
13	Hall	4436 Bear Creek Pike	80238	CO
14	Hilson	36 Albee Square Apt 25B	72034	AR
15	Youman	9315 Hwy 6 East	22801	VA
16	Ward	84 Rose Circle SW	95492	CA
17	Meares	654 Bloomfield Lane	30215	GA

Figure 12.2.24

b. To remove duplicate entries, we use the Remove Duplicates tool in Excel. As before, you'll first need to select the data. On the Data tab, in the Data Tools section, select Remove Duplicates. A window will appear in which you will select the columns that might contain duplicates. Ensure all columns are selected and then click OK. An alert window will appear informing you that two duplicates were removed and that 14 unique values remain. Figure 12.2.25 shows the list of submissions with duplicates removed. Notice that a few cells are still highlighted in the state column because several people live in the same state. These duplicates are fine since we were only concerned about entire entries being duplicated.

	A	B	C	D
1	**Last Name**	**Street Name**	**Zip Code**	**State**
2	Fletcher	3028 W 6th St	30062	GA
3	Ward	84 Rose Circle SW	95492	CA
4	Grisham	4726 Sarah Ln	38655	MS
5	Allgood	427 Camden Drive NE	32765	FL
6	Meares	654 Bloomfield Lane	30215	GA
7	Headlee	190 Elizabeth Street NE	46077	IN
8	Schutz	20 Colonial Circle	30075	GA
9	Dobson	4469 E. Paradise Village Pkwy	37214	TN
10	Edlin	81 Paseo Manuel Girona	10468	NY
11	Jureidini	57 Midway Circle	37217	TN
12	Hutcheson	38 Bunker Hill	78705	TX
13	Hall	4436 Bear Creek Pike	80238	CO
14	Hilson	36 Albee Square Apt 25B	72034	AR
15	Youman	9315 Hwy 6 East	22801	VA

Figure 12.2.25

Example 12.2.6

Identifying Issues within Data

Figure 12.2.26 is a partial spreadsheet from several months of data on daily trending YouTube videos in the United States. The data includes the video id, trending date, video title, channel title, tags, and views. Examine the spreadsheet and identify any issues that might need to be addressed before analysis can begin.

	A	B	C	D	E	F
1	video_id	trending_date	title	channel_title	tags	views
2	2kyS6SvSYSE	17.14.11	WE WANT TO TALK ABOUT OUR MARRIAGE	CaseyNeistat	SHANtell martin	748374
3	1ZAPwfrtAFY	17.14.11	The Trump Presidency: Last Week Tonight with John Oliver (HBO)	LastWeekTonight	last week tonight trump presiden	2418783
4	puqaWrEC7tY	17.14.11	Nickelback Lyrics: Real or Fake?	Good Mythical Morning	rhett and link|"gmm"|"good myt	343168
5	d380meD0W0M	17.14.11	I Dare You: GOING BALD!?	nigahiga	ryan|"higa"|"higatv"|"nigahiga"|	2095731
6	gHZ1Qz0KiKM	17.14.11	2 Weeks with iPhone. X	iJustine	ijustine|"week with iPhone X"|"i	119180
7	39idVpFF7NQ	17.14.11	Roy Moore & Jeff Sessions Cold Open - SNL	Saturday Night Live	SNL|"Saturday Night Live"|"SNL S	2103417
8	nc99ccSXST0	17.14.11	5 Ice Cream Gadgets put to the Test	CrazyRussianHacker	5 Ice Cream Gadgets|"Ice Cream"	$817,732.00
9	jr9QtXwC9vc	17.14.11	The Greatest Showman | Official Trailer 2 [HD] | 20th Century FOX	20th Century Fox	Trailer|"Hugh Jackman"|"Michell	826059
10	TUmyygCMMGA	17.14.11	Why the rise of the robots won‚Äôt mean the end of work	Vox	vox.com|"vox"|"explain"|"shift c	256426
11						
12	9wRQljFNDW8	17.14.11	Dion Lewis' 103-Yd Kick Return TD vs. Denver! | Can't-Miss Play | NFL	NFL	NFL|"Football"|"offense"|"defen	81377
13	VifQlJit6A0		(SPOILERS) 'Shiva Saves the Day' Talked About Scene Ep. 804 | The W	amc	The Walking Dead|"shiva"|"tiger	104578
14	5E4ZBSInqUU	17.14.11	Marshmello - Blocks (Official Music Video)	marshmello	marshmello|"blocks"|"marshmel	687582

Figure 12.2.26

Solution

	A	B	C	D	E	F
1	video_id	trending_date	title	channel_title	tags	views
2	2kyS6SvSYSE	17.14.11	WE WANT TO TALK ABOUT OUR MARRIAGE	CaseyNeistat	SHANtell martin	748374
3	1ZAPwfrtAFY	17.14.11	The Trump Presidency: Last Week Tonight with John Oliver (HBO)	LastWeekTonight	last week tonight trump presiden	2418783
4	puqaWrEC7tY	17.14.11	Nickelback Lyrics: Real or Fake?	Good Mythical Morning	rhett and link|"gmm"|"good myt	343168
5	d380meD0W0M	17.14.11	I Dare You: GOING BALD!?	nigahiga	ryan|"higa"|"higatv"|"nigahiga"|	2095731
6	gHZ1Qz0KiKM	17.14.11	2 Weeks with iPhone. X	iJustine	ijustine|"week with iPhone X"|"i	119180
7	39idVpFF7NQ	17.14.11	Roy Moore & Jeff Sessions Cold Open - SNL	Saturday Night Live	SNL|"Saturday Night Live"|"SNL S	2103417
8	nc99ccSXST0	17.14.11	5 Ice Cream Gadgets put to the Test	CrazyRussianHacker	5 Ice Cream Gadgets|"Ice Cream"	$817,732.00
9	jr9QtXwC9vc	17.14.11	The Greatest Showman | Official Trailer 2 [HD] | 20th Century FOX	20th Century Fox	Trailer|"Hugh Jackman"|"Michell	826059
10	TUmyygCMMGA	17.14.11	Why the rise of the robots won‚Äôt mean the end of work	Vox	vox.com|"vox"|"explain"|"shift c	256426
11						
12	9wRQljFNDW8	17.14.11	Dion Lewis' 103-Yd Kick Return TD vs. Denver! | Can't-Miss Play | NFL	NFL	NFL|"Football"|"offense"|"defen	81377
13	VifQlJit6A0		(SPOILERS) 'Shiva Saves the Day' Talked About Scene Ep. 804 | The W	amc	The Walking Dead|"shiva"|"tiger	104578
14	5E4ZBSInqUU	17.14.11	Marshmello - Blocks (Official Music Video)	marshmello	marshmello|"blocks"|"marshmel	687582

Figure 12.2.27

a. Line 11 is an empty line and needs to be deleted.

b. Cell B13 is missing data. Since this is a date and all other dates are the same, the best assumption would be that it is the same date.

c. Cell C4 has extra leading spaces that need to be removed.

d. Cells D6 and D7 also contain leading spaces to be removed.

e. Column E lists any tags the video received. Since the tags are all contained in one cell separated by the | character, we can use the Text to Column feature to separate them into different columns.

f. The number of views in cell F8 is listed as currency and needs to be changed to a general number.

✔ Skill Check 12.2.3

Identify the rows in the following spreadsheet that have potential data issues that need to be resolved.

	A	B	C	D
1	Customer ID	Job	Credit Amount	Risk
2	38476531	2	30062	Low
3	3902j894	1	95492	High
4	37957209	2	38655	Low
5	33875429	3	32765	Low
6	30972548		30215	High
7	31147639	1	46077	Low

Figure 12.2.28

OBJECTIVE 3 Data Parsing

Data wrangling also includes arranging a datafile into a structured format. This process is called *data parsing*. Like many terms, data parsing has several meanings. In data science, it can mean everything from breaking the data into manageable sections to the sophisticated use of artificial intelligence to analyze text. Instead of having to decipher a jumbled mess of values, we can parse the data, making it possible to make sense of it by detecting distinctions and patterns. In fact, we have already done a bit of data parsing at the beginning of this section when we used the Text to Columns tool in Excel. With that tool, we took data and separated it into manageable pieces. In the next few examples, we'll look at how to parse data by pulling specific pieces of information from a string of text, creating tables to quickly isolate certain types of data, and joining two or more data sets together.

Let's begin with extracting specific data from a string of text in a spreadsheet. In Excel, the **MID** function allows you to extract particular information from a text string. For instance, suppose you would like to pull out only the area codes in a set of phone numbers or identify the last names in a list of clients' full names. By using the MID function, you can determine the number of characters that you would like to retrieve from anywhere in the string of text. This command is more nuanced than splitting the data using the Text to Columns feature in that it keeps the data in its original format while pulling out only what you need.

Example 12.2.7

Extracting Text from a String in Excel

Figure 12.2.29 shows the credit card transactions for a kitchen renovation project completed by Jenn's Interior Solutions. For accounting to accurately keep track of spending on the project, both the date and the vender need to be separated into their own columns. (The full data set is available for download from math.hawkeslearning.com in Microsoft Excel format.)

a. Use the MID function to parse out the date for each transaction.

b. Use the MID function along with the FIND tool to identify the vender in each transaction.

◢	A	B	C
1	**Credit Card Transactions**		
2			
3	**Description**	**Date**	**Vendor**
4	11/27/20 THE HOME DEPOT, 6986 ATLANTA GA		
5	11/27/20 Crate&barrel Cb2 Nod, Il		
6	11/28/20 AMZ Mktp Us*12345XYZ, WA		
7	11/30/20 WWW COSTCO COM, WA		
8	12/10/20 ETSY.COM, NEW YORK		

Figure 12.2.29

Solution

a. To parse the date from each description, we will use the **MID(text, start_num, num_chars)** function, where "text" is the text string to be parsed and can either be entered directly or as a cell reference, "start_num" is the position of the first character you want to extract in the text, and "num_chars" specifies the number of characters you want to extract. All three of these values are required in the formula.

In the Description column, the date is the first piece of information provided, which means start_num = 1. Each date has the format DD/MM/YY, which has 8 characters (including the slashes), so num_chars = 8. The cell reference will change for each instance of the formula, but since the first description is contained in cell A4, we'll start with text = A4.

Start by typing "=MID(A4,1,8)" in cell B4 and press Enter.

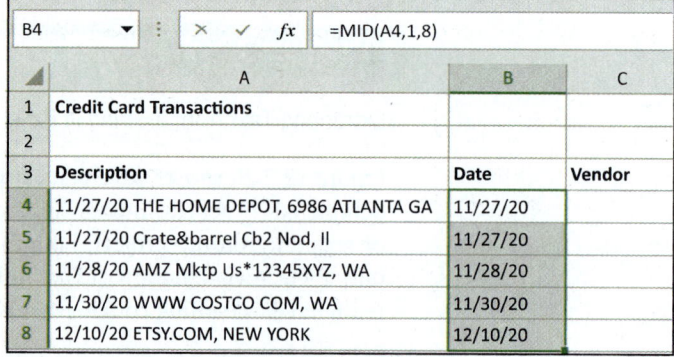

B4	▼	⋮	× ✓ fx	=MID(A4,1,8)		
	A				B	C
1	Credit Card Transactions					
2						
3	Description				Date	Vendor
4	11/27/20 THE HOME DEPOT, 6986 ATLANTA GA				11/27/20	
5	11/27/20 Crate&barrel Cb2 Nod, Il					
6	11/28/20 AMZ Mktp Us*12345XYZ, WA					
7	11/30/20 WWW COSTCO COM, WA					
8	12/10/20 ETSY.COM, NEW YORK					

Figure 12.2.30

To parse the date from the remaining transactions, drag the green square at the bottom-right corner of cell B4 down through cell B8.

B4	▼	⋮	× ✓ fx	=MID(A4,1,8)		
	A				B	C
1	Credit Card Transactions					
2						
3	Description				Date	Vendor
4	11/27/20 THE HOME DEPOT, 6986 ATLANTA GA				11/27/20	
5	11/27/20 Crate&barrel Cb2 Nod, Il				11/27/20	
6	11/28/20 AMZ Mktp Us*12345XYZ, WA				11/28/20	
7	11/30/20 WWW COSTCO COM, WA				11/30/20	
8	12/10/20 ETSY.COM, NEW YORK				12/10/20	

Figure 12.2.31

b. To pull out the vendor, we can once again use the MID function. However, unlike the dates in part a., we cannot guarantee that the name of each vendor contains the same number of characters. Instead of inserting the number of characters to extract, we can tell the computer where to find the end of the string using the **FIND(find_text, within_text, [start_num])** function. In this function, "find_text" is the character or string of text we need to find, "within_text" is the text we are searching in, and "start_num" tells the function where to start. The only optional argument is "start_num," which is considered to be 1 if the value is omitted from the function.

Since each vendor name has a comma after it, we will use the FIND function to locate the comma, thus finding the end of the string that we want to extract. We will use "," for find_text. Again, the cell reference will change for each transaction description, but the first reference is in A4, so we'll use A4 for within_text. We'll omit the value for start_num. So for the first transaction description, we will use "FIND(",",A4)". This function returns the location of the comma in cell A4. Excel returns a value of 24 for FIND(",",A4), which means that the comma is the 24th character in cell A4. We will use this finding as an internal value in the MID function.

We know that the name of the vendor starts on character 10, because the date is 8 characters long and is followed by a space. We now also know that the name of the vendor ends on whatever value is found by FIND(",",A4). So for the MID function, the number of characters we are extracting, or the length of the string we are extracting, will be the value found by FIND(",",A4) minus 10. This means that we will use FIND(",",A4)-10 for the num_chars argument in the MID function and we will use 10 for start_num.

Now we can use the MID function. In cell C4, type "=MID(A4,10,FIND(",",A4)-10)".

C4		× ✓ fx	=MID(A4,10,FIND(",",A4)-10)	
	A		**B**	**C**
1	Credit Card Transactions			
2				
3	Description		Date	Vendor
4	11/27/20 THE HOME DEPOT, 6986 ATLANTA GA		11/27/20	THE HOME DEPOT
5	11/27/20 Crate&barrel Cb2 Nod, Il		11/27/20	
6	11/28/20 AMZ Mktp Us*12345XYZ, WA		11/28/20	
7	11/30/20 WWW COSTCO COM, WA		11/30/20	
8	12/10/20 ETSY.COM, NEW YORK		12/10/20	

Figure 12.2.32

To complete the vendor column, drag the green square at the bottom-right corner of cell B4 down through cell C8.

C4		× ✓ fx	=MID(A4,10,FIND(",",A4)-10)	
	A		**B**	**C**
1	Credit Card Transactions			
2				
3	Description		Date	Vendor
4	11/27/20 THE HOME DEPOT, 6986 ATLANTA GA		11/27/20	THE HOME DEPOT
5	11/27/20 Crate&barrel Cb2 Nod, Il		11/27/20	Crate&barrel Cb2 Nod
6	11/28/20 AMZ Mktp Us*12345XYZ, WA		11/28/20	AMZ Mktp Us*12345XYZ
7	11/30/20 WWW COSTCO COM, WA		11/30/20	WWW COSTCO COM
8	12/10/20 ETSY.COM, NEW YORK		12/10/20	ETSY.COM

Figure 12.2.33

Next, let's look at a technique that can quickly combine and organize data in a spreadsheet into summary calculations. A **pivot table** is a way to group information according to specified characteristics and quickly run summary calculations on the groups, such as finding the sum, count, average, or standard deviation. The table arranges (or *pivots*) the data in the manner you choose.

Suppose you need to analyze rental car inventory and you have the data given in Figure 12.2.34. Instead of trying to sort the data in the current layout, a pivot table can be used to quickly rearrange the table into a more useful format.

For instance, the following pivot table in Figure 12.2.35 uses the style of car as the row labels while the columns indicate the number of cars there are for each model year. Notice you can easily see that there are 12 SUVs—six that are 2021 models and six that are 2022 models.

	A	B	C	D
1		Rental Car Inventory		
2	No. of Seats	Style	Year	Color
3	4	Economy	2022	White
4	5	Sedan	2022	Blue
5	5	Sedan	2021	Black
6	5	Sedan	2022	Red
7	5	SUV	2021	White
8	5	SUV	2021	Red
9	5	Truck	2020	Green
10	7	SUV	2022	Grey
11	5	Sedan	2022	Black
12	5	Sedan	2022	White
13	5	Sedan	2021	White
14	7	SUV	2022	Slate
15	7	Van	2022	Black
16	7	Van	2022	White
17	5	Sedan	2022	White
18	5	Sedan	2022	Black
19	15	Van	2021	Grey
20	5	SUV	2022	Black
21	5	Sedan	2022	Blue
22	8	SUV	2022	Black
23	5	SUV	2022	Sunset
24	5	SUV	2021	Silver
25	4	Sporty Car	2022	Black
26	4	Truck	2022	Black
27	7	SUV	2021	White
28	7	SUV	2022	Silver
29	5	Sedan	2021	White
30	4	Sporty Car	2022	Black
31	5	Sedan	2021	Black
32	5	SUV	2021	Silver
33	5	SUV	2022	White

Figure 12.2.34: Spreadsheet with Rental Car Inventory Data

Count of No. of Seats	Column Labels ▾			
Row Labels ▾	2020	2021	2022	Grand Total
Economy			1	1
Sedan		4	7	11
Sporty Car			2	2
SUV		6	6	12
Truck	1		1	2
Van		1	2	3
Grand Total	**1**	**11**	**19**	**31**

Figure 12.2.35: Pivot Table with Style as Row Labels

Alternatively, the pivot table in Figure 12.2.36 uses the style of car, separated by model year, as the row labels with the number of cars in the column. Notice again that it is just as easy to identify the years of SUV vehicles. However, in this configuration you no longer can just glance to identify how many 2022 vehicles there are in total; you will need to do some addition on your own.

> 💬 **Helpful Hint**
>
> Pivot tables can be thought of as a summary of filtering the original format of the data set. In the pivot tables in Figures 12.2.35 and 12.2.36, the data values in the rows are the count of how many items remain in the No. of Seats column when the original data is filtered by Style and Year. For instance, if the Style column is filtered to "Sedan" and the Year column is filtered to "2021", then 4 items will remain in the No. of Seats column.

Row Labels ▾	Count of No. of Seats
− Economy	1
2022	1
− Sedan	11
2021	4
2022	7
− Sporty Car	2
2022	2
− SUV	12
2021	6
2022	6
− Truck	2
2020	1
2022	1
− Van	3
2021	1
2022	1
Grand Total	**31**

Figure 12.2.36: Pivot Table with Style and Year as Row Labels

While both pivot tables display exactly the same information, depending on your needs, it might be beneficial to use one version instead of the other.

In Excel, a pivot table is created by selecting the data range you want to analyze and then clicking on the PivotTable button in the Insert tab. In the "Create PivotTable" window that appears, you'll choose where you want your pivot table to appear (in the same spreadsheet or a new one). Notice that in Figure 12.2.37, we selected to place the pivot table in cell E3 of the existing spreadsheet.

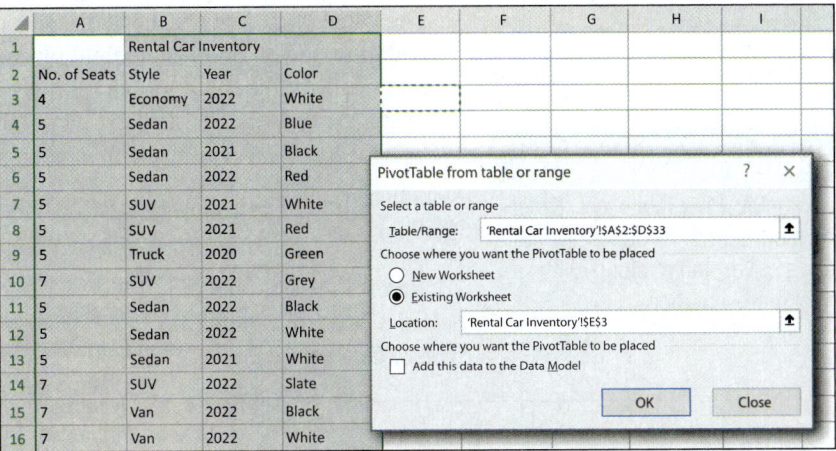

Figure 12.2.37

Once the pivot table spreadsheet opens, the PivotTable Fields pane will open at the side. There are five main areas within the default view of the pane, which are shown in Figure 12.2.38.

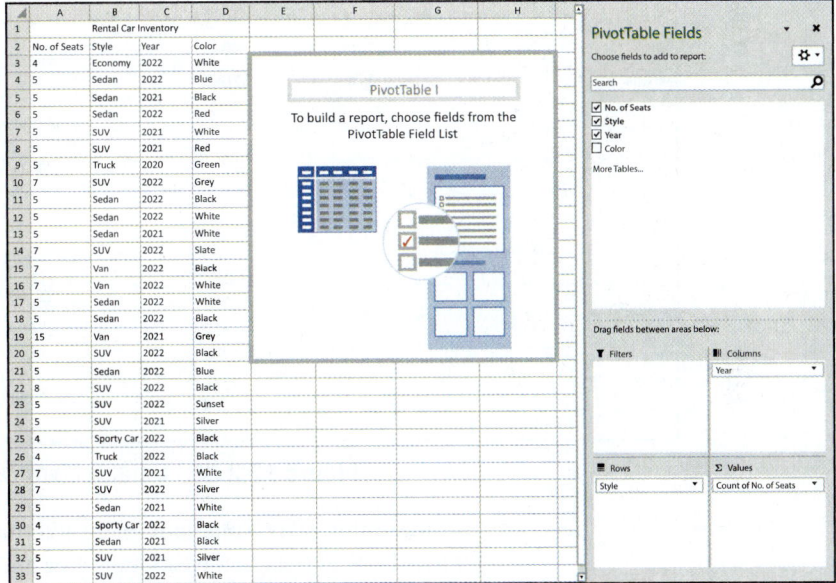

Figure 12.2.38: Spreadsheet with PivotTable Fields

- The top portion of the PivotTable Fields pane contains the list of fields to choose from, which are the column titles of the selected data. Once a field is selected, that field will automatically be moved to one of the four areas depending on the data type: Filters, Columns, Rows, or Values. You can also drag and drop the field names to whichever of these four areas works best for your purposes.

- The Filters area specifies the fields that the pivot table will use as a basis for sorting the data. While it can be left blank, this area provides the ability to filter the results in your table.

- The Columns area is where the fields that the pivot table will use as column labels are listed.

- The Rows area contains the fields that the pivot table will use as row labels. The rows are the main way the pivot table breaks down the data.

📈 **TECH TIP**

Along with the PivotTable tool, Excel has a tool called PivotCharts that creates a pivot table along with a graphical representation.

- The Values area contains the fields that will be used to determine the data values within the pivot table. If the field placed in this area consists of numbers, the pivot table will default to finding the sum of those values. If the field placed in this area consists of text, the pivot table will default to finding the count of those values. Excel allows you to change these defaults if needed.

Note that if no field is placed in the Columns area, then the name of the fields in the Values area will be used as the column labels.

To recreate the pivot table in Figure 12.2.35, we first need to select three fields: No. of Seats, Style, and Year. Excel will place these fields into areas based on the type of data they contain.

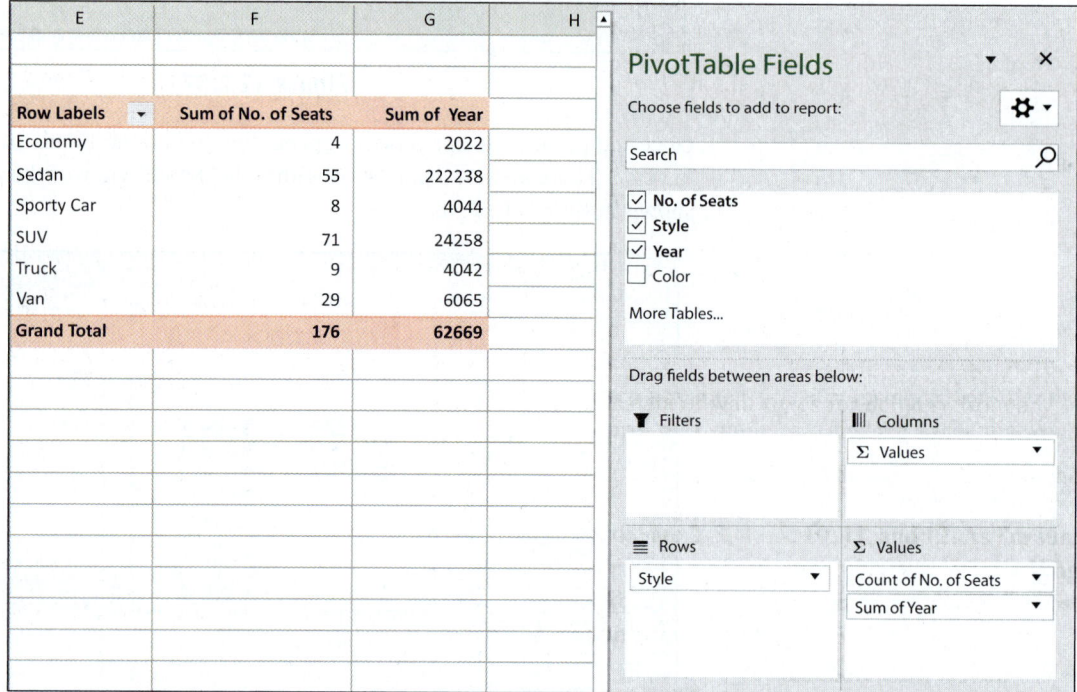

Figure 12.2.39: Default Pivot Table after Selecting Fields

To make this pivot table look more like the one in Figure 12.2.35, we need to move the Year field to the Columns area and make one other change.

Notice that in Figure 12.2.39 the grand total of the Sum of No. of Seats column is 176. This value refers to the total number of seats in all of the vehicles, but we want to refer to the total number of vehicles. The total is not what we're expecting because the Values area is set to find the *sum* of the No. of Seats. We need to change this to the *count* of No. of Seats, or the pivot table will not display the results we want. The default calculation of Sum can be changed by clicking on the drop-down arrow to the right of the field name and selecting Value Field Settings, which brings up the **Value Field Settings** window. In this window, we can select a different field summary. Select Count and click OK.

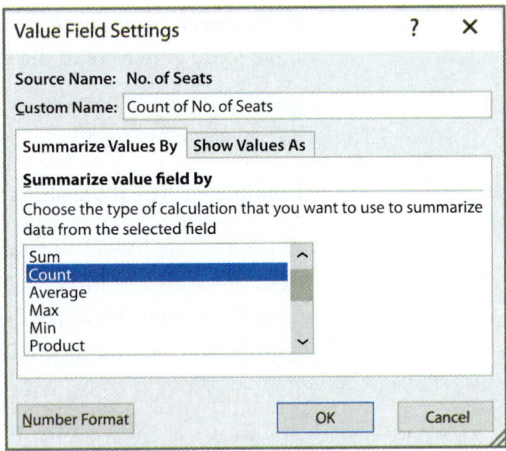

Figure 12.2.40: Value Field Settings Window

Our pivot table now matches the pivot table in Figure 12.2.35.

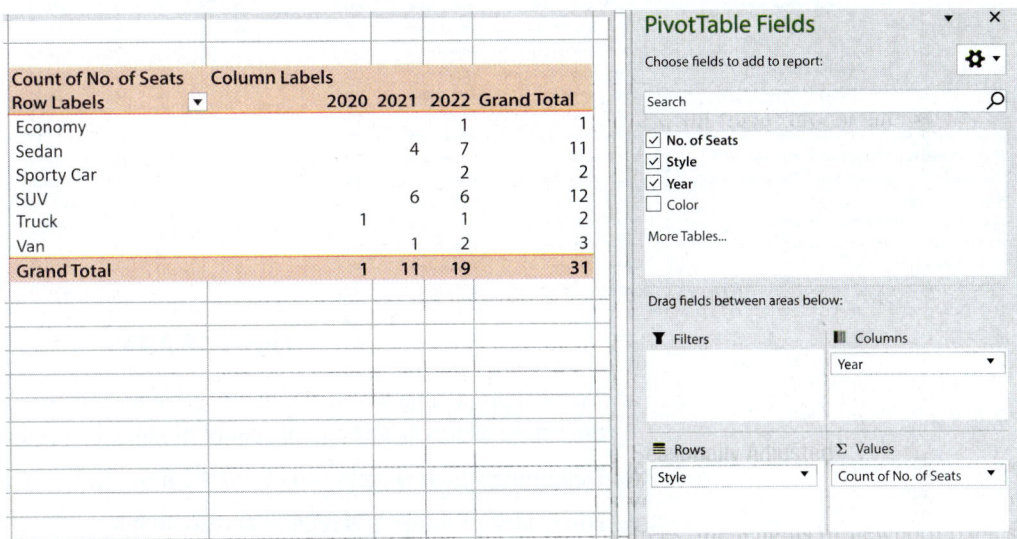

Figure 12.2.41: Pivot Table with PivotTable Fields Panel

To recreate the pivot table in Figure 12.2.36, the only thing we need to do is relocate the Years field to the Rows area.

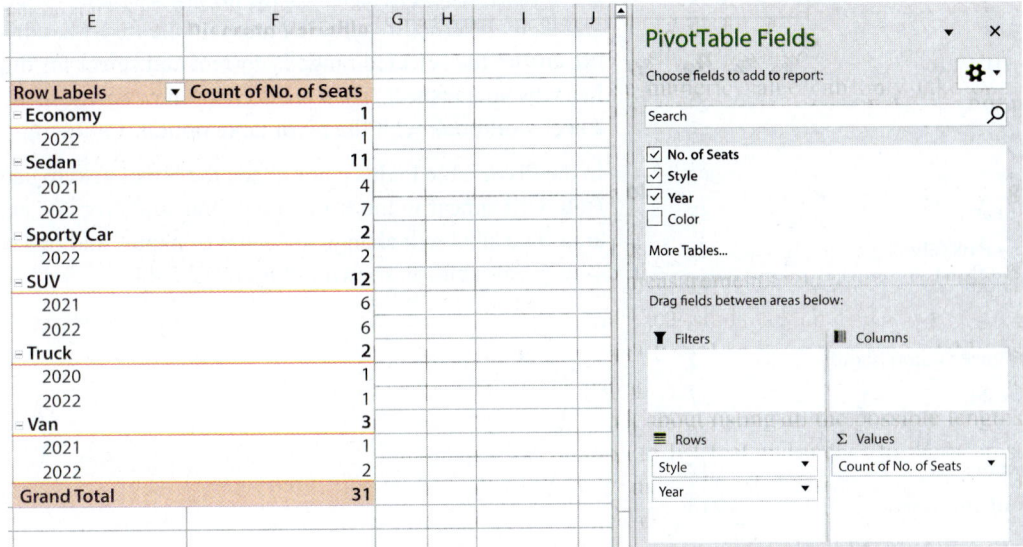

Figure 12.2.42: Pivot Table with PivotTable Fields Panel

If we decide we want the year to be the main row label in the pivot table, all we have to do is rearrange the fields in the Rows area. Try this for yourself.

Example 12.2.8

Creating a Pivot Table in Excel

Airbnb began listing vacation rental properties in 2008. Figure 12.2.43 shows the first 19 listings of 48,879 along with characteristics for accommodations in New York City, New York, for 2019. (The full data set is available for download from math.hawkeslearning.com in Microsoft Excel format.)

	A	B	C	D	E	F	G	H	I	J	K
1	AirBnB ID	name	host_id	neighborhood	room_type	price	minimum_nights	number_of_reviews	last_review	reviews_per_month	availability_365
2	2539	Clean & quiet apt home by the park	2787	Kensington	Private room	149	1	9	10/19/18	0.21	365
3	2595	Skylit Midtown Castle	2845	Midtown	Entire home	225	1	45	5/21/19	0.38	355
4	3647	THE VILLAGE OF HARLEM....NEW YORK!	4632	Harlem	Private room	150	3	0			365
5	3831	Cozy Entire Floor of Brownstone	4869	Clinton Mill	Entire home	89	1	270	7/5/19	4.64	194
6	5022	Entire Apt: Spacious Studio/Loft by central park	7192	East Harlem	Entire home	80	10	9	11/19/18	0.1	0
7	5099	Large Cozy 1 BR Apartment in Midtown East	7322	Murray Hill	Entire home	200	3	74	6/22/19	0.59	129
8	5121	BlissArtsSpace!	7356	Bedford-Stuyvesan	Private room	60	45	49	10/5/17	0.4	0
9	5178	Large Furnished Room Near B'way	8967	Hell's Kitchen	Private room	79	2	430	6/24/19	3.47	220
10	5203	Cozy Clean Guest Room - Family Apt	7490	Upper West Side	Private room	79	2	118	7/21/17	0.99	0
11	5238	Cute & Cozy Lower East Side 1 bdrm	7549	Chinatown	Entire home	150	1	160	6/9/19	1.33	188
12	5295	Beautiful 1br on Upper West Side	7702	Upper West Side	Entire home	135	5	53	6/22/19	0.43	6
13	5441	Central Manhattan/near Broadway	7989	Hell's Kitchen	Private room	85	2	188	6/23/19	1.5	39
14	5803	Lovely Room 1, Garden, Best Area, Legal rental	9744	South Slope	Private room	89	4	167	6/24/19	1.34	314
15	6021	Wonderful Guest Bedroom in Manhattan for SINGLES	11528	Upper West Side	Private room	85	2	113	7/5/19	0.91	333
16	6090	West Village Nest - Superhost	11975	West Village	Entire home	120	90	27	10/31/18	0.22	0
17	6848	Only 2 stops to Manhattan studio	15991	Williamsburg	Entire home	140	2	148	6/29/29	1.2	46
18	7097	Perfect for Your Parents + Garden	17511	Fort Greene	Entire home	215	2	198	6/28/19	1.72	321
19	7322	Chelsea Perfect	18946	Chelsea	Private room	140	1	260	7/1/19	2.12	12
20	7726	Hip Historic Brownstone Apartment with Backyard	20950	Crown Heights	Entire home	99	3	53	6/22/19	4.44	21

Figure 12.2.43

a. Create a pivot table in Excel to summarize the data using the neighborhood of the listing as row labels and the count of name as values.

b. Determine which neighborhood had the most rentals during 2019.

c. How many of the 48,879 rentals were available to rent all 365 days of the year, possibly indicating that they were full-time rentals?

Solution

a. Open the Excel spreadsheet with the Airbnb data. Select the columns of data and then click the PivotTable button on the Insert tab. In the window that pops up, ensure the selected range is correct and select for the pivot table to appear in a new spreadsheet. Then press OK. A new spreadsheet will open in the same Excel workbook with the pivot table options on the right side.

In the PivotTable Fields pane, check the boxes next to name and neighborhood. Both will appear in the Rows area. Drag and drop the name field to the Values area. The label will change to "count of name" and two columns will appear in your pivot table, as shown in Figure 12.2.44.

Row Labels ▼	Count of name
Allerton	42
Arden Heights	4
Arrochar	21
Arverne	77
Astoria	900
Bath Beach	17
Battery Park City	69
Bay Ridge	141
Bay Terrace	6
Bay Terrace, Staten Island	2
Baychester	7
Bayside	39
Bayswater	17
Bedford-Stuyvesant	3713

Figure 12.2.44

b. To find the neighborhood with the most rentals in 2019, we need to sort the Count of name column from smallest to largest. Select any cell in the Count of name column, then go to the Data tab and click the button for Sort Largest to Smallest, as shown in Figure 12.2.45. After sorting, it's clear that Williamsburg had the most rentals in 2019.

Row Labels	Count of name
Williamsburg	3918
Bedford-Stuyvesant	3713
Harlem	2657
Bushwick	2463
Upper West Side	1970
Hell's Kitchen	1957
East Village	1852
Upper East Side	1798
Crown Heights	1564
Midtown	1545
East Harlem	1117
Greenpoint	1115
Chelsea	1113
Lower East Side	911

Figure 12.2.45

c. To determine how many rentals were available to rent all 365 days of the year, we need to add another field to our pivot table. In the PivotTable Fields pane, check the box next to availability_365. This field will be added to the Values area as "sum of availability_365". (You'll notice that a column is added to the table with the sum of the availability of all properties in each neighborhood, which is not helpful!) Drag and drop the availability_365 field to the Filter area. You'll notice a filter appear above the pivot table. Click the drop-down, then type "365" in the search bar, and press Enter. The data will be filtered to only include properties that are available for rent all year long, as shown in Figure 12.2.46.

availability_365	365

Row Labels	Count of name
Hell's Kitchen	91
Bedford-Stuyvesant	87
Midtown	84
Harlem	70
Bushwick	55
East Flatbush	43
Williamsburg	42
Upper West Side	40
Upper East Side	38
Murray Hill	36
Crown Heights	30
Chelsea	28
Financial District	25
East Harlem	22
Greenpoint	22
Flushing	21
Ridgewood	21

Figure 12.2.46

The sales data for the first quarter is shown in the pivot table.

a. Which salesperson had the most sales in each month?

b. Who had the most sales in the first quarter?

Sum of Order Amount	Column Labels			
Row Labels	January	February	March	Grand Total
Albertson, Dan	$ 925.00	$ 1,375.00	$ 350.00	$ 2,650.00
Benton, William	$ 829.00	$ 520.00	$ 2,350.00	$ 3,699.00
Canton, Marcel	$ 1,520.00	$ 1,055.00	$ 900.00	$ 3,475.00
Dunlap, Kathy	$ 1,200.00	$ 1,460.00	$ 920.00	$ 3,580.00
Edmond, Lanny	$ 475.00	$ 1,250.00	$ 680.00	$ 2,405.00
Fields, Lucy	$ 1,395.00	$ 400.00	$ 600.00	$ 2,395.00
Goode, Alex	$ 740.00	$ 300.00	$ 400.00	$ 1,440.00
Grant, Joshua	$ 550.00	$ 1,100.00	$ 1,825.00	$ 3,475.00
Harold, Becca	$ 660.00	$ 1,800.00	$ 1,000.00	$ 3,460.00
Isling, Lauren	$ 1,210.00	$ 1,000.00	$ 750.00	$ 2,960.00
Grand Total	$ 9,504.00.00	$10,260.00	$ 9,775.00	$29,539.00

Figure 12.2.47

Sometimes, you need to combine the information in two spreadsheets into one. This process is called *joining two data sets*. Excel uses the function **LOOKUP** to join data sets by a common unique identifier. The key here is that you must have at least one *unique* identifier that appears in both data sets. Things like social security numbers, student ID numbers, or individual customer numbers all work because they are unique to the individual. Typically, descriptors like "name" or "address" won't work as an identifier because there is no guarantee that these attributes will all be unique.

Example 12.2.9

Joining Data Sets

The spreadsheet in Figure 12.2.48 is comprised of two tables containing order and customer information for an Etsy store. Determine the identifier that should be used to join these data tables together using the LOOKUP function in Excel.

	A	B	C	D	E	F	G	H	I
1		Etsy Order					Customer Information		
2	Online Request	Cust. ID	Item			Cust. ID	Last Name	Email	Order Date
3	100	4062	spoons			2519	Adair	4youonly@example.com	2/13/22
4	101	2948	copper pot			1357	Alley	kplly@example.com	1/25/22
5	102	1357	necklace			4611	Brent	batmanB@example.com	1/15/22
6	103	4324	cheese board			4991	Canton	ccanton@example.com	2/1/22
7	104	2757	note cards			4062	Clarret	cl12@example.com	2/3/22
8	105	1774	ceramic bowl			1774	Groppe	11gobig@example.com	1/30/22
9	106	1917	tumbler			2757	Lee	sword@example.com	2/2/22
10	107	2519	candle			2019	Lincoln	studentnow@example.com	2/10/22
11	108	4969	spoons			2948	Long	clemlong@example.com	3/1/22
12	109	2019	plates			4324	Roberts	goldilocks@example.com	1/14/22
13	110	4991	cheese knife			4969	Yawn	84runner88@example.com	2/10/22
14	111	4611	wallet			1917	Young	turn88@example.com	1/30/22

Figure 12.2.48

✔ **Skill Check 12.2.5**

Suppose you were given multiple datasets on student enrollment for your institution. Identify at least one unique identifier that could be used to join data in an Excel spreadsheet.

Solution

In order to join datasets using the LOOKUP function in Excel, we need a unique identifier for each row of data. Both tables contain the common identifier "Cust. ID", which can be used for merging tables.

Example 12.2.10

Using LOOKUP in Excel

KMJ Carpets gives a quarterly bonus to the Regional Sales Manager with the highest sales. You have received a spreadsheet from the sales team containing the totals for each salesperson and a spreadsheet from the HR department identifying the specific region for each salesperson. Each unique employee ID number is given in both spreadsheets. You've combined the two spreadsheets into one document. The spreadsheet from the sales team is titled "Sales Report" and the spreadsheet from HR is titled "Sales Regions." (The full data sets are available for download from math.hawkeslearning.com in Microsoft Excel format.)

	A	B	C
1	KMJ Carpets	Quarterly Sales Report	
2		December 2021	
3	Employees		
4	Last Name	ID	Net Sales
5	Allen	5980	$12,239.00
6	Clark	4294	$10,730.00
7	Davis	6797	$ 8,302.00
8	King	4741	$ 4,179.00
9	Lee	4520	$15,381.00
10	Lopez	4291	$18,704.00
11	Martinez	2709	$ 2,714.00
12	Scott	1525	$12,933.00
13	Smith	7717	$15,620.00
14	Smith	2951	$ 3,015.00
15	White	8357	$12,231.00
16	Wilson	6714	$ 4,302.00
17	Wright	8345	$ 1,989.00

Figure 12.2.49: Sales Regions Spreadsheet

	A	B	C	D
1	Employee			
2	Last Name	ID	Start Date	Region
3	Scott	1525	1/2/1999	SOUTHEAST
4	Martinez	2709	4/2/2016	SOUTHEAST
5	Smith	2951	8/25/2022	EAST
6	Lopez	4291	2/26/2004	NORTH
7	Clark	4294	4/3/2011	NORTH
8	Lee	4520	8/17/2008	SOUTHEAST
9	King	4741	11/26/2006	EAST
10	Allen	5980	10/22/2020	NORTH
11	Wilson	6714	4/27/2012	SOUTHEAST
12	Davis	6797	5/11/2022	EAST
13	Smith	7717	8/15/2007	NORTH
14	Wright	8345	3/25/2016	EAST
15	White	8357	7/13/2020	EAST

Figure 12.2.50: Sales Regions Spreadsheet

a. Add a Region column to the Sales Report spreadsheet using the LOOKUP function.

b. Use a pivot table to determine which regional sales manager will win the bonus.

Solution

a. Notice that the employees are listed in alphabetical order in the Sales Report spreadsheet, but they are listed by employee ID number in the Sales Regions spreadsheet. Therefore, a simple copy and paste from one sheet to the next will not work. Notice that using last names doesn't work either since there are multiple employees with the last name "Smith."

We will use the **LOOKUP(lookup_value, lookup_vector, [result_vector])** function to pull data from the Sales Regions spreadsheet and place it into the Sales Report spreadsheet. In the LOOKUP function, lookup_value is the value we're looking for in the Sales Regions spreadsheet (this can be a cell reference), lookup_vector is the area of the data table that we're searching for the lookup_value in, and result_vector is the area of the data table that we will pull a value from.

The lookup_vector will be a range from the "Sales Regions" spreadsheet; it will be in the form '[spreadsheet_name]'![beginning_cell]:[ending_cell]. We want to search the ID column, so the beginning_cell = $B$4 and the ending_cell = $B$16. This gives us lookup_vector = 'Sales Regions'!$B$3:$B$15. (Note that the use of the $ in the range tells Excel that we want to lock these values if we copy and paste the formula to other cells. We use them here because the ranges in the lookup functions need to remain the same.)

The result_vector has the same format. We want to return the sales region for the specific salesperson, so we have result_vector = 'Sales Regions'!$D$3:$D$15.

Type "Region" in cell D4 of the Sales Report spreadsheet to label the column. In cell D5, type "=LOOKUP(B5,'Sales Regions'!$B$3:$B$15,'Sales Regions'!$D$3:$D$15)" and press Enter. The LOOKUP function will return the sales region for Allen. (Verify this yourself by checking the ID in the Sales Regions spreadsheet).

D5	▼	:	× ✓ *fx*	=LOOKUP(B5,'Sales Regions'!$B$3:$B$15,'Sales Regions'!$D$3:$D$15)					
	A	**B**	**C**	**D**	**E**	**F**	**G**	**H**	**I**
1	KMJ Carpets	Quarterly Sales Report							
2		December 2021							
3	Employees								
4	Last Name	ID	Net Sales	Region					
5	Allen	5980	$12,239.00	NORTH					
6	Clark	4294	$10,730.00						
7	Davis	6797	$ 8,302.00						
8	King	4741	$ 4,179.00						
9	Lee	4520	$15,381.00						
10	Lopez	4291	$18,704.00						
11	Martinez	2709	$ 2,714.00						
12	Scott	1525	$12,933.00						
13	Smith	7717	$15,620.00						
14	Smith	2951	$ 3,015.00						
15	White	8357	$12,231.00						
16	Wilson	6714	$ 4,302.00						
17	Wright	8345	$ 1,989.00						

Figure 12.2.51

To complete the column, drag the bottom-right corner of cell D4 down through cell D17. The results should appear as in Figure 12.2.52. If your results return errors, check your formula to ensure you included the dollar signs in your cell references to lock the lookup region.

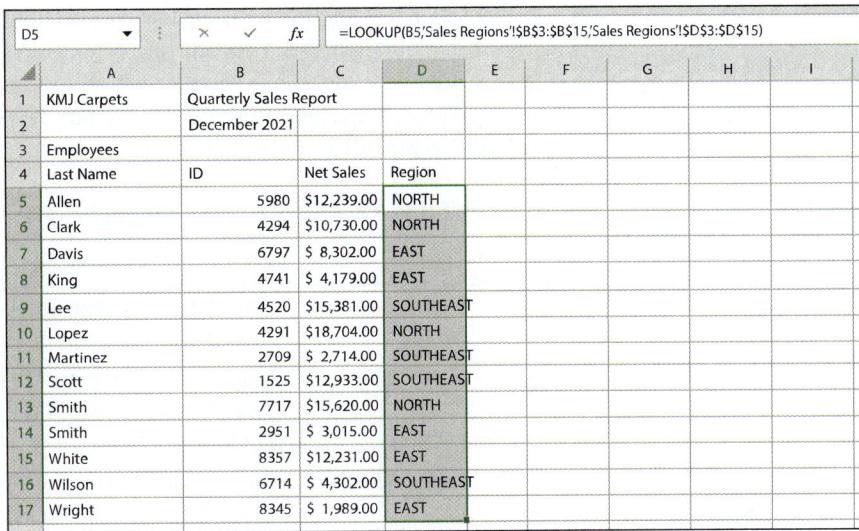

Figure 12.2.52

b. To determine which region sales manager will receive a bonus, we need to determine the total amount of sales in each region. While we could do this by hand, we will employ the use of a pivot table.

Select the data for the Net Sales and the Region, including those headers. This is the data we want to analyze. Click Insert > Pivot Table. This time, we'll insert the pivot table in the same spreadsheet. In the Create PivotTable box, select the radio button next to Existing Spreadsheet and type F4 for the location. Then click OK. The pivot table tool will appear on the right side of the spreadsheet.

Select the check boxes next to Region and Net Sales. If necessary, drag and drop Region to the Rows area and Net Sales to the Values area. The complete pivot table will appear as in Figure 12.2.53.

	A	B	C	D	E	F	G
1	KMJ Carpets	Quarterly Sales Report					
2		December 2021					
3	Employees						
4	Last Name	ID	Net Sales	Region		Row Labels ▼	Sum of Net Sales
5	Allen	5980	$12,239.00	NORTH		EAST	29716
6	Clark	4294	$10,730.00	NORTH		NORTH	57293
7	Davis	6797	$ 8,302.00	EAST		SOUTHEAST	35330
8	King	4741	$ 4,179.00	EAST		Grand Total	122339
9	Lee	4520	$15,381.00	SOUTHEAST			
10	Lopez	4291	$18,704.00	NORTH			
11	Martinez	2709	$ 2,714.00	SOUTHEAST			
12	Scott	1525	$12,933.00	SOUTHEAST			
13	Smith	7717	$15,620.00	NORTH			
14	Smith	2951	$ 3,015.00	EAST			
15	White	8357	$12,231.00	EAST			
16	Wilson	6714	$ 4,302.00	SOUTHEAST			
17	Wright	8345	$ 1,989.00	EAST			

Figure 12.2.53

We can see from the pivot table that the North region made the most sales with a total of $57,293. As a result, the sales manager of the North will receive a bonus this quarter.

Finally, it is important to note that data cleaning and parsing shouldn't be considered a one-time job, but a reoccurring process. As you get to know your data, it often sparks new ideas and ways in which to transform the information. Once in the analysis stage, you may need to return to wrangling in order to parse the data differently after you've noticed an interesting insight or a stubborn roadblock.

Table 12.2.2: Summary Table of Excel Tools and Functions for Cleaning

	What it does	Location/Syntax
Text to Columns	Takes text in one or more cells and splits it across multiple cells.	Found on the Data Tab.
TRIM	Removes all spaces from a text string except for single spaces between words.	TRIM(text)
Find and Replace	Locates particular number of text string and replaces it with something else.	Found on the Home Tab, in the Editing section
MID	Returns a specific number of characters from a text string, starting at the position you specify, based on the number of characters you specify.	MID(text, start_num, num_chars)
FIND	Locates one text string within a second text string and returns the number of the starting position of the first text string from the first character of the second text string.	FIND(find_text, within_text, [start_num])
LOOKUP	Searches a row or column for a value.	LOOKUP(lookup_value, lookup_vector, [result_vector])

Skill Check Answers

1. 9 columns **2.** pet **3.** Rows 3 and 6 **4. a.** January: Marcel Canton; February: Becca Harold; March: William Benton; **b.** William Benton **5.** Answers could include student IDs or SSN numbers

12.2   Exercises

✔ CONCEPT CHECK

1. Data wrangling is the process of acquiring _____ data, cleaning it, organizing it, and transforming it so that the data is _____.

2. In Excel, the _____ function allows you to extract a defined number of characters from a string of text.

3. In Excel, the _____ function can find information in an array of data based on a specified identifier.

4. True or False: Excel reads a space as a character.

5. True or False: Pivot tables can summarize the information in a large data set into table.

For each data file snippet, determine the delimiter that would be used to separate the data into columns and determine how many columns the data would be separated into.

6. Date,Time,PurchaseTotal,PaymentMethod
   01072021,1132,$523.45,Credit
   02072021,1623,$723.11,Paypal
   02072021,1704,$312.77,Credit

7. CustID;Age;Birthdate;Membership;AutoRenew;Flavor
   021789;44;07161976;Premium;Yes;Raspberry
   022893;30;01011990;Basic;Yes;Chocolate
   023113;35;03301985;Premium;No;Vanilla

8. ItemName Size Inventory Price LastReordered
   PinkShirtSS S 24 $18.99 07/23/21
   PinkShirtLS M 15 $19.99 03/12/21
   GreenShirtSS L 12 $18.99 05/15/21

9. Title,Author,Page Count,Word Count,Star Rating,Copies Sold (Millions),List Price
   A Grand Scheme,Lynn Minning,277,73682,4.2,1.73,$15.99
   The Princess Pickle,Susie Roadings,298,82546,3.8,2.77,$16.50
   The Scout,Allison Feist,308,81620,4.3,3.85,$14.99

Describe the inconsistencies in the data in each spreadsheet and explain the steps that should be taken to clean up the data.

10.

	A	B	C	D	E
1	Date	Time	Customer Name	Email Address	Subscription
2	11-10-2022	0732	Emmet Murphy		Premium
3	10.12.2022	1157	Susan Franklin	Sfranks@emails.com	Basic
4	12-10-2022	2005	Kyla Jones	no@no.no	Basic
5					
6	13-10-2022	1053	Ed	ed@edssite.co	Premium
7	10.13.2022	1442	Stephen Moore	SMoore@yahoos.com	Basic

11.

	A	B	C	D	E
1	Chapter	Status	Page Count	Approved	Editor
2	1	Finalized	72	No	Susan
3	2	Draft	55	No	Kyle
4	3	Editing	58	No	Antonio
5	3	Editing	58	No	Antonio
6	5	Draft	73		Susan
7	6	Finalized	60	Yes	Kyle
8	7	Finalized	63	Yes	Antono
9	8	Draft	59	No	Susan

12. Which of the following data will Excel find if you run the command "Find *ing" and select the "Match case" option?

    a. swing

    b. JUMPING

    c. Ingress

    d. ingested

13. Which of the following data will Excel find if you run the command "Find s?n"?

    a. sand

    b. spin

    c. sin

    d. stain

Determine the output of each Excel function.

14. MID("Once upon a time, there lived a lovely princess",6,6)

15. MID("The sum of twelve and twenty is thirty-two.",12,10)

16. Suppose the string "11/20/22 The Pillow Store $76.34" is in cell A5. What will the function MID(A5,10,FIND("$",A5)-10) return?

17. Suppose the string "051121 Okra Express $76.34" is in cell B3. What will the function MID(B3,8,FIND("$",B3)-8) return?

Use the provided pivot table to answer the questions.

18. The following pivot table displays a partial inventory of a clothing store, where the columns indicate the season the item of clothing is intended for.

Style and Color ▾	Fall	Spring	Winter	Grand Total
⊟ Cardigan		28		28
Blue		8		8
Ivory		10		10
Pink		9		9
Purple		1		1
⊟ Coat	14		14	28
Black	6		5	11
Blue	4		4	8
Green	4		5	9
⊟ Jacket	23	22		45
Black	12	7		19
Blue	6	7		13
Gray	5	8		13
Grand Total	37	50	14	101

    a. What is the total inventory of each type of clothing?

    b. Which seasons are the clothing items intended for?

    c. What color cardigans are in stock?

    d. How many items in stock are intended for spring?

19. The following pivot table displays a summary of houses that are listed for sale by a real estate agent, where the columns indicate the number of bathrooms the house has.

Style and Bedrooms	1	2	2.5	Grand Total
⊟ Cape Cod	16	24		40
2	4			4
3	12	18		30
4		6		6
⊟ Colonial	3	24	5	32
2	2	1		3
3	1	13		14
4		10	5	15
⊟ Farmhouse	6	14		20
2	2			2
3	4	12		16
4		2		2
⊟ Ranch	10	18		28
2	2			2
3	8	12		20
4		6		6
Grand Total	35	80	5	120

a. What styles of house are listed as for sale by this real-estate agent?

b. How many Cape Cod style houses listed for sale have three bedrooms?

c. How many houses listed for sale have two bathrooms?

d. How many farmhouses are listed for sale?

20. The following spreadsheet is comprised of two tables containing order and customer information for an online art store. Determine the identifier that should be used to join these data tables together using the LOOKUP function in Excel.

	A	B	C	D	E	F	G
1	Order Information				Customer Information		
2	Online Request	ID	Item		ID	Name	Sign-up Date
3	202	1939	Kittens		1159	Abigail Berry	01-09-2020
4	203	1321	Flowers		1752	Adam Diaz	02-12-2019
5	204	1551	Springtime		1099	Ali Richardson	14-04-2020
6	205	1304	Kittens		1337	Daniel King	10-06-2018
7	206	1882	Ducks		1919	Jade Davis	12-02-2020
8	207	1099	Nighttime		1939	James Knight	05-04-2019
9	208	1182	Dusk		1321	Liam Smith	17-05-2020
10	209	1159	Eclipse		1551	Martha MacDonnell	02-11-2018
11	210	1529	Sunset		1529	Mira Ward	29-03-2021
12	211	1752	Flowers		1882	Olivia Skywell	11-08-2019
13	212	1337	Nighttime		1182	Omar Torres	12-02-2021
14	213	1919	Eclipse		1304	Skylar Davis	25-12-2021

21. The following partial spreadsheet is comprised of two tables containing book inventory and author information for an online bookstore. Determine the identifier that should be used to join these data tables together using the LOOKUP function in Excel.

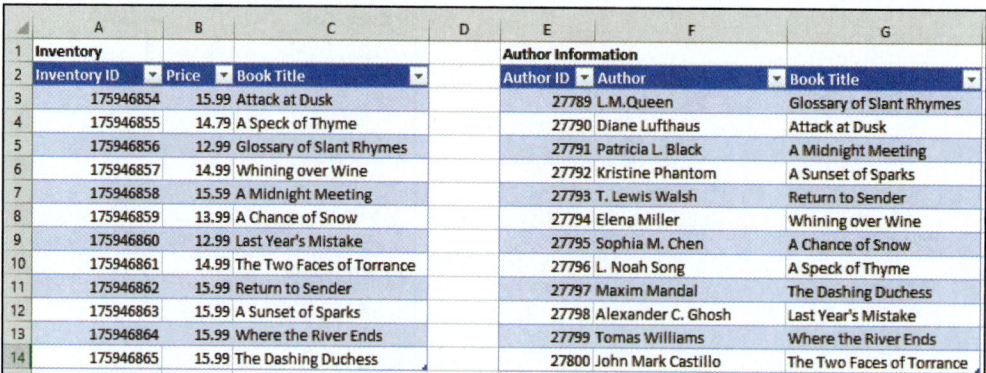

	A	B	C	D	E	F	G
1	Inventory				Author Information		
2	Inventory ID	Price	Book Title		Author ID	Author	Book Title
3	175946854	15.99	Attack at Dusk		27789	L.M.Queen	Glossary of Slant Rhymes
4	175946855	14.79	A Speck of Thyme		27790	Diane Lufthaus	Attack at Dusk
5	175946856	12.99	Glossary of Slant Rhymes		27791	Patricia L. Black	A Midnight Meeting
6	175946857	14.99	Whining over Wine		27792	Kristine Phantom	A Sunset of Sparks
7	175946858	15.59	A Midnight Meeting		27793	T. Lewis Walsh	Return to Sender
8	175946859	13.99	A Chance of Snow		27794	Elena Miller	Whining over Wine
9	175946860	12.99	Last Year's Mistake		27795	Sophia M. Chen	A Chance of Snow
10	175946861	14.99	The Two Faces of Torrance		27796	L. Noah Song	A Speck of Thyme
11	175946862	15.99	Return to Sender		27797	Maxim Mandal	The Dashing Duchess
12	175946863	15.99	A Sunset of Sparks		27798	Alexander C. Ghosh	Last Year's Mistake
13	175946864	15.99	Where the River Ends		27799	Tomas Williams	Where the River Ends
14	175946865	15.99	The Dashing Duchess		27800	John Mark Castillo	The Two Faces of Torrance

✏ **WRITING & THINKING**

22. Explain why it would be useful to have two different pivot tables of the same data set in a presentation.

12.2 PROJECT

PIVOTING AN INVENTORY

One aspect of data wrangling is making sure the raw data is cleaned up and organized into a form that can be easily analyzed. This organized data should also be easy to work with to answer the questions being asked about the data. A pivot table is a tool that can quickly summarize a large data set. Pivot tables also have the ability to present several completely different views of the data.

Suppose the inventory manager of a garden center was asked to create a report on the current inventory of blueberry bushes. She wasn't provided a specific question to answer, so she created the following two pivot tables to spark conversation about the current inventory.

Age (in Years) ▾	Count of Blueberry Size
─ Bluecrop	8
1	1
2	2
3	5
─ Bluegold	16
1	3
2	6
3	7
─ Hardyblue	8
1	4
2	4
─ Jersey	5
1	2
2	1
3	2
─ Legacy	12
1	8
2	3
3	1
─ Patriot	18
1	5
2	6
3	7
Grand Total	**67**

Harvest Season and Sun Exposure ▾	Count of Age (in Years)
─ **Early**	**18**
─ Full Sun	18
Patriot	18
─ **Late Season**	**17**
─ Full Sun	12
Legacy	12
─ Full to Partial Sun	5
Jersey	5
─ **Midseason**	**32**
─ Full Sun	8
Bluecrop	8
─ Full to Partial Sun	24
Bluegold	16
Hardyblue	8
Grand Total	**67**

1. Describe the information presented in the pivot table on the left.

2. Describe the information presented in the pivot table on the right.

3. What information is displayed in both tables?

4. How many different varieties of blueberry bushes does the garden center sell? What varieties do they have?

5. Which blueberry bush varieties require full sun?

6. Write a question that the pivot table on the left would answer.

7. Write a question that the pivot table on the right would answer.

8. Suppose the inventory manager is given the following question right before the meeting: "What is the age of each blueberry bush in stock and when is the harvest season for each?" Describe how she could create a pivot table to answer this question.

OBJECTIVES

1. Create scatter plots.

2. Calculate correlation coefficients.

3. Identify statistical significance of experimental results.

4. Calculate the regression line for a data set.

Figure 12.3.1: John Snow Memorial Pump

Once data is in an accessible format, it's time to delve into the data looking for patterns—some of which may appear obvious and some of which are more hidden. Data exploration is the creative side of data science that intertwines with the statistical aspect. Imagine that you are handed a big stack of photographs of your family ancestors from the last 100 years and are asked to categorize them without being told what the categories are. How do you go about deciding on the categories? The categorization process begins by noticing similarities as well as differences between the photos and posing questions to yourself. For instance, you might decide to sort by date, or particular people, or places, or even specific events. This is the idea behind exploring your data. Getting to know your data helps you decide how broad or narrow a category will be. It might take several tries to decide on categories that are manageable, but the important thing is to have an open mind and to be inquisitive.

In the mid-1800s, a London physician name John Snow did just that. He pushed back against the accepted Victorian mindset on the cause of cholera—a disease that could turn deadly within hours if left untreated, even among perfectly healthy people. The belief at the time was that bad air or vapors—not bacteria—were the cause of cholera. In contrast, Dr. Snow believed bacteria contaminating water was the source of the disease. He believed that sewage dumped into the river or into cesspools near town wells could contaminate water supplies, leading to a rapid spread of disease. When a large outbreak occurred in London, he spent many, many hours mapping every case that was documented, looking for a connection. The results pointed to a single source of water for the area: the Broad Street pump. By visualizing the data, he standardized a mapping tool that revolutionized the science of epidemiology. He used his maps to narrow in on the common source of water connecting hundreds of cases of cholera. To this day, scientists working to combat disease outbreaks often ask, "Where is the handle to this pump?" in reference to Snow's work.

In that same spirit, we tackle data exploration by always asking "What can we learn from the data?" As data analysts (that is, researchers who use data to answer a question), we use statistical methods to separate "data noise" from actual trends in the data.

OBJECTIVE 1 Scatter Plots

Just as John Snow did with his disease map, it is helpful to first look at the data visually. The most common graph used to show the relationship between two quantitative variables is called a scatter plot. A **scatter plot** is a graph on the coordinate plane that contains one point for each pair of data. The horizontal axis (x-axis) represents one variable, and the vertical axis (y-axis) represents the other. Unlike a line graph, the points on a scatter plot are not connected. The reason the points are not connected is that each point represents a separate data pair. Scatter plots do not necessarily represent a change in the value of a variable over time as a line graph does.

📖 SCATTER PLOT

A **scatter plot** is a graphical display of data that is used to show how variables might relate to one another. The most common type of scatter plot is created by plotting two variables on the coordinate plane, where each piece of data is represented by a point.

Figure 12.3.2 shows five different scatter plots. Notice the data in scatter plots A, B, and C have visible relationships, while the data in D and E don't display any obvious patterns.

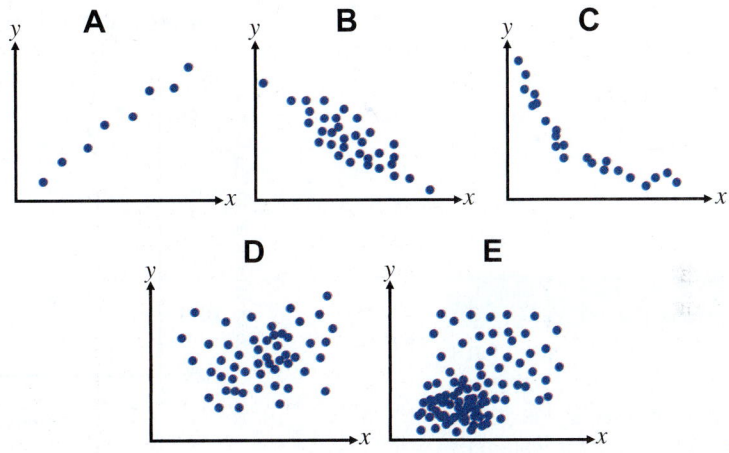

Figure 12.3.2

Helpful Hint

Quantitative variables are counts or measurements. See Section 11.1 for more on types of variables.

The importance of scatter plots like the ones in Figure 12.3.2 is that they help us to identify trends in the data. For example, do the points fall in a linear pattern, a curved pattern, or no pattern at all? If the trend seems to follow the pattern of a straight line, as in scatter plots A and B, there is said to be a linear relationship between the variables. Scatter plot C has a visible pattern; it's just not linear.

Example 12.3.1

Creating a Scatter Plot to Identify Trends in Data

Providers of cloud computing offer services such as email hosting, data storage, and online office applications—all available over the internet. Consider the following data collected from a sample of small businesses regarding the number of years they have been in business and the approximate percentage of their workload that is in the cloud, whether public or private. Use the data in Table 12.3.1 to create a scatter plot that shows the relationship between the number of years in business and the percentage of workload in the cloud. Remember to label both axes as well as the graph itself.

Table 12.3.1: Sample of Cloud Usage by Small Businesses

Number of Years in Business	1.25	8.0	0.5	3.75	5.0	12.25	3.5	2.25
% of Workload in the Cloud	88%	25%	100%	75%	50%	10%	25%	80%

Solution

When creating a scatter plot, the first step is to determine which variable each axis represents. We will let the *x*-axis represent the number of years the business has been in operation and the *y*-axis represent the percentage of cloud usage.

By Hand

To create the scatter plot, graph each year-percentage pair as a point on the graph. For instance, the first business has been around 1.25 years and has 88% of their workload in the cloud. This is represented by the point (1.25, 88). When we plot each of these points, we obtain the following scatter plot.

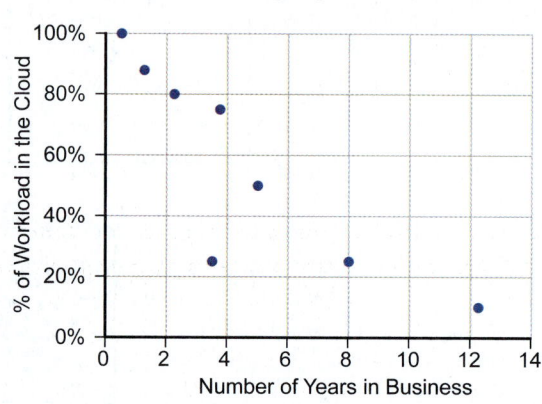

Figure 12.3.3

> **Helpful Hint**
>
> Figure 12.3.3 appears to have an outlier, a data value that is extreme compared with the rest of the data values. See Section 11.3 for more on outliers.

TI-83/84 Plus

To create a scatter plot using a TI-83/84 Plus CE calculator, start by turning on STAT PLOT. Press [2nd] then [y=]. Select Plot1 and highlight On. Press [enter] and then press [2nd] [mode] to return to the home screen.

Now press [y=] and clear any functions listed.

Press [stat] and select Edit.... Enter the scatter plot values in L1 and L2. Be sure you are entering the number of years in business in L1 and the percentage of cloud usage in L2. Once all the values are entered, press [zoom] and select ZoomStat. The scatter plot will appear as in Figure 12.3.5.

Notice that the points fall in a linear pattern that moves downward from left to right. This type of pattern can be described as a linear relationship with a negative slope.

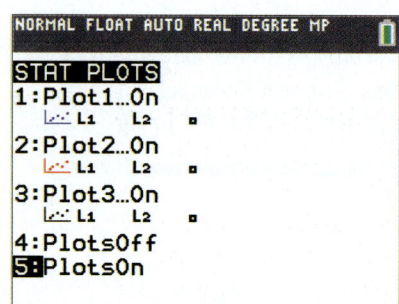

Figure 12.3.4

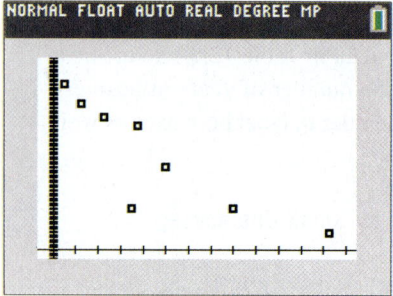

Figure 12.3.5

When there appears to be a linear relationship between data on a scatter plot, we say the variables are **correlated**. Both the direction and the position of the points (that is, how close the points are to lying in a straight line) tell us how the data are correlated. If the trend follows a line that has a positive slope (that is, as the values of one variable increase, so do the values of the other variable), then the two variables are said to have a *positive linear correlation*. If the trend follows a line that has a negative slope (that is, as the values of one variable increase, the values of the other variable decrease), then the two variables are said to have a *negative linear correlation*.

📖 LINEAR CORRELATION

Positive Linear Correlation

When data are plotted in a scatter plot and there appears to be an upward trend (that is, as one variable increases, the other variable increases as well), we say there is a **positive linear correlation** between the variables.

Negative Linear Correlation

When data are plotted in a scatter plot and there appears to be a downward trend (that is, as one variable increases, the other decreases), we say there is a **negative linear correlation** between the variables.

Example 12.3.2

Identifying Correlations

Consider the relationship between the following variables and what kind of correlation might show up in a scatter plot of the data. Decide if the variables would likely have a positive linear correlation, a negative linear correlation, or no linear correlation.

a. The number of cigarettes smoked and the probability of lung cancer

b. The amount of credit card debt incurred by college freshmen and their IQ score

c. The number of minutes spent on social media sites by college students and their first semester grades

Solution

a. As the number of cigarettes smoked increases, so does the chance of lung cancer. Thus, the scatter plot is likely to have upward-trending data points. The variables would have a positive linear correlation.

b. The scatter plot for these variables would likely contain a wide range of credit card debt and a wide range of IQ scores. It would be unlikely that there is a linear relationship between these two variables. Thus, they have neither a positive linear correlation nor a negative linear correlation.

c. As the number of minutes (or hours) spent on social media sites increases, a student's grades are likely to decrease. This would result in a downward-trending scatter plot, which would indicate a negative linear correlation between the amount of time spent on social media sites and grade point average.

> ✔ **Skill Check 12.3.1**
>
> Identify two variables that would likely have a positive linear correlation.

OBJECTIVE 2 Correlation Coefficient

Let's consider a new parameter that measures the strength of the linear relationship between two variables; that is, how strongly one of the variables is linearly dependent on the other. The strength of the linear relationship is determined by how closely the points in the scatter plot resemble a straight line. Thus, the stronger the relationship, the more the plot looks like a straight line. And the weaker the relationship, the more scattered the points are in the plot. Figure 12.3.6 demonstrates the difference between a strong correlation and a weak correlation.

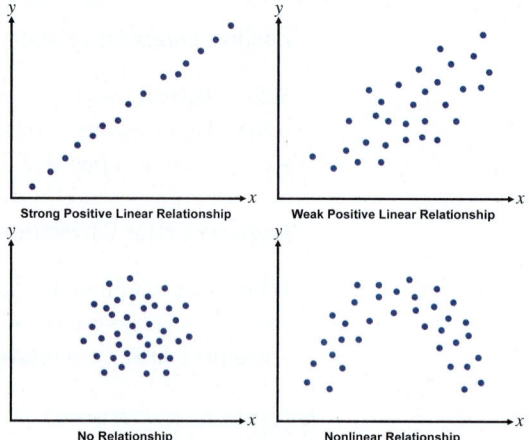

Figure 12.3.6: Strength of a Linear Relationship

The parameter that measures the strength of the linear relationship between the two variables is called the **Pearson correlation coefficient**. The correlation coefficient for a sample is denoted by r and is always a value between -1 and 1, inclusive. The stronger the correlation, the closer the correlation coefficient is to either -1 or 1. If there is a very strong positive linear correlation, r will be close to 1. If there is a strong negative linear correlation, r will be close to -1. The closer r is to 0, the less correlation there is between the variables. A correlation coefficient of 0 means that there is no linear relationship between the variables at all.

The formula for calculating the correlation coefficient r is complicated and is tedious to compute by hand. Since technology provides a more convenient way of determining the correlation, we will use calculator and spreadsheet methods in our calculations.

> **💬 Helpful Hint**
>
> The correlation coefficient is usually rounded to the nearest ten thousandth.

> **📖 PEARSON CORRELATION COEFFICIENT**
>
> The **Pearson correlation coefficient** is the parameter that measures the strength of a linear relationship between two quantitative variables. The correlation coefficient for a sample is denoted by r. It always takes a value between -1 and 1, inclusive.
>
> $$-1 \leq r \leq 1$$

Example 12.3.3

Calculating the Correlation Coefficient

The following is a small sample of data collected from male participants in a 161 km trail ultramarathon. The survey collected the Body Mass Index (BMI) of each participant and their age. A table of the data, along with the scatter plot of the data, is given. Find the correlation coefficient r between the variables and interpret your results.

Table 12.3.2: BMI vs. Age of Men in a 161 km Trail Ultramarathon

BMI	23.4	25.7	23.3	25.9	24.1	23.0	22.4	21.6	23.4	24.8	24.3	24.7
Age	53	47	28	52	60	34	39	36	52	54	53	53

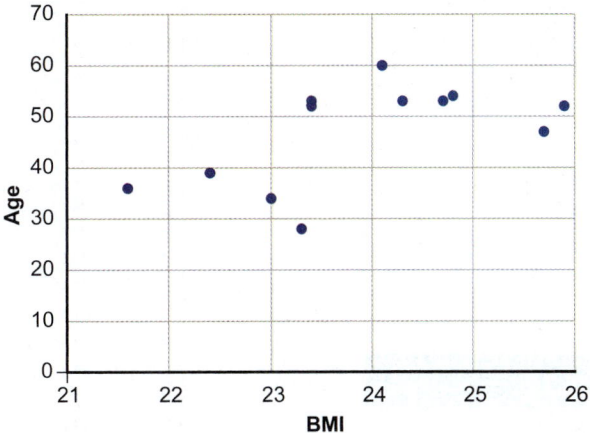

Figure 12.3.7: BMI vs. Age of Men in a 161 km Trail Ultramarathon

Solution

TI-83/84 Plus

To calculate the Pearson correlation coefficient, we need to input the data in the calculator. Begin by pressing stat and select Edit.... Enter the values for the BMIs in L1 and the values for the ages in L2. Then press stat, choose CALC and select LinReg(ax+b). Be sure that L1 is listed for Xlist:, L2 is listed for Ylist:, and FreqList: is left blank. Highlight Calculate and press enter. The output should appear as shown in Figure 12.3.8.

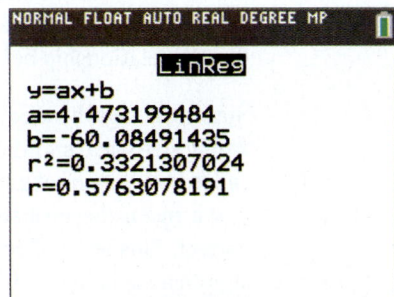

Figure 12.3.8

Microsoft Excel

To calculate the Pearson correlation coefficient in Excel, we will use the **PEARSON(array1, array2)** function, where "array1" refers to the cells containing the x-values and "array2" refers to the cells containing the y-values.

Enter labels for your columns in cells A1 and A2, then enter the values for BMI in column A and the values for age in column B. In an empty cell, type "=PEARSON(A2:A13,B2:B13)" and press Enter. The value of r will appear in the cell.

Figure 12.3.9

✓ **Skill Check 12.3.2**

Calculate the correlation coefficient for the following data.

Table 12.3.3

Age (in Years)	34	43	49	51	65
Bone Density (in mg/cm²)	946	875	804	723	691

Notice that the value of r is positive, which matches with the positive slope of the scatter plot. Since $r \approx 0.5763$, we know that there is a weak positive linear correlation between a male participant's BMI and his age.

OBJECTIVE 3 Significance

So far we have discussed the linear relationship between two variables. However, we have not determined if that relationship occurred by chance or if a given relationship is statistically significant. The term *statistically significant* indicates that the relationship between two variables is unlikely to have occurred by chance.

Statistical significance can be considered strong or weak. As you can imagine, strong statistical significance helps support the fact that our results are real and not caused by luck or chance. In statistics, when we refer to the **level of confidence**, c, we mean the probability that the assertions made about the data are in fact correct. This level of confidence is tied to the **level of significance**, α. The level of significance is the probability that we are wrong in our assertions made about the data. The level of confidence c and the level of significance α are related in that their values add to one (that is, $c + \alpha = 1$).

📖 **LEVEL OF CONFIDENCE AND LEVEL OF SIGNIFICANCE**

Level of Confidence

The **level of confidence**, c, is the probability that the assertions made about the data are correct.

Level of Significance

The **level of significance**, α, is the probability that the assertions made about the data are incorrect.

$$c + \alpha = 1$$

Often, statisticians reference the level of significance in their reporting, leaving the reader to infer the level of confidence. For example, suppose a level of significance for a correlation is given as $\alpha = 0.05$. This means there is a 5% probably the variables are actually not related at all even though we think there is a correlation between the two variables. However, it also means that the level of confidence is $1 - \alpha$, or 0.95, implying that there is a 95% probability that the variables are correlated.

If we know the level of significance and the sample size of a set of data, we can look up the critical value for the Pearson correlation coefficient to determine how large r needs to be for the relationship to be statistically significant. The critical value conveys the minimum value of r that is needed to conclude there is a statistically significant relationship. The most common levels of significance used to test for correlation are $\alpha = 0.05$ and $\alpha = 0.01$. Table C in Appendix A lists critical values for the Pearson correlation coefficient at both of these levels of significance. You need to know the level of significance α along with the sample size n to use these tables. A portion of Table A is shown in Table 12.3.4.

Table 12.3.4: Critical Values of the Pearson Correlation Coefficient

n	$\alpha = 0.05$	$\alpha = 0.01$
4	0.950	0.990
5	0.878	0.959
6	0.811	0.917
7	0.754	0.875
8	0.707	0.834
9	0.666	0.798
10	0.632	0.765
11	0.602	0.735
12	0.576	0.708

📖 **STATISTICALLY SIGNIFICANT**

If $|r|$ is greater than the critical value listed in Pearson's Correlation table, then r is **statistically significant**, which means the relationship is unlikely to have occurred by chance.

Example 12.3.4

Identifying Statistical Significance

Use the critical values in Table 12.3.5 to determine if the correlation between BMI and age from the previous example is statistically significant. Recall that $r \approx 0.5763$. Use a 0.05 level of significance.

Table 12.3.5: BMI vs. Age for Men in a 161 km Trail Ultramarathon

BMI	23.4	25.7	23.3	25.9	24.1	23.0	22.4	21.6	23.4	24.8	24.3	24.7
Age	53	47	28	52	60	34	39	36	52	54	53	53

Solution

There are twelve pairs of data in Example 12.3.3, so the sample size, n, is 12. We are told to use a 0.05 level of significance, so $\alpha = 0.05$. In Table 12.3.6, the critical value is found where the row for $n = 12$ intersects the column for $\alpha = 0.05$.

Table 12.3.6: Critical Values of the Pearson Correlation Coefficient

n	$\alpha = 0.05$	$\alpha = 0.01$
4	0.950	0.990
5	0.878	0.959
6	0.811	0.917
7	0.754	0.875
8	0.707	0.834
9	0.666	0.798
10	0.632	0.765
11	0.602	0.735
12	0.576	0.708
13	0.553	0.684
14	0.532	0.661

Thus, the critical value is 0.576. Comparing this critical value to the absolute value of the correlation coefficient we found for the data in Example 12.3.3, we have $|0.5763| > 0.576$, and thus $|r| >$ critical value. So the linear relationship between the variables is statistically significant at the 0.05 level of significance. Therefore, we have enough evidence to conclude that a linear relationship exists between BMI and age for male ultramarathoners.

✔ **Skill Check 12.3.3**

Determine whether the correlation coefficient is statistically significant at the specified level of significance assuming that the scatter plot of the data shows a linear pattern.

$r = -0.499$, $\alpha = 0.01$, $n = 26$

When we do find a data set that proves to have a statistically significant r, even at the 0.01 level of significance, we need to be careful to not assign causation to the situation. When two variables are statistically correlated, we are often tempted to infer that one thing caused the other to happen. For instance, the line graph in Figure 12.3.10 shows that the per capita consumption of mozzarella cheese correlates with the number of civil engineering doctorates awarded.[1] However, we instinctively know that changing one of these values does not have an effect on the other.

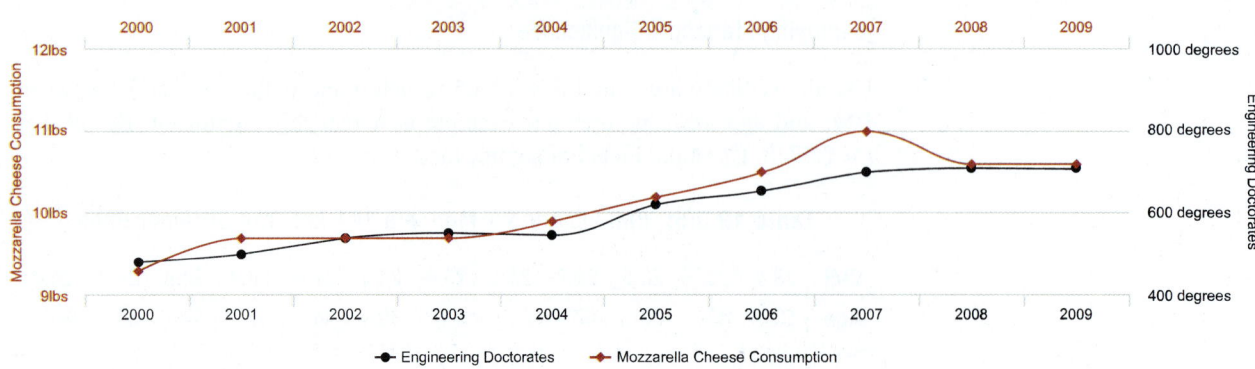

Per Capital Consumption of Mozzarella Cheese
Correlates with
Civil Engineering Doctorates Awarded

Figure 12.3.10

---

1   Spurious Correlations, Tyler Vigen, https://tylervigen.com/spurious-correlations

Other statistically significant situations may not be so immediately obvious in their absurdity and may involve a third factor. For example, we might find that the number of watermelons consumed is positively correlated with the number of drownings that occur. This in no way means that eating a watermelon causes you to drown. Instead it might indicate that a third factor—the time of year—is playing a roll. During the hotter months of the year, more people eat watermelons (which are in season) while at the same time the number of drownings increases simply because more people go swimming when it's hot outside. Can you think of a third factor that might be lying behind the correlation of cheese and civil engineering in Figure 12.3.10?

OBJECTIVE 4 Regression Lines

Once we've established that there actually is a statistically significant correlation between two variables x and y, we can predict what value y might take when given a specific value of x. In other words, we can model the behavior between the two variables. To make predictions, we use the **regression line**, or *line of best fit*. The regression line is a particular line that most closely "fits" the data points on the scatter plot. For instance, consider the scatter plot shown in Figure 12.3.11, which shows the birth weight for a newborn male (in grams) versus gestational age (in completed weeks). There are several lines shown that fall among the points on the scatter plot, but which one is the best fit?

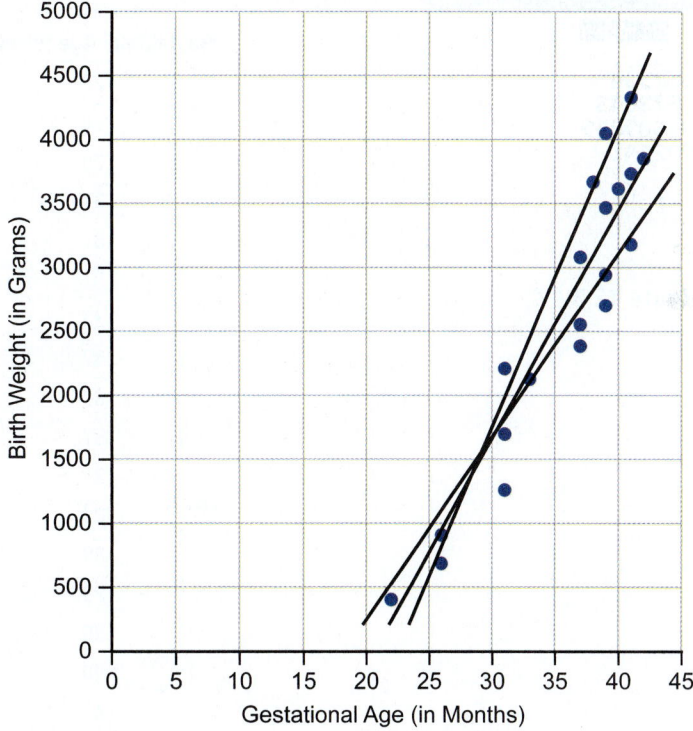

Figure 12.3.11: Birthweight and Gestational Age for Newborn Males

There is a particular line called the regression line, or the *line of best fit*, and it is the line for which the average variation from the data is the smallest. The line $\hat{y} = ax + b$ is determined by its slope a and its y-intercept b, where $\hat{y}$ is the predicted value of the y variable. Note that calculators and statistical software often omit the "hat" on the y in the equation of a regression line, as in the following forms: $y = mx + b$, $y = ax + b$, or $y = a + bx$.

> ### 📖 REGRESSION LINE
>
> The **regression line**, also known as the **line of best fit**, is the line for which the average variation from the data is the smallest. It can be represented by
>
> $$\hat{y} = ax + b,$$
>
> where a is the slope of the line and b is the y-intercept.

Let's stop and note that before ever using a regression model to predict the value of the y-variable, we must first ensure that the relationship between the two variables is indeed linear and statistically significant. Only then should a regression model be used for prediction purposes. If r fails to be statistically significant, then the results are not meaningful enough to use for predictions.

Table 12.3.7 shows the data that was used to make the scatter plot in Figure 12.3.11. Both the figure and the table display each newborn male's birth weight (in grams) versus gestation age (in completed weeks). Figure 12.3.12 shows the output screen for a TI 83/84 plus calculator listing the value for r calculated from the data, as we did in Example 12.3.3.

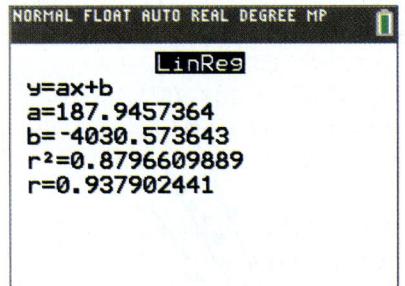

Figure 12.3.12

Table 12.3.7: Birth Weight and Gestational Age for Newborn Males

Gestational Age (Weeks)	Birth Weight (Grams)
22	401
26	908
26	686
31	1259
31	1698
31	2209
33	2127
37	2384
37	2552
37	3080
38	3665
39	2701
39	4049
39	3465
39	2942
40	3613
41	4328
41	3179
41	3733
42	3851

Using the Pearson Correlation Coefficient Table of Critical Values in Appendix A Table C, we can confirm that r is most definitely statistically significant. In fact, as long as the sample size is bigger than 5, $r \approx 0.9379$ is significant at both the 95% and 99% confidence levels. (Check this for yourself.)

Knowing that r is statistically significant allows us to utilize the regression line to make predictions. Notice that along with r, the calculator also computes the values of the slope, a, and the y-intercept, b, as shown in Figure 12.3.12. Given these values, we can write the regression line by substituting the calculated values $a = 187.946$ and $b = -4030.574$ (both rounded to the thousandths place) into $\hat{y} = ax + b$ to get the regression line.

$$\hat{y} = 187.946x - 4030.574$$

> **Helpful Hint**
>
> Excel can also quickly calculate the necessary variables for the regression line, as we will demonstrate in Example 12.3.5.

Now we can predict the weight of a newborn male born between 22 and 42 gestational weeks. Suppose we wish to know the estimated weight of a newborn male at 40 gestational weeks. We can substitute $x = 40$ into the formula and simplify, as follows.

$$\hat{y} = 187.946x - 4030.574$$
$$\hat{y} = 187.946(40) - 4030.574$$
$$\hat{y} = 3487.266$$

Based on our regression model, we can predict that a newborn male born at 40 weeks will weigh approximately 3487 grams.

> **Helpful Hint**
>
> A prediction should **not** be made with a regression model if any of the following statements are true.
>
> - The data do not fall in a linear pattern when graphed on a scatter plot.
>
> - The correlation coefficient is not statistically significant.
>
> - You wish to make a prediction about a value outside the range of the sample data.
>
> - The population is different than that from which the sample data were drawn.

As with all predictions, you should be careful not to get too carried away. We've already stated that you should only use the regression line for predictions if you find r to be statistically significant. You should also make sure that you are only predicting for values that are within the range and population of the original sample data. In other words, don't try to predict something that is either way out of the scope of the original data or not from the same type of population the sample was drawn from. For instance, it would not be wise to try to predict a newborn male's weight for a gestational age of 19 weeks. Nor would it be reasonable to apply these results to newborn females, which are a different population.

Example 12.3.5

Predictions Using the Regression Line

The following data set gives the teen birth rates (per 1000) during one year in fourteen of the northeastern states along with the percent of the state's population who are living with incomes below the federally defined poverty level. Use the data, along with its scatter plot, to answer the following questions.

Tables 12.3.8: State Poverty Levels vs. Teen Birth Rate (per 1000)

State	Poverty %	Birth Rate (ages 15–19)
Maine	12.0	11.1
New Hampshire	6.6	8.0
Vermont	9.2	8.8
Massachusetts	10.0	7.2
New York	11.9	11.7
Pennsylvania	11.4	14.1
Connecticut	10.4	8.3
Rhode Island	10.1	11.5
New Jersey	9.1	10.3

State	Poverty %	Birth Rate (ages 15–19)
Delaware	8.1	16.7
Maryland	7.8	14.1
Ohio	12.4	18.9
Virginia	10.1	14.3
West Virginia	16.5	25.4

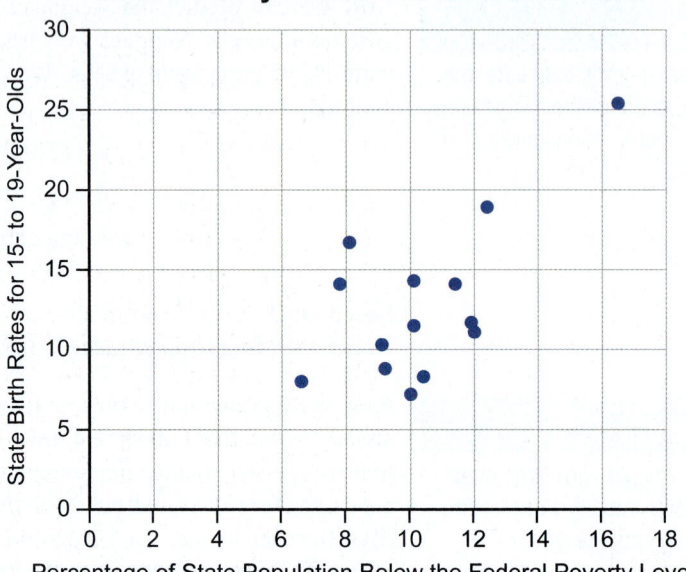

Figure 12.3.13

a. Determine if there is a statistically significant relationship between state poverty levels and teen birth rates at the 0.05 level.

b. If the relationship is significant, find the linear regression line in the form $\hat{y} = ax + b$ for these data.

c. If appropriate, predict the teen birth rate for a northeastern state that has 11% of its population with incomes below the federal poverty level.

d. Discuss why it is or is not appropriate to use this regression line to predict the birth rate of the state of California.

e. Discuss why it is or is not appropriate to use this regression line to predict the birth rate for a northeastern state where only 1.5% of the population has incomes below the federal poverty level.

Solution

a. Begin by letting the variable x represent the poverty percentages in each state and the variable y represent the teen birth rates. We can see from the scatter plot that the points do fall in somewhat of a linear pattern. In order to determine if there is a statistically significant relationship, we must calculate the value of the correlation coefficient, r.

TI-83/84 Plus

Begin by pressing [stat] and selecting Edit.... Enter the values for the poverty percentages in L1 and the values for birth rates in L2. Then press [stat] and choose CALC and select LinReg(ax+b). Be sure that L1 is

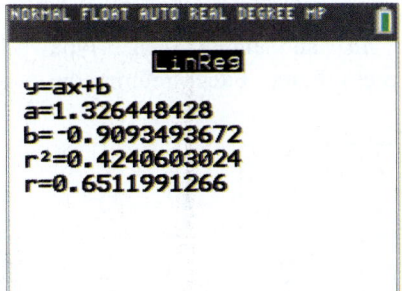

Figure 12.3.14

listed for `Xlist:`, L2 is listed for `Ylist:`, and `FreqList:` is left blank. Highlight `Calculate` and press enter. The output should appear as shown in Figure 12.3.14.

Microsoft Excel

To calculate the Pearson correlation coefficient, we will again use the **PEARSON(array1, array2)** function, where "array1" refers to the cells containing the x-values and "array2" refers to the cells containing the y-values.

Enter labels for your columns in cells A1 and B1, then enter the values for the poverty percentages in column A and the values for the birth rates in column B. In an empty cell, type "=PEARSON(A2:A15,B2:B15)" and press Enter. The value of r will appear in the cell.

E2	▼	× ✓ fx	=PEARSON(A2:A15,B2:B15)		
	A	B	C	D	E
1	Poverty %	Birth Rate (15-19)			
2	12	11.1		r =	0.651199
3	6.6	8			
4	9.2	8.8			
5	10	7.2			
6	11.9	11.7			
7	11.4	14.1			
8	10.4	8.3			
9	10.1	11.5			
10	9.1	10.3			
11	8.1	16.7			
12	7.8	14.1			
13	12.4	18.9			
14	10.1	14.3			
15	16.5	25.4			

Figure 12.3.15

Using Table C in Appendix A, we find that with a sample size of 14 and a 0.05 level of significance, the critical value is 0.532. The absolute value of the correlation coefficient, $|r| \approx 0.6512$, is greater than 0.532, so there is enough evidence at the 0.05 level of significance to conclude that the correlation between the two variables is statistically significant.

b. From part a. we know it is now appropriate to consider the linear regression model.

TI-83/84 Plus

To determine the coefficients in the linear regression model, we will once again press stat, choose `CALC` and select `LinReg(ax+b)`. Figure 12.3.14 shows that `a = 1.326` and `b = -0.909`.

Microsoft Excel

We need to calculate the slope and y-intercept separately. To calculate the slope, we can use the Excel formula **SLOPE(known_ys, known_xs)**. (Notice that this formula uses the y-values in column B and then the x-values in column A.) In part a., we entered the x-values into column A and the y-values into column B. In an empty cell, type "=SLOPE(B2:B15,A2:A15)" and press Enter. Excel returns the value 1.326448.

Similarly, to calculate the y-intercept, we can use the Excel formula **INTERCEPT(known_ys, known_xs)**. In an empty cell, type "=INTERCEPT(B2:B15,A2,A15)" and press Enter. Excel returns the value − 0.90935.

	A	B	C	D	E
1	Poverty %	Birth Rate (15-19)			
2	12	11.1		r =	0.651199
3	6.6	8		a =	1.326448
4	9.2	8.8		b =	-0.90935
5	10	7.2			
6	11.9	11.7			
7	11.4	14.1			
8	10.4	8.3			
9	10.1	11.5			
10	9.1	10.3			
11	8.1	16.7			
12	7.8	14.1			
13	12.4	18.9			
14	10.1	14.3			
15	16.5	25.4			

Figure 12.3.16

Thus, the equation of the regression line, in the form $\hat{y} = ax + b$, is as follows.

$$\hat{y} = 1.326x - 0.909$$

c. To make a prediction, substitute the value $x = 11$ into the regression line and solve for $\hat{y}$.

$$\hat{y} = 1.326x - 0.909$$
$$\hat{y} = 1.326(11) - 0.909$$
$$\hat{y} = 13.677$$

Thus, we would predict that a northeastern state with 11% of its population at an income less than the federal poverty level would have a teen birth rate of 13.677 births per 1000.

d. Predictions should only be made for states that fall into the original population. California is not in the northeast of the United States and cannot be assumed to have the same characteristics. So it is not appropriate to use the regression line for predictions in this case.

e. It is not meaningful to predict the value of y for this percentage level because the value of $x = 1.5$ is outside the range of the original data. The original data only considered poverty levels between 6.6% and 16.5%, so we should only predict teen birth rates for poverty-level percentages within this range.

✔ **Skill Check 12.3.4**

The average temperature in Florida has a statistically significant correlation to the monthly precipitation in inches in the state. The regression line $\hat{y} = 0.225x - 15.42$ models this relationship when the temperature is between 75 and 90. Use the line to predict the monthly precipitation if the temperature is 80 degrees.

Skill Check Answers

1. Correct responses suggest that as one variable increases, so does the other. For example, the number of candy bars consumed on a daily basis and weight gain; or the distance between two locations and the length of time it takes to drive between the two. **2.** $r \approx -0.9396$ **3.** Yes, r is statistically significant. **4.** 2.58 inches

12.3 Exercises

✔ CONCEPT CHECK

1. A scatter plot is a graph that shows the _____ between two variables.

2. When data points in a scatter plot appear to follow an upward trend, there is a _____ linear correlation between the variables.

3. Once the relationship between two variables is determined to be statistically significant, a _____ can be fit to the data set to model the situation.

4. True or False: The Pearson correlation coefficient is always a positive value.

5. True or False: The level of confidence is the probability that assertions made about the data are incorrect.

🚀 PRACTICE

Use the linear regression model $\hat{y} = ax + b$ to predict the y-value for each value of x.

6. $\hat{y} = 28.01x + 17.83$

   a. $x = 21$

   b. $x = 31$

   c. $x = 40$

7. $\hat{y} = -16.5x + 230.55$

   a. $x = 5$

   b. $x = 13$

   c. $x = 35$

In each scatter plot, determine whether there appears to be a positive linear correlation, a negative linear correlation, or no linear correlation.

8.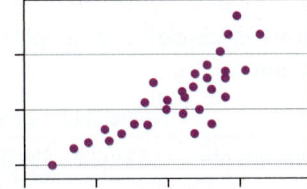

9.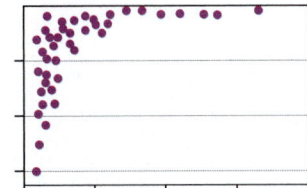

10.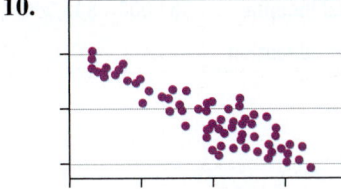

11.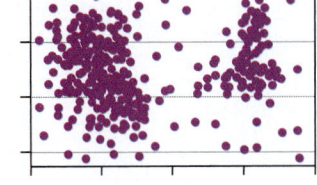

12.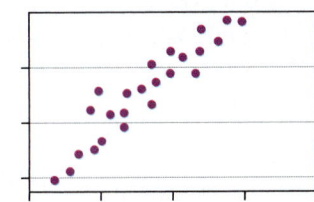

13.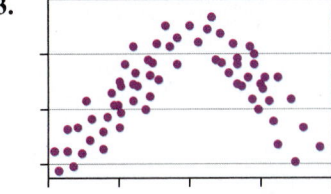

Consider each set of variables and predict whether the variables would have a weak negative relationship, a strong negative relationship, a weak positive relationship, a strong positive relationship, or no relationship at all.

14. Body weight and hours of exercise per week

15. A person's height and their self-esteem

16. Vision ability and IQ

17. Number of hours spent studying for a test and the grade on the test

Determine whether each correlation coefficient is statistically significant at the specified level of significance for the given sample size.

18. $r = 0.703$, $\alpha = 0.01$, $n = 12$

19. $r = 0.403$, $\alpha = 0.05$, $n = 25$

20. $r = 0.378$, $\alpha = 0.05$, $n = 29$

21. $r = 0.809$, $\alpha = 0.01$, $n = 8$

For each data set, find the following.

a. Estimate the correlation in words as positive, negative, or no correlation.

b. Calculate the correlation coefficient r. Round your answer to the nearest thousandth, if necessary.

c. Determine whether r is statistically significant at the 0.01 level of significance.

22. The following table gives the number of hours a student watches TV per week and his or her overall GPA.

Hours of TV Per Week and Overall GPA

TV Hours	20	10	25	15	14	13	21	9	5
GPA	2.0	2.46	2.3	2.9	3.0	3.2	3.5	3.3	3.7

23. The following table gives a sample of annual income and number of years of education.

Annual Income and Years of Education

Annual Income	$21,000	$39,000	$40,000	$39,500	$42,000	$55,500
Years of Education	12	12	14	16	16	16
Annual Income	$61,000	$45,000	$100,000	$142,000	$240,000	$205,000
Years of Education	17	16	16	20	22	21

24. The following table shows the diastolic blood pressure reading and the stress test score for 20 adults.

Blood Pressure Reading and Stress Test Score

Stress Test Score	Diastolic Blood Pressure Reading
51	67
59	66
62	71
63	76
64	73
68	77
71	77
70	76
72	80
82	82
78	79
79	83
83	81
84	83
88	85
87	90
89	82
91	80
90	86
90	88

25. The following table shows the heights of identical twins in centimeters.

Heights of Identical Twins

Sibling 1	Sibling 2
110.5	109.5
116.6	115.6
122.6	121.6
128.2	127.4
133.5	133.5
138.8	140.2
145.0	146.7
152.3	151.9
159.6	155.0
165.1	156.6
168.3	157.1
169.9	157.6
170.7	158.0

🚀 APPLICATIONS

Use the provided table to create a scatter plot of the data. Remember to label both axes as well as the graph itself. Describe whether the scatter plot indicates there appears to be a positive linear correlation, negative linear correlation, or no correlation.

26. Allied Technologies is trying to determine how well the different locations are utilizing the TFS system to track their work. The following table shows the data collected regarding the number of employees and the approximate percentage of their workload that had been accounted for in TFS.

Sample of TFS Usage by Allied Technologies

Number of Employees	12	8	15	16	9	11	7	13
% of Workload Accounted for in TFS	76	64	85	88	67	73	61	79

27. A group of hikers decided to record the temperature at various heights along their climb. Consider the following data they collected regarding their height above sea level and the temperature.

Sample of Temperatures at Different Elevations

Height Above Sea Level (m)	500	800	1000	2000	1500	3000	1200	2500
Temperature (°C)	26	25	24	14	20	10	23	15

28. Casandra made several trips from college to home. She often brought friends which meant additional luggage as well. She has been keeping track of her fuel consumption. The following table shows the approximate additional weight per trip and the fuel consumption in miles per gallon.

Sample of Fuel Consumption per Trip

Additional Weight in Vehicle (lb)	500	1000	750	600	850	900
Fuel Consumption (Miles per Gallon)	45	42	43	44	42	43

29. Melissa started exercising for 30 minutes per day and she recorded her weight every morning for a week, as shown in the following table.

Sample of Weight per Day

Day	1	2	3	4	5	6
Weight (lb)	178	177	178	176	179	177

30. The following table gives the data for the number of months students have taken piano lessons and the number of songs they can play from memory.

Number of Months and Number of Songs

# of Months	0	5	3	10	22	19	30	15	8	12
# of Songs	1	4	0	5	9	12	10	0	4	9

a. Determine the regression line $\hat{y} = ax + b$. Round the slope and y-intercept to the nearest thousandth.

b. Determine if the regression equation is appropriate, at the 0.05 level of significance, to use for making predictions. If so, answer part c.

c. If a student has taken eight months of piano lessons, make a prediction for the number of songs they can play from memory, if appropriate.

31. The following table shows students' test grades on the first two tests in an introductory literature class.

Test Grades in Introductory Literature Class

Test 1 (x)	61	45	71	81	89	55	84	91	95	59	77	88
Test 2 (y)	67	79	68	80	87	68	87	90	97	71	77	74

a. Determine the regression line $\hat{y} = ax + b$. Round the slope and y-intercept to the nearest thousandth.

b. Determine if the regression equation is appropriate, at the 0.05 level of significance, to use for making predictions. If so, answer part c.

c. If a student scored a 70 on his first test, make a prediction for his score on the second test, if appropriate.

32. The following shows the results on evaluations measuring self-esteem and perceived family support from 10 adolescents.

Self-Esteem and Perceived Family Support Evaluation Results

Self-Esteem	30	31	31	28	27	26	15	32	27	33
Family Support	13	13	19	21	8	4	10	12	7	17

a. Determine the regression line $\hat{y} = ax + b$. Round the slope and y-intercept to the nearest thousandth.

b. Determine if the regression equation is appropriate, at the 0.05 level of significance, to use for making predictions. If so, answer part c.

c. If an adolescent had a self-esteem score of 22, make a prediction for his perceived family support score, if appropriate.

33. A medical equipment company wishes to show that a new device works with the same degree of accuracy and precision as an earlier model to perform an electrocardiogram. One of the measurements tested was the change in radio electric waves during a cardiac cycle. The following results were collected from both healthy adults and those with cardiovascular problems.

Change in Radio Electric Waves During Cardiac Cycle

# of 5 mm Squares Between R Waves	
Old	**New**
2	2
3	3
4	4.5
3	3
6	6
4	4.5
3	3
5	5
3	3.5
2	2
6	6
4	4
6	6
5	5
3	3
2	2

a. Determine the regression line $\hat{y} = ax + b$. Round the slope and y-intercept to the nearest thousandth.

b. Determine if the regression equation is appropriate, at the 0.01 level of significance, to use for making predictions. If so, answer part c.

c. If the old machine had a reading of 5.5, make a prediction for the new machine reading, if appropriate.

12.3 PROJECT

CORRELATION: STORKS AND BABES

The Theory of the Stork[2] states that the number of out-of-hospital births in an area decreases as the number of storks in the area decreases and increases as the number of storks in the area increases. In the project for Section 5.2, you were introduced to a data set collected over a period of ten years consisting of pairs of storks in Brandenburg (the countryside around Berlin) and out-of-hospital births in the city of Berlin. That data set is reproduced below. In this project, we will determine the actual line of best fit for the data and determine whether there is a correlation between the number of pairs of storks and the number of out-of-hospital births in the area.

Year	Pairs of Storks (x)	Out-of-Hospital Births (y)
1990	970	900
1991	850	790
1992	980	780
1993	1210	890
1994	1280	960
1995	1270	1080
1996	1360	1070
1997	1120	1250
1998	1320	1130
1999	1370	1200

1. Create a scatter plot using this data set.

2. Find the equation of the regression line (line of best fit) for the data set and sketch the line on the scatter plot created in part 1.

3. Calculate the Pearson correlation coefficient for the data set. Then, check for significance at both the 0.05 and 0.01 levels. Is there a statistically significant correlation between stork pairs and out-of- hospital deliveries at both levels of significance?

4. If there are different conclusions for the different levels of significance in part 3, discuss potential causes and implications of this difference. If the conclusion is the same, did you find it surprising considering the scatter plot?

5. Use the equation of the regression line found in part 2 to predict the number of out-of-hospital births when during a given year 1100 pairs of storks were observed.

6. Use the equation of the regression line found in part 2 to predict the number of out-of-hospital births during a given year when 0 pairs of storks are observed.

7. Are the results obtained in parts 5 and 6 reasonable? Explain why or why not.

---

2   Thomas Höfer, Hildegard Przyrembel, and Silvia Verleger, "New Evidence for the Theory of the Stork," *Paediatric and Perinatal Epidemiology*, Volume 18 (January 2004): 88–92.

12.4 DATA STORYTELLING

OBJECTIVES

1. Learn how to properly communicate results.
2. Create word clouds.
3. Create bubble charts.
4. Read infographics.
5. Read data dashboards.

Fun Fact

British chef Jamie Oliver kicked off his 2014 TED talk, "Teach Every Child about Food," with this powerful opening line:

"Sadly, in the next eighteen minutes, when I do our chat, four Americans that are alive will be dead from the food they eat."

Watch the full TED talk here: hawkes. biz/OliverTEDTalk

OBJECTIVE 1 Communicating the Results

The last pillar of data science is communicating your results. It's been said that your findings are only as useful as your ability to communicate them. The best data scientists help their audience, client, or colleagues *see* the data and understand what it reveals. Remember, it's not your job as the data scientist to avoid the truth of the data. It's your job to deliver the truth of the data in the best way possible. For instance, you may need to explain your statistical insights to a business manager who needs to be convinced to spend money or change processes, or you may need to explain the results to a programmer who doesn't think statistically, to a fellow expert, or to the general public. In essence, to explain the results, you need to be a good storyteller.

Telling the story of data is a powerful skill no matter the medium: blog posts, emails, power points, technical papers, conference presentations, or one-on-one conversations. As you begin to pull together your ideas, you should focus on the message you want to send. A good story evokes a strong response and has a lasting impact. Think about the setting (the time and place) over which the data story exists. Does the setting of the data impact its story? Is it important to the narrative?

For a presentation to be effective, it needs to have

- a well-defined purpose,
- strategic organization,
- clear transitions between points,
- strong supporting details, and
- an awareness and inclusion of the audience.

Effective presentations also have rhetorical appeals to logic and reasoning, authority and trustworthiness, and emotions. If a presentation is missing any of these elements, the story you are telling about the data might not come across as clearly or strongly as possible.

No matter the size of the audience, effective presentations have several essential elements that you should be aware of. When people listen to presentations, they tend to remember the beginning and ending most clearly. As a result, you must create a compelling introduction that will show the audience the purpose of the presentation and get them excited about the journey you're going to take them on. Like a map, this introduction will show the audience where they're going and help them see the precise benefits they will gain from what they're about to hear.

In addition to clearly stating the purpose of the presentation, the introduction sets the audience's expectations. Start by greeting the audience and briefly tell them who you are. Use an appeal to credibility (*ethos*) to establish your authority by highlighting your past work experience or knowledge of the subject. You can also appeal to emotion (*pathos*) to reassure the audience that you understand them and their needs. Consider mentioning the presentation's time frame; not only will this help the audience know what to expect, it will also encourage you to stay on track. It's helpful to remember that the average attention span is constantly shrinking.

Fun Fact

Professor David Spiegelhalter, Chair of the Winton Centre for Risk and Evidence Communication at the University of Cambridge, won the 2020 Faraday Prize for his ability to vividly illustrate key insights from the disciplines of statistics and probability to the public at large and to key decision makers. He urges people to think about their emotional reactions to numbers.

✔ Skill Check 12.4.1

Brainstorm two appropriate engagement activities for the following audiences:

a. A club's fundraising committee

b. A women's motivational conference

c. A training session for new sales associates

Helpful Hint

Eye contact between the speaker and the audience is very powerful. Hold your head up and look into the eyes of your audience—just try not to stare at a single person.

✔ Skill Check 12.4.2

Think of at least two gestures not listed in Table 12.4.1 and explain what the gestures might suggest.

Another important goal of the introduction is to engage the audience. Keep in mind that people like to feel included, not lectured. One way to do this is by adding a funny image to your presentation (one that is relevant to the topic) to make the audience laugh and to lighten the mood. To actively engage the audience, consider asking a question or taking a poll during the presentation. You can also add a plot twist: once you've got their attention, take them on a different path.

> "Increasing sales by 35% is a great result, but that's not what we're going to talk about."

You might reenact the situation—take them back in time and show them the path that your brain followed to get you to the results. The goal is to put the audience in your shoes and let them get excited with you.

> "It's the Fall of 2021 and this is where we were."

It's also important to make sure everyone is on the same page. Audiences often have a mixed level of technical or subject-based knowledge, and you need to speak to all of them at once.

> "Just so we're all together, when I use the term bad cholesterol, I mean your LDL cholesterol."

Lastly, some of the best ways to get your audience excited about what you're talking about is to be passionate about your findings and smile! It's contagious.

In addition to compelling introductions and clear organization, presentations have a few more tools to help people listen to, understand, and remember information. Here, we'll focus on two of those tools: nonverbal signals and visual aids.

During presentations, your movements, gestures, and facial expressions are nonverbal signals that are visible to the audience, and they affect your message. Subconscious tics, like rocking back and forth or crossing your arms, are distracting and may even hurt your credibility by suggesting that you are not confident in what you are presenting. Channel your energy into gestures that express and emphasize your ideas. Consider the following commonly recognized gestures to reinforce communication.

Table 12.4.1: Common Nonverbal Gestures

Gesture	What It Suggests
Hold up a finger, or raise fingers sequentially	An important point or list of reasons, items, or examples
Starting with hands in front of the body, move hands in opposite directions	Opposing or contrasting ideas
Sweep one hand across the audience	Inclusivity
Place one hand over heart	Personal belief or emotion
Arms extended to the sides with palms facing up	Openness, humility, or trustworthiness

Good visual tools, like pictures, graphs, or charts add strength to your presentation. The best visuals help enhance your story and emphasize the main points to your audience. However, these visuals should *support* your point; they should not *be* your point. They are not the main thing that the audience should focus on—you are! Have you ever sat through a presentation where the speaker just read the slides out loud to you? I'm sure you thought, "I can read. Tell me something that isn't on the slide." The text in the visual portion of presentations should be limited to very short summaries or bullet points that are elaborated on by the oral portion of the presentation.

Today, it is common for people to post or tweet about what they hear and see in a presentation. This social-media traffic is a great way to expand the impact of what you have to say. It is worth thinking about what you might say that is "tweetable," and make sure that if someone posts a picture of one of your slides that it tells the story that you are trying to tell.

Overly elaborate visual aids, flashy images, or unrelated figures can end up distracting and confusing the audience. Try to keep what you display clean and easy to read by using the same fonts, complementary colors, and the same style of graphics. With visuals, the old adage "less is more" certainly applies. Minimal visuals keep your audience focused on your words. Relate the problem and how you approached it, gently walking your audience through the process without all the data cleaning and stats that were involved. Even if people don't understand something, they can see it with good visuals. You don't have to acknowledge every obvious anomaly in the data. Leaving some mystery is good for the end of the presentation so that there is an easy transition into the question-and-answer portion of your talk.

Lastly, a strong conclusion is just as important as a compelling introduction. You have taken your audience on a journey, and you need to make sure that at the end they feel that they have arrived at the destination and understand what they should take away from it. Try to find a way to make them continue to think about what you had to say, even when they go on with the rest of their day. Below are some dos and don'ts to follow for a conclusion.

DO:

- ✅ Reflect on the impact of your data and how it may affect others. Share the evidence that what you did works. Include practical applications.

- ✅ Consider using a creative ending strategy, such as leaving your audience with a final thought or question.

- ✅ Consider referencing the introduction of your story and tie it to the conclusion.

DON'T:

- ❌ Over present or stop abruptly because you don't know how to end the presentation.

- ❌ Explicitly state the point of your narrative.

- ❌ Summarize the entire story you've just told.

Now that we have some basics about how to organize your presentation, we will turn to ways to represent your data's story. Throughout the remainder of this section, we are going to explore a variety of approaches to data visualization—ways to visually make what your data has to say come alive. Used in the right way, each approach can go a long way to help your audience gain profound insight into the data. It is important to note that, as well as communicating the data's story to an audience, data visualizations can also be invaluable tools to you throughout the entire process—helping you explore your data once it has been suitably cleaned.

> 💬 **Helpful Hint**
>
> Don't turn your back to the audience and read from the screen. Doing so suggests that you don't know what it says or that you've lost your place.

OBJECTIVE 2 Word Clouds

As we have seen, data can come in all shapes and sizes: numerical, chronological, logical, and of course textual. We will begin our tour of data visualizations with a way to visualize a passage of text, such as a speech, or opinions from a series of interviews, or even the contents of a book.

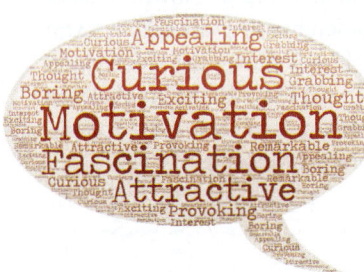

Figure 12.4.1

Word clouds, or *tag clouds*, represent word counts in a passage of text using a picture format. Each word is counted and then represented by how frequently it appears. The font size or color of the word visually signifies its frequency in the body of the text—the bigger and bolder the word, the more it occurs. Because word clouds are so quick and easy to create using free software, they make a great way to immediately envisage data in documents, surveys, or evaluations.

The benefits of using word clouds to tell a story spring from the positive emotional response and engagement that they provide that a page of text simply cannot. Word clouds stimulate curiosity and result in questions that lead to further investigation. Word clouds made from open-ended questionnaires pick up on common likes or dislikes as well as themes that might otherwise be missed, because similarities begin to emerge.

Websites like WordArt.com and WordClouds.com provide platforms that make the mechanics of creating word clouds stress-free. You simply upload your text in one of several formats and personalize your image. Your job as the storyteller is to choose the best way to speak visually through the word cloud. For instance, suppose you want to convey to your audience the most prominent context from a large piece of text. You could show a list of words from the text or even read through sections of the text. However, you would likely lose your audience to boredom. In contrast, well-constructed word clouds can convey that same information in a much more engaging way.

Later in this section, we will consider some more weighty texts, but for now let's begin with something more familiar. Consider the word clouds in Figure 12.4.2. All three designs are made from the first chapter of text in *Alice's Adventures in Wonderland* by Lewis Carroll. Take a moment and decide which of the three word clouds does the best job conveying the content of the chapter.

Figure 12.4.2: Word Clouds of Chapter 1 of Alice's Adventures in Wonderland

Although it is a nice idea, the book graphic is harder to read because of the light colors that were chosen. The tree is also nice, but a little more difficult to discern what words have the same weight in the text because of its shape. If you chose the bottom cloud as the best choice, we would agree that it does the best at visually describing Chapter 1 of the book and is far more intriguing than a simple list of words.

When used correctly, word clouds can be insightful and meaningful. However, there are shortcomings with them that need to be noted. Sometimes they are simply extra fluff and provide little insight into the data. Figure 12.4.3 shows the word cloud created from a set of real estate data. Because the tool used to make the graphic was looking for text strings in the file, the only information it picked up was the column headings listed in Row 1. Obviously, this is not a useful word cloud.

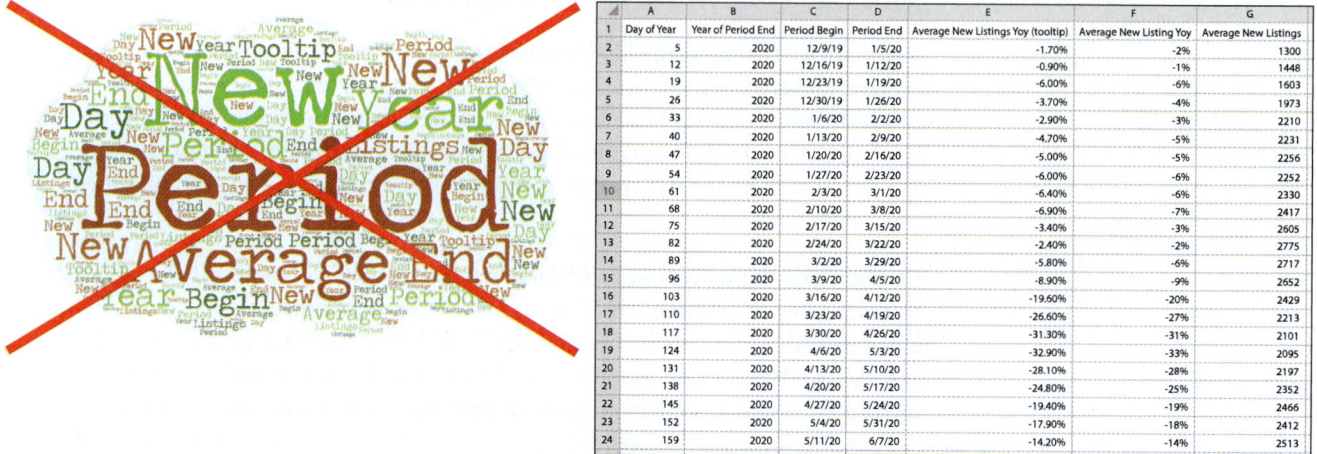

	A	B	C	D	E	F	G
1	Day of Year	Year of Period End	Period Begin	Period End	Average New Listings Yoy (tooltip)	Average New Listing Yoy	Average New Listings
2	5	2020	12/9/19	1/5/20	-1.70%	-2%	1300
3	12	2020	12/16/19	1/12/20	-0.90%	-1%	1448
4	19	2020	12/23/19	1/19/20	-6.00%	-6%	1603
5	26	2020	12/30/19	1/26/20	-3.70%	-4%	1973
6	33	2020	1/6/20	2/2/20	-2.90%	-3%	2210
7	40	2020	1/13/20	2/9/20	-4.70%	-5%	2231
8	47	2020	1/20/20	2/16/20	-5.00%	-5%	2256
9	54	2020	1/27/20	2/23/20	-6.00%	-6%	2252
10	61	2020	2/3/20	3/1/20	-6.40%	-6%	2330
11	68	2020	2/10/20	3/8/20	-6.90%	-7%	2417
12	75	2020	2/17/20	3/15/20	-3.40%	-3%	2605
13	82	2020	2/24/20	3/22/20	-2.40%	-2%	2775
14	89	2020	3/2/20	3/29/20	-5.80%	-6%	2717
15	96	2020	3/9/20	4/5/20	-8.90%	-9%	2652
16	103	2020	3/16/20	4/12/20	-19.60%	-20%	2429
17	110	2020	3/23/20	4/19/20	-26.60%	-27%	2213
18	117	2020	3/30/20	4/26/20	-31.30%	-31%	2101
19	124	2020	4/6/20	5/3/20	-32.90%	-33%	2095
20	131	2020	4/13/20	5/10/20	-28.10%	-28%	2197
21	138	2020	4/20/20	5/17/20	-24.80%	-25%	2352
22	145	2020	4/27/20	5/24/20	-19.40%	-19%	2466
23	152	2020	5/4/20	5/31/20	-17.90%	-18%	2412
24	159	2020	5/11/20	6/7/20	-14.20%	-14%	2513

Figure 12.4.3: Housing Data and Corresponding Word Cloud

Other factors can also affect the observer's visual interpretation of data displayed in a word cloud. For instance, consider the weight of the words in Figure 12.4.4.

Figure 12.4.4: Word Cloud Projecting the Wrong Tone

The words *helpful*, *quiet*, and *attractive* have similar weights, but they appear to be different sizes because of the length of the words and their random placements in the shape. Notice that the word *attractive* does not carry the same weight as *quiet* but has gained a place of prominence in the word cloud simply by its positioning. The shape of the word cloud itself can also have an impact on what it is communicating. Although the words in Figure 12.4.4 are all positive in nature, the word cloud itself is projecting sadness.

Example 12.4.1

Interpreting Word Clouds

Bill Clinton's first State of the Union address, delivered on February 17, 1993, was one of the most-watched State of the Union addresses in history according to the Nielsen TV ratings, pulling in nearly 67 million viewers. The flag word cloud in Figure 12.4.5 is made from the text of his speech.

Figure 12.4.5: Word Cloud of President Clinton's Speech

a. Based on the word cloud, what do you think was the main policy focus that President Clinton laid out in his speech?

b. How might you modify the word cloud to help make those main points more prominent?

c. Compare the word cloud of President Clinton's speech to the word cloud in Figure 12.4.6 of the second State of the Union speech from President George W. Bush, who became president after Clinton. In his speech, President Bush was addressing the September 11th terrorist attacks on the World Trade Center in New York. Can you recognize the different topics that were addressed in January 2002?

Figure 12.4.6: Word Cloud of President Bush's Speech

Solution

a. The most prominent words are *American*, *people*, *will*, *must*, and *new*. However, other weighty words are to do with jobs, tax cuts, healthcare, and the deficit. The focus may have been on improving the economy.

b. Removing some of the common but less informative words like *will*, *can*, and *must* would make the policy intentions more pronounced.

c. The words *will* and *American* are still very prominent, but there is more focus on security and terror in the world since the September 11th terrorist attacks occurred only 4 months before.

Example 12.4.2

Creating Word Clouds

Create a word cloud of the text of *Little Women* by Louisa May Alcott, downloaded from Project Gutenberg website (gutenberg.org). Use your word cloud to deduce who is the main protagonist, as well as supporting characters.

Solution

The first step in creating a word cloud of *Little Women* is to obtain the full text from Project Gutenberg. To do this, go to Gutenberg.org and search for "Little Women". After selecting the correct book from the results, click on the link for the text in the Plain Text UTF-8 format. This will open the .txt file in a separate browser window. (If you wanted to download the text, you would need to right-click on the link and select "Save link as.") Note that the link is "https://www.gutenberg.org/cache/epub/514/pg514.txt". This will be useful in the next step.

Next, go to wordart.com and click on the "Create Now" link. In the new browser window that opens, remove any existing words from the "Words" area, then click the "Import" button. In the pop-up box, select the "Web" option and enter the Web URL of the .txt file of Little Women in the box, as shown in Figure 12.4.7.

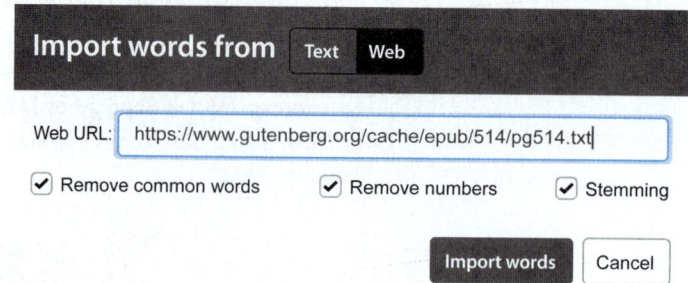

Figure 12.4.7

Ensure that the boxes next to "Remove common words," "Remove numbers," and "Stemming" are checked. Then, click "Import words." Once the words are selected, click "Visualize." The resulting word cloud will look similar to Figure 12.4.8. (Note that the visualization is randomized, so your word cloud will look different.)

Figure 12.4.8

Judging from the word cloud, Louisa May Alcott's main character in the book Little Women is Jo. Other characters include Meg, Laurie, Amy, and Beth.

Figure 12.4.9

OBJECTIVE 3 Bubble Charts

Throughout this book, we have looked at a variety of ways to represent numerical data and the relationship between two variables with pie charts, bar charts, and scatter plots. A **bubble chart** is an extension of the scatter plot. It allows the analysis of data points that each consist of three pieces of information. Two of the variables are plotted along the x- and y-axes, while the size of the dot is determined by the third variable. Sometimes, by making the color of each bubble have a significance, a fourth dimension can also be represented.

Figure 12.4.10 shows the bubble chart that the World Bank used in 2018 to highlight the relationship between basic sanitation services (proper toilets or clean water for washing or drinking), childhood growth, and the size of the economy's population.[3] The x-axis represents the percentage of the population that has access to basic sanitation services, and the y-axis measures the percentage of children under the age of 5 who have stunted growth. The size of the bubble corresponds to the economy's population. Notice that they went a step further and used color to denote seven geographical areas of the world. Based on your knowledge of populations across the world, can you guess what country is represented by the largest of the blue circles from South Asia? How about the largest of the gold circles from the East Asia and Pacific area? If you guessed India and China, you'd be correct.

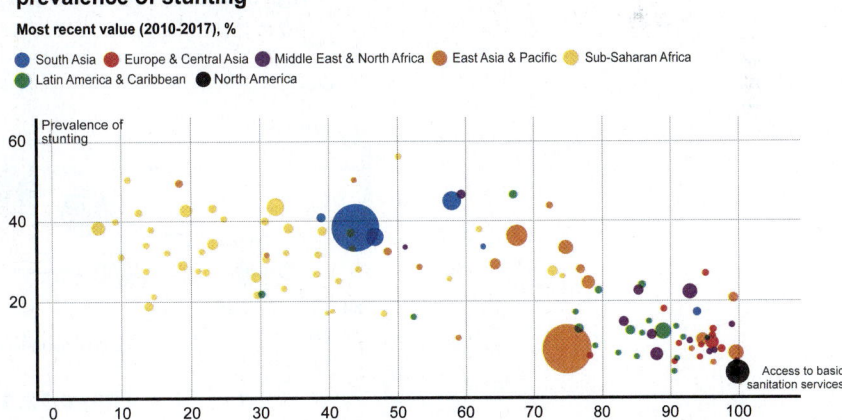

Figure 12.4.10

Since bubble charts are a newer type of visual than scatter plots, our brains have a bit more work to do to focus in on the implications of a third variable in a graph. In 2006, a Swedish statistician and physician introduced some of the most famous bubble charts as a way to visually tell the evolving story of our world, its economy, and its future. Telling his audience that he had an urgency to communicate the story that massive amounts of data was pointing to, Hans Rosling held captive those in attendance by using bubble charts along with his distinctive ability to tell a story. He led people through his charts, helping them to integrate the significance of the size of bubble as he went along.

3   "World Development Indicators," The World Bank, last accessed November 18, 2021, https://databank.worldbank.org/sanitation-and-stunting/id/96d4f71b.

It's important to fight the urge to use a bubble chart just to add pizzazz to a presentation. Make sure the addition of the third (or fourth) variable actually adds to your story and doesn't detract from it. Too many bubbles can result in bubble clusters, which can result in a lack of clarity.

Example 12.4.3

Interpreting a Bubble Chart

Figure 12.4.11 shows a bubble chart made from data shown in Table 12.4.2. The table lists the top ten NBA players with the highest average points per game between 1996 and 2020 according to data.world.

Table 12.4.2: Top Ten NBA Players between 1996 and 2020

Name	Total Games	Total Points	Total Assists
Allen Iverson	914	23,823	5484
Anthony Davis	519	12,430	1181
Damian Lillard	606	14,733	3962
James Harden	824	20,727	5161
Kevin Durant	849	22,824	3502
Kobe Bryant	1346	32,573	6407
LeBron James	1256	33,993	9346
Luka Dončić	123	3050	904
Michael Jordan	306	7742	1285
Trae Young	139	3371	1202

Top Ten NBA Players between 1996 and 2020

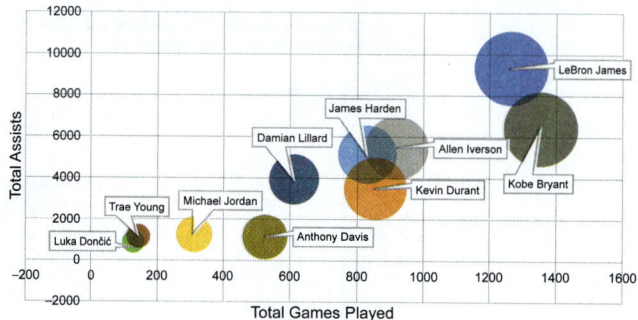

Figure 12.4.11

a. What does the size of each bubble represent?

b. Given that Michael Jordan began his NBA career in 1984 and retired for good in 2003, (retiring twice in between by sitting out four of those years) do you think this bubble chart fairly represents his career?

c. What can you deduce from the bubble chart about the relationship between the number of games played and the number of points a player had, as well as the number of assists and the number of points?

Solution

a. From the bubble chart, we can see that the horizontal and vertical axes represent the number of games played and the number of assists for each player. Thus, the size of the bubbles represents the total number of points each player scored.

b. Because Michael Jordan began his career in the NBA in 1984, these data only contain about half of his games, career total points and assists. Therefore, we can say that this bubble chart does not represent his whole career fairly.

c. The number of points a player attained is represented by the size of the bubble. In general, each player's bubble gets bigger as you move along the x-axis. The bubbles also increase in size as you move up the y-axis. Thus, the more games a player played, the more points he scored. Similarly, a player with more assists tended to score more points. The exception to this second point is with Anthony Davis. The size of his bubble, indicating the number of points he scored, is larger in comparison to the three other players with similar numbers of assists.

Example 12.4.4

Interpreting a Bubble Chart

Data for six-month's worth of loans made in a certain county in Illinois is displayed in the bubble chart in Figure 12.4.12. The x-axis shows the size of the loan, the y-axis shows the number of loans made at that level, and the size of the bubbles represents the total amount of money the banks needed to fund the loans at that level.

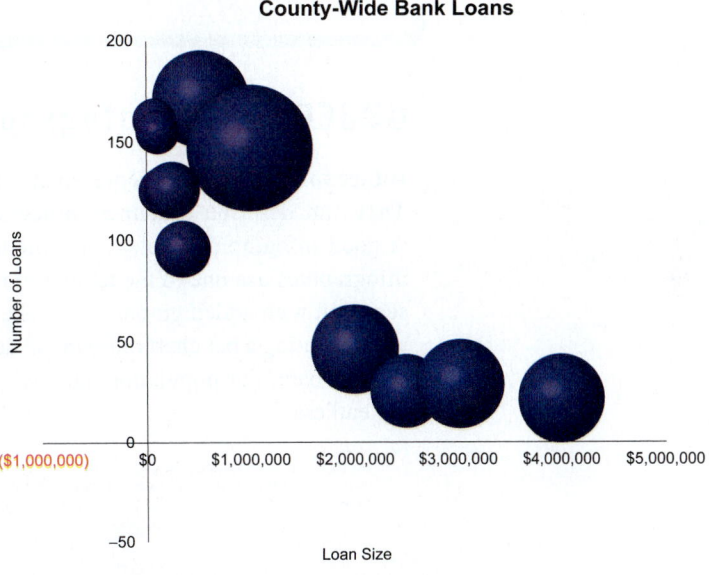

Figure 12.4.12

Use the bubble chart to answer the following questions.

a. What is the largest loan that was made?

b. What level of loan was made most during the six months?

c. Which level of loan required the most funds from the bank during this time?

d. Why are the answers to questions a. and b. different?

e. If you were presenting the answers to questions a. through d. to a local civic group, would you use the bubble chart in your presentation?

f. A bank is planning to open a new branch in this county. Based on the bubble chart, what types of loans should the bank expect to have applications for? How would this influence the staffing of the bank and marketing targets?

Solution

a. The bubble that is farthest to the right is at $4,000,000, so the largest loan made during this period was for $4,000,000.

b. The bubble that is highest along the y-axis is the most frequently made loan. Its x-value is around 500,000, so we know that loans at the $500,000 level were made the most during this period.

c. Because the bubble at the million-dollar level is the largest, we know that million-dollar loans required the most funds during this time.

d. Although the largest loan was for $4,000,000, there were fewer of these loans granted. There were many more loans at the $1,000,000 level, which accounts for their larger portion of the overall net funds.

e. A local civic group would likely be less interested in the exact figures of the loans, and thus the overall picture the bubble chart gives is a good resource to use for the presentation.

f. Loans at or below the $1,000,000 level account for both the most loans granted and the largest portion of overall net funds. Therefore, the bank should expect most applications to be for loans at or below $1,000,000, with few loans above $2,000,000. The bank should train employees to process loans in this range and focus marketing campaigns aimed at attracting customers needing loans up to $1,000,000.

OBJECTIVE 4 Infographics

Infographics are visual representations of any kind of information, not just data. They can describe timelines, processes, comparisons, as well as information. A good infographic is clear and memorable with concise text. Often, effective infographics use one of the textual or numerical graph types that we have already seen, but with added graphics to make the message more engaging. For instance, when creating a bar chart of financial data, you can use stacks of money rather than colored bars. For population data, why not use lines of people? The possibilities are endless.

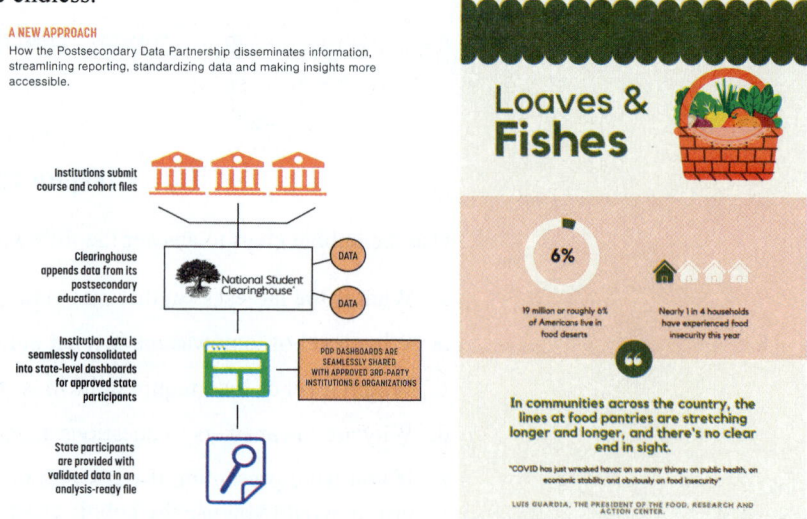

Figure 12.4.13: Examples of Infographics[4]

---

4   Clearinghouse Infographic courtesy of Complete College America

When creating an infographic, you need to consider who your audience is, what the audience needs to learn from the graphic, and why the information is relevant to them. Remember, you are telling the story of data by taking a complex data set and translating it into an easily understood, engaging picture. The good news is you don't have to be a designer to pull together an effective infographic. Free software such as Adobe Spark or Snappa have thousands of templates to choose from. Excel and Word also have built-in tools like SmartArt that you can customize yourself.

Example 12.4.5

Interpreting Infographics

The portion of an infographic[5] shown in Figure 12.4.14 is from the University of Southern California Rossier's online teaching degree program. In it they compare the amount of money spent on education to education performance metrics for the United States and eleven other countries.

a. Who spends the most overall on education among the 12 countries?

b. Who spends the most per school-aged child among the 12 countries?

c. Where does the US rank in math test scores among the 12 countries?

d. Do you think this is an effective infographic? Why or why not?

Solution

a. The size of the circle indicates that out of the twelve countries shown, the United States spends the most on education.

b. Based on the bar graphs in the first line, we can see that the United States also has the highest per-child spending amount.

c. Math test scores rankings are shown on the fourth line of bar charts. The United States is the tenth out of the twelve countries for math test scores.

d. The infographic is both eye-catching and memorable. The use of lines joining the data to its respective country communicates a sense of complexity that is almost certainly involved in the data story. However, it is a very busy graphic, which might turn some users away. The use of the total amount spent on education as the main category could be questioned since the United States is the wealthiest country in the list. The per-student spending amount would have been a better metric to use at the top.

OBJECTIVE 5 Data Dashboards

A **data dashboard** is fundamentally a compilation of dynamic graphs all in one location. It allows users the ability to visually intake a wide range of information all at once, providing a means to both monitor and analyzing relevant datasets. The best dashboards are informative in an interactive and intuitive way. Dashboards are highly customizable; departments, companies, or any particular group can choose which goals or processes to monitor in order to advance improvements or generate new ideas.

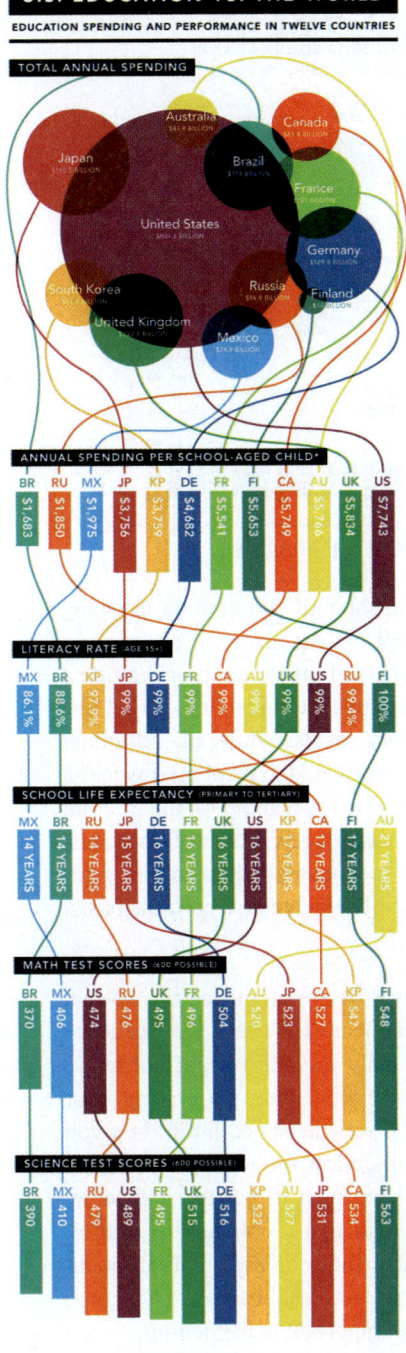

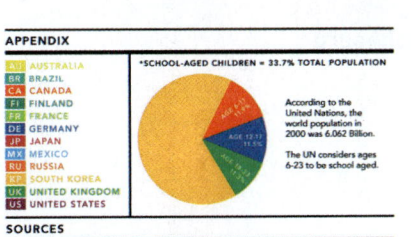

Figure 12.4.14

---

5   Rossier Staff, "U.S. Education Spending and Performance vs. The World [INFOGRAPHIC]", USCROssier, February 9, 2011, https://rossieronline.usc.edu/blog/u-s-education-versus-the-world-infographic/.

Figure 12.4.15: Examples of Data Dashboards

As the examples in Figure 12.4.15 illustrate, a dashboard can contain as many or as few metrics as desired. A dashboard can tell the data story at a birds-eye view and encourage drilling down to specific pockets of information or provide a detailed analysis of particular data sets. Management-level personnel might use dashboards to track overall revenue and inventory levels, while an individual sales dashboard might keep a sales representative motivated to meet their weekly, monthly, quarterly, and yearly goals. A business that relies heavily on the internet could use a dashboard to display web analytics like webpage visits, traffic sources, click-throughs, or bounce rates. When dashboards are *dynamic*, they provide real-time information as the data updates. Other dashboards update on a periodic basis: hourly, daily, monthly, and so on.

The main goal of a dashboard is to raise awareness of important metrics so that appropriate and timely action can be taken. A dashboard isn't like a noticeboard where you share anything interesting. It should only include things that really matter to your business. Think about what you see when you are driving your car: if the speedometer shows that you are driving over the speed limit, then you need to slow down; if the fuel gauge says that you are low on gas, then you need to stop at a gas station; and a check engine light means a trip to the mechanic. Data dashboards function similarly; they may signal that a company is getting dangerously low on inventory or that a particular location is especially popular. No matter the information to be conveyed, the best data-dashboards are designed to help the user know at a glance when something important is happening.

Characteristics of a Highly Effective Data Dashboard

- Aligns with goals
- Uses clear and intuitive visuals
- Places numbers into context
- Has a logical layout
- Tells a story
- Keeps data current

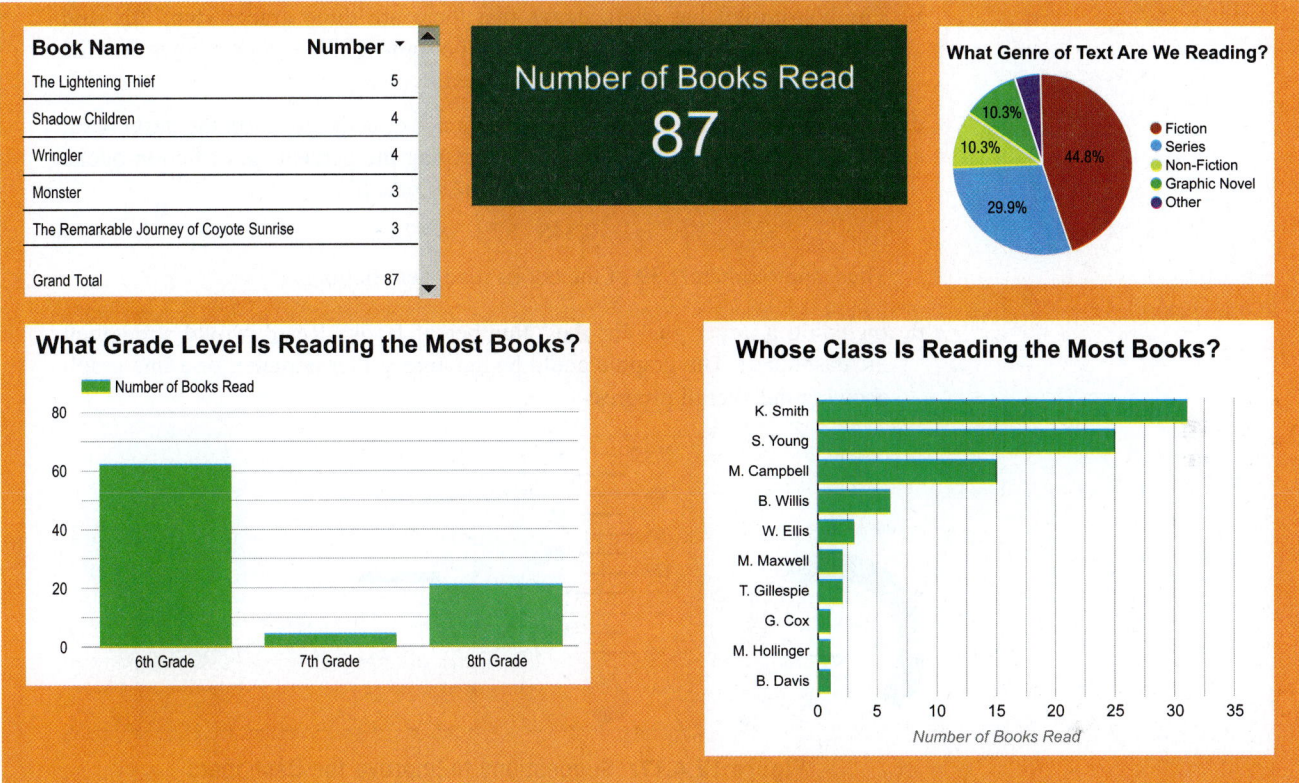

Example 12.4.6

Reading Dashboards

As part of the push to increase student reading outside of class assignments, Peace Middle School put together the following reading dashboard for their students.

Figure 12.4.16

a. Describe what each of the five graphics displays and any connections that exist between the graphs.

b. How many grades levels are represented in the dashboard? How many teachers are represented?

c. How many fiction books were read to date?

d. How could you change the graphic for "Number of Books Read" so that it tells a better story?

e. Does the dashboard give you a good sense of where the school is on their progress for reading? List any ways that you might make the dashboard better and explain your reasoning.

Solution

a. There are five graphics, and each displays a different type of data.

Top Left: This graphic displays the book titles that students are reading along with the number of times a student has read the book. The "Grand Total" at the bottom will be the same as the middle figure, "Number of Books Read."

Top Middle: The grand total matches the sum of the bars in the bottom left graph as well as the sum of the bars in the bottom right graph.

Top Right: The pie chart displays the percentages of each genre of book read.

Bottom Left: This bar chart displays of the numbers of books read by each grade level.

Bottom Right: This bar graph displays the numbers of books read by the students, grouped by teacher.

b. The bottom-left bar chart shows that there are three grades (6th, 7th, and 8th) represented in the dashboard. The bottom-right bar graph shows that the classrooms of 12 teachers are represented.

c. Although the number of fiction books is not directly given on the dashboard, we can calculate this number by multiplying the percentage of fiction books read, 44.8 %, by the total number of books read, 87.

$$0.448 \cdot 87 = 38.976$$

Thus, approximately 39 of the books read were fiction.

d. Including a target goal for the "Number of Books Read" would strengthen the dashboard. The graphic could be a gauge, a thermometer, or a line graph showing the overall progress.

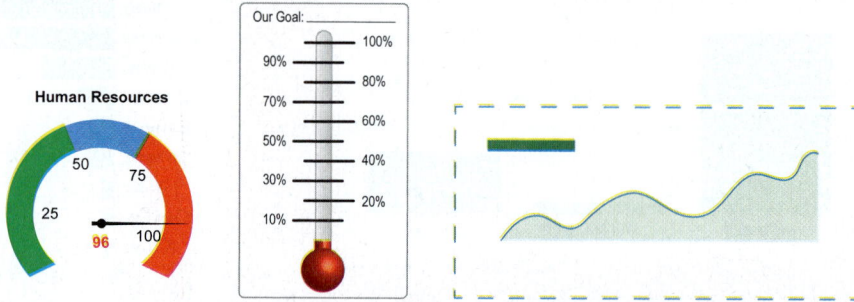

Figure 12.4.17: Suggestions to Improve the Dashboard

e. Overall, the dashboard gives a good snapshot of which students in general are reading, and which grade levels and classes could be doing a better job. All of the different graphs are very easy to read, and the dashboard does not feel overcrowded. However, there is nothing on the dashboard which gives a sense of how this rate of book reading compares to other years, or what goals each grade is trying to attain. Having a graphic such as the one described in part d. could help in this area. There is also nothing to indicate how current the data in the charts are or how many students there are in each grade. Lastly, the bar chart showing breakdown by teacher does not provide any indication what grade level each teacher teaches. It could be helpful as a link between the two bar charts.

Skill Check Answers

1. Answers will vary. For example, **a.** Tell a story about past fundraising experiences **b.** Tell a funny story about unmotivated people **c.** Do an icebreaker activity **2.** Answers will vary. **a.** Scratching head suggests confusion or uncertainty. **b.** Raising eyebrows suggests shock or disbelief. **c.** Bowing suggests gratitude or respect. **d.** Sitting down (with good posture) suggests comfort/ease or focus on something else. **3.** Most used word is State. Next three highest used words are People, Laws, and Right; No, you cannot tell how many times each word was used in a word cloud.

12.4   Exercises

✔ CONCEPT CHECK

1. A _____ illustrates how often common words are used within a text.

2. Bubble charts are an extension of _____.

3. _____ can describe timelines, processes, comparisons, and information.

4. Data dashboards are a compilation of _____ in one location.

🚀 APPLICATIONS

The following is a word cloud created from the book *Frankenstein; or, the Modern Prometheus* by Mary Wollstonecraft (Godwin) Shelley. Use the word cloud to answer the questions.

5. Based on this word cloud, what would you describe as the theme of the book?

6. How might you modify the word cloud to better illustrate that it is a horror fiction novel?

7. Explain why the words *Frankenstein*, *monster*, or *creature* are not prominent in the word cloud.

The following is a word cloud created from the book *Pride and Prejudice* by Jane Austen. Use the word cloud to answer the questions.

8. Based on the word cloud, who are the main characters of the book?

9. What type of novel would you guess *Pride and Prejudice* is based on the word cloud?

10. Explain why *Mr* is the most prominent word in the word cloud?

For each book, go to https://www.gutenberg.org/ to find a copy of the text and then create a word cloud.

11. *Great Expectations* by Charles Dickens

12. *Pygmalion* by George Bernard Shaw

The following bubble chart summarizes the sales success for a sales team during the first quarter of a year. The size of the bubbles represents the total sales, in dollars, made during the quarter. Use the bubble chart to answer the questions.

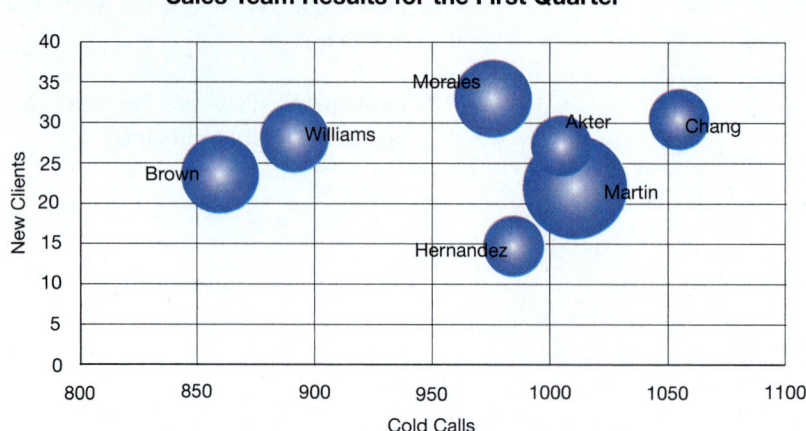

13. Which salesperson made the most cold calls during the quarter?

14. Which salesperson made the most sales during the quarter?

15. Which salesperson acquired the most new clients during the quarter?

16. Compare the success of Williams and Akter.

The following bubble chart summarizes the success of social media posts by an online store in regards to shares and resulting website visits. The size of the bubbles represents the total number of sales made as the result of the post. Use the bubble chart to answer the questions.

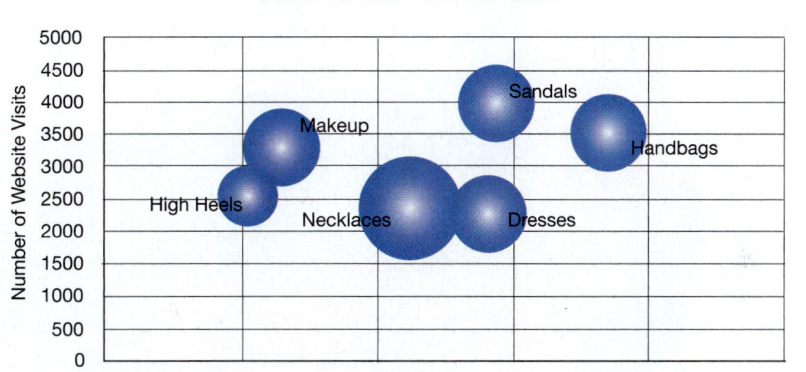

17. Which social media post topic was the most successful in terms of sales?

18. Which social media post topic was the most successful in terms of shares?

19. Which social media post was least successful in terms of website visits?

20. Based on this bubble chart, can a conclusion be drawn that a high number of shares or a high number of website visits translates into a high number of sales? Explain your reasoning.

The following infographic shows the marriage rates in the United States by state compared to the national average as reported by the United States Census Bureau. Use the infographic to answer the following questions[6].

State Marriage and Divorce Rates Compared to National Rate: 2018

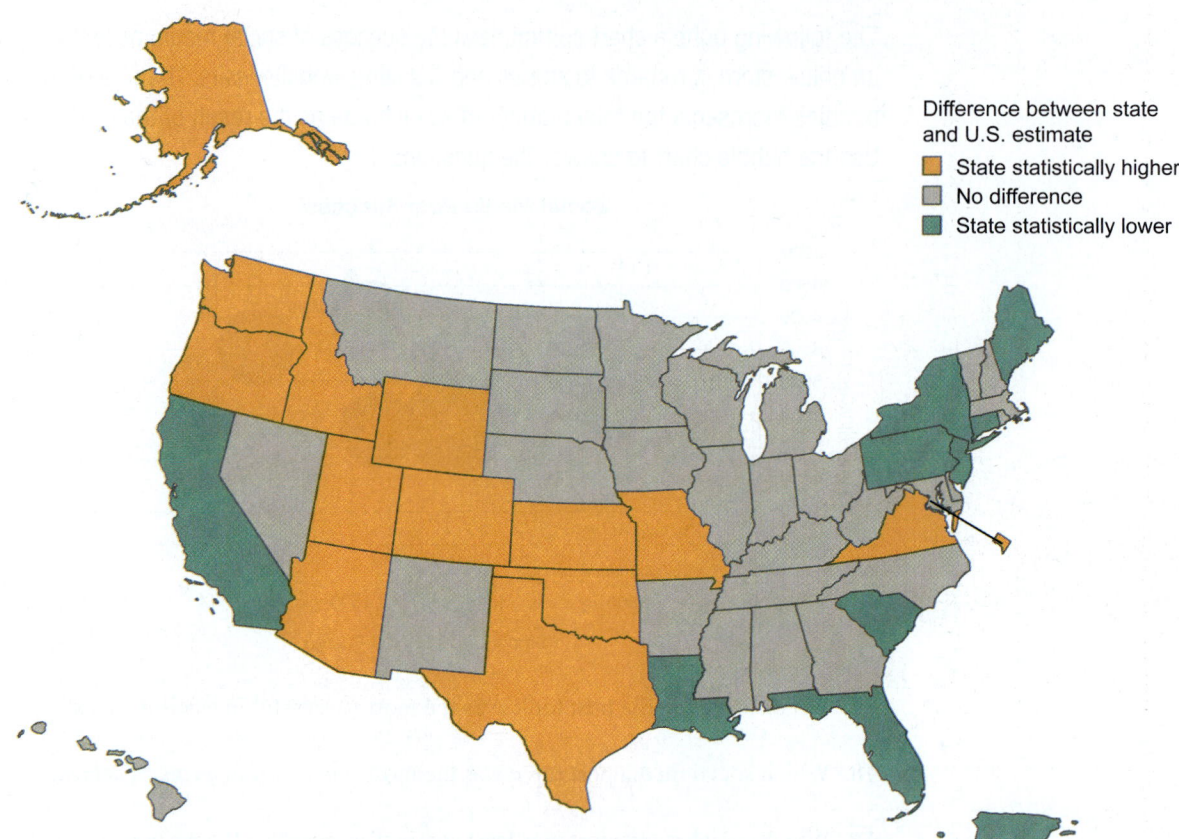

21. Does the infographic convey that the marriage rate in most states in the United States is higher than, lower than, or no different than the national average?

22. Can the infographic be used to illustrate specific differences in marriage rates among different individual states in the United States? Explain why or why not.

23. Suppose a campaign wanted to use this infographic to show that marriage rates are lower in popular tourist destinations. What additional information could be added to tell this story?

A sales firm has the following dashboard displayed in their office to show success during the current quarter. Use the dashboard to answer the following questions.

24. Describe the main information displayed in the dashboard.

25. According to the Weekly Sales table, which customer base made up the majority of sales during the quarter? Explain your reasoning.

6   U.S. Marriage and Divorce Rates by State: 2008 & 2018," United States Census Bureau, January 15, 2020, https://www.census.gov/library/visualizations/interactive/marriage-divorce-rates-by-state.html.

26. How could you change the Top Sales portion so that it tells a better story?

27. List ways that the dashboard could be improved to tell a better story of how well the company is doing overall.

Chapter 12 Exercises

1. For the following research question, identify a weakness and then give three stronger, alternative questions: "Which SPF of sunscreen is best for overall skin health?"

2. The owner of a shoe store wants to know how to optimize the window display to draw more customers into the store. Identify three questions that a data scientist who is unfamiliar with the specific domain knowledge indicated can ask their team to help strengthen their research process.

3. For the following data file snippet, determine the delimiter that would be used to separate the data into columns and determine how many columns the data would be separated into.

   ```
   Date,PurchaseTotal,PaymentMethod,AccountMade,Subscription
   11012021,$663.85,Paypal,yes,montly
   11072021,$627.91,Paypal,no,none
   11112021,$442.23,Credit,yes,biweekly
   ```

4. The following pivot table displays the inventory of electronic devices available to be borrowed from a local library.

Row Labels	Available	Checked Out	Grand Total
− Kids' Tablet	2	12	14
Needs Update		1	1
Up to Date	1	11	13
− Laptop	11	16	25
Needs Update		2	2
Up to Date	9	14	23
− Tablet	15	16	31
Up to Date	15	16	31
Grand Total	**26**	**44**	**70**

   a. What is the total inventory for each device that can be borrowed from the library.

   b. How many devices need to be updated?

   c. How many devices are currently checked out?

   d. How many total devices are in the inventory?

In each scatter plot, decide whether there appears to be a positive linear correlation, a negative linear correlation, or no linear correlation.

5.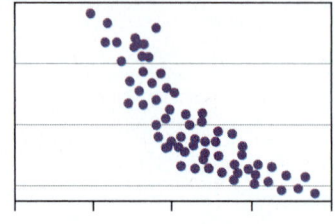

6.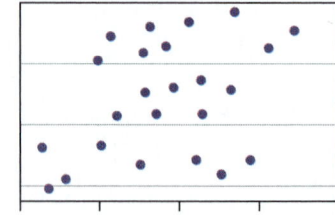

Consider each set of variables and predict whether the variables would have a weak negative relationship, a strong negative relationship, a weak positive relationship, a strong positive relationship, or no relationship at all.

7. Score on the ACT and first semester GPA in college

8. A person's eye color and their self-esteem

Determine whether each correlation coefficient is statistically significant at the specified level of significance for the given sample size.

9. $r = 0.913$, $\alpha = 0.01$, $n = 15$ **10.** $r = 0.525$, $\alpha = 0.05$, $n = 50$

For each data set, find the following.

a. Estimate the correlation in words as positive, negative, or no correlation.

b. Calculate the correlation coefficient r. Round your answer to the nearest thousandth, if necessary.

c. Determine whether r is statistically significant at the 0.01 level of significance.

11. The following table gives the number of hours a student watches TV per week and his or her overall GPA.

Hours of TV per Week and Overall GPA

TV Hours	30	10	28	12	11	13	21	9	5
GPA	1.8	2.46	2.4	3.1	3.4	3.2	3.5	3.3	3.7

12. The following table shows the annual salary of 10 adults and their corresponding years of education.

Adult Annual Salary and Education

Salary ($)	Years of Education
125,000	19
100,000	20
40,000	16
35,000	16
41,000	18
29,000	12
35,000	14
24,000	12
50,000	16
60,000	17

Use the linear regression model $\hat{y} = ax + b$ to predict the *y*-value for each value of *x*.

13. $\hat{y} = 25.33x + 353.16$

    a. $x = 8$

    b. $x = 12$

    c. $x = 18$

Solve

14. The following table shows students' test grades on the first two tests in an introductory literature class.

Test Grades on the First Two Tests in an Introductory Literature Class

Test 1 (*x*)	81	55	81	90	79	45	84	91	82	69	77	88
Test 2 (*y*)	79	68	83	87	77	55	87	90	87	71	77	74

    a. Determine the regression line $\hat{y} = ax + b$. Round the slope and *y*-intercept to the nearest thousandth.

    b. Determine if the regression equation is appropriate at the 0.05 level of significance to use for making predictions. If so, answer part c.

    c. If a student scored a 70 on his first test, make a prediction for his score on the second test, if appropriate.

15. Describe how you might design a word cloud of a horror novel to accurately convey the tone of the book.

16. Describe the benefits of using a bubble chart instead of a scatterplot to display data.

CHAPTER 12 PROJECT

Telling a Story with Data

As you have learned throughout this chapter, data can be acquired in a variety of ways and then cleaned up, analyzed, and used to tell a story. Whether the story told about the data is true or is strongly backed by the data is something that should always be considered. While visualizations of data can help simplify the story that is being told, these visualizations can often be misleading. Consider the following data set, which shows the youth literacy rate, poverty level, access to electricity, birthrate, and mobile cellular subscriptions worldwide over a 15-year span.

Year	Youth Literacy Rate (%)	Poverty Level (%)	Access to Electricity (%)	Birthrate (per 1000 people)	Mobile Cellular Subscriptions (per 100 people)
2003	87.882	24.7	80.009	20.859	22.22
2004	88.343	22.9	80.16	20.703	27.292
2005	88.475	21	80.161	20.57	33.766
2006	88.764	20.3	81.251	20.422	41.564
2007	88.667	19.1	82.205	20.341	50.263
2008	89.451	18.4	82.284	20.232	59.375
2009	89.523	17.6	82.765	20.036	67.496
2010	89.567	16	83.3	19.809	76.162
2011	89.783	13.9	82.115	19.628	83.716
2012	90.337	12.9	84.746	19.51	87.929
2013	90.674	11.4	85.031	19.298	92.461
2014	90.993	10.7	85.553	19.21	96.053
2015	91.03	10.1	86.579	18.957	97.421
2016	91.447	9.7	87.73	18.942	100.72
2017	91.629	9.3	88.617	18.621	102.869

Source: The World Bank|Data, accessed December 18, 2021, https://data.worldbank.org/.

1. Suppose the following graph was given as part of a presentation on world youth literacy rates. What does the graph seem to imply?

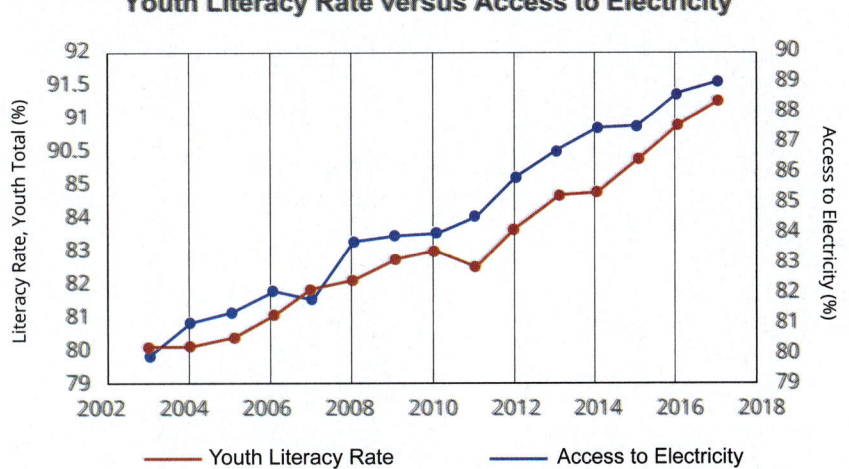

2. Explain how the graph might be misleading.

3. Create a graph that shows the data graphed using the same vertical scale. Does this graph change the story that is told by the two data sets? If so, describe how. If not, explain why.

4. Find the Pearson correlation coefficient between the youth literacy rate and access to electricity rounded to nearest ten thousandth. With a level of significance of $\alpha = 0.01$, is the relationship statistically significant?

5. Based on your findings in part 4, if you wanted to show the relationship between youth literacy rates and access to electricity, would you use the graph from part 1 or the graph from part 3 in a presentation? Which graph tells the better story about the connection? Explain your reasoning.

6. Calculate the Pearson correlation coefficient (rounded to the nearest ten thousandth) for the youth literacy rate compared to each of the remaining categories (that is, compared to poverty level, birthrate, and mobile cellular subscriptions). Determine which of these relationships are statistically significant at a level of significance of $\alpha = 0.01$.

7. Using the data provided in the table, create an infographic illustrating how youth literacy rates are related to poverty level, access to electricity, birthrate, and mobile cellular subscriptions. Your infographic should contain at least 4 different elements that clearly explain the relationships between youth literacy rates and each of other categories of the data set. (**Hint:** Search the internet for infographics to get ideas on how to structure your infographic.)

8. Is the infographic you created intentionally misleading in anyway? If so, explain how. Explain your reasoning for including misleading information.

CHAPTER 12 REVIEW

Section 12.1 The Science of Data

Definitions

Scientific Method for Data Science

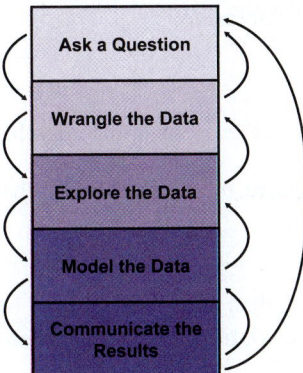

Figure 12.1.2

Domain Knowledge: Domain knowledge, or *business insight*, implies an understanding of how a particular business works in general and how a company's data can be used to achieve aspirational business goals. Regardless of the domain you are working in, there are some important things to keep in mind when forming the question(s) to be answered.

- Use specific well-defined terms and concepts.

- Avoid subjective judgements.

- Focus on a specific area of concern.

- Ask questions within the data you have access to.

Section 12.2 Data Wrangling

Definitions

Data Wrangling: Data wrangling, or *data munging*, is the process of acquiring raw data, cleaning it, organizing it, and transforming it so that the data is user-friendly.

Text String: A **text string**, or *string*, is a group of characters used as data.

Section 12.3 Data Exploration

Definitions

Scatter Plot: A **scatter plot** is a graphical display of data that is used to show how variables might relate to one another. The most common type of scatter plot is created by plotting two variables on the coordinate plane, where each piece of data is represented by a point.

Positive Linear Correlation: When data are plotted in a scatter plot and there appears to be an upward trend (that is, as one variable increases, the other variable increases as well), we say there is a **positive linear correlation** between the variables.

Negative Linear Correlation: When data are plotted in a scatter plot and there appears to be a downward trend (that is, as one variable increases, the other decreases), we say there is a **negative linear correlation** between the variables.

Pearson Correlation Coefficient: The **Pearson correlation coefficient** is the parameter that measures the strength of a linear relationship between two quantitative variables. The correlation coefficient for a sample is denoted by r. It always takes a value between -1 and 1, inclusive.

$$-1 \leq r \leq 1$$

Level of Confidence: The **level of confidence**, c, is the probability that the assertions made about the data are correct.

Level of Significance: The **level of significance**, α, is the probability that the assertions made about the data are incorrect.

$$c + \alpha = 1$$

Statistically Significant: If $|r|$ is greater than the critical value listed in Pearson's Correlation table, then r is **statistically significant**, which means the relationship is unlikely to have occurred by chance.

Regression Line: The **regression line**, also known as the **line of best fit**, is the line for which the average variation from the data is the smallest. It can be represented by

$$\hat{y} = ax + b,$$

where a is the slope of the line and b is the y-intercept.

Section 12.4 Data Storytelling

Definitions

Word Clouds: Word clouds, or *tag clouds*, represent word counts in a passage of text using a picture format. Each word is counted and then represented by how frequently it appears. The font size or color of the word visually signifies its frequency in the body of the text—the bigger and bolder the word, the more it occurs.

Bubble Charts: A **bubble chart** is an extension of the scatter plot. It allows the analysis of data points that each consist of three pieces of information. Two of the variables are plotted along the x- and y-axes, while the size of the dot is determined by the third variable. Sometimes, by making the color of each bubble have a significance, a fourth dimension can also be represented.

Infographics: Infographics are visual representations of any kind of information, not just data. They can describe timelines, processes, comparisons, as well as information. A good infographic is clear and memorable with concise text.

Data Dashboards: A **data dashboard** is fundamentally a compilation of dynamic graphs all in one location. It allows users the ability to visually intake a wide range of information all at once, providing a means to both monitor and analyzing relevant datasets.

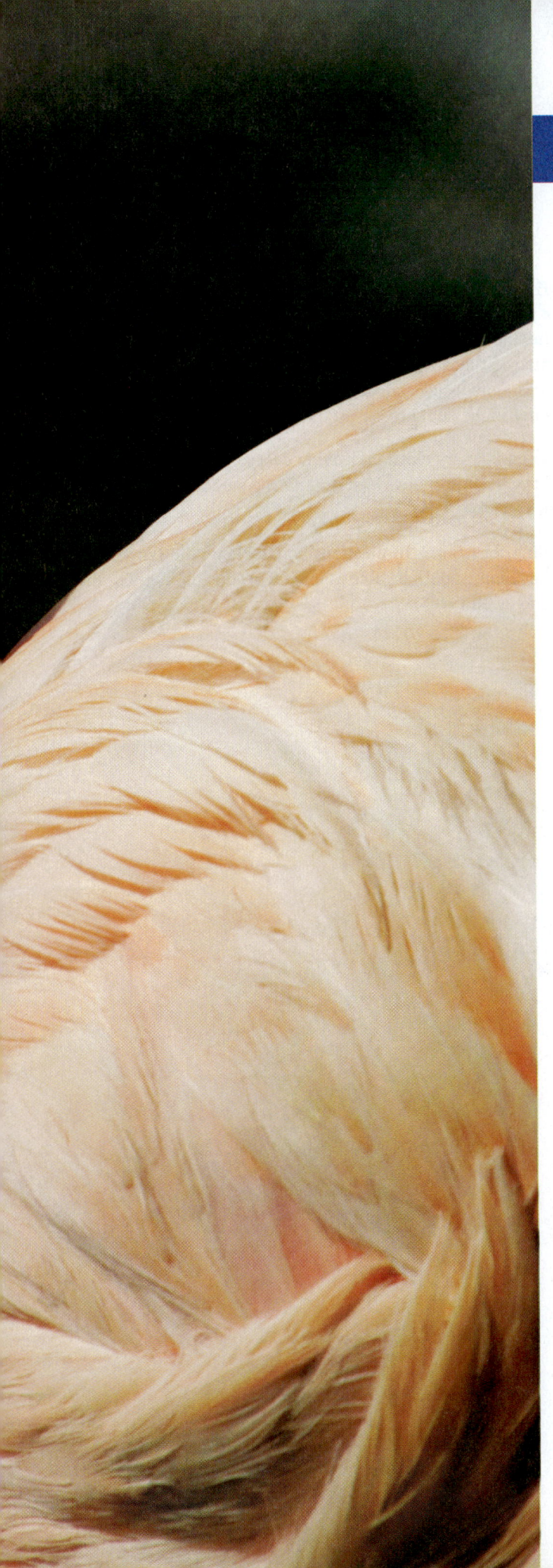

CHAPTER 13

Voting and Apportionment

🔖 SECTIONS

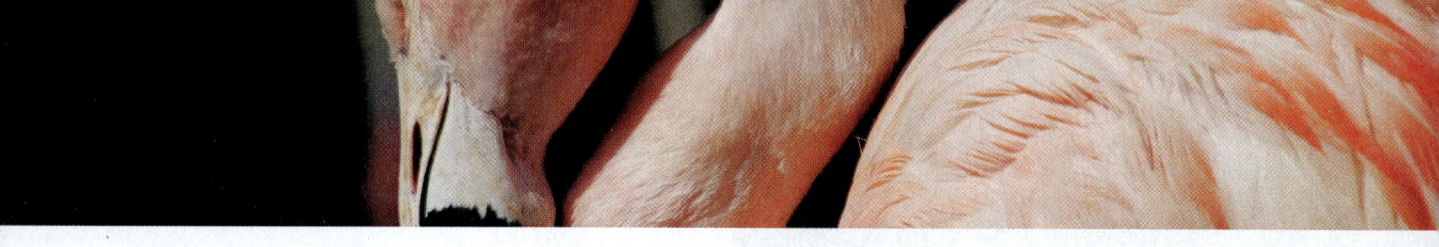

Introduction

In October 2021, the National Basketball Association (NBA) released their picks for the 75 Greatest Players in NBA History to be named to their 75th anniversary team. The picks were voted on by media members, current players, former players, coaches, and team executives. Due to a tie, the team actually features 76 players. Interestingly, the list includes all of the players that were named to a similar list that was compiled before the 1996–97 season to honor the 50 greatest players of all time. Once the list was announced, fans immediately took to social media to express either their delight in the outcomes or their disbelief in the "unfairness" of the choices.

At the end of the day, a winner (or winners) has to be chosen in every election, leaving everyone else behind. And just like the choices for the NBA anniversary team, controversy almost always surrounds the outcomes of a voting process. Someone is likely to feel the process wasn't fair.

In this chapter, we'll look at different methods of deciding the winner of elections and evaluate the fairness of each approach. The next time you cast a vote and view the results, you'll be able to judge for yourself how fair the election was and what "fair" really means when it comes to elections.

OBJECTIVE 1 Preference Tables

With a click-bait title like "Nine Ways to Pick a Contest Winner So Everyone Is Happy," an online article implies that there are not only lots of ways to choose a winner, but that a winner can be chosen in such a way so that everyone is happy with the outcome. While the first is certainly true, the second point is still up for debate. The truth is that no matter the arena, choosing a winner is a more nuanced process than it sounds. There are many ways to decide the outcome of elections or choose winners, and the idea of "fair" is a difficult concept to enact. Throughout this chapter, we will study different methods that have been used throughout history, and many that are still used today, to choose winners and how those processes attempt to heighten some aspect of fairness.

Typically, in elections, we are only concerned with determining an overall winner. But sometimes, the desired outcome is a ranking of the candidates; that is, first place, second place, and so on. We begin our discussion of how winners can be chosen by looking at this second option—ranking candidates in an election. Let's consider a scenario in which four candidates are competing against one another in an election: Russo, Satou, Tremblay, and Williams. Voters have been asked to indicate their rankings of the candidates on what is called a **preference ballot**. This type of ballot allows the voter to give each candidate a ranking.

📖 PREFERENCE BALLOT

A **preference ballot** is a ballot that allows a voter to rank the items in order of preference from most preferred to least preferred.

Often we are asked to indicate only our top candidate on a ballot, but a preference ballot gives us the opportunity to voice how we rank all of the candidates. This information allows the comparison of different ways of tallying the votes.

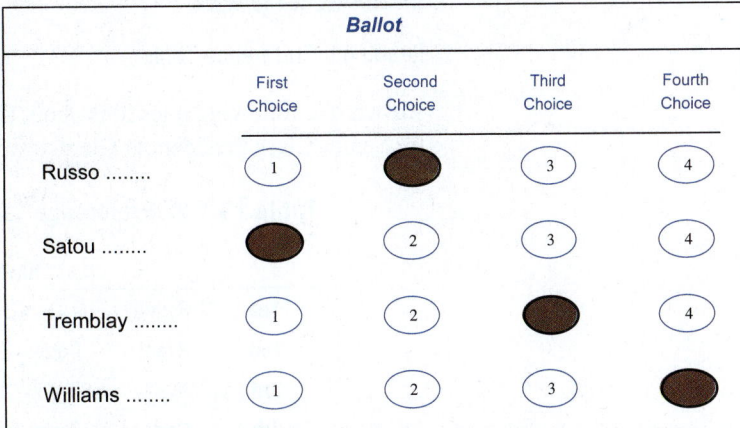

Figure 13.1.1: Preference Ballot

In order to compile the results of all of the voter rankings from the ballots, a **preference table** is used. A preference table summarizes how often each possible ranking occurred in the voting process. By convention, if a particular ranking receives no votes, it is not included in the preference table.

📖 **PREFERENCE TABLE**

A **preference table** summarizes all of the individual preference ballots in an election by tallying the number of ballots with the same order of ranking.

Since there are four candidates in our election example, let's consider the maximum size that the preference table might be. From our previous discussion on probability, we know that if there are n distinct items to place in order, then there are $n!$ ways in which to order the items. So, there are $4! = 4 \cdot 3 \cdot 2 \cdot 1 = 24$ unique ways to order the four candidates on our preference ballot. Potentially, there could be 24 columns in the preference table. In our election example, only four different rankings were chosen by the voters. Therefore, the preference table for this election will have four columns. The following preference table shows how often each particular ranking occurred among the 776 voters in our election.

Table 13.1.1: Preference Table for Candidates

	Rankings			
1st	Williams	Tremblay	Satou	Russo
2nd	Satou	Satou	Williams	Williams
3rd	Tremblay	Williams	Tremblay	Satou
4th	Russo	Russo	Russo	Tremblay
Total Votes	132	210	167	267

Notice that the bottom row of the table indicates how many individual voters ranked the candidates in the particular order listed in each column. For example, we know from the first column that 132 people ranked Williams first, Satou second, Tremblay third, and Russo fourth. In order to explore different methods of choosing a winner in an election, we'll refer back to this preference table of our four candidates throughout the section.

Example 13.1.1

Reading a Preference Table

Answer the following questions about the preference table shown for the election for Senior Class President at Clarkstown High School.

Table 13.1.2: Preference Table for Senior Class President

	Rankings				
1st	Sydney	Ava	Ava	Carley	Carley
2nd	Ryan	Carley	Carley	Zaire	Sydney
3rd	Ava	Ryan	Sydney	Ryan	Ava
4th	Carley	Zaire	Ryan	Ava	Zaire
5th	Zaire	Sydney	Zaire	Sydney	Ryan
Total Votes	15	29	6	24	1

a. How many different rankings were possible for the election?

b. How many students voted in the election?

c. Which candidate had the most first-place votes?

d. How many students thought that the order of candidates should be Ava, Carley, Sydney, Ryan, then Zaire?

Solution

a. Because there were five candidates on the ballot, we can use factorials to find the number of possible rankings.

$$5! = 5 \cdot 4 \cdot 3 \cdot 2 \cdot 1 = 120$$

Therefore, there were 120 possible rankings of the five candidates in the election.

b. To find the total number of students that voted in the election, we add the numbers in the row labeled "Total Votes."

$$15 + 29 + 6 + 24 + 1 = 75$$

Thus, there were 75 students who voted in the election.

c. To find the candidate with the most first-place votes, we need to count the number of votes each candidate received where they were ranked in first place. By looking across the 1st row, we see that there was only one ranking order that placed Sydney first. At the bottom of that column, we know that this particular ranking received 15 votes. So Sydney received 15 first-place votes overall.

Table 13.1.3: Preference Table for Senior Class President

	Rankings				
1st	Sydney	Ava	Ava	Carley	Carley
2nd	Ryan	Carley	Carley	Zaire	Sydney
3rd	Ava	Ryan	Sydney	Ryan	Ava
4th	Carley	Zaire	Ryan	Ava	Zaire
5th	Zaire	Sydney	Zaire	Sydney	Ryan
Total Votes	15	29	6	24	1

There were two rankings in which Ava received first-place votes: the second column of rankings and the third. Therefore, Ava had a total of 29 + 6 = 35 first-place votes.

Table 13.1.4: Preference Table for Senior Class President

	Rankings				
1st	Sydney	Ava	Ava	Carley	Carley
2nd	Ryan	Carley	Carley	Zaire	Sydney
3rd	Ava	Ryan	Sydney	Ryan	Ava
4th	Carley	Zaire	Ryan	Ava	Zaire
5th	Zaire	Sydney	Zaire	Sydney	Ryan
Total Votes	15	29	6	24	1

Carley was also ranked first in two rankings, which are shown in the last two columns of the table. Her first-place rankings total 24 + 1 = 25.

Table 13.1.5: Preference Table for Senior Class President

	Rankings				
1st	Sydney	Ava	Ava	Carley	Carley
2nd	Ryan	Carley	Carley	Zaire	Sydney
3rd	Ava	Ryan	Sydney	Ryan	Ava
4th	Carley	Zaire	Ryan	Ava	Zaire
5th	Zaire	Sydney	Zaire	Sydney	Ryan
Total Votes	15	29	6	24	1

Notice that there are no rankings that place either Ryan or Zaire in first. Therefore, Ava received the most first place rankings with a total of 35 votes.

d. The number of students who prefer the order of ranking to be Ava, Carley, Sydney, Ryan, and then Zaire is found in the third column of rankings. The total votes for this column is 6, so there were 6 students who chose this ranking.

> ✔ **Skill Check 13.1.1**
>
> Which candidate in Example 13.1.1 had the most rankings for fifth place?

Summarizing the results from preference ballots is the first step toward determining a winner in an election. The remainder of this section focuses on five different methods that have been used historically to choose winners: the majority rule decision, the plurality method, the Borda count method, the plurality with elimination method, and the pairwise comparison method. In each case, we'll use the preference ballot for the four candidates we introduced earlier in order to highlight how each different method can produce varying outcomes, even while the vote count stays the same.

OBJECTIVE 2 Majority Rule Decision

Using the **majority rule decision** to declare a winner means that the winner is supported by a majority of the voters; that is, more than 50% of the voters rank a single candidate in first place. When there are only two choices, the majority rule will almost always have a winner. The only possible time the majority rule will not produce a winner when there are only two candidates is when both candidates receive exactly 50% of the votes. To avoid this stalemate, many voting organizations allow the chair of the committee to only have a vote in order to break a tie. When there are more than two candidates, there is no guarantee that a winner can be named with the majority rule decision. Let's return to our example of the four candidates: Russo, Satou, Tremblay, and Williams. Suppose they were running for office in a voting system that uses a majority rule decision to choose its leaders. Who would win?

Example 13.1.2

Using the Majority Rule Decision

Use the majority rule decision to determine a winner in the election results from Satou, Williams, Tremblay, and Russo. The preference table is reprinted as Table 13.1.6 for ease of reference.

Table 13.1.6: Preference Table for Candidates

	Rankings			
1st	Williams	Tremblay	Satou	Russo
2nd	Satou	Satou	Williams	Williams
3rd	Tremblay	Williams	Tremblay	Satou
4th	Russo	Russo	Russo	Tremblay
Total Votes	132	210	167	267

Solution

In order to be declared a winner with the majority rule decision, a candidate must have more than 50% of the first-place votes. The first thing to determine is the total number of voters so we can calculate the number needed for a majority. To find the total number of voters, we add the values in the row labeled "Total Votes."

$$132 + 210 + 167 + 267 = 776$$

There were 776 votes cast in this election. To have a majority, a candidate needs more than half of the votes, or more than

$$\frac{776}{2} = 388 \text{ votes.}$$

So a minimum of 389 votes are needed for a majority.

Looking across the first row, we can see that each candidate has only one ranking where they were placed first. The number of first place rankings are as follows.

Russo:	267 first-place votes
Satou:	167 first-place votes
Tremblay:	210 first-place votes
Williams:	132 first-place votes

This means that no candidate obtained the minimum of 389 first-place votes needed for a majority. Therefore, using the majority rule decision, we cannot declare an immediate winner. Instead, an alternative method for choosing a winner will be needed.

OBJECTIVE 3 Plurality Method

What happens when the majority rule doesn't produce a winner, as we just saw in Example 13.1.2? A popular alternative method is the *plurality method*. The **plurality method** states that the candidate with the most first-place votes wins—majority or not. This method is often confused with the majority rule decision. When there are only two candidates, the plurality method is equivalent to the majority rule decision. However, if there are more than two choices, the winner of the plurality method does not need to have a majority of votes to win. If our same four candidates were running for office in India, where a plurality method is used to select a winner, which candidate would win?

Example 13.1.3

Using the Plurality Method

Use a plurality method to determine a winner with the four candidates: Satou, Williams, Tremblay, and Russo. The preference table is reprinted as Table 13.1.7.

Table 13.1.7: Preference Table for Candidates

	Rankings			
1st	Williams	Tremblay	Satou	Russo
2nd	Satou	Satou	Williams	Williams
3rd	Tremblay	Williams	Tremblay	Satou
4th	Russo	Russo	Russo	Tremblay
Total Votes	132	210	167	267

Solution

When using the plurality method, the only consideration is who has the most first-place votes. The other rankings play no role in the process here. Recall each candidate's first-place votes from Example 13.1.2.

Russo:	267 first-place votes
Satou:	167 first-place votes
Tremblay:	210 first-place votes
Williams:	132 first-place votes

Therefore, Russo wins the election using the plurality method with 267 first-place votes.

We know that in the plurality method, only the first-place votes count, and all other rankings are irrelevant. Let's take a closer look at the last example. Russo was the winner with the most first-place votes, but look at the other rankings he received—all of the other sets of rankings placed him in fourth place. So although he won with 267 votes, the rest of the voters (132 + 210 + 167 = 509 voters) think he should be in last place. In other words, although he has more first-place votes than any other candidate, a majority of voters definitely think he should not be the winner. Do you think this poses any issues? Let's turn our attention to a method of vote counting that actually takes all of the rankings into account.

OBJECTIVE 4 Borda Count Method

The **Borda count method** assigns each ranking a specific number of points based on how many candidates are in the election. A first-place ranking receives the most points, second place gets one fewer, and so on until the last place, which receives one point. For instance, in our election with four candidates, first-place votes receive four points, second-place votes receive three points, third-place votes receive two points, and fourth-place votes receive one point. Points are then totaled for each candidate. The winner is the candidate with the most points overall. One of the benefits in using the Borda count method is that all of the preferential rankings of each voter play a role in the process. This method is often used in polls that rank sports teams or players. In fact, the Heisman Trophy, given to the most outstanding college football player in the United States each year, uses the Borda count method to choose a winner. In the Republic of Slovenia, the Italian and Hungarian national communities each elect one member to the National Assembly by using the Borda count method. Suppose our four candidates were running to be a member for the Hungarian community. Who would be chosen to represent the Hungarian community in the Slovenia National Assembly?

Example 13.1.4

Using the Borda Count Method

Use the Borda count method to determine a winner with our four candidates. The preference table is reprinted as Table 13.1.8.

Table 13.1.8: Preference Table for Candidates

	Rankings			
1st	Williams	Tremblay	Satou	Russo
2nd	Satou	Satou	Williams	Williams
3rd	Tremblay	Williams	Tremblay	Satou
4th	Russo	Russo	Russo	Tremblay
Total Votes	132	210	167	267

Solution

The Borda count method requires that we assign points for each ranking. Because there are four candidates in our election, each ranking will receive the following number of points.

1st place = 4 points   3rd place = 2 points

2nd place = 3 points   4th place = 1 point

To help keep the information straight, the calculations for each candidate are organized in a table. For instance, Russo was ranked fourth in the first three rankings, so each of those votes gets 1 point. He then gets four points for each of the 267 votes that had him ranked in first place. Added together, this gives Russo a total of 1577 points.

Table 13.1.9: Borda Count Tally

	Rankings				Total Points
Russo	4th 1(132) = 132	4th 1(210) = 210	4th 1(167) = 167	1st 4(267) = 1068	1577
Satou	2nd 3(132) = 396	2nd 3(210) = 630	1st 4(167) = 668	3rd 2(267) = 534	2228
Tremblay	3rd 2(132) = 264	1st 4(210) = 840	3rd 2(167) = 334	4th 1(267) = 267	1705
Williams	1st 4(132) = 528	3rd 2(210) = 420	2nd 3(167) = 501	2nd 3(267) = 801	2250
Total Votes	132	210	167	267	

The final column in the table calculates each candidate's point total as follows.

Russo receives	$132 + 210 + 167 + 1068 = 1577$ points.
Satou receives	$396 + 630 + 668 + 534 = 2228$ points.
Tremblay receives	$264 + 840 + 334 + 267 = 1705$ points.
Williams receives	$528 + 420 + 501 + 801 = 2250$ points.

When we take into account the ranking preferences of all the voters, we have Williams as the overall winner with 2250 total points. Recall from the previous example that he received the fewest first-place votes.

Determining the results of elections with the Borda count method may produce a completely different result than using the plurality method. Russo, who won the election when using the plurality method, is actually in last place using the Borda count method. It might be surprising to you that precisely the same set of votes produces two different winners depending on the counting method used. This is not to say that one method is necessarily better than the other, but rather that it's important to choose a method that best fits the situation at hand and one that will represent the purpose of the election. It's also a wise decision to decide upon a method of choosing a winner *before* the elections occur and not after the votes have been cast.

Before you rush to decide on one of these methods for electing a president of your local club, let's consider two other variations.

OBJECTIVE 5 Plurality with Elimination Method

The plurality method that we discussed earlier is a one-vote, winner-take-all method of naming a winner. However, a variation on this method, called **plurality with elimination**, requires the winner to have a majority of the votes and uses a series of eliminations, if necessary, to choose the winner. This method can be done by running several consecutive elections, often called runoffs, or by using a preference ballot to begin with. After the first count of votes, if a candidate has a majority of votes, then he is named the winner. However, if no single candidate has a majority of votes, then the candidate with the least amount of first-place votes is removed from the ballot, and a runoff election is held with the remaining candidates.

If a preference ballot is used in the first election, there is no need for a second voting process. The losing candidate is simply removed from the preference ballots and then the rankings for each voter are adjusted accordingly. For instance, suppose the candidate who is removed from the ballot after the first count was originally ranked first by a voter. Once he is removed, the voter's second place choice is then moved up a ranking to first place, the third place moves to second place, and so on. Once this type of adjustment happens for each ballot, the votes are recounted. In each round, only first-place votes are counted to eliminate the candidate with the fewest first-place votes. This method of determining a winner is used to elect the leader of the Labour Party in the United Kingdom. Suppose our four contestants were running for Labour Party leader. Who would be appointed?

Example 13.1.5

Using the Plurality with Elimination Method

Use a plurality with elimination to determine a winner with the four candidates. The preference table is reprinted as Table 13.1.10.

Table 13.1.10: Preference Table for Candidates

	Rankings			
1st	Williams	Tremblay	Satou	Russo
2nd	Satou	Satou	Williams	Williams
3rd	Tremblay	Williams	Tremblay	Satou
4th	Russo	Russo	Russo	Tremblay
Total Votes	132	210	167	267

Solution

In the plurality with elimination process, only first-place votes are considered in each counting cycle. By looking at the preference table, we know that each of the candidates has the following number of first-place votes. Recall that a majority of the votes in this election is 389.

Russo:	267 first-place votes
Satou:	167 first-place votes
Tremblay:	210 first-place votes
Williams:	132 first-place votes

We already know that Russo would win in a plurality contest with his 267 votes, but since he doesn't have the required 389 votes for a majority, an elimination must happen. Since Williams has the least amount of first-place votes with 132, he is eliminated, as shown in Table 13.1.11.

Table 13.1.11: Plurality with Elimination Cycle 1

	Rankings			
1st	~~Williams~~	Tremblay	Satou	Russo
2nd	Satou	Satou	~~Williams~~	~~Williams~~
3rd	Tremblay	~~Williams~~	Tremblay	Satou
4th	Russo	Russo	Russo	Tremblay
Total Votes	132	210	167	267

Now, the preference table must be adjusted accordingly. In the first column, Satou will now be ranked first, Tremblay second, and Russo third. In the second rankings column, Tremblay and Satou remain in first and second place, but Russo will move into third place. In the third rankings column, both Tremblay and Russo move up one place. Finally, in the last rankings column, Satou and Tremblay move up one place. The new preference table for the second cycle is shown in Table 13.1.12. Note that the total votes in the bottom row will stay the same in each cycle.

Table 13.1.12: Plurality with Elimination Cycle 1 Simplified

	Rankings			
1st	Satou	Tremblay	Satou	Russo
2nd	Tremblay	Satou	Tremblay	Satou
3rd	Russo	Russo	Russo	Tremblay
Total Votes	132	210	167	267

Now, we recount the number of first-place votes for each candidate and determine if anyone has the required 389 votes for a majority.

Satou now has two rankings that place him in first place, which gives him a total of

$$132 + 167 = 299 \text{ votes.}$$

Tremblay and Russo remain the same with 210 and 267 first-place votes, respectively.

Although Satou is now the leader with 299 first place votes, he does not have a majority, so we must remove a candidate and continue with another elimination. This time it is Tremblay who has the least amount of first-place votes, so he is eliminated.

Table 13.1.13: Plurality with Elimination Cycle 2

	Rankings			
1st	Satou	~~Tremblay~~	Satou	Russo
2nd	~~Tremblay~~	Satou	~~Tremblay~~	Satou
3rd	Russo	Russo	Russo	~~Tremblay~~
Total Votes	132	210	167	267

Adjusting the table in the same manner as before, we have the following preference table with the two remaining candidates, Satou and Russo.

Table 13.1.14: Plurality with Elimination Cycle 2

	Rankings			
1st	Satou	Satou	Satou	Russo
2nd	Russo	Russo	Russo	Satou
Total Votes	132	210	167	267

Counting the votes, we see that Satou now has $132 + 210 + 167 = 509$ first-place votes to Russo's 267. Therefore, Satou now wins with a majority of the first-place votes.

OBJECTIVE 6 Pairwise Comparison Method

The last type of counting we will consider is the **pairwise comparison method**. In this method, each candidate is "paired" with every other candidate in a head-to-head vote count. The candidate with the most number of votes in each comparison receives one point. If there is a tie in the comparison count, both candidates receive $\frac{1}{2}$ of a point. After all pairs of comparisons are made, the candidate with the most points overall is declared the winner. The pairwise comparison method is helpful for making decisions involving fairly complicated criteria. It allows the comparison between each and every option available to play a role in the decision-making process. This method of decision making is often used to choose leaders in organizations, assess risks about a certain decision, or even to select the most qualified candidate for a job.

How many head-to-head comparisons need to be made when using this method? Multiplying the total number of candidates, n, by the number of other candidates in the race they could be paired with, $n - 1$, is a good place to start. However, if we use the product $n(n - 1)$, we'll end up with duplicates because a head-to-head matchup between Candidate A and Candidate B is the same as a matchup between Candidate B and Candidate A. To account for the duplicates, we need to divide the product $n(n - 1)$ in half. The following formula will give us the number of head-to-head comparisons needed for the **pairwise comparison method**.

> **Helpful Hint**
>
> We can think of the number of pairwise comparisons as the combination of choosing two candidates from a pool of n candidates, or $_nC_2$.

📖 **NUMBER OF PAIRWISE COMPARISONS**

The **number of pairwise comparisons** that must be made if there are n candidates is

$$\frac{n(n-1)}{2}.$$

Suppose our original four candidates are running for a seat on a board in an organization that uses the pairwise comparison method to decide elections. Who would have been elected to the Board of Trustees? We will use the pairwise comparison method in Example 13.1.6 to examine this question.

Example 13.1.6

Using the Pairwise Comparison Method

Use the pairwise comparison method to determine a winner with our four candidates. The preference table is reprinted as Table 13.1.15.

Table 13.1.15: Preference Table for Candidates

	Rankings			
1st	Williams	Tremblay	Satou	Russo
2nd	Satou	Satou	Williams	Williams
3rd	Tremblay	Williams	Tremblay	Satou
4th	Russo	Russo	Russo	Tremblay
Total Votes	132	210	167	267

Solution

In the pairwise comparison method, each candidate is pitted against every other candidate to see who received the highest ranking. We can use the formula to determine how many comparisons will need to be made in our tally. Since there are four candidates, $n = 4$. Substituting this into the formula, we have the following.

$$\text{Number of Comparisons} = \frac{4(4-1)}{2} = \frac{12}{2} = 6$$

In order to help us keep track of all six comparisons, we'll create a table where each candidate has a column and a row. Since there is no need for comparisons between a candidate and himself, an X is placed in the cells that appear on the diagonal. Notice that each head-to-head comparison is listed twice in the chart; that is, a comparison between Russo and Satou (Row 1, Column 2) is the same as a comparison between Satou and Russo (Row 2, Column 1). Since there is no need to duplicate the comparisons, we've grayed out the cells that are duplicates. Check for yourself that this leaves us with the required six comparisons.

Table 13.1.16: Pairwise Comparison Grid

	Russo	Satou	Tremblay	Williams
Russo	X			
Satou		X		
Tremblay			X	
Williams				X

In each head-to-head cell, list the candidate with the higher ranking from each of the four original rankings and the number of votes he received from that ranking. For the comparison in the Russo vs. Satou cell, we can create a table to compare the rankings for the two candidates.

Table 13.1.17: Head-to-Head Comparison for Russo and Satou

	Rankings			
Russo	4th	4th	4th	1st
Satou	2nd	2nd	1st	3rd
Total Votes	132	210	167	267

Using Table 13.1.17, we can see that Satou is ranked higher in the first three columns, so he receives all of the votes for those rankings. The last 267 votes go to Russo in this head-to-head comparison. The same is done for each of the head-to-head matchups, and the results are shown in Table 13.1.18.

Table 13.1.18: Completed Pairwise Comparison Grid

	Russo	Satou	Tremblay	Williams
Russo	X	S = 132; S = 210; S = 167; R = 267	T = 132; T = 210; T = 167; R = 267	W = 132; W = 210; W = 167; R = 267
Satou		X	S = 132; T = 210; S = 167; S = 267	W = 132; S = 210; S = 167; W = 267
Tremblay			X	W = 132; T = 210; W = 167; W = 267
Williams				X

Once all of the votes have been awarded based on the different rankings, the overall winner in each head-to-head matchup is given 1 point. If there is a tie in the votes, each candidate gets $\frac{1}{2}$ of a point. Table 13.1.19 shows who received the point for each matchup.

Table 13.1.19: Pairwise Comparison Grid of Winners

	Russo	Satou	Tremblay	Williams
Russo	X	S = 509; R = 267 [Satou]	T = 509; R = 267 [Tremblay]	W = 509; R = 267 [Williams]
Satou		X	S = 566; T = 210; [Satou]	W = 399; S = 377; [Williams]
Tremblay			X	W = 566; T = 210; [Williams]
Williams				X

We can see that Russo received 0 points, Satou received 2 points, Tremblay received 1 point, and Williams received 3 points. Therefore, Williams is the winner when using the pairwise comparison method.

As we've seen, each of the methods for determining a winner is just as viable as the others, but the choice of voting method can affect the outcome of the election. Table 13.1.20 displays the named winner from our preference ballot based on the method of counting. Tremblay was the only candidate not to win in any of the scenarios, while Williams is named winner in two of the options. Stop and consider which method you would choose to name the winner in our original election.

Table 13.1.20: Comparison Table of Winners for Different Counting Methods

Candidate	Majority Rule Decision	Plurality	Borda Count	Plurality with Elimination	Pairwise Comparison
Russo		Winner			
Satou	No Winner			Winner	
Tremblay					
Williams			Winner		Winner

Although no counting method is inherently superior to any other, some methods fit the needs of different situations better than others. This decision is left up to those designing the election. Regardless of the chosen method, the decision of how to choose the winner should always be made before the first ballot is cast. Table 13.1.21 gives a summary of the five election counting methods we've discussed.

Table 13.1.21: Summary of Election Counting Methods

Majority Rule Decision	The winner must have more than 50% of the votes to win.
Plurality	The candidate with the most votes wins. No majority required.
Borda Count	Voting results are organized in a preference table and each ranking is assigned a specific number of points based on how many candidates are in the election. The candidate with the most ranking points is declared the winner.

Plurality with Elimination	A series of runoff elections, or eliminations, where the candidate with the least amount of first-place votes is removed from the ballot each round. A winner is declared when a candidate has a majority of the first-place votes.
Pairwise Comparison	Every candidate is compared head-to-head with the other candidates. In each pair of comparisons, the candidate with the greater number of higher rankings is given a point. The candidate with the most points after all head-to-head comparisons are made is the winner.

Skill Check Answer

1. Sydney, with 53 votes that gave her a fifth place ranking.

13.1   Exercises

✔ CONCEPT CHECK

1. The _____ method of choosing a winner in an election is a single vote with a winner-take-all approach. The winner need not have a majority of the votes to win; she just needs the most votes.

2. The _____ method for elections pits each candidate against all of the others individually in order to choose a winner.

3. Setting up an election so that voters are required to rank all the candidates requires use of a _____.

4. In order to give weight to each of the different rankings for candidates, the _____ attributes each ranking a certain number of points, and the candidate with the most points wins the election.

5. In a _____ election, the winner must receive more than 50% of the votes.

💡 PRACTICE

Create a preference table to show the results for each set of preference ballots.

6. Candidates in the election are: Maria, Daniel, Eisa.

   Maria, Eisa, Daniel: 29

   Eisa, Daniel, Maria: 43

   Eisa, Maria, Daniel: 33

   Daniel, Maria, Eisa: 17

   Maria, Daniel, Eisa: 3

   Daniel, Eisa, Maria: 21

7. The candidates are: Lars, Noah, Stephen, Oliver, Jack.

   Stephen, Oliver, Jack, Lars, Noah: 212

   Noah, Oliver, Jack, Lars, Stephen: 133

   Oliver, Stephen, Jack, Noah, Lars: 543

   Jack, Stephen, Lars, Noah, Oliver: 24

   Oliver, Noah, Lars, Jack, Stephen: 179

   Noah, Lars, Stephen, Jack, Oliver: 8

   Stephen, Lars, Oliver, Noah, Jack: 201

   Stephen, Jack, Lars, Noah, Oliver: 11

8. The candidates are: Jones (J), Brown (B), Wang (W), Cohen (C), Diaz (D).

   JBDCW, CWJDB, WBDJC, BDCWJ, JBDCW, JBDCW, CWJDB, CWJDB, WBDJC, JBDCW, BWJDC, WBDJC, CWJDB, BWJDC, BWJDC, BJWCD, WBDJC, BWJDC, JBDCW, BJWCD, WBDJC, BWJDC, JBDCW, JBDCW, DBWCJ

9. Candidates in the election are: Nguyen (N), Dey (D), Smith (S), Mori (M), Abbadi (A).

   MADNS, NDSMA, DSMAN, SMAND, SMAND, NDSMA, MANDS, DSMAN, SAMND, NDSMA, SMAND, NDSMA, MANDS, SAMND, NDSMA, MANDS, MANSD, SDMAN, MADNS, MANDS, MANDS, SMAND, NDSMA, DSMAN, SAMDN, MADNS, SAMDN, SMAND, SMAND, NDSMA

Use the given preference table to answer each question.

10. Answer the following questions about the given preference table.

Preference Table for Candidates

	Rankings				
1st	A	C	C	E	E
2nd	B	E	E	D	A
3rd	C	B	A	B	C
4th	E	D	B	C	D
5th	D	A	D	A	B
Total Votes	153	129	63	404	111

   a. How many possible unique rankings are there of the candidates in the election?

   b. How many people voted in the election?

   c. How many voters place Candidate C in first place?

   d. Which candidate wins using the plurality method?

   e. How many votes would a candidate need to have a majority?

   f. Do any of the candidates have a majority of first-place votes?

11. A wedding planning website asked visitors to rank the following six all-time favorite wedding songs. The song choices are *Nothing Compares to You* by Sinead O'Connor, *Love Me Tender* by Elvis Presley, *Close to You* by Maxi Priest, *When You Say Nothing at All* by Ronan Keating, *Just the Way You Are* by Billy Joel, and *You Made Me Love You* by Al Jolson. Use the preference table to answer the questions.

Preference Table for Favorite Wedding Songs

	Rankings			
1st	Presley	Priest	O'Connor	Keating
2nd	Jolson	O'Connor	Presley	Joel
3rd	Keating	Joel	Priest	Presley
4th	Priest	Jolson	Keating	O'Connor
5th	Joel	Presley	Joel	Jolson
6th	O'Connor	Keating	Jolson	Priest
Total Votes	1320	2010	3167	2697

a. How many possible unique rankings are there of the six wedding songs?

b. How many people participated in the website questionnaire?

c. Which song was ranked number one by the most voters?

d. How many votes are required for a song to have a majority of first-place votes?

e. Do any of the songs have a majority of first-place votes?

Complete each preference table.

12. The student programming board is deciding which band to bring to campus for a student concert. The possibilities are Matisyahu, Lady A, and The Black Keys. Complete the following preference table so that The Black Keys win the election using the majority rule.

Preference Table for Student Concert

	Rankings		
1st	The Black Keys	Lady A	Matisyahu
2nd	Lady A	Matisyahu	The Black Keys
3rd	Matisyahu	The Black Keys	Lady A
Total Votes	?	9	2

13. A hospital needs a new doctor. The hiring committee wants to get feedback from the current staff. They have the nursing staff and resident doctors rank the candidates in order of their preference. Complete the following preference table so that Perez is the staff's top choice using the plurality method.

Preference Table for New Doctor

	Rankings			
1st	Ross	Porter	Walker	Perez
2nd	Perez	Walker	Ross	Porter
3rd	Walker	Perez	Perez	Ross
4th	Porter	Ross	Porter	Walker
Total Votes	?	21	22	36

14. Phonetex is trying to decide which new feature to include in an advertising campaign: 12 megapixel camera, longer battery life, voice-activated commands, or reception quality. The company surveyed its current customers and asked them to rank the four features in order of preference. Complete the following preference table so that voice-activated commands is the favorite among current customers using the plurality with elimination method.

Preference Table for New Features

	Rankings			
1st	Camera	Battery Life	Voice-Activated	Voice-Activated
2nd	Reception Quality	Reception Quality	Reception Quality	Camera
3rd	Battery Life	Voice-Activated	Camera	Reception Quality
4th	Voice-Activated	Camera	Battery Life	Battery Life
Total Votes	44	11	?	56

15. Rhone County is electing a new county executive. The candidates are McMillian, Witt, and Longsdon. Complete the following preference table so that McMillian wins the election using the Borda count method.

Preference Table for Candidates

	Rankings		
1st	?	Witt	McMillian
2nd	?	McMillian	Longsdon
3rd	McMillian	Longsdon	Witt
Total Votes	601	632	650

16. *College Living Magazine* is compiling a list of the top five features of a college or university that students used when choosing an institution for higher education. They asked students to rank the importance of the following features of colleges and universities: size of the school, strength of the degree program, location, and number of student activities. Complete the following preference table so that location is the top influencer among students if the pairwise method of comparison is used.

Preference Table for Most Important Features

	Rankings				
1st	Size	Location	Activities	Size	Strength
2nd	Activities	Strength	Size	Location	Activities
3rd	Strength	Size	Location	Strength	Location
4th	Location	Activities	Strength	Activities	Size
Total Votes	100	?	55	105	89

Calculate the number of pairwise comparisons that must be made in each election.

17. In the history club on campus, five students are running for the office of president.

18. Running for president of the National Association of College Students are the following students: L. Mayer, R. Schartz, M. Gillmore, L. Bilbo, R. Rosevear, K. Colletta, T. Yawn, S. Oswald, and K. Smith.

◆ APPLICATIONS

19. The following table shows the number of votes for first, second, and third places that the top five college football players received in 2011 for the Heisman Trophy—College Football's highest award. Determine the number of total points that each player would receive using the Borda count method if a first place vote receives 3 points, a second place receives 2 points, and third place receives 1 point.

Preference Table for 2011 Heisman Trophy Candidates

Player	School	1st	2nd	3rd
Robert Griffin III	Baylor	405	168	136
Andrew Luck	Stanford	247	250	166
Trent Richardson	Alabama	138	207	150
Montee Ball	Wisconsin	22	83	116
Tyrann Mathieu	Louisiana State	34	63	99

Source: Heisman Trophy. "2011 - 77th Award Robert Griffin III Baylor University." https://www.heisman.com/heisman-winners/robert-griffin-iii/

20. In professional baseball, the Baseball Writers' Association of America hands out two Most Valuable Player (MVP) awards each year. One for the American League and one for the National League. The following table shows the voting breakdown for the top 10 players in the National League. Determine the number of total points that each player received using the Borda count method, given that a first place vote receives 14 points, a second place receives 9 points, third place receives 8 points, and on down to 1 point for tenth place.

National League MVP 2011												
Player	Team	1st	2nd	3rd	4th	5th	6th	7th	8th	9th	10th
Ryan Braun	Milwaukee Brewers	20	12								
Matt Kemp	Los Angeles Dodgers	10	16	6							
Prince Fielder	Milwaukee Brewers	1	4	11	9	1	3		2		1
Justin Upton	Arizona Diamondbacks	1		8	11	6	3	1	1		1
Albert Pujols	St. Louis Cardinals			1	6	11	6	4		2	
Joey Votto	Cincinnati Reds			4	3	2	8	3	3	4	1
Lance Berkman	St. Louis Cardinals			1	2	6	3	7	2	4	3
Troy Tulowitzki	Colorado Rockies						3	4	8	5	4
Roy Halladay	Philadelphia Phillies			1		1	1	6	2		3
Ryan Howard	Philadelphia Phillies				1	3	1	1		1	3

Source: Baseball Writers' Association of America. "2011 NL MVP: Ryan Braun Slugs His Way to Award." https://bbwaa.com/11-nl-mvp/

21. In an attempt to choose a list of the "Best Places to Live in America," readers of a magazine were asked to rank eight towns. The following preference table summarizes the results of the survey.

Preference Table for Best Places to Live in America					
	Rankings				
1st	Louisville, CO	Middleton, WI	Leesburg, VA	Leesburg, VA	Hanover, NH
2nd	Papillion, NE	Louisville, CO	Middleton, WI	Milton, MA	Leesburg, VA
3rd	Hanover, NH	Liberty, MO	Milton, MA	Louisville, CO	Milton, MA
4th	Milton, MA	Milton, MA	Liberty, MO	Papillion, NE	Solon, OH
5th	Liberty, MO	Papillion, NE	Hanover, NH	Solon, OH	Middleton, WI
6th	Solon, OH	Solon, OH	Papillion, NE	Hanover, NH	Papillion, NE
7th	Middleton, WI	Leesburg, VA	Solon, OH	Middleton, WI	Louisville, CO
8th	Leesburg, VA	Hanover, NH	Louisville, CO	Liberty, MO	Liberty, MO
Total Votes	345	231	211	202	16

a. Determine a winner using the plurality method.

b. Determine a winner using the plurality with elimination method.

c. Which method do you think chooses the winner that best reflects the votes, or would you choose a different method? Explain your answer.

22. The following preference table summarizes the election results for city mayor of Clamptonville.

Preference Table for Candidates

	Rankings			
1st	Costa	Little	Little	Torres
2nd	Braugh	Allen	Costa	Braugh
3rd	Little	Torres	Braugh	Allen
4th	Torres	Braugh	Allen	Little
5th	Allen	Costa	Torres	Costa
Total Votes	1231	542	1001	654

a. Determine the winner for city mayor using the Borda count method.

b. Determine the winner for city mayor using the plurality with elimination method.

c. Determine the winner with the pairwise method of comparison.

d. Suppose that Braugh withdrew from the race before the votes were counted. Candidates ranked below him in any of the orders simply move up a ranking. Who would be declared the winner using the Borda count method with the new preference table? Is this a different winner than in part **a.**?

23. A university's Department of Business is electing a new chairman. As part of the process, each candidate gives a brief question and answer session with the business students. The students are then asked to rank the candidates in order of preference. The following table shows the results of the student votes.

Preference Table for Candidates

	Rankings					
1st	B	A	C	C	D	B
2nd	A	B	D	B	A	A
3rd	D	C	B	D	B	C
4th	C	D	A	A	C	D
Total Votes	43	12	13	21	51	9

a. Determine the student preference for chairman by using the pairwise method of comparison.

b. Determine the student preference for chairman by using the plurality with elimination method.

c. Determine the student preference for chairman by using the plurality method.

d. If you were chosen to represent the students by announcing a preference for chairman, which candidate would you recommend? How would you defend your choice based on the answers to **a.–c.**?

24. The Tennessee Department of Transportation is considering a new bus route between its state capital, Nashville, and one of the other four largest cities: Knoxville, Memphis, Chattanooga, and Clarksville. Each of the 99 State Representatives was asked to rank the cities in order of preference for the new route. Their votes are summarized in the following table.

Preference Table for New Bus Route

	Rankings					
1st	Chattanooga	Memphis	Chattanooga	Knoxville	Clarksville	Memphis
2nd	Memphis	Knoxville	Clarksville	Memphis	Knoxville	Knoxville
3rd	Clarksville	Clarksville	Knoxville	Chattanooga	Memphis	Chattanooga
4th	Knoxville	Chattanooga	Memphis	Clarksville	Chattanooga	Clarksville
Total Votes	23	13	9	33	12	9

a. Which city would receive the new bus route using the pairwise method of comparison?

b. Which city received the most first-place votes? Is this a majority?

c. Which city would receive the new bus route using the plurality with elimination method?

d. Which city would receive the new bus route using the Borda count method?

e. Explain which method you think is most "fair" in choosing the new bus route.

25. In the 1968 United States Presidential Election, three candidates shared most of the November 5 popular vote, although there was a small portion of votes for other candidates. The following table displays the breakdown of the popular vote along with the electoral vote for the election.

1968 US Presidential Election Results

Presidential Candidate	Popular vote		Electoral Vote
	Count	**Percentage**	
Richard Milhous Nixon	31,783,783	43.40%	301
Hubert Horatio Humphrey	31,271,839	42.70%	191
George Corley Wallace	9,901,118	13.50%	46
Eugene McCarthy	25,634	0.00%	0
Other	217,624	0.30%	0
Total Votes	**73,199,998**		**538**

Sources: 1968 Presidential General Election Results, http://www.uselectionatlas.org. Electoral Votes for President and Vice President, http://www.archives.gov.

a. Who won the popular vote by plurality? Was it a majority?

b. Who won the electoral vote by plurality? Was it a majority?

c. Compare the outcomes of the popular vote between Nixon and Humphrey to that of the electoral vote for the two candidates.

d. Suppose Wallace had dropped out of the election. Is it possible that Humphrey would have won the presidency? Explain your answer.

13.1 PROJECT

WHO WON? IT DEPENDS ON HOW YOU TALLY THE VOTES

In Section 13.1, you learned about five different methods for tallying votes from a preference table and explored how the different methods sometimes lead to different outcomes. Now it's time to start thinking about who might prefer which method in which situation.

As a recap, here are the methods to tally votes from a preference table.

- The **majority rule decision** means that the winner is supported by a majority of the voters; that is, more than 50% of the voters rank a single candidate in first place.

- The **plurality method** states that the candidate with the most first-place votes wins—majority or not.

- The **Borda count method** assigns each ranking a specific number of points based on how many candidates are in the election.

- The **plurality with elimination method** requires the winner to have a majority of the votes and uses a series of eliminations, if necessary, to choose the winner.

- The **pairwise comparison method** pairs each candidate with every other candidate in a head-to-head vote count.

Let's compare two of the methods by analyzing the pros and cons of each. Here are some reflection questions to guide your responses.

    a. In what situation might a person prefer this method?

    b. Who stands to benefit from this method?

    c. Who is least likely to benefit from this method?

    d. What is a real-world example where this method is used? (This may require some internet research.)

1. Select any two of the methods for tallying votes from a preference table.

    Method 1: _____

    Method 2: _____

2. Describe an advantage of your chosen Method 1.

3. Describe a downside of your chosen Method 1.

4. Describe an advantage of your chosen Method 2.

5. Describe a downside of your chosen Method 2.

6. Give an example where a person would prefer one of your chosen methods over the other. State clearly which method is preferred from that person's perspective and why.

OBJECTIVES

1. Understand fairness criteria as related to Arrow's Impossibility Theorem.

2. Determine if the Condorcet criterion is violated.

3. Determine if the majority criterion is violated.

4. Determine if the monotonicity criterion is violated.

5. Determine if the irrelevant alternatives criterion is violated.

6. Determine if the dictator criterion is violated.

OBJECTIVE 1 Fairness Criteria

As you were reading through the different ways to choose a winner in the previous section, maybe one or more of the methods made you think to yourself, "That's not very fair." But someone hearing your thoughts might also respond, "Not fair to whom?" So how do we begin to define *fairness*? For instance, is it fair that if a person has a majority of the votes, then that person should always win the election under any circumstances? Or is it more fair that the candidate who is rejected by the fewest voters is declared the winner?

Throughout history, scholars have proposed a variety of conditions as standards or benchmarks for measures of fairness to help ensure electoral integrity. Five of the most natural conditions have become widely accepted yardsticks of fairness. They are the Condorcet criterion, the majority criterion, the monotonicity criterion, the irrelevant alternatives criterion, and the dictator criterion. These criteria have become a means by which we can evaluate election models. Although some seem more obvious than others, each strives to make the outcome of an election as fair as possible.

It might surprise you to know that none of the election counting methods we discussed in the previous section meet all five fairness criteria. In fact, Kenneth Arrow, an American economist and social choice theorist, demonstrated that it is impossible for a non-dictatorship to satisfy all the criteria and produce a perfect election. In his PhD thesis, *A Difficulty in the Concept of Social Welfare* (1950), and later in his book, *Social Choice and Individual Values*, Arrow outlined and proved a theorem that states that if there are three or more choices on a ballot, there *cannot* be a voting method that will satisfy all five fairness criteria. His theory is called **Arrow's Impossibility Theorem**.

In this section, we will consider each of the five criteria for fairness. The examples of the election methods discussed in the previous section will be used to show how they each violate at least one of the fairness criteria and, in turn, support Arrow's Impossibility Theorem.

OBJECTIVE 2 The Condorcet Criterion

The first of the conditions is referred to as the **Condorcet criterion**. It is named for the French mathematician Nicolas de Condorcet. It simply states that if a candidate wins the head-to-head comparison against *every other* candidate, then that candidate should also win the overall election in a fair voting system. By design, elections using the pairwise comparison method adhere to the Condorcet criterion. However, the plurality method, the Borda count method, and the plurality with elimination method all have the possibility of violating this criterion. That doesn't mean that these counting methods will always violate the Condorcet criterion, but there is not a guarantee that they won't.

📖 CONDORCET CRITERION

The **Condorcet criterion** states that if a candidate wins the head-to-head comparison against every other candidate, then that candidate should also win the overall election in a fair voting system.

Example 13.2.1

Condorcet Criterion and the Plurality Method

Three students are in an election for president of the National Society of Collegiate Scholars on campus. Members were asked to rank the three candidates. The results of the membership votes are shown in the following preference table.

Table 13.2.1: Preference Table for National Society of Collegiate Scholars

	Rankings			
1st	Charles	Charles	Andrew	Bethany
2nd	Bethany	Andrew	Charles	Charles
3rd	Andrew	Bethany	Bethany	Andrew
Total Votes	30	31	65	21

a. Determine the winner if the society used the plurality method for determining the winner.

b. Determine the winner if the society used the pairwise comparison method to determine the winner.

c. Does the plurality method adhere to the Condorcet criterion for this election? Explain your answer.

Solution

a. In order to find a winner using the plurality method, we need to see which candidate has the most first-place votes. We can read across the first row to find the total number of first-place votes each candidate received as shown.

Andrew: 65 first-place votes

Bethany: 21 first-place votes

Charles: 30 + 31 = 61 first-place votes

Therefore, Andrew is the winner if the plurality method is used to elect the president of the National Society of Collegiate Scholars.

b. Using the pairwise comparison method requires that we determine the winner of each head-to-head matchup between students. Using the formula we found earlier, we know that there are three comparisons required.

$$\text{Number of Comparisons} = \frac{3(3-1)}{2} = \frac{6}{2} = 3$$

Therefore, using a comparison table to keep track of the head-to-head comparison, we have the following.

Table 13.2.2: Completed Pairwise Comparison Grid

	Andrew	Bethany	Charles
Andrew	X	B = 30; A = 31 A = 65; B = 21	C = 30; C = 31 A = 65; C = 21
Bethany		X	C = 30; C = 31 C = 65; B = 21
Charles			X

Table 13.2.3: Pairwise Comparison Grid of Winners

	Andrew	Bethany	Charles
Andrew	X	B = 51; A = 96 [Andrew]	C = 82; A = 65 [Charles]
Bethany		X	C = 126; B = 21 [Charles]
Charles			X

Adding each candidate's votes together in the comparisons, we see that Andrew received 1 point while Charles received 2 points. Therefore, Charles is the winner using the pairwise comparison method. Notice that Charles won the head-to-head comparison against both of the other candidates, Andrew and Bethany.

c. In the pairwise comparison of the candidates, Charles wins the head-to-head comparisons against all other candidates. However, the plurality method chose Andrew as the winner. This means that for this election, the Condorcet criterion of fairness is violated by the plurality method.

OBJECTIVE 3 The Majority Criterion

The **majority criterion** is probably the most obviously named. It states that if a candidate receives a majority of votes in an election, that candidate should win. Although it seems that this should always be true, this is not the case. Of the counting methods discussed in the previous section, the Borda count is the only method that can violate the majority criterion. That is, there are situations where the Borda count method does not favor a candidate who would win by the majority rule decision.

📖 MAJORITY CRITERION

The **majority criterion** states that if a candidate receives a majority of the votes in an election, that candidate should win.

Example 13.2.2

Majority Criterion and the Borda Count Method

In an election for chairman of the board of directors for a major company, shareholders were given the opportunity to rank the top five nominees. The results are posted in Table 13.2.4.

Table 13.2.4: Preference Table for Chairman Of The Board

	Rankings			
1st	H. Beridze	M. Gruber	T. Taylor	T. Taylor
2nd	L. Wright	R. Jensen	H. Beridze	H. Beridze
3rd	M. Gruber	H. Beridze	M. Gruber	L. Wright
4th	T. Taylor	T. Taylor	R. Jensen	M. Gruber
5th	R. Jensen	L. Wright	L. Wright	R. Jensen
Total Votes	2300	3100	4000	2200

a. Determine the winner of the board elections using the majority rule decision.

b. Determine the winner of the board elections using the Borda count method.

c. Does the Borda count method satisfy the majority criterion for this election? Explain your answer.

Solution

a. To find a winner using the majority rule decision, we simply count the number of firstplace votes for each candidate. They are as follows.

H. Beridze:	2300 first-place votes
M. Gruber:	3100 first-place votes
R. Jensen:	0 first-place votes
T. Taylor:	4000 + 2200 = 6200 first-place votes
L. Wright:	0 first-place votes

In order to win an election with a majority rule decision, a candidate needs more than half of the votes. In this case, there are $2300 + 3100 + 4000 + 2200 = 11,600$ votes available. Half of the votes would be $\dfrac{11,600}{2} = 5800$ votes. So a minimum of 5801 votes are needed for a majority.

Therefore, T. Taylor is the winner by the majority rule decision with 6200 votes.

b. To determine a winner using the Borda count method, each ranking will receive the following number of points.

1st place = 5 points

2nd place = 4 points

3rd place = 3 points

4th place = 2 points

5th place = 1 points

Table 13.2.5 shows the point calculations. The final column is the sum of each candidate's points.

Table 13.2.5: Borda Count Tally for Chairman of the Board

	Rankings				Total Points
H. Beridze	1st $5(2300) = 11{,}500$	3rd $3(3100) = 9300$	2nd $4(4000) = 16{,}000$	2nd $4(2200) = 8800$	45,600
M. Gruber	3rd $3(2300) = 6900$	1st $5(3100) = 15{,}500$	3rd $3(4000) = 12{,}000$	4th $2(2200) = 4400$	38,800
R. Jensen	5th $1(2300) = 2300$	2nd $4(3100) = 12{,}400$	4th $2(4000) = 8000$	5th $1(2200) = 2200$	24,900
T. Taylor	4th $2(2300) = 4600$	4th $2(3100) = 6200$	1st $5(4000) = 20{,}000$	1st $5(2200) = 11{,}000$	41,800
L. Wright	2nd $4(2300) = 9200$	5th $1(3100) = 3100$	5th $1(4000) = 4000$	3rd $3(2200) = 6600$	22,900
Total Votes	2300	3100	4000	2200	

From the final column, it is clear that H. Beridze is the winner with 45,600 points using the Borda count method.

c. Using the Borda count method, H. Berdize would be declared the winner. However, T. Taylor received a majority of the votes and would win using the majority rule method. Therefore, the Borda count method does not satisfy the majority rule criterion in this election.

OBJECTIVE 4 The Monotonicity Criterion

The third condition is often referred to as the **monotonicity criterion**. It considers elections that involve runoffs or have multiple stages. This criterion states that if a candidate wins the first round of an election, and then only gains support and does not lose any support in the next round, the candidate should also win the second round. For example, if Jensen were ranked second in your list originally, then in round two Jensen must be ranked in either first or second place while the order of the other candidates is unchanged.

> 📖 **MONOTONICITY CRITERION**
>
> The **monotonicity criterion** states that if a candidate wins an early round of an election and only gains support and does not lose support in subsequent rounds, then that candidate should win the election.

Example 13.2.3

Monotonicity Criterion and the Plurality with Elimination Method

The city council for the town of Whitman voted to elect a new vice president. All council members were asked to rank the four candidates in order of preference. Show that by using the plurality with elimination method, the monotonicity criterion is violated. Table 13.2.6 shows the results of the vote.

Table 13.2.6: Preference Table for Whitman City Council Vice President

	Rankings			
1st	Clarke	Roberts	Green	Green
2nd	Green	Clarke	Roberts	White
3rd	Roberts	White	Clarke	Clarke
4th	White	Green	White	Roberts
Total Votes	6	5	4	2

Solution

In order to show that the monotonicity criterion is violated using the plurality with elimination method, we need to first find the winner of the election using the plurality with elimination method. Recall that when using this method, if no candidate has a majority of the first-place votes, the candidate with the least amount of first-place votes is eliminated in each round. A majority is more than half the number of votes. In this case, there were $6 + 5 + 4 + 2 = 17$ votes. Half of 17 is 8.5, so a majority consists of at least 9 votes. The following is a list of the number of first-place votes received by each candidate.

Clarke:	6 first-place votes
Green:	$4 + 2 = 6$ first-place votes
Roberts:	5 first-place votes
White:	0 first-place votes

Therefore, since no candidate had a majority of votes, White is eliminated with the least amount of first-place votes.

Table 13.2.7: Plurality with Elimination Cycle 1

	Rankings			
1st	Clarke	Roberts	Green	Green
2nd	Green	Clarke	Roberts	~~White~~
3rd	Roberts	~~White~~	Clarke	Clarke
4th	~~White~~	Green	~~White~~	Roberts
Total Votes	6	5	4	2

Note that White did not have any first-place votes, so none of our vote counts change when he is eliminated. So the candidate with the next smallest number of first-place votes, Roberts, is eliminated next.

Table 13.2.8: Plurality with Elimination Cycle 2

	Rankings			
1st	Clarke	~~Roberts~~	Green	Green
2nd	Green	Clarke	~~Roberts~~	White
3rd	~~Roberts~~	White	Clarke	~~Clarke~~
Total Votes	6	5	4	2

That leaves Green and Clarke as the only remaining candidates. Table 13.2.9 shows the standings without White and Roberts.

Table 13.2.9: Plurality with Elimination Cycle 2 Simplified

	Rankings			
1st	Clarke	Clarke	Green	Green
2nd	Green	Green	Clarke	Clarke
Total Votes	6	5	4	2

We can now compare Green and Clark and determine the winner. Clarke wins in the first two columns of rankings, but Green wins in the final two. Therefore, Clarke has 11 first-place votes and Green has six. Clarke is the winner using the plurality with elimination method.

In order to show that the monotonicity criterion is violated using this method, we need to show that a different winner is produced if some of the council members change their votes in such a way that the winner, Clarke, increases his ranking while the order of the other candidates stays the same. Suppose that the two voters in the last column of the original preference table changed their rankings to place Clarke first. You can see that their rankings for the other candidates stay in the same preference order (Green, White, then Roberts).

Table 13.2.10: Whitman City Council Vice President Original Election

	Rankings			
1st	Clarke	Roberts	Green	Green
2nd	Green	Clarke	Roberts	White
3rd	Roberts	White	Clarke	Clarke
4th	White	Green	White	Roberts
Total Votes	6	5	4	2

Table 13.2.11: Whitman City Council Vice President Second Election

	Rankings			
1st	Clarke	Roberts	Green	Clarke
2nd	Green	Clarke	Roberts	Green
3rd	Roberts	White	Clarke	White
4th	White	Green	White	Roberts
Total Votes	6	5	4	2

We will once again use the plurality with elimination method. Each candidate has the following number of first-place votes.

Clarke:	6 + 2 = 8 first-place votes
Green:	4 first-place votes
Roberts:	5 first-place votes
White:	0 first-place votes

Once again, White is eliminated without affecting any of the first-place votes. Therefore, we look to find the candidate with the next lowest number of first-place votes. This time, Green is eliminated with only four votes. Table 13.2.12 shows both White and Green eliminated.

Table 13.2.12: Plurality with Elimination Cycles 1 and 2 Simplified

	Rankings			
1st	Clarke	Roberts	Roberts	Clarke
2nd	Roberts	Clarke	Clarke	Roberts
Total Votes	6	5	4	2

Now we see that Clarke has $6 + 2 = 8$ votes and Roberts has $5 + 4 = 9$ votes. That gives Roberts a majority and therefore the title of Whitman City Council Vice President in this second voting process.

Notice that Clarke won in the first voting. He then gained two first-place votes in the second process, but ultimately lost the election to Roberts. This illustrates that by using the plurality with elimination method, it is possible to violate the monotonicity criterion.

Example 13.2.3 used the same method, plurality with elimination, to produce a different winner after voters were allowed to change their votes, illustrating that the plurality with elimination method can violate the monotonicity criterion. It is the only counting method discussed in the previous section that has a possibility of violating the monotonicity criterion.

OBJECTIVE 5   The Irrelevant Alternatives Criterion

The fourth condition considers cases where candidates drop out of an election. The **irrelevant alternatives criterion** states that if a candidate wins an election, then that same candidate would win the election even if at least one candidate withdraws from the election. This criterion is the most stringent of all the criteria because it may be violated in four of the five voting methods; the majority rule decision is the only method that never violates the irrelevant alternatives criterion.

> 📖 **IRRELEVANT ALTERNATIVE CRITERION**
>
> The **irrelevant alternatives criterion** states that if a candidate wins an election, then that same candidate would win the election even if at least one candidate withdraws from the election.

Example 13.2.4

Irrelevant Alternatives Criterion and the Pairwise Method of Comparison

As part of the process of choosing the next chancellor of the university, faculty members were asked to rank the four finalists in order of preference. Table 13.2.13 summarizes their votes.

Table 13.2.13: Preference Table for Chancellor

	Rankings				
1st	Khan	Khan	Patel	Mason	Mason
2nd	Patel	Lewis	Lewis	Lewis	Khan
3rd	Mason	Patel	Mason	Khan	Lewis
4th	Lewis	Mason	Khan	Patel	Patel
Total Votes	166	290	235	123	236

a. Use the pairwise comparison method to determine the favorite candidate among the faculty.

b. Before a chancellor could be named, Patel took a job elsewhere and Lewis decided the university was not the right fit for him after visiting the campus. Consequently, Patel and Lewis withdrew their names from the application pool. Determine the faculty favorite after Patel and Lewis are removed from the ballot.

c. Based on your solutions to parts a. and b., does this election comply with the irrelevant alternative criterion?

Solution

a. The pairwise comparison method requires that we determine the winner of each head-to-head matchup between the candidates. Using the formula, we know that there are six comparisons required.

$$\text{Number of Comparisons} = \frac{4(4-1)}{2} = \frac{12}{2} = 6$$

Using a table to keep track of the head-to-head comparisons, we have the following.

Table 13.2.14: Completed Pairwise Comparison Grid for Chancellor

	Patel	Mason	Lewis	Khan
Patel	X	P = 166; P = 290; P = 235; M = 123; M = 236	P = 166; L = 290; P = 235; L = 123; L = 236	K = 166; K = 290; P = 235; K = 123; K = 236
Mason		X	M = 166; L = 290; L = 235; M = 123; M = 236	K = 166; K = 290; M = 235; M = 123; M = 236
Charles			X	K = 166; K = 290; L = 235; L = 123; K = 236
Khan				X

Table 13.2.15: Pairwise Comparison Grid of Winners for Chancellor

	Patel	Mason	Lewis	Khan
Patel	X	P = 691; M = 359 [Patel]	P = 401; L = 649 [Lewis]	K = 815; L = 235 [Khan]
Mason		X	M = 525; L = 525 [Tie]	K = 456; M = 594; [Mason]
Charles			X	K = 692; L = 358 [Khan]
Khan				X

> **💬 Helpful Hint**
>
> Remember that in a head-to-head comparison, a tie gives each candidate $\frac{1}{2}$ of a point.

b. Adding each candidate's votes together in the comparisons, we see that Patel received 1 point, Mason received $1\frac{1}{2}$ points, Lewis received $1\frac{1}{2}$ points, and Khan received 2 points. Therefore, Khan is the winner using the pairwise method of comparison. When both Patel and Lewis withdraw from the race, only Mason and Khan remain. We know from the head-to-head comparison in Table 13.2.15, that when these two face one another, Mason is the winner. Therefore, Mason is the winner using the pairwise comparison method after Patel and Lewis withdraw.

c. Using the pairwise comparison method both before and after Patel and Lewis withdrew resulted in two different winners. Therefore, this method did not satisfy the irrelevant alternative criterion in this election.

OBJECTIVE 6 The Dictator Criterion

The fifth and final criterion is more about the voter rather than the method of counting the votes. The **dictator criterion** states that no single voter is allowed to decide the outcome of an election. Simply put, in order to be fair, an electoral system must be a democracy and not a dictatorship.

> **📖 DICTATOR CRITERION**
>
> The **dictator criterion** states that no single vote is allowed to decide the outcome of an election.

The following table summarizes each of the voting methods discussed previously versus the fairness criteria used in democratic elections.

Table 13.2.16: Voting Methods and Fairness Criteria

	Condorcet Criterion	Majority Criterion	Monotonicity Criterion	Irrelevant Alternatives Criterion	Dictator Criterion
Majority Rule Decision	Never violates	Never violates	Never violates	Never violates	May violate
Plurality	May violate	Never violates	Never violates	May violate	Never violates
Borda Count	May violate	May violate	Never violates	May violate	Never violates
Plurality with Elimination	May violate	Never violates	May violate	May violate	Never violates
Pairwise Comparison	Never violates	Never violates	Never violates	May violate	Never violates

As our table shows, it would appear that using a majority rule as an election method might in fact be the perfect voting system because it never violates any of the four fairness criteria in a democracy. However, the one "fly in the ointment," so to say, is exactly what we saw in the previous section—using the majority rule doesn't always produce a winner. This is why Arrow requires that valid voting systems always produce a distinct winner.

13.2   Exercises

✔ CONCEPT CHECK

1. The _____ states that if a candidate wins the head-to-head comparison—as in the pairwise comparison method—against every other candidate, then that candidate should also win the overall election in a fair voting system.

2. If a candidate receives a majority of votes in an election but does not win, the _____ has been violated.

3. The _____ requires that if Candidate X wins an election, then X would also win a second election if each voter were allowed to reorder the candidates in such a way that X only increases in ranking while the order of the other candidates remained the same.

4. If the winner of an election is changed when one or more candidates drops out of the election, the _____ has been violated.

5. The _____ is the only voting method that can violate the majority criterion.

🚀 **APPLICATIONS**

Use the given preference table to answer each question.

6. A group of students were asked to listen to auditions by four bands for the homecoming dance and then rank the bands. The following preference table summarizes their selections.

Preference Table for Homecoming Bands

	Rankings				
1st	Band B	Band A	Band C	Band D	Band A
2nd	Band D	Band D	Band B	Band A	Band C
3rd	Band C	Band B	Band A	Band B	Band B
4th	Band A	Band C	Band D	Band C	Band D
Total Votes	23	15	16	11	5

a. Which band is the plurality winner?

b. Which candidate wins the election using a pairwise method of comparison?

c. Does this election, using the plurality method, satisfy the Condorcet criterion? Explain your answer.

7. Susan, Courtney, and Jade ran for Freshman SGA representative. The following preference table shows the results of the race.

Preference Table for SGA Representatives

	Rankings		
1st	Susan	Courtney	Susan
2nd	Courtney	Susan	Jade
3rd	Jade	Jade	Courtney
Total Votes	32	55	16

a. Which candidate is the winner using the Borda count method?

b. Which candidate is the winner using the pairwise method of comparison?

c. Does the Borda count method satisfy the Condorcet criterion in this situation? Why or why not?

8. Work colleagues are trying to decide on which restaurant to use for their end-of-year party. Since they don't all agree on a favorite, they decide to rank the four possibilities and decide on a restaurant that way. The results are shown in the preference table.

	Preference Table for Favorite Restaurants				
	Rankings				
1st	Tanvi's	O'Reilly's Pub	Ralphie's	Tanvi's	Salty's
2nd	Ralphie's	Salty's	Tanvi's	O'Reilly's Pub	Ralphie's
3rd	Salty's	Tanvi's	Salty's	Salty's	O'Reilly's Pub
4th	O'Reilly's Pub	Ralphie's	O'Reilly's Pub	Ralphie's	Tanvi's
Total Votes	8	3	4	11	9

a. Which restaurant is the winner using the plurality with elimination method?

b. Which restaurant is the winner using the pairwise method of comparison?

c. Does the plurality with elimination method satisfy the Condorcet criterion if the colleagues chose a restaurant using plurality with elimination? Why or why not?

9. Jack is scheduling classes for next semester and has three possible classes for a particular time slot. He surveys students to see which class has the most interest out of Electricity and Magnetism, Heat and Optics, and Mechanics.

	Preference Table for Classes			
	Rankings			
1st	H & O	Mechanics	H & O	E & M
2nd	E & M	H & O	E & M	Mechanics
3rd	Mechanics	E & M	Mechanics	H & O
Total Votes	81	132	36	13

a. Which class has the most interest using the majority rule?

b. Which class has the most interest using the Borda count method?

c. Does the Borda count method satisfy the majority criterion if Jack uses this method to fill the available time slot? Explain your answer.

10. The history department is adopting a new textbook for the Freshman Introduction class. Instructors in the history department were asked to rank the top four textbooks in order of usability for the class. The preference table shows the outcome of their rankings.

Preference Table for History Textbooks

	Rankings			
1st	Excursions Through Time	Thinking Historically	A Look in Time	Through the Ages
2nd	Thinking Historically	A Look in Time	Thinking Historically	A Look in Time
3rd	A Look in Time	Excursions Through Time	Through the Ages	Thinking Historically
4th	Through the Ages	Through the Ages	Excursions Through Time	Excursions Through Time
Total Votes	7	3	7	4

a. Use the plurality with elimination method to choose a new textbook for adoption.

b. After some discussion in the department, the seven instructors whose rankings appear in the first column decide to change their votes so that the textbooks titled *A Look in Time* and *Thinking Historically* swap rankings. Create a new preference table to show this change and then use the table to decide on a textbook using the plurality with elimination method once again.

c. Is the monotonicity criterion satisfied? Explain your answer.

11. Charles Beauregard High School sponsors a senior class trip each year. This year, the students can choose between trips to Charleston, SC; Jacksonville, FL; New Orleans, LA; Chicago, IL; or Atlanta, GA. The students were asked to rank their choices in order to choose a destination.

Preference Table for Senior Class Trip

	Rankings			
1st	New Orleans, LA	Charleston, SC	Atlanta, GA	Chicago, IL
2nd	Chicago, IL	New Orleans, LA	New Orleans, LA	Charleston, SC
3rd	Atlanta, GA	Atlanta, GA	Chicago, IL	Atlanta, GA
4th	Charleston, SC	Jacksonville, FL	Charleston, SC	New Orleans, LA
5th	Jacksonville, FL	Chicago, IL	Jacksonville, FL	Jacksonville, FL
Total Votes	375	348	289	115

a. Which city would be the senior trip destination using the plurality with elimination method?

b. After consideration of travel time, the senior class sponsors decided to eliminate Chicago, IL, from the options. Which city is now the winning destination using plurality with elimination?

c. Does the plurality with elimination method satisfy the irrelevant alternatives criterion? Explain your answer.

12. A nonprofit organization is looking to use social networking sites to increase interest. They poll donors to decide where to focus their resources. The donors are asked to rank the following sites by how often they use them.

Preference Table for Social Networking Sites

	Rankings			
1st	Facebook	Pinterest	Facebook	Twitter
2nd	Twitter	Facebook	Twitter	Facebook
3rd	Google+	Twitter	LinkedIn	LinkedIn
4th	LinkedIn	LinkedIn	Pinterest	Google+
5th	Pinterest	Google+	Google+	Pinterest
Total Votes	23	40	44	35

   a. Which site is given the higher ranking using the pairwise method of comparison?

   b. If the nonprofit decides to eliminate Pinterest and Google+ from the list and adjust the votes accordingly, which site would then be the top place for the nonprofit to focus on using the pairwise method?

   c. Does the pairwise method satisfy the irrelevant alternatives criterion for this preference table? Why or why not?

13. The men of the fraternity Chi Rho on campus are electing a new president. The candidates are M. Jones, H. Kennedy, and T. Parchment. The summary of the rankings of the candidates from the members is given in the chart.

Preference Table for Chi Rho Presidential Candidates

	Rankings			
1st	H. Kennedy	M. Jones	H. Kennedy	T. Parchment
2nd	M. Jones	T. Parchment	T. Parchment	H. Kennedy
3rd	T. Parchment	H. Kennedy	M. Jones	M. Jones
Total Votes	8	14	10	11

   a. Name the fraternity's next president using the Borda count method.

   b. Because of unexpected family issues, M. Jones had to withdraw from the university on short notice. Create a new preference table to show this change and then use the table to determine the next president using the Borda count method.

   c. Is the irrelevant alternatives criterion satisfied with this change? Explain your answer.

14. For the end of year banquet, the program committee needs to decide on the type of cuisine. They have four choices: Italian, French, Mexican, or Chinese. The following preference table summarizes the rankings of the members surveyed. Determine if the Condorcet criterion is satisfied if the committee uses the Borda count method to choose a cuisine.

Preference Table for Banquet Cuisine

	Rankings				
1st	French	Mexican	Italian	Italian	Mexican
2nd	Italian	Italian	Mexican	Chinese	French
3rd	Mexican	Chinese	French	Mexican	Chinese
4th	Chinese	French	Chinese	French	Italian
Total Votes	4	21	14	11	9

15. In preparation for their Oscar party, a sorority decides to have an election of their own. In the Best Actor category are George Clooney, Johnny Depp, Colin Firth, and Zac Efron. The rankings from the sorority members are shown in the following preference table. Determine if the majority criterion will be satisfied if the sorority uses the Borda count method to choose the winner for Best Actor.

Preference Table for Best Actor

	Rankings				
1st	Johnny Depp	Zac Efron	Johnny Depp	George Clooney	Colin Firth
2nd	Colin Firth	Colin Firth	George Clooney	Colin Firth	Zac Efron
3rd	Zac Efron	George Clooney	Zac Efron	Johnny Depp	Johnny Depp
4th	George Clooney	Johnny Depp	Colin Firth	Zac Efron	George Clooney
Total Votes	26	16	16	15	9

16. *Pretty People* magazine wants to name the number one activity for a first date amongst singles. They enlist the help of an online dating site that asks members to rank five activities in order of their preference as part of their sign-up process. The five activities include: movie, nice dinner, picnic, sporting event, and concert. The following preference table displays the results of the rankings. Determine if the Condorcet criterion is satisfied if the magazine uses the plurality method to name the top activity.

Preference Table for First Date Activities

	Rankings					
1st	Nice Dinner	Picnic	Picnic	Sporting Event	Sporting Event	Movie
2nd	Picnic	Sporting Event	Sporting Event	Concert	Nice Dinner	Nice Dinner
3rd	Concert	Concert	Nice Dinner	Picnic	Movie	Concert
4th	Movie	Nice Dinner	Concert	Nice Dinner	Concert	Picnic
5th	Sporting Event	Movie	Movie	Movie	Picnic	Sporting Event
Total Votes	40	34	21	22	19	20

17. United Way asks its volunteers to rank the nonprofit organizations it serves in order to give out the annual award for Local Volunteer of the Year. This year's organizations include the Breast Cancer Foundation, Community Garden, Big Brother/Big Sister, YMCA, and Manna Café. Complete the preference table so that Community Garden is the winner using the Borda count method, but the Condorcet criterion is violated.

Preference Table for Local Volunteer of the Year Award

	Rankings		
1st	MC	CG	BB/BS
2nd	BB/BS	BCF	MC
3rd	CG	BB/BS	CG
4th	YMCA	YMCA	BCF
5th	BCF	MC	YMCA
Total Votes	?	16	10

18. An online dating site asks new users to rank certain traits in order of importance when they are matched with another user. The site uses the top choice as the first criterion for matching people together. The following preference table summarizes the rankings for appearance, personality, income, profession, and height. Complete the preference table so that personality is the top trait if the site uses the Borda count method to count the votes and the majority criterion is not violated.

Preference Table for Important Traits

	Rankings			
1st	?	Personality	Income	Appearance
2nd	?	Profession	Personality	Income
3rd	Profession	Appearance	Appearance	Personality
4th	Height	Height	Profession	Height
5th	Income	Income	Height	Profession
Total Votes	45	56	33	48

19. College dining is considering opening a national fast food franchise on campus. Their choices have been narrowed down to Subway, Taco Bell, McDonald's, KFC, or Pizza Hut. Students were asked to rank these establishments in order of preference, and the results are shown in the preference table. If college dining uses the plurality method to choose the students' top choice of Subway, complete the preference table so that the irrelevant alternatives criterion is violated when KFC is removed from the race.

Preference Table for New Franchise on Campus

	Rankings			
1st	Subway	Taco Bell	KFC	Subway
2nd	McDonald's	Subway	Taco Bell	Taco Bell
3rd	Taco Bell	Pizza Hut	Subway	Pizza Hut
4th	KFC	McDonald's	Pizza Hut	McDonald's
5th	Pizza Hut	KFC	McDonald's	KFC
Total Votes	56	60	?	45

20. Sigma Air is exploring the possibility of establishing a new hub in one of the following cities: Cheyenne, WY; Little Rock, AR; Amarillo, TX; and Portland, OR. They surveyed their Gold and Platinum members to ask their preference for the new hub. Complete the preference table of the rankings so that using the plurality with elimination method Amarillo is the winner, and the Condorcet criterion is satisfied.

Preference Table for New Hub Location

	Rankings		
1st	Amarillo, TX	Portland, OR	Cheyenne, WY
2nd	Portland, OR	Little Rock, AR	Amarillo, TX
3rd	Cheyenne, WY	?	Little Rock, AR
4th	Little Rock, AR	?	Portland, OR
Total Votes	70	34	37

21. The following preference table summarizes the outcome of an election.

Preference Table for Candidates

	Rankings			
1st	Luke	Luke	Lauren	Blake
2nd	Hannah	Blake	Blake	Luke
3rd	Lauren	Lauren	Hannah	Lauren
4th	Blake	Hannah	Luke	Hannah
Total Votes	31	9	23	41

a. Which candidate is the plurality winner?

b. Which candidate wins the election using a pairwise method of comparison?

c. Does this method satisfy the Condorcet criterion? Why or why not?

22. Every year, the Oviedo Little League gives out a Coach of the Year award. The board of directors selects three names to put on the ballot, and parents send in their votes. The results are shown in the following preference table.

Preference Table for Coach of the Year

	Rankings			
1st	Hall	McKee	Price	Hall
2nd	McKee	Price	McKee	Price
3rd	Price	Hall	Hall	McKee
Total Votes	7	62	21	11

a. Which coach would win using the majority rule?

b. Which coach would win using the Borda count method?

c. Does the Borda count method satisfy the majority criterion

23. A high school math teacher is trying to pick a name for her puppy. She's narrowed it down to four choices and has decided to let the senior class vote. Determine if the majority criterion will be satisfied if the teacher uses the Borda count method to choose the winner.

Preference Table for Puppy Name

	Rankings				
1st	Paladin	Knight	Knight	Paladin	Paladin
2nd	Ranger	Paladin	Whiskey	Ranger	Whiskey
3rd	Whiskey	Ranger	Paladin	Whiskey	Knight
4th	Knight	Whiskey	Ranger	Knight	Ranger
Total Votes	26	73	68	51	25

24. Suzie's family has a reunion every other year. On the last night of each reunion, family members can propose a location for the next reunion. Family members rank the top four proposals.

Preference Table for Reunion Location

	Rankings			
1st	Charleston	San Diego	Dallas	New Orleans
2nd	Dallas	Charleston	San Diego	Dallas
3rd	San Diego	New Orleans	Charleston	Charleston
4th	New Orleans	Dallas	New Orleans	San Diego
Total Votes	6	5	4	3

a. Use the plurality with elimination method to choose where the next reunion should be held.

b. After some discussion among the cousins, the 3 family members whose rankings appear in the rightmost column decide to change their votes so that Charleston is their first pick, but the rest of their ranking keeps the same order. Create a new preference table to show this change and then use the table to decide on a location using the plurality of elimination method once again.

c. Is the monotonicity criterion satisfied? Explain your answer.

13.2 PROJECT

WHAT'S FAIR? DEPENDS ON WHO YOU ASK

In Section 13.2, we discussed five different conditions for determining the fairness of an election. Now it's time to explore how voting procedures and ideas about fairness impact elections in the real world. In this project, we'll consider the fairness of ranked-choice voting, which is equivalent to the plurality with elimination method that is used with preference ballots.

As a recap, here are the fairness criteria.

- The **Condorcet criterion** states that if a candidate wins the head-to-head comparison against every other candidate, then that candidate should also win the overall election in a fair voting system.

- The **majority criterion** states that if a candidate receives a majority of votes in an election, that candidate should win.

- The **monotonicity criterion** states that if a candidate wins an early round of an election and only gains support and does not lose support in subsequent rounds, then that candidate should win the election.

- The **irrelevant alternatives criterion** states that if a candidate wins an election, then that same candidate would win the election even if at least one candidate withdraws from the election.

- The **dictator criterion** states that no single vote is allowed to decide the outcome of an election.

1. Find an article online from a reputable source explaining some of the downsides of ranked-choice voting. Record your source and two or three key takeaways from the article.

2. Find an article online from a reputable source explaining some of the advantages of ranked-choice voting. Record your source and two or three key takeaways from the article.

3. Find an article online from a reputable source detailing a voting procedure in a state or municipality that uses a vote-tallying method other than majority rule. Record your source and two or three key takeaways describing the method. Reflecting on what you've learned about fairness in voting, state one downside and one advantage of the method described.

13.3 APPORTIONMENT

OBJECTIVES

1. Calculate the standard divisor and standard quota.

2. Allocate resources using Hamilton's method.

3. Allocate resources using Jefferson's method.

4. Allocate resources using Webster's method.

5. Allocate resources using the Huntington-Hill method.

6. Identify weaknesses in apportionment methods.

As we saw in the previous section, the concept of complete fairness is not attainable in elections. Instead, we make informed choices before votes are cast by choosing a method of counting that best fits the situation, all the while knowing the potential shortcomings of each particular method.

The same type of dilemma rears its head when we want or need to distribute things among a group in the most fair-minded way possible. For instance, consider the allotment of funding for public higher education institutions in a state. The state legislature often allocates billions of dollars of its overall budget to the university system to be shared among multiple institutions. The question is then, how should the funds be distributed among the schools? One possibility is that the school with the largest student population gets the most money. Another method might allocate the same amount of funds to all of the institutions regardless of the student population. A third option would be to base the size of the portions allocated to each institution on the number of credit hours each student is enrolled in. Regardless of the method, you can imagine that each of these ways of distributing the money would likely garner both support and opposition from residents across the state, each arguing for the same end goal—*fairness*. This concept of distributing items in the fairest way possible is called **apportionment**.

📖 **APPORTIONMENT**

Apportionment is a method of fairly dividing resources or items among individuals or groups.

OBJECTIVE 1 The Standard Divisor and the Standard Quotient

In the United States, we use apportionment to determine how many representatives each state is allowed in the federal government. Congress is comprised of 535 voting members: 100 from the Senate and 435 from the House of Representatives. In the Senate, the distribution of 100 votes among the states is easy; each of the fifty states has two Senators. However, in the House of Representatives, the 435 voting members are apportioned by state population. Each time the census is conducted, there is a new debate on how congressional seats are distributed throughout the states, based on shifting populations.

As it turns out, it is not always straightforward to determine a method that apportions available resources in a manner that ensures all members within a population receive a fair share. However, there are conventional ways to approach the problem. The first step in all of the approaches is to determine a **standard divisor**. The standard divisor for apportionment is the average number of people per item to be apportioned. It is found by dividing the total population size by the number of items to be apportioned. For example, the standard divisor for the US House of Representatives is the average number of people represented per congressional seat.

📖 **STANDARD DIVISOR**

The **standard divisor** for apportionment is the average number of people per item to be apportioned.

$$\text{Standard Divisor} = \frac{\text{Total Population}}{\text{Number of Items to Be Apportioned}}$$

Example 13.3.1

Finding the Standard Divisor

At the beginning of each school year, the student body is given 9 delegates on the governing board. The delegates are divided among the four classes (freshmen, sophomores, juniors, seniors) according to student population. Calculate the standard divisor that will be used to assign the delegates.

Table 13.3.1: Student Body Class Size

Freshmen	Sophomores	Juniors	Seniors
224	267	135	169

Solution

To calculate the standard divisor, divide the total number of students in the school by the number of delegates to be shared out.

$$\text{Standard Divisor: } \frac{224 + 267 + 135 + 169}{9} = \frac{795}{9} \approx 88.3333$$

Thus, each student delegate will represent approximately 88.3333 students.

Now that we can determine roughly how many people each item represents (the standard divisor), we can begin to hand out the items based on a group's population. For instance, if representatives are being apportioned with a standard divisor of 100, it means that for every 100 people, a single representative is given. A group of 500 people would get $\frac{500}{100} = 5$ representatives. A standard divisor of 75 means that a group of 500 people should get $\frac{500}{100} \approx 6.6667$ representatives. This number of approximate representatives, or items, each subgroup is assigned is known as the **standard quota.**

📖 **STANDARD QUOTA**

The **standard quota** for apportionment is the average number of items to be apportioned to each subgroup.

$$\text{Standard Quota} = \frac{\text{Subgroup Population}}{\text{Standard Divisor}}$$

Example 13.3.2

Finding the Standard Quota

Use the information from Example 13.3.1 to determine the standard quota of representatives for each student class on the school governing board. The class sizes are reprinted for you here.

Table 13.3.2: Student Body Class Size

Freshmen	Sophomores	Juniors	Seniors
224	267	135	169

Solution

In order to find the standard quota for each subpopulation, we need to divide each class size by the standard divisor, which was calculated in Example 13.3.1 as 88.3333.

Standard Quotas

Freshmen: $\dfrac{224}{88.3333} \approx 2.5359$

Sophomores: $\dfrac{267}{88.3333} \approx 3.0226$

Juniors: $\dfrac{135}{88.3333} \approx 1.5283$

Seniors: $\dfrac{169}{88.3333} \approx 1.9132$

✔ **Skill Check 13.3.2**

Three classes have student populations of 87, 52, and 89, respectively. Determine the standard quota for each class given that the standard divisor is 13.4.

Apportionment issues arise when the standard quota (that is, the number of items each subgroup receives) is a fraction, as we saw in Example 13.3.2. We cannot assign the freshmen 2.5359 representatives on the governing board. In the case of the US House of Representatives, it also doesn't make sense to talk about a state having 4.7333 representatives. Because of this issue, at times we need to use the *lower quota* or *upper quota* for each subgroup. The **lower quota** is the standard quota rounded *down* to the nearest whole number. The **upper quota** is the standard quota rounded *up* to the nearest whole number. For instance, the lower quota for a state with 4.7333 representatives is 4 and the upper quota is 5.

However, even rounding quotas to whole numbers does not eliminate all problems. Using lower quotas might result in leftover resources not being allocated, while using upper quotas could overpromise but under deliver by allocating more resources than are actually available. Historically, there have been four methods used by Congress to tackle this issue of apportionment: Hamilton's method, Jefferson's method, Webster's method, and the Huntington-Hill method. We'll spend the remainder of this section looking at each method.

OBJECTIVE 2 Hamilton's Method

In the spring of 1791, Congress passed a bill to apportion the seats in the House of Representatives using a method introduced by Alexander Hamilton, now known as Hamilton's method (or the *method of largest remainders*). This method uses the lower quotas for each subgroup to allocate resources. It then distributes any leftover resources according to how much each standard quota was rounded. The standard quota rounded down the most is first in line for a leftover resource, then the standard quota with the next largest rounding, and so on until all of the resources are allocated. For instance, suppose there are thirteen representatives to divide among four classes. Table 13.3.3 shows each class along with their standard quota, lower quota, and fractional remainder.

Table 13.3.3: Class Quotas

Class	Standard Quota	Lower Quota	Fractional Remainder
A	1.1127	1	0.1127
B	3.9345	3	0.9345
C	2.6889	2	0.6889
D	4.8022	4	0.8022

Based on the lower quotas, $1 + 3 + 2 + 4 = 10$ resources are allocated, leaving three representatives still to be assigned. Hamilton's method now looks back at the fractional remainders of the standard quotas. We can sort the classes by their remainders from greatest to least, as shown in Table 13.3.4.

Table 13.3.4: Class Quotas Sorted by Fractional Remainder

Class	Standard Quota	Lower Quota	Fractional Remainder
B	3.9345	3	0.9345
D	4.8022	4	0.8022
C	2.6889	2	0.6889
A	1.1127	1	0.1127

This means that class B incurred the biggest rounding effect with the lower quotas. We allocate the resources going down this order until there are none left. So classes B, D, and C all receive one additional representative.

💬 **Math Milestone**

Before it was ever used, Hamilton's method of apportionment was vetoed by President Washington in 1791, becoming the first veto by a US President. It was later adopted in 1850 and used until 1910.

📖 **HAMILTON'S METHOD OF APPORTIONMENT**

1. Calculate the standard divisor.
2. Calculate the standard quota for each subgroup.
3. Calculate the lower quota for each subgroup.
4. Assign each subgroup the number of resources based on its lower quota.
5. Assign any remaining resources based on the fractional remainder of the standard quotas, in order from largest to smallest.

Example 13.3.3

Applying Hamilton's Method

A new suburb of Chicago is seeking a charter to become a city of its own. The suburb will have 25 council members divided among six zones. Use Hamilton's method to determine how the council members will be apportioned between the six zones. The populations for each zone are given in Table 13.3.5.

Table 13.3.5: Suburb Population by Zone

Zone	Population
1	2897
2	4538
3	3362
4	14,003
5	8450
6	12,339

Solution

Begin by calculating the standard divisor. Add all of the subpopulations together and divide by the number of members to be apportioned, which is 25.

Standard Divisor:

$$\frac{2897 + 4538 + 3362 + 14,003 + 8450 + 12,339}{25} = \frac{45,859}{25} = 1823.56$$

Next, calculate the standard quota for each subgroup. We will extend the table to help us keep track of the calculations.

Table 13.3.6: Suburb Population by Zone

Zone	Population	Standard Quota
1	2897	$\frac{2897}{1823.56} \approx 1.5887$
2	4538	$\frac{4538}{1823.56} \approx 2.4885$
3	3362	$\frac{3362}{1823.56} \approx 1.8436$
4	14,003	$\frac{14,003}{1823.56} \approx 7.6789$
5	8450	$\frac{8450}{1823.56} \approx 4.6338$
6	12,339	$\frac{12,339}{1823.56} \approx 6.7664$

To calculate the lower quota for each zone, we round each standard quota down. Thus, we have the following lower quotas.

Table 13.3.7: Suburb Population by Zone

Zone	Population	Standard Quota	Lower Quotas
1	2897	1.5887	1
2	4538	2.4885	2
3	3362	1.8436	1
4	14,003	7.6789	7
5	8450	4.6338	4
2	12,339	6.7664	6

If we allocate the lower quotas for each zone, we apportion $1 + 2 + 1 + 7 + 4 + 6 = 21$ council seats. That leaves four seats to hand out. Hamilton's method states that we give the four leftover seats to the four subgroups with the largest fractional remainders in their standard quotas.

We will reorder the zones according to their fractional remainder (greatest to least) to help us see which zones get the extra seats.

Table 13.3.8: Suburb Population by Zone

Zone	Population	Standard Quota Sorted by Fractional Remainder
3	3362	1.8436
6	12,339	6.7664
4	14,003	7.6789
5	8450	4.6338
2	4538	2.4885
1	2897	1.5886

Therefore, Zones 3, 6, 4, and 5 each get 1 additional council seat. So the council seats are apportioned as follows.

Table 13.3.9: Suburb Population by Zone

Zone	Population	Standard Quota	Lower Quota	Distribution of Remaining Council Seats	New Apportionment
1	2897	1.5886	1		1
2	4538	2.4885	2		2
3	3362	1.8436	1	1	2
4	14,003	7.6789	7	1	8
5	8450	4.6338	4	1	5
6	12,339	6.7664	6	1	7
Total	**45,589**		**21**		**25**

This means that Zone 1 receives 1 seat, Zone 2 receives 2 seats, Zone 3 receives 2 seats, Zone 4 receives 8 seats, Zone 5 receives 5 seats, and Zone 6 receives 7 seats. Notice that the number of council seats among the zones is now $1 + 2 + 2 + 8 + 5 + 7 = 25$, which is the correct number of seats.

OBJECTIVE 3 Jefferson's Method

George Washington vetoed Hamilton's method of apportionment before it was ever used to apportion seats in the US Congress, and another method of apportioning representatives was sought out. Thomas Jefferson proposed a method that tried to remedy the potentially unfair rounding situation in Hamilton's method by using a *modified divisor*. The new method begins just the same as Hamilton's did. However, if the standard divisor does not distribute all of the seats, a new **modified divisor** is chosen by trial and error until the modified lower quotas equal the number of seats to be apportioned.

Math Milestone

The Jefferson method of apportionment was used to apportion US House representatives were apportioned using the Jefferson Method for 40 years, from 1790 to 1830.

📖 JEFFERSON'S METHOD OF APPORTIONMENT

1. Calculate the standard divisor.

2. Calculate the standard quota for each subgroup.

3. Calculate the lower quota for each subgroup.

4. Assign each subgroup the number of resources based on its lower quota.

5. If there are remaining resources to be distributed, chose a modified divisor by trial and error until the sum of the lower quotas equals the number of resources to be apportioned.

It should be noted that whenever the sum of the quotas found is smaller than the number of items, then the modified divisor should be made *smaller* to increase the sum. Conversely, if the sum of quotas is larger than the number of items, then modified divisor should be made *larger* to decrease the number of items allocated to each subgroup.

Example 13.3.4

Applying Jefferson's Method

A university has 18 scholarships to be apportioned among 225 math majors, 417 history majors, and 308 computer science majors. Use Jefferson's method to determine how the scholarships should be apportioned among the three major groups.

Solution

Begin by calculating the standard divisor. Add all of the subpopulations together and divide by the number of scholarships to be apportioned, 18.

$$\text{Standard Divisor: } \frac{225+417+308}{18} \approx 52.7778$$

Next, steps 2 and 3 of Jefferson's method say to calculate the standard quota and lower quota for each subgroup.

Table 13.3.10: Standard and Lower Quotas

Major	Standard Quota	Lower Quota
Math	$\frac{225}{52.7778} \approx 4.2632$	4
History	$\frac{417}{52.7778} \approx 7.9010$	7
Computer Science	$\frac{308}{52.7778} \approx 5.8358$	5
Total Scholarships		16

This gives out 16 scholarships—two short of the number needed. Jefferson's method tells us that we should modify the divisor either up or down and find the modified quotas. Because we need larger quotas, we want a smaller divisor. Rounding the standard divisor down to 52, we now calculate the modified quotas as shown in Table 13.3.11. Remember this is a trial-and-error process, so we could technically choose any number for the modified divisor.

Table 13.3.11: Modified Quotas and Modified Lower Quotas with a Modified Divisor of 52

Major	Modified Quota	Modified Lower Quota
Math	$\frac{225}{52} \approx 4.3269$	4
History	$\frac{417}{52} \approx 8.0192$	8
Computer Science	$\frac{308}{52} \approx 5.9232$	5
Total Scholarships		17

Notice that we have not apportioned all 18 scholarships. So we need adjust the modified divisor once again. Let's use 51 for our divisor this time. The new modified quotas are calculated in Table 13.3.12.

Table 13.3.12: Modified Quotas and Modified Lower Quotas with a Modified Divisor of 51

Major	Modified Quota	Modified Lower Quota
Math	$\frac{225}{51} \approx 4.4118$	4
History	$\frac{417}{51} \approx 8.1765$	8
Computer Science	$\frac{308}{51} \approx 6.0392$	6
Total Scholarships		17

✔ **Skill Check 13.3.3**

Use Hamilton's method to apportion the scholarships in Example 13.3.4.

Finally, we have the correct number of scholarships apportioned. Four math majors, eight history majors, and six computer science majors will receive scholarships.

OBJECTIVE 4 Webster's Method

Jefferson's method was thought to be the fairest method of apportionment until the 1820 census, when a flaw was discovered that could possibly award additional seats in Congress to a state even though the state population decreased or vice versa. After a repeat of this anomaly in the next census in 1830, Jefferson's method was replaced with one proposed by Senator Daniel Webster of Massachusetts. The difference in Webster's method once again aimed at how to distribute the remaining resources if the standard quotas did not equal the number of resources to be apportioned. His method states that the initial assignments are found by rounding the standard quotas to the nearest integer, not simply rounding down as in Jefferson's method and Hamilton's method. If the fractional part of the standard quota is 0.5 or more, round up; otherwise, round down.

Math Milestone

Webster's method was used by the US House of Representatives after the 1840 census. From 1850 to 1900, Congress finally decided to adopt Hamilton's original method. Webster's method was later readopted and used from 1910 to 1940.

📖 WEBSTER'S METHOD OF APPORTIONMENT

1. Calculate the standard divisor.
2. Calculate the standard quota for each subgroup.
3. Round each quota to the nearest integer.
4. Assign each subgroup the number of resources based on the rounded quota.
5. If there are remaining resources to be distributed, chose a modified divisor by trial and error until the sum of the rounded quotas equals the number of resources to be apportioned.

Example 13.3.5

Applying Webster's Method

A computer tech firm has three divisions with 135, 98, and 132 employees, respectively. A total of 11 administrative assistants must be allocated to the three divisions according to their size. Use Webster's method to determine how many administrative assistants should be allocated to each division.

Solution

Begin by calculating the standard divisor. Add all of the subpopulations together and divide by the number of assistants to be apportioned, which is 11.

$$\text{Standard Divisor: } \frac{135+98+132}{11} \approx 33.1818$$

Steps 2 and 3 in Webster's method say to find the standard quotas for each subgroup and round to the nearest integer. The calculations are shown in Table 13.3.13.

Table 13.3.13: Standard Quotas and Rounded Quotas for Administrative Assistants

Division Population	Standard Quota	Rounded Quota
135	$\frac{135}{33.1818} \approx 4.0685$	4
98	$\frac{98}{33.1818} \approx 2.9534$	3
132	$\frac{132}{33.1818} \approx 3.9781$	4
Total Assistants		11

✔ Skill Check 13.3.4

Use Hamilton's method to apportion the administrative assistants in Example 13.3.5.

We can see from the bottom row that we have allocated all eleven administrative assistants to the respective divisions. The first division has 4 administrative assistants, the second division has 3, and the third division has 4.

OBJECTIVE 5 The Huntington-Hill Method

The last method of apportionment that we will consider is the Huntington-Hill method. The Huntington Hill method is credited to the chief statistician of the Census Bureau, Joseph Hill, after the 1940 census and was revised by Harvard mathematician Edward Huntington. The Huntington-Hill method is similar to Webster's method in that it uses the same procedure to determine the standard quota, with one exception: the determining factor of whether quotas should be rounded up or down is based on the **geometric mean** of the upper quota and lower quota. The geometric mean is found by multiplying two numbers together and taking the square root.

> **📖 GEOMETRIC MEAN**
>
> The **geometric mean** of any two numbers m and n is $\sqrt{m \cdot n}$.

The Huntington-Hill method states that if the geometric mean of the integers above and below the standard quota is less than the standard quota, then the standard quota gets rounded up. If the standard quota is less than the geometric mean, then it gets rounded down.

For instance, suppose that a company allocating new administrators calculates a standard quota of 5.4823 administrators for a single department. Rounding down to 5 is arithmetically correct, but an argument could be made for rounding up instead. The Huntington-Hill method, however, determines rounding based on the geometric mean. The geometric mean of a standard quota 5.4823 is $\sqrt{m \cdot n} = \sqrt{5 \cdot 6} \approx 5.4772$. Since the standard quota is greater than the geometric mean (that is, $5.4923 > 5.4772$), we round up to 6.

> **📖 THE HUNTINGTON-HILL METHOD OF APPORTIONMENT**
>
> 1. Calculate the standard divisor.
>
> 2. Calculate the standard quota for each subgroup.
>
> 3. Calculate the geometric mean of the standard quota for each subgroup.
>
> 4. If the standard quota is less than the geometric mean, round the quota down. If the standard quota is greater than the geometric mean, round the quota up.
>
> 5. Assign each subgroup the number of resources based on the rounded quota.
>
> 6. If there are remaining resources to be distributed, chose a modified divisor by trial and error until the sum of the rounded quotas equals the number of resources to be apportioned.

Example 13.3.6

Applying the Huntington-Hill Method

After a census, 100 seats in parliament need to be re-apportioned among 5 counties with the following populations. Use the Huntington-Hill method to apportion the seats.

Table 13.3.14: Population per County

County	Population
1	35,589
2	17,425
3	3658
4	11,457
5	6871
Total	75,000

Solution

Begin by calculating the standard divisor. The total population is given in the table as 75,000.

$$\text{Standard Divisor: } \frac{75,000}{100} = 750$$

Step 2 of the Huntington-Hill method says to find the standard quotas for each subgroup. The calculations are shown in Table 13.3.15.

Table 13.3.15: Standard Quotas by County

County	Population	Standard Quota
1	35,589	$\frac{35,589}{750} \approx 47.452$
2	17,425	$\frac{17,425}{750} \approx 23.2333$
3	3658	$\frac{3658}{750} \approx 4.8773$
4	11,457	$\frac{11,457}{750} \approx 15.276$
5	6871	$\frac{6871}{750} \approx 9.1613$
Total	75,000	

The next step is to take the geometric mean of each standard quota. This is the square root of the product of the lower quota and the upper quota. We then compare the standard quota to the geometric mean. If the standard quota is less than the geometric mean, we round it down. If the standard quota is greater than the geometric mean, we round it up. The calculations are shown in Table 13.3.16.

Table 13.3.16: Rounded Quotas per County

County	Population	Standard Quota	Geometric Mean	Rounded Quota
1	35,589	47.452	$\sqrt{47 \cdot 48} \approx 47.4974$	47
2	17,425	23.2333	$\sqrt{23 \cdot 24} \approx 23.4947$	23
3	3658	4.8773	$\sqrt{4 \cdot 5} \approx 4.4721$	5
4	11,457	15.276	$\sqrt{15 \cdot 16} \approx 15.4919$	15
5	6871	9.1613	$\sqrt{9 \cdot 10} \approx 9.4868$	9
Total	75,000			99

Notice that the only county where the quota was rounded up was County 3. This rounding leaves one seat unassigned. Therefore, we need to modify the divisor and repeat steps 3 through 5. Since we need larger quotas, we will use a smaller divisor, say 749. The calculations with the modified divisor are shown in Table 13.3.17.

Table 13.3.17: Rounded Modified Quotas per County

County	Population	Modified Quota	Geometric Mean	Rounded Quota
1	35,589	$\frac{35,589}{749} \approx 47.515$	$\sqrt{47 \cdot 48} \approx 47.4974$	48
2	17,425	$\frac{17,425}{749} \approx 23.264$	$\sqrt{23 \cdot 24} \approx 23.4947$	23
3	3658	$\frac{3658}{749} \approx 4.884$	$\sqrt{4 \cdot 5} \approx 4.4721$	5
4	11,457	$\frac{11,457}{749} \approx 15.296$	$\sqrt{15 \cdot 16} \approx 15.4919$	15
5	6871	$\frac{6871}{749} \approx 9.174$	$\sqrt{9 \cdot 10} \approx 9.4868$	9
Total	75,000			100

> ✔ **Skill Check 13.3.5**
>
> Use Hamilton's method to apportion the scholarships in Example 13.3.6.

Using the modified divisor of 749, County 1 gains the additional seat, making all the seats apportioned among the counties. This means that County 1 receives 48 seats, County 2 receives 23 seats, County 3 receives 5 seats, County 4 receives 15 seats, and County 5 receives 9 seats.

OBJECTIVE 6 Apportionment Weaknesses

Just as we saw with voting methods, perfect "fairness" is unachievable using apportioning methods. Throughout history, as the United States has grown and congress has had to grapple with how to fairly assign representation to each state, these methods have revealed a set of *paradoxes*—statements that seemingly contradict themselves. The shortcomings noticed in Hamilton's Method of apportionment have become known as the **Alabama paradox**, the **population paradox**, and the **new states paradox**.

The **Alabama paradox** occurs when an increase in the number of items being apportioned causes a subgroup to lose an item. This paradox, which opposes common sense, was first observed after the 1880 census. The chief clerk at the time decided to calculate the apportionments for each state if the House of Representatives consisted of 275 members. He also did this for every possible house size up to 350 members—computing them all by hand! He discovered that if the House of Representatives had 299 seats, Alabama would get eight of those seats, but if the House of Representatives had 300 seats, Alabama would only get seven seats. Thus, the idea of increasing the number of items to be apportioned and a subgroup actually losing an item is called the Alabama paradox.

📖 **THE ALABAMA PARADOX**

The **Alabama paradox** occurs when an increase in the number of items to be apportioned causes a subgroup to lose an item.

Example 13.3.7

Illustrating the Alabama Paradox

Due to growth in student population, a school district was given money to hire 10 new middle school teachers for their three middle schools: Brown Middle School, Peachtree Middle School, and MLK Middle School. The teachers will be assigned to the schools based on their student populations using Hamilton's method of apportionment. The student populations are shown in Table 13.3.18.

Table 13.3.18: Middle School Student Population

Middle School	Student Population
Brown Middle	203
Peachtree Middle	588
MLK Middle	600
Total	1391

a. Use the Hamilton method to apportion the 10 teachers.

b. Show that the Alabama paradox occurs if the district is able to hire 11 teachers instead of 10.

Solution

a. To apportion the 10 teachers we need to begin by finding the standard divisor.

$$\text{Standard Divisor: } \frac{1391}{10} = 139.1$$

Using Hamilton's method, we then find each school's standard quota and round down to assign the teachers. Any unassigned teachers are given out in order of the largest fractional remainders. The calculations are shown in Table 13.3.19.

Table 13.3.19: Hamilton's Method of Apportionment with 10 Teachers

Middle School	Student Population	Standard Quota	Lower Quota	Distribution of Unassigned Teachers	Apportionment
Brown Middle	203	$\frac{203}{139.1} \approx 1.4594$	1	1	2
Peachtree Middle	588	$\frac{588}{139.1} \approx 4.2272$	4		4
MLK Middle	600	$\frac{600}{139.1} \approx 4.3134$	4		4
Total	1391		9		10

Using Hamilton's method, Brown Middle School receives the additional teacher because their fractional remainder is the largest. Thus, Brown Middle School will receive 2 new teachers while Peachtree Middle School and MLK Middle School will each receive 4 new teachers.

b. We need to calculate the standard divisor with 11 teachers this time.

$$\text{Standard Divisor: } \frac{1391}{11} = 126.4545$$

Table 13.3.20 shows the calculations using Hamilton's method with 11 teachers apportioned among the three middle schools.

Table 13.3.20: Hamilton's Method of Apportionment with 11 Teachers

Middle School	Student Population	Standard Quota	Lower Quota	Distribution of Unassigned	Apportionment
Brown Middle	203	$\frac{203}{126.4545} \approx 1.6053$	1		1
Peachtree Middle	588	$\frac{588}{126.4545} \approx 4.6499$	4	1	5
MLK Middle	600	$\frac{600}{126.4545} \approx 4.7448$	4	1	5
Total	1391		9		11

With eleven teachers to distribute, Brown Middle School now only receives one new teacher while Peachtree Middle and MLK Middle both receive five new teachers. This means that the Alabama Paradox will occur in the school district if there are eleven teachers instead of ten to apportion—Brown Middle gets fewer teachers even though there are more teachers to distribute.

Around the 1900s, Hamilton's method was dealt another blow, this time by what would come to be known as the **population paradox**. The US government decided to add 30 seats to the House of Representatives, increasing the number to 386. At the time, Virginia's population was growing much faster than the population of Maine. However, after using Hamilton's method to apportion the seats, Virginia lost a seat while Maine gained a seat. This paradox observes the anomaly of faster growth leading to loss of apportionment.

📖 THE POPULATION PARADOX

The **population paradox** occurs when subgroup A loses an item to subgroup B when the rate of growth of the population of subgroup A is greater than the rate of the growth in subgroup B.

Example 13.3.8

Illustrating the Population Paradox

Parish council representatives are to be distributed among the five parishes they represent according to each parish's population. Table 13.3.21 shows the current apportionment of the 17 representatives along with the population growth that has occurred since the last apportionment.

Table 13.3.21: Apportionment for Parish Representatives and Population Growth

Parish	Former Population	Apportionment	Percent Increase	Current Population
Eddy	13,877	7	0.8%	13,988
Longly	5080	2	12.7%	5725
Martyn	8109	3	3.0%	8352
Meeds	7592	4	9.0%	8275
Viant	1100	1	17.5%	1293
Total	35,758	17		37,633

Use Hamilton's method to reapportion the 17 representatives with the new populations, showing that the population paradox occurs.

Solution

Begin by calculating the standard divisor with 17 representatives over the increased population size of 37,633. Thus, the standard divisor is $\frac{37.633}{17} \approx 2213.7059$. Table 13.3.22 shows the calculations using Hamilton's method with 17 representatives apportioned among the five parishes.

Table 13.3.22: Hamilton's Method of Apportionment for Parish Representatives with Population Increases

Parish	Percent increase	Population	Standard Quota	Lower Quota	Distribution of Remaining Representatives	New Apportionment
Eddy	0.8%	13,988	$\frac{13,988}{2213.7059} \approx 6.3188$	6		6
Longly	12.7%	5725	$\frac{5725}{2213.7059} \approx 2.5862$	2	1	3
Martyn	3.0%	8352	$\frac{8352}{2213.7059} \approx 3.7729$	3	1	4
Meeds	9.0%	8275	$\frac{8275}{2213.7059} \approx 3.7381$	3	1	4
Viant	17.5%	1293	$\frac{1293}{2213.7059} \approx 0.5841$	0		0
Total		37,633		14		17

Now we can compare the old distribution to the new distribution in light of the population growth to see if the population paradox occurred. Table 13.3.23 shows the comparison.

Table 13.3.23: Apportionment for Parish Representatives and Population Growth

Parish	Former Apportionment	Percent Increase	New Apportionment
Eddy	7	0.8%	6
Longly	2	12.7%	3
Martyn	3	3.0%	4
Meeds	4	9.0%	4
Viant	1	17.5%	0
Total	17		17

Notice that Viant Parish grew by 17.5%, yet lost their only seat, while Longly Parish and Martyn Parish both gained a seat but grew at slower rates of 12.7% and 3.0%, respectively. This shows that the population paradox occurred with the Hamilton method of apportionment.

Another potential flaw in Hamilton's method was discovered in 1907 when Oklahoma became a state. At the time, Oklahoma's population dictated it should have five seats in the House of Representatives, bringing the total number of seats from 386 to 391. The idea was that these new seats would simply be added, leaving the number of seats unchanged for the other states. However, when the apportionment was recalculated, Maine actually gained a seat (four instead of three) while New York lost a seat (from 38 to 37). This idea that adding a subgroup, along with its appropriate number of items, causes other subgroups to experience a change in their apportionment is known as the **new states paradox.**

📖 THE NEW STATES PARADOX

The **new states paradox** occurs when the addition of a new subgroup, with a corresponding increase in the number of available items, can cause a change in the apportionment of items among the other subgroups.

Example 13.3.9

Illustrating the New States Paradox

A college campus has a student leadership board consisting of 13 representatives from the 3 different branches of student life: academic, social, and service. The members of the board are apportioned based on the student enrollment in the organizations of each of the different branches. The calculations using the Hamilton method, with a standard divisor of 894.7692, are shown in Table 13.3.24.

Table 13.3.24: Hamilton's Method of Apportionment for Student Leadership Board with 3 Branches

Student Life Branch	Population	Standard Quota	Lower Quota	Apportionment
Academic	3106	3.4713	3	4
Social	5649	6.3134	6	6
Service	2877	3.2154	3	3
Total	11,632			13

The student population has voted to add representation for pre-professional organizations to the leadership board. The standard divisor is 894.7692, which means that there should be 1 representative for every 894.7692, students in pre-professional organizations. Since the new branch has 1363 students, we find $\frac{1363}{894.7692} \approx 1.5233$, and round that down to 1 so that this new branch adds one representative to the board. Show that the new states paradox occurs with this addition.

Solution

Begin by calculating the new standard divisor with 14 representatives, which includes the new one for the pre-professional branch. The population will increase by 1363 students as well. Thus, the standard divisor is $\frac{11,632+1363}{14} \approx 928.2143$. Table 13.3.25 shows the calculations using Hamilton's method with 14 representatives apportioned among the four student life branches.

Table 13.3.25: Hamilton's Method of Apportionment for Student Leadership Board With 4 Branches

Student Life Branch	Population	Standard Quota	Lower Quota	Apportionment
Academic	3106	3.3462	3	3
Social	5649	6.0859	6	6
Service	2877	3.0995	3	3
Pre-Professional	1363	1.4684	1	2
Total	12,995			14

Notice that once the pre-professional group and its one representative is added in, the apportionment actually assigns the new branch two representatives, while the academic branch loses one. Because the addition of a new subgroup changed the apportionment of another subgroup, the new states paradox has occurred.

Hamilton's method of apportionment managed to be used for quite a long time in the United States House of Representatives before these paradoxes forced it to be replaced with some of the other methods we've looked at. However, although they avoided the paradoxes, these new methods created problems of their own. Occasionally, they violate what is known as the quota rule. The **quota rule** states that any fair apportionment method should assign every subgroup either its lower quota or its upper quota. Because the methods of apportionment introduced after Hamilton's all use a modified divisor, sometimes the modified quotas are different from the lower or upper quotas for that subpopulation.

> 📖 **THE QUOTA RULE**
>
> The **quota rule** states that any fair apportionment method should assign every subgroup either its lower quota or its upper quota.

Does a method exist that fairly apportions and does not violate any of the paradoxes or the quota rule? The short answer is no, although our examples in the text seem to suggest that it is possible because they adhere to the quota rule. However, in 1983, two mathematicians Michel Balinski and Peyton Young proved that there cannot be a perfect apportionment method. They proved that any apportionment method that does not violate the quota rule must produce paradoxes and any method that does not produce paradoxes must violate the quota rule. Although no perfect method for apportionment exists, we can make an informed decision about the most appropriate apportionment method, understanding the potential for flaws for paradoxes while striving to adhere to the quota rule.

Skill Check Answers
1. 25, 125 **2.** 6.4925; 3.8806; 6.6418 **3.** Math 4; History 8; Computer Science 6
4. Div. Pop 135 = 4; Div. Pop 98 = 3; Div. Pop 132 = 4
5. County 1 = 48; County 2 = 23; County 3 = 5; County 4 = 15; County 5 = 9

13.3 Exercises

✔ CONCEPT CHECK

1. A _____ represents the number of items that will be apportioned to each subgroup.

2. The Huntington-Hill method of apportionment uses the _____ to determine if the standard quota should be rounded up or down.

3. The _____ occurs when an increase in the number of available items causes a group to lose an item—even though populations remain the same.

4. The _____ states that subgroup A can lose an item to subgroup B even when the rate of growth of the population of subgroup A is greater than in subgroup B.

5. The _____ happens when the addition of a new subgroup, with a corresponding increase in the number of available items, can cause a change in the apportionment of items among the other subgroups.

💡 PRACTICE

Use the given table to solve each problem.

Student Enrollment in the University of California System		
Campus	Enrollment	Full Time Equivalent (FTE)
Berkeley	33,558	14,161
Davis	32,290	20,883
Irvine	28,000	12,558
Los Angeles	37,221	28,292
Merced	2700	799
Riverside	20,956	4689
San Diego	25,938	18,274
San Francisco	18,140	4174
Santa Barbara	21,016	6081
Santa Cruz	15,012	4597

6. The University of California system consists of 10 campuses and has a budget of $2.6 billion in state funding. How much money in state funding would each campus receive if the state divided the money equally among the campuses (round to nearest dollar)?

7. If the state of California wanted to allocate the money equally based on total student enrollment, how much would be appropriated per student, rounded to the nearest cent?

8. Based on the result from Exercise 2, how much money would the campus of Los Angeles and Merced receive based on the number of students enrolled on each campus?

9. If the state apportions the funds based on FTE, how much money will the campuses of Davis and Santa Cruz receive (round to nearest dollar)?

10. As part of a "green" initiative, the state is wanting to apportion 500 new electric vehicles to their university system campuses. The state decides to apportion these vehicles based on the number of students at each university.

   a. Find the SD.

   b. Find the SQ for the San Francisco and Irvine campuses.

11. Supposed the state decides to apportion the 500 electric vehicles based on FTE.

   a. Find the SD.

   b. Find the SQ for the Riverside and Santa Barbara campuses.

12. Use the Hamilton method to apportion the 500 electric vehicles to all 10 campuses based on the number of students.

13. Use the Hamilton method to apportion the 500 electric vehicles to all ten campuses based on FTE. Compare your results with the apportionments from Exercise 7 and determine if the apportionments are different when the apportionment basis is different.

The given table shows the number of students enrolled in history, liberal arts math, and English during the fall and spring semesters. Use it to solve each problem.

Course Enrollment per Semester

Subject	Fall	Spring
History	1902	1922
Liberal Arts Math	14,200	14,200
English	3898	3938

14. If there are 197 full time teaching positions available for apportionment among the three departments based on course enrollment, answer each of the following questions.

   a. Find the number of teaching positions that should be apportioned to each department in the fall using Jefferson's method.

   b. Find the number of teaching positions that should be apportioned to each department in the spring using Jefferson's method.

   c. Does Jefferson's method create an example of the population paradox?

   d. Find the number of teaching positions that should be apportioned to each department in the fall using Hamilton's method.

   e. Find the number of teaching positions that should be apportioned to each department in the spring using Hamilton's method.

   f. Does Hamilton's method create an example of the population paradox?

   g. Find the number of teaching positions that should be apportioned to each department in the fall using the Huntington-Hill method.

   h. Find the number of teaching positions that should be apportioned to each department in the spring using the Huntington-Hill method.

   i. Does the Huntington-Hill method create an example of the population paradox?

15. An English teacher at a high school can teach six classes. There are 35 students enrolled in English I, 43 in English II, and 48 in English III.

    a. Find the SD and SQ to determine how many sections of each course should be offered?

    b. Use the Jefferson method to determine the apportionment of students to the courses to determine the number of sections needed per course. determine the number of sections needed per course.

16. Repeat Exercise 15 b. using the Webster method.

17. Repeat Exercise 15 b. using the Hamilton method.

18. Repeat Exercise 15 b. using the Huntington-Hill method.

19. A county is divided into four districts with the populations of: Northern: 5500, Southern: 6350, Eastern: 3470, and Western: 1950. There are 16 seats on the county board to be apportioned.

    a. Use the Jefferson method to apportion the board seats.

    b. Use the Huntington-Hill method to apportion the board seats.

    c. Compare the apportionments for the Jefferson and Huntington-Hill methods and determine if there is a difference in how the seats are apportioned.

20. A biology department uses 25 graduate assistants in teaching its undergraduate courses. The enrollments for each of the courses that these students teach is as follows. How many graduate assistants should be assigned to each course using the Jefferson method?

Enrollment per Course

Course	Enrollment
Survey of Biology	450
Zoology	200
Cell Biology	175
Plant Biology	280

21. Use the Hamilton method to round each of the following numbers to a whole number while preserving the total.

$$12.65 + 3.48 + 2.57 + 4.39 + 1.91 = 25$$

22. Use Hamilton's method to round each of the following numbers to a whole number while preserving the total.

$$32.61 + 58.37 + 55.02 + 23.11 + 54.89 = 224$$

23. Suppose there are 76 faculty members in the sciences, 86 in the humanities, and 16 in the professional and trade schools. An 11-person faculty committee is to be formed.

    a. Use Hamilton's method to determine the allocation of committee members based on department size.

    b. Use Jefferson's method to determine the allocation of committee members based on department size.

    c. Use the Huntington-Hill method to determine the allocation of committee members based on department size.

    d. Use Webster's method to determine the allocation of committee members based on department size.

24. Suppose Learn-A-Lot University has enrollments on its three campuses as follows. There are 40 police officers to be distributed among these campuses based on enrollment. Use Hamilton's method to apportion the police officers.

Enrollment per Campus			
Campus	1	2	3
Enrollment	10,170	9150	680

25. If the number of police officers to be apportioned increases by 1 to 41 (see Exercise 24) determine the apportionment of officers using Hamilton's Method and show that the Alabama paradox occurs when the number of officers increases by 1.

26. Suppose a country has six states with populations as given in the table. There are 250 seats in the House of Representatives for this country. Use Webster's method to apportion the representatives.

Population by State	
State	**Population**
A	1646
B	6936
C	154
D	2091
E	685
F	988
Total	**12,500**

27. Suppose a college homecoming planning committee has 17 members. The makeup of the committed is to be based on the size of the classes: Freshman = 422, Sophomore = 356, Junior = 321, and Senior = 288.

    a. Find the number of members from each class to be apportioned to the committee using Webster's method.

    b. Find the number of members from each class to be apportioned to the committee using Jefferson's method.

    c. Find the number of members from each class to be apportioned to the committee using the Huntington-Hill method.

13.4 WEIGHTED VOTING SYSTEMS

OBJECTIVES

1. Identify parts of a weighted voting system.

2. Determine the division of power within a weighted voting system.

3. Calculate power indices for weighted voting systems.

Fun Fact

The US Electoral College, as defined in the Constitution, is a process to elect the President of the United States. The Founding Fathers established it as a compromise between Congress choosing a President or a popular vote among qualified citizens.

OBJECTIVE 1 Weighted Voting Systems

Not all voting systems are based on a "one person, one vote" system like the ones we've looked at so far in this chapter. Some are **weighted voting systems**, meaning participants have differing numbers of votes. For instance, in the United States, we elect the President using a weighted voting system called the electoral college. In this system, each state plus Washington, D.C., is given a certain number of votes based on its population size. In 2020, there were 538 electoral votes distributed among the 50 states and Washington, D.C. California, being the most populous state, had 55 electoral votes, while Alaska, one of the least-populated states, had the guaranteed minimum of 3 votes. This shows that in the electoral college, the influence of a state on the election is *weighted* according to the number of citizens who live there.

Weighted voting systems are also commonly used in situations that involve shareholders. Votes are distributed according to the number of shares that a shareholder owns—the more shares you own, the more votes you have.

In weighted voting systems, the voters are often called **players (P)**, or *participants*, and the **weight (w)** each player holds is the number of votes they control. In order to win, a candidate or proposal must have a **quota (q)** of votes. The quota is the minimum number of votes needed to win, which is at least half, but no more than the total number of votes that exist. If the quota requires a minimum of half of the total votes for a proposal to pass, the quota is called a *simple majority*. If more than half of the votes are required, the quota is called a *super majority*.

The standard notation for weighted systems is to identify the quota, followed by the weight each player carries. For example, the system represented by [10: 4, 5, 7] requires a quota of 10 votes. There are three players that share the 4 + 5 + 7 = 16 votes. Player 1 has 4 votes, player 2 has 5 votes, and player 3 has 7 votes.

WEIGHTED VOTING SYSTEM

A **weighted voting system** consists of n players (P_n) each controlling a number of votes (w), called their **weight**. It is described using the following notation.
$$[q: w_1, w_2, \ldots, w_N]$$

Here, the **quota (q)** is the minimum number of votes needed to win.

Example 13.4.1

Identifying the Parts of a Weighted Voting System

The board of directors for a sporting footwear company makes its decisions using a weighted voting system represented by [25: 8, 6, 5, 3, 3, 3, 2, 2, 1, 1, 1, 1].

a. Determine the number of players in the voting system.

b. How many total votes are available among all of the players?

c. What is the quota for a proposal to win the board's approval?

d. What is the weight of P_3?

Solution

a. To determine the number of players, we need to count the number of weights that are listed following the quota and colon. Thus, this is a weighted voted system with 12 players $P_1, P_2, P_3, ..., P_{12}$.

b. To determine the number of total votes available, we add up the weights of all of the players.

Total number of votes: $8 + 6 + 5 + 3 + 3 + 3 + 2 + 2 + 1 + 1 + 1 + 1 = 36$

c. The quota is found at the beginning of the list, before the colon. Thus, the quota is 25 votes.

d. The player's weights are listed in order, thus the weight of P_3 is 5; that is, player 3 has 5 votes.

Example 13.4.2

Representing a Weighted Voting System

The library council has four voting members representing the district's main areas: North, South, East, and West. The north district has four votes, the south district has three, the east district has two, and the west district has one vote. A minimum of six votes is required to pass any proposal on the council.

a. Use the notation for weighted voting systems to represent the library council's voting system.

b. Determine how many total votes are in the system.

Solution

a. Let the districts, North, South, East, and West, be represented by P_1, P_2, P_3, and P_4, respectively. Their corresponding weights are $w_1 = 4$, $w_2 = 3$, $w_3 = 2$, and $w_4 = 1$. We are also told that the quota for any vote of the library council is 6. Thus, the system can be represented as [6: 4, 3, 2, 1].

b. The total number of votes is found by adding all of the weights together.

Total Number of Votes: $4 + 3 + 2 + 1 = 10$

Example 13.4.3

Determining the Quota for a Weighted Voting System

Consider the voting system $[q: 4, 3, 2, 2, 1, 1]$

a. What is the largest value that the quota can be?

b. What is the smallest value that the quota can be?

c. What is the value of the quota if at least two-thirds of the votes are required to pass a proposal?

Solution

a. The quota value can be no more than the total number of votes possible. To find this, we add together the weights of all of the players.

$$\text{Total Number of Votes} = 4 + 3 + 2 + 2 + 1 + 1 = 13$$

Thus, the largest value the quota can be is 13, which is the total number of votes.

b. The quota must be *at least* half of the total possible votes, or $\frac{13}{2} = 6.5$ votes.

We have to round this up since rounding down would be too few votes. Thus, the smallest value that the quota can be is 7. This is called a simple majority.

c. To determine a quota of two-thirds of the votes, we need to multiply the fraction by the total number of possible votes.

$$\frac{2}{3} \cdot 13 \approx 8.67$$

Again, we must round up to the nearest whole number in order to have enough votes. Thus, if two-thirds of the votes are required, the quota is 9. Because this is more than half of the votes, this is called a *super majority*.

✔ **Skill Check 13.4.1**

Given the voting system represented by [6: 5, 3, 1], determine the number of players, the quota, and the total number of votes.

OBJECTIVE 2 Power within a Weighted Voting System

A player's *power* is their ability to influence decisions in an election. If a player can single-handedly pass a proposal or block one from passing, we would all agree that player had significant power. If a player has the ability to pass a proposal based on their vote alone (that is, they have control of a number of votes that is greater than or equal to the quota), then we say that player is a **dictator**. The combined weight of the other players does not outweigh the dictator's share of votes and thus they cannot overrule the dictator. For instance, consider the system [13: 14, 5, 4, 3]. Player 1 is a dictator with more votes than the quota requires. All of the other players' votes combined cannot override the votes of player 1.

> **Helpful Hint**
>
> The difference between a dictator and a player with veto power is that a proposal is guaranteed to pass if a dictator votes in favor of it.

We also describe power in terms of being able to stop a vote from passing. If a player has enough votes to keep a proposal from passing, we say the player has **veto power**. In other words, a vote cannot pass without their support. This does not mean that the proposal is guaranteed to pass with the support of the player, it means that the proposal will not pass without their support. For example, for a vote to pass in the system [6: 5, 3, 1], the proposal must have the support of player 1. This is because player 2 and player 3 do not have enough combined votes to meet the quota without player 1. Thus, player 1 has veto power. Notice that although player 1 has veto power, that player is not a dictator. Player 1 does not have enough votes to guarantee that the vote will pass; the player only has enough power to stop it from passing. While not all weighted systems have a player with veto power, it is possible for more than one player to have veto power.

On the other end of the power spectrum is the **dummy player**. A dummy player is one who does not have enough votes to assert any effect on the outcome of a proposal. Consider the system [10: 6, 5, 1]. Player 3 has no effect on the outcome of any proposal. Player 1 and player 2 must both be in support of the proposal for it to pass, but don't need the support of player 3. As you might have suspected, when there is a dictator, all other players are dummy players, because regardless of their weight, they essentially have no power at all.

> 📖 **PLAYER POWER IN A WEIGHTED VOTING SYSTEM**
>
> **Dictator**
>
> A **dictator** is a player with the power to pass a proposal single-handedly; that is, the dictator has at least as many votes as the quota.
>
> **Veto Power**
>
> When a player's votes are needed in order to pass a proposal, that player is said to have **veto power**.
>
> **Dummy Player**
>
> A **dummy player** is a player who does not have enough votes to have an effect on the outcome of a proposal.

Example 13.4.4

Determining Player Power

In each of the voting systems, determine which voters are dictators, are dummy players, or have veto power.

a. $[11: 5, 5, 3, 1]$ **b.** $[8: 9, 3, 2, 1, 1]$ **c.** $[7: 4, 3, 2]$

Solution

a. There are no dictators since no player has more than the quota of 11 votes.

There are no dummy players in this voting system. A proposal will need the support of both players 1 and 2, along with the support of either player 3 or player 4.

Players 1 and 2 have veto power, because without the support of both of them, a proposal cannot pass.

b. Player 1 is a dictator since they have 9 votes which is more than the quota of 8 votes.

Because there is a dictator, all other players (P_2, P_3, P_4, P_5) are dummy players.

Only the dictator, player 1, has veto power.

c. There are no dictators since no player has more than the quota of 7 votes.

Player 3 is a dummy player since Players 1 and 2 must both be in support of a proposal for it to pass, and they don't need the support of player 3.

Players 1 and 2 both have veto power since without their support a proposal cannot pass.

As you might imagine, voting power can sometimes lead to partnerships, even unpredictable ones. Players often decide to join forces and vote together, creating a *coalition*. A **coalition** is formed when any set of players decides to vote the same way. Recall from Chapter 2 that the number of subsets in a set of n objects is 2^n. We can use this fact to determine the number of possible coalitions in a voting system. A voting system with n players has $2^n - 1$ possible coalitions. We subtract one from the number of subsets because the empty set with no players in it is not considered a coalition. We commonly denote the players in a coalition by listing their player number (P_n) between curly brackets {}.

If the combined weight of a coalition has enough votes to meet the quota, it is a *winning coalition*; if not, it is a *losing coalition*. For instance, in the voting system [6: 4, 3, 1, 1], a coalition consisting of P_1 and P_2 is a winning coalition because it has a total of $4 + 3 = 7$ votes, which is enough to pass a proposal. The coalition consisting of P_1, P_3, and P_4 is also a winning coalition. (Check this for yourself.) A coalition consisting of all of the players in the voting system is considered a winning coalition.

> ### 💬 Helpful Hint
>
> A coalition can consist of a single player. For instance, in a system with three players, P_1, P_2, and P_3, there are $2^3 - 1 = 7$ coalitions, which are $\{P_1\}$, $\{P_2\}$, $\{P_3\}$, $\{P_1, P_2\}$, $\{P_1, P_3\}$, $\{P_2, P_3\}$, and $\{P_1, P_2, P_3\}$.

> ### 📖 COALITION
>
> A **coalition** is a group of players joining together to vote the same way. If the combined weight of the coalition is at least the quota, it is a *winning coalition*. In a voting system with n players, there are $2^n - 1$ possible coalitions. A coalition consisting of players P_1, P_2, and P_3 is denoted by $\{P_1, P_2, P_3\}$.

Example 13.4.5

Determining Voting Coalitions

The Homeowner Association (HOA) board of a condominium consists of owners who vote on building issues. The voting system is represented by [10: 4, 4, 3, 2].

a. Determine the number of possible coalitions that could be formed in the HOA board.

b. Name the two winning coalitions that do not consist of all of the players.

Solution

a. There are four players in the HOA voting system. Therefore, there are $2^4 - 1 = 15$ possible coalitions.

b. The winning coalitions must have a combined weight of at least 10. In order to reach ten votes, players 1 and 2 must be included in the coalition, but their votes aren't enough to win on their own; either player 3 or player 4 must also be in a winning the coalition. Thus, the two winning coalitions are $\{P_1, P_2, P_3\}$ and $\{P_1, P_2, P_4\}$.

> ### ✔ Skill Check 13.4.2
>
> Given the voting system [18: 8, 7, 6, 4, 4, 4, 2], determine if the coalition $\{P_3, P_4, P_5, P_6\}$ is a winning coalition.

Within a coalition, players can have a certain type of power. If a player were to leave a winning coalition and cause it to become a losing coalition, we define that player as a **critical player**. In other words, a critical player has the power to change the effectiveness of a coalition. For instance, in the voting system [7: 4, 3, 2, 2], consider the winning coalition $\{P_1, P_2, P_4\}$. Their combined weight is $4 + 3 + 2 = 9$. Players 1 and 2 are considered critical players because if they left the coalition, the combined weight would drop below the quota of 7. However, player 4 is not critical because if they were to leave, the coalition would still have a combined weight of 7.

📖 **CRITICAL PLAYER**

If a player in a winning coalition were to leave and cause the coalition to become a losing coalition, that player is considered a **critical player**.

Example 13.4.6

Determining Critical Players

a. In the voting system [12: 7, 6, 2, 1], determine which voters are critical players in the coalitions $\{P_1, P_2\}$ and $\{P_1, P_2, P_3\}$.

b. In the voting system [20: 12, 5, w, 1], what is the smallest value of w that would make player 3 a critical player in the coalition $\{P_1, P_2, P_3\}$?

Solution

a. The coalition $\{P_1, P_2\}$ has a combined weight of $7 + 6 = 13$. Both players are critical in that their combined weight would drop below the quota of 12 if either left.

The coalition $\{P_1, P_2, P_3\}$ has a combined weight of $7 + 6 + 2 = 15$. If player 1 or player 2 leaves the coalition, the combined weight would fall below the quota, so both are critical. Note that player 3 is not critical because the coalition would still have a combined weight of 13 without them.

b. For player 3 to be critical, the minimum combined weight of the coalition would need to be equal to the quota, which is 20. This means that the combined votes can be represented by the equation $12 + 5 + w = 20$. Solving for w, we get the following.

$$12 + 5 + w = 20$$
$$17 + w = 20$$
$$w = 3$$

Thus, player 3 would need a minimum of 3 votes to be a critical player in the coalition.

✔ **Skill Check 13.4.3**

Determine if player 4 is a critical player for the coalition $\{P_1, P_2, P_4\}$ in the voting system [27: 10, 9, 8, 8, 5]

OBJECTIVE 3 Power Indices

Weight, veto power, dictator status, dummy player, and critical player are all terms used to describe how much power an individual player might have in a voting system. However, how do we merge all of these descriptions together to describe the overall power each player might have? There are two different methods: the Banzhaf Power Index and the Shapley-Shubik Power Index. We will look at the Banzhaf Power Index first.

The **Banzhaf Power Index** rates the overall power of a player based on the number of times the player is a critical player in a winning coalition. This index is a fraction that compares the number of times a specific player is a critical player in any possible coalition to the number of times *any* player could be a critical player in *any* coalition. Essentially, it asserts that the more often a player is critical, the more power they have. The list of each player's power index is called the **Banzhaf Power Distribution**.

For instance, consider the voting system [5: 3, 2, 2]. Which player has the most power? We know that player 1 has the most votes, but what leverage does that provide? A Banzhaf Power Distribution can give us more insight. The distribution is $P_1 = 60\%$, $P_2 = 20\%$, and $P_3 = 20\%$. From this perspective, player 1 obviously has much more power than the other two players. In fact, player 1 has more power than both of the others combined according to the Banzhaf Power Index. Because of the Banzhaf Power Index, we are able to quantify the amount of power each voter has.

📖 CALCULATING THE BANZHAF POWER INDEX FOR PLAYER P_n

Step 1: List all winning coalitions for the voting system.

Step 2: Identify the critical players for each winning coalition.

Step 3: Count the total number of times each player is a critical player.

Step 4: Add together the critical player counts found in Step 3.

Step 5: Calculate the Banzhaf Power Index using the following fraction.

$$\frac{\text{Number of Times Player } P_n \text{ Is a Critical Player}}{\text{Total Number of Times All Players Are Critical Players}}$$

Example 13.4.7

Determining the Banzhaf Power Index

a. Determine the Banzhaf Power Index for each player in the voting system [10: 6, 4, 2].

b. In the voting system [11: 6, 5, 4, 1], how much more power does player 2 have than player 3 according to the Banzhaf Power Index?

Solution

a. Follow the five steps to calculate the Banzhaf Power Index for each player in the voting system [10: 6, 4, 2].

Step 1: List all winning coalitions for the voting system.

The winning coalitions are $\{P_1, P_2\}$ and $\{P_1, P_2, P_3\}$.

Step 2: Identify the critical players for each winning coalition.

Table 13.4.1: Winning Coalitions in the Voting System [10: 6, 4, 2]

Winning Coalition	Critical Players
$\{P_1, P_2\}$	P_1, P_2
$\{P_1, P_2, P_3\}$	P_1, P_2

Step 3: Count the total number of times each player is a critical player.

Table 13.4.2: Critical Players in the Voting System [10: 6, 4, 2]

Player	Number of Times Critical
P_1	2
P_2	2

Step 4: Add together the critical player counts found in Step 3.

There are 4 total critical player counts.

Step 5: Calculate the Banzhaf Power Index using the fraction

$$\frac{\text{Number of Times Player } P_n \text{ Is a Critical Player}}{\text{Total Number of Times All Players Are Critical Players}}.$$

Table 13.4.3: Banzhaf Power Index for the Voting System [10: 6, 4, 2]

Player	Number of Times Critical	Power Index
P_1	2	$\frac{2}{4} = 50\%$
P_2	2	$\frac{2}{4} = 50\%$

Note that even though players 1 and 2 have a different number of votes (6 and 4, respectively), they have equal power indices.

b. The list of critical players in the winning coalitions in the voting system [11: 6, 5, 4, 1] is as follows.

Table 13.4.4: Winning Coalitions in the Voting System [11: 6, 5, 4, 1]

Winning Coalition	Critical Players
$\{P_1, P_2\}$	P_1, P_2
$\{P_1, P_2, P_3\}$	P_1, P_2
$\{P_1, P_2, P_4\}$	P_1, P_2
$\{P_1, P_3, v_4\}$	P_1, P_3, P_4
$\{P_1, P_2, P_3, P_4\}$	P_1

Now we can count the number of times each player is critical and the total number of critical player counts in order to represent each as a fraction. Player 1 is critical 5 times, player 2 is critical 3 times, and players 3 and 4 are each critical 1 time. So the total number of critical player counts is $5 + 3 + 1 + 1 = 10$. Now we can find the Banzhaf Power Index for each player.

Table 13.4.5: Banzhaf Power Index for the Voting System [11: 6, 5, 4, 1]

Player	Number of Times Critical	Power Index
P_1	5	$\frac{5}{10} = 50\%$
P_2	3	$\frac{3}{10} = 30\%$
P_3	1	$\frac{1}{10} = 10\%$
P_4	1	$\frac{1}{10} = 10\%$

Thus, we see that in the given voting system, player 2 has three times more power than player 3.

Notice that in the Banzhaf Power Index, *leaving* a coalition is a potentially powerful move and a player is critical if the loss of their votes causes a coalition to change from a winning one to a losing one. In contrast, the **Shapley-Shubik Power Index** considers *joining* a coalition to be the potentially powerful move and a player is considered to be a **pivotal player** if their additional votes change the coalition from a losing one to a winning one.

In this index, the order in which players join a coalition is key. Coalitions where the order of players is important is called a **sequential coalition**. We use the notation $\langle P_1, P_2, P_3, ..., P_N \rangle$ where P_1 is the first player to join, P_2 is the second player to join, and so on. Consequently, the coalition $\langle P_1, P_2, P_3 \rangle$ is different from the coalition $\langle P_2, P_3, P_1 \rangle$.

> **Helpful Hint**
>
> There is only one pivotal player in any sequential coalition.

📖 SEQUENTIAL COALITION AND PIVOTAL PLAYER

Sequential Coalition

A **sequential coalition** is a coalition where the order in which N players join is important. It is denoted as $\langle P_1, P_2, P_3, ..., P_N \rangle$, where P_1 is the first player to join, P_2 is the second player to join, and so on.

Pivotal Player

A **pivotal player** is the player who joins a sequential coalition and causes the coalition to change from a losing coalition to a winning coalition.

Example 13.4.8

Determining the Pivotal Player

Wilson, Clinton, & Niese Law Firm has the following voting system [8: 7, 6, 4, 3, 3, 2]. Determine the pivotal player in each of the sequential coalitions listed.

a. $\langle P_2, P_4, P_3, P_6, P_5, P_1 \rangle$

b. $\langle P_6, P_4, P_5, P_3, P_2, P_1 \rangle$

c. $\langle P_2, P_1, P_3, P_4, P_5, P_6 \rangle$

Solution

The notation for each sequential coalition tells us the order in which the players join the coalition. To determine the pivotal player, we need to keep track of the weight of the coalition as the players join and note when the coalition changes from a losing one to a winning one; that is, when the weight of the coalition is more than the quota. We will use a table in each instance to keep track of the votes. At the top of each table we have included the voting system for your reference.

Table 13.4.6: System [8: 7, 6, 4, 3, 3, 2] Sequential Coalition $\langle P_2, P_4, P_3, P_6, P_5, P_1 \rangle$

Coalition	Coalition Weight	Status of Coalition
$\langle P_2 \rangle$	6	Losing
$\langle P_2, P_4 \rangle$	$6 + 3 = 9$	Winning

Player 4 is the pivotal player in this coalition.

Table 13.4.7: System [8: 7, 6, 4, 3, 3, 2] Sequential Coalition $\langle P_6, P_4, P_5, P_3, P_2, P_1 \rangle$

Coalition	Coalition Weight	Status of Coalition
$\langle P_6 \rangle$	2	Losing
$\langle P_6, P_4 \rangle$	$2 + 3 = 5$	Winning
$\langle P_6, P_4, P_5 \rangle$	$2 + 3 + 3 = 8$	Winning

Player 5 is the pivotal player in this coalition.

Table 13.4.8: System [8: 7, 6, 4, 3, 3, 2] Sequential Coalition $\langle P_2, P_1, P_3, P_4, P_5, P_6 \rangle$

Coalition	Coalition Weight	Status of Coalition
$\langle P_2 \rangle$	6	Losing
$\langle P_2, P_1 \rangle$	$6 + 7 = 13$	Winning

Player 1 is the pivotal player in this coalition.

> ✔ **Skill Check 13.4.4**
>
> Determine the pivotal player for the coalition $\langle P_3, P_4, P_2, P_1 \rangle$ in the voting system [4: 3, 1, 1, 1].

Now that we've established how to identify the pivotal player in a sequential coalition, we can calculate the Shapley-Shubik Distribution for a voting system. This power index only considers coalitions that consist of *all* N players in the system. Let's pause and consider how many sequential coalitions are possible in a voting system. Recall from Chapter 10 that the number of ways n distinct objects can be arranged is $n! = (n)(n-1)(n-2)\cdots(2)(1)$. Thus, the total number of sequential coalitions of N players is $N!$. For example, if a system has 3 players, then there are $3! = 6$ sequential coalitions, and they are as follows.

$$\langle P_1, P_2, P_3 \rangle \quad \langle P_1, P_3, P_2 \rangle \quad \langle P_2, P_1, P_3 \rangle \quad \langle P_2, P_3, P_1 \rangle \quad \langle P_3, P_1, P_2 \rangle \quad \langle P_3, P_2, P_1 \rangle$$

The Shapley-Shubik index measures power as the proportion of the number of times a player is the pivotal player in a sequential coalition to the number of sequential coalitions (consisting of all N players) that are possible. Again, this index asserts that the more times a player is pivotal, the more power that player holds. Let's once again consider the voting system [5: 3, 2, 2]. The Shapley-Shubik Power distribution for this system is $P_1 \approx 66.67\%$, $P_2 \approx 16.67\%$, and $P_3 \approx 16.67\%$. By this power index, player 1 has even more power over the other two voters than the Banzhaf index indicated. Used together, these indices paint a convincing picture of the power allocation.

CALCULATING THE SHAPLEY-SHUBIK POWER INDEX FOR PLAYER P_n

Step 1: List all $N!$ sequential coalitions that contain all N players in the voting system.

Step 2: Identify the pivotal player for each sequential coalition.

Step 3: Count the total number of times player P_n is a pivotal player.

Step 4: Calculate the Shapley-Shubik Index using the following fraction.

$$\frac{\text{Number of Times Player } P_n \text{ Is a Pivotal Player}}{N!}$$

Example 13.4.9

Determining the Shapley-Shubik Power Distribution

A charitable organization has a volunteer board of directors with the voting system [7: 4, 3, 2]. Determine the Shapley-Shubik Power Distribution for the voting system.

Solution

Step 1: List all $N!$ sequential coalitions that contain all N players in the voting system.

There are three players, so there are $3! = 6$ sequential coalitions containing all three players. They are as follows.

$$\langle P_1, P_2, P_3 \rangle \quad \langle P_1, P_3, P_2 \rangle \quad \langle P_2, P_1, P_3 \rangle \quad \langle P_2, P_3, P_1 \rangle \quad \langle P_3, P_1, P_2 \rangle \quad \langle P_3, P_2, P_1 \rangle$$

Step 2: Identify the pivotal player for each sequential coalition.

By listing the sequential coalitions with the weight of each player, we can determine the pivotal player in each coalition by noticing when the weight of the coalition reaches at least the quota of 7.

Table 13.4.9

Sequential Coalition and Player Weights	Pivotal Player
$\langle P_1, P_2, P_3 \rangle$ 4, 3, 2	P_2
$\langle P_1, P_3, P_2 \rangle$ 4, 2, 3	P_2
$\langle P_2, P_1, P_3 \rangle$ 3, 4, 2	P_1
$\langle P_2, P_3, P_1 \rangle$ 3, 2, 4	P_1
$\langle P_3, P_1, P_2 \rangle$ 2, 4, 3	P_2
$\langle P_3, P_2, P_1 \rangle$ 2, 3, 4	P_1

Step 3: Count the total number of times player P_n is a pivotal player.

The number of times each player is pivotal is as follows: $P_1 = 3$, $P_2 = 3$, and $P_3 = 0$. Notice, that player 3 is a dummy player because there in not a coalition in which he is a pivotal player.

Step 4: Calculate the Shapley-Shubik Index.

$$\frac{\text{Number of Times Player } P_n \text{ Is a Pivotal Player}}{N!}$$

Player 1: $\dfrac{3}{6} = 0.5 = 50\%$

Player 2: $\dfrac{3}{6} = 0.5 = 50\%$

Player 3: $\dfrac{0}{6} = 0 = 0\%$

The power distribution indicates that player 1 and player 2 have equal weight on the board, while player 3 has no influence.

✔ **Skill Check 13.4.5**

Determine the Shapley-Shubik Power distribution for the voting system [8: 4, 3, 3, 2].

Example 13.4.10

Determining the Power Distributions of a Voting System

Consider the voting system [9: 6, 4, 2, 1] for the local government council board.

a. Determine the Banzhaf Power Distribution for the council.

b. Determine the Shapley-Shubik Power Distribution for the council.

c. Discuss the power allocation among the council members.

Solution

a. The list of critical players in the winning coalitions is as follows.

Table 13.4.10: Critical Players in the Winning Coalitions for the Voting System [9: 6, 4, 2, 1]

Winning Coalition	Critical Players
$\{P_1, P_2\}$	P_1, P_2
$\{P_1, P_2, P_3\}$	P_1, P_2
$\{P_1, P_3, P_4\}$	P_1, P_3, P_4
$\{P_1, P_2, P_3, P_4\}$	P_1

Now we can count the number of times each player is critical and add these values together to find the total number of critical player counts. We then use those values to find the Banhaf Power Index for each voter.

Table 13.4.11: Banzhaf Power Index for the Voting System [9: 6, 4, 2, 1]

Player	Number of Times Critical	Banzhaf Power Index
P_1	4	$\frac{4}{8} = 50\%$
P_2	2	$\frac{2}{8} = 25\%$
P_3	1	$\frac{1}{8} = 12.5\%$
P_4	1	$\frac{1}{8} = 12.5\%$

b. There are four players, so there are 4! = 24 sequential coalitions containing all four players. They are organized in the table below, with the pivotal player identified in each.

Table 13.4.12: Pivotal Players for the Voting System [9: 6, 4, 2, 1]

Sequential Coalition with Player Weights	Pivotal Player	Sequential Coalition with Player Weights	Pivotal Player
$\langle P_1, P_2, P_3, P_4 \rangle$ 6, 4, 2, 1	P_2	$\langle P_3, P_2, P_1, P_4 \rangle$ 2, 4, 6, 1	P_1
$\langle P_1, P_2, P_4, P_3 \rangle$ 6, 4, 1, 2	P_2	$\langle P_3, P_2, P_4, P_1 \rangle$ 2, 4, 1, 6	P_1
$\langle P_1, P_3, P_2, P_4 \rangle$ 6, 2, 4, 1	P_2	$\langle P_3, P_1, P_2, P_4 \rangle$ 2, 6, 4, 1	P_2
$\langle P_1, P_3, P_4, P_2 \rangle$ 6, 2, 1, 4	P_4	$\langle P_3, P_1, P_4, P_2 \rangle$ 2, 6, 1, 4	P_4
$\langle P_1, P_4, P_2, P_3 \rangle$ 6, 1, 4, 2	P_2	$\langle P_3, P_4, P_2, P_1 \rangle$ 2, 1, 4, 6	P_1
$\langle P_1, P_4, P_3, P_2 \rangle$ 6, 1, 2, 4	P_3	$\langle P_3, P_4, P_1, P_2 \rangle$ 2, 1, 6, 4	P_1

Sequential Coalition with Player Weights	Pivotal Player	Sequential Coalition with Player Weights	Pivotal Player
$\langle P_2, P_1, P_3, P_4 \rangle$ 4, 6, 2, 1	P_1	$\langle P_4, P_1, P_3, P_2 \rangle$ 1, 6, 2, 4	P_3
$\langle P_2, P_1, P_4, P_3 \rangle$ 4, 6, 1, 2	P_1	$\langle P_4, P_1, P_2, P_3 \rangle$ 1, 6, 4, 2	P_2
$\langle P_2, P_3, P_1, P_4 \rangle$ 4, 2, 6, 1	P_1	$\langle P_4, P_3, P_1, P_2 \rangle$ 1, 2, 6, 4	P_1
$\langle P_2, P_3, P_4, P_1 \rangle$ 4, 2, 1, 6	P_1	$\langle P_4, P_3, P_2, P_1 \rangle$ 1, 2, 4, 6	P_1
$\langle P_2, P_4, P_1, P_3 \rangle$ 4, 1, 6, 2	P_1	$\langle P_4, P_2, P_1, P_3 \rangle$ 1, 4, 6, 2	P_1
$\langle P_2, P_4, P_3, P_1 \rangle$ 4, 1, 2, 6	P_1	$\langle P_4, P_2, P_3, P_1 \rangle$ 1, 4, 2, 6	P_1

Now we can count the number of times each player is pivotal and divide this by $N!$, or 24 in this case.

Table 13.4.13: Shapley-Shubik Power Index for the Voting System [9: 6, 4, 2, 1]

Player	How Many Times Pivotal	Shapley-Shubik Power Index
P_1	14	$\frac{14}{24} \approx 58.33\%$
P_2	6	$\frac{6}{24} \approx 25\%$
P_3	2	$\frac{2}{24} \approx 8.33\%$
P_4	2	$\frac{2}{24} \approx 8.33\%$

Compare the power indices of each player using the two different methods.

Table 13.4.14: Power Indices for the Voting System [9: 6, 4, 2, 1]

Player	Banzhaf Power Index	Shapley-Shubik Power Index
P_1	50%	58.33%
P_2	25%	25%
P_3	12.5%	8.33%
P_4	12.5%	8.33%

Note that player 1 has more power using the Shapley-Shubik Index. Player 2's power is the same with either method. Players 3 and 4 have equal power using either method with a lower power using the Shapley-Shubik Index. Notice that players 2, 3, and 4 combined have equal power to player 1 using the Banzhaf Power Index.

13.4   Exercises

✔ CONCEPT CHECK

1. A group of players joining together to vote the same way is called a _____.

2. A _____ is a player who causes a winning coalition to become a losing coalition if they were to leave.

3. A player with the power to form a winning coalition by themself is called a _____.

4. The total number of sequential coalitions is found from the following formula: _____.

💡 PRACTICE

5. Five partners start a business, with each owning the following number of shares: P_1 owns 9 shares, P_2 owns 6 shares, P_3 owns 5 shares, P_4 owns 3 shares, and P_5 owns 2 shares. If one share is equal to one vote and they use a simple majority quota, represent this weighted voting system.

6. Using the information in Exercise 5, if the quota is a two-thirds majority, represent the weighted voting system.

7. Consider the weighted voting system $[25: 10, 7, 4, 4, 2, 2, 1, 1, 1]$.

   a. What is the quota for this voting system?

   b. What is the weight of P_5?

   c. If only the last six voters vote for a motion, does the motion pass?

   d. If P_1 and P_2 vote against a motion, does the motion pass?

8. Consider the weighted voting system $[q: 10, 7, 4, 4, 2, 2, 1, 1, 1]$.

   a. What is the smallest quota that requires a majority for this voting system?

   b. What is the largest quota for this voting system?

9. In the weighted voting system $[q: 20, 19, 15, 8, 4, 2]$, if a two-thirds majority of votes is needed, what is q?

10. Which players in the voting system $[20: 7, 7, 3, 3, 2]$ have veto power?

11. In the weighted voting system $[q: 18, 15, 12, 9, 6, 3]$, what is the largest value of q so that no voter has veto power?

12. In the weighted voting system $[q: 10, 7, 4, 4, 2, 2]$, if every voter has veto power, what is q?

13. Consider the weighted voting system $[42: 22, 16, 10, 6, 4]$. Does this voting system contain a dictator? Does this system contain a dummy player?

14. Consider the weighted voting system $[20: 12, 10, 3]$. Does this voting system contain a dictator? Does this system contain a dummy player?

15. Consider the weighted voting system $[42: 20, 16, 10, 6, 4]$.
    a. How many voters are in the system?
    b. What is the quota for the system?
    c. Is the coalition $\{P_2, P_3, P_4, P_5\}$ a winning coalition?
    d. What are all of the winning coalitions?
    e. Are there any critical players?
    f. Do any of the players have veto power?

16. Consider the weighted voting system $[42: 20, 16, 10, 6, 4]$. Determine the Banzhaf Power Index for each player.

Consider the weighted voting system [30: 12, 10, 9, 5, 4, 4]. Determine the pivotal player in each of the sequential coalitions listed.

17. $\langle P_1, P_3, P_2, P_6 \rangle$

18. $\langle P_4, P_2, P_3, P_6, P_5, P_1 \rangle$

19. Consider the weighted voting system $[7: 4, 2, 2, 2, 2]$. Find the Banzhaf Power Index for each voter. Are any of the voters dictators? Do any of the voters have veto power?

20. Find the Banzhaf Power Index for each player in the voting system [10: 7, 5, 3, 2].

21. Consider the weighted voting system, $[100: 80, 60, 30, 20]$. Calculate the Shapley-Shubik Power Index for each player in this system. Are any of the voters dictators? Do any of the voters have veto power?

🚀 **APPLICATIONS**

22. The United Nations Security Council consists of fifteen members of the United Nations; five permanent member countries and ten nonpermanent member countries. In order for a vote to pass, nine members, including all five of the permanent members, must be in agreement. An equivalent weighted voting system is $[39: 7, 7, 7, 7, 7, 1, 1, 1, 1, 1, 1, 1, 1, 1, 1]$. According to our previous calculations, this means there are 15! or about 1.3 trillion permutations for the members.[1]

    a. Without computing all possible coalitions, determine when a nonpermanent member would be a pivotal voter.

    b. What is the Shapley-Shubik Power Index for each nonpermanent member?

    c. What is the Shapley-Shubik Power Index for the combined nonpermanent members?

    d. What is the Shapley-Shubik Power Index for each permanent member?

    e. What is the Shapley-Shubik Power Index for the combined permanent members?

23. There are 435 members of the United States House of Representatives.

    a. If a simple majority is needed to pass a bill, what constitutes a winning coalition?

    b. If a two-thirds majority is needed to ratify an amendment, what constitutes a winning coalition?

24. In the weighted voting system $[43: 25, 20, 19, 17]$, determine each of the following.

    a. List all permutations in which P_1 is a pivotal player.

    b. List all permutations in which P_2 is a pivotal player.

    c. Calculate the Shapley-Shubik Power Index for each player.

    d. Calculate the Banzhaf Power Index for each player.

    e. Is there a player with veto power?

25. If the quota in Exercise 23 is increased to 50, how would the Shapley-Shubik Power Index change?

---

1   United Nations, http://www.un.org

26. A company has five shareholders and a total of 500 shares. The quota for passing a measure is the number of votes where shareholders own 251 or more shares. The number of shares owned by each shareholder is as follows: $S_1 = 200$, $S_2 = 123$, $S_3 = 120$, $S_4 = 40$, and $S_5 = 17$. Suppose there is an investor S6 that wants to buy shares and currently owns no shares:

   a. What are the winning and losing coalitions? Compute the number of votes needed to make a losing coalition a winning coalition.

   b. How many shares can S_1 sell to S_2 without causing any of the winning coalitions listed in part **a.** to lose or any of the losing coalitions in part **a.** to win?

   c. How many shares can S_1 sell to S_5 without causing any of the winning coalitions listed in part **a.** to lose or any of the losing coalitions in part **a.** to win?

   d. How many shares can S_1 sell to S_6 without causing any of the winning coalitions listed in part **a.** to lose or any of the losing coalitions in part **a.** to win?

13.4 PROJECT

POWER DYNAMICS AT PLAY

In Section 13.4, you learned about weighted voting systems and the power that each player in the system has. In this project, you'll investigate power dynamics. Power dynamics are where weighted voting systems get interesting. Players may have opposing viewpoints that motivate them to desire different outcomes. Each player wants to win and wants to know the likelihood of winning.

Consider a small company with a small number of shareholders who disagree about the direction the company should take. Each member is likely to be acutely aware of how much their vote counts and with whom they need to align to be part of a winning coalition. Let's look at a few power dynamics at play.

1. Consider a scenario where player 1 is a dictator and is interested in selling some shares to another player, but wants to remain a dictator after the sale. Is this possible in the voting system [20: 25, 5, 3, 1]? Explain why this is or isn't possible. If it is possible, how many shares can player 1 sell?

2. Consider a scenario where player 1 and player 4 rarely vote in the same manner, while player 2 and player 4 often vote the same. In the voting system [14: 12, 7, 3, 2], why would player 4 be glad to see player 1 sell two shares to player 2? (**Hint:** Consider the possible winning coalitions in each voting system.)

3. Using the same scenario as part 2, use the Banzhaf Power Index to describe in words how player 4's power changes if player 1 sells two shares to player 2.

Chapter 13 Exercises

Create a preference table to show the results for each set of preference ballots.

1. The candidates are: Smith (S), Patel (P), Harvey (H), Knight (K), Jordan (J).

Preference Voting Results

HKPJS	PJKSH	HKPJS	SHJPK	SHKPJ
SHJPK	KJPHS	PHSJK	PJSHK	PJSHK
JPHKS	SHJKP	KPJSH	KPJSH	SHJPK
PJKSH	SHJPK	PJSHK	HKPJS	SHJPK
SHKPJ	HKPJS	PJKSH	KJPHS	KJPHS
PJKSH	KJPHS	HKPJS	KJPHS	SHJPK
KJPHS	PJSHK	SHKPJ	SHJKP	KJPHS

2. Candidates in the election are: Lawson, Eric, Misty.

   Misty, Eric, Lawson: 27

   Lawson, Misty, Eric: 41

   Misty, Lawson, Eric: 31

   Eric, Misty, Lawson: 15

   Eric, Lawson, Misty: 5

   Lawson, Eric, Misty: 19

Calculate the number of pairwise comparisons that must be made in each election.

3. Seven students are running for senior class president.

4. Running for local councilman: J. Pitts, K. McMillian, J. Wallace, D. McLaughlin, K. Smith, and T. Knight.

Use the given preference table to answer each question.

5. Answer the following questions about the given preference table.

Preference Table for Candidates

		Rankings			
1st	A	E	C	C	A
2nd	D	B	A	A	E
3rd	B	C	B	E	C
4th	C	A	D	B	D
5th	E	D	E	D	B
Total Votes	221	212	109	84	167

a. How many possible unique rankings are there of the candidates in the election?

b. How many people voted in the election?

c. How many voters place Candidate A in first place?

d. Which candidate wins the using plurality method?

e. How many votes would a candidate need to have a majority?

f. Do any of the candidates have a majority of first-place votes?

6. A photography club held a contest where the pictures had to be taken solely with a cell phone. The club members were asked to rank the following entries in order of preference. The preference table summarizes the results of the photo contest.

1. 2. 3.

4. 5.

Preference Table for Best Photo

	Rankings			
1st	3.	3.	2.	5.
2nd	1.	5.	4.	4.
3rd	2.	4.	1.	3.
4th	4.	1.	3.	2.
5th	5.	2.	5.	1.
Total Votes	54	83	65	92

a. Determine the photo winner using the Borda count method.

b. Determine the winner using the plurality with elimination method.

c. Determine the winner with the pairwise method of comparison.

d. Suppose that Picture 4 was thrown out for violating the rules. Pictures ranked below Picture 4 simply move up a ranking. Which picture would be declared the winner using the Borda count method with the new preference table? Is this a different winner than in part **a.**?

7. In the Campus Sing-Off, each contestant is ranked by the general student body via the Internet, and by all student organizations sponsoring a contestant. The top 10 entries receive votes in each ballot. The votes are distributed in the following manner.

1st place = 12 points	2nd place = 10 points	3rd place = 8 points
4th place = 7 points	5th place = 6 points	6th place = 5 points
7th place = 4 points	8th place = 3 points	9th place = 2 points
10th place = 1 point		

The following table shows how the student organization ranked each contestant in the first semifinal round. The contestants are listed by their last names and sponsor organizations.

Voting Results for First Semifinal																		
Harris; SGA				1st							3rd							
Icin; Social Work Club	6th		6th	6th		7th	6th	2nd	7th	8th	3rd	2nd	10th	7th	9th	10th	5th	
Green; ΑΛΠ	2nd	6th			3rd	1st	8th	3rd		8th	4th	1st	7th	6th		10th	6th	8th
Lawson; FCA	9th													7th				
Albert; GSA	1st	8th	2nd	7th		7th	1st	2nd	6th	6th	2nd	2nd	4th	9th	2nd	1st	1st	7th
Roman; ΓΒΘ	4th	7th	3rd		6th		9th	7th		3rd	5th	5th	10th	3rd	8th	6th	2nd	1st
Switzer; Galois Club		9th		4th	8th	9th			10th	10th	9th			3rd	8th	3rd		
Belton; Hispanic Culture Center				7th		9th	10th			9th								10th
Finley; History Club		4th		5th	10th		10th	10th		9th			3rd		1st			
Issen; NAEA				10th		6th			8th			10th	8th	5th	6th	4th		9th
San Marie; NBS	7th		9th		2nd												8th	
Centus; NTSS	5th	1st	1st	8th	5th	4th		8th	10th	2nd				4th			4th	3rd
Dennis; ΩΨΦ		3rd	10th	3rd		8th	2nd		3rd		7th	7th		8th	10th		5th	
Russan; ΦΑ	3rd	5th	4th	1st		5th	3rd	1st	1st	1st	9th	4th	1st		4th	2nd	9th	4th
Hunter; ΦΜΑ				4th	3rd	5th	5th	7th			6th	6th			7th			
Austin; ΣΓΡ		10th				6th	9th											
Molda; Student Design Group	8th		5th	9th	7th	2nd	4th		5th	5th	6th	8th	5th	1st	9th	5th	7th	2nd
Irene; Voices of Praise	10th	2nd	8th	2nd			4th	4th	4th	1st			9th	2nd	5th	3rd		6th

a. Determine the number of total points each contestant received in the first semifinal round from the student organizations.

b. Which top 10 contestants did the organizations think should move on to the second semifinal?

8. Online voting for the best student music video was available for two weeks. Students were asked to rank the videos in order of preference. The following table shows the results of the student votes.

Preference Table for Best Student Music Video

	Rankings					
1st	Video 2	Video 1	Video 2	Video 3	Video 3	Video 4
2nd	Video 1	Video 2	Video 1	Video 4	Video 2	Video 1
3rd	Video 4	Video 3	Video 3	Video 2	Video 4	Video 2
4th	Video 3	Video 4	Video 4	Video 1	Video 1	Video 3
Total Votes	143	212	99	43	121	151

a. Determine the winning video by using the plurality method.

b. Determine the winning video by using the pairwise method of comparison.

c. Determine the winning by using the plurality with elimination method.

d. If you were determining a winning video, which video would you declare the winner based on the online votes? How would you defend your choice based on the answers to **a.–c.**?

9. In the 1992 United States Presidential Election, three candidates shared most of the November 3 popular vote, although there was a small portion of voters for other candidates. The following table displays the breakdown of the popular vote along with the electoral vote for the election.

1992 US Presidential Election Results

Presidential Candidate	Popular Vote		Electoral Vote
	Count	**Percentage**	
George H. Bush	39,104,545	37.45%	168
Bill Clinton	44,909,889	43.01%	370
Ross Perot	19,742,267	18.91%	0
Other	669,958	0.63%	0
Totals	**104,426,659**	**100%**	**538**

Source: Federal Election Commission. "Federal Elections 92. Election Results for the US President, the US Senate, and the US House of Representatives." June 1993. http://www.fec.gov/pubrec/fe1992/federalelections92.pdf

a. Who won the popular vote by plurality? Was it a majority?

b. Who won the electoral vote by plurality? Was it a majority?

c. Suppose Perot had dropped out of the election. Is it possible that Clinton would have lost the presidency? Explain your answer.

10. The student government is deciding which author to bring to campus next year for the annual Books to Tables talk. The possibilities are Marci Shimoff, coauthor of *Chicken Soup for the Woman's Soul*, Peter Legge, author of *How to Soar with the Eagles*, and Tania Aebi, author of *Maiden Voyage*. Complete the following preference table so that Tania Aebi wins the election using the majority rule.

Preference Table for Author for Books to Tables Talk

	Rankings		
1st	Tania Aebi	Peter Legge	Marci Shimoff
2nd	Peter Legge	Marci Shimoff	Tania Aebi
3rd	Marci Shimoff	Tania Aebi	Peter Legge
Total Votes	?	5	4

11. The Campus Housing Welcome Back committee is deciding what movie to show free to students the first week of school. The movie options are *Django Unchained*, *Zero Dark Thirty*, *Silver Linings Playbook*, and *Life of Pi*. Students were asked to rank the movies by preference. The following preference table summarizes their selections.

Preference Table for Movie Showing

Movie	Rankings						
1st	Zero	Zero	Django	Silver	Life	Django	Django
2nd	Django	Life	Life	Zero	Django	Silver	Zero
3rd	Life	Silver	Zero	Django	Zero	Zero	Silver
4th	Silver	Django	Silver	Life	Silver	Life	Life
Total Votes	30	23	15	16	11	15	27

a. Which movie is the plurality winner?

b. Which movie wins using a pairwise method of comparison?

c. Does using the plurality method satisfy the Condorcet criterion? Explain your answer.

12. An online poll asked readers to rank the top three law schools in the country. The following preference table shows the results after two days of online voting.

Preference Table for Top Law Schools

	Rankings		
1st	Stanford	Yale	Stanford	Harvard
2nd	Harvard	Stanford	Harvard	Yale
3rd	Yale	Harvard	Yale	Stanford
Total Votes	84	132	39	13

a. Which law school would win using the majority rule if voting closed today?

b. Which law school would win using the Borda count method if voting closed today?

c. Does the Borda count method satisfy the majority criterion for these poll results? Explain your answer.

13. The Association of Marketing Professionals is choosing a city for its next conference. The committee making the decision asks the association's members to rank the top five cities they would like to visit. The following preference table gives the results of the member survey. Would the plurality with elimination method satisfy the irrelevant alternatives criterion if the committee decided to eliminate Jacksonville, FL, from the options after the vote? Explain your answer.

Preference Table for Conference Location

	Rankings			
1st	Boston, MA	Nashville, TN	San Diego, CA	New Orleans, LA
2nd	New Orleans, LA	Boston, MA	Boston, MA	Nashville, TN
3rd	San Diego, CA	San Diego, CA	New Orleans, LA	San Diego, CA
4th	Nashville, TN	Jacksonville, FL	Nashville, TN	Boston, MA
5th	Jacksonville, FL	New Orleans, LA	Jacksonville, FL	Jacksonville, FL
Total Votes	**375**	**348**	**289**	**115**

14. The Chamber of Commerce is electing a new president. The candidates are N. Pitts, A. Palmer, and J. Layne. The summary of the rankings of the candidates from the members is given in the table. Is the irrelevant alternatives criterion satisfied using the Borda count method if N. Pitts has to withdraw from the ballot after votes are cast because he moved? Explain your answer.

Preference Table for Chamber of Commerce President

	Rankings			
1st	A. Palmer	N. Pitts	A. Palmer	J. Layne
2nd	N. Pitts	J. Layne	J. Layne	A. Palmer
3rd	J. Layne	A. Palmer	N. Pitts	N. Pitts
Total Votes	**12**	**15**	**13**	**16**

15. *Living Day-to-Day* magazine wants to name the number-one feature homebuyers want in a house. They ask their readers to rank eight home features in order of importance. The eight home features are open-concept homes, smaller homes, outdoor living spacings, neutral decor, modern kitchens, linen closest & smart storage options, energy-efficient fixtures & appliances, and two-car garage with organization. The following preference table displays the results of the rankings. Determine if the Condorcet criterion is satisfied if the magazine uses the plurality method to name the top feature.

Preference Table for Homebuyers' Number One Feature

Feature	Rankings					
1st	Outdoor	Smaller	Open	Kitchen	Smaller	Outdoor
2nd	Open	Outdoor	Smaller	Outdoor	Efficient	Kitchen
3rd	Kitchen	Kitchen	Outdoor	Efficient	Garage	Storage
4th	Neutral	Efficient	Neutral	Open	Open	Efficient
5th	Efficient	Storage	Kitchen	Storage	Outdoor	Garage
6th	Garage	Garage	Storage	Garage	Kitchen	Neutral
7th	Storage	Neutral	Efficient	Smaller	Storage	Open
8th	Smaller	Open	Garage	Neutral	Neutral	Smaller
Total Votes	**421**	**234**	**331**	**472**	**253**	**119**

16. Driving Divas asks its readers to rank the following car features in order of "must haves" in a new car. The car features are remote keyless entry, OnStar system, antilock brakes (ABS), electronic stability/skid-control system, telescoping steering wheel/adjustable pedals, and rear-seat DVD player. Complete the preference table so that rear-seat DVD player is the winner using the Borda count method, but the Condorcet criterion is violated in a head-to-head matchup with antilock brakes.

Preference Table for Must-Have Car Feature

Feature	Rankings			
1st	DVD	Keyless	Adjustable	Stability
2nd	Adjustable	?	DVD	ABS
3rd	ABS	?	OnStar	OnStar
4th	Keyless	Stability	ABS	Keyless
5th	OnStar	OnStar	Stability	Adjustable
6th	Stability	Adjustable	Keyless	DVD
Total Votes	35	44	28	20

17. In his next newspaper column, Tom plans to publish his results from a study on the top six qualities that make a great leader. The following preference table summarizes the rankings for the traits. Complete the preference table so that "sense of humor" is the top trait if the site uses the Borda count method to count the votes, and the majority criterion is not violated.

Preference Table for Top Leadership Qualities

Quality	Rankings			
1st	?	Honesty	Honesty	Communication
2nd	?	Sense of Humor	Communication	Sense of Humor
3rd	Confidence	Ability to Delegate	Sense of Humor	Honesty
4th	Commitment	Commitment	Confidence	Ability to Delegate
5th	Ability to Delegate	Confidence	Ability to Delegate	Commitment
6th	Communication	Communication	Commitment	Confidence
Total Votes	159	66	42	50

Use the given table to solve each problem.

Population per County

County	Population
A	8578
B	9878
C	10450
D	7565
E	4563
F	6347

18. Suppose a state decides to apportion 250 new highway patrol officers on the basis of number of residents in a county.

   a. Find the standard divisor, SD.

   b. Find the standard quota, SQ, for county A and county B. Round to three decimal places.

19. Use the Hamilton method to apportion the 250 patrol officers based on the population of the county.

Use the given table to solve each problem.

Student Enrollment in the University of Texas System

Campus	Enrollment	Full Time Equivalent (FTE)
Arlington	33,439	20,594
Austin	51,112	46,402
Brownsville	13,836	9398
Dallas	18,684	12,537
El Paso	18,160	16,271
San Antonio	30,968	23,198
Tyler	5064	4810
Permian Basin	16,266	2696
Pan American	21,016	15,494

20. The University of Texas system consists of nine campuses and has a budget of $13.1 billion. How much money in funding would each campus receive if the system divided the money equally among the campuses (round to nearest dollar)?

21. If the state of Texas wanted to allocate the money equally based on total student enrollment, how much would be appropriated per student?

22. Based on the result from Exercise 21, how much money would the campuses of Arlington and Pan American receive based on the student enrollment on each campus.

23. If the state apportions the funds based on FTE, how much money will the campuses of Austin and Dallas receive (round to nearest dollar)?

24. The University of Texas system wants to apportion 825 new electric vehicles to their campuses. The state decides to apportion these vehicles based on the student enrollment at each university.

  a. Find the SD. Round to three decimal places.

  b. Find the SQ for the Brownsville and Tyler campuses. Round to three decimal places.

25. Suppose the state decides to apportion the 825 electric vehicles based on FTE.

  a. Find the SD. Round to three decimal places.

  b. Find the SQ for the San Antonio and Permian Basin campuses. Round to three decimal places.

26. Use the Jefferson method to apportion the 825 electric vehicles to all 10 campuses based on the number of students.

27. Use the Hamilton method to apportion the 825 electric vehicles to all 10 campuses based on FTE. Compare your results with the apportionments from Exercise 26 and determine if the apportionments are different when the apportionment basis is different.

Solve each problem.

28. A science teacher at a high school can teach five classes. There are 27 students enrolled in Physical Science, 24 in Biology, and 37 in Anatomy.

  a. Find the SD and SQ of each course. Round to three decimal places.

  b. Use the Jefferson method to determine the apportionment of students to the courses to determine the number of sections needed per course.

29. Repeat Exercise 28 b. using the Huntington-Hill method.

30. Repeat Exercise 28 b. using the Hamilton method.

31. An English department uses 19 graduate assistants in teaching its undergraduate courses. The enrollments for each of the courses these students teach is as follows. Using Webster's method, how many graduate assistants should be assigned to each course?

Students per Course	
Course	Students
Comp I	675
Comp II	455
Survey of Lit	197
Poetry	385

32. A state consists of six counties: A, B, C, D, E, and F.

The senate for the state is to have 30 members apportioned to the counties based on the old populations using Webster's method. The new populations are also given.

Population by County

County	Old Population	New Population
A	8578	9244
B	9878	9166
C	10,450	10,580
D	7565	10,254
E	4563	6680
F	6347	5993

a. Apportion the members of the senate using the old populations.

b. Apportion the members of the senate using the new populations.

c. Apportion the members of the senate using the Huntington-Hill method and the old populations.

d. Apportion the members of the senate using the Huntington-Hill method and the new populations.

e. Does the population paradox occur when calculating the apportionments using the Huntington-Hill method and the old and new populations?

33. The city of Hillsboro wants to allocate $55 million to youth sports and activity programs. The current participation rates are listed in the table.

Participants per Activity

Activity	Number of Participants
Swimming	245
Library	2595
Baseball	1150
Softball	978
Theater	753
Computer Hobbies	477
Music	1658

a. Find how much the city should allocate to each activity based on the number of participants using the Jefferson method.

b. Find how much the city should allocate to each activity based on the number of participants using Webster's method.

c. Find how much the city should allocate to each activity based on the number of participants using the Huntington-Hill method.

d. Find how much the city should allocate to each activity based on the number of participants using the Hamilton method.

34. Three groups, A, B, and C, have the number of members shown in the table. If Jefferson's method is used to apportion the representatives of the groups on a panel, does the Alabama paradox occur if the number of representatives is increased from 40 to 41?

Members per Group	
Group	**Members**
A	3340
B	4500
C	8875

35. Using the table in Exercise 34, if Webster's method is used to apportion the representatives of the groups on a panel, does the Alabama paradox occur if the number of representatives is increased from 40 to 41?

36. What is the quota for $[8, 5, 3, 2]$ in a simple majority?

37. What is the quota for $[8, 5, 3, 2]$ for a two-thirds majority?

38. Consider the weighted voting system $[25: 9, 7, 4, 3, 2, 2, 1]$.

    a. What is the quota for this voting system?

    b. What is the weight of P_5?

    c. If only the last six voters vote for a motion, does the motion pass?

    d. If P_1 and P_2 vote against a motion, does the motion pass?

    e. What are the possible winning coalitions for this voting system?

39. Six partners in a law firm have the following voting system $[96: 34, 25, 18, 18, 12, 4]$.

    a. How many voters are in the system?

    b. What is the quota for the system?

    c. What are all of the winning coalitions?

    d. Are there any critical players?

    e. Do any of the players have veto power?

    f. Is there a dummy player?

    g. Determine the Banzhaf Power Index for each player.

40. In the weighted voting system $[q: 10, 7, 4, 4, 2, 2]$, if every voter has veto power, what is q?

41. A voting system is represented by $[61: 35, 30, 24, 21]$.

    a. List all sequential coalitions in which the voter with weight 35 is pivotal.

    b. List all sequential coalitions in which the voter with weight 30 is pivotal.

    c. Calculate the Shapley-Shubik Power Index for each player in the system.

    d. Are any of the voters dictators?

    e. Do any of the voters have veto power?

42. A professional football team is getting ready for the amateur player draft. The head coach, general manager, head scout, and team physician will use the voting system $[8: 5, 4, 3, 2]$ to make decisions regarding players to draft.

   a. List all sequential coalitions in which P_1 is a pivotal player.

   b. List all sequential coalitions in which P_2 is a pivotal player.

   c. List all of the winning coalitions.

   d. Calculate the Shapley-Shubik Power Index for each player.

   e. Calculate the Banzhaf Power Index for each player.

   f. Is there a player with veto power?

CHAPTER 13 PROJECT

Person of the Year

It's your turn to run an election! You are going to set up and run an election to name the most influential person of the year. You will choose four well-known figures to have on your ballot, collect votes, and then name a winner.

Step 1: Develop the ballot.

1. Choose at least four living people to place on your preference ballot for *Most Influential Person of the Year*. Create a preference ballot so that voters can rank each of the personalities. Be sure that the people on the ballot are well-known enough that the voters will recognize them.

2. Who do you expect to be the winner of your election?

Step 2: Decide on how you will choose your winner.

3. Which fairness criteria do you think is the most important in your election? Why?

4. Based on your answer above, choose two counting methods from those covered in Chapter 13 that you will use to determine a winner: Majority Rule, Plurality, Borda Count, Plurality with Elimination, or Pairwise Comparison.

5. Do you anticipate that both of your chosen methods will produce the same winner? Why or why not?

Step 3: Collect data.

6. Ask at least 30 voters to rank the personalities in order of the most influential to the least. Try not to influence your voters in any way as you collect their choices. List the results in a preference table.

7. What are some factors that might have affected the way people voted in your election?

Step 4: Analyze your results.

8. Determine a winner using each of the methods you chose in Step 2.

Step 5: Discuss your results.

9. Did both counting methods produce the same winner?

10. If the counting methods you chose produced different winners, which method do you think best reflects the "will of the people"? Why?

11. Who do you think would disapprove the most with your results? Why?

12. What do you think would change if there were more candidates in your election? What if there were fewer candidates?

▤ CHAPTER 13 REVIEW

Section 13.1 How to Determine a Winner

Definitions

Preference Ballot: A **preference ballot** is a ballot that allows a voter to rank the items in order of preference from most preferred to least preferred.

Preference Table: A **preference table** summarizes all of the individual preference ballots in an election by tallying the number of ballots with the same order of ranking.

Formula

Number of Pairwise Comparisons: The number of pairwise comparisons that must be made if there are n candidates is

$$\frac{n(n-1)}{2}.$$

Section 13.2 Flaws in Voting Methods

Definitions

Condorcet Criterion: The **Condorcet criterion** states that if a candidate wins the head-to-head comparison against every other candidate, then he should also win the overall election in a fair voting system.

Majority Criterion: The **majority criterion** states that if a candidate receives a majority of the votes in an election, that candidate should win.

Monotonicity Criterion: The **monotonicity criterion** states that if a candidate wins an early round of an election and only gains support and does not lose support in subsequent rounds, then that candidate should win the election.

Irrelevant Alternative Criterion: The **irrelevant alternatives criterion** states that if a candidate wins an election, then that same candidate would win the election even if at least one candidate withdraws from the election.

Dictator Criterion: The **dictator criterion** states that no single vote is allowed to decide the outcome of an election.

Section 13.3 Apportionment

Definitions

Apportionment: **Apportionment** is a method of fairly dividing resources or items among individuals or groups.

The Alabama Paradox: The **Alabama paradox** occurs when an increase in the number of items to be apportioned causes a subgroup to lose an item.

The Population Paradox: The **population paradox** occurs when subgroup A loses an item to subgroup B when the rate of growth of the population of subgroup A is greater than the rate of the growth in subgroup B.

The New States Paradox: The **new states paradox** occurs when the addition of a new subgroup, with a corresponding increase in the number of available items, can cause a change in the apportionment of items among the other subgroups.

The Quota Rule: The **quota rule** states that any fair apportionment method should assign every subgroup either its lower quota or its upper quota.

Formulas

Standard Divisor: The **standard divisor** for apportionment is the average number of people per item to be apportioned.

$$\text{Standard Divisor} = \frac{\text{Total Population}}{\text{Number of Items to Be Apportioned}}$$

Standard Quota: The **standard quota** for apportionment is the average number of items to be apportioned to each subgroup.

$$\text{Standard Quota} = \frac{\text{Subgroup Population}}{\text{Standard Divisor}}$$

Geometric Mean: The **geometric mean** of any two numbers m and n is $\sqrt{m \cdot n}$.

Procedures

Hamilton's Method of Apportionment

1. Calculate the standard divisor.
2. Calculate the standard quota for each subgroup.
3. Calculate the lower quota for each subgroup.
4. Assign each subgroup the number of resources based on its lower quota.
5. Assign any remaining resources based on the fractional remainder of the standard quotas, in order from largest to smallest.

Jefferson's Method of Apportionment

1. Calculate the standard divisor.
2. Calculate the standard quota for each subgroup.
3. Calculate the lower quota for each subgroup.
4. Assign each subgroup the number of resources based on its lower quota.
5. If there are remaining resources to be distributed, chose a modified divisor by trial and error until the sum of the lower quotas equals the number of resources to be apportioned.

Webster's Method of Apportionment

1. Calculate the standard divisor.
2. Calculate the standard quota for each subgroup.
3. Round each quota to the nearest integer.
4. Assign each subgroup the number of resources based on the rounded quota.
5. If there are remaining resources to be distributed, chose a modified divisor by trial and error until the sum of the rounded quotas equals the number of resources to be apportioned.

The Huntington-Hill Method of Apportionment

1. Calculate the standard divisor.
2. Calculate the standard quota for each subgroup.
3. Calculate the geometric mean of the standard quota for each subgroup.
4. If the standard quota is less than the geometric mean, round the quota down. If the standard quota is greater than the geometric mean, round the quota up.

5. Assign each subgroup the number of resources based on the rounded quota.

6. If there are remaining resources to be distributed, chose a modified divisor by trial and error until the sum of the rounded quotas equals the number of resources to be apportioned.

13.4 Weighted Voting Systems

Definitions

Weighted Voting System: A **weighted voting system** consists of n players (P_n) each controlling a number of votes (w) called their **weight**. It is described using the following notation.

$$[q: w_1, w_2, \ldots, w_N]$$

Dictator: A **dictator** is a player with the power to pass a proposal single-handedly; that is, the dictator has at least as many votes as the quota.

Veto Power: When a player's votes are needed in order to pass a proposal, that player is said to have **veto power**.

Dummy Player: A **dummy player** is a player who does not have enough votes to have an effect on the outcome of a proposal.

Coalition: A **coalition** is a group of players joining together to vote the same way. If the combined weight of the coalition is at least the quota, it is a *winning coalition*. In a voting system with n players, there are $2^n - 1$ possible coalitions. A coalition consisting of players P_1, P_2, and P_3 is denoted by $\{P_1, P_2, P_3\}$.

Critical Player: If a player in a winning coalition were to leave and cause the coalition to become a losing coalition, that player is considered a **critical player**.

Sequential Coalition: A **sequential coalition** is a coalition where the order in which N players join is important. It is denoted as $\langle P_1, P_2, P_3, \ldots, P_N \rangle$, where P_1 is the first player to join, P_2 is the second player to join, and so on.

Pivotal Player: A **pivotal player** is the player who joins a sequential coalition and causes the coalition to change from a losing coalition to a winning coalition.

Procedures

Calculating the Banzhaf Power Index for Player P_n

Step 1: List all winning coalitions for the voting system.

Step 2: Identify the critical players for each winning coalition.

Step 3: Count the total number of times each player is a critical player.

Step 4: Add together the critical player counts found in Step 3.

Step 5: Calculate the Banzhaf Power Index using the following fraction.

$$\frac{\text{Number of Times Player } P_n \text{ Is a Critical Player}}{\text{Total Number of Times All Players Are Critical Players}}$$

Calculating the Shapley-Shubik Power Index for Player P_n

Step 1: List all $N!$ sequential coalitions that contain all N players in the voting system.

Step 2: Identify the pivotal player for each sequential coalition.

Step 3: Count the total number of times player P_n is a pivotal player.

Step 4: Calculate the Shapley-Shubik Index using the following fraction.

$$\frac{\text{Number of Times Player } P_n \text{ Is a Pivotal Player}}{N!}$$

CHAPTER 14

Graph Theory

⬛ SECTIONS

Introduction

You may have heard of the River Crossing Puzzle. It goes something like this: A farmer has a chicken, a fox, and some grain. He needs to take everything across the river, but he can only transport one item at a time. He can't leave the fox with the chicken because the fox will eat the chicken. Nor can he leave the chicken with the grain. So how does the farmer get everything safely across the river? While we probably wouldn't use graph theory to solve this problem due to its simplistic nature, if we added more constraints or wanted to show that a solution is the only solution, graph theory would be extremely useful.

In 1735, Swiss mathematician Leonhard Euler found himself faced with a similar problem: whether it was possible to walk through the city of Königsberg and cross each of the seven bridges once and only once. Euler eventually wrote a paper proving mathematically that it was impossible. His abstract representation of the problem became the beginnings of the field of graph theory.

While graph theory may have a humble beginning, its usefulness has flourished into endless applications because graphs can be used to make connections between almost anything. Some common business applications are optimization problems such as garbage collection and mail delivery. At a more personal level, social graphs can connect you with other people, places, and things you interact with online and in real life. These contact networks are useful in tracking infectious diseases such as Covid-19 and can also be fundamental in solving criminal investigations. Similarly, flight networks connect places of departure or destination with airports, aircrafts, and cargo weights. Even cryptocurrencies such as IOTA use storage systems where individual items connect to each other.

You utilize graph theory technology almost every day, whether you are aware of it or not. You do this when you use navigation systems that provide directions to find the shortest route for you to take, or when you surf the web, or when you use social media networks such as Twitter and Facebook. Just think, all this is possible because of a simple problem about bridges!

OBJECTIVES

1. Identify parts of a graph.

2. Determine if a graph is connected or disconnected.

3. Use the properties of edges and vertices.

4. Determine a vertex covering for a graph.

5. Determine the chromatic number of a graph.

OBJECTIVE 1 Graphs

In the world of mathematics, the field of graph theory is a relative newcomer. However, as one of the fastest-growing and most applicable branches of mathematics, graph theory permeates throughout academic disciplines such as computer science, linguistics, biology, chemistry, physics, cartography, and operations research. Even social media apps utilize graph theory to help users find potential friends by identifying the activities and demographics that they have in common.

Figure 14.1.1: Social Media Friends

So, what is graph theory? Graph theory is the study of graphs—an abstract way to model different kinds of networks, systems, or structures like the one seen in Figure 14.1.1. Unlike Cartesian graphs of functions, in graph theory the actual position of the points on a graph is unimportant. What is important is whether the points are interconnected and, if so, how they are connected. Graphs remove all extraneous information and only record what is important for the given problems—the connections. To study graphs of this type, we first need vocabulary that describes their different features.

In the field of graph theory, a **graph** consists of a set of points called *vertices* and a set of lines called *edges*. The edges associate particular pairs of the vertices together. We can illustrate a graph by representing the vertices as dots and the edges as lines that join two vertices together, just as in Figure 14.1.1. For instance, if you think of your friends on your favorite social media platform as the set of vertices (dots), then the edges (lines) represent their connection to you and their connections to one another.

📖 GRAPH

A **graph** consists of a set of vertices and a set of edges that join pairs of vertices together. Graphs are generally represented with a capital letter, such as *G*.

Example 14.1.1

Identifying Parts of a Graph

In graph A, identify the vertices and edges.

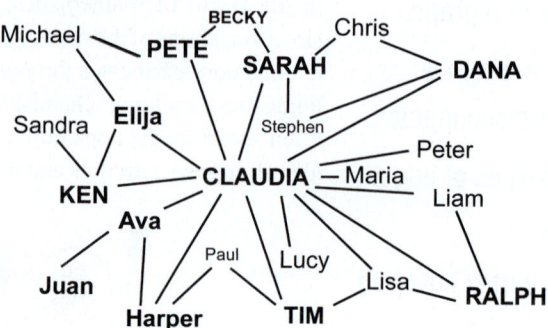

Figure 14.1.2: Graph A

Solution

In graph A, the vertices are represented by the names and the edges are the lines joining the names to one another.

As you can imagine, when working with graphs, we need a way to distinguish each vertex. A distinct lower case letter or number is commonly assigned to each vertex for identification. Figure 14.1.3 shows the five vertices in graph G labeled with the letters u, v, w, x, and y.

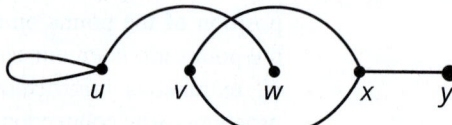

Figure 14.1.3: Graph G

Notice in graph G, that even though the edge joining vertices u and w crosses an edge joining v and x, the point where they cross is not a vertex. In graph theory, not every place where edges meet is a vertex.

Similarly, when we refer to the edges, a common practice is to denote them by the two vertices at each end of the edge. For instance, in graph G, the edge between vertices x and y is labeled xy, as shown in Figure 14.1.4. Note that when labeling edges using vertices, the order in which the vertices are labeled is irrelevant. In other words, the edge between the vertices x and y can be labeled either xy or yx.

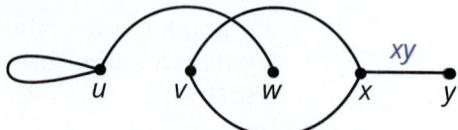

Figure 14.1.4: Graph G

How do we distinguish between the two edges joining vertices v and x in G? We cannot refer to them both as vx. The notation using the ends of the edges is no longer sufficient to distinguish between edges. When multiple edges exist between a pair of vertices, we name the edges with unique identifiers to set them apart. For instance, one could label the edges in G as illustrated in Figure 14.1.5.

$$e_1 = uu$$

$$e_2 = uw$$

$$e_3 = vx \text{ (drawn on top in Figure 14.1.5)}$$

$$e_4 = vx \text{ (drawn on bottom in Figure 14.1.5)}$$

$$e_5 = xy$$

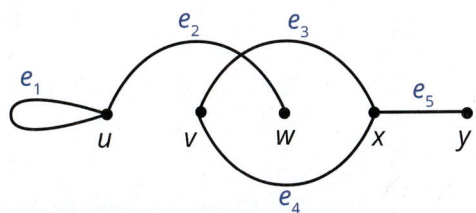

Figure 14.1.5: Graph G

Notice that both ends of edge e_1 are the same vertex, u. We call an edge like this a loop.

> **Helpful Hint**
>
> A graph can be thought of as a bunch of plastic balls connected with a number of strings. The essential information about the graph is which balls are connected via strings. The balls can either be in a big clump or spread out in any way, but it's still the same graph.

> 📖 **LOOP**
>
> A **loop** is an edge that has the same vertex for both of its ends.

Example 14.1.2

Identifying Parts of a Graph

For the given graphs, label each vertex and edge. Identify any loops.

a.

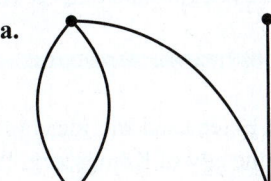

Figure 14.1.6: Graph K

b.

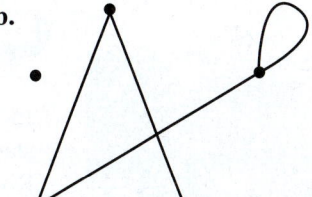

Figure 14.1.7: Graph L

Solution

a. First, label the four vertices in graph K with lowercase letters. We'll use a, b, c, and d. Note that the assignment of the letters is not critical; they can be in any order.

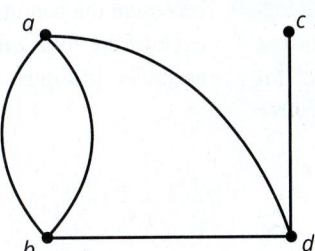

Figure 14.1.8: Graph K

Second, since the graph has two edges between the vertices a and b, it is not sufficient to use the ends as labels. Instead, use distinct identifiers in order to distinguish between the edges. We will use f_1, f_2, f_3, f_4, and f_5.

Although graph K might appear to have a loop on the left-hand side of it, there is not an edge that begins and ends at the same vertex. Therefore, there are no loops in graph K.

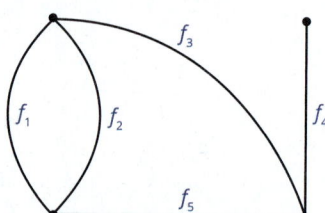

Figure 14.1.9: Graph K

b. In graph L, label the five vertices with lowercase letters. We use v, w, x, y, and z. Since there is at most one edge between any pair of vertices, we can label the edges using either format. This time, we will use the pairing method to label the edges. Note that there are only five vertices and not six, remembering that there is not a vertex every time edges intersect.

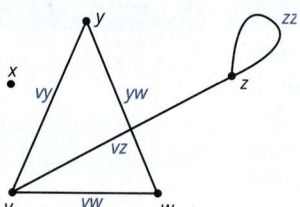

Figure 14.1.10: Graph L

Notice that edge zz begins and ends at vertex z and is, therefore, a loop.

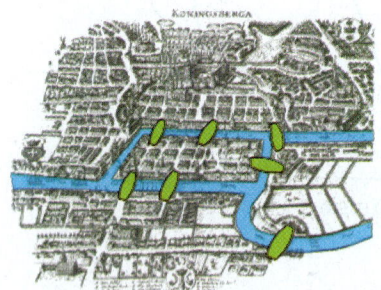

Figure 14.1.11

In 1735, Leonhard Euler used the ideas of vertices and edges to settle a long-standing dispute in the city of Königsberg, Prussia. The city was set on the Pregel River and included two islands that were connected to the mainland and each other by seven bridges. The townspeople of Königsberg disagreed on whether it was possible to visit all parts of the city and cross each of its seven bridges once and only once. They were not allowed to cross half a bridge and then turn around and cross the other half of the bridge from the other side, however they did not have to begin and end the journey in the same spot. Figure 14.1.11 shows a map of the city with the bridges highlighted.[1]

Founding an entire new branch of mathematics, Euler's genius idea was to represent the islands and bridges as a graph. He used a different vertex to represent each of the four land masses and an edge to represent each of the seven bridges, as shown in Figure 14.1.12.

---

1   Image courtesy of Bogdan Giuşcă.

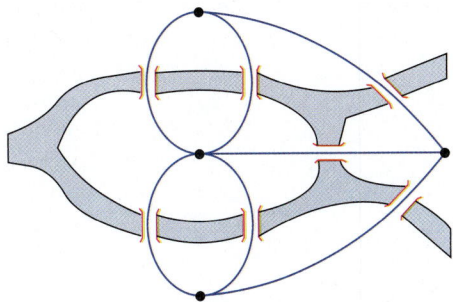

Figure 14.1.12: Graph of the Königsberg Bridge Problem

Routes around the city then became sequences of vertices and the edges that connected them. In order to settle the argument, Euler proved that a *walk* could not be found that visited every land mass and crossed every bridge precisely once. A **walk** in graph theory is a way to describe a particular sequence of vertices and edges in a given order.

> 📖 **WALK**
>
> A **walk** in a graph is a finite list of alternating vertices and connecting edges that begins and ends with a vertex.

For instance, in Figure 14.1.13, the walk $v, e_4, x, e_5, y, e_5, x$ is highlighted in graph G. Notice that a walk is not a "jump" between vertices. Each vertex must be connected by an edge. For instance, u, v, w is not a walk.

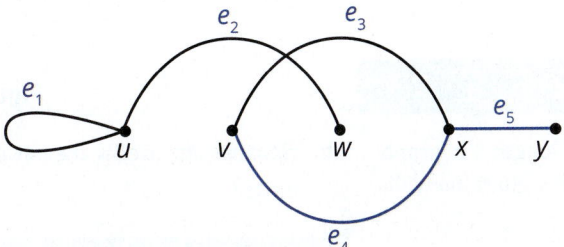

Figure 14.1.13: A Walk in Graph G

The *length of a walk* is the number of edges traversed in the walk. To complete the walk v, x, y, x, you have to traverse the edge from v to x, the edge from x to y, and then finally, the edge from y to x. Thus, the walk v, x, y, x is of length 3. If a walk starts and ends at the same vertex, we say the walk is **closed**. We could make our walk closed in Figure 14.1.13 if we continue the walk back to the starting point v. The closed walk would then be $v, e_4, x, e_5, y, e_5, x, e_4, v$. Notice that there is no restriction on the number of times a vertex or edge appears in a walk.

Example 14.1.3

Identifying a Walk

Use graph B to answer the following questions. Notice that since there is only one edge between each pair of vertices, we do not need to label the edges.

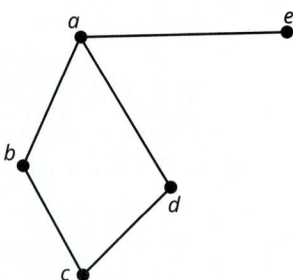

Figure 14.1.14: Graph *B*

a. Identify a walk of length 2 that starts at vertex *a* and ends at vertex *c*.

b. Is the walk you just found a closed walk?

Solution

a. There are two possible walks that start at vertex *a*, end at vertex *c*, and have a length of 2. The walk *a*, *b*, *c* is highlighted in blue and the walk *a*, *d*, *c* is highlighted in red. Either is an acceptable answer.

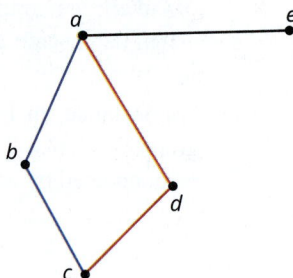

Figure 14.1.15: Graph *B*

b. Because the walks do not start and end at the same vertex, neither walk is closed.

> ✔ **Skill Check 14.1.1**
>
> Identify a walk of length 3 in graph *B* in Example 14.1.3. Start the walk at vertex *c*.

Sometimes restrictions are placed on walks. For instance, in the Königsberg bridge problem, every land mass (vertex) had to be visited while crossing every bridge (edge) only once. When we restrict a walk such that no edges or vertices can be repeated, we call it a **path**. The walk v, e_4, x, e_5, y is a path in graph *G*, as is shown in Figure 14.1.16.

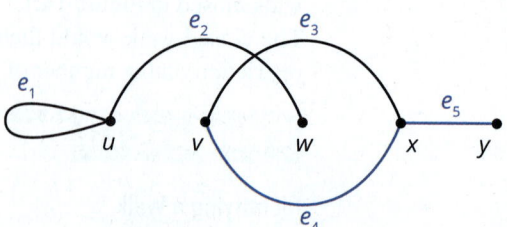

Figure 14.1.16: Graph *G*

> 📖 **PATH**
>
> A **path** is a walk with no repeated edges or vertices.

Example 14.1.4

Identifying a Path

Identify a path of length 3 starting at vertex a and ending at vertex d in graph C.

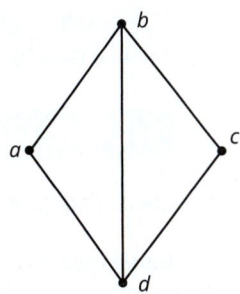

Figure 14.1.17: Graph C

Solution

Since the path must be of length 3 and start at vertex a and end at d, we cannot go straight to d. We need to move from a to b. From there, if we move to d next, we still only have a path of length 2. Instead, we need to move from b to c and then to d.

This walk a, b, c, d, is a walk of length 3 and only crosses each vertex once and never duplicates an edge. Therefore, it is the required path.

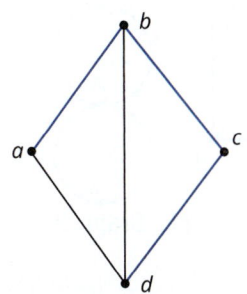

Figure 14.1.18: Graph C

OBJECTIVE 2 Connected and Disconnected Graphs

Remember that the definition of a graph has no specifications as to where the vertices lie on the page nor the shape of the edges joining them. The edges can be drawn as curves or lines, short or long, etc. The definition of a graph simply describes which pairs of vertices are joined.

For instance, in Figure 14.1.19, we've drawn our original graph G on the top, but on the bottom is a different view of graph G, this time without any edges crossing. Notice that both views of the graph have exactly the same vertices with the same edges joining them. If any of the vertices or edges or their connections were different, then the graphs would not be different views of the same graph.

You might be wondering if this second view of G is actually one graph or two. It certainly has two pieces, called *components*; however, it is still only one graph. Notice that although you can move across edges between vertices u and w, and also between vertices v, x, and y, there is no way to reach vertex v from vertex w. If it is possible to reach any vertex of a graph by moving along edges from any other vertex in the graph, then that graph is called **connected**. As we stated, G is in fact one graph, but the graph is **disconnected**.

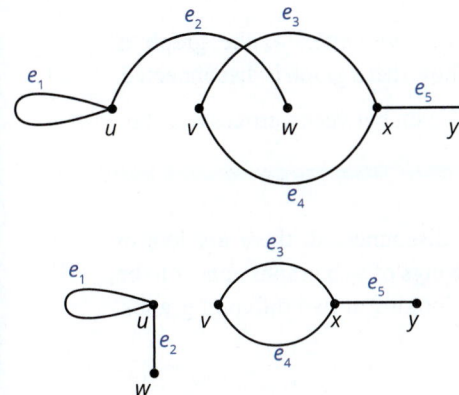

Figure 14.1.19:
Two Different Views of Graph G

CONNECTED AND DISCONNECTED GRAPHS

A graph is **connected** if there is at least one path connecting each pair of distinct vertices.

A graph is **disconnected** if it has at least one pair of vertices that is not connected by a path.

Example 14.1.5

Identifying Connected and Disconnected Graphs

Label each graph as connected or disconnected. Explain your answer.

a.

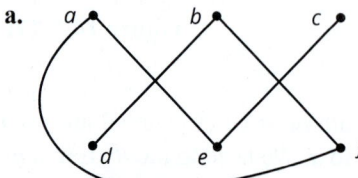

Figure 14.1.20: Graph *A*

b.

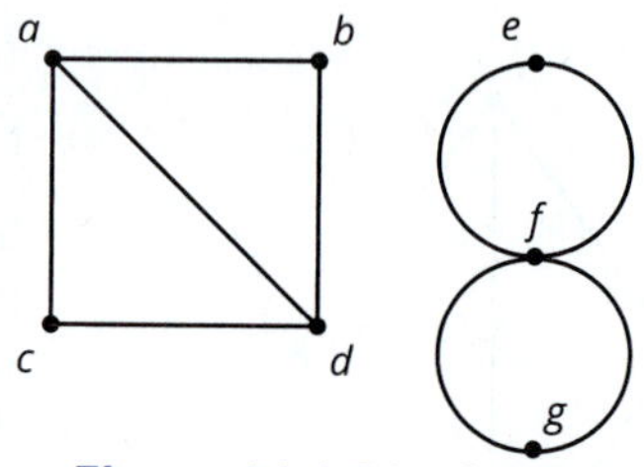

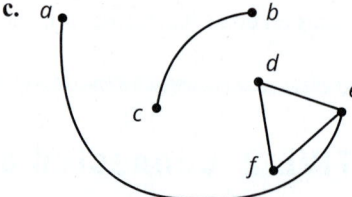

Figure 14.1.21: Graph *B*

c.

Figure 14.1.22: Graph *C*

Solution

a. Even though each vertex is not directly connected with every other vertex by single edges, this graph is connected because there are paths between each pair of vertices.

b. Because there is no path connecting vertex *b* to vertex *e*, the graph is disconnected. Only one example is needed to show that a graph is disconnected.

c. This graph is disconnected because there is no path between vertices *a* and *c*.

Regardless of whether a graph is connected or disconnected, there are lots of ways in which the graph can be drawn. If the drawings of two graphs appear to be different, it doesn't necessarily mean that we are looking at two different graphs. Example 14.1.6 illustrates this point.

Example 14.1.6

Identifying Different Views of the Same Graph

Decide whether the pairs of graphs shown could be different views of the same graph or not. Explain your answer.

a.

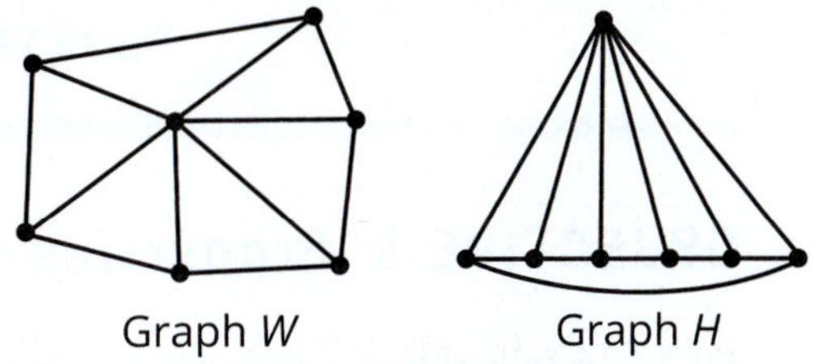

Graph *W*   Graph *H*

Figure 14.1.23

b.

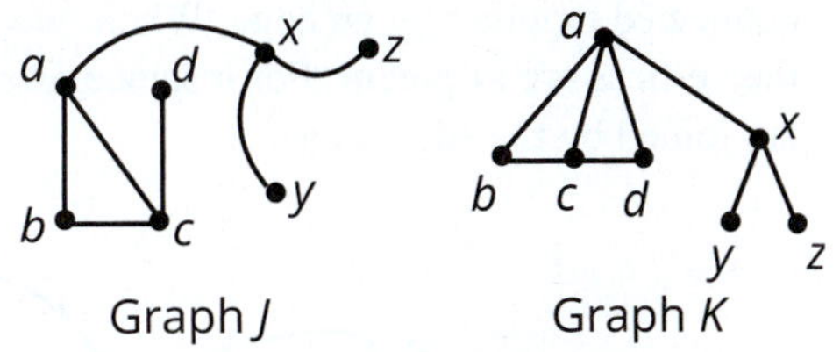

Graph *J*   Graph *K*

Figure 14.1.24

c.

Graph *M*   Graph *N*

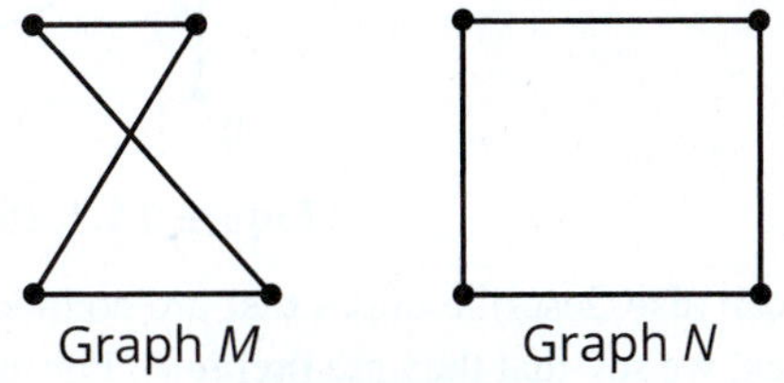

Figure 14.1.25

Solution

a. In both graphs *W* and *H*, there are seven vertices. If you think of numbering the vertices 1 through 7, you can check that the edges join the same vertices in both diagrams. Check this for yourself in the following figure. Therefore, these are different views of the same graph.

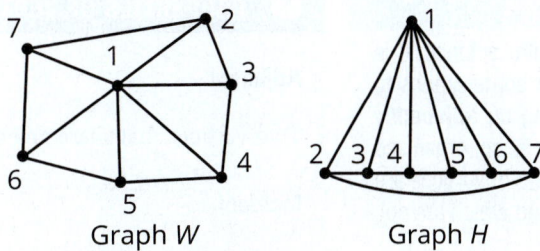

Graph *W*   Graph *H*

Figure 14.1.26

b. Graphs *J* and *K* are not the same. Graph *J* does not include an edge between the vertices *a* and *d*; however, graph *K* does. Therefore, graphs *J* and *K* cannot be different views of the same graph.

c. Both graphs *M* and *N* contain four vertices. If you label the vertices of each graph carefully, you can see that the edge sets of Graph *M* and Graph *N* are identical. Therefore, these are different views of the same graph.

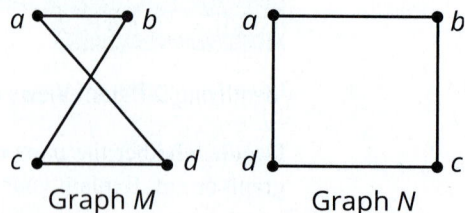

Graph *M*   Graph *N*

Figure 14.1.27

OBJECTIVE 3 Properties of Edges and Vertices

Figure 14.1.28 shows a diagram of graph *G*, which we have been examining throughout the section. We can describe the placement of vertices that are connected together by an edge. When two vertices are joined by an edge, we say the vertices are **adjacent**. For instance, vertices *x* and *y* are adjacent because they are joined by the edge e_5.

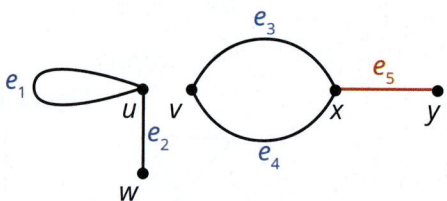

Figure 14.1.28: Graph *G*

We can also describe edges that are connected together. When two edges share a vertex, we say that they are **incident**. For instance, in Figure 14.1.28, edges e_1 and e_2 are incident because they share the vertex *u*.

The **degree** of a vertex is the number of edges that are incident to that vertex. We denote the degree of the vertex by *d*(vertex). For instance, in graph *G*, the degree of vertex *v* is $d(v) = 2$ and the degree of vertex *y* is $d(y) = 1$. A loop contributes 2 to the degree count. In graph *G*, vertex *u* has two edges that are incident to it, and one of them is a loop. Therefore, the degree of *u* is three; that is, $d(u) = 3$.

> ✔ **Skill Check 14.1.2**
>
> Identify all pairs of adjacent vertices and all pairs of incident edges in Figure 14.1.28.

> 💬 **Fun Fact**
>
> In the field of computer science, it is particularly useful for some graphs to be described by using an **adjacency matrix**. This matrix is an organized list to show which pairs of vertices are connected together and which are not.

> 📖 **PROPERTIES OF VERTICES AND EDGES**
>
> **Adjacent**
>
> Two vertices that share an edge are **adjacent**.
>
> **Incident**
>
> Two edges that share a vertex are **incident** to one another. Also, a vertex is **incident** to the edges that have that vertex as an endpoint.
>
> **Degree**
>
> The **degree** of a vertex denoted *d*(*u*), is the number of edges that are incident to *u*.

Example 14.1.7

Identifying Parts of a Graph

Use Graph A to solve the following problems.

a. Determine the degree of each of the vertices.

b. Name the pairs of vertices that are adjacent.

c. Name the pairs of edges that are incident.

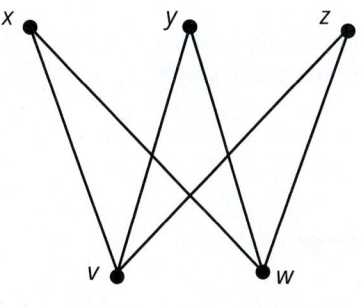

Figure 14.1.29: Graph A

Solution

a. To determine the degree of each vertex, we simply count the number of edges incident to that vertex. Therefore,

$$d(x) = 2; \; d(y) = 2; \; d(z) = 2; \; d(v) = 3; \text{ and } d(w) = 3.$$

b. Two vertices are adjacent if they are ends of the same edge. Therefore, all the following pairs of vertices are adjacent.

x, v	y, w
x, w	z, v
y, v	z, w

c. Two edges are incident if they share a common vertex. Therefore, all of the following sets of edges are incident.

xv, xw	xv, yv, zv
yv, yw	xw, yw, zw
zv, zw	

As Euler demonstrated with the bridges of Königsberg, it is often helpful to visually represent a graph given certain features or constraints. Sometimes this will involve finding the right visual representation, or drawing, of the given vertices and edges. Sometimes this will involve creating a graph that has a required structure. The next two examples demonstrate both of these possibilities.

Example 14.1.8

Drawing a Graph Given the Edges

a. Draw a five point star; that is, draw a graph with five vertices, v_1, v_2, v_3, v_4, and v_5, whose edges are $v_1 v_3$, $v_3 v_5$, $v_5 v_2$, $v_2 v_4$, and $v_4 v_1$.

b. Determine the degree of each vertex.

c. Redraw the graph without edges crossing.

d. Determine the degree of each vertex in the new drawing.

Solution

a. Begin by placing the five vertices on the page and labeling them v_1 through v_5, as shown in Figure 14.1.30. Remember, the placement of the vertices is not important.

Now, connect the vertices with the given edges. For example, edge v_1v_3 connects vertex v_1 with vertex v_3. Figure 14.1.31 shows the completed graph.

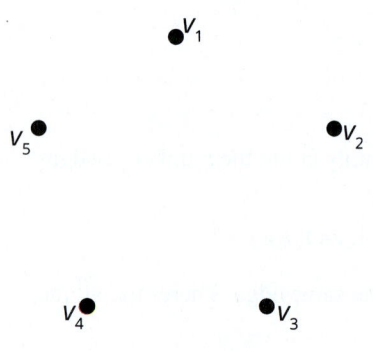

Figure 14.1.30

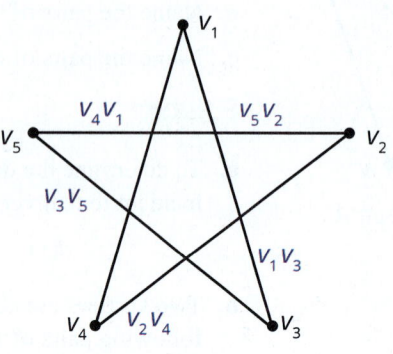

Figure 14.1.31

b. To determine the degree of each vertex, count the number of edges incident to each vertex.

$$d(v_1) = 2; \; d(v_2) = 2; \; d(v_3) = 2; \; d(v_4) = 2; \text{ and } d(v_5) = 2$$

c. To redraw the graph without edges crossing, think of "pulling" edge v_1v_3 to the outside of the graph, as shown in Figure 14.1.32.

Next, do the same with edge v_3v_5.

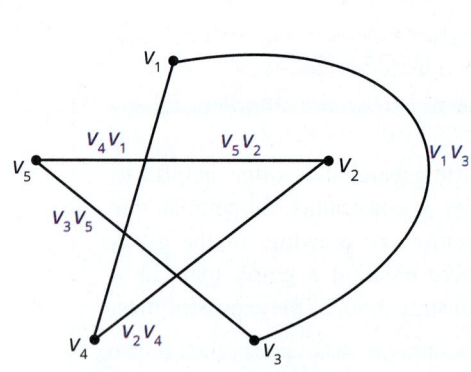

Figure 14.1.32

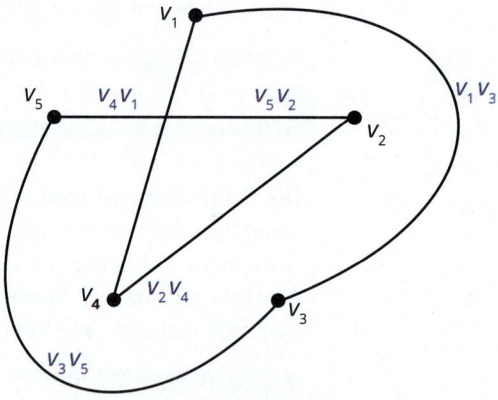

Figure 14.1.33

The last thing to do is to pull vertices v_2 and v_4 out of the middle. When you do, think of flipping their position, as shown in the following graph. Check for yourself that the edges are all still the same in this new layout.

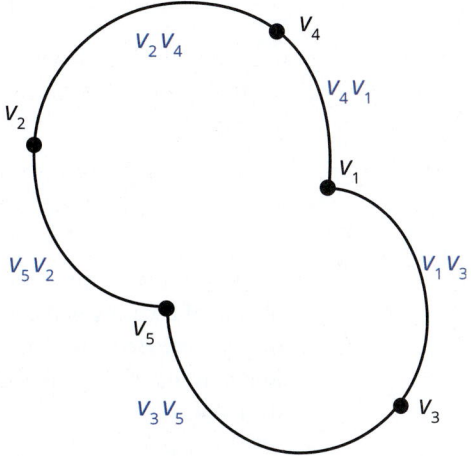

Figure 14.1.34

The graph is now drawn without any edges crossing. You can see that the "star" now looks round.

d. The degree of each vertex does not change when the graph is redrawn. Each vertex still has the same edges connected to it and, hence, the same degree. (This will always be the case with any graph.) Therefore, all vertices in this graph still have degree 2.

Example 14.1.9

Drawing a Graph without Edges Predetermined

A research chemist is analyzing the structure of compounds of aluminum, iron, and oxygen. Each aluminum atom makes three chemical bonds, each oxygen atom makes two chemical bonds, and each iron atom makes six chemical bonds. Create a possible form for a molecule of two aluminum atoms, two oxygen atoms, and one iron atom where no atom makes a chemical bond to itself.

Solution

To create the structure of the molecule, we will represent the atoms and their bonds as a graph. Each vertex will represent an atom and each edge will represent a chemical bond that joins one atom with another. So we have to draw a graph with no loops that has two vertices of degree 3, two vertices of degree 2, and one of degree 6.

Unlike the previous example, there are many graphs that meet the requirements given since there are no restrictions on which vertices must be joined together by edges. We will show one possible form for a molecule of this type.

First notice that there must be a total of five vertices, one for each atom. Place the vertices on the graph.

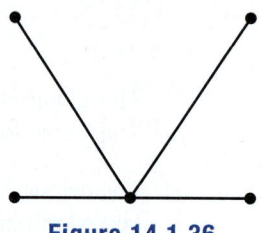

Figure 14.1.35

Let's begin with the vertex with the highest degree of degree 6. Recall that the degree of a vertex is the number of edges that are connected to it, not the number of vertices. Because there are no loops allowed, the vertex must have six edges joined to it. We'll join each of the other four vertices to it, giving us a vertex of degree 4.

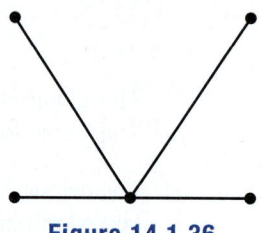

Figure 14.1.36

To get to degree 6, we'll need to draw two more edges somewhere. We'll choose to give the bottom two vertices on either side another edges with our central vertex, as shown.

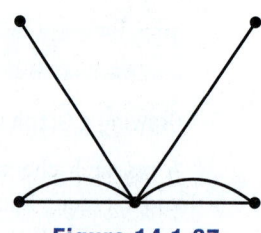

Figure 14.1.37

Let's stop and determine what the graph looks like so far. We have a graph with five vertices: one of degree 6, two with degree 2, and two with degree 1. Recall that we need two vertices of degree 3, two vertices of degree 2, and one of degree 6.

Notice that we have more than one possible ways to complete our graph here. One way is to join each vertex of degree 1 to a vertex of degree 2, which increases each of their degrees by one, giving us the desired degree count. Our final graph for the molecule's form is as follows.

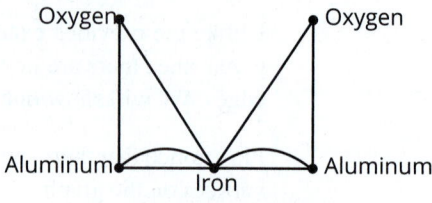

Figure 14.1.38

✔ **Skill Check 14.1.3**

Draw a different graph with the same restrictions as in Example 9: no loops, having two vertices of degree 3, two vertices of degree 2, and one of degree 6.

OBJECTIVE 4 Vertex Coverings

Graphs can also be used to optimize resources in real-life situations. For instance, suppose you need to hire security guards to protect precious art work in a museum. You'd like to have enough guards so that every room in your museum can be seen by at least one guard at all times. At the same time, you want to be prudent with your money and not have an excess of guards. You need to establish how many security guards you need to hire and where they should be deployed. Or suppose you need to place Wi-Fi hotspots all over a building. Once again you would like to minimize the cost of covering the building with wireless internet and, hence, both their number and placement are critical.

Both of the scenarios described can be modeled with graphs to help find optimal solutions. To do this, we need to introduce a new graph structure—a **minimum vertex cover**. First, a **vertex cover** is a group of vertices with the property that every vertex in the graph is either in the group or adjacent to at least one of the vertices in the group. We can optimize the covering by choosing the smallest group of vertices that will cover the graph. This we call a minimum vertex cover.

📖 VERTEX COVERINGS

Vertex Cover

A **vertex cover** is a set of vertices A so that every vertex in the graph is either in A or adjacent to a vertex in A.

Minimum Vertex Cover

When a vertex cover A is as small as possible, A is called a **minimum vertex cover**.

To see how a vertex cover can be useful, let's consider the problem involving the museum guards. Remember that we want to hire the fewest (minimum) number of guards that will be able to guard all the rooms at once. To begin, we need to know the layout of the museum. Suppose we have a blueprint for the museum building, as shown in Figure 14.1.39.

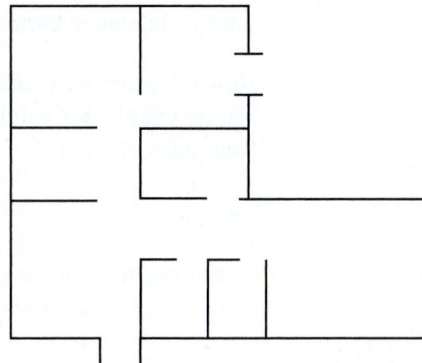

Figure 14.1.39: Musuem Blueprint

The first step is to begin by representing the blueprint with a graph. Let each room be represented with a vertex. If the rooms share a door, then the vertices are adjacent. In other words, they will have an edge connecting them. Since there are ten rooms, the graph will have ten vertices. Remember that the position of the vertices makes no difference; but to make things easier, we will layout the vertices in the same manner as the blueprint.

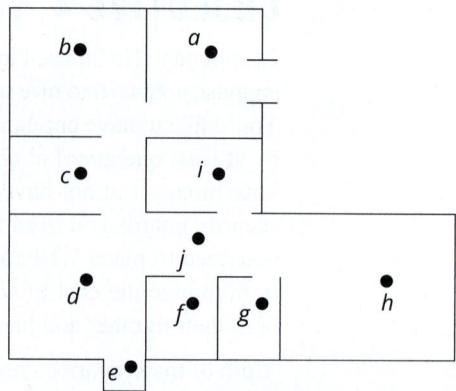

Figure 14.1.40: Musuem Blueprint Graph with Vertices

Using the blueprint as a guide, join the vertices together with an edge if they share a doorway, as shown in Figure 14.1.41.

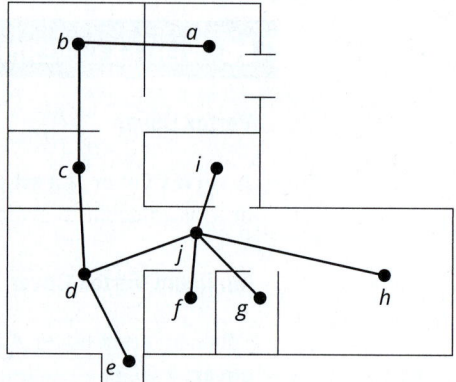

Figure 14.1.41: Museum Blueprint Graph with Vertices and Edges

Now that the building is represented by a graph, we can find a solution to the guard problem.

Example 14.1.10

Finding a Minimum Vertex Cover

Using the graph we created in Figure 14.1.41, determine the minimum number of security guards that would be needed if a guard can secure their room and all the rooms adjacent to it.

Solution

We can begin by placing a guard in room j, represented by a blue G. That way, rooms d, f, g, h, and i are all covered since they are adjacent to j.

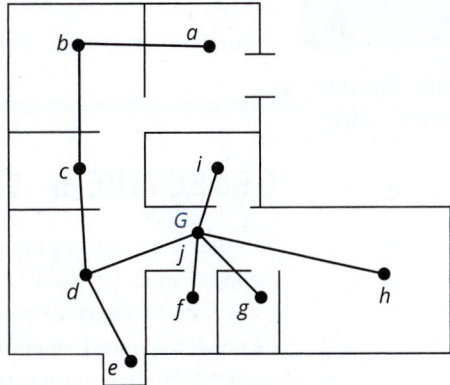

Figure 14.1.42

Then, we have the remaining four rooms *a*, *b*, *c*, and *e* that need a guard. Since these four rooms are connected in a row, this will require a minimum of two guards. One guard in room *b* who can cover room *b* as well as *a* and *c*, and one guard in either room *d* or *e*.

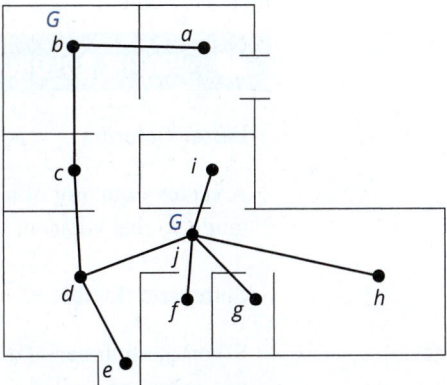

Figure 14.1.43

Here's where we can use our best judgment for the scenario. Although room *d* is secured by the guard in room *j*, the guard in room *j* might be a little overworked with all the rooms assigned to his position. We could technically place the final guard in room *e* and accomplish our objective. However, placing the guard in room *d* might be a more sensible solution, helping to ease the job of the guard in room *j*. Unless, of course, room *e* contains an incredibly valuable jewel. Either way, the minimum number of guards needed to cover every room in the museum is three.

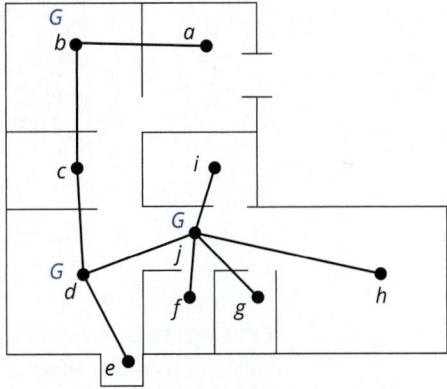

Figure 14.1.44

✔ **Skill Check 14.1.4**

Find another minimum vertex covering for the museum graph from Example 14.1.10.

The set of vertices that form the guards' post assignments, $A = \{b, d, j\}$, is a minimum vertex cover for the museum graph.

OBJECTIVE 5   Chromatic Numbers

Another way to use graphs to optimize resources is by categorizing the vertices. Suppose that we want to give cellular towers a frequency band so that nearby towers don't interfere with one another. How many different frequencies would be needed to cover an entire area? Or, suppose we want to paint a geometric mural such that no two bordering shapes have the same paint color. How many different paints would need to be purchased?

In both of these cases, a vertex coloring can be used to optimize the solutions. A **vertex coloring** of a graph is an assignment of colors to the vertices in such a way that no two adjacent vertices have the same color. In mathematics, the "colors" for vertex coloring might be colors, letters, or numbers, depending on which is most convenient.

> 📖 **VERTEX COLORINGS**
>
> **Vertex Coloring**
>
> A **vertex coloring** of a graph is an assignment of colors to the vertices of the graph so that adjacent vertices receive different colors.
>
> **Chromatic Number**
>
> When the number of colors used in a vertex coloring is as small as possible, this number is called the **chromatic number** of a graph and is denoted $\chi(G)$, read "chi of G."

The field of designing maps, called cartography, illustrates how vertex colorings can be utilized. Consider Figure 14.1.45, which shows a map of the eleven counties in a state.

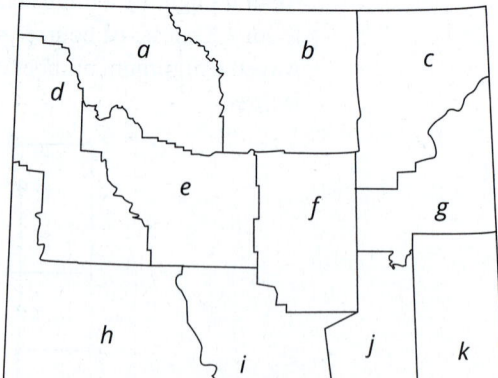

Figure 14.1.45: State Counties

Let's represent the map using a graph where the counties are vertices and an edge exists if counties share a border. Once again, begin by representing each of the eleven counties with a vertex, as shown in Figure 14.1.46.

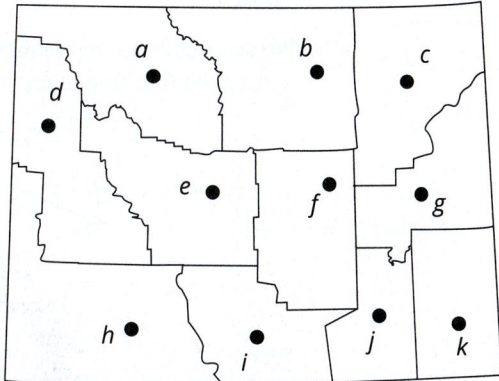

Figure 14.1.46: State Counties Graph with Vertices

Joining each neighboring county with an edge if they share a border gives us the following graph. Check this for yourself.

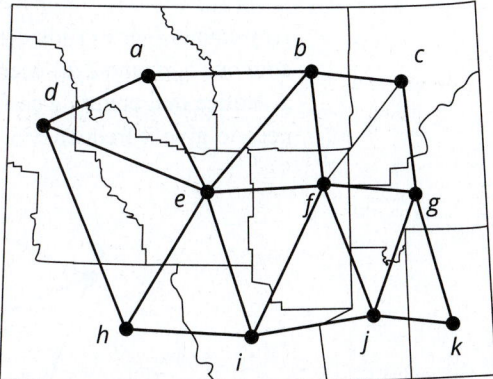

Figure 14.1.47: State Counties Graph with Vertices and Edges

Now that the counties are represented by a graph, we can use the graph to find a vertex coloring. Example 14.1.11 looks at how to find a vertex coloring and the chromatic number of the graph.

Example 14.1.11

Finding a Vertex Coloring

A wireless communications service provider needs to place cell towers in all counties of the state shown in Figure 14.1.47 in such a way that the frequencies for each tower do not interfere with each other. A vertex in the graph represents a county and an edge represents a shared border between counties.

When two cell towers are placed in adjacent counties, they interfere with each other's signal unless they are on different frequencies. The provider could place a different frequency in each county and have 11 frequencies for the state. However, to minimize their costs, the provider needs to use as few frequencies as possible. Find the chromatic number of the graph and determine the best arrangement of the frequencies of the cell towers for the provider.

Solution

We can represent the frequencies of a cell tower by using a number. Let's begin by putting the first frequency in county *a*, as shown with the number 1 on the graph.

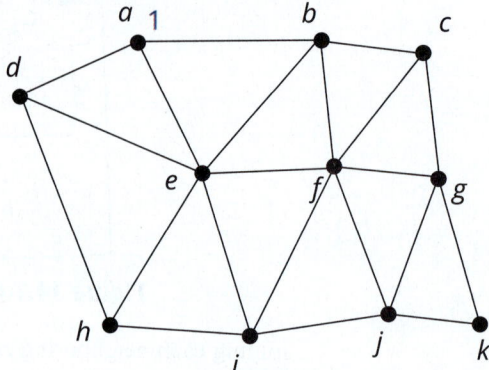

Figure 14.1.48

Any county (vertex) adjacent to county *a* is not allowed to use frequency 1. So, the counties *b*, *d*, and *e* all need new frequencies. Start by giving county *b* frequency 2. Notice that county *e* can't be on 1 or 2 since it connects to both *a* and *b*, so we need to give it frequency 3. Here's how the graph looks so far.

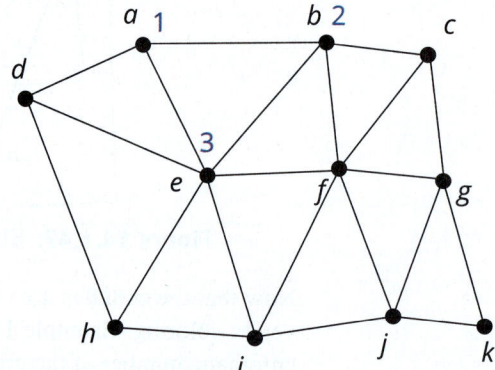

Figure 14.1.49

Returning to county *d*, we know that it cannot be the same as either of the adjoining counties *a* and *e*, which have frequencies 1 and 3 respectively. Since frequency 2 is available, we'll use it again, as we are trying to use as few frequencies as possible.

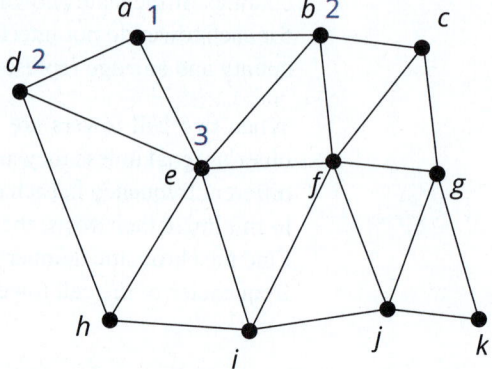

Figure 14.1.50

Continue in the same manner until all vertices are numbered.

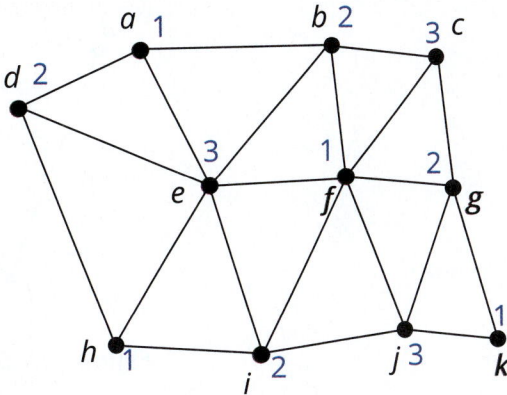

Figure 14.1.51

You can see that we needed three different frequencies to cover all the counties without interference. Therefore, the county graph has a chromatic number of 3; that is, $\chi(G) = 3$.

Think about the graph created in Figure 14.1.41 for the placement of museum guards and the graph created in Figure 14.1.47 for the placement of cell phone towers. Both graphs are shown again in Figure 14.1.52.

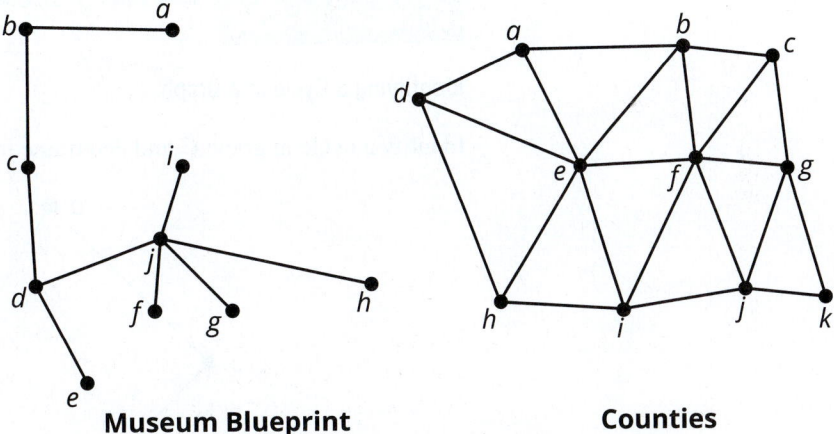

Museum Blueprint   **Counties**

Figure 14.1.52: Graphs for Museum Blueprint and State Counties

Although both graphs describe the structure of their underlying scenarios, the blueprint graph on the left is a little simpler than the county graph on the right. The difference is that the county graph has lots of interlocking **cycles**, which are walks that start and end at the same vertex. For instance, among the many cycles in the state counties graph are a, d, e, shown in red and c, f, e, h, i, j, k, g shown in blue in Figure 14.1.53.

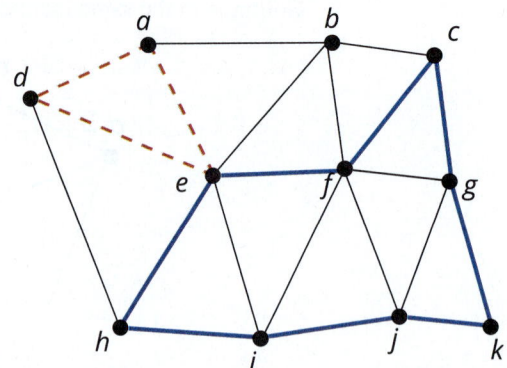

Figure 14.1.53: Two Cycles in the Counties Graph

Notice that although a cycle starts and ends at the same vertex, we do not list it twice when referring to the cycle.

> 📖 **CYCLE**
>
> A **cycle** is a walk that starts and ends at the same vertex and has no edges or vertices repeated except for the starting vertex.

The length of a cycle is the number of edges or number of vertices in the cycle. For instance, the cycle with dashed lines (a, d, e) in Figure 14.1.53 is a cycle of length 3 and the blue cycle (c, f, e, h, i, j, k, g) is a cycle of length 8. Notice that in a cycle, the number of edges is the same as the number of vertices.

Example 14.1.12

Identifying a Cycle in a Graph

Identify a cycle in graph G and determine its length.

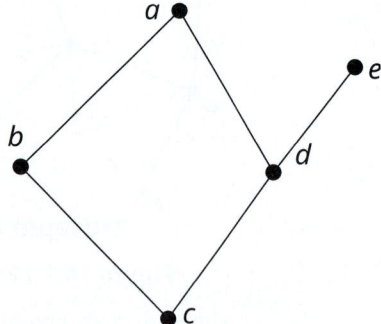

Figure 14.1.54: Graph G

Solution

The only cycle in graph G is the cycle among the vertices a, b, c, and d. To refer to the cycle we can begin at any of the vertices and list them in order as we trace the cycle. For instance, beginning with vertex b, the cycle is b, c, d, a. Beginning with the vertex a, the same cycle is a, d, c, b. Changing the direction or the starting vertex does not create a different cycle. Because the cycle has four vertices and four edges, the cycle has length 4.

✔ **Skill Check 14.1.5**

Identify a cycle of length 6 in the following graph.

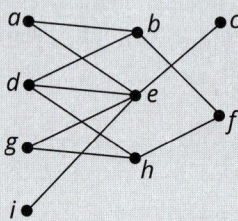

Figure 14.1.55

Skill Check Answers

1. Any of the following: c, b, a, e; c, b, a, d; c, b, a, b; c, b, c, b; c, d, a, e; c, d, a, b; c, d, a, d; c, d, c, d 2. Adjacent vertices: x and y; u and w; v and x; Incident edges: e_1 and e_2; e_3 and e_4; e_3 and e_5; e_4 and e_5

3. Answers will vary. For example,

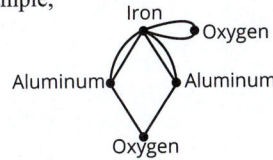

4. $B = \{b, e, j\}$

5. Answers will vary.
   For example, a, b, d, h, g, e; g, e, a, b, f, h; b, f, h, d, e, a; d, e, g, h, f, b..

14.1 Exercises

✔ CONCEPT CHECK

1. The _____ of a vertex is the number of edges that are incident to that vertex.

2. A/an _____ consists of a set of vertices and a set of edges that connect pairs of vertices.

3. A/an _____ associates a pair of vertices in a graph.

4. The end of an edge in a graph is called a/an _____.

5. A/an _____ is an edge whose ends are the same vertex.

6. Two edges are said to be _____ if they share a vertex.

7. The _____ of a graph is the minimum number of colors needed in a vertex coloring of a graph.

8. Two vertices are said to be _____ if they share an edge.

9. A sequence of vertices such that each vertex is adjacent to the next in the list, the last vertex is adjacent to the first, and there are no repeated edges or vertices, except for the starting vertex, is called a/an _____.

💡 PRACTICE

List the vertices and edges of each graph.

10.

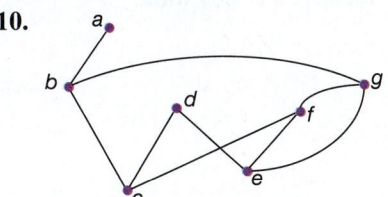

11.

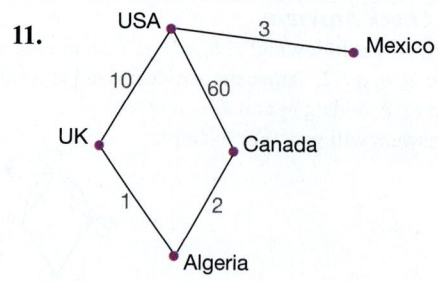

12.

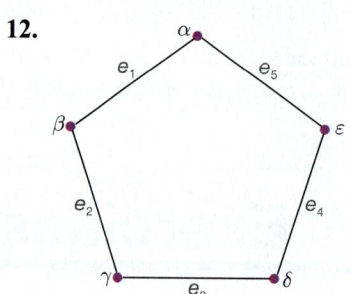

13.

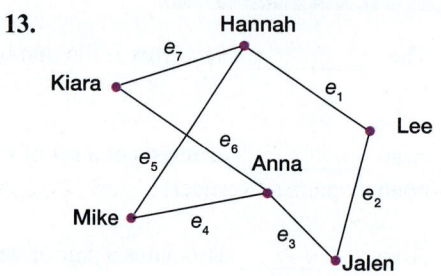

14.

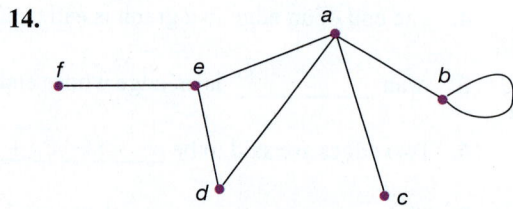

15. Use the graph to answer the following.

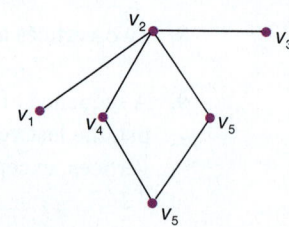

    a. Identify a walk of length two that starts at vertex v_1 and ends at vertex v_3.

    b. Is the walk in part a. closed?

    c. Identify a closed walk.

16. Use the graph to answer the following.

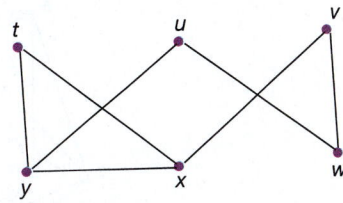

    a. Identify a walk of length 3 that starts at vertex u and ends at vertex v.

    b. Identify a walk of length 5 that starts at vertex t and ends at vertex t.

    c. Are the walks in parts a. and b. closed?

17. An **Euler trail** in a graph G is a walk that contains every edge in G precisely once. Each of the following graphs contains an Euler trail, identify it.

    a. 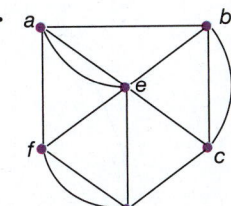   **b.**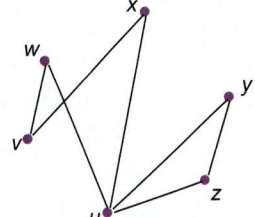

18. Identify two paths in the following graph. What is the length of each path?

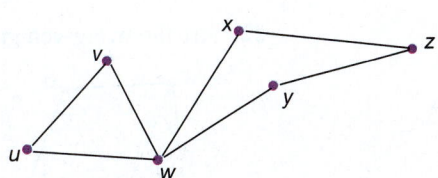

19. The length of a path is the number of edges in the path. Identify a path of the specified length in the graph.

    a. Path of length 2

    b. Path of length 3

    c. Path of length 6

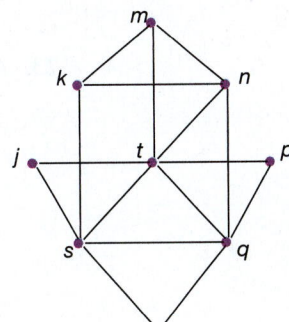

20. Which of the given graphs is connected?

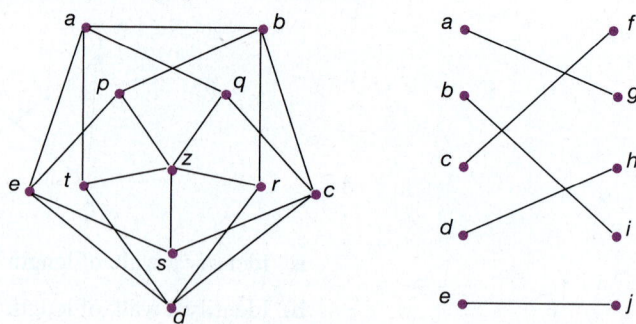

21. Which of the given graphs is connected?

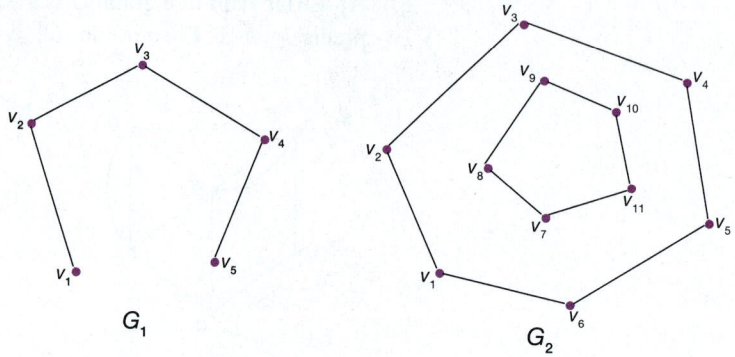

22. Are the two given graphs the same graph? Explain.

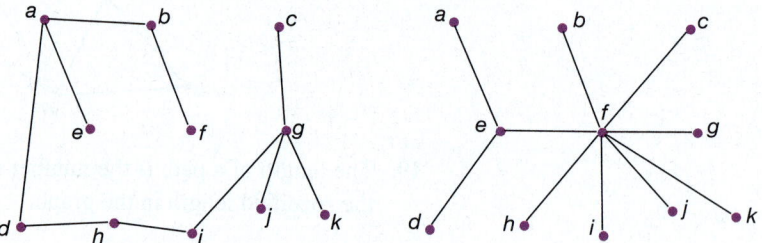

23. Are the two given graphs the same? Explain.

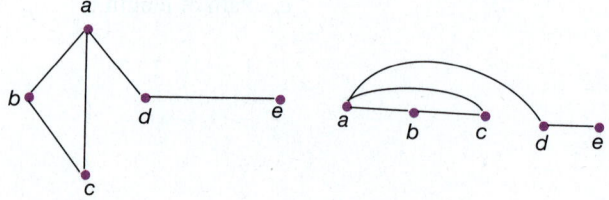

24. Draw a graph that has five vertices r, s, t, u, v, and the edges rt, ru, rv, st, su, and sv.

25. Draw a graph with vertices u, v, w, x, y, z, and edges ux, uv, uy, vx, vy, vw, wy, wz, wx, with no crossings.

26. Draw a graph that has 7 vertices a, b, c, d, e, f, g, and edges ab, ac, ae, bc, cd, de, ef, fg, and gg.

27. Draw a graph with vertices α, β, π, σ, ω, and edges $\alpha\beta$, $\beta\pi$, $\pi\sigma$, $\sigma\omega$, $\alpha\omega$, $\beta\omega$, and $\alpha\sigma$, with no edges crossing.

28. Find a vertex coloring of the graph. What is the chromatic number of that graph?

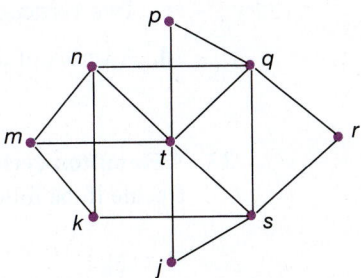

29. Identify a cycle in the following graph and determine its length.

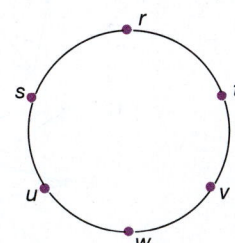

30. Identify a cycle of the required length in the graph.

    a. Cycle of length 4

    b. Cycle of length 6

    c. Cycle of length 8

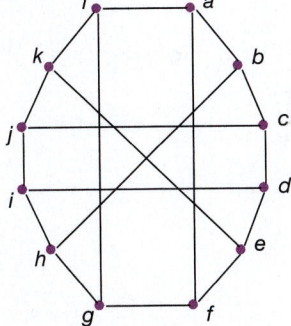

31. Identify the following features in the graph. In some cases, more than one answer is correct.

    a. An edge

    b. Two adjacent vertices

    c. Two incident edges

    d. Two vertices that are not adjacent

    e. A vertex of degree 4

    f. A cycle of length 5

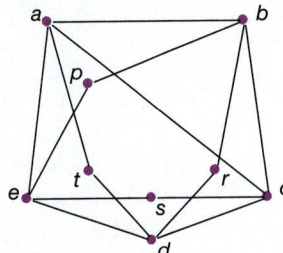

32. Identify the following features in the given graph. In some cases, more than one answer is correct.

    a. Two adjacent vertices

    b. Two incident edges

    c. Two vertices that are not adjacent

    d. A vertex of degree 2

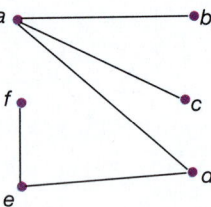

33. A **Hamilton cycle** in a graph G is a cycle that contains every vertex in G. Decide if the following graphs contain Hamilton cycles.

    a.

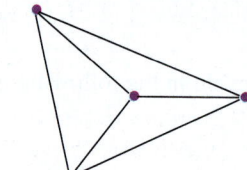

    b.

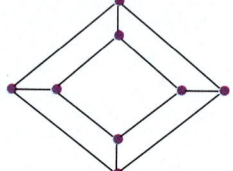

    c.

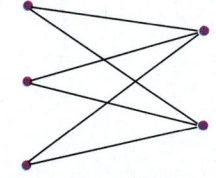

🚀 **APPLICATIONS**

34. The graph shows a blueprint of rooms in a museum. Find a way to place guards in the rooms so that every room either has a guard or shares a doorway with a room that has a guard and the minimum number of guards is used.

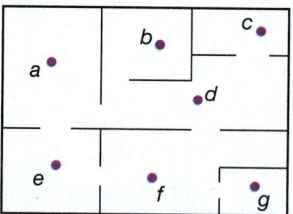

35. The graph shows a blueprint of rooms in a house. Find a way to place security cameras so that every room either has a camera or shares a doorway with a room that has a camera, and the minimum number of cameras is used.

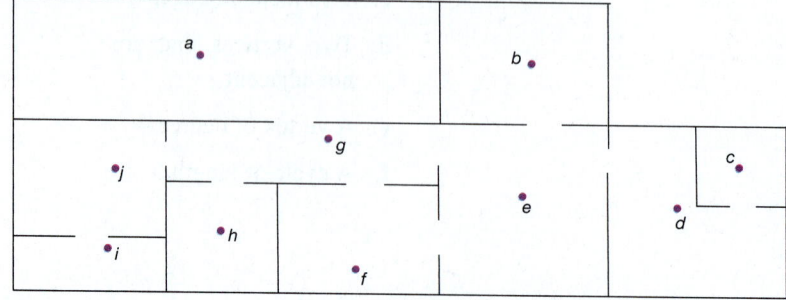

36. Use a vertex coloring to assign colors to the counties of Vermont so that adjacent counties have different colors. What is the minimum number of colors required? Find a vertex coloring using the minimum number of colors.

37. Use a vertex coloring to assign colors to the counties of Connecticut so that adjacent counties have different colors. What is the minimum number of colors required? Find a vertex coloring using the minimum number of colors.

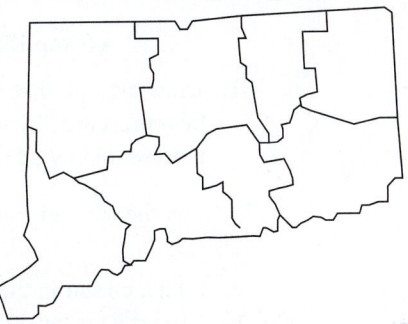

✎ WRITING & THINKING

38. Draw two different connected graphs, each with four vertices and four edges.

14.1 PROJECT

THE KÖNIGSBERG BRIDGES

In Section 14.1, you learned about the famous Königsberg bridge problem that asks whether it is possible to start at one of the land masses in the city and cross every bridge exactly once. Let's investigate a slight modification of the problem: is it possible to start your journey on one of the land masses, cross every bridge exactly once, and return to the *original* land mass?

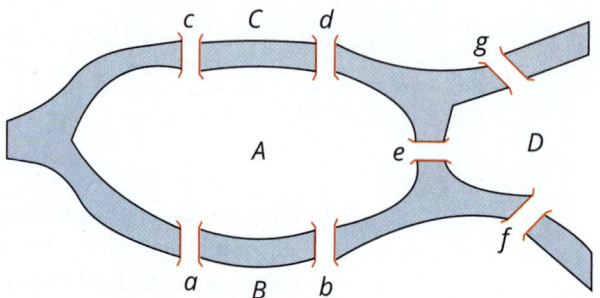

A Simplified Diagram of Königsberg and Its Bridges

1. Draw a graph that models the situation. The land masses A, B, C, and D will be represented by vertices while the bridges a, b, c, d, e, f, and g will be represented by edges. Determine the degree of each vertex.

In graph theory, a **circuit** is a walk that starts and ends at the same vertex.

2. Find a circuit in the graph that was created in part 1.

We define an **Euler circuit** as a circuit that uses each edge exactly once. In other words, an Euler circuit starts at a vertex, uses every edge exactly once, and then returns to the same vertex.

3. Rephrase our version of the Königsberg bridge problem in terms of Euler circuits.

4. How many Euler circuits are in the graph?

Notice that in order to produce an Euler circuit, you must enter a vertex using one edge and leave that vertex using a different edge—that is, the edges that meet at a vertex must come in pairs. This indicates that a connected graph will have an Euler circuit when all of its vertices have even degrees.

5. Use this fact to justify why the modified Königsberg bridge problem does or does not have a solution.

6. Draw a graph that represents 4 land masses and 7 bridges that would have an Euler circuit and explain your answer.

7. Could you remove one of the Königsberg bridges to create an Euler circuit? If yes, which bridge should be removed? If not, explain why.

14.2 TREES

OBJECTIVES

1. Identify trees.

2. Find a spanning tree in a graph.

3. Determine the number of leaves on a tree.

OBJECTIVE 1 Introduction to Trees

Some of the simplest graphs that we can consider have no cycles. Graphs without cycles prove to be convenient models for many real-life scenarios: family trees, finding shortest routes between cities, maximizing flow of pipes in a network, or minimizing cost for laying cable between states. All of these instances can be modeled using a type of graph called a **tree**.

📖 TREE

A connected graph with no cycles is called a **tree**.

Remember that a connected graph is one in which there is a path between any two given vertices. In a tree, however, because there are no cycles, there is only one distinct path between any two vertices.

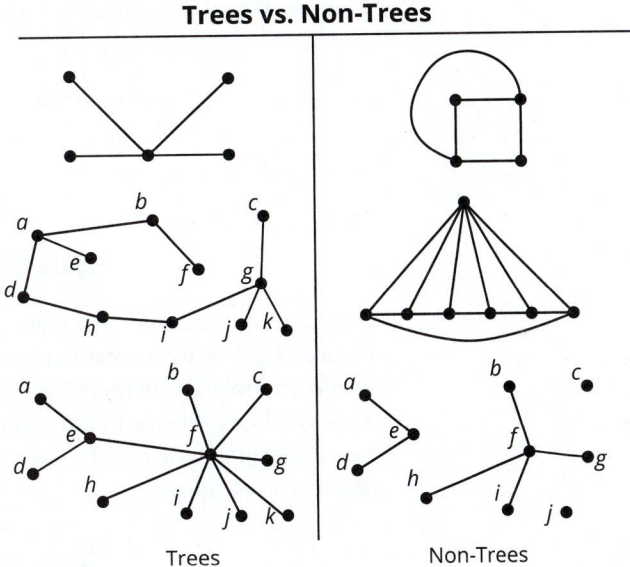

Figure 14.2.1

Example 14.2.1

Identifying Trees

Determine if each of the following graphs is a tree.

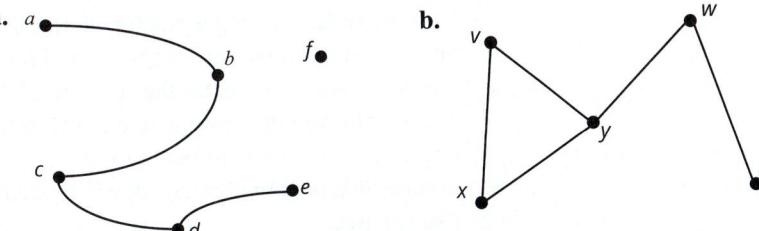

a.

b.

Figure 14.2.2: Graph G_1 **Figure 14.2.3:** Graph G_2

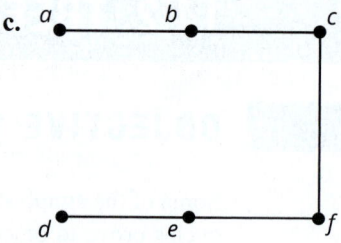

Figure 14.2.4: Graph G_3

Solution

a. Although G_1 does not have a cycle in it, it is not a connected graph because there is not a path between vertex f and other vertices. Therefore, it is not a tree.

b. G_2 is not a tree because there is a cycle among the vertices v, x, and y.

c. G_3 is a tree because it is a connected graph without any cycles.

Consider the set of vertices in Figure 14.2.5.

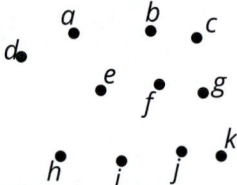

Figure 14.2.5: Set of Vertices

Suppose you are told to create a connected graph using the vertices from Figure 14.2.5 with no restrictions on the number of edges in the graph. How would you join the vertices? As you can imagine, there are an infinite number of possibilities. Figure 14.2.6 shows diagrams of two graphs, both of which are connected graphs using the vertices. They are just two of the many possible graphs you could create.

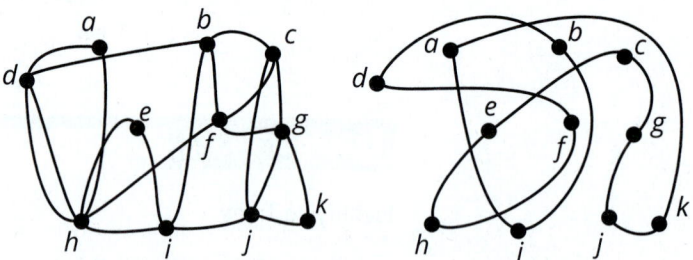

Figure 14.2.6: Two Possible Connected Graphs

Now, consider creating a graph with the same vertices, still having no restrictions on the number of edges, but with the knowledge that each edge you draw costs you $10. Would you draw the same graph? If the goal is to spend as little money as possible, you certainly would never join a pair of vertices with more than one edge. Most of us would begin to think of a way to find a connected graph of the vertices that uses the fewest edges. In other words, we would look to find a graph that is a tree.

Figure 14.2.7 shows two of the many connected graphs that could be drawn with the fewest number of edges; that is, they are trees. Notice that each of these would cost you the same amount of money to draw—$100.

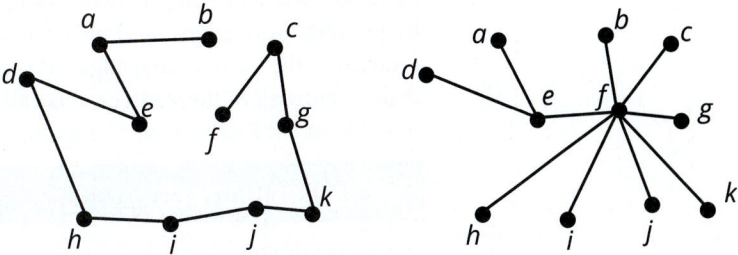

Figure 14.2.7: Two Possible Trees

Because both graphs are trees, neither one contains a cycle. If there were a cycle, any one of the edges of the cycle could be removed without disconnecting the graph. Therefore, a tree connects vertices with the fewest possible edges. In fact, the number of edges in a tree is always one fewer than the number of vertices. The following theorem states this principle.

🔑 TREE THEOREM

Let graph T be a tree on v vertices. Then graph T has $v - 1$ edges.

Example 14.2.2

Determining the Number of Edges Needed for a Tree

Determine the number of edges needed to connect the given vertices so that the resulting graph is a tree.

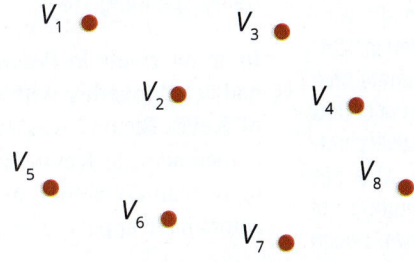

Figure 14.2.8

Solution

To determine the number of edges needed to construct a tree from the given vertices, we simply need to subtract 1 from the number of vertices. Since there are 8 vertices, the number of edges needed is $8 - 1 = 7$.

✔ **Skill Check 14.2.1**

Construct two different trees with the vertices in Example 14.2.2.

OBJECTIVE 2 Spanning Trees

By modeling situations with graphs, we can use trees to find ways to maximize or minimize efficiency in a network. We can create a tree within an existing graph by pruning away edges until only a tree remains. This tree is a *subgraph*; that is, a subset of the vertices and edges of the original graph. When a subgraph is a tree that contains all of the vertices, it is called a **spanning tree**. For many graphs, it is possible to have more than one spanning tree.

> **SPANNING TREE**
>
> A **spanning tree** is a subgraph of a connected graph, which is itself connected and contains all the vertices of the original graph, but has no cycles.

The key to a graph being a tree is that a tree doesn't contain any cycles. So, to find a spanning tree in a graph, one needs to identify the cycles in the graph and delete enough edges to break the cycles while still keeping the graph connected. The following process gives a method for finding a spanning tree.

> **STEPS FOR CONSTRUCTING A SPANNING TREE**
>
> 1. In a connected graph G, identify a cycle. If more than one cycle exists, choose one at random.
>
> 2. Choose an edge from the cycle selected and delete it from the graph.
>
> 3. While the graph contains a cycle, repeat Steps 1 and 2.

Example 14.2.3

Using Spanning Trees

In an interview in *Premier Magazine* in 1994, Kevin Bacon commented that he had acted together with almost everyone in Hollywood. The game "Six Degrees of Kevin Bacon" was invented based on this comment. The goal is to link any named actor to Kevin Bacon by beginning with the named actor and naming no more than six actors, ending with Kevin Bacon, where each successive pair of actors have appeared in a movie together.

Find a spanning tree for the following graph, in which the vertices are actors, and two actors are joined by an edge if they appeared together in a movie, where the edges are labeled with the movie.

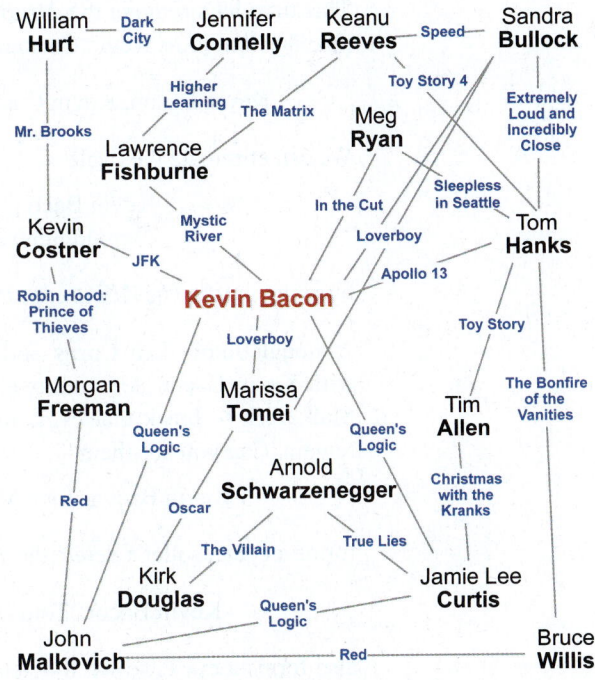

Figure 14.2.9

Solution

The algorithm says that we have to identify cycles one by one, then eliminate edges until no more cycles remain. There are lots of cycles to choose from here, but let's begin with the following cycle.

> Kevin Bacon, Sandra Bullock, Marisa Tomei

Do you see this cycle on the graph? We are free to delete any of the edges of this cycle to construct the spanning tree. Let's remove the edge *Loverboy* joining Sandra Bullock and Marisa Tomei.

The next cycle we'll consider is the following.

> Kevin Bacon, Sandra Bullock, Tom Hanks

Remove the edge *Loverboy* connecting Kevin Bacon and Sandra Bullock.

Staying with Sandra Bullock and Tom Hanks, let's remove the edge *Toy Story 4* between Tom Hanks and Keanu Reeves, which is in the cycle Tom Hanks, Keanu Reeves, Sandra Bullock.

Now were starting to make some progress. Look at the following cycle.

> Kevin Bacon, Lawrence Fishburne, Keanu Reeves,
> Sandra Bullock, Tom Hanks

We can remove *Speed* to eliminate the cycle.

The next cycle we'll consider is the following.

> Kevin Bacon, Lawrence Fishburne, Jennifer Connelly, William Hurt, Kevin Costner

This time let's remove the *Mr. Brooks* edge. While we're at it, let's also remove the edge that joins Kevin Costner to Morgan Freeman. That eliminates the cycle

Kevin Bacon, Kevin Costner, Morgan Freeman, John Malkovich.

We can eliminate the cycle

Kevin Bacon, Marisa Tomei, Kirk Douglas,
Arnold Schwarzenegger, Jamie Lee Curtis

by removing the *The Villain* edge that joins Kirk Douglas to Arnold Schwarzenegger.

Although Jamie Lee Curtis and John Malkovich were both in *Queen's Logic* with Kevin Bacon, let's remove the edge that joins Jamie Lee Curtis and John Malkovich to break that cycle of length three. Now, there are three cycles that remain. Can you see them?

Kevin Bacon, John Malkovich, Bruce Willis, Tom Hanks

forms a cycle, so let's delete the *The Bonfire of the Vanities* edge.

Kevin Bacon, Tom Hanks, Tim Allen, Jamie Lee Curtis

also forms a cycle, so we'll delete the *Toy Story* edge. Finally, there is one cycle remaining:

Kevin Bacon, Meg Ryan, Tom Hanks.

We will remove this cycle by deleting the *In the Cut* edge. Now, we have the following spanning tree.

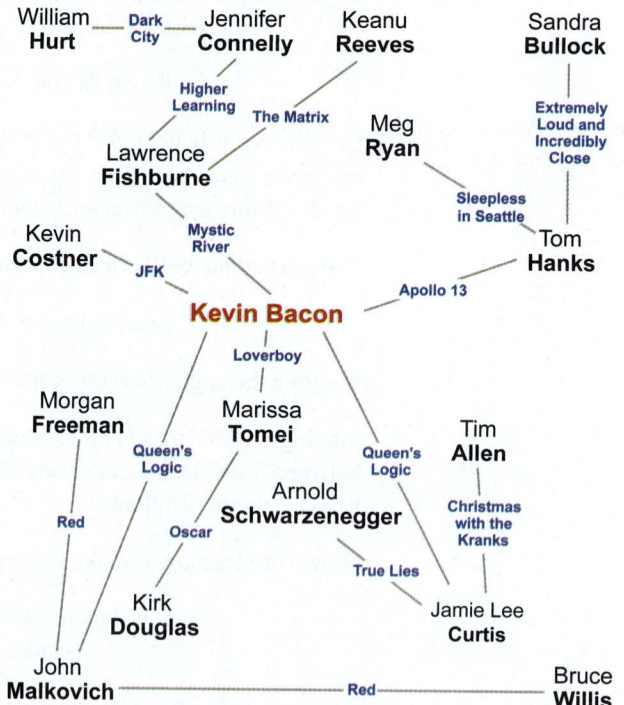

Figure 14.2.10

> ✔ **Skill Check 14.2.2**
>
> Find another spanning tree for the Kevin Bacon graph in Example 14.2.3 by deleting a different set of edges.

How does the spanning tree help to play the Six Degrees of Kevin Bacon Game? By using the algorithm to find a spanning tree where Kevin Bacon is always at most six vertices away from any other actor, you can play the Six Degrees of Kevin Bacon Game with all the actors in the graph. For example, according to the spanning tree in Example 14.2.3, Kirk Douglas is 2 degrees from Kevin Bacon.

We often assign values, or **weights**, to edges of a graph. We thought about this earlier when it cost us $10 an edge to connect vertices. We can assign weights to the edges to represent not only costs, but distances, time, numbers of people, volumes, or any other measurement. There are occasions when we want to minimize the spanning tree that we choose in order to find the most efficient route. One way to do this would be to list out all the possible weighted spanning trees and then choose the one with the smallest values. You can imagine that this would take a great deal of time and resources, and it is not practical with a large graph, such as a graph of the US interstate system. A GPS (Global Positioning System) navigation system uses weighted graphs to determine shortest routes and does this quickly. The following algorithm can be used to find a minimum-weight spanning tree.

STEPS FOR CONSTRUCTING A MINIMUM-WEIGHT SPANNING TREE

1. Consider each edge in the graph in order of descending weight.

2. If the edge being considered is part of a cycle, remove it. If not, it must remain in the graph, and you can move on to the next edge.

3. Repeat Steps 1 and 2 until all edges have been considered.

Example 14.2.4

Constructing a Minimum-Weight Spanning Tree

SpeedFirst Telecommunications is going to run its fiber optic cable in a new development. The edges in the given graph represent the possible ways that the cable can be run. The weight of each edge is the length of the cable needed, in meters. Find the minimum spanning tree using the algorithm to obtain the most cost-efficient way for SpeedFirst to run the cable. What is the minimum amount of cable that SpeedFirst will need to run in the new development?

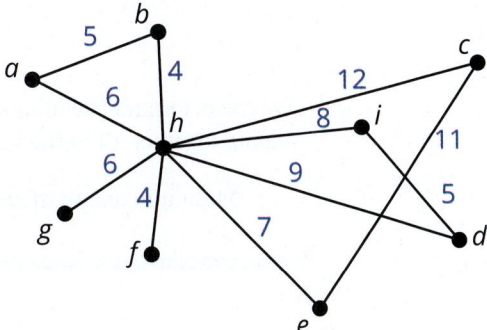

Figure 14.2.11

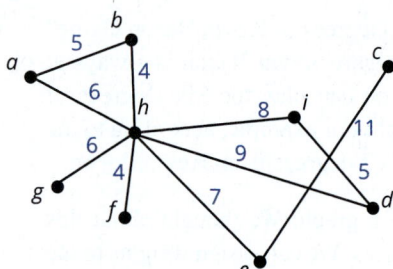

Figure 14.2.12

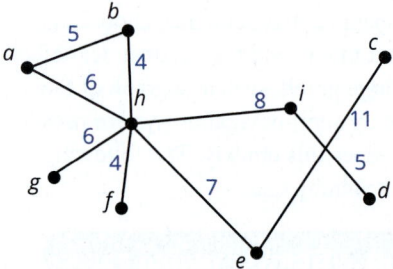

Figure 14.2.13

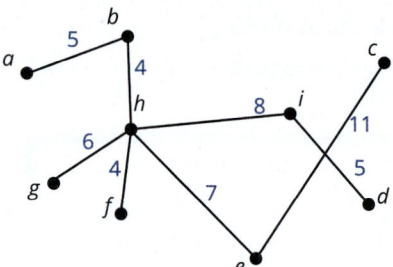

Figure 14.2.14

Solution

We begin with the edge with the largest weight, which is the edge $hc = 12$. Since it has the highest weight and is part of cycle h, c, e, we can remove it, as shown in Figure 14.2.12.

Next, look at edge $ec = 11$. This edge is no longer part of a cycle since the edge hc was removed, so this edge must remain in order to connect vertex c to the graph.

The next highest weighted edge is $hd = 9$. This edge is part of the cycle h, d, i, so we can remove this edge. Figure 14.2.13 shows the graph as it stands now.

Edge $hi = 8$ has the next highest weight. This is not part of a cycle since the edge hd was removed, so this edge remains to connect vertex i.

Next, we consider $he = 7$. This edge is not part of a cycle since the edge hc was removed, so it must remain to connect the vertex e.

Look at edges $ah = 6$ and $gh = 6$. Edge ah is part of cycle a, h, b, so it can be removed. Edge gh is not part of a cycle, so it must remain to connect the vertex g, as shown in Figure 14.2.14.

Next, we consider edges $ab = 5$ and $id = 5$. Both of these edges must remain to connect vertices a and d since they are not part of any cycles once the other edges were removed.

Finally, we look at edges $bh = 4$ and $fh = 4$. Again, neither of these edges are part of any cycles, so they must remain to connect vertices b and f.

We have considered all edges in G, so we are done. Thus, our minimum-weight spanning tree is the following.

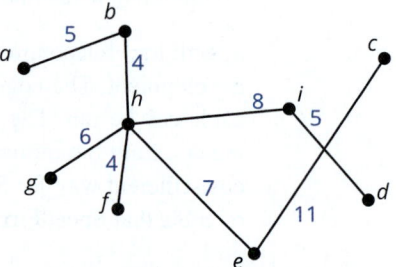

Figure 14.2.15

We can determine the minimum length of cable that SpeedFirst will need to run by adding together all of the weights in the minimum spanning tree.

Minimum length of cable = $4 + 4 + 5 + 5 + 6 + 7 + 8 + 11 = 50$ meters.

OBJECTIVE 3 Leaves on a Tree

No discussion of trees is complete without talking about leaves. In graph theory, a vertex of degree one is called a **leaf**. We can determine the number of leaves on a tree by inspecting each vertex.

📖 LEAF

A vertex of degree 1 in a tree is called a **leaf**.

Example 14.2.5

Determining the Number of Leaves on a Tree

Determine the number of leaves on the tree T.

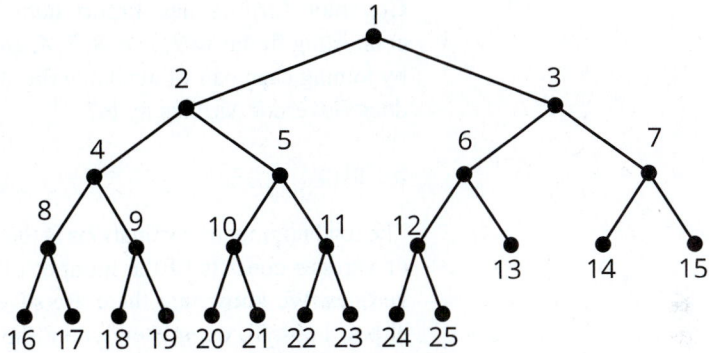

Figure 14.2.16: Tree T

Solution

To find the leaves in T, we need to look for vertices of degree 1. These will be vertices with only one edge connected to them. We've circled the leaves in Figure 14.2.17

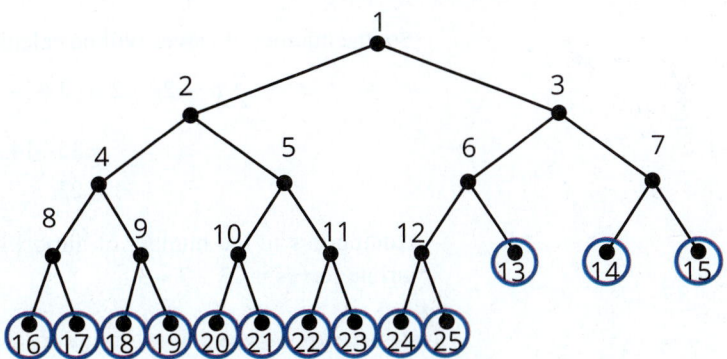

Figure 14.2.17: Tree T

There are 13 leaves on tree T.

As you can imagine, if the graph is very large, inspecting each vertex one at a time would get tedious. Instead, we can determine the number of leaves in a tree by knowing the degrees of all the non-leaf vertices; that is, the vertices with degree 2 or more.

$f(x)$ NUMBER OF LEAVES ON A TREE

If a tree has k vertices with degrees $d_1, d_2, \ldots, d_k$, each greater than 1, then the **number of leaves on the tree** is $\sum\limits_i d_i - 2k + 2$.

Example 14.2.6

Determining the Number of Leaves on a Tree

Governor Airlines has airport hubs in seven cities. Those seven cities have connecting flights to 2, 3, 6, 8, 7, 4, and 5 cities, respectively. If the graph formed by joining each pair of available flights is a tree, how many different destinations does Governor Airlines fly to?

Solution

The total number of destinations is the number of vertices in the tree. That number of vertices consists of the number of leaves on the tree as well as the internal vertices. We know that there are seven internal vertices representing the seven airport hubs. To completely count the number of destinations, we need to know the number of leaves.

The formula for the number of leaves is

$$\sum\limits_i d_i - 2k + 2,$$

where k is the number of vertices with degree greater than 1.

So, the number of leaves will be calculated as follows.

$$\sum\limits_i d_i - 2k + 2 = \left(2 + 3 + 6 + 8 + 7 + 4 + 5\right) - 2\left(7\right) + 2$$

$$= 35 - 14 + 2$$

$$= 23$$

Adding this to the number of airport hubs, the total number of destinations the airline serves is $23 + 7 = 30$.

Skill Check Answers

1.

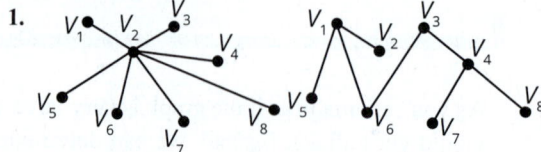

2. Answers will vary.

14.2 Exercises

✔ CONCEPT CHECK

1. A tree is a _____ graph with no cycles.

2. If graph T is a tree on v vertices, then graph T has $v - 1$ _____.

3. A _____ is a subgraph of a connected graph, which is itself connected and contains all the vertices of the original graph but has no cycles.

4. To find a spanning tree, start by identifying any _____ in the graph and delete enough edges to break the _____ while keeping the graph connected.

5. A vertex of degree _____ in a tree is called a leaf.

💡 PRACTICE

Determine if each graph is a tree.

6.

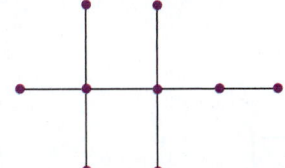

7.

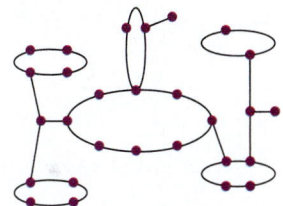

8.

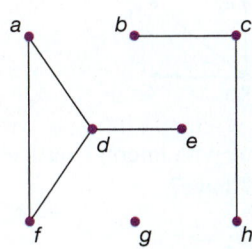

9.

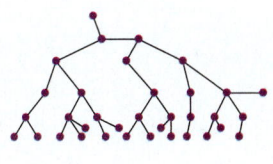

Determine the number of edges needed to form a tree in each graph.

10.

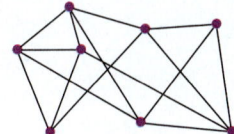

11.

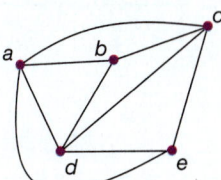

12.

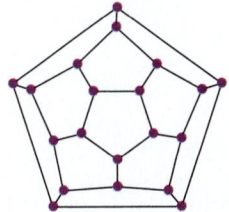

13.

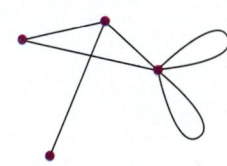

Find a spanning tree in each graph.

14.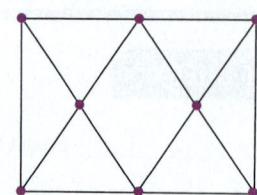

15.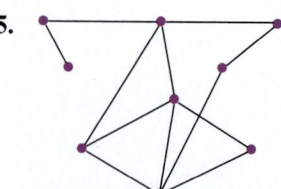

16.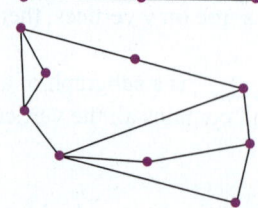

17.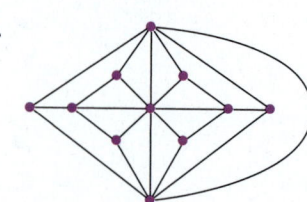

Find the number of leaves in each tree.

18.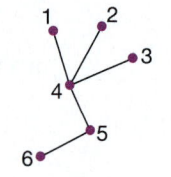

19.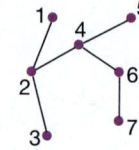

20.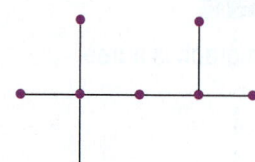

21.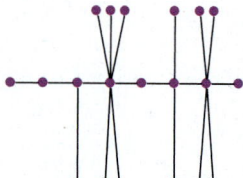

22.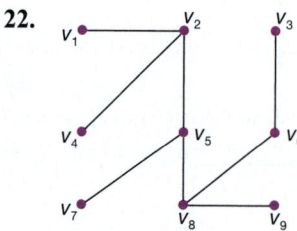

23.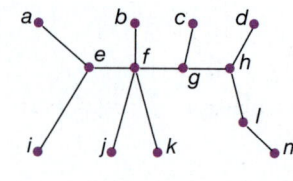

24. Let T be a tree with internal vertices of degree 4, 5, 6, 2, and 3. How many leaves does T have?

🚀 **APPLICATIONS**

25. The sidewalks at a university are laid out according to the following graph. Each corresponding edge is labeled with the length of that section of sidewalk. If all of the sidewalks are covered with snow, which sections of sidewalks should be cleared to connect all sections of the university while clearing as little snow as possible?

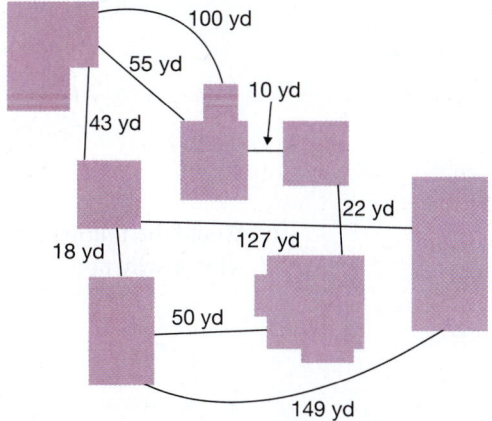

26. A traveling salesman has to visit a number of destinations by a route that is as short as possible. Let G be a graph in which the vertices are the destinations, the edges represent roads that join the destinations, and the weights represent the lengths of the roads.

   a. If the minimum spanning tree has a weight of 100 miles, show that the salesman must drive at least 100 miles to visit all of the destinations.

   b. Show that he needs to drive no more than 200 miles to complete his route.

27. Eurojet Hovercrafts has five hubs connecting European destinations on both sides of the English Channel. If the routes form a tree, with the five hubs having routes to 4, 5, 7, 3, and 6 ports respectively, how many cities are served by Eurojet Hovercrafts?

✏️ **WRITING & THINKING**

28. What is the chromatic number of a tree? (**Hint:** Recall from Section 14.1 that the chromatic number of a graph is the number of colors used in a minimum vertex coloring.)

14.2 PROJECT

ALGEBRA TREES

In this project, you will explore the use of trees to represent algebraic expressions and other mathematical formulas. Writing expressions in this way can be useful when trying to create a computer program that performs complicated calculations. For example, if we want to represent the expression $x + y$ we could use the following tree.

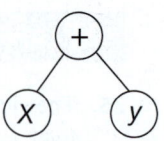

A slightly more complicated expression, such as $2(x + y)$, would result in a tree with more levels.

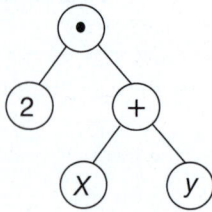

1. Would the following tree also represent the expression $2(x + y)$? Explain why or why not.

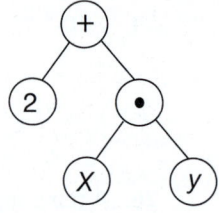

2. Write an algebraic expression that can be represented by the tree below?
   Hint: Start from the leaves and move your way up.

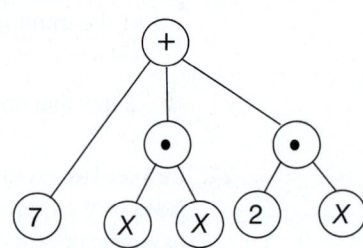

3. Explain how the order of operations are related to this type of tree.

4. The formula to convert a temperature from degrees Fahrenheit to degrees Celsius is given by the equation Using only the operations $C = \dfrac{5}{9}(F - 32)$.
   ⊕, and ⊙ draw a tree to represent the right-hand side of the equation.

5. Draw a tree for the same formula from part 4 where this time you are allowed to also use the operation of division and all the numbers involved must be integers.

14.3 MATCHINGS

OBJECTIVES

1. Determine if a graph is bipartite.

2. Find a matching in a bipartite graph.

OBJECTIVE 1 Bipartite Graphs

Each year, close to 26,000 medical students medical students graduate from US medical schools and seek to move onto the next step in their education—*residency*. They dream of the perfect program for their needs and begin searching for their match. Each student has a number of residency programs that they aspire to be a part of and each applies to some of those programs. In a perfect world, each student would be matched up with a program that they applied to. The question is, with this many students can it be done?

This is an example of a situation that appears in a surprisingly wide number of settings. High school students pairing up with prom dates, students enrolling in classes, and telephone connections on a switchboard are all examples of things that have to be paired, or *matched* up.

To begin to analyze this problem, we'll use the medical residency problem and translate it into the language of graph theory. These types of problems require a graph with two groups of vertices. One group, the left-hand vertices, will represent the medical students with one vertex for each student. The other group, the right-hand vertices, will represent the residency programs with one vertex for each residency position. We call a graph in two parts like this a **bipartite graph**.

📖 BIPARTITE GRAPH

A **bipartite graph** is a graph in which the vertices can be partitioned into precisely two subsets so that every edge joins a vertex in one subset to a vertex in the other subset.

Notice that no vertices on the same side of the graph are joined together in a bipartite graph. In the case where every vertex on the left is joined with every vertex on the right, we call the graph a *complete bipartite* graph.

Example 14.3.1

Identifying Bipartite Graphs

Determine if the following graphs are bipartite.

a.

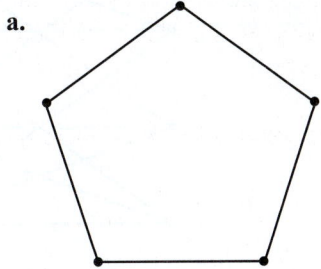

Figure 14.3.1

b.

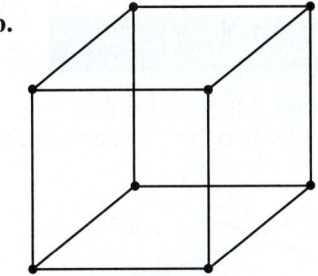

Figure 14.3.2

Solution

a. To determine if the graph is bipartite, we need to show that the vertices can be divided into two groups. Let's assume there are two groups, the left-side group and the right-side group. Begin by labeling the vertices either L for the left-hand side or R for the right-hand side. We will start at the top and label the vertices alternately as we go clock-wise around the graph. If vertices are adjacent, they cannot have the same label.

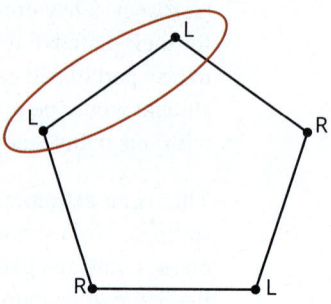

Figure 14.3.3

Notice that we end up with two adjacent vertices both labeled L. Therefore, this graph is not bipartite.

b. We can carry out the same process of labeling vertices with either L or R for the cube. We've also numbered each vertex to help keep track of them.

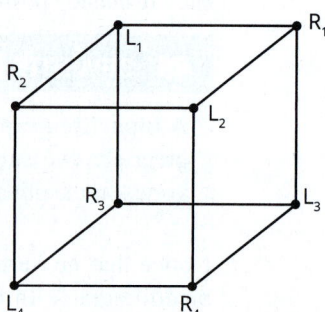

Figure 14.3.4

Notice that no adjacent vertices have the same label. Therefore, this graph is bipartite. Figure 14.3.5 is an alternate drawing showing the vertices on their left and right sides.

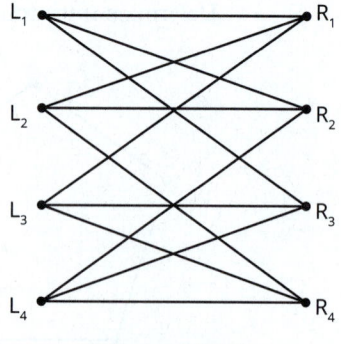

Figure 14.3.5

✔ **Skill Check 14.3.1**

Determine if the following graph is bipartite.

Figure 14.3.6

OBJECTIVE 2 Matchings of Bipartite Graphs

Let's go back to our residency example. To draw the bipartite graph for the students, we place an edge joining each *student* vertex to each *residency* vertex for which the student is an applicant. Assume that each residency has only one position available. Figure 14.3.7 shows a portion of this.

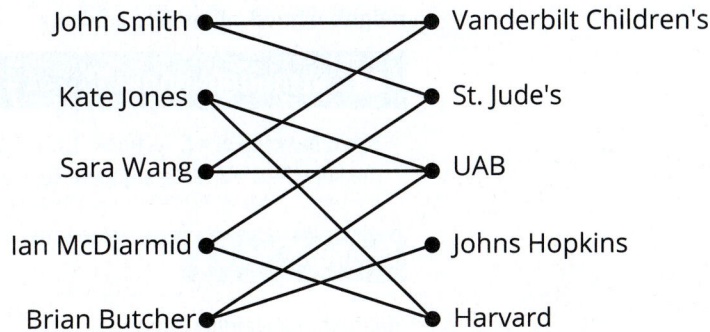

Figure 14.3.7: Diagram of a Bipartite Graph of Residency Students

In order to pair each student with one of their residency program choices so that every student has a spot in a particular residency, we need to find a *matching* in the graph. A **matching** is a subset of edges in the graph so that each vertex involved is incident with one matching edge, just like the dashed edges in Figure 14.3.8 In other words, a vertex never has two edges incident to it in a matching.

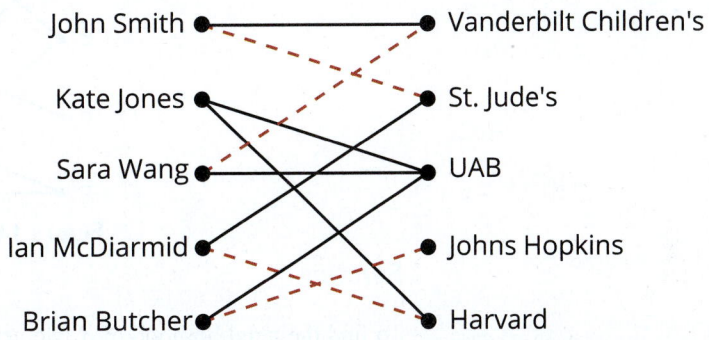

Figure 14.3.8: A Matching for Residency Students

📖 **MATCHING**

A **matching** is a subset of edges in a graph so that each vertex is incident with at most one edge.

A *matching from the left into the right* in a bipartite graph pairs each vertex on the left with exactly one vertex on the right. This does not imply that every vertex on the right is paired with a vertex on the left, but it does mean that every vertex on the left is paired.

In our matching, the edges pair the students with the places of residency. Since every vertex is incident with an edge in the matching, every student has a residency position. Moreover, every student is assigned to precisely one position and every position assigned has only one student.

This all sounds fine in principle, but you can imagine that finding a matching in a bipartite graph that is even remotely large can be very laborious to do by trial and error. So, how in general might we go about creating a matching or even deciding if one exists? First let us think about existence. What kind of situations would indicate that it would be impossible for a matching to exist?

Let's imagine that in our residency example, Vanderbilt Children's is not taking applications this year after all. Now these five students have only applied to four different residency positions. In graph theory terms, the four residencies are the **neighborhood** of the five students.

📖 NEIGHBORHOOD

If we have a set of vertices A, then the **neighborhood** of A, denoted $N(A)$, is the set of all vertices adjacent to a vertex in A.

Example 14.3.2

Identifying a Neighborhood

Let A be the set of vertices $\{d, e\}$ in the Graph G below. Identify the neighborhood of A.

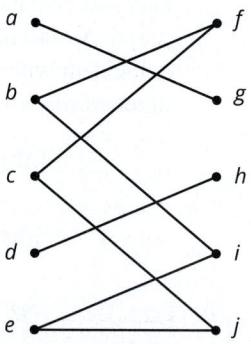

Figure 14.3.9: Graph G

Solution

To find the neighborhood of A, we need to find all the vertices on the right-hand side that are adjacent to the vertices in A. Vertex d is adjacent to vertex h on the right-side and vertex e is adjacent to both vertices i and j.

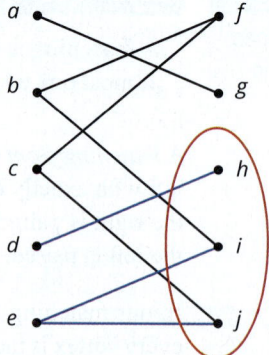

Figure 14.3.10: Graph G

So, $N(A) = \{h, i, j\}$.

✔ **Skill Check 14.3.2**

Let B be the set of vertices $\{a, b\}$ from the graph in Example 14.3.2. Identify the neighborhood of B.

It is clear that there is a problem trying to find a matching with unequal numbers of vertices in the two sets of vertices. In our residency example, we can certainly begin to match students up, but there is no way to match them all—there just aren't enough chosen residency positions in the neighborhood for each residency applicant to have a place all to their own. For example, consider the situation in Figure 14.3.11.

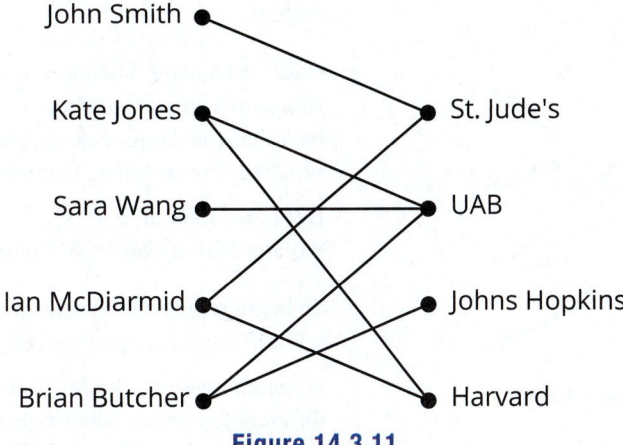

Figure 14.3.11

Of course, there is nothing magical about the numbers five and four. Any situation where the number of residency positions is less than the number of applicants is going to create the same problem. In fact, no matter what the bipartite graph represents, we will not be able to find a matching if a set of vertices has a neighborhood smaller than itself. What is surprising, perhaps, is that this is the only possible barrier to finding a matching. This fact is described in a theorem by Philip Hall with the unusual name of the *Marriage Theorem*.

> 🔑 **HALL'S MARRIAGE THEOREM**
>
> Let G be a bipartite graph. Then there is a matching of the left-hand vertices into the right-hand vertices if and only if for every subset of left-hand vertices A, the number of vertices in $N(A)$ is at least as large as the number of vertices in A.

Example 14.3.3

Determining If a Matching Is Possible

Use Hall's Marriage Theorem to determine if a matching is possible for the following graphs.

a.

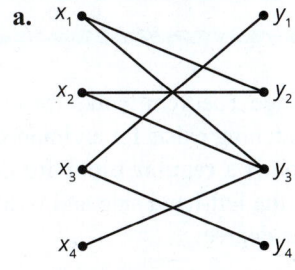

Figure 14.3.12

b. Three top prizes are to be given to the three highest salespeople—a new phone, a flight voucher, and a gift card of choice. John, Mo, and Richard have their sights set on the prizes they would prefer to win. John would like either the new phone or the gift card. Mo prefers the flight voucher, while Richard's top pick is the new phone. Is there a way for each salesperson to receive a prize they prefer if there is only one of each prize?

Solution

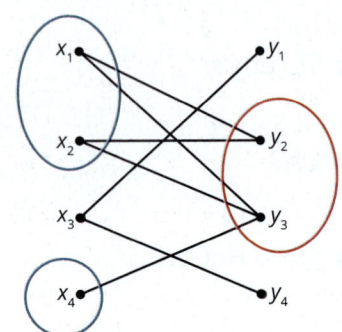

Figure 14.3.13

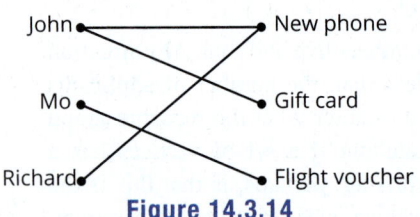

Figure 14.3.14

a. Hall's Marriage Theorem says that for there to be a matching, whichever subset of vertices we choose on the left, their neighborhood on the right must be at least as large. For several subsets of vertices on the left this property is satisfied. For instance, consider $A = \{x_3\}$, which has $N(A) = \{y_1, y_4\}$.

But if we look at $A = \{x_1, x_2, x_4\}$, the neighborhood is $N(A) = \{y_2, y_3\}$, which violates Hall's Marriage Theorem and shows that a matching is not possible.

b. To begin represent the salesmen and prizes as a graph with the salesmen on the left and the prizes on the right, as in Figure 14.3.14.

A matching from the left into the right is a way for each salesman to win a different top prize. You can probably see that this can be done by looking at the graph, but let's see how Hall's Marriage Theorem also establishes this.

To use Hall's Marriage Theorem we need to consider each possible subset of salesman vertices and check the size of the neighborhood. We have seven subsets to work through. We will deal with them one at a time.

1. $A = \{John\}$ $N(A) = \{New\ Phone,\ Gift\ Card\}$

2. $A = \{Mo\}$ $N(A) = \{Flight\ Voucher\}$

3. $A = \{Richard\}$ $N(A) = \{New\ Phone\}$

4. $A = \{John,\ Mo\}$ $N(A) = \{New\ Phone,\ Gift\ Card,\ Flight\ Voucher\}$

5. $A = \{John,\ Richard\}$ $N(A) = \{New\ Phone,\ Gift\ Card\}$

6. $A = \{Mo,\ Richard\}$ $N(A) = \{New\ Phone,\ Flight\ Voucher\}$

7. $A = \{John,\ Mo,\ Richard\}$ $N(A) = \{New\ Phone,\ Gift\ Card,\ Flight\ Voucher\}$

As we can see, for each subset the number of elements in $N(A)$ is greater than or equal to the number of elements in A. Therefore, Hall's Marriage Theorem ensures that there is a matching, and a way for each salesman to win a prize they prefer.

Since Hall's Marriage Theorem works for every bipartite graph, it allows us to guarantee that a matching exists for an important special case of bipartite graphs. We say that a graph is a **regular bipartite graph** if there are the same number of vertices in both the left-hand side and right-hand side of the graph, and every vertex has the same degree.

📖 REGULAR GRAPH AND REGULAR BIPARTITE GRAPH

Regular Graph

A **regular graph** is a graph where every vertex has the same degree.

Regular Bipartite Graph

A **regular bipartite graph** is a regular graph with the same number of vertices on both the left-hand side as the right-hand side.

Example 14.3.4

Identifying a Regular Bipartite Graph
Determine whether each graph is a regular bipartite graph.

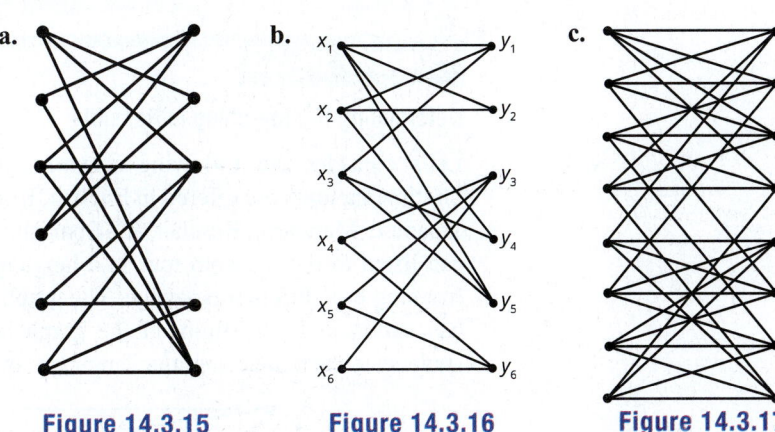

a. Figure 14.3.15   **b.** Figure 14.3.16   **c.** Figure 14.3.17

Solution

a. A regular bipartite graph always has the same number of vertices on the left and right-hand sides. This graph has six vertices on the left-hand side and five on the right. Therefore, this graph is not a regular bipartite graph.

b. This graph has six vertices on the left and right-hand sides. In a regular bipartite graph, every vertex has the same degree. Among lots of other differences, vertex x_1 has degree 4, but vertex x_2 has degree 3. Therefore, this graph is not a regular bipartite graph.

c. This graph has eight vertices on each side, and every vertex has degree 4. Therefore, it is a regular bipartite graph.

Let's see how applying Hall's Marriage Theorem shows that every regular bipartite graph has a matching. Suppose that in a regular bipartite graph G, every vertex has degree 5. Choose a set A of left-hand vertices; say k of them. Each of those k vertices is adjacent with five right-hand vertices, but of course, they can overlap in all kinds of ways. We need to show that the neighborhood of these k vertices has at least k vertices itself.

Let's say that there are l right-hand vertices in the neighborhood of A. For Hall's Marriage Theorem to apply, we need to show that k is $\leq l$. Now, let's count the edges between A and its neighborhood. There are $5k$ edges "coming out" of the k vertices in A on the left and going into the l vertices on the right. Because G is a regular bipartite graph, we know the l vertices also only have five edges each some of which must come from A. However, not all of the edges have to come from A. (Remember we chose a subset of vertices from the graph, so l could have edges coming from vertices outside of A.) We know that

$$5k \leq 5l, \text{ which means } k \leq l.$$

Hence, the condition of Hall's Theorem is satisfied for a regular graph; no matter how big k is, we must have a matching.

> 🔑 **REGULAR BIPARTITE GRAPH THEOREM**
>
> A regular bipartite graph has a matching.

Example 14.3.5

Determining If a Matching Is Possible

Let's consider the following scenario. At the British Museum in London, multimedia tours are offered in English, Korean, Arabic, French, German, Italian, Japanese, Mandarin, Russian, and Spanish. A group of ten students wonder if they can listen to the museum tour in a language they understand with each student listening to a different language. The graph shows which languages each student can understand. Determine if the graph has a matching that would allow the students to each have a unique language for their tour.

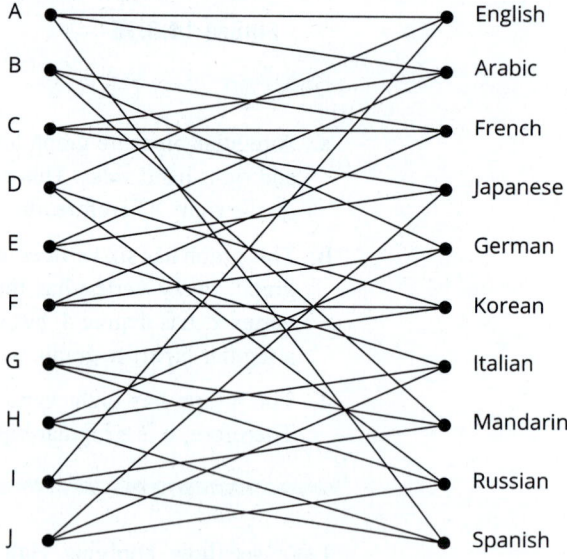

Figure 14.3.18

Solution

One way to determine if the graph has a matching is to show it is a regular bipartite graph. To do this, we need to check for two things:

1. All vertices must have the same degree.

2. The number of vertices on the left-hand side must be the same as the number of vertices on the right-hand side.

First to determine the degree of each vertex, we need to count the number of edges incident to each vertex on both sides of the graph.

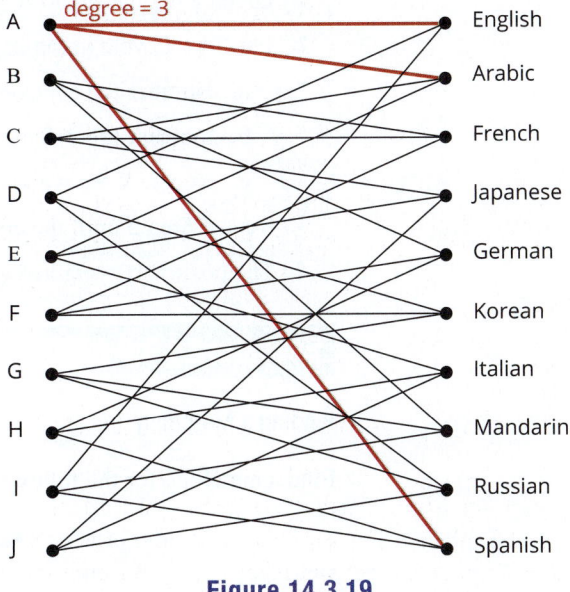

Figure 14.3.19

We see that all vertices have degree 3, so the first criterion is met.

We can count the number of vertices on each side of the graph to make sure there are equal numbers in each.

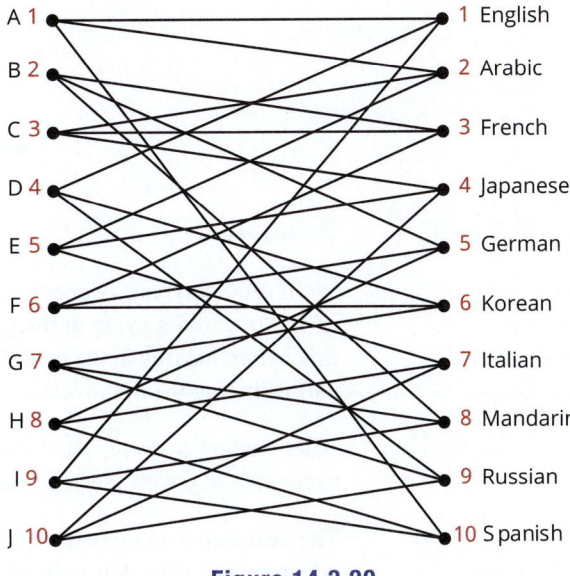

Figure 14.3.20

Since this graph is a regular bipartite graph, we know from the theorem that it has a matching.

Of course, knowing that a matching exists and finding it are two different things. To find a matching in a regular graph, we can use Schriver's Algorithm.

> ⧉ **SCHRIVER'S ALGORITHM**
>
> **Steps for Finding a Matching in a Regular Bipartite Graph**
>
> 1. Given a regular bipartite graph G, give every edge in G a weight of 1.
> 2. Let C be a cycle in the edges of positive weight.
>   a. Number the edges of C successively in turn.
>   b. Increase the weight of the even-numbered edges by 1.
>   c. Decrease the weight of the odd-numbered edges by 1.
> 3. Repeat Step 2 until the edges with positive weight contain no cycle.
> 4. The positively weighted edges form a matching.

Example 14.3.6

Finding a Matching

Find a matching for the following regular bipartite graph.

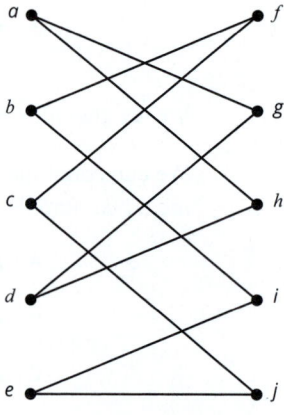

Figure 14.3.21

Solution

We'll use Schriver's Algorithm to find a matching. Let every edge have a weight of 1. Now, find a cycle in the graph. Take a moment and see if you can find one first before looking at the next Figure that shows the cycle. Sometimes the hardest part is finding the first cycle.

After finding a cycle, label the edges consecutively. Figure 14.3.22 shows the cycle a, h, d, g with the edges labeled in order, not by weight.

The next step is to increase the weights of the even numbered edges 2 and 4 by 1, which means they will now each have a weight of 2. We then decrease the odd numbered edges 1 and 3 by 1, which means they have a weight of 0 and can no longer be considered in the algorithm because only positive weighted edges are considered. In Figure 14.3.23, the edges with positive weights are shown and the edges with zero weights are removed since they are not part of the matching.

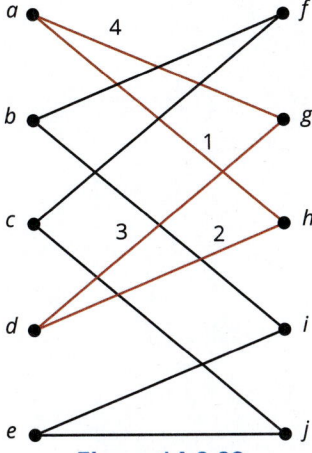

Figure 14.3.22

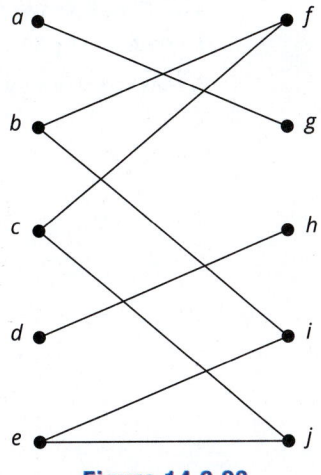

Figure 14.3.23

We begin again at Step 2 by finding a cycle from the edges with positive weights. The only remaining cycle is *f, b, i, e, j, c*. Label the cycle with consecutive numbers, as shown in Figure 14.3.24.

Once again, increase the weight of the even numbered edges by 1 and decrease the odd edges by 1. The only remaining positive weighted edges give us the following graph.

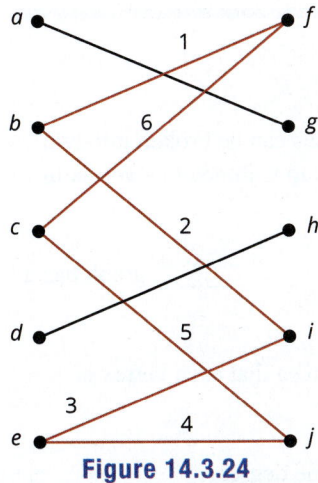

Figure 14.3.24

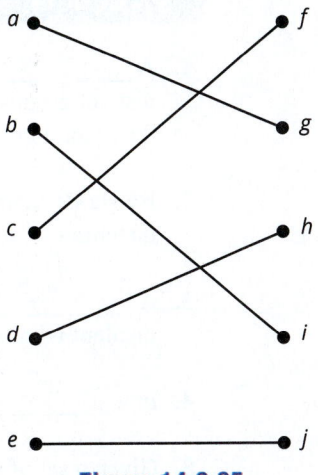

Figure 14.3.25

Because there are no more cycles, we can stop. The remaining edges form the matching we required.

✔ **Skill Check 14.3.3**

Find a different matching for the graph in Example 14.3.6.

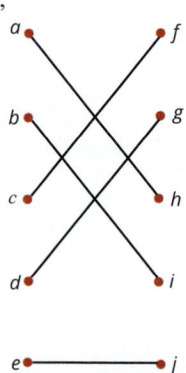

14.3   Exercises

✔ CONCEPT CHECK

1. A _____ is one in which the vertices can be broken into two distinct groups where each vertex in the first group is joined to a vertex in the second group.

2. By Hall's Marriage Theorem, we know that a _____ graph has a matching.

3. A _____ is a subset of edges in a graph so that each vertex is incident with only one edge.

4. In a _____, every vertex has the same degree.

5. Given a set of vertices A, the _____ of A is the set of all vertices adjacent to a vertex in A.

💡 PRACTICE

Determine if each graph is bipartite. Justify your answer.

6.

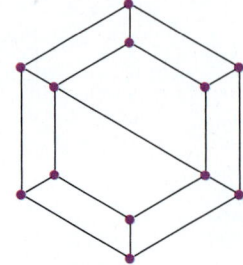

7.

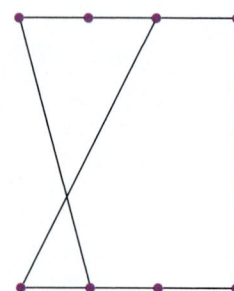

Use Graph *Q* to identify the neighborhood of each set.

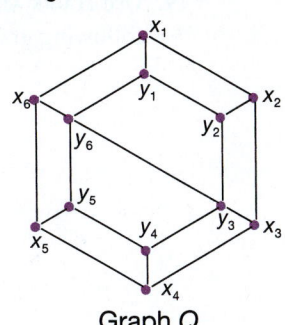

Graph *Q*

8. $A = \{x_1, y_1\}$

9. $B = \{$the set of odd numbered x's$\}$

10. $C = \{y$ vertices$\}$

11. $D = \{$all vertices in $Q\}$

Determine if each graph has a matching. Justify your answer.

12.
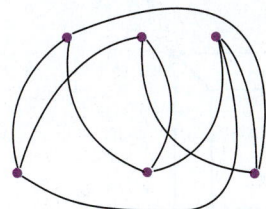

13.
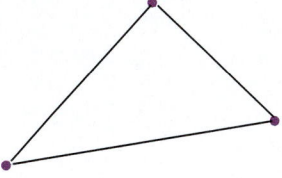

14.
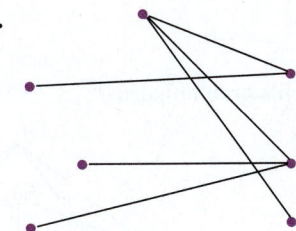

15.
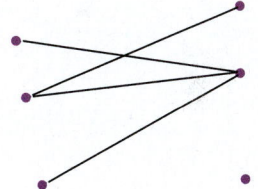

16.
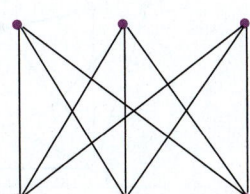

17.
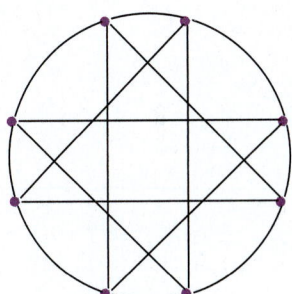

18.
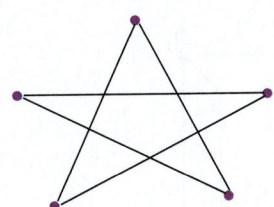

19. Use Hall's Marriage Theorem to determine if there is a matching in the following graph.

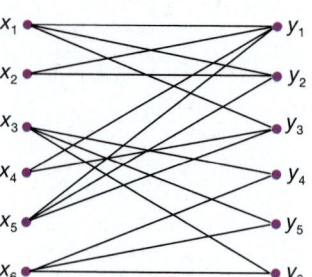

Determine if each graph is regular bipartite.

20.

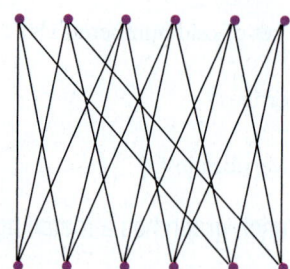

21.

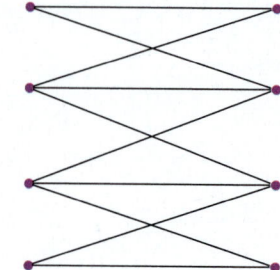

22.

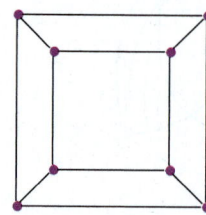

23.

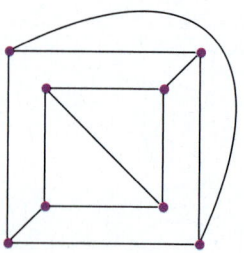

24. Is Grötzsch's graph bipartite?

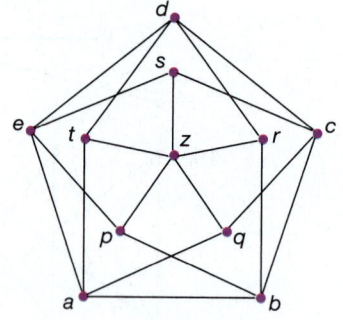

Find a matching in each regular bipartite graph.

25.

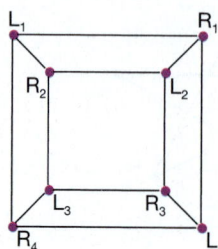

26.

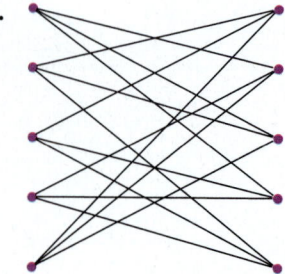

Form a bipartite graph from the information in each table. Find a matching for each scenario, if possible.

27.

Teacher	Courses
Mr. Hall	Political Science, American History
Ms. Cutlidge	English I, English II, English III, Advanced Writing, Computer Science
Mrs. Roseview	English I, English II, English III, Advanced Writing
Mr. Burden	Algebra II, Precalculus, Calculus, PE
Mr. Smith	American History, US History, Economics, Political Science
Mrs. Jones	Political Science, US History, PE, Computer Science
Ms. Rodriguez	Algebra I, Algebra II, Geometry, Calculus

28.

Volunteers	Positions Willing to Serve In
Kerri	greeter, usher, refreshments
Luke	parking attendant, usher
Ron	sound board, video, lights
Jose	lights, parking attendant, greeter
Cho	nursery, sound board, video
Mia	refreshments, nursery, greeter
Margaret	parking attendant, refreshments, lights
Lenton	video, nursery, greeter

29. Use Schriver's Algorithm to find a matching of the British Museum example given in the text. We know from the text that it is a regular bipartite graph, and hence does have a matching.

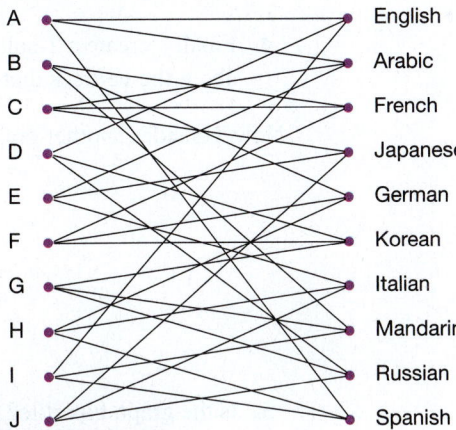

WRITING & THINKING

30. Let graph G be a bipartite graph with the same number of vertices on each side. Suppose that for every set of vertices B on the right-hand side, the number of vertices in $N(B)$ on the left-hand side is at least as large as the number of vertices in B. Explain why graph G has a matching from the left-hand side to the right-hand side.

31. An edge coloring of a graph G is an assignment of colors to the edges, so that incident edges have different colors. Explain why edges of the same color form a matching.

14.3 PROJECT

THE CHROMATIC NUMBER OF BIPARTITE GRAPHS

Recall from Section 14.1 that a *vertex coloring* of a graph is an assignment of colors to the vertices of that graph such that adjacent vertices have different colors. The chromatic number of a graph is the minimum number of colors needed to produce a vertex coloring. In this activity, you will investigate the chromatic number of bipartite graphs.

Consider the following graph.

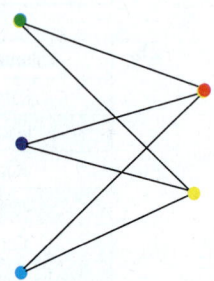

1. Is the graph a bipartite graph? If so, does it have a matching? Explain why or why not.

2. The current vertex coloring uses 5 different colors. Create a vertex coloring using only 4 colors. Explain why this is possible.

3. Modify your 4-color vertex coloring to obtain a 3-color vertex coloring.

4. Finally, create a 2-color vertex coloring for the graph. What do you notice about the vertices that have same colors?

Now, consider another graph.

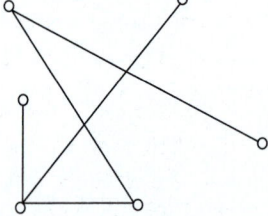

5. Is the graph bipartite? Explain why or why not.

6. What is the chromatic number of this graph? (**Hint:** Start with a 6-color vertex coloring and remove one color at a time as we did before.) What similarities do you notice between your final vertex coloring and the one you found in part 4?

7. Is the chromatic number of a bipartite graph always 2? Explain why or why not.

14.4 PLANAR GRAPHS

OBJECTIVES

1. Identify parts of a planar graph.

2. Use Euler's formula for planar graphs.

3. Determine whether a graph is planar by identifying complete subgraphs.

4. Determine whether a graph is planar by finding graph minors.

OBJECTIVE 1 Introduction to Planar Graphs

Recall from the definition of a graph that not every place in a graph drawing where two edges meet is a vertex. A line between two vertices indicates that those vertices are joined by an edge. The edges can interweave with one another in all kinds of ways. However, there are situations when we want to restrict ourselves to graph drawings where the edges never cross one another, and only meet at vertices. For instance, consider the graph that represents the circuit board in Figure 14.4.1.

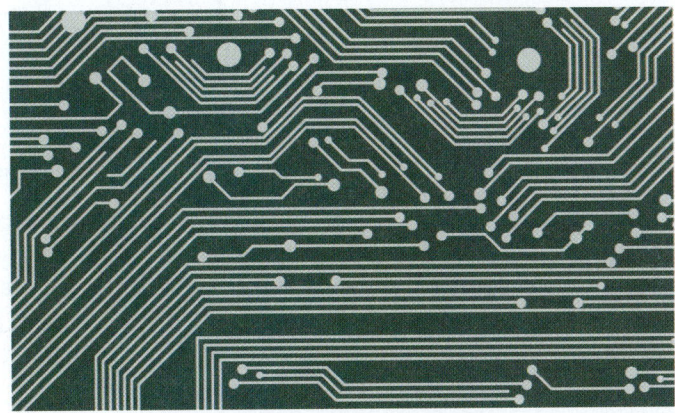

Figure 14.4.1: Circuit Board

The vertices are the components and the edges are the connections that join the components together. If the connections cross in the wrong place, it could cause the circuit board to malfunction. At first glance it might seem that circuit boards can be arbitrarily complicated, having any design you want, but that turns out not to be true. Although they can be very complicated, the design is limited to those graphs that can be drawn on a page with edges that never meet other than at the vertices. When two edges meet other than at a vertex, we call this an edge crossing. Chances are that you didn't think that when you first saw a circuit board, did you?

This restriction of not allowing edges to cross can sometimes cause exasperation. For instance, no matter how hard we try, we cannot have five vertices all mutually linked together (that is, with all possible edges drawn between them) without having at least two of the edges crossing. Try it for yourself. We will explore why you are doomed to failure later. In this section, we examine graphs that do not have any edges crossing when drawn on a plane. This type of graph is known as a **planar graph**.

💬 **Think Back**

Recall that a plane is a flat surface without thickness, depth, or boundaries; like an infinite piece of paper.

📖 **PLANAR GRAPHS**

Graphs that can be drawn on a plane without edges crossing are called **planar graphs**.

Planar Graphs vs. Nonplanar Graphs

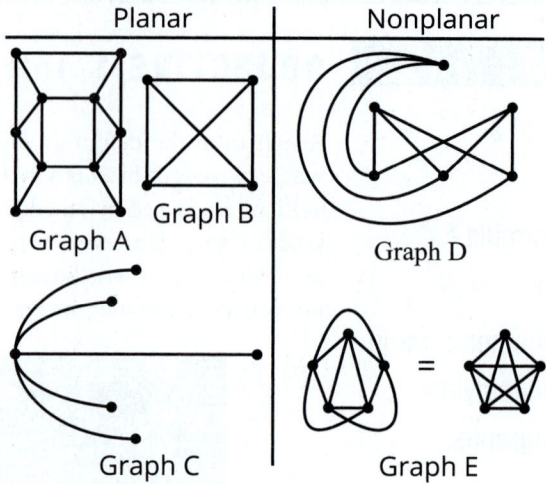

Figure 14.4.2

Figure 14.4.2 shows both planar graphs and nonplanar graphs. At first glance, the planar graph B might appear to be nonplanar. However, if we redraw the graph, we see that it can be drawn without any edges crossing, as in Figure 14.4.3.

Fun Fact

The website https://www.jasondavies.com/planarity/ has a browser-based game related to untangling planar graphs!

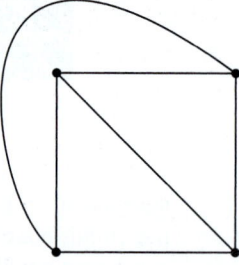

Figure 14.4.3: Alternate View of Graph B

What this tells us is that we cannot merely depend on drawings to determine if a graph is planar. However, we can use some algorithms and theorems to help us classify a graph as planar or not. We will begin with *Euler's formula*, which captures the structure of a planar graph, To explain the formula, we first need some new vocabulary.

✔ **Skill Check 14.4.1**

Draw the following planar graph without any edges crossing.

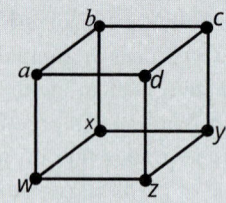

Figure 14.4.4

If you have a planar graph, the graph has another feature that we can describe in addition to the vertices and edges. The graph breaks the plane into certain areas, which we call **faces**.

📖 **FACE**

In a planar graph, a **face** is a region inside a cycle of edges or the infinite exterior region on the outside of the graph. We denote the number of faces of a graph by f.

A planar graph always has at least one face, which is the exterior face. For instance, in Figure 14.4.5, the exterior face on the tree T is shaded. Notice that an exterior face extends indefinitely.

If there is an interior face in a planar graph, it is bounded by a cycle. Figure 14.4.6 shows the graph H. There are three faces in graph H: one surrounded by the cycle A, B, C, D, which we've shaded orange, one surrounded by the cycle C, D, E,

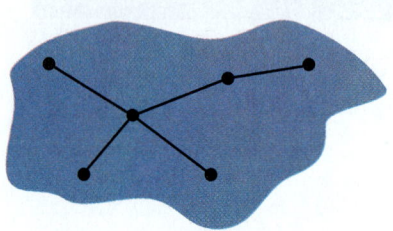

Figure 14.4.5: Faces of Tree T

which we've shaded blue, and don't forget the exterior face on the outside the graph, which we've shaded in light green. Since graph H has 3 faces, $f = 3$.

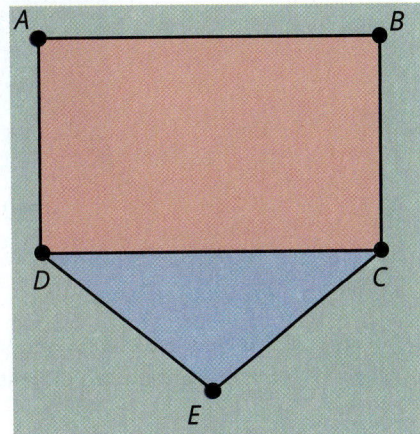

Figure 14.4.6: Faces of Graph H

You can think about faces in a "crafty" manner. If you were to cut the graph H out of this textbook by cutting along it's edges, the sections of paper you get (the orange piece, the blue piece, and the larger light green piece) would represent the three faces of the graph. Thinking in this way, although it has no cycles, even a tree has one face, as we saw in Figure 14.4.5.

Example 14.4.1

Determining the Number of Faces in a Planar Graph

Two drawings of the same graph J are shown. Verify that both drawings have the same number of faces.

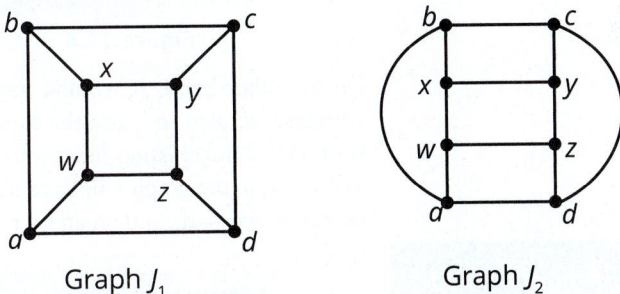

Graph J_1 Graph J_2

Figure 14.4.7

Solution

We can count each of the faces by counting the contained spaces in the graph as well as the exterior face. We've numbered the spaces for each graph in the following figure.

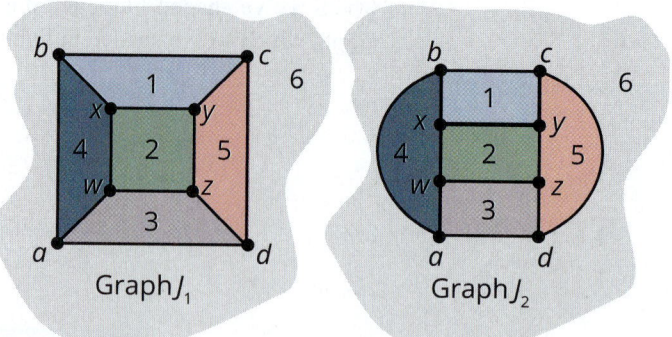

Figure 14.4.8

You can see that $f = 6$ in both diagrams. In fact, graph J will always have six faces no matter how we draw it.

Notice that if we were to remove any edge from a cycle in a graph, we decrease the number of faces by one. For instance, if we remove the edge CD from graph H in Figure 14.4.9, we create one large interior face.

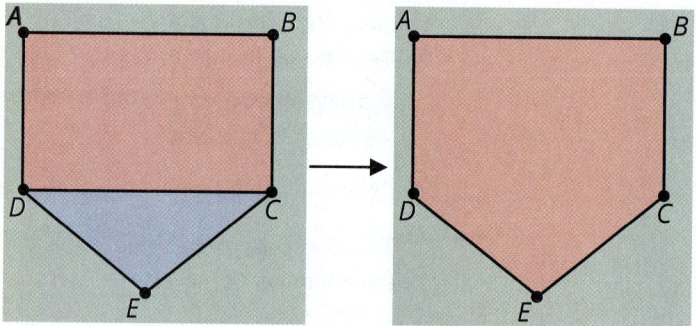

Figure 14.4.9: Graph H with an Edge Removed

On the other hand, if we add any edge to a graph joining two of the existing vertices, we also increase the number of faces by one since that new edge will split one of the existing faces into two parts. For instance, if we add the edge AC to the graph, the graph will have four faces because the new edge splits the orange face into two parts and creates a new cycle A, B, C, as Figure 14.4.10 shows.

✔ **Skill Check 14.4.2**

a. What would happen if we removed the edge CE from the original graph H in Figure 14.4.6?

b. What would happen if we add the edge BD to the original graph H in Figure 14.4.6?

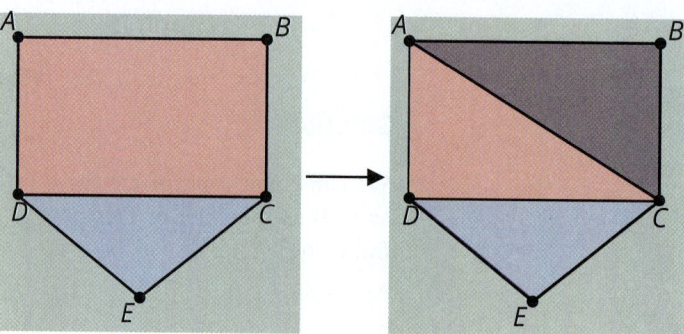

Figure 14.4.10: Graph H with an Edge Added

OBJECTIVE 2 Euler's Formula

Let's think for a minute about trees, which we looked at in Section 13.2. For a tree, we know the relationship between the numbers of edges, vertices, and faces. We've already seen in Figure 14.4.5 that every tree, since it has no cycles, has precisely one face. Because a tree has no cycles, it has no way to "surround" a specific area of the plane in order to make a face. Therefore, it has only the exterior face. So for any tree, $f = 1$.

On the other hand, we also know from Section 13.2 that the number of edges in a tree is one fewer than the number of vertices. If we denote the number of edges in a graph as e and the number of vertices in a graph as v, then we can write this as $e = v - 1$. Putting both v and e on the same side of the equation, we have $v - e = 1$. So, for a tree, we can add these together to see the following.

$$(f) + (v - e) = 1 + 1$$
$$f + v - e = 2$$

We now know from the previous discussion, that each time we add an edge without adding any extra vertices, we also add a face. So this equation relating f, v, and e holds true for any planar graph. The mathematician Euler first proved this and it is now referred to as Euler's formula in graph theory.

$f(x)$ EULER'S FORMULA

If G is a connected planar graph with v vertices, e edges, and f faces, then

$$v + f - e = 2.$$

Let's use Euler's formula in an example.

Example 14.4.2

Verifying Euler's Formula for a Graph

Confirm Euler's formula for the planar graph G.

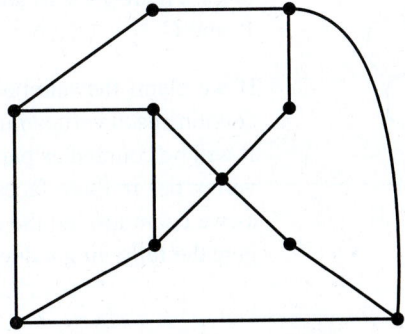

Figure 14.4.11: Graph G

Solution

We can see that G has 10 vertices and 14 edges. All that remains is to find the number of faces. It has 5 internal faces, as indicated in Figure 14.4.12.

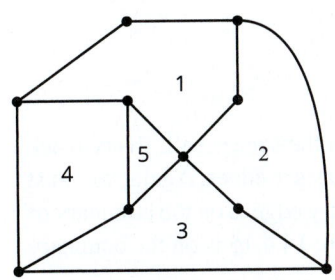

Figure 14.4.12: Graph G

However, we must also remember to include the external face. That makes a total of 6 faces.

Now we have that $v = 10$, $e = 14$, and $f = 6$. Substituting these values into Euler's formula, we can confirm that the number of vertices plus the number of faces minus the number of edges is equal to 2.

$$v + f - e = 10 + 6 - 14 = 2$$

Example 14.4.3

Applying Euler's Formula

A soccer ball traditionally consists of hexagonal white faces and pentagonal black faces. Suppose that we use a box cutter and cut out one of the white, hexagonal faces from the ball to make a hole. Now, imagine that we could stretch the ball covering outward at the hole we made, so that it lies completely flat in a plane as shown. How many black pentagonal faces are there?

Solution

Remember that Euler's formula says that the number of vertices plus the number of faces minus the number of edges in a planar graph must equal 2. To use Euler's formula, we need to find a way to express the number of vertices, edges, and faces in a soccer ball, and then substitute them into the formula.

Let's tackle the vertices first. On the soccer ball, each black pentagon has five vertices, and each white hexagon has six vertices. We can count the total number of vertices on the entire ball one face at a time by counting 5 for every black pentagon and 6 for every white hexagon. If we let P represent the number of pentagonal faces, we can multiply it by 5. Similarly, we'll let H represent the number of hexagonal faces and multiply it by 6. Then, the total number of vertices can be represented by the sum $5P + 6H$.

However, notice that every vertex on the soccer ball is a vertex of three adjacent faces. Figure 14.4.14 shows how vertex v is adjacent to the three faces labeled A, B, and C.

If we claim the number of vertices on the soccer ball is $5P + 6H$, we end up counting each vertex multiple times. Look at vertex v in the previous Figure again. It will be counted as part of each of the three faces in which it lies. In fact, every vertex lies in three faces, which means every vertex will be counted three times as we count around the faces. So, we must divide the number of vertices by 3 to give the following value for v.

$$v = \frac{5P + 6H}{3}$$

We can do the same to count the number of edges on the soccer ball. Every black pentagon has five edges, and every white hexagon has six edges. Again, we must be careful not to over count the edges. Notice that every edge is on the boundary of two faces. For instance, the edge highlighted in Figure 14.4.15 is on the boundary of the face labeled A as well as the face labeled B.

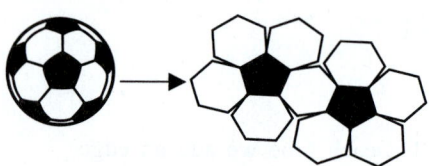

Figure 14.4.13

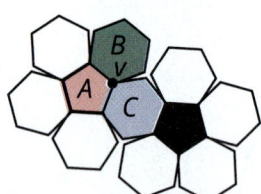

Figure 14.4.14

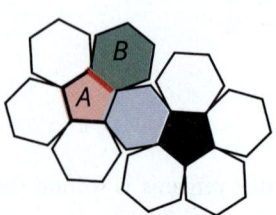

Figure 14.4.15

So, the number of edges is given by the following.

$$e = \frac{5P + 6H}{2}$$

Now that we've found expressions for the number of vertices and edges, we just need to express the number of faces. The total number of faces will be the number of black pentagons P plus the number of white hexagons H. (Recall that we cut out one of the white hexagonal faces, which counts as the exterior face. As a result, we do not need to count an exterior face.) Substituting these expressions into the formula we have the following.

$$v + f - e = 2$$
$$\frac{5P + 6H}{3} + (P + H) - \frac{5P + 6H}{2} = 2$$

Next, simplify this equation by removing the fractions and combining like terms together.

$$\frac{5P + 6H}{3} + (P + H) - \frac{5P + 6H}{2} = 2$$
$$6\left(\frac{5P + 6H}{3} + (P + H) - \frac{5P + 6H}{2}\right) = 6(2)$$
$$10P + 12H + 6P + 6H - (15P + 18H) = 12$$
$$10P + 12H + 6P + 6H - 15P - 18H = 12$$
$$(10P + 6P - 15P) + (12H + 6H - 18H) = 12$$
$$P = 12$$

This means that no matter how many hexagons there are, there must always be precisely 12 black pentagons on the soccer ball.

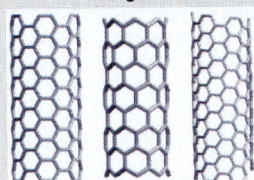
The answer to Example 14.4.3 may be somewhat surprising to you. It is actually confirming that no matter what size soccer ball you want to make, you will always need precisely 12 pentagons. Since hexagons are tessellating shapes, meaning that they can cover a surface without any gaps or overlaps, we can add more hexagons to the ball cover, but we can't ever add more pentagons. If we tried to, a consequence of Euler's formula means that the configuration would no longer fit as a ball cover no matter how hard you try. This applies to any ball made of pentagons and hexagons.

One consequence of Euler's formula is that, given a fixed number of vertices, if we restrict ourselves to having at most one edge between each pair of vertices, there is a limit to how many edges can be added while keeping the graph planar. Euler's formula puts an upper limit on the number of total edges in a planar graph. In order for a graph to be planar, the number of edges it can have is at most $3v - 6$, where v is the number of vertices in the graph.

f(x) COROLLARY OF EULER'S FORMULA

A planar graph G with v vertices has at most $3v - 6$ edges. That is, $e \leq 3v - 6$.

Example 14.4.4

Applying the Corollary of Euler's Formula

If graph G has 13 vertices, what is the greatest number of edges that graph G can have and be a planar graph?

Solution

The corollary of Euler's Formula says that if graph G is to be a planar graph, then the number of edges can be at most $3v - 6$. Since we know that graph G has 13 vertices, $v = 13$. Therefore, in graph G the number of edges, e, can be at most

$$e \leq 3(13) - 6$$
$$e \leq 39 - 6$$
$$e \leq 33$$

✔ **Skill Check 14.4.3**

What is the maximum number of edges that a five-vertex graph can have and still be planar? (**Hint:** Draw a graph to help you find the answer.)

Example 14.4.5

Establishing That a Graph Is Not Planar Using Euler's Corollary

Use the corollary of Euler's formula to establish that graph G is not planar.

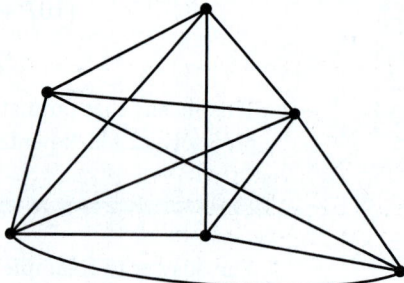

Figure 14.4.17: Graph G

Solution

To use the corollary of Euler's formula, we need to count the number of edges and the number of vertices in graph G. There are six vertices (labeled alphabetically) and 13 edges (labeled numerically) as labeled in the following graph.

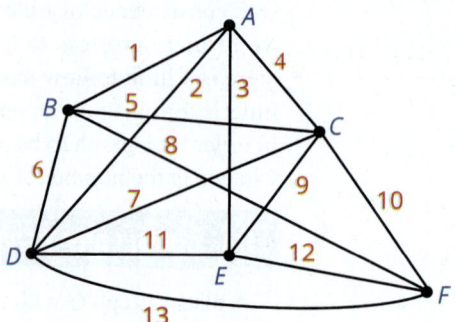

Figure 14.4.18: Graph G

So $v = 6$ and $e = 13$. Now we can substitute these values into the corollary to see if graph G meets the restriction to be planar.

$$e \overset{?}{\le} 3(v) - 6$$

$$13 \overset{?}{\le} 3(6) - 6$$

$$13 \overset{?}{\le} 18 - 6$$

$$13 \nleq 12$$

The number of edges is too large, therefore it is not possible for G to be planar.

OBJECTIVE 3 Complete Graphs

We've now looked at several characteristics of planar graphs. To summarize, if a graph is planar, it will meet all of the following characteristics.

- It can be drawn with no edges crossing.

- It has definable faces.

- Euler's formula relates the number of vertices, faces, and edges by $f + v - e = 2$.

- The number of edges is always at most $3v - 6$.

Not all graphs meet these criteria, so not all graphs are planar. We will now introduce two important families of graphs and analyze when they are planar. The first of these families are complete graphs. A **complete graph** is defined to be a graph in which all vertices are connected to all other vertices. We denote a complete graph by K_n, where n is the number of vertices in the graph.

📖 COMPLETE GRAPHS

A **complete graph** K_n is the graph on n vertices in which every vertex is joined to every other vertex by a single edge.

> 💬 **Fun Fact**
>
> Complete graphs are represented by the capital letter K. Some sources claim that the K stands for the German word for complete, *komplett*, while others say that the notation gives honor to the graph theorist Kazimierz Kuratowski.

Example 14.4.6

Drawing Complete Graphs

Draw the complete graph K_6.

Solution

The complete graph K_6 is the graph with six vertices, each joined to every other vertex by a single edge. Begin by arranging the six vertices in a circle.

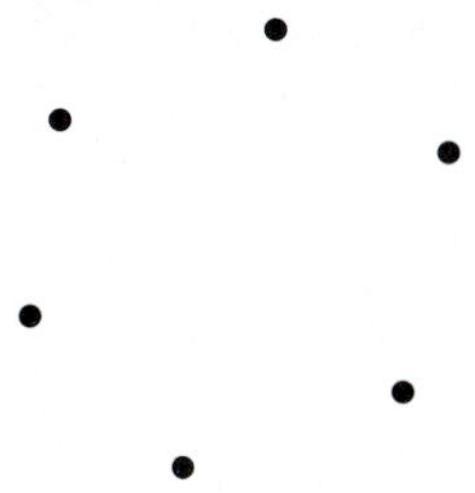

Figure 14.4.19

We will use a systematic way to join each vertex to every other vertex so that each pair of vertices has a single edge between them. We started in the upper left corner and joined the vertex to each of the others in a clockwise motion, as shown in Figure 14.4.20. Moving in a clockwise motion around the circle of vertices, we draw edges to join each vertex to each other vertex.

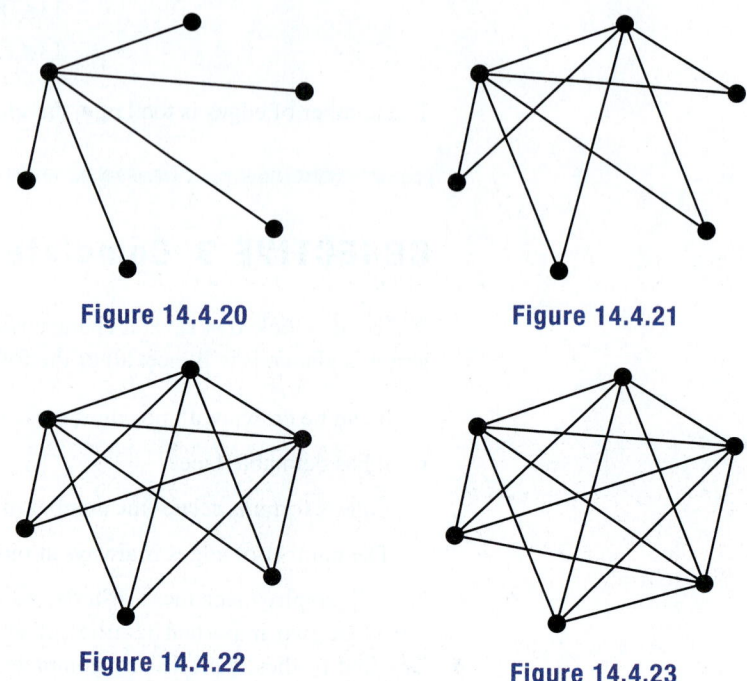

Figure 14.4.20   **Figure 14.4.21**

Figure 14.4.22   **Figure 14.4.23**

The complete K_6 graph is shown in Figure 14.4.24.

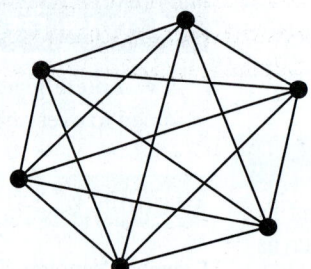

Figure 14.4.24: Complete K_6

✔ **Skill Check 14.4.4**

Draw the complete graph K_7.

The complete graphs K_3 and K_4 are both planar, as you can see in Figure 14.4.25.

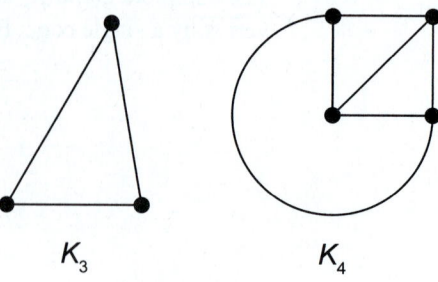

K_3 K_4

Figure 14.4.25

What about the complete graph K_5? Consider the following graph, K. Graph K has five vertices that all have degree 4. In other words, every vertex is joined to every other vertex as shown in Figure 14.4.26.

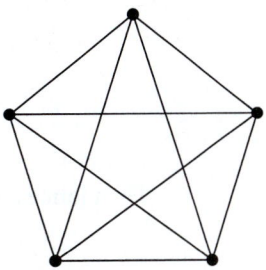

Figure 14.4.26: Graph K

If graph K is planar, it will meet all the characteristics. However, if we can find one characteristic it doesn't satisfy, we know it's not planar. Let's see if graph K satisfies the corollary of Euler's formula. We know from Figure 14.4.26 that $v = 5$ and $e = 10$. Verify these values for yourself. If we substitute these values into the inequality, we have the following.

$$e \overset{?}{\leq} 3(v) - 6$$
$$10 \overset{?}{\leq} 3(5) - 6$$
$$10 \overset{?}{\leq} 15 - 6$$
$$10 \nleq 9$$

Since the inequality simplifies to a false statement, it is impossible for graph K to be planar. Thus, K_5 is not planar.

The second family of graphs is the complete bipartite graphs. A **complete bipartite graph** is a bipartite graph where each vertex on the left is joined to every vertex on the right with a single edge. We denote the complete bipartite graph with m vertices on the left-hand side and n vertices on the right-hand side by $K_{m,n}$.

📖 COMPLETE BIPARTITE GRAPH

A **complete bipartite graph** $K_{m,n}$ is the bipartite graph with m vertices on the left-hand side and n vertices on the right-hand side. Each vertex on the left is joined to every vertex on the right with a single edge.

Example 14.4.7

Drawing Complete Bipartite Graphs

a. Draw the complete bipartite graph $K_{2,2}$.

b. Is the complete bipartite graph $K_{2,2}$ planar?

Solution

a. The complete bipartite graph $K_{2,2}$ has two vertices on the left-hand side and two vertices on the right-hand side of the graph. Begin by arranging the four vertices, two on each side.

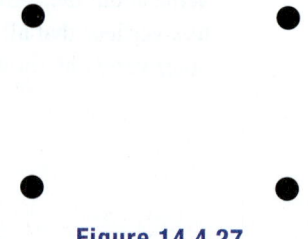

Figure 14.4.27

Then join each vertex on the left with both vertices on the right.

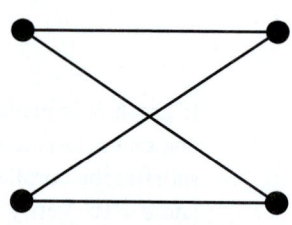

Figure 14.4.28

b. We can establish that $K_{2,2}$ is planar by drawing it so that the edges do not cross. If there is no way to do that, then $K_{2,2}$ is not planar. Notice that if we draw one of the crossing edges on the outside of the vertices, we can see that $K_{2,2}$ is planar.

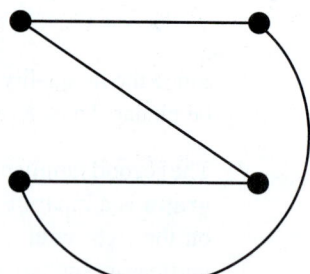

Figure 14.4.29

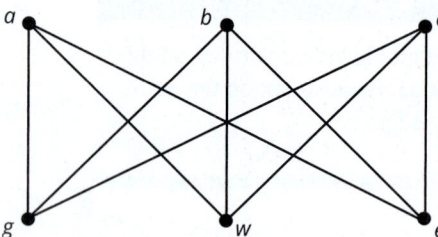

Figure 14.4.30: Graph $K_{3,3}$

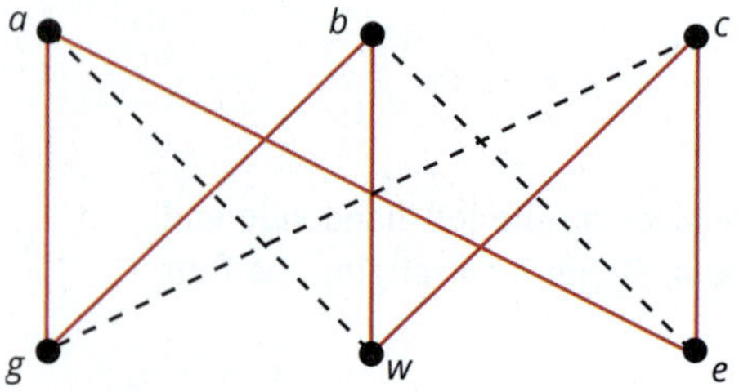

Figure 14.4.31: Cycle in Graph U

Let's consider whether the complete bipartite graph $K_{3,3}$ is planar. The graph $K_{3,3}$ is a regular bipartite graph where each vertex has degree 3. It has six vertices, three on the top, and three on the bottom.

Let's check to see if $K_{3,3}$ satisfies $e \leq 3v - 6$ with $v = 6$ and $e = 9$.

$$e \overset{?}{\leq} 3(v) - 6$$

$$9 \overset{?}{\leq} 3(6) - 6$$

$$9 \overset{?}{\leq} 18 - 6$$

$$9 \leq 12$$

It does satisfy the inequality. However, that doesn't guarantee that $K_{3,3}$ is planar. In Figure 14.4.30, $K_{3,3}$ certainly has edges crossing. However, can we draw $K_{3,3}$ without the crossings? Notice that the vertices a, g, b, w, c, e form a cycle of length 6. We've highlighted this cycle in Figure 14.4.31.

In any planar drawing of the graph, this cycle needs to be disentangled. We show this in Figure 14.4.32.

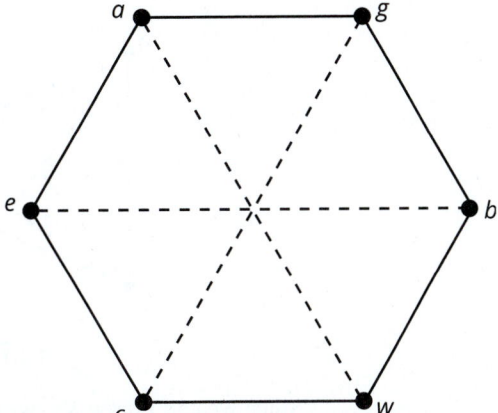

Figure 14.4.32: Alternate View of Graph U

That leaves us to uncross the edges shown by the dotted lines in Figure 14.4.32: aw, be, and gc. Notice that we can't possibly draw more than one of those edges "inside" the cycle without creating a crossing; and similarly we can only draw one edge "outside" the cycle. Since there are three edges remaining, that still leaves one edge left to draw that has to cross something either inside the cycle or outside. So, it is impossible to draw $K_{3,3}$ in the plane without crossings. Therefore, graph $K_{3,3}$ is not a planar graph.

Example 14.4.8

Classifying Graphs as Planar or Not

Decide whether graph F is planar or not.

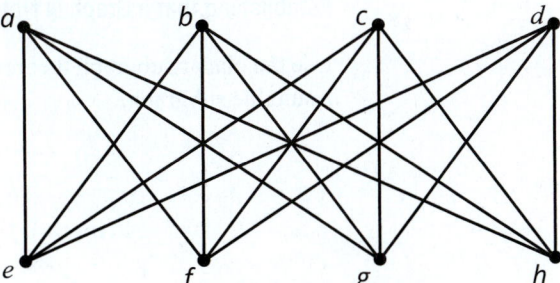

Figure 14.4.33: Graph F

Solution

Graph F has $v = 8$ vertices and $e = 16$ edges.

Let's first check that this satisfies the corollary of Euler's formula.

$$e \overset{?}{\leq} 3(v) - 6$$

$$16 \overset{?}{\leq} 3(8) - 6$$

$$16 \overset{?}{\leq} 24 - 6$$

$$16 \leq 18$$

It does satisfy the inequality. However, that doesn't guarantee that graph F is planar. Notice that graph F has a subgraph on a, b, c, e, f, g that is the same as $K_{3,3}$, which we saw is not planar.

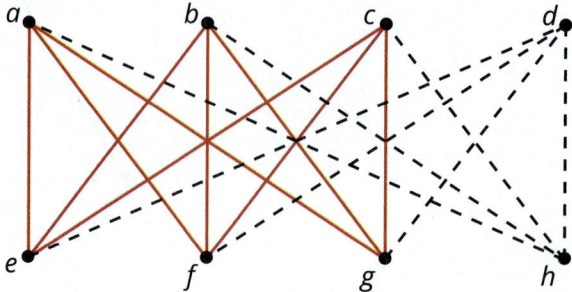

Figure 14.4.34: Graph F

Since we can't draw $K_{3,3}$ without edges crossing, then graph F cannot be planar.

The logic that we employed to determine that graph F was not a planar graph is logic that can be applied much more broadly. In fact there is a theorem that states that if a graph contains a subgraph that is not planar, then the whole graph is not planar. Consequently since we have shown that neither K_5 nor $K_{3,3}$ is planar, any graph that contains either K_5 or $K_{3,3}$ cannot itself be planar.

🔑 PLANAR SUBGRAPH THEOREM

Graph G is not planar if any subgraph of G is not planar.

Example 14.4.9

Establishing that a Graph Is Not Planar Using Subgraphs

Use the planar subgraph theorem to show that the graph L is not planar by finding a suitable subgraph.

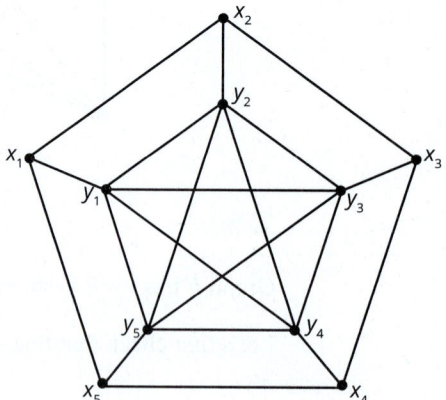

Figure 14.4.35: Graph L

Solution

In order to show that graph L is not planar, it is enough to find either a K_5 or $K_{3,3}$ subgraph. Let's consider the five vertices in the middle of the graph labeled $y_1, y_2,$

y_3, y_4, and y_5. If they form the graph K_5, meaning each vertex is connected to all other vertices, then the graph is not planar.

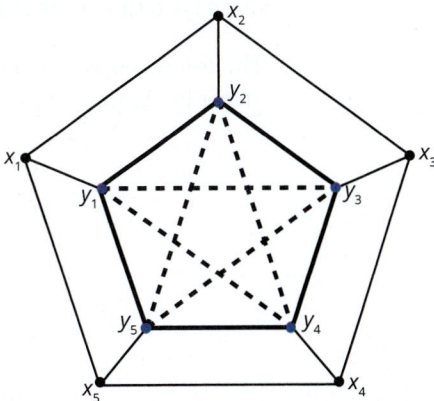

Figure 14.4.36: Graph L

Notice that each of the vertices y_1 through y_5 have degree 4 and are all connected to each other. Check this for yourself. The existence of this subgraph is sufficient to say that graph L is not planar.

OBJECTIVE 4 Graph Minors

To close this section, we will improve on the previous theorem in a way that allows us to completely characterize planar graphs. To do that, we need one more graph theory operation called **contracting**. Given a graph G and an edge $e = uv$, the new graph H, obtained by shrinking e until the vertices u and v are the same vertex, is called "G contract e" and written G/e. In other words, contracting an edge means that it disappears in such a way that its end vertices become one. We illustrate this in Figure 14.4.37.

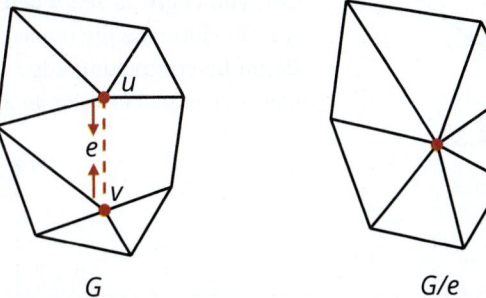

G   G/e

Figure 14.4.37: Contracting an Edge in a Graph

If G is planar then G/e is also planar. We say that a graph that is obtained from G by a sequence of edge or vertex deletions and edge contractions is a *minor* of G. We can now finally characterize planar graphs completely. Since we know that neither $K_{3,3}$ nor K_5 are planar, no planar graph can have one of these as a minor. In fact, that is the only barrier to planarity.

🔑 PLANAR GRAPH THEOREM

G is planar if and only if G has neither $K_{3,3}$ nor K_5 as a minor.

Example 14.4.10

Classifying Graphs as Planar or Not

The following is a famous graph called the Grötzsch graph. Show that it is not planar by finding a K_5 minor of the graph.

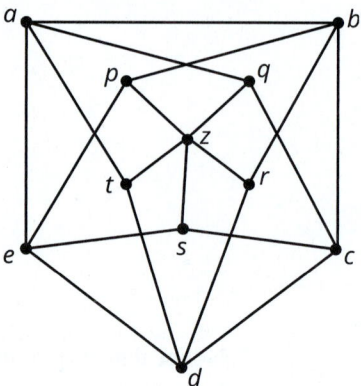

Figure 14.4.38: Grötzsch Graph

Solution

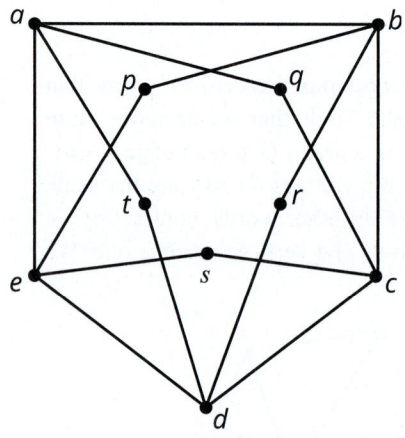

Figure 14.4.39

In order to show that the Grötzsch graph is not planar, we need to show that it contains a K_5 minor. Without stronger algorithms that are beyond the scope of this text, we need to simply "find" a minor by a process of deleting vertices, deleting edges, or using edge contractions. Remember that looking for a K_5 minor means that we are looking for five vertices that are all connected to each other. Obviously, there are too many vertices in the original graph, so we'll begin by deleting the middle vertex z. When we delete a vertex from a graph, all edges connected to it also get deleted because they no longer have a vertex at both of their ends. Figure 14.4.39 shows the Grötzsch graph after deleting z.

Can you begin to see where the five vertices in the minor might come from? If we can eliminate the inner vertices, we might just have the graph we are seeking. Begin by contracting edge aq so that the vertices a and q become one. We now have one edge between aq and c as the following graph shows.

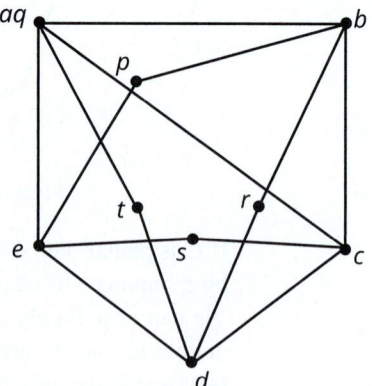

Figure 14.4.40

In the same manner, we can contract the edge pb as well as edge sc. Figure 14.4.41 shows the contracted graph so far.

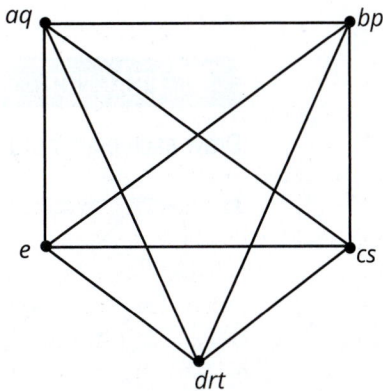

Figure 14.4.41

What remains is to contract edges *td* and *rd*. Once this is done, we have the following graph of five vertices of degree 4 all connected together. In other words, a K_5 minor of the Grötzsch graph.

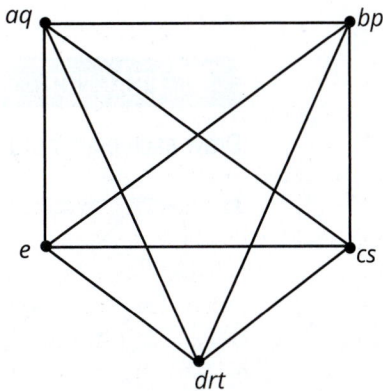

Figure 14.4.42

Math Milestone

The Grötzsch graph is the smallest triangle-free graph with chromatic number 4. It is named after German mathematician Herbert Grötzsch, born in 1902 in Halle, Germany. He died in 1993, a week before his 91st birthday.

Hence, we have shown that the Grötzsch graph is not planar since it contains a K_5 minor.

Skill Check Answers

1. Answers will vary. Two possibilities are as follows.

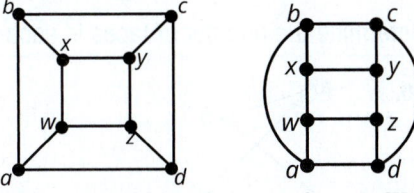

2. a. There would then only be two faces. **b.** We'd split the orange face in two, creating four faces in total. **3.** Nine edges **4.**

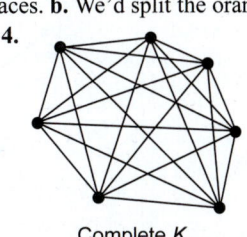

Complete K_7

14.4   Exercises

1. Graphs that can be drawn on a plane without _____ are called planar graphs.

2. In a planar graph, a _____ is a region inside a cycle of edges or the infinite exterior region on the outside of the graph.

3. Euler's formula states that if G is a connected planar graph with v vertices, e edges, and f faces, then _____.

4. A planar graph G with v vertices has at most _____ edges.

5. A _____ graph is defined to be a graph in which each vertex is connected to all other vertices.

Draw each planar graph without any crossing edges.

6.

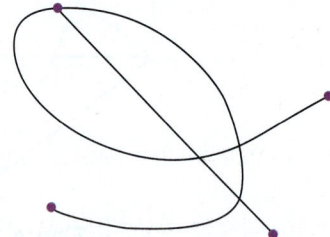

Graph *K*

7.

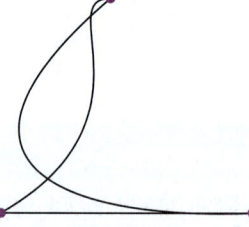

Graph *S*

8.

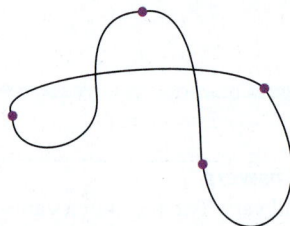

Graph *N*

9.

Determine the number of faces in each planar graph.

10.

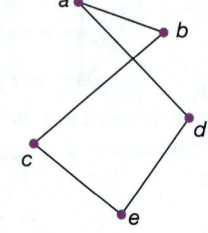

Graph *N*

11.

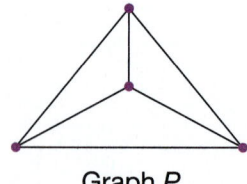

Graph *P*

12.

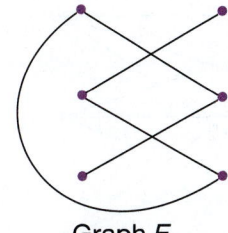

Graph *E*

Verify Euler's formula for each graph.

13.

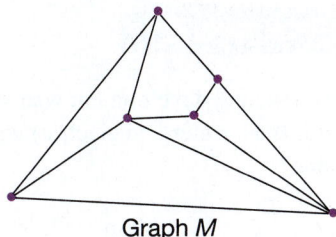

Graph *M*

14.

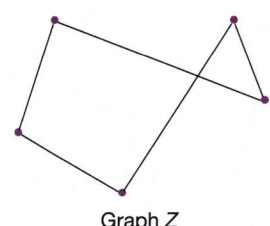

Graph *Z*

15.

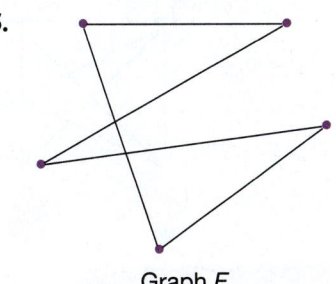

Graph *F*

16.

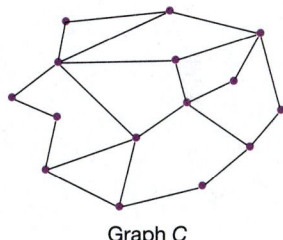

Graph *C*

Use Euler's formula and the corollary of Euler's formula.

17. If a graph G has 9 vertices, what is the greatest number of edges that graph G can have and be a planar graph?

18. If a graph G has 16 vertices, what is the greatest number of edges that graph G can have and be a planar graph?

19. If G is a connected planar graph with 8 vertices and 10 edges, then how many faces does G have?

20. If G is a connected planar graph with 10 edges and 6 faces, then how many vertices does G have?

Determine if each graph is planar.

21.

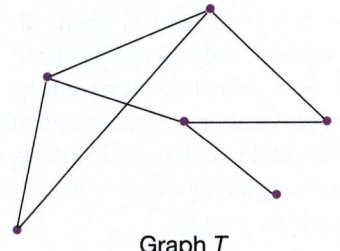

Graph *T*

22.

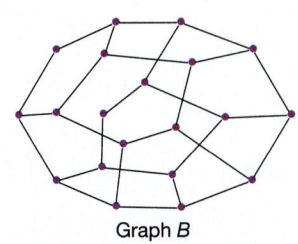

Graph *B*

23.

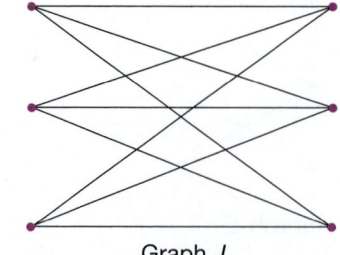

Graph *J*

24.

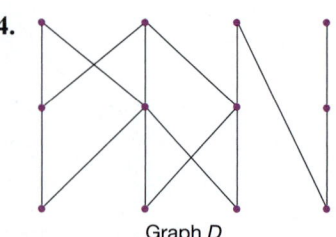

Graph *D*

25. A new housing development wants to connect three houses to their three utilities (gas, water and electric) without their cables/pipes crossing. Can it be done?

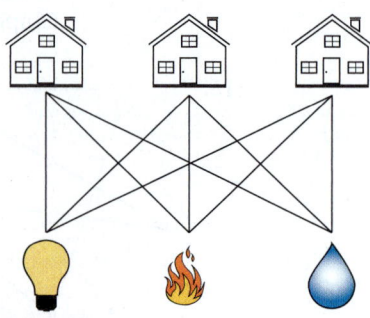

26. Show that a planar graph must have a vertex of degree at most 5.

27. Find the chromatic number of K_3, K_5, and K_n.

28. Find the chromatic number of $K_{5,5}$ and $K_{k,k}$.

29. Show that any planar graph can also be drawn on a sphere.

30. What happens if you draw the graph of a cube on a sphere and then flatten the faces so that they are the same size?

14.4 PROJECT

PLANAR POLYHEDRONS

In this section, you learned about planar graphs and how their numbers of v vertices, e edges, and f faces are related by Euler's formula: $v + f - e = 2$. You may have also learned about vertices, edges, and faces in geometry as part of the study of convex polyhedrons. A convex polyhedron is a solid made up of flat polygonal faces joined at their edges and vertices with the additional property that any line segment joining any two points on the surface of the polyhedron stays on or inside the polyhedron.

A cube is an example of a convex polyhedron.

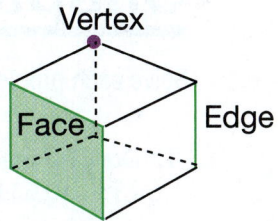

1. How many faces, edges, and vertices does a cube have?

2. Show that a cube satisfies Euler's formula.

The fact that a cube satisfies Euler's formula isn't a coincidence. We can turn a cube into a planar graph without changing the number of faces, edges, or vertices. The process is illustrated as follows.

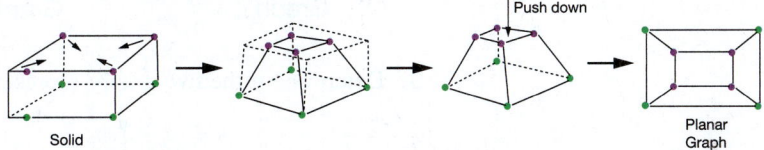

3. Draw the planar graph that corresponds to the tetrahedron, illustrated as follows. (**Hint:** In this case, all you must do is push down on the top vertex.)

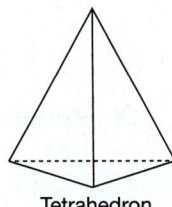

Tetrahedron

4. Show that a tetrahedron satisfies Euler's formula.

We can turn any convex polyhedron into a planar graph using a process like the one described for the cube. That is, Euler's formula $v + f - e = 2$ is satisfied for all convex polyhedrons.

5. One of your classmates claims that they have constructed a convex polyhedron out of two triangles, two squares, six pentagons, and five octagons. Explain why such a convex polyhedron cannot exist. (**Hint:** Each vertex of a convex polyhedron must border at least three faces, and the sum of the degrees of all the vertices in a graph is equal to twice the number of edges.)

Chapter 14 Exercises

Solve each problem.

1. Draw a graph that has six vertices (v_1, v_2, v_3, v_4, v_5, v_6) and the edges v_1v_4, v_1v_5, v_1v_6, and v_2v_3. Is the graph connected?

2. Determine if the three graphs represent the same graph.

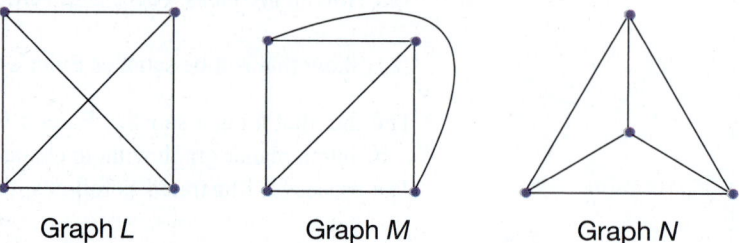

    Graph *L*   Graph *M*   Graph *N*

3. Determine if the two graphs represent the same graph.

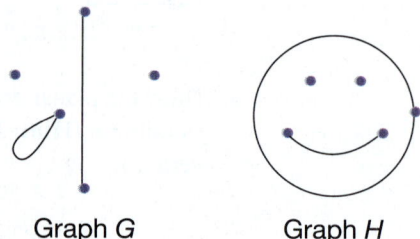

    Graph *G*   Graph *H*

4. Use the following graph to answer the questions below.

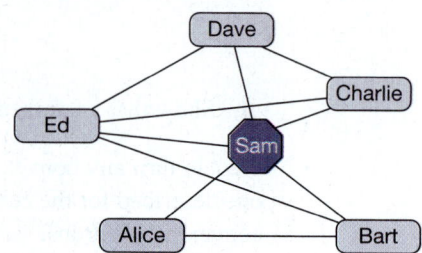

  a. What is the degree of "Sam"?

  b. Identify a path of length 5. If no such path exists explain why.

  c. Identify a path of length 6. If no such path exists explain why.

  d. Identify a cycle of length 5.

5. Find a vertex coloring of graph *W*. What is the chromatic number of that graph?

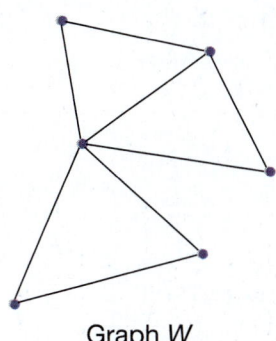

Graph *W*

6. Use a vertex coloring to assign colors to the regional electricity companies in England and Wales so that companies whose regions share a border have different colors. What is the minimum number of colors required? Find a vertex coloring using the minimum number of colors.

Regional Electricity Companies in England and Wales

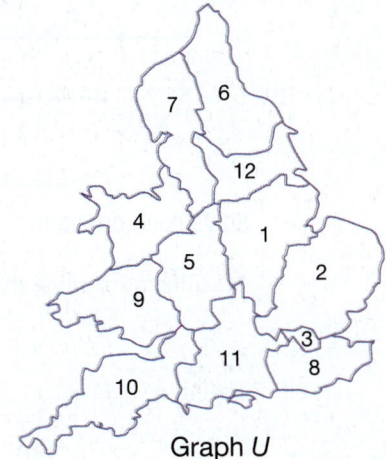

Graph *U*

7. Determine if the graph formed by the subway map is a tree.

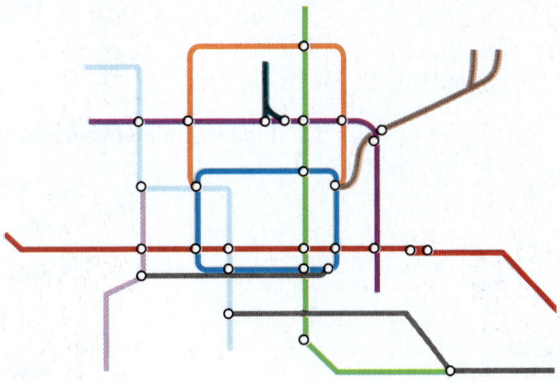

8. Determine whether the graph is a tree.

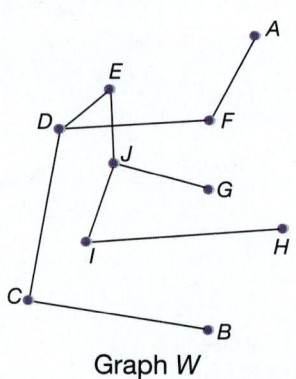

Graph *W*

Determine the number of edges needed to form a tree in each graph.

9.

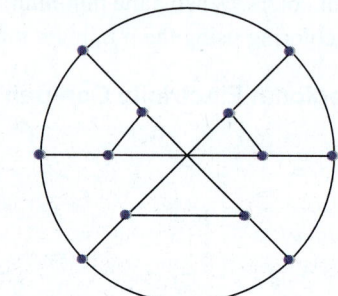

10.

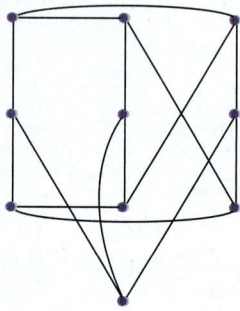

Solve each problem.

11. Find a spanning tree for the following graph.

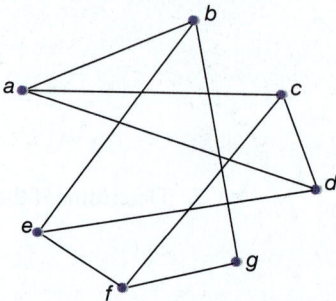

12. Find a minimum-weight spanning tree in the following graph.

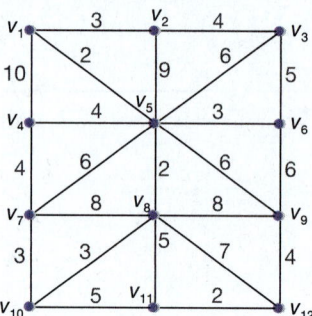

13. A national internet grid is needed to connect the cities in the following graph with optical fiber in the most cost-effective way possible. The weights on the edges represent the lengths of cable that would have to be used to connect each pair of cities, when that can be achieved.

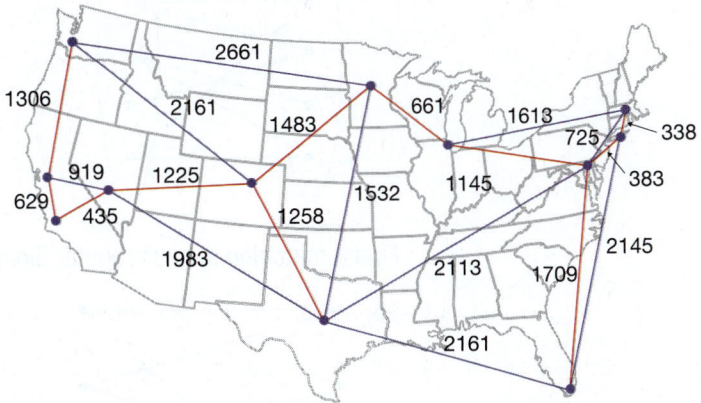

a. The proposed connections are represented by the red edges. Check that this graph is a spanning tree. Explain why a spanning tree is a suitable way to lay the optical fiber.

b. Is the proposed set of connections a minimum-weight spanning tree?

14. Find the number of leaves in the given trees.

a.

b.

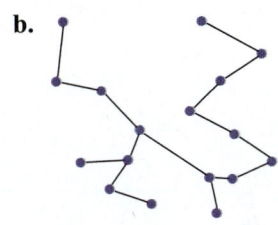

Use the given graph to identify the neighborhood of each set.

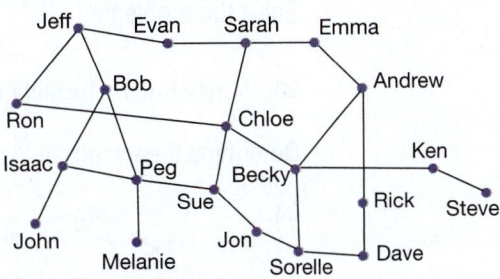

15. $A = \{\text{Bob, Evan}\}$

16. $A = \{\text{Sorelle, Peg, Evan}\}$

Determine if each graph has a matching. Justify your answer.

17.

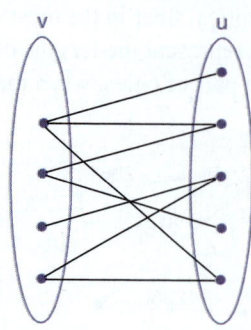

18.

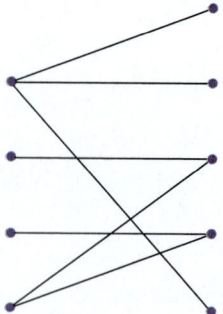

Find a matching in each regular bipartite graph.

19.

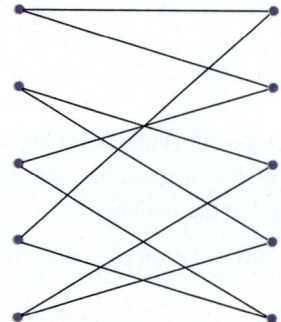

20.

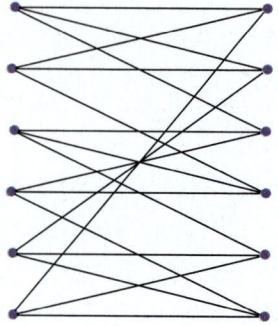

Determine the number of faces in each planar graph.

21.

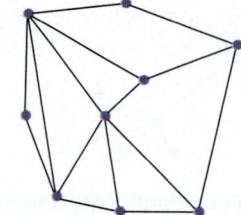

22.

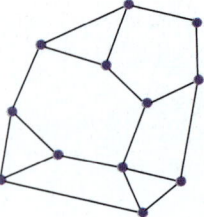

Solve the problem.

23. Verify Euler's formula for the graphs in Exercises 21 and 22.

Determine if each graph is planar.

24.

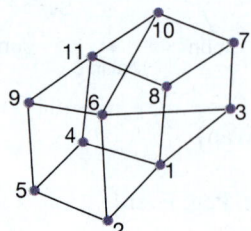

25.

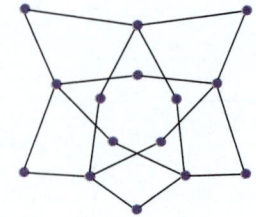

CHAPTER 14 PROJECT

How Connected are You?

Merriam-Webster defines social media as "forms of electronic communication through which users create online communities to share information, ideas, personal messages, and other content." It's hard to live in the modern world and not belong to some type of social media community (such as Facebook, Snapchat, Twitter, Instagram, Pinterest… the list is long).

In this project, you will use graphs to visualize and think about your community in social media. You are going to create a graph using a social media site where both parties agree to be connected, such as Facebook or LinkedIn. A different type of graph (that is, directed graphs, which are beyond the scope of this chapter) is created when you consider one-sided connections such as "following" someone on Instagram or Twitter.

As many people often have very large communities on social media, your graph will likely be a subgraph of your entire community. Using yourself as the first vertex, you will create a graph that consists of at least 15 people as the vertices, but the more you can include, the better! If two people are connected via your social media, their vertices will be joined with an edge.

1. Before you begin, anticipate why it is important for you to use a social media where both parties agree to be connected when constructing your graph. Describe your predictions here.

On a separate piece of paper, create graph G of your social media community. Use a minimum of 15 people to whom you are "connected" and whom are also "connected" to you. The more connections you include, the better idea you will get of your community. Begin with yourself as the first vertex. Using the person's name, label each of the vertices. If any two people are connected via your social media, their vertices should be joined with an edge. If you don't have a social media account where connections are two-way, you will need to ask a friend who has one if you can create their graph of connections and answer the questions based on their graph.

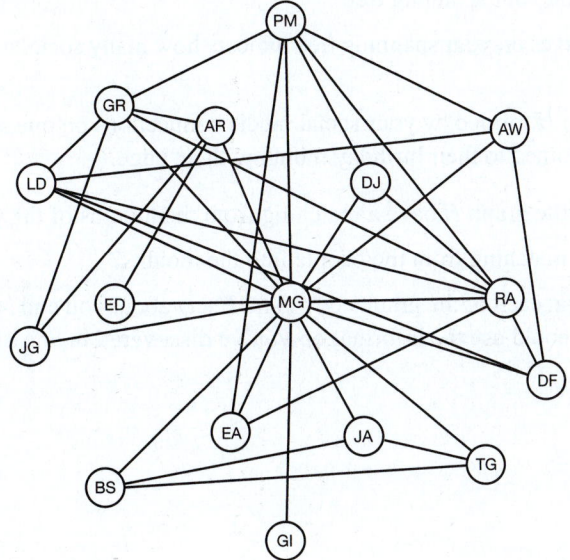

Answer the following questions about your graph G.

2. What was the most difficult part of constructing G?

3. How many vertices does G have?

4. How many edges does G have?

5. Is G connected? Why or why not?

6. Is G connected if you remove your vertex from the graph?

7. Suppose G becomes a disconnected graph upon removing your vertex, what does that mean in social terms about your groups of connections?

Now create a new graph on a separate piece of paper in which you remove your vertex from the graph G. Use this new graph F to answer the following questions.

8. What can you say about the neighborhoods of the vertices in F?

9. Which vertex in F has the largest degree? What is its degree?

10. What can you say in terms of social media about the person with the largest vertex degree in F?

11. Find the largest complete subgraph in F.

12. What does this subgraph represent in terms of social media?

13. Find the chromatic number χ of F.

14. Find a vertex coloring of F with χ colors.

15. What can you say about a group of people in F whose vertices all have the same color in your coloring?

16. What would it mean in terms of the social media if F has a high chromatic number?

17. What would it mean in terms of the social media if F has a low chromatic number?

18. Find a minimum vertex cover of F.

19. Why might a minimum vertex cover be useful to know if friends of yours were using social media to organize a surprise party for you?

20. Find a spanning tree of F or, if F is not connected, find a spanning tree of the largest connected subgraph of F.

21. How long is the longest path in your spanning tree?

22. What does the longest path represent in terms of your social media connections?

23. How many leaves are on your spanning tree?

24. Does the number of leaves on your spanning tree indicate how many social media connections you have? Why or why not?

Consider the bipartite graph H formed by your social media connections on one side and the months of the year on the other side. A person is joined to their birthday month with an edge.

25. When would this bipartite graph H have a matching from the months of the year into the people?

26. When would H have a matching from the people into the months?

27. Finally, think about what each of the graphs G, F, and H say about you and social media. Describe at least one practical way that you could use the information you've discovered in one of the the three graphs.

CHAPTER 14 REVIEW

Section 14.1 Introduction to Graph Theory

Definitions

Graph: A **graph** consists of a set of vertices and a set of edges that join pairs of vertices together. Graphs are generally represented with a capital letter, such as G.

Loop: A **loop** is an edge that has the same vertex for both of its ends.

Walk: A **walk** in a graph is a finite list of alternating vertices and connecting edges that begins and ends with a vertex.

Path: A **path** is a walk with no repeated edges or vertices.

Connected Graph: A graph is **connected** if there is at least one path connecting each pair of distinct vertices.

Disconnected Graphs: A graph is **disconnected** if it has at least one pair of vertices that is not connected by a path.

Adjacent: Two vertices that share an edge are **adjacent**.

Incident: Two edges that share a vertex are **incident** to one another. Also, a vertex is incident to the edges that have that vertex as an endpoint.

Degree: The **degree** of a vertex denoted $d(u)$, is the number of edges that are incident to u.

Vertex Cover: A **vertex cover** is a set of vertices A so that every vertex in the graph is either in A or adjacent to a vertex in A.

Minimum Vertex Cover: When a vertex cover A is as small as possible, A is called a **minimum vertex cover**.

Vertex Coloring: A **vertex coloring** of a graph is an assignment of colors to the vertices of the graph so that adjacent vertices receive different colors.

Chromatic Number: When the number of colors used in a vertex coloring is as small as possible, this number is called the **chromatic number** of a graph and is denoted $\chi(G)$, read "chi of G."

Cycle: A **cycle** is a walk that starts and ends at the same vertex and has no edges or vertices repeated except for the starting vertex.

Section 14.2 Trees

Definitions

Tree: A connected graph with no cycles is called a **tree**.

Spanning Tree: A **spanning tree** is a subgraph of a connected graph, which is itself connected and contains all the vertices of the original graph, but has no cycles.

Leaf: A vertex of degree 1 in a tree is called a **leaf**.

Procedures

Steps for Constructing a Spanning Tree

1. In a connected graph **G**, identify a cycle. If more than one cycle exists, choose one at random.

2. Choose an edge from the cycle selected and delete it from the graph.

3. While the graph contains a cycle, repeat Steps 1 and 2.

Steps for Constructing a Minimum-Weight Spanning Tree

1. Consider each edge in the graph in order of descending weight.

2. If the edge being considered is part of a cycle, remove it. If not, it must remain in the graph, and you can move on to the next edge.

3. Repeat Steps 1 and 2 until all edges have been considered.

Theorem

Tree Theorem: Let graph T be a tree on v vertices. Then graph T has $v-1$ edges.

Formula

Number of Leaves on a Tree: If a tree has k vertices with degrees $d_1, d_2, \ldots, d_k$, each greater than 1, then the **number of leaves on the tree** is $\sum_i d_i - 2k + 2$.

Section 14.3 Matchings

Definitions

Bipartite Graph: A **bipartite graph** is a graph in which the vertices can be partitioned into precisely two subsets so that every edge joins a vertex in one subset to a vertex in the other subset.

Matching: A **matching** is a subset of edges in a graph so that each vertex is incident with at most one edge.

Neighborhood: If we have a set of vertices A, then the **neighborhood** of A, denoted $N(A)$, is the set of all vertices adjacent to a vertex in A.

Regular Graph: A **regular graph** is a graph where every vertex has the same degree.

Regular Bipartite Graph: A **regular bipartite graph** is a regular graph with the same number of vertices on both the left-hand side as the right-hand side.

Theorems

Hall's Marriage Theorem: Let G be a bipartite graph. Then there is a matching of the left-hand vertices into the right-hand vertices if and only if for every subset of left-hand vertices A, the number of vertices in $N(A)$ is at least as large as the number of vertices in A.

Regular Bipartite Graph Theorem: A regular bipartite graph has a matching.

Procedure

Schriver's Algorithm: Steps for Finding a Matching in a Regular Bipartite Graph

1. Given a regular bipartite graph G, give every edge in G a weight of 1.

2. Let C be a cycle in the edges of positive weight.

    a. Number the edges of C successively in turn.

    b. Increase the weight of the even-numbered edges by 1.

    c. Decrease the weight of the odd-numbered edges by 1.

3. Repeat Step 2 until the edges with positive weight contain no cycle.

4. The positively weighted edges form a matching.

Section 14.4 Planar Graphs

Definitions

Planar Graphs: Graphs that can be drawn on a plane without edges crossing are called **planar graphs**.

Face: In a planar graph, a **face** is a region inside a cycle of edges or the infinite exterior region on the outside of the graph. We denote the number of faces of a graph by f.

Complete Graphs: A **complete graph** K_n is the graph on n vertices in which every vertex is joined to every other vertex by a single edge.

Complete Bipartite Graph: A **complete bipartite graph** $K_{m,n}$ is the bipartite graph with m vertices on the left-hand side and n vertices on the right-hand side. Each vertex on the left is joined to every vertex on the right with a single edge.

Formulas

Euler's Formula: If G is a connected planar graph with v vertices, e edges, and f faces, then

$$v + f - e = 2.$$

Corollary of Euler's Formula: A planar graph G with v vertices has at most $3v - 6$ edges. That is, $e \le 3v - 6$.

Theorems

Planar Subgraph Theorem: Graph G is not planar if any subgraph of G is not planar.

Planar Graph Theorem: G is planar if and only if G has neither $K_{3,3}$ nor K_5 as a minor.

A   Standard Normal Distribution

Numerical entries represent the probability that a standard normal random variable is between $-\infty$ and z, where $z = \dfrac{x - \mu}{\sigma}$.

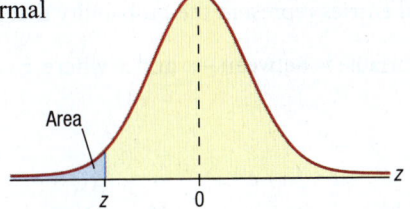

Area

z	0.09	0.08	0.07	0.06	0.05	0.04	0.03	0.02	0.01	0.00
−3.4	0.0002	0.0003	0.0003	0.0003	0.0003	0.0003	0.0003	0.0003	0.0003	0.0003
−3.3	0.0003	0.0004	0.0004	0.0004	0.0004	0.0004	0.0004	0.0005	0.0005	0.0005
−3.2	0.0005	0.0005	0.0005	0.0006	0.0006	0.0006	0.0006	0.0006	0.0007	0.0007
−3.1	0.0007	0.0007	0.0008	0.0008	0.0008	0.0008	0.0009	0.0009	0.0009	0.0010
−3.0	0.0010	0.0010	0.0011	0.0011	0.0011	0.0012	0.0012	0.0013	0.0013	0.0013
−2.9	0.0014	0.0014	0.0015	0.0015	0.0016	0.0016	0.0017	0.0018	0.0018	0.0019
−2.8	0.0019	0.0020	0.0021	0.0021	0.0022	0.0023	0.0023	0.0024	0.0025	0.0026
−2.7	0.0026	0.0027	0.0028	0.0029	0.0030	0.0031	0.0032	0.0033	0.0034	0.0035
−2.6	0.0036	0.0037	0.0038	0.0039	0.0040	0.0041	0.0043	0.0044	0.0045	0.0047
−2.5	0.0048	0.0049	0.0051	0.0052	0.0054	0.0055	0.0057	0.0059	0.0060	0.0062
−2.4	0.0064	0.0066	0.0068	0.0069	0.0071	0.0073	0.0075	0.0078	0.0080	0.0082
−2.3	0.0084	0.0087	0.0089	0.0091	0.0094	0.0096	0.0099	0.0102	0.0104	0.0107
−2.2	0.0110	0.0113	0.0116	0.0119	0.0122	0.0125	0.0129	0.0132	0.0136	0.0139
−2.1	0.0143	0.0146	0.0150	0.0154	0.0158	0.0162	0.0166	0.0170	0.0174	0.0179
−2.0	0.0183	0.0188	0.0192	0.0197	0.0202	0.0207	0.0212	0.0217	0.0222	0.0228
−1.9	0.0233	0.0239	0.0244	0.0250	0.0256	0.0262	0.0268	0.0274	0.0281	0.0287
−1.8	0.0294	0.0301	0.0307	0.0314	0.0322	0.0329	0.0336	0.0344	0.0351	0.0359
−1.7	0.0367	0.0375	0.0384	0.0392	0.0401	0.0409	0.0418	0.0427	0.0436	0.0446
−1.6	0.0455	0.0465	0.0475	0.0485	0.0495	0.0505	0.0516	0.0526	0.0537	0.0548
−1.5	0.0559	0.0571	0.0582	0.0594	0.0606	0.0618	0.0630	0.0643	0.0655	0.0668
−1.4	0.0681	0.0694	0.0708	0.0721	0.0735	0.0749	0.0764	0.0778	0.0793	0.0808
−1.3	0.0823	0.0838	0.0853	0.0869	0.0885	0.0901	0.0918	0.0934	0.0951	0.0968
−1.2	0.0985	0.1003	0.1020	0.1038	0.1056	0.1075	0.1093	0.1112	0.1131	0.1151
−1.1	0.1170	0.1190	0.1210	0.1230	0.1251	0.1271	0.1292	0.1314	0.1335	0.1357
−1.0	0.1379	0.1401	0.1423	0.1446	0.1469	0.1492	0.1515	0.1539	0.1562	0.1587
−0.9	0.1611	0.1635	0.1660	0.1685	0.1711	0.1736	0.1762	0.1788	0.1814	0.1841
−0.8	0.1867	0.1894	0.1922	0.1949	0.1977	0.2005	0.2033	0.2061	0.2090	0.2119
−0.7	0.2148	0.2177	0.2206	0.2236	0.2266	0.2296	0.2327	0.2358	0.2389	0.2420
−0.6	0.2451	0.2483	0.2514	0.2546	0.2578	0.2611	0.2643	0.2676	0.2709	0.2743
−0.5	0.2776	0.2810	0.2843	0.2877	0.2912	0.2946	0.2981	0.3015	0.3050	0.3085
−0.4	0.3121	0.3156	0.3192	0.3228	0.3264	0.3300	0.3336	0.3372	0.3409	0.3446
−0.3	0.3483	0.3520	0.3557	0.3594	0.3632	0.3669	0.3707	0.3745	0.3783	0.3821
−0.2	0.3859	0.3897	0.3936	0.3974	0.4013	0.4052	0.4090	0.4129	0.4168	0.4207
−0.1	0.4247	0.4286	0.4325	0.4364	0.4404	0.4443	0.4483	0.4522	0.4562	0.4602
−0.0	0.4641	0.4681	0.4721	0.4761	0.4801	0.4840	0.4880	0.4920	0.4960	0.5000

B   Standard Normal Distribution

Numerical entries represent the probability that a standard normal random variable is between $-\infty$ and z, where $z = \dfrac{x - \mu}{\sigma}$.

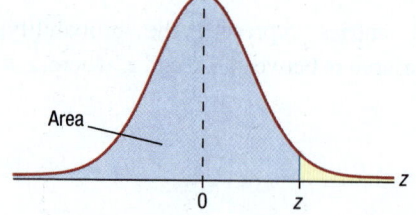

Area

z	0.00	0.01	0.02	0.03	0.04	0.05	0.06	0.07	0.08	0.09
0.0	0.5000	0.5040	0.5080	0.5120	0.5160	0.5199	0.5239	0.5279	0.5319	0.5359
0.1	0.5398	0.5438	0.5478	0.5517	0.5557	0.5596	0.5636	0.5675	0.5714	0.5753
0.2	0.5793	0.5832	0.5871	0.5910	0.5948	0.5987	0.6026	0.6064	0.6103	0.6141
0.3	0.6179	0.6217	0.6255	0.6293	0.6331	0.6368	0.6406	0.6443	0.6480	0.6517
0.4	0.6554	0.6591	0.6628	0.6664	0.6700	0.6736	0.6772	0.6808	0.6844	0.6879
0.5	0.6915	0.6950	0.6985	0.7019	0.7054	0.7088	0.7123	0.7157	0.7190	0.7224
0.6	0.7257	0.7291	0.7324	0.7357	0.7389	0.7422	0.7454	0.7486	0.7517	0.7549
0.7	0.7580	0.7611	0.7642	0.7673	0.7704	0.7734	0.7764	0.7794	0.7823	0.7852
0.8	0.7881	0.7910	0.7939	0.7967	0.7995	0.8023	0.8051	0.8078	0.8106	0.8133
0.9	0.8159	0.8186	0.8212	0.8238	0.8264	0.8289	0.8315	0.8340	0.8365	0.8389
1.0	0.8413	0.8438	0.8461	0.8485	0.8508	0.8531	0.8554	0.8577	0.8599	0.8621
1.1	0.8643	0.8665	0.8686	0.8708	0.8729	0.8749	0.8770	0.8790	0.8810	0.8830
1.2	0.8849	0.8869	0.8888	0.8907	0.8925	0.8944	0.8962	0.8980	0.8997	0.9015
1.3	0.9032	0.9049	0.9066	0.9082	0.9099	0.9115	0.9131	0.9147	0.9162	0.9177
1.4	0.9192	0.9207	0.9222	0.9236	0.9251	0.9265	0.9279	0.9292	0.9306	0.9319
1.5	0.9332	0.9345	0.9357	0.9370	0.9382	0.9394	0.9406	0.9418	0.9429	0.9441
1.6	0.9452	0.9463	0.9474	0.9484	0.9495	0.9505	0.9515	0.9525	0.9535	0.9545
1.7	0.9554	0.9564	0.9573	0.9582	0.9591	0.9599	0.9608	0.9616	0.9625	0.9633
1.8	0.9641	0.9649	0.9656	0.9664	0.9671	0.9678	0.9686	0.9693	0.9699	0.9706
1.9	0.9713	0.9719	0.9726	0.9732	0.9738	0.9744	0.9750	0.9756	0.9761	0.9767
2.0	0.9772	0.9778	0.9783	0.9788	0.9793	0.9798	0.9803	0.9808	0.9812	0.9817
2.1	0.9821	0.9826	0.9830	0.9834	0.9838	0.9842	0.9846	0.9850	0.9854	0.9857
2.2	0.9861	0.9864	0.9868	0.9871	0.9875	0.9878	0.9881	0.9884	0.9887	0.9890
2.3	0.9893	0.9896	0.9898	0.9901	0.9904	0.9906	0.9909	0.9911	0.9913	0.9916
2.4	0.9918	0.9920	0.9922	0.9925	0.9927	0.9929	0.9931	0.9932	0.9934	0.9936
2.5	0.9938	0.9940	0.9941	0.9943	0.9945	0.9946	0.9948	0.9949	0.9951	0.9952
2.6	0.9953	0.9955	0.9956	0.9957	0.9959	0.9960	0.9961	0.9962	0.9963	0.9964
2.7	0.9965	0.9966	0.9967	0.9968	0.9969	0.9970	0.9971	0.9972	0.9973	0.9974
2.8	0.9974	0.9975	0.9976	0.9977	0.9977	0.9978	0.9979	0.9979	0.9980	0.9981
2.9	0.9981	0.9982	0.9982	0.9983	0.9984	0.9984	0.9985	0.9985	0.9986	0.9986
3.0	0.9987	0.9987	0.9987	0.9988	0.9988	0.9989	0.9989	0.9989	0.9990	0.9990
3.1	0.9990	0.9991	0.9991	0.9991	0.9992	0.9992	0.9992	0.9992	0.9993	0.9993
3.2	0.9993	0.9993	0.9994	0.9994	0.9994	0.9994	0.9994	0.9995	0.9995	0.9995
3.3	0.9995	0.9995	0.9995	0.9996	0.9996	0.9996	0.9996	0.9996	0.9996	0.9997
3.4	0.9997	0.9997	0.9997	0.9997	0.9997	0.9997	0.9997	0.9997	0.9997	0.9998

C   Critical Values of the Pearson Correlation Coefficient

n	$\alpha = 0.05$	$\alpha = 0.01$
4	0.950	0.990
5	0.878	0.959
6	0.811	0.917
7	0.754	0.875
8	0.707	0.834
9	0.666	0.798
10	0.632	0.765
11	0.602	0.735
12	0.576	0.708
13	0.553	0.684
14	0.532	0.661
15	0.514	0.641
16	0.497	0.623
17	0.482	0.606
18	0.468	0.590
19	0.456	0.575
20	0.444	0.561
21	0.433	0.549
22	0.423	0.537
23	0.413	0.526
24	0.404	0.515
25	0.396	0.505
26	0.388	0.496
27	0.381	0.487
28	0.374	0.479
29	0.367	0.471
30	0.361	0.463
35	0.334	0.430
40	0.312	0.403
45	0.294	0.380
50	0.279	0.361
55	0.266	0.345
60	0.254	0.330
65	0.244	0.317
70	0.235	0.306
75	0.227	0.296
80	0.220	0.286
85	0.213	0.278
90	0.207	0.270
95	0.202	0.263
100	0.197	0.256

Note: r is statistically significant if $|r|$ is greater than or equal to the value given in the table.

Tax Forms

Form 1040 (2020)

Form **1040**	Department of the Treasury—Internal Revenue Service (99) **U.S. Individual Income Tax Return**	**2020**	OMB No. 1545-0074	IRS Use Only—Do not write or staple in this space.

Filing Status
Check only one box.

☐ Single ☐ Married filing jointly ☐ Married filing separately (MFS) ☐ Head of household (HOH) ☐ Qualifying widow(er) (QW)

If you checked the MFS box, enter the name of your spouse. If you checked the HOH or QW box, enter the child's name if the qualifying person is a child but not your dependent ▶

Your first name and middle initial	Last name		Your social security number
If joint return, spouse's first name and middle initial	Last name		Spouse's social security number

Home address (number and street). If you have a P.O. box, see instructions.		Apt. no.	**Presidential Election Campaign**

Check here if you, or your spouse if filing jointly, want $3 to go to this fund. Checking a box below will not change your tax or refund.

City, town, or post office. If you have a foreign address, also complete spaces below.	State	ZIP code

Foreign country name	Foreign province/state/county	Foreign postal code

☐ You ☐ Spouse

At any time during 2020, did you receive, sell, send, exchange, or otherwise acquire any financial interest in any virtual currency? ☐ Yes ☐ No

Standard Deduction

Someone can claim: ☐ You as a dependent ☐ Your spouse as a dependent
☐ Spouse itemizes on a separate return or you were a dual-status alien

Age/Blindness **You:** ☐ Were born before January 2, 1956 ☐ Are blind **Spouse:** ☐ Was born before January 2, 1956 ☐ Is blind

Dependents (see instructions):
If more than four dependents, see instructions and check here ▶ ☐

(1) First name Last name	(2) Social security number	(3) Relationship to you	(4) ✔ if qualifies for (see instructions):	
			Child tax credit	Credit for other dependents
			☐	☐
			☐	☐
			☐	☐
			☐	☐

Attach Sch. B if required.

1	Wages, salaries, tips, etc. Attach Form(s) W-2		**1**	
2a	Tax-exempt interest	2a	**b** Taxable interest	**2b**
3a	Qualified dividends	3a	**b** Ordinary dividends	**3b**
4a	IRA distributions	4a	**b** Taxable amount	**4b**
5a	Pensions and annuities	5a	**b** Taxable amount	**5b**
6a	Social security benefits	6a	**b** Taxable amount	**6b**
7	Capital gain or (loss). Attach Schedule D if required. If not required, check here ▶ ☐		**7**	
8	Other income from Schedule 1, line 9		**8**	
9	Add lines 1, 2b, 3b, 4b, 5b, 6b, 7, and 8. This is your **total income** ▶		**9**	
10	Adjustments to income:			
a	From Schedule 1, line 22	10a		
b	Charitable contributions if you take the standard deduction. See instructions	10b		
c	Add lines 10a and 10b. These are your **total adjustments to income** ▶		**10c**	
11	Subtract line 10c from line 9. This is your **adjusted gross income** ▶		**11**	
12	**Standard deduction or itemized deductions** (from Schedule A)		**12**	
13	Qualified business income deduction. Attach Form 8995 or Form 8995-A		**13**	
14	Add lines 12 and 13		**14**	
15	**Taxable income.** Subtract line 14 from line 11. If zero or less, enter -0-		**15**	

Standard Deduction for—
• Single or Married filing separately, $12,400
• Married filing jointly or Qualifying widow(er), $24,800
• Head of household, $18,650
• If you checked any box under *Standard Deduction,* see instructions.

For Disclosure, Privacy Act, and Paperwork Reduction Act Notice, see separate instructions.   Cat. No. 11320B   Form **1040** (2020)

Form 1040 (2020)   Page **2**

			16	
16	**Tax** (see instructions). Check if any from Form(s): **1** ☐ 8814 **2** ☐ 4972 **3** ☐ _____		16	
17	Amount from Schedule 2, line 3		17	
18	Add lines 16 and 17		18	
19	Child tax credit or credit for other dependents		19	
20	Amount from Schedule 3, line 7		20	
21	Add lines 19 and 20		21	
22	Subtract line 21 from line 18. If zero or less, enter -0-		22	
23	Other taxes, including self-employment tax, from Schedule 2, line 10		23	
24	Add lines 22 and 23. This is your **total tax** ▶		24	
25	Federal income tax withheld from:			
a	Form(s) W-2	25a		
b	Form(s) 1099	25b		
c	Other forms (see instructions)	25c		
d	Add lines 25a through 25c		25d	
26	2020 estimated tax payments and amount applied from 2019 return		26	

• If you have a qualifying child, attach Sch. EIC.
• If you have nontaxable combat pay, see instructions.

27	Earned income credit (EIC)	27		
28	Additional child tax credit. Attach Schedule 8812	28		
29	American opportunity credit from Form 8863, line 8	29		
30	Recovery rebate credit. See instructions	30		
31	Amount from Schedule 3, line 13	31		
32	Add lines 27 through 31. These are your **total other payments and refundable credits** ▶		32	
33	Add lines 25d, 26, and 32. These are your **total payments** ▶		33	

Refund

Direct deposit? See instructions.

34	If line 33 is more than line 24, subtract line 24 from line 33. This is the amount you **overpaid**		34	
35a	Amount of line 34 you want **refunded to you.** If Form 8888 is attached, check here ▶ ☐		35a	
▶ b	Routing number ☐☐☐☐☐☐☐☐☐ ▶ c Type: ☐ Checking ☐ Savings			
▶ d	Account number ☐☐☐☐☐☐☐☐☐☐☐☐☐☐☐☐☐			
36	Amount of line 34 you want **applied to your 2021 estimated tax** ▶	36		

Amount You Owe

For details on how to pay, see instructions.

37	Subtract line 33 from line 24. This is the **amount you owe now** ▶		37	
	Note: Schedule H and Schedule SE filers, line 37 may not represent all of the taxes you owe for 2020. See Schedule 3, line 12e, and its instructions for details.			
38	Estimated tax penalty (see instructions) ▶	38		

Third Party Designee

Do you want to allow another person to discuss this return with the IRS? See instructions ▶ ☐ **Yes.** Complete below. ☐ **No**

Designee's name ▶	Phone no. ▶	Personal identification number (PIN) ▶

Sign Here

Under penalties of perjury, I declare that I have examined this return and accompanying schedules and statements, and to the best of my knowledge and belief, they are true, correct, and complete. Declaration of preparer (other than taxpayer) is based on all information of which preparer has any knowledge.

Joint return? See instructions. Keep a copy for your records.

Your signature	Date	Your occupation	If the IRS sent you an Identity Protection PIN, enter it here (see inst.) ▶
Spouse's signature. If a joint return, **both** must sign.	Date	Spouse's occupation	If the IRS sent your spouse an Identity Protection PIN, enter it here (see inst.) ▶

Phone no.	Email address

Paid Preparer Use Only

Preparer's name	Preparer's signature	Date	PTIN	Check if: ☐ Self-employed
Firm's name ▶			Phone no.	
Firm's address ▶			Firm's EIN ▶	

Go to *www.irs.gov/Form1040* for instructions and the latest information.   Form **1040** (2020)

Schedule 1 (Form 1040)

SCHEDULE 1 (Form 1040) Department of the Treasury Internal Revenue Service	**Additional Income and Adjustments to Income** ▶ Attach to Form 1040, 1040-SR, or 1040-NR. ▶ Go to *www.irs.gov/Form1040* for instructions and the latest information.	OMB No. 1545-0074 **2020** Attachment Sequence No. **01**

Name(s) shown on Form 1040, 1040-SR, or 1040-NR	Your social security number

Part I — Additional Income

1	Taxable refunds, credits, or offsets of state and local income taxes	**1**	
2a	Alimony received	**2a**	
b	Date of original divorce or separation agreement (see instructions) ▶ _____		
3	Business income or (loss). Attach Schedule C	**3**	
4	Other gains or (losses). Attach Form 4797	**4**	
5	Rental real estate, royalties, partnerships, S corporations, trusts, etc. Attach Schedule E	**5**	
6	Farm income or (loss). Attach Schedule F	**6**	
7	Unemployment compensation	**7**	
8	Other income. List type and amount ▶ _____	**8**	
9	Combine lines 1 through 8. Enter here and on Form 1040, 1040-SR, or 1040-NR, line 8	**9**	

Part II — Adjustments to Income

10	Educator expenses	**10**	
11	Certain business expenses of reservists, performing artists, and fee-basis government officials. Attach Form 2106	**11**	
12	Health savings account deduction. Attach Form 8889	**12**	
13	Moving expenses for members of the Armed Forces. Attach Form 3903	**13**	
14	Deductible part of self-employment tax. Attach Schedule SE	**14**	
15	Self-employed SEP, SIMPLE, and qualified plans	**15**	
16	Self-employed health insurance deduction	**16**	
17	Penalty on early withdrawal of savings	**17**	
18a	Alimony paid	**18a**	
b	Recipient's SSN . . . . ▶		
c	Date of original divorce or separation agreement (see instructions) ▶ _____		
19	IRA deduction	**19**	
20	Student loan interest deduction	**20**	
21	Tuition and fees deduction. Attach Form 8917	**21**	
22	Add lines 10 through 21. These are your **adjustments to income**. Enter here and on Form 1040, 1040-SR, or 1040-NR, line 10a	**22**	

For Paperwork Reduction Act Notice, see your tax return instructions.   Cat. No. 71479F   Schedule 1 (Form 1040) 2020

Getting Started with Microsoft Excel (Desktop)

The Basics of Microsoft Office Excel 365

Microsoft Excel is a spreadsheet program that allows users to track and analyze data. Spreadsheets such as those created with Microsoft Excel are widely used in the business world to perform various tasks such as accounting, budgeting, billing, reporting, planning, and tracking. The instructions provided in this Appendix are intended for non-mobile users.

When you open Excel, you will see various tabs along the top such as File, Home, Insert, Page Layout, Formulas, Data, etc. The numerous commands and options available when working within Excel can be found under these tabs. At the bottom, there is a tab labeled Sheet1. The icon to the right of the Sheet1 tab creates a new tab with a blank spreadsheet. Tabs can be renamed by double-clicking on them and entering new text. Figure C.1 shows what a new workbook looks like when Excel is first opened.

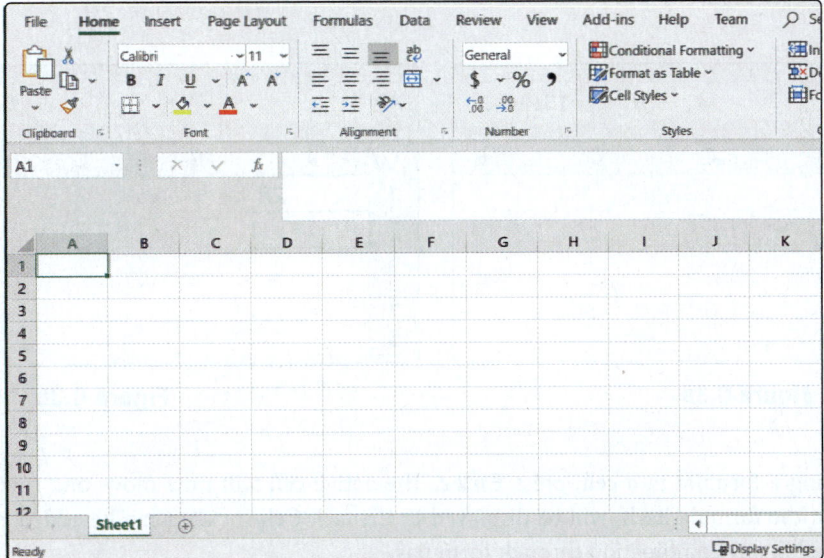

Figure C.1

Cells

Cells in Microsoft Excel may contain numerical data and/or text. The cells' locations are described by their positions in terms of columns and rows. Columns are listed from left to right across the top of the worksheet and are labeled with letters. Rows are listed from top to bottom along the left side of the worksheet and are labeled with numbers. A column letter followed by a row number describes the active cell position. For example, **B2** would be referring to the cell found in the second row of Column B (see Figure C.2). A thick solid border will outline the active cell. To help you identify the cell address, the column and row headers are also highlighted. Another way to identify the active cell's address is to look at the window just above the header for Column A.

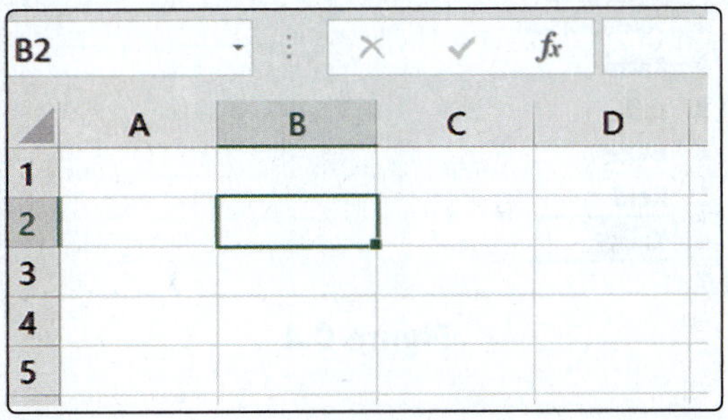

Figure C.2

The column width can be altered by placing your cursor between two column labels; when it turns into a vertical line with arrows pointing to the left and right, you can then click and drag to the left or right to make the column width larger or smaller. The same can be done with row heights. Alternatively, double-clicking between two column or row labels will autofit the row height or column width to the data it contains.

To change the active cell, move the mouse to the desired cell and click. The border will now be surrounding the new cell and the address will have changed in the active cell address window. The arrow keys can also control navigation of the active cell.

To the right of the cell address window is the formula window (labeled with f_x). This displays the contents of the active cell. The contents of the cell can be edited either within the formula window or the cell itself. Thus, as you enter data in the active cell, it is displayed within the cell and in the formula window at the top of the worksheet.

Suppose you wanted to sum the numbers 15, 10, and 3. To do this, enter the formula **=15+10+3** into cell **A1**. The equal sign, "=", is the command to start a formula in Excel. Notice that the formula is seen within in the cell as well as in the formula window (see Figure C.3a). When you press **Enter**, you will see that the active cell moves to cell **A2**, and the solution of 28 is now displayed in cell **A1** (see Figure C.3b).

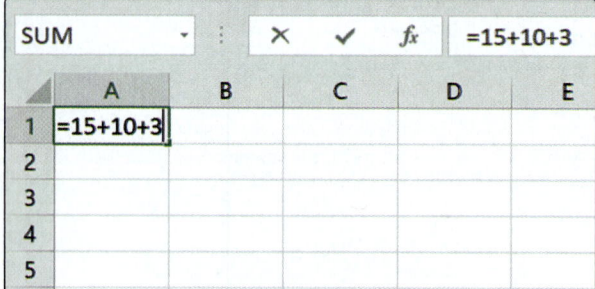

Figure C.3a

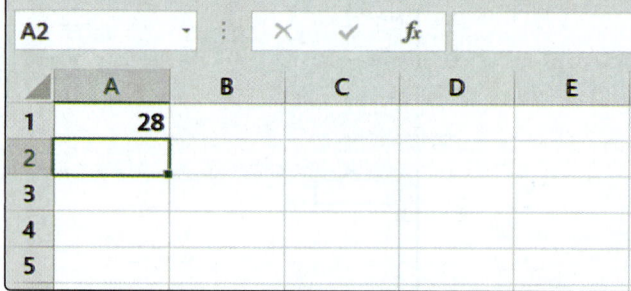

Figure C.3b

When you finish entering a formula in a cell, press **Enter**. The active cell will now move one row down. The answer to the formula, rather than the formula itself, will be displayed in the cell. Cells or data can be added, subtracted, multiplied, divided, and manipulated in any combination through formulas.

Filling Cells

Excel will try to anticipate a reoccurring word or phrase within a column. After you have entered some text in a column, if you type the first letter of a previous cell again, Excel will fill in the rest of the word or phrase. To accept the automatic completion of the word, just hit the **Enter** key or click in a new cell. If you do not want to use the automatic completion of the word, simply keep typing in the cell and the new word or phrase will appear. Under the Review tab, there is an option for Excel to spell check your spreadsheet.

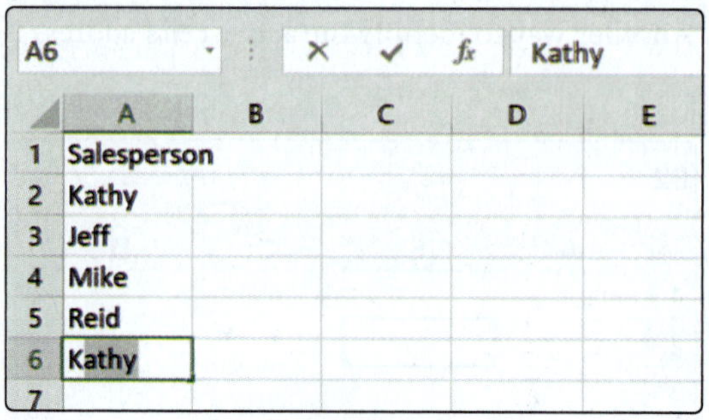

Figure C.4

Excel has a helpful tool when entering a series of data. Try entering 1 into cell **A1** and 2 into cell **A2**. Then highlight both of the cells. You will see a small square at the bottom right of the highlighted section.

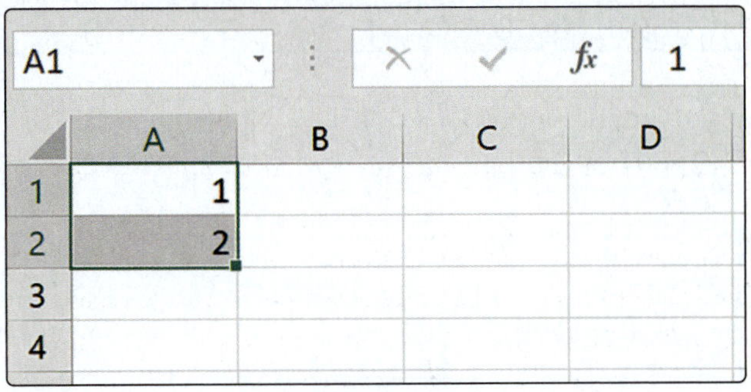

Figure C.5

If you move your cursor over this square, the cursor should change into a narrow plus sign. Now, click the left mouse button and drag the mouse down to include cells **A3** through **A10**. When you release the mouse button, the cells should now be filled with values from 1 to 10. We will refer to this as *filling cells*. Also notice that at the bottom of the screen, Excel gives some summary statistics about the highlighted cells. The average, count of the number of values, and sum are displayed for any highlighted cells.

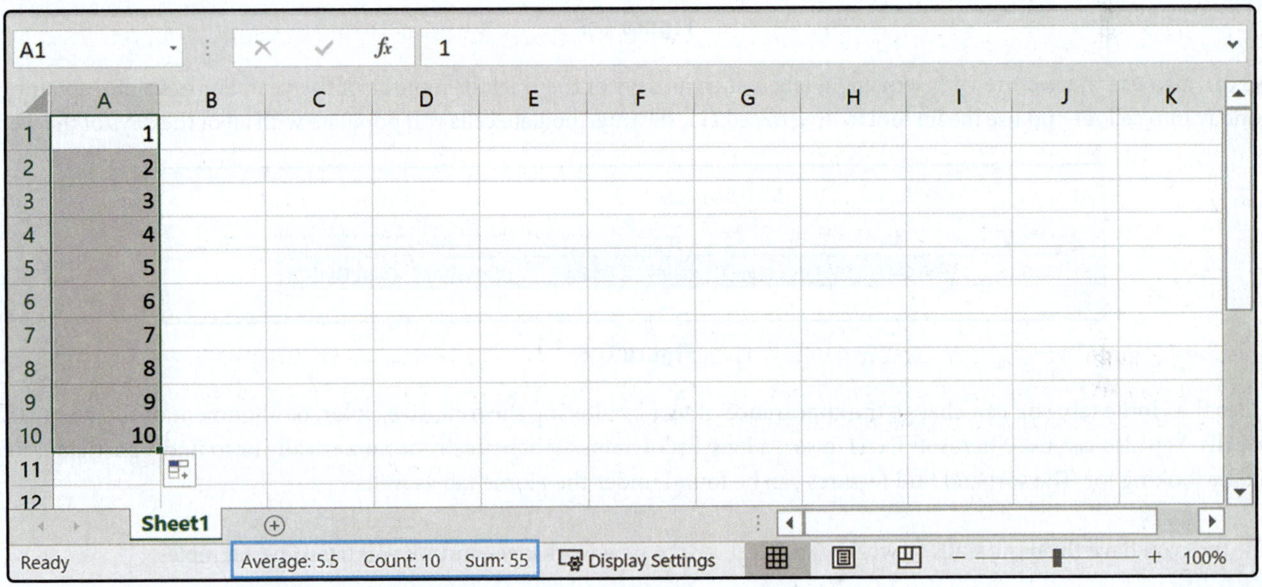

Figure C.6

The fill tool will work for other (nonsequential) values as well. Excel finds the relationship between two data points and replicates this when the tool is used. As an example, try this with multiples of 5. Enter 5 in cell **A1** and 10 in cell **A2**. Highlight cells **A1** and **A2** again and put your cursor over the square in the bottom right of the highlighted section. Click the left mouse button and drag down to cell **A10**. Release the button. Now the values should have changed to 5 through 50, in increments of 5.

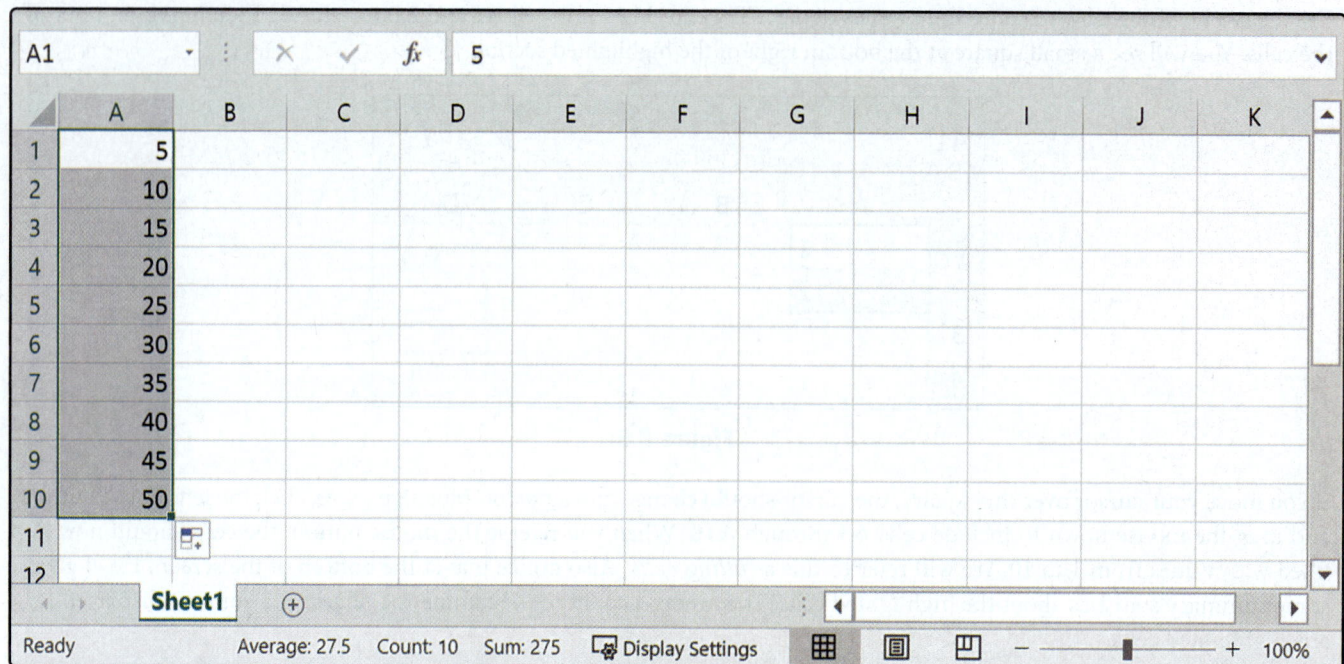

Figure C.7

You can also use this feature with sequential labels such as days of the week or months of the year. For example, if you type Monday into cell **A1** and use the fill tool to drag to cell **G1**, the intermediate cells will populate with all of the days of the week.

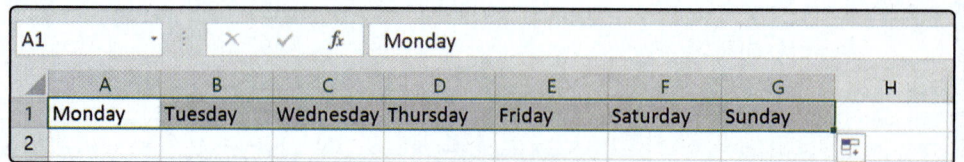

Figure C.8

Under the Home tab, you can change the appearance of text by altering the font, size, color, or alignment of the text within the cells. You also can use Microsoft Excel to sort a long list of items alphabetically or numerically or to find a particular item you are looking for. The sort and find features can be found under the Home tab as well.

Now that you have the basic skills to work with Excel, start a new worksheet and we will try some examples.

Formulas and Addressing Using Excel

Enter the labels Checking Balance and Savings Balance in cells **A1** and **B1**, respectively. You will notice that the label you typed in cell **A1** is cut off. We need to resize the columns to allow for the entire label. Move your cursor to the line that separates the column labels for A and B. Your cursor will change to an arrow pointing off to the left and right. You can click and drag the cursor to the right to increase the column width, or simply double-click to have the width autofit the label. Repeat this for Column B.

Next, select the column labels A and B and you should see both columns highlighted. Under the Home tab there is an area labeled Number. The buttons in this area format any numbers you enter in the selected cells. With columns A and B highlighted, click on the $ button. Now all of the values put in these columns will be formatted as currency.

Next, we will use the fill tool to get some values to work with. Start by entering 100 in cell **A2** and 200 in cell **A3**. Now move to cell **B2**, enter 1000, and enter 2000 in cell **B3**. Highlight cells **A2** through **B3** by clicking in **A2** and dragging down and to the right to include **B3**. With those cells highlighted, put your cursor over the square in the bottom right of the highlighted section. When your mouse changes to a narrow plus sign, click and drag down to row 11. When you release the mouse button, the cells should fill with data and end with $10,000.00 in cell **B11**. You should have a screen that looks like the following.

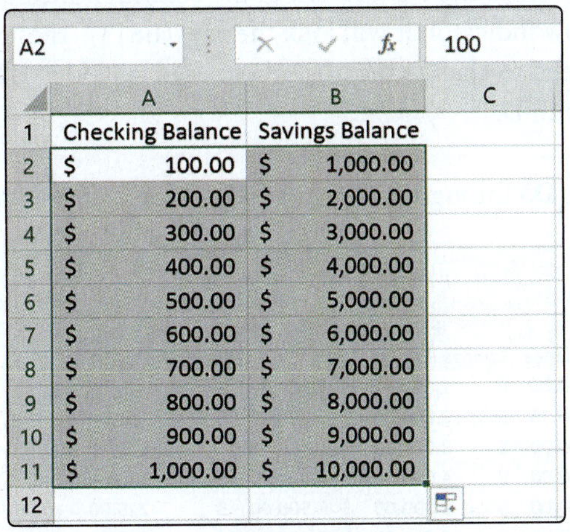

Figure C.9

Formulas can be applied to manipulate data between cells. We will see, however, that special attention needs to be paid to the copying of formulas. We will continue working with the above worksheet with the checking and savings balances while we try some formulas.

We want to see what your total bank balance would be if you had the amount listed in a particular row in your checking and savings accounts. In cell **C2**, type = to indicate you are entering a formula, and then click in cell **A2**, type +, and click in cell **B2**. Press **Enter**. You have just added cells **A2** and **B2** together, and the sum ($1,100.00) is displayed in cell **C2**. Under the Home tab, the Clipboard section contains buttons allowing you to cut, copy, and paste cells. With cell **C2** selected, either click on the **copy** button or use the keyboard shortcut **Ctrl+c** to copy the formula. Highlight cells **C3** through **C11**. Under the Home tab, press the **Paste** button or use the keyboard shortcut **Ctrl+v** to paste the formula into these cells. Notice that the value $1,100.00 that was displayed in **C2** was not copied, but the formula that made up that value. The formula changed for each of the rows and substituted the new row value in the formula. This is called *relative addressing*. The formula changes relative to its cell address. The last cell should show the value $11,000.00. If the value is not visible, the column needs to be resized.

Suppose we wanted to see what the annual interest would be on our savings account at the levels listed in Column B. Create a new label in cell **A13** called Interest Rate. Now in cell **B13**, input the interest rate 0.045. We will need to format the cell for percentages rather than currency. To do this, highlight cell **B13**, and under the Home tab in the Number area, select the **%** button next to the **$** button that we used to originally format the cells. The number will most likely be displayed as 5%. We can adjust the number of decimals displayed by clicking on the button, also located in the Number section under the Home tab. This button increases the number of decimal places displayed by one decimal place each time it is clicked. (Notice that there is a similar button next to the one mentioned that decreases the number of places displayed.) The value should now display as 4.5%.

Now we want to use this value to compute interest for each level of savings. Label cell **D1** Savings Interest. Resize the cell to fit the label. In cell **D2**, we will put our formula. Type = to start the formula, and then select cell **B13** to get the percentage rate. Type * to indicate multiplication, and select cell **B2**. Upon pressing **Enter**, the resulting value will be displayed ($45.00). Thus, the annual interest gained on $1,000.00 in savings at 4.5% APR is $45.00.

Since we have the formula, we can copy it to the rest of the column and get the interest income for each of the savings levels. With cell **D2** selected, click in the lower right-hand corner of the cell and drag down to **D3**. Is this the result you expected? Probably not. Remember the rules of relative addressing. As you copy the formula down one row, the formula values change by one row. So in cell **D3**, the formula is pulling from cell **B14** for the interest rate and cell **B3** for the principal amount. You will notice that cell **B14** is empty. Thus, Excel computes this formula as being equal to zero. However, want to use cell B13 for each of these formulas in Column D. So how do we lock in the address of cell **B13**? We use *absolute addressing*. The **$** symbol is the key to locking the position in the formula. If the $ is in front of the column indicator of a cell reference, it will lock the column ($B13), and if it is in front of the row indicator, it will lock the row (B$13). These can be used together to lock a cell reference to a specific cell. In cell **D2**, we need to change the formula to read =**$B$13*B2**. This will "lock" cell B13 into the formula when it is copied throughout Column D.

Go back and copy the new formula into cells **D3** through **D11** using the fill tool. The results should now look like Figure C.10.

D11			f_x	=B$13*B11	
	A	**B**	**C**	**D**	**E**
1	Checking Balance	Savings Balance		Savings Interest	
2	$ 100.00	$ 1,000.00	$ 1,100.00	$ 45.00	
3	$ 200.00	$ 2,000.00	$ 2,200.00	$ 90.00	
4	$ 300.00	$ 3,000.00	$ 3,300.00	$ 135.00	
5	$ 400.00	$ 4,000.00	$ 4,400.00	$ 180.00	
6	$ 500.00	$ 5,000.00	$ 5,500.00	$ 225.00	
7	$ 600.00	$ 6,000.00	$ 6,600.00	$ 270.00	
8	$ 700.00	$ 7,000.00	$ 7,700.00	$ 315.00	
9	$ 800.00	$ 8,000.00	$ 8,800.00	$ 360.00	
10	$ 900.00	$ 9,000.00	$ 9,900.00	$ 405.00	
11	$ 1,000.00	$ 10,000.00	$11,000.00	$ 450.00	
12					
13	Interest Rate		4.5%		
14					

Figure C.10

Excel can compute a multitude of calculations on data values in the spreadsheet. If you press the f_x button next to the formula window, you can explore all of the functions and formulas that Excel has to offer.

Charts

Suppose you have the following data about ticket sales from your county fair. Adult tickets are $20 and child tickets are $12. Ticket sales for the week are listed in Figure C.11.

C8			f_x	7740	
	A	**B**	**C**	**D**
1		Adult	Child	
2	Sunday	$ 10,120.00	$ 5,760.00	
3	Monday	$ 9,040.00	$ 3,600.00	
4	Tuesday	$ 8,380.00	$ 3,360.00	
5	Wednesday	$ 7,620.00	$ 3,132.00	
6	Thursday	$ 7,900.00	$ 2,520.00	
7	Friday	$ 10,560.00	$ 4,944.00	
8	Saturday	$ 13,260.00	$ 7,740.00	
9				

Figure C.11

From these data, you might see how the ticket sales change daily and wish to compare child and adult ticket sales. This information might be easier to understand if it were graphically displayed on a chart. Enter the data as seen above in a new worksheet and use a bar chart for this particular example.

To create a chart, we need to select the data to be graphed. We want to chart the ticket sales for both the adults and children. In a bar chart, the dollar amount will determine the height of the bar. The days of the week should also be included since they are the labels for the bars.

Highlight the data in Columns A, B, and C from rows 1 to 8. (Note that Excel can interpret the first row or column in a set of data as labels and not part of the data.) Under the Insert tab, there is a Charts group of icons.

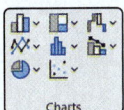

Figure C.12

Choose Column (in the upper left corner), and then select the top-left graph (the first one listed under the 2-D column heading). This is a clustered column graph. Excel creates a side-by-side bar graph based on the highlighted data. After the chart is created, a set of tabs labeled Chart Design and Format appear. Additionally, three icons appear to the right of the chart: Chart Elements, Chart Styles, and Chart Filters. These tabs and icons can be used to edit the chart that has been made.

Click on the chart title, type a new title of "Ticket Sales", and press **Enter**. Using the Chart Elements icon, you can edit the chart appearance by adding axis titles, data labels, or gridlines. Microsoft Excel makes it easy to create the chart you want.

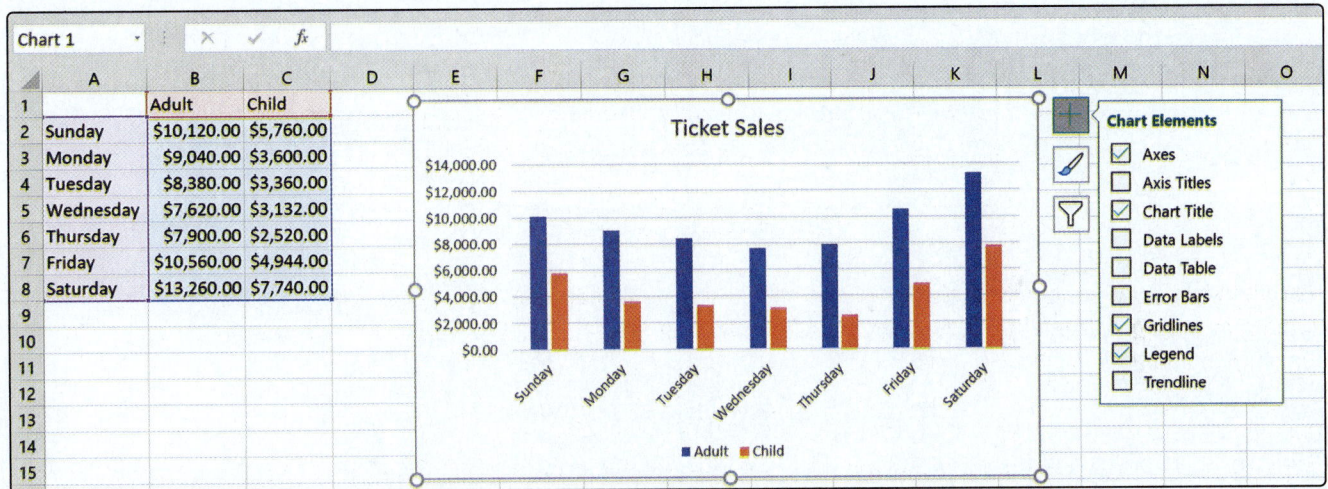

Figure C.13

Excel can create most types of charts, ranging from pie charts to line graphs to scatter plots. These types of charts are indicated by the menus seen when an icon in the Charts group is clicked on.

Answer Key

Chapter 1: Critical Thinking and Problem Solving

1.1 Exercises

1. Inductive   **3.** False

5. Deductive reasoning

7. Deductive reasoning

9. Deductive reasoning

11. Inductive reasoning

13. 9, 11, 13; arithmetic; common difference = 2

15. −5, −10, −15; arithmetic; common difference = −5

17. $\frac{1}{16}, \frac{1}{32}, \frac{1}{64}$; geometric;

common ratio = $\frac{1}{2}$

19. 100,000, 1,000,000, 10,000,000; geometric; common ratio = 10

21.

Geometric; common ratio = 2. (A figure with 24 sides)

23. Answers will vary. There are seven days in a week, so if the first Tuesday of the month was the 2nd (an even day), then seven days later that Tuesday's date would be the 9th (an odd day). Therefore, not every Tuesday in the month is an even day.

25. Answers will vary.

5 − 3 = 2. The difference between any two odd numbers is always an even number. Therefore, if the difference between two numbers is even, the numbers do not have to both be even.

27. Answers will vary.

If $a = 5$, $b = 1$, and $c = 2$ then $a > b$ and $a > c$, but $b < c$.

29. b   **31.** b

33. 10 and 15

 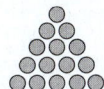

35.

9	3	4	5	6	8	1	2	7
8	2	6	7	1	4	5	9	3
1	5	7	9	2	3	4	6	8
2	7	8	1	5	9	3	4	6
6	4	1	3	8	7	2	5	9
3	9	5	6	4	2	7	8	1
5	6	3	4	9	1	8	7	2
7	8	9	2	3	5	6	1	4
4	1	2	8	7	6	9	3	5

37. a. 1,333,332   **b.** 1,333,332

c. Inductive reasoning

1.2 Exercises

1. order of magnitude

3. True

5. Answers may vary. Ten thousands

7. Answers may vary. Tens

9. b. Knowing the average weight of a newborn Asian elephant can help determine how large it must grow during the gestation period.

11. a. Knowing the number of employees who work in the office can be helpful if you can also estimate the amount of time a single person spends in a meeting.

13. $540,000   **15.** $7875

17. Answers will vary. Need to provide food, drinks, cake or cupcakes, decorations, and games.

19. Answers will vary. Need to lease or buy a building, to purchase all kitchen equipment, to purchase dining tables and chairs, and to set aside money to cover employee paychecks for 6 months.

21. Answers may vary.

a. $1120;   **b.** $7840

23. Answers may vary.

a. 5400 pounds

b. 64,800 pounds

25. Approximately 29,450 ears of corn per acre and 2,945,000 ears of corn for the season.

27. Approximately 7254 sprinkles

29. 810 tacos

31. 319 jelly beans

33. Answers will vary. For example,

$120/hr · 100 hr
+ 2($90/hr · 50 hr)
+ 3($120/hr · 100 hr) = $57,000

35. Answers will vary.
Bills come to about
$180 + $140 + $180 = $500 per month. So leftover income will be about $1700 − $500 = $1200.

37. Answers will vary.
($25,000 − $7500) ÷ 4 = $4375

39. Answers will vary.
9 · $5 + 6 · $30 = $225

41. 60 · 160 will be the most precise. Answers will vary. For example, these numbers are rounded to the nearest 10 instead of 100, so there will be less rounding error.

1.3 Exercises

1. True   **3.** False

5. 20 pieces of candy

7.

16	2	3	13
5	11	10	8
9	7	6	12
4	14	15	1

9.
③
② ⑥
⑤ ④ ①

11.
①
⑧ ⑤
③ ⑨
⑦ ⑥ ② ④

13.

4	3	8
9	5	1
2	7	6

15. a. $72 \cdot 1$; $36 \cdot 2$; $24 \cdot 3$; $18 \cdot 4$; $9 \cdot 8$; $8 \cdot 9$; $4 \cdot 18$; $3 \cdot 24$; $2 \cdot 36$; $1 \cdot 72$

b. $9 \cdot 8$ or $8 \cdot 9$

17. Marcus should purchase the miniature racetrack with cars, the fire station playset, the dinosaur expedition playset, and either the toy soldiers set or the construction truck set for a total of either $84.66 or $88.66.

19. Yes, Sofía can do the aquarium, art museum, and wine tasting for $148.

21. 405,450   **23.** 122,500

25. 140 total games

27.

29. Three 41¢ stamps and six 8¢ stamps

31. 6 packs of fudge swirls and 7 packs of extra chocolate chunk

33. 41 and 22   **35.** 4

Chapter 1 Exercises

1. Inductive reasoning

3. Answers will vary. Most Super Bowls have been played in January. For example, Super Bowl XXXVII was played on January 26, 2003.

5. Answers will vary. The sum of two odd numbers is always even, so if the sum is odd, then the numbers cannot both be odd. For example, $4 + 5 = 9$, and 4 is even, not odd.

7. 108, 216, 648; neither

9. $12,345,678 \cdot 8 + 8 = 98,765,432$

11. Answers may vary. Tens

13. Approximately 4544 people

15. 45 at $35 and 50 at $55

17.
    4
   2 3
   9 7
  5 1 8 6

19. 4; 7; 10; 13; 49

Chapter 2: Set Theory

2.1 Exercises

1. False; the set may be an infinite set.

3. True

5. True

7. True   **9.** True

11. $B = \{$Nebraska, Nevada, New Hampshire, New Jersey, New Mexico, New York, North Dakota, North Carolina$\}$

13. Answers will vary.

15. $F = \{$Monday, Tuesday, Wednesday, Thursday, Friday$\}$

17. $B = \{3, 6, 9, 12, 15\}$

19. $D = \{$Wyoming, Nebraska, Kansas, New Mexico, Oklahoma, Utah, Arizona$\}$

21. $|X| = 10$

23. $|Y| = 43$, as of printing

25. Neither   **27.** Neither

29. $H = \{x \mid x \in \mathbb{N}, x \le 50\}$

31. $K = \{x \mid x \in U, x \text{ is an athlete}\}$

33. $M = \{3x \mid x \text{ is a positive integer}\}$

35. $P = \{x^2 \mid x \text{ is a whole number}\}$

37. Not the empty set

39. Not the empty set

41. $A' = \{$c, d, f, g, h, i, j, l, m, n, o, p, q, r, u, v, w, x, y, z$\}$

43. $A' = \{$c, d, f, g, h, i, j, l, m, n, o, p, q, r, u, v, w, x, y, z, A, B, C, D, E, F, G, H, I, J, K, L, M, N, O, P, Q, R, S, T, U ,V, W, X, Y, Z$\}$

45. 4   **47.** 6

49. 15

51. Yes; they have the same elements, just in a different order.

53. Yes; P, R, and S are all equivalent. They all contain 5 elements. Q is not equivalent to any of them.

55. No; they have different elements.

57. B is equivalent to C since they each have 5 elements. A and D are not equivalent to the other sets.

2.2 Exercises

1. Venn diagram **3.** element

5. True

7. $A' = \{$Alex, Georgina, Charles, Olivia, Richard, Karl, Rhonda, Matthew$\}$

9. $M = \{5,10,15,20,25\}$, $N = \{12,24\}$, $P = \{2,4,9,16,25\}$

11. $M' = \{1, 2, 3, 4, 6, 7, 8, 9, 11, 12, 13, 14, 16, 17, 18, 19, 21, 22, 23, 24\}$

13. $Q = \{2, 4, 6, 8, 10, 12, 14, 16, 18\}$
$R = \{2, 4, 6, 8, 10, 12, 14, 16, 18\}$
$S = \varnothing$

15. $R' = \{1,3,5,7,9,11,13,15,17,19\}$

17. Yes

19. $\varnothing$, $\{$*The Card Players* by Cézanne$\}$, $\{$*No. 5 1948* by Pollock$\}$, $\{$*Woman III* by de Kooning$\}$, $\{$*The Card Players* by Cézanne, *No. 5 1948* by Pollock$\}$, $\{$*The Card Players* by Cézanne, *Woman III* by de Kooning$\}$, $\{$*No. 5 1948* by Pollock, *Woman III* by de Kooning$\}$, $\{$*The Card*

Players by Cézanne, *No. 5 1948* by Pollock, *Woman III* by de Kooning}.

21. ∅, {lemon}, {lime}, {lemon, lime}

23. ∅, {four}, {vier}, {cuatro}, {quatre}, {four, vier}, {four, cuatro}, {four, quatre}, {vier, cuatro}, {vier, quatre}, {cuatro, quatre}, {four, vier, cuatro}, {four, vier, quatre}, {four, cuatro, quatre}, {vier, cuatro, quatre}, {four, vier, cuatro, quatre}

25.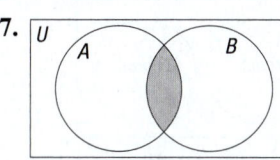

27. **a.**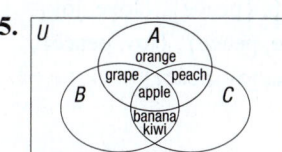

b.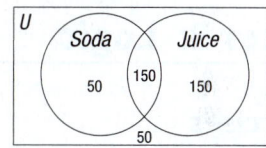

29.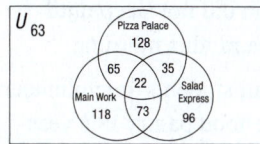

31. Proper subsets are *B* and *C*.

33. ∅, {2}, {4}, {6}, {2, 4}, {2, 6}, {4, 6}

35. ∅, {cloudy}, {rainy}, {sunny}, {cloudy, rainy}, {cloudy, sunny}, {rainy, sunny}

37. ∅

39. $2^8 = 256$ subsets and $2^8 - 1 = 255$ proper subsets

41. 5   **43.** 8

45. California   **47.** 22,825

49. Because you have to eliminate the set itself (where you choose all 6 keys) and you have to eliminate the empty set (where you don't choose any keys), all proper non-empty subsets: $2^6 - 2 = 62$.

2.3 Exercises

1. intersection   **3.** complement

5. $A \cap B'$

7.

9.

11.

13. $A \cap B = \{4\}$ **15.** $|A \cap B| = 1$

17. $X \cup Y = \{1, 3, 5, 13, 21\}$

19. $|X \cup Y| = 5$

21. We have $(X \cup Y)' = \{2, 8, 34\}$. Since $X' = \{2, 8, 34\}$ and $Y' = \{1, 2, 5, 8, 13, 34\}$, we have $X' \cap Y' = \{2, 8, 34\}$. Therefore, $(X \cup Y)' = X' \cap Y'$.

23. $H \cap R = \{r, i, h\}$

25. $|H \cap R| = 3$

27. $D \cup N = \{d, a, n, c, e, o, t\}$

29. $|D \cup N| = 7$

31. We have $(D \cap N)' = \{d, u, c, a, t, i, o\}$. Since $D' = \{u, t, i, o\}$ and $N' = \{d, u, c, a, i\}$, $D' \cup N' = \{d, u, c, a, t, i, o\}$. Therefore, $(D \cap N)' = D' \cup N'$.

33. $B \cup (C \cap P)$ $= \{b, i, o, g, r, a, p, h, y, c\}$

35. We have $C \cup P = \{c, h, i, p, o, t, l, e, r, a, y\}$, so $(C \cup P)' = \{g, b\}$. Similarly, $C' = \{y, r, g, a, b\}$ and $P' = \{o, g, h, t, b, l, e\}$, so $C' \cap P' = \{g, b\}$. Therefore, $(C \cup P)' = C' \cap P'$.

37. $S \cup (M \cap T) = \{s, n, o, w, m, e, l, t, d\}$

39. We have $M \cap T = \{o, d, e\}$, so $(M \cap T)' = \{m, l, t, w, n, s\}$. Similarly, $M' = \{t, w, n, s\}$ and

$T' = \{m, l, n, s\}$ so $M' \cup T' = \{m, l, t, w, n, s\}$. Therefore, $(M \cap T)' = M' \cup T'$.

41. Will, David, Kim, Barbara, Alden, Morgan, Ali, Holly, Jessica, Jeff, Kent

43. 11   **45.** 32

2.4 Exercises

1. four   **3.** True

5.

7. 50 students

9. **a.** 21   **b.** 1   **c.** 20

11. **a.** $\left|(A \cup B \cup C)'\right| = 3$. This is the number of people who did not like any part of the meal;

b. $|B \cup C| = 69$. This is the number of people who liked the main course or dessert or both.

13. 600 students were surveyed.

15. **a.** 62   **b.** 17   **c.** 5

17. **a.** 33   **b.** 111   **c.** 24

19. 92; After accounting for each of the patients in categories A, B, and Rh, there were 92 of the 200 patients not accounted for. Therefore, those 92 must be O−.

Chapter 2 Exercises

1. False; the set containing 3 is not an element of the set.

3. True   **5.** True

7. False; the cardinal number of the empty set is 0.

9. $A = \{2, 4, 6, 8, 10, 12\}$

11. $C = \{x \mid x \in \mathbb{R}, 100 < x < 1000\}$

13. $|A| = 2, |B| = 3$

15. Proper subsets of A: $\varnothing$, $\{$Felix$\}$, $\{$Amber$\}$

17. No, A has 2 elements and B has 3 elements.

19. Subsets of G: $\varnothing$, $\{$I$\}$, $\{$II$\}$, $\{$III$\}$, $\{$I,II$\}$, $\{$I,III$\}$, $\{$II, III$\}$, $\{$I, II, III$\}$; Subsets of F: $\varnothing$, $\{$love$\}$, $\{$joy$\}$, $\{$peace$\}$, $\{$love, joy$\}$, $\{$love, peace$\}$, $\{$joy, peace$\}$, $\{$love, joy, peace$\}$

21. No, they contain different elements.

23. 255

25.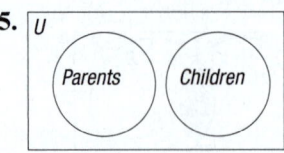

Universal set will vary.

27.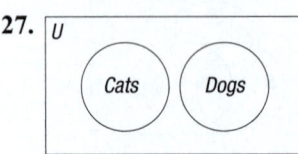

Universal set will vary.

29. $\{1, 3, 5\}$

31. $\{7\}$ **33.** 3

35. $\{$m, e, d, i, c, a, l, n, k$\}$

37. $\{$n, k$\}$

39. $\{$b, f, h, j, o, p, q, r, s, v, w, x, y, z$\}$

41. $\{$a, b, c, d, e, f, g, h, i, j, k, l, m, n, o, p, q, r, s, t, u, v, w, x, y, z$\}$

43. $|A \cap B| = 2$

45. a.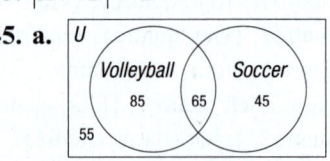

b. 85   **c.** 45   **d.** 55

47. 33%

Chapter 3: Logic

3.1 Exercises

1. paradox   **3.** and, or

5. True   **7.** Yes

9. No   **11.** No

13. No   **15.** Yes

17. Yes

19. 329 is not the number of people who applied for the same job I did.

21. Austin did not sleep until 8:00 a.m. this morning.

23. Not all students are volunteering at the food pantry this year.

25. $\sim m \vee \sim n$ **27.** $\sim b \vee a$

29. $t \Rightarrow \sim r$ **31.** $w \Rightarrow z$

33. $\sim a \Leftrightarrow b$

35. Driving makes me smile or it is sunny.

37. If it is sunny, then I grill more often than I bake.

39. If it is not sunny, then I grill no more than I bake.

41. If I do not have 1000 Facebook friends, then the home page of my website does not have a bounce rate of less than 20%.

43. The home page of my website does not have a bounce rate of less than 20% if and only if my video did not reach 1000 views on YouTube.

45. The moon is not full or I don't know if it's cloudy or bright outside.

47. The moon is full if and only if I do know if it's cloudy or bright outside.

49. If I have not lost my glasses, then I do know if it's cloudy or bright outside.

3.2 Exercises

1. truth values

3. both a and b are false

5. True

7.

w	z	~w	~z	~w ∨ ~z
T	T	F	F	F
T	F	F	T	T
F	T	T	F	T
F	F	T	T	T

9.

m	n	~n	m ⇔ ~n
T	T	F	F
T	F	T	T
F	T	F	T
F	F	T	F

11.

s	t	s ⇔ t	~(s ⇔ t)
T	T	T	F
T	F	F	T
F	T	F	T
F	F	T	F

13.

p	r	q	p ∨ r	(p ∨ r) ⇒ q
T	T	T	T	T
T	T	F	T	F
T	F	T	T	T
T	F	F	T	F
F	T	T	T	T
F	T	F	T	F
F	F	T	F	T
F	F	F	F	T

15.

m	n	r	m ∧ n	(m ∧ n) ⇔ r
T	T	T	T	T
T	T	F	T	F
T	F	T	F	F
T	F	F	F	T
F	T	T	F	F
F	T	F	F	T
F	F	T	F	F
F	F	F	F	T

17.

a	b	c	~a	~b	~b⇔c	(~b⇔c) ⇒ ~a
T	T	T	F	F	F	T
T	T	F	F	F	T	F
T	F	T	F	T	T	F
T	F	F	F	T	F	T
F	T	T	T	F	F	T
F	T	F	T	F	T	T
F	F	T	T	T	F	T
F	F	F	T	T	F	T

19. *a*: I plan to go to the movies.; *b*: I plan to go out to eat this weekend.; $a \wedge b$

a	b	a∧b
T	T	T
T	F	F
F	T	F
F	F	F

21. *a*: My calculator is working correctly; *b*: I need a new calculator; *c*: My calculator calculates that one plus one equals three; $(\sim a \wedge b) \Leftrightarrow c$

a	b	c	~a	~a∧b	(~a∧b)⇔c
T	T	F	F	F	F
T	T	F	F	F	T
T	F	T	F	F	F
T	F	F	F	F	T
F	T	T	T	T	T
F	T	F	T	T	F
F	F	T	T	F	F
F	F	F	T	F	T

23. *a*: Meg gets gas.; *b*: Meg's car will break down.; *c*: Meg will miss her exam.; $\sim a \Rightarrow (b \wedge c)$

a	b	c	~a	b∧c	~a⇒(b∧c)
T	T	T	F	T	T
T	T	F	F	F	T
T	F	T	F	F	T
T	F	F	F	F	T
F	T	T	T	T	T
F	T	F	T	F	F
F	F	T	T	F	F
F	F	F	T	F	F

25. Not a tautology

27. Tautology

29. Not a tautology; *a*: The bird is blue; *b*: The bird is green; *c*: The bird is from here; $(a \vee b) \Rightarrow \sim c$

a	b	c	~c	a∨b	(a∨b)⇒~c
T	T	T	F	T	F
T	T	F	T	T	T
T	F	T	F	T	F
T	F	F	T	T	T
F	T	T	F	T	F
F	T	F	T	T	T
F	F	T	F	F	T
F	F	F	T	F	T

31. Tautology; *a*: The animal is a penguin; *b*: The animal wears a bow tie; $(a \Rightarrow b) \Leftrightarrow (\sim b \Rightarrow \sim a)$

a	b	~a	~b	a⇒b	~b⇒~a	(a⇒b)⇔ (~b⇒~a)
T	T	F	F	T	T	T
T	F	F	T	F	F	T
F	T	T	F	T	T	T
F	F	T	T	T	T	T

3.3 Exercises

1. Logically equivalent

3. conjunction

5. False

7. None are logically equivalent.

p	q	p⇒q	~(p⇒q)	p∨q	~(p⇒q) ⇒ p∨q
T	T	T	F	T	T
T	F	F	T	T	T
F	T	T	F	T	T
F	F	T	F	F	T

p	q	~p	~q	~p∧~q	p∨q	~(p∨q)	(~p∧~q) ⇔~(p∨q)
T	T	F	F	F	T	F	T
T	F	F	T	F	T	F	T
F	T	T	F	F	T	F	T
F	F	T	T	T	F	T	T

9. $p \wedge (\sim p \vee q)$ and $p \wedge q$ are logically equivalent.

p	q	~p	~p∨q	p∧ (~p∨q)	p∧q	[p∧(~p∨q)] ⇔(p∧q)
T	T	F	T	T	T	T
T	F	F	F	F	F	T
F	T	T	T	F	F	T
F	F	T	T	F	F	T

11. Logically equivalent

m	n	~n	n⇒m	m∨~n
T	T	F	T	T
T	F	T	T	T
F	T	F	F	F
F	F	T	T	T

13. Logically equivalent

r	s	~s	r⇒s	r∧~s	~(r⇒s)
T	T	F	T	F	F
T	F	T	F	T	T
F	T	F	T	F	F
F	F	T	T	F	F

15. Logically equivalent

p	q	~p	q∧~p	p∨q	p∨(q∨~p)
T	T	F	F	T	T
T	F	F	F	T	T
F	T	T	T	T	T
F	F	T	F	F	F

17. The dog will not wear a bow tie or will wear a hat, or the dog will not have his picture taken.

19.

p	q	~p	~q	p∨q	~(p∨q)	~p∧~q
T	T	F	F	T	F	F
T	F	F	T	T	F	F
F	T	T	F	T	F	F
F	F	T	T	F	T	T

21. $p \wedge q$ **23.** $\sim(\sim p \wedge \sim q)$

25. $a \wedge \left(\sim(c \vee d)\right)$

$a \wedge (\sim c \wedge \sim d)$

$a \wedge \sim c \wedge \sim d$

27. $(w \vee z) \wedge \sim(w \wedge z)$

$(w \vee z) \wedge (\sim w \vee \sim z)$

29. I leave my computer alone for 15 minutes, and it does not go to sleep.

31. I am in Charleston, South Carolina, and I am not on Eastern Standard Time.

33. Statements *a* and *c* are logically equivalent to the given conditional statement.

35. It is not true that in Biology the nucleotide bases adenine and thymine do not pair together in DNA or the bases pair together in RNA.

37. Converse: If I get an A in Biology, then I got an 89 on the final. Inverse: If I do not get an 89 on the final, then I will not get an A in Biology. Contrapositive: If I do not get an A in Biology, then I did not get an 89 on the final. Biconditional: I will get an 89 on the final if and only if I get an A in Biology.

39. Converse: If I smile, then I have seen a puppy. Inverse: If I do not see a puppy, then I will not smile. Contrapositive: If I do not smile, then I have not seen a puppy. Biconditional: I have seen a puppy if and only if I smile.

41. $p \Rightarrow \sim q$ **43.** $q \Rightarrow p$

3.4 Exercises

1. premise, conclusion

3. sound   **5.** True

7. Premise: We stop burning fossil fuels today. Conclusion: There is enough carbon dioxide in the atmosphere that temperatures will continue to rise for a few hundred years.

9. Premise: Fast food is easily available in grocery shops, gas stations, and dispensers everywhere. Conclusion: Fast food obesity has strikingly increased.

11. Premise: A man is struck down by a heart attack in the street. Conclusion: Americans will care for him whether or not he has insurance.

13. Premises: Penguins are black and white.; Some old TV shows are black and white. Conclusion: Some penguins are old TV shows.

15. Premises: All potatoes have skin; I have skin. Conclusion: I must be a potato.

17. Inductive, invalid

19. Inductive, invalid

21. Deductive, valid

23. Valid   **25.** Invalid

27. Valid

29. Missing piece: The hand sanitizer was bigger than 3.4 fl oz.

31. Missing piece: My guitar is tuned to an open D tuning.

33. Missing piece: Emma is over 16.

35. Missing piece: You did not buy a new car.

37. a. Premise 1: If a movie is extremely popular and financially successful, then it is considered a blockbuster, Premise 2: The 2009 movie *Avatar* was not a blockbuster, Conclusion: *Avatar* was not popular nor financially successful.

b. p: A movie is extremely popular and financially successful; q: It is considered a blockbuster; $((p \Rightarrow q) \wedge \sim q) \Rightarrow \sim p$

c. The argument is a tautology, and therefore is valid.

p	q	$\sim p$	$\sim q$	$p \Rightarrow q$	$(p \Rightarrow q)$ $\wedge \sim q$	$((p \Rightarrow q) \wedge$ $\sim q) \Rightarrow \sim p$
T	T	F	F	T	F	T
T	F	F	T	F	F	T
F	T	T	F	T	F	T
F	F	T	T	T	T	T

d. The argument is not sound because premise 2 is untrue. In fact, the movie *Avatar* is one of the highest-grossing films of all time.

39. False dilemma

41. Post hoc, ergo propter hoc

43. Ad hominem

45. Dicto simpliciter

Chapter 3 Exercises

1. The puppy could keep her eyes open after 10:00.

3. Not all houses have fireplaces (or at least one house does not have a fireplace).

5.

a	b	c	$\sim b$	$a \vee \sim b$	$(a \vee \sim b) \Rightarrow c$
T	T	T	F	T	T
T	T	F	F	T	F
T	F	T	T	T	T
T	F	F	T	T	F
F	T	T	F	F	T
F	T	F	F	F	T
F	F	T	T	T	T
F	F	F	T	T	F

7. *i*: I go to the movies. *y*: You go to the movies. *k*: Kathy goes to the movies.

9. Converse: If I enroll in the next course, then my grade in this course will be an A. Inverse: If my grade in this course is not an A, then I cannot enroll in the next course. Contrapositive: If I cannot enroll in the next class, then my grade in this course was not an A. Biconditional: My grade in this course will be an A if and only if I can enroll in the next course.

11.

p	q	$p \Rightarrow q$	$p \wedge (p \Rightarrow q)$	$(p \wedge (p \Rightarrow q)) \Rightarrow q$
T	T	T	T	T
T	F	F	F	T
F	T	T	F	T
F	F	T	F	T

13. a.

15. Valid argument

17. Valid argument

19. Premise: Criminals support global warming; Conclusion: If you support global warming, you are a criminal

21. False dilemma

Chapter 4: Ratios, Percentages, Rates, and Proportionality

4.1 Exercises

1. fraction

3. percentage

5. False **7.** 20%

9. 28.77%

11. one-sixth of 138

13. One-third of 93

15. 96

17. **a.** 10 **b.** 14

19. **a.** 4.5 **b.** 7.5

21. Answers will vary; possible answers must include 2 : 5 and may include 48 : 120

23. Answers will vary; possible answers must include 5 : 15 : 9 and may include 50 : 150 : 90

25. **a.** $\dfrac{\$750}{\$2800} = \dfrac{15}{56}$

 b. $\dfrac{\$800}{\$2800} = \dfrac{2}{7}$

27. $\dfrac{520}{950} = \dfrac{52}{95}$

29. $\dfrac{1}{100}$ are left-handed

31. $\dfrac{89}{100}$ do not report having a disability

33. 16.25%

35. 30%

37. Toffee with a 76.2% delicious rating.

39.

State	Proportion of Eligible Voters who Voted	# of Eligible Voters	# of Votes Cast
Illinois	67.02%	9,027,082	6,049,950
Minnesota	79.96%	4,118,462	3,293,122
Tennessee	59.81%	5,124,867	3,065,183
Louisiana	64.61%	3,373,932	2,179,897

41. Hindi: 957,670,980 people; Christian: 2,257,367,310 people

43. 246 nonherbivores

45. City A: 2467 citizens; City B: 1898 citizens

47. Utah: 64,000 people; South Dakota: 17,700 people

49. 69 were not there to learn about online courses

51. **a.** $\dfrac{3}{2}$ **b.** 8 drops

53. **a.** $\dfrac{3}{8}$ **b.** 3 cups

55. 50,000 : 415,000; 465,000 elephants

57. 12,748 : 10,473; 23,221 students

59. $\dfrac{1}{2}$ or 1 : 2 or 1 to 2

61. $\dfrac{18}{1.3}$ or 18 : 1.3 or 18 to 1.3

63. **a.**

City	# of Billionaires	Population	Unit Ratio
Mumbai, India	38	20,000,000	1 : 526,316
Los Angeles, USA	44	4,000,000	1 : 90,909
Beijing, China	57	20,400,000	1 : 357,895
London, UK	66	9,300,000	1 : 140,909
Moscow, Russia	73	12,500,000	1 : 171,233

 b. All five cities have a lower ratio of billionaires within the population than does New York City. Note that London has the most similar city population size while Los Angeles has the most similar ratio of billionaires within the population. For Mumbai with its population of 20,000,000 to have the same ratio of billionaires as New York City, it would need 269 billionaires.

65. 60 locals

67. 1092 uninsured cars

4.2 Exercises

1. list price **3.** 1.04

5. original amount

7. 56; 400% increase

9. 323; 68% decrease

11. 19.22; 31% decrease

13. $16.86 **15.** $1.23

17. $13.27 **19.** $1.43

21. $2.66 **23.** $38.49

25. $629.85 **27.** $243.20

29. False **31.** $79.52

33. $54.36

35. approximately $34.90

37. $189.88 **39.** $124.20

41. $8.81 **43.** $523.67

45. $416.15 **47.** $345.60

49. $8340 **51.** $95,400

53. 20% decrease

55.

Change Between	Absolute Change	Percentage Change
1901 and 2001	207,518,075	267.48% increase
1901 and 2011	235,097,446	303.02% increase
2001 and 2011	27,579,371	9.67% increase

57. $21,360 **59.** $0.48

61. 37.04 million

63. Company Z has the better absolute change but both companies have the same percentage growth.

4.3 Exercises

1. rate

3. rate of change

5. false

7. $\dfrac{\$4800}{15 \text{ credit hours}} = \dfrac{\$320}{1 \text{ credit hour}}$

9. $\dfrac{15¢}{1 \text{ min}}$ **11.** $\dfrac{\$1473.00}{1 \text{ month}}$

13. $\dfrac{65 \text{ hours}}{3 \text{ weeks}}$

15. $\dfrac{\$3732}{\$32,750} \approx 11.4\%$

17. $\dfrac{3932}{29,070} \approx 13.5\%$

19. b.   **21.** a.

23. 13.5 grams   **25.** 340 milliliters

27. 56.28 inches   **29.** 2484 words

31. 25.17 mpg   **33.** a.

35. $18.59/day

37. $0.19/mile or 19.12¢/mile

39. $0.75/pound or 74.75¢/pound

41. $1.25/eggplant; 0.80 eggplants/$1

43. a. US Dollars to Japanese Yen
$$= \dfrac{0.009121 \text{ USD}}{1 \text{ JPY}}$$
b. Japanese Yen to US Dollars
$$= \dfrac{109.642453 \text{ JPY}}{1 \text{ USD}}$$
c. 4.56 USD

45. a. Shortly after January 2020

b. Approximately −0.8% per year

c. January 2018 to January 2019 had very little change

d. Shortly after January 2020 to January 2021

4.4 Exercises

1. 1   **3.** diagonal

5. False

7. Approximately 0.28 ounces per inch

9. 120 km/hr

11. 2400 calories/lb

13. $120/day

15. Approximately 1.89 liters per hour

17. Approximately 62 miles per hour

19. 122 centimeters/year

21. 72 ounces/day

23. 5.96 kilometer/liter

25. 0.537 kilometers/minute

27. 9 cups/day

29. 15.14 liters/minute

31. a. $54 per hour

b. Approximately $25.14 per hour

c. Approximately $45.88 per hour

33. a. 50 drops/minute

b. 45 drops/minute

c. 40 drops/minute

35. Approximately 8.66×10^{-5} inches

4.5 Exercises

1. proportional

3. solving a problem

5. origin   **7.** $k = 5$

9. $k = 11$

11. a. 14 N/m;   **b.** 3 meters

13. a. 25 N/m;   **b.** 9 meters

15. $\dfrac{1}{48}$ **17.** $\dfrac{1}{87}$

19. 26.6 inches   **21.** 26.5 inches

23. 2.6 inches

25. 22 indoor swimmers

27. 2.5 boxes

29. 384 chocolate hearts

31. 120 T-shirts   **33.** $6.91

35. $4.50

37. 250 cans

39. No, the values do not increase at the same rate each time.

41. No, the graph of the values doesn't go through the origin.

43. Yes; $k = 400$

45. No, the graph of the values doesn't go through the origin.

Chapter 4 Exercises

1. 360 people   **3.** $\dfrac{3}{7}$

5. 68 games   **7.** $11.70

9. $647.06   **11.** 25%

13. $\dfrac{42 \text{ students}}{3 \text{ advisors}} = \dfrac{14 \text{ students}}{1 \text{ advisors}}$

15. $5.99 for a package of 8, since that is about $0.75 each instead of $1.43 each.

17. 30.72 ounces

19. a. Approximately 0.21 mm per minute;

b. Approximately 0.85 mm per minute;

c. Approximately 0.004 mm per minute

21. 2.5 cups/container

23. 5.42 feet

25. $\dfrac{1}{25}$

27. 184 inches

Chapter 5: Algebra: Equations, Inequalities, and Functions

5.1 Exercises

1. one

3. infinitely many

5. ordered pairs   **7.** $x = \dfrac{8}{7} = 1\dfrac{1}{7}$

9. $x = 3$ **11.** No solution

13. $z = 10.9$ **15.** $y = -3$

17. Infinitely many solutions

19. $k = gt - v$ **21.** $r = \dfrac{I}{Pt}$

23. $h = \dfrac{V}{\pi r^2}$ **25.** −3

27. 856   **29.** 12

31. 1.08

33. Domain: all real numbers; Range: all real numbers greater than or equal to 5.

35. 20 attempts   **37.** $47 per ticket

39. $x = 4.9$

41. a. $x = \dfrac{3}{8}$ **b.** $x = \dfrac{3}{4}$ **c.** $x = \dfrac{5}{8}$

43. 1.75 hours

45. a. False. The minimum and maximum values are not the same for each graph. The range of the 2018 graph is $[82,100]$. The range of the 2019 graph is $[81,98]$. The range of the 2020 graph is $[80,99]$.

b. True. All three graphs have a domain of the months January through December.

c. True. The range of the 2019 graph is $[81,98]$, which falls within the range of the 2020 graph, $[80,99]$.

47. Domain: nonnegative integers less than or equal to 150; Range: Real numbers greater than or equal to $7000

5.2 Exercises

1. slope   **3.** slope-intercept

5. Parallel   **7.** 0

9. $\dfrac{1}{2}$

11. $m = -4$; y-intercept: $(0,0)$

13. $m = \dfrac{1}{2}$; y-intercept: $\left(0, \dfrac{3}{2}\right)$

15. $m =$ undefined; y-intercept: none

17. $y = -6x + 28$

19. $y = 8.4x - 1.2$

21. $m = -\dfrac{5}{4}$; $b = \dfrac{9}{2}$; $\left(\dfrac{13}{5}, \dfrac{5}{4}\right)$

23. $m = -3.8$; $b = 1.4$; $(13, -48)$

25. $y = -\dfrac{2}{3}x + 3$ **27.** $y = -\dfrac{1}{4}x + \dfrac{3}{2}$

29. $y = \dfrac{1}{8}x + \dfrac{17}{2}$ **31.** $y = -\dfrac{5}{3}x - 5$

33.

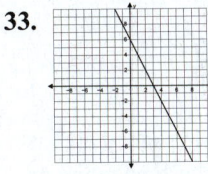

35.

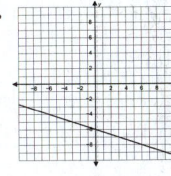

37.

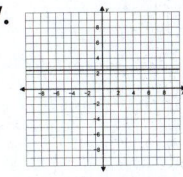

39. a. $y = -0.051x + 17,545.544$

b. $14,995.54

c. The calculated price is slightly higher. Other factors might influence the price of the car, such as body damage.

41. a. $-0.065x + 18,784.24$

b. $15,534.24;

c. The value calculated with this equation is about $500 more than the value calculated in Exercise 39. Answers will vary but might include that 50,000 is inside the range of values that should be used in the equation.

43. a. $y = 0.016x + 8.1$;

b. $16.10

c. The actual price of $16.50 is rather similar to the estimated one. The relationship of size to price is approximately linear.

45. a. $y = 0.022x + 6.375$

b. $17.38;

c. The price of $17.38 is a bit farther from the real price than $15.95. However, it is still close enough to the real price for the same reasons.

47. The break-even point is (8,13) so the business will start to earn profit after 8 months.

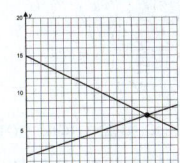

49. The break-even point is (16,7) so the mass of the element should be 16 g per 100 g of the alloy.

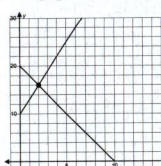

5.3 Exercises

1. never intersect, intersect at infinitely many points, intersect at exactly one point

3. money spent, money collected

5. cost   **7.** No

9. No   **11.** (2,1)

13. Infinitely many solutions

15. $(-1,-3)$

17. $(0,1)$

19. Infinitely many solutions

21. (2,1)

23. $P(3000) = 778.25

25. Textbook cost, $x = 140$; Calculator cost, $y = 30$

27. Non-student tickets, $x = 542$; Student tickets, $y = 181$

29. Mutual fund, $x = 8500; CD, $y = 1500

31. $C(x) = 270 + 50x$; $R(x) = 140x$; the break-even point is (2,420) so the family consists of no more than three people.

33. a. $C(x) = 31.90x + 100,900$

b. $R(x) = 37.50x$

c. $P(x) = 5.60x - 100,900$

d. Break-even point is 18,017.8571; when 18,018 pairs of shoes are made and sold, both cost and revenue are $675,675.

35. a. $C(x) = 18.26(0.25)x + 8.16x$
$= 12.725x$

b. $R(x) = 18.26x$

c. $P(x) = 18.26x - 12.725x$
$= 5.535x$

d. $309.96;

e. Answers will vary. Nation-wide studies show that drivers disagree strongly on whether insurance, car maintenance and depreciation are real factors in their earnings.

37. a. $C(x) = 56,000 + 12x$

b. $R(x) = 40x$

c. $P(x) = 28x - 56,000$

e. $x = 2000$

f. $28,000

5.4 Exercises

1. equal

3. test point

5. not included

7. a. In the solution set;

b. Not in the solution set;

c. In the solution set

9. a. In the solution set;

b. Not in the solution set;

c. Not in the solution set

11.
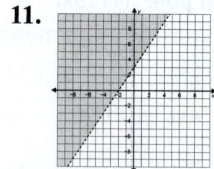

13.
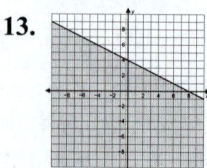

15.
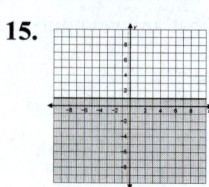

17.
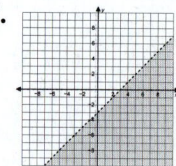

19.

21.

23.

25. Open circle

27. Closed circle

29.

31.

33.

35. a. $8x + 12y > 1200$

b. No, because
$8(94) + 12(26) - 1200 = -136 < 0$

c.
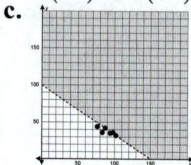

d. Likely no, because most points do not lie in the region satisfy-ing the inequality for positive profit.

37. a. $0.75x + 1.25y \geq 250$

b.
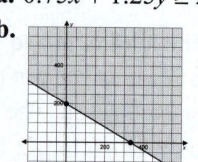

c. No **d.** 200;

e. Any whole number solution; for example, (200, 200).

39. a. $0.8x - 4.5y \geq 1500$

b.
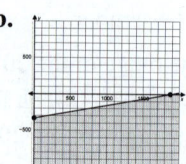

c. 1875 (the x-intercept)

d. Any whole number solution where $x > 1875$; for example, (2050, 2);

e. Although it is a solution to the inequality, it is not possible to have a negative number of broken vases.

41. a. $x + y \leq 13$; $15x + 8.5y > 60$

b.
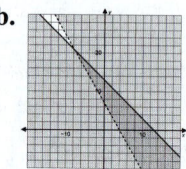

c. 4 hours of tutoring

d. 7.5 hours

e. Any point in the solution set where x is a whole number and y is a multiple of 0.5.

43. Let x be the number of short ads and y be the number of long ads.

$$\begin{cases} 215x + 380y \leq 4000 \\ 110x + 200y \geq 1400 \end{cases}$$

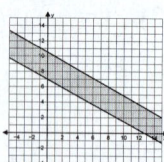

Any point in the shaded area is a combination of how many of each type of ad the newspaper could place to meet the goal. For example, (2, 8), (5, 6), (13, 2), etc.

45. Let x be the number of hours spent producing a regular glue and y be the number of hours spent producing a glue with increased adhesiveness.

$$\begin{cases} 43x + 165y \le 2800 \\ \quad\quad x + y \ge 14 \end{cases}$$

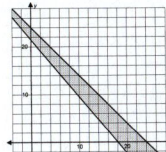

Any point in the shaded area is a combination of hours the factory could spend making each type of glue to meet the goal. For example, (4, 18), (10, 12), (20, 2), etc.

5.5 Exercises

1. linear inequalities

3. limitation

5. feasible region

7. $z = 80{,}000 + 2000x - 200y$

9. $z = 10x + 6y$

11. $z = 1.5a + 1.25b + 2.1c$

13. Point c. $(-5, -270)$ minimizes the objective function with the value $z = -48{,}850$.

15. Point b. $(0, 400)$ maximizes the objective function with $z = 118{,}000$.

17. Point b. $(0, 405)$ maximizes the objective function with $z = 166{,}050$.

19. The maximum value of z is 60 and occurs at the point $(5, 15)$.

21. The minimum value of z is 93 and occurs at the point $(9, 3)$.

23. The minimum value of z is -4 and occurs at the point $(0, 4)$.

25. a. $z = 157x + 423y$

   b. $6x \le y$

   c. $2x + 3y \le 180$

   d. $x \ge 0, y \ge 0$

   e. 0 apple trees and 60 peach trees

   f. $25,380

27. a. $z = 2.50a + 2.00b$

   b. $6a + 9b \ge 132$

   c. $4a + 5b \ge 82$

   d. $3a + b \ge 45$

   e. $a \ge 0; b \ge 0$

   f. 13 packages from Company A and 6 packages from Company B

   g. $44.5

29. The company should buy 8 of the 30-seat buses and 7 of the 50-seat buses to have the maximum number of seats 590. Objective function: $z = 30x + 50y$; Constraints: $45x + 60y \le 780, x \le 9$, $y \le 7, x \ge 0, y \ge 0; z(8, 7) = 590$

31. To minimize the cost, an expectant mother should take 1 Baby Bliss Prenatal Pill and 3 Natural Gift Prenatal pills. Objective function: $z = 0.4x + 0.34y$; Constraints: $x \ge 0, y \ge 0$, $0.1x + 0.1y \ge 0.4$, $200x + 400y \ge 1000$, $9x + 6y \ge 27$; Test points: $z(5,0) = \$2, z(3, 1) = \$1.54$, $z(1, 3) = \$1.42, z(0, 4.5) = \$1.53$.

5.6 Exercises

1. $ax^2 + bx + c = 0$

3. standard form

5. vertex

7. $x = -1, 7$

9. $x = \dfrac{3}{4} \pm \dfrac{\sqrt{17}}{4}$

11. $x = -\dfrac{1}{3} \pm \dfrac{\sqrt{7}}{3}$

13. $x = -2 \pm \sqrt{6}$

15. $x = -3, \dfrac{1}{2}$

17. 117 feet; 42 feet

19. 47.27 feet; 3.44 seconds

21. 8.84 seconds

23. Between inner cables: from -2 to 2, distance is 4 ft; Between outer cables: from -4 to 4, distance is 8 ft.

25. Distance: 2 meters; Length: 8 meters

27. a. 7 meters **b.** 4 meters

29. a. 6 centimeters

   b. 9 centimeters

31. a. Normal conditions: 4.5 meters, When windy: 3.92 meters. When rainy: 2.33 meters

   b. Normal conditions: 2.8 meters, When windy: 2.22 meters, When rainy: 0.63 meters

   c. Normal conditions: 14 seconds, When rainy: 10.58 seconds

   d. The ball falls the fastest when it is rainy, followed by the windy weather, and then the normal conditions. However, at the beginning of the fall, the ball falls slightly slower in the windy weather.

33. a. $N(t) = N_0 e^{2t}$

   b. $A_l(t) = 10{,}000 + 243t$

   c. 9.84×10^9 cells/L

   d. 20.8 minutes

35. a. $h(t) = -5t^2 + 3.9t + 6$

   b. $d(t) = 2.32t$

   c. 5.92 meters;

   d. 1.55 seconds

   e. 6.76 meters;

   f. 0.39 seconds

5.7 Exercises

1. $f(x) = ab^x$

3. Frequency

5. 10

7. 16,341 people

9. a. Approximately 1066 people

   b. Approximately 1211 people

   c. Approximately 1498 people

11. pH $= 3.52$; acidic

13. $[H+] = 0.001$

15. 100 times more

17. One octave: 440 Hz; two octaves: 880 Hz; three octaves: 1760 Hz

19. Two octaves higher

21. 438.948 Hz

23. a. Brother: $B(x) = 50 + 5x$;
   Sister: $S(x) = 50(1.1)^x$

   b. Brother: $65; Sister: $66.55

   c. 1 day;

   d. Brother, because after being 1 day late paying back the money, you would need to pay your sister more money.

25. $I = 1015.5$
   $\approx 3,162,277,660,168,379$

27. 8.70

29. Rather get the pennies.

Chapter 5 Exercises

1.

x	$f(x) = x^2 + 2x - 3$
−3	0
−2	−3
0	−4
1	−3
2	0
3	12

3. indep var = miles driven; dep var = cost of rental; $y = 0.25x + 35$ for $x \geq 0$; $97.50

5.
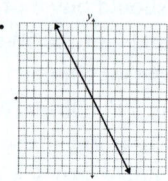

7.
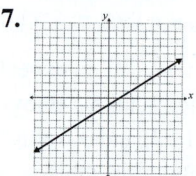

9. $m = \dfrac{3}{4}$; y-intercept $= (0, -3)$

11. a. $C(x) = 5400 + 1400x$

   b. $R(x) = 2000x$

   c. $P(x) = 600x - 5400$

   d. $x = 9$

   e. $6600

13.
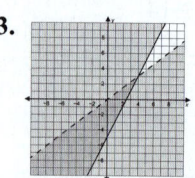

15. The maximum value of z is 95 and occurs at the point $(10, 15)$.

17. a. 19 miles per gallon

   b. 30 miles per gallon

   c. About 44 miles per gallon

19. a. About 24,367 people

   b. May 2220

21. About 6.85 seconds

Chapter 6: Finance

6.1 Exercises

1. borrowing

3. principal

5. False   **7.** $1000

9. $233.75   **11.** $126

13. $3712

15. a. $5484.47 **b.** $1984.47

17. a. $18,325.20 **b.** $12,675.20

19. a. $65,758.59 **b.** $50,758.59

21. a. $387,717.95

   b. $380,417.95

23. a. 5.12% **b.** 5.12% **c.** 5.13%

   d. 7.76% **e.** 7.78% **f.** 7.79%

25. $360   **27.** $12,500

29. $402.30   **31.** $722.44

33. $13,960.32 **35.** $16,518.97

37. The second bank, its value after 7 years is $14,079.92, as opposed to $13,947.11 for the first.

39. a. $5287.23;

   b. About 4 years

41. $16,912.57 **43.** $1398.84

45. About 3.5 years

47. About 6.67 years, or 6 years and 8 months

49. 2.14%   **51.** 1.66%

53. $88,956.99

55.

First Bank of Lending Loan APY	
Loan Amount	APY
< $20,000	11.73%
$20,000–$99,000	9.30%
≥ $100,000	5.88%

6.2 Exercises

1. present value **3.** shareholder

5. True   **7.** $7574.87

9. $18,621.74 **11.** $26,701.34

13. $17,229.90 **15.** $10.22

17. $31.39   **19.** $10,071.48

21. $48,733.54 **23.** $2925.07

25. $782.48

27. a. $17.91 **b.** $39.39 **c.** $91.19

29. a. $317.41

   b. $57,133.80; $I = $17,866.20

31. $1673   **33.** $1776

35. a. $6012;

   b. $288,576 deposited and $211,424 in interest

37. $471.17   **39.** $5877.10

41. $107,395.23

43. a. $87,971.89;

   b. $40,000 deposited and $47,971.89 in interest

45. a. $56,556.06; **b.** $163,566.51;
   c. $72,846.51; **d.** $545.22

47. a. NKE; **b.** −$0.74; **c.** $173.78;

   d. Decreased by $1.46;

   e. About 1.6 billion shares;

   f. $77

49. While there are no guarantees, the rule of thumb was developed based on the historical data of the interest rates on different types of saving strategies. Therefore, unless the market experiences a crisis more severe than in the past, the four percent rule of thumb will likely ensure that the funds will last 30 years.

6.3 Exercises

1. down payment

3. collateral   **5.** open-ended

7. a. $331.15   **b.** $7947.60

9. a. $501.81   **b.** $6021.72

11. a. $168.89   **b.** $6080.04

13. a. Please see math.hawkeslearning.com for full answer.;
   b. $215,761.92; 163 payments; the 164th payment is the first one where more than half of the payment goes to the principal.

15. $1168.97   **17.** $880.29

19. a. $960.10;   **b.** $172,818

21. a. $2599.60   **b.** $124,780.80
   c. $10,780.80

23. a. $125.36   **b.** $3008.64
   c. $358.64

25. $506.65   **27.** $361.63

29. The five-year loan with an APR of 7.5% ($400.76 monthly payment, $4045.60 in interest paid)

31. Option A: PMT =$319.61, Total cost = $7670.64; Option B: PMT = $424.76, Total cost = $7645.68

33. Option A: PMT =$360.44, Total amount =$30,276.96; Option B: PMT =$496.21, Total amount =$29,772.60

35. a. $32,200;   **b.** $197,800

37. $44,262.71

39. a. $823.33;   **b.** $1124.58

41. $324,776.49

43. a. Approximately 114 payments;
   b. Approximately 27 payments;
   c. $1260

45. 260%   **47.** 910%

49. a. $8718.72;   **b.** $9146.16;
   c. The better credit score would save $427.44.

51. Option 1

53. When the interest rate is fixed and the loan has a long term in a country with a high inflation rate, the future worth of the total sum of regular payments may be less than the present worth of the loan.

6.4 Exercises

1. Form 1040   **3.** deductions

5. False   **7.** $58,133.55

9. $65,913.23   **11.** $32,866.32

13. a. $78,490.88 **b.** $72,440.88

15. $2778.50   **17.** $12,334.00

19. $52,040.00   **21.** $34,057.10

23. $67,023.00   **25.** $65,459.40

27. $51,966.50

29. The taxpayer received a federal tax refund of $3895.59 − $3763.14 = $128.45.

31. a. $26.10   **b.** $111.60
   c. $344.29   **d.** $1455.71

33. $81,048   **35.** $47,956

37. $96,045   **39.** $58,851

6.5 Exercises

1. income, expenses

3. prorate   **5.** $38.16

7. $67.80   **9.** $40

11. $41.38

13. a. $4500   **b.** $1500

15. a. $2425   **b.** $808.33

17. Necessities: $2560; Wants: $1536; Savings: $1024

19. Necessities: $1236; Wants: $741.60; Savings: $494.4

21. a. $2132   **b.** $132

23. a. $4245.50   **b.** $1125.50

25. $730   **27.** $561

29. $506

31. Answers will vary. A budget can help you get a solid understanding of the amount of money you make and how you spend your money. A budget can also help prevent you from overspending each month.

Chapter 6 Exercises

1. $390

3. a. $7061.72; $7451.05; $390,674.35
   b. $1561.72; $2696.05; $383,374.35

5. a. $24,839.67 **b.** $24,854.70

7. a. $38,481.59 **b.** $8481.59

9. a. $153,926.38
   b. $33,926.38

11. $2387.67

13. a. $330.61
   b. $71,411.24; $I = $63,588.76

15. a. $1150   **b.** 97.5%

17. $243,932.24

19. The 5-year loan, because the 4-year loan has a higher payment than you can afford ($638.39/month) and the 6-year loan has you paying more interest ($6246 instead of $4775).

21. a. $251.70   **b.** $6040.80

23. $57,592.27

25. a. $4937.50;   **b.** $1645.83

27. $1916/month

Chapter 7: Numeration and Measurement Systems

7.1 Exercises

1. 10, 2, 3   **3.** 20

5. $\left(2\cdot10^3\right)+\left(3\cdot10^2\right)+\left(5\cdot1\right)$ or
$\left(2\cdot10^3\right)+\left(3\cdot10^2\right)+\left(0\cdot10^1\right)+\left(5\cdot1\right)$

7. $\left(6\cdot10^2\right)+\left(2\cdot10^1\right)$
$+\left(1\cdot1\right)+\left(9\cdot10^{-1}\right)$

9. $\left(8\cdot10^1\right)+\left(2\cdot10^{-2}\right)$ or
$\left(8\cdot10^1\right)+\left(0\cdot1\right)+\left(0\cdot10^{-1}\right)+\left(2\cdot10^{-2}\right)$

11. 951   **13.** 300,062

15. 10,200.07   **17.** 46

19. $\left(11\cdot60^1\right)+\left(55\cdot1\right)=715$

21. $\left(48\cdot60^2\right)+\left(8\cdot60^1\right)+\left(20\cdot1\right)$
$=173,300$

23. ⟨ ▽▽

25. ▽   ⟨⟨⟨   ▽▽▽
▽▽▽
▽▽▽

27. ▽   ⟨ ▽   ⟨

29. 8

31. $\left(10\cdot20\right)+\left(4\cdot1\right)=204$

33. $\left(3\cdot20^3\right)+\left(0\cdot20^2\right)+\left(3\cdot20\right)$
$+\left(3\cdot1\right)=24,063$

35. ▬▬   **37.** ●●●●
●●
▬▬

39. ●●●
●
▬▬

41. Answer will vary depending on the current year. Babylonian

2022: ⟨⟨⟨ ▽▽▽   ⟨⟨⟨ ▽▽
⟨
2023: ⟨⟨⟨ ▽▽▽   ⟨⟨⟨ ▽▽▽
⟨
2024: ⟨⟨⟨ ▽▽▽   ⟨⟨⟨ ▽▽▽
⟨   ⟨ ▽

Mayan

2022: ▬
●
●●

2023: ▬
●●●

2024: ▬
●
●●●●

43. Answers will vary. Possible answers might include that it's an advantage to have a symbol for zero. One disadvantage is that it takes up a lot of space to write.

45. Answers will vary. Encourage students to think about the fact that without any a place holder, the numbers 1 and 60 look the same. One way to remove the ambiguity is to consider the context of how the number is being used. For instance, 60 of anything (for example, tablespoons or pints) doesn't seem to make sense in a recipe. Likewise, a length of 1 might seem too small for a distance to travel.

7.2 Exercises

1. hieroglyphics   **3.** M

5. changed   **7.** 6050

9. 700,000   **11.** 1,405,003

13. ℰ∩ ∩ I I I I I I I I I

15.

17. ⚏ 𓏏 𓏏 𓏏 ∩ ∩

19. 444   **21.** 301

23. 8555   **25.** 90,809

27. $\rho\ \varsigma\ \varsigma$ **29.** , $\alpha o\eta$

31. $\eta\ \upsilon\kappa\delta$ **33.** 48
M

35. 651   **37.** 975

39. XLII   **41.** CLXIII

43. CDXCI   **45.** 54

47. 1008   **49.** 25,575

51. 一百零三   **53.** 七十五

55. 二千零二

57. Answers will vary depending on the current year. For example,

2022: , $\beta\kappa\beta$

2023: , $\beta\kappa\gamma$

2024: , $\beta\kappa\delta$

59. Answers will vary depending on the current year. For example,

2022: 二千零二十二

2023: 二千零二十三

2024: 二千零二十四

61. a. 52;

b. Answers will vary. Can include the look of the number or the fact the Super Bowl is in the year after the main portion of the football season;

c. Answers will vary.

63. Possible answers may include: Drawing the numerals takes up a lot of space and time. The formation of some of the numerals is complex. A new character is needed for each power of 10.

65. Answers will vary.

7.3 Exercises

1. 10   **3.** letters

5. 1000s, 8s   **7.** 624

9. 5460   **11.** 513

13. 1752

15. $BB_{16}=187$ and $603_2=192$; Thus, 603_2 is larger.

17. 1101001_3 **19.** 40_{12}

21. 1010111100_2 **23.** $21E_{16}$

25. $3B_{16}$ **27.** $72C_{16}$

29. $B038_{16}$ **31.** 111101_2

33. 111111101101_2

35. 111101000010010101_2

37. 144_7 **39.** 1100_2

41. 51103_6 **43.** 273_9

45. $7E5_{16}$ **47.** 1202_5

49. 220_3 **51.** 1032_4

53. 148_{16}

55.

+ Base 8	0	1	2	3	4	5	6	7
0	0	1	2	3	4	5	6	7
1	1	2	3	4	5	6	7	10
2	2	3	4	5	6	7	10	11
3	3	4	5	6	7	10	11	12
4	4	5	6	7	10	11	12	13
5	5	6	7	10	11	12	13	14
6	6	7	10	11	12	13	14	15
7	7	10	11	12	13	14	15	16

57.

× Base 4	0	1	2	3
0	0	0	0	0
1	0	1	2	3
2	0	2	10	12
3	0	3	12	21

59.

+ Base 7	0	1	2	3	4	5	6
0	0	0	0	0	0	0	0
1	0	1	2	3	4	5	6
2	0	2	4	6	11	13	15
3	0	3	6	12	15	21	24
4	0	4	11	15	22	26	33
5	0	5	13	21	26	34	42
6	0	6	15	24	33	42	51

61. Base 6   **63.** Base 8

7.4 Exercises

1. 1   **3.** liter

5. are   **7.** 156 in.

9. 7 lb 8 oz   **11.** 10 pt

13. 255 minutes   **15.** 18,480 ft

17. 64 fl oz   **19.** 1.5625 mi²

21. 0.23 km   **23.** 1.23 m

25. 0.356 kL   **27.** 1200 g

29. 72 600 dm²

31. 6 272 000 cm²

33. 99 000 000 cm²

35. 3.45 a   **37.** 104 fl oz

39. A king-size mattress is 16 inches wider.

41. 32.5 gallons

43. 1.36 km

45. $2.40

47. Trey can put 22 tacks along each chair (or approximately 22.25 tacks)

49. 0.9 hectares   **51.** 19 000 cm³

53. meters or kilometers

55. millimeters or centimeters

57. 0.0000006 kg, 0.61mg, 100.6 mg, 1021 g

7.5 Exercises

1. dimensions

3. approximations

5. 2.54   **7.** 241.5 km

9. 45.75 meters   **11.** 29.64 acres

13. 52 km²   **15.** 1.225 oz

17. 1190.7 g   **19.** 1.584 gal

21. 6.6667 tbsp

23. 29,976,308 km

25. 0.5167 mi/min

27. 354.375 g/L

29. 0.31 lb/ft   **31.** 13.56 yd²/lb

33. 350.6 °F   **35.** 135 °C

37. 77 °F   **39.** −11.111 °C

41. 100 cm, 100 in., 10 ft, 100 yd, 100 m, 10 km, 10 mi

43. 1206.11 lb/ft³ **45.** 0.19 L/sec

47. 62 miles per hour

49. 13 min

51. 4 fl oz; 374 °F

53. 19.4 °F

55. 12.125 gallons; $452.02

57. 5.45 mi/hr

59. No, they would be underbaked or not cooked at all because of the oven temperature. Also, there is twice as much cinnamon as called for, so they might not taste good.
350 g ≈ 12.25 oz
400 g ≈ 14 oz
22.5 mL ≈ 1.5 tbsp
140 °C ≈ 284 °F

Chapter 7 Exercises

1. $\left(1\cdot10^{2}\right)+\left(7\cdot10^{1}\right)+\left(5\cdot10^{0}\right)+\left(8\cdot10^{-1}\right)+\left(3\cdot10^{-2}\right)$

3. 1449   **5.** 7206

7. 120,052   **9.** 847

11. 1047   **13.** 6039

15. 76   **17.** $15B_{16}$

19. 1004_{5} **21.** 2112_{3}

23. 216 oz   **25.** 2400 min

27. 152 400 cL

29. 57 750 000 m²

31. 56.68 m   **33.** 15.9 qt

35. 29.444 °C   **37.** 0.504 g/mL

Chapter 8: Number Theory

8.1 Exercises

1. 6   **3.** False

5. Composite   **7.** Prime

9. Composite   **11.** Composite

13. Composite

15. $2 \cdot 2 \cdot 2 \cdot 2$

17. $2 \cdot 3 \cdot 3 \cdot 3 \cdot 3$

19. $5 \cdot 5 \cdot 5 \cdot 5$ **21.** $3 \cdot 5 \cdot 7$

23. 7   **25.** 45

27. 36   **29.** 35

31. 7   **33.** 1

35. 8   **37.** 6

39. Yes   **41.** No

43. No   **45.** No

47. a. 8 groups   **b.** 4 violinists
c. None

49. 4 boxes

51. a. 12 ft

    b. Troop A:
15 ft × 12 ft;
Troop B: 17 ft × 12 ft

53. All prime numbers smaller than $\sqrt{1291} \approx 35.9305$: 2, 3, 5, 7, 11, 13, 17, 19, 23, 29, and 31

8.2 Exercises

1. check-sum digit

3. False   **5.** True

7. True   **9.** True

11. True   **13.** 1

15. 0   **17.** 8 (mod 10)

19. 3 (mod 4)   **21.** 5 (mod 12)

23. 2   **25.** 1

27. 1:00 p.m.   **29.** 12:00 a.m.

31. 1200   **33.** 0700

35. 4 (mod 7)   **37.** 0 (mod 3)

39. 3   **41.** 10

43. 0

45. No, the check-sum digit should be 1.

47. Valid

49. No, the check-sum digit should be 6.

51. 7   **53.** 7

55. 0   **57.** 5

59. 1   **61.** 0

63. Yes

65. No, the check-sum digit should be 5.

8.3 Exercises

1. Fermat's Little Theorem

3. False

5.
$$\begin{aligned}
2^{23} - 2 &= 2^{16} \cdot 2^7 - 2 \\
&\equiv 9 \cdot 13 - 2 \,(\text{mod } 23) \\
&\equiv 2 - 2 \,(\text{mod } 23) \\
&\equiv 0 \,(\text{mod } 23)
\end{aligned}$$

7.
$$\begin{aligned}
3^{79} - 3 &= \left(3^4\right)^{19} \cdot 3^3 - 3 \\
&\equiv (2)^{19} \cdot 27 - 3 \,(\text{mod } 79) \\
&\equiv 2^{17} \cdot 2^2 \cdot 27 - 3 \\
&\equiv 11 \cdot 4 \cdot 27 - 3 \,(\text{mod } 79) \\
&\equiv 3 - 3 \,(\text{mod } 79) \\
&\equiv 0 \,(\text{mod } 79)
\end{aligned}$$

9.
$$\begin{aligned}
2^{17} - 2 &= 2^5 \cdot 2^5 \cdot 2^5 \cdot 2 \cdot 2 - 2 \\
&\equiv (15 \cdot 15)(15 \cdot 4) - 2 \,(\text{mod } 17) \\
&\equiv 4 \cdot 9 - 2 \,(\text{mod } 17) \\
&\equiv 2 - 2 \,(\text{mod } 17) \\
&\equiv 0 \,(\text{mod } 17)
\end{aligned}$$

11.
$$\begin{aligned}
4^{31} - 4 &= 4^3 \cdot 4^3 \cdot 4^3 \cdot 4^3 \cdot 4^3 \cdot 4^3 \cdot 4^3 \cdot 4^3 \\
&\quad \cdot 4^3 \cdot 4^3 \cdot 4 - 4 \\
&\equiv 2^5 \cdot 2^5 \cdot 4 - 4 \,(\text{mod } 31) \\
&\equiv 1 \cdot 1 \cdot 4 - 4 \,(\text{mod } 31) \\
&\equiv 4 - 4 \,(\text{mod } 31) \\
&\equiv 0 \,(\text{mod } 31)
\end{aligned}$$

13.–31. Answers will vary.

8.4 Exercises

1. prime   **3.** False

5.
$$\begin{aligned}
3^{2(13-1)(7-1)+1} - 3 \\
= 3^{2(12)(6)+1} - 3 \\
= 3^{145} - 3 \\
= \left(3^6\right)^{24} \cdot 3 - 3 \\
\equiv (1)^{24} \cdot 3 - 3 \,(\text{mod } 91) \\
\equiv 1 \cdot 3 - 3 \,(\text{mod } 91) \\
\equiv 3 - 3 \,(\text{mod } 91) \\
\equiv 0 \,(\text{mod } 91)
\end{aligned}$$

7. 49   **9.** 33

11. 39   **13.** 6

15. 64   **17.** 38

19. 12   **21.** 31

23. $d = 151$

25. $p = 17$, $q = 11$, and $d = 59$

Chapter 8 Exercises

1. 4 groups   **3.** 170

5. GCD = 21; because the GCD is not 1, the numbers are not relatively prime.

7. 0   **9.** 2

11. No, the check-sum digit should be 5.

13. 1   **15.** 3

17. 4,459,580   **19.** 12345

Chapter 9: Geometry

9.1 Exercises

1. circumference

3. tessellation

5. $600   **7.** 25 bags

9. 77.04 cm^2 **11.** 37.68 in.^{2}

13. 246 ft^2 **15.** $21.60

17. 15 boxes

19. 38 shrubs; 165 flowers

21. Rotate wheels forward 6.4 rotations; turn 90° left; rotate wheels forward 1.9 rotations; turn 90° left; rotate wheels forward 1.1 rotations.

23. 13 inches   **25.** $641.30

27. $37.15

29. $C = 18.84$ in.; $A = 28.26$ in.^{2}

31. $11.42   **33.** 18 m^2

35. 9.29 in.^{2} **37.** 4 m^2

9.2 Exercises

1. three; two   **3.** triangular

5. False   **7.** Volume

9. Volume   **11.** Surface area

13. 120 ft^3

15. a. 270 m^3 **b.** 18 m^3

17. 162 cubes   **19.** 96 ft²

21. 486 in.²

23. a. 14,657,414.63 mi²

   b. 3,517,779,524 mi²

25. a. 144π ≈ 452.39

   b. 49 sheets

27. a. 33,912 ft³

   b. 27,129.6 bushels

   c. 169.56 min (about 2.8 hours)

29. a. Vase A: 490.87 cm³
   Vase B: 502.65 cm³

   b. Vase B holds 11.78 cm³ more
   than Vase A.

31. a. 169.65 cm³

   b. About 86 minutes

33. a. 85.3 ft³

   b. About 3.2 ft³

35. 0.32 m³

37. Answers will vary. At a height of
20 cm and a volume of 400 cm³,
the radius should be 2.5 cm.

39. The radius cannot be smaller than
2.3 ft.

41. a. $SA = 65.81$ m²

   b. Cost: $477.13

9.3 Exercises

1. 180

3. sine; cosine; tangent

5. False   **7.** Acute

9. Acute   **11.** Obtuse

13. 56°   **15.** 81°

17. $x = 4$

19. $m\angle 1 = 96°$ **21.** $m\angle 3 = 58°$

23. 47° and 43°   **25.** 84° and 96°

27. No, $\dfrac{360°}{140°} \approx 2.57$

29. $\sin A = 0.38$;
   $\cos A = 0.92$;
   $\tan A = 0.42$;
   $\sin B = 0.92$;
   $\cos B = 0.38$;

31. Law of Cosines

33. Law of Cosines

35. $\sqrt{5} \approx 2.24$

37. a. 106.5 in. or 8.9 ft;

   b.
Maximum Cabinet Height

$$< \sqrt{\text{Ceiling Height}^2 - \text{Cabinet Depth}^2}$$

39. Calvin will need to make a right
triangle with sides measur-
ing 60 feet and 71.5 feet with a
hypotenuse measuring 93.3 feet to
get the 50° angle. He would then
need to measure 30 feet along
the hypotenuse to mark the next
corner.

41. 148.25 feet   **43.** 47.57 feet

45. a. 5.99 ft   **b.** 14.83 ft

47. 61.6°   **49.** 38.7°

51. 49.7 feet tall   **53.** 75 mi

55. Answers will vary. You can create
the following tessellation.

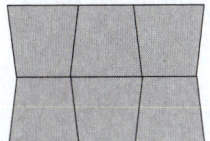

Chapter 9 Exercises

1. 145.96 in.²   **3.** $528

5. 48.33 in.³, 42.96 in.³, 31.24 in.³

7. a. 1.13 in.³

   b. 3.75 in. by 4.5 in. by 2.75 in.

9. 3.75 in.

11. 13° and 77°   **13.** 78° and 102°

15. a. 9.54 ft   **b.** 15.26 ft

17. 117.57 ft

Chapter 10: **Probability**

10.1 Exercises

1. trial

3. tree diagram

5. False

7. {RCS, RCM, RCN, RLS, RLM,
RLN, WCS, WCM, WCN, WLS,
WLM, WLN, SCS, SCM, SCN,
SLS, SLM, SLN}

7. cont.

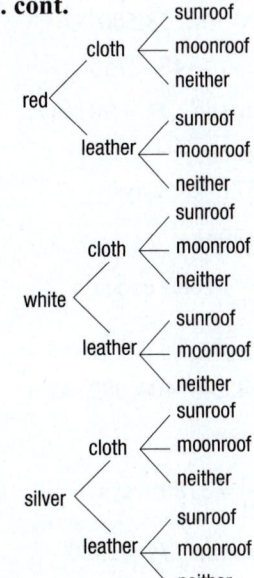

9. {AAAA, AAAN, AANA, AANN,
ANAA, ANAN, ANNA, ANNN,
NAAA, NAAN, NANA, NANN,
NNAA, NNAN, NNNA, NNNN}

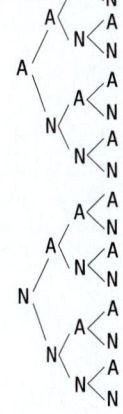

11. $\{11, 22, 33, 44, 55, 66, 77, 88, 99\}$

13. $\{$WFS, WFR, WFC, WFM, WDS, WDR, WDC, WDM, BFS, BFR, BFC, BFM, BDS, BDR, BDC, BDM, PFS, PFR, PFC, PFM, PDS, PDR, PDC, PDM$\}$

15. $\{$PCM, PCN, PGM, PGN, LCM, LCN, LGM, LGN$\}$

17. $\{$COT, COS, CMT, CMS, POT, POS, PMT, PMS, NOT, NOS, NMT, NMS$\}$

19. Classical

21. Experimental

23. Classical

25. $\dfrac{70}{200} = 0.35$

27. a. $\dfrac{82}{500} = 0.164$

  b. $\dfrac{405}{500} = 0.81$

29. $\dfrac{2}{15} \approx 0.133333$

31. $\dfrac{38}{115} \approx 0.330435$

33.

$\dfrac{6}{12} = 0.5$

35. $\dfrac{18}{38} \approx 0.473684$

37. $\dfrac{11}{321} \approx 0.034268$

39. $\dfrac{64}{213} \approx 0.300469$

41. $\dfrac{1}{3} \approx 0.333333$

43. $\dfrac{1}{4} = 0.25$

45. a. $\dfrac{8}{19} \approx 0.421053;$

  b. $\dfrac{38}{38} = 1;$

  c. $\dfrac{15}{19} \approx 0.789474$

47. Answers will vary. With classical probability, the entire sample space is known. With experimental probability, only part of the sample space can be determined by observing or collecting data. For both types of probability, the probability is calculated by dividing the size of the event space (or number of times the event occurs) by the size of the sample space (or number of times the experiment is performed).

10.2 Exercises

1. multiply

3. product

5. True

7. 362,880

9. 1

11. 75,600

13. 210

15. 6188

17. 840

19. 24

21. 30

23. 6

25. 3

27. $\dfrac{1}{6}$

29. 1099

31. $9 \cdot 10 \cdot 10 = 900$

33. $5 \cdot 2 \cdot 2 \cdot 3 = 60$

35. $7 \cdot 6 \cdot 5 \cdot 4 \cdot 3 \cdot 2 \cdot 1 = 5040$

37. $16 \cdot 14 \cdot 7 \cdot 12 = 18,816$

39. $9 \cdot 10 \cdot 10 \cdot 5 = 4500$

41. a. 67,600 **b.** 58,500

43. 10,000   **45.** 2730

47. Permutation; $_{10}P_2 = 90$ ways

49. Permutation; $_{84}P_3 = 571,704$ ways

51. Permutation; $\dfrac{8!}{3!2!2!1!} = 1680$ codes

53. Combination; $_{123}C_6 = 4,249,404,082$ ways

55. Permutation; $\dfrac{9!}{1!4!2!2!} = 3780$ ways

57. Permutation; $_{12}P_3 = 1320$

59. Combination; $_8C_4 = 70$

61. Permutation; $_{20}P_7 = 390,700,800$

63. Permutation; $\dfrac{10!}{6!4!} = 210$

65. Permutation; $\dfrac{12!}{4!4!2!2!} = 207,900$

67. Answers will vary; should be $_{\text{(number of class members)}}C_3$

10.3 Exercises

1. complement

3. against, for

5. False

7. $\dfrac{3}{1000} = 0.003$

9. $\dfrac{1}{26 \cdot 25 \cdot 24} = \dfrac{1}{15600}$
  ≈ 0.000064

11. $\dfrac{1}{_{48}C_2} \approx 0.000887$

13. a. Permutation

  b. $10^4 = 10,000$

  c. $\dfrac{10}{10,000} = 0.001$

15. $\dfrac{1}{_{20}C_5} = \dfrac{1}{15,504} \approx 0.000064$

17. $\dfrac{1}{_{10}P_5} = \dfrac{1}{30,240} \approx 0.000033$

19. $\dfrac{1}{_{52}C_2} = \dfrac{1}{1326} \approx 0.000754$

21. $\dfrac{1}{_6P_6} = \dfrac{1}{720} \approx 0.001389$

23. a. $_{46}C_4 = 163,185$

  b. $\dfrac{_4C_4}{163,185} = \dfrac{1}{163,185} \approx 0.000006$

  c. $\dfrac{_{41}C_4}{163,185} = \dfrac{101,270}{163,185} \approx 0.620584$

25. a. $10^4 = 10,000$

  b. $_{10}P_4 = 10 \cdot 9 \cdot 8 \cdot 7 = 5040$

  c. $\dfrac{1}{4!} = \dfrac{1}{24} \approx 0.041667$

  d. $\dfrac{1}{3 \cdot \left(\dfrac{4!}{2!1!1!}\right)} = \dfrac{1}{36} \approx 0.027778$

  e. Repeat a digit

27. $\dfrac{_3C_2}{_{14}C_2} = \dfrac{3}{91} \approx 0.032967$

29. a. $_{11}C_5 = 462;$

   b. $\left(_6C_3\right)\left(_5C_2\right) = (20)(10) = 200;$

   c. $\dfrac{200}{462} \approx 0.4329$

31. $1 - \dfrac{5}{26} \approx 0.807692$

33. $1 - \dfrac{(5803.7 + 15,004.0)}{137,147.0}$

   $\approx 1 - 0.151718$

   ≈ 0.848282

35. $1 - \dfrac{6}{36} \approx 0.833333$

37. 75%

39. $\dfrac{3,874,000}{26,233,000} \approx 0.147677$

41. $\dfrac{1}{7}$ 43. $\dfrac{2}{5}$

45. $\dfrac{5}{7}$

47. a. 0.822   b. $80.24

49. a. $\dfrac{8}{9} \approx 0.8889$, $\dfrac{9}{10} = 0.9$,

   $\dfrac{14}{15} \approx 0.9333$, $\dfrac{16}{17} \approx 0.9412$,

   $\dfrac{20}{21} \approx 0.9524$

   b. Dustin Johnson was most likely to win.

51. The set of even numbers larger than 0.

53. The complement consists of players aged 8 or 9, so there are nine players in the complement of *A*.

55. All employees who have worked there for less than or equal to five years.

10.4 Exercises

1. subtracting   3. independent

5. False

7. Mutually exclusive

9. Mutually exclusive

11. These are not independent events. Because we are not allowed to repeat characters, choosing a character omits it from being chosen the next time.

13. Yes

15. $\dfrac{24}{36} = \dfrac{2}{3} \approx 0.666667$

17. $\dfrac{16}{52} = \dfrac{4}{13} \approx 0.307692$

19. $\dfrac{9}{30} = 0.3$

21. $\dfrac{10}{19} \approx 0.526316$

23. $\dfrac{13}{36} \approx 0.361111$

25. $\dfrac{22}{52} \approx 0.423077$

27. $\dfrac{8}{36} = \dfrac{2}{9} \approx 0.222222$

29. $\dfrac{8}{52} = \dfrac{2}{13} \approx 0.153846$

31. a. $\dfrac{102}{137} \approx 0.744526$

   b. $\dfrac{76}{137} \approx 0.554745$

33. 0.46   35. $\dfrac{9}{16} = 0.5625$

37. $\dfrac{1}{2} \cdot \dfrac{1}{2} = \dfrac{1}{4} = 0.25$

39. $\dfrac{13}{52} \cdot \dfrac{4}{52} = \dfrac{52}{2704} \approx 0.019231$

41. $\dfrac{3}{20} \cdot \dfrac{5}{19} \approx 0.039474$

43. $\dfrac{6}{12} \cdot \dfrac{2}{11} \approx 0.090909$

45. $\dfrac{6}{10} = \dfrac{3}{5} = 0.6$

47. $\dfrac{3}{51} \approx 0.058824$

49. 0.666667

51. $\dfrac{31}{35} \approx 0.8857$

53. 0.8077   55. 0.1774

10.5 Exercises

1. successes, failures

3. True   5. Yes; binomial

7. No; there is no fixed number of trials.

9. No; the probability of each success does not remain the same.

11. No; there are three outcomes instead of two.

13. No; there are more than two possible outcomes.

15. Yes; binomial

17. 0.2734   19. 0.2592

21. 0.2330   23. 0.0289

25. 0.3331   27. 0.1876

29. 0.2447   31. 0.9417

33. 0.0345   35. 1.29×10^{-12}

37. 0.1445   39. 0.3223

41. 0.5331   43. 0.4086

45. 0.0706

10.6 Exercises

1. probability experiment

3. False

5. Expected value = 2.15

7. Expected value = 16

9. $2.33

11. Expected value = 1

13. 1.201 strings

15. a. $1300   b. $200

17. $115

19. a. $7.20   b. $576

21. $23,500

23. a. Expected value = −$9.73

   b. Lose $243.25

25. a. Expected value = −$40

   b. Lose $600

27. Your sister's offer. The expected value from her offer is $10, while the expected value from your mother's offer is only $7.

29. a. Expected value = 1.14 points

  b. Expected value = 1.03 points

  c. Yes, it is since he is expected to make fewer points with that strategy.

Chapter 10 Exercises

1. Experimental

3. The sample space is the set of $14 \cdot 14 = 196$ ordered pairs. where the first of each pair one of the 14 possible outcomes on the first spin and the second is one of the 14 possible outcomes for the second spin. That is, $250 would be paired with each of the 14 possible outcomes on the second spin, as would $300, $350, and so on.

5. $\dfrac{150}{1237} \approx 0.121261$; higher than expected based on the classical probability of about 0.083333 that we calculated in Exercise 4b.

7. $\dfrac{5!}{2!1!1!1!} = 60$

9. a. $\dfrac{{}_5C_4}{{}_6C_4} = \dfrac{5}{15} \approx 0.333333$

  b. $6! = 720$

  c. $\dfrac{5!}{6!} = \dfrac{120}{720} \approx 0.166667$

  d. $\dfrac{2 \cdot 5!}{6!} = \dfrac{1}{3} \approx 0.333333$

11. $\{H1, H2, H3, H4, H5, H6, T1, T2, T3, T4, T5, T6\}$

13. $\dfrac{10}{100} \cdot \dfrac{10}{100} = \dfrac{1}{100} = 0.01$

15. $\dfrac{50}{100} \cdot \dfrac{25}{99} = \dfrac{25}{198} \approx 0.126263$

17. $\dfrac{25}{100} \cdot \dfrac{24}{99} + \dfrac{10}{100} \cdot \dfrac{9}{99} + \dfrac{15}{100} \cdot \dfrac{14}{99}$

  $+ \dfrac{25}{100} \cdot \dfrac{24}{99} = \dfrac{5}{33} \approx 0.151515$

19. $\dfrac{15}{36} = \dfrac{5}{12} \approx 0.416667$

21. $\dfrac{13}{36} \approx 0.361111$

23. $\dfrac{1}{52} \approx 0.019231$

25. 0.0745   **27.** 0.237668

29. 0.3504

31. a. $2409.09   **b.** $4000

33. –$48   **35.** 4.2 days

Chapter 11: Statistics

11.1 Exercises

1. counts, measurements, observations

3. bias

5. Population of interest: women with PCOS; Variables of interest: insulin level, symptoms of PCOS

7. Population of interest: all people; Variables of interest: noise level, cortisol level

9. Sample statistic

11. Sample statistic

13. Population: cancer patients; Sample: 675 cancer patients; Sample statistics: 76% and 63%.

15. Population: Middle-aged women; Sample: 36,000 health records of women; Sample statistics: 18% and 11%; Population parameter: 40%

17. Population: All coffee consumers; Sample: 6195 customers who complete the survey; Parameter: 77%; Statistics: 45%, 32%, and 23%

19. Observational study

21. Observational study

23. Experiment

25. Systematic sampling

27. Cluster sampling

29. Convenience sampling

31. Random sampling

33. Random sample, since there are not clear divisions in the sample group of 1200 patients. Possible biases include age, race, and stage of development of the skin disease in the patients. Answers will vary.

35. Convenience sample, since the population is all children with autism, this is a very large group that may be difficult to study. It may be best to start with a convenience sample to get an idea of where to start. A possible bias to consider is that the effects may be different on children at different levels of functionality, so it may be best to make sure that a range of children are represented. Answers will vary.

37. a. Explanatory variable: type of antiviral agent; Response variable: viral load;

  b. Treatment: new antiviral agent;

  c. Treatment group: Group A; Control group: Group B;

  d. Single-blind. Yes, because the viral load is not a subjective measure.

39. a. Explanatory variable: vitamin C; Response variable: iron level

  b. Treatment: iron + vitamin C

  c. Treatment group: group that takes iron with vitamin C; Control group: group that takes only iron

  d. Neither single-blind nor double-blind. Yes, because iron level is not a subjective measure.

41. Answers will vary. For example, income level, health of residents, happiness of residents, average annual temperature, number of activities in the area.

43. Both begin with separating the population into groups, but cluster sampling surveys every member of certain randomly chosen groups, whereas stratified sampling takes a random sample from each group.

11.2 Exercises

1. frequency distribution

3. frequency histogram

5. True

7. a.

Class	Freq.	**b.** Rel. Freq.
0–9	0	0%
10–19	6	17.1%
20–29	8	22.9%
30–39	5	14.3%
40–49	4	11.4%
50–59	3	8.6%
60–69	5	14.3%
70–79	4	11.4%

c. 11.4%   **d.** 0%

e. 20–29 calls

9.

Class	Freq.	Rel. Freq.
1800–2199	1	$\frac{1}{15} \approx 7\%$
2200–2599	4	$\frac{4}{15} \approx 27\%$
2600–2999	6	$\frac{6}{15} = 40\%$
3000–3399	3	$\frac{3}{15} = 20\%$
3400–3799	1	$\frac{1}{15} \approx 7\%$

11. a. Relative frequency; the amount spent on a particular category as a percentage of Sofie's total expenses

b. Housing and Utilities

c. $2946.30

13. a.

Hours	Number of Students	Relative Frequency
0.0–4.4	9	16.4%
4.5–8.9	14	25.5%
9.0–13.4	21	38.2%
13.5–17.9	11	20%

b. 38.2%

c. 41.8% or 41.9%

15. a. iPhone **b.** 19% **c.** 28%

d. No, because the percentage calculated in part **c.** is not above 50%.

17. a. Approximately 6 hours

b. Approximately 7 hours

c. There is essentially no difference in the number of hours wives spend sleeping.

d. A pair of side-by-side bar graphs: one for husbands and one for wives.

19. a. 106 **b.** 43 **c.** 33%

d. 750–799 words and 850–899 words

e. Yes, because the percentage of essays with 700 or more words is 51.9%. It is close enough to 50%.

21. a. Although it's easy enough to see the peak, the way the horizontal tick marks are displayed on the graph makes it difficult to easily identify the date of the peak. A good guess would be somewhere around June 13th, 2021

b. Based on the graph shown, it is true that the average usage time has decreased during this period. However, we need to consider other factors along with a larger window of time.

23. The class widths are not equal.

25. A pie chart is an inappropriate way to display this data. Possible answers could include reference to the fact that not all states are represented in the graph. The pie chart gives percentages based on only these 8 states. The percentages in the pie chart make it seem that the percentage of women who had mammograms is much less than it really is. It could appear that Massachusetts contains 14% of all women age 50 and older who reported mammograms. Furthermore, the percentages add up to more than 100% because each percentage is the total for each category, which means this graph is not a representation of a whole.

11.3 Exercises

1. sum, divided **3.** left

5. True

7. Mean: $19.59; median: $18.35; mode: $32.00; range: $21.92; standard deviation: $8.739; unimodal

9. Mean: 310; median: 310; mode: no mode; range: 0; standard deviation: 0

11. Mean: 22.9; median: 24.5; mode: 24, 26, 35; range: 29; standard deviation: 9.0; multimodal

13. Mode   **15.** Mode

17. 86

19. At least 1040 g each

21. Yes

23. No, percentiles give the relative position in terms of percentages. Without knowing more information about the size of the sample, we cannot compute the size of the sample.

25. Georgia

27. The child weighs less than the median weight.

29. a. Min = $21,590, Q_1 = $25,030, Q_2 = $33,385, Q_3 = $35,960, Max = $38,680

b. 15%

c. [$21,590,$38,680]

31. Min = 20, Q_1 = 56, Q_2 = 84, Q_3 = 114, Max = 168

33. a. Min: $100,960; Q_1: $120,900; Q_2: $145,850; Q_3: 159,635; Max: $182,500

b. 75%   **c.** $81,540

35. Min: 310; Q_1: 510; Q_2: 705; Q_3: 1225; Max: 3700

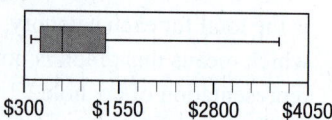

37. No, you should not operate with means. We cannot assume that all five classes are the same size. For instance, suppose one section has 100 students with a mean of 72 and another section has only 11 students, but with a mean of 79. We cannot give equal emphasis to the two means. You will not get an accurate answer.

39. Amelia scoring an 82 is just her raw score; it means she got 82% of the questions on the test correct. Amelia scoring in the 82nd percentile tells you that she scored at least as well as 82% of the people who took the test. It tells you more about her relative performance on the test.

41. Yes, the five-number summary can be put back together in numerical order: 9, 11, 13.5, 17.5, 19. So, $Q_1 = 11$.

11.4 Exercises

1. discrete

3. standard deviations

5. True

7. a. X3 has the largest mean and X1 has the smallest.

b. X3 has the largest standard deviation and X1 has the smallest.

c. BMW X1 generally costs less than BMW X2.

9. a. $z = 0.67$ **b.** $z = -0.93$

c. $z = 1$ **d.** $z = 0$

11. a. $z = -2.11$ **b.** $z = 1.81$

c. $z = 0.43$

13.

	z	x	μ	σ
a.	−0.87			
b.			277.12	
c.		130.30		
d.				8.99

15. 92.79%   **17.** 1.32%

19. 15.15%   **21.** 99.90%

23. 50.00%   **25.** 95%

27. 28.64%   **29.** 27.13%

31. 33.81%   **33.** 98.80%

35. $0.378 \approx 0.38$ **37. a.** −2.1

39. a. 5.16%   **b.** 95.76%

c. 99.94%   **d.** 88.88%

41. a. 68%   **b.** 280 and 748

c. Below 163 and above 865

43. a. 99.7%

b. 495 mL and 499 mL

c. Bottles that contain at least 501 mL of water are the top 2.5%. Bottles that contain at most 493 mL of water are the bottom 2.5%.

45. Her second race

47. Sofia is taller when compared to her class.

49. 17.92%   **51.** 27.14%

53. 25.25%

55. $z = 0$, since half the z-scores lie above this value, and half the z-scores lie below this value.

11.5 Exercises

1. point estimate

3. interval estimate

5. $18   **7.** $(16.30, 19.70)$

9. a. About $(820, 1010)$. We can be 95% confident that the mortality rate in 2010 was from 820 to 1010 deaths per 100 000 population.

b. Although the sample mean for 2012 falls below 900, the confidence interval still contains 900. Therefore, we cannot conclude with 95% confidence that the actual value is less than 900.

c. Because these are 95% confidence intervals, it is not possible to be 100% certain for any of the years. However, the interval for 2016 is completely below 850, so it is the most likely year to have the mortality below 850.

d. Although the sample mean for 2016 is smaller than that of 2015, their confidence intervals overlap. Therefore, we cannot say that there is a statistically significant difference between the two years' mortality rates.

11. We can be 90% confident that the percentage of patients whose pain is completely relieved by the anesthetic drug is between 76.2% and 81.4%.

13. $(14.6, 15.4)$; The professor can be 90% confident that the mean amount of time that her students spend studying is between 14.6 and 15.4 hours per week.

15. $(18.6, 19.8)$; The writer can be 95% confident that the mean computer usage time for American households is between 18.6 and 19.8 hours per week.

17. $(691, 749)$; We can be 98% confident that the mean amount of money that homeowners spend on lawn service each year is between $691 and $749.

19. $(21.1, 24.1)$; The physical therapist can be 99% confident that the mean recovery time for patients using the new therapy after ACL surgery is between 21.1 and 24.1 days.

21. $(0.60\%, 17.09\%)$; With 95% confidence, we can say that the percentage of all faculty members at the community college who know sign language is between 0.60% and 17.09%.

23. $(14.41\%, 31.31\%)$; We can say with 95% confidence that the percentage of all kindergartners who say pancakes are their favorite breakfast food is between 14.41% and 31.31%.

25. $(51.70\%, 68.30\%)$; With 95% confidence, the percentage of all students at that college who do not regularly check their campus e-mail accounts is between 51.70% and 68.30%.

27. a.

Margin of Error vs. Sample Size, 95% Confidence

b. The graph shows us that as the sample size increases, the margin of error gets smaller. This means that with the same level of confidence, we can create a tighter confidence interval, which gives a better approximation of the population parameter.

c. A confidence interval for a population parameter is constructed using a sample statistic and a margin of error. Since we are only given the margin of error and not the sample statistics, we do not have enough information to construct the confidence intervals.

29. A margin of error tells how many percentage points the result of the poll may differ from the true result for the whole population. Thus, from the estimated sample

parameters, we can conclude that in 95% of such surveys, the percentage of US adults who find dating is now harder than it was 10 years ago will be from 44.9% to 49.1%. Similar reasoning is used for the other parameters.

31. a. $68\% \pm 2.2\% = (65.8\%, 70.2\%)$

b. 95% Confidence Interval for the Population Percentage:
Rep/Lean Rep, Too Much:
$74\% \pm 3.3\% = (70.7\%, 77.3\%)$
Dem/Lean Dem, Too Much:
$64\% \pm 3.0\% = (61.0\%, 67.0\%)$
The confidence intervals for the two groups do not overlap. Therefore, we can say that we are 95% confident that Republicans are more likely to say technology companies have too much power than Democrats.

33. a. 2.7%

b. With 95% confidence, the percentage of US households owning a pet is from 56.6% to 62.0%.

Chapter 11 Exercises

1. Population: all US college freshmen; Sample: 203,967 freshmen who responded; Sample statistic: 27.6% describe themselves as "liberal;" Sample statistic: 20.7% describe themselves as "conservative;" Sample statistic: 47.4% describe themselves as "middle of the road."

3. Not necessarily; it is convenient, but may not be representative of the campus as a whole.

5. a. Answers will vary. For example, choose one store and survey people as they enter the store; choose one mall entrance and survey people as they enter the mall; set up a booth in the middle of the mall and ask

people to participate as they walk by, etc.

b. Answers will vary.

7. a. Explanatory variable: the drug for insomnia; Response variable: side effects

b. Treatment: the drug for insomnia

c. Treatment group: group that was taking the drug; Control group: group that was taking placebo

d. Single-blind. No, because the researchers can be biased when interpreting the patients' answers along with unintentionally pushing them for certain answers.

9. a. 74%   **b.** 5940   **c.** 8100

d. Yes, it is categorical data. However, it may not include ALL types of jobs, in which case the graph could be misleading.

11. Mean = 3.22; median = 3.21; mode = 3.29; range = 0.44; stdev = 0.13; min = 3.05; $Q_1 = 3.07$; med = 3.21; $Q_3 = 3.29$; max = 3.49

13. a. Min = 322, $Q_1 = 326$, $Q_2 = 335$, $Q_3 = 342$, Max = 345

b. 40%   **c.** [322, 345]

15. a. $z_1 = 6.25$ **b.** $z_2 = -1.5$

c. $z_3 = 12.5$

17. Hasef: $z = 0.889$; Kimberly: $z = 0.885$; Hasef's score was better.

19. 38.59%   **21.** 99.93%

23. 12.10%   **25.** 99.94%

27. 95.45%   **29.** 13.59%

31. True; $z > 0$, so it is higher than the mean.

33. a. 40.13%   **b.** 45.03%

c. 14.84%

Chapter 12: Data Science

12.1 Exercises

1. Domain knowledge

3. False

5. Answers will vary.

  a. Does the use of caffeine affect the productivity rates of night-shift workers?

  b. Data can be collected by surveying employees of local businesses that have a night-shift.

  c. A scatterplot can be used to see if there is a trend with productivity and caffeine usage between employees who do and do not consume caffeine.

  d. A bar graph can be created using data for employees who do not consume caffeine versus employees who do.

  e. A social media post could be used.

7. Answers will vary.

  a. What are the most commonly purchased items during the Thanksgiving holiday and how much of each are purchased?

  b. Previous years' sales data could be used along with latest trends in Thanksgiving meals along with decorations.

  c. The sales data across multiple years can be analyzed to see if sales of different items are consistent or are increasing or decreasing.

  d. Line graphs can be used to show sales trends across the years.

  e. An infographic illustrating the trends (whether increasing, decreasing, or steady) for various popular products along with a list of suggested items to keep in stock for the Thanksgiving season.

9. Answers will vary. A weakness is that the question is too vague; it could refer to software, features such as the camera, or price. Alternative questions include "Which main features should the next iteration of the flagship smartphone include?", "Which features do users wish to see in the next iteration of the flagship smartphone?", and "What are the market trends for main features of smartphones?"

11. Answers will vary. A weakness is that the question does not define what "better" means. Alternative questions include "Does flossing before brushing your teeth result in less overall plaque growth between dental cleanings than brushing before flossing?," "Does flossing before brushing your teeth result in less overall gum inflammation between dental cleanings than brushing before flossing?", and "Does flossing before brushing your teeth result in fewer cavities per year than brushing before flossing?"

13. Answers will vary. A weakness is that the question is too vague; it does not define what kind of focus is under consideration nor does it mention a frequency of meditation. Alternative questions include "Does meditating for 20 minutes a day increase a person's ability to focus on work?", "Does meditating for 30 minutes each morning increase a person's decision-making ability?", "Does meditating for 20 minutes in the middle of the workday increase an employee's ability to focus on tasks during the afternoon?"

15. Answers will vary. Questions include "What is the most popular time of day for people to visit the bakery?", "What are the most popular bakery items?", and "What is the age group of the typical customer?"

17. Answers will vary. Questions include "What do employees enjoy most about working for the company?", "What type of training is offered to new employees?", "and "What do employees who leave within a year of being hired have in common?"

19. Answers will vary. Questions include "How many people visit the museum per day?", "What are the main points of congestion in the current lobby?", and "What are the restrictions with redesigning the lobby?"

12.2 Exercises

1. raw, user-friendly

3. LOOKUP

5. True

7. Delimiter is semicolon, six columns

9. Delimiter is a comma, seven columns

11. Lines 4 and 5 seem to be duplicates, Chapter 5 doesn't have an Approved status, and "Antonio" seems to be misspelled in cell E8. The duplicate line can be removed with the Remove Duplicates tool, the missing approval and misspelled name can be fixed with Find and Replace.

13. a. and c.

15. twelve and

17. Okra Express

19. a. Cape cod, colonial, farmhouse, and ranch

  b. 30   c. 80   d. 20

21. Book Title

12.3 Exercises

1. possible relationship

3. regression line

5. False

7. **a.** 148.05   **b.** 16.05

 c. −346.95

9. No linear correlation

11. No linear correlation

13. No linear correlation

15. No linear relationship

17. Strong positive relationship

19. Yes   21. No

23. **a.** Positive   **b.** $r = 0.897$

 c. Yes

25. **a.** Positive   **b.** $r = 0.981$

 c. Yes

27. There appears to be a negative linear correlation.

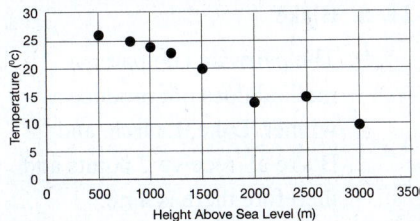

Sample of Temperatures at Different Elevations

29. There appears to be no correlation.

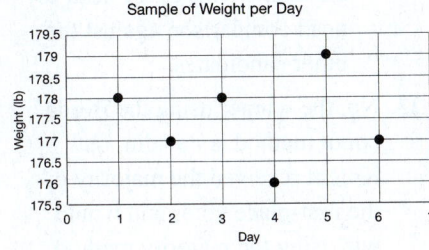

Sample of Weight per Day

31. **a.** $\hat{y} = 0.418x + 47.535$

 b. Yes, $r = 0.695$

 c. 76.8

33. **a.** $\hat{y} = 0.993x + 0.121$

 b. Yes, $r = 0.990$

 c. 5.6

12.4 Exercises

1. word cloud

3. Infographics

5. Answers will vary. The theme is about individuals, friends, and feelings.

7. Answers will vary. There is less of a focus on the monster than on the writer's feelings.

9. Answers will vary. A love story

11. Answers will vary.

13. Chang   15. Morales

17. Necklaces

19. Necklaces and dresses seem to tie for least successful.

21. No different than the national average

23. Answers will vary. Information about tourism numbers in 2018 could be added to the map for states that show statistically lower marriage rates compared to the national rate.

25. Existing customers make up a majority of the sales because their bar is consistently higher.

27. Answers will vary. The corresponding data from the same quarter in the previous year could also be displayed so there is a benchmark for comparison. Additional information could be added to detail the success of individual products being sold.

Chapter 12 Exercises

1. Answers will vary. A weakness is that the question is too vague and doesn't explain what "best" means for the overall skin health. Alternative questions include "Which sunscreen blocks the most UVA rays over a period of 2 hours?", "Which sunscreen causes the least amount of irritation to sensitive skin in the form of rashes or clogged pores?", "Which sunscreen has the longest lasting protection against UVB rays?"

3. Delimiter is a comma, five columns

5. Negative linear correlation

7. Weak positive relationship

9. Statistically significant

11. **a.** Negative   **b.** $r = -0.708$

 c. Not statistically significant

13. **a.** 555.8   **b.** 657.12

 c. 809.1

15. Answers will vary. Dark colors could be used. An eerie shape related to the story could be formed by the words.

Chapter 13: Voting and Apportionment

13.1 Exercises

1. plurality

3. preference ballot

5. majority rule

7.

Rankings								
1st	Stephen	Noah	Oliver	Jack	Oliver	Noah	Stephen	Stephen
2nd	Oliver	Oliver	Stephen	Stephen	Noah	Lars	Lars	Jack
3rd	Jack	Jack	Jack	Lars	Lars	Stephen	Oliver	Lars
4th	Lars	Lars	Noah	Noah	Jack	Jack	Noah	Noah
5th	Noah	Stephen	Lars	Oliver	Stephen	Oliver	Jack	Oliver
# Votes	212	133	543	24	179	8	201	11

9.

Rankings										
# Votes	3	7	3	6	5	2	1	1	2
1st	Mori	Nguyen	Dey	Smith	Mori	Smith	Mori	Smith	Smith
2nd	Abbadi	Dey	Smith	Mori	Abbadi	Abbadi	Nguyen	Dey	Abbadi
3rd	Dey	Smith	Mori	Abbadi	Nguyen	Mori	Abbadi	Mori	Dey
4th	Nguyen	Mori	Abbadi	Nguyen	Dey	Nguyen	Smith	Abbadi	Mori
5th	Smith	Abbadi	Nguyen	Dey	Smith	Dey	Dey	Nguyen	Nguyen

11. **a.** 720 **b.** 9194

   c. *Nothing Compares to You* by Sinead O'Connor

   d. 4598 **e.** No

13. Any number smaller than 36

15. Witt must be in second place and Longsdon in first place in column one

17. 10

19. Robert Griffin III: 1687; Andrew Luck: 1407; Trent Richardson: 978; Montee Ball: 348; Tyrann Mathieu: 327

21. **a.** Leesburg, VA

   b. Louisville, CO

   c. Answers will vary.

23. **a.** Candidate B

   b. Candidate B

   c. Candidate B

   d. Candidate B; he or she won using all methods discussed.

25. **a.** Nixon; no **b.** Nixon; yes

   c. Nixon won both the popular vote and the electoral vote. The popular vote is much closer than the electoral vote shows.

   d. No. Even with all of Wallace's electoral votes, Humphrey would only have had 237 electoral votes. A majority of 270 electoral votes were required to win the election.

13.2 Exercises

1. Condorcet criterion

3. monotonicity criterion

5. Borda count method

7. **a.** Susan **b.** Courtney

   c. Courtney won all head-to-head comparisons, so the Condorcet criterion applies and is not satisfied.

9. **a.** Mechanics

   b. Heat and Optics

   c. No, the Borda count method declares Heat and Optics the winner, but Mechanics has a majority of first-place votes, so the majority criterion is not satisfied.

11. **a.** New Orleans, LA

   b. New Orleans, LA

   c. Yes, New Orleans is the preferred destination both before and after Chicago was taken out of the selected options, so the irrelevant alternatives criterion is satisfied.

13. **a.** H. Kennedy

   b. T. Parchment

Rankings				
1st	H. Kennedy	T. Parchment	H. Kennedy	T. Parchment
2nd	T. Parchment	H. Kennedy	T. Parchment	H. Kennedy
# Votes	8	14	10	11

   c. No, Kennedy was the winner before Jones pulled out of the election, but not after, so the Borda count method does not satisfy the alternative criterion here.

15. The majority criterion is satisfied because Johnny Depp wins using both the Borda count method or the majority method.

17. Any x such that x is less than or equal to 11

19. Any x between 42 and 100, inclusive

21. **a.** Blake

   b. The pairwise comparison method does not produce a winner. Luke, Lauren, and Blake all receive 2 points and therefore there is a tie.

   c. The Condorcet criterion does not apply here since no candidate wins every head-to-head comparison against the other candidates.

23. No; the winner using the Borda count method is Paladin, but Knight received the majority of the first-place votes and would win using the majority method.

13.3 Exercises

1. standard quota

3. Alabama paradox

5. new states paradox

7. $11,071.79

9. Davis: $474,166,057; Santa Cruz: $104,378,746

11. a. 229.016

  b. Riverside: 20.475;
  Santa Barbara: 26.553

13. The apportionments are different.

Campus	Electric Vehicles
Berkeley	62
Davis	91
Irvine	55
Los Angeles	124
Merced	3
Riverside	20
San Diego	80
San Francisco	18
Santa Barbara	27
Santa Cruz	20

15. a. SD = 21; SQ: I = 1.667;
  II = 2.048; III = 2.286

  b. They should offer two sections
  of each course with about 21
  students per section.

17. They should offer two sections
  of each course with about 21
  students per section.

19. a. Northern: 5; Southern: 6;
  Eastern: 3; Western: 2

  b. Northern: 5; Southern: 6;
  Eastern: 3; Western: 2

  c. No, the apportionments are the
  same.

21. $13 + 3 + 3 + 4 + 2 = 25$

23. a. Sciences: 5; humanities: 5;
  professional and trade
  schools: 1

  b. Sciences: 5; humanities: 5;
  professional and trade
  schools: 1

  c. Sciences: 5; humanities: 5;
  professional and trade
  schools: 1

  d. Sciences: 5; humanities: 5;
  professional and trade
  schools: 1

25. Campus 1: 21 police officers;
  campus 2: 19 police officers;
  campus 3: 1 police officer;
  The Alabama paradox occurs
  because when the total number of
  police officers to be apportioned
  increased, the number of police
  officers appointed to campus 3
  decreased.

27. a. Freshmen: 5 members;
  sophomore: 4 members;
  junior: 4 members;
  senior: 4 members

  b. Freshmen: 5 members;
  sophomore: 4 members;
  junior: 4 members;
  senior: 4 members

  c. Freshmen: 5 members;
  sophomore: 4 members;
  junior: 4 members;
  senior: 4 members

13.4 Exercises

1. coalition   **3.** dictator

5. $[13: 9, 6, 5, 3, 2]$

7. a. 25   **b.** 2

  c. No   **d.** No

9. 46   **11.** 45

13. No dictator; no dummy player

15. a. 5   **b.** 42   **c.** No

  d. $\{P_1, P_2, P_3\}, \{P_1, P_2, P_4\},$
  $\{P_1, P_2, P_3, P_4\},$
  $\{P_1, P_2, P_4, P_5\},$
  $\{P_1, P_2, P_3, P_5\},$
  $\{P_1, P_2, P_3, P_4, P_5\}$

  e. P_1 and P_2 are critical in all winning coalitions.

  f. P_1 and P_2 have veto power.

17. P_2

19. $\text{BPI}(P_1) = \dfrac{10}{26} \approx 0.385 = 38.5\%;$

  $\text{BPI}(P_2) = \text{BPI}(P_3) = \text{BPI}(P_4)$

  $= \text{BPI}(P_5) = \dfrac{4}{26} \approx 0.154 = 15.4\%;$

  no dictators; none have veto
  power

21.

  $\text{SSPI}(P_1) = \dfrac{12}{24} = 0.5 = 50\%;$

  $\text{SSPI}(P_2) = \text{SSPI}(P_3) = \text{SSPI}(P_4)$

  $= \dfrac{4}{24} \approx 0.167 = 16.7\%;$

  no dictators; none have veto
  power

23. a. Any coalition of at least 218
  members

  b. Any coalition of at least 291
  members

25. P_1: 25%; P_2: 25%; P_3: 25%;
  P_4: 25%

Chapter 13 Exercises

1.

	Rankings									
1st	H	P	S	S	K	P	P	J	S	K
2nd	K	J	H	H	J	H	J	P	H	P
3rd	P	K	J	K	P	S	S	H	J	J
4th	J	S	P	P	H	J	H	K	K	S
5th	S	H	K	J	S	K	K	S	P	H
# Votes	5	4	6	3	7	1	4	1	2	2

3. $\dfrac{7(6)}{2} = 21$

5. a. $5! = 120$ **b.** 793

  c. 388   **d.** Candidate A

  e. 397   **f.** No

7. a.

Harris; SGA	20
Icin; Social Work Club	73
Green; $\Delta\Lambda\Pi$	86
Lawson; FCA	6
Albert; GSA	132
Roman; $\Gamma B \Theta$	89
Switzer; Galois Club	37
Belton; Hispanic Culture Center	10
Finley; History Club	38
Issen; NAEA	33
San Marie; NBS	19
Centus; NTSS	82
Dennis; $\Omega\Psi\Phi$	56
Russan; ΦA	130
Hunter; ΦMA	45
Austin; $\Sigma\Gamma P$	8
Molda; Student Design Group	92
Irene; Voices of Praise	88

  b.

1	Albert; GSA	132
2	Russan; ΦA	130
3	Molda; Student Design Group	92
4	Roman; $\Gamma B \Theta$	89
5	Irene; Voices of Praise	88
6	Green; $\Delta\Lambda\Pi$	86
7	Centus; NTSS	82
8	Icin; Social Work Club	73
9	Dennis; $\Omega\Psi\Phi$	56
10	Hunter; ΦMA	45

9. a. Clinton; no   **b.** Clinton; yes

   c. No; answers will vary. For example, although it is possible that the popular vote may have changed, the electoral vote would not (since Perot did not receive any electoral votes) and Clinton would still win the election.

11. a. *Django Unchained*

   b. *Zero Dark Thirty*

   c. No; the Condorcet criterion is not satisfied because the plurality method chose *Django Unchained* as the winner, but *Zero Dark Thirty* won the head-to-head comparisons against every other candidate.

13. The irrelevant alternatives criterion would be met using the plurality with elimination method because even if Jacksonville, FL were removed, the winner is still Boston, MA.

15. The Condorcet criterion is satisfied because outdoor living spaces wins using the plurality method and also wins all head-to-head matchups with the other categories.

17. The missing ranking for honesty is 2nd and the missing ranking for sense of humor is 1st.

19.

County	Officers
A	45
B	52
C	55
D	40
E	24
F	34

21. $62,816.18

23. Austin: $4,014,970,172; Dallas: $1,084,773,995

25. a. 183.515 students/vehicle

   b. San Antonio: 126.409 vehicles; Permian Basin: 14.691 vehicles

27. Yes, the apportionments are different.

Campus	Allocation
Arlington	112
Austin	253
Brownsville	51
Dallas	68
El Paso	89
San Antonio	126
Tyler	26
Permian Basin	15
Pan American	85

29.

Course	Allocation
Phys Sci	2 Sections
Biology	1 Section
Anatomy	2 Sections

31.

Course	Allocation
Comp I	8 Grad Stud
Comp II	5 Grad Stud
Surv of Lit	2 Grad Stud
Poetry	4 Grad Stud

33. a.

Activity	Allocation
Swimming	$1,715,249
Library	$18,167,643
Baseball	$8,051,171
Softball	$6,846,996
Theater	$5,271,767
Comp Hob	$3,339,485
Music	$11,607,689

b.

Activity	Allocation
Swimming	$1,715,249
Library	$18,167,643
Baseball	$8,051,171
Softball	$6,846,996
Theater	$5,271,767
Comp Hob	$3,339,486
Music	$11,607,688

c.

Activity	Allocation
Swimming	$1,715,249
Library	$18,167,643
Baseball	$8,051,171
Softball	$6,846,996
Theater	$5,271,767
Comp Hob	$3,339,486
Music	$11,607,688

d.

Activity	Allocation
Swimming	$1,715,249
Library	$18,167,643
Baseball	$8,051,171
Softball	$6,846,996
Theater	$5,271,767
Comp Hob	$3,339,486
Music	$11,607,688

35. No   **37.** 13

39. a. 6   **b.** 96

   c. $\{P_1, P_2, P_3, P_4, P_5, P_6\}$, $\{P_1, P_2, P_3, P_4, P_5\}$, $\{P_1, P_2, P_3, P_4, P_6\}$

   d. Yes, all players are critical in some coalition.

   e. Yes, P_1, P_2, P_3, P_4

   f. No

   g. $\text{BPI}(P_1) = \text{BPI}(P_2)$

$= \text{BPI}(P_3) = \text{BPI}(P_4) = \dfrac{3}{14}$

$\approx 0.214; = 21.4\%$

$\text{BPI}(P_5) = \text{BPI}(P_6)$

$= \dfrac{1}{14} \approx 0.071 = 7.1\%$

41. a. $\langle P_2, P_1, P_3, P_4 \rangle, \langle P_2, P_1, P_4, P_3 \rangle,$ $\langle P_3, P_4, P_1, P_2 \rangle, \langle P_4, P_3, P_1, P_2 \rangle,$ $\langle P_2, P_3, P_1, P_4 \rangle, \langle P_3, P_2, P_1, P_4 \rangle,$ $\langle P_2, P_4, P_1, P_3 \rangle, \langle P_4, P_2, P_1, P_3 \rangle$

   b. $\langle P_1, P_2, P_3, P_4 \rangle, \langle P_1, P_2, P_4, P_3 \rangle,$ $\langle P_3, P_4, P_2, P_1 \rangle, \langle P_4, P_3, P_2, P_1 \rangle,$ $\langle P_1, P_3, P_2, P_4 \rangle, \langle P_3, P_1, P_2, P_4 \rangle,$ $\langle P_1, P_4, P_2, P_3 \rangle, \langle P_4, P_1, P_2, P_3 \rangle$

   c. $\text{SSPI}(P_1) = \text{SSPI}(P_2)$

$= \dfrac{8}{24} = \dfrac{1}{3} \approx 0.333; = 33.3\%$

$\text{SSPI}(P_3) = \text{SSPI}(P_4)$

$= \dfrac{4}{24} = \dfrac{1}{6} \approx 0.167 = 16.7\%$

   d. No   **e.** No

Chapter 14: Graph Theory

14.1 Exercises

1. degree   **3.** edge

5. loop

7. chromatic number

9. cycle

11. Vertices: USA, UK, Algeria, Canada, Mexico; edges: 10, 1, 2, 60, 3

13. Vertices: Hannah, Lee, Jalen, Anna, Mike, Kiara; Edges: e_1, e_2, e_3, e_4, e_5, e_6, e_7

15. a. v_1, v_2, v_3

b. No, it does not start and end at the same vertex.

c. Answers will vary. The walk must start and end with the same vertex.

17. a. Answers will vary. For example, $a, b, c, b, e, a, e, c, d, e, f, d, f, a$.

b. Answers will vary. For example, u, w, v, x, u, y, z, u.

19. a. Answers will vary. For example, k, m, n.

b. Answers will vary. For example, r, q, t, n.

c. Answers will vary. For example, j, t, m, k, s, q, p.

21. G_1 is connected; G_2 is disconnected

23. Yes. Explanations will vary. For example, both graphs consist of the same vertices and edges.

25. Answers will vary. For example,

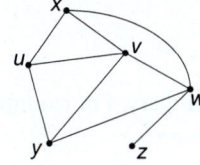

27. Answers will vary. For example,

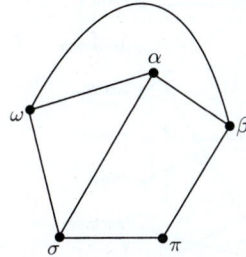

29. Answers will vary because the cycle may be written beginning with any of the vertices. However, the cycle must include all vertices in their correct order. The first vertex in the cycle list should not be repeated at the end of the list. For example, r, t, v, w, u, s is of length 6.

31. a. Answers will vary. For example, ab.

b. Answers will vary. For example, a and e.

c. Answers will vary. For example, ae and de.

d. Answers will vary. For example, c and r.

e. Answers will vary. For example, a.

f. Answers will vary. For example, a, b, c, d, e.

33. Answers will vary. For example,

a.

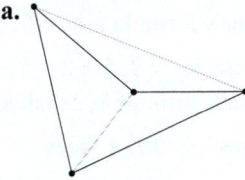

b.

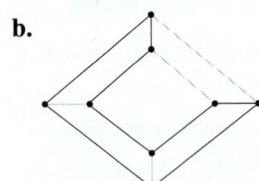

c. This graph doesn't have a Hamilton cycle.

35. Answers will vary. Cameras can be placed in rooms i, g, b, and d or in rooms j, g, b, and d.

37. Minimum number of colors = 3;

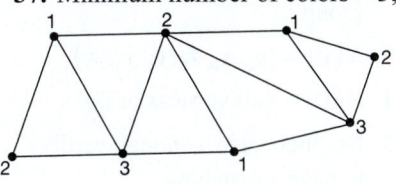

14.2 Exercises

1. connected

3. spanning tree

5. one

7. No, the graph has cycles in it, and is therefore not a tree.

9. Yes, the graph is a tree.

11. 4   **13.** 3

15. Answers will vary.

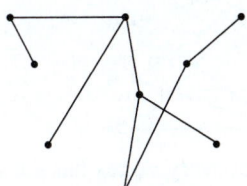

17. Answers will vary.

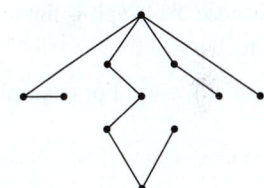

19. 4   **21.** 14

23. 8

25.

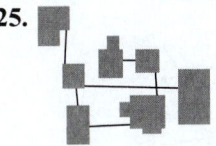

27. 22 cities

14.3 Exercises

1. bipartite graph

3. matching

5. neighborhood

7. No, the vertices cannot be divided into two groups without edges joining vertices in the same group.

9. $N(B) = \{x_2, x_4, x_6, y_1, y_3, y_5\}$

11. $N(D) = \{$all vertices in $Q\}$

13. No, there aren't enough vertices to have a matching.

15. No, there is a vertex without an edge, and hence it cannot have a matching.

17. The graph is regular bipartite, so it has a matching.

19. No, if $A = \{x_1, x_2, x_4, x_5\}$, $N(A)$ has only three vertices.

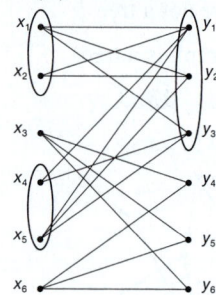

21. No, not every vertex has the same degree.

23. No, because the graph is not bipartite.

25. Answers will vary. For example,

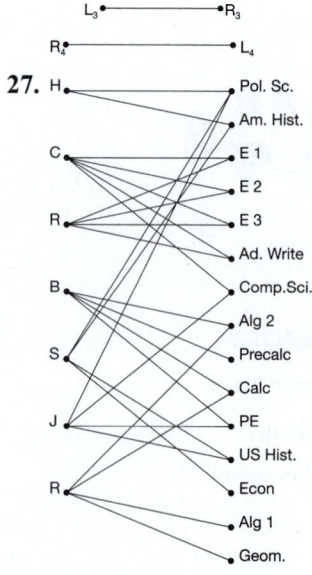

27.

No matching possible.

29. Answers will vary.

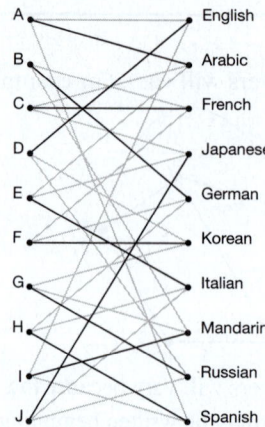

A — English
B — Arabic
C — French
D — Japanese
E — German
F — Korean
G — Italian
H — Mandarin
I — Russian
J — Spanish

31. No two edges of the same color can share a vertex. Therefore, every edge of the same color has a distinct pair of vertices it joins together

14.4 Exercises

1. edges crossing

3. $v + f - 3 = 2$

5. complete

7.

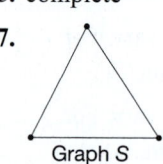

Graph S

9.

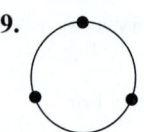

11. 4

13. $v = 6, e = 10, f = 6; 6 + 6 - 10 = 2$, so Euler's formula is satisfied.

15. $v = 5, e = 5, f = 2; 5 + 2 - 5 = 2$, so Euler's formula is satisfied.

17. 21 edges   **19.** 4 faces

21. Yes, the graph is planar.

23. No, the graph is not planar (it is $K_{3,3}$).

25. No, the graph is not planar (it is $K_{3,3}$).

27. $\chi(K_3) = 3; \chi(K_5) = 5; \chi(K_n) = n$

29. Any planar drawing when drawn on a sphere will have its exterior face changed from an infinite face to a finite one.

Chapter 14 Exercises

1. No, the graph is not connected.

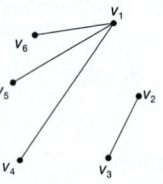

3. Yes, they are the same graph.

5. Chromatic number = 3; Answers will vary. For example,

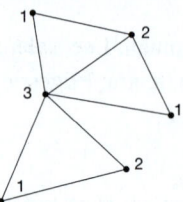

7. No, the graph is not a tree because it contains at least one cycle.

9. 11

11. Answers will vary. For example,

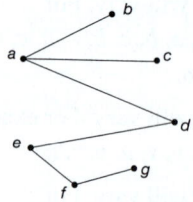

13. a. Yes, it is a spanning tree. A spanning tree will ensure that all cities are connected with optical fiber.

   b. Yes, it is a minimum-weight spanning tree.

15. $N(A) = \{$Jeff, Isaac, Peg, Sarah$\}$

17. The graph has a matching.

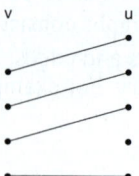

19. Answers will vary. For example,

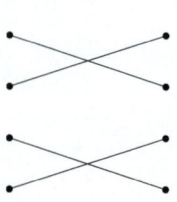

21. 8

23. #21: $f + v - e = 2$

$8 + 9 - 15 \overset{?}{=} 2$

$17 - 15 \overset{?}{=} 2$

$2 = 2$

#22: $f + v - e = 2$

$8 + 12 - 18 \overset{?}{=} 2$

$20 - 18 \overset{?}{=} 2$

$2 = 2$

25. No, the graph is not planar because it contains K_5 as a minor.

Index